Molecular Biology

Molecular Biology

A.V.S.S. Sambamurty

Alpha Science International Ltd.
Oxford, U.K.

A.V.S.S. Sambamurty
Professor in Botany (Retd.)
Sri Venkateswara College,
Delhi University, Delhi, India

Alpha Science International Ltd.
7200 The Quorum, Oxford Business Park North
Garsington Road, Oxford OX4 2JZ, U.K.

www.alphasci.com

ISBN 978-1-84265-414-9

Printed in India

DEDICATED
TO
LORD SRI VENKATESWARA

Preface

Molecular Biology deals with the entire "cell" in a three-dimensional way, viz., the structural details, the functional details and their regulatory mechanisms. It is an exhaustive treatment of the subject covering 41 chapters, several appendices and glossaries. Molecular Biology is taught in schools, colleges and universities. As such it is a fundamental science at all levels of study.

The book starts with the chapter on Chemistry of Life, which deals with the chemical nature of the bond, the structure and function of nucleic acids, carbohydrates, proteins, enzymes and lipids. These constitute the crux of the entire Molecular Biology. In the next four chapters introduction and tools, techniques in Molecular Biology, Bioenergetics, Metabolism, Glycolysis and Tricarboxylic Acid Cycle and their regulatory mechanisms are discussed.

Next few chapters are focussed on certain organelles like Mitochondria, Chloroplasts, Golgi Apparatus, Lysosomes, Peroxisomes and Microbodies in a detailed way. Cell structures like Plasmamembrane, Nucleus, DNA, RNA and Genetic Code are discussed in detail.

From chapters 21-41 the entire Molecular Genetics, Cytogenetics, Classical Genetics and Genomics are discussed. Special topics like Genomics, Immunology and Immunity, Human Genetics, Human Molecular Genetics and Cancer Genetics have been treated with up to date literature giving a glimpse of molecular aspects of human beings. It is hoped that this book will cater to the needs of M.Sc. and B.Sc. students of Molecular Biology, Genetics, Pharmacy, Biotechnology, Medicine, Biochemistry, Botany and Zoology.

I thank my daughter Ramaa Sambamurty for help in various ways and Mrs Bhawna Sharma in computer typing of the book. I also thank Narosa Publishing House for their neat and speedy publication.

Acknowledgements

The author greatly acknowledges the following Publishers and associated authors for permitting to use some of their tables and illustrations in this book:

Van Nostrand Co., N.Y., USA (*Avers, C.J.*)

Prentice Hall, Englewood Cliffs, N.J., USA (*Brock, T.M.*)

Chapman & Hall, London, UK (*Brown, T.A.*)

Holden Day, San Fransisco, USA (*Drake, D.W.*)

Jones & Bartlett Inc., Boston, USA (*Friefelder, D.*)

John Wiley & Sons, N.Y., USA (*Gardner, E.J. et al*)

E.L.B.S., London, UK (*Hayes, W*)

Benjamin/Cummings Publishing Co., California, USA (*Watson J.D. et al*)

Rastogi Publications, Meerut, India (*Gupta P.K.*)

Freeman, W.H. & Co., N.Y., USA (*Watson, J.D. et al*)

Pearson Education, Singapore, China (*Watson, J.D. et al*)

Contents

CHAPTER 1

Chemistry of Life

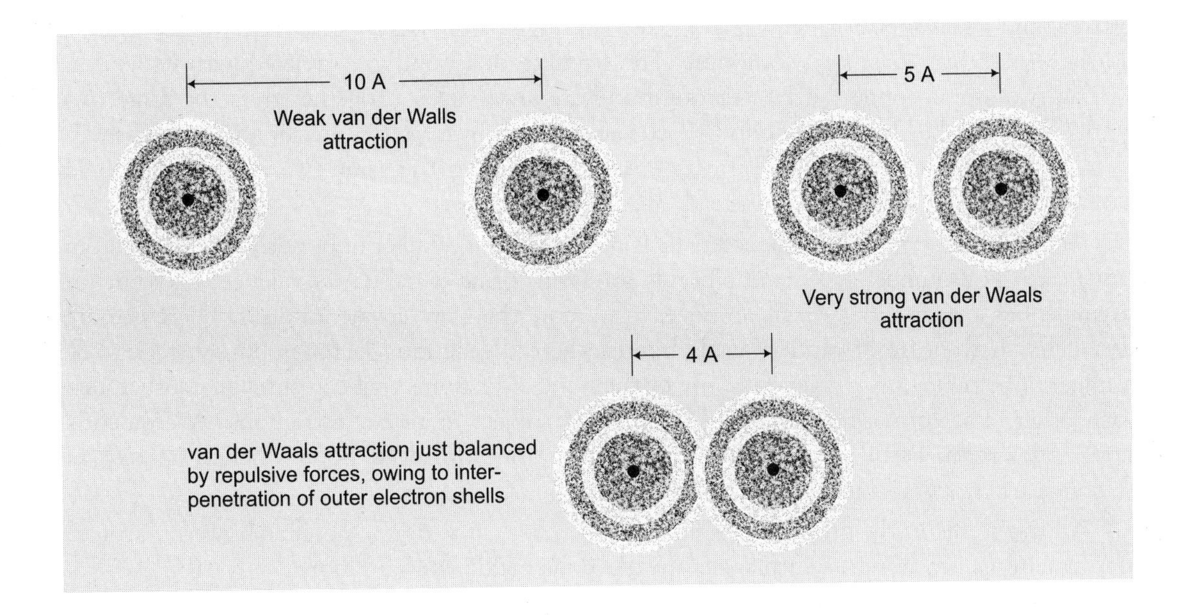

In this chapter on the 'Chemistry of life', an introduction to the chemical phenomena essential to life processes like nature of the chemical bond, bond energy, thermodynamics, structure and function of proteins, enzymes, carbohydrates, glycolysis, citric acid cycle, have been discussed. These will help the students to understand the biochemistry of reactions involved in molecular genetics. All reactions are atom specific in gene activity involving DNA, RNA, etc.

SECTION A

THE NATURE OF THE CHEMICAL BOND

Characteristics of Chemical Bond

A chemical bond is an attractive force that holds atoms together. Aggregates of definite size are called molecules. Originally, it was thought that only covalent bonds hold atoms together in molecules but now weaker attractive forces are known to be important in holding together many macromolecules. For example, the four polypeptide chains of hemoglobin are held together by the combination of several weak bonds. It is thus customary also to call weak positive interactions chemical bonds, even though they are not strong enough, when present singly, to effectively bind two atoms together.

Bond Length

Chemical bonds are characterized in several ways. A most obvious characteristic of a bond is its strength. Strong bonds almost never fall apart at physiological temperatures. This is why atoms united by covalent bonds always belong to the same molecule. Weak bonds are easily broken, and when they exist singly, they exist fleetingly. Only when present in ordered groups do the weak bonds exist for a long time. The strength of a bond is correlated with its length, so that two atoms connected by a strong bond are always closer together than the same two atoms held together by a weak bond. For example, two hydrogen atoms bound covalently to form a hydrogen molecule (H;H) are 0.74 A apart, whereas the same two atoms, when held together by vander Waals forces instead, are held 1.2 A apart.

Another important bond characteristic is the maximum number of bonds that a given atom can make. The number of covalent bonds an atom forms is called its valence. Oxygen, for example, has a valence of two. It can never form more than two covalent bonds. There is more variability in the case of vander Waals bonds where the limiting factor is purely steric. The number of possible bonds is limited only by the number of atoms that can simultaneously touch each other. The formation of hydrogen bonds is subject to more restrictions. A covalently bonded hydrogen atom usually participates in only one hydrogen bond whereas an oxygen seldom participates in more than two hydrogen bonds.

All the atoms in the grey must lie in the same plane. Rotation is possible however, around the remaining two bonds, which make up the polypeptide configurations.

The angle between two bonds originating from a single atom is called the BOND ANGLE. The angle between two specific covalent bonds is always approximately the same. For example, when a carbon atom has four single bonds they are directed tetrahedrally (bond angle = 100). In contrast, the angles between weak bonds are much more variable.

Bonds differ also in the freedom of rotation they allow. Single covalent bonds permit free rotation of bond atoms (Fig. 1.1) whereas double and triple bonds are quite rigid. For example, the carbonyl (C=O) and imino (N=H) groups bound together by the rigid peptide bond must lie in the same plane (Fig. 1.2), because of the partial double bond character of the peptide bond. Much weaker ionic bonds show completely opposite behaviour they impose no restrictions on the relative orientations of bonded atoms.

Chemical Bonds Explained by Quantum Mechanics

The nature of the forces strong as well as weak that give rise to chemical bonds remained a mystery to chemists until the quantum theory of the atom (quantum mechanics) was developed in the 1920s. Then, for the first time the various empirical laws about how chemical bonds are

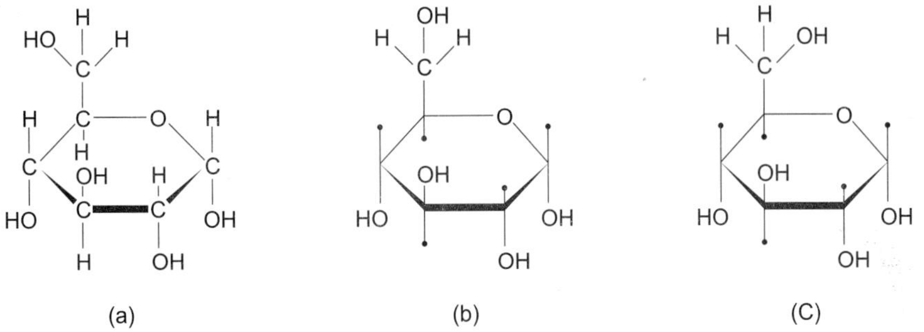

Fig. 1.1 Rotation about the C_5—C_6 bond in glucose. This carbon-carbon is a single bond, and so any of the three configurations (a), (b), (c) may occur.

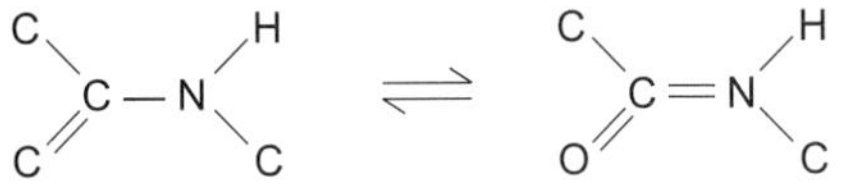

Fig. 1.2 The planar shape of the peptide bond. Shown is a portion of an extended polypeptide chain. Almost no rotation is possible about the peptide bond because of its partial double bond character.

formed were put on a firm theoretical basis. It was realized that all chemical bonds, weak as well as strong, were based on electrostatic forces. Quantum mechanical explanations were provided not only for the covalent bonding by the sharing of electrons, but also for the formation of weaker bonds.

Chemical Bond Formation Involves a Change in the Form of Energy

The spontaneous formation of a bond between two atoms always involves the release of some of the internal energy of the unbonded atoms and its conversion tom another energy form. The stronger the bond the greater the amount of energy which is released upon its formation. The bonding reaction between two atoms A and B is thus described by

$$A + B - AB + \text{energy}$$

where AB represents the bonded aggregate. The rate of the reaction is proportional to the frequency of collision between A and B. The unit most commonly used is the calorie, the amount of energy required to raise the temperature of 1 gram of water from 14.5 to 15.5°C. Since thousands of calories are usually required in the breaking of a mole of chemical bonds most chemical changes in chemical reactions are expressed in kilocalories per mole.

Atoms joined by chemical bonds, however, do not remain together forever. There also exist forces, which breed chemical bonds. By far the most important of these forces arises from heat energy. Collisions with fast moving molecules or atoms can break chemical bonds. During a collision some of the kinetic energy of moving molecule is given up as it pushes apart two bonded atoms. The faster a molecule is moving (the higher the temperature), the greater the probability that, upon collision, it will break a bond. Hence, as the temperature of collection of molecules increases, the stability of their bonds decreases. The breaking of a bond is thus always indicated by the formula

$$AB + \text{energy} - A + B$$

The amount of energy that must be added to break a bond is exactly equal to the amount which was released upon its formation. This equivalence follows from the first law of thermodynamics, which states that energy (except as it is interconvertible with mass can be made nor destroyed.

Equilibrium between Bond Making Bond Breaking

Every bond is thus a result of the combined actions of bond making (arising from electrostatic interactions) and bond breaking forces. When equilibrium is reached in a closed system, the number of bonds forming per unit time will equal the number breaking. Then the proportion of bonded atoms is described by the following mass action formula:

$$K_{eq} = \frac{Conc^{AB}}{Conc^{A} \times Conc^{B}}$$

where K is the equilibrium constant, and conc, conc and conc are the concentrations of A, B and AB in moles per liter respectively. Whether we start with only free A and B with only the molecule AB or with a combination of AB and free A and B at equilibrium proportions of A, B, and AB will reach the concentration given by K_{eq}.

The Concept of Free Energy

There is always a change in the form of energy as the proportion of bonded atoms move toward the equilibrium concentration. Biologically, the most useful way to express this energy change is through physical chemists, concept of free energy, G. Here we shall not give a rigorous description of free energy, nor show how it differs from other forms of energy. Free energy is the energy that has the ability to do work.

The second Law of Thermodynamics tells us that a decrease of free energy (ΔG is negative) always occurs in spontaneous reactions. When equilibrium is reached, there is no further change in the amount of free energy ($\Delta G = 0$). The equilibrium state for a closed collection of atoms is, thus, that state that contains the least amount of free energy.

The free energy lost as equilibrium is approached is either transformed into heat or used to increase the amount of entropy. The amount of entropy is a measure of the amount of disorder. The greater the disorder, the greater the amount of entropy. The existence of entropy means that many spontaneous chemical reactions do not proceed with an evolution of heat. For example, in the dissolving of NaCl in water, heat is absorbed. There is, nonetheless, a net decrease in free energy because of the increase of disorder of the Na^{+} and Cl^{-} ions as they move from a solid to a liquid phase.

K_{eq} is Exponentially Related to ΔG

It is obvious that the stronger the bond, and hence the greater the change in free energy (ΔG), which accompanies its formation, the greater, the proportion of atoms that must exist in the bonded form. This common sense idea is quantitatively expressed by the physical chemical formula:

$$\Delta G = -RT \ln K_{eq} \quad \text{or} \quad K_{eq} = e^{-\Delta G/RT}$$

where, R is the universal gas constant, T the absolute temperature, e = 2.718, in the logarithm of K to the base *e* and K_{eq} is the equilibrium constant.

Insertion of the appropriate values of R (= 1.987 cal/deg/mole) and T (= 298 at 25°C) tells us (Table 1.1) that ΔG values as low as 2 kcal/mole can drive a bond forming reaction to virtual completion, if all reactants are present at molar concentrations.

Covalent Bonds are Very Strong

The G values accompanying the formation of covalent bonds from free atoms such as hydrogen or oxygen are very large and negative in sign, usually –50 to –110 kcal/mole. Application of equation, G = –Rt in Keg or Keg = eG/RT tells us that Keg of the bonding reaction will be correspondingly large and so the concentration of hydrogen or oxygen atoms existing unbound will be very small. For example, G value of-100 kcal/mole tells us that, if we start with 1 mole /liter of the reacting atoms, only one in 10 atoms will remain unbound when equilibrium is reached.

Weak Bonds have Energies between 1 and 7 Kcal/Mole

The main types of weak bonds important in biological systems are the van der Waals bonds, hydrogen bonds, and ionic bonds. Sometimes, the distinction between a hydrogen bond and an ionic bond is arbitrary. The weakest bonds are the van der Waals bonds. These have energies, (1 to 2 kcal/mole) that are only slightly greater than the kinetic energy of heat motion. The energies of hydrogen and ionic bonds range between 3 and 7 kcal/mole.

In liquid solutions, almost all molecules are forming a number of weak bonds to nearby atoms. All molecules are able to form van der Waals bonds: hydrogen and ionic bonds can also form between molecules (ions) which have a net charge or in which the charge is unequally

Table. 1.1 The numerical relationship between the equilibrium constant and G at 25°C.

K_{eq}	*G kcal/mole*
0.001	4.089
0.01	2.726
0.1	1.363
1.0	0
10.0	–1.363
100.0	2.726
1000.0	–4.089

distributed. Some molecules thus, have the capacity to form several types of weak bonds. Energetic considerations, however, tell us that molecules always greater tendencies to form the stronger bond.

Weak Bonds Constantly Made and Broken at Physiological Temperatures

The energy of the strongest weak bond is only about 10 times larger than the average energy of kinetic motion (heat) at 25°C (0.6 kcal/mole). Since there is a significant spread in the energies of kinetic motion many molecules with sufficient kinetic energy, to break the strongest weak bond, always exist at physiological temperatures.

Enzymes not Involved in Making (Breaking) of Weak Bonds

The average lifetime of single weak bond is only a fraction of a second. Cells do not need a special mechanism to speed up the rate at which weak bonds are made and broken. Correspondingly, enzymes never participate in weak reactions of weak bonds.

Distinction between Polar and Nonpolar Molecules

All forms of weak interactions are based on attractions between electric charges. The separation of electric charges can be permanent or temporary, depending upon the atoms involved. For example, the oxygen molecule (O;O) has a symmetric distribution of electrons between its two oxygen atoms, and so each of its two atoms are uncharged. In contrast, there is a non-uniform distribution of charge in water (H;O;H) where the bond electrons are unevenly shared (Fig. 1.3). They are held more strongly by the oxygen atom, which thus carries a considerable negative charge, whereas the two hydrogen atoms together have an equal amount of positive charge. The center of the positive charge is on one side of the center of the negative charge. A combination of separated positive and negative charges is called an ELECTRIC DIPOLE MOMENT. Unequal electron sharing reflects dissimilar affinities of the bonding atoms for electrons. Atoms which have tendency to gain electrons are called electronegative atoms. Electropositive atoms have a tendency to give up electrons.

Molecules such as H_2O, which have a dipole moment, are called POLAR MOLECULES. NONPOLAR MOLECULES are those with no effective dipole moments. In methane (CH_4), for example, the carbon and hydrogen atoms have similar affinities for their shared electron pairs and so neither the carbon nor the hydrogen atom is noticeably charged.

The distribution of charge in a molecule can also be affected by the presence of nearby molecules particularly if the affected molecule is polar. This effect may cause a nonpolar molecule to acquire a slight polar structure. If the second molecule is not polar, its presence will still alter the nonpolar molecule establishing a fluctuating charge distribution. Such induced effects, however, give rise to a much smaller separation of charge than is found in polar molecules, thus resulting in smaller interaction energies and correspondingly, weaker chemical bonds.

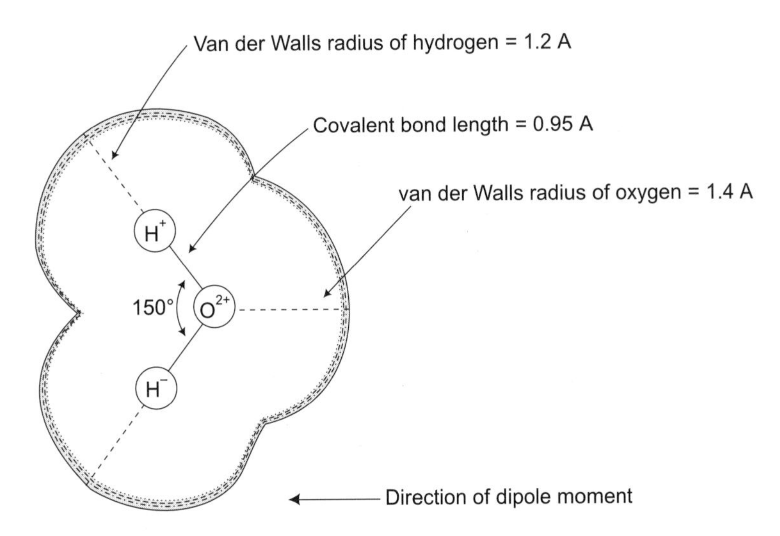

Fig. 1.3 The structure of a water molecule.

Van Der Waals Forces

Van der Waals bonding arises from a nonspecific attractive force originating when two atoms come close to each other. It is based not upon the existence of permanent charge separations but rather upon the induced fluctuating charges caused by the nearness of molecules. It, therefore, operates between all types of molecules polar as well as nonpolar. It depends heavily upon the distance interacting groups, since the bond energy is inversely proportional to the sixth power of between distance (Fig. 1.4).

There also exists a more powerful van der Waals repulsive force, which comes into play at even shorter distances. This repulsion is caused by the overlapping of the outer electron shells of the atoms involved. The van der Waals attractive and repulsive forces balance at a certain distance specific for each type of atom. This distance is the so called van der Waals radius (Table 1.2). The van der Waals bonding energy between two atoms separated by the sum of their van der Waals radii increases with the size of the respective atoms. For two average atoms it is only about 1 kcal/mole which is just slightly more than the average thermal energy of molecules at room temperature (0.6 kcal/mole).

This means that van der Waals forces are an effective binding force at physiological temperatures only when several atoms in a given molecule are bound to several atoms in another molecule. Then the energy of ineteraction is much greater than the dissociating tendency resulting from random thermal movements. In order for several atoms to interact effectively, the molecular fit must be precise, since the distance separating any two interacting atoms must not be much greater than the sum of van der Waals radii. The strength of interaction rapidly approaches zero when this distance is only slightly exceeded. Thus the strongest type of van der Waals contact arises when a molecule contains a cavity exactly complementary in shape to a protruding group of another molecule. This type of situation

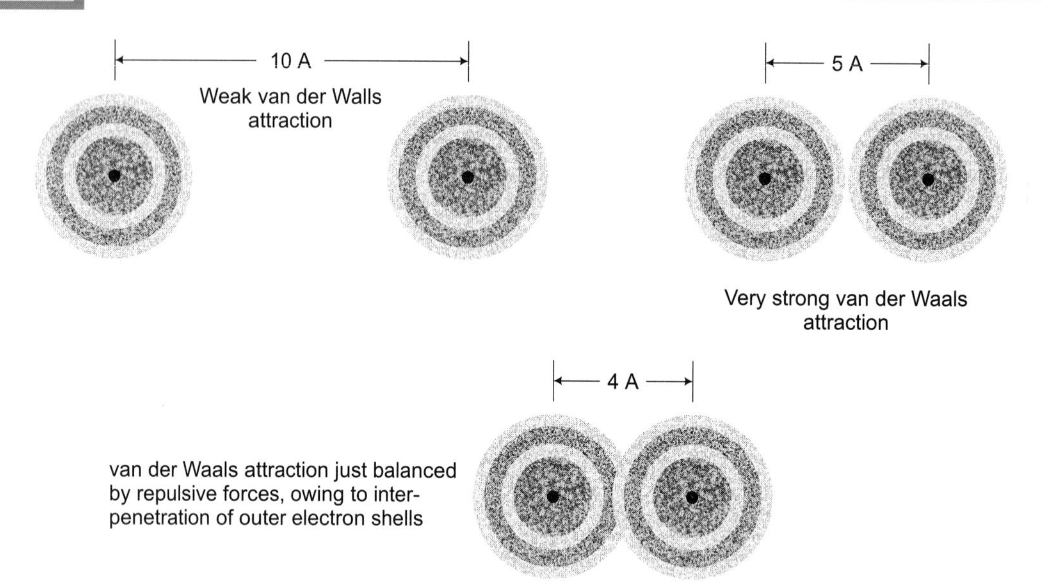

Fig. 1.4 Diagram illustrating van der Waals attraction and repulsion forces in relation to electron distribution of monoatomic molecules on the inert rare gas argon.

Table 1.2 Van der Waals radii of the atoms in biological molecules.

Atom	*van der Waals radius, A*
H	1.2
N	1.5
O	1.4
P	1.9
S	1.85
CH_3 group	2.0
Half thickness of aromatic molecule	1.7

thought to exist between antigen and its specific anibody. In this instance, the binding energies sometimes can be as large as 10 kcal/mole, so that antigen-antibody complexes seldom fall apart. Many polar molecules are seldom affected by van der Waals interactions since such molecules can acquire a lower energy state (lose more energy) by forming other types of bonds.

Hydrogen Bonds

A hydrogen bond arises between a covalently bound hydrogen atom with some positive charge negatively charged covalently bound acceptor atom (Fig. 1.5). For example, the hydrogen atoms of the imino group (N-H) are attracted by the negatively charged keto oxygen atoms (C = O) Sometimes the hydrogen bonded atoms belong to groups with a unit of charge (e.g., NH_3 + or COO-). In other cases, both the donor hydrogen atoms and the negative acceptor atoms have less than a unit of charge.

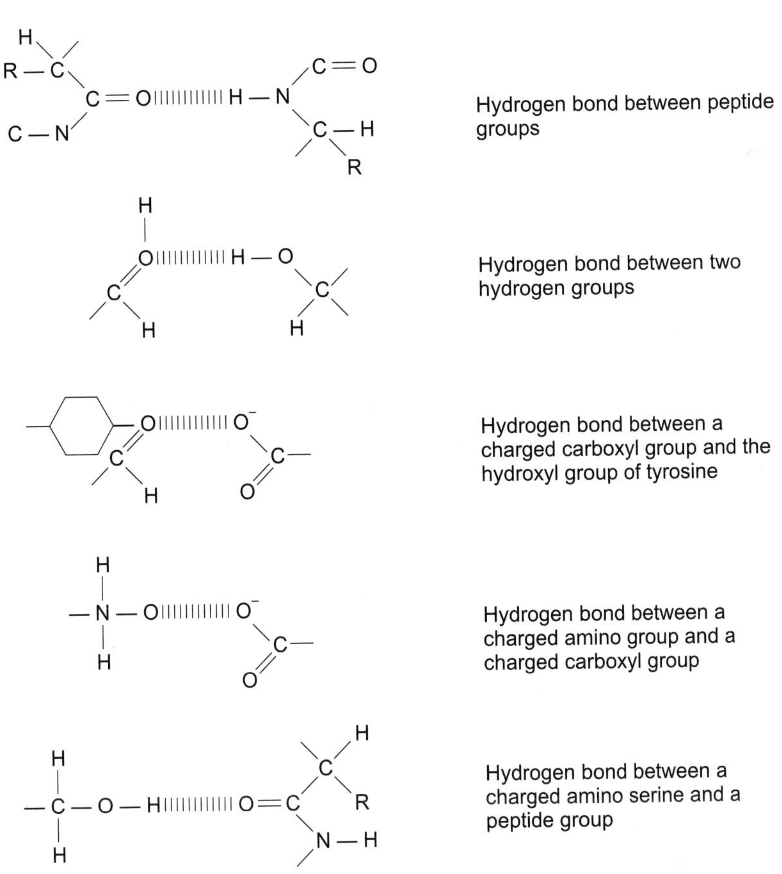

Hydrogen bond between peptide groups

Hydrogen bond between two hydrogen groups

Hydrogen bond between a charged carboxyl group and the hydroxyl group of tyrosine

Hydrogen bond between a charged amino group and a charged carboxyl group

Hydrogen bond between a charged amino serine and a peptide group

Fig. 1.5 Examples of hydrogen bonds in biological molecules.

Table 1.3 Approximate bond lengths of biologically important hydrogen bonds.

Bond	*Approximate bond length, A*
O—H———O	2.70 ± .10
O—H———O	2.63 ± .10
O—H———N	2.88 ± .13
N—H———O	3.04 ± .13
N—H———O	2.93 ± .10
N—H———N	3.10 ± .13

The biologically most important hydrogen bonds involve hydrogen atoms covalently bound to oxygen (O-H) or nitrogen atoms (N-H). Likewise, the negative acceptor atoms are usually nitrogen or oxygen. Table 1.3 lists some of the most important hydrogen bonds. Bond energies range between 3 and 7 kcal/mole, the stronger bonds involving the greater charge differences between the donor and acceptor atoms. Hydrogen bonds are thus weaker than covalent bonds, yet considerably stronger than van der Waals bonds. A hydrogen bonds therefore, will hold two atoms closer together than the sum of their van der Waals radii, but not so close together as covalent bond would hold them.

Hydrogen bonds, unlike van der Waals bonds, are highly directional. In optimally strong hydrogen bonds the hydrogen atom points directly at the acceptor atom (Fig. 1.6). If it points indirectly the bond energy is much less. Hydrogen bonds are also much more specific than van der Waals bonds, since they demand the existence of molecules with complementary donor hydrogen and acceptor groups.

Fig. 1.6 Directional properties hydrogen bonds. In (a) the vector along the covalent O—H bond points directly at the acceptor oxygen, thereby forming a strong bond. In (b) the vector points away from the oxygen atom, resulting in a much weaker bond.

Water Molecules form Hydrogen Bonds

Under physiological conditions, water molecules rarely ionize to form H^- and OH^+ ions. Instead, they exist as polar H–O–H molecules both the hydrogen and oxygen atoms form strong hydrogen bonds. In each H_2O molecule the oxygen atom can bind two eternal hydrogen atoms, whereas each hydrogen atom can bind to one adjacent oxygen atom. these bonds are directed tetrahedrally (Fig. 1.7) so that, in its solid and liquid forms, each water molecule tends to have four nearest neighbours, one in each of the four directions of a tetrahedrally. In ice the bonds to these neighbours are very rigid and the arrangements fixed. Above the melting temperature (0°C) the energy of thermal motion is sufficient to break the hydrogen bonds and to allow the water molecules to change their nearest neighbours continually. Even liquid form, however, at a given instant most water molecules are bound by four strong hydrogen bonds.

The Uniqueness of Molecular Shapes: The Concept of Selective Stickiness

Even though most cellular molecules are built up from only a small number of groups, such as OH, NH_2, and CH_3, great specificity exists as to which molecules tend to lie next to each other. This is because each molecule has unique bonding properties. One very clear demonstration comes from the specificity of stereoisomers. For example, proteins are always constructed from L-amino acids, never from their mirror images, the D-amino acids (Fig. 1.8). Though the D-and L-amino acids have identical covalent bonds their bonding properties to asymmetric molecules are often very different. Thus most enzymes are specific for L-amino acids. If an L-amino acid is able to attach to a specific enzyme the D-amino acid is unable to bind.

The general rule exists that most molecules in cells can make good "weak" bonds with only a small number of other molecules. This is partly because all molecules in biological systems

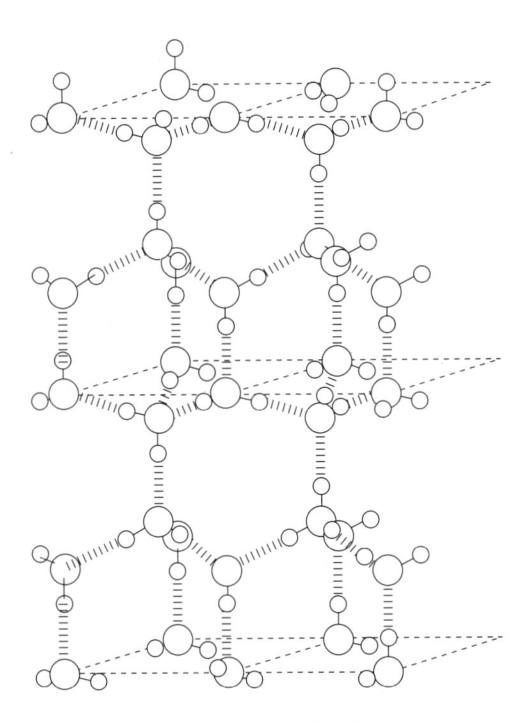

Fig. 1.7 Schematic diagram of lattice formed by H_2O molecules. the energy gained by forming specific hydrogen bonds (|||||) between H_2O molecules favors the arrangement of the molecules in adjacent tetrahedrons. Oxygen atoms are indicated by large circles, and hydrogen atoms by small circles. Although the rigidity of the arrangement depends upon the temperature of the molecules, the pictured structure is nevertheless, predominant in water as well as in ice.

exist in an aqueous environment. The formation of bond in a cell depends not only upon whether two molecules bind well to each other, but also upon whether the bond will permit their water solvent to form the maximum number of good hydrogen bonds.

DNA can Form a Regular Helix

At first glance, DNA looks even more unlikely to form a regular helix as does an irregular polypeptide chain. DNA not only has an irregular sequence of side groups, but in addition, all its side groups are hydrophobic, both the purines (adenine and guanine) and the pyrimidines (thymine and cytosine), even though they contain polar C=O and NH_2 groups, are quite insoluble in water because their flat sides are completely hydrophobic.

Nonetheless, DNA molecules usually have regular helical configurations. This is because most DNA molecules contain two polynucleotide strands that have complementary structures (see chapter 2 for more details). Both internal and external secondary bonds stabilize the structure. The two strands are held together by hydrogen bonds between pairs of complementary purines and pyrimidines (Fig. 1.9) Adenine (amino) is always hydrogen bonded to thymine (keto), where as guanine (keto) is hydrogen bonded to cytosine (amino). In addition, virtually all the surface atoms in the sugar an phosphate groups form bonds to water molecules.

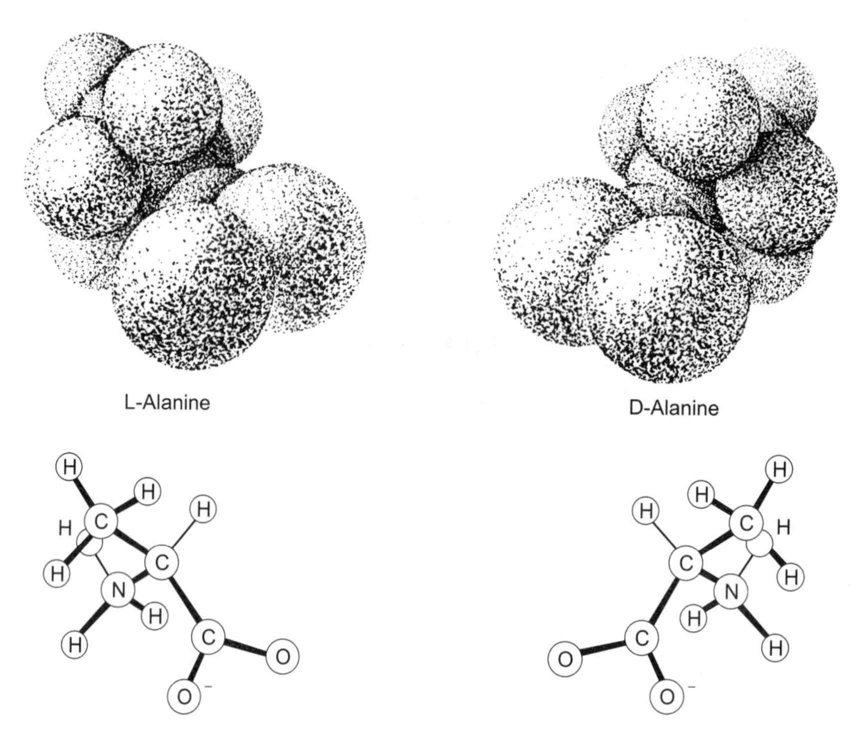

Fig. 1.8 The two stereoisomes of the amino acid alanine.

The purine-pyrimidine base pairs are found in the center of the DNA molecule. This arrangement allows their flat surfaces to stack on top of each other and so limits their contact with water. This stacking arrangement would be much less satisfactory if only one chain were present.

A single chain could not have a regular backbone because its pyrimidines are smaller than the purines, and so the angle of helical rotation would have to vary with the sequence of bases. The presence of complementary base pairs in double-helical DNA makes a regular structure possible, since each base pair is of the same size.

DNA Molecules are Stable at Physiological Temperatures

The double-helical DNA molecule is very stable at physiological temperatures, for two reasons. First, disruption of the double helix breaks the regular hydrogen bonds and brings the hydrophobic purines and pyrimidines into contact with water. Second, individual DNA molecules have a very large number of weak bonds, arranged so that most of them cannot break without the simultanous breaking of many others. Even though thermal motion is constantly breaking apart the terminal purine-pyrimidine pairs at the ends of each molecule, the two chains do not usually fall apart because the hydrogen bonds in the middle are still intact (Fig. 1.10). Once a break occurs the most likely next event is the reforming of the same hydrogen bonds to restore the original molecular configuration. Sometimes, of course, the first breakage is

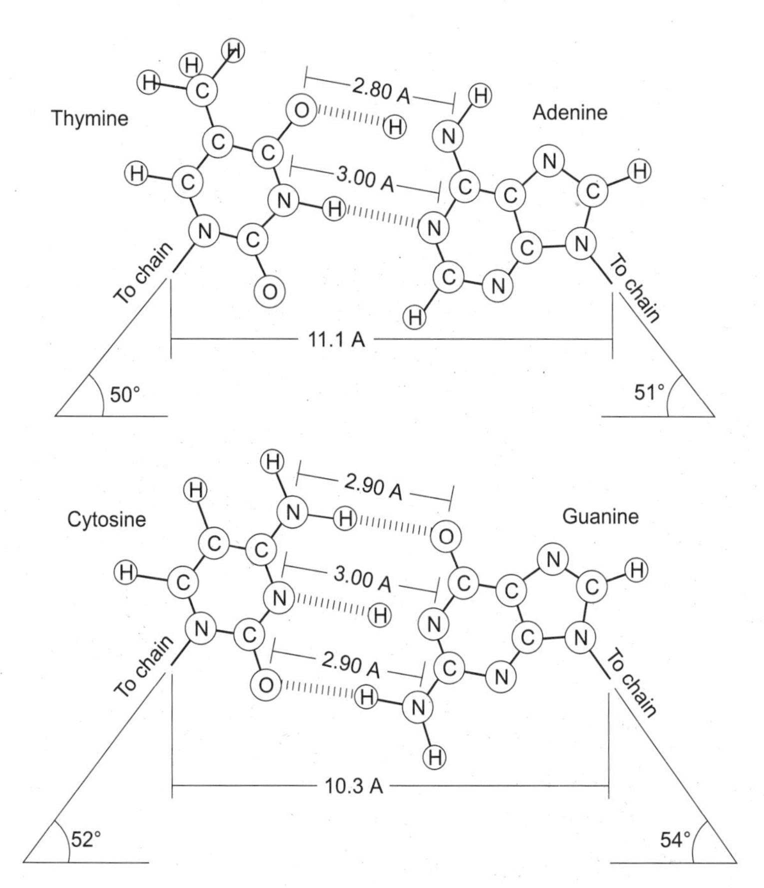

Fig. 1.9 The hydrogen-bonded base pairs in DNA. Adenine is always attached to thymine by two hydrogen bonds, where as guanine always bonds to cytosine by three hydrogen bonds. The obligatory pairing of the smaller pyrimidine with the larger Purine allows the two sugar-phosphate backbones to have identical helical configurations. All the hydrogen bonds in both base pairs are strong. since each hydrogen atom points directly at its acceptor atom (nitrogen or oxygen).

followed by a second one, and so forth. Such multiple breaks, however, are quite rate so that double helixes held together by more than ten nucleotide pairs are very stable at room temperature.

The same principle also governs the stability of most protein molecules. Stable protein shapes are never due to the presence of just one or two weak bonds, but must always represent the cooperative result of a number of weak bonds.

Ordered collections of hydrogen bonds become less and less stable as their temperature is raised above physiological temperatures. At physiologically abnormally high temperatures, the simultaneous breakage of several weak bonds is more frequent. After a significant number have broken a molecule usually loses its original form (the process of denaturation) and assumes an inactive (denatured) configuration.

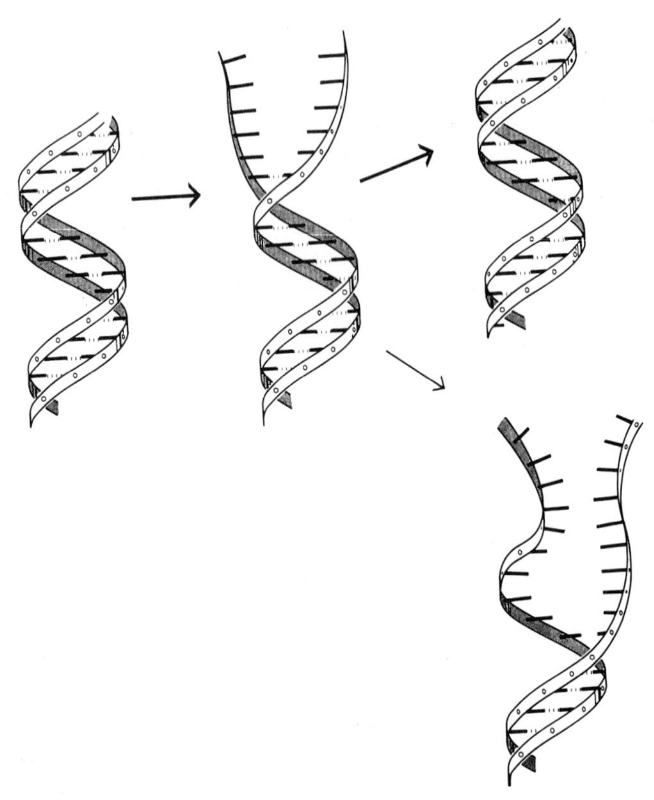

Fig. 1.10 The breaking of terminal hydrogen bonds in DNA by random thermal motion Because the internal hydrogen bonds continue to hold the two chains together the immediate reforming the broken bonds is highly probable. Also shown is the very rare alternative: the breaking of further hydrogen bonds, and the consequent disentangling of the chains.

BOND ENERGIES AND THERMODYNAMICS

Food Molecules are Thermodynamically Unstable

There is great variation in the amount of free energy possessed by specific molecules. This is a consequence of the fact that all covalent bonds do not have the same bonds. As an example, the covalent bond between oxygen and hydrogen is considerably stronger than the bonds between hydrogen and hydrogen or oxygen and oxygen. The formation of an O-H bond at the expense of O–O or H–H will thus release energy. Energetic considerations tell us that a sufficiently concentrated mixture of oxygen and hydrogen will be transformed into water.

A molecule thus possesses a larger amount of free energy if linked together by weak covalent bonds than if it is linked together by strong bonds. This idea seems almost paradoxical at first glance, since it means that the stronger the bond the less energy it can give off. But the notion automatically makes sense when we realize that an atom that has formed a very strong bond has already lost a large amount of free energy in this process. Therefore, the best food molecules

(molecules which donate energy) are those molecules that contain weak covalent bonds and are, thereby thermodynamically unstable.

For example, glucose is an excellent food molecule, since there is a great decrease in free energy when it is oxidized by O_2 to yield CO_2 and H_2O. On the contrary, CO_2 is not a food molecule in animals, since in the absence of the energy doner ATP, it cannot spontaneously be transformed to more complex organic molecules even with the help of specific enzymes. CO_2 can be used as a primary source of carbon in plants only because the energy supplied by light quanta during photosynthesis results in the formation of ATP.

DISTINCTION BETWEEN DIRECTION AND RATE OF A REACTION

The chemical reactions by which molecules are transformed into other molecules which contain less free energy do not occur at significant rates at physiological temperatures in the absence of a catalyst. This is because even a "weak covalent bond" is, in reality, very strong and is only rarely broken by thermal motion within a cell. In order for a covalent bond to be broken in the absence of a catalyst, energy must be supplied to push apart the bonded atoms. When the atoms are partially apart, they can recombine with new partners to form stronger bonds. In the process of recombination, the energy released is the sum of the free energy supplied to break the old bond plus the difference in free energy between the old and the new bond. The energy that must be supplied to break the old covalent bond in a molecular transformation is called the activation energy. The activation energy is usually less than the energy of the original bond because molecular rearrangements generally do not involve the production of completely free atoms. Instead, a collision between the two reacting molecules is required, followed by the temporary formation of a molecular complex (the activated state). In the activated state, the close proximity of the two molecules makes each other's bonds more laible, so that less energy is needed to break a bond than when the bond is present in a free molecule.

Most reactions of covalent bonds in cells are, therefore, described by

$$A\text{–}B + C\text{–}D — A\text{–}D + C\text{–}B — \Delta G$$

The mass action expression for such reaction is

$$K_{eg} = \frac{Conc^{A\text{-}D} \times Conc^{C\text{-}B}}{Conc^{A\text{-}B} \times Conc^{C\text{-}D}}$$

where $conc^{A\text{-}B}$, $Conc^{C\text{-}D}$, etc, are the concentrations of the several reactants in moles per liter. Here also, the value of K_{eq} is related to ΔG by equatation

$$\Delta G = \text{-RT in } K_{eq}$$

Since energies of activation are generally between 20 and 30 kcal/mole, activated states practically never occur at physiological temperatures. High activation energies should thus be considered barriers preventing spontaneous rearrangements of cellular covalent bonds.

ENZYMES LOWER ACTIVATION ENERGIES

Enzymes are absolutely necessary for life because they lower activation energies. The function of enzymes is to speed up the rate of the chemical reactions requisite to cellular existence by lowering the activation energies of molecular rearrangements to values that can be supplied by the heat of motion. When a specific enzyme is present, there is no longer an effective barrier preventing the rapid formation of the reactants possessing the lowest amount of free energy. Enzymes never affect the nature of an equilibrium: They merely speed up the rate at which it is reached Thus, if the thermodynamic equilibrium is unfavorable for the formation of a molecule, the presence of an enzyme can in no way bring about its accumulation.

The need for enzymes to catalyze essentially every cellular molecular rearrangement means that knowledge of the free energy of various molecules cannot by itself tell us whether an energetically feasible rearrangement will, in fact occur. The rate of the reactions must always be considered. Only if a cell possesses a suitable enzyme will the corresponding reaction be important.

A Metabolic Pathway is Characterized by a Decrease in Free Energy

Thermodynamics tells us that all biochemical pathways must be characterized by a decrease in free energy. This is obviously the case for degradative pathways, in which thermodynamically unstable food molecules are converted to more stable compounds, such as CO_2 and H_2O, with the evolution of heat. All degradative pathways have two primary purposes: (1) to produce the small organic molecules, and (2) to conserve a significant fraction of the free energy of the original food molecule in a form that can do work, by coupling some of the steps in degradative pathways with the simultaneous formation of molecules that can store free energy (high-energy molecules) like ATP.

Not all the free energy of a food molecule is coverted into the free energy of high-energy molecules. If this were the case, a degradative pathway would not be characterized by a decrease in free energy. No driving force would exist to favor the breakdown of food molecules. Instead, we find that all degradative pathways are characterized by a conversion of at least one half the free energy of the food molecule into heat or entropy. For example, it is now believed that, in cells, approximately 40% of the free energy of glucose is used to make new high-energy compounds, the remainder being dissipated into heat energy and entropy.

High-energy Bonds Hydrolyze with Large Negative ΔG's

A high-energy molecule contains a bond (S) whose breakdown by water (hydrolysis) is accompained by a large decrease in free energy (5 kcal/mole or more). The specific bonds whose hydrolysis yields these large negative ΔG's are called high- energy bonds. Both these terms are in a real sense, misleading, since it is not the bond energy but the free energy of hydrolysis that is high. Nonetheless, the term high-energy bond is generally comployed, and, for convenience we shall continue this usage by marking high-energy bonds with the symbols.

The most important high-energy compound is ATP. It is formed from inorganic phosphate Ⓟ and ADP, using energy obtáined either from degradative reactions or from the sun (Photosynthesis). There are, however, many other important high-energy compounds. Some are directly formed during degradative reactions; others are formed using some of the free energy of ATP. Table 1.4 lists the most important types of high-energy bonds. All involve either phosphate or sulfur atoms.

Peptide Bonds Hydrolyze Spontaneously

The formation of a dipeptide and a water molecule from two amino acids requires a ΔG of 1 to 4 kcal/mole, depending upon which amino acids are being bound. This ΔG value decreases progressively if amino acids are added to longer polypeptide chains; for an infinitely long chain the ΔG is reduced to 0.5 kcal/mole. This decrease reflects the fact that the free charged NH_3^+ and COO^- groups at the chain ends favor the hydrolysis (breakdown accompanied by the uptake of a water molecule) of nearby peptide bonds.

These positive ΔG values by themselves tell us that polypeptide chains cannot form from free amino acids. In addition, we must take into account the fact that water molecules have a much, much higher concentration (generally 100) than any other cellular molecule. All equilibrium reaction in which water participates are thus strongly pushed in the direction that consumes water molecules. This is easily seen in the definition of equilibrium constants. For example, the reaction forming a dipeptide.

$$\text{amino acid (a)} + \text{amino acid (b)} - \text{dipeptide (a-b)} + H_2O$$

has the following equilibrium constant:

$$K_{eq} = \frac{\text{Conc (a)} \times \text{Conc (b)}}{\text{Cons (a - b)} \times \text{Conc } (H_2O)}$$

where concentrations are given in moles/liter. Thus for a given K_{eq} value (related to ΔG by the formula ΔG = –RT in K), a much greater concentration of H_2O means a correspondingly smaller concentration of the dipeptide. The relative concentrations, are, therefore, very important. In fact, a simple calculation shows that dydrolysis may often proceed spontaneously even when the ΔG for the nonhydrolytic reaction is –3 kcal/mole.

Thus, in theory, proteins are unstable and, given sufficient time, will spontaneously degrade to free amino acids. On the other hand, in the absence of specific enzymes, these spontaneous rates are too slow to have a significant effect on cellular metabolism. That is, once a protein is made, it remains stable unless its degradation is catalyzed by a specific enzyme.

WATER AS A SOLVENT OF LIFE

Physical Properties of Water

The physical properties of water differ distinctly from those of other solvents. For example, water as a hydride of oxygen (H_2O) has a higher melting point, heat of vaporization, and surface

Table 1.4 Important classes of high-energy bonds.

Class	*Molecular*	*DG of reaction, kcal/mole*
Pyerophosphate	(P)~(P) pyrophosphate	(P)~(P) ⇌ (P) + (P) ΔG = –6
Nucleoside diphosphates	Adenosine — (P)~(P) (ADP)	ADP ⇌ AMP + (P) ΔG = –6
Nucleoside triphosphates	Adenosine — (P)~(P)~(P)	ATP ⇌ ADP + (P) ΔG = –7 ATP ⇌ AMP + (P)~(P) ΔG = –8
Enol phosphates	O^- O C C — O ~ (P) CH_2 Phosphoenol pyruvate (PEP)	PEP ⇌ pyruvate + (P) ΔG = –12
Amino acyl adenylates	Adenosine (P) ~ O R C — C — NH_3 O H	AMP ~ AA ⇌ AMP + AA ΔG = –7
Guanidinium phosphates	O H_2C — C O N H H_3C C — N ~ (P) NH Creatine phosphate	Creatine ~ P ⇌ Creatine + P ΔG = –8
Thioesters	O H_3C — C S-CoA Acetyl-CoA	Acetyl CoA CoA-SH + acetate ΔG = –8

tension than do the comparable hydrides of sulphur (H_2S) and nitrogen (NH_3) Table 1.5. Such properties are indicative of strong intermolecular forces in liquid water. It is the electrical dipolar nature of the water molecule, that accounts for these forces. In a water molecule, the highly electronegative oxygen atom attracts the bonding, electrons from each of the two hydrogen

Table 1.5 Some physical properties of the hydrides of oxygen, sulfur, and nitrogen

Hydride	*Melting point*	*Boiling point*	*Heat of Vaporization (cal/g)*
H_2O	0	100	540
H_2S	–85	–60	132
NH_3	–78	–33	327

atoms, polarizing the bonds. A partial positive ($^+$) region is created around each of the two resultant hydrogen nuclei. These partial positive regions and the partial negative ($^-$) region around the oxygen atom make water a dipolar molecule. It is this polar nature of water that allows for electrostatic attraction between its molecules. The result of such an attraction is called a hydrogen bond, which occurs between the oxygen atom of one water molecule and a hydrogen atom of another.

Because of the almost tetrahedral arrangement of the oxygen electrons (bond angle 104.5°) each water molecule can potentially hydrogen bond with four other molecules studies on liquidies water reveal an average of 3.4 hydrogen bonds per molecule. The bonding properties which allow water to bind to itself make water a relatively structured solvent and account for its strong internal cohesion as a liquid. Several rare physical properties of water offer certain biological advantages, two of which are important for the maintenance of the constant internal temperature required by many organisms. One property is water's high heat of vaporization (the number of calories absorbed when a gram or liquid is vaporized). The other property is the high specific heat capacity of water (the number of calories required to raise the temperature of one gram of substance 1°C). As a result, water absorbes heat well and, therefore, is of great value in helping to keep an organism's temperature constant.

The fact that water achieves its maximum density at 4°C is also important to many biological systems. The solidification of water at O°C to form ice produces a less dense phase which floats. Because ice does not sink, bodies of water do not freeze from the bottom upward; this permits aquatic organisms to remain in their normal environment during winter Also once frozen, bodies of water thaw more readily because the coldest water is near the surface, exposed to sun and atmosphere.

WATER AS A SOLVENT

Water is an excellent solvent for ionic compounds, such as salts, because the attraction between the ionic components of the molecules and the water dipoles is sufficient to overcome the attraction between the ions themselves. Nonionic polar compounds, such as sugars and simple alcohols, are also very soluble in water. Polar functional groups, such as the hydroxyl group, of nonionic compounds readily make hydrogen bond with water molecules, dispersing the compounds among the water molecules.

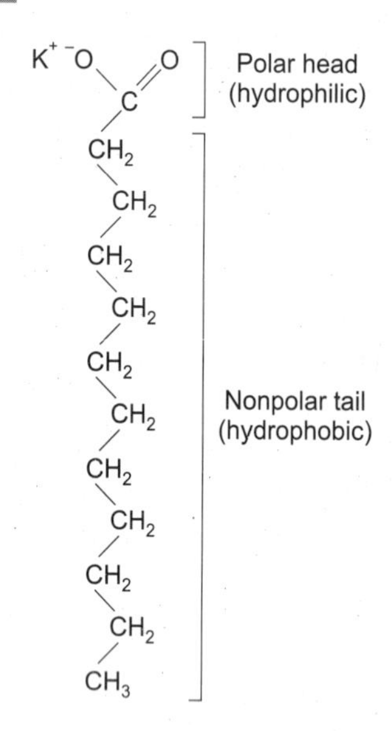

Fig. 1.11 Potassium laurate, a 12-carbon fatty acid.

An interesting phenomenon occurs when amphipathic molecules, possessing both polar (hydrophilic) and nonpolar (hydrophobic) groups are dispersed in water. Salts of fatty acids (Fig. 1.11) are examples of amphipathic molecules because of their polarhead (carboxylate group) and nonpolar tail (hydrocarbon chain). In a dilute aqueous solution, such, salts form micelles. These are aggregations of molecules with polar carboxylate ions on the exterior and nonpolar hydrocarbon chains in the interior, which create an internal hydrophobic environment. Cleansing by scap (alkali salts of fatty acids) is accomplished by the entrapment of water-insoluble dirt and grease in the hydrophobic interior of soap micelles. Aside from this practical consideration, micelle formation is important for an understanding of organized biological systems because amphipathic molecules are involved in the formation and structure of membranes.

HYDROGEN ION CONCENTRATION OF BIOLOGICAL SYSTEMS

The aqueous environments of biological systems have hydrogen ion (H^+) concentrations that remain remarkably constant. Maintenance of appropriate H^+ concentration is vitally important for the life of any organism because biochemical reactions are highly sensitive to fluctuations in the concentration of this ion.

In the dynamics of metabolism, the presence and production of many biomolecules continuously affect the amount of H^+ present. If there were no mechanisms to control alterations in H^+ concentration, the effective coordination of the many reactions that constitute metabolism would be rapidly lost. Life processes would then cease. Therefore, an understanding of the mechanisms that strictly control H^+ environment is essential for an appreciation of the firmly regulated aqueous systems that life requires.

DISSOCIATION OF WATER

Water itself contributes hydrogen ions to biological systems because it ionizes to a very slight extent to yield a hydrogen ion and a hydroxyl ion (OH^-). The H^+ does not exist as such in aqueous solution but, rather, in association with a water molecule (proton transfer) to form a hydronium ion (H_3O^+).

$$H_2O + H_2O == H_3O^+ + OH^-$$

However, for convenience and by convention, the dissociation of H_2O is usually written

$$H_2O = = H^+ + OH^-$$

The dissociation can be expressed according to the concepts of the Law of Mass Action, as follows:

$$K_{eq} = (H^+)\ (OH^-)/(H_2O)$$

This expression states that, at equilibrium, the mathematical product of the concentrations (signified by) of the products of dissociation (H^+ and OH^-) divided by the concentration of undissociated water (H_2O) is equal to value known as the equilibrium constant (K_{eq}). From this expression, the concentration of H^+ in pure water can be calculated if the K_{eq} is known.

The equilibrium constant for the dissociation of H_2O at 25°C has been accurately determined to be 1.8×10^{-16} M (Moles/liter), a value that reflects very slight dissociation. The concentration of H_2O in pure, undissociated water is 55.5 M, obtained by dividing 1,000 g (weight of 1 liter of H_2O) by 18 g (gram molecular weight of H_2O) This value of 55.5 m is generally accepted as the concentration of undissociated water (H_2O) in the above equilibrium equation, because dissociation is so slight. It is also usually regarded as the concentration of H_2O in dilute aqueous solutions. Since the actual concentration does not differ significantly from that of pure H_2O.

When the values for K_{eq} and (H_2O) are substituted into the above equation, the following ion product is obtained.

$$(H^+)\ (OH^-) = (1.8 \times 10^{-16})\ (55.5) = 1.0 \times 10^{-16}\ M$$

Since (H^+) = (OH^-) when water dissociates, the following expression is derived:

$$(H^+)^2 = 1.0 \times 10^{-14}\ M$$

$$(H^+) = 1.0 \times 10^{-7}\ M$$

Thus, the (H^+) of pure water is 1.0×10^{-7} M. Although the value is expressed only to two significant figures, it is accurate enough for most purposes.

P^H Scale

Hydrogen ion concentrations are routinely expressed as pH values. The pH expresses (H^+) as a logarithmic function and is defined as the negative logrithm of the hydrogen ion concentration.

$$pH = \log 1/(H^+) = {}^-\log (H^+)$$

Although pH is an expression of hydrogen ion activity (as measured by pH meters) no distinction is usually made between ion activity and concentration. As seen in Table 1.6 pH values offer a convenient means of stating widely varying (H^+) in small positive numbers. The pH scale of 0 to 14 accommodates H^+ concentrations of 1 M to 1×10^{14} M. A P^H value of 7 (H^+) of pure water) is considered the neutral pH. Increases in (H^+) (pH values smaller than 7)

Table 1.6 The pH scale

H^+ (M)	pH	
1.0	0	↑
0.1	1	
0.01	2	
0.001	3	Acidic
1×10^{-4}	4	
1×10^{-5}	5	
1×10^{-6}	6	↓
1×10^{-7}	7	Neutral
1×10^{-8}	8	↑
1×10^{-9}	9	
1×10^{-10}	10	Basic
1×10^{-11}	11	
1×10^{-12}	12	
1×10^{-13}	13	
1×10^{-14}	14	↓

produce acidic conditions, and decreases (pH values larger than 7) result in basic, or alkaline, conditions. In the use of pH values, it is important to remember that the numbers represent a logarithmic function and that therefore a decrease or increase of one pH unit, e.g., $pH^7 - pH^6$, represents a teniold difference in (H^+).

SECTION B

NUCLEIC ACIDS

Introduction

In 1869 Fredrick Miescher, a young Swiss studying with the eminent Hoppy-Seyler in Germany, isolated nuclei from white blood cells and found that they contained a hitherto unknown phosphaterich substance which he named *nuclein*.

When nuclein was established to be acidic in nature, its name was changed to nucleic acid, Research on these bimolecules in the first decades of this century revealed that they, like proteins, are polymers. The monomeric unit of a nucleic acid is called a *nucleotide*; thus nucleic acids are also referred to as *Polyucleotides*. A nucleotide can be hydrolyzed to a nitrogenous base, a five carbon sugar, and phosphoric acid. *Additionlly* it was discovered that there are two types of nucleic acid; ribo nucleic acid (RNA) and deoxyribonucleic acid (DNA). RNA differs chemically from DNA in that it has ribose as its pentose and DNA has 2-deoxyribose. RNA also has the nitrogenous base. Uracil, whereas DNA has thymine, a methylated derivative of uracil. This latter distinction is not absolute as once thought, since it is now known that certain RNAs also contain thymine. Continued studies on nucleic acids revealed that the nucleotide units are linked to one another by phosphodiester bonds forming macromoleculer structures which, in the case of

$$
\begin{array}{c}
\text{O} \\
\| \\
-\text{O}-\text{P}-\text{O}^- \\
| \\
\text{O}_-
\end{array}
$$

Phosphodiester linkage

DNA, can have moleculer weights of billions. Both types of nucleic acids are present in all plants and animals. Viruses also contain nucleic acids; unlike a plant or animals, a virus has either RNA or DNA, but not both.

Although the chemistry of nucleic acids was seriously studied after their discovery, 75 years passed before the biological significance of these macromolecules was realized. The suggestion made by Avery and his colleagues, in 1944, that DNA is genetic material was the first specific biological role proposed for nucleic acid. The announcement, although belatedly appreciated, laid the cornerstone for a new and productive era of biological research on nucleic acids. As for RNA, it was not until 1957 that a specific cellular function for this nucleic acid (RNA involvement in protein synthesis) was established. (It should be noted, however, that RNA had been identified earlier as the genetic material of some viruses. The emergence of moleculer biology emphasized the biological eminence of both DNA (the gene) and RNA, whose different cellular species have prominent roles in protein synthesis (gene expression). Understanding the chemistry and the biochemical functioning of nucleic acids has been one of the most meaningful achievements of twentieth century science.

LEVELS OF NUCLEIC ACID STRUCTURE

Nitrogenous Bases (Fig. 1.12)

Nucleic acids contain two classes of nitrogenous bases, purines and pyrimidines. The two purines of RNA and DNA are adenine and guanine, which chemically are amino- and oxy-substituted purine molecules. With respect to pyrimidines cytosine is common to both RNA and DNA, whereas uracil occurs in RNA, with thymine being its counterpart in DNA. The three bases are substituted pyrimidine molecules Because of the tautomerism exhibited by purines and pyrimidines, a pH-dependent equilibrium exists between keto (lactam) and enol (lectim) forms of the bases. The two tautomeric structures for cytosine are depicted in Fig. 1.12. At physiological pH, the keto forms of the bases predominate.

The above five nitrogenous bases were once believed to account for the total base composition of the animal and plant nucleic acids. It is now known that other bases called modified bases, also occur in polynucleotide structures. For example, transfer RNAs (tRNAs) are a class of nucleic acids which contain a significant percentage of modified bases in their

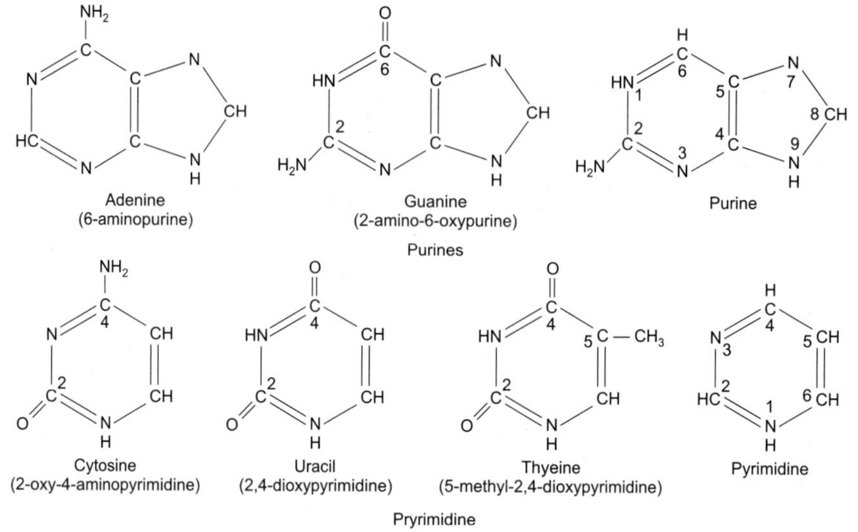

Adenine
(6-aminopurine)

Guanine
(2-amino-6-oxypurine)

Purine

Purines

Cytosine
(2-oxy-4-aminopyrimidine)

Uracil
(2,4-dioxypyrimidine)

Thyeine
(5-methyl-2,4-dioxypyrimidine)

Pyrimidine

Pryrimidine

(A)

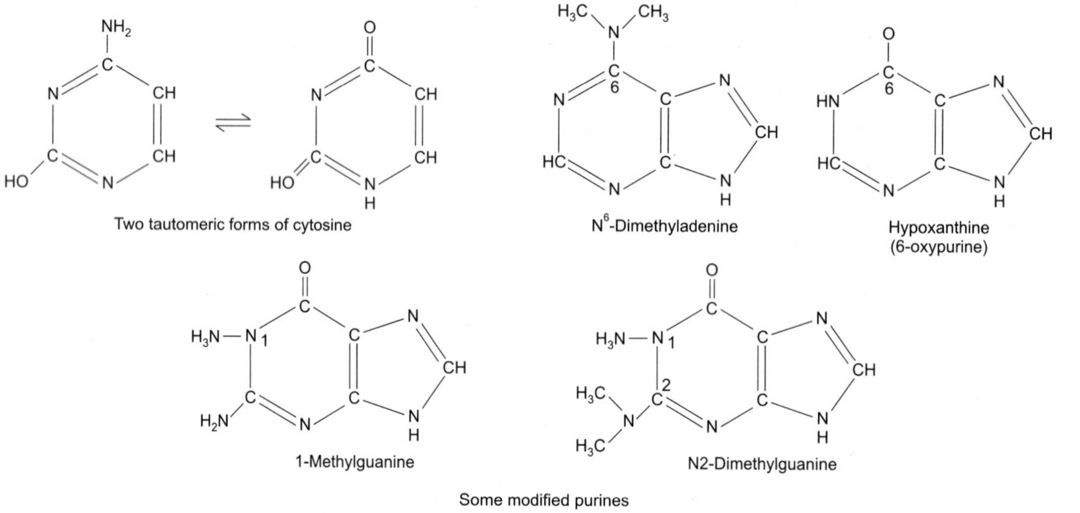

Two tautomeric forms of cytosine

N^6-Dimethyladenine

Hypoxanthine
(6-oxypurine)

1-Methylguanine

N2-Dimethylguanine

Some modified purines

(B)

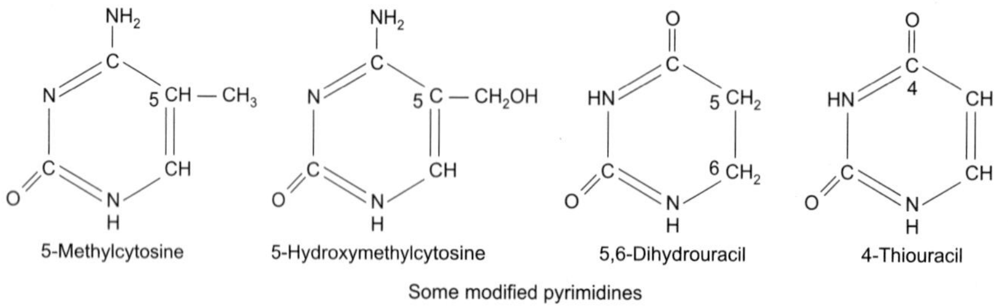

5-Methylcytosine

5-Hydroxymethylcytosine

5,6-Dihydrouracil

4-Thiouracil

Some modified pyrimidines

(C)

Fig. 1.12

structure Fig. 1.12 depicts some of the modified purines which have been identified in nucleic acids. Methylation is the most common form of purine modification. Methylation of purines (particularly of adenine) in DNA is now known to occur in the genetic material of microorganisms and animals, and it is belived that plant genomes will also be shown to have methylated purines. Some naturally accouring forms of modified pyrimidines are shown is Fig. 1.12. 5-Methylctosine is a common component of plant and animals DNA; in fact, up to 25 per cent of the cytosyl residues of plant genomes are methylated. The DNA of T-even bacteriophages of *Escherichia coli* has no cytosine but instead has 5-hydroxymethylcytosine and its glucoside derivatives.

Evidence has now been obtained that methylation of sperm and oocyte DNA during gametogenesis (production of sperm and ova) is responsible for the differential imprinting essential for successful embryonic development. An embryo diploid for the male or female set of chromosomes does not develop only the combination of a male and a female genetic complement can bring about embryonic development. Thus, the particular methylation patterns of sperm and ova genetic material suggest that the paternal and maternal genetic contribution are distinctive, e.g., in gene expression, and that each is imprinted to remember its parental origin throughout the development and life of an organism. The evolving scientific scenario, therefore, implicates the methylated bases, once called minor bases, as major determinants in the genetic regulation of the life processes.

Nucleosides (Fig. 1.13)

Nucleosides are compounds that have a purine or pyrimidine covalently bonded to D-ribofuranose (ribonucleosides) or to 2-deoxy-D-ribofuranous (deoxyribonucleosides) in an N-B-glycosidic linkage. Bonding involves the hemiacetal group of C-l' of the pentose and the N-9 nitrogen atom of a purine or N-1 of a pyrimidine. In chemical nomenclature, the carbon atoms of a sugar in a nucleoside are identified by primed numbers to distinguish them from the atoms of the nitrogenous base. One nucleoside found in tRNAs that has a different linkage is pseudouridine which has the C-1 of ribose attached directly to the C-5 of uracil. The uracil moiety at physiological pH has one oxy group in the keto form and the other in the enol form.

Table 1.7 lists the common names of the major nucleosides of RNA and DNA. Thymidine is the historical name of the deoxyribonucleoside of thymine, long thought to be the only naturally occurring nucleoside of that base. Now that thymine has been identified in tRNAs, ribothymidine and deoxythymidine are the preferred and least confusing name. 3' or 5'- carbon of a deoxyribonucleotide. Naturally occurring nucleotides are commonly 5'- monophosphates (Fig. 1.7) listed the teo trival names and abbreviations used for each of the 5'- nucleotide of the five major bases of RNA and DNA; their classification as acids is an alternative older nomenclature.

The di- and triphosphate derivatives of monophoshonucleosides also occur naturally. Nucleosides triphosphates are substrates.

Table 1.7 Trivial names of nucleosides:

Base	*Ribonucleoside*	*Deoxyribonucleside*
Adenine	Adenosine	Deoxyadenosine
Guanine	Guanosine	Deoxyguanosine
Uracil	Uridine	Deoxyuridine
Cytosine	Cytidine	Deoxycytidine
Thymine	Ribothymidine or thymine ribonucleoside	Deoxythymidine or thymidine

Table 1.8 Nucleoside 5'-monophosphates of RNA and DNA

Ribonucleoside 5'-phosphates	*Deoxynucieoside 5'-phosphate*
Adenosine 5'-monophoshate 5'-Adenylic acid, AMP	Deoxyadenosine 5'-monophosphate 5'-Deoxyadenylic acid, damp
Guanosine 5'monophsphate 5'-Guanylic and GMP	Deoxyguanosine 5'-Monophosphate 5'Deoxyguanlic acid, dGMP
Cytidine 5'-monophosphate 5'- cytidylic acid, CMP	Doxycytidine 5'-monophoshate 5'-Deoxycytidylic acid, Dcmp
Uridine 5'-monophosphate 5'-Uridylic acid, UMP	Deoxythymine 5'-monophosphate 5'-Deoxythymidylic acid, dTMP

Nucleotides (Fig. 1.13)

Nucleotides are phosphate (phosphoric acid) esters of nucleosides. There are classes of nucleotides since the phosphate ester can be at the 2'-, 3'-, or 5'- carbon of ribonucleotide or at the for the synthesis of nucleic acids and the triphosphates of ribonucleosides (ATP, GTP, CTP, and UTP) furnish the energy needed for many biochemical reactions. Adenosine 5'mono-, di-, and triphosphates are a critically important group of biomolecules because of their key roles in the conservation and utilization of chemical energy in all biological systems. As shown in the phosphorus atoms of ATP are identified as X, B, and Y. Ribonucleotides of adenine also occur in the structure of certain coenzymes, such as FAD and NAD+. Another nucleotide of adenine, cyclic AMP is an intracellular mediator of the physiological action of many hormones and a metabolic regulator of some biological processes in *E. coli*. Ribonucleotides of other nitrogenous bases and their derivatives are also used for intermediary metabolic purposes, e.g., uridine 5'-diphosphate linked to glucose (UDPG) is a glucose donor thus, as a class of biomolecules, nucleotides are utilized for a variety of biologically distinct functions. This is especially true of adenine-containing molecules because they, like glucose and its derivatives, are extensively involved in many aspects of the chemical process of life.

Use of Nucleoside Analogues as Drugs (Fig. 1.14)

In medical therapy against certain diseases use is made structural analogues of nucleosides as drugs. For example, in seeking a therapy for the treatment of acquired immune deficiency syndrome (AIDS) patients two nucleoside analogues, 3'-azidodeoxyyhymidine (AZT) AND 2',

Adenosine
(9,β-o-ribofuranosyl adenine)

Deoxythymidine
(1,β-2′-ribofuranosyl thymine)

(A) A ribo- and a dexoyribonucleoside

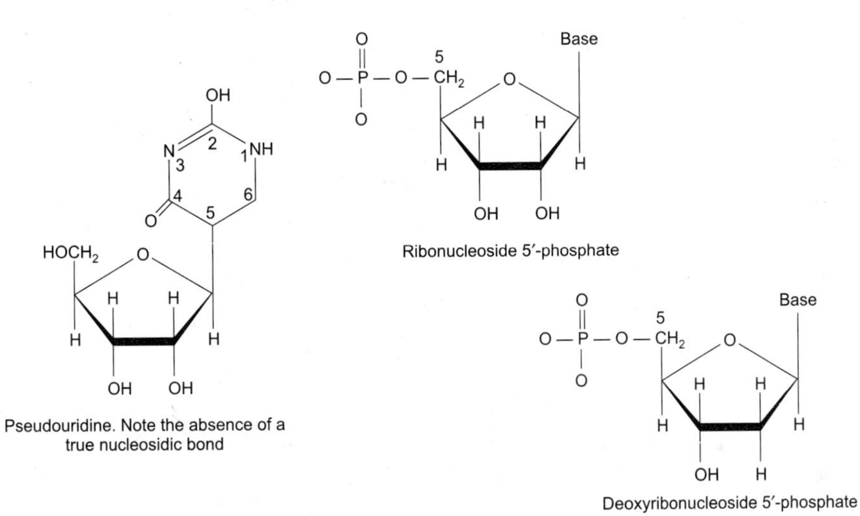

Pseudouridine. Note the absence of a true nucleosidic bond

Ribonucleoside 5′-phosphate

Deoxyribonucleoside 5′-phosphate

(B) General structures of a ribo- and a deoxyribonucleotide

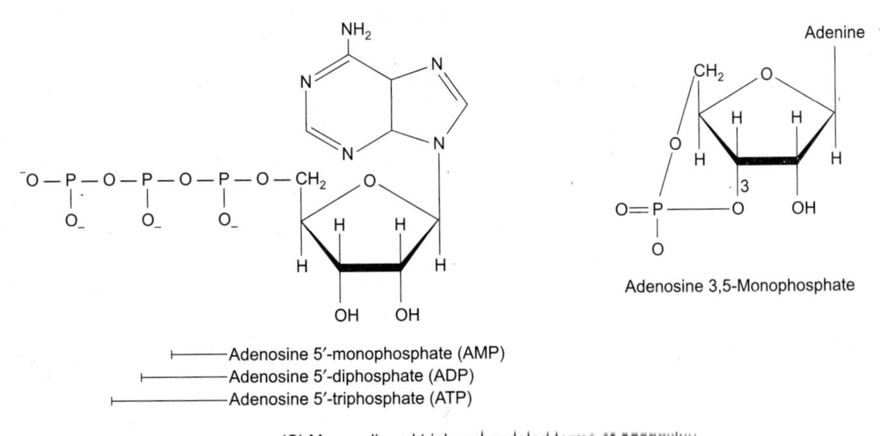

Adenosine 3,5-Monophosphate

(C) Mono-, di, and triphosphorylated forms of adenosine

Fig. 1.13

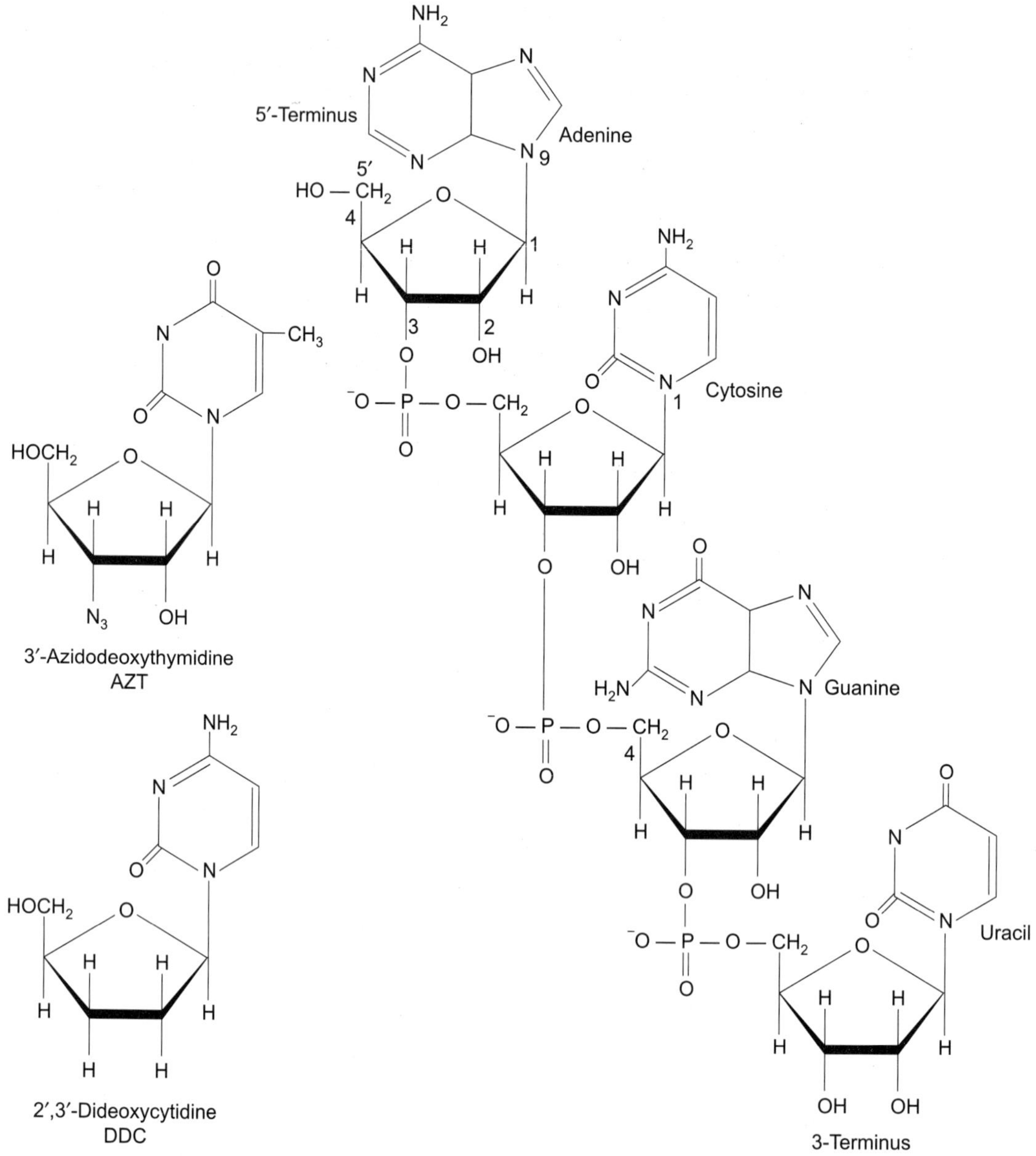

5′-Terminus

Adenine

Cytosine

Guanine

Uracil

3-Terminus

3′-Azidodeoxythymidine
AZT

2′,3′-Dideoxycytidine
DDC

Fig. 1.14 A tetranucleotide with 3′- and 5′ hydroxyl termini.

3'- dideoxycytidine (DDC) are being used. The disease is caused by the human immunodeficiency virus (HIV) which is an RNA virus that relies on a specific enzyme, an RNA-dependent DNA polymerase, for its replication. When given to patients, AZT and DDC are converted into their triphosphate forms, which can then compete with dTTP and dCTP respectively as substrates for DNA synthesis. When incorporated the analogues terminate DNA synthesis because their lack of a 3'-hydroxyl group prevent continued elongation of the DNA molecule being synthesized because the DNA -synthesizing enzyme of HIV is considerably more sensitive to AZT and DDC inhibition than are the analogues enzymes of its host cell, these nucleoside analogues (if their effect are not serious) offer hope as an effective treatment (not a cure) for AIDS.

Linked to glucose (UDPG)is a glucose donor thus, as a class of biomolecules, nucleotides are utilized for a variety of biologically distinct functions. This is especially true of adenine-containing molecules because they, like glucose and its derivatives, are extensively involved in many aspects of the chemical process of life.

Polynucleotide Structure (Fig. 1.14)

Figure 1.14 shows the structure of a tetranucleotide fragment of RNA containing the four major nitrogenous bases of the nucleic acid. The individual nucleotides are joined by phosphodister bonds which involve the 3,'- and 5'- carbon atoms of the ribosyl residues and are called 3'-5'-phosphodiester linkages. The some 3'-5'-diester linkage occurs between the deoxyribosyl residues of DNA. Like proteins, many nucleic acids being linear polymers have two termini one called the 5'-terminus (or 5'-end) and the other the 3'-terminus. As illustrated in the ribosyl moiety at the 5'-terminus has a free C-5' (not in phosphodiester linkage) and the ribosyl residues at the 3-terminus a free C-3. It is not unusual however for cellular polynucleotides to have the terminal C-5 in a monophosphate ester linkage.

For convenience polynucleotide are usually depicted or written in an abbreviated from. The abbreviated structure of the tetranucleotide illustrated in is shown in the same figure. The tetranucleotide can also be written 5'-ApCpGpU-3'-or 5'-A-C-G-U-, i.e., forms that emphasize the sequence and content the nitrogenous bases of the molecule. To indicate a 5'-phosphate group, the same tetranucleotide would be written 5'-pApCpGpU-3'- or 5'-pA-C-G-U-3'; a" p" following the "U" of the tetranucleotide would designate a 3'-phosphate group.

Until the 1950s tetranucleotide with the four different bases in varying arrangement, such as the one shown in were believed to be the basic repeating unit of RNA structure and the analogous tetradexoynucleotide the basic unit of DNA. The concept that DNA of chromosomes possessed the genetic information of the cell was not seriously considered because the macromolecule's assumed monotonous structure of repeating tetranucleotide was thought incapable of having the chemical versatility required to account for the thousand of diverse hereditary traits. Proteins were the candidate of choice because they have a greater number of different monomeric units (amino acids) which can be arranged in an almost endless variety of sequences. Thus because of tetranucleotide hypothesis, it was difficult to many scientists to accept the idea of DNA as genetic material when it was first proposed in 1944 by Avery and his colleagues. The subsequent finding the DNA bases are not arranged in repetitive tetranucleotide sequences but occur in many different sequences aided in dispelling doubts about the biological function of the nucleic acid.

Structure of DNA (Fig. 1.15 & 1.16)

Elucidation of the double-helical structure of DNA by Watson and Crick was made feasible by several pertinent finding of other scientists. That most DNA possess a common and possibly by the same three-dimensional structure was suggested by the similarity in X-ray diffraction patterns obtained by Maurice Wilkins (Nobel Prize, 1962) and Rosalind Franklin on DNA from different sources. Their finding further indicate a helical structure containing two or more poly-deoxyribonucleotide. Independently, Erwin Chargaff had provided the crucial observation that, in DNA obtained from a wide variety of organisms, the molar ratio of adenine to thymine and that of guanine to cytosine were close to unity. These results indicated that a specific relationship must exist between the two bases within each of the ratios. The results of titration studies had also suggested that hydrogen bonding between bases of the deoxynucleotide was responsible for the association of the long DNA strands of the molecule. Watson and Crick using these available data to construct models of DNA, proposed what is now confirmed as the correct structure, the double-helical in the structure, two strands of DNA are would together about a common axis in a right-handed double-helix. The spirals of the helix are the deoxyribosy residues linked by phosphodiester bonds, and the bases if the deoxyribonucleotides project perpendicularly into the center of the helix. The helix is maintained by hydrogen bonding between bases of the two DNA strand. The hydrogen bonding specifically occurs between an adenine of one strand and a thymine of the other and, similarly, between guanine and cytosine. The two DNA strand are thus complementary because its complementary hydrogen-bonding base on the other matches, every base of one strand. The two strand are also antiparallel, meaning that the 3;-5'-diphosphodiester linkages are in opposite directions in the two strands, as indicated.

As illustrated in Fig. 1.17 three forms of DNA, called A-DNA, B-DNA and Z-DNA, have been described. The A and B forms are right-handed helices, and the most recently described form, Z- DNA, is a left-handed helix. In the A form, the base pairs are about 20 inclined with the axis of the helix, with base pairing occurring every 2.7 A. In the B form the base pairs lie about perpendicular (2 tilt) to the helix axis and occur every 3.4 A. The width of the both forms is 20 A. Because of the differences in basepair spacing, B-DNA is longer and thinner than A-DNA. In solution DNA assumes the B form and, under conditions of dehydration, the A form. Weather regions of A-DNA exist in cells is still uncertain. However double- stranded RNA, e.g., viral double-stranded RNA molecules assumed an A-like structure because the 2'-hydroxyl groups of the ribosyl moieties of RNA prevent formation of the B form.

Formation of Z-DNA the third form, is facilitated by sequences of alternation purines and pyrimiaines, e.g., GCGCGCGC. The base-pair spacing in this left-handed helical version of DNA is 3.7 A, and the bases are about 6 inclined with the axis of the helix.

The width of the Z forms 16 to 17 A. Methylation of cytosyl residues in alternation CG sequences helps to bring about the transition of B-DNA to Z-DNA because the added hydrophobic methyl groups stabilize the Z-DNA structure. Only small segments of a cell's DNA

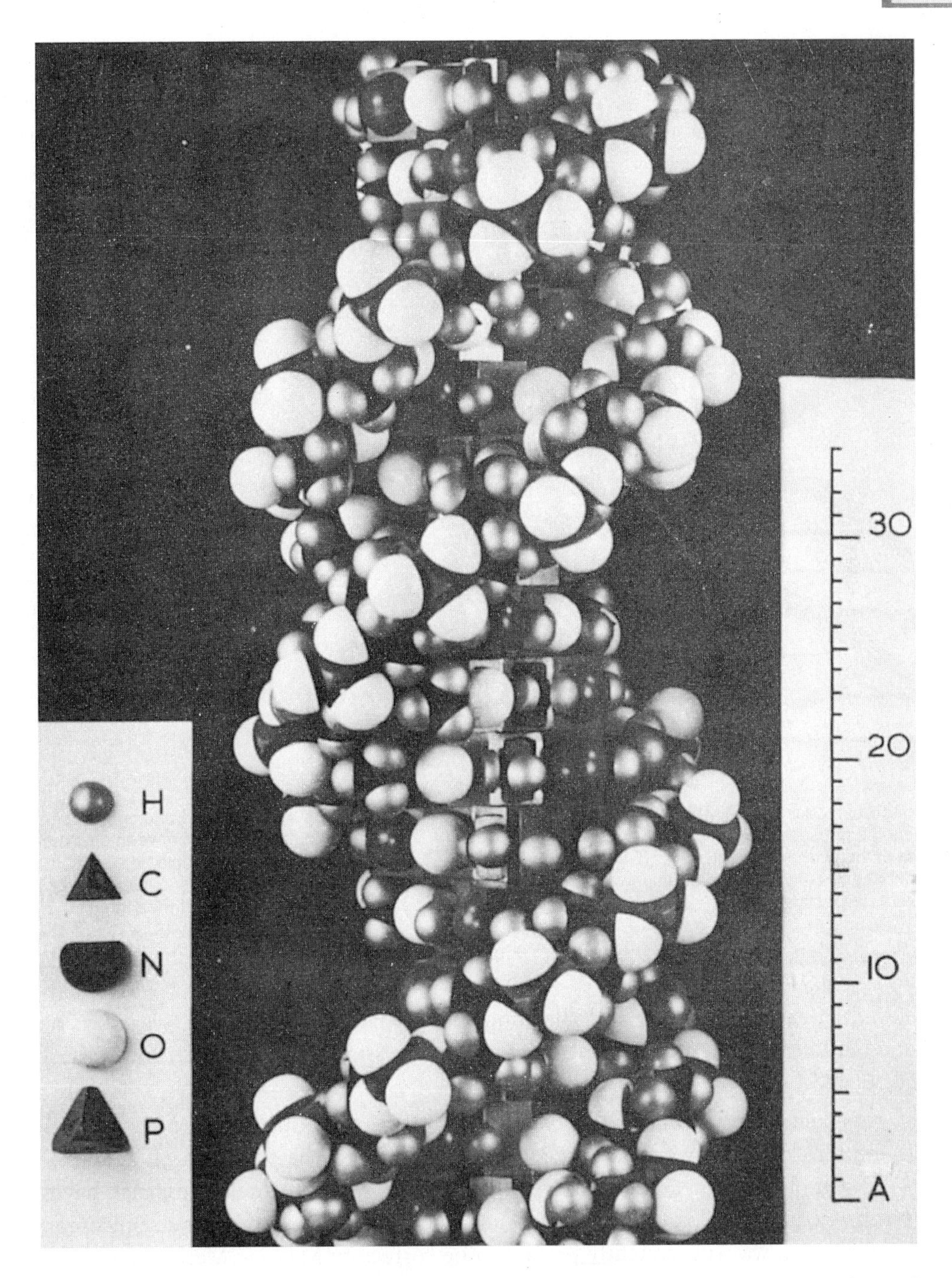

Fig. 1.15 A space-filling model of double-helical DNA. the size of the circles reflects the van der Waals radii of the different atoms (Courtesy of M.H.F. Wilkins).

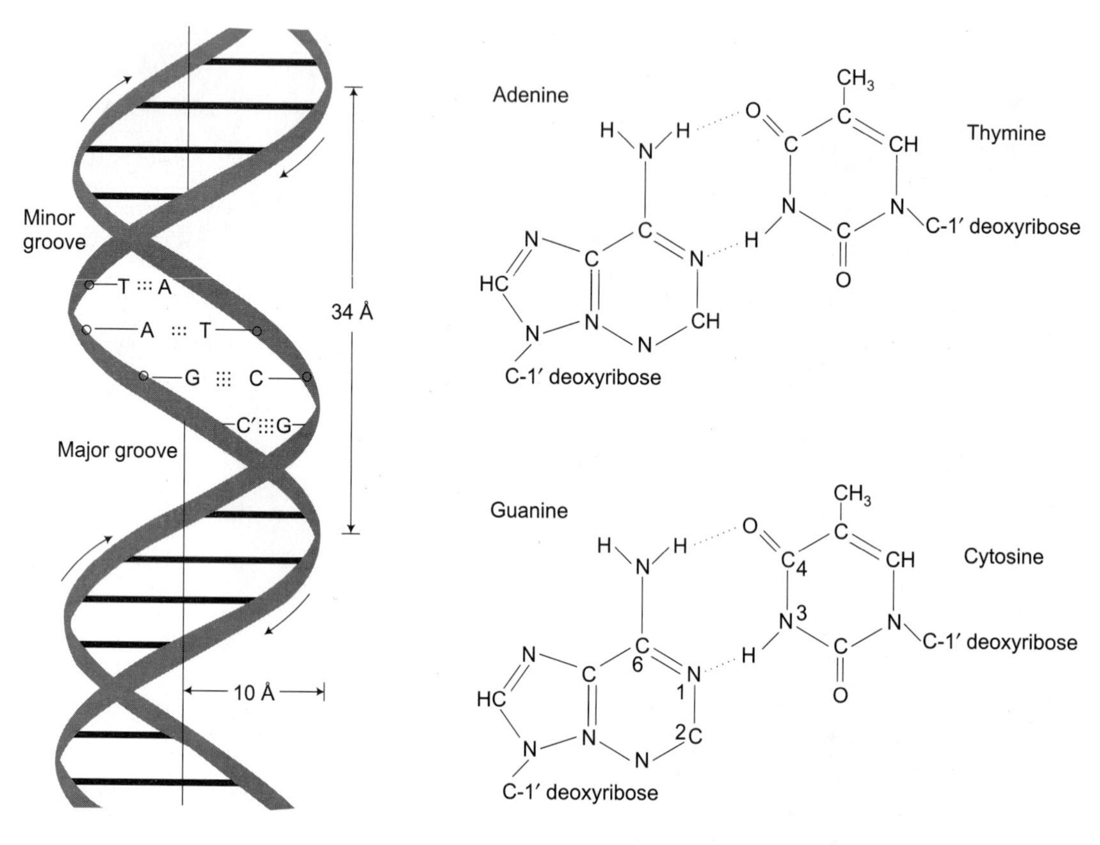

Double-helical model of DNA. The two ribbons represent the phosphate (P) and deoxyribose (D) moieties of the two DNA strands. A ::: T and G ::: C represent adenine-thymine and guanine-cytosine base pairs, respectively.

Base pairing by hydrogen bonding between adenine and thymine and between guanine and cytosine in a DNA double helix. An adenine-thymine pair has two hydrogen bonds and a guanine-cytosine pair has three.

Fig. 1.16

are believed to exist as Z-DNA, and the function of this particular DNA structure remains enigmatic (although a role in the control of gene expression, dominated by methylation, has been postulated.)

The complementary nature of the two strands suggested a mode of DNA, replication to Watson and Crick. They proposed that replication occurs by the synthesis of complementary strands on each of the two parent strands yield two daughter DNA molecules, each having one newly synthesized and one parental strand in its structure. This proposal for a semiconservative mode of replication was subsequently proved to be correct by Mathew Meselson and Franklin Stahl.

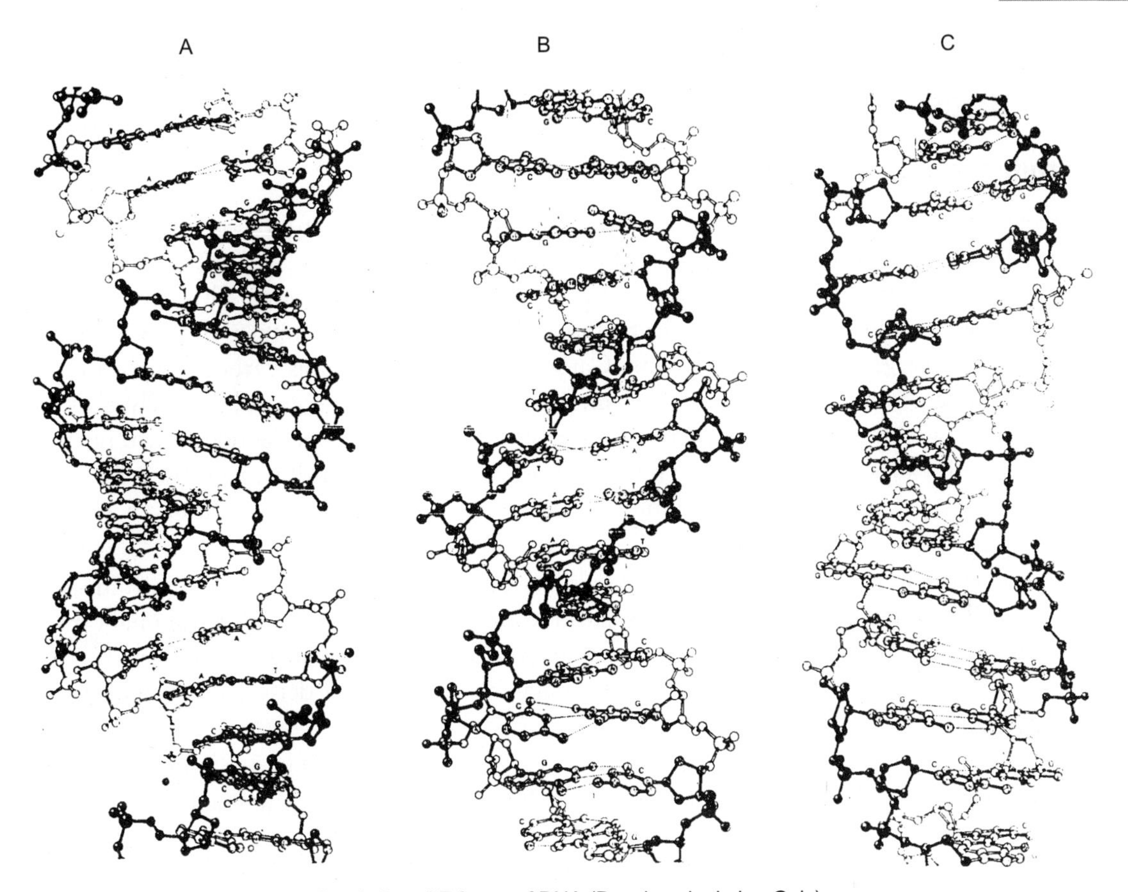

The A, B and Z forms of DNA (Drawings by Irving Geis)

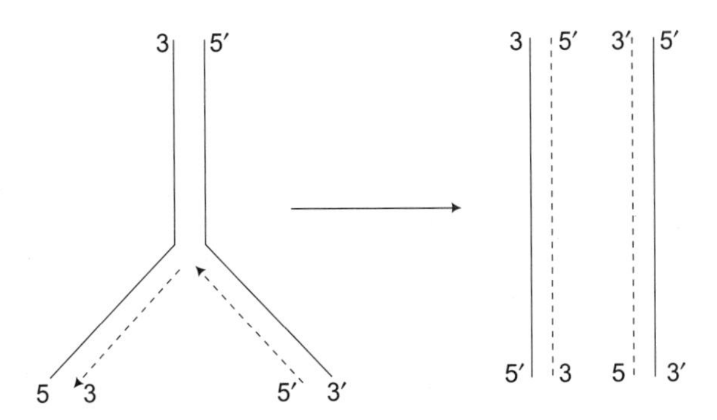

Diagram of the semiconservative replication of DNA. Solid lines represent the two parental strands of DNA; the dashed red lines the newly replicated (daughter) strands.

Fig. 1.17

SECTION C

CARBOHYDRATES

Introduction

Carbohydrates, on the basis of mass, are the most abundant class of biomolecules in nature. More commonly known as sugar, carbohydrates are the major end products of the photosynthetic incorporation of inorganic carbon (CO_2) into living matter. This conversion of solar energy into the chemical energy of biomolecules makes carbohydrates the primary source of metabolic energy for living organisms. Carbohydrates also serve as carbon source for the synthesis of other biomolecules and as polymeric storage forms of energy. Additionally, they are components of many structure ane cellular secretory material and of nucleotide which, in turn, are also used a variety of function. Thus, in living systems, carbohydrates are used for many different purposes and are an outstanding example of the functional versatility a class of biomolecules can possess.

Carbohydrates are defined as polyhydroyaldehydes or polyphdroxyketons and derivatives. A carbohydrates is an aldehyde (-CHO) if the carbonyl oxygen is associated with a terminal carbon atom and a ketone (= C= O) if the carbonyl oxygen is bonded to an internal carbon. This definition avoids classification by empirical formula and encompasses derivatives such as deoxy- and amino-sugars.

In nature carbohydrates occur as monosaccharides (individual or simple sugars) oligosaccharides, and polysaccharides. Oligosaccharides are generally defined as molecules containing two to ten monosaccharide units and polysaccharide are larger polymeric carbohydrates some of which have molecular weights of several millions. It is within the context of these three classifications that the broad subject of carbohydrates is presented.

MONOSACCHARIDES (Fig. 1.18 & 1.19)

Nomenclature and Fischer Projections

A monosaccharide is identified by the number of carbon atoms it contains by its carbonyl functional group, e.g., aldose if an aldehyde and ketone. The smallest carbohydrates are generally considered the three carbon sugars, glyceraldehydes (the aldotriose) and dihydroxyacetone (the ketotriose). In case of glyceraldehydes there are two stereoisomers the D- AND L-forms stereochemically, all sugars can be related to one of these two isomers, in the classification of monosaccharides having more asymmetric carbon atom, the symbols D and L always refer to the configuration of the asymmetric carbon most distal to the carbonyl carbon (analogous to the one in glyceraldehyde). In biological systems, the D- forms of sugars predominate.

Number of carbon atoms in a sugar	Name
3	Triose
4	Tetrose
5	Pentose
6	Hexose
7	Heptose

1CHO
H — 2C — OH
3CH_2OH

D-Glyceraldehyde
an aldotriose

1CH_2OH
2C = O
3CH_2OH

Dihydroxyacetone
a ketotriose
(no asymmetric carbon)

Carbon sugars

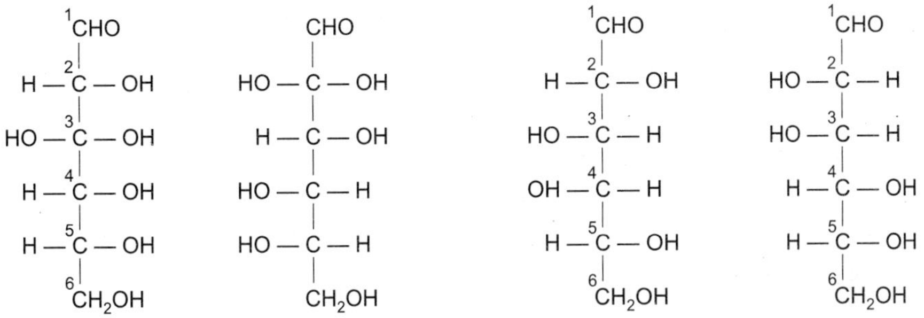

1CHO
H — 2C — OH
HO — 3C — OH
H — 4C — OH
H — 5C — OH
6CH_2OH

D-Glucose
asymmetric carbon atom

CHO
HO — C — OH
H — C — OH
HO — C — H
HO — C — H
CH_2OH

L-Glucose

Enantiomers of glucose, an aldehexose

1CHO
H — 2C — OH
HO — 3C — H
OH — 4C — H
H — 5C — OH
6CH_2OH

D-Glactose

1CHO
HO — 2C — H
HO — 3C — H
H — 4C — OH
H — 5C — OH
6CH_2OH

D-Mannose

Two naturally occuring aldohexoses

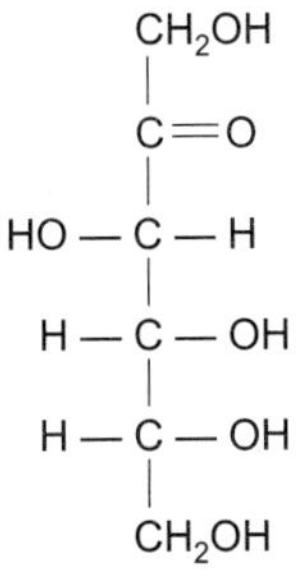

CH_2OH
C = O
HO — C — H
H — C — OH
H — C — OH
CH_2OH

D-Glucose
asymmetric carbon atom

A 2-ketohexose

Fig. 1.18

The systematic nomenclature for monosaccharides becomes cumbersome for larger sugars because they have two or more asymmetric centers and hence, increased numbers of stereoisomers. Thus trivial, or common names of these carbohydrates are generally used. Glucose (also called dextrose) the most prevalent organic compound in nature, is an aldohexose containing four asymmetric carbons and is therefore one of 16 possible streroisomers (2). In a Fischer projection of a sugar, A D or L assignment is based on the asymmetry at the penultimate

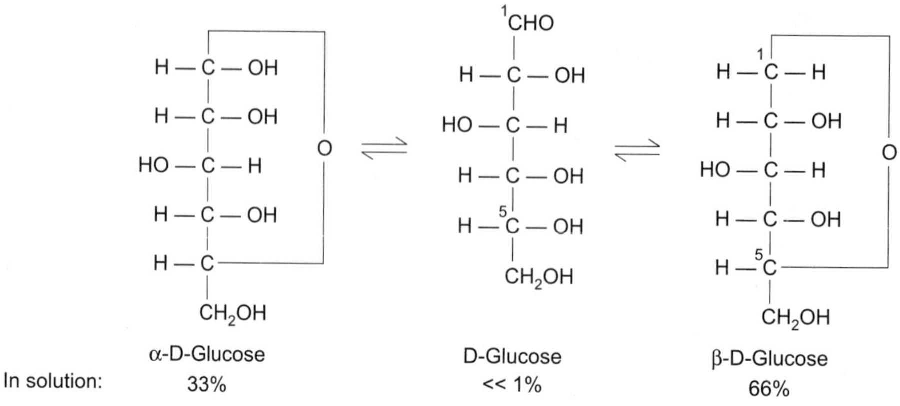

	α-D-Glucose	D-Glucose	β-D-Glucose
In solution:	33%	<< 1%	66%

Mutarotation of glucose (as Fischer projections)

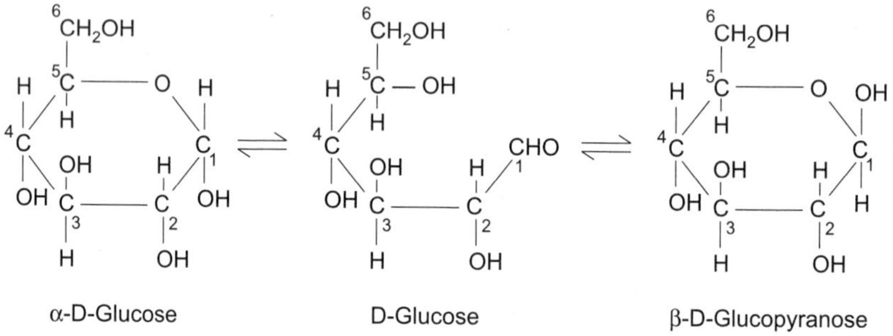

α-D-Glucose ⇌ D-Glucose ⇌ β-D-Glucopyranose

Mutarotation of glucose (as Haworth projections)

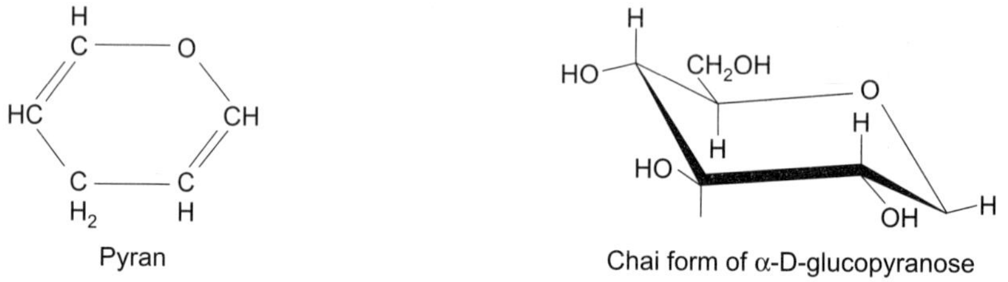

Pyran

Chai form of α-D-glucopyranose

Fig. 1.19

carbon atom of the molecule, which is C-5 in an aldohexose. As in the formula for D-glyceraldhyde the hydroxyl group is written to the right of C-5 to designate the D form of glucose. The enantiomer (mirror image isomer) of D-glucose which has the opposite configuration at each of the four asymmetric centers.

D-glucose and D-mannose, two other aldohexose frequently found in living organisms, are also stereoisomers of D-glucose. As in the case of D-glucose, the placement of the hydroxyl group of C-5 identifies these sugars as D-isomers. D-glucose and D-galactose are also referred to as epimers because two monosaccharides differ only in the configuration at a single carbon atom (C-5). D-glucose and D-mannose are also epimers because the differences is again limited by the asymmetry at a single carbon atom (C-2). However, D-galactose and D-mannose are not epimers, since they: differ in the asymmetry at both C-2 and C-4. D-fructose sometimes called levulose is common biological sugar and it is one of eight isomeric 2-ketohexoses, having three asymmetric carbons.

Closed Ring Structure

As indicated in x- or B- forms of D-glucose are interconvertible; this spontaneous phenomenon is called mutarotation, and it occur with anomers of many monosaccharides. With respect to D-glucose either anomer when dissolved in water will slowly undergo mutarotation until an equilibrium mixture consisting of about one third x-D-glucose and two third B-D-glucose is attained.

Haworth Projections

A more representative projection of the closed-ring structure of monosaccharides is that first suggestion by–Walter H. Haworth (Nobel Prize 1937) in 1925. Illustates the Haworth projection of X- and B-D-glucose. Abbreviated versions of this type of projections for D-glucose either would not show the carbons except C-6 or additionally, would omit the hydrogen's and indicate hydroxyl groups by short lines. Haworth projections more closely approximate the predominant "chair" structure of glucose and other hexose sugars that exist in solution. Because the six membered ring of glucose is similar to that of pyran, it is called the pyranose form of glucose; he anomers are referred to as either X- or B-D-glucopyranose.

OTHER MONOSACCHARIDES OF IMPORTANCE (Fig. 1.20 & 1.21)

Both types of nucleic acid, DNA and RNA have specific pentoses in their structure. D-ribose is the sugar component of RNA and 2-deoxy-D-ribose that of RNA. Deoxyribose is a deoxy sufar one that has a hydrogen atom substituted for a hydroxyl group at one or more of its carbon atoms. Deoxy sugars provide an example of one of several distinct classes of modified monosaccharides, which have specific structure biological functions. Another class of modified monosaccharides is that of the amino sugars, e.g., D-glucosamine which have an amino group substituted for hydroxyl group. N-acetyl-D-glucosamine is a common derivative of glucosamine. Amino sugar frequently occur in large quantities in structural materials. Chitin of the exoskeletons of invertebrates, e.g., crabs and lobsters is a linear polymer of N-acetyl-D-glucosamine. Other substituted monosaccharides include phosphorylated forms of sugars, e.g., D-glucose I-phosphate and D-glucose 6-phosphate. Phosphorylated sugars are often utilized for interacellular metabolic processes.

D-Ribose β-D-Ribofuranose 2-Deoxy-D-ribose α-D-Deoxy-D-ribofuranose

D-Ribose and D-deoxyribose

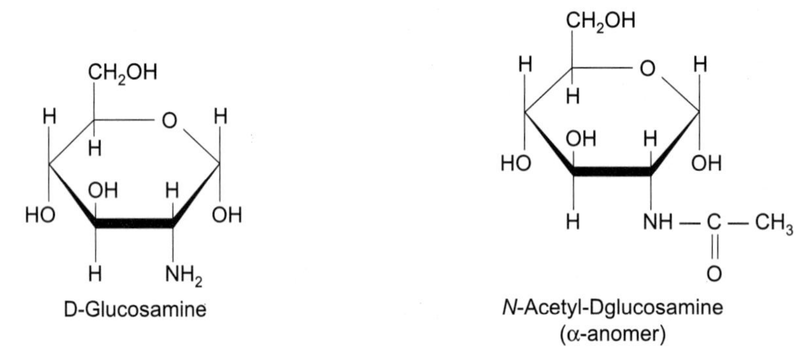

D-Glucosamine

N-Acetyl-Dglucosamine (α-anomer)

Two modified monosaccharides

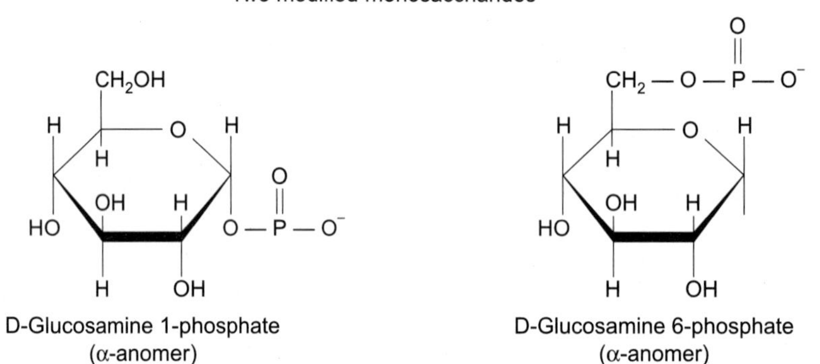

D-Glucosamine 1-phosphate (α-anomer)

D-Glucosamine 6-phosphate (α-anomer)

Two phosphorylated forms of glucose

Fig. 1.20

Oxidized (acidic) and reduced (alcoholic) forms of sugars are also present in living systems. Uronic acids, such as glucuronic acid are often constituents of acid mucopolysaccrides, high moleculer weight compounds frequently found in large quantities in structural tissues such as bone, cartilage and skin.

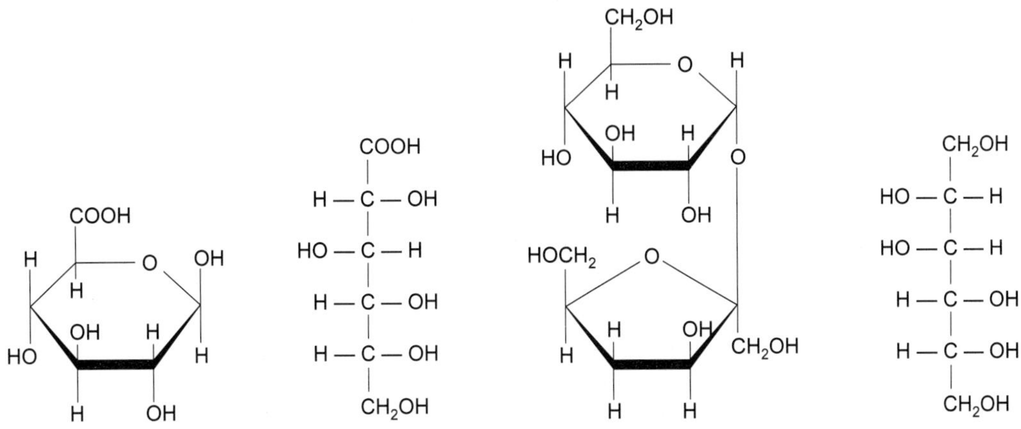

D-Glucuronic acid

A uronic acid

D-Gluconic acid

An aldonic acid

O-α-D(glucopyranosul-(1-2)-
β-D-fructofuranoside

Sucrose

D-Mannito

An alditol

Fig. 1.21

OLIGOSACCHARIDES

In nature the most abundant oligosaccharides are the disaccharides sucrose and lactose. Sucrose (table sugar) occurs in plant. Where it is synthesized from D-glucose and D-fructose. A glycosidic linkage between the anomeric C-1 of A-D-glucose and the anomeric C-2 of B-D-fructose joins the two monosaccharides by an oxygen bridge producing an X-(1-2) linkage. Lactose the carbohydrate of mammalian milk is composed of D-galactose and D-glucose. In this disaccharide the glycosidic linkage between the anomeric C-1 of B-D-galactose and the nonanomeric C-4 of D-glucose is B-(1-4).

The synthesis of lactose by lactose synthetase a heterogeneous dimer, is a novel example of the modification of catalytic specificity by dimmer formation, (a form of allosteric conformational change). One of the two protomers is an enzyme (galactosyl transferase). That occurs widely in animal tissues, mammary glands during pregnancy, by itself it catalyzes the following reaction;

UDP-galactose + N- acetylglucosamine –N-acetyllactosamine + UDP

UDA is uridine diphosphate, which serves as a molecular carrier of carbohydrates in certain enzymatic reactions. For milk production the second protomer of lactose synthesis A-lactalbumin, is synthesized specifically in mammary tissue and this protein's interaction with galactosyl transferac alters the substrate specificity so that the dimeric enzyme catalyzes the synthesisod lactose in the presence of glucose.

UDP-galactose + glucose –lactose + UDP

A-Lactalbumin occurs only in mammary tissue, hence lactose is unique to mammalian milk.

GLYCOLYSIS

Introduction

One of nature's most primitive means of extracting chemical energy from organic compounds is glycolysis, the anaerobic conversion of a molecule of glucose into two lactic acid molecules (Fig. 1.23). The series of chemical reactions constituting the glycolytic pathway is believed to have evolved in the reduced environment (no oxygen) in which primordial life arose. Glycolysis represents a successful attempt by early forms of life to transform some of the chemical energy of glucose, produced by photosynthetic processes, into other forms containing utilizable metabolic energy e.g., solar-glucose- ATP. The continuing is evident by the fact nearly all forms of life, aerobic and anaerobic, utilize glycolysis. In fact, aerobic oxidation of carbohydrates, a subsequent evolutionary development, requires that sugars first be degraded anaerobically to pyruvate by glycolysis. All carbohydrates, from monosaccharides to polysaccharides, need to be routed metabolically through glycoysis to yield their chemical energy to biological systems.

Alcoholic fermentation, whereby a molecule of glucose is converted into two ethanol and two carbon dioxide molecules is the same anaerobic catabolic process as glycolysis except for the final stages, which produce different end products. In glycolysis, the last metabolic intermediate, pyruvate, is reduced to CO_2. The biological studies on fermentation initiated early in the nineteenth century culminated in the present century with a detailed biochemical description of the anaerboic utilization of carbohydrates. By 1940 the accumulated results of interrelated studies on yeast (alcoholic fermentation) and muscle (glycolysis) provided biochemistry with its first major enzymatic pathway. Glycolysis, for many years, was called the Embden Meyerhof pathway Gustav Embden and Otto Meyerhof (Nobel Prize 1922) were German biochemists who, among a number of other distinguished scientists, made significant contributions toward the molecular elucidation of glycolysis. As a developing science, biochemistry accrued significant benefits from these pioneering studies on carbohydrate metabolism, because they introduced many new techniques and concepts into research on enzymes and intermediary metabolism.

Reaction of the Glycolytic Pathway

The conversion of D-glucose into two lactate molecules involves 11 enzymatic steps. If glycogen, the polysaccharide storage form of carbohydrates in animals, supplies the glucose for glycolysis,

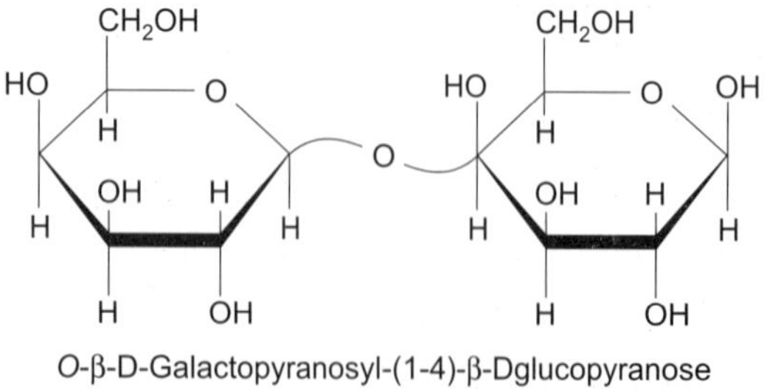

O-β-D-Galactopyranosyl-(1-4)-β-Dglucopyranose

Fig. 1.22

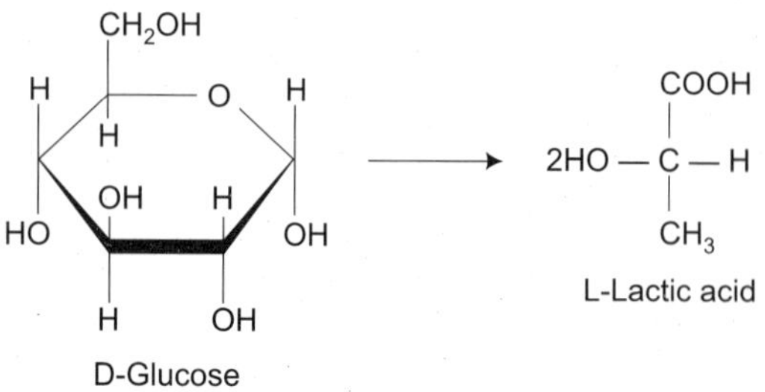

D-Glucose → L-Lactic acid

Fig. 1.23

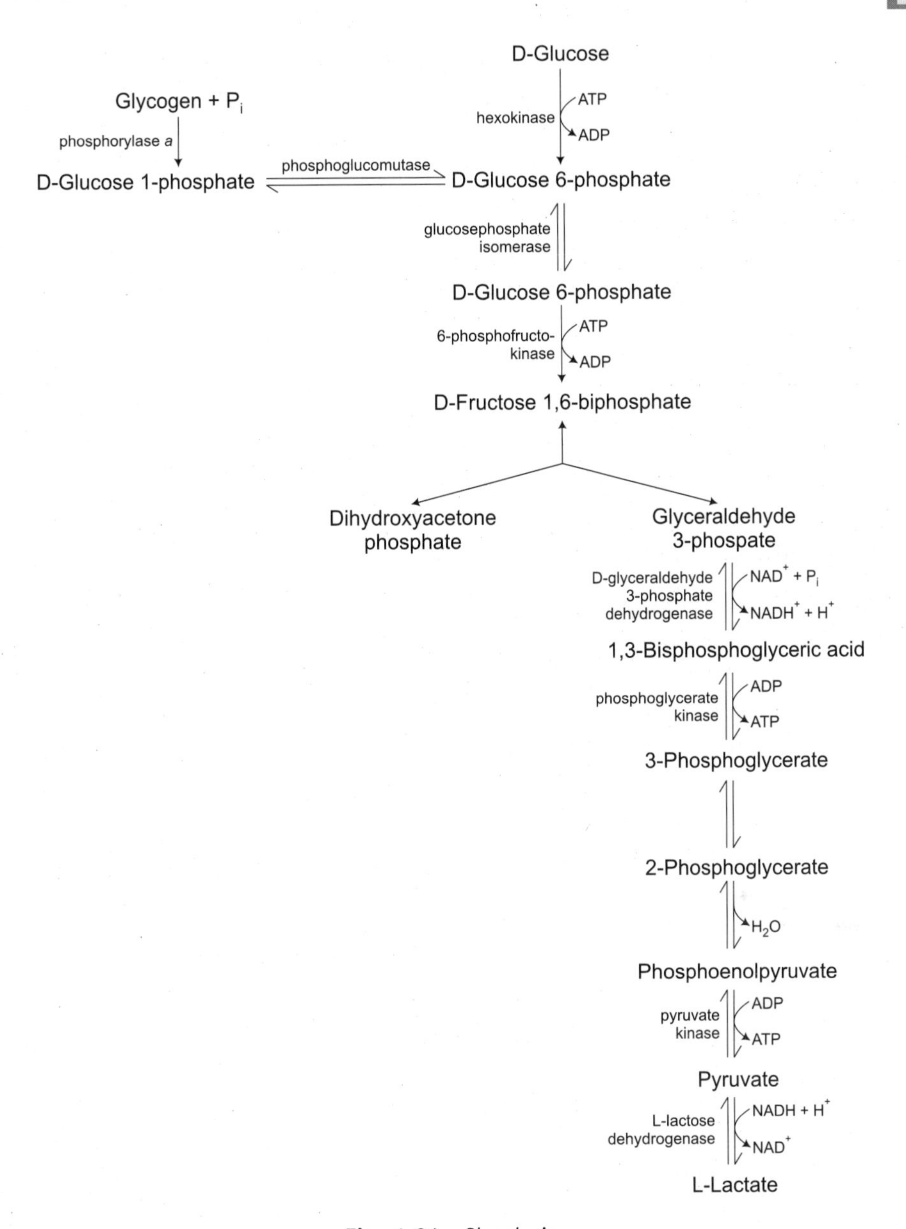

Fig. 1.24 Glycolysis.

then 12 reactions are involved. Elucidation of the pathway entailed the systematic identification and characterization of each enzyme to establish its role in the anaerobic metabolism of glucose. As the individual steps of glycolysis were described, they were fitted together, like pieces of a puzzle; this ultimately led to a biochemical description of the entire process.

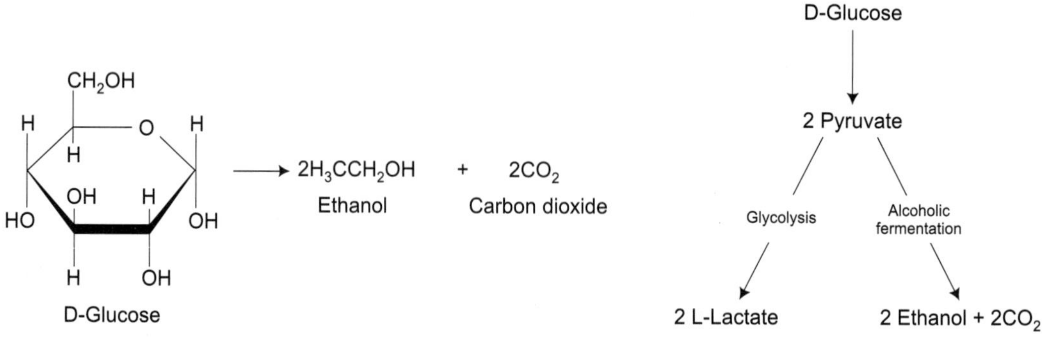

D-Glucose $\longrightarrow$ $2H_3CCH_2OH$ (Ethanol) + $2CO_2$ (Carbon dioxide)

Alcoholic fermentation

D-Glucose → 2 Pyruvate → Glycolysis → 2 L-Lactate; Alcoholic fermentation → 2 Ethanol + $2CO_2$

Difference between glycolysis and alcoholic fermentation

Fig. 1.25

Pentose Phosphate Pathway (Fig. 1.26)

Another fundamentally important pathway of carbohydrate metabolism in most forms of life is the pentose phosphate pathway. The physiological importance of this pathway does not include the generation of ATP (like glycolysis) but, rather, the production of essential NADPH (reducing power) and the synthesis and/or interconversion of various sugars, including the production of glycolytic intermediates. Also known as the hexose monophosphate or phosphogluconate pathway, the cystosolic pentose phosphate pathway was elucidated about the middle of this century by several noted biochemists, including Fritz Lipmann, Efraim Racker, Bernard Horecker, and their colleagues.

The pathway commences with the oxidation of glucose 6-phosphate to 6-phosphoglucono-8-lactone; the reaction is catalyzed by glucose6-phosphate dehydrogenase, which requires $NADP^+$ and Mg^{2+} as cofactors. The enzyme was the first $NADP^+$ -linked dehydrogenase described by Otto Warburg in the early 1930s. In the second reaction, lactonase cleaves the lactone ring to yield the oxidized derivative of glucose 6-phosphogluconate. In the next reaction, 6-phosphogluconate dehydrogenase, which also requires NADP+ AND Mg^{2+}, catalyzes the production of CO_2 and pentose D-ribulose 5-phosphate O_2. The pentose can serve as a substrate for two different reactions, one of which is catalyzed by phosphopentose isomerase to produce D-ribose 5-phosphate.

In some metabolic systems, the pathway's principal function is to produce NADPH and ribose; hence, it terminates after the synthesis of the pentose. The ribose synthesized is used for the production of nucleotides which, in turn, serve as components of RNA and DNA and of coenzyme. The production of NADPH is also a key metabolic aspect of the pathway, since the coenzyme is selectively used for reductive biosynthesis and for furnishing reducing power in biological systems. For example, in humans, NADPH is needed for the normal maintenance of red blood cells, where it serves to generate reduced glutathione, a deficiency of which results

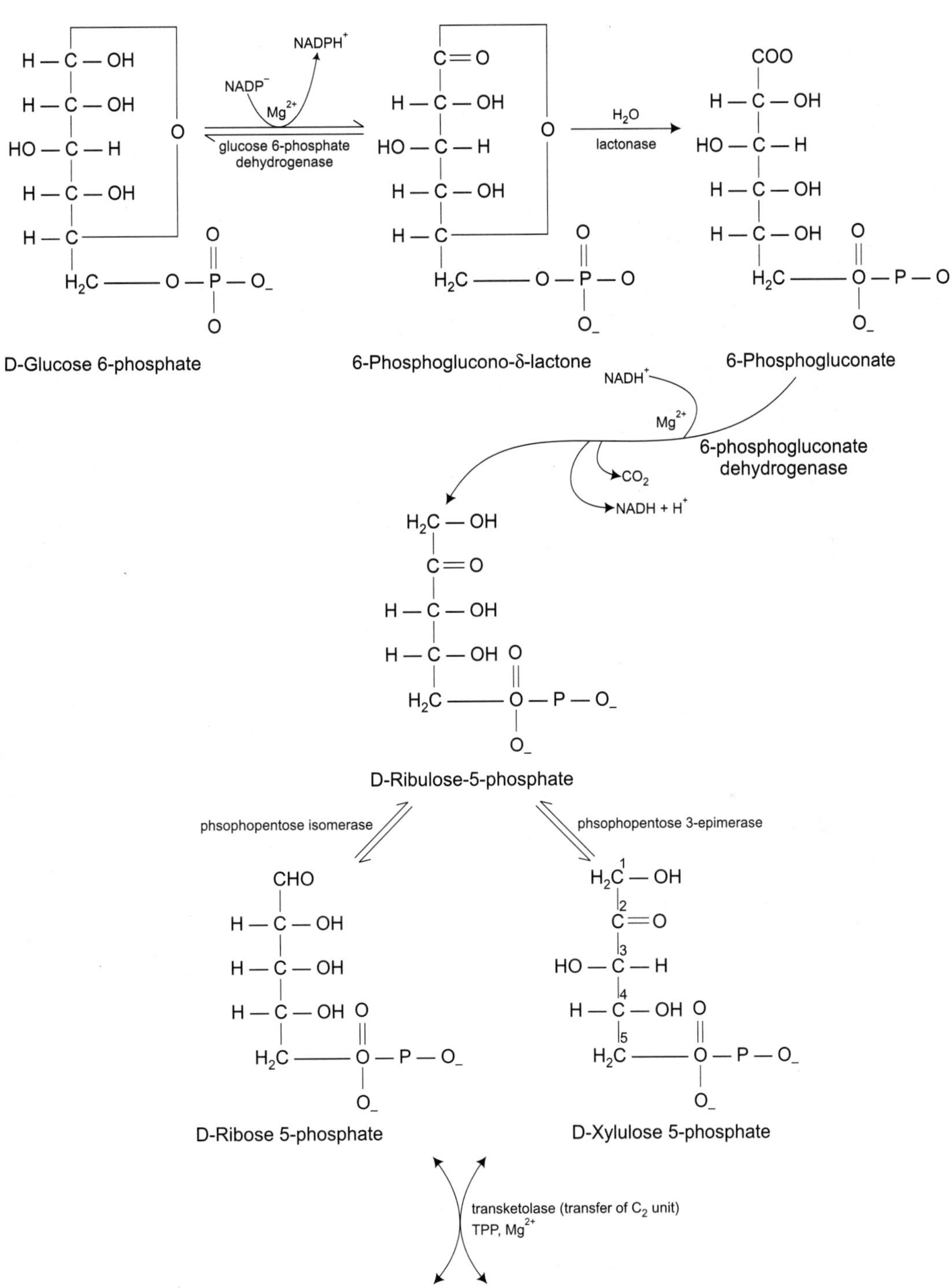

D-Glucose 6-phosphate

$NADP^-$ Mg^{2+} $NADPH^+$

glucose 6-phosphate dehydrogenase

6-Phosphoglucono-δ-lactone

H_2O

lactonase

6-Phosphogluconate

$NADH^+$ Mg^{2+}

6-phosphogluconate dehydrogenase

CO_2

$NADH + H^+$

D-Ribulose-5-phosphate

phsophopentose isomerase

phsophopentose 3-epimerase

D-Ribose 5-phosphate

D-Xylulose 5-phosphate

transketolase (transfer of C_2 unit)
TPP, Mg^{2+}

Fig. 1.26 Contd.

Fig. 1.26 Contd.

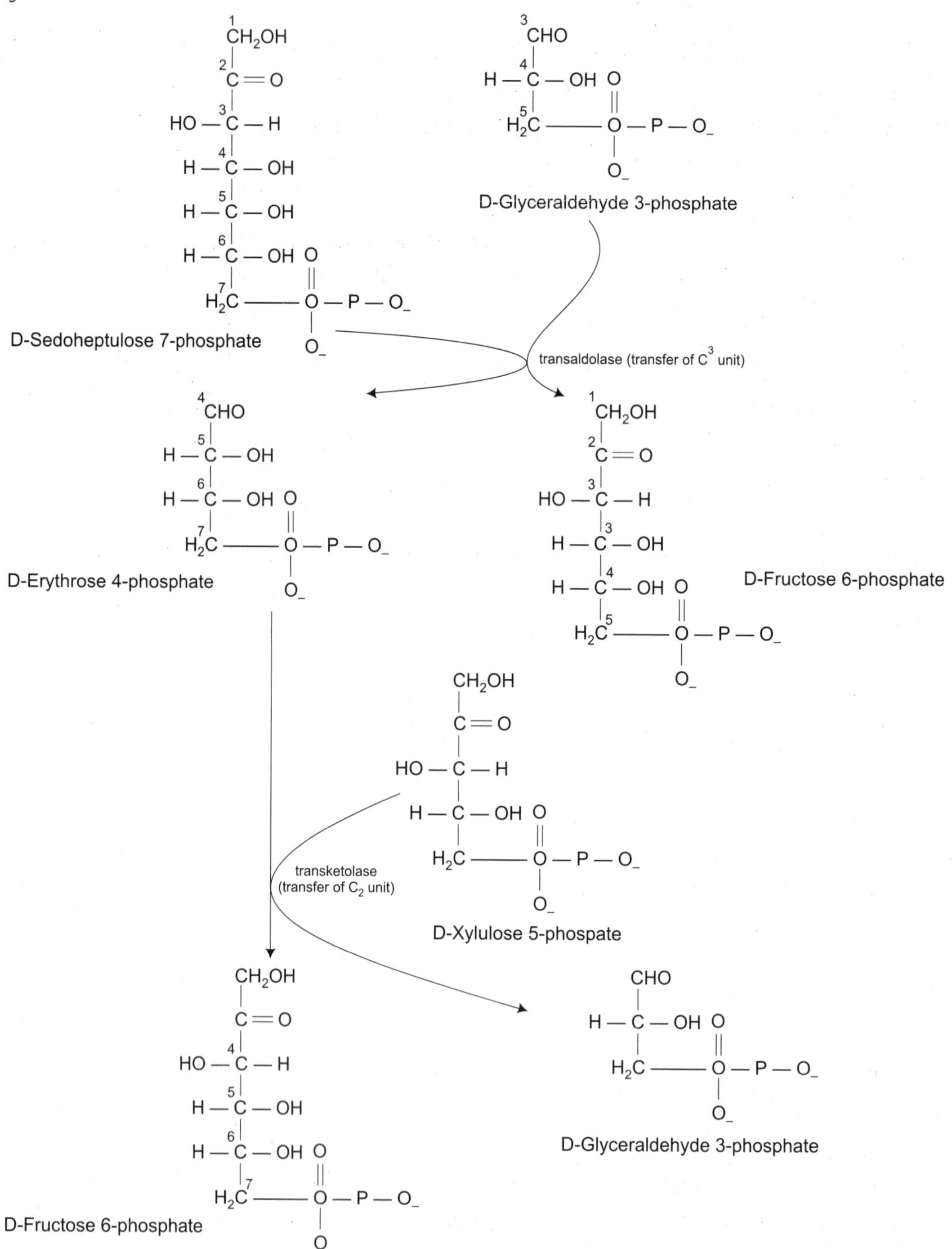

Fig. 1.26

in hemolysis of the cells. The most prevalent human inborn error is glucose 6-phosphate-dehydrogenase deficiency (first reaction of the pathway); this impaired ability to synthesize NADPH is usually manifested clinically as acute hemolytic anemia, induced by drugs (e.g., pamaquine, an antimalarial agent) or by toxic substances (e.g., accidental ingestion of mothballs by an afflicted child). Glucose 6-phosphate-dehydrogenase deficiency is an X-linked recessive trait, most frequently observed in regions of the world where malaria is endemic. It is estimated that 100,00,000 people in the world have the deficiency. Like sickle-cell anemia, a deficiency of the dehydrogenase's activity is believed to protect an individual against malaria and to be another example of an inborn error that, from a populational viewpoint, can be advantageous in a particular environment.

The three other enzymes of the pathway are phosphopentose 3-epimerase, transketolase, and transaldolase. The epimerase, the second reaction that utilizes D-ribulose 5-phosphate as a substrate, converts that pentose (by epimerization of C-3) to d-xylulose 5-phosphate. The transketolase reaction, using D-ribose 5-phosphate and D-xylulose 5-phosphate as substrates, catalyzes the transfer of the glycoaldehyde (CH2OH-co-) unit from xylulose to ribose to produce D-sedoheptulose 7-phosphate (seven-carbon sugar) and D-glyceraldehde 3-phosphate (three-carbon intermediate of glycolysis). Thiamin pyrophosphate (TPP) and Mg^{2+} are required for the transketolase reaction. In the transaldolase reaction, the dihydroxyacetone moiety (CH_2OH-CO-CHOH-) of sedaheptulose is transferred to glyceraldehyde to yield D-fructose 6-phosphate and D-erythrose 4-phosphate (a tetrose). If D-xylulose 5-phosphate then donates- two carbons to D-erythrose 5-phosphate, via the transketolase reaction, D- fructose 6-phosphate and D-glyceraldehyde 3-phosphate are generated. Thus, as summarized in figure , the pentose phosphate pathway can serve to convert pentose into glycolytic intermedia for anaerobic and/ or aerobic respiration. As noted, one glyceraldehydes and two fructose intermediates of glycolysis can be produced from three phosphorylated pentoses.

As seen in Fig. 1.27, the pentose phosphate pathway provides a biochemical route by which three-, four-, five-, six-, and seven-carbon carbohydrates can be synthesized or interconverted. The principal enzymes in this versatile exchange of carbon atoms between carbohydrates are transaldolase (transfer of a C3 unit) and transketolase (transfer of C2 unit) with the latter enzyme being able to utilize various 2-keto and aldose sugar phosphates as donor and acceptor substrates, respectively. The pathway is also of prime importance in the photosynthetic process.

1. D-Xylulose 5-PO_3^{2-} + D-ribose 5-PO_3^{2-} $\rightleftharpoons$ D-sedoheptulose 7-PO_3^{2-} + D-glyceraldehy 3-PO_3^{2-}

2. D-Sedoheptulose 7-PO_3^{2-} + D-glyceraldehyde 3-PO_3^{2-} $\rightleftharpoons$ D-fructose 6-PO_3^{2-} + D-erythrose 4-PO_3^{2-}

3. D-Xylulose 5-PO_3^{2-} + D-ertythrose 4-PO_3^{2-} $\rightleftharpoons$ D-fructose 6-PO_3^{2-} + D-glyceraldehye 3-PO_3^{2-}

Net overall: 2 D-Xylulose 5-PO_3^{2-} + D-ribose5-PO_3^{2-} $\rightleftharpoons$ 2 D-fructose 6-PO_3^{2-} + D-glyceraldehyde 3-PO_3^{2-}

2 (5C) + (5C) 2(6C) + 3C

Fig. 1.27 Production of three-and six-carbon intermediates of glycolysis from pentoses.

AEROBIC SYNTHESIS OF ATP TRICARBOXYLIC ACID CYCLE (TCA)

Introduction

The aerobic oxidation of glucose of CO_2 and H_2O, called respiration, was a critically important evolutionary advancement because it allowed organisms to greatly enhance their ability to extract the chemical energy of the sugar for use in life processes. Although the anaerobic pathway of glycolysis is successful in releasing some of the free energy content of glucose, the amount is only a small fraction of the total energy that can be made available by complete oxidation of the hexose to CO_2 and H_2O The values for the two modes of glucose oxidation are:

Anaerobic: Glucose $\rightarrow$ 2 lactate (ΔG° = -47.0 kCal mol^{-1})

Aerobic: Glucose + $6O_2 \rightarrow 6CO_2 + 6H_2O$ (ΔG° = -686.0 kCal mol^{-1})

Thus, the two molecules of lactate (end product of glycolysis) still contain about 93 percent of the free energy content of the glucose molecule.

The aerobic oxidation of glucose requires the participation of three metabolically interrelated processes (tricarboxylic acid cycle, electron transport, and oxidative phosphorylation), all of which take place in the mitochondria. Whereas glycolysis occurs in the cytosol of eucaryotic cells, aerobic oxidation is a highly specialized cellular phenomenon limited to mitochondria (or the plasma membrane of prokaryotes). Mitochondria are often referred to as the energy factories of the cell. All biomolecular fuels (carbohydrates, lipids, and amino acids) are oxidized in the mitochondria, where the bioenegetic processes for the generation of chemical energy are concentrated.

In aerobic oxidation, the pyruvate produced by glycolysis is not converted into lactate but, rather, into CO_2 and acetyl-Co A; the latter is then utilized by the tricarboxylic acid cycle (TCA). The TCA cycle completes the oxidation of the four remaining carbon atoms of glucose molecule, i.e., two acetyl-CoA to four CO_2 The subsequent production of water, to complete the production of six H_2O per oxidized glucose, involves electron transport, which consists of a series of reactions that transfer the electron pairs derived from the oxidation of the sugar to oxygen to form water. The flow of electron through electron transport is an evolutionary development that ensures a release of electron energy, some of which is conserved through the synthesis of ATP by oxidative phosphorylation; a process physically coupled to the electron transport system.

The elucidation of the TCA cycle was one of the classic achievements of biochemistry during the first half of this century. As was true for the studies on glycolysis, a number of distinguished scientists, beginning in the 1910s, aided in uncovering the molecular details of an oxidative process in anima tissues that proved to be the central oxidative pathway for energy production. The accumulated findings about the oxidation of specific di-and tricarboxylic acids to CO_2 by molecular oxygen prompted the English biochemist Hanz A. Krebs (Nobel Prize, 1953) to study this process of aerobic oxidation. One intriguing observation, first noted by Albert Szent-Gyorgyi (Nobel Prize, 1937), was that when small amounts of added organic acids were oxidized by minced tissues, oxygen consumption was much larger than the quantity needed to oxidize the exogenous acid. Szent-Gyogyi correctly surmised that some endogenous molecule was also being oxidized. Krebs brilliantly deduced and established the relationship between the utilization of endogenous pyruvate and the oxidation process and in 1937, proposed the citric acid cycle, the original name of the TCA cycle. The cycle, including later modifications, is diagrammed in Fig. 1.28. As illustrated, the "catalytic nature" of the cycle is maintained by the continuous regeneration of oxaloacetate. Theoretically, only one molecule of oxaloacetate (or of another intermediate of the cycle) is required for the oxidation of many molecules of pyruvate.

Production of Acetyl – CoA from Pyruvate

The formation of acetyl-CoA is a prerequisite for the entrance of the carbon atoms of pyruvate, produced by all carbohydrates being utilized for energy, into the TCA cycle. As implied by the requirements of six cofactors (Fig. 1.28), acetyl-CoA production is a complicated enzymatic process catalyzed by an aggregate of enzymes called the pyruvate dehydrogenase complex.

SECTION D

PROTEINS — THEIR STRUCTURE AND FUNCTION

Introduction

A eucaryotic or prokaryotic cell contains thousand of different proteins. The most abundant class of biomolecules in cell. Because each species of life possesses a chemically distinct group of proteins millions of different proteins exist in the biological world. The genetic information contained in chromosomes determines the protein composition of an organism and, in this manner, endows members of a species with their macromolecular uniqueness.

As is true of many bimolecular proteins exhibit functional versatility and are therefore utilized in a variety of biological roles. A number of important functions performed by proteins are listed in Table 1.9. The prominent position occupied in biological systems was correctly surmised by the Dutch chemist Gerardus J. Muldur, Who in 1838 introduces the term protein, derived from the Greek word proteins, meaning first of importance.

Mulder commented:

"There is present in plants and animals a substance which ...is without doubt the most important of the known substances in living matter, and without it, life would be impossible on our planet. This material has been named protein"

Although biologically active proteins are macromolecules that range in molecular weight from about 6,000 (single protein chain) to several millions (proteins complexes. All are polymers composed of covalently linked amino acids. The number, chemical nature, and sequences order or amino acids in a protein chain determine the distinctive structure and characteristic chemical behavior of each protein. For this reason an appreciation of the chemical properties of amino acids is a prerequisite for the understanding of biomolecular explanations of the ways proteins function in their biological roles.

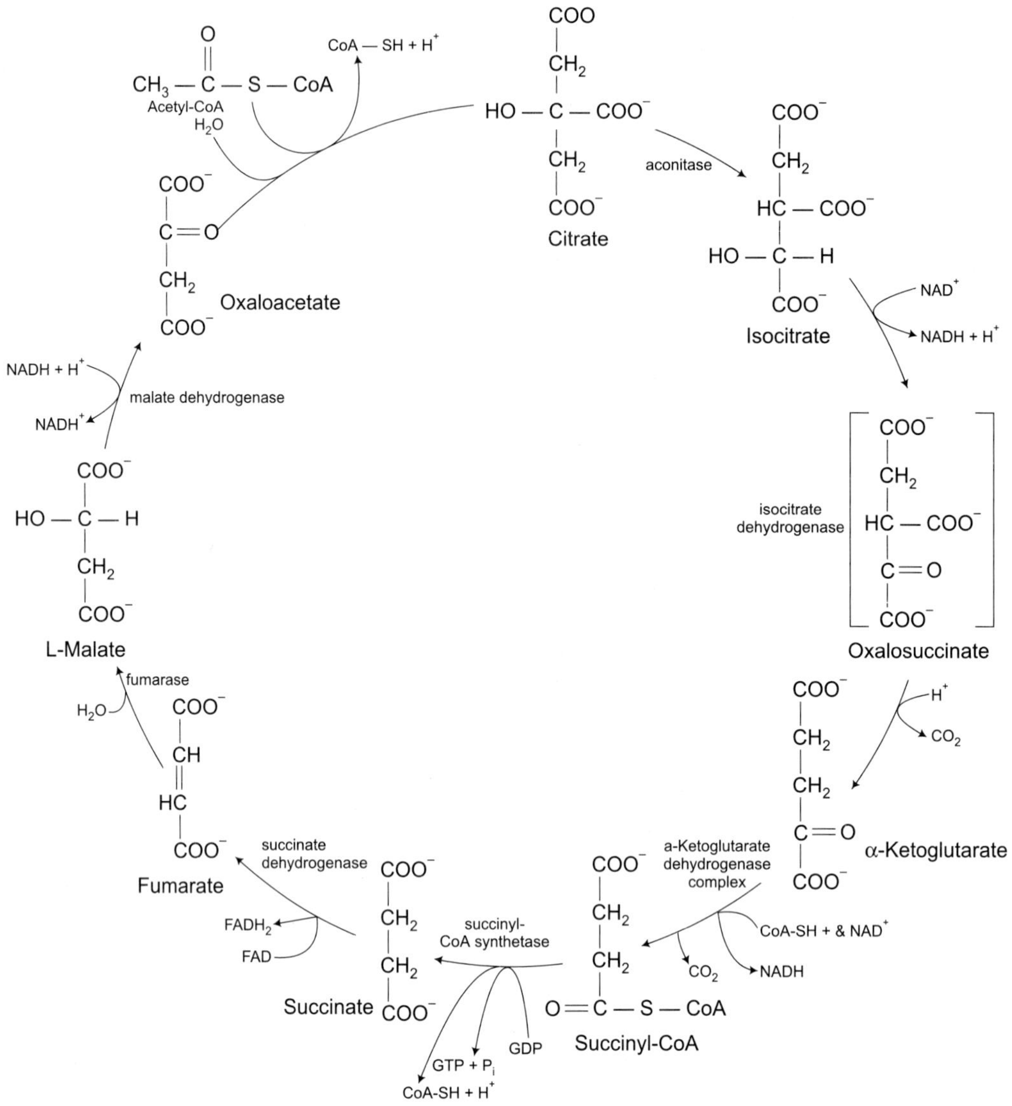

Fig. 1.28 The tricarboxyic acid cycle (TCA). Utilized for energy, into the TCA cycle. As implied by the requirements of six cofactors acetyl-CoA production is a complicated enzymatic process catalyzed by an aggregate of enzymes called the *pyruvate dehydrogenase complex.*

Two General Properties of Amino Acids

Amino acids are small biomolecules (average MW of about 135) having the, general structure depicted in Figure, all A-amino acids are organic acids (A-COOH) containing both an amino (NH_2) group and a hydrogen atom bonded to the A-carbon. They differ from one another by the chemical composition of their R groups (side Chains). Their A-COOH and A-NH_2 groups

Table 1.9 Some biological functions of proteins.

Function	*Examples*
Enzymatic Activity	Glycolate oxicase of glyoxymes Alcohol dehydrogenase in alcoholic fermentation
Transport	Hemoglobin-oxygen transport in blood (vertebrates) Ceruloplasmin-copper transport in blood
Storage	Ferritin-iron storage (spleen) Casein- amino acid storage (milk)
Structure	Collagen-fibrous connective tissue (cartilage, bones, tendons)
Contraction	Myosin-thick filament in skeletal muscle Actin-thin filament in skeletal muscle
Protection	Antibodies-interact with foreign protein Fibrinogen-protein required for blood clotting
Hormonal activity	Insulin-regulator of glucose metabolism Growth hormone-required for bone growth
Toxins	Snake venom-hydrolytic (degradative) enzymes *Clostridium botulinum* toxins-lethal bacterial food toxins

are ionized in solution at physiological pH, with the deprotonated carboxyl group bearing a negative, resonance-stabilized charge and the protonated amino group a positive charge (Fig. 1.29), An amino acid in its dipolar state is called a zwitterions. The dissooiable A-COOH and A-NH_3^+ groups are responsible for the two characteristic pKa' values of A-amino acids. The ionization of the amino acid alanine is illustrated in Figure. At low PH values, molecules of alanine bear a net charge of + 1 because both functional groups are protonated, e.g., at pH, 99 percent are positively charged.

Another distinct property of A-amino acids is the asymmetry of their A-carbon atom. Because four different substituent groups are bonded to this atom, all amino acids except glycine exhibit stereoisomerism. The two optical isomers of alanine are shown in (Fig. 1.30). The D and L nomenclature and structural formulas were adopted by biochemists from those proposed for glyceraldehydes and the Fischer projection of an A-amino acid metric carbon for the D-isomer and the left for the L-isomer. In biochemistry, structural formulas of amino acids are often written not as Fischer projections but in a manner best suited for the particular topic being presented. Figure depicts several structural formulas commonly used. Almost all biological functions involving amino acids have strict requirements for L- isomers; however, there is

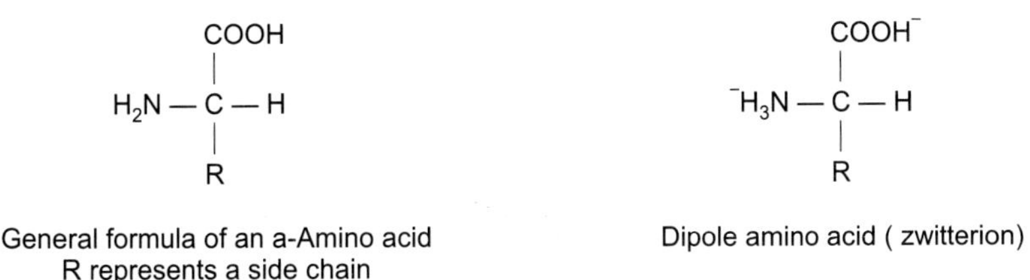

Fig. 1.29

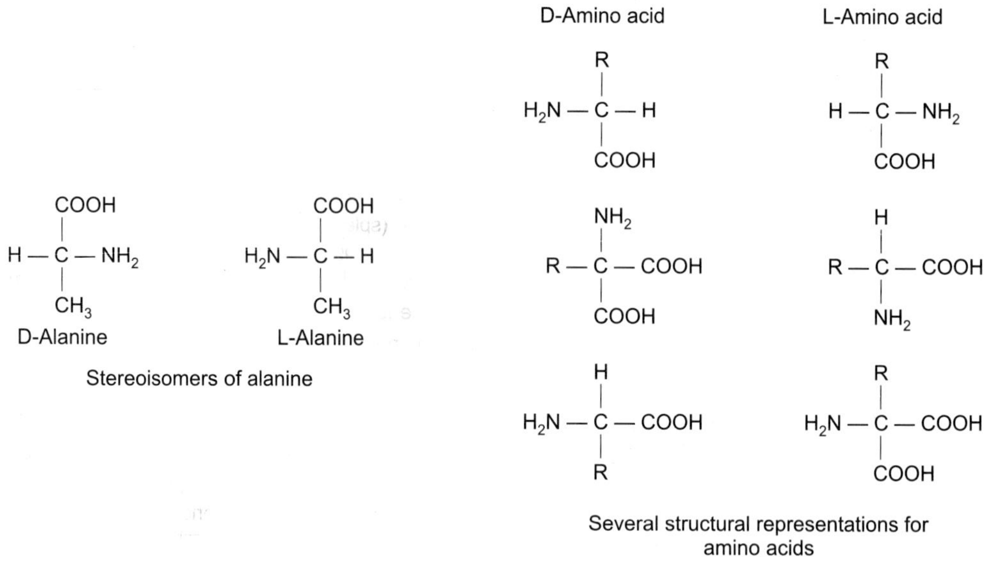

COOH | H—C—NH$_2$ | CH$_3$
D-Alanine

COOH | H$_2$N—C—H | CH$_3$
L-Alanine

Stereoisomers of alanine

D-Amino acid

R | H$_2$N—C—H | COOH

NH_2 | R—C—COOH | COOH

H | H$_2$N—C—COOH | R

L-Amino acid

R | H—C—NH$_2$ | COOH

H | R—C—COOH | NH$_2$

R | H$_2$N—C—COOH | COOH

Several structural representations for amino acids

Fig. 1.30

limited biological use of D-amino acids, since they are present in some biological materials, e.g. certain bacterial cell walls and some antibiotics. A few amino acids possess two asymmetric carbons and, of the four optical isomers, only one is usually biological active.

AMINO ACID RESIDUES OF POLYPEPTIDES

Of over 100 naturally occurring amino acids, only 20 are utilized in polypeptide biosynthesis. Francis Crick coined the term magic 20 to distinguish this set of amino acids required by all living species for protein synthesis. More then 20 amino acids have been identified in protein synthesis. More than 20 amino acids have been identified in protein structures; however in all cases chemical modification of certain of the "magic 20" amino acids) occurring after incorporation into a polypeptide) accounts for these additional residues. For example, certain incorporated praline residues can be converted into hydroxyproline residues a some serine residues into phos-phoserine residues.

The solubility and ionization properties of R groups and influential traits of amino acids and collectively they contribute greatly to the native three-dimensional structures of individual polypeptides. It is on the basis of these two characteristics of their R groups that the 20 amino acids are classifies into four categories, as follows.

Nonpolar Groups (Fig. 1.31)

Eight of the amino acids have nonpolar side chains and, as a group, display varying degrees of hydrophobicity. Four (alanine, valine, leucine and isoleucine) have aliphatic noncyclic R groups (Structural formulas of the amino acids are shown as the ionized forms that predominate at

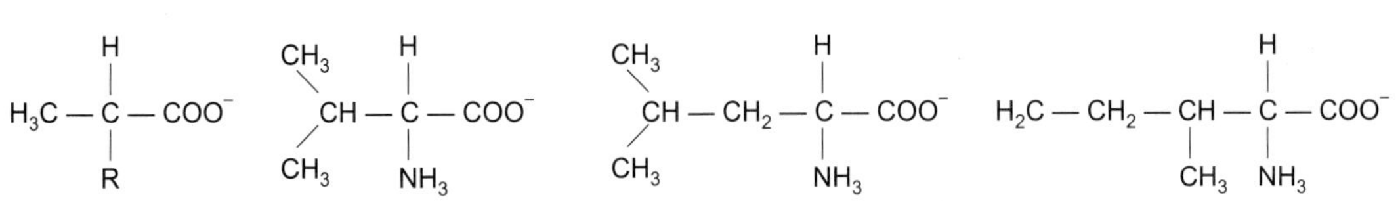

Alanine
Ala, A

Valine
Val, A

Leucine
Leu, L

Isoleucine
Ille, Ii

Fig. 1.31 Alanine and three branched-chain amino acids.

physiological pH. Conventional three and one letter abbreviations are also listed). Alanine is the least hydrophobic of the eight because of its small methyl side chain. Valine, Leucine and isoleucine are called the branched-chain amino acids because of the branching in their aliphatic R groups.

The fifth "amino" acid of this group is proline, which has an aliphatic heterocyclic structure that includes both the R group and the a-nitrogen atom. Thus, proline differs from the other 19 amino acids since it is an imino acid, having an imino (= NH) rather than an amino (-NH_2) group. The sixth amino acid, methionine has a sulphur atom in its nonpolar side chain and is one of two sulpfur-containing amino acids incorporated into proteins. The remaining two amino acids, phenylalanine and trytophan have water-insoluble aromatic rings in their structures. Phenylalanine has phenyl group in its side chain and tryptophan an indole group (a condensed ring composed of benzene and pyrrole); both are considered aromatic amino acids.

Uncharged Polar R Groups (Fig. 1.32 & 1.33)

This class contains seven amino acids which are relatively hydrophilic because of the polar functional groups in their side chains. Glycine, the nonasymmetric amino acid (Figure is sometimes considered nonpolar. However, glycine's small R group (a hydrogen atom) exerts essentially no effect on the hydrophilicity of the molecule. Three amino acids (serine, threonine, and tyrosine) are hydroxylated, and the OH groups contribute to their polarity. Tyrosine, like phenylalanine and tryptophan, is also an aromatic amino acid. Cysteine, the other sulphur-containing amino acid (Figure, is polar because of its sulfhydrl (-SH) group. Often in protein structures, two cysteinyl residues are covalently linked to each other through oxidation of their sulfhydryl form of cysteine is called cystine. Asparagine and glutamine are derived from aspartic acid and glutamic acid and each has a polar amide group in its side chain.

Negatively Charged (Acidic) Polar R Groups (Fig. 1.34)

Both aspartic acid and glumatic acid have a second carboxyl group, which is fully ionized (negatively charged) at physiological pH (Fig. 1.34). This ionization contributes significantly to the polarites of their side chains. Aspartic acid (HA), when its -COOH group dissociates to COO^-, is called aspartate (A-) and, similarly, glumatic acid becomes glutamate. Both are referred to as acidic amino acids because they denote H^+ when placed in solution.

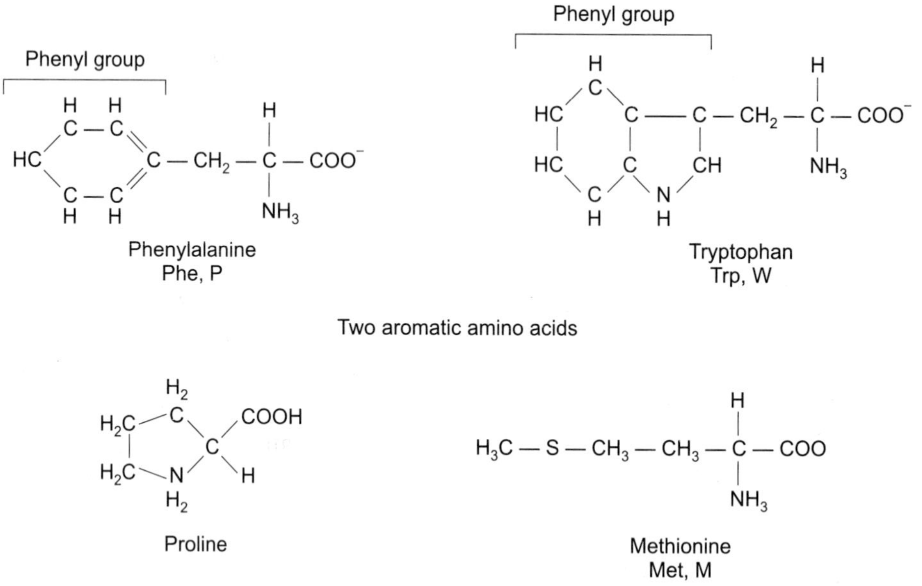

Phenyl group

HC_6H_4—C—CH_2—$C(H)(NH_3)$—COO^-

Phenylalanine
Phe, P

Phenyl group

Tryptophan
Trp, W

Two aromatic amino acids

Proline

$H_3C — S — CH_3 — CH_3 — C(H)(NH_3) — COO$

Methionine
Met, M

Fig. 1.32

Positvely Charged (Basic) Polar R Groups (Fig. 1.34)

Lysine and arginine, two of the three amino acids, have R groups, that are positively charged at physiological pH (Figure). Ionic charges are provided by protonation of the amino group of the e-carbon of lysine and of the guanidinium group of arginine. The third amino acid, histidine, has an imidazolium R group with a pK value of 6.0 and, therefore, is less than 10 percent protonated at pH 7. Histidine's basic properties are clearly marginal. Of the 20 amino acids, histidine is the only one whose isoionic point of about 7.6 is near physiological pH.

Allo Forms of Amino Acids (Fig. 1.35)

Isoleucine and threonine are examples of amino acids having two asymmetric carbons. Among the four possible stereoisomers of each, only the L-isomer is used for protein synthesis. The structural formulas for the isoleucine isomers are illustrated. The allo nomenclature identifies the other isomers (In addition to the D-and L-forms), which are also enantiomers (mirroe images). L-isomlucine and L-allo-isolucine have the same configuration at the A-carbon but opposite configurations at the second asymmetric carbon (B C), the same is true for D-isolucine and D-allo-isolucine. If a compound has more than one asymmetric carbon, the stereo isomers that are not mirror images of each othera re called diastereomers. The case of isolucine, for example, L-isolucine is a enantiomer of D-isolucine and diasteroemer of L-and D-allo-isolucione.

X-ray Crystalography in Determining Protein Structure

In the 1950s, elucidation of the A-helical abd B-pleated sheet structures of proteins and the double helix of the nucleic acid DNA was made possible by x-ray crystallographic studies. The

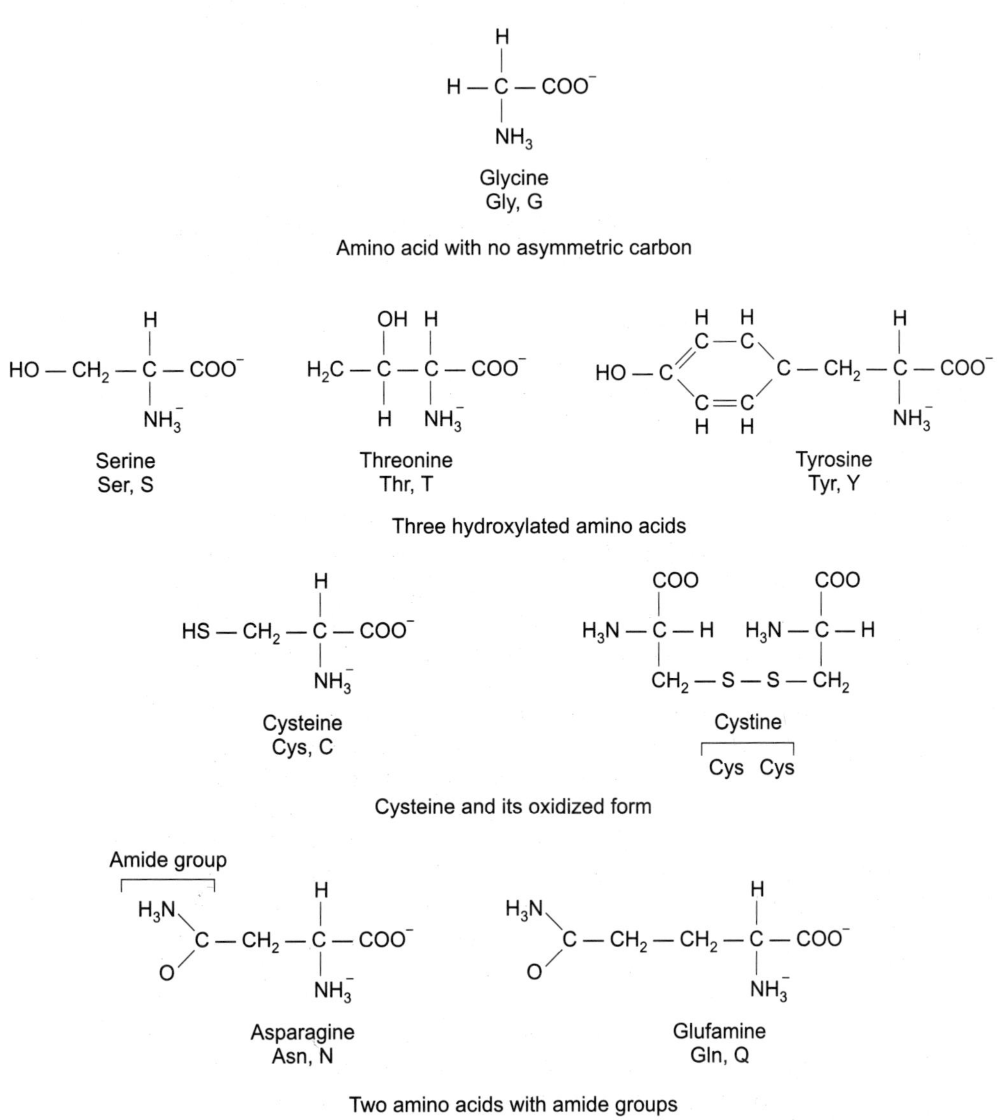

Glycine
Gly, G

Amino acid with no asymmetric carbon

Serine
Ser, S

Threonine
Thr, T

Tyrosine
Tyr, Y

Three hydroxylated amino acids

Cysteine
Cys, C

Cystine
Cys Cys

Cysteine and its oxidized form

Amide group

Asparagine
Asn, N

Glufamine
Gln, Q

Two amino acids with amide groups

Fig. 1.33

value of x-ray crystallography to the study of the more complicated tertiary and quaternary conformations of proteins was established by the definitive determination of the structures of two oxygen-carrying proteins, myoglobin by John Kendrew and hemoglobin by Max Perutz. Twenty-three years of dedicated effort by Perutz (who began the study when ha was a graduate student) were required to obtain the structure of hemoglobin. Both scientists shared the Nobel Prize 1962 for their monumental contributions to science.

The basic experimental design in x-ray crystallographic studies of protein involves the passage of an x-ray beam through a protein crystal and the detection of its diffraction

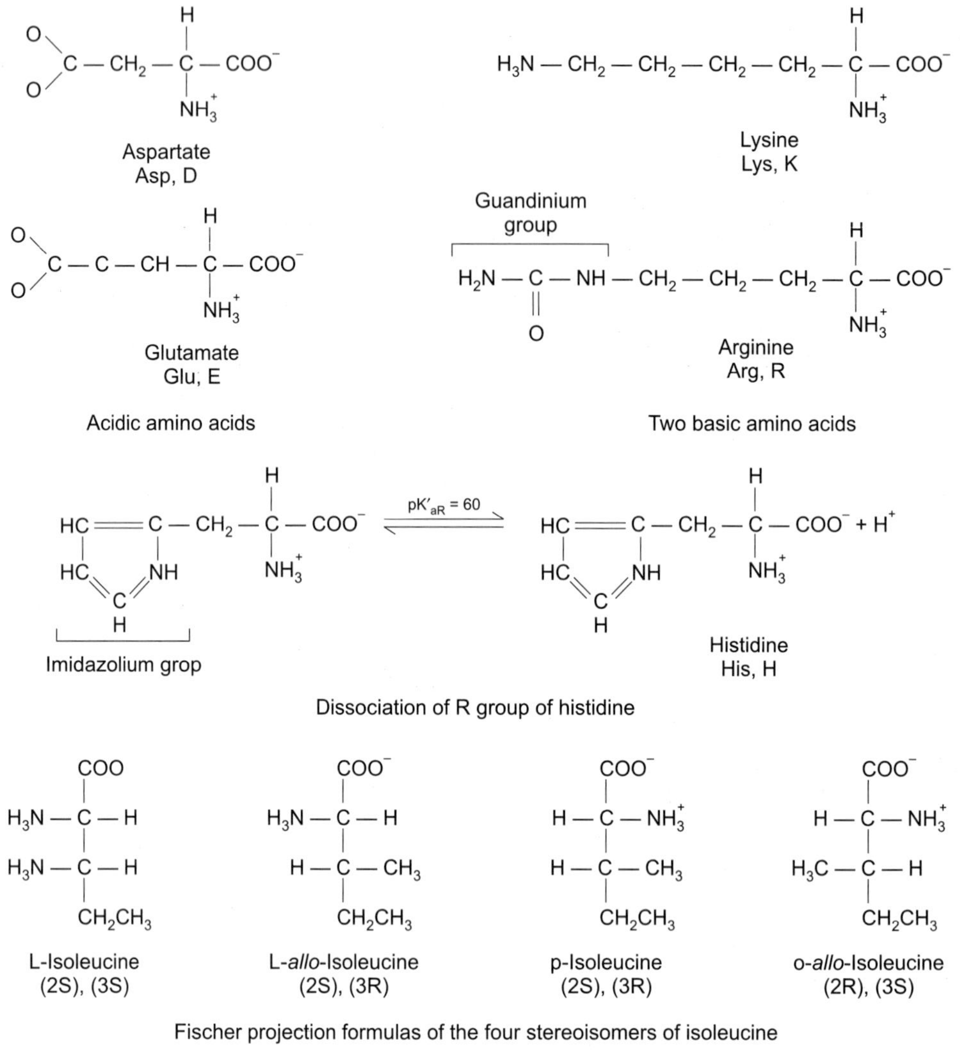

Aspartate
Asp, D

Lysine
Lys, K

Guandinium group

Glutamate
Glu, E

Arginine
Arg, R

Acidic amino acids

Two basic amino acids

$pK'_{aR} = 60$

Imidazolium grop

Histidine
His, H

Dissociation of R group of histidine

L-Isoleucine
(2S), (3S)

L-*allo*-Isoleucine
(2S), (3R)

D-Isoleucine
(2S), (3R)

D-*allo*-Isoleucine
(2R), (3S)

Fischer projection formulas of the four stereoisomers of isoleucine

Fig. 1.34

(scattering) pattern (Fig. 1.36). Electrons of the individual atoms of the macromolecule responsible for the diffraction, with each electron, contributing to the overall pattern (Fig. 1.37). The intesnsity of the blackened emulsion spot on the film is proportional to the number of the electrons in the atoms responsible for that scatter point, i.e., the heavier the atoms the greater the intensity. These intensities, which collectively indicate the locations of the atoms fixed in space in molecular structure, are the basic data used to construct a three-dimensional image of the protein. Separate x-ray diffractions are obtained with the protein containing one or two heavy atoms. Eg., lead or uranium, to provide high density areas in the diffraction patterns by which the phase of the scattered beams can be determined. Through the

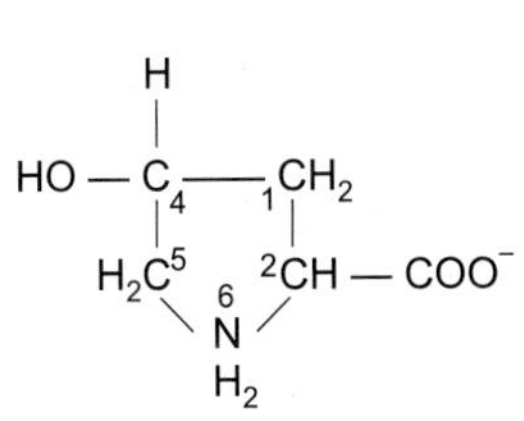

4-Hydroxyproline 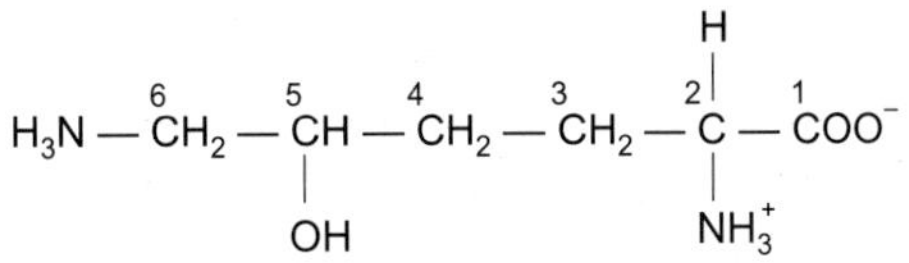 5-Hydroxylysine

Hydroxylated forms of proline and lysine

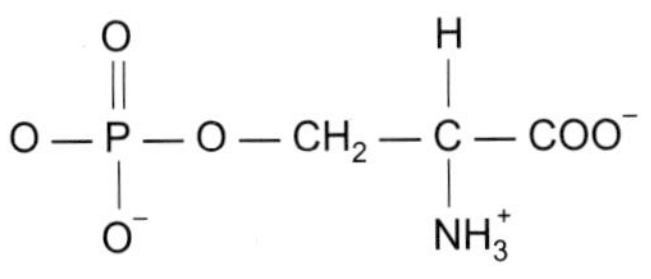

Phosphoserine 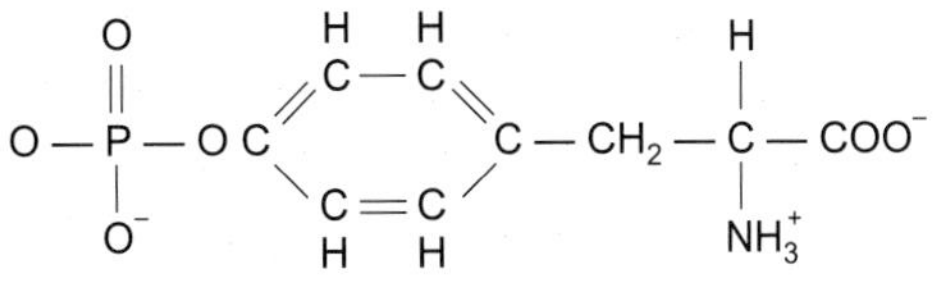 Phosphotyrosine

Two phophorelated amino acids

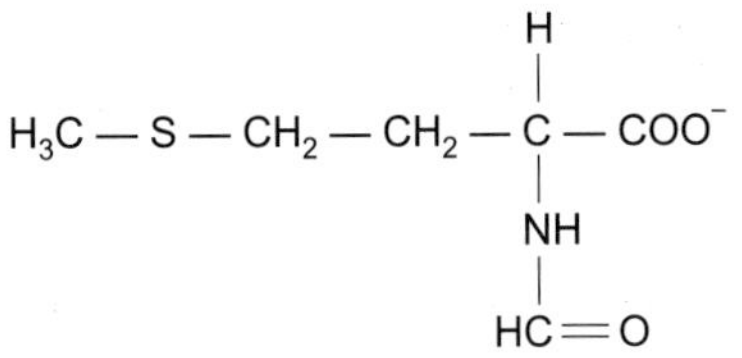

A formylated amino acid

Fig. 1.35

use of sophisticated mathematical analysis (Fourier series) and computers, two-dimensional electron-density patterns of parallel sections of the molecules are constructed which, when stacked one upon another, yield a three-dimensional electron distribution. In x-ray analysis, greater resolution enhances the details of the individual atomic arrangements in the protein structure. Resolution of 6 and 1.4 A for the myoglobin.

Schematic of the diffraction of any-ray beam by its passage through a crystal. Between 1912 and 1915. William Henry Bragg and his son, William Lawrence Bragg, developed the technique of x-ray diffraction by determining the crystalline structure of NaCl.As the joint 1915 Nobel price recipients in physics, the Braggs became the only father-son combination to receive the award and W. Lawrence, who was 25 years old at the time the youngest scientist to be so honored.

A view through several consecutive sections of the three-dimensional electrondensity map of the plant protein concanavalin A. Regions of equal elect on density are drawn as contours in each section and consecutive sections, 0.5 A apart are stacked to represent the three-dimensional electron density. The dots represent atomic positions The contours on the sections in the back of the stack appear fainter than thickness of the plastic sheets on which the map is plotted.

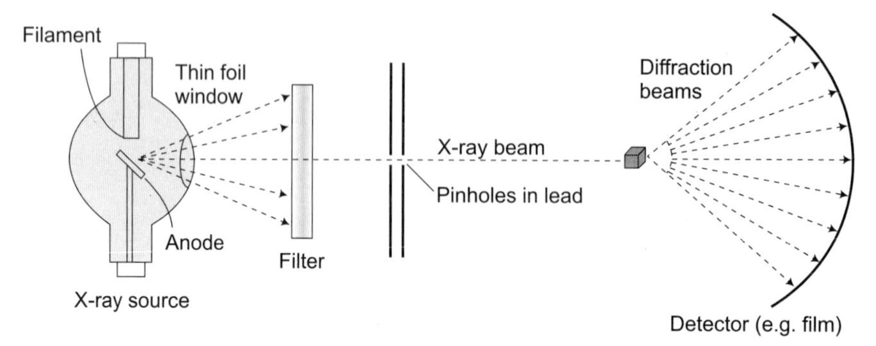

Schematic of the diffraction of an x-ray beam by its passage through a crystal

Between 1912 and 1915, William Henry Bragg and his son, William Lawrence Bragg, developed the technique of x-ray diffraction by determining the crystalline structure of NaCl. As the joint 1915 Nobel prize recipients in physics, the Braggs became the only father-son combination to receive the award and W. Lawrence who was 25 years old at the time, the youngest scientist to be so honoured

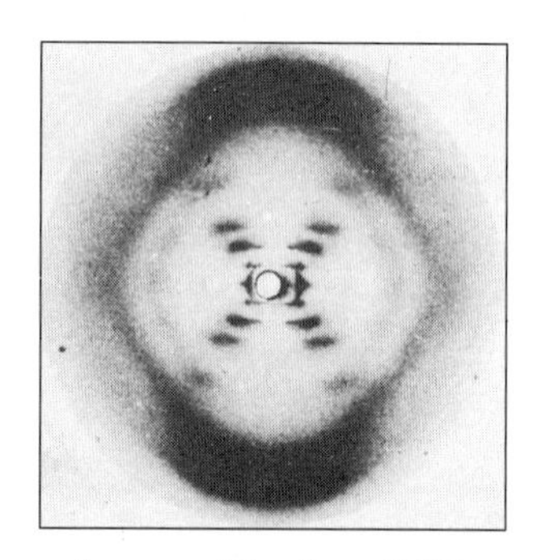

An x-ray diffraction of DNA

Fig. 1.36

Allostery: Regulation of a Proteins Functon by Changing its Shape

The binding of either small large molecules (ligands) to a protein can cause a substancial change in the conformation of that protein. Such ligands-induced conformational changes can have a variety of effects, from increasing, the affinity of the protein for a second ligand, to switching the enzymatic activity of a protein on or off. This is known as allosteric regulation and is prevalent control mechanism in biological system. “Allostery” means “other shape,” and the basic mechanism is as follows: a ligand binding at one side on a protein changes the shape of that protein. As a result of that change, an active site, or another binding site, elsewhere on the protein is altered in a way that increases or decreases its activity. Examples of proteins controlled in this way range from metabolic enzymes to transcriptional regulatory proteins.

The ligand (the allosteric effector) is very often a small molecule-a sugar or an amino acid. But allosteric regulation of a given protein can also be mediated by the binding of another protein and a very similar effect can, in some cases, be triggered by enzymatic modification of a single amino acid residue within the regulated protein. We will see examples of allosteric regulations by all three mechanisms in this section.

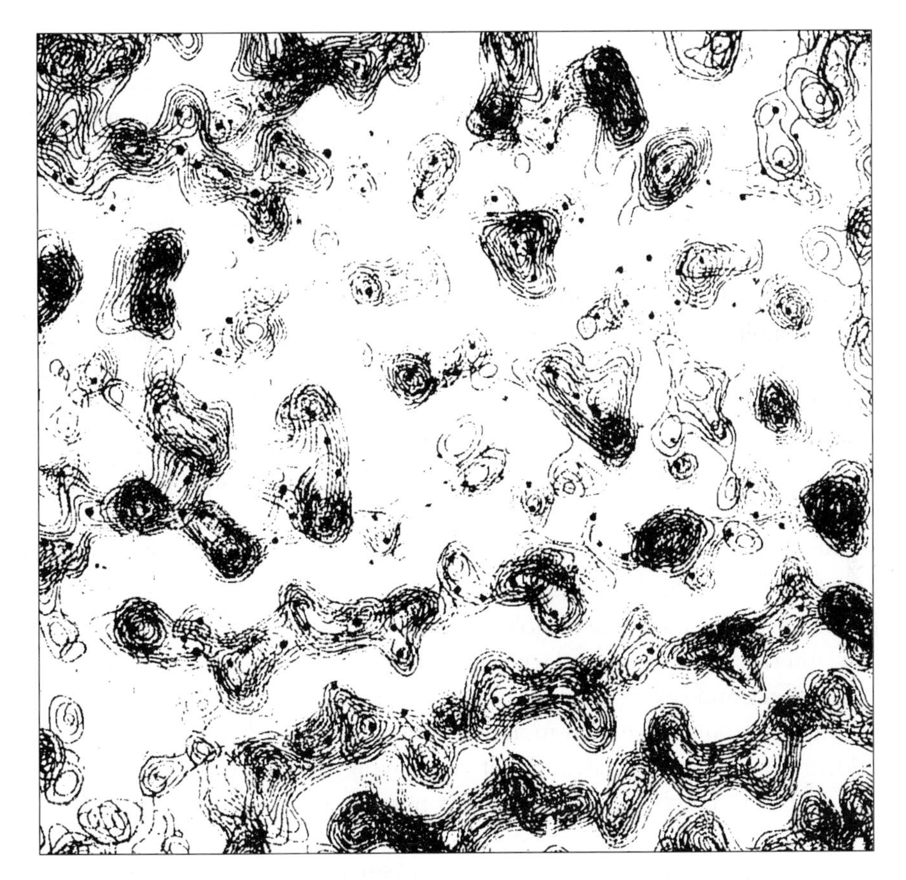

Fig. 1.37 A view through several consecutive sections of the three-dimensional electron-density map of the plant protein concanavalin A. Regions of equal electron density are drawn as contours in each section and consecutive sections, 0.5 A apart. are stacked to represent the three-dimensional electron density. The dots represent atomic positions. The contours on the sections in the back of the stack appear fainter than those in the front because of the thickness of the plastic sheets on which the map is plotted.

SUMMARY

DNA, RNA, and proteins are all polymers, each composed of a defined set of subunits joined by covalent bonds. For example DNA is made of chains of nucleotides, and proteins are chains of amino acids. The three dimensional shape of each such polymer is further determined by multiple weak, or secondary, interaction between those subunits. Thus, in the case of DNA hydrogen bonds and stacking interactions between the basis of nucleotides account for the double-helical character of that molecule. Likewise, the stable three dimensional structure of a given protein requires multiple weak interaction between (non adjacent) amino acids within the polypeptide chain. In this chapter we looked at how those weak interactions determine the shape of molecules and the interactions between and among them, particularly proteins.

There are multiple levels to the structural organizations of a protein. The initial covalent linkage of the amino acid is the primary structure. Each amino acid is linked to the next by a peptide bond. Secondary structure is formed by interactions between amino acid typically found rather near each other in the primary structure of the protein. The A helix and B sheet are examples of secondary structural elements. The tertiary structure of a protein is the final three dimensional shape of a polypeptide chain and is determined by the arrangement of the various elements of secondary structure in an energetically favorable way. For many proteins there is another level of structural organization-the quaternary structure.

This refers to multimerization of individual polypeptide chains into dimmer or higher-order structure. Many proteins work as multimers-hemoglobin is a tetramer, for example, and many DNA binding proteins work as dimmers.

Many native proteins contain several discrete folded sections (domains) that are stable by themselves and which aride from a continuous amino acid sequences. Combinations of such domains account for a large variety of all known proteins. The number of truly unique domains is probably only a few hundred. Each domain is often associated with a specific functional activity for example DNA binding.

The specific shape of each macromolecule restricts the number of other molecules with which it can interact strong secondary interactions between molecules demand both a complementary (lock-and-key) relationship between the two bonding surfaces and the involvement of many atoms. Although molecules bound together by only one or two secondary bounds frequently fall apart, a collection of these weak bounds can result in a quite stable complex. The fact that double-helical DNA does not fall apart spontaneously shows just how stable such complexes can be all though complexes held together by multiple weak bound are not observed to fall apart spontaneously, there assemble can occur spontaneously, with the correct bounds forming in a step by step manner (the principal of self assembly).

The binding of specific protein to specific sequence along DNA molecules also involves the formation of weak bounds; usually hydrogen bounds between group on DNA basis and appropriate acceptor or donor group on protein most regulatory proteins use An helix to recognized specific DNA sequence. That recognization helix fits into the major groove of DNA, and the amino acid in the helix contact the edges of basis in a sequence-specific manner. These contacts are stabilized by the binding energy of the specific interaction. DNA binding proteins also contain regions that allow non specific bonding to the DNA backbone. These non specific backbone interaction permit linear diffusion along DNA, allowing proteins to reach there specific target sequence more quickly. A few proteins use B sheets (rather than A helices) to be recognized specific DNA sequence and interactions with the minor grooves but these are much less common.

Proteins perform many functions, such as catalysis or DNA binding. These activities are commonly regulated by the binding of small ligands or other proteins to the protein in question, or through enzymatic modifications of residues within that proteins. These ligands, or modifications, often regulate protein function through allostery. That is the ligand binds (or the modification targets) a site on the protein separate from the region of that protein that mediates its main function (the active site of an enzyme, DNA-binding domain, etc.,) This binding or modification triggers a change in thee shape of the protein which increases or decreases the activity of the active site, or DNA-binding domain, essentially switching the activity on or off.

In other cases, a protein may be controlled by modification or binding of a second protein in ways that do not involve, allostery. For example, modification can create a site on a protein that is recognized by a second protein. Such protein-protein interactions can recruit proteins to particular locations or substrates, and in that way control their activity.

SECTION E

ENZYMES

General Catalytic Properties of Enzymes

Enzymes are remarkably effective catalysis, responsible for the thousands of coordinated chemical reactions involved in biological processes of living systems. Like an inorganic catalyst, an enzyme accelerates the rate of a reaction by lowering the energy of activation required for the reaction to occur. However, unlike a simple organic catalyst, an enzyme lowers the energy of activation by replacing a large activation barrier with multiple lower barriers. This property is illustrated in Fig. 1.38 where, as noted, the difference between the energy coordinates of the enzymatic and nonenzymatic reactions is the amount of energy (G) needed for the reactant (substrate) to attain the activated state. Note that for both reactions, the difference between the free energy of A (reactant) and B (product), called the G of the reaction, is the same.

As a catalyst, an enzyme is not destroyed in the reaction and therefore remains unchanged and is reusable. An outstanding feature of an enzyme as a catalyst is a subtrate specificity, which determine its biological substrate (absolute substrate specificity); others have a broader specificity and utilize a group of structurally similar biomolecules as substrates (relative group specificity). Glucose 6-phosphatase is an example of an enzyme with specificity for D-glucose 6-phosphate; it catalyzes the hydrolytic removal of the phosphate moiety. Conversely, acid and alkaline phosphates can act upon various phosphorylated substrates and therefore exhibit broad specificity. Another critical biological feature of enzymatic reactions is that their substrate and catalytic specificities ensure synthesis of only specific biomolecular products without the concomitant production of by-products, in contrast to many reactions of organic chemistry.

Some enzymes are referred to as simple proteins because they require only their protein structure for catalytic activity. Other enzymes are conjugated proteins since they each require a non-protein component, called a cofactor, for activity. The enzymes that need metal cofactors e.g., are called metaloenzymes. Other enzymes have organic biomolecules, coenzymes, as cofactors. The coenzyme forms of B-vitamins, for example, are required for many reactions. If a cofactor (metal ion or coenzyme) is firmly bound to an enzyme, it is called a prosthetic group, e.g., the B-vitamin biotin in its coenzyme form is covalently bonded to a lysyl residue of the

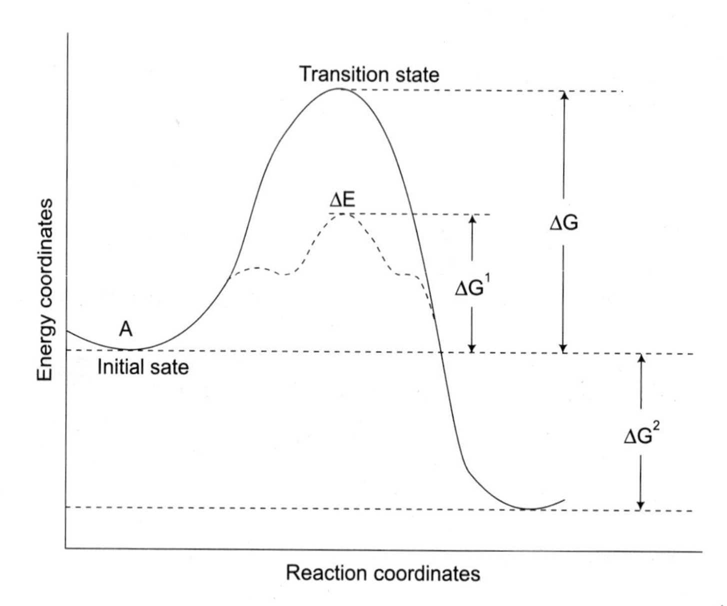

Fig. 1.38 Energy diagram for a reaction A → B. A is the reactant and B the product. $A^{\ddagger}_{NE}$ and $A^{\ddagger}_{E}$ are the activated complexes in the nonenzymatic and enzymatic reactions, respectively $\Delta G^{\ddagger}_{NE}$ and $\Delta G^{\ddagger}_{E}$ are the free energies of enzymatic reactions, respectively. *DG* is the overall free-energy change of the reaction.

enzyme pyruvate carboxylase. The complex composed of an apoenzyme (protein portion of an enzyme) and its cofactor (s) is called a holoenzyme.

Classification of Enzymes by Catalytic Function

Berzelius was correct in his 1837 prediction that thousands of catalytic processes occur in biological system. A little over a century later, Beadle and Tatum's one gene-one enzyme hypothesis emphasized that concept that each of these thousands of reactions is catalyzed by a specific enzyme, a proposal that is now accepted as generally true. In the 1960s, the International Union of Biochemistry (IUB) established a Commission on Enzyme Nomenclature to adopt a systematic classification and nomenclature for the ever-increasing number of enzymes being identified and described. The Commission, identifying enzymes by, the types of reactions they catalyzed, defined six major classes (Table 1.10); examples of each class are listed in Table 1.11. Oxidation-reduction reactions are catalyzed by oxidoreductases, and reactions involving the transfer of a group utilize transferases. Hydrolases employ H_2O to cleave covalent bonds, and this class includes enzymes that degrade polysaccharides, proteins, and nucleic acids. Lyases cleave or remove groups from compounds by electron rearanements (eliminations) and therefore create double bonds in one of the products. If a lyase reaction is reversible, a group may be added to a double bond (deaminase reaction in Table 1.11). Isomerases catalyze internal arrangements within a substrate and therefore do not involve the addition or removal of groups. Ligases catalyze the formation of various types of covalent bonds to synthesize biomolecules and require an input of chemical energy, provided by hydrolysis of

Table 1.10 Six major classes of enzymes.

Class number	*Enzymes*	*Catalyitc function*
1	Oxidoredu tases	Oxidation-reduction reactions
2	Transferases	Group-transfer reactions
3	Hydrolases	Hydrolytic reactions
4	Lyases	Reactions involving elimination of a group by the cleaving of a bond (leaving a double bond) or the addition of a group to a double bond
5	Isomerases	Reactions involving isomerizations
6	Ligases	Reactions joining together two molecules coupled with the hydrolysis of high-energy pyrophosphate bonds

biomolecules such as nucleoside triphosphates, e.g., ATP. As an energy donor ATP, is hydrolyzed by some ligased to yield AMP and PPl (inorganic pyrophosphate) and by other ligases to yield ADP and Pl (inorganic phosphate).

In the official nomenclature, every enzymes is distinctly identifiable by its formal name and by a four-component number. Alcoholdehyreogenase is identified in scientific reports as Alcohal: NAD+oxidoreducatse, E. C. 1.1.1.1. The first number 1 refers to class I (oxidoreducatse); the second, to the type of group oxidized (I = alcohol); The third to the oxidizing agent (l = the coenzyme NAD+); and the fourth, to the specific reaction(l = alcohol dehydrogenase.). Because official names are often lengthy, the trivial names of enzymes are generally used after initial identification.

Enzymes Assays

To study an enzyme, it is necessary to have an assay by which to measure its catalytic activity. Assays are designed to measure the rate of product formation or the rate of substrate disappearance. An assay that measures product formation is preferred because it involves a direct measurement, ass opposed to the indirect measurement obtained from an assay relying on the determination of substrate disappearance. Frequently the amount of product produced in a given length of time is measured in a fixed-time assay. The manner in which the amount of product is determined depends on its chemical and physical properties. If the product is colored or can undergo a reaction to produce a colored solution then absorbance of light at an appropriate wavelength can be measured (colorimetric assay) and related to the concentration of product at the time of sampling Spectrophotometric assays are especially useful because the progress of the reaction can be monitored continuously in kinetic assays. The activities of enzymes that utilize NAD+ or NADH as coenzymes are assayed spectrophotometrically because NADH, but not NAD+ has an absorbance peak at a wavelength of 340 nanometers (nm), which is an the ultraviolet region of the spectrum. Hence, increases or decreases in absorbance, which measure appearance or disappearance, respectively, of NADH can be conveniently monitored in a spectrophotometer. A radioactive assay is another commonly used method for monitoring enzymatic activity. By using sabstrate labeled with accurate determinations can be made of minute quantities of isolated, radioactive product.

Table 1.11 Rpresentative enzymes of the six major classes and their reactions.

	Enzyme class		*General*
1.	Oxidoreductases	Type of oxidation	
	Alcohol dehydrogenase	Alcohol to aldehydle	H_3CCH_2OH ® $H_3C\overset{O}{\overset{\|}{C}}H + 2H^+$
	Succinate dehydrogenase	Double bond information	$^-OOCCH_2CH_2COO^- \rightarrow {}^-OOC\overset{H}{C}{=}\underset{H}{C}COO^- + 2H^+$
2.	Transferases	Group transferred	
	Phosphotransferase	Phosphoryl	$RO-\underset{O^-}{\overset{O}{\overset{\|}{CH}}}-O + HOR' \rightarrow ROH + O-\underset{O^-}{\overset{O}{\overset{\|}{P}}}-OR'$
	Aminotransferase	Amino	$R-\underset{NH_3^+}{CH}-COO^- + R'-\overset{O}{\overset{\|}{C}}-COO^- \rightleftharpoons R-\overset{O}{\overset{\|}{C}}-COO^- - R'-\underset{NH_3^+}{CH}COO^-$
3.	Hydrolases	Bond hydrolysed	
	Peptidase	Peptide	$R-\overset{O}{\overset{\|}{C}}-\underset{H}{N} + R'-HOH \rightarrow R-\overset{O}{\overset{\|}{C}}-O + {}^+H_3N-R'$
	Phosphatatase	Monophosphate ester	$R-O-\underset{O^-}{\overset{O}{\overset{\|}{P}}}-O^- + HOH \rightarrow R-\overset{O}{\overset{\|}{O}}H + HPO_4^{2-}$
4.	Lyases	Group removed	
	Decarboxylase	Carbon dioxide	$R-\underset{NH_3^+}{\overset{O}{\overset{\|}{C}}}-COO^- \rightarrow R-\underset{NH_3^+}{CH_2} + O{=}C{=}O$
	Deaminase	Ammonia	$R-CH_2\underset{NH_2}{C}HR' \rightleftharpoons RCH{=}CHR' + NH_3$
5.	Isomerase	Group Isomerized	
	Epimerase	C-3 of a five-carbon sugar	D-Fructose 5-phosphate $\rightleftharpoons$ D-xylulose 5-phospate
	Racemase	α-Carbon substituents	D-Alanine $\rightleftharpoons$ D-alanine
6.	Ligases	Covalent bond formed	
	Acetyl-CoA synthetase	C-S	Acetate + CoA—SH + ATP $\rightleftharpoons$ acetyl—S—CoA + AMP + PP_i
	Pyruvate carboxylase	C-C	Pyruvate + CO2 + H2O + ATP $\rightleftharpoons$ oxaloacetate + ADP + P_i

Optimal conditions are also determined for an enzymatic assay, including determination of the optimum pH, temperature, and ionic strength, for the reaction. Because enzymes are globular proteins, most are thermolabile and being to denature (indicated by loss of enzymatic activity) at temperatures between 45 and 50°C. Lonic strength and pH are also important parameters because they specify the charges of amino acid residues that may influence the three-dimensional structure of an enzyme and thus, its catalytic activity. The appropriate concentrations of enzyme and of cofactors (if needed) to use in a particular volume of assay mixture are also established empirically.

Enzyme Reaction Rates

When increasing concentrations of substrate are used in a series of assays, standardized by the above criteria, a plot of V (velocity, or reaction rate) vs. substrate concentration often produces a hyperbolic curve. Velocity is usually expressed as units, e.g., amoles of product produced or substrate transformed per minute, or specific activity (units per milligram of protein). An analysis of the plot reveals reaction characteristics of the enzyme. The hyperbolic curve analogous to the oxygen-dissociation curve of myoglobin, is representative of the reaction kinetics of a noncooperative enzyme. The plot shows that the velocity increases with substrate concentration until maximum v (Vmax) is approached asymptotically, after which larger concentrations of substrate do not significantly enhance the reaction rate. In the lower region of the curve, the reaction approaches first-order kinetics, meaning that v is a direct function of substrate concentration because the active sites of the enzyme molecules are not saturated. At the plateau at the upper portion of the plot, the reaction approaches zero-order kinetics because the active sites of all the enzyme molecules are saturated and the reaction rate is therefore independent of further increases in substrate concentration. For the intermediate portion of the curve, as the enzyme approaches substrate concentration. Routine enzyme assays are designed to follow zero-order kinetics to avoid the influence of substrate concentration on reaction velocity. Under such conditions, measured rates are directly proportional to the concentration of the enzyme itself.

The substrate concentration needed for half-maximum velocity (1/2 V max) is called the Km value (michealis constant) Table 1.12 and is expressed in units of substrate concentration 9 moles per liter or M). Km may be considered an approximate measure of the affinity of an enzyme for its substrate; the lower the Km values of 1 × 10 M and 1 × 10 M, the one with the 10 M value requires a substrate concentration 100-fold greater to attain its 1/2 V max than that required by the one with the 10 M value. The Km value is a characteristic property of an enzyme and, as seen in a wide range of values is found among enzymes and their substrates.

Enzyme Inhibition

Substances that specifically decrease the rate of enzymatic reactions are called inhibitors, and, in enzymology, inhibitory phenomena are studied seriously because of their importance to many different areas of research. For many years inhibitors have been classified as either competitive or noncompetitive, terms, which represent two general types of reversible inhibition.

Table 1.12 K_M Values for some enzymes.

Enzme	*Substrate K_m*	*(M)*
Catalase	H_2O_2	2.5×10^{-2}
β-Galactosidase	Lactose	4×10^{3}
β-lactamase	Benzylpenicilin	5×10^{5}
Glutamate dehydrogenase	α-Ketoglutarate	2×10^{3}
	NH_4^-	5.7×10^{2}
	NADH	1.8×10^{5}
Hexokinase	Glucose	1.5×10^{-4}
	Fructose	1.5×10^{-3}

Competitive Inhibition

As implied by its name, a competitive inhibitor has been classically envisioned as a compound that competes with a natural substrate of an enzyme for the active (substrate-binding sire. Such an inhibitor is almost always structurally similar to the natural substrate and, by mimicry, binds to the enzyme and precludes catalytic activity. Competitive inhibition is reversible and can be overcome by increasing substrate concentration. The effectiveness affinities the enzyme has for the substrate and inhibitor.

Oxaloacetate and malonate are competitive inhibitors of succinate dehydrogenase, which catalyzes the conversion of succinate into fumarate and the substrate and these former two inhibitors have similar structures. The pharmaceutical industry relies heavily on the concept of competitive inhibition in the synthesis of drugs. Sulfanilamide, a sulfa drug (Fig. 1.39) is an antibacterial agent because it effectively competes with p-aminobenzoic acid, which is required for synthesis of the essential metabolite folic acid by some pathogens. Sulfanilamide can inhibit the bacterial enzyme selectively because humans require the B-vitamin folic acid (cannot synthesis the vitamin) from p-aminobenzoic acid and, therefore, do not have the comparable enzyme. Much of the chemical therapy used to combat cancer utilizes compounds structurally designed to compete with substrates needed for the enzymatic processes of DNA replications in cancer cells, and this manner, cell division can be prevented.

Catalytic Effectiveness of Enzymes

Enzyme are usually much more effective as catalysis than are the inorganic or organic catalysts commonly used in chemistry. One of the most catalytically potent enzymes, carbonic anhydrase, has a turnover number of 600,000 per second (Table 1.13). This means that when an enzyme molecule is fully saturated with its substrate (CO_2), 600,00 substrate molecules are transformed into product (H_2CO_3) in 1 second. Not all enzymes however are that active, and the turnover number of most enzyme range from 1 to 10 per second for their natural substrates. Note that the turnover number, or catalytic coefficient (K_{cat}), of an enzyme is its V max per unit enzyme concentration; thus $K_{cat} = V_{max}/[E]_t$. Although biochemists have learned much about mechanisms of enzyme catalysis by chemical studies on active sites and by sequences and x-ray

NH_2 / COOH

p-Aminobenzoic acid

NH_2 / SO_2NH_2

Sulfanilamide (sulfa drug)

Fig. 1.39

Table 1.13 Turnover numbers of some enzmes.

Enzyme	*Turnover number/sec*
Carbonic anhydrase	600,00
Acetylcholinesterase	25,000
Lactate dehydrogenase	1,000
b-Galactosidas	208
Phosphoglucomutase	21
Tryptophan synmetase	2

crystallographic analyses of enzyme, satisfactory explanations to account for the amazingly high rates of catalysis exhibited by most enzymes are still being sought.

Acid Base Catalysis

The principal of acid base catalysis, well identified for organic chemical reactions, are useful in explaining fundamental chemical mechanisms of some enzymes. Two types of acid-base catalysis general and lewis are of importance to studies on enzyme mechanisms. General acid-base catalysis relies on the Bronsted-Lowry concept of acids (proton donors) and bases (proton acceptors). In this type of catalysis, a specific group on enzyme is viewed as an acceptor or donor of protons. Amino, carboxyl, sulfhydryl, imidazolium, and phenolic (tyrosine) groups of enzymes can serve as acids and or bases. Lewis acid-base catalysis stresses the lewis concept of acids and bases to explain certain enzymatic mechanisms. Lewis acid are defined as electron-pair acceptors; thus they constitute electrophilic (electron-seeking) groups. Conversely, Lewis bases are electron-pair donors, nucleophilic (nucleus-seeking) groups. Certain metal ions, e.g, Mg^{2+}, Mn^{2+} and Fe^{3+} are electrophilic. Hydroxly (e.g, serine) imidazole (histidine) carboxylate (glutamate or aspartate), and sulfhydryl (cysteine) groups are nucleophilic because each has an electron pair which could intract with an electron-deficient center of a substrate molecule. Such reactions of nucleophilic and electrophilic groups are important to certain catalytic mechanisms. Histidine is versatile in acid base catalysis because its imidazole group can serve as a Lewis base (strong nucleophile) and, in its protonated form, the imidazoliumion can behave as a Bronsted-Lowry acid. An illustrative example of acid-base catalysis is the mechanism proposed for the peptidase chymotrypsin.

Activation of Chymotrypsinogen

Chymotrypsin is one of several proteolytic digestive enzymes synthesized in the pancreas as zymogens, inactive precursors of enzymes. The prefix pro or suffix ogen added to an enzyme's name designates, the zymogenic state. Chymotrypsinogen, other zymogens, and digestive enzymes, including ribonucleases and lipases, are stored in lipid-protein membranes (called zymogen granules) in a mammalian pancreas and, when needed for digestive purposes, are secreted into a duct that leads into the duodenum. Where they are activated. The biosynthesis of proteolytic enzymes as zymogens and their storage in membranes are necessary precautions to avoid enzymatic hydrolysis of pancreatic tissue (acute pancreatitis).

Table 1.14 Some pancreatic zymogens.

Zymogen	*Active enzyme*
Chymotrypsinogen	Chymotrypsin
Trypsinogen	Trypsin
Procarboxy-peptidases	Carboxy-peptidases
Proelastase	Elastase

Chymotrypsinogen is composed of a single polypeptide chain with 245 amino acid residues and five disulfide bridges. The activation of the zymogen to produce chemotrypsin first involves the selective cleavage by trypsin of the peptide bond linking Arg 15 and lle 16 ,yielding chymotrypsin. This single hydrolysis of a peptide bond converts the inavtive form of the enzyme into active chymotrypsin, which then proceeds to cleave selectively other chymotrypsin molecules to produce chymotrypsin, the stable form of the active enzyme. Chymotrypsin catalyzes the removal of two dipeptides (residues 14 and 15 residues 147 and 148) from chymotrypsin in the production of chymotrypsin. As a result, chymotrypsin, which has a molecular weight of 25,000 consists of three chains (designated A, B, and C); the chains remain covalently linked by disulfide bonds.

SUMMARY

Enzymes are the remarkably efficient biological catalysts responsible for the thousands of reactions needed for the chemical processes of life. All enzymes are proteins and can be classified either as simple proteins, whose catalytic activities rely on the enzyme's polypeptidyl structure, or as conjugated proteins, which require a nonprotein components (s) for activity. The nonprotein components of enzymes, called cofactors, include metal ions, e.g., Mg^{2+}, and organic biomolecules (coenzymes), e.g., B-vitamins. If a cofactor is tightly bound to an enzyme, it is called a prosthetic group. An enzyme is classified by the type of reaction it catalyzes, and the six major classes of enzymes are oxidoreductases, transferases, hydrolases, lyases, isomerases, and ligases.

In an enzyme-catalyzed reaction, increasing the substrate concentration increases the reaction rate (v). However, after the active sites of the enzyme molecules are saturated with substrate, the reaction rate become independent of substrate concentration (zero-order reaction); the maximal velocity theoretically attainable under these conditions is called Vmax. The substrate concentration at which the reaction rate is half that of Vmax is called Km (Michaelis constant). Each enzyme displays a characteristic Km of its substrate (s). The relationship between substrate concentration, v, Vmax, Km, and the initial reaction rate of an enzyme reaction is mathematically expressed in the Michaelis-Menten equation. The Km and Vmax values of an enzymatic reaction are often determined by plotting the reciprocals of (S) and v values (Linerweaver-Burk plot). Ideally, the Linerweaver-Burk plot yields a straight line with a slope equal to K_m/V_{max} and the l/v and l/S intercepts are equal to l/V_{max} and $-1/K_m$, respectively.

The three general types of inhibition of enzymatic activity are called competitive, noncompetitive, and irreversible. Competitive and noncompetitive inhibition is reversible phenomena. A competitive inhibitor lowers the rate of an enzyme-catalyzed reaction by competing with the substrate for the active site to form an enzyme-inhibitor complex; this type of inhibition can be reversed by increasing the substrate concentration. A noncompetitive inhibitor interacts not with the active site but, instead, with another region of the enzyme or enzyme-substrate complex. An irreversible inhibitor usually inactivates an enzyme by covalently bonding to an essential group in its active site.

An isozyme is an enzyme that occur in multiple forms within an organism. An isozyme pattern may very during the development of an organism and/or in different tissues. Like hemoglobin, many enzymes are allosteric, and their activities are regulated by positive and/or negative effectors which increase and/or decrease the reaction rates, respectively. Allosteric enzymes are generally key metabolic enzyme, and control of their activities is critical for the coordination of the metabolic activities of a living cell.

The principal of acid-base catalysis that apply to organic reactions have proved useful in explaining the catalytic mechanisms, of a number of enzymes. The proposed catalytic mechanism for chymotrypsin is illustrative of the biochemical importance of general acid-base catalysis (involving Bronsted-Lowry acids and bases) and Lewis acid base catalysis (involving electrophilic and/or nucleophilic groups). Chymotrypsin is a digestivepeptidase which is synthesized in the pancreas as a zymogen (inactive precursor of an enzyme), called chymotrypsinogen. After the zymogen is transported into the duodenum, four selective hydrolytic cleavages convert chymotrypsinogen into A-chymotrypsin, the stable, active form of the enzyme. A-chymotrypsin is a serine protease, i.e., has a highly nucleophilic serly oxygen (essential for catalysis) at its active site. Ser 195 is a strong nucleophilie (Lewis base) because of the hydrogen bonding that occur between the following three residues: Asp 102, His 57, and Ser 195. The two-step cleavage of a peptide bond by chymotrypsin is initiated by a nucleophilic attack on the carbonyl carbon of the substrate by the oxygen atom of Ser 195, resulting in the formation of an acyl-enzyme intermediate and the first product, an amine. A nucleophilic attack by the oxygen atom of a water molecule on the acyl carbon of acyl-enzyme intermediate than occur to yield the second product, an acid, and the restored enzyme.

Comparision of the results of the detailed biological and chemical studies on the four serine proteases chymotrypsin, trypsin elastase, and subtilisin provides evidence that the first enzyme share a comman ancestral gene-an example of divergent evolution. Although the fourth protease, subtilisin, also possesses an active-site serine and a hydrogen bonded network involving an aspartyl, a histidyl, and a seryl residue, the enzyme appears to have evolved independently of the other three. The evolution of two distinct types of serine proteases with the same catalytic mechanism is representative of independent convergent evolution.

The specialized catalytic capabilities of enzyme have been used for industrial and medical purposes for many years and continue to increase in their commercial importance. The current development of immobilized enzymes is introducing new industrial applications for these biological catalysts.

SECTION F

LIPIDS

Lipids are a chemically heterogeneous collection of molecules insoluble in water but soluble in nonpolar (organic) solvents such as ether, chloroform, and benzene; it is because of their similar solubility properties that they are usually considered together. Like the carbohydrates, the lipids serve two major roles in cells: (1) They occur as constituents of certain structural components of cells, particular membranous organelles; and (2) they may be stored in cell as reserve energy source. The most common cell lipids include the **fatty acids, neutral fats, glycerophosphatides, sphingolipids, plamalogens, glycolipids, and steroids.**

Fatty Acids

The saturated fatty acids consist of long hydrocarbon chains terminating in a carboxyl group and conform to the following general formula: CH_3–$(CH_2)_n$–COOH. In nearly all naturally occurring fatty acids, n is an even number, usually 14 (i.e. palmitic acid) or 16 (stearic acid). In unsaturated fatty acids, two or more of the carbon atoms of the hydrocarbon chain are linked by double bonds. The two most common unsaturated fatty acids are oleic acids and linoleic acid (Fig. 1.40)

Fatty acid molecules contain both hydrophilic and hydrophobic parts. In their dissociated states (not shown in the formula of Fig. 1.40), the carboxyl ends the' molecules are soluble in water, while the long hydrocarbon chains repel water. As a result, fatty acids form monomolecular layers on water, with their polar ends facing into the water and their hydrophobic ends directed away from the water (Fig. 1.41). This phenomenon is shared by lipids other than fatty acids (see below) and contributes to the suitability of lipids in membrane formation.

Saturated region ... Saturated region

$$CH_3 — (CH_3)_7 — CH{=}CH — (CH_2)_7 — COOH$$

Oleic acid

Saturated region ... Saturated region

$$CH_3 — (CH_2)_4 — CH{=}CH — CH_2 — CH{=}CH — (Ch_2)_7 — COOH$$

Linoleic acid

Fig. 1.40 The unsaturated fatty acids—*oleic* and *linoleic* acid.

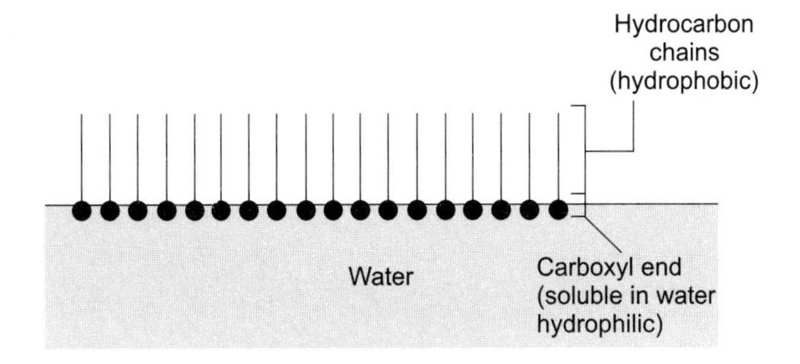

Fig. 1.41 As the result of possessing hydrophilic and hydrophobic regions, fatty acids from monomolecular layers on water.

Neutral Fats

The neutral fats or triglycerides are esters of glycerol and three fatty acids and have the general formula shown in Fig. 1.42. In this formula n, n, and n" may be the same number of different numbers, and the three fatty acids may be saturated and/or unsaturated. Unlike the fatty acids, the neutral fats are entirely nonpolar. Neutral fats, which are used by the cell as source of energy, represent the major type of stored lipid, and most of the recovered in the soluble phase of disrupted cells takes this form.

Glycerophosphatides

The major members of this group of lipids are derivatives pf phosphatidic acid. Phosphatidic acid is similar to a triglyceride except that one of the fatty acids is replaced by a phosphate group (Fig. 1.43). Phosphatidic acis and its derivatives are present in cell membranes, where they play anactive role in membrane function, in addition to serving as structural constituents. The most common derivatives of phosphatidic acis are phosphatidyl choline (also called lecithin), ethanolamine (also called cephalin), phosphatidyl serine, and phosphatidyl inositol (Fig. 1.44). Free rotation about the single bonds of the glycerol backbone allow the hydrophilic phosphate

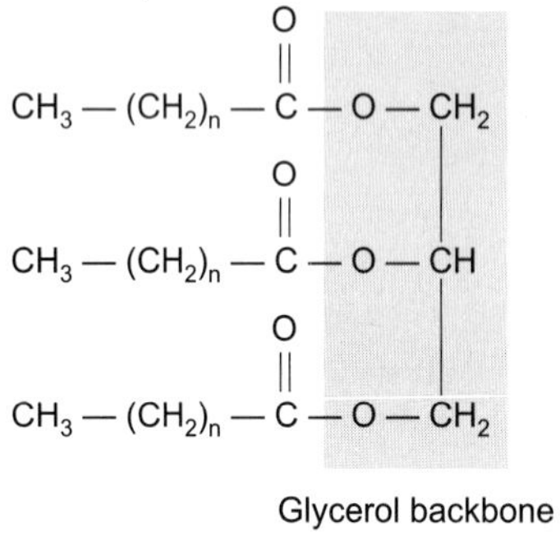

Fig. 1.42 Generalized formula for a neutral fat.

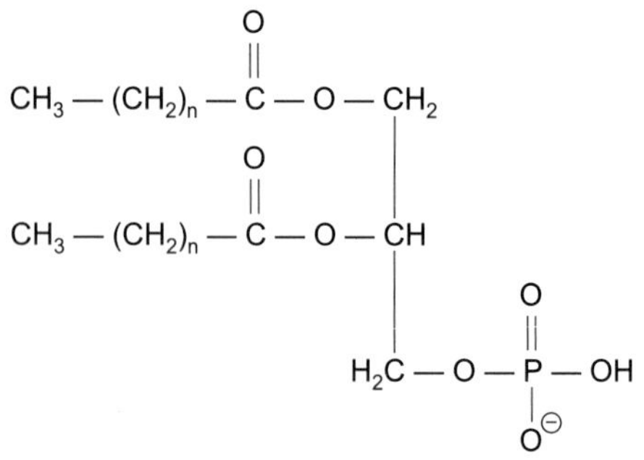

Fig. 1.43 Phosphatidic acid.

and its derivatives to face away from the hydrophobic portion, as shown in the structural formula of Fig. 1.44. Consequently, glycerophosphatides (and other lipids discussed below) are amphipathic-that is, one end of the molecule is hydrophobic (i.e., the end containing the hydrocarbon chains), while the other end is extremely hydrophilic as a result of the charged nature of the dissociated phosphate group and other substituents. These properties are important to the organization of glycerophosphatides in cell membrance.

Plasmalogens

Plasmalogens are a special class of lipids especially abundant in the membranes of nerve and muscle cells. Plasmalogens are also found in blood. A plasmalogens is similar to a glycerophosphatide except that an unsaturated ether is substituted at one of the glycerol postitions (Fig. 1.45).

Sphingolipids

Sphingolipids are derivates of sphingosine, an amino alcohol possessing a long, unsaturated hydrocarbon chain (Fig. 1.46b). In sphingomyelin, the amino group of the sphingosine skeleton is linked to a fattyacid, and the hydroxyl group is esterified to phosphorylcholine.

Glycolipids

The glycolipids are sugar-containing lipids similar to sphingosine. In glycolipids, the amino group of the sphingosine skeleton is acylated by a fatty acid (as in sphingomylin), but the hydroxyl group is associated with one or more monosaccharides. The simplest glycolipids are the cerebrosides which as their name suggests, are abundant in brain cells. However, cerebrosiders are also present in kidney, liver, and spleen cells. The sugar of a crerbroside is either glucose or galactose (Fig. 1.47). Gangliosiders are glycolipids found in cell membranes and are especially abundant in the membranes of matter cells of the brain. A chain of several sugar molecules (including galactose, glucose, and neuraminic acid) composes the carbohydrate portion of the molecule.

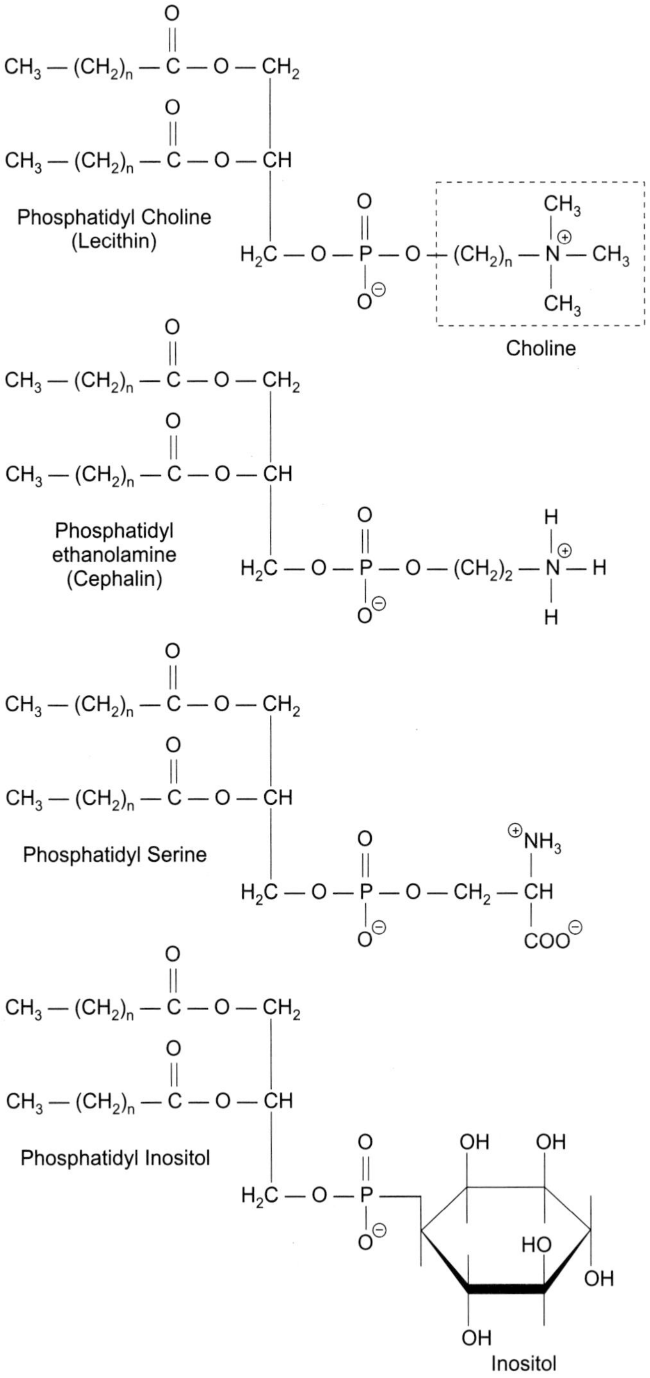

Phosphatidyl Choline (Lecithin)

$$CH_3-(CH_2)_n-\overset{O}{\overset{\|}{C}}-O-CH_2$$
$$CH_3-(CH_2)_n-\overset{O}{\overset{\|}{C}}-O-CH$$
$$H_2C-O-\overset{O}{\overset{\|}{P}}(O^{\ominus})-O-(CH_2)_n-\overset{\oplus}{N}(CH_3)_3$$

Choline

Phosphatidyl ethanolamine (Cephalin)

$$CH_3-(CH_2)_n-\overset{O}{\overset{\|}{C}}-O-CH_2$$
$$CH_3-(CH_2)_n-\overset{O}{\overset{\|}{C}}-O-CH$$
$$H_2C-O-\overset{O}{\overset{\|}{P}}(O^{\ominus})-O-(CH_2)_2-\overset{\oplus}{N}H_3$$

Phosphatidyl Serine

$$CH_3-(CH_2)_n-\overset{O}{\overset{\|}{C}}-O-CH_2$$
$$CH_3-(CH_2)_n-\overset{O}{\overset{\|}{C}}-O-CH$$
$$H_2C-O-\overset{O}{\overset{\|}{P}}(O^{\ominus})-O-CH_2-CH(^{\oplus}NH_3)(COO^{\ominus})$$

Phosphatidyl Inositol

$$CH_3-(CH_2)_n-\overset{O}{\overset{\|}{C}}-O-CH_2$$
$$CH_3-(CH_2)_n-\overset{O}{\overset{\|}{C}}-O-CH$$
$$H_2C-O-\overset{O}{\overset{\|}{P}}(O^{\ominus})-\text{Inositol}$$

OH OH HO OH OH

Inositol

Fig. 1.44 Phosphatidyl choline (lecithin), phosphatidyl othanolamine (cephalin). phospatidyl serine, and phosphtidyl insoitol.

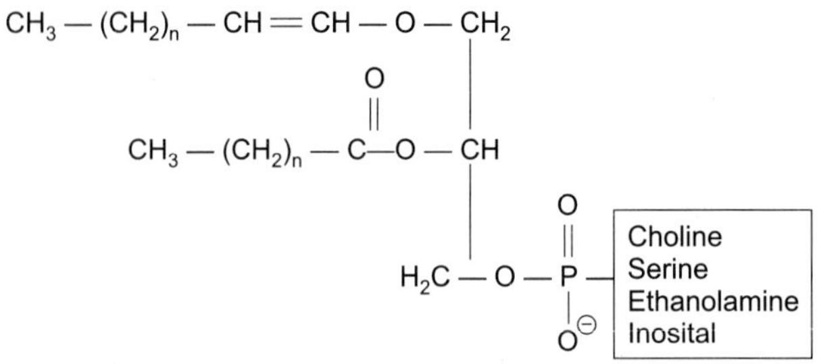

$$CH_3-(CH_2)_n-CH=CH-O-CH_2$$

Choline
Serine
Ethanolamine
Inosital

Fig. 1.45 Gdeneralized formula for a plasmalogen.

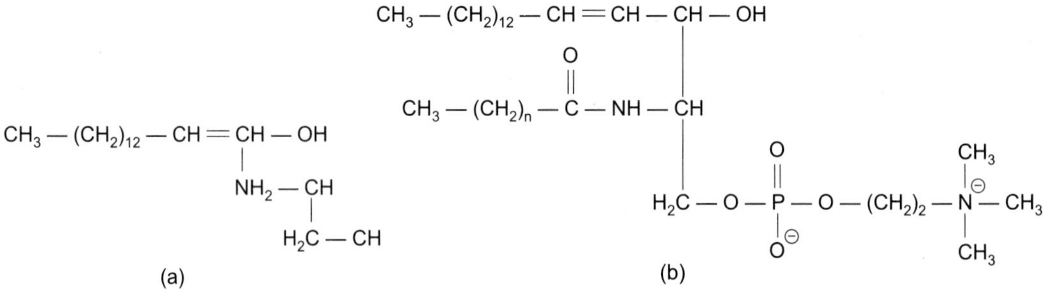

Fig. 1.46 The sphingolipds sphingosine (a) and sphingomylin (b).

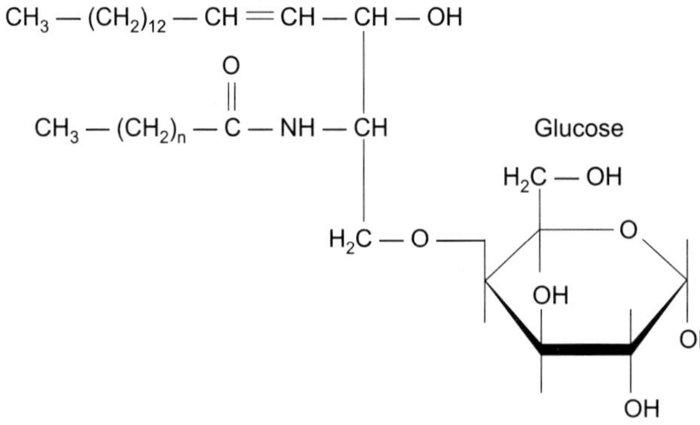

Glucose

Fig. 1.47 A glcycolipid. the molecule shown is a cefrebroside and contains only one sugar, in this instance glucose corebrosldes may also contain galactose in gangliosides the hexose is replaced by a chain of sugars.

Steroids

The steroids are a physiologically important class of complex lipids consisting of a system of fusedcyclohexane and cyclopentance rigs. All are derivatives of perhydrocyclo-pentanophenanthrene, which consists of three fused cyclohexane rings (in a nonlinear arrangement) and a terminal cyclopentane ring (Fig. 1.48a).

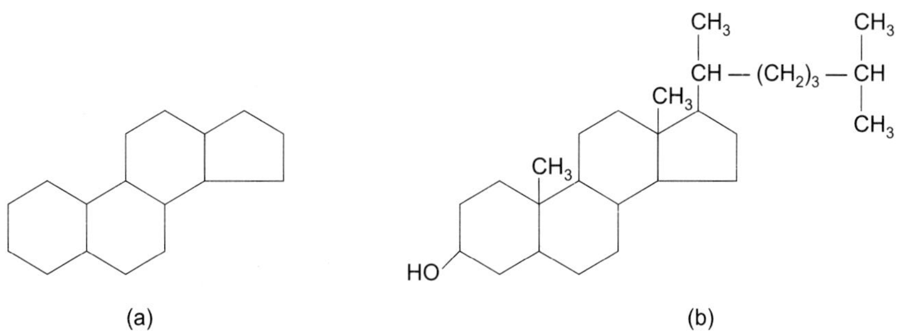

Fig. 1.48 Perthydrocyclopentanophenathrene (a) and one of its derivatives, cholesterol (b).

Steroids have widely different physiological properties. The properties of the steroid derivatives are determined by the group attached to the basic skeleton. Some are hormones (estrogen, progesterone, corticosterone, etc.), some are vitamins (i.e., vitamin D), and other are regular constituents of subcellular structures. Probably the best known steroid and an important constituent of the plasma membranes of certain animal cells is cholesterol (Fig. 1.48b).

Table 1.15 shows the distribution of lipids in the four major cell fractions obtained when liver tissue is dispersed and serially centrifuged. The distribution is believed to be fairly representative at least of animal cells. It should be noted that the mitochondrial and microsomal fractions of the cells are riched in lipid. Most of the lipid in these fractions is phospholipid (i.e., glycerophosphatides, etc.) serving as structural constituents of the mitochondrial and microsomal membranes, where is combined with membrane protein. Phospholipids occur in similar quantities in the chloroplasts of plant cells. Most of the lipids found in the nonsedimenting soluble phase of the cell (or "cytosol") are neutral fats and are in rapid metabolic turnover. The neutral fat content of the cytosol fractions of tissues specialized for fat storage (e.g. adipose tissue) is much higher. We will return to a consideration of lipids in conjunction with several chapters that deal with the structure and function of the major cellular organelles.

SUMMARY

The **lipids** are a heterogeneous collection of macromolecules soluble in nonpoar solvents that play two major roles in cells and tissue: they are sources of reserve energy and structural

Table 1.15 Lipid content of rat liver cell fractions.

Fraction	*Total lipid % of dry weight*	*Percentage of total lipid*		
		phospholypid	*Sterol*	*Neutral fat*
Nuclei	16	90	5	31
Mitochondria	21	90	6	1
Microsomes	32	90	6	0
Soluble	7	30	4	70

constituents of cellular membranes. The simplest lipids, the saturated and unsaturated fatty acids, consist of unbranched hydrocarbon chains teminating in carboxyl groups. In the neutral facts, these carboxyl groups are esterified to the three carbon atoms of glycerol and represent the most abundant form of stored lipid. Neutral facts are in rapid metabolic turnover. Glycerophosphatides, in which one fatty acid chain of a neutral fat is replaced by the polar phosphate group or a phosphate derivative, are amphipathic and, as such, play an important part in the structure and organization of cell membranes. The most physiologically diverse lipids are the steroids, formed by a series of fused; ringed hydrocarbons. Steroids serve as hormones, as vitamin, and occasionally as constituents of cell membranes.

CHAPTER 2

Introduction to Molecular Biology

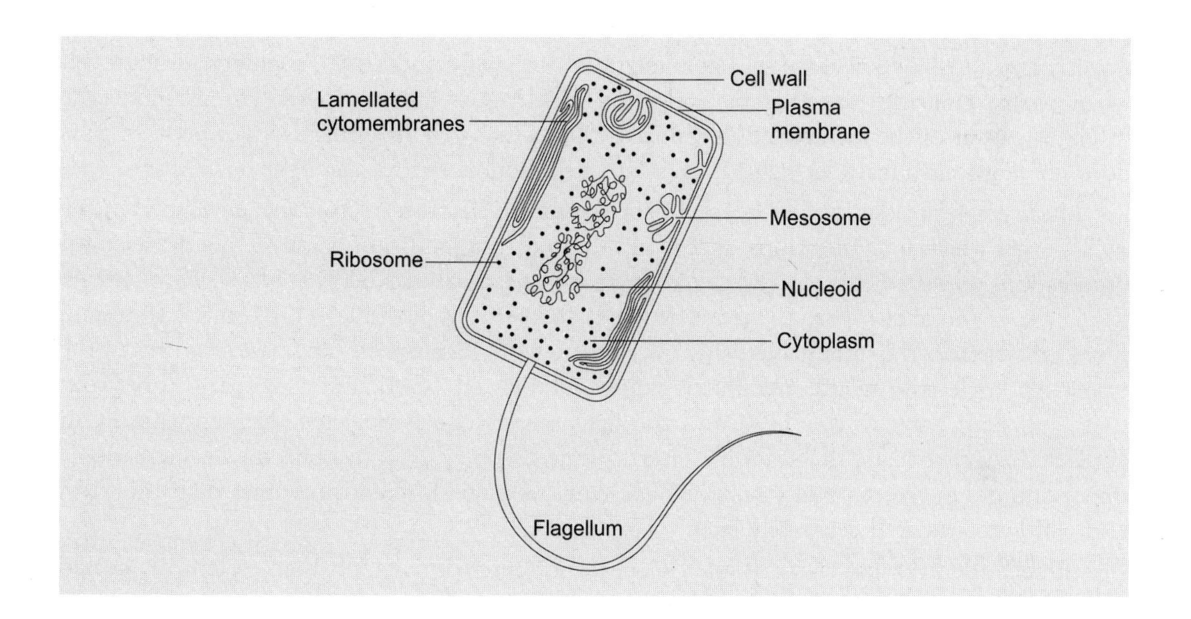

MICROSCOPY

Fundamentals of Light Microscopy and Transmission Electron Microscopy

Until the 1940s, most of our knowledge concerning the structure and organization of cells was obtained by light microscopy, and major structure and organelles, including the cell wall, nuclei, chromosomes, chloroplasts, mitochondria, vacuoles, centrioles, flagella, and cilia, had been described.

The smallest distance, d, between two points resolvable as separate points when viewed through lenses is given by the relationship.

$$d = \frac{0.6}{n \sin} \quad (1)$$

In this equation, is the wavelength of the light (radiation) employed to illuminate the specimen; n is the refractive index of the air or liquid between the specimen and the liner and is the aperture angle. The product, n sin is called the lens numerical aperture, and for a good microscope lens, it would be about 1.4.

Equation 1.1 also shows that the resolving power of s microscope varies with the wavelength of the source of illumination. The human eye cannot directly detect light having a wavelength of less than about 400 nm (see Table 2.1 for metric measurements). Therefore, in the case of the light microscope, the maximum resolving power is about 0.6(400/1.4), or about 0.17 μ, that is, points less than about 0.2 μ apart cannot be distinguished as separate points by light microscopy (in practice, the limit is closer to .05 μ). Using glass optic of the finest quality, it is possible to observe cells at a magnification of about 2000X. Resolution is improved when source emitting rays that have shorter wavelengths are employed. For example, the resolving power of the ultraviolet light microscope (which requires quartz optics because glass does not transmit ultraviolet light) is approximately double that of the light microscope.

Much greater resolution has been obtained with the electron microscope, developed in the 1930s, with which magnifications of several hundred thousands are possible. The wavelength of radiations used with the electron microscope is typically about 0.005 nm (0.05 A), although resolution of the order of an angstrom or less is theoretically possible. This is many thousand times greater than that attainable using microscope with glass optics. The basic features of the transmission electron microscope (often simply abbreviated TEM), and a comparison between the component parts of the TEM and the light microscope is depicted diagrammatically in Fig. 2.1. In recent years, the scanning electron microscope (SEM) has become an increasingly important tool of the cell biologist. The SEM employs quite different principles than the TEM and will be considered separately later.

In both the light and electron microscopes, the source of radiation is an electrically heated tungsten filament. In the light microscope, the light emitted from the glowing filament is focused by a condenser onto the specimen to be observed. In the transmission electron microscope, the condenser focuses electron electrons emitted by the excited tungsten atoms into a boam and directs this onto the specimen. While the condenser of a light microscope consists of one or a few glass lenses, the condenser of the electron **microscope** consists of

Table 2.1 Metric measurements of size.

1 meter (m)	= 39.4 inches (in)
1 meter (m)	= 100 centimeters (cm)
1 centimeter (cm)	= 10 millimeters (mm)
1 millimeter (mm)	= 1000 micrometers (urn) or microns (u)
1 micrometer (um)	= 1000 nanometers (nm) or millimicrons (mu)
1 nanometer (nm)	= 10 angstroms (A)

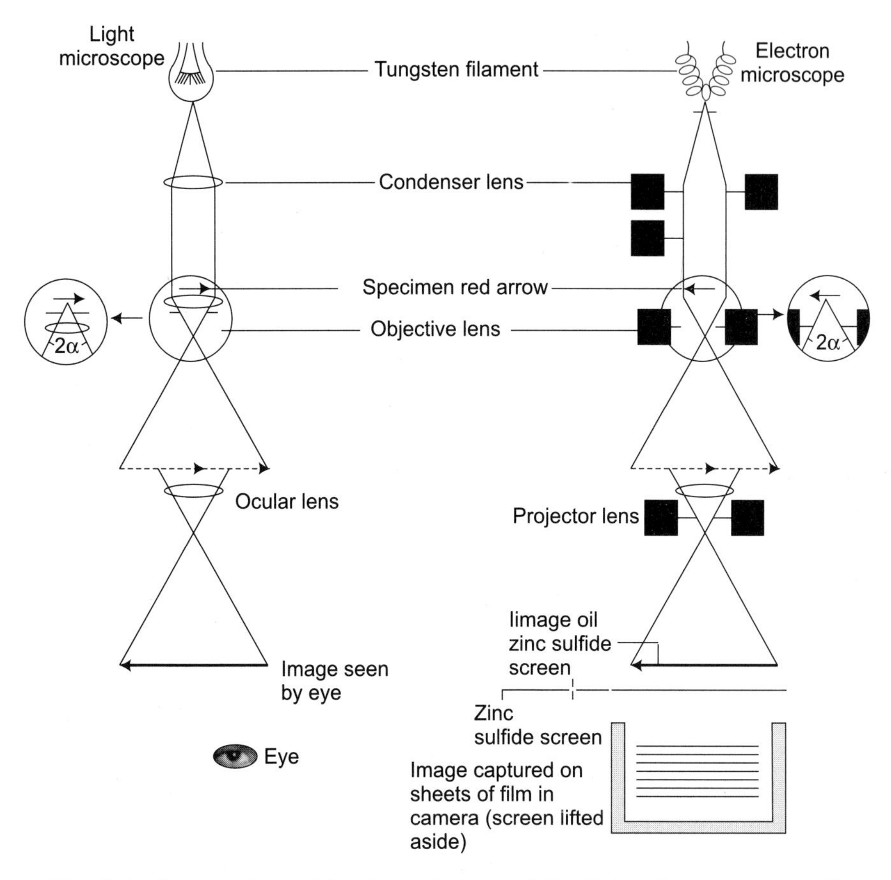

Fig. 2.1 A comparison of the basic features of the light microscope and TEM.

several large, circular electromagnets. Indeed all "lenses" of the electron microscope are electromagnets. In both microscopes, the radiation passes through the specimen and is then refocused by the objective lenses. The last lens of the light microscope is the ocular, through which the image may be viewed with the eye. The image of the electron microscope is viewed after is magnetic projection onto a zinc sulfide screen. The molecules of the screen excited by the impining electron and emit visible light during their return to ground state. Alternatively, the image may be captured on photographic film housed in a special camera mounted below the movable zinc sulfide screen.

The lenses of the light microscope have a fixed focal length and are focused by moving them nearer to or further from the specimen. In the electron microscope, focusing is accomplished by manipulating the amount of current flowing through the windings of the series of electromagnet lenses. This alters the electromagnetic force through which the electron beam must pass. The column through which the electron beam passes must be evacuated of air. If the vacuum is inadequate, the electrons would be scattered by collisions with gas molecules. Consequently, the specimen, the filament, the electromagnets, and the zinc sulfide screen are all mounted within a sealed compartment connected to a vacuum pump. In order to avoid-excessive scattering or absorption of elections by the specimen itself, the material to be examined must be cut into extremely thin sections. Two special forms of light microscopy warrant further description because of their widespread as use and special applications: phase-contrast microscopy and fluorescence microscopy.

PHASE-CONTRAST MICROSCOPY

Although most regions of an unstained cell are transparent, they may have different refractive indexes. Consequently, light rays travel through these regions at different velocities and may be refracted or bent to different extents. The phases of light rays that pass directly through an object and those that pass across its edges (i.e., at the in terrace where the refractive index changes) will necessarily be altered. The phase increases the contrast between the object in focus and its surroundings. In the phase-contrast microscopes, the phases of light rays entering the object are shifted by an annular diaphragm below the condenser. The phases of rays passing through and around the object are shifted again by a phase plate in the objective lens. The result is a striking increase in the contrast of the object as certain regions appear much brighter (owing to additive effects of rays brought into phase), while other regions appears much darker (owing to the cancelling effects of rays shifted further out of phase).

Fluoresence Microscopy. Certain chemical substances emit visible light, when they are illuminated with ultraviolet light. The effect is termed fluorescence and is put to use in the fluorescence microscope in which ultraviolet light rays are focused on the specimen. Some cellular components (e.g., cytoplasm, mitochondria, and certain granules) possess a natural fluorescence and appear in various colors. Other, nonfluorescing structures can be made of fluoresce by staining them with fluorescent dyes (fluorochromes). One of the most contemporary uses of fluorescence microscopy involves the preparation of antibodies that will bind to specific cellular proteins. The antibodies are first complexed with fluorescein (a fluorescent dye), and the fluorescein-labeled antibody is then applied to the cells. Cell structures containing the specific proteins capable of binding the fluorescein-labeled antibody are caused to fluoresce when examined with the fluorescence microscope, dramatically revealing their detail.

Preparation of Materials for Microscopy

The preparation of biological material for examination with either the light microscope or the transmission electron microscope involves a series of physical and chemical manipulations that include (1) fixation, (2) embedding, (3) sectioning, and (4) mounting.

Fixation. One notable advantage of the light microscope is the ability to observe whole, living cells. It is also possible to employ "vital stains," which improve contrast but do not interfere with normal cell activity. More frequently, however, the cells are first killed and fixed. The fixation step is intended to preserve the structure of the material by preventing the growth of bacteria in the sample and by precluding postmortem changes. Formaldehyde and osmium tetroxide (OsO_4) are examples of fixatives most often employed for light microscopy. OsO_4 has a very high electron density, and since this gives contrast to the resulting image, OsO_4 has also found widespread use as a fixative in electron microscopy. Other popular fixatives include potassium permanganate and glutaraldehyde. After fixation for the required length of time, the samples are dehydrated by successive exposures to increasing concentrations of alcohol or acetone.

Embedding. Cells or tissues to be examined by light microscopy are usually embedded in warm, liquid paraffin wax. The wax, which both surrounds the tissue and infiltrates it, hardens upon cooling, thereby supporting the tissue externally and internally. The resulting solid paraffin block is then trimmed to the appropriate shape before being sectioned. The ultrathin sections required for electron microscopy necessitate the use of harder embedding and infiltrating materials, such as methacrylate, Epon, or Vestopal. These initially are in liquid form and are poured into small molds containing pieces of the fixed tissue; upon heating, they polymerize into hard plastics.

Sectioning. The trimmed blocks containing the embedded samples are sectioned using a microtome. In this instrument, the block is sequentially swept over the blade of a knife that cuts the block into a series of thin sections forming a ribbon. Between each stroke, the block is advanced a short distance toward the knife. For light microscopy, the microtome knives are usually constructed of polished steel and can provide sections several microns thick. The sections for electron microscopy must be much thinner (typically 100 to 500 A) and require more elaborate microtomes (called ultramicrotomes), which either diamond knives or knives prepared by fracturing plate glass are used in place of polished steel.

Mounting. Sections prepared for light microscopy are mounted on glass slides and may be stained with dyes of various colors that specifically attach to different molecular constituents of the cells. Sections to be examined with the electron microscope are generally not stained (no colors are seen with the electron microscope), although contrast may be improved by "poststaining" with electron-dense materials such as uranyl acetate, uranyl nitrate, and lead citrate. The sections are mounted on copper "grids" (small disks perforated with numerous openings) that have been coated with a thin (sometimes monomolecular) film of carbon. That grid supports the film, which in turn supports the thin section. Thus the beam of electrons must pass through the spaces of the grid, the supporting film, and the section before striking the fluorescent screen.

Specialized Application of Transmission Electron Microscopy

Shadow Casting. In shadow casting, the sample (usually containing small particles such as viruses or macromolecules) is spread on a coated grid which is then placed in an evacuated chamber. A chromium and platinum wire is heated until the metal is vaporized, and the vapor is deposited onto the sample at a precise angle. The metal piles up in front of the sample particles but leaves clear areas behind them. If the resulting electron photomicrographs are printed in reverse, the areas containing the electron-dense metal that had piled up against the particles appear light, while the electron-transparent areas behind the particles appear as dark shadows. Because the vaporized metal atoms tend to be projected in a straight line, the shadows are cast at precise angles. In this manner, the general shape and profile of a particle may be discerned.

Negative Staining. In the negative-staining procedure, the sample (again small particles such as viruses or macromolecules) is surrounded by an electron-dense material, such as phosphotungstic acid, that permeates the open superficial interstices of the sample. When the excess material is removed, the sample particles appear as light (i.e., electron-transparent) areas surrounded by a dark background.

Freeze Fracturing. Freeze fracturing is a technique in which the tissue is first fractured (i.e., cracked) along planes of natural weakness that run through each cell. These planes generally occur between the two layers of molecules that comprise the limiting membrane around the cells various vesicular organelles. The tissue to be freeze-fractured is first impregnated with glycol and then frozen at about–130°C in liquid Freon. The frozen tissue is transferred to an evacuated chamber containing a microtome and steel knife (also maintained at about–100°C using liquid nitrogen). The microtome knife is used to produce a fracture plane across the tissue. When the plane of the fracture intersects the membrane of a vesicular structure (e.g., nucleus, mitochondrion, vacuole, etc.), the membrane is split along its center, producing two "half-membranes". These are called the E half and P half. The E half formerly faced the cell's external phase and the P half faced the internal phase (cytosol). One surface of each half-membrane is the original membrane surface while the other surface is the newly exposed fracture face. The vacumm of the chamber is then used to sublimate water on the cut surface to a depth of several hundred angstroms. New membrane faces exposed by sublimation are termed Es and Ps. An electron-dense combination of metal (usually platinum) and carbon is then deposited on the cut surface at an angle and piles up in front of and behind projections form the surface, as well as in pits and depressions. Additional carbon is added to form an electron-transparent backing.

The shadowed and coated tissue is removed from the chamber, and the tissue itself is either floated off or dissolved away, thereby leaving only the carbon-platinum "replica". The replica is trimmed to the proper size, placed on a grid, and examined with the transmission electron microscope. Note that the replica is actually a template like impression of the distribution of particles in the original specimen. The electron beam readily passes through portions of the replica containing the carbon but is absorbed by the areas containing the platinum. The resulting images, which have a three-dimensional impact, are considerably different from those obtained with sectioned material (Fig. 2.2a, b).

THE SCANNING ELECTRON MICROSCOPE

Scanning electron microscopy has become an increasingly popular technique since in introduction as a biological tool in the 1960s. With this technique, the surface topography of a specimen may be examined to considerable detail. At the present time, resolution id of the order of 50 A, so that most specimens examined are the sizes of whole cells or clusters of cells. The organization of the scanning electron microscope (SEM) is shown in Fig. 2.3 and is

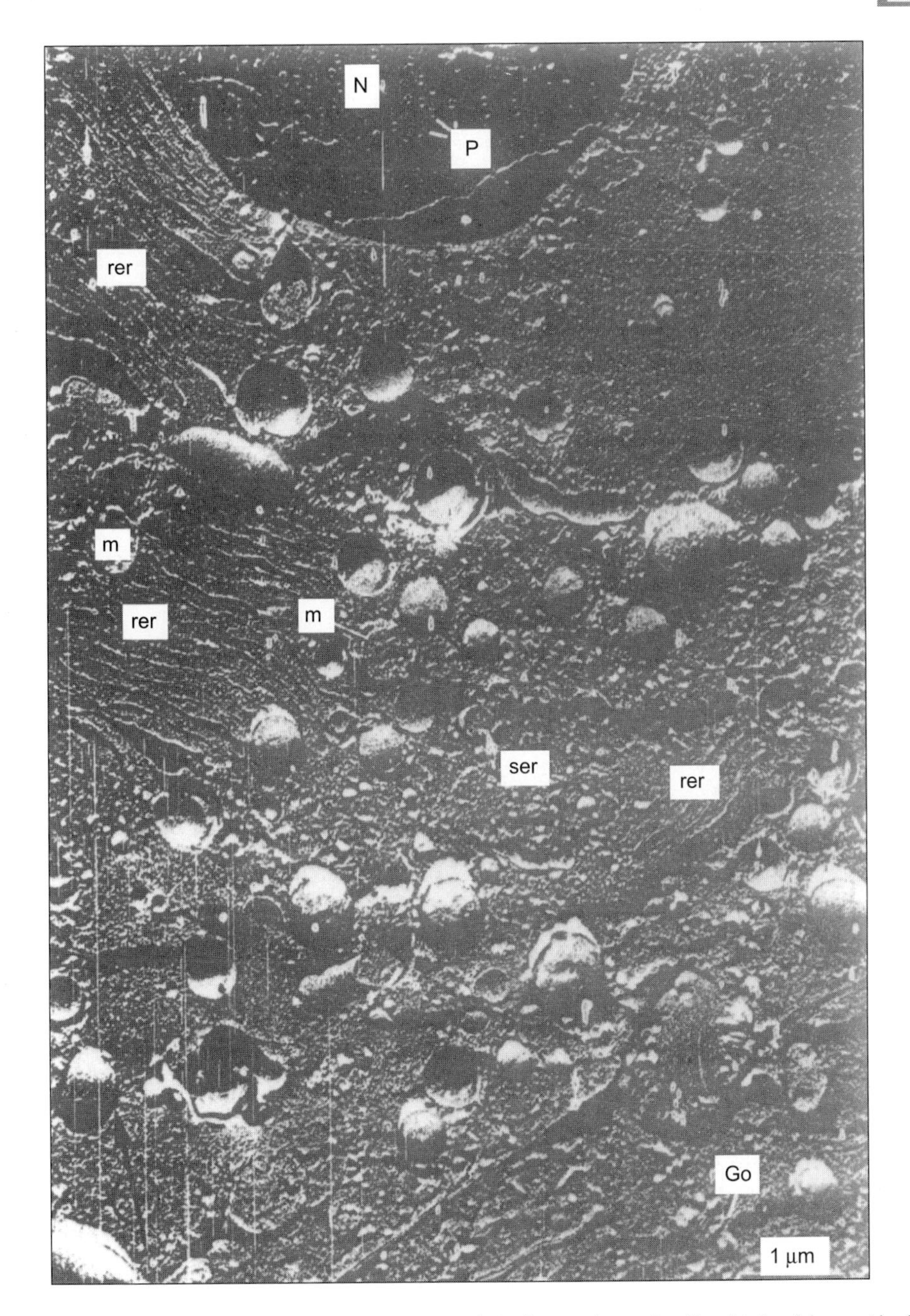

Fig. 2.2a A comparison of electron photomicrographs of similar regions of cells obtained by sectioning and by freeze-fracturing (a and b). Micrographs of liver tissue (N, nucleus; P, pores; rer rough endoplasmic reticulum; ser, smooth endoplasmic reticulum; Go, Golgi body; m, mitochondria; Mt, microbody; BC, bile canaliculus; Gl, glycogen) (c and d). Cortical regions of egg cells (pm, plasma membrane; cg, cortical granule). The direction of shadowing is indicated by the circled arrow (courtesy fo Dr. E.G. Pollock).

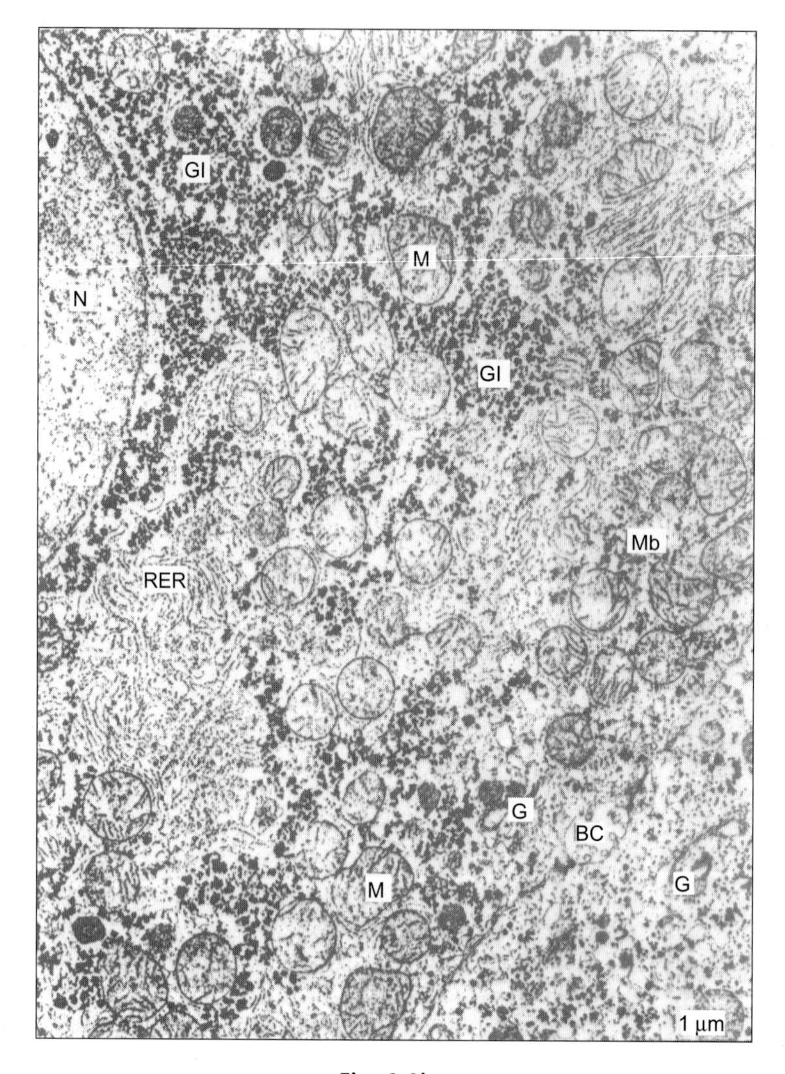

Fig. 2.2b

basically similar to the TEM. However, instead of the electron beam passing through (i.e., being "transmitted" by) the specimen, the interaction of the electrons of the beam (called "primary" electrons) with the surface of the specimen causes the emission of "secondary" electrons from the surface. The beam rapidly scans back and froth over the surface of the specimen, thereby producing bursts of secondary electrons. Greater numbers of secondary electrons are produced when the beam strikes projections from the specimen surface than when the beam enters a pit or depression in the surface. Hence, the number of secondary electrons produced at each point on the specimen surface, as well as the direction in which scattering occurs, depends upon the surface topography. Therefore, there are quantitative and qualitative differences in the secondary

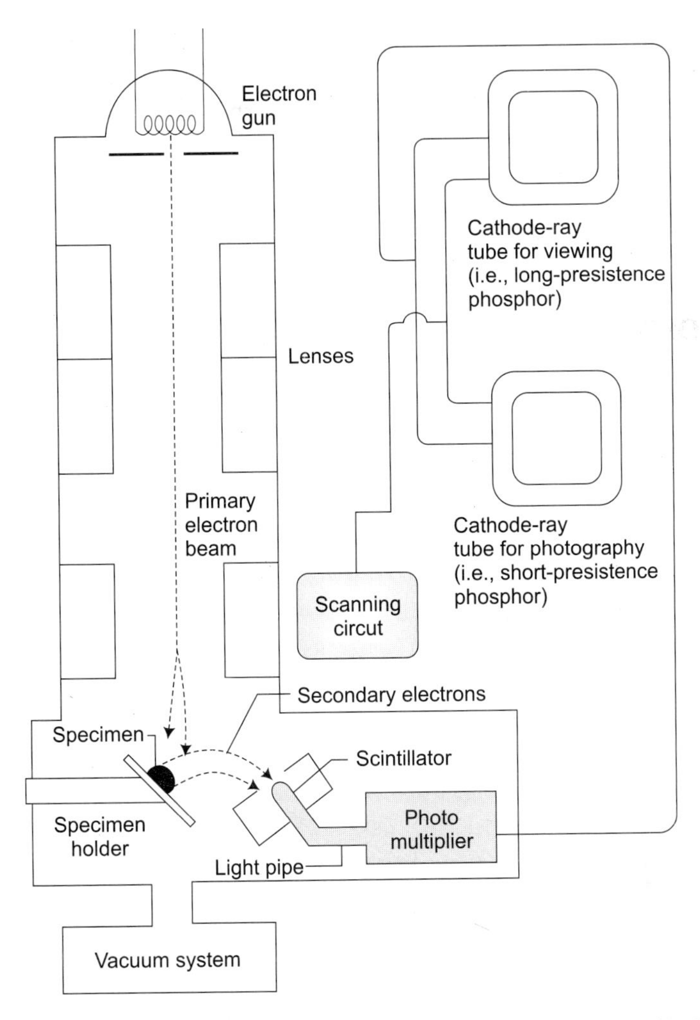

Fig. 2.3 Essential components of scanning electron microscope (SEM).

electron bursts produced by the scanning electron beam. These ultimately give rise to an image in the following way.

Secondary electrons ejected at each point on the specimen surface are accelerated toward a positively charged scintillator located at the side of the specimen. Light scintillations created upon impact of these electrons with the scintillator are conducted by a light guide to the photocathode of a photomultiplier tube. Electrical pluses produced in the photomultiplier tube are then amplified, and the resulting signal is released to a cathode-ray tube. The result is an image much like that of a television, consisting of light and dark spots. The scanning of the specimen surface by the primary electron beam is synchronized with the projection of a beam on the television screen in such a way that each portion of the specimen is reproduced in corresponding region of the television image.

Samples to be examined are usually coated first with a metal (typically a gold-palladium alloy), forming a layer about 500 A thick, and then are affixed to a supporting disk that is placed in the beam path. The metal coating efficiently reflects the primary electrons of the beam and also produces large numbers of secondary electrons.

Since the specimen being examined with the SEM can be rotated, it is possible to obtain views from different angles. This provides additional information of the material being studied Fig. 2.4 and 2.5.

STEREO MICROSCOPY (STEREOSCOPY)

True three-dimensional (i.e., stereoscopic) images of the specimen being studied can be obtained if one photomicrograph is taken as through the specimen were being viewed with the left eye only, and a second is obtained representing the right eye view. (The two views are obtained by a minor tilling of the sample in the horizontal plane). When the two micrograph are placed side by side and the stereo pair is viewed through the appropriate pair of lenses (called "stereo viewers"). A striking three-dimensional is seen revealing details and geometric relationship that cannot be discerned from a single photomicrograph. Stereo views of the surface topograph of tissues and cells are readily obtained with specimen prepared for SEM study.

The internal organization of cells is revealed in three dimensions by high-voltage transmission electron stereoscopy. In this procedure, cells are placed or cultured on a conventional grid, are then fixed and dehydrated, and the grid and cells are sandwiched between layers of carbon. The samples are examined in a TEM in which the accelerating voltage is great enough to penetrate the entire thickness of the cell (about 1 million volts). The cells are photographed at various tilt angles to produce the stereo pairs needed for the three-dimensional image.

The viewing of stereomicrographs may present some difficulties, especially for the novice. Generally, fewer problems are encountered with SEM stereoscopic views, since the objects in the photomicrographs are opaque (i.e., certain object are clearly in front of others). Transmission stereomicrographs are more difficult to assimilate and interpret because most of the objects are translucent. However, no other procedure provides direct images of the three-dimensional morphology of the cell's interior. A single stereo pair can reveal the entire population of mitochondria, lysosomes, or other organelles (see below) distributed through the cell.

CELL STRUCTURE

Cell Structure: A Preview

Free-living cells and the cells of multicellcular organisms are subdivided into two major classes-eukaryotes (i.e., "true nucleus") and prokaryotes (i.e., "before nucleus"). In eukaryotes the constituents of the cell nucleus (chromosomes, DNA, etc.) are separated from the rest of the cell

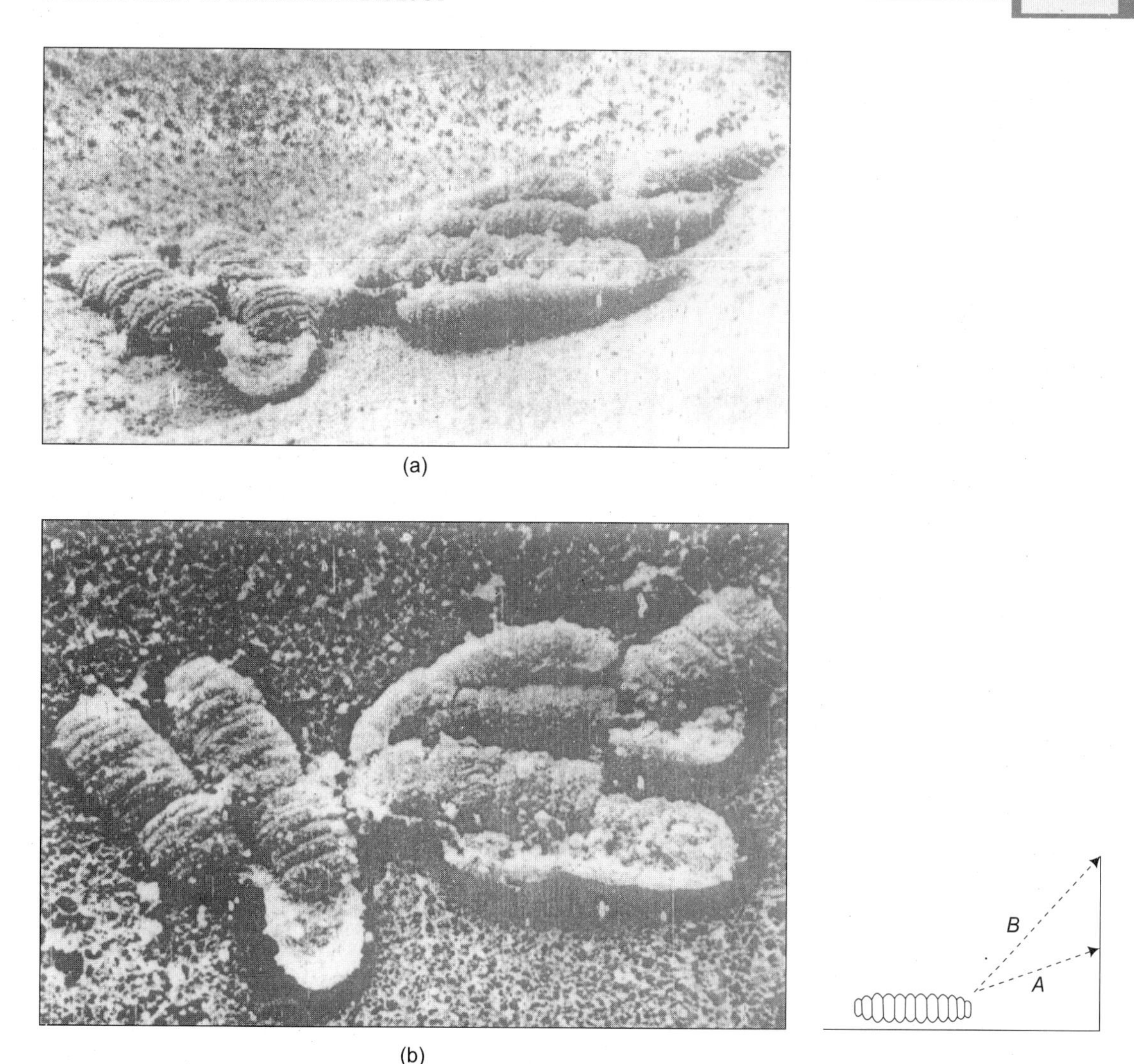

Fig. 2.4 Scanning electron micrographs of chains of bacterical cells (*Simonsiella*) viewed from two (i.e., a and b) different aangles showing the varying perspectives attainable. Magnification 4000 × (Courtesy of Drs. D.A. Kuhn, J. Pangborn and J.R. Woods).

by a boundary membrane. Whereas in prokaryotes these materials are not separated. Although the presence or absence of a true nucleus is the most obvious distinction between eukaryotic and prokaryotic cells. It will soon become clear than these two groups of cells also differ in many other important respects. Essentially all animal and plant cells are eukaryotic, whereas prokaryotic cells include bacteria, blue-green algae, and the so-called pleuropneumonia-like organisms (PPLO) or mycoplasmas.

EUCARYOTIC CELLS: THE GENERALIZED ANIMAL CELL

Animal cells vary considerably in size, shape, organelle composition and physiological roles. Consequently, there is no "typical" ell that can serve as an example of all animal cells. There are however a number of cell structures common to the majority of animal cells that are similar or identical in organization. These structure are depicted in the generalized animal cell diagrammed in Fig. 2.5 and described briefly in the following sections. They are dealt with in greater detail in later chapters that are individually devoted to the structure and functions of cell organelles. Fig. 2.6 is an electron photomicrograph of an animal cell containing many of the structures to be discussed.

The Plasma Membrane. The contents of the cell (cytoplasm and cytoplasmic organelles) are separated from the external surrounding by a limiting membrane, the plasma membrane also called cell membrane or plasmalemma), which is composed of protein, lipid, and carbohydrate. This structure regulates the passage of materials between the cell and its surroundings and in some tissues is involved in intercellular communication (e.g., nerve tissue). In some tissue cells, portion of the plasma membrane is modified to from a large number of fingerlike projections called microvilli because of their resemblance to the much larger villi of the small intestine. The microvilli greatly increase the surface area of the cell and provide for the more quantitative passage of materials across the plasma membrane. When a large number of the cells are in close contact with one another (as, for example, in a tissue), it is not unusual to observe special from of junctions between opposing plasma membranes. These take the from of tight junctions, desmosomes, and gap junctions.

The plasma membrane should not be thought of as uniform or as having the same composition over its entire surface. Instead, the composition and organization vary in different regions of the membrane. Some areas of the plasma membranes of adjoining cells in the tissue; other areas face the bile channels (bile canaliculi) into which substances are secreted by the liver cell. Still other portions of the plasma membrane face the epithelial lining of capillaries from which substances are absorbed. Each of these regions of the plasma membrane is differently composed and differently organized and, in fact, is continually undergoing change and reorganization.

The Endoplasmic Reticulum and Ribosomes. With in the cytoplasm of most animal cells in an extensive network of branching and anastomosing membrane-limited cannels or eisternae collectively called the endoplasmic reticulum (Fig. 2.7). the membranes of the endoplasmic reticulum (usually abbreviated ER) divide the cytoplasm into two phases; the luminal phase and the hyaloplasmic phase or cytosol. The luminal phase consists of the material enclosed within the cisternae of the endoplasmic reticulum, while the cytosol surrounds the ER membranes.

In the cytosol are large numbers of small particles called ribosomes. These particles are distributed either along the hyaloplasmic surface of the endoplasmic reticulum ("attached" ribosomes) or free in the hyaloplasma ("free" ribosomes). There is some evidences that the free ribosomes are interconnected by fine filaments. Endoplasmic reticulum with associated ribosomes is called rough ER (RER).

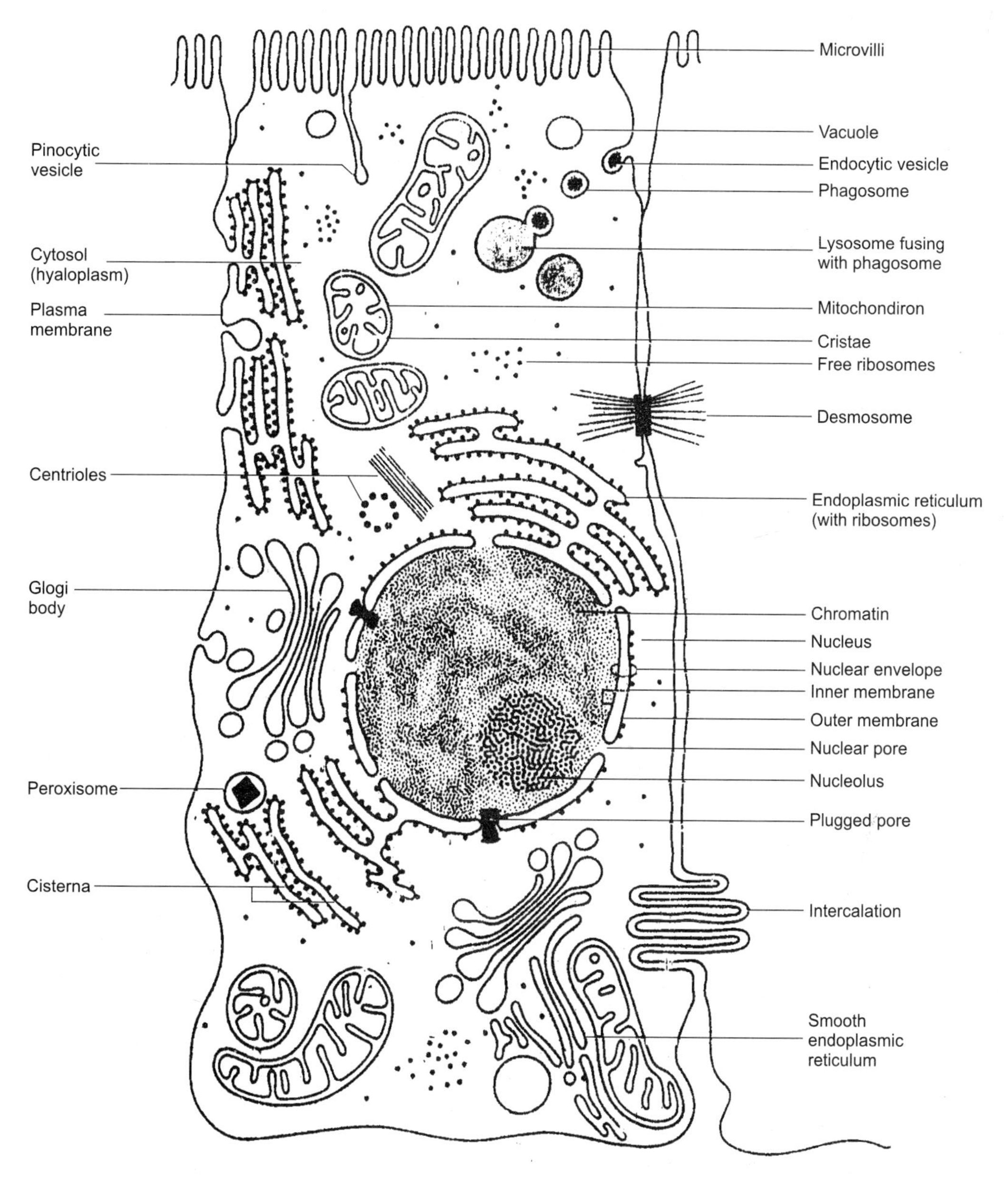

Fig. 2.5 The generalized animal cell.

Whereas smooth ER (SER) is devoid of attached ribosome. Ribosomes carry out the synthesis of the cell's proteins. Certain potions of he endoplasmic reticulum may from a continuum with the plasma membrane and the nuclear envelop.

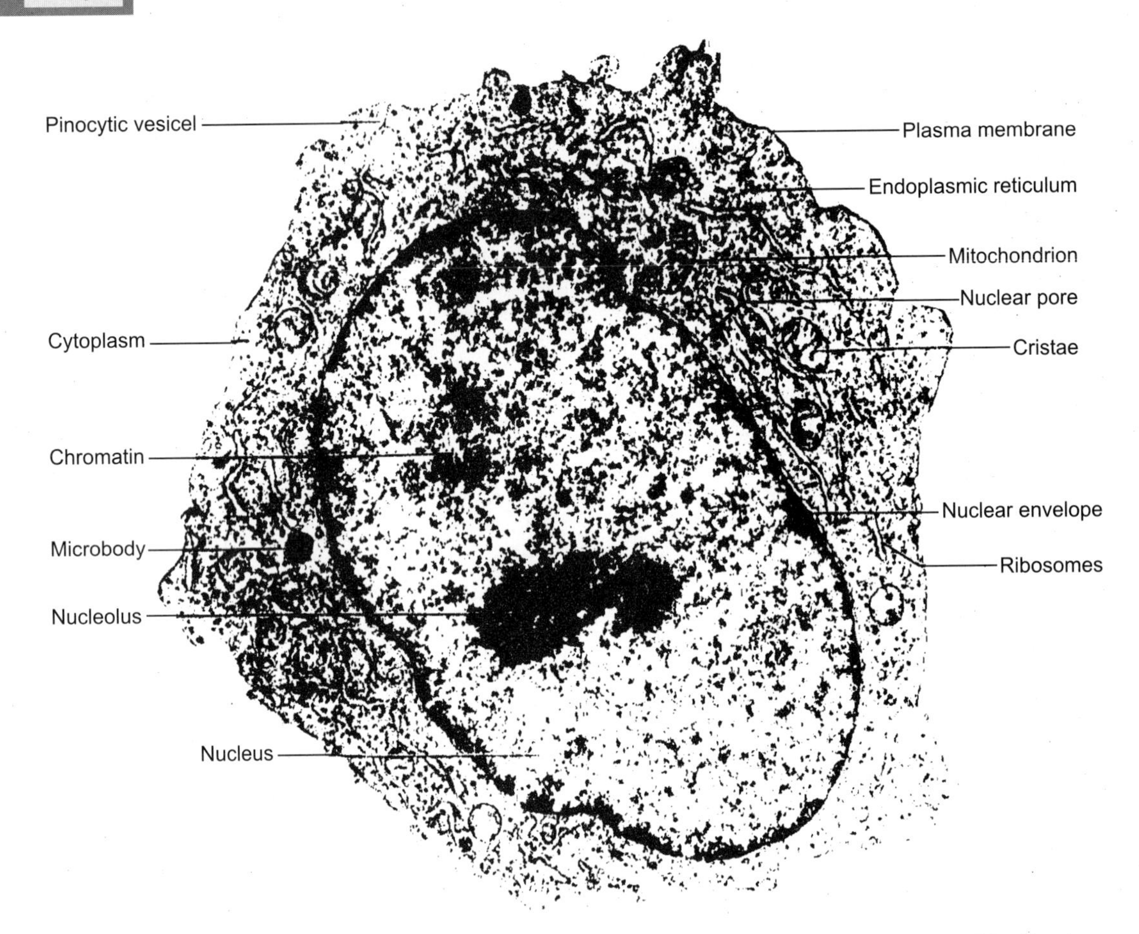

Fig. 2.6 Electron micrograph of whole animal cell. Magnification, 13,000 × (Courtesy of R. Chao).

Mitochondria. With in the cytoplasm are numerous vesicular organelles called mitochondria (Figs. 2.8 to 2.10). Each mitochondrion is bordered by a double membrane. The outer membrane is smooth and continuous, and the inner membrane displays numerous in folding called cristae (or crystal membranes). These greatly increase the surface area of the inner membrane. The space between neighboring cristae is called the mitochondrial matrix and often contains crystal-like inclusions. Mitochondria are engaged in numerous metabolic functions in the cell, including energy-producing phases of carbohydrate and fit metabolism (called respiration) and prophyrin bio-synthesis.

The Golgi Apparatus. The golgi apparatus also called golgi body or golgi complex, consists of a unique network of cisternae similar to and possibly continuous with the luminal phase of the endoplasmic reticulum. The cisternae of a Goigi body are often stacked together in parallel rows, and this state the body is referred to as a dictyosome. The Golgi apparatus is frequently surrounded by vesicles of various sizes that apparently are discharged from the margins of the

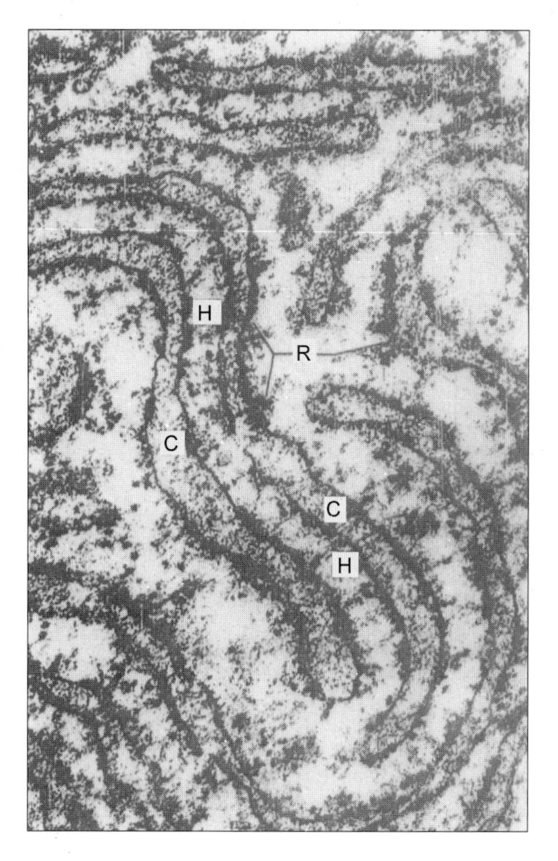

Fig. 2.7 Endoplasmic reticulum.

main body of the organelle (Fig. 2.10). Varieties of functions are ascribed to the Golgi apparatus, including secreoty activity (especially the secretion of enzymes) and the prolifereations of additional membranes for the cell.

Lysosomes. Many cells contain vesicular structures that are generally smaller than mitochondria and are called lysosomes (Fig. 2.13). The lysosomes and are bounded by a single membrane and contain quantities of various hydrolytic enzymes capable of digesting protein, nucleic acid, polysaccharide, and other materials. Under normal conditions, these enzymes are inactive and isolated from the cytoplasm. However, if the lysosomal membranes are ruptured, the ruptured, the released enzymes can degrade the cell. Lysosomes are believed to be responsible for the intracellular digestion of food particles ingested by the cell, the scavenging of worn and therefore poorly functioning organelles, and a number of other cell functions.

Peroxisomes and Glyoxysomes. Many cells contain small numbers of peroxisomes and/or glyoxysomes. These small organelles, which are bounded by a single membrane, contain a number of enzymes whose functions are related to the metabolism of hydrogen peroxide and glyoxylic acid.

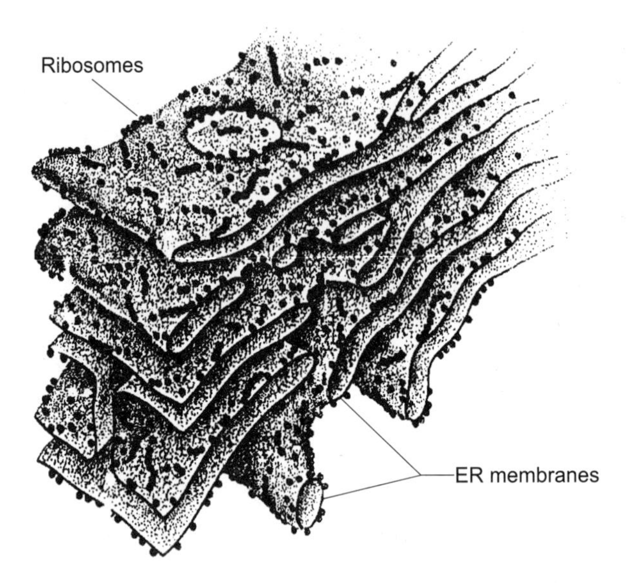

Fig. 2.8 Ribosomes EM.

The Nucleus. The nucleus is a relatively large structure frequently but not always located near the center of the cell. The contents of the nucleus are separated from the cytoplasm by two membranes that together form the nuclear envelope. At various positions, the outer membrane of the envelop (membrane 1) fuses with the inner membrane (membrane 2) to form pores. Nuclear pores provide a measure of continuity between the cytosol and the contents of the nucleus. Occasionally, the nuclear pores are plugged by a dense material. The outer nuclear side and may from continuities with the membranes of the endoplasmic reticulum. Since the latter may be continuous with the plasma membrane, the perinuclear space (i.e., the space between the inner and membrane, of the nuclear envelop) corresponds to the luminal phase and may be considered external to the cell (see Fig. 2.10).

The nuclear envelop and the pores that pentetrate it are dramatically revealed in freeze-fracture preparations (Fig. 2.11).

The cytosol-contacting half of the outer nuclear membrane is fractured away, exposing the inner half of that membrane (i.e., face EF of Fig.). Also fractured away are pieces of the inner nuclear membrane (the half-membrane that faced the cytosol (i.e., PF of Fig. 2.11). The nuclear pores penetrate both membranes and in Fig. 2.10 appear to be non randomly distributed in the nuclear envelop.

The nucleus contains the genetic machinery of the cell (chromosomal DNA, histones, etc.), using either the light microscope or the electron microscope, the nucleus often revels one or more dense, granular structure called nucleoli. Nucleoli are not bounded by a membrane and appear to be formed in part from localized concentrations of ribosomal materials.

Flagella and Cilia. Many free-living cells (such as protozoa and other microorganisms) possess locomotor organelles that project from the cell surface. In animals cells these are either flagella

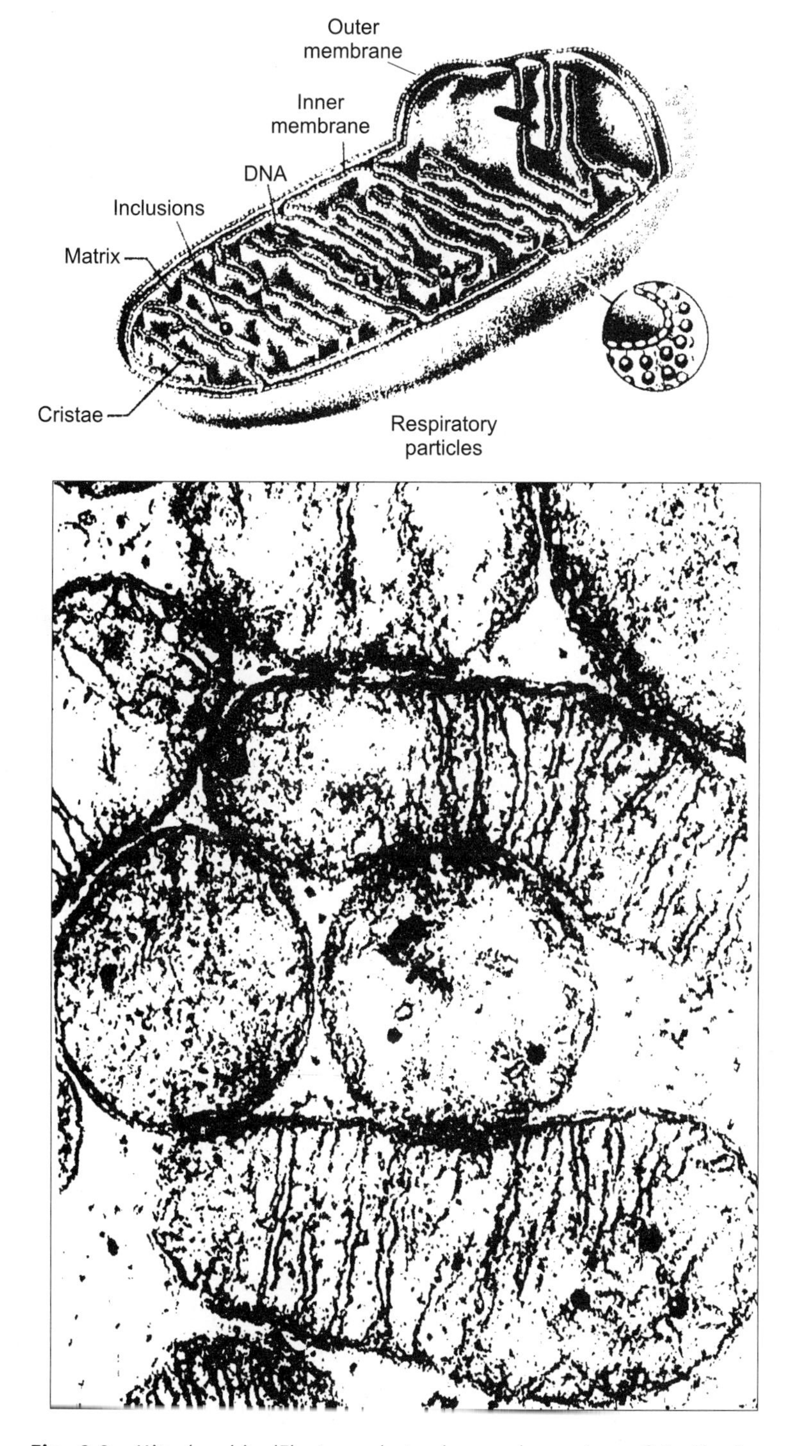

Fig. 2.9 Mitochondria (Electron photomicrograph courtesy of R. Chao).

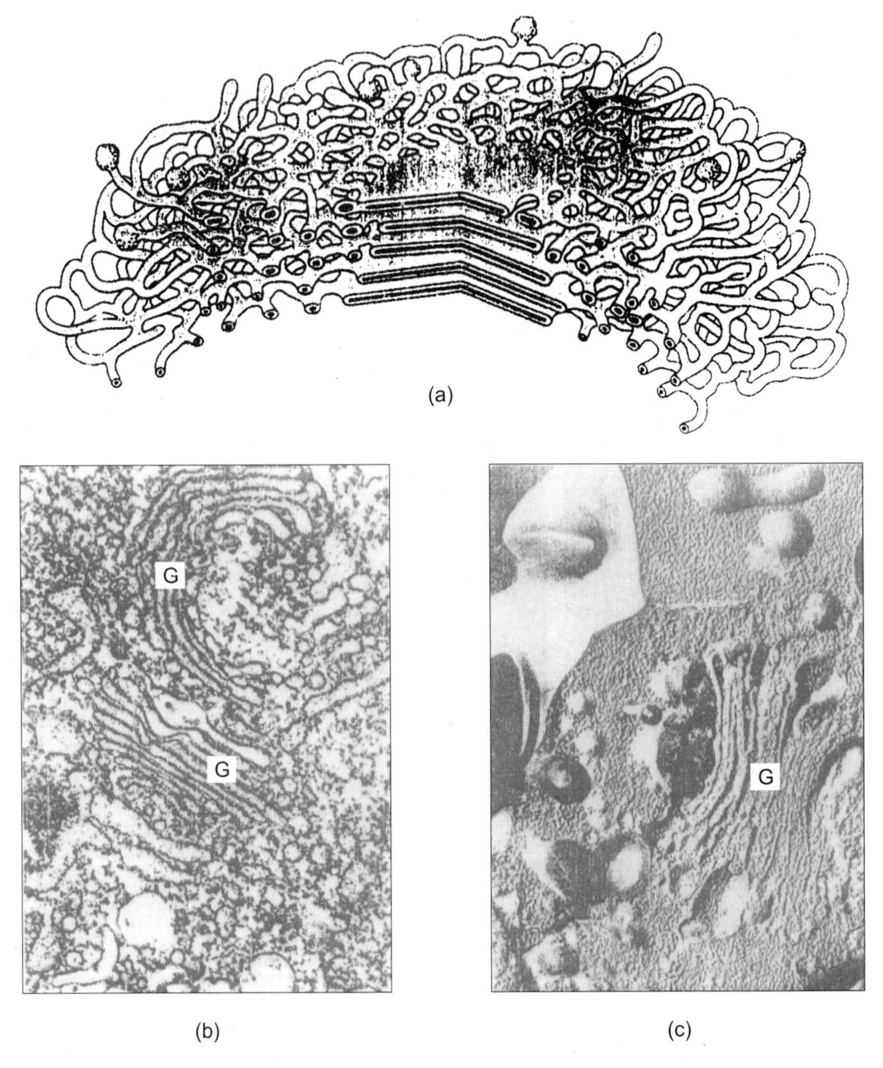

Fig. 2.10 The golgi apparatus, (a) Diagram depicting the three-dimensional relationship among the cisternae, channels, and peripheral vesicles. (b) Typical appearance of Golgi bodies (G). In thin sections. (c) Appearance of golgi bodies in freeze-fractured cells (Photomicrographs courtesy of Dr. E.G. Pollock).

or cilia (Fig. 2.12). The tissue cells of multicultural animals may also contain cilia, but they are employed here to advance a substrate across the cell surface (such as mucus in the respiratory tract or the egg cell during its passage through the oviduct and not for self-locomotion. The organelles are called cilia when they are short but present in large numbers and are called flagella when long but few in number. Each cilium or flagellum is covered by an extension of the plasma membrane. Internally, these organelles contain a specific array of microtubules which run from the basal plate toward the tip of the structure. This array consists of two central microtubles and nine pairs of peripheral (outer) microtubles (Fig. 2.12).

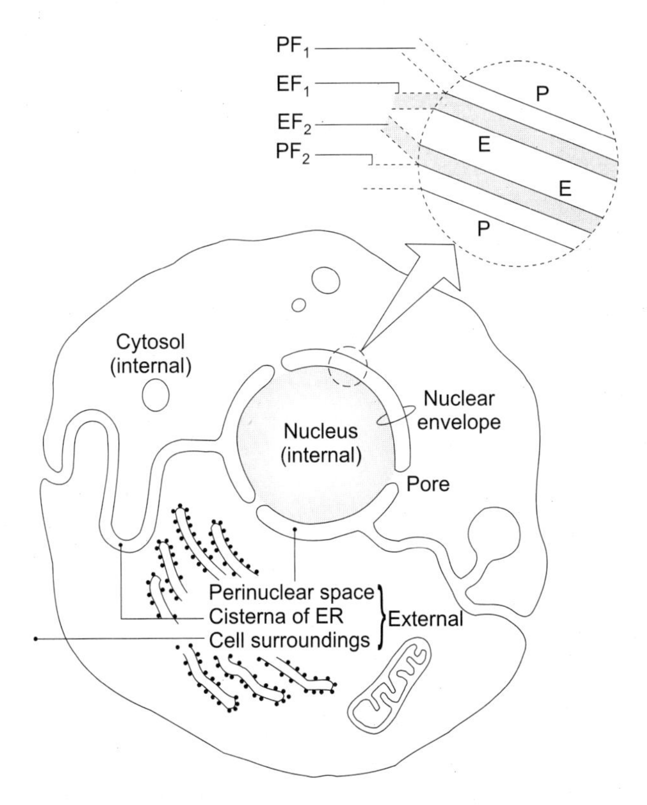

Fig. 2.11 "Internal" and "external" phases of the cell. Region of the nuclear envelope is enlarged to illustrate the relationship among various faces of the inner and outer nuclear membranes.

Other structural elements commonly found in animal cells include microfilaments, which may participate in intracellular movement and communication. These may be scattered through the cytoplasm, but may are located just under the plasma membrane and may be anchored to it. Basal bodies are found at the base of locomoter organelles and may give rise to the microtubules of these structures. In animal cells, pairs of centrioles are observed near the cell nucleus and may involved in the mechanics of cell division.

EUCARYOTIC CELLS: THE GENERALIZED PLANT CELL

All the organelles described in the preceding section as regular constitutes of animal cell are also found in similar form in may plant cells. Several other organelles are unique to plant issues and include the cell wall, plasmodesmata, chloroplasts and large vacuoles. A generalized plant cell are depicted in Fig. 2.13.

The Cell Wall. The cell wall is a thick polysaccharide-containing structure immediately surrounding the plasma membrane Fig. 2.13. In multi cellular plants, the plasma memberance of neighboring cells are separated by these walls and adjacent plant cells have their walls fused

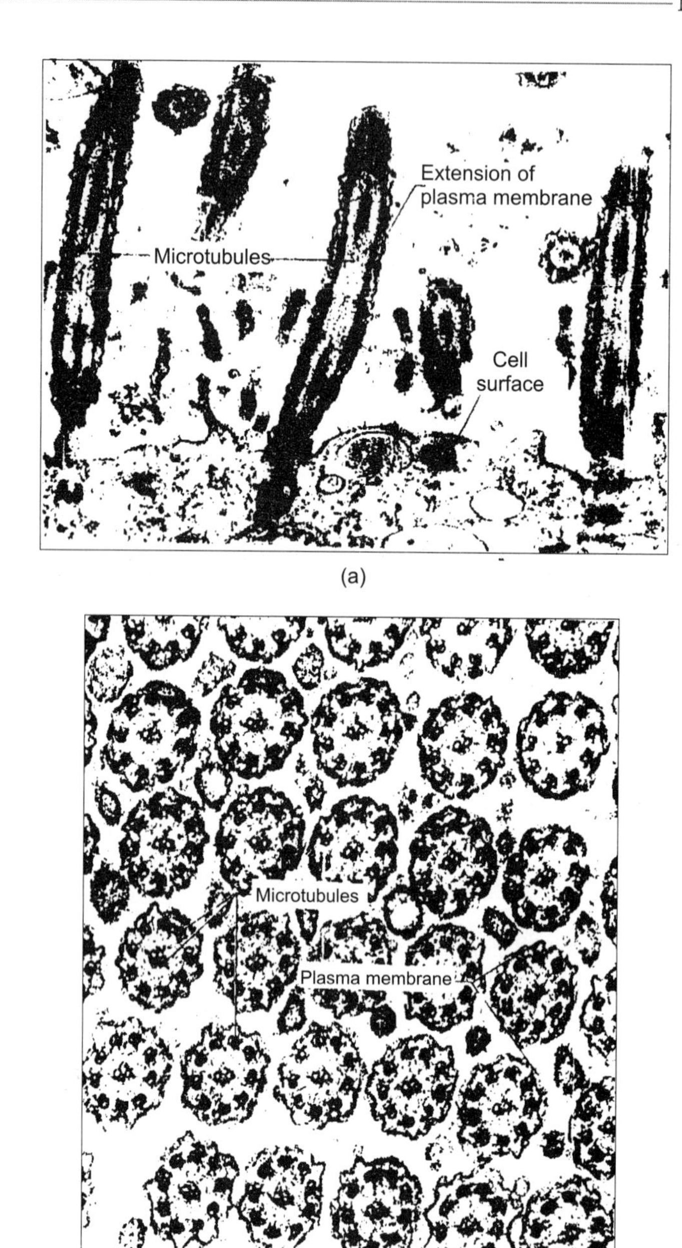

Fig. 2.12 (a) Electron photomicrograph showing tangentially sectioned cilia projecting from the cell surface (b). Cross section through nuberous cilia, with long axis perpendicular to plane of the page. Note the arrangements of microtubules. Magnifications; (a) 25,000 × (b) 40,000 × (Photomicrographs courtesy of R. Chao).

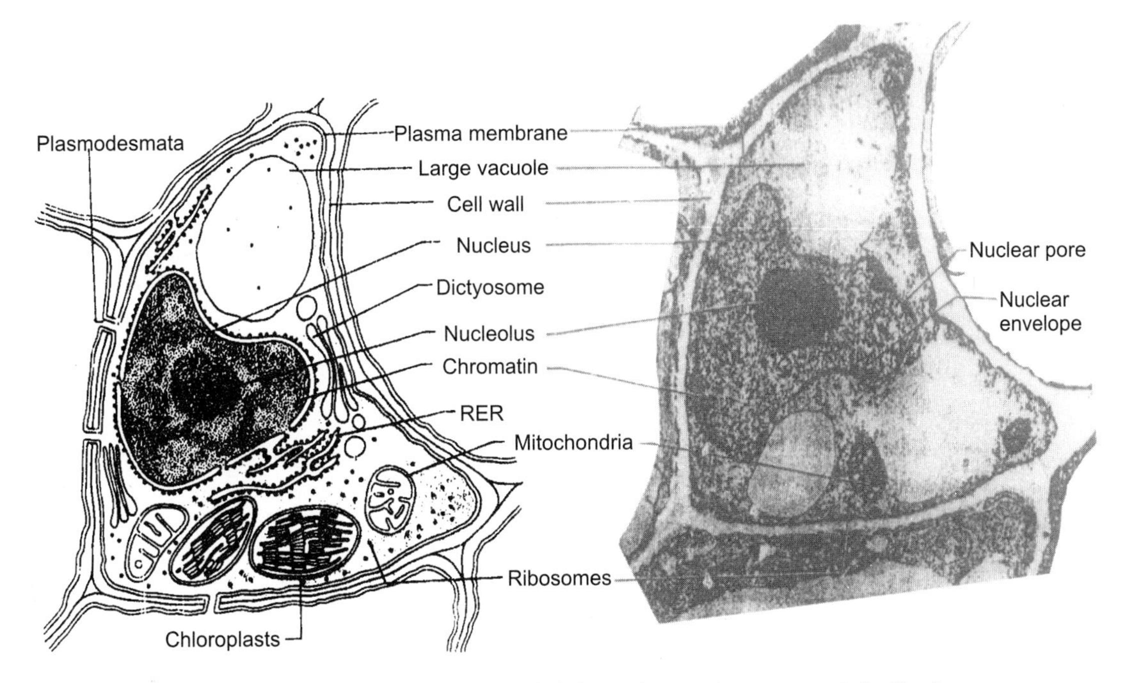

Fig. 2.13 The generalized plant cell (Photomicrograph courtesy of R. Chao).

together by a layer of material called the middle lamella. The cell wall serves both protective and a supportive function for the plant. The degree to which the cell wall may be involved in the regulation of the exchange of materials between the plant cells and its surrounding is difficult to assess but is most likely restricted to micro molecules of considerable size. As in animal cells most of the regulation of exchanges between the cytoplasm and the extra cellular surrounding of plant cell is a function of the plasma membrance.

Plasmodesmata. At intervals the plant cell wall may be interrupted by cytoplasmic bridges between one cell and its neighbor Fig. 2.14. These bridges are called plasmodesmata and represent regions in which channel-like extensions of the plasma memberance of neighboring cells merge. The channels serve in intercellular circulation of materials.

Chloroplasts. The ability to use light as source of energy for sugar synthesis from water and carbondioxide distinguished animal cells and certain plant cells. This process, termed photosynthesis, is carried on in the chloroplasts Fig. 2.15. These organelles are commonly ovoid structures bounded by an outer membrane but also containing a number of internal membrances. Internally the chloroplasts consist of a serious of membrance arranged in lamellae (parallel sheets) and supported in a homogeneous matrics called the Stroma. The membrance are arranged as thin sacs (called Thylakoids) that contain chlorophyll and may be stacked on top of another forming structures called grana. Lamellar membranes containing the grana are called straoma lamellae Fig. 2.14.

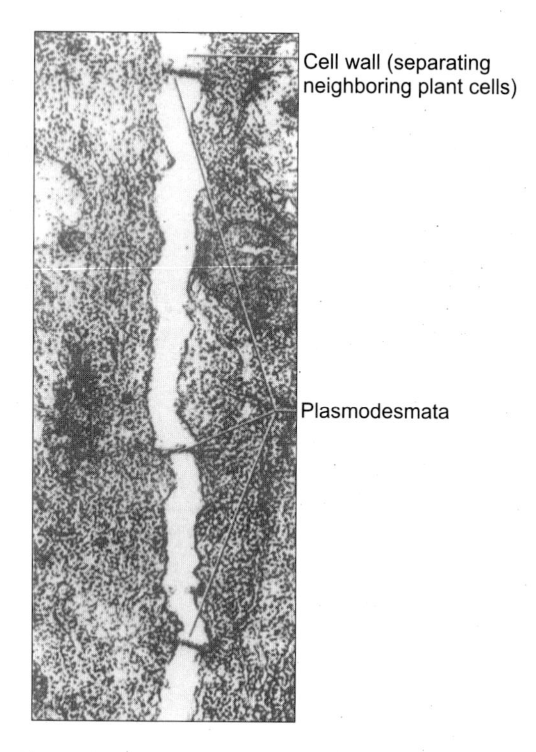

Fig. 2.14 Plasmodesmata (Photomicrograph courtesy of N. Herzog).

Vacuoles. All though vacuoles are present in both animal and plant cells, they are particularly large and abandon in plant cells Fig. 2.13, often occupying a major portion of the cell volume and forcing the remaining cell structures into a thin peripheral layer. These vacuoles are bounded by single memberance and are formed by the coaleseence of smaller vacuoles during the plant's growth and development. Vacuoles serve as sites for the storage of water and product or metabolic intermediates.

PROCARYOTIC CELLS: BACTERIA

The bacteria are structurally distinct from eucaryotic microorganisms such as protozoa and contains a number of unique cellular organelles. The typical bacterial cells is about the size of mitochondrion of an animal or plant cell, and in view of the small size, it is to be expected that the organelles of bacteria would be correspondingly smaller. The generalized structure of a bacterium is shown in Fig. 2.16.

The Bacterial Cell Wall. The bacterial cell is enclosed within a wall that differs chemically from the cell wall of plants in that it contains protein and lipid as well as polysachride. It contains of a particular "mucopeptide" (a protein-carbohydrate complex) has been the basis of the histo-chemical classification of bacteria, being high in the so-called "Gram-Positive" bacteria

Fig. 2.15 Chloroplast of sugar beet (Electron photomicrograph courtesy of Dr. W. Laetsch).

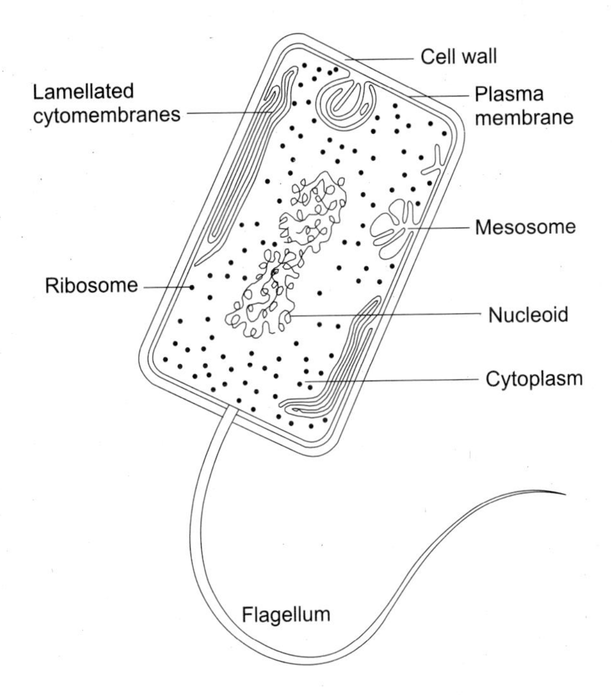

Fig. 2.16 The generalized bacterium.

(such as the Bacillus subtils) and low in the gram-negative bacteria (such as *Escherichia coli* and the *Simonsiella*). In some bacteria the cell wall is surrounded by an additional structure called a capsule. The cell wall and capsule confer shape and form to the bactrerium and also act as an osmotic barrier.

Mesosmes. In foldings of the plasma memberane of gram-positive bacteria give rise to structures called Mesosmes (or chondrioids) these structures appear to have many properties similar to those of the mitochondria of animal and plant cells yet they are not thought to have the sum physiological role. Mesosmes may be absent in gram-negative bacteria in which the plasma membrane itself process respiratory, activity. Intrusions of the plasma membrane form the photosynthetic organelles (chromatophores) of the photosynthetic bacteria.

Lamellate Cytomembranes. In some bacteria, there is a lameller arrangement of membrane within the cytoplasm that may arise from the plasma membrane, however there is not structure comparable to the endosplasmic reticulum of animal and plant cells. Whereas bacteria are often packed with ribosomes, these are free in the cytoplasm and are not attached to membranes. Although bacterial ribosomes, like the ribosomes of eucaryotic cells, are the sites of protein synthesis, considerable differences exist between the organelles of these two groups. Lamellate membranes are particularly abundant in the autotrophic bacteria, which support their growth through photosynthesis or similar processes.

Nucleoids. In bacteria the contents of the nucleus are not separated from the cytoplasm by membranes, although the nuclear material may be confined to a specific region of the cell. The nuclear "area" of the bacterium is sometimes referred to as nucleoid. During bacterial cell division, the nuclear material are distributed tot eh daughter cells without formation of observable chromosomes. Nucleoli are not present in the nucleus.

Bacterial Flagella. Many bacteria contain one or more flagella employed for cellular locomotion. These organelles arise from a small basal granule in the cytoplasm and penetrate the plasma membrane and cell wall. Bacterial flagella are smaller than those of animal and plant cells and are simpler in organization, containing a single filament of globular proteins (called flagellin) surrounded by a sheath.

Some bacteria, such as *E. coli*, *P. mirabilus*, and *B. subtilis*, occur as separate, individual cells. However, in a number of groups, the daughter cells remain attached following division, so that chains (e.g. streptococci) or filaments are formed. An example of a filamentous genus is *Simonsiella*, which colonizes the mucosal epithelial surface of the mouth. The individual cells of some filamentous bacterial reveal a dorsal-ventral differentiation; that is, the ventral surface (which is *Simonsiella*) attaches to and glides along the epithelium) is structured differently than the dorsal surface (which faces away from the epithelium). This is apparent not only in the scanning electron micrographs of whole filaments and in transmission electron micrographs of thin sections through the filaments (Fig. 2.17), which reveal an internal differentiation. Individual Cells of a filament exhibit features common to single-cell (i.e., nonfilamentous) forms like *E. coli* and *B. subtilis*.

PROCARYOTIC CELL: BLUE-GREEN ALGAE

The blue-green algae more closely resemble bacteria than they do algae and often are classified with the bacteria. Although most blue-green algae are blue-green in color, they do occur in a

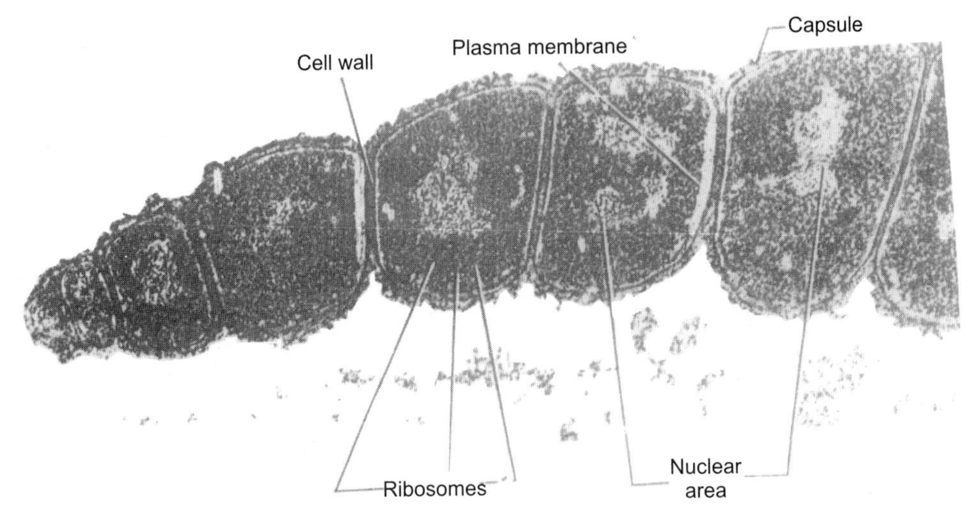

Fig. 2.17 The filamentous bacterium *Simonsiella.* (a) Section through several cells of a filament revealing internal organization and dorsal-ventral differentiation.

wide range of colors. Their name may be attributed to the fact that the first species recognized as members of the group were blue-green. The blue-green algae are photosynthetic prokaryotes and occur as individual cells, as small clusters or colonies of cells, or as long, filamentous chains (Figs. 2.18 and 2.19). Blue-green algae lack locomotor organelles, and a gelatinous sheath replaces the capsule typical of bacteria. The photosynthetic apparatus consists of lamellae lined with pigment granules sometimes referred to as phycobilosomes.

The PPLO (i.e., pleuropneumonia-like organism) or mycoplasmas, which cause a number of diseases in humans and other animals, are the smallest (i.e., about 0.1 U in diameter) of all free-living cells. They are smaller even than some of the larger viruses. The PPLO is bounded at its surface by a membrane composed of protein and lipid, but internally cell's compositions is more or less, diffuse. The only microscopically discernible features within the cell are its genetic complement, which consists of a double helical strand of circular DNA, and a number of ribosomes (Fig. 2.20). The PPLO appears to contain the bare minimum of structural organization required for a viable, freeliving cell and may represent a form intermediate between viruses and bacteria. The relative sizes of typical eucaryotic cells, bacteria, PPLOs and viruses are compared in Fig. 2.21, which dramatizes the differences that exist.

VIRUSES

So far, the descriptions in this chapter have been restricted to various kinds of cells. In this section, we are concerned with the organization and activities of viruses, but it should be emphasized at the outset that viruses are not cells and that it is debatable whether viruses constitute living systems. Viruses are described here because of their intimate association with cells and because of their contributions to our understanding of certain cellular phenomena. It

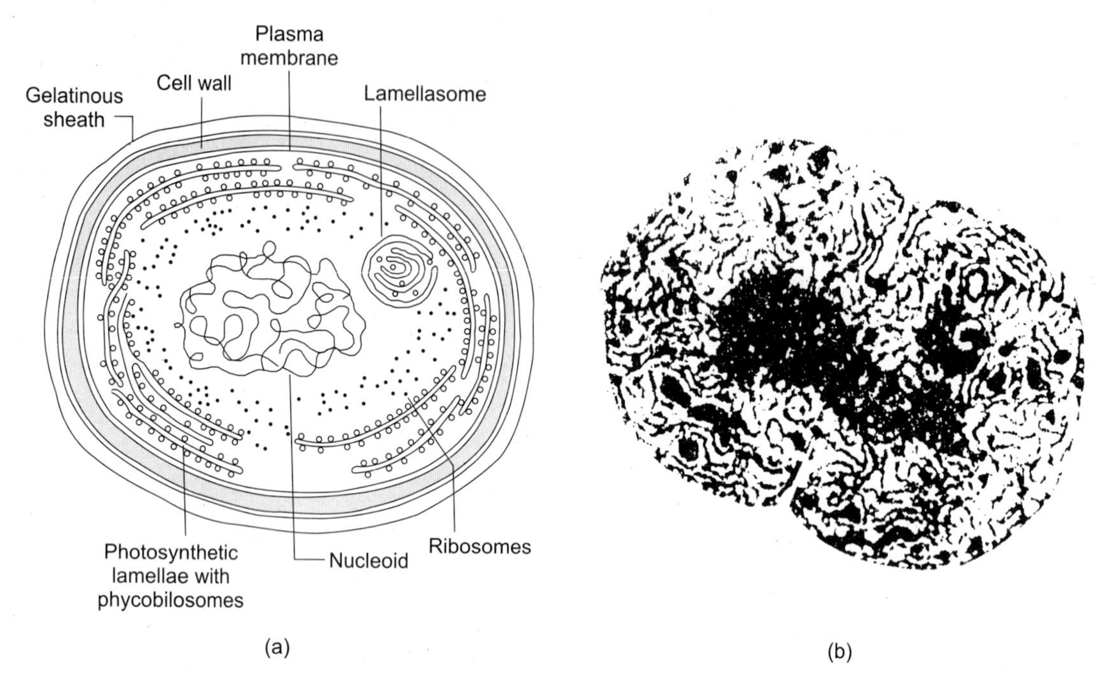

Fig. 2.18 (a) Generalized blue-green alga cell. (b) Electron photomicrograph of dividing *Anabaena*. Note the formation of the septum. Magnification, 15,000 × (Courtesy of Drs. H.W. Beams and R.G. Kessel).

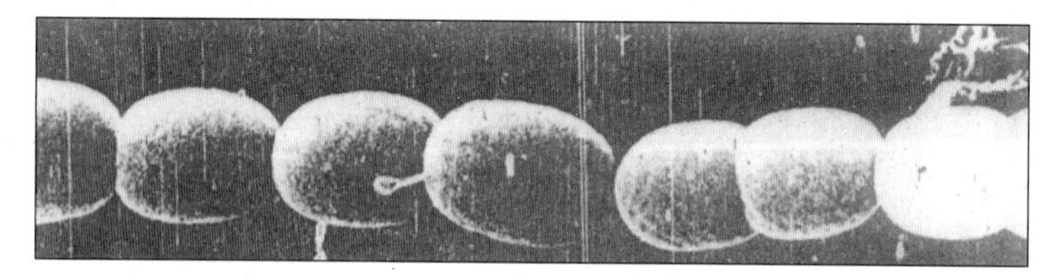

Fig. 2.19 Scanning electron micrograph of a filament of the blue-green alaga *Anabaena* (compare with the thin section of Figure 1.40b). Maginification, 6000 × (Courtesy of Drs. F.A. Elserling and S. Elpert).

will become apparent as we deal in later chapters with the structure and interactions of nucleic acids and proteins and with gene expression that much of our present-day understanding is based on studies initiated with viruses.

STRUCTURE OF VIRUSES

Although all viruses or virions are extremely small, they are diverse in size and in organization. Generally, viruses range in diameter (or length) from about 20 nm to about 200 nm. Thus the largest viruses are actually larger than the smallest cells. However, even the smallest of cells (bacteria, PPLOs, etc) are subject to infection by viruses. Among these viruses that attack animal

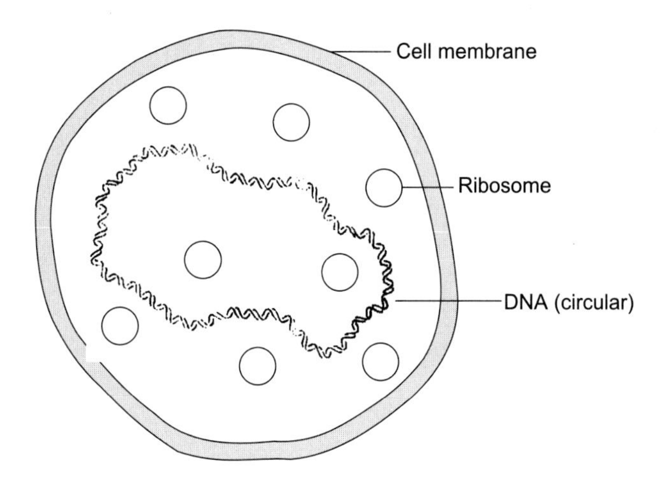

Fig. 2.20 Structure of PPLO.

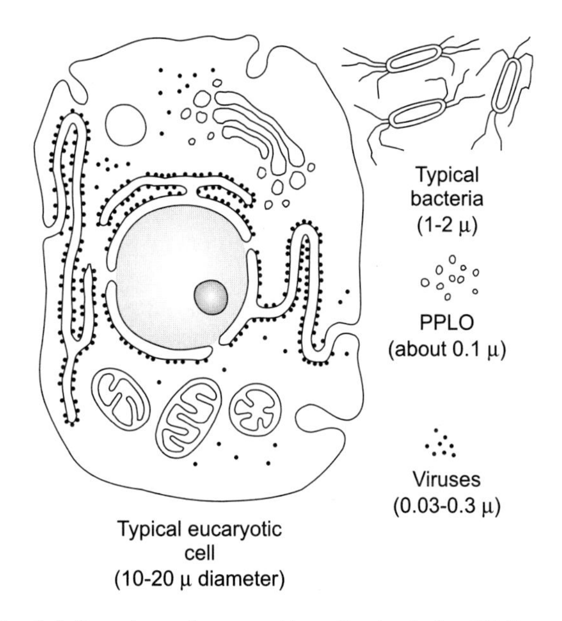

Fig. 2.21 Relative sizes of eucaryotic cells, bacteria, PPLO, and viruses.

cells, the most notorious are the viruses that cause disease in humans. Smallpox, chicken pox, rabies, poliomyelitis, mumps, measles, influenza, hepatitis, and the common cold are all produced by viruses. Even certain leukemias and cancers are of viral origin.

Most virions are either rod-shaped or quasi-spherical and contain a nucleic acid core surrounded by a specific geometric array of protein molecules that form a coat or capsid (Fig. 2.22). The proteins of the capsid are arranged to form either a helical pattern (when the virus is rod-like) or an isometric pattern (when the virus is globular). In the latter state, the virus

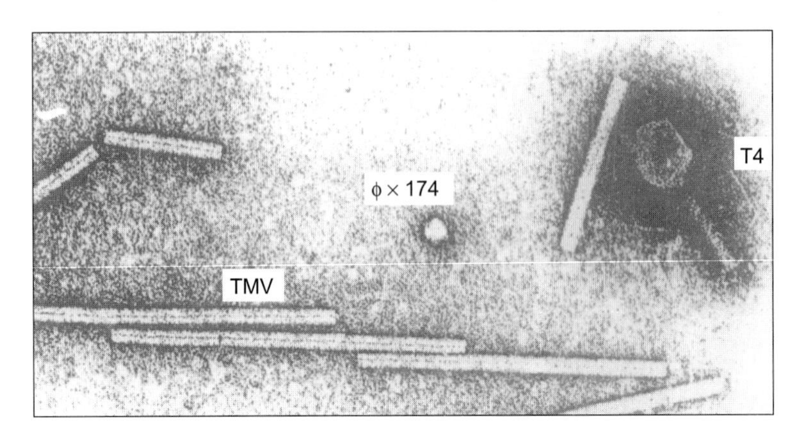

Fig. 2.22 Viruses (a) Negatively stained preparation of a mixture of tobacco mosaic virus (TMV) and the T4 and $\phi \times 174$ bacteriophages.

appears much like a polyhedron. The viruses causing chicken pox, mononucleosis, fever blisters, and cold are examples of virions having polyhedral capsids. Helical capsids are more common among viruses that infect plant cells and bacteria. The tobacco mosaic virus (TMV), which infects the leaves of the tobacco plant, is among the most extensively studied viruses and exhibits the helical capsid pattern.

In many animal viruses and in some plant viruses, a lipoprotein envelope surrounds the capsid (e.g. influenza virus, herpesvirus, and smallpox virus). Among the largest and most complex virions are those that attack bacteria (i.e., the bacteriophages). Most extensively studies among these are the T2, T4, and T6 (i.e., the "T-even") bacteriophages (Fig. 2.22). These bacteriophages have a tail-like structure emerging from the capsid (Fig. 2.23). The tail is enclosed in a sheath of proteins arranged in a helical pattern, while the head of the virus is polyhedral. The end of the tail frequently reveals specialized structures (Fig. 2.23) involved in attachment to the surface of the host cell (Fig. 2.24A).

In the free or isolated state, viruses exhibit no metabolism and are incapable of proliferation. Proliferation of viruses requires a host cell and in its simplest and most direct form takes the following pattern. One or more viruses attach to specific sites on the surface of the host cell and insert their core nucleic acid into the host. Release of the core nucleic acid fro a virus can be achieved experimentally; Fig. 2.24 dramatically reveals the uncoiled DNA molecules released by a T4 bacteriophage. Once inside the host cell, the viral nucleic acid redirects the metabolism of the host so the new viral protein and new viral nucleic acids are formed. These viral components combine in the host to form large numbers of new virions that egress from the cell by disruption of its plasma membrane (i.e., cell lysis). The cycle of infection then repeats itself. The proliferative cycle of a virus is best understood for the bacteriophages and is depicted diagrammatically in Fig. 2.25.

The virion envelope that characterizes many animal viruses is acquired from a portion of the plasma membrane of the host cell as the virus emerges.

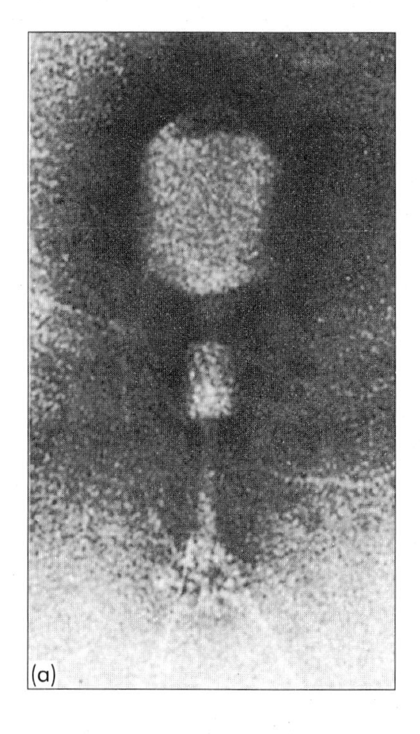

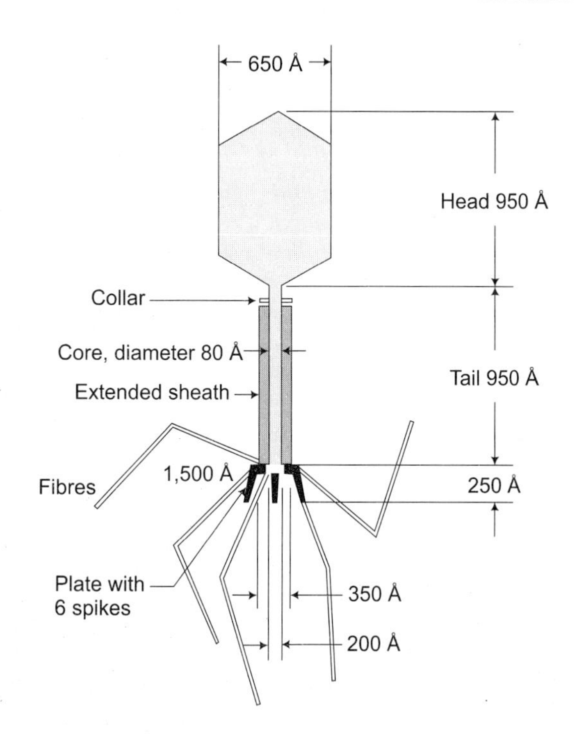

Fig. 2.23 The detailed structure of the T-even phage particle [After E. Kellenberger, *Adv. Virus Research* 8, 1 (1962)].

On some occasions and only for certain viruses, the injected nucleic acid does not cause proliferation and release of new virions. Instead, the injected nucleic acid is incorporated into the host's genetic material, and the host cell continues to function in its normal manner. However, duplication of the host's genetic material prior to cell division is accompanied by duplication of the incorporated viral nucleic acid. Several generations of cell may be produced, each containing a copy of the viral nucleic acid. Viruses exhibiting this phenomenon are called temperate viruses, because they do not cause the death of the immediate host. Viruses that engage only in the cycle described earlier and that kill the host cell are called virulent viruses. The dormant viral nucleic acid within the host is referred too as a provirus, and the infected cell is said to be lysogenic, because sooner or later, in one of the generations of the host cell, the provirus nucleic acid will begin to direct the replication of new virion and the this in turn will lead to cell lysis and release of new infective virus particles.

CLASSIFICATION OF VIRUSES AND THE NATURE VIRAL NECLEIC ACIDS

The classification of viruses poses certain problems, and several different approaches have been used. One method is to classify the virus according to the type of host cell. Hence these are animal viruses, plant viruses and bacterial viruses (i.e., bacteriophage). This method is not

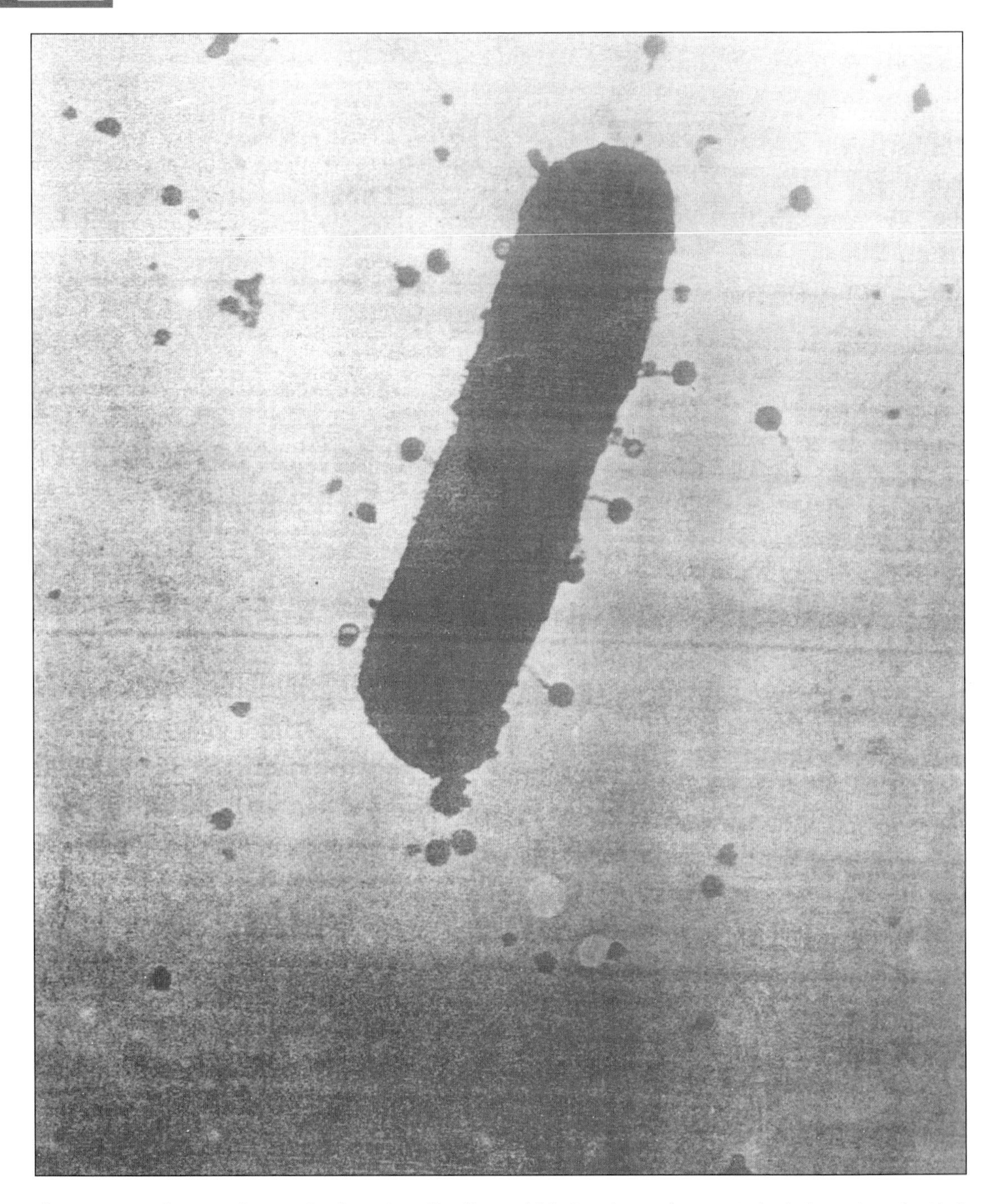

Fig. 2.24a Eelctrom micrograph of an *E. coli* cell to which T5 phages have attached themselves by their tail [After T.F. Anderson, *Cold Spring Harbor Symp. Quant. Biol.* 18, 197 (1953)].

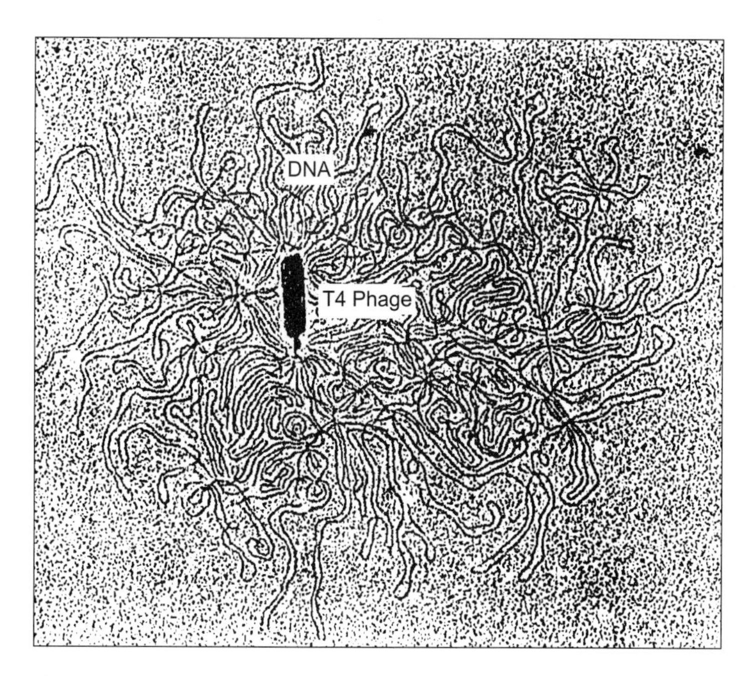

Fig. 2.24b DNA molecule released from a "giant" T4 bacteriophage particle. The single DNA molecule is more than 150 μ long. Magnification, 20,000 × [Courtesy of Dr. F.A. Eiserling].

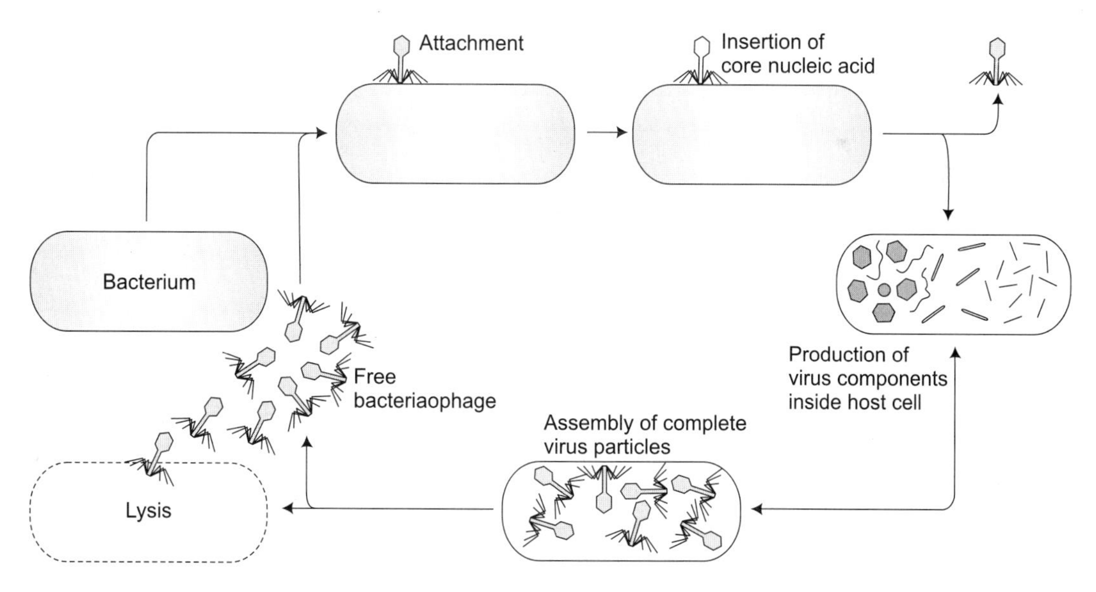

Fig. 2.25 Proliferative cycles of the bacteriophage.

always satisfactory; for example a few viruses infect both animals and plants. Another method employs comparisons of virus morphology (e.g. capsid shape and geometric symmetry), an interesting approach is, to classify viruses according to the nature of their genetic material. Even the beginning student of biology becomes quickly aware of the functional relationship between DNA, RNA, and proteins in cells. Genetic information stored in molecules of DNA is transcribed or copied into a corresponding RNA molecules (called a messenger), and the messenger then directs the synthesis of a specific protein. The protein then acts as an enzymes or structural component of the cell. In some viruses. However, the genetic information comprising the virion core is RNA and not DNA. Among the RNA viruses are TMV and the virion causing polio, mumps, measles, influenza, and colds. While RNAs of cells are single-stranded molecules viral RNAs can be single stranded (e.g., TMV, RNA) or double stranded (e.g., the reoviruses. Like eucaryotic and prokaryotic cells, the genetic information of many viruses is encoded in a DNA core. The DNA viruses include the t-even bacteriophages and those that cause chickenpox, herpes blisters, infections mononucleosis, and shingles. However, the DNA may be double-stranded linear molecules (e.g., T5 and T7 bacteriophages), double-stranded circular molecules (polyoma and SV 40 viruses), single-stranded linear molecules (the parvoviruses), or single-stranded circular molecules (× 174 bacteriophage). Whatever form the nucleic acid takes, it includes the genetic information for the synthesis of the variety of proteins that either become components of new viruses or are involved in the redirection of the host cell's metabolism.

It was noted at the beginning of this discussion that while viruses themselves are not cells, the study of viruses has yielded a wealth of information about cells. Research with certain viruses has provided crucial and it times astounding information about the chemistry of and interactions among nucleic acids and proteins- TMV. Studies involving the tobacco mosaic virus (Fig. 2.22), began nearly a century ago and represent the stating point in the field of virology, in the 1950s TMV was at the focal point of research that verified that nucleic acids and not proteins compose the genetic apparatus and that genetic information can be encoded in RNA a swell as in DNA.

0 × 174, studies with the 0 × 174 bacteriophage which infects *E. coli* revealed that the information for viral proliferation can be encoded in a single strand of DNA and does not require a double strand (i.e., double-helix). Moreover, the structure of 0 × 174 DNA has now been completely analyzed, and the entire base sequence is known. A most astounding finding yielded by these studies is that the coding sequences for several of the virus, proteins are included within sequences for other proteins. That is, certain genes overlap, similar finding are being reported for other viruses including similar virus 40 (SV 40), that possess double-stranded DNA and have both temperate and virulent phases.

Reoviruses. Reoviruses (e.g. Rous sarcoma virus, avian leukemia virus, and other Cancer-causing viruses) have RNA as their core nucleic acid, but unlike X174, TMV, and other RNA viruses, replication within the host cells requires that the inserted RNA be used for the preliminary synthesis of DNA, following which transcription and translation take the conventional pattern. The reoviruses have demon stranded that transcription can take place in the reverse direction, that is, from RNA to DNA. Our current understanding of the molecular mechanisms for strong or expressing genetic information is based to a large degree on studies using viruses. Some of the concepts discussed above only in introductory terms are treated more fully in later chapters.

SUMMARY

In the 300 years that followed the introduction of microscopy to biological science, the concept evolved that the cell is fundamental unit of structure and function in all living things. This notion is referred to as the cell doctrine. Microscopy remains one of the cell biologist's most important and powerful research tools as new variations of light and electron microscopy have appeared notably freeze-fracture transmission electron microscopy, scanning electron microscopy, and stereoscopic electron microscopy. With these tools the detailed structure and organization of nearly all sub cellular organelles have been revealed.

Free-living cells and the cells of multicellular organisms are divided into two classes: eukaryotes (nearly all animal and plant cells) and prokaryotes (bacteria, blue-green algae, mycopasmas, etc.). Eucaryotic cells are characterized by a number of discrete organelles-especially the nucleus, mitochondria, golgi bodies, lysosomes, peroxisomes, rough and smooth endoplasmic reticulum, and in plant cells, chloroplasts and cell wall. Like eucaryotic cells many prokaryotes possess a limiting (plasma membrane and ribosomes, but lack the true nucleus and other discrete organelles characteristic of eukaryotes.

Although viruses are not cells, they are intimately associated with cells, and their study has made invaluable contributions to our understanding of cell function. In the isolated state, viruses are incapable of metabolizing or proliferating. Proliferation requires preliminary infection of a hast cell, which takes the form of insertion of the viral nucleic acid (DNA or RNA). Following this, the metabolism of the host is redirected to make the components of the virus. Assembly of new virus particles within the host is ultimately followed by their egress to begin a new cycle of infection.

Most of the remaining chapters in the book are devoted to a detailed description of the biochemistry, structure, and physiological functions of cells and their organelles.

CHAPTER 3

Techniques in Molecular Biology

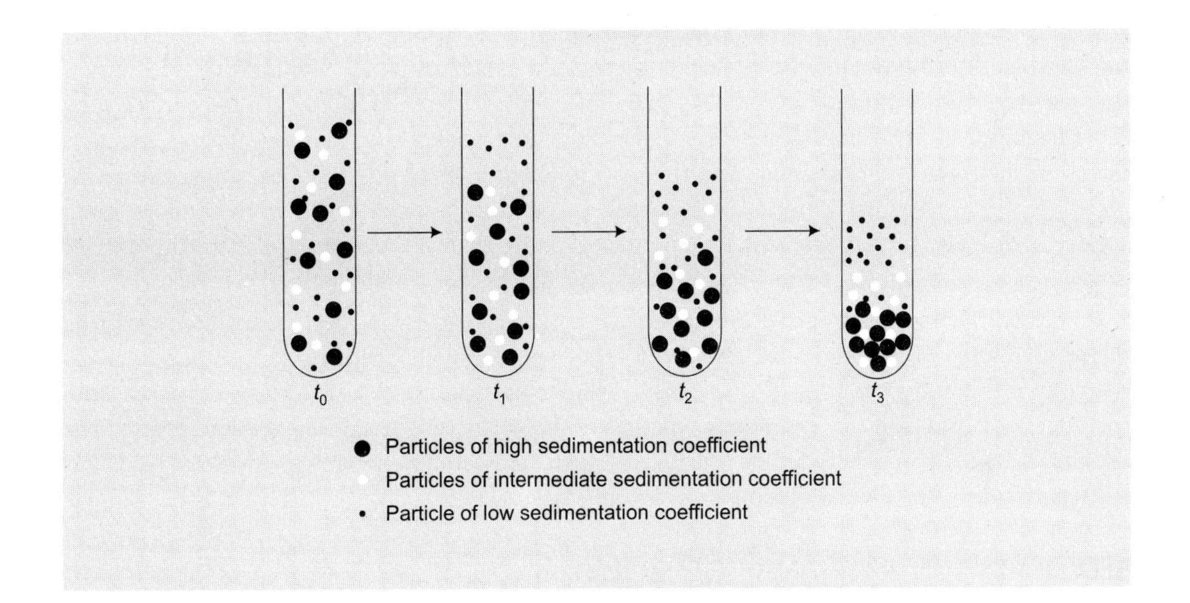

DISRUPTION OF CELLS

Cells can be disrupted by chemical means. For example, enzymes that specially degrade the components of the cell wall or cell membrane may be added to the tissue or cell suspension. Alternately, proteolytic or lipolytic agents that dissolve the membrane may be used, some cells are sufficiently fragile that they are readily disrupted by successive freezing and thawing. Erythrocytes (red blood cells) and certain others cells may be broken by the osmotic pressure created within them when they are placed in distilled water or hypotonic solution.

After cells have been disrupted, the goal is to separate and isolate the structures that have been released. Since cellular organelles and other constituents vary in size shape, and density, they settle through the liquid in which they are suspended at different rates. Consequently, disrupted cells are more often fractionated by some form of **centrifugation** than by any other method. Indeed, centrifugation has become one of the most widely employed procedures in cellular research and one of the most important tools of the cell biologist. In the following section, we consider the principles and applications of centrifugation.

CENTRIFUGATION (Fig. 3.1 & 3.2)

Theory of Centrifugation

If a container is filled with a liquid suspension of particles of varying size and density, the particles will gradually settle to the bottom of the container under the influence of gravity. The rate at which settling occurs can be greatly increased by increasing the gravitational effect upon the particles. This is the underlying principle for isolating large or dense particles by centrifugation. A tube containing the suspension of particles (e.g., a tissue homogenate) is placed in the rotor of a centirifuge and then is rotated at high speed. The resulting acceleration greatly increases the gravitational pull on the suspended particles, causing their more rapid sedimentation to the bottom of the tube along paths that are perpendicular to the axis of rotation (i.e., along radii of the circle being swept out by the rotating tube) this relation.

The sedimentation of particles by centrifugation is, in effect, a method for concentrating them; therefore one of the major physical forces opposing such concentration is diffusion. In the case of the sedimentation of cells or sub-cellular particles such as nuclei or mitochondria, the effects of diffusion are essentially nil. However, when centrifugation is employed for the sedimentation of much smaller particles (such as cellular proteins, nucleic acids or polysaccharides), the effects of diffusion become significant.

SEDIMENTATION RATE AND COEFFICIENT

The RCF or "g force' applied to particles during centrifugation may readily by calculated and is independent of the physical properties of the articles being sedimented. However, a particle's sedimentation rate at a specified RCF depends upon the properties of the particles itself. Also, since the RCF varies directly with x. it is clear that the sedimentation rate changes with changing distance from the axis of rotation. (for particles setting under the influence, of the earth's gravity alone, these sedimentation rate becomes constant). The instantaneous sedimentation rate of a particle during centrifugation is determined by three forces: (1) Fc (i.e., the centrifugal force), (2) Fb, the buoyant force of the medium, and (3) f the frictional resistance to the particle's movement.

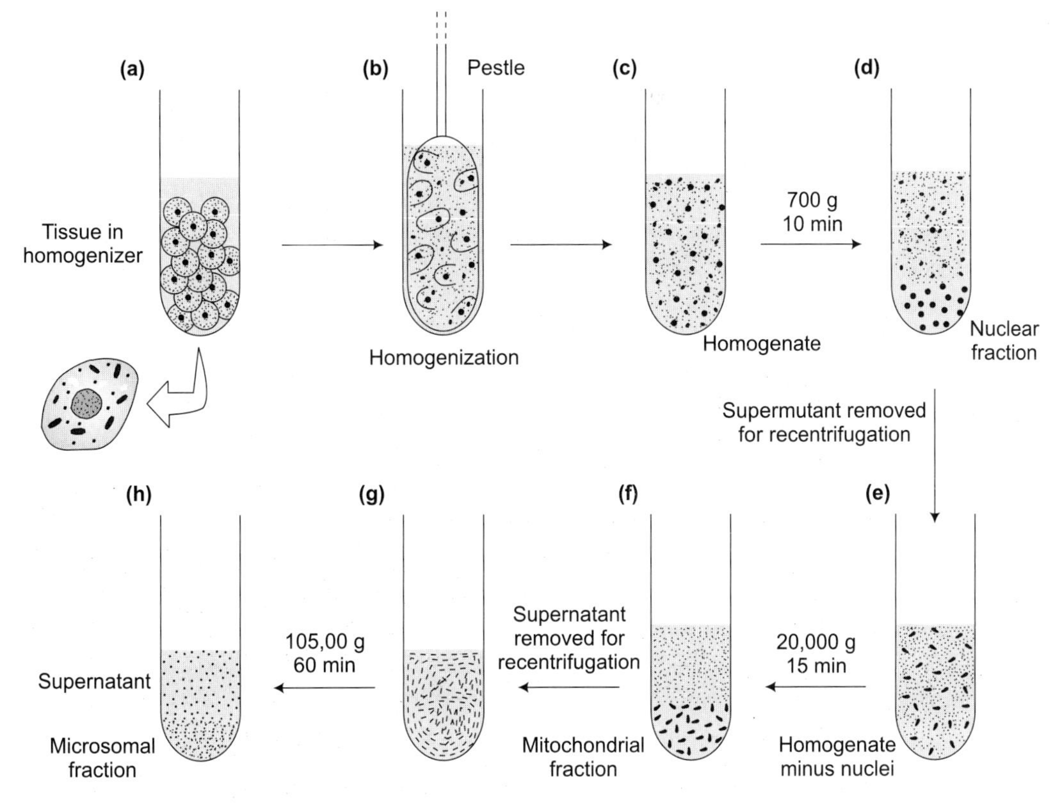

Fig. 3.1 Fractionation of liver by differential centrifugation, Nuclei (pale color), mitochondria (solid color) lysosomes (white), microsomes (black) and cytosol (grey).

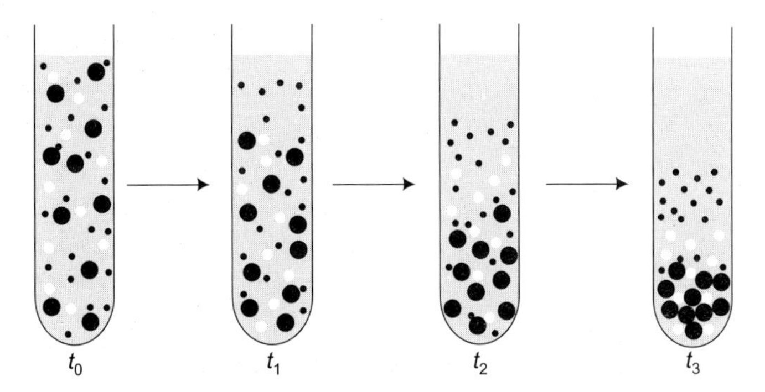

Fig. 3.2 Effects of differential centrifugation. At time t_0,all particles are uniformly distributed through centrifuge tube. During centrigation (times t_1, t_2 and t_3) all particles sediment according to their sedimentation coefficients. Although particles of highest sedimentation coefficient quickly reach the bottom of the tube, cross-contamination by smaller particles is unavoidable.

THE ANALYTICAL ULTRACENTRIFUGE

The sedimentation coefficient of a particle may be experimentally determined in an instrument known as an analytical ultracentrifuge. This instrument is equipped with an optical system (called Schlieren optics) that permits visual observation of the sedimenting particles as one or more moving boundaries formed between regions of the particle suspension having different refractive indexes. The measured rate at which these boundaries move under specified conditions is used to determine the sedimentation coefficient, and the number of boundaries formed is an index of the heterogeneity of the sample.

The first analytical ultracentrifuges were designed and built by the Nobel Prize-wining biochemist T. Svedberg in the 1920s. Following Svedberg's pioneering work, the analytical ultracentrifuge was advanced to its present status as one of the more important tools of the molecular biologist, principally through contributions of E. G. Pickels and J. W. Beams.

The rotor spins in the ultracentrifuge typically contains two compartments. Into one compartment is placed a 'reference cell" containing the sample-free solvent, while the other receive the cell containing the sample to be analyzed. The interior of each cell is sector shaped and bounded above and below by parallel quartz windows to permit light from below the rotor to pass through the reference and sample during rotation. As the particles sediment, boundaries are formed at the trailing edges of each particulate species. When these boundaries pass in front of the optical system, the resulting change in refractive index.

The sedimentation coefficients a of many cellular macromolecules such as proteins, polysaccharides, and nucleic acids fall in the range 1 × 10 second to 200 × 10 seconds (i.e., the dimensions of s are seconds). For convenience, a unit called the Svedberg unit (after T. Svedberg) and abbreviated s is used to describe sedimentation coefficients and is equal to the constant 10 seconds. Thus, most cellular proteins have sedimentation coefficients between 1 and 200 S. the sedimentation coefficients of a number of cell constituents are listed in Table 3.1.

DIFFERENTIAL CENTRIFUGATION

During centrifugation, particles sediment through the medium in which they are suspended at rates relates to their size and density. Differences in the sedimentation coefficients of the various sub cellular particles provide the means for their effective separation. Differential centrifugation, a technique introduced to cellular research in the early 1940s by the noted biologist Albert Claude, is one of the classical producers for isolating subculture particles and involves the stepwise removal of classes of particles at increasing RCF.

The material to be fractionated is subjected first to low-speed centrifugation in order to sediment the largest (or densest) Particles present. Following this, the unsedimented material (called the supernatant) is centrifuged at a higher speed to sediment particles of somewhat smaller size (and/or lower density). The sequence is repeated several times until all particles have been sedimented and the sediments then used for further experimentation and analysis.

Table 3.1 Some representative sedlmentation coefficients.

Particle	*Sedimentation coefficlent (s)*
Proteins	
Cytochrome *c*	1.7
Hemoglobin	4.1
Fibrinogen	7.6
Hemocyanin	59
Cell organelles	
Nucleosomes	11
Ribosomes	70 to 80
Membrane fragments	10^2 to 10^4
Plasma membranes	up to 10^5
Lysosomes	4×10^3 to 2×10^4
Peroxisomes	4×10^3
Mitochondria	1×10^4 to 7×10^4
Chloroplasts	10^5 to 10^6
Nuclei	10^6 to 10^7

Tissue fractionation begins with the disruption of the cells and the preparation of a subcellular particle suspension or homogenate. Cells are most frequently disrupted using the shear forces generated by special grinders, blender, pressure cell, or insonators. Chemical procedures involving lytic agents or osmotic pressure are also used in certain instances.

Once the tissue homogenate is prepared, the method of choice for separating subcellular organelles and particles is centrifugation. Centrifugal fractionation may involve one or a combination of different approaches. In differential centrifugation, gross differences in the sedimentation rates of certain subcellular particles are used to produce a series of particulate sediments at successively higher "g forces". Different families of particles may also be prepared on the basis of size and/or density differences using the density gradient approach. A variety of centrifugal devices are used to effect the purification of subcellular particles including conventional swinging-bucket and fixed-angle rotors and the more sophisticated vertical and zonal rotors.

Heterogeneous mixtures of very large particles, such as whole cells, may be fractionated by centrifugal elutriation or simply by unit gravity sedimentation. Electronic sorting, countercurrent distribution, and electrophoresis are alternative methods. When particularly large quantities of cells or particles are to be harvested, continuous-flow centrifugal procedures are generally employed.

Dialysis and Ultra Filtration

Semipermeable membranes, such as those prepared from celiophane or collodion (cellulose nitrate) may be used to separate solutes on the basis of molecular weight differences. In dialysis,

Table 3.2 Methods for the isolation and characterization of cellular macromolecules.

Method	*Principal impelling factor(s)*	*Principal retarding or opposing factor(s)*	*Separation depends primarily upon*
Countercurrent distribution	Mechanical	Solubility	Differential partition
Dialysis/ultrafiltration	Osmotic effects, concentration gradients, hydrodynamic force	Molecular sieve effects	Molecular size
Ultracentrifugation	Centrifugal force	Friction, bouyancy, diffusion	Molecular size, shape effective density
Electrophoresis			
Moving-boundary	Electrostatic force	Friction, diffusion	Molecular ionic properties
Zone	Electrostatic force	Friction, diffusion, molecular sieve effects	Molecular ionic properties
Discontinous	Electrostatic force, Kohlrausch function	Friction diffusion, molecular sieve effects	Molecular ionic properties and molecular size
Immunoelectrophoresis	Electrostatic force	Diffusion, molecular sieve effects	Molecular ionic properties, biological activity
Isoelectric focusing	Electrostatic force	Diffusion	Molecular ionic properties
Paper chromatography	Hydrodynamic force	Association/dissciation effects, diffusion	Adsorption/partition differences
Thin-layer chromatography	Hydrodynamic force	Association/dissociation effects, diffusion	Adsorption/partition differences
Ion-exchange chromatography	Hydrodynamic force	Electrostatic forces	Molecular ionic properties
Affinity chromatography	Hydrodynamic force	Molecular affinity	Biological activity
Gel filtration	Hydrodynamic force	Molecular sieve effects	Molecular size
Gas Chromatography	Gas pressure	Diffusion	Adsorption/partition differences

the solute mixture is place in a bag formed from tubular sheets of the semipermeable membrane, and the bag is immersed in an aqueous medium (usually distilled water). Molecules larger than the pores of the membrane are confined to the tubing, while smaller molecules diffuse into the surrounding liquid (Fig. 3.3). Semipermeable membranes can be treated chemically or physically in order to alter the sizes of the pores so that solutes of varying molecular weight are rendered permeable. Generally, dialysis is used with unmodified membranes to quickly separate low-molecular-weight solutes (i.e., molecular weight of less than 5000) such as salts, sugars, and amino acids from proteins and polysaccharides present in the sollution. However, high molecular-weight solutes do differentially penetrate membranes having large pore sizes (Table 3.3).

In ultrafiltration, force is used to drive the smaller molecules along with solvent through the semipermeable membrane. As a result, not only are permeable and impermeable molecules separated but the impermeable species is also simultaneously concentrated (Fig. 3.4).

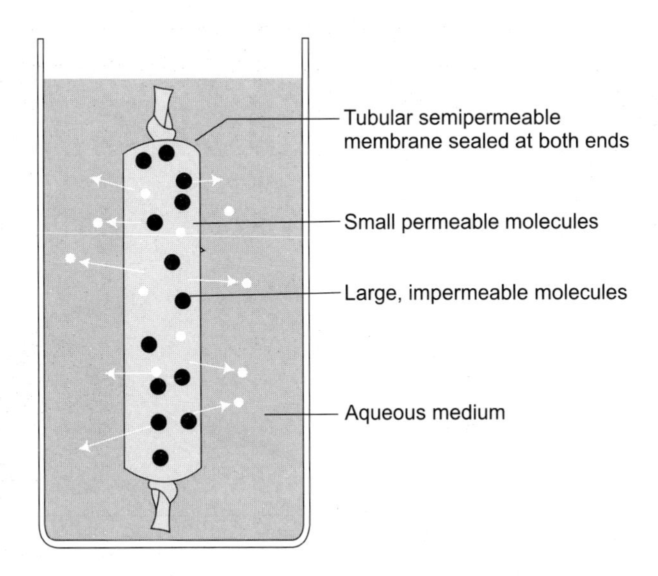

Fig. 3.3 Dialysis using a semipermeable membrane.

Table 3.3 Relationship between molecular weight and permeability to modified cellophane membranes.

Solute	*Molecular Weight*	$T^{\circ}_{1/2}$ *(min)*
Tryptophan	204	4
Bacitracin	1,422	15
Cytochrome *c*	12,000	60
Ribonuclease	13,600	120
Lysozyme	14,000	138
Trypsin	20,000	240
Chymotrysin	25,000	300
Pepsin	35,500	4,800

[a]$T^{\circ}_{1/2}$ is the amount of time required for one-halft of the solute to permeate the membrane.

Ultracentrifugation

Centrifugation can be employed not only for the separation of cells, subcellular organelles, and other particulate constituents of cells but also for molecular separations. Since its initial development in the 1920s by Svedberg, the analytical ultracentrifuge has been used repeatedly to evaluate the heterogeneity or purity of molecular constituents extracted from cells and to estimated molecular sizes on the basis of sedimentation rate. Physical separations (as opposed to analytical studies) became possible with the development of ultracentrifuges and preparation, rotors capable of generation an RCF in excess of 500,000 gm. Using forces of this magnitude, true separations.

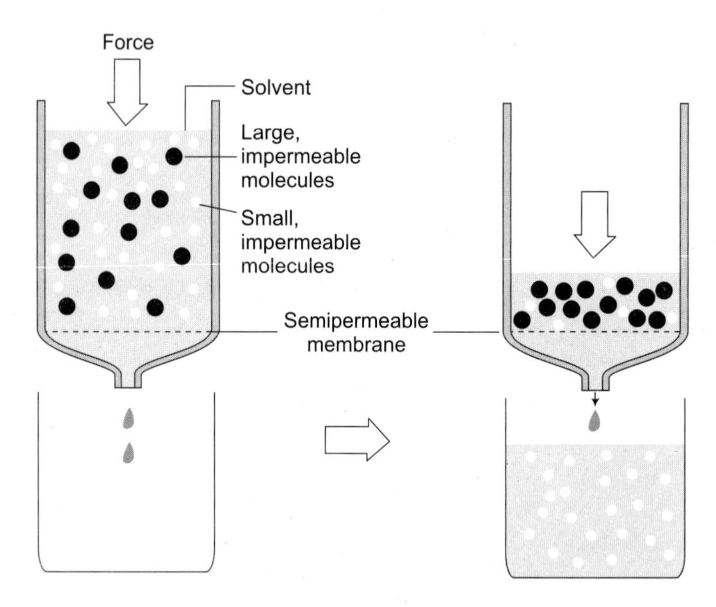

Fig. 3.4 Ultrafilltration through a semipermeable membrane using hydrodynamic force.

Table 3.4 Isoelectric points of some proteins.

Protein	*Isoelectric pH*
Lysozyme	11.0
Cytochrome *c*	10.6
Ribonuclease	9.5
Normal human hemoglobin	7.1
Myoglobin	7.0
Horse hemoglobing	6.9
Transferring	5.9
Fibrinogen	5.8
Insulin	5.4
Beta lactoglobulin	5.1
Urease	5.0
Plasma albumin	4.8
Egg albumin	4.6
Haptoglobin	4.1
Pepsin	1.0

Electrophoresis

The term electrophoresis originally described the migration of charged particles through a liquid or semisolid medium under the influence of an electrical potential. More recently, the word has come to refer to any technique by which molecules are separated from one another in electrical potential gradients on the basis of differences in their net charges regardless of the conducting medium. The method is employed most often for the separation of different proteins, since the side chains of many of the constituent amino acids exist in a dissociated form and contribute some number of positive and negative charges to the macromolecule. However, other molecules such as carbohydrates can also be separated by electrophoresis after being complexed with inorganic ions or other charged groups.

The fundamental principles of electrophoresis are quite simple. If two electrodes are inserted into a solution containing an electrolyte and a suspension of macromolecules of varying net charge, the macromolecules will be accelerated toward the electrode of opposite sign with a force proportional to the magnitude of the chare on the macromolecule and the strength of the applied electrical potential. Since each of the migrating particles will encounter frictional resistance to its movement, each will soon attain some maximum velocity of migration. Therefore, if a mixture of these macromolecules is exposed to a constant field strength, the maximum velocities attained will be different for particles of differing net charge. In addition to net charge and field strength, several other factors influence the rate of electrophoretic migration of proteins, including molecular size and shape, pH, and the nature of the medium through which migration occurs. These factors will be considered later.

The rate at which a protein migrates toward one or the other electrodes during electrophoresis is called its electrophoretic mobility (usually expressed in square centimeters per second per volt) and is dependent on the relative numbers of positively charged and negatively charged amino acid side chains. Whether or not a particular amino acid side chain carries a charge is, is turn, determined in part by the pH of the protein solution. As a pH is lowered (i.e., the concentration of H^+ is increased), negatively charged groups such as the secondary COO^- of aspartic and glutamic acid and the O^- of tyrosine become protonated, thereby neutralizing these negative charges. At the same time, some secondary amino groups such as those of lysine and arginine may accept additional protons, thereby increasing the number of positive charges associated with the protein. In contrast, as the pH is raised (i.e., the concentration of OH^- is increased), protons are dissociated from these chains and make the protein more negative. These relationships are shown in Fig. 3.5. Therefore, the numbers and types of charges associated with the amino acid side chains of a protein are determined by pH. At low pH, proteins tend to carry more positive than negative side chains and therefore, posses a net positive charge and migrate toward the cathode (the negative electrode) during eletrophoresis. At high pH, negatively charged side chains predominate, and the protein migrates toward the anode.

It follows from the above discussion that for every protein there will be a pH at which the number of positive and negative charges will be equal. If electrophoresis is carried out at this pH, no migration occurs, since the protein has no net charge. Above this pH, the net charge on the protein becomes increasingly negative, and its electrophoretic mobility toward the cathode increases. The relationships between pH and electrophoretic mobility for the proteins egg albumin and plasma beta lactoglobulin are shown in Fig. 3.6. The pH at which a protein possesses no electrophoretic mobility is called the isoelectric point and is a characteristic of each protein (Table 3.4). Since pH markedly influences electrophoretic mobility, it is important to maintain a constant electrolyte pH during electrolysis.

Paper Chromatography

Paper chromatography is a technique in which a mixture of solutes is separated into discrete zones on a sheet of filter paper on the basis of (1) differences in solute partition between a stationary aqueous phase tightly bound to the cellulose fibers of the paper and a mobile organic liquid phase passing through the sheet by capillary action (i.e., liquid-liquid partition) and (2) differences in solute adsorption to the cellulose fibers and dissolution in the mobile liquid (i.e., solid-liquid partition). Although the separation is based upon a combination of both phenomena, liquid-liquid partition differences are the more significant. The principles of the technique were set down in the 1940s by R. Consden, A. H. Gordon, A. J. Martin, and R. L. Synge and are not unlike those that are in effect during countercurrent distribution.

In practice, a rectangular sheet of filter paper is saturated with the aqueous phase and allowed to air dry, and the sample is applied near the end of the sheet as a narrow zone. The sheet is then suspended in a closed chamber in which the air has been saturated with the vapors of the mobile organic phase, the edge of the paper immersed in a bath containing the mobile phase (Fig. 3.1). Capillary action causes the mobile phase to slowly percolate through the paper from one end (the end containing the mixture to be separated) to the other. Movement of the liquid may be downward (descending chromatography) or upward (ascending chromatography). The solute mixture differentially partitions itself between the flowing solvent and the stationary phase time and time again as the solvent front advances toward the edge of the paper. Usually, the solvent front is allowed to migrate through the paper until it has almost reached the other end, at which time the sheet is removed and dried and the solute zones located by the appropriate chemical or physical means. However, in certain instance, where the solutes trail far behind the solvent front, it is desirable to allow the solvent to run off the edge of the paper sheet (descending chromatography only) so that maximum resolution of the solutes is achieved.

The rate of movement of a solute during paper chromatography is usually expressed as a dimensionless term Rf, where Rf is the ratio of the distance traveled by the solute front. Naturally, the Rf can only be calculated in those instances when the solvent is not allowed to leave the end of the paper sheet.

Greater resolution of the solutes may be obtained using two-demensional paper chromatography (Fig. 3.1): after chromatography using a particular solvent system in one direction along the paper sheet, the sheet is dried and rotated 90°, and another solvent system is used to chromatograph the solutes a second time. In this manner, solutes not fully separated by partition in the first solvent may be completely separated using the second solvent (and vice versa). Paper chromatography can also be combined with zone electrophoresis to provide two-dimensional analysis of a solute mixture, a technique known as "fingerprinting".

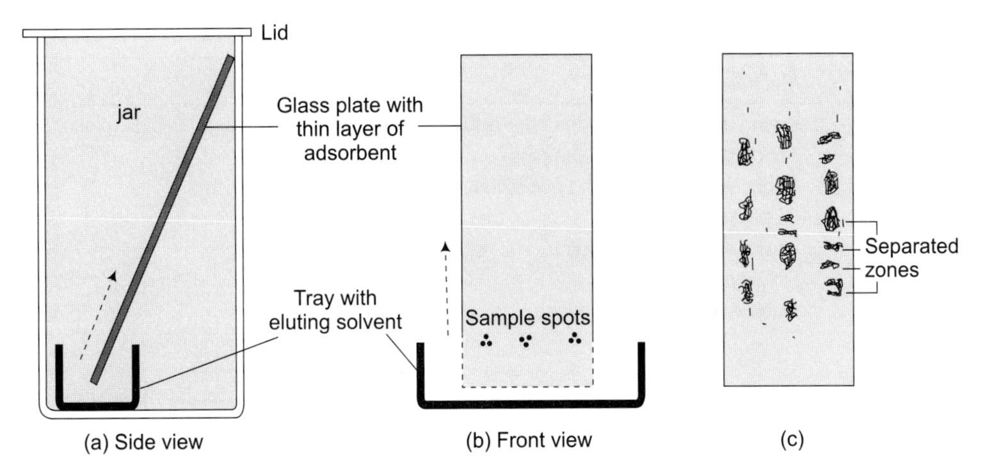

Fig. 3.5 Thin-layer chromatography (a and b). Glass plate containing thin layer of adsorbent and samples is immersed in tray containing shallow layer of eluting solvent. Capillary action causes solvent to ascend through adsorbent (arrows) which separates sample spots into series of zones (c).

Thin-Layer Chromatography

Thin-layer chromatography (abbreviated TLC) is an especially valuable method for rapidly separating unsaturated and saturated fatty acids, triglycerides, phospholipids, steroids, peptides, nucleotides and numerous other biological substances. In effect, TLC is a modification of paper chromatography in which the sheets of filter paper are replaced by glass plates covered with a thin, uniform layer of absorbent. The essential features of the technique may be described as follow an aqueous slurry of the selected absorbent is uniformly spread over a glass plate to produce a thin layer and is then dried. Following this, the sample (usually prepared in a volatile solvent) is applied near one end of the long axis of the plate as a spot or thin line. When the sample dries, the plate is supported vertically so that the end near the sample zone is immersed in a tray containing a shallow layer of the eluting solvent (usually an organic solvent of low polarity). Capillary action causes the solvent to ascend slowly through the layer of adsorbent and as in paper chromatography, the solutes become distributed along the plate on the basis of differential partition between the stationary and mobile phases (Fig. 3.3). Although ascending TLC is the most common, descending and horizontal separations may also be carried out. TLC separations are vary rapid, rarely exceeding 20 to 30 minutes.

After the separation has been achieved the glass plate is removed from the tray of solvent and allowed to dry. Zones containing colored substances can be detected directly, and others may be identified if they contain compounds that fluoresce when exposed to ultraviolet light. Many zones may also be rendered visible by spraying the plate with certain reagent dyes or stains. Adsorbent may also be scraped off various regions of the plate and the separated molecules eluted from the adsorbent particles. Table 3.5 lists some frequently used adsorbents and their applications.

Table 3.5 Adsorbents used for thin-layer chromatography.

Asorbent	*Materials separated*
Silica gel	Amino acids, polypeptides fatty acids steroids, phospholipids, glycolipids, plasma lipids
Alumina	Amino acids, steroids, vitamins
Kieselgulr	Oligosaccharides, amino acids, fatty acids, triglycerides steroids
Celite	Steroids
Cellulose powder	Amino acids, nucleotides
Hydroxylapatite	Polypeptides, Proteins
Polyethlenimine	Nucleotides oligonucleotides

Ion-exchange column chromatography. Proteins and other macromolecules may also be separated by the technique known as ion- exchange chromatography. The separation is carried out in glass columns packed with grains of an ion-exchange resin (polymers to which numerous ionzable groups have chemically added). Resins bearing negative charges are called cation exchangers and positively charged resins anion exchangers. As a solution of ions is passed through the column, the ions compete with each other for the charged sites on the resin. Consequently, rate of movement of any ion through the column depends on its affinity for the resin sites, it; degree of ionization, and the nature and concentration of competing ions in the solution. The differential rates of movement of ions through the column is the basis for protein and nucleic acid separations, since these molecules possess a variety of positively and negatively charged groups. Some ion exchangers used for protein and nucleic acid separations are listed in Table 3.6.

Affinity chromatography

Affinity chromatography is a novel form of column chromatography in which the molecules (principally proteins and nucleic acids) to be isolated from the sample under study are retarded in their passage through the column by their specific biological reaction with the column matrix. It is the biological nature of he interaction between the sample and the column matrix

Table 3.6 Some commonly used ion exchangers.

	Polymer	*Functional (Ionic) Group*
Anion exchangers		
Diethylaminoethylcellulose	Cellulose	$- O - (CH_2)_2 - N^+ H - (C_2H_5)_2$
Dowex-2	Polystyrene-divinyl-benzene	$- N^+ - (CH_3)_2C_2H_5OH$
Polylysine-Kieselguhr (PLK)	Polylysine	$- (CH_2)_4 - NH_3$
Amberlite IRA-400	Polystyrene-divinyl-benzene	$- N + (CH_3)_3$
Cation exchangers		
Carboxymethylccllulose	Cellulose	$- O - CH_2 - COO$
Hydroxylapatite	–	$- PO_4 \equiv$
Amberlite XE-64	Polymethacrylate	$- COO$
Sephadex SE	Dextran	$- (CH_2)_2 - SO_3$

that distinguishes this form of chromatography from others. The specificity may take the form of an antigen-antibody reaction, the complex of an enzyme with its substrate or inhibitor, hydrogen bounding between complementary polynucleotide, and so on.

The column is packed with porous gel particles (usually agarose, a linear polymer of the monosaccharide galactose) to which ligands having a high affinity for specific biological components have been covalently coupled. When the sample is applied to the column, a only the constituents having a high affinity for the ligand are bound, while other compounds are rapidly eluted. The ultimate desorptior, and isolation of the bound species is achieved by significantly altering the pH and/or ionic strength of the eluent (Fig 3.6).

Affinity chromatography has been used with great success for the isolation of specific immunoglobulin from antisera. This is achieved by first coupling the gel with specific an tigenic materials.

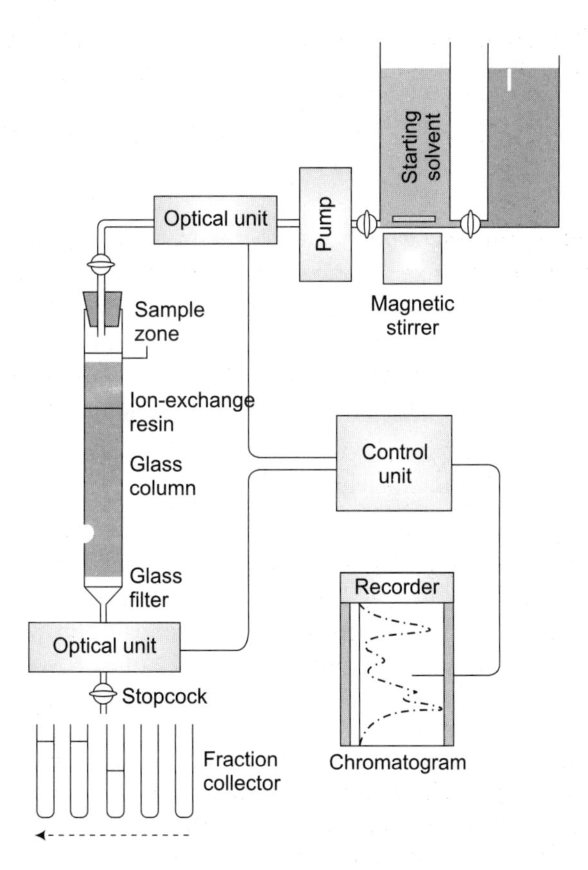

Fig. 3.6 Diagram of components used in ion-exchange column chromatography. Gradlent elution method is shown. The solvent is pumped through the optical unit of an ultraviolet light absorption monitor before enterinq the column. Flow out of the column is also monitored and the two absorbances compared in proteins. The absorbances are plotted on a strip-chart recorder. Peaks in the tracing (called a chromatogram) correspond to protein zones emerging from the column.

Gel Filtration (Molecular Sieving)

Gel filtration is a method for separating molecules on the basis of size (i.e., molecular weight) differences. The procedure is carried out using glass columns packed with nonionic, porous gel particles. Since the gels do not possess ionic groups, separation of a mixture of macromolecules does not involve the temporary formation of bound with the gel. The most commonly used gels are cross-linked dextrans produced by reacting dextrans with epichloronydrin) in which the degree of cross-linkage determines the average pore size of the gel. Gel particles may be produced with a variety of pore sizes.

Gas Chromatography

Gas chromatography is a special form of column chromatography in which a gas is used as the mobile phase (instead of a liquid) and either a liquid or a solid is used as the stationary phase. When a liquid is used as the stationary phase, the technique is called gas-liquid chromatography (GLC), and separations are based primarily on differences in the partition of the molecules in the sample between the stationary liquid and the moving gas. In gas-solid chromatography (GSC), separations result from the differential adsorption of sample molecules to the stationary phase as they are carried through the column by the gas. Of the two method, GLC is by far, the method most often employed.

The basic components of the gas chromatography are the source of gas, the sample introduction chamber, the chromatographic column, the detector, and the recorder (Fig. 3.19), the gas (usually nitrogen carbon dioxide, helium, or argon) is contained within a high pressure cylinder connected to the column through metal tubing. A valve, pressure gauge, and flowmeter are used to accurately regulate the flow of gas. The sample is introduced into the flow of gas using a microsyringe and needle inserted through a self-sealing diapharagm in the sample chamber. The chamber itself is enclosed within a heating block so that the sample (if it is not already in a gaseous form) will immediately be vaporized upon introduction into the chamber and will be swept into the column by the gas. Gas chromatograph column are made of glass, copper, or stainless steel tubing and are also enclosed in an oven; since they may be several feet long, they are often twisted to from a spiral.

The selection of packing material for the column depends upon whether the separation is to be based upon partition or adsorption. For GLC, the column is packed with an inert solid such as kieselguhr, which is impregnated and lightly coated with a liquid of low volatility (so that it will not be eluted from the column at the operating temperature used). For GSC, the stationary phase is an adsorbent such as charcoal or silica gel. Different molecules in the sample will be carried through the column at different rates, depending on their adsorption or partion characteristics, and will emerge from the end of the column at different times. Located near the exit of the column and also housed within an oven is the detector that monitors the composition of the emerging gas relays electrical electrical signals proportional to the amounts of separated components to a strip-chart recorder. The separated components are thus recorded as a series pf peaks in the chromatogram tracing. The most common form of detector is the flame ionization chamber in which the components are successively mixed with hydrogen and air and burned in a high-voltage field. Migration of the ionized fragments in this field creates a current registered by the recorder. Most chromatographic separations are analytical, and the technique is so sensitive that minute quantities of sample are required. However, separated components may also be collected as condensate in tubes as they emerge from the heated column and are rapidly cooled.

Gas chromatography may be used to separate lipids, oligosaccharide, and amino acids after their preliminary conversion to volatile derivatives (many lipids may be chromatographed without conversion). Gas chromatography has been used with great success for the separation of different fatty acids. Depending upon whether a nonpolar or a polar liquid phase is used, fatty acids may be separated according to boiling point and size or degree of saturation.

The methods described in this chapter for separating, isolating and studying macromolecules and their constituents are widely used in cellular research. Student interested in pursuing this project can find a more comprehensive discussion of these methods in the books and articles listed at the end of the chapter, together with descriptions of other separation methods that are used less often but are also important.

SUMMARY

Over the years, a variety of analytical and preparative techniques have been developed for separating, analyzing and isolating the macromolecular constituents of cells and tissues. Older methods such as salting in, salting out, and isoelectric precipitation relied on differences in the solubility of the molecular species under investigation. Differences in molecular size are used to achieve separations in such diverse approaches as dialysis, ultrafiltration, gelfiltration (molecular sieving), and ultracentrifugation. Differences in the net electrostatic charges of molecules are used in various types of electrophoetic separations, including moving-boundary electrophoresis, zone electrophoresis, and disc elctrophoresis. Biological activities of the molecules are used to advantage in such techniques as immunoelectrophoresis and affinity chromatography. Many chromatographic techniques utilize partition differences to achieve separations, the most

Table 3.7 Phusical Biological and Effective Half-Lives of Some Radioisotopes

			Half-life	
Element	*Radioisotope*	*Physical*	*Biological*	*Effective*
Hydrogen	^{3}H	12.3 yr	19 days	19 days
Carbon	^{14}C	5.570 yr	180 days	180 days
Phophorous	^{32}P	14.3 days	3 yr	14.1 days
Calcium	^{45}Ca	164	73 yr	163 days
Iron	^{59}Fe	45.1 days	3.4 yr	27.1 days
Cobalt	^{60}Co	5.3 yr	8.1 days	8.0 days
Iodine	^{134}I	8.1 days	156 days	7.7 days
Radium	^{226}Ra	1.620 yr	104 days	104 days

popular formats being thin-layer chromatography, anion and cation exchange chromatography, and gas chromatography. Because such molecular parameters as size, charge, and solubility are used as the basis for effecting a separation, information concerning the unique physical and chemical properties of the species under investigation is simultaneously acquired.

RADIOACTIVE ISOTOPES AS TRACERS IN CELL BIOLOGY

Isotopes are chemical elements that have the same atomic number (i.e., the number of protons in the nucleus of the atom) but different atomic masses (i.e., the sum of the number of protons and neutrons in the nucleus). Certain isotopes are unstable and undergo spontaneous disintegrations (transmutations) accompanied by the emission of particulate and sometimes also electromagnetic radiations. These atoms are said to be radioactive and are called radioisotopes or radionuclides; their presences may readily be detected by instruments sensitive to their radiations. Generally, an organism cannot distinguish between the stable and radioactive forms of the same element so that both are metabolized in an identical manner. It is for this reason that radioisotopers have proven extremely useful to biologists, since these elements may conveniently by employed as tracers. That is, the fate of a given element (or molecule) in an organism (or even in an individual cell) may be studied by introducing the radioactive form of that element and following the uptake and subsequent localization of the radioactivity.

If the radioisotope is initially a part of a larger molecule, then the fate of all or part of that molecule may similarly be followed. The use of radioisotopically labeled compounds is particularly described when the compound to be administered is a normal constituent of the cell or organism and would be impossible to distinguish from stable molecules already present. Many organic and inorganic compounds of biological may now be obtained that have one or more specific atomic positions occupied by radioisotopes. Because of the extremely high sensitivities of most radiation detectors, the quantities of radioisotopes employed in tracer studies are small enough to preclude significant damage to cell constituents by the radiation.

Advantages of the Radioisotope Technique

Results obtained from experiments involving the use of radioisotopers are quantitative, since the amount of radioactivity present and available for detection is directly proportional to the radioisotope content. Moreover, numerous biological studies carried out routinely using radioisotopes can only be performed with great difficulty or are virtually impossible without them. Some examples may be cited to illustrate the value of the radioisotope technique.

Properties of Radioactive Isotopes

Most radiations emitted by radioisotopes are the result of charges in the arrangement of unstable atomic nuclei. Whether or not a given atomic nucleus is stable depends in turn upon the numbers of neutrons (N) and protons (Z) that it contains: the relationship between nuclear stability and the neutron. It should be noted that for he lighter elements, nuclear stability when N = Z, whereas in the stable heavier elements, the number of neutrons exceeds the number of protons, with the neutron excess increase with atomic number. Isotopes with N and Z numbers outside of the stable region undergo spontaneous changes in which nuclear neutrons and protons ae interconvert. These nuclear transmutations are accompanied by the emission of particulate and electomagenetic radiation.

Types of Radiation Emitted by radioisotopes

The most common types of nuclear radiations are alpha particles, positive and negative beta particles, and gamma rays. Alpha particles, which consist of two protons and two neutrons and are therefore identical to helium nuclei, are emitted by radioisotopes of high atomic number such as uranium, polonium, thorium, and radium; that is,

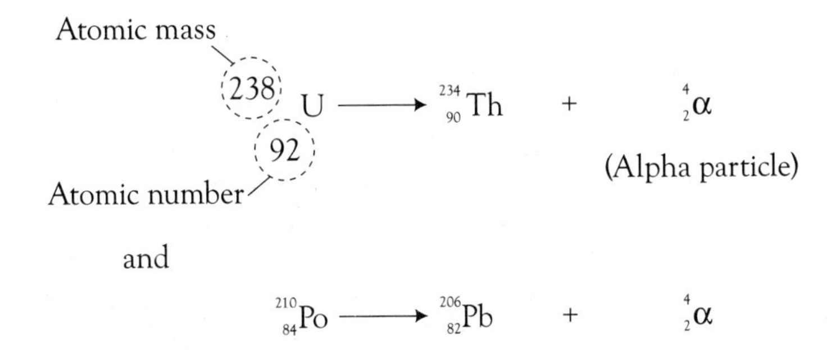

Alpha-emitting radioisotopes are rarely used as tracers in biological studies.

Positive beta particles, also called positrons, are emitted from nuclei in which the N : Z ratio is below that which is stable. The nuclear transmutation involves the conversion of a proton into a neutron, positron and neutrino, the latter two being ejected from the nucleus. The positron possesses a unit positive charge and is equal in mass to an electron, while the neutrino has neither mass nor charge. The isotope “C decays by positron emission:

$$^{11}_{0}C \rightarrow {}^{11}_{5}B + {}^{0}_{1+1}e + \text{Neutrino}$$

Unlike alpha and beta radiation, gamma radiation is not particulate but is electromagnetic.

Energy of Radiation and Its Interaction with Matter

The energy of radiation is measured in electron volts (abbreviated ev), one electron volt being the kinetic energy acquired by an electron in a potential difference of one volt. Beta particles and gamma rays emitted by radioisotopes often used as tracers have energies ranging from about, 1 × 10 to 4 × 10 ev ot 0.01 to 4.0 Mev (1 Mev equals 1 million electron volts). This energy range may be compared with the energy of chemical bound, which is of the order of a few electron volts. The kinetic energy acquired beta particles and gamma rays interact with matter by ionizing and exciting atoms in their path. In the case of beta particles, ionization results from the repulsion of orbital electrons by the negatively charged particle. As a result, beta particles tracking through matter produce a wake of electrons and positively charged ions (called ion pairs). The number of ion pairs produced per unit path length, called the specific ionization, is not constant but increases as the beta particle slows down. Also, the path of the beta paricle is not liner but is quite erratic as a result of its repulsion by the orbital electrons of atoms with which it interacts. High-energy beta particles may traverse the linear equivalent of 1 to 2 m in air. Since gamma radiation is electromagnetic, its probability of interacting with matter is less than that for beta particles. Consequently, gamma rays have a much lower specific ionization and a much longer and also liner path length.

Half-life

The number of atoms in a sample of radioisotope that disintegrate during a given time interval decreases logarithmically with time and is unaffected by chemical and physical factors that normally alter the rates of chemical processes (i.e., temperature, concentration, pressure, etc). radioactive decay is therefore a classical example of first-order reaction. A convenient term used to described the rate of decay of a radioisotope is the physical half-life, T-that is, the amount of time required to reduce the amount of radioactive material to one-half its previous value. Each radioactive decays at a characteristic rate and therefore has a unique half-life. For example, the amount of radioactivity arising from a sample of Fe is reduced to one-half its original value in 45.1 days, to one-fourth in 90.2 days, to one-eight in 135.3 days and so on. The amount of decay occurring in the course of a tracer experiment must be taken into account when radioisotopes of short physical half-life such as Na, P, Cl, K, and I are used. Of course, this is not a problem in experiments involving H and C.

Autoradiography

Any discussion of the uses of radioisotopes as tracers in cell biology would be incomplete without at least a brief description of autoradiography. The materials and equipment used in this technique differ significantly from those employed in the methods previously described. In autoradiography, the biological sample containing the radioisotope is placed in close contact with a sheet or film of photographic emulsion. Rays emitted by the radioisotope enter the photographic emulsion and expose it in a manner similar to visible light. After some period of time (usually several days to several weeks), the film is developed and the location of the radioisotope in the original sample determined from the exposure spots on the film. Unlike the methods described earlier, autoradiography is generally not employed as a quantitative technique (although it can be under certain conditions); instead, it is used to determine the specific region of localization of a radioactive tracer. The method is most often used with histological sections to determine the precise location of a labeled compound in a tissue or in a cell and may applied either at the light microscope or electron microscope level. Autoradiography has also been successfully employed in conjunction with electrophoresis, chromatography (i.e., thin-layer chromatography paper chromatography, etc), and other molecular fractionation methods for identifying zones containing labeled compounds.

Autoradiography may be employed with large but thin slices of tissues or organs (gross autoradiography) or with smaller pieces sectioned and prepared for light or electron microscopy.

For example, the deposition of calcium in bone has been studied using Ca by cutting thin, flat longitudinal slice through bone and placing them against large sheets of photographic film. When used in conjunction with light microscopy, the paraffin sections are first mounted on slides which are then coated with the photographic emulsion and stored in a lighttight, usually lead-lined box (in order to minimize background exposure that result from the effects of cosmic radiation and other sources of radiation). Several days or weeks later, after photochemical development, the sections are conventionally stained to better visualize the biological material and the distribution of radioisotope determined microscopically from the location of dark exposure spots (called "grains") on the section.

Autoradiography may also be used with thin sections of tissues prepared for electron microscopy. This producer involves preliminary examination and photography of the thin section followed by coating, the grid with a very thin layer of photographic emulsion containing silver halide crystals of particularly small size. After several days the grid is developed and again examined and photographed with the electron microscope in order to identify those regions of the original section that contain clusters of metallic sliver grains and therefore contain the radioactive tracer.

SUMMARY

Certain atomic isotopes are unstable and emit particulate (and sometimes electromagnetic) radiations. A number of radioactive isotopes, especially H, C, P, S, and Fe are particularly valuable to cell biologists as tracers of metabolism and function. With radioisotopes, it is possible to follow ionic or molecular flux under conditions of zero net exchanges to simplify chemical analyses, to measure pool sizes, and to determine precursor-product relationships.

The most common types of additions are alpha particles, beta particles, and gamma rays. Only negative beta-particles-emitting isotopes and gamma-ray-emitting isotopes are normally used by biologists. The energy acquired by a beta particle during nuclear transmutation may vary over a specific range, whereas gamma rays have discrete energy value. Beta particles and gamma rays interact with matter by ionizing or exciting atoms in their path, and it is this property that is used in instruments designed to detect and measure the quantity to radioisotope present in a sample. The simplest and most common detector is the Geiger-Muller counter and is used principally with tracers emitting beta particles of intermediate or high energy. Solid scintillation counters are used with isotopes emitting gamma rays and liquid scintillation counters, with isotopes emitting weak beta rays. H and C are used as tracers more often then any other isotopes, and since these emit very weak beta particles, the liquid scintillation approach is a particularly important technique in cell biology. In the typical liquid scintillation counter, the sample is mixed with a scintillation fluid in a glass vial. The energies of the beta rays emitted by the sample are ultimately converted to flashes of light which are detected by photomultiplier tubes and counted.

Autoradiography is a special modification of the radioisotopes tracer technique most often used in conjunction with light and /or electron microscopy. Biological samples (usually tissue sections) containing the tracer are placed in contact with special photographic films or emulsions. The emitted rays expose the film, producing grains, the numbers and distributions of which within a cell or tissue provide quantitative and qualitative information about the movement and/ or localization of metabolic intermediates or products.

CHAPTER 4

Bioenergetics

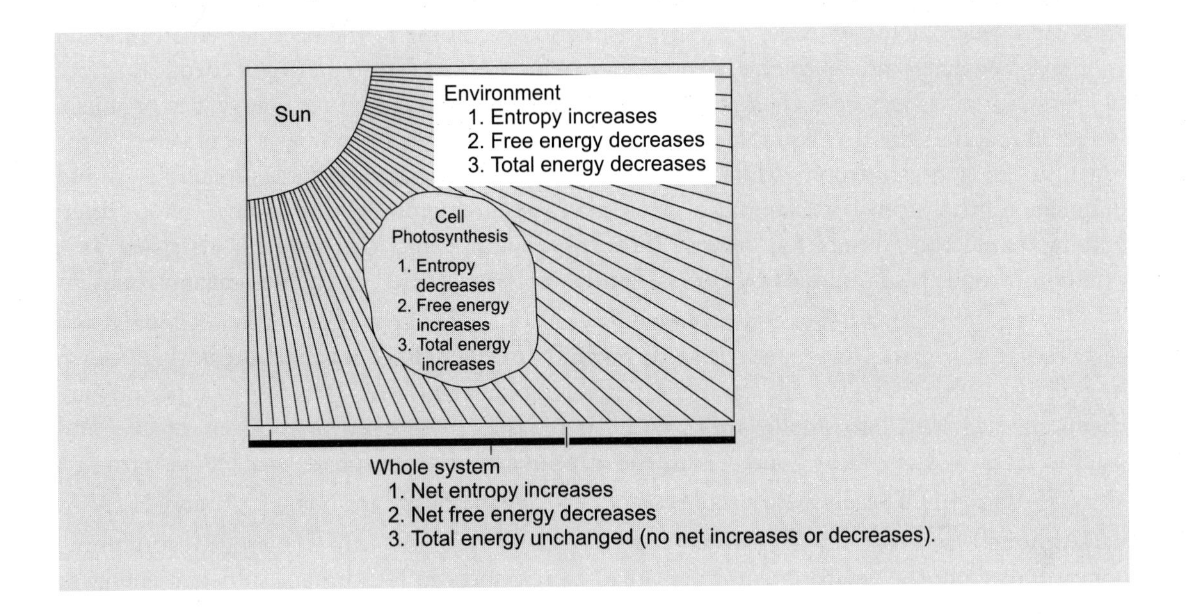

In the study of physiology and biochemistry, cells are frequently thought of as machines in which all events are explainable as one or more chemical reactions. As the result of fluid dynamics, as electrical fluxes across partitions, or as the absorption or emission of light. In other words, the principles of physics and chemistry, for cell actions conform to the laws of the physical science.

This chapter discusses those laws of physics and chemistry that are important for an understanding of cell metabolism. Including the breakdown of some molecules (catabolism) and the synthesis of others (anabolism). The absorption of light aw in vision and photosynthesis, propagation, of electrical charges such as nerve impulses and fluid dynamics as they relate to diffusion and osmotic phenomena.

ENERGY

All the chemical reactions and physical events in cells are related to energy changes. For example, the synthesis of new membranes in a growing cell requires or consumes energy. That energy must be ultimately absorbed from the environment in some form, such as light energy or chemical energy, and they transformed by the cell into forms that can provide the energy for the membrane synthesizing reactions. Depending upon the source of the environment's energy cells and organisms are divided into two basic groups. If the organism required organic chemicals (foods) from the environment, it is classified as a heterotroph. If the organism can survive with only inorganic compounds and energy sources such a slight, it is called an autotroph. Microbiologists make a further distinction according to the need for environmental energy or raw materials. By their classification system, an organism requiring a chemical source of energy supply is termed a chemotroph, but if only light is required for energy, the organism is a phototroph. A further requirement for organic raw materials for synthesis would classify the organism as a organotroph, while a requirement for only inorganic raw materials would designate a lithotorph. For example, a green plant requires only light for energy (phototroph) and inorganic compounds for growth (lithotroph) and could therefore be classified as a photolithotroph. Most animals by this system would be classified as chemo-oraganotrophs.

The primary sources of energy and raw materials for heterotrophs are proteins, lopids, and carbohydrates. Organisms remove these compounds from environment and break them down in progressive stages called the catabolic reactions of metabolism. As these compounds are chemically degraded into smaller units, their energy is both released in the form of heat and used to form new chemical bonds, as in the attachment of a phosphate to ADP to form

ATP (Fig. 4.1) The ultimate primary products of catabolism are NH_3, CO_2 and H_2O.

Although autotrophic organisms absorb CO_2, H_2O and small nitrogen-containing compounds from the environment, these small compounds do not contain sufficient energy to maintain the organisms. Consequently, autotrophs absorb energy in the form of light, and using the light energy, they synthesize simple organic acids from CO_2 and water (photosynthesis) as well as phosphorylate ADP and synthesize amino acids from the organic acids by incorporating NH_4 (aminacion). From these simpler molecules more complex molecules such as proteins, lipids and polysaccharides are formed (Fig. 4.2) however, the synthesis of these molecules requires further energy consumption which cell supplies from its pool of ATP, and with each change, energy in the form of heat is lost from the cell.

In general, autotrophic organisms not only contain the enzymes for the anabolic reactions just described but also possess catabolic enzyme systems similar to those of the heterotrophs and produce ATP upon the breakdown of carbohydrates, lipids and proteins. Heterotrophs have anabolic enzyme systems that require ATP and are capable of synthesizing macromolecules much like the autotrophs, but they are unable to carry out photosynthesis. Cells tend to cycle compounds internally, and compounds are also cycled between organisms (Fig. 4.3) with each change, there is a corresponding change in energy.

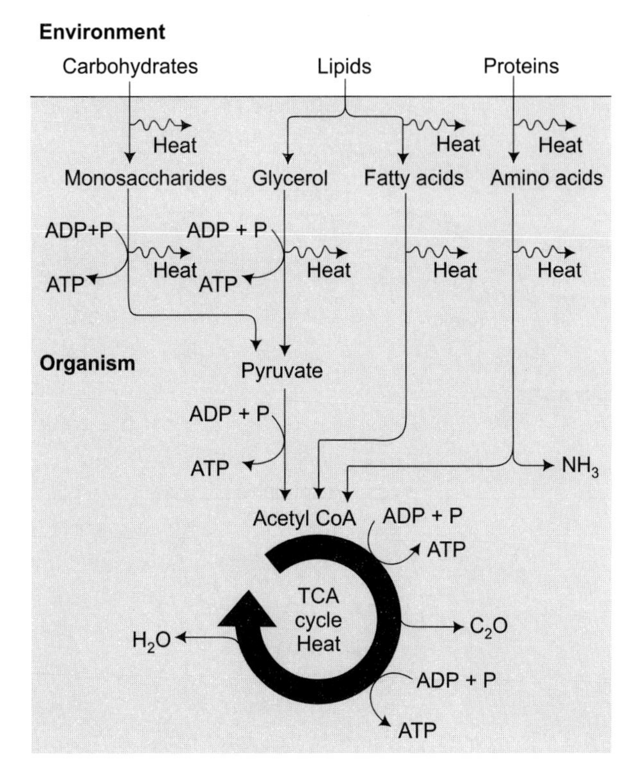

Fig. 4.1 Energy flow in living systems during catabolic reaction.

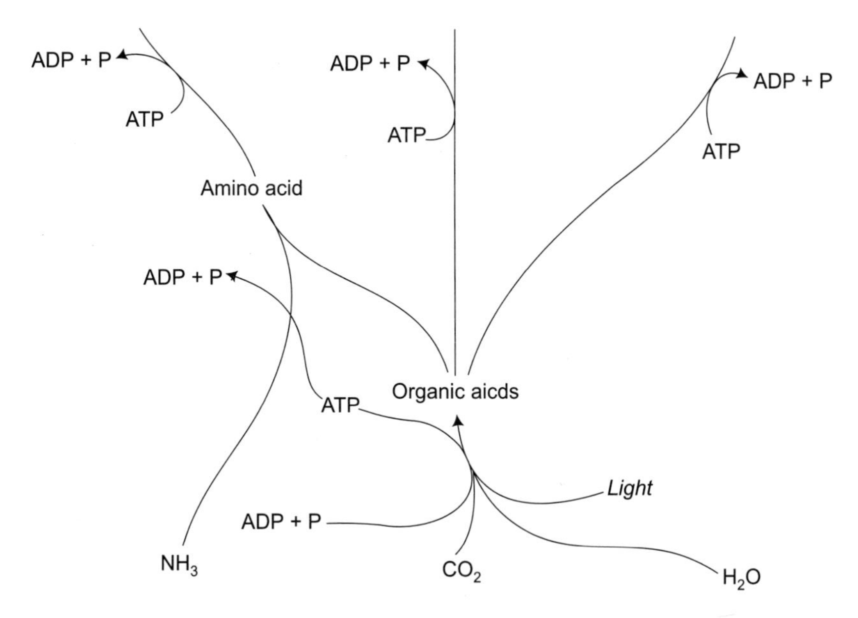

Fig. 4.2 Energy flow in living systems during anabolic reaction.

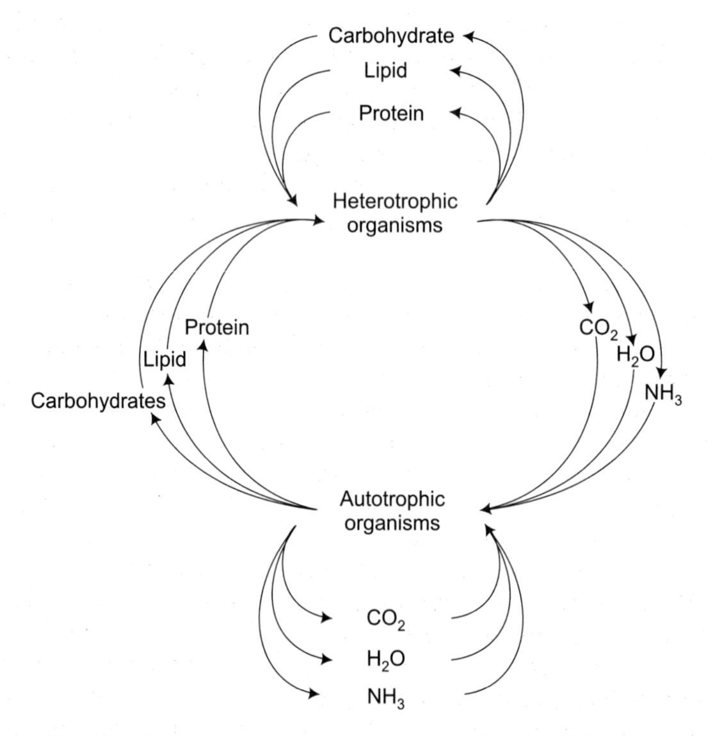

Fig. 4.3 The cycles produced by the interchange of foodstuff.

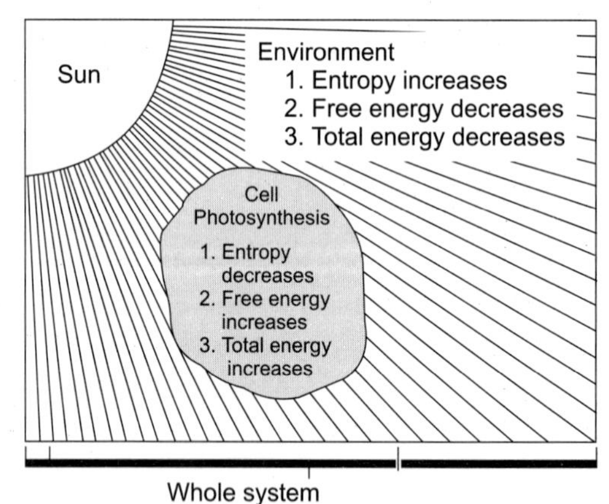

Whole system
1. Net entropy increases
2. Net free energy decreases
3. Total energy unchanged (no net increases or decreases).

Fig. 4.4 Relationship of entropy and free energy between various parts of a system and the net changes in energy.

THE LAWS ABOUT ENERGY AND ENERGY CHANGES

The laws that relate to energy in general, and that therefore apply to cellular energy as well, are the two laws of thermodynamics. The first law is concerned with the conservation of energy and requires no modifying statements when applied to biological systems. The first law states the following.

ENERGY CANNOT BE CREATED OR DESTROYED BUT CAN BE CONVERTED FROM ONE FORM TO ANOTHER

This definition applies to cells, to organelles, and even to single chemical reactions. In practice, the measurement of energy in a cell or similar units is difficult, since energy may escape from the cell into the surrounding environment during the measurement. Similarly, energy may be gained from the environment; for example a photosynthesizing cell absorbs energy in the form of light. The escape or absorption of energy by a body should not be confused with the destruction or creation of energy, which by the firs law of thermodynamics does not occur. Usually the term "system" is used to delimit the matter in which the observer is interested in theoretical discussions, energy is assumed not to enter or leave the system. In practical measurements the system is isolated as much as possible, and exchange of energy between the system and the surroundings is measured.

Applied to the cell, the first law of thermodynamics indicates that the cell has finite amount of energy. That energy may be (1) potential, as in the bonds of molecules or in the pressure-volume relationships within the cell membranes (2) electrical, as in the distribution of charges across membranes or (3) kinetic, as in the thermal activity of molecules. The first law also states that these energies may be converted from one form into another; in other words, the potential chemical energy could decrease and, in so doing, be converted into electrical or kinetic energy. However, the total energy within the cell when isolated as a system remains constant. It is not possible for the cell to make energy out of matter or to convert energy into matter or to destroy any energy. For example, in the breakdown or a polysaccharide into CO_2 and H_2O, some of the potential energy in the carbohydrate is transferred to ADP and phosphate to make potential energy in the form of ATP, and some of the energy is converted to kinetic energy (which is of little practical use to the cell); however, none of the energy is lost or destroyed, and all the energy originally in the polysaccharide should be accountable for in other forms within the system.

The second law of thermodynamics explains the direction of energy changes:

IN ALL PROCESSES OF ENERGY CHANGE, THE ENTROPY OF THE SYSTEM INCREASES UNTIL EQUILIBRIUM IS ACHIEVED

Entropy is an expression of the energy in a system that is of no value for performing work and that cannot be converted into useful of reactive forms of energy. For example, a molecule of sucrose at 25°C requires a certain amount of internal energy just to be at that temperature and to maintain the thermal agitation of the molecule as well as the oscillations of the component atoms and their particles. This is unavailable energy, since it cannot be given up without a change in temperature. This unavailable energy changes with a change in temperature and is thus expressed by the paired symbols TS, or T for absolute temperature and S, which is called the entropy.

In catabolism of sucrose in a cell, many molecules of energy - rich ATP may be formed. Although, superficially it may appear that useful energy has increased in the form of ATP, within the whole cell (system) the total amount of useful energy has decreased and the unavailable energy has increased. Some of the potential energy that was present in the sucrose has been converted into useful potential energy in the ATP, but some has also been converted into kinetic energy, which raises the temperature of the cell and the entropy (TS). In addition, when a large molecule such as sucrose is catabolized to many smaller molecules the energy for agitation of the many small molecules is greater than that for the large molecules, thus, contributing to the increase in entropy.

Table 4.1 Standard Free Energy Changes of Common Biochemical Reactions at pH 7.0 and 25°C.

Reaction	*ΔG° kcal mol^{-1}*
Hydrolysis:	
Acid anhydrids:	
Acetic anhydride + H_2O → 2 acetate	– 21.8
Pyrophosphate + H_2O → 2 phosphate	– 8.0
Esters:	
Ethyl acetate + H_2O → ethanol + acetate	– 4.7
Glucose-6-phosphate + H_2O → glucose + phosphate	– 3.3
Amides:	
Glutamine + H_2O → glutamate + NH_4^+	– 3.4
Glycylglycine + H_2O → 2 glycine	– 2.2
Glycosides:	
Sucrose + H_2O → glucose + fructose	– 7.0
Maltose + H_2O → 2 glucose	– 4.0
Esterification:	
Glucose + phosphate → glucose-6-phosphate + H_2O	+ 3.3
Rearrangement:	
Glucose-1-phosphate → glucose-6-phosphate	– 1.7
Fructose-6-phosphate → glucose-6-phosphate	– 0.4
Elimination:	
Malate → fumarate + H_2O	+ 0.75
Oxidation:	
Glucose + $6O_2$ → $6CO_2$ + $6H_2O$	– 686
Palmitic acid + $23O_2$ → $16CO_2$ + $16H_2O$	– 2338

Source. From A. L. Lehninger Biochemistry (2nd ed.). Worth Publisher New-York. 1975, p. 397.

Table 4.2 Standard Redox Potentials at pH 7.0 and 25°-37° C.

Reductant	*Oxidant*	*E'_o(volts)*
Pyruvate	Acetate + $2H^+$ + $2e^-$	– 0.70
Acetaldehyde	Acetate + $2H^+$ + $2e^-$	– 0.58
H_2	$2H^+$ + $2e^-$	– 0.42
$NADH_2$	NAD^+ + $2H^+$ + $2e^-$	– 0.32
Ethanol	Acetaldehyde + $2H^+$ + $2e^-$	– 0.197
Lactate	Pyruvate + 2H + $2e^-$	– 0.185
Succinate	Fumarate + 2H + $2e^-$	– 0.031
Ubiquinol	Ubiquinone $2H^+$ + $2e^-$	+ 0.10
2 Cytochrome $b_{(ox)}$	2 Cytochrome $b_{(red)}$ + $2e^-$	+ 0.030
2 Cytochrome $c_{(ox)}$	2 Cytochrome $c_{(red)}$ + $2e^-$	+ 0.254
2 Cytochrome $a_{3(ox)}$	2 Cytochrome $a_{3(red)}$+ $2e^-$	+ 0.385
H_2O	1/2 O_2 + $2H^+$ + $2e^-$	+ 0.816

E'_o = 0.0 volts

Suggestions that cells can decrease entropy by photosynthesis are misleading. Although it is true that photosynthesis can convert small molecules with little potential energy (CO and HO) into large molecules with considerable potential energy (sugars) and that the entropy of the cell does decrease, "high-grade" energy (i.e., light) was absorbed from the surrounding environment (outside the system). If the light energy consumed is included as part of the system, then the net energy changes will reveal an increase in entropy and a decrease in useful energy. These relationships are expressed in Fig. 4.

Not all the energy is changed in a system each time there is a chemical reaction (or similar phenomenon). For example, when sucrose is hydrolyzed to glucose and fructose, much of the potential energy of the original source still remains in the resulting glucose and fructose molecules. The change in energy, rather than total energy, is a commonly used measure. The change in total energy, H, is called the enthalpy and is composed of two factors: (1) the change in useful energy (or free energy) capable of performing work, ΔG; and (2) the entropy factor T S, which is the absolute temperature times the change in entropy. For example,

$$\Delta H = \Delta G + T\Delta S \quad (1)$$

Since the change in entropy is difficult to determine while the important aspect of reactions in cells is the change in amount of useful or free energy (G), equation (1) is more frequently written

$$\Delta G = \Delta H - T\Delta S \quad (2)$$

The change in free energy (G) can also be defined as the total amount of free in the product minus the total amount of free in the reactions; that is,

$$G = G_{Prod} - G_{React} \quad (3)$$

A reaction that has a negative G value (i.e., the products have less energy than the reactants) will occur spontaneously. A reaction with a positive G value requires an input of energy from some outside source. The hydrolysis of sucrose,

$$\text{Sucrose} + H_2O \rightarrow \text{glucose} + \text{fructose}$$

has a negative value. If 5 moles of sucrose are mixed with water the reaction proceeds spontaneously and the ΔG can be determined. The value is, of course, greater than if 4 or 2 moles of sucrose were used. The G is dependent upon the concentration of reactants and products. Since the ΔG value very with concentration, it is difficult to compare changes in free energy between various reactions. As a result, the concept of standard free energy, ΔG, was established by convention. This value represents the change in free energy when the reactions are present in 1.0 molar concentrations (1.0 molar is frequently substituted) at 25°C and a pressure of 1.0 atmosphere. G is the change in standard free at pH 7.0 a value more useful in biological, systems.

The ΔG value for he hydrolysis of sucrose is -7.0 kcal; that is, the reaction will proceed spontaneously and release 7 kcal of free energy if 1 mole of sucrose is mixed with 1000 ml of H_2O. However, in cells concentrations of reactions and products vary and the actual ΔG should not be confused with the ΔG°.

G is calculated from the equilibrium constant.

$$G = -RT \text{ In } K \qquad (4)$$

$$= -2.303RT \log K \qquad (5)$$

where R is the gas constant (1.98. cal degree-1 mole -1). T is the absolute temperature (degree Kelvin), and K is the equilibrium constant, Table 4.1 lists a number of G values for common reactions. The equilibrium constant is determined analytically.

In the reaction $A + B \rightarrow C + D$

$$K = \frac{[C][D]}{[A][B]} \text{ at equilibrium} \qquad [6]$$

[A] and [B] are the concentrations of reactants, and [C] and [D] are concentrations of products, if the equilibrium constant is 1.0, then the $\Delta G = 0$, if the equilibrium constant is less than 1.0, then the G is positive (e.g,. + 2.73 Kcal mole for K = .01), and the reaction is said to be endergonic because it requires an additional supply of energy to make it proceed. A K value greater than 1.0 would indicate a negative ΔG (–1.36 kcal mole for K = 10), and the reaction would be called exergonic because energy is spontaneously released in the reaction.

LIGHT AND CHEMICAL TRANSDUCTION

The conversion of energy from one to another is called transduction. Light photons are converted into potential chemical bond energy is such seemingly diverse processes as photosynthesis and vision, Chemical bond energy is transformed into the light energy in bioluminescence, such as the light emitted by fireflies and the light emitted by microorganisms in the sea when agitated by a passing boat.

Natural electromagnetic radiations comprise more than the spectrum of colors perceived by the human eye. The range of natural radiations is described by the electromagnetic spectrum (Table 3). Each of these radiations mat be envisioned as a stream of moving packets of energy called photons. The streams of photons move in a wave-like manner as illustrated in Fig. 4.5. the distances from the crest of one wave to the crest of the next (or from a point on one wave to an equivalent point on an adjacent wave) is defined as the wavelength. Among the differences between radiations in the electromagnetic spectrum are differences in wavelengths. At one extreme are the long radio waves or television waves with a wavelength of 1 to 1m. At the opposite end of the spectrum are the gamma rays with wavelength as short as 0.0001 × 10 m (10 nm.).

INTRACELLULAR PHOSPHATE TURNOVER

In all cells, most of the more significant energy changes and reaction coupling occur with ADP and ATP. This is true in the cytoplasm, mitochondria, ribosomes, nucleus, and other organelles. The cell has a number of the nucleotidephosphate pools. Although

$$\text{ADP} + \text{phosphate} \rightarrow \text{ATP}$$

Exchanges are most common.

$$\text{ATP} \rightarrow \text{AMP} + \text{pyrophosphate}$$

Reactions occur in lipid metabolism as does

$$\text{CTP} \rightarrow \text{CDP} + \text{phosphate}$$

Reactions. Uridine triphosphate (UTP) is utilized in polysaccharide synthesis; guanosine triphosphate is required in protein synthesis. They deoxy derivatives, dATP . dGTP. dUTP, and dCTP, are all used in DNA synthesis.

Regenerations of each of the triphosphates is catalyzed by the relatively nonspecific nucleotide diphosphate kinase:

$$\text{CDP} + \text{ATP} \rightarrow \text{CTP} + \text{ADP}$$

$$\text{UDP} + \text{ATP} \rightarrow \text{UTP} + \text{ADP}$$

$$\text{GDP} + \text{ATP} \rightarrow \text{GTP} + \text{ADP}$$

$$\text{dCDP} + \text{ATP} \rightarrow \text{dCTP} + \text{ADP}$$

$$\text{dADP} + \text{GTP} \rightarrow \text{dATP} + \text{GDP}$$

The total concentration of nucleotides in cells in a steady state is relatively constant. In a relatively inactive steady-state cell, most of the adenylate system is in the form of ATP. The major catabolic pathways operate at a level that keeps the system filled with phosphates. When cellular activity increases. The level of ATP decreases and the ADP + ATP level increases. The change in levels of these compounds initiates an acceleration in reaction sequences, such as glycolysis, which generates ATP. In a similar manner, when activity decrease, the level of ATP will quickly increase, causing in turn the slowing ATP generating systems. The mechanism of regulation is through allosteric enzymes that are modified by ATP, ADP and AMP.

Although ATP hydrolysis is a major source of energy in the cell and pools of ATP of ATP are relatively predictable, ATP does not form a substantial energy pool. The amount of ATP, in the cell will last only a short time during periods of increased activity. Most cells contain additional reservoirs of energy-rich compounds that can quickly be converted into ATP. Skeletal muscle, smooth muscle, and never cells, contain phosphocreatine reservoirs. When ATP levels fall, creatine kinase is activated, catalyzing the reaction,

$$\text{Phosphoreatine} + \text{ADP} \rightarrow \text{creatine} + \text{ATP}$$

$$(\text{G} = -3.0 \text{ kcal mole})$$

phosphoarginine and polymetaphosphate act in a similar way in invertebrate and bacterial cells, respectively.

REDOX COUPLES

In addition to the exchange of energy between substrates in a reaction, energy can also be changes during electron transfer between oxidants and reductants, or redox reactions. An oxidants or oxidizing agent is a substance that loses electrons to a reductant or reducing agent. Thus, a reductant is a substance that absorbs electrons from anoxidant. Substances have various potentials to retain or lose electrons. Hydrogen is known to dissociate.

$$\text{H} \rightarrow 2\text{H} + 2e^-$$

Table 4.3 Wavelengths of the radiations of the electromagnetic spectrum.

Type of radiation	*Photon Energy (ergs × 10^{-12})*	*Wavelength (nin°)*
Television waves	$2.0 \times 10^{10} - 2.0 \times 10^{8}$	$10^{2} - 10^{11}$
Radar waves	$2.0 \times 10^{-8} - 2.0 \times 10^{-5}$	$10^{6} - 10^{2}$
Radio waves	$2.0 \times 10^{-11} - 2.0 \times 10^{-5}$	$10^{6} - 10^{12}$
Infrared (far)	0.005 – 0.99	$2 \times 10^{3} - 4 \times 10^{3}$
Infrared (near)	0.99 – 2.5	$780 - 2 \times 10^{3}$
Red light	2.5 – 3.2	620 – 780
Orange light	3.2 – 3.4	590 – 620
Yellow light	3.4 – 3.6	545 – 590
Green light	3.6 – 4.1	490 – 545
Blue light	4.1 – 4.6	430 – 490
Violet light	4.6 – 5.1	390 – 430
Ultraviolet (long)	5.1 – 6.6	300 – 390
Ultraviolet (short)	6.6 – 9.9	200 – 300
Ultraviolet (very short)	9.9 – 132	15 – 200
X-rays (soft)	$99.3 - 2.0 \times 10^{4}$	0.1 – 20
X-rays (hard)	$2.0 \times 10^{4} - 3.9 \times 10^{6}$	0.0005 – 0.1
Gamma rays	$1.4 \times 10^{4} - 19.9 \times 10^{6}$	0.0001 – 0.14

[a]1 nm (nanometer) = 10^{-9} (meter); 1 nm = 1 mμ (millimieron); 1 nm = 10 Å (angstrom).

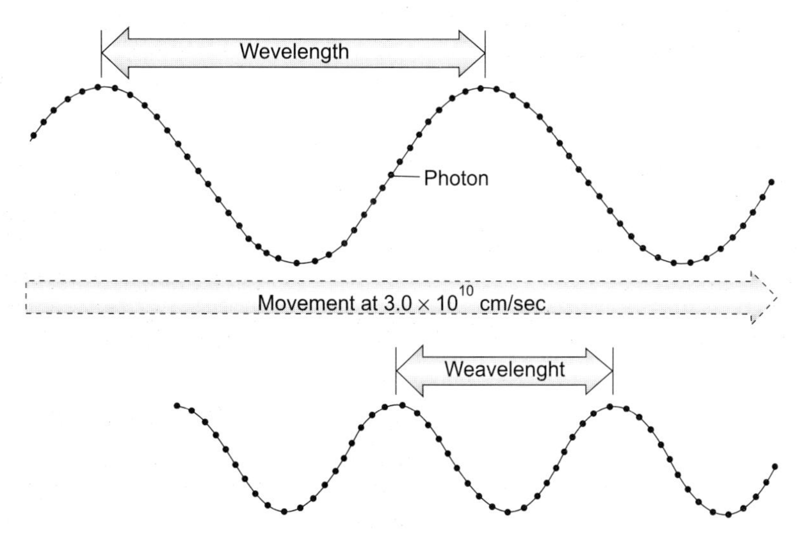

Fig. 4.5 Particulate, wavelike characteristics of electromagnetic radiations.

And is used as a standard against which other substances may be compared for their ability to absorb electrons from hydrogen (positive potential) or lose electrons to hydrogen (negative potential). The measurements are made using an electrode standardized against hydrogen. By placing the electrode in a solution of the substance to be measures, the standard electrode potential, E, can be determined in volts. Oxidizing agents (substances capable of absorbing electrons from hydrogen) have positive potentials (Table 4.2). E values represent potentials at pH 0 and 25°C; more commonly, measurements are made at other pH values, especially pH small, the nanometer (10 m) is most frequently used unit (1 nm = 1 mu = 10 A). Each of the photons in these radiations travels at the same speed, namely 3 × 10 cm sec.

The radiations in the electromagnetic spectrum also differ in the amount of energy each contains. Although radiations behave as though each is composed of discrete packets of energy called quanta, the energy in a quantum is related to the wavelength, as defined by Planck's low:

$$q = hv = hc \qquad (8)$$

where q is the quantum in ergs, h is Planck's constant [6.624 × 10 erg-second), v is the frequency of the light wave c is the speed of light (3.0 × 10 cm/sec), and wavelengths contain more energy than those with longer wavelengths.

In addition to the electromagnetic radiation, particulate radiations carry energy and affect biological systems. Alpha rays and beta rays are considered particulate radiations because they consist of streams of energy containing particles that move at comparable but varying speeds. An alpha ray consists of a stream of helium nuclei, and a beta ray is composed of a stream of electrons.

Radiations are absorbed at random in a substance by individual molecules or atoms. Individual molecules (atoms) will absorb the energy only in units of 1 quantum. This one

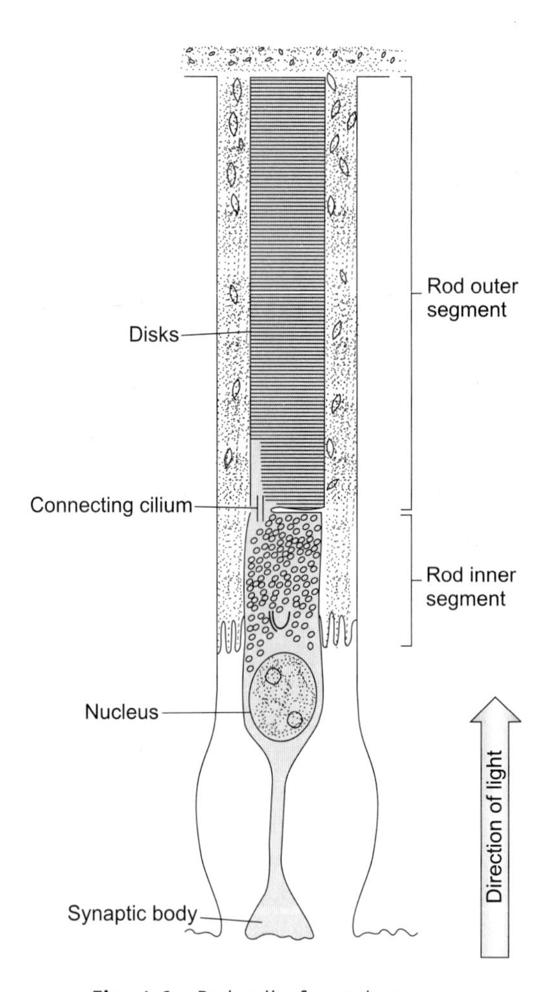

Fig. 4.6 Rod cell of vertebrate eye.

molecules (atom) one quantum proportion is known as the Einstein-Stack law or the primary reaction. The energy of the electromagnetic radiation is absorbed by an electron in the molecule. The energy from the radiation causes the electron either to be transferred to an outer orbital at a higher energy level or to spin at a faster rate. Since the allowable orbitals and speeds of electrons fall within limited ranges and whole quanta of energy must be absorbed at a time, not all molecules absorb all radiations. Only those radiations containing the specific energy per quantum to raise an electron to a higher defined energy state or to increase the spin rate to allowable limits will be absorbed by specific molecules. Other radiations will be diffracted or completely pass through the substance. Very high energy radiations such as X-rays and gamma rays may cause an electron to be displaced from the molecules, thereby forming an ion (ionization).

Electrons raised to higher energy levels are very unstable and return to their original, stable orbitals or ground state in a fraction of a second (generally, less than 10 seconds). The energy

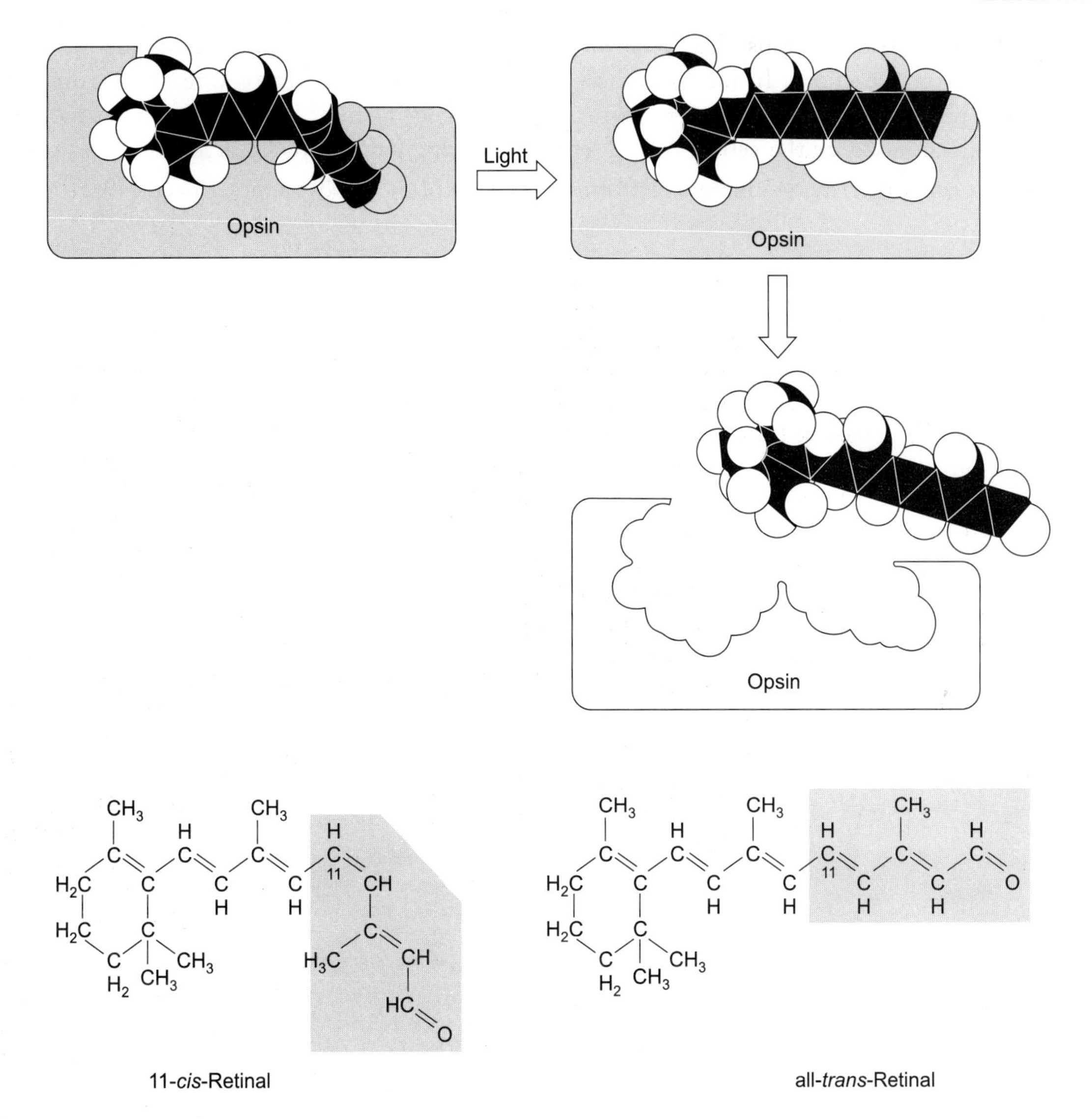

Fig. 4.7 Light-induced change in rhodopsin.

released by the molecules in returning to the ground state is a secondary reaction, which may be (1) the emission of light or fluorescence, (2) the emission of heat, or (3) the activation of another molecules. It is this third possibility that induces biological reactions such as those

SUMMARY

Energy changes regulate the chemical and physical translations in cells. Autotrophic cells derive their energy from light. Heterotrophic cells derive energy from catabolism of proteins, lipids, carbohydrates. The reactions that break down large molecules, thereby yielding energy, are called exergonic reactions. In most instances, energy released from exergonic reactions is used to attach phosphate to ADP by an energy-rich bond, thereby forming ATP, ATP and its energy-rich bonds are used by the cell to "drive" energy-requiring (endergonic) reactions.

These and other energy changes conform to the laws of thermodynamics. Specifically, these are:

First low: Energy cannot be created or destroyed but can be converted from one form into another.

Second law: in all processes involving energy changes, the entropy of the system in creases until equilibrium is achieved.

The energy changes in cellular reactions obey these laws and be expressed by the relationship.

$$G = H - TS$$

where the changes in useful, free energy is equal to the change in total energy less the entropy factor or the product of the absolute temperature and the change in entropy.

Two reactions tied together by energy exchange in which the energy released from the exergonic reaction drives the endergonic reacting are said to be coupled. The coupling can be mediated by ATP or by an equilibrium shift in which the exergonic reaction consumes the product of the endergonic reaction. Energy can be exchanged during electron transfer between oxidants and reductants (redox reaction) as well as between substrates.

CHAPTER 5

Metabolism

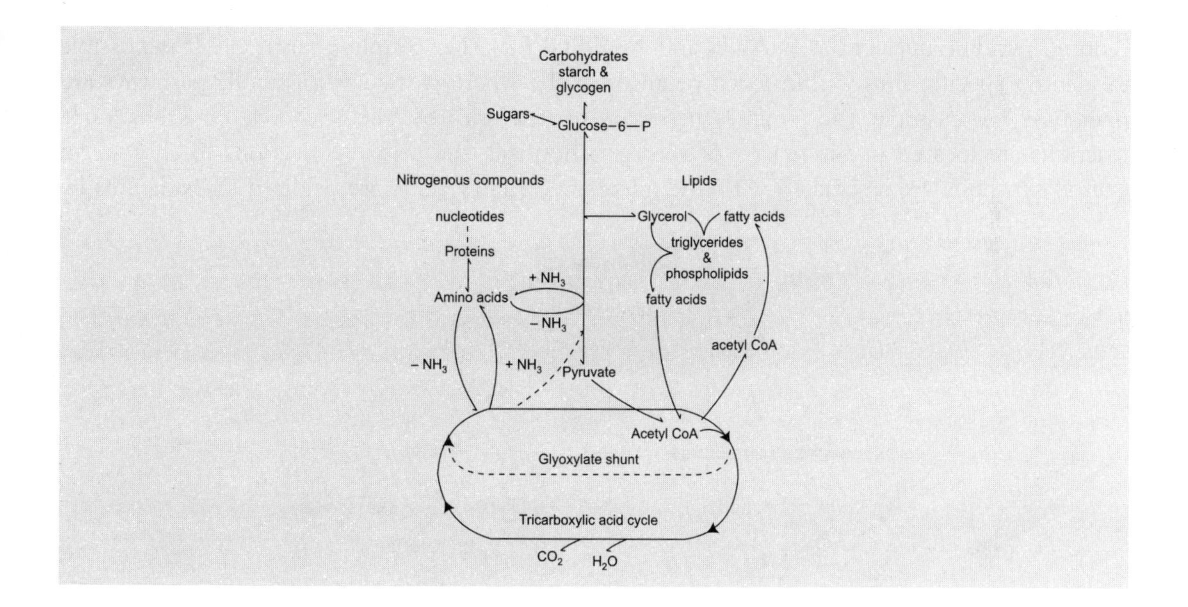

The metabolic processes in cells include all the individual chemical reactions and sequences of reactions that convert various substrates into product. The individual reactions may be spontaneous and energy yielding (the exergoinc reactions, or they may be energy consuming (endergonic reactions). Commonly the primary reactions are converted into products by means of a sequence of reactions, each reaction enzymatically catalyzed, and each subsequent reaction in the sequences requiring the products of the prior reaction as a substrate.

The overall reaction sequences may be classified as catabolic (i.e., degradative) if the ultimate products of the reaction sequences are considered to be subunits or parts of the initial substrate. Alternatively, if the products are a result of the combining of two or more different substrates, the sequence is considered to be anabolic (i.e., synthetic). A sequence of reaction is usually referred to as a metabolic pathway or simply a pathway. Some pathways are common to all living organisms or cell. Some pathways function as very active reaction sequences from which less active pathway may branch or join. The more active pathways are usually referred to as central pathways of metabolism.

Figure 5.1 diagrammatically shows some of the relationship between the catabolic and anabolic pathways of the major groups of compounds Intermediates in the break down of carbohydrates can be diverted to lipid synthesis or to the formation of nitrogen compounds such as nucleotides and amino acids. Lipids in microbial and plant (and to a limited extent animal) cells can be converted into carbohydrates and nitrogen compounds. Likewise, nitrogen compounds, once identified, can be converted into lipids or carbohydrates. All the compounds may be broken down, their catabolism acting as source of energy for ATP synthesis or to provide reduced pyridine nucleotides (NADH and NADPH) for other coupling reactions. It is possible to identify specific sites within a cell or an organelle where particular metabolic pathways are operative. For example, the enzymes necessary for the tricarboxylic acid cycle or Krebs' cycle reactions are located in the matrix or the mitochondria; the primary reactions in cholesterol (sterol) synthesis are associated with the micorsomes; fatty acids are oxidized (B-oxidation) by

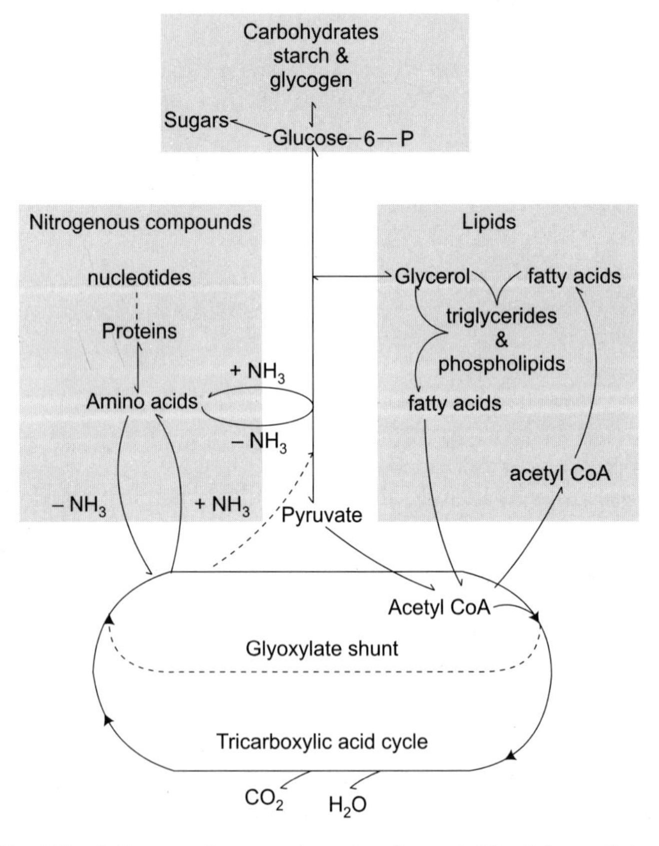

Fig. 5.1 Pathways of conversion of major metablic Intermediates.

reaction sequences in the mitochondria and are synthesized in the fluid of the cytoplasm (cytosol); proteins are formed on the ribosomes.

The intermediates, as well as the products of a pathway, may be drawn off and used in other pathways. For example, in the breakdown of carbohydrates, a large number of intermediate compounds are formed before the ultimate products. CO_2 and H_2O, are formed. It is possible for some of the intermediates to be "pulled away" from the catabolic process and used in the formation of fatty acids. Other intermediates may be used in the formation of amino acids. A number of natural mechanisms have been found that enable the cell to regulate the activity of the pathways. These metabolic control mechanisms are discussed in the next chapter.

In this chapter, some of the major pathways that are common to most cells are discussed. For convenience the major pathways of carbohydrate metabolism will be discussed first followed by those of lipid and nitrogen metabolism. Although described separately, it should be remembered that in the cell various pathways may be operative at the same time, and one pathway may influence the rate of metabolic reactions of another pathway.

ANALYSIS OF METABOLIC PATHWAYS

Our current knowledge of metabolic pathways was assembled form a variety of observations and experiments. Initially, pathways are identified by observing the consumption of reactants and accumulation of products. For example, the consumption of sugar and the production of carbon dioxide and alcohol during alcoholic fermentation is a pathway that has been known for centuries; i.e.

$$C_6H_{12}O_6 \rightarrow 2CO_2 + 2C_2H_5OH \text{ (ethyl alcohol)} \quad (1)$$

By analysis, it is possible to measure the amount of sugar consumed and the amount of CO_2 and ethyl alcohol produced. From such an analysis, it is learned that all the carbon in the sugar is converted to these two products. Therefore, one may conclude that no other products are formed from the reactants and that the reaction can be balanced as written in equation.

Quantitative analysis of reactants and products does not reveal the steps or individual reactions that bring about the overall reaction; nor does such an experiment reveal coupled reactions such as the formation of ATP during fermentation.

MARKER AND TRACER TECHNIQUES

A frequently used technique for identifying the steps in a pathway is to incorporate a radioactive isotope (or in rare instances a "heavy" isotope) as one or more of the atoms in a substrate (Table 5.1) such a substrate is said to be labeled with a tracer. The labeled substrate is then fed to the cell, and metabolism is allowed to proceed for a period or time. Afterward the reaction is stopped by heating, addition of enzymatic inhibitors, or some other means and the cell is extracted to remove potential intermediates; By chromatographic or related techniques the components are separated, their isotopic activity measured from their radioactivity and they are

Table 5.1 Isotopes commonly used as tracers.

Isotopes[a]	*Type of radiation*	*Half-life*
$^{2}_{1}H$ ("deuterium")	None	Stable
$^{3}_{1}H$ ("tritium")	β-particles	12.3 years
$^{13}_{6}C$	None	stable
$^{14}_{6}C$	β-particles	5.57 × 10^3 years
$^{15}_{7}N$	None	stable
$^{18}_{8}O$	None	Stable
$^{24}_{11}Na$	β-particles; γ-rays	15.0 hours
$^{32}_{15}Na$	β-particles	14.3 days
$^{15}_{16}S$	β-particles	87.2 days
$^{36}_{17}Cl$	β-particles	3.0 × 10^5 years
$^{42}_{19}K$	β-particles; γ-rays	12.5 hours
$^{45}_{20}Ca$	β-particles	164 days
$^{59}_{26}Fe$	β-particles; γ-rays	45.1 days
$^{131}_{53}I$	β-particles; γ-rays	8.1 days

chemically identified. Those components with labeled atoms are then studied, and possible reaction sequences are postulated to account for the formation of the marked intermediates.

Because in a reaction sequence such as

$$\text{Substrate} \rightarrow B^* \rightarrow C^* \rightarrow D^* \rightarrow E^* \rightarrow \text{product}^* \quad (2)$$

the reaction forming B must occur before the reaction forming C, which in turn must occur before reaction forming C, which in turn must occur before reaction D and so forth, the time factor can be used in the elucidation of the sequence. The labeled substrate (*) is given to the cell at a designated time, and the reaction is allowed to proceed for only a short interval before all reactions are stopped. In such a short time span, one or a few, but not at all, intermediates formed will contain the labeled compound. Separation, identification, and analysis of these intermediates would proceed as before. The experiment is then repeated with labeled substrate fed to fresh cells, and the experiment is allowed to proceed for a somewhat longer period of time before being halted for analysis.

The experiment is repeated as many times and for as long duration as is necessary to reveal each of the successive intermediates formed.

$$^*\text{Substrate} \rightarrow B^* \rightarrow C \quad D \quad (3)$$

When it is suspected that a substrate may be split into two or more products atoms at various positions in the substrate (* and +) may be specifically labeled so that each of the products formed may be followed and identified. Because some intermediates are not formed in sufficient amounts to easily allow extraction and identification, additional nonlabeled, intermediates may be added. This material mixes with the labeled intermediate being formed and provides sufficient "carrier" for extraction and analysis; in effect it traps some of the labeled compound. The technique is known as isotopic trapping.

$$\text{Substrate} \rightarrow B^* \rightarrow C^* \rightarrow D^* \rightarrow E^* \rightarrow \text{product}^* \qquad (4)$$

Carrier added: C (added at the C* step)

Another method for clarifying the role of an intermediate is called overloading and involves two parallel experiments. In one experiment, the substrate is labeled, and the amount of label appearing in the product is followed (i.e. via reaction). In the second experiment, the substrate is labeled as before, but a significant amount of unlabeled intermediate (i.e. C in reaction) is also added to the system. If C is indeed an intermediate in the pathway, then the amount of label in the product formed during the second experiment should be less than the measured in the product during the first experiment.

ENZYME TECHNIQUES

There are relatively few metabolic reactions in cells that are not catalyzed by enzymes. Thus, one would normally expect to be able to identify an enzyme for each metabolic reaction postulated or identified in a cell. In studies of the cell enzymes, the cells are disrupted and a cell-free extract prepared. After supplying the cell-free extract with selected substrates and cofactors, enzymatic activity can be demonstrated by measuring the rate of appearance of a product. The cell-free extract can be fractioned by centrifugation into its components, thereby isolating the mitochondria, cell membranes, ribosomes, other organelles, or the cytosol. Each of these fractions can also be tested for enzymatic activity. In addition, using a variety of chromatographic techniques applicable to protein isolations, the enzymes can be specifically purified. Ultimately, an enzyme in its pure crystalline state may be obtained.

ENZYME PRODUCTION AND INHIBITION

A third approach to determining metabolic reaction sequences is by the use of specific enzymes inhibitors or genetic mutant cells that fail to produce a specific enzyme. Numerous enzyme inhibitors are known that block specific reaction steps. In the conversion of precursor A to product E, that is,

$$A \xrightarrow{1} B \xrightarrow{2} C \xrightarrow{3} D \xrightarrow{4} E \qquad (5)$$

an inhibitor that blocks the enzyme of reaction 2 would prevent the formation of final product E. But one should still be able to make two additional observations: (1) intermediate B should still be formed from A, since step 1 is not inhibited; in fact, B may even be found to accumulated in excess; and (2) addition of an exogenous source of C or D should allow for the continued production of final product E. The identification of an inhibitor for a metabolic pathway thus provides a tool for finding the intermediate before the block and also allow for the testing of suspected intermediates that come after the block.

In a similar manner, a cell mutant that lacks the genetic information for producing a specific enzyme may serve as a test organism for studying the steps of the pathway involving that enzyme.

CARBOHYDRATE METABOLISM

The steps in the catabolism of sugars are known in great detail. For the most part, cells decompose carbohydrates by similar metabolic pathways whether they are plant cells, animal cells, or bacterial cells. There are alternative methods of oxidizing carbohydrates, but the central pathway found in most cells is that outlined in Fig. 5.2. Polysaccharides such as starch and glycogen are cleaved into their component saccharide units by the addition of inorganic phosphate (phosphorolysis) through the action of phosphorylase enzymes (reaction 6). Oligosaccharides are hydrolyzed first into their component monosaccharide units, and then the monosaccharides such as glucose are phosphorylatd by enzymatic reaction with ATP (reactions 8 and 9). The phosphorylaed sugars from each of these sources are sequentially subjected to the enzymatic reactions outlined in reactions 10 through 18, producing the intermediate pyruvate. This central pathway is frequently referred to as glycolysis.

GLYCOLYSIS

The major features of glycolysis are as follows:

1. The saccharides are phosphorylated. In the case of monosaccharides such as lucose, fructose, fructose and the like, 2 moles of ATP per ole of monosaccharide re utilized. Glycogen and starch require only 1 mole of ATP per mole of glucose-quivalent, since inorganic phosphate is acquired during phosphosrolysis.
2. The six-carbon sugar disphosphate is split by aldolase (-12), producing 2 three-carbon units glycerol-dehyde-3-phosphate and dihydroxyacetone phosphate; he latter subsequently forms a second mole of glyceralde hyde-3-phosphate (-13).
3. A major oxidation and phosphorylation of the substrate is catalyzed by glyceraldehydes-3-phosphate dehydrogenase. Two moles of hydrogen are removed per mole of substrate and reduce 2 moles of the coenzyme NAD. In the same reaction inorganic phosphate is bound to the acid.

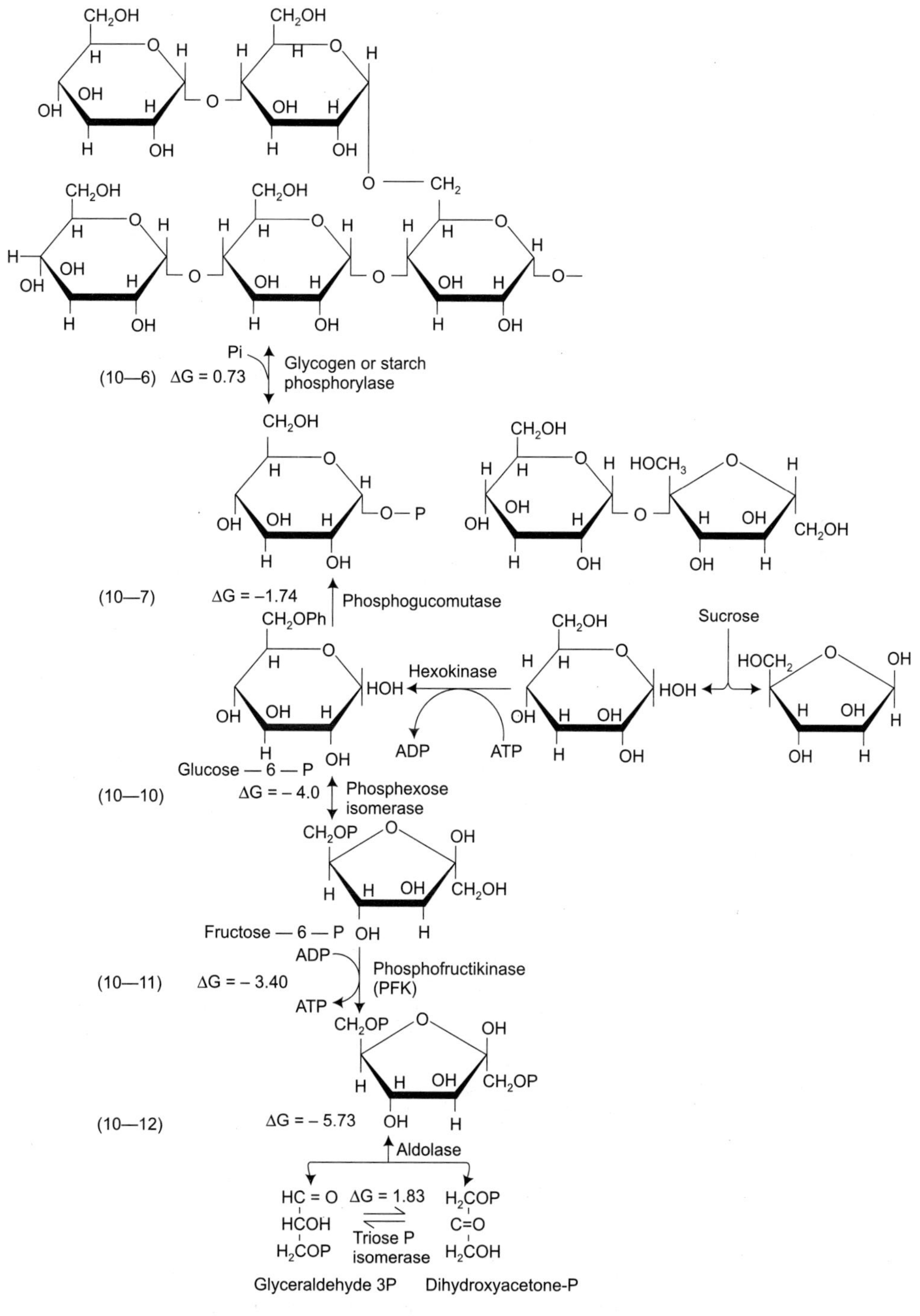

(10—6) ΔG = 0.73 Pi Glycogen or starch phosphorylase

(10—7) ΔG = –1.74 Phosphogucomutase

Sucrose

Hexokinase

ADP ATP

Glucose — 6 — P

(10—10) ΔG = – 4.0 Phosphexose isomerase

Fructose — 6 — P

(10—11) ΔG = – 3.40 ADP ATP Phosphofructikinase (PFK)

(10—12) ΔG = – 5.73 Aldolase

ΔG = 1.83 Triose P isomerase

Glyceraldehyde 3P Dihydroxyacetone-P

Fig. 5.2 Contd.

Fig. 5.2 Contd.

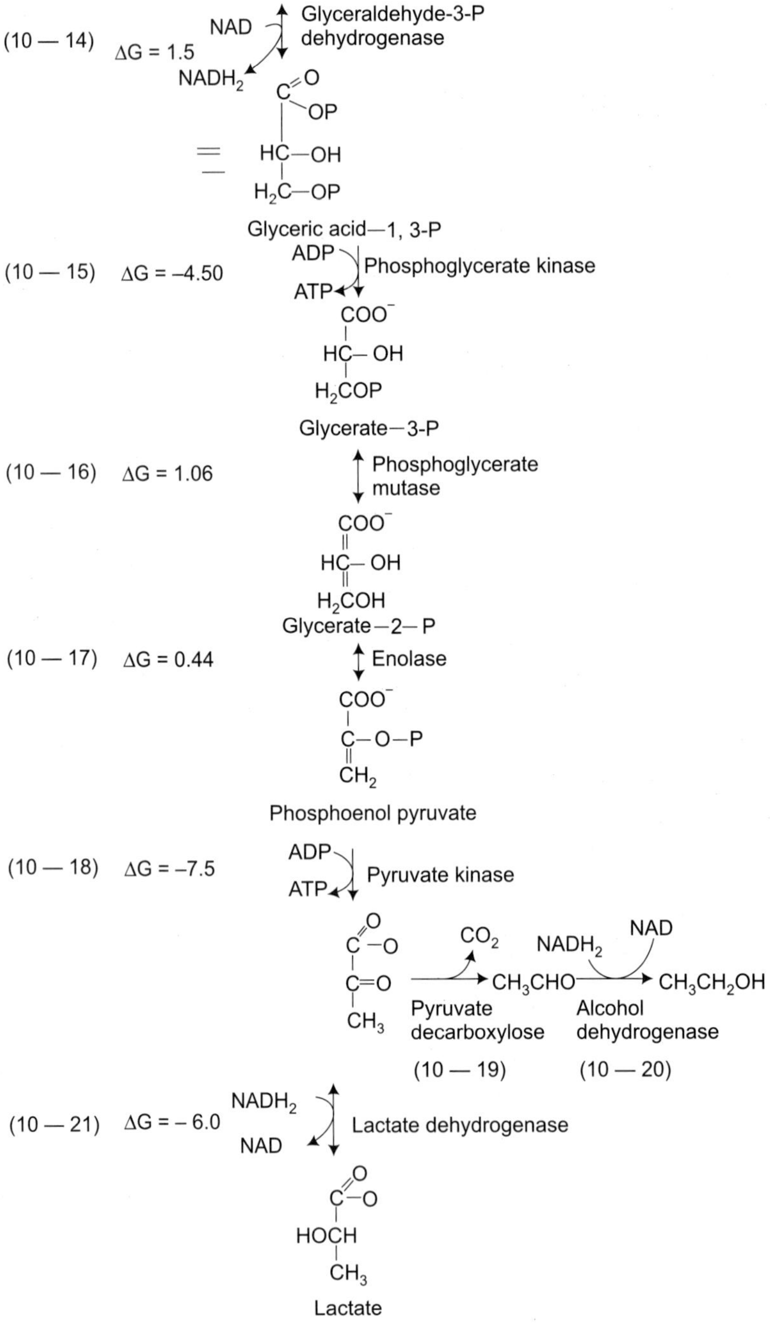

(10 — 14) $\Delta G = 1.5$ NAD → $NADH_2$ — Glyceraldehyde-3-P dehydrogenase

C(=O)OP
|
HC—OH
|
H_2C—OP

Glyceric acid—1, 3-P

(10 — 15) $\Delta G = -4.50$ ADP → ATP — Phosphoglycerate kinase

COO^-
|
HC— OH
|
H_2COP

Glycerate—3-P

(10 — 16) $\Delta G = 1.06$ Phosphoglycerate mutase

COO^-
||
HC— OH
||
H_2COH

Glycerate—2—P

(10 — 17) $\Delta G = 0.44$ Enolase

COO^-
|
C—O—P
||
CH_2

Phosphoenol pyruvate

(10 — 18) $\Delta G = -7.5$ ADP → ATP — Pyruvate kinase

C(=O)—O
|
C=O
|
CH_3

→ (CO_2) CH_3CHO — Pyruvate decarboxylose (10 — 19)

CH_3CHO → ($NADH_2$ → NAD) CH_3CH_2OH — Alcohol dehydrogenase (10 — 20)

(10 — 21) $\Delta G = -6.0$ $NADH_2$ → NAD — Lactate dehydrogenase

C(=O)—O
|
HOCH
|
CH_3

Lactate

Fig. 5.2 Central pathways for carbohydrate metabollsm.

4. In the final reactions of glycolysisthe intermediates are dephosphorylated by reaction with ADP. For each mole of monosaccharide oxidized to puruvate, 2 moles of ATP are consumed and 4 moles of ATP are produced, resulting in a net production of 2 moles of ATP per mole of monosaccharide. Note that in the case of glycogen or starch-three is a net production of 3 moles of ATP per mole of glucose-equivalent.

Pyruvate does not accumulate in very large amount in cells. Instead, it is conveted into other products. The enzymes that react with pyruvate very with the type of organism and with the nature of the environment. The more common fates of pyruvate are (1) its fermentation to ethylalcohol and carbon dioxide in cells such as yeast, (2) its anaerobic conversion into lactate in cells such as muscle, and (3) its conversion into acetate in the mitochondria of most organisms living under aerobic conditions.

ANAEROBIC RESPIRATION AND FERMENTATION

In glycolysis there is no specific requirement for oxygen. Oxidation reactions do occur, such as the removal of two hydrogenes from glyceraldehyde-3-phosphate, and NAD is reduced to NADH, but oxygen per se is not consumed. In the absence of oxygen, pyrucate may be reduced to a variety of different compounds. Alcoholic fermentation (reactions 10-19 and 10-20) is a common pathway in microorganisms and is of industrial importance. In these last two steps, 1 mole of Co is given off per mole of pyruvate (i.e., 2 moles of CO_2 per mole in monosaccharide) and NADH is reoxidized to NAD in the final step, producing ethanol. The stoichiometry and cyclic action of NAD are important; in the oxidation of each mole of glyceraldehydes-3-phosphate 9 reaction 10-14), a mole of NAD is reduced to NADH; a mole of NADH is reoxidized to NAD during the conversion of a mole of acetaldehyde to ethanol (reaction 10-20). The levels of NAD and NADH in the cell are relatively small. If the NAD reduced in the earlier reaction was not reoxidized, this central pathway would soon be blocked at the glyceraldehydes-3-phosphate step by the lack of sufficient NAD. ATP and ADP are also cycled between the ATP-requiring reactions in the early steps of glycolysis and the ATP-producing reactions in the later steps. Cells contain pools of adenylates (ATP, ADP, and AMP), and these compounds are drawn from the pools when needed and then readded to the pools. Although the stoichiometry indicates that two ATP molecules are consumed for every four molecules of ATP produced in fermentation or glycolysis, the earlier reactions may proceed by drawing ATP from the pool, while the latter reactions return ATP to the pool.

Another common fate of pyruvate that occurs in the absence of oxygen is its conversion to lactate. This is a normal process in muscle cell when they are deprived of oxygen and in many plant and bacterial cells living under anaerobic conditions (without air or, specifically, without oxygen). In these instances, a single enzyme and reaction reduce pyruvate to lactate (reaction-21). The NADH is reoxidized to NAD in equimolar amounts to that consumed in the earlier glyceraldehydes oxidation step.

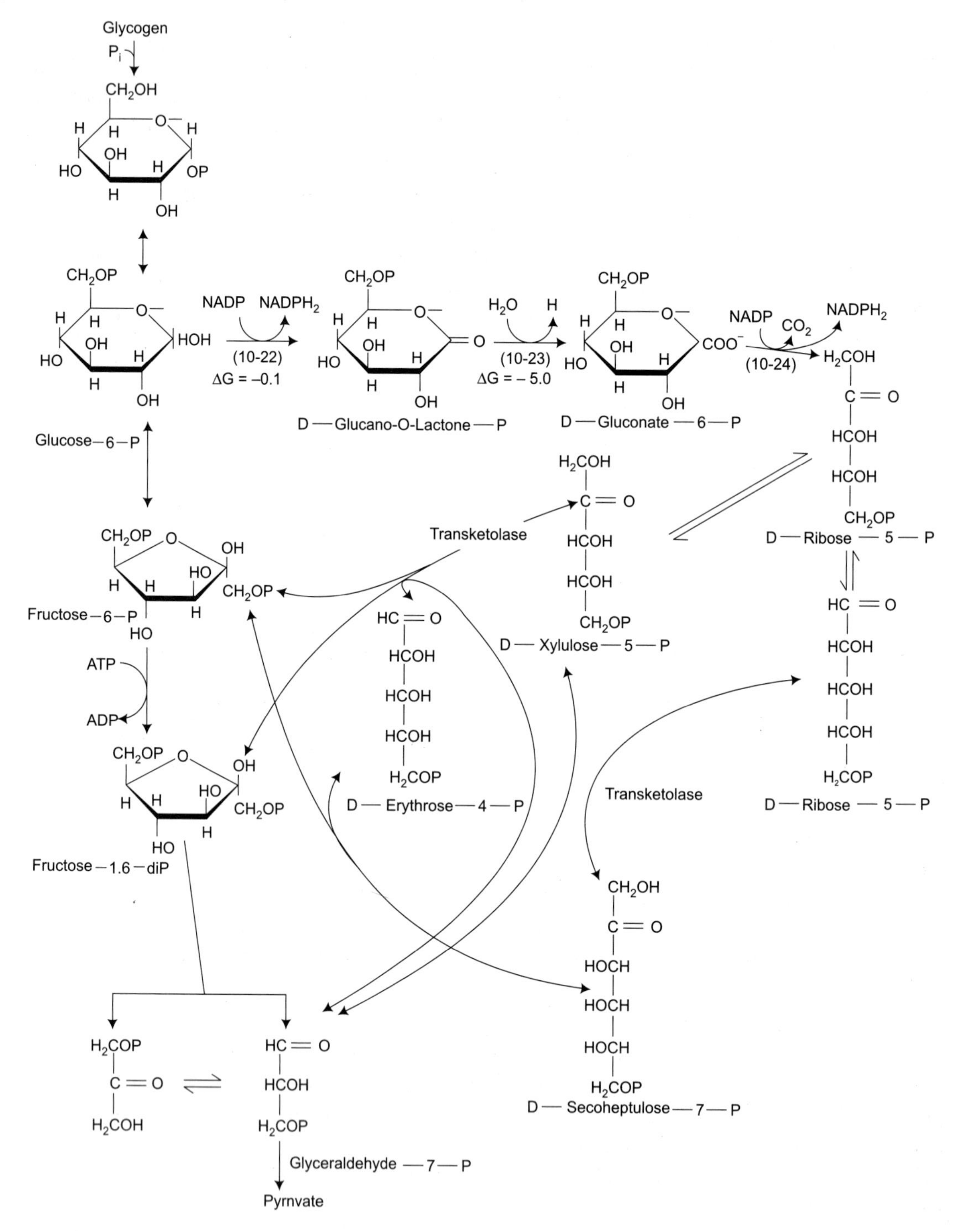

Fig. 5.3 Phosphogluconate pathway (Hexose monophosphate shunt; pentose phosphate shunt).

OXIDATION OF PYRUVATE

Except in the case of the strict anaerobes (organisms that cannot live in the presence of oxygen), when oxygen is present, most cells will oxidize pyruvate rather than reduce it to lactate, ethanol, or other compounds.

Energetically, there is an advantage for the cell to oxidize is also decarboxylation step that degrades the hexose to a pentose. Through the action of an isomerase and an epimerase, three pentose phosphates may be formed: ribose-5- phosphate, ribulose-5- phosphate, and xylulose-5-phosphate. The compounds may be incorporated into nucleic acids or they may be acted upon by a set of transaldolases and transketolases. The transketolases catalyze the transfer of 2-carbon moieties from compounds such as the pentose phosphate could have a 2-carbon potion of the molecules transferred to xylulose-α5-phosphate, and sedo-heptulose-7-phosphate and glyceraldehydes-3-phosphate would result. In a some what similar action, transaldolases catalyze the transfer of 3-carbon moieties; for example, sedoheptulose-7-phosphate, thus forming erythrose-4-phosphate and fructose-6-phosphate. Because of the action of these enzymes 3-, 4-, 5-, 6-, and 7- carbon sugar phosphates are interconvertible, and the intermediates of the phosphogluconate pathway may all be channeled into glycolysis for subsequent breackdown. Further conversions may occur in the mitchondria in the tricarboxylic acid enzymes and terminal respiration compounds to from CO_2 and HO_2. Fig. 5.4 shows the major reaction steps for the routing of hexoses through the phosphogluconate pathway and into the glycolytic pathway.

The enzymes and reactions of the phosphogluconate pathway are also found in the stroma of chloroplasts and are used there to regenerate pentoses in photosynthesis.

THE GLYOXYLATE PATHWAY

The glyoxylate pathway is essentially a bypass of the CO_2-evolving steps of the tricarboxylic acid cycle. The enzymes for this bypass are commonly found implants and are generally absent in the tissues of animals. The enzymes are localized in organelles called glyoxysomes. Fig. 5.5 summarizes the key reaction of the bypass. Two moles of acetate are required for each turn of the pathway rather than the 1 mole required in the tricarboxylic acid cycle. The first mole of acetate (acetyl CoA) condenses with oxaloacetate to from citrate and then isocitrate. The inducible enzyme isocitric lyase then catalyzes the conversion to succinate and glyoxylate. The glyoxylate then reacts the pyruvate to CO_2 and water rather than to reduce it to lactate. In the production of lactate, there is an overall change in standard free energy (G') of –47.0 kcal per mole of glucose and a net production of 2 moles of ATP. However, in the breakdown of a mole of glucose through glycolysis and further oxidation to CO_2 and water the G' = -680.0 kcal per mole of glucose, and a net of 36 to 38 moles of ATP are produced.

All the enzymes concerned with the oxidation of pyruvate are found in the mitochondria, either in the central matrix or associated with the mitochondrial membranes. The stages of

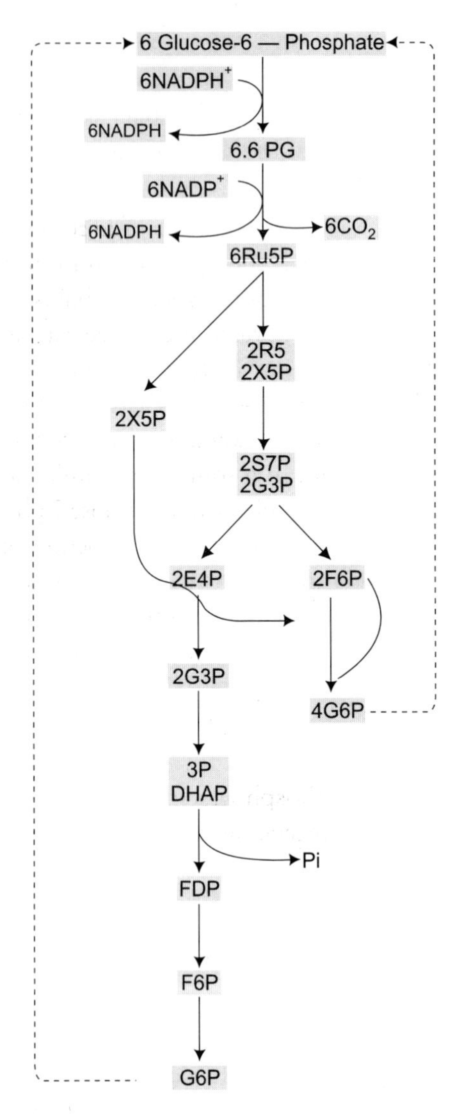

Fig. 5.4 Balance sheet and diagram illustrating how the phosphogluconate intermediates formed from 6 moles of glucose 6 phosphate can be recycled back to glycolysis or 5 moles of glucose 6-phosphate. The other mole-equivalent becomes CO_2.

Key: 6PG = 6—Phosphogluconate; Ru5P = ribulose 5— phophate; R5P = ribose 5—Phosphate; X5P = xylulose 5—phosphate; S7P = sedoheptulose 7—phosphate; G3P = glyceraldehydes 3—phosphate; G6P = glucose 6—phosphate; E4P = erthrose 4—phosphate.

pyruvate oxidation-(1) formation of acetyl-CoA, (2) the tricarboxylic acid cycle (Krebs cycle) reactions, and (3) electron transport and oxidative phosphorylation-are closely associated with the structure of the mitochondrion.

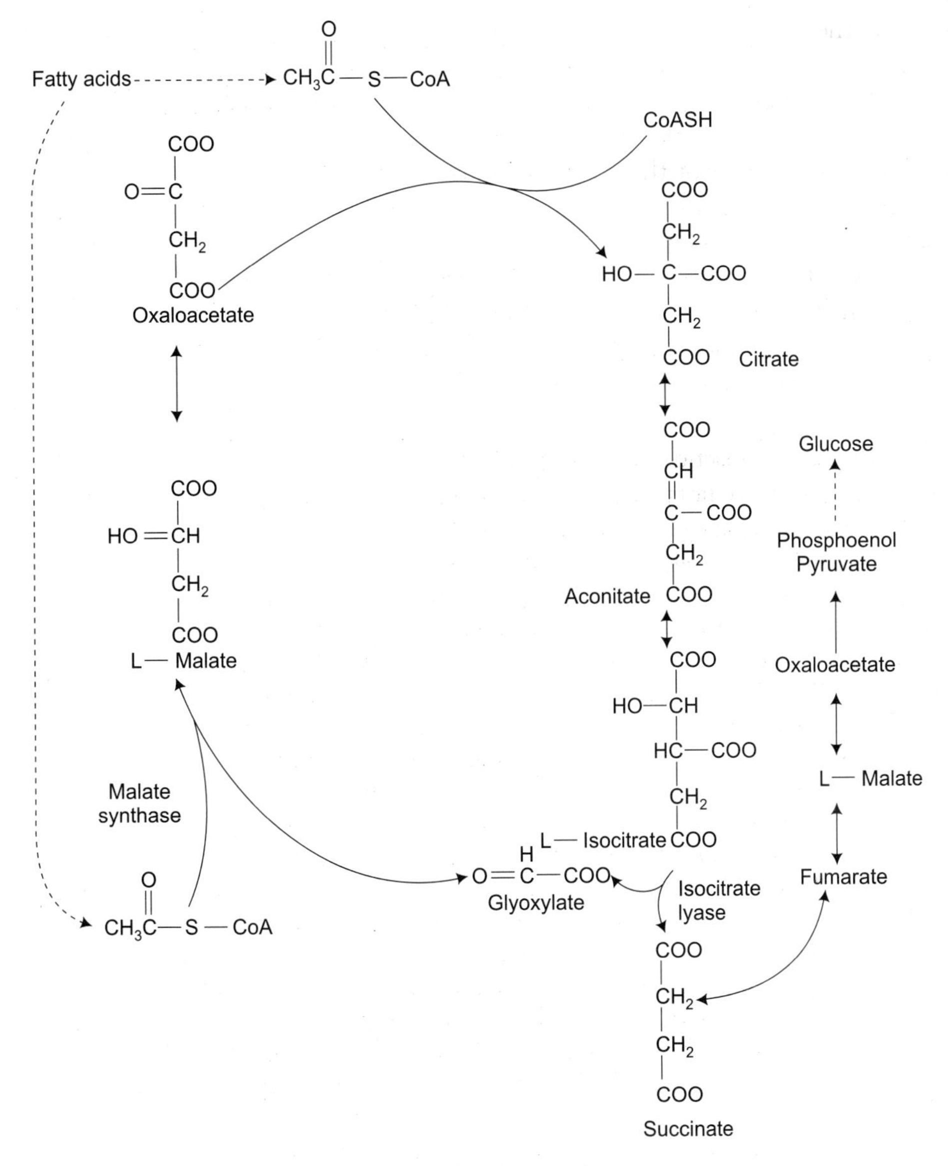

Fig. 5.5 Glyoxylate pathway.

OTHER PATHWAY OF CARBOHYDRATE CATABOLISM PHOSPHOGLUCONATE PATHWAY

This alternative oxidative pathway, also called the pentose phosphate pathway or the hexose monophosphate shunt, occurs, implants and in most animal tissues. However, its activity in comparison with glycolysis is usually lower and varies considerably from tissue to tissue. The pathway serve most cells as the primary means of (1) converting hexoses into those pentoses necessary for the synthesis of nucleotides and nucleic acids, (2) degrading pentoses so that they may be catabolized in the glycoytic pathway, and (3) generating reduced pyridine nucleotides (NADPH) in the cytosol for synthetic such as fatty acid synthesis, steroid synthesis, and amino synthesis. In animal tissues, the latter functions occur extensively in the liver, in the mammary glands, and in the cortex of the adrenal glands. It has been demonstrated that 20% of the hexose metabolized by the mammary gland occurs via the phosphogluconate pathway. In heart or skeletal muscle, little synthesis of fatty acids (or. other related substances) occurs, and correspondingly little phosphogluconate pathway activity is observed. Fig. 5.3 illustrates the major reaction steps of the phosphogluconate pathway. Reactions 22 and 24 describe the two oxidative steps that produce the reduced NADPH required in the synthesis reactions. Step 24 with a second mole of acctate (acetyl CoA) to produce malate, which is converted back to oxaloacetate.

This bypass set of reactions enables plants and microbes to transform acetyl CoA derived from the breakdown of fatty acids and convert it into carbohydrate. The excess succinate produced by the glyoxylate pathway can be chan-need back up the glycolytic pathway for the formation of sugar and subsequently polysaccharides. To achieve the reversal of the glycolytic pathway, certain bypass reactions are needed to make synthesis of sugar energetically possible.

GLUCONEOGENESIS

Glucose can be synthesized by reversing most of the reaction steps of glycolysis. Different initial reactions are utilized in different types of cell and tissues in order to get various substrates started back up the pathway. Also the carbohydrate finally produced vary with conditions and with types of cells. Photosynthetic organisms can reduce CO through a reversal of the glycoytic reactions after an initial fixation with ribulose-diphosphate. Liver cells can regenerate glucose from lactate by reversing the glycolytic reactions. Almost all cells can transaminate or deaminate key amino acids into triboxylic acid cycle intermediates and, by conversion into phosphoenol pyruvate, reverse glycolysis to produce glucose. Plant and bacterial cells as described above can oxidize fatty acids to acetyl CoA and, via the glyoxylate cycle, form tricarboxylic acid intermediates which can then be converted to phosphenol pyryvate and thereby start reverse glycolysis. Fig. 5.6 diagrams the major initial reactions and features of gluconeogenesis. In the reversal of glycolysis, there are three reactions that occur as alternates to those found in the catabolic sequence:

1. Pyruvate is not directly converted to phosphoenol pyruvate (the energetics of that particular reaction are not favorable: G = +7.5 kcal/mole); instead, the pyruvate is converted first to tricarboxylic acid cycle intermediates within the mitochondria, which then pass into the cytosol, where phosphoenol pyruvate carboxykinase, together with GTP, converts these to phosphoenol pyruvate (G = +1.0 kcal/mole).
2. The dphosphorylation of fructose-1, 6-diphosphate is not coupled to ATP by the catabolic sequence enzyme (phosphofructokinase) but instead is acted upon by fructose diphosphatase to yield fructose-6-phosphate and inorganic phosphate.

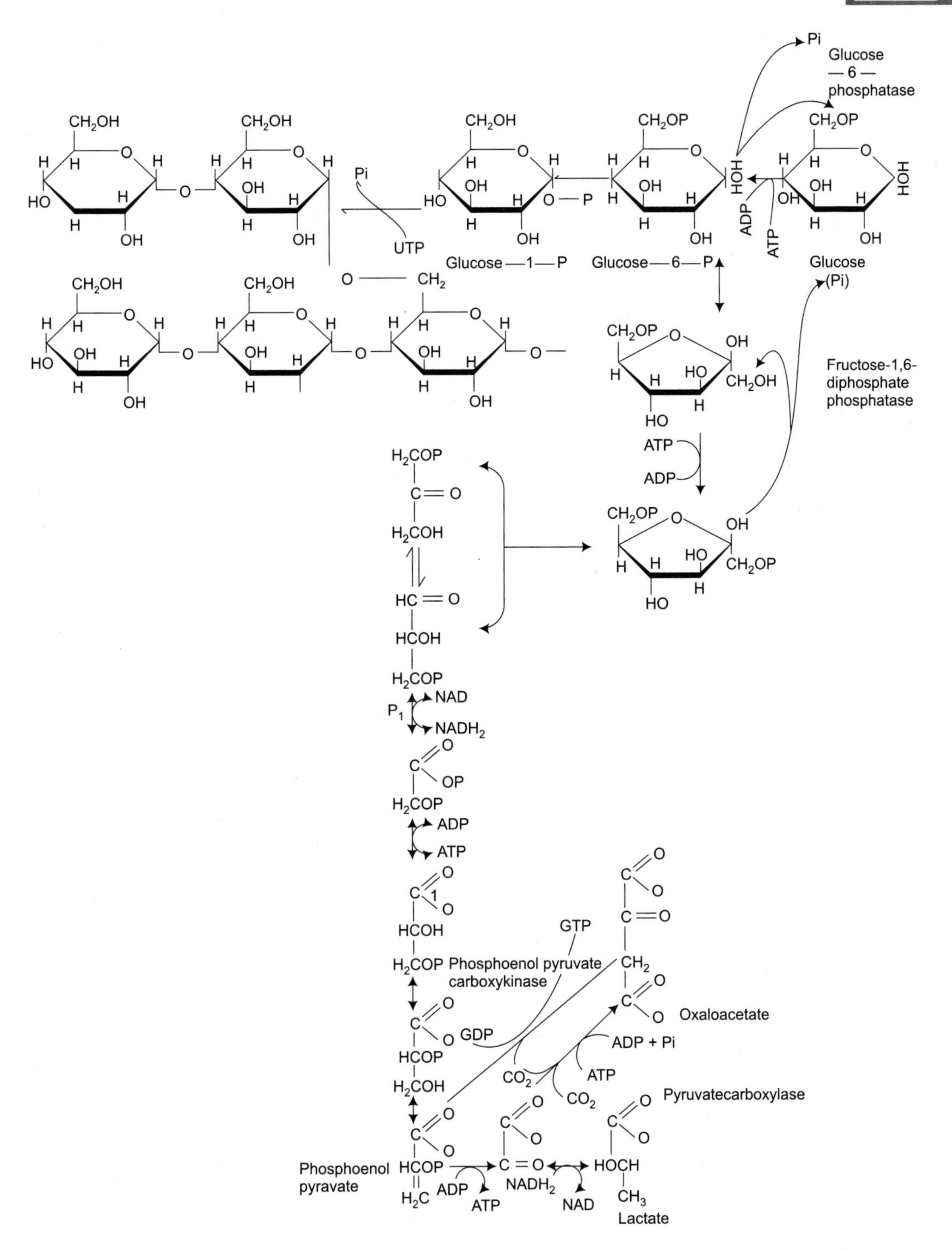

Fig. 5.6 Pathways of gluconeogenesis.

3. In some tissues such as liver and kidney the glucose-6 phosphate may be broken down to glucose and inorganic phosphate by glucose-6-phosphate rather than by the kinase.

Synthesis of Glycogen and Starch

The synthesis of glycogen and starch follow basically the same reaction steps beginning with glucose-6-phosphate: however, different enzymes are involved, and usually different nucleoside triphosphate sugars are formed. In both pathways (Fig. 5.7), glucose-6- phosphate is first converted to glucose-1-phosphate, which the reacts with UTP in the case of glycogen formation in animals and ATP in starch formation in plants. The resulting pyridine diphosphate-glucose component then becomes the glucosyl donor reacting with the preexisting polysaccharide chain (usually called the "primer" molecule). The glycogen or starch molecule is thereby lengthened

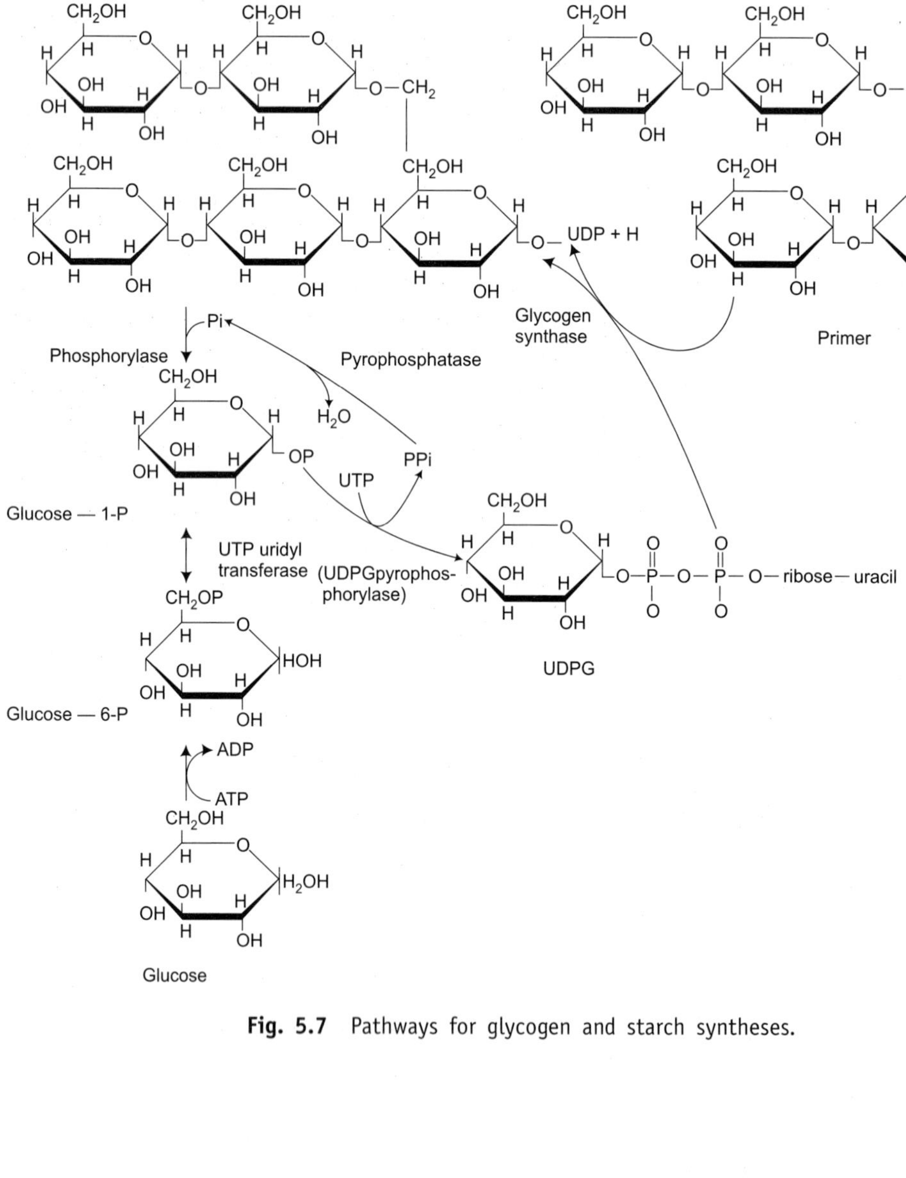

Fig. 5.7 Pathways for glycogen and starch syntheses.

by one glucose unit at a time through an x, 1 → 6 linkages necessary to form the branches that occur in the glycosidic chains of glycogen and starch.

LIPID METABOLISM

Triglycerides

The synthesis of fatty acids and their ultimate incorporation into triglycerides follows a pathway that is significantly different from that which results in the catabolic breakdown of these compounds. Basically, the linear fatty acid molecule is built up two carbon units at a time by the consumption of acetate in the form of acetyl CoA (Fig 5.8) there are six enzymatic steps to add each 2-carbpn unit to the chain.

1. The priming reaction, in which acetyl CoA is bound first to the nonenzymatic acyl carrier protein, (ACP) and then transferred to the enzyme ACP-acyltransferase
2. The malonyl reaction, in which a second acetyl CoA converted to malonyl CoA by incorporation of HCO_3 is bound to the acyl carrier protein
3. The condensation reaction, in which the malonyl group loses CO_2, becoming an acetyl group which then attached to the acetyl group on the enzyme
4. A first reduction step involving $NADPH_2$
5. A dehydration step
6. A second reduction step involving $NADPH_2$

The resulting fatty acid, which is four carbons in length, can now act as the primer and repeat the sequence. Each cycle through the sequence adds two carbon units to the growing chain length. Desaturation of the fatty acids occurs by additional enzymatic steps. The acetyl CoA for the synthesis is primarily generated in the mitochondria by decarboxylation of pyruvate, by oxidative degradation of some amino acids, or by oxidation of the other fatty acids. The mitochondrial acetyl CoA must be converted into other molecules to escape from the mitochondria into the cytosol, where acetyl CoA is reformed and incorporated into the fatty acids. The triglycerides are formed by the condensation of the fatty acyl CoA molecules onto dihydroxy-acetone-phophate (produced by the reactions of glycosis) or glycerol-3-phophate (formed through the reduction of dihydroxyacetone phosphate by $NADH_2$)

Degradation of triglycerides occurs initially in the cytosol by hydrolysis to glycerol and fatty acids; the glycerol then enters the glycolytic pathway. Fatty acids must be activated and transported into the matrix of the mitochondria, where they undergo the degradative and oxidative steps outlined in Fig. 5.9. The enzymatic sequence of steps 3 to 6 (Fig. 5.9) repeats, removing two carbon units at a time until breakdown is complete and only lcetyl CoA or malonyl CoA remains. This process is known as B-oxidation, since it is the bond at the second carbon atom that is cleaved. Other oxidations are known, for example, -oxidations occurs in some germinating seeds producing CO_2, w-oxidation occurs in the liver of mammals, but its role is not understood.

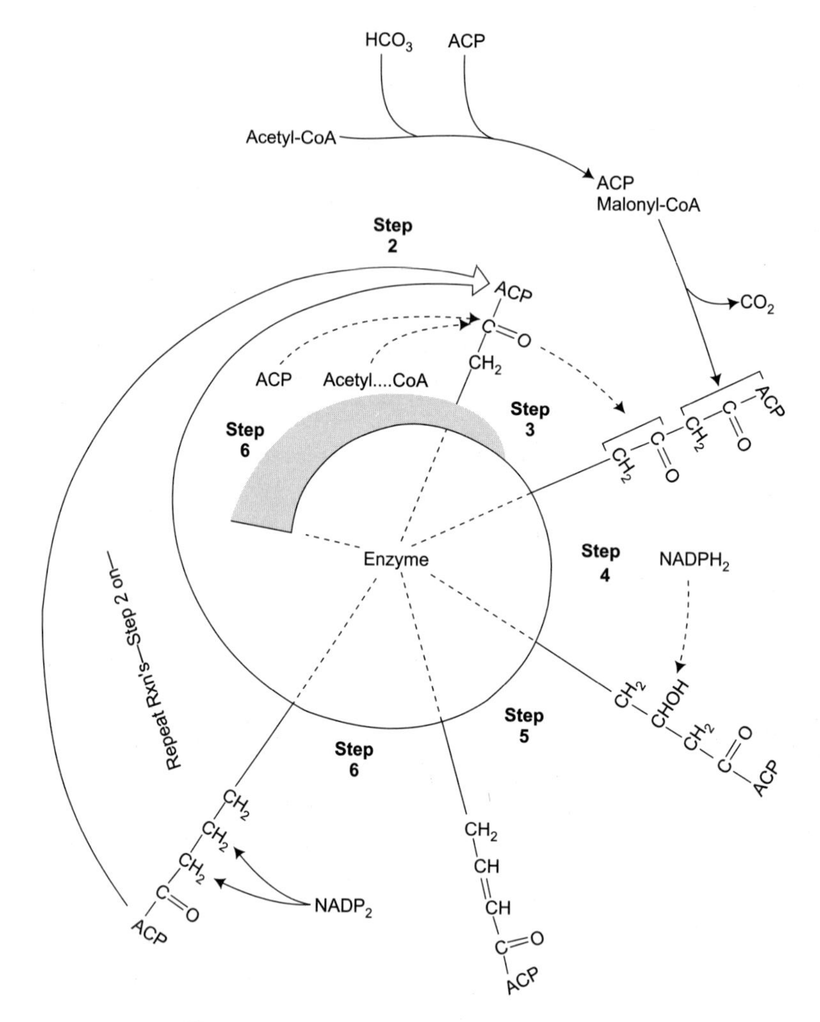

Fig. 5.8 Fatty acid biosynthesis (see text).

The synthesis and degradation of most other lipids such as phospholipids and sterols are known in detail but will not be considered here. Synthesis of cholesterol occurs by he progressive buildup of 2-carbon fragments (i.e. acetyl CoA units) and involves about 25 enzymatic steps.

NITROGEN METABOLISM

Cells require a variety of nitrogen compounds for survival. Central to the requirement for nitrogen is the formation of amino acids. Amino acids are necessary not only for the synthesis of proteins but also as the primary source of nitrogen in the synthesis of the nucleotide building blocks of nucleic acids. Not all cells or organisms are able to synthesize all the amino acids

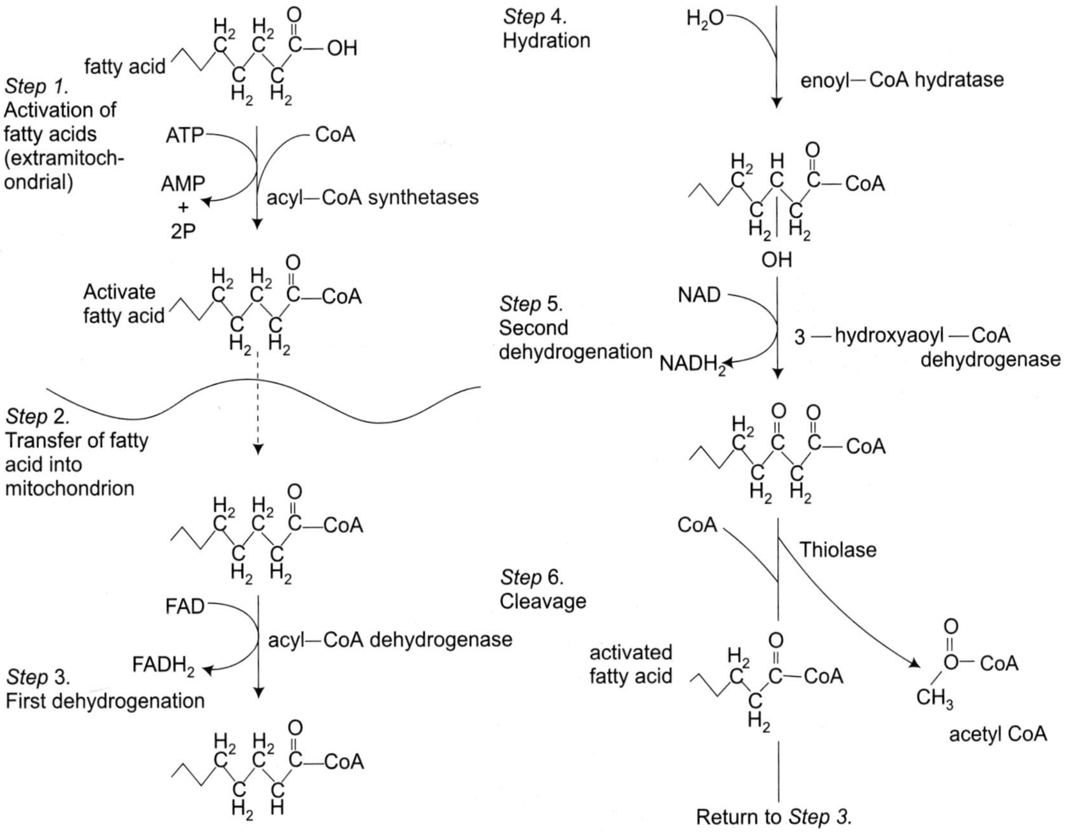

Step 1. Activation of fatty acids (extramitochondrial)
fatty acid
ATP, CoA → AMP + 2P
acyl—CoA synthetases
Activate fatty acid

Step 2. Transfer of fatty acid into mitochondrion

Step 3. First dehydrogenation
FAD → $FADH_2$
acyl—CoA dehydrogenase

Step 4. Hydration
H_2O
enoyl—CoA hydratase

Step 5. Second dehydrogenation
NAD → $NADH_2$
3—hydroxyaoyl—CoA dehydrogenase

Step 6. Cleavage
CoA
Thiolase
activated fatty acid
acetyl CoA
Return to Step 3.

Fig. 5.9 Oxidation of fatty acids.

necessary to their existence. For example, the cells of humans beings are able to make only 10 of the 20 amino acids required for human protein synthesis. The other 10 are referred to as the essential amino acids and are obtained from plant or microbial sources in the diet. Nitrogen for the synthesis of the nonessential amino acids of humans and other higher animals is obtained from ammonium ions, since the enzyme systems for utilizing nitrate, nitrite, or atmospheric nitrogen are not present. However, many microbes, leguminous plants, and a few other plants can use atmospheric nitrogen. Most higher plants can make the amino acids needed for protein synthesis by using ammonia, nitrate, and nitrite as sources of nitrogen.

Each of the 20 different amino acids required for protein synthesis is synthesized by a different system of enzymes. Several utilize intermediates of the glycolytic and tricarboxylic acid cycle pathways. The decomposition of amino acids involves pathways different from the synthetic pathways but the products after deamination are usually further metabolized by the glycolytic of the tricarboxylic acid cycle reactions.

CANCER CELL METABOLISM

Caner cells have a distinct type of metabolism, one that is different from most normal tissues. While cancer cells posses all the enzymes necessary for glycolysis, the tricarboxylic acid cycle reactions, and terminal respiration, the rates of utilization of these pathways are distinctive. Less oxygen is consumed by cencerous tissue than by normal tissue, but significantly more glucose is consumed. Instead o converting most of the glucose to CO_2 and H_2O through terminal respiration as normal tissue does, cancerous tissue converts large amount of glucose to lactate, even though oxygen is available. This phenomenon is called aerobic glycolysis. The lactate produced by cancerous tissue can be converted back into glucose in the liver, but this resynthesis process generally consumes six molecules of ATP per mole lactate (or glucose) compared to a net generation of only two molecules per glucose during its break down in the cancerous cells. In effect the cancerous cell tens to be a "parasite" of the other, normal tissues of the organism.

FUNCTIONS OF METABOLIC PATHWAYS

The sequences of metabolic reactions called metabolic pathways serve a number of functions. First and most obvious is the formation of an end product needed by the cell of organism. This product may be used as part of the structure of the cell, or in some cases when secreted from the cell. It may be incorporated into an extracellular structural part of an organism. The proteins in the matrix of cartilage and the proteins of the middle lamellae of the plant cell walls are examples of such secretions. The end product may also function in the cell as regulatory agent for other reactions. Enzymes and hormones ar such end products, End products may also become storage or reserve compounds such as starch, glycogen, and certain lipids.

Second, a metabolic pathway may function to provide energy-rich compounds such as ATP, GTP, and UTP for other energy-requiring reactions. Metabolic pathways rarely provide these compounds as end products of reaction sequences but instead produce them at one or more of the intermediate steps of a pathway. For example, in glycolysis, ATP is produced at two intermediate steps (Fig. 5.2; reactions 5.15 and 5.18). Other energy-related compounds such as NADH2 and NADPH2 are also products of intermediate reactions and may be utilized directly in other reactions or oxidized to provide energy in the form of ATP.

Third, intermediates in metabolic pathways may be drawn upon or utilized as substrates for other metabolic sequences. For example, the tricarboxylic acid cycle (Krebs cycle) is a sequence of reactions that effectively oxidizes acetyl CoA to CO2, $NADH_2$, and GTP, but many of the intermediates are drawn off and utilized for other purposes. Acetyl CoA itself is used in lipid syntheses, and alpha-ketoglutarate and oxaloacetate are used in amino acid syntheses. Metabolic reactions that serve in both a catabolic (energy-producing) and an anabolic (biosynthetic) function are called amphibolic pathways. Amphibolic pathways, which may have their intermediates diverted to other reaction sequences, frequently include specialized reactions that replenish these intermediates. This specialized enzymatic reactions are called anapleurotic reactions. The importance of anapleurotic reactions is apparent in the tricarboxylic acid cycle. To oxidize 1 mole of acetyl CoA, 1 mole of oxaloacetate is required. At the end of the sequence, 1 mole of oxaloacetate is produced and is therefore available to react with additional acetyl CoA. If, however some of the intermediates of the cycle are diverted into side reactions to form amino acids or other compounds 1 mole of acetyl CoA will not result in the entire production of 1 mole of oxaloacetate, and the capability of oxidizing further acetyl CoA will be reduced. Anaplcurotic reactions, which regenerate intermediates of the cycle from external compounds, could reestablish the full capacity of the cycle. The "Wood-Werkman" reaction in bacteria (Carbon dioxide is bound to pyruvate to form oxaloacetate) is an example of such an anapleurotic mechanism. The glyoxylate pathway described earlier is also an anapleurotic mechanism, effectively forming the tricarboxylic acid cycle intermediates, succinate, and malate.

CALCULATIONS OF ENERGY CHANGE

In most instances, there is sufficient knowledge about the major metabolic pathways of the cell so that the specific enzymatic reactions in which ATP (GTP or other related compounds) is formed or consumed are known. By inspection of a metabolic chart, such as that in figure 10-2, one should be able to calculate in number of moles of ATP consumed or produced and the net change for any reactant-to-product sequence. For example, if 1 mole of sucrose is oxidized to pyruvate, 4 moles of ATP would be consumed.

1 mole in reaction 10-8a

1 mole in reaction 10-8b

2 moles in reaction 10-11

Total = 4 moles (consumed)

and 8 moles of ATP would be produced;

4 moles in reaction 10-15

4 moles in reaction 10-18

Total = 8 moles (produced)

Therefore, in the glycolytic oxidation of 1 mole of sucrose, there is a net production of 4 moles of ATP. These "paper calculations" provide the theoretical amounts of ATP expected from the glycolytic sequence of reactions. Experimentally, these numbers are approached but not always obtained, for laboratory conditions do not always provide advantageous conditions, intermediates may be drawn into other reactions (especially in whole, cell preparations), co-factors may not be in the proper concentration, and so on.

The efficiency of a metabolic pathway is usually determined by an analysis of the change in free energy, AG*. For example, in the conversion of glucose to lactate (Fig. 5.2), the enzymatic sequence may be considered to be composed of the example reaction.

$$\underset{\text{(Glucose)}}{C_6H_{12}O_6} \rightarrow \underset{\text{(Lactate)}}{2C_3H_6O_3}$$

And the endergoni reaction

$$2ADP + 2P \quad 2ATP + H_2O \rightarrow$$

The standard free energy change (AG*) for the catabolic reactions is determined by adding the values for each of the reactions.

A summation of the net ADP/ATP reactions would be

ENERGY

AG*(kcal mole)

-7.3

-7.3

+14.6, i.e, (7.3 × 2)

+14.6. i.e. (7.3 × 2)

Total + 14.6 kcal mole-1

The reactions of glycolysis and the fermentation of glucose to lactate thus produce much more free energy than they consume in the formation of ATP. The efficiency of these pathways would therefore be

$$14.6/44.04 \times 100 \text{ or } 33.1\%$$

Attempts to measure the energetic of glycolysis and fermentation in intact cells (i.e., in vivo) have produced striking results. In red blood cells (which are ideal in such studies, since this cell derives most of its energy from glycolysis and from fermentation to lactate), the efficiency is about 53%, this is much higher than that expected from the calculations above. To determine this in vivo efficiency, the steady-state concentrations of all the glycolytic intermediates are measured, and from these values, the actual equilibrium constants are calculated and the AG values determined. The AG values (rather than the AG* values) reveal the greater efficiency of red blood cells. Skeletal muscle cells also reveal an efficiency level higher than that anticipated from calculations and summations of the type carried out above. Therefore while AG* values provide figures or easy comparison under defined circumstances (i.e., pH 7.0: 25* C), the differences in substrate concentration, pH, and other factors may bring about extreme variations in efficiency of enzymatic conditions under natural conditions, that is in vivo.

SUMMARY

Cells break down compounds by sequences of enzyme reactions (**catabolism**) to obtain energy or to form smaller molecules that can be used to build other, larger molecules.

Sequences of enzyme reactions also proved the mechanism for the synthesis of new molecules (**anabolism**). These reaction sequences, called metabolic pathways, are frequently associated with specific organelles. The enzymes that catalyze the reactions are often compartmentalized within, between, or on membranes of the organelles.

In general, carbohydrates, lipids, and proteins can be catabolized to yield energy and a pool of small molecules, and these may either be used in the synthesis of new macromolecules or be excreted from the cell. The products of catabolism in one organelle may be transported to other organelles for further catabolism or anabolism. Reaction sequences and their intracellular location have been determined by **marker** and radioactive tracer techniques, studies of enzyme activity and inhibition, and through the use of mutant cells.

Some of the more common pathways in cells are listed below:

1. **Glycolysis**–The catabolism of monosaccharides to pyruvate.
2. **Fermentation**–The catabolism of monosaccharides to products such as ethanol and CO_2 in the absence of air (anaerobic conditions).
3. **Oxidation of pyruvate (Krebs cycle)**–Catabolism to CO_2 and water in the presence of oxygen.
4. **Phosphogluconate pathway** (pentose phosphate or hexose monophosphate shunt)–pathway for formation and/or catabolism of pentoses.
5. **Glyoxylate pathway**–Conversion of acetyl CoA to carbohydrates.
6. **Gluconeogensis**–Synthesis of glucose from simple acids.
7. **Glycogen and starch synthesis.**
8. **Fatty acid synthesis**–Pathway for formation of fatty acids from acetyl CoA.
9. **B-oxidation**–Pathway for breakdown of fatty acids.
10. **Protein synthesis**–Mechanism for formation of proteins from amino acids.

CHAPTER 6

Metabolic Regulation

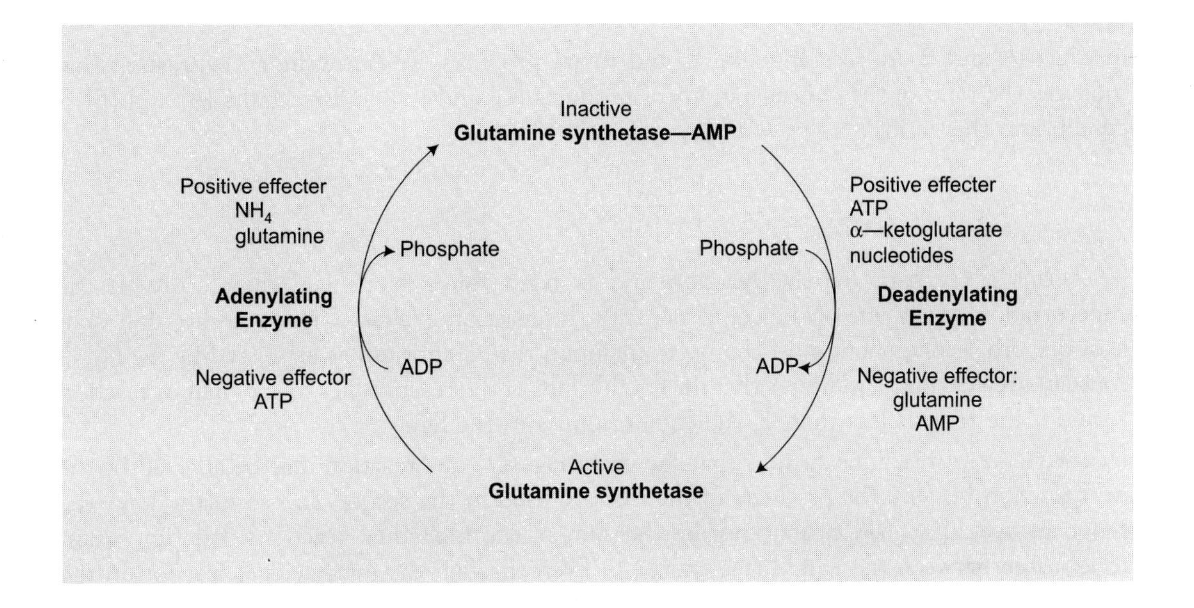

The diverse metabolic reactions and reaction sequences in cells have been briefly described and outlined in the preceding three chapters. From these descriptions, it is clear that a substrate can be enzaymatically converted into a great variety of intermediates and products. Although there are a number of essentially unidirectional reactions, the metabolic network of reversible and cyclical reaction sequences provides for the possible conversion of almost any metabolite into any other metabolite. In a superficial comparison, one could visualize the metabolic pathway as a breaching and connecting network of water pipes in which the water can be caused to flow between any two points in the network under suitable conditions. For example. Simply an excess of water (metabolite) in one part of the system could cause flow to the other parts of the system. In vivo and in vitro experiments have indicated that the "flow" of metabolites through metabolic pathways is not as free and uncontrolled as this pipeline-network analogy. Metabolic conversions are controlled or regulated by a variety of mechanisms that cannel metabolites into needed compounds or into stable reserve products and prevent energetically wasteful conversions.

Cells have evolved a diverse set of regularly mechanisms. Individual reactions may be controlled by one or more processes from simple mass action to complex hormonally controlled enzyme systems. The more common of these processes is described in this chapter.

REGULATION BY MASS ACTION

For any reversible reaction, such as

$$[A] + [B] \rightarrow [C] + [D] \quad (1)$$

in which A and B are reactants and C and D are products, we can write an expression that indicates the ratio of the concentrations of products [C] and [D], and reactants [A] and [B] at equilibrium this ratio,

$$K_{eq} = \frac{[C][D]}{[A][B]}$$

Forms a constant for the reaction and is fixed for a particular temperature. If the concentration of any one of the components of the reaction is altered, the concentration of at least one other component must change to maintain the equilibrium as expressed by the K_{eq} A constant greater than 1 indicates that the equilibrium of the reaction lies to the right in reaction 1, and if the ratio is less than 1, the equilibrium is to the left.

In a sequence of reactions such as that shown in Fig. 6.1, one reaction may be affected by the next reaction because the products of the first are used in the second and so forth. Thus the K for an overall sequences may not be the sum of the individual reactions. (the important relationship between the K and free energy, G Even through the intermediate reaction in the sequences in Fig. 6.1 has a K less than 1 (with an equilibrium that lies to the left), the sequence may still proceed when glucose and ATP are introduced, since the formation of glucose-6-P will necessitate a conversion to fructose-6-P to maintain the K.

A limited degree of regulation of alternative metabolic pathway is also achieved by the law of mass action at branch points in a pathway. Although the equilibrium is to the right in each of the reactions in Fig. 6.2, the equilibrium is "further" to the right in the reaction forming fructose-6-P, and more product would be formed from reactant along that, branch than in the branch that forms glucose-1-P. However, in glucose metabolism, both products are part of a continuing sequences of reactions, and subsequent reactions would affect the net K_{eq} and could alter the overall direction of conversion.

REGULATION BY ENZYME ACTIVITY

Regulation of metabolism is most commonly controlled at the cellular level by altering the activities of the enzymes or by altering the number of enzyme molecules present. The regulation

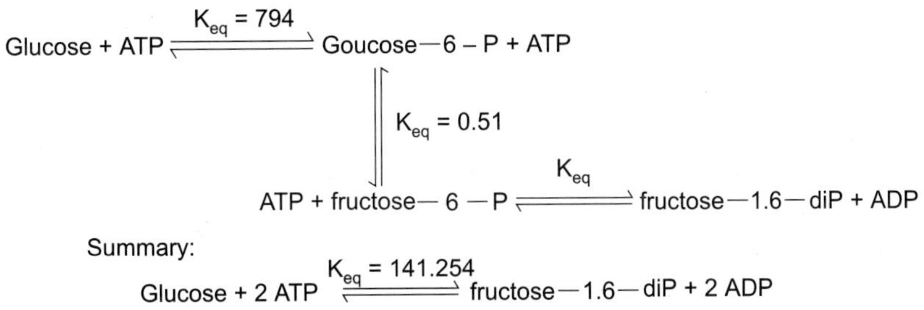

Glucose + ATP
K_{eq} = 794
Goucose—6 – P + ATP
K_{eq} = 0.51
ATP + fructose— 6 —P
K_{eq}
fructose—1.6—diP + ADP
Summary:
Glucose + 2 ATP
K_{eq} = 141.254
fructose—1.6—diP + 2 ADP

Fig. 6.1 Example of linear metabolic pathway.

Glucose — 1 — P
K_{eq} = 19
Glucose + ATP
K_{eq} = 871
Glucose —6—P
K_{eq} = 0.51
Fructose — 6 — P

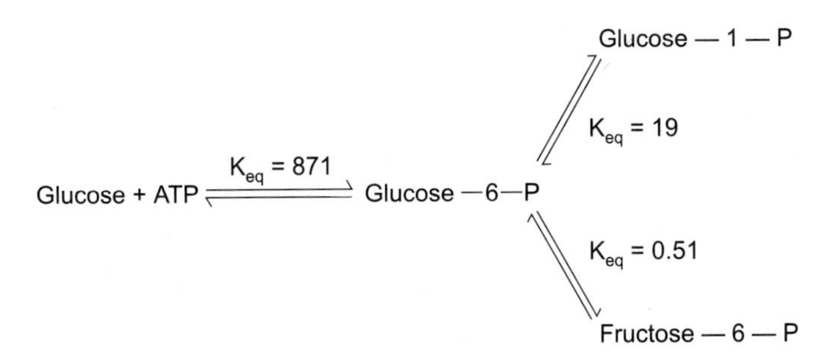

Fig. 6.2 Example of branching metabolic pathway.

by altered activity can be brought about by (1) changes in substrate concentration, (2) allosteric effectors such as AMP or glucose-6-phosphate, (3) irreversible covalent bond modification such as the hydrolytic activation of pancreatic zymogens, (4) reversible covalent bond modifications such as the phosphorylase enzyme activation, and (5) noncolvalent modifications brought about by the action of one enzyme on another.

SUBSTRATE CONCENTRATION EFFECTORS

The activity of an enzyme (i.e. the rate of product formation) increases as a hyperbolic function as the substrate concentration is raised (Fig. 6.3) until a maximum reaction velocity (V_{max}) is achieved. However, excessively high substrate concentration may actually reduce enzyme activity. Each enzyme, subject to experimental conditions, has a characteristic maximum velocity, as exemplified by the maximum velocities of the glycolytic enzymes in brain tissue shown in Table 6.1. If the substrate in this tissue were in excess, then on e would except that aldolase with the lowest maximum velocity would be the rate-limiting reaction in the sequences. In vivo studies have indicated that enzyme are rarely saturated by substrate. In experiments with mouse brain tissue it has been shown that hexokinase phosphoglucoisomerase and aldolase, under normal conditions, function at a substrate concentration somewhat equal to or greater than the K_m (Michaelis-Menten constant). Small changes in substrate concentration do not significantly alter the rate of metabolism through this part of the glycolytic pathway. Of

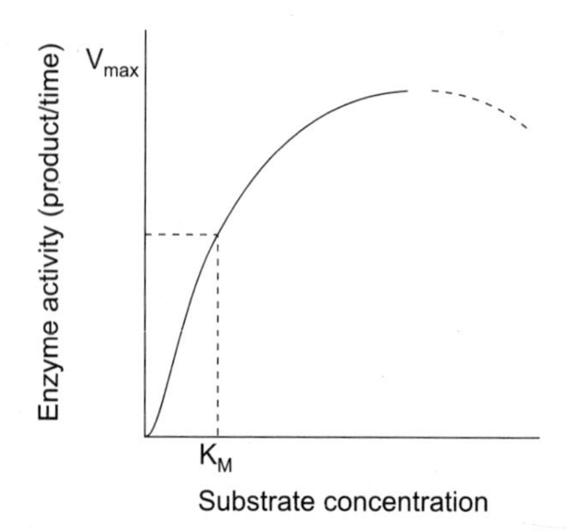

Fig. 6.3 Hyperbolic effect of substrate concentration on enzyme activity.

Table 6.1 Maximum Activity of Glycolytic Enzymes in Mouse Brain Tissue under Anoxic Conditions.

Enzyme (in Order of Glycolysis)	*V_{max} (mmoles/Kg/min)*
Hexokinase	15.2
Phosphoglucoisomerase	154.0
Phosphofructokinase	26.7
Aldolase	7.6
Glyceraldehyde-3-phosphate dehydrogenase	96.0
Phosphoglycerate kinase	750.0
Phosphoglycerate mutase	145.0
Enolase	36.0
Pyruvate kinase	95.0
Lactic dehydrogenase	129.0

Source: Copyright © American Society of Biological Chemists, Inc., *J. Biol. Chem. 239*, 31 (1964.)

greater regulatory importance here would be the amount of enzyme present. The last six enzymes in mouse brain glycolysis function at substrate levels significantly below the K_m. Small changes in the substrate concentration at these levels directly and significantly alter rate of enzyme activity. Therefore, under anoxic conditions the rate of production of lactate will not appreciably increase by supplementing the tissue with glucose. However, supplements of compounds such as glycerol that enter the glycolytic pathway below the level of aldolase will cause an increase in lactate production.

The limiting factor for the rate of an enzyme catalyzed reaction in vivo is the affinity between the enzyme and the substrate. At branch points along metabolic pathways, two enzymes compete for the same substrate. The enzyme with the lower K_m value will react more rapidly with the substrate at low substrate concentrations, while the enzyme with the higher Km value will be more active at high substrate concentrations (Fig 6.4). Thus, when present at low levels, a substrate may be channeled primarily into one pathway, while the major direction of metabolism may shift to other pathways at higher substrate levels.

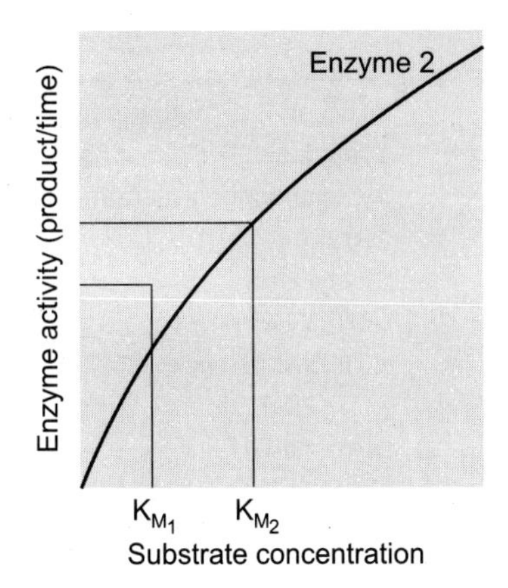

Fig. 6.4 Kinetics of two enzymes with different K_m at varying substrate concentrations.

ALLOSTERIC EFFECTORS

The regulatory effects of substrate concentrations and the mass action factor generally influence all reactions in metabolic pathway; therefore, these mechanisms are not very specific. There are, however, a number of mechanisms by which specific reactions in a pathway can be regulated. In one such mechanism, enzymes catalyzing specific reactions of the pathway are influenced by the type and amount of certain regulatory metabolites present. These enzymes are called allosteric enzymes because their of specific metabolites to a site on the protein other than the active site. Some of the more common regulatory metabolites are listed in Table 6.2. The binding of the regulatory metabolite to the allosteric enzyme may (1) cause an inhibitory or a stimulatory response in enzyme activity, and (2) cause a change in the level of activity of the pathway that is proportional to the concentration of the regulatory metabolite.

In general, inhibitory allosteric effects are caused by the accumulation of the product of a reaction sequence, with the resultant binding of the product to an enzyme at or near the beginning of the sequence. Where branches occur in a pathway, the end product usually affects one of the enzymes at the branch point (Fig. 6.5). This type of inhibition is called end-product inhibition or feedback inhibition. The end-product metabolite is called an effectors (or modulator), and if its effect is inhibitory, it is termed negative effectors. Positive effectors also occur; generally, these are either the original substrate of a metabolic pathway or an early intermediate in the sequence and serve to increase the activity of an enzyme further along the metabolic pathway. This regulatory mechanism is called feedforward stimulation, in contrast to feed back inhibition. An allosteric enzyme may be affected by one modulator, in which case it is said to be monovalent, or by two or more modulators, in which case it is polyvalent.

The regulation of amino acid synthesis in *Eschrichia coli* provides a clesr example of control of divergent metabolic pathways by feedback inhibition. An outline of the metabolic pathways for the synthesis of three amino acids is shown in Fig. 6.6. Lysine, methionine, and threonine are each synthesized from aspartate, and each may be utilized in protein synthesis. Without

Table 6.2 Common regulatory metabolites.

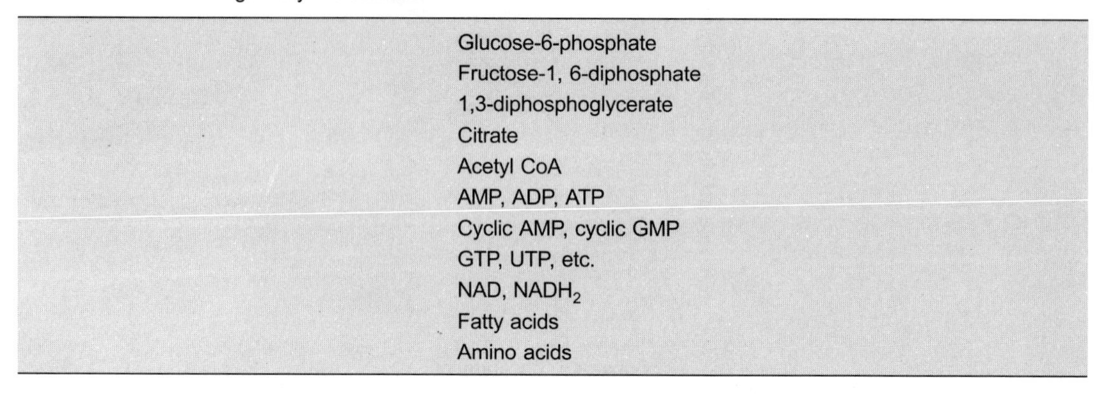

Glucose-6-phosphate
Fructose-1, 6-diphosphate
1,3-diphosphoglycerate
Citrate
Acetyl CoA
AMP, ADP, ATP
Cyclic AMP, cyclic GMP
GTP, UTP, etc.
NAD, $NADH_2$
Fatty acids
Amino acids

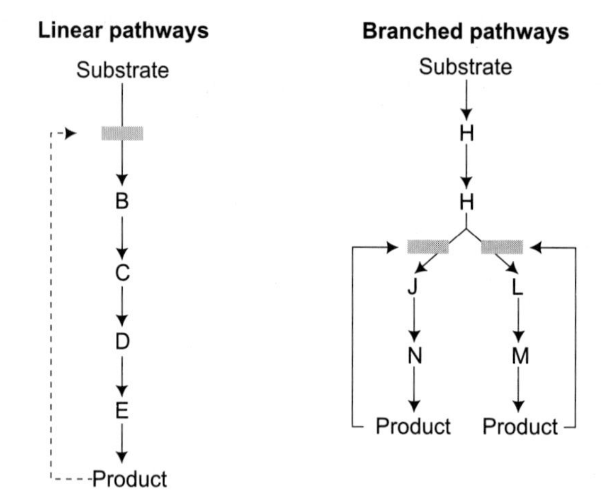

Fig. 6.5 Feedback inhibition mechanisms (___) for linear and branched pathways.

metabolic controls the consumption or utilization of anyone of these amino acids would stimulate the pathways and cause unneeded synthesis of the unused amino acid as well as the one utilized. Such an unregulated system would consume vital resources and energy; both factors should have survival implications to the organism and. evolutionary consequences to the species. However, in *E. coli*, the allosteric regulatory mechanisms are most effective. The accumulation of each amino acid provides a feedback inhibition of the first enzyme in the specific branch of the pathway leading to the synthesis of that amino acid. In Fig. 6.6, this negative effect is shown by the dotted lines. Moreover, an additional level of regulation is achieved through effects on the enzyme aspartokinase, which catalyzes the phosphorylation of as partake (Fig. 6.6). This enzyme exists in three forms or isozymes. The presence of the three isozymes is symbolized in Fig. 6.6 by using separate arrows to show the conversion of aspartate to aspartylphosphate. One of he isozymes is specifically and completely inhibited by threonine; the second (which is present only in small amounts) is specifically inhibited by homoserine; the

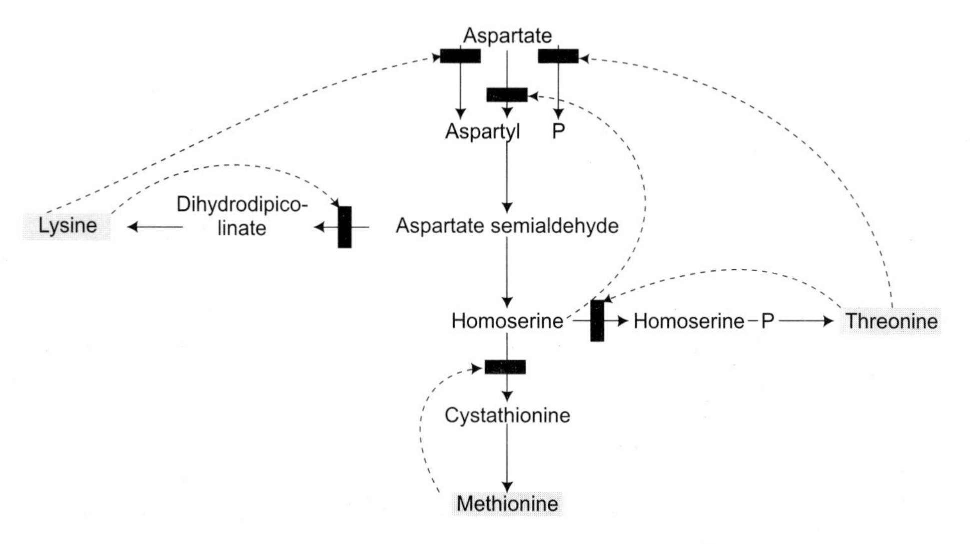

Fig. 6.6 Allosteric feedback control of amino acid synthesis ___ = negative feedback).

third isozymeis specifically inhibited by lysine. In addition, the latter isozyme is repressed by lysine. (Repression is a regulatory mechanism that reduce the number of enzyme molecules in the cell.

COVALENT BOND

Modification of Enzyme Activity

A number of enzymes are synthesized in what is called an inactive (or zymogen) form and must be covalently modified to become active. The modification may be irreversible, as is the case with the hydrolytic modification of zymogens such as pepsinogen (to form pepsin) and trypsinogen (to form trypsin). Other enzymes maybe reversibly covalently activated and deactivated. The reversible activation and deactivation of glutamine synthetase is a well-studied example (Fig. 6.7). This enzyme catalyzes the conversion of glutamate to glutamine,

$$\text{Glutamate} + NH_3 + \text{ATP} \rightarrow \text{glutamine} + \text{ADP} + \text{phosphate}$$

And the transfer of a glutamyl group to hydroxylamine,

$$\text{Glutamine} + \text{NHOH} \rightarrow \text{glutamyl-NHOH} + NH_3$$

Interestingly, it was found that glutamine synthetase was inactivated when treated with ammonia for 10 minutes. Inactivation was found to be the result of covalent adenylation, as shown in Fig. 6.7. the active and inactive forms of the enzyme are spectroscopic ally different, have different metal specificities (Mg^{++} and Mn^{++}), and are themselves acted upon by two different enzymes (adenylating and deadenylating) which change their activities. The latter enzymes are further regulated by glutamıne and ATP.

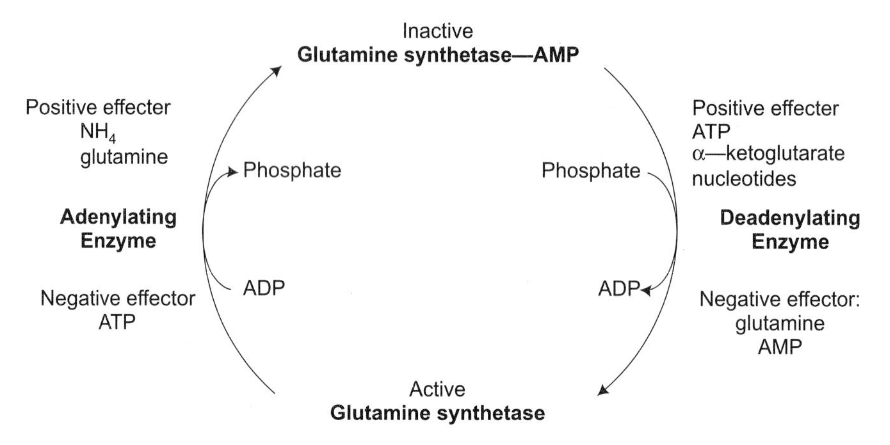

Fig. 6.7 Regulation of glutamine synthetase by activation and inactivation.

REGULATION BY NUMBER OF ENZYME MOLECULES

Isozymes (Isoenzymes)

In a number of instances, enzymes that catalyze specific reactions have been found to exist in multiple forms in tissues and organisms. These isozymes or isoenzymes are coded for by different genes and therefore have different amino acid complements. Isozymes can usually be separated from on another by gel electrophoresis (Chapter 1).

Of the tissue extracts. One of the most exhaustively studied of the known isozymes is lactic dehydrogenase which catalyzes one of the teminal reactions in glycolysis; that is,

$$\text{Pyruvate} + \text{NADH} \rightarrow \text{lactate} + \text{NAD}$$

In rates and in a number of other vertebrates, this enzyme is present in five forms. Each of he five forms has a molecular weight of about 134.000 Daltons, consists of four polypeptide chains of about 33,500 Daltons each, and catalyzes the same reaction. Each of the four polypeptides may be of two types, usually referred to as M and H. In rat skeletal muscle tissue, the predominant form of the isozymes contains four polypeptides of M type; in contrast, in rat heart muscle, the predominant form contains four polypeptides of the H type. The other isozyme form are made up of three H and One M, two M, and one H and three M polypeptides (usually abbreviated H M, H M, and H M). Some of these forms predominate in other tissues, although each tissue has some of each isozyme (Fig. 6.8).

The M isozyme is prevalent in embryonic tissue and in skeletal muscle tissue. It has a low k for pyruvate and high V_{max} for converting pyruvate to lactate. It is well adapted to these tissues, which are frequently deprived of oxygen, and must depend upon 1he breakdown of glucosssse to lactate for energy. The isozyme is likewise beneficial to the regulation of metabolism in heart muscle. This isozyme has a high K and a low V_{max} for pyuvate-to-lactate conversions and is inhibited by excess pyuvate. Heart muscle is primarily aerobic converting pyuvate to CO and HO rather than to lactate. The lactic dehydrogenase enzyme is most active during emergency conditions when the oxygen supply is low.

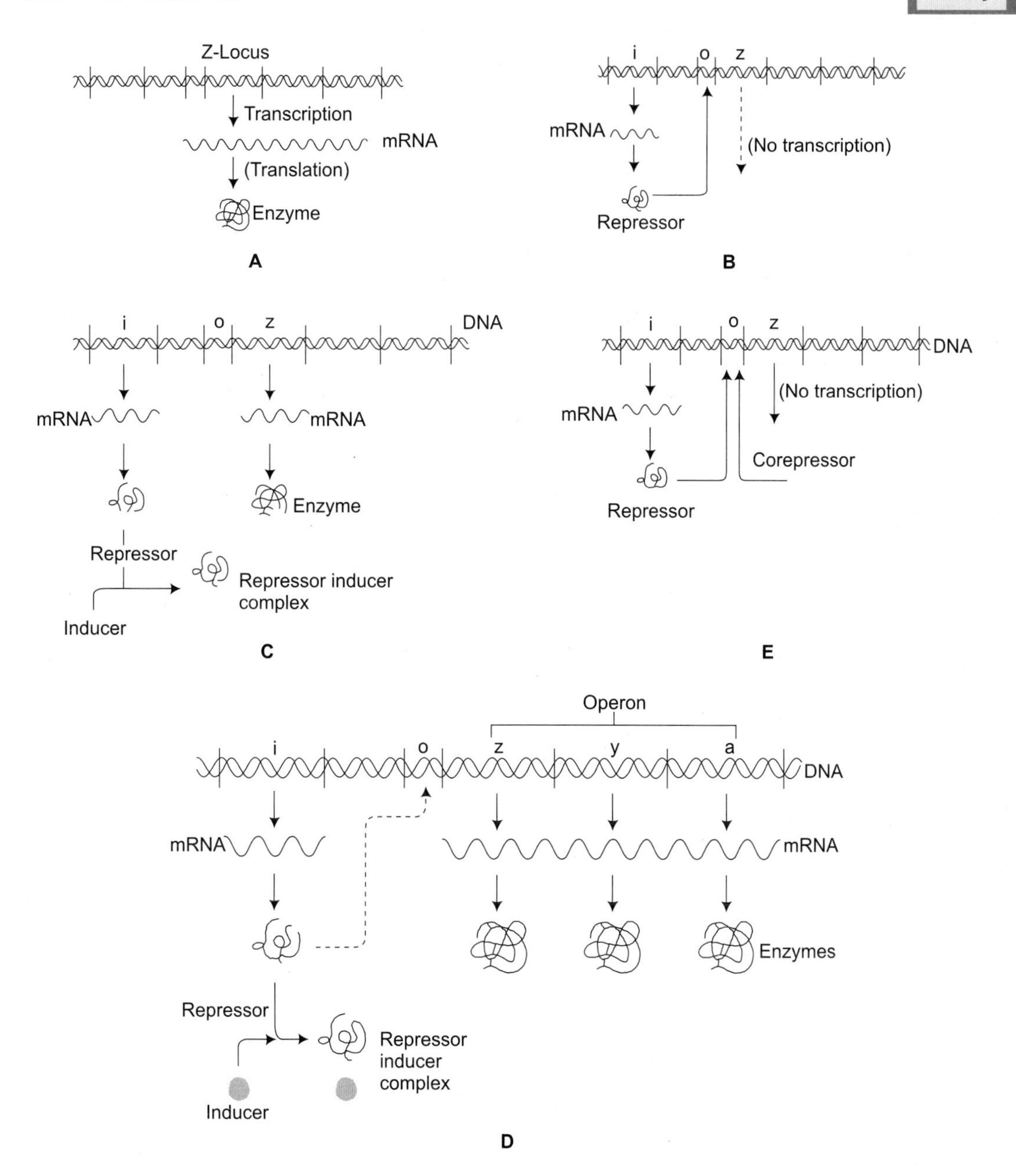

Fig. 6.8

REGULATION OF ENZYMES SYNTHESIS (Fig. 6.8)

The control of an enzyme-catalyzed reaction through the regulation of the number of enzyme molecules available in the cell may be achieved at the various biochemical steps leading from transcription of the DNA to mRNA to translation of the mRNA into polypeptides. Two levels of control based on these steps are usually recognized; these are transcriptional control mechanisms and translation control mechanisms. Most of the present understanding of these control mechanisms stems from work with prokaryotic organisms, for the regulation of gene expression in eucaryotic cells is much more complex.

CONSTITUTIVE AND INDUCED ENZYMES

Work with microorganisms has indicated that enzymes fall into two categories with respect to their occurrence and number in cells. Those that appear to always be present and that occur in relatively constant concentrations are called constitutive enzymes. The enzymes of the glycolytic pathway in microbes are usually constitutive. The second type of enzyme may be found lacking in cells or be present only in small amounts, whereas upon introduction of a specific metabolic, usually a substrate, these enzymes quickly increase in concentration. Since their synthesis appears to be induced by the presence of the substrate, they are called inducible enzymes.

One of the first thoroughly studied inducible enzymes was B-galactosidase. Wild-type *E. coli* cells normally metabolize glucose and will metabolize only glucose even if lactose is also present. The enzymes for glucose metabolism are all constitutive and are thus present, while the enzyme needed to initiate lactose metabolism, B-galatosidase, is present in only minor amounts-according to one study, no more than five copies per cell. If wild-type *E. coli* cells are placed in a growth medium containing only lactose as the carbon source, they are at first unable to utilize this disaccharide. Soon, the cells repond by synthesizing B-galactosidase and the lactose in thus hydeolyzed to glucose and galactose and the resulting sugars metabolized by glycolysis. A number of B-galactosides besides lactose are able to act as inducers; these include methyl B-galactoside and allolactose. Actually, the application of any of these induces initiates the synthesis of not one but three enzymes in *E. coli*: (1) B-galactoside permease, an enzyme formed in the plasma membrane that promotes the rate of transfer of B-galactosides across the membrane even against a concentration gradient; (2) B-thiogalactosides; acetyltransferase, used in the metabolism of galactosides; and (3) B-galactosidase, the key enzyme for initiating lactose decomposition. When, as in this case, induction can be brought about by a single agent and result in the appearance of several enzymes, the process is known as coordinate induction.

ENZYME REPRESSION

The presence of a specific substance any inhibit the synthesis of an enzyme or sequence of enzymes in a metabolic pathway; this process is called enzyme repression (or in the case of the repression of a sequence of enzymes, coordinate repression). *E. coli* has all the enzyme systems necessary to synthesize the 20 amino acids from organic acids and NH+. if no other nitrogen source is present. However, if one of the amino acids is introduced exogenously, the synthesis of the enzymes in the pathway leading to that amino acid will be inhibited, and the number of these enzymes quickly becomes reduced.

CATABOLIC REPRESSION (Fig. 6.8A-E)

Catabolic repression is a specific type of repression of enzyme synthesis in which a catabolite, usually glucose functions, to repress the formation of enzyme that would allow the decompostition of other substrates. For example, glucose represses the formation of B-galactosidase even when lactose (an inducer of this enzyme) is present. Glucose is even known to repress the formation of constitutive enzyme. A common phenomenon in microbes is the suppression of aerobic respiration and electron transport at high glucose concentrations, even in the presence of ample oxygen. Under these conditions, the cells utilize the glycolytic and fermentative pathways.

REPRESSION AND TRANSCRIPTION

The relationship between induction and repression of enzymes in microbes were clarified by the studies of Monod and Jacob and their colleagues at the Pasteur Institute in France. These investigators showed that the mechanism of regulation is tied to the transcription of the DNA code into mRNA. Today, it is clear that portions of DNA are coded with specific information on the sequencing of amino acids to form a specific enzyme (or other protein). These segments of the DNA are termed the structural genes and designated the z-locus. Their encoded information is transcribed first into mRNA molecules, and these are then translated into polypeptide chains: because they found mutants of *E. coli* that contained B-galactosidase even in the absence of an inducer (just as through B-galactosides was a constitutive enzyme). Monod and Jacob concluded that a locus of the DNA other than the structural gene (i.e. the i-locus) must be responsible for inhibition in the absence of inducer. This i-locus is called the regulatory gene. When this gene undergoes mutation, it can no longer inhibit the structural gene, and therefore the structural gene is expressed as what appears to be a constitutive enzyme. Mutants of this kind are called constitutive mutants.

Pardee, Monod and Jacob found that if the DNA containing a normal i-locus and z-locus is introduced into the cell containing a mutated i-locus, the cells behave as a normal cell with inducible B-galactosidase. In other words, the normal regulatory gene (i-locus) regulated the structural genes (z-loci) of both normal and mutated DNA. Thus, they proposed the existence of a product of the regulatory gene, a repressor substance, which diffused to other sites in the cell. The product of the regulatory gene, which has been found to be a small polypeptide, was postulated of diffuse to a site called the o-locus, which is next to or near the structural gene.

The existence of this o-locus has been substantiated by the finding of mutants defective in DNA at this point. The o-locus or operator controls the transcription of the structural gene. In wild-type cells in the absence of an inducer, the regulatory gene is transcribed and translated into a repressor protein that diffuses to the operator, where it binds and inhibits the operator. With the operator locus inhibited, transcription of the structural gene does not occur. It is suggested that when an inducer such as lactose is added, it combines with the repressor protein to from a repressor-inducer complex that can no longer bind to and inhibit the operator locus.

COMPARTMENTALIZATION

A final regulatory mechanism that is most evident in eukaryotes is the physical separation of groups of enzymes be cellular membrane boundaries. Selected groups of enzymes are compartmentalized in organelles. For example, the enzymes of the tricarboxylic acid cycle are physically separated from those of glycolysis by their confinement within the mitchondria. The enzymes of he "dark reactions" of photosynthesis (which function in basically the same manner as many of those of glycolysis) are physically isolated in the stroma of the chloroplast and are not associated with the enzymes of glycolysis that occur in the cytosol.

Many of these "isolated" enzyme sequences use substrated and/or cofactors produced by enzymes confined in other parts of the cell. Regulation of the transport of these compounds across the membranes from one cell compartment to another affords yet another level of control of metabolism.

SUMMARY

The direction and ate of metabolic pathways and individual enzymatic reactions may be controlled by one or more regulatory mechanisms:

1. **Regulation by mass action:** the rate and direction of a reaction is changed by addition of substrates or removal of products.
2. **Enzyme activity:** the rate of is altered by inhibition or activation of an enzymes.
3. **Allosteric effectors:** metabolites that attach to enzymes at a site other than that occupied by the substrate may stimulate or inhibit activity.
4. **Bond modification:** zymogens are activated by altering covalent bonds.
5. **Isozymes:** metabolic regulation by means of multiple forms of an enzyme. Different forms have different activities under varying conditions.
6. **Enzyme induction:** Regulation at transcriptional and translational levels of enzyme synthesis.
   (a) **Transcriptional:** control through a repressor-inducer complex affecting an operator gene segment of DNA and controlling the expression of structural genes (operon theory).
   (b) **Translational:** rare, possible a mechanism for controlling amounts of different enzymes regulated by the same operon.
7. **Hormones:** hormone-stimulated induction of enzymes.
8. **Compartmentalization:** isolation of enzymes and therefore metabolic pathways in organelles. Changes in permeability of membranes amt separates from enzymes.

THE GROWING PROBLEM OF ANTIBIOTIC RESISTANCE

Not too long ago, it was widely believed that human health would no longer be threatened by serous bacterial infections. Bacterial diseases such as tuberculosis, pneumonia gonorrhes, and dozens of others would be stopped dead in their tracks by administration of any one of a number of antibiotics-compounds that would selectively kill bacteria without harming the human host in which they grow. It has become painfully evident in the past decade or so that the announcement of the death of infectious bacteria was premature. Bacteria that were once susceptible to a variety of antibiotics are becoming increasingly resistant to these drugs. The development of bacterial resistance provides an excellent example of natural selection; the widespread as use of these drugs has killed off susceptible cells, leaving the rare resistant individuals so survive and repopulate the ranks. The result has been a marked upswing in the incidence and virulence of a number of disease, including pneumonia, tuberculosis, and a host of diseases caused by streptococcal and staphylococcal bacteria. Many infectious disease specialists are predicting that the problem will become more acute in the coming years and that fatalities from once curable diseases will increase sharply. Here, we will look briefly at the mechanism of action of antibiotics-particularly those that target enzymes, which is the subject of this chapter-and the development of bacterial resistance.

PABA 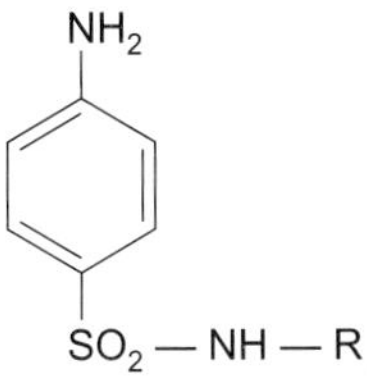 **Sulfa drugs** 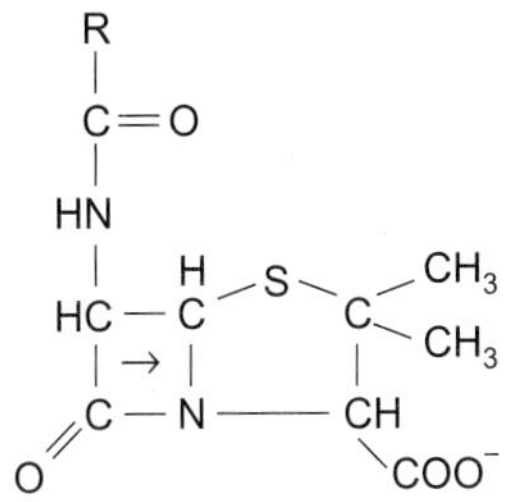 **Penicillin**

Antibiotics work because they are able to target bacterial activities without affecting those of eucaryotic cells. Several types of targets in bacterial cells have proven most vulnerable. These include:

1. Enzymes involved in the formation of the bacterial cell wall. Penicillin and its derivatives are structural analogues of the substrates of a family of transpeptidases that catalyze the final cross-linking reactions that give the cell wall its protective properties. If these reactions do not occur, the cell wall falls apart. Penicillin is an irreversible inhibitor of the transpeptidases; the antibiotic fits into the active site of the enzyme, forming an irreversible complex that cannot be dislodged. Vancomycin, an antibiotic that bacteria have proven least able to develop resistance to, inhibits an enzyme that acts at an early stage in cell wall formation.
2. Components of the system by which bacteria duplicate, transcribe, and translate their genetic information. Although prokaryotic and eukaryotic cells have a similar system for storing and utilizing genetic information, there are many basic differences between the two types of cells that pharmacologists can take advantage of. Rifamycin, for example, is an antibiotic that selectively inhibits the bacterial RNA polymerase, the enzyme that transcribes DNA into RNA. Similarly, streptomycin and the tetracyclines bind to prokaryotic ribosomes, but not eukaryotic ribosomes.

3. enzymes that catalyze metabolic reactions that occur specifically in bacteria. Sulfa drugs, for example, are effective antibiotic because they closely resemble the compound p-aminobenzoic acid (PABA).

Which bacteria convert enzymatic ally to the essential coenzyme folic acid. Since humans lack a folic acid-synthesizing enzyme, they must obtain this essential coenzyme in their diet, and consequently, sulfa drugs have no effect on human metabolism.

Bacteria become resistant to antibiotics through a number of distinct, mechanisms, many of which can be illustrated using penicillin as an example. Like most antibiotic, penicillin is a natural compound, that is, a compound normally produced by a living organism, in this case a mold. Penicillin protects the mold from bacterial pathogens in the same way that it protects (or once protected) humans. Bacterial cells have probably been exposed to penicillin like compounds for hundreds of millions of years, thus it shouldn't be surprising that they have evolved weapons to defend themselves against such compounds.

Pencillin is a B-lactam; that is, it contains a characteristic 4-membered B-lactam ring (shown by arrow).

As early as 1940, researchers discovered that certain bacteria posess an enzyme called B-lactamase (or penicillinase) that is able to split open the lactam ring, rendering the compound harmless to the bacterium. At the time when pencillin was first introduced as an antibiotic during World War II, none of the major disease-causing bacteria possessed a gene for B-lactamase. This can be verified by examining the genetic material of bacteria descended from laboratory cultures that were started in the preantibiotic era. Today, the B-lactamase gene is present in a wide variety of infectious bacteria, and the production of B-lactmase by these cells is the primary cause penicillin resistance.

The widespread occurrence of the B-lactamase gene illustrates how readily genes can spread from one bacterium to another, not only among the cells of a given species, but between species. There are several ways this can happen, including conjugation in which DNA is passed from one bacterial cell to another; transduction, which a bacterial gene is carried from cell to cell by a virus; and transformation, in which a bacterial cell is able to pick up naked DNA from its surrounding medium. Pharmacologists have attempted to counter the spread of B-lactamase by synthesizing penicillin derivatives that are more resistant to the hydrolytic enzyme. As might be expected, natural selection quickly produces bacteria whose B-lactamase can split the new forms of the antibiotic. As noted by Julian Davies. "a single base change in a gene encoding a bacterial B-lactamase may render unless $100 million dollars worth of pharmaceutical research effort." One approach the t has met with limited-success is to treat patients with two separate drugs; a penicillin-like antibiotic to inhibit the transpeptidase and a separate enzyme inhibitor (e. g., clavulanic acid) to inhibit the B-lactamase.

Not all penicillin-resistant bacteria have acquired a B-lactamase gene. Some are resistant because they possess modifications in their cell walls that block entry of the antibiotic; others are resistant because they are able to selectively export the antibiotic once it has entered the cell; still others are resistant because they possess modified transpeptidases that fail to bind the antibiotic. Bacterial meningitis, for example is caused by the bacterium *Neisseria meningitides*, which has yet to display evidence it has acquired B-lactamase. Yet these bacteria are becoming resistant to penicillin because their transpeptidases are losing affinity for the antibiotics. Comparison of the genes encoding the resistant transpeptidase with the genes encoding the corresponding enzymes in susceptible strains (isolated from cultures begun in the preantibiotic era) reveals major difference in nucleotide sequences. These findings indicate that the bacterial cells are not becoming drug resistant as the result o genetic mutation, which would produce small genetic changes, but rather by acquiring new genes from another species.

CHAPTER 7

The Plasma Membrane

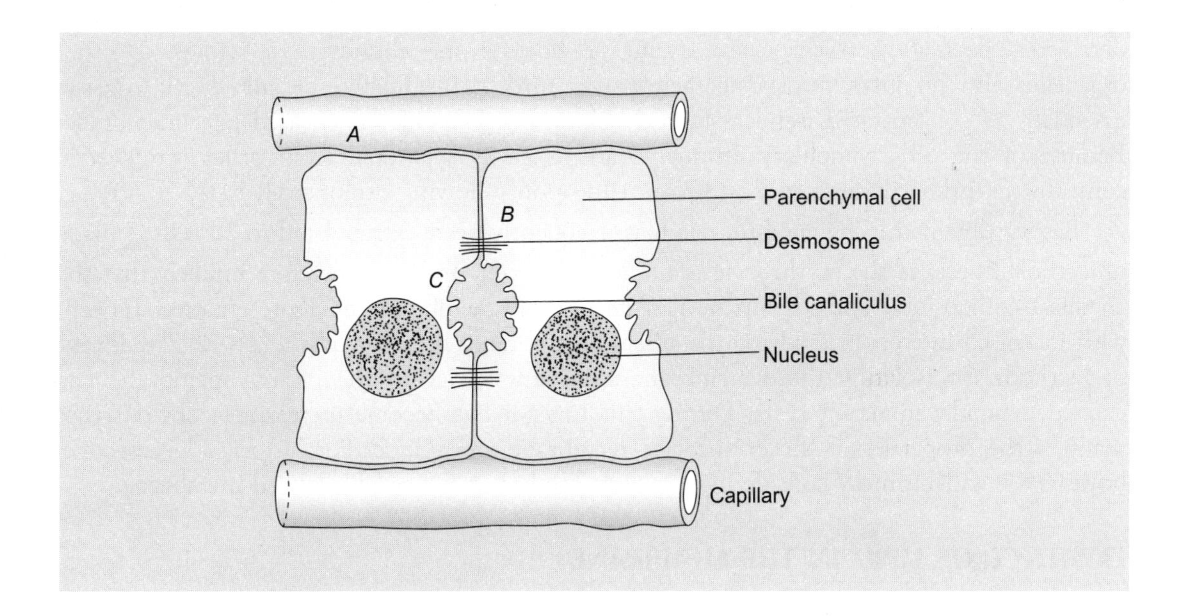

The plasma membrane delimits the cell, physically separating the cytoplasm from the surrounding cellular environment. This implies that all substances either entering or exiting cell must pass through the plasma membrane. Only rarely does the plasma membrane play a passive role in the exchange of molecules between the cell and its surroundings. Instead, the flux of substances may actually be facilitated by continuous molecular changes within the membrane. In many instances, transport through the membrane is achieved by the active participation of carrier molecules within the membrane and incurs the expenditure of large amount of chemical energy. The cellular ingestion (or excretion) of some material is associated with gross movement and separations of fragments of the membrane from main body. Stages of this activity can be seen and studied with the electron microscope (sometimes also the light microscope).

In this chapter, we consider the structure and chemical organization of the plasma membrane and the mechanisms by which the transport of materials across the membrane may be achieved. As will become evident in subsequent chapters, much of the information presented here can be directly extrapolated to the membrane the encase cell organelles as well as to other cytomembranes.

EARLY STUDIES ON THE CHEMICAL ORGANIZATION OF THE PLASMA MEMBRANE

Among all animal and plant cells, none has been more extensively studied than the mammalian erythrocyte or red blood cell. The erythrocyte has long been the favorite of investigators studying the plasma membrane because relatively pure membrane preparations are so easily obtained. The mature erythrocyte contains no nucleus, mitochondria, ribosomes, or other organelles and no intracytoplasmic membrane; instead this highly specialized cell consists essentially of concentrated (semicrystalline) solution of hemoglobin encased in a membrane. Because of the cell's simplicity, its membrane is easily separated from other cytoplasmic constituents (primarily hemoglobin) by centrifugation following osmotic cell lysis.

Results obtained using erythrocytes have frequently been extrapolated to all cells. This is unfortunate because the erythrocyte is not a typical cell, and it is therefore unlikely that the chemical composition and organization of its limiting membrane are representative. Indeed, with increased interest in studying the plasma membranes of other cells and with the advent of methods for isolating these membranes, our knowledge of the plasma membrane has expanded rapidly in recent years. During this time, It has become increasingly obvious that many of the properties of the erythrocyte membrane are unique. For historical perspective, however we will begin by considering the early studies of the red blood cell membrane.

EXISTENCE OF LIPID IN THE MEMBRANE

As early as 1899, E. Overton recognized that the boundary of animal and plant cells was "impregnated" by lipid material. Overton's conclusion were based on exhaustive studies of the rates of penetration of more than 500 different chemical compounds into animal and plant cells. In general, compounds soluble in organic solvents entered the cells more rapidly than compounds soluble in water. These differences were attributed to the "selective solubility" of the membrane; that is, lipid soluble materials would pass into the cell by dissolving in the corresponding lipid elements that made up the membrane. Overton suggested that cholesterol and lecithins might be among the lipid constituents of the plasma membrane, a suggestion that was later substantiated chemically. The pioneering studies of Overton at the turn of the century set the stage of Gorter, Grendel, Cole, Danielli, Harvey, Davson, and others who attempted to determine the specific manner in which the lipid might be organized with in the membrane.

THE LANGMUIR TROUGH

One of the most valuable instruments used to study the behavior of lipid films is the langmuir trough (Fig. 7.1).

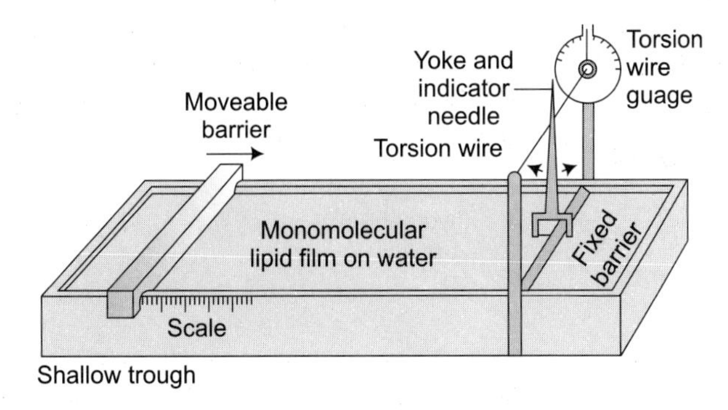

Fig. 7.1 The Langmuir trough.

If lipid containing hydrophilic groups (such as the catboxyl groups of fatty acids or the phosphate groups of phospholipids) is dissolved in a highly volatile solvent and several drops are then carefully applied to the surface of water, the lipid spreads out to form a thin, monomolecular film in which the hydrophilic parts of each molecules project into the water surface while the hydrophobic parts are directed up, away from the water (Fig. 7.2). In 1917, I. Langmuir introduced a clever technique for measuring the specific minimum surface area occupied by a monomolecular film of lipid and the force necessary to compress all the lipid molecules into this area. His device, known as the Langmuir trough or Langmuir film balance, has been used extensively over the past several decades in connection with physical measurements of membrane lipids.

Langmuir himself employed this device to study the behavior of the surface films formed by a variety of organic compounds, and for this work, together with his electronics innovations, he received the Nobel Prize in 1932. Others have applied the same technique in specific studies of membrane lipids.

GORTER AND GRENDEL'S BIMOLECULAR LIPID LEAFLET MODEL

In 1925, E. Gorter and F. Grendel published the results of their studies on the organization of lipid in the membrane of the red blood cell. Their studies were carried out using blood from a variety of mammals, including dogs, sheep, rabbits, guinea pigs, goats, and humans, and all yielded essentially the same results. The lipid present in accurately measured quantities of washed red blood cells was extracted with acetone and the acetone was then evaporated, leaving the lipid as a residue. This residue was redissolved in benzene and spread in a Langmuir trough to form a tightly packed monomolecular layer. The surface area occupied by extracted lipid was then measured.

Gorter and Grendel determined the numbers of red cells present in each sample of blood analyzed and estimated the total surface area of the cells by multiplying the cell number by the average surface area per cell. (The surface area of the erythrocyte was estimated using the relationship proposed by Knoll that for red blood cells the surface area = 2d, being the diameter

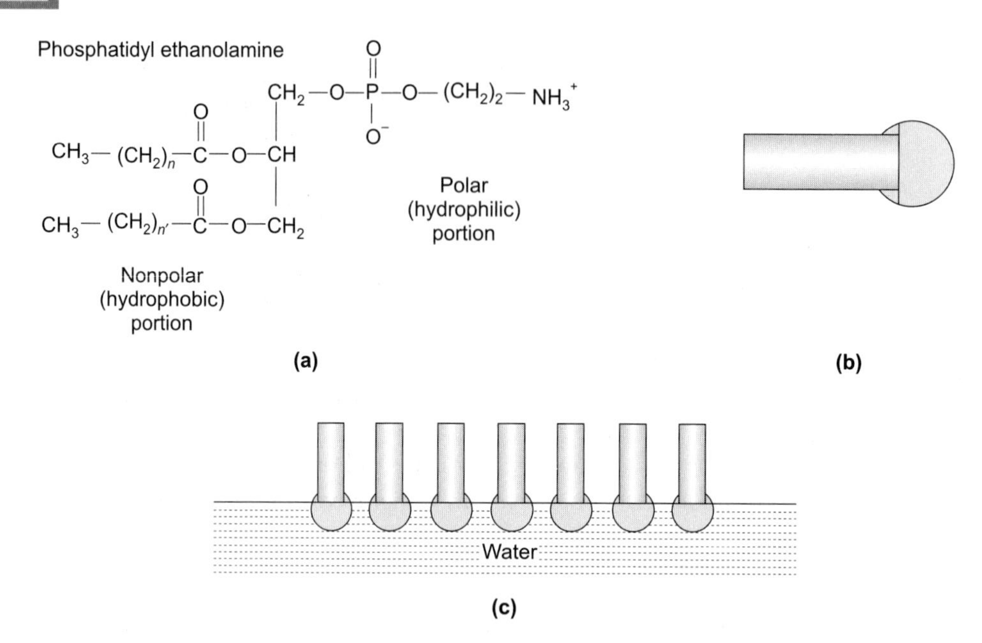

Fig. 7.2 Formation of a monomolecular lipid film on water. Phosphatidyl ethanolamine (a) represents a typical lipid molecule possessing polar (hydrophilic) and non-polar (hydrophobic) regions. These regions of the phospholipid molecule are depicted diagrammatically in (b). When spread on water, the hydrophilic part of each lipid project into the water surface, while the hydrophobic parts are directed up, away from the water (c).

of the cell determined microscopically. This estimate was subsequently shown to be an error. By dividing the total surface area occupied by a monomolecular layer of membrane lipid extracted from these cells by the total cell surface area the number of the lipid layers present in the membrane was obtained. The value varied between 1.8 and 2.2, leading Gorter and Grendel to propose that the cell membrane was formed by a bimolecular lipid sheet. They further suggested that the polar ends of the lipid molecules of the one layer were directed outward (from the cell) toward the surrounding plasma, while the polar ends of the lipid molecules coming the other layer were directed inward toward the cell hemoglobin. Thus the nonpolar and extremely hydrophobic ends of the lipid molecules in each layer would face one another (Fig. 7.3).

In the past 10 years, the results of Gorter and Grendel have been reexamined under improved conditions by a number of investigators. The validity of Gorter and Grendel's bimolecular lipid leaflet model depends on the assumptions that (1) all the erythrocyte lipids are in the plasma membrane, (2) all of this lipid was extracted using their acetone procedure, and (3) the average surface area of the cells was accurately estimated. The first assumption has been verified-all the erythrocyte lipid is in the plasma membrane. However, it is now clear that Gorter and Grendel extracted only 70 to 80% of the total lipid. This error would seriously alter the predicted ratio of lipid film area to cell surface area if it were not the fact that Gorter and

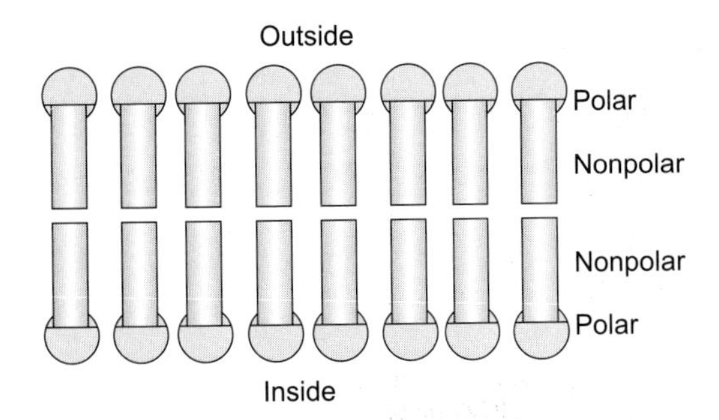

Fig. 7.3 Bimolecular lipid leaflet model for the structure of the red blood cell membrane proposed by Gorter and Grendel.

Grandel also underestimated the red blood cell surface area by a comparable amount. The two errors cancelled each other out, so that the ratio of 2.1 is still obtained.

THE DANIELLI-DAVSON MEMBRANE MODEL

A consistent observation made for cell membranes that was not explained by the bimolecular lipid leaflet model was the very low surface tension of the cell membrane. In 1935, J. F. Danielli and E.N. Harvey proposed that oil droplets and other lipid inclusions in cells were bonded at their surface by an organized layer of the lipid and a layer of protein. It was postulated that the protein, which consisted of a monomolecular layer of hydrated molecules, faced the aqueous cytoplasm and simultaneously interacted with the polar portions of the lipid layer. The nonpolar portions of the lipid layer faced the hydrophobic oil phase of the droplet interior (Fig. 7.4). In this structure, the natural surface activity of the protein would account for the low interfacial tension of the droplet membrane. Shortly thereafter, Danielli and H. Davson suggested that the plasma membrane itself might be composed of two such lipid-protein bilayers–one facing the interior of the cell and the other facing the external milieu. This arrangement is shown in Fig. 7.5. Danielli and Davson proposed that such a membrane would exhibit selective permeability, using capable of distinguishing between molecules of different size and solubility properties and also between ions of different charge. Between the outer and inner protein layers. The modified Danielli-Davson membrane model is shown in Fig. 7.6. In this arrangement, the association between the surface proteins and the bimolecular lipid leaflet would be mainted primarily by electrostatic interactions between the polar ends of each lipid molecule and charged amino acid side chains of the polypeptide layers. Either electrostatic or van der Waals bonds could bind others groups to the outer protein surface.

It should be kept in mind that the model shown in Fig. 7.6 was formulated before the plasma membrane was first seen, since the use of the electron microscope to study the organization of the plasma membrane began around 1957. Nevertheless, even the thickness of the membrane was estimated to be about 100 A, based on the known lengths of extended phospholipids molecules (about 35 A) and the thickness of several layers of pleated sheet polypeptide or a single layer of the alpha-helix form (10-15 A).

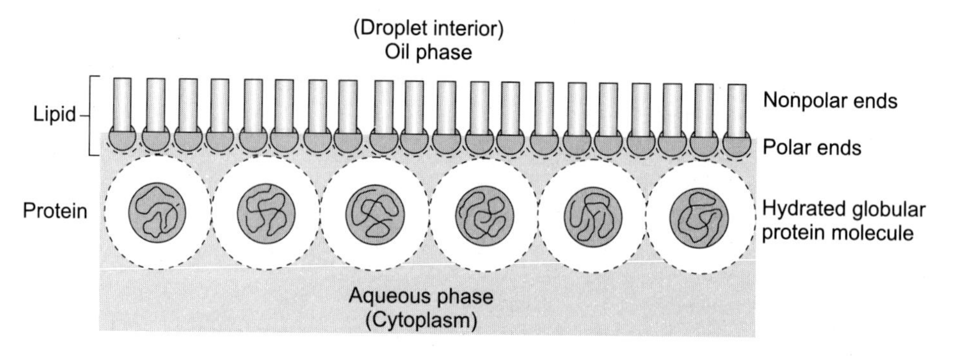

Fig. 7.4 Danielli and Harvey's 1935 model of the protein-lipid bilayer formed at the interface between a cell oil droplet and aqueous cytoplasm.

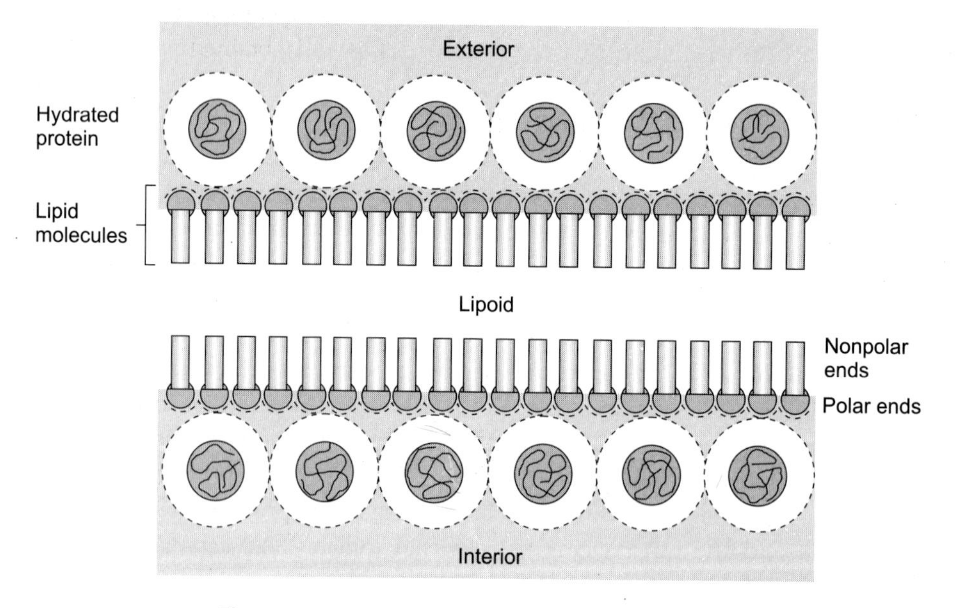

Fig. 7.5 Danielli-Davson membrane model (1935).

ROBERTSON'S UNIT MEMBRANE

In the late 1950s, electron microscopy provided additional information about the structure of the plasma membrane. J.D. Robertson was a pioneer in this area, showing that membranes fixed with osmium tetroxide revealed a characteristic trilaminar appearance consisting of two parallel outer dark (osmiophilic) layers and a central light (osmiophobic) layer (Fig. 7.7). The osmiophilic layers typically measured 20-25 A in thickness and the osmiophobic layer measured, 25-35 A, yielding a total thickness of 65-85 A. This value compared favourably with the thickness predicted on the basis of chemical studies. However, the thickness of the osmiophilic outer layer was much grater than that predicted for the outer protein coats, while the thickness

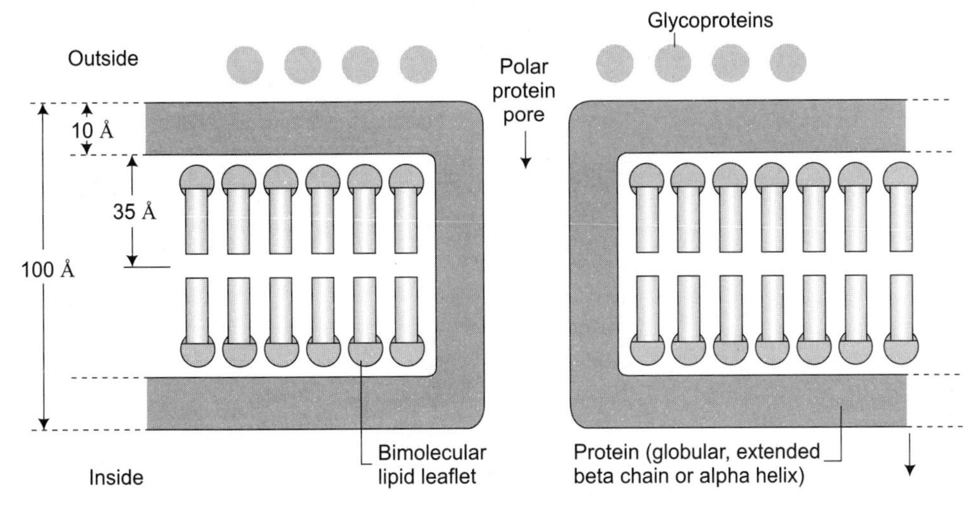

Fig. 7.6 Modified Donielli-Davson membrne model (1950).

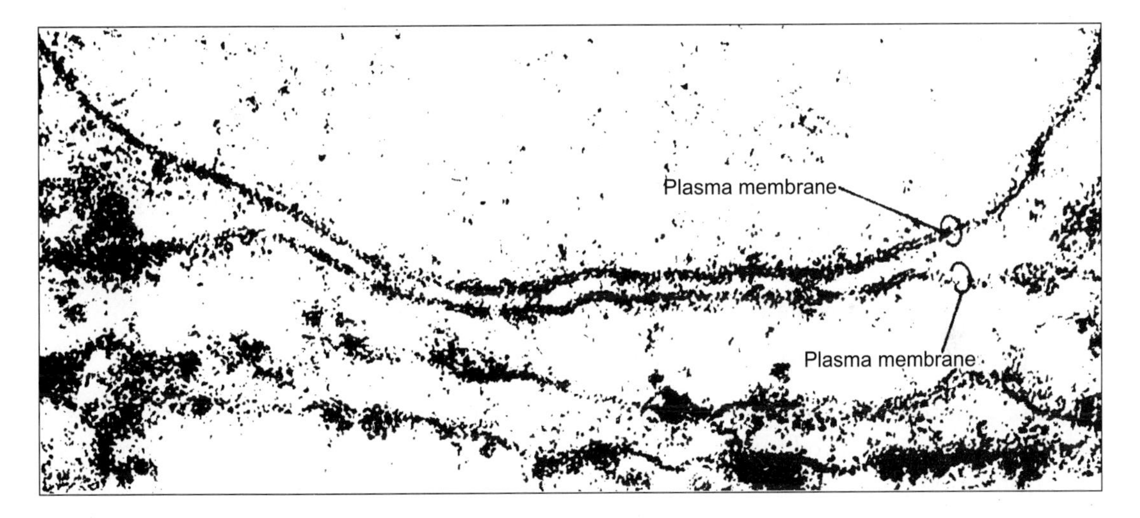

Fig. 7.7 Trilaminar appearance of the plasma membrane (Electron photomicrograph courtesy of R. Chao).

of the central osmiophobic layer was too small to be the bimolecular lipid leaflet (see Fig 7.8). To account for these apparent discrepancies, Robertson suggested that the dark layers might be produced by osmium ions binding to both the polar amino acid side chains of the protein and the polar ends of the phospholipids molecules, while the light central layer represented only the nonpolar fatty acid chains of each phospholipid that would not bind osmium. In some cells, the outer dark line was thicker than the internal dark line, and this was presumed to be due to the binding of additional osmium ions by adsorbed glycoproteins or other osmiophilic molecules.

Robertson and others demonstrated that the trilaminar pattern was characteristic of many other cellular membranes, including the endoplasmic reticulum, the membrane of mitochondria,

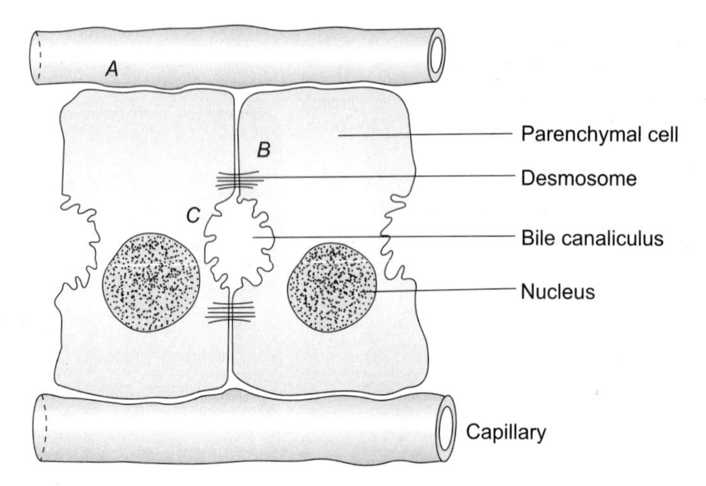

Fig. 7.8 Three faces of the liver cell. (A) Juxtaposition of plasma membrane and sinusold capillary membrane. (B) Juxtaposition with neighboring cell. (C) Juxtaposition with bile Canaliculus . Each face contains specific membrane constituents and properties.

chloroplasts, and Golgi bodies. In view of the underlying unity in the appearance of nearly all cell membranes studied, Robertson proposed his now famous unit membrane model. According to Robertson, the unit membrane consisted of a bimolecular lipid leaflet sandwiched between outer and inner layers of protein organized in the pleated sheet configuration. Such an arrangement was presumed to be basically the same in all cell membranes. While acknowledging specific chemical differences between membranes (i.e., the particular molecular species that make up each membrane differ), Robertson proposed that the pattern of molecular organization was fundamentally the same.

The Fluid-Mosaic Model of Membrane Structure

At the present time, the most widely accepted model of membrane structure is the fluid-mosaic model (an expression introduced by S.J. Singer and G. Nicolson to describe both the properties and organization of the membrane). According to this model (Fig. 7.9), the membrane contains a bimolecular lipid layer, the surface of which is interrupted by proteins. Some proteins are attached at the polar surface of the lipid (i.e., the peripheral, or extrinsic, proteins), while others penetrate the bilayer or span the membrane entirely (i.e., the integral, or intrinsic, proteins). The peripheral proteins and those parts of the integral proteins that occur on the outer membrane surface frequently contain chain of sugars (i.e., they are glycoproteins). The sugar chains are believed to be involved in a variety of physiological phenomena including the adhesion of cells to their neighbors. Membrane lipid is primarily phospholipids, although quantities of neutral lipids may also be present. Some of the lipid at the outer surface is complexed with carbohydrate to form glycolipid.

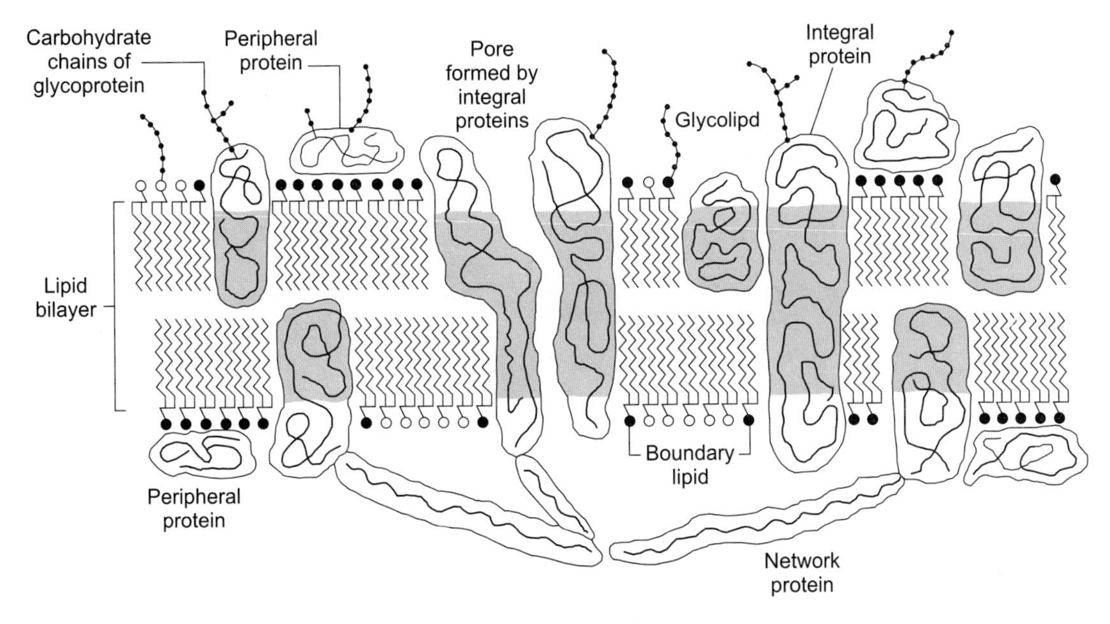

Fig. 7.9 The *fluid mosaic* model of membrane structure.

Freeze-Fractured Membranes

The fluid mosaic model of membrane structure is beautifully supported by the visual evidence provided when freeze-fractured membranes are examined with the transmission electron microscope. D.Branton, who pioneered this field, showed that membranes rapidly frozen at the temperature of liquid nitrogen and cut or chipped with a microtome blade readily fracture along specific planes. When the plane of the fracture intersects the plane of the membrane, the membrane is split along the center of the lipid bilayer, producing two "half-membranes" called the E half and the P half. The E half is that portion of the membrane that faced the cell exterior, while the P half corresponds to the portion that faced the protoplasm (cytosol). One side of each half-membrane is the original membrane surface, called E and P faces, while the other side is the newly exposed fracture face, called the E fracture face (EF) and p fracture face (PF). The fracture faces are extremely delicate and are not examined directly. Instead, a thin film of platinum and carbon is evaporated onto the surface of the fracture faces to produce a replica which is then examined by transmission electron microscopy.

In many instance, before the replica is made, water (as well as other volatile materials) near the fracture surfaces is eliminated by sublimation (i.e., by carefully raising the temperature of the sample). This step, which used to be called "freeze etching," exposes additional surface features of the fracture face.

Electron micrographs of freeze-fractured cells show the membranes to be covered by numerous small particles. There is convincing evidence that the particles are membrane proteins (e.g., they disappear when the membranes are first treated with proteolytic enzymes).

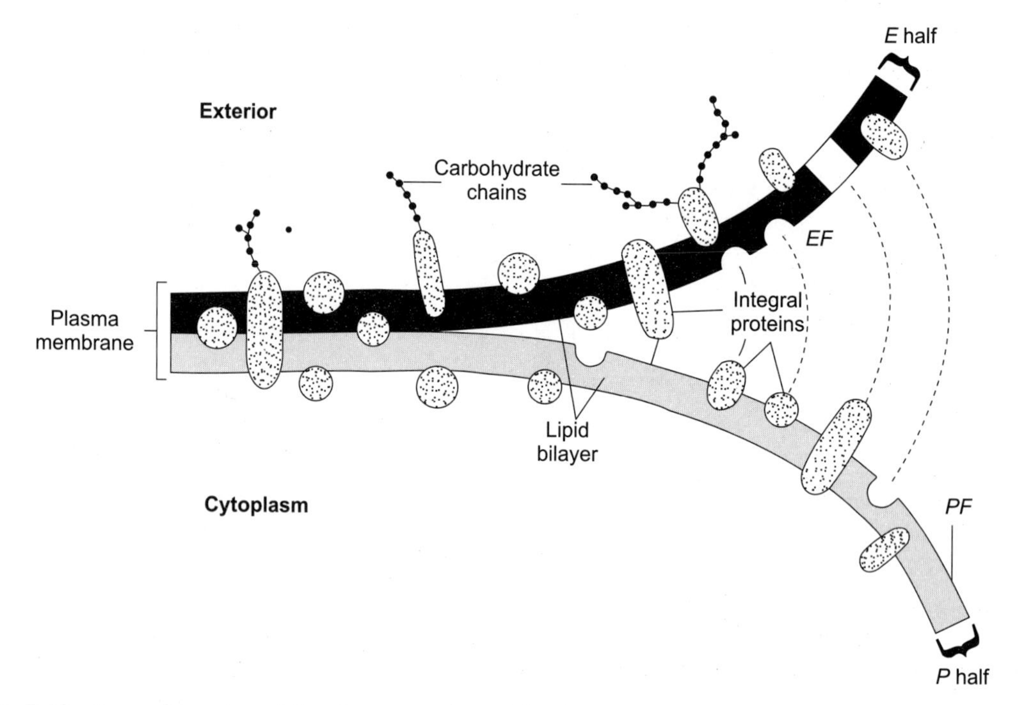

Fig. 7.10 Freeze-franturing of the plasma membrane. The fracture plane occurs at the center of lipid bilayer and passes over (or under) the integral mambrane proteins.

This suggests that the plane of fracture passes around the protein molecules rather than through them. This relationship is depicted in Fig. 7.10. The relatively uniform background apparent in the fracture face corresponds to the surface of one-half of the lipid bilayer.

MEMBRANE PROTEINS

Peripheral (Extrinsic) Proteins

Peripheral or extrinsic membrane proteins are generally loosely attached to the membrane and are more readily removed than are the integral proteins. Peripheral proteins are rich in amino acids with hydrophilic side chains that permit interaction with the surrounding water and with the polar surface of the lipid bilayer. Peripheral proteins on the cell's exterior membrane often contain chains of sugars.

Integral (Intrinsic) Proteins

Integral or intrinsic membrane proteins contain both hydrophilic and hydrophobic regions. Those portions of the protein that are buried in the lipid bilayer are rich in amino acids with hydrophobic side chains. The latter are believed to form hydrophobic bonds with the fatty acid tails of the membrane phospholipids. Portions of integral proteins that project outward from the lipid bilayer are rich in hydrophilic amino acids; those projecting from the outer membrane surface may contain carbohydrate chains.

Integral proteins that span the membrane. It was M. Bretscher who first demostrated the existence of integral proteins that span the entire membrane. In a series of elegant experiments, Brestscher showed that radioactive ligands specific from membrane proteins of the erythrocyte were bound in smaller quantities to intact cells than to disrupted cells. Disruption for the cells was shown to expose portions of the membrane proteins previously facing the cell interior, thereby allowing additional radioactive ligand to associate with the protein.

Integral proteins that span the entire membrane contain outer regions that are hydrophilic and a central region that is hydrophobic. Carbohydrate associated with the hydrophilic region facing the cell's surroundings is believed to play a role in maintaining the orientation of the protein within the membrane. The hydrophilic sugars, together with the hydrophilic side chains of amino acids in the outer region of the protein, effectively prevent reorientation of the protein in the direction of the hydrocarbon core of the lipid bilayer.

Asymmetric Distribution of Membrane Proteins

The outer and inner regions of the cell membrane do not contain either the same or equal amount the various peripheral and integral proteins. For example, the outer half of the erythrocyte membrane contains far less protein than does the inner half. In addition, various membrane proteins may be present in significantly different quantities; the membrane of some cells contains a hundred times as many molecules of one protein species as another. Moreover, regardless of absolute quantity, all copies of a given membrane protein species have exactly the same orientation in the membrane. This is in stark contrast with the more uniform distribution of the various membrane lipids. The differential distribution of proteins in the various regions of the plasma membrane within a single cell was described earlier in connection with liver cells and intestinal epithelium. This irregular distribution of membrane proteins is known as membrane asymmetry. Not only plasma membranes but also membranes of the endoplasmic reticulum and vesicular organelles (e.g., mitochondria) ar asymmetric.

Mobility of Membrane Proteins

When cells are grown in culture, there is an occasional fusion of one cell with another to form larger cell. The frequency of cell fusion can be greatly increased by adding sendai virus to the cell cullture. In the presence of this virus, even different strains of cells can be induced to fuse, producing hybrid cells, or heterokaryons. D. Frye and M. Edidin utilized this phenomenon do demonstrate that membrane proteins may not maintain fixed positions in the membrane but may move about laterally through the bilayer. Frye and Edidin induced the fusion of human and mouse cells to form heterokaryons and, using fluorescent antibody labels, followed the distribution of human and mouse membrane proteins in the heterokaryon during the time interval that followed fusion. At the onset of fusion, human and mouse membrane proteins were respectively restricted to their "halves" of the hybrid cell, but in less than an hour both protein types became uniformly distributed of the membrane (Fig. 17.1). The distribution of the membrane proteins was not dependent on the availability of ATP and was not prevented by metabolic inhibitors, indicating that lateral movement of proteins in the membrane occurred by diffusion.

Not all membrane proteins are capable of lateral diffusion. G. Nicolson and others have obtained evidence suggesting that some integral proteins are restrained within the membrane by a network of protein lying just under the membrane's inner surface. This network may, in turn, be associated with a system of micro-filaments and microtubules in the cytosol (i.e., the "cystoskeleton").

Enzymatic Properties of Membrane Proteins

Membrane proteins have been shown to possess enzymatic activity. Table 7.1 lists some of the enzymes that are now recognized constituents of the plasma membrane. To this list of proteins must be added receptor proteins (such as the insulin-binding sites of the liver cell membrane) and structural proteins. Recent evidence suggests that glycosyl transferees may also be present on the membrane's outer surface, where they add sugars to the ends of oligosaccharides associated with membrane protein or lipid.

Isolation and Characterization of Membrane Proteins

Because of the relative ease with which they may be pirifies, the plasma membranes of erythrocytes provided much of the early information on the chemistry of proteins (and lipids) present in membranes. Now however, plasma membranes can be obtained from many cell types in a reasonably uncontaminated state using various forms of density gradient centrifugation.

Table 7.1 Enzymes present in the plasma membrane.

Adenosine triphosphatase (Mg^{++} stimulated)	Monoglyceride lipase
Adenosine triphosphatase (Mg^{++}, Na^{+}, K^{+} stimulated)	Triglyceride lipase
Adenosine triphosphatase (Mg^{++}, Ca^{++} stimulated)	Acetylcoenzyme A synthetase
Nucleoside diphosphate phosphatase	Invertase
Nucleoside triphosphate pyrophosphatase	Maltase
5' Nucleotidase	Isomaltase
Adenylcyclase	Lactase
Protein kinasw	Trehalase
Acetylphosphatase (K^{+} stimulated)	Furanase
Alkaline nitrophenyl phosphatase	Cellobiase
Acid nitrophenyl phosphatase	UDP glycosidase
NAD pyrophosphatase	Nitrophenyl glycosidase
Alkaline glycerophosphatase	Collagen glycosyl transferase
Alkaline phosphdiesterase	Leucyl β-naphthyl amidase
NAD glycohydrolase	NADH dehydrogenase
Cholesterol esterase	Nucleoside kinase
Phosphatidyl inositol kirase	Triosephosphate dehydrogenase
Diglyceride kinase	Phosphoribos isomerase
Phosphatidate phosphatase	Xanthine oxidase
Sphingomyelinase	

Table 7.2 Lipids present in the plasma membrane.

Plasma Membrane of	*Major Lipids Present*	*Protein Lipid (wt/wt)*
Liver cell	Cohlesterol, phosphatidyl choline, phosphatidyl ethanolamine, phosphatidyl serine, sphngomyelin	1.0-1.4
Intestinal epithelial cell	Cholesterol, phosphatidyl choline, phosphatidyl ehanolamine, phosphyatidyl serine, sphingomyelin	4.6
Erythrocyte	Phosphatidyl inositol; cholesterol, phosphatidyl choline, phosphatidyl ethoanolamine, phosphatidyl sernine sphingomyelin	1.6-1.8
Myelin	Cholespterol, cerebrodides, phosphatidyl ethanolamine, phosphatidyl choline	0.25
Gram-positive bacteria	Diphosphatidyl glycerol, phosphatidyl glycerol, phosphatidyl	2.0-4.0

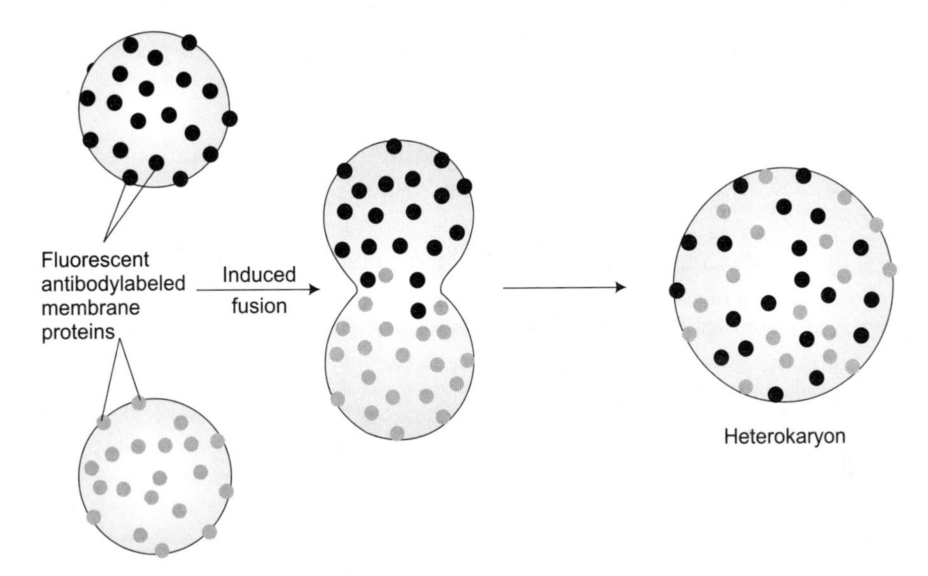

Fig. 7.11 Movement of proteins in the plasma membrane. When fiourescent antibody labeling of the membrane proteins of different cells is followed by Sendai virus induced cell fusion the proteins are soon observed to distribute throughout the membrane of the heterokaryon.

Nevertheless, the individual protein constituents of the membrane are not so easily extricated for individual study because of their high degree of insolubility. Varying degrees of success in extracting proteins from the plasma membrane have been achieved using sodium dodecyl sulfate (SDS) and Triton X-IOO (two organic detergents) and concentrated solutions of urea, n-butanol, and ethylene diamine tetraacetic acid (EDTA). These chemicals have a disaggregating effect on membranes, causing the release of many of the membrane proteins by dissociating the bonds that link the proteins together or to other.

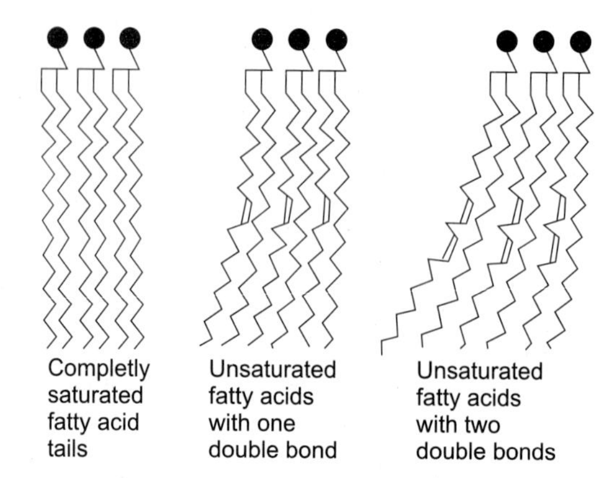

Fig. 7.12 Disruption of the orderly stacking of phospholipids having saturated fatty acid tails by unsaturated fatty acids having one or more double bonds.

Membrane Lipids

Much more is known about the specific lipid composition of cell membranes because the lipids are more readily extracted from the membranes using a variety of organic solvents. Once extracted from isolated membranes, the lipids may be separated and identified using chromatographic or other produces. Nearly all the membrane studied so far appear to contain the same types of lipid molecules. Phospholipids such as phosphatidyl ethanolamine, phosphatidyl serine, phosphatidyl inositol, phosphatidyl choline (lecithin), and sphingomyelin are the most common constituents, but cholesterol may also be present. Table 7.3 lists the most common lipids found in a variety of cell membranes and also shows their protein-to-lipid weight rations; the latter varies considerably.

Membrane Carbohydrate

It has already been noted that carbohydrate is present in the plasma membrane as short, sometimes branched chains of sugars attached either to exterior peripheral proteins (forming glycoproteins) or to the polar ends of phospholipid molecules in the outer lipid layer (forming glycolipid). No membrane carbohydrate is located at the interior surface.

Table 7.3 Distribution of lipids in the erythrocyte membrane.

	Interior Lipid Monolayer (%)	*Exterior Lipid Monolayer (%)*
Total	50	50
Sphingomyclin	6	20
Phosphatidyl choline	9	23
Phosphatidyl ethanolamine	25	6
Phosphatidyl serine	10	0
Phosphatidyl inositol	0	0

The oligosaccharide chains of the membrane are formed by various combinations of six principal sugars: *D-galactose, D-mannose, L-fucose, N-acetylneuraminic acie (also called sialic acid), N-acetyl-D-glucosamine and N-acetyl-D-galactosamine* (see chapter 4 and 5 for chemical structures). All of these may be derived from glucose.

Possible functions of membrane carbohydrate

Several roles have been suggested for the present on the outer surface of the plasma membrane. One possibility is that because they are highly hydrophilic, the sugars help to orient the glycoproteins (and glycolipids) in the membrane so that they are kept in contact with the external aqueous environment and are likely either to rotate toward the interior or to diffuse transversely.

Certain plasma transport proteins, hormones, and enzymes are glycoproteins, and in these molecules, carbohydrate is important to physiological activity. It would therefore not be inappropriate to expect that in certain glycoporteins of the plasma membrane the carbohydrate moiety is basic to either enzymic or some other activity.

Surface carbohydrate is clearly responsible for the various human blood types (e.g., ABO types, MN types, etc) and other tissue types. That is, the sugar sequence and the arrangement of the sugar chains in the membranes of blood cells of an individual with type A blood differ from those of an individual with type B blood, and so on. The carbohydrate is responsible for cell type specificity and is therefore fundamental to the specific antigenic properties of cell membranes. These antigenic properties are linked in some manner to the body's immune system and the ability of that system to distinguish between cells that should be present in the organism (i.e., native cells) and foreign cells. Foreign cells (such as bacteria, other microorganisms, transplanted tissue, or transfused blood) may be recognized as foreign because their membrane glycoproteins contain different carbohydrate markers than those present in the individual's own tissues. Such a situation triggers the immune response. In contrast, an individual's won cell membrane carbohydrate organization is recognized as being native (referred to as 'recognition of self') and does not normally trigger an immunological response. Of course, neither does blood transfusion or tissue transplantation if the carbohydrate organization in the membrane of the "donor's" and "recipient's" cells is the same. Cell-specific membrane carbohydrate organization is considered further below in connection with the actions of lectins and antibodies.

Oppenheimer, Roseman, Roth, and others have clearly implicated surface carbohydrate in the adhesion of a cell to its neighbors in a tissue; presumably, the carbohydrate acts as an adhesive maintaining the integrity of the tissue by linking neighboring cells together. **Contact inhibition**, the phenomenon in which cells grown in culture stop dividing when they touch one another (thereby limiting the growth of the population), may possibly be attributable to a mechanism triggered by interaction of carbohydrates on neighboring cells.

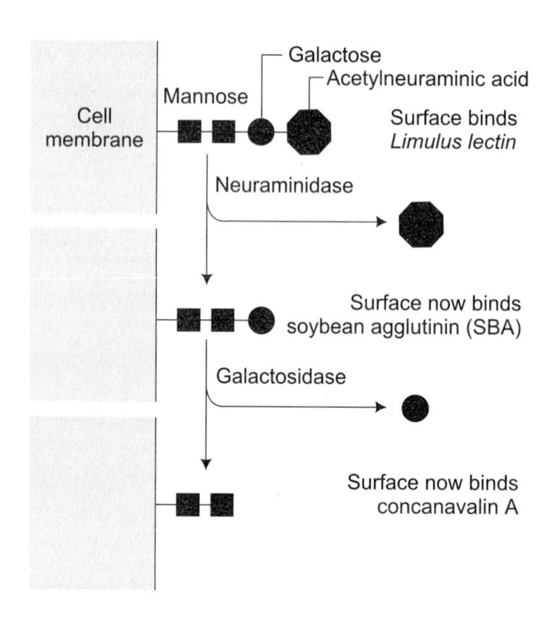

Fig. 7.13 Structure of surface carbohydrate is establlished by successive treatments with lectins and terminal suar-cleaving enzymes. In this illustration, binding of *Limulus* lectin by th cells indicates th prsence of terminal acetylneuraminic acid groups. Thier removal using *neuraminidase* is followed by binding of soybean agglutinin, indicating that the acetylneuraminic acid was linked either to galactose or acetylglucosamine. Which of these two alternative exists is revealed by sensitivity to the specific enzyme added in the next round. In the illustration, it is *galactosidase* that now allows concanavalin A binding, thereby indicating that the teminus was galactose. Removal of galactose and binding of Con A indicates that the next sugar is mannose.

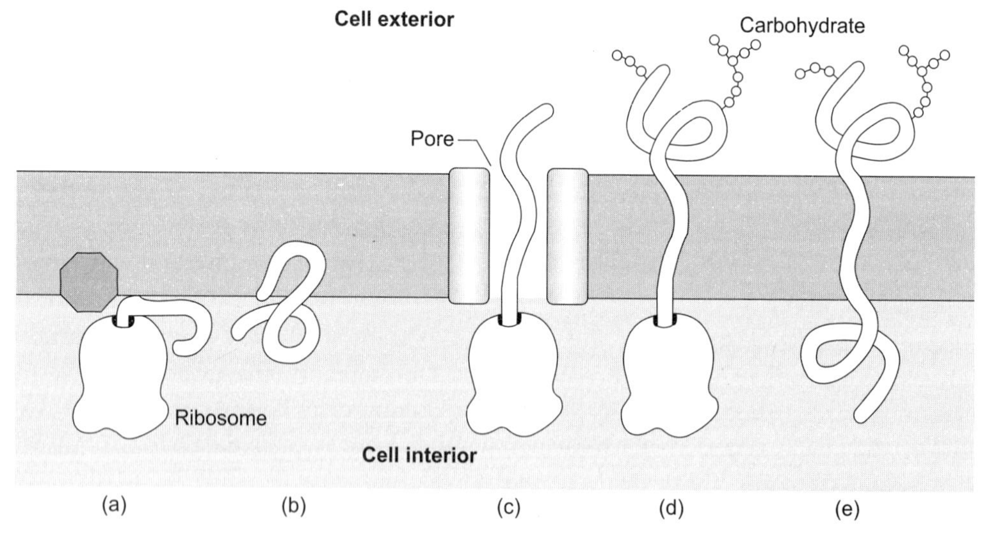

Fig. 7.14 Origin of membrane protein asymmetry, (a and b) Synthesis of interior integral protein. (c, d and e) Synthesis of integral protein spanning the membrane. Colared regions of polypeptide chains represent hydroyphabic sections. Geometric areas of the membrane represent a ribosome attachment site and a channel for protein extrusion. see text for details.

LECTINS, ANTIBODIES, ANTIGENS AND THE PLASMA MEMBRANE

Lectins

Lectins are a special class of proteins (found principally in plants, especially legumes, and also in some invertebrates) that have a high affinity for sugars and combine with them in much the same manner as an enzyme combines with its substrate or an antibody combines with an antigen. Because the interaction of the lectin with sugar is specific (see Table 7.4), lectins can be used to map the distribution of sugar son the cell surface.

Following their discovery in plants some 90 years ago by H. Stillmark, the lectins were for some time called phytohemagglutinins because of their ability to cause the agglutination of red blood cells. However, lectins will aggultinate many kinds of cells, including bacteria. Lectin molecules contain two or more sugar-binding sites, and when large numbers of lectins bind simultaneously to sugars on the surfaces of separate cells (thereby cross-linking the cells), the result is agglutination. It should be noted that binding of the lectin to the cell surface sugars can occur without ensuing agglutination if no cross-linking takes place. The presence and extent of cross-linking is dependent on the balance of lectin concentration and the numbers of surface sugars. In this respect, the lectin-sugar interaction is much like that of an antibody-antigen reaction (see below). However, unlike antibodies, which chemically are very similar proteins (Chapter 4), lectins are of diverse structure, organization, and size.

Lectins will bind free sugars as well as sugars attached to cell membranes. Consequently, lectin-induced cell agglutination can be blocked by preliminary addition of the appropriate free sugar to a suspension of cells.

Lectins have been used to verify that the plasma membranes of malignant cells and normal cells differ. Malignant cells are much more readily agglutinated by lectins than the normal cells from which they are derived; that is, the malignant cells can be caused to agglutinate at much lower lectin concentrations than are required to agglutinate normal cells. It has been found that the increased agglutinability of the malignant cells results from increased glycoprotein mobility in the lipid bilayer of the plasma membrane. Since the malignant cell membrane is more fluid, lectins are able to cluster the glycoproteins in the membrane (i.e., draw them together) and thereby make it possible to form greater numbers of cross-bridges.

Table 7.4 Some lectins and their sugar specificities.

Lectin	*Sugar Specificitiy*
Concanavalin A (Con A) (from jack beans)	D-Mannose
What germ agglutinin (WGA)	*N*-Acetyl-*D*-glucosamine
Ricinis Communis agglutinin (RCA)	D-Galactose
Soybean agglutinin (SBA)	*N*-Acetyl-D-galactosamine
Lima bean lectin	*N*-Acetyl-D-galactosamine
Limulus lectin	*N*-Acetylneuraminic acid

How lectins can be used to determine the composition of the sugar chains of surface carbohydrate is illustrated in Fig. 17.3. The lectins bound by unmodified cell membranes establish the choice of terminal sugar; these may then be cleaved from the carbohydrate using the specific enzyme. The newly exposed terminai sugars are now examined for their lectin-binding characteristics, following which another enzymatic sugar removal is carried out. Repetition of this sequence of treatments progressively reveals the order of sugars.

Antigens and Antibodies

An antigen may be defined as any molecule that has the capacity to stimulate antibody production by the immune system of higher animals. Typically, antigens are glycoproteins in the membranes of the cell or in other particles foreign to the animal. For example, the antigens present in the membranes of bacterial cells or in viruses act to stimulate antibody production by the immune system of the infected animal. The antibodies or immunoglobulin (see chapter 4) produced in response to the presence of the antigen combine with the antigen to form a complex, and this is followed by a series of reactions in which the antigen-bearing agents (e.g., the bacteria) are destroyed.

Antibodies are synthesized by lymphocytes, a subpopulation of white blood cells produced either in the bone marrow (B-lymphocytes) or in the thymus gland (T-lymphocytes). The reaction between antibody and antigen is very specific, a particular antibody combining with only one type of antigen. An enormous variety of B- and T-lymphoctes are present in the body's tissues, each capable of manufacturing only a single antibody type (and therefore capable of reacting with a single type of antigen of foreign cell).

Some of the antibodies manufactured by a lymphocyte are maintained in its plasma membrane. In the presence of the corresponding antigen, a reaction takes place on the surface of the lymphocytes that acts as a stimulus to the surface of the lymphocyte that acts as a stimulus to the production and secretion of additional quantities of antibody. The surface reactions also trigger lymphocyte proliferation, so that even larger quantities of antibody became available. Following the initial reaction on the lymphocyte cell surface, subsequent antigen-antibody reactions may not involve lymphocytes directly. Instead, the secreted antibodies react with either free antigens or more likely with antigens in the invading cells membranes. The involvement of either lymphocyte membranes of foreign cell membranes (or both) in the antigen-antibody reaction makes these molecules especially valuable tools for studying the properties of the cell surface.

It should be clear that the surface antibodies of the lymphocyte of one animal can serve as antigens if these lymphocyte are transferred to the bloodstream of another animal. The transferred lymphocytes will be treated much like any other foreign cell and serve to stimulate the production of anti-immunoglobulin antibodies (AIA). AIA has been especially useful in probing the distribution of glycoproteins in the cell membrane.

Intercellular Junctions and Other Specializations of the Plasma Membrane

The plasma membranes of neighboring cells in a tissue frequently exhibit specialized junction regions believed to play roles in cell-to-cell adhesion and in intercellular transport. The most common of these junctions are (1) **tight junctions** (zonula occludens), (2) **intermediate junctions on belt desmosomes** (also called terminal bars or zonula adherens), (3) **spot desmosomes** (macula adherens), (4) **gap junctions (nexuses),** and (5) **plasmodesmata** (Figs. 7.15 and 7.16).

In tight junctions, the plasma membranes of the neighboring cells fuse at one or more points. The tight junctions generally occur in the same circumferential regions of the cell, so that they give rise to belts of fusion points with neighboring cells. The belt obliterates the intercellular space and acts as a barrier to the flow of materials between the cell surfaces. Internally, the belts of tight junctions are reinforced by a network of fine filaments radiatng into the cytoplasm. The tight junctions between cells are formed by two interdigitating rows of

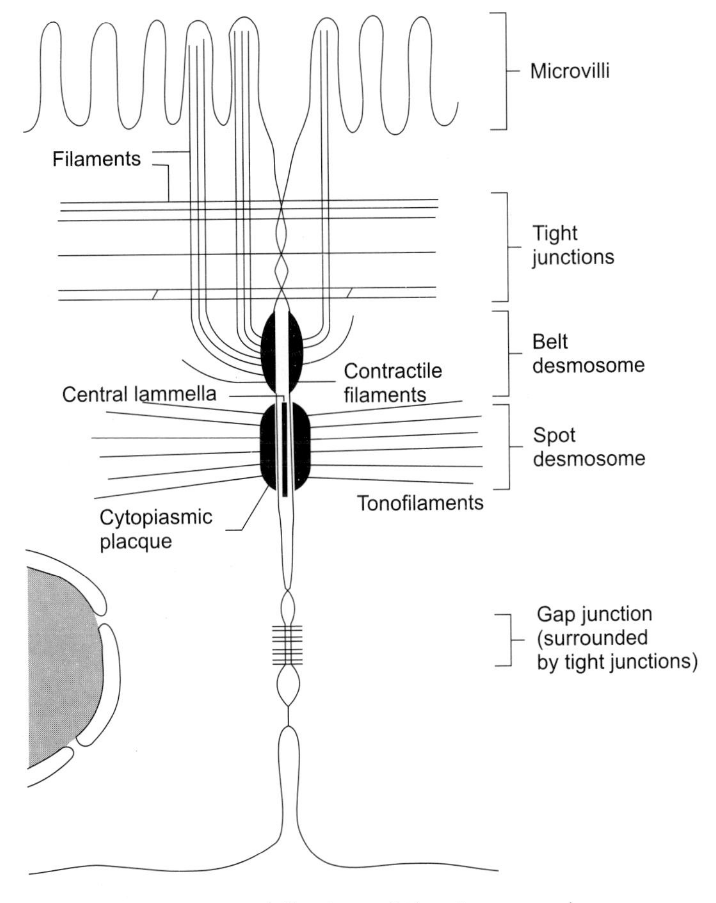

Fig. 7.15 Specializations of the plasma membrane.

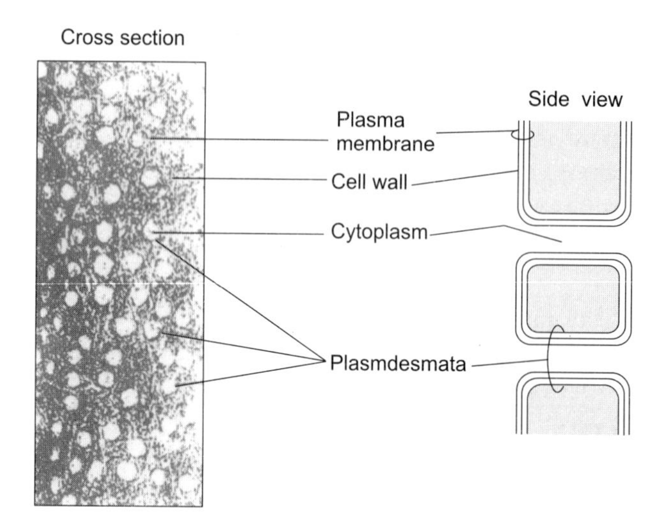

Fig. 7.16 Plasmodesmata.

membrane particles (probably integral proteins), one row contributed by each cell. The number of rows of particles (usually called sealing strands) and the extent to which they interconnect to form a network vary from one type of tissue to another. It is thought that sealing strands also act to deter the movements of other proteins within the membrane. In this way, the differential distribution of certain membrane proteins is maintained, providing for the functional specialization of different faces of the cell (see earlier).

Intermediate junctions, or belt desomosomes, are girdles of contractile filaments (i.e., they contain actin) attached to the interior surface of he plasma membrane. The girdle of contractile filaments interweaves with another web of filaments the extends to the microvilli (Fig. 7.15).

Unlike tight junctions and belt desmosomes, spot desmosomes (macula adherens) do not form a belt around the cell. Instead they are discrete, button like attachment points scattered over the opposing membrane surfaces. In the region of the spot desmosome the adjacent cell membranes are strictly parallel, somewhat thicker, and separated by an intracellular space of about 300 A. This gap characteristically is granular in appearance and contains a central dense band (the central lamella). The cytoplasm adjacent to the plasma membrane is divided into two regions: a lucid zone that lies immediately next to membrane and a neighboring dense band called the cytoplasmic placque. Microfilaments called tonofilaments arise in this region, radiate into the cell, and may be linked to other spot desmosomes. Spot desmosomes are believed to be the strongest points of attachment between neighboring cells.

In plant tissues, the cytoplasm of neighboring cells may be connected through numerous narrow channels that peneirate the fibrous cell wall separating the cells. The channels, called plasmodesmata, are formed by extensions of the plasma membranes of the cells (Fig. 15-32), and are much larger than the channels of the nexus. Plasmodesmata provide for the direct exchange of materials between neighboring cells in the tissue.

Passive Movements of Materials through Cell Membranes

In this section, we consider some of the fundamental principles the govern the passive movements of water, ions, and various other molecules through cell membranes. The term passive is intended to denote that the movement of the substance through the membrane is not associated with any chemical or metabolic activities in the membrane. Passage through the membrane in such factors as the concentration gradient across the membrane and the chemical and physical relationships between the membrane and substances inside and outside the cell. Later, we will direct our attention to movements though the membrane that are accompanied by chemical changes metabolic activity, or gross molecular rearrangements within the membrane itself.

Osmosis and Diffusion across Membranes

Substances that are able to pass through membranes are said to be permeable to the membrane. Nearly all plasma membranes are **permeable** to water. If water (or some other solvent) is the only substances that can pass through the membrane, the membrane is said to be **semi-permeable.** Membranes that display a gradation of permeability to water and dissolved solutes (i.e., membranes that permit water to pass through more readily than salts, sugars, etc,) are said to be **selectively permeable.**

Water molecules are continuously moving into the out of the cell through the plasma membrane. Such movements are generally not discernible as changes in cell size or shape because the flux in each direction is the same. When the concentrations of solutes inside and outside the cell differ, the water flux in one direction may be greater than in the other directions, and the cell may swell or shrink. Water moves from a region of low solute concentration to one of higher solute concentration in order to establish concentration equilibrium. The movement of water (or some other solvent) in response to such a solute concentration gradient is known as **osmosis**.

Osmosis may readily be demonstrated using an artificial membrane such as cellophane, which is permeable to water small molecules such as salts, sugars and amino acids but is impermeable to larger molecules such as proteins. If a cellophane bag filled with a concentrated salt solution is connected to a length of vertical glass tubing and is then immersed in container of distilled water, water will pass into the bag by osmosis. The entry of water into the cellophane bag will cause the water level of the glass tubing to rise (Fig. 7.16). Salts permeate cellophane membranes more slowly than water, so that some time elapses before the salt molecules pass out of the bag into the surrounding water. The movement of solute molecules (in this case, salt molecules) from a region of high concentration 9 inside the bag) to one of lower concentration (outside the bag) occurs by the process of **diffusion.** In the case illustrated in Fig. 7.17, the initial movement of water into the cellophane bag is followed by the outward diffusion of alt from the bag. As the solute concentration inside the bag decreases, the liquid level in the glass tubing again falls as water molecules leave the bag by osmosis. The movements of salt and water molecules continue until the salt concentrations inside and outside the bag are equal.

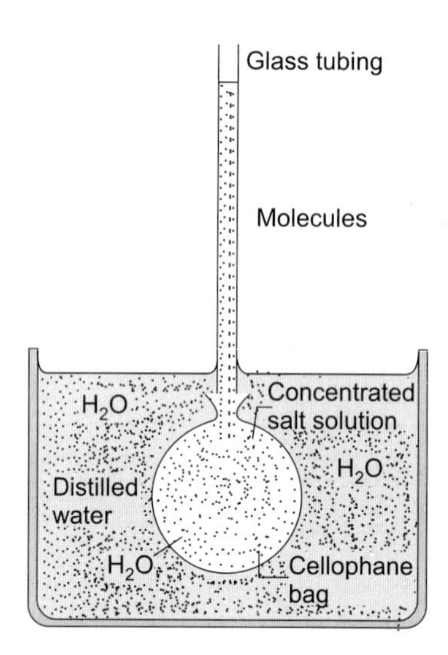

Fig. 7.17 Osmosis of water into a cellophane bag containing a conscentrated salt solution. Because the solute cncentration inside the bag is greater than outside the bag (i.e.,the solute concentration outside the bag is zero), water enters through the cellophane membrane by osmosis. The indlux of water causes the water level in the attached glass tubing to rise.

Consider a case in which the cellophane bag is filled with a solution containing an impermeable solute. As in the previous instance, water will enter the bag by osmosis, causing the liquid level in the glass tubing to rise. Since the solute is impermeable, it cannot diffuse from the bag, and concentration equilibrium across the membrane cannot be achieved. consequently, water will continue to enter the bag and rise in the glass tubing until a height is reached at which the pressure at the base of the water column is just great enough to prevent any further water will remain at this level indefinitely. The pressure that is created inside the bag by the impermeable solute and that supports the column of water is called osmotic pressure. Its value can be approximated by measuring is expressed of the water column. Usually, osmotic pressure is expressed in millimeters of mercury (i.e., mm Hg) rather than in inches of water. Devices used to measure osmotic pressure are called **osmometers** (Fig. 7.18).

Osmosis and Diffusion Across Cell Membranes

Cellular phenomena associated with osmosis and diffusion across the plasma membrane are readily demonstrated using red blood cells, sea urchin eggs, or certain plant cells. The plasma in which the red blood cells are normally suspended contains same concentration of impermeable salt (0.15 M NaCI) as the reythrocyte cytoplasm (0.15 M KCl) normal plasma is said to be isotonic to be red cell. If the plasma is dilute with water. Its salt concentration will decrease, and the plasma will become hypotonic to the red blood cell. Any suspending medium containing an impermeable solute concentration that is lower than the corresponding solute concentration in the cells suspended in that medium is considered hyotonic. In the case of hypotonic plasma, water will enter the red cells by osmosis, causing the cells to swell. The same

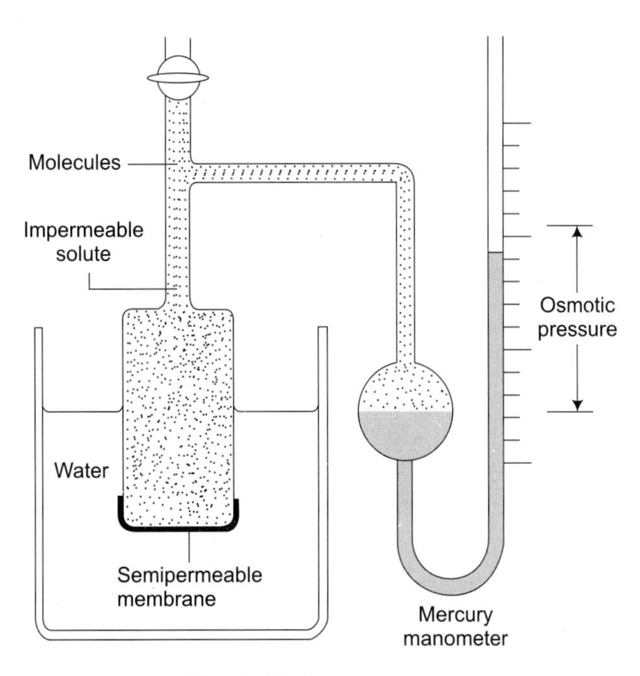

Fig. 7.18 An osmometer.

effect can be produced by placing red blood cells in any hypotonic solution. Just how much water will enter the cell depends upon hoe hypotonic the suspending medium is. For example, if the red cells are suspended in plasma containing one-half the normal salt concentration, water will enter the cells until they swell to twice their original volume. This will reduce the internal salt concentration to one-half its former value bringing the internal and external salt concentrations into equilibrium (Fig. 7.19).

If the cells are suspended in a solution of even greater hypotonicity, then proportionately more water will have to enter the cell to reduce the internal salt concentration to that outside the cell. Obviously, cells can tolerate only a certain amount of swelling before the membrane ruptures. Spilling the cell contents into the surrounding medium; this is called osmotic lysis. Red blood cells lyse when suspended in very dilute salt solutions or in distilled water (Fig. 7.21). In the specific case of the red blood cell, this phenomenon is called hemolysis, since hemoglobin from the red blood cell is released into the suspended medium. Other animal cells behave in a similar manner in appropriately phpotonic media.

Plant cells generally do not lyse even when placed in distilled water because cell swelling is limited by the rather inflexible cellulose cell wall. In hypotonic solutions, plant cells well as water enters the cytoplasmic vacuoles, by osmosis. This forces the cytoplasm to the margins of the cell wall. Under these conditions, the plant tissue becomes turgid (Fig. 7.20).

Equal concentration of impermeable salts and nondissociating (nonionizing) molecules (such as sucrose and other sugars do not exhibit the same osmotic effects. For example, 0.15 M NaCl exerts twice the osmotic pressure as does 0.15 M sucrose. This is because 0.15 M NaCl

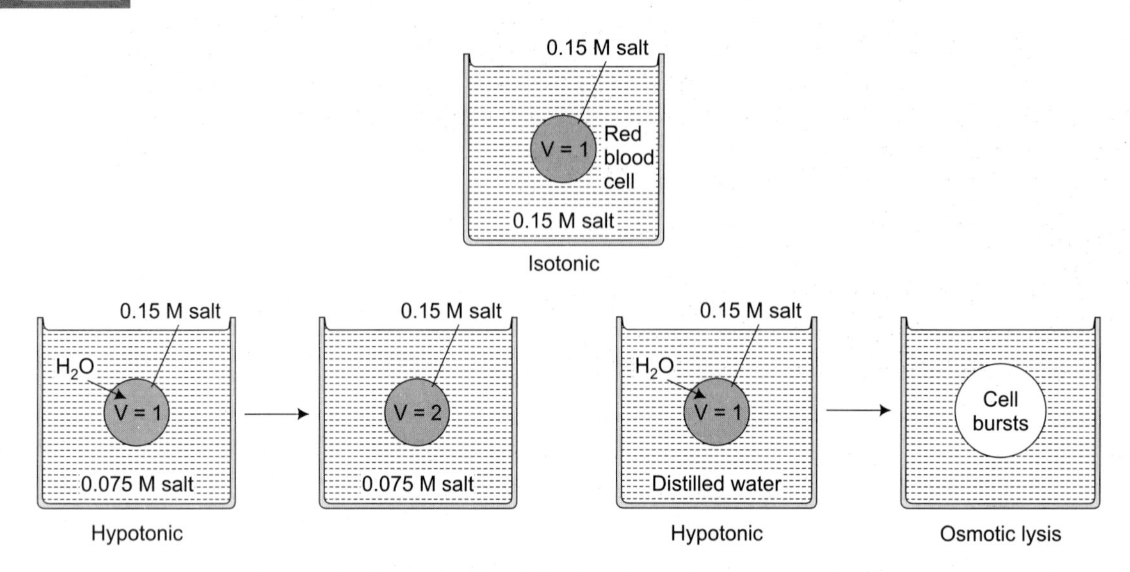

Fig. 7.19 Behavior of red blood ceels in hypotonic solutions.

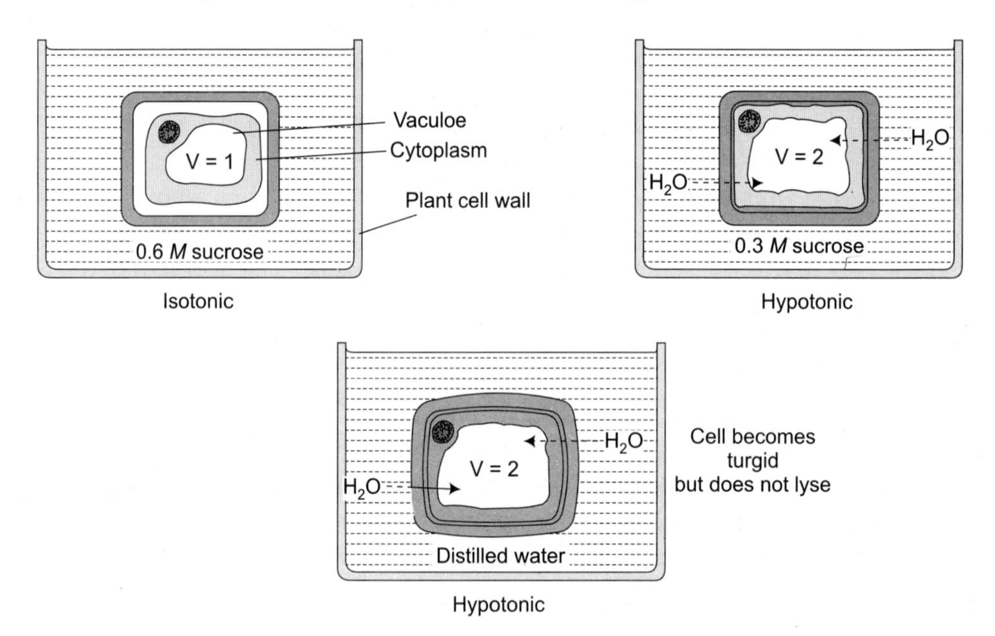

Fig. 7.20 Behavior of plant cells in hypotonic solutions.

undergoes dissociation in water to produce twice the number of particles (i.e., ions) per cubic centimeter as 0.15 M sucrose. Thus, 0.15 M Na Cl, 0.15 M KCl, 0.10 M CaCl, and 0.3 M sucrose would all exert the same osmotic pressure, since they produce the same particle concentrations in water.

Sucrose is impermeable to most cell membranes, including the membrane of the red blood cell. Therefore, both 0.15 M NaCI and 0.3 M sucrose are isotonic to red blood cells. Solutions that contain higher concentrations of impermeable solute than are found inside cells are said to be hypertonic (Fig. 7.21). Water moves by osmosis from the cells into the medium until the concentrations of solute inside and outside the cell are the same. The shrinkage associated with such water loss is called **crenation.** Other animal cells behave in a similar manner.

When plant cells are placed in hypertonic media, they undergo plasma lysis; that is water passes from the cyto.

The Gibbs-Donnan Effect

Proteins behave like ions because the k. groups of their amino acids may bear positive or negative charges. The net ionic charges of the protein molecule depend on the relative numbers of positive and negative side groups. Most soluble proteins behave like anions because they more negative than positive sites. Unlike many others ions, proteins are generally too large to permeate cell membranes. The Gibbs-Donnan effect (after J.W. Gibbs and F.G. Donnan) describes the effect the protein ions have on the equilibrium distributions of small ions across a selectively permeable membrane. When a selectively permeable membrane separates an electrolyte solution containing proteins from one that lacks proteins, the concentrations reached at equilibrium for each permeable ionic species will not be the same on both sides of the having the same sign as the protein will be lower on the side of the membrane containing the protein (i.e., inside the cell), while the concentration of small ions of opposite sign will be higher. The Gibbs-donnan effect does not influences the equilibrium distributions of nonionized substances such as glucose, urea and the like.

Active Transport

During diffusion (passive or facilitated), substances pass through the plasma membrane until some sort of equilibrium is achieved. The equilibrium may be of the Gibbs-Donnan variety or may be simple concentration equilibrium. Both involve an inter play between the concentration of soluble solute inside and outside the cell. Cells can also accumulate solutes in quantities far in excess of that expected by any of the above mechanisms if the solute is rendered insoluble

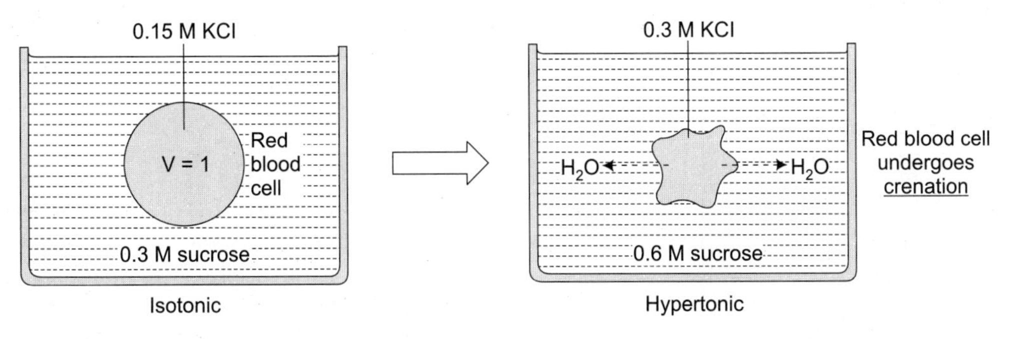

Fig. 7.21 Crenation of red blood cells in hypertonic solutions.

once it has entered the cell, since insoluble materials do not contribute to concentration gradients. Alternatively, once inside the cell, a solute may enter a metabolic pathway and be chemically altered, thereby reducing the concentration of that particular solute and allowing additional solute permeation. In all the cases we have so far considered, solute permeation of the of the membrane hinges on the presence of a concentration gradient, with the solute moving in the direction of the gradient.

Substances can also move through the plasma membrane into or out of the cell against a concentration gradient. This requires the expenditure of energy on the part of the cell, and is called active transport. Active transport ceases when cells are (1) cooled to very low temperatures (such as 2-4 C), (2) treated with metabolic poisons such as cyanide or iodoacetic acid, or (3) deprived of a source of energy. The best understood and most exhaustively studied instances of active transport are those that involve the movements of sodium and potassium ions across the plasma membranes of erythrocytes, nerve cells, and *Nitella* cells and that result in an ionic concentration gradient across the cell membrane. The mechanism that establishes and maintains these gradients appears to be basically similar in all of those cells and can be illustrated with the erythrocyte. The formation of cytoplasmic vesicles from the plasma membrane and the consequent entrapment within these vesicles of materials formerly in the cell surroundings is called endocytosis. Several different kinds of endocytosis have been described including pinocytosis, rhopheocytosis, and phagocytosis. Movements of materials from the cell into its surroundings by the fusion of of cytoplasmic vesicles with the plasma membrane constitute exocytosis. As seen in Fig. 7.23, endocytosis and exocytosis are variations of the same fundamental phenomenon, different from one another in the *direction* of bulk solute movement.

Endocytosis

Pinocytosis. Using time-lapse photography to study tissue culture cells, W.H.Lewis in 1931 described a phenomenon in which small amount of culture medium were trapped in invaginations of the plasma membrane and then pinched off to form small cytoplasmic vesicles. Since the entire process appeared much like some form of organized cell drinking, Lewis termed the phenomenon pinocytosis (pinos means "I drink" in greek). Lewis' observations with tissue culture cells were confirmed in 1934 by S.O.Mast and W.L.Doyle studying amoebae in which pinocytosis is readily observed with the light microscope. Using electron microscopy, it became

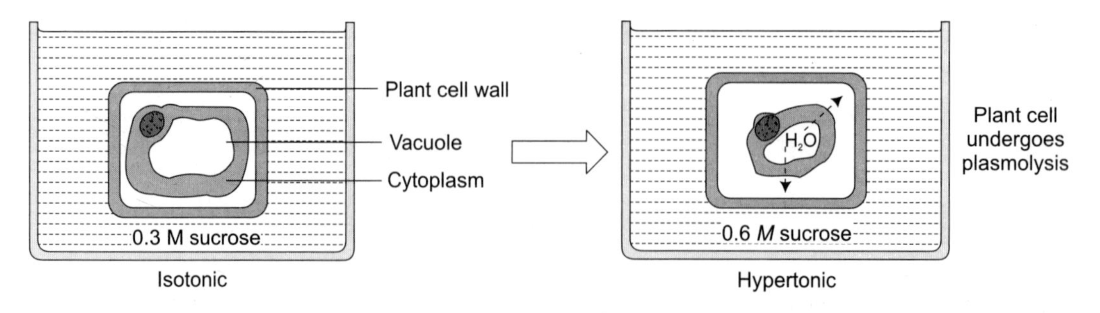

Fig. 7.22 Plasmolysis of plant cells in hypertonic solution.

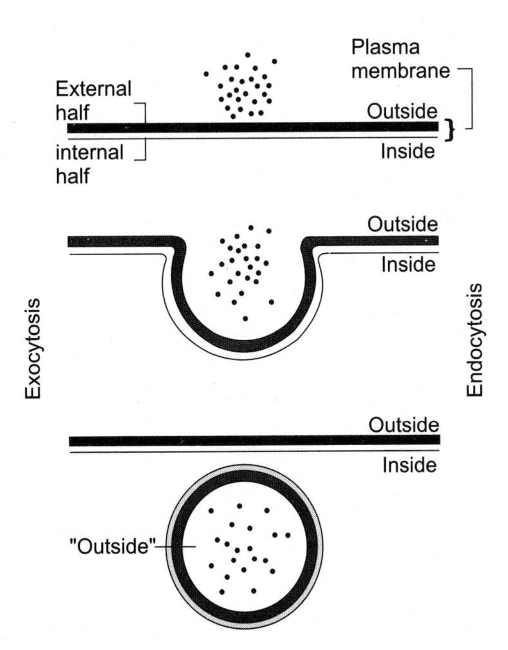

Fig. 7.23 Relationship of external and internal halvos of the plasma membrane to the inner and outer faces of vesicles in exocy, osis and endocytasis.

clear in the 1950's that pinocytosis is a common phenomenon occurring at intervals in many different kinds of cells including leukocytes, kidney cells, intestinal epithelium, liver macrophages, and plant root cells.

Pinocytosis is induced by the presence of appropriate concentrations of proteins, amino acids, or certain ions in the medium surrounding the cell. The first step in the process involves a simple binding of the inducer substance to specific receptor sites on the cell membrane. This is followed by invagination of the membrane to form either small pinocytic vesicles or channels (Fig. 7.24). Although binding of the inducer is not inhibited by cyanide or low temperature, the formation of pinocytic vesicles is, and it is therefore dependent on cell metabolism. Formation of pinocytic and other endocytic vesicles, is believed to involve the contractions of intracellular microfilaments whose ends are anchored in the plasma membrane.

Pinocytic vesicles (which usually are less than 1 nm in diameter) detach from the cell membrane and migrate toward the interior of the cell. In the cell interior, these may fragment into smaller vesicles or coalesce to form larger ones. Unless the vesicles are "tagged" by inducing pinocytosis in the presence of radioactive tracers, they soon become indistinguishable from other vacuoles in the cell.

Although pinocytosis is induced by the presence of specific substances in the cell surroundings, other materials are also enclosed by the pinocytic vesicles including water, salts, and so on. These substances, together with the inducer molecules, may enter the cytosol from the vesicle by diffusion, active transport, or related transport mechanisms.

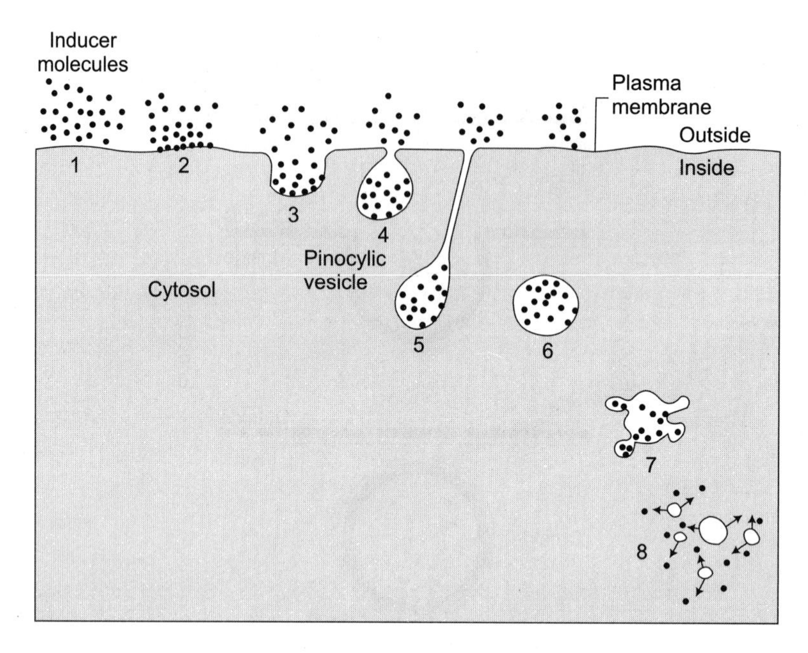

Fig. 7.24 Stages of pinocytosis: 1-2, binding of Inducer molecules to plasma membrane: 3-5, invagination of the membrane: 6-8, detachment from plasma membrane and fragmentation into smaller vesicles.

Rhopheocytosis

Rhopheocytosis is a bulk transport mechanism in which small quantities of cytoplasm, together with their inclusions, are transferred from one cell to another. Rhopheocytosis was first demonstrated in bone marrow tissues by M.Bessis. In the marrow, maturing red blood cells (erythroblasts) are attached to reticuleondothelial cells to form large numbers of erythroblast islands (Fig. 7.25). The reticuleondothelial cells of the marrow contains large quantities of iron derived from the breakdown of hemoglobin from red blood cells. In the reticuleondotheial cell, hemoglobin iron is converted to ferritin-a high molecular-weight protein containing up to 23% iron by weight. Because of their high density, clusters of ferritis in reticuloendothelial cells are readily identified with the electron microscope, which has therefore been employed to study the fate of these molecules. During the maturation of red blood cells, ferritin granules, together with small amounts of cytoplasm, are transferred from reticuloendothelial cells to erythroblasts by the simultaneous invagination and evagination of their adjacent plasma membranes. Once inside the erythroblast, ferritin iron (and iron derived directly from the circulating blood plasma) passed into the cytosol and is used in the synthesis of new hemoglobin molecules.

Phagocytosis

Phagocytosis which was first described by E.Metchnikoff in the late nineteenth century is similar to pinocytosis but involves the engulfment of much larger quantities of particulate material. For example, entire ciliates, rotifers or other microscopic organisms may be phagocytosed by an

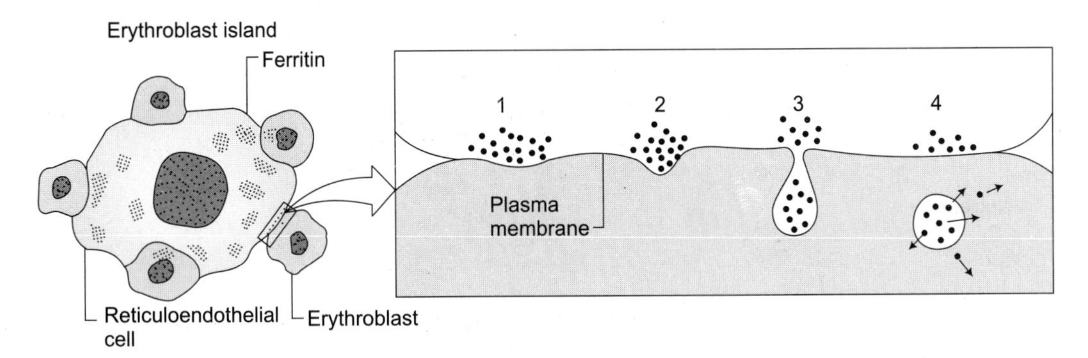

Fig. 7.25 Rhopheocytosis in the bone marrow: stages 1-4 show successive evagination and invagination of abjacent plasma membranes transferring ferritin and some cytplasm from the reticuloendothelial cell to the erythroblasi.

amoeba and enclosed within one or more vacuoles called phagosomes. *Food vacuoles*, or *food cups* (see Fig. 7.26). During phagocytosis, the "prey" may be temporarily immobilized by secretions from the phagocytic cell. The phagocytosis of ciliates by amoebae is characterized by the flowing of the amoeba's cytoplasm into footlike projections (pseudopodia) that gradually encircle and fully encapsulate the ciliate. Using a similar mechanism, certain white blood cells phagocytose hundreds of bacteria. The removal and destruction of old red blood cells in the liver, spleen, and bone marrow by reticuloendothelial cells in these organs also occur by phagocytosis. Following phagocytosis, the phagosomes fuse with primary lysosomes digest the engulfed material, converting it to a form that may be transported across the vacuolar membranes.

Exocytosis

Exocytosis is the mechanism by which large quantities of material enclosed within a cell vacuole are transferred to the cell surroundings by fusion of the vacuole with the plasma membrane. In a sense, the process is the reverse of pinocytosis or phagocytosis, the contents of the vacuole being emptied into the extra cellular space. The best understood form of exocytosis is secretion. When the secretory vesicle touches the plasma membrane, lipids in both membranes are moved aside, making the membranes more fluid. After the vesicle's contents have been discharged to the outside, the vesicle membrane is incorporated into the plasma membrane.

SUMMARY

The plasma membrane delimits the cell and activity participates in the movement of materials between the cell and its surroundings. Early studies on the chemistry of the plasma membrane focused upon the content and organization of its lipids resulting in the bimolecular lipid leaflet model. This concept was subsequently modified to account for the membrane proteins, resulting first in the unit membrane model and more recently in the fluid mosaic model. The latter model appears to be more consistent with biochemical studies and electron microscopic analyses of plasma membranes. Both the proteins and lipids of the membrane are asymmetrically

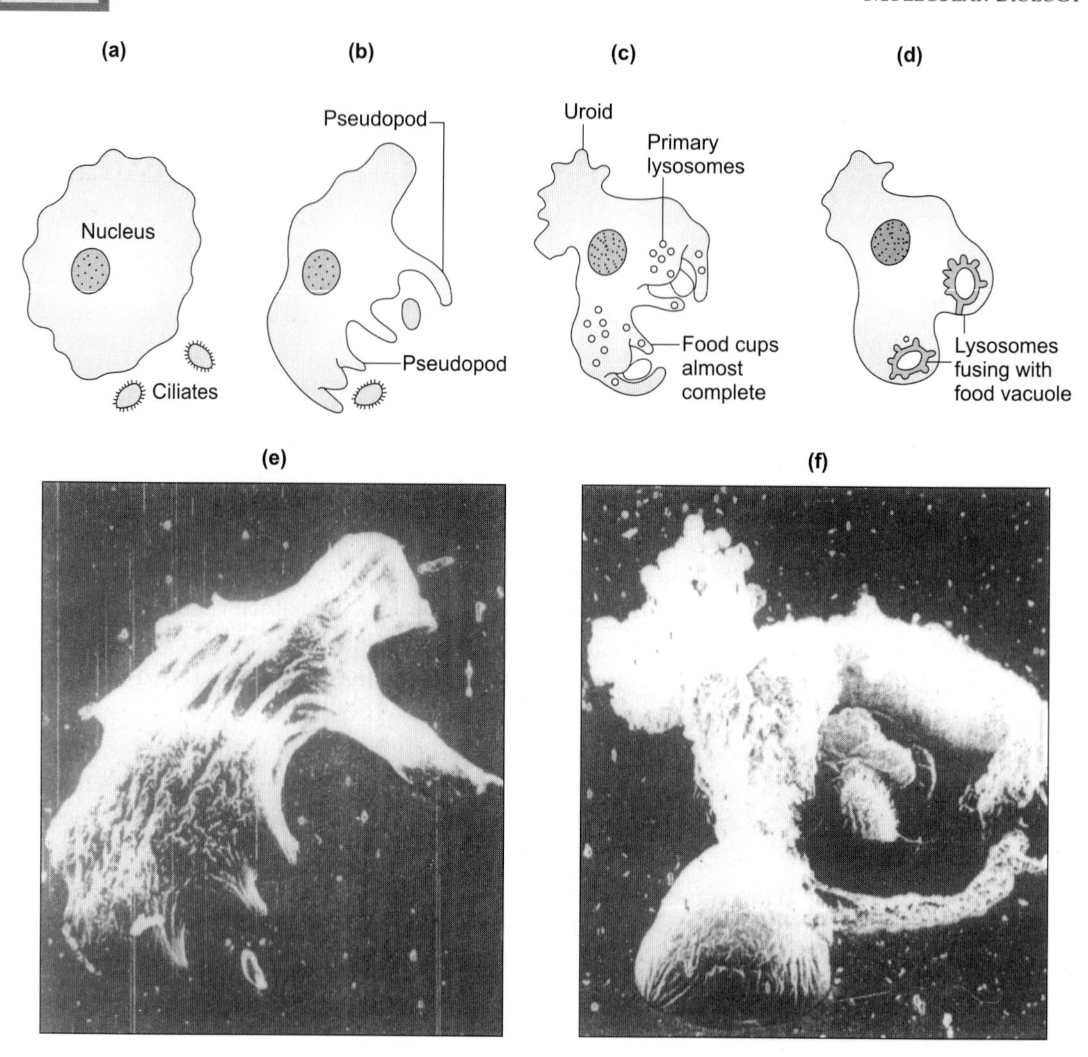

Fig. 7.26 Phagocytosis. In the presence of phagocytosable material (a), the cell forms pseudopodia (b) that entrap line prey in vacuoles (c and d). The fustion of primary lysosomes with these vacuoles is followed by digestion of the prey. Scanning electron micrographos are an amoeba prior to (e) and during (f) phagocytosis and correspond to stages b and d of the diagram (Photomicrograph kindly provided by Dr. K.W. Jeon).

distributed across the inner and outer halves. In some cells, the proteins in the inner half of the membrane may be anchored in position by a cytoskeletal network. Carbohydrate chains associated with protein (and lipid) on the outer membrane surface play roles in maintaining membrane organization, in transmembrane transport, in cell-to-cell adhesion, and in providing membrane antigenic properties. In tissues, the plasma membranes of neighboring cells exhibits specialized junction regions including tight junctions, desmosomes, gap junctions, and plasmodesmata. Such junctions are important maintaining tissue integrity and in regulating the passage of substances across a tissue and from cell to cell.

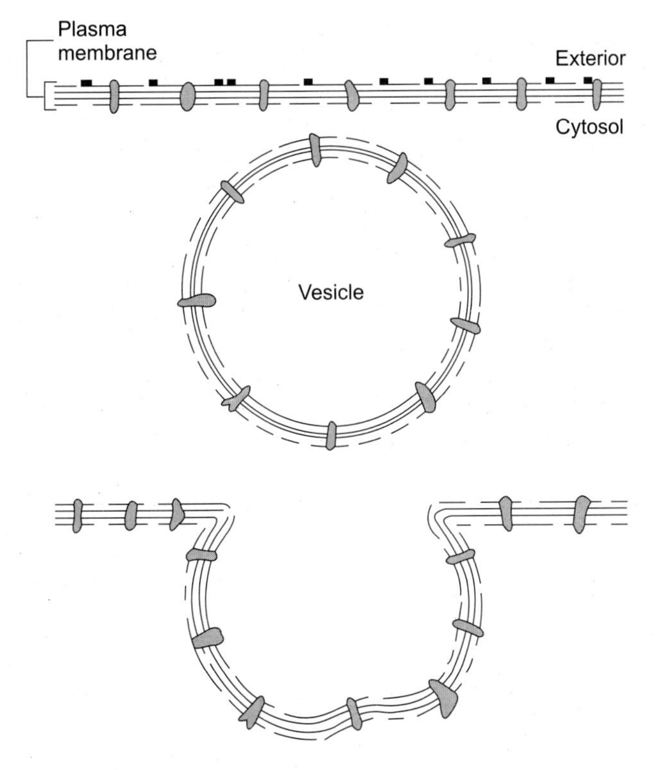

Fig. 7.27 Fusion of exocytic vesicle with the plasma membrane.

The movement of materials across the cell membrane may be achieved by passive and active mechanisms. The passive movements of solutes (by diffusion) and water (by osmosis) across the membrane occur principally as the result of concentration gradients. In some cases, diffusion through the membrane is facilitated by carrier molecules processing the properties of enzymes. Substances can also move through the plasma membrane against a concentration gradient; this consumes cellular energy (i.e., ATP is hydrolyzed in the process) and is termed active transport. For individual ions and small molecules, active transport for individual ions and small molecules, active transport is effected by membrane-associated enzymes acting as pumps; however, transport into and out of cells (called endocytosis and excytosis) can occur in bulk through gross movements of the plasma membrane and takes the form of pinocytosis, rhophcocytosis, and phagocytosis.

PLASMA MEMBRANE-SYNOPSIS

Plasma membranes are remarkably thin, delicate structures, yet they playa key role in many of a cell's most important functions. The plasma membrane separates the living cell from its environment; it provides a selectively permeable barrier that allows the exchanges of certain substances, while preventing the passage of others; it contains the machinery that physically transports substances from one side of the membrane to another; it contains receptors that bind to specific ligands in the external external space and relay the information to the cell's internal compartments; it mediates interactions with other cells; it provides a framework in which components can be organized; it is a site where energy it transuded from one type to another.

Membranes are lipid protein assemblies in which the components are held together in a thin sheet by noncovalent bonds. The membrane is held together as a cohesive sheet by a lipid bilayer consisting of a bimolecular layer of amphipathic lipids whose polar head groups face outward and hydrophobic fatty acyl tails face inward. Included among the lipids are phosphoglycerides, such as phospholipids sphingomyclin and the carbohydrate-containing cerebrosides and gangliosides (glycolipids); and cholesterol. The proteins of the membrane can be divided into three groups; integral proteins that penetrate into and through the lipid bilayer with portions exposed on both the cytoplasmic and extracellular membrane surfaces; peripheral proteins that are present wholly outside that lipid bilayer, but are noncovalently associated with either the polar head groups of the lipid bilayer or with the surface of an integral protein; and lipid-anchored proteins that are outside the lipid bilayer but covalently linked to a lipid that is part of the bilayer. The transmembrane segments of integral proteins occur typically as an ∝helix, which may be predominantly hydrophobic if they simply span the bilayer or amphipathic if they line an internal aqueous channel.

Membranes are highly asymmetric structures whose two leaflets have very different properties. As examples, all of the membrane's carbohydrate chains face away from the cytosol; many of the integral proteins bear sites on their extracellular surface that interact with extracellular ligands and sites on their internal surface that interact with peripheral proteins that form part of an inner membrane skeleton; and the phospholipids content of the two halves of the bilayer is highly asymmetric. The organization of the proteins within the membrane is best revealed in freeze-fracture replicas in which the cells are frozen, their membranes are split through the center of the bilayer by a fracture plane, and the exposed internal faces are visualized by formation of a metal replica.

The physical state of the lipid bilayer has important consequences for the lateral mobility of both phospholipids and integral proteins. The viscosity of the bilayer and the temperature at which it undergoes phase transition depends upon the degree of unsaturation and the length of the fatty acyl chains of the phospholipids. Maintaining a fluid membraneis important for many cellular activities, including signal transduction, cell division, and formation of specialized membrane regions. The lateral diffusion of proteins within the membrane was originally demonstrated by cell fusion and can be quantitated by techniques that follow the movements of proteins tagged with fluorescent compounds or electron-dense markers. Measurement of the diffusion coefficients of integral proteins suggests that most are subject to restraining influences that inhibit their mobility. Proteins may be restrained by association with other integral proteins or with peripheral proteins located on either membrane surface. Because of these various types of restraint, membranes are able to achieve a considerable measure of organizational stability in which particular membrane regions are differentiated from one another.

The erythrocyte plasma membrane contains two major integral proteins, band 3 and glycophorin, and a well-defined inner skeleton composed or peripheral proteins. Each band 3 subunit spans the membrane at least a dozen times and contains an internal channel through which bicarbonate and chloride anions are exchanged. Glycophorin is a heavily glycosylated protein of unknown function containing a single transmembrane domain consisting of a hydrophobic helix. The major component of the membrane skeleton is the fibrous protein spectrin, which interacts with other peripheral proteins such as ankyrin to provide support for the membrane and restrain the diffusion of its integral proteins.

The plasma membrane is a selectively permeable barrier that allows solute passage by several mechanisms, including simple diffusion through either the lipid bilayer or membrane channels, facilitated diffusion, and active transport. Diffusion is an energy-independent process in which a solute moves down and electrochemical gradient, dissipating the free energy stored in the gradient. Small inorganic solutes, such as O_2, CO_2, and H_2O, penetrate the lipid bilayer readily, as do solutes with large partition coefficients (high lipid solubility). Ions and polar organic solutes, such a sugars and amino acids, require special transporters to enter or leave the cell.

Water moves by osmosis directly through the lipid bilayer of a semipermeable membrane from a region of lower solute concentration (the hypotonic compartment) to a region of higher solute concentration (the hypertonic compartment). Osmosis plays a key role in a multitude of physiological activities. In plants, for example, the influx of water generates turgor pressure against the cell wall that helps support nonwoody tissues. Ions diffuse through a plasma membrane by means of special protein-lined channels that are often specific for particular ions. Ion channels are usually gated and controlled by either voltage or chemical ligands, such as neurotransmitters.

Facilitated diffusion and active transport involves integral membrane proteins that combine specifically with the solute to be transported. Facilitative transporters act without the input of energy and are capable of moving solutes down a concentration gradient in either direction across the membrane. They are thought to act by changind conformation, which alternately exposes the solute binding site to the two sides of the membrane. The glucose transporter is a facilitative transporter whose presence in the plasma membrane is stimulated by increasing levels of insulin. Active transporters require the input of energy and move ions and solutes against a concentration gradient. P-type active transporters, such as Na+K + ATPase, are driven by the transfer of a phosphate group from ATP to the transporter, changing its affinity toward the transported ion. Secondary active transport a second solute against a gradient. For example, the active transport of glucose across the apical surface of an intestinal epithelial cell is driven by the co-transport of Na^+ down its electrochemical gradient.

The resting potential across the plasma membrane is due largely to the limited permeability of the membrane of K^+ and is subject to the dramatic change. The resting potential of a typical nerve or muscle is about-70 mV (inside negative). When the membrane of an excitable cell is depolarized past a threshold value, events are initiated that lead to the opening of the gated Na^+

channels and the influx of Na^+, which is measured as a reversal in a voltage across the membrane. Within milliseconds after they have opened, the Na^+ gates close and gated potassium channels open, which leads to an efflux of K^+ and a restoration of the resting potential. The series of dramatic changes in membrane potential following depolarization constitutes an action potential.

Once an action has been initiated, it becomes self-propagating. Propagation occurs because the depolarization that accompanies an action potential at one site on the membrane is sufficient to depolarize the adjacent membrane, which initiates an action potential at that site. In a myelinated axon, an action potential at one node in the sheath is able to depolarize the membrane at the next node, allowing the action potential to hop rapidly from node to node. When the action potential reaches the terminal knobs of an axon, the calcium gates in the plasma membrane open, allowing an influx of Ca^{2+}, which triggers the fusion of the membranes of neurotransmitter-containing secretory vesicles with the overlying plasma membrane. The neurotransmitter diffuses across the synaptic cleft where it binds to receptors on the postsynaptic membrane, including either the depolarization or hyperpolarization of the target cell.

CYSTIC FIBROSIS: THE CLINICAL SIGNIFICANCE OF MEMBRANE TRANSPORT

On the average, 1 out of every 25 persons of Northern European descent carries one copy of the gene that causes cystic fibrosis. These people, in other words, are heterozygous at this genetic locus. Since they show no symptoms of the mutant gene, most heterozygotes are unaware they are carriers (unless their DNA has been subjected torecent screening procedures). Consequently, approximately 1 out of every 2500 infants in this Caucasian population ($1/25 \times 1/25 \times 1/4$) are homozygous recessive at this locus and born with cystic fibrosis (CF).

Cystic fibrosis is a disease of abnormal fluid secretion. Although a variety of organs are affected, including the intestine, pancreas, sweat glands, and reproductive tract, the respiratory tract usually. exhibits the most severe effects. Victims of CF produce thickened, sticky mucus that is vary hard to propel out of the airways leading from the lungs. As a result, these individuals typically suffer from chronic lung infections, which progressively destroy pulmonary function.

In 1984, cells cultured from CF patients were shown to exhibit an abnormally low efflux of chloride ions, suggesting that the defect may lie in a gene encoding an ion channel or transporter. Since the movements of water out of epithelial cells by osmosis is directly affected by the movement of salts, it is not unexpected that a decrease in chloride efflux would lead to an increase in the concentration and hence viscosity of bodily secretions.

The gene responsible for cystic fibrosis was isolated in 1989. For the first time, victims of cystic fibrosis saw a glimmer of light at the end of a long, dark tunnel; there was hope that a cure for their crippling disease might be developed. Once the sequence of the CF gene was determined and the amino acid sequence of the corresponding polypeptide was deduced, it was apparent that the polypeptide was an ABC transporter. Like the other transporters in this superfamily, the polypeptide contains two domains situated within the lipid bilayer each of which contains six membrane spanning segments, and two nucleotide-binding domains (NBDs) that project into the cytoplasm. Unlike the other membranes of the superfamily, the protein involved in CF includes a regulatory (R) domain containing several serine residues that can be phosphorylated by a proteinkinase (called PKA) that is activated by the second messenger, cyclic AMP. The protein was named cystic fibrosis transmembrane conductance regulator (CFTR), an ambiguous term that reflected the fact that researchers weren't sure of its precise function, whether it was a chloride channel (unlike other membranes of the superfamily, which were active transporters) or a protein that somehow regulated chloride conductance through other membrane channel. The problem was solved only after the protein was solved only after the protein was purified, incorporated into artificial lipid bilayers, and shown to act as a cyclic AMP-regulated chloride channel. The opening of the gate apparently requires two independent events: the cyclic AMP dependent phosphorylationof the R domain and the binding of ATP to the nucleotide-binding domains The reason for dual control over CI- conductance is unclear.

In the past few years, researchers have isolated approximately 200 different mutations that give rise to cystic fibrosis, and the effect of each of these alterations on protein structure and function has been studies As a result, more is now known about the molecular basis of the disease than the physiological basis for the accompanying tissue and organ pathology. Seventhly percent of the alleles responsible for cystic fibrosis in the United States contain the same genetic alteration (designated 508) they are all missing three base pairs of DNA that encode a phenylalanine at the 508th position within one of the nucleotide-binding domains of the CFTR polypeptide. Subsequent research has revealed that CFTR polypeptides lacking this particular amino acid fail to be processed normally within the membranes of the ER and Golgi complex and, in fact, never reach the surface of epithelial cells. As a result, FC patients that are homozygous for the A508 allele completely lack the CFTR chloride channel and have severe form of the disease. Other CF patients with less severe forms have mutant alleles that encode a CFTR that is able to reach the surface of cells, but mediates a reduced chloride conductance. The mildest forms are characterized by infertility, with little or no damage to major organs.

Ever since the isolation of the gene responsible for CF, the development of a cure by gene therapy-that is, by replacement of the defective gene with a normal version-has been a major goal of CF researchers. Cystic fibrosis is a good candidate for gene therapy because the worst symptoms of the disease are caused by cells that line the airways and, therefore, are accessible to substances that can be delivered by inhalation of an aerosol. To date, clinical trials have begun using two different types of delivery systems. In one set of trials, the normal CFTR gene is incorporated into the DNA of a defective adenovirus, a type of virus that normally causes upper respiratory tract infections. The recombinant virus particles are then allowed to infect the cells of the airway, delivering the normal gene to the genetically deficient cells. In the other trials, the DNA encoding the normal CFTR gene is contained within liposomes When introduced into the lungs and airways, the liposomes fuse with the plasma membranes of the epithelial cells, delivering their DNA contents into the cytoplasm. To date, both approaches have met with success in temporarily correcting the genetic defect in cells of the respiratory tract. Liposomes may have the advantage over viruses as delivery system in that they are less likely to stimulate a destructive immune response within the patient following repeated treatments.

CHAPTER 8

The Cell Nucleus

The structure and function of the nucleus are discussed in this chapter. Our present knowledge of nuclear structures and their functions has been derived principally through the application of modern techniques, especially electron microscopy and radioactive tracer methods. The functions of the nucleus are primarily concerned with the replication and transcription of its nucleic acids, which ultimately results in regulatory control of the cell by proteins.

The nucleus of the eucaryotic cell is delimited by a pair of membranes called the nuclear envelope. The outer and inner membrances of this envelope are continuous with each other only around the margins of large pores, which penetrate both membranes. Elsewhere, the two membranes are separated by an intermembrane space. Ribosomes may be attached to the outer surface (or cytosol side) of the outer nuclear membrane (which also may fold out into the cytoplasm), but ribosomes are not considered to be nuclear structures.

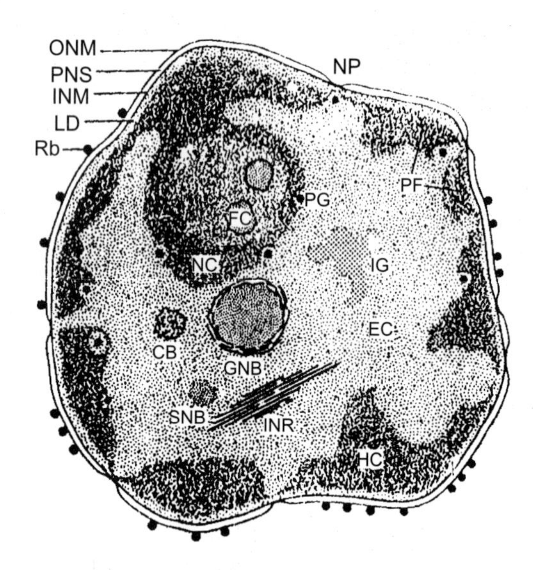

Fig. 8.1 Diagram of a section through the nucleus, locating the main structures and defined regions. CB = colled body, EC = euchromatin, FC = fibrillar canter of nucleolus, GNB = granular nuclear body, HC = heteochromatin, IG = interchromatin granules, INM = inner nuclear membrane, INR = intranuclear rodiet, LD = lamina densa, NC = nucleolus associated chramatin, NP = nuclear pare, ONM = outer nuclear membrane, PF = perinuclear space, Rb = ribosome, SNB = simple nuclear body.

Within the nucleus are a number of structures and defined regions (Fig. 8.1). the structures and their functions are described below.

CHROMATIN

During mitosis and meiosis chromosomes condense by compaction of their constituent DNA and protein. As a result, they are easily stained so that they can be studied during nuclear division. During interphase the chromosomes are not condensed, and it is difficult, if not impossible, to distinguish individual chromosomes. However, there are regions of the nucleus called chromocenters that so stain deeply even during interphase. In 1928, E. Heitz identified these chromocenters as portions of chromosomes that remain condensed throughout the cell cycle.

Chromosome material that stains with basic days is called chromatin. The location and intensity of the staining permits differentiation of two different forms, or states, of chromatin, portions of the chromosomes that stain lightly with basic dyes and are associated with relatively uncondensed chromosomes are termed euchromatin, Darkly staining regions associated with condensed porteions of interphase (and early prophase) chromosomes are called heterochromatin. In additions, their addition, there is usually some condensed chromation around the nucleous, called perinucleolar chromatin, and some inside the nucleolus, called intranucleolar chromatin. The perinucleolar and intranucleolar chromatin. The perinucleolar and intranucleolar chromatin appear to be connected and jointly are referred to as nucleolar chromatin.

Dispersed chromatins are found. Although heterochromatin has been shown to contain very few, if any, structural genes, it has been identified with gene expression. When "euchromatic" genes of a known function in organisms such as *Drosophila* are relocated into a position adjacent to the heterochromatin, their phenotypic expression is modified. The exact role of the heterochromatin is not known, but because it is found in all nuclei, it is presumed to have a basic function in gene expression. During prophase in nuclear that contain few chromosomes, it is possible to identify the heterochromatic regions of specific chromosomes. The heterochomatin is also characterized by its high content of repelitive DNA sequences, and its replication in the cell cycle occurs at a later time than euchromatin. Euchromatin is believed to be the material containing the structural genes and can be expressed when decomposed during interphase. Euchromatin that, temporarily has been physiologically inactivated by condensation so that its gene content cannot be exoressed is some times called facultative heterochromatin, as opposed to perpetually condensed constitutive heterochromatin.

STRUCTURE AND COMPOSITION OF CHROMATIN

Electron-microscopic techniques have not yet been sufficiently perfected to allow clear observation of the structure of chromatin in the interphase nucleus. However, in certain pathological state, the heterochromatin is loosened, and a fibrillar substructure is observed. When a well-developed lamina densa is present, the fibrillar heterochromatin takes on a special granular appearance near the inner membrane. The special association has led investigators to speculate that there is a connection between the chromosomes and the nuclear membrane. The fibrils vary in size; most appear to be about 100 A in diameter, but a number are thicker (about 150-250 A). nucleolar chromatin is also fibrillar in substructure, with fiber diameters a of about 100A.

As noted above, chromatin is any. chromosome material that stains with basic stains. In terms of chemical composition, this includes DNA and they closely associated basic proteins, among which the histones are the most common (Table 8.1). Basic dyes probably stain nuclear RNA as well, but most investigators consider chromatin to be composed of DNA and histones.

Table 18.1 Chemical composition of typical chromatin.

	Content Rilative to DNA (%)				
Source of Chromatin	*DNA*	*Histone*	*Nonhistone protein*	*RNA*	*Template Activity of Chromatin*
Pea vegetative bud	1.0	1.30	0.10	0.11	6
Pea embryonic axis	1.0	1.03	0.29	0.26	12
Pea growing cotyledon	1.0	0.76	0.36	0.13	32

Source: From J. Bonner and J.E. Varner, *Plant Biochemistry*, Academic Press, New York, 1976.

The weight ratio in chromatin of histone to DNA varies from about 0.8 to 1.3, with an average of about 1.1. The ratio varies not only with the species of organism but also with the tissue. It appears that histones are associated with the chromatin of all eucaryotic organisms except fungi, which resemble prokaryotic organisms in this respect.

The ultra structure of chromatin is thought to involve a repeating pattern of bodies that are associations of DNA and histone. These bodies, called nucleosomes (Fig. 8.2), are composed of an octamer of histones around which about 200 base pairs of DNA are coiled. The octamer consists of our pairs of histones: two copies each of H2A and H2B, which are rich in lysine, and two copies each of H3 and H4, which are rich in arginine. H1, another histone, is frequently associated with the nucleosome and may function in the coiling of the nucleosome chain but is not a part of the nucleosome. The number of base pairs of DNA associated with each nucleosome of the chain varies with organism and tissue from which the nucleosomes are extracted. The range is about 140 to 240 base pairs per nucleosome. Both thin and thick chromatin filaments have been described by investigators. The thin chromatin filament (about 100A in diameter) is probably a linear array of the nucleosome. The thick (300 A diameter) chromatin filament appears to be a spiral of the thin filament with a 100 A pitch and a 100 A hole down the central core.

SITES OF DNA REPLICATION

The sites of DNA replication in the nucleus have been determined by Autoradiography and electron microscopy of cells, synchronized in their growth cycle through the action of certain drugs. Although there is some concern that the drugs used to induce synchrony may alter the process, studies made with randomly growing cells tend to support the findings made with synchronously growing cells. Once the cells are synchronized, tritiated thymidine (a precursor in DNA synthesis) is added to the growth medium at selected times, and the uptake and sites of deposition of the laleled thymidine are followed.

By use of these techniques, replication has been shown to begin at random sites in the dispersed chromatin (euchromatin) during the S phase of the growth cycle.

By the late S phase, replication shifts to the periphery (heterochromatic areas) of the nucleus. The site and time of replication of nuclear DNA are not yet fully under stood. Some evidence indicates that rDNA replication occurs during the late S phase. However, tritiated thymidine is incorporated into nucleolar chromatin even after short pluses of the label. The observation the rDNA replication occurs later in the S phase could be associated with the more prevalent condensed nucleolar chromatin.

SITES OF TRANCRIPTION

Our understanding of the mechanism of transcription at the molecular level in the nucleous is fairly detailed. Site-specific localization of the process is also rapidly becoming clear. Initiation

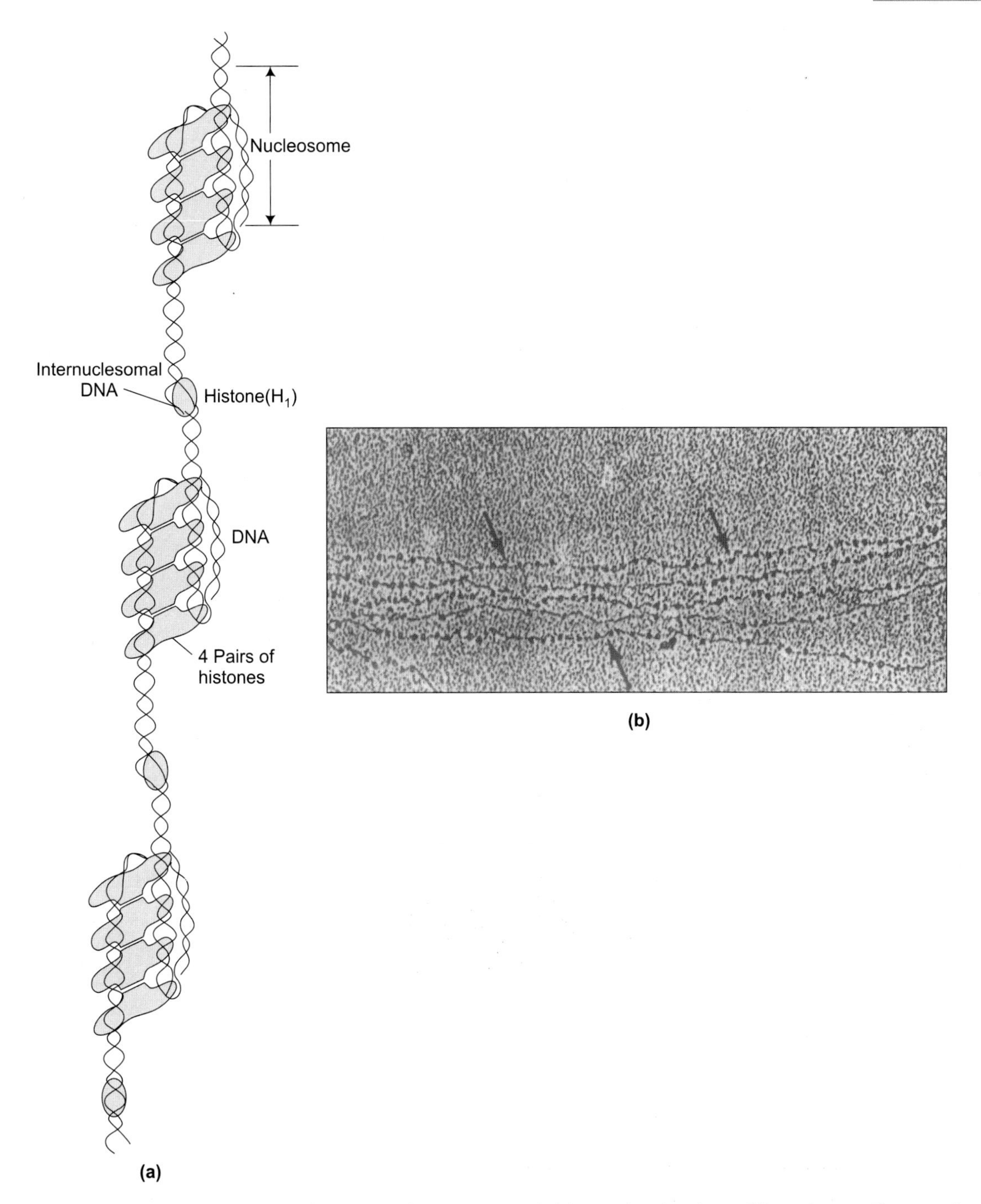

Fig. 8.2 (a) Diagram of part of a chain of nucleosomes held together by the coilling of DNA. Shape and size of nucleosomes are not to scale and represent only the basic concept of structure. (b) Electron photomcrograph of nuleosomes (arrows) (Conrtesy of Dr. F. Puvion-Dutilleul).

of transcription in the nucleolus is at the border between the intranucleolar chromatin and the remaining (granular) portion of the nucleolus.

When tritiated uridine is introduced, the label first appears in this border region. It than accumulates in the fibrillar region between the intranucleoar chromatin and the granular region and than moves into the granular region. The fibers of the fibrillar regions are believed to be precursors of the granules of the granular regions. Dispersal of the label in the granular regions is fairly uniform. The nucleolar fibers may be correlated with the RNA that sediments at 45 S and the granules, with the RNA that sediments at 23 S.

The site of nuclear transcription outside the nucleolus is not at all certain. Some evidence suggests hat all areas of the nucleus may undergo transcription, including both condensed and decondensed (dispersed) chromatin. However, there is other evidence indicating that transcription begins in the perichromatin region (i.e., the region between the peripheral condensed chromatin and the dispersed chromatin.

THE NUCLEOLUS

The nucleolus was one of the first subcellular organelles to be identified by microscopy. It was initially described by Fontana in 1774. Light microscopy using basic dyes has shown nucleoli to vary greatly in size, shape and number depending on the physiological state of the cell. Lymphocytes, which have little ribosome synthesis, have nucleoli reduced to crescent-shaped structures. Cells activity synthesizing ribosomes are enlarged, reveling "Giant-size" nucleoli. Nucleoli have been described with vacuoles and with dense areas called nucleolini. The electron microscope has shown that the nucleolous is mud of granular and fibrillar elements, as well as the intranucleolar and perinucleolar chromatin described above. The nucleolus is not bordered by a membrane. It is clear that primary function of the nucleolus is the synthesis of the rRNA species found in the large and small subunits of the ribosomes and the packing of these rRNAs with ribosomal proteins to from peribosomal particles. It has been suggested that they nay function in the formation of polysomes, but their interaction with mRNA.is not known.

THE NUCLEAR ENVELOPE

The presence of a membrane separating the nuclear material from the cytoplasm is one of the characteristics distinguishing eucaryotic organisms from prokaryotic organisms. The existence of a "membrane" delimiting the nucleus was first demonstrated by O. Hertwig in 1893. However, little interest was addressed to this "membrane" until studies with the electron microscope revealed that it was not simply a single membrane, but rather a double membrane in which the outer membrane had features that clearly distinguished it from the inner membrane. The two membranes lie close together, one surrounding the other. The two membranes fuse together at the nuclear pores but elsewhere are separated by the perinuclear space. The folded appearance at the nuclear pores and the perinuclear space between the membranes aptly justifies the term nuclear envelope to describe the entire structure.

The functions of the nuclear envelop are diverse. Clearly, it acts to compartmentalize the cytoplasm and nucleoplasm. Mitochondira, Golgi bodies, vacuoles, lysosomes, chloroplasts, and other cytoplasmic or ganele usually do not enter the nucleus, and the chromosomes and nucleoli rarely exit into the cytoplasm. However, ribosome precursors from the nucleolus and mRNA and tRNA molecules do leave the nucleus through the nuclear pores. Also molecules enter the nucleus from the cytoplasm, and a number of nucleocytiplasmic effects result.

The nuclear envelope is not just a physical barrier, but also functions in connection with cellular activities on each side. Ribosomes may be attached to the outer membrane, which there by takes on a structural appearance like rough endoplasmic reticulum. Elctron micrographs reveal that the outer membrane is continuous with the endoplasmic reticulum, and there is evidence that the endoplasmic reticulum forms as an outgrowth of the outer nuclear membrane. In some plants, the network of cytoplasmic membranes extending from the nuclear membrane is continuous with membranes of chloroplasts. In effects, the chlroplast is anchored to the nuclear envelope, and channels (cisternae) between the two structures provide for the exchange of genetic information (Fig. 8.3).

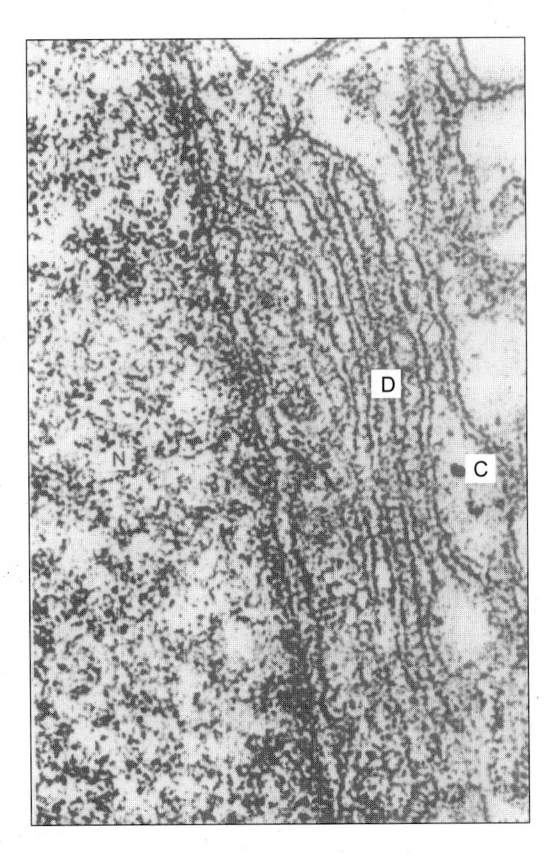

Fig. 8.3 Relationship between the nuclear envelope and developing Golgi apparatus (D). N nucleus; C, cisternae (Phatomicorgrapgh courtesy of Dr. E.G. Pollock).

The inner membrane of the nuclear envelope has a function associated with the chromosome. in the interphase nuclei of many cells; the heterochromatin is closely appressed to the inner membrane. This associated does not occur at or near the nuclear pores; the chromosomes seem to be firmly attached to the interpore sections. The purpose of the connections is not known, but suggestions being studies currently include (1) a role in chromosome replication, (2) a role in orienting the chromosomes during interphase and early prophase of mitosis and meiosis, and (3) a site of information of the nuclear envelope.

It seems reasonable to expect that some materials are exchanged between the nucleoplasm and the cytoplasm through open nuclear pores. Such exchanges presumably involve large molecular complexes such as ribonucleoprotein particles. In additions to this route of exchange, there arc others. For example, small ions and molecules readily exchange between the two sols by permeating the nuclear membranes, and except for a few special cases, electrical potential differences across the nuclear envelope are negligible. Larger molecules and particles may pass through the membrane by, formation of small pockets and vesicles that transverse the envelope and empty on the other side. It is also possible that small sections of the entire envelope evaginate or break away and undergo dissolution (as is known to occur during mitosis and meiosis).

Pores are not unique to nuclear membranes, occurring also in the plasma membrane and the membranes of the endoplasmic reticulum. However, the pores of the nuclear membrane possess a special character. The number of nuclear pores varies widely among cell types but, in general, ranges from 1 to about 60 pores per square micrometer. Unlike pores in the plasma membrane, the unclear pore is a complex of structures. The orifice is essentially circular, with occasional evidence of a polygonal shape. The inner pore diameter is markedly constant for a given cell type but may yield variable values depending on the technique used to fix and examine the cell. The usual range is 600-1000 A, with most nuclear envelope having pore orifice diameters of about 725 A.

The inner and outer rims of the pore are made up of nonmembranous material, which gives the pore the appearance of lips (Fig. 8.4) see in freeze-etched preparation, these appear as rings and are therefore called annuli (sing. Annulus). The annulus is not uniform but has eight granular subunits symmetrically placed about it. The lumen of the pore frequently contains distinct substructures. Some are conical projections from the sides the center, while others are fingerlike or fibrous.

CHROMOSOMES

Chromosomes were first described by numerous investigators in the period between 1875and 1880, and the name "chromosome" was introduced by Walderyer around 1880. These structures readily take up basic dyes and are easily seen between the poles of a cell during division. The material composing the chromosome, chromatin, is highly condensed during the division stages, so that the chromosomes can readily be counted and individually described when spread on a microscope slide. The number of chromosomes per nucleus varies greatly

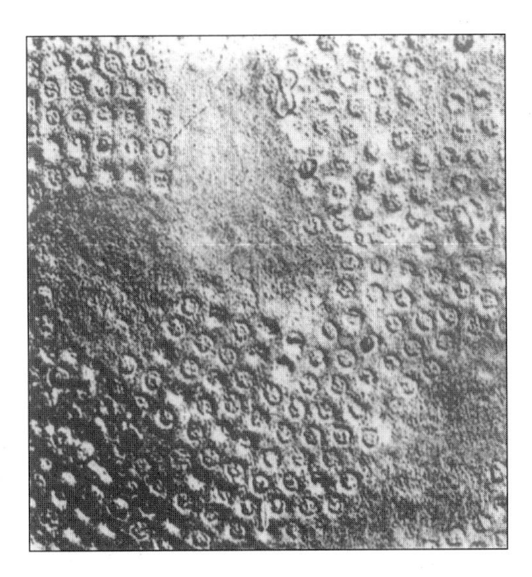

Fig. 8.4 Freeze fracture photomicrograph of the surfaces of the nuclear envelope and nuclear pores; (Courtesy of Dr. E.G. Pollock).

among the various animal and plant species, but for each specific species a constant number of chromosomes, each exhibiting specific shape and size at metaphase (Fig. 8.5) can be identified.

Chromosome shape and size during the stages of nuclear division (mitosis and meiosis) and during interphase. Each chromosome has two arms, one on each side of the primary constriction or centromere (also called kinetochore). The metaphase chromosome has doubled so that it appears to have two sets of arms (Fig. 8.6). Secondary constrictions associated with nucleoli are observed in some chromosomes and are called nucleolar-organizing regions (NOR). Tertiary constrictions are seen in nearly all chromosomes. Although their significance is not under stood, they help to distinguish one chromosome from another. During interphase the chromatin is spread out (decondensed), so that structure of the chromosome is virtually impossible to observe.

ULTRASTRUCTURE OF THE CHROMOSOME

The condensed chromosomes visible as distinct bodies during mitosis are composed of an organized array to chromatin fibers, but in their decondensed state give rise to a highly filamentous network. Each chromatin fiber is believed to contain one double-heix molecule of DNA. The average diameter of the chromatin fiber is about 100 A, whereas the diameter of the DNA double helix is only 20 A. The difference is due to supercoiling of the DNA molecule and the presence of large quantities of proteins (primarily histones), some enzymes, and RNA in the fiber.

Chromatin fibers fall into two classes, based on their diameter. Type A chromatid fibers are narrow strands of high electron density, whereas type B fibers are thicker and show light

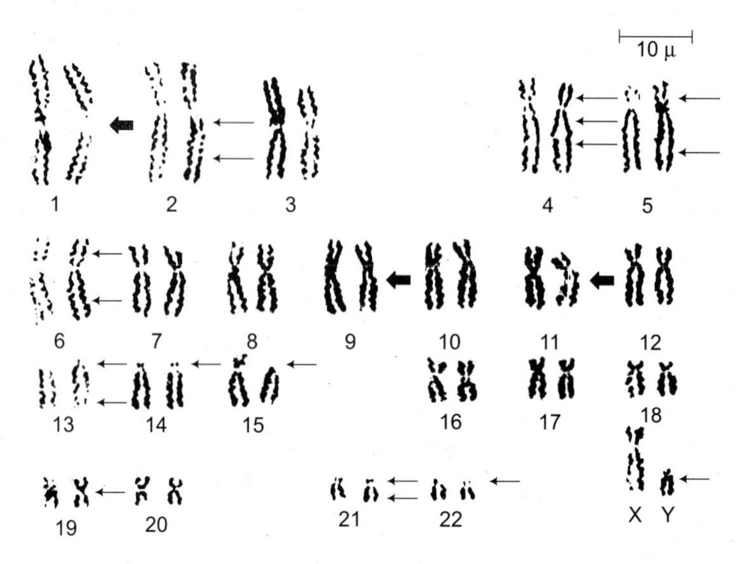

Fig. 8.5 Melaphse chromosomes of a male human cell as seen with the light microscope. The colchicine treatment used in the preparation of the chromosomes and the centromere (or primary constriction) create the charactenristic "X" appearance of the aims (Courtesy Y. Ohnukl, copyright 1968 Springer-verlag, Chromosome 25, 416).

diagonal cross-striation, indicating that this chromatin may be supercoiled. Chromatin with varying thicknesses, presumably alternating area of type A and B fibers, have also been described.

The packing ratio of interphase chromatin fibers averages about 56.1 (the packing ratio is the length of DNA divided by the length of the chromatin fiber.). The packing ratio of metaphase chromatids are usually greater than 100: 1, thus, it is estimated that during prophase the chromatin fiber undergoes a two-fold condensation, since the total amount of DNA remains uncharged. Since the chromatin fiber also twists and. folds, the chromosome appears to shorten much more than two fold.

In addition to its condensation, the total mass of the chromosome doubles during prophase. The increase is caused by the accumulation of more non-DNA constituents, most specifically nonhistone proteins. The increase is reflected by measurements of type B fibers of interphse and metaphase chromosomes. Interphase type B have diameter of 230-250 A, whereas metaphase type B fibers have diameter that usually exceed 300A.

The folding fibers that accompanies chromosome condensation during prophase does not appear to be regular or consistent. The fibers are twisted in an apparently random manner, producing configurations that are probably not duplicated in other chromosomes (not even homologous chromosomes) or the same chromosome during successive mitotic divisions.

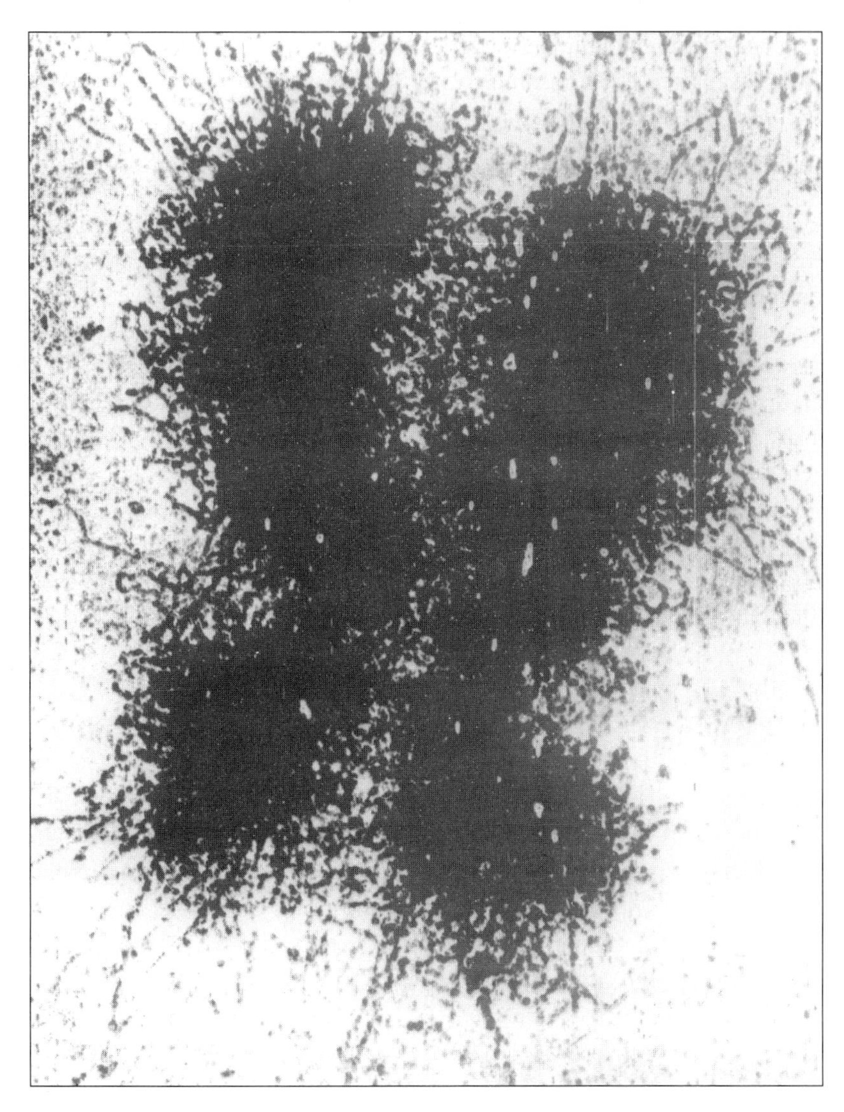

Fig. 8.6 Electron photomicrograph of human metaphase chromosome (Courtesy Gunther F. Bahr, Armed forces institute of pathology).

METAPHASE CHROMOSOME-CHROMOSOME ASSOCIATIONS

Although the chromosomes arranged on the equatorial plate at metaphase are usually described as random and independent, it has been known for some time that certain specific associations occur. Thee may be size assortment on the spindle with the long chromosomes on the outside and the short on the inside. Certain chromosomes are known to group together or have arms directed toward one another. G. Huskins has shown that human chromosomes are bound together by chromosome-to chromosome "connectives". The connectives are composed of DNA and protein and are strong enough to hold the chromosomes together even when teased out of cell. Fibers linking separate chromosomes have been visualized using electron microscopy.

POLYTENE CHROMOSOMES

In the salivary glands of dipteran files, of which the genus *Drosophila* has been most extensively studied, and in certain other tissues as well, the interphase nucleus is characterized by extremely large chromosomes. These so-called giant chromosomes are actually several hundred parallel and tightly packed copies of the same chromosome (Fig. 8.7). Because of their multiple structure, they are called polytene chromosomes. Along the length of the polytene chromosome, disks or bands of variably staining intensities can be distinguished. The identification of the genetic content of these bands (and interband regions) has played a major part in the genetic analysis of dipteran organisms.

Some of the band of giant chromosomes appear to be swollen or "puffed", the specific bands exhibiting puffing varying from one tissue to another even within the same organism. It is now well established that the puffed bands of these interphase chromosomes correspond to regions in which the DNA is being activity transcribed into mRNA.

BACTERIAL AND VIRAL CHROMOSOMES

The "chromosomes" of bacteria and viruses are single nucleic acid molecules, in most cases, they consist of a double helix of DNA, but in certain viruses they consist of only a single strand of DNA or RNA. The DNA may be linear or circular, depending on the organism.

Bacteria. The chromosomes of *E. coli* and *B. sublilis* have been studier more extensively than those of any other bacteria. In both species the chromosomes is a single, circular DNA molecule more than a millimeter long. The circular molecule is packed into a dense mass less than 1 μ in diameter. The circular DNA can be released intact by proper osmoitc treatment of the cell. The bacterial chromosome is attached either to the plasma membrane or to an in folding of the membrane called a mesosome. Attachment to a membrane is necessary for separation of daughter chromosomes, since these cells do not form a spindle. Within the nucleoid, the DNA appears to be folded back and forth in a semiparallel manner, rather than being coiled and supercoiled, although DNA from osmotically exploded cells has a tendency to coil upon release.

Viruses. The tobacco mosaic virus (TMV) was the first to be crystallized and its structure determined in detail. TMV is a cylindrical tube 300 A long and 180 A in diameter (Fig. 8.8). Its "chromosome" consists of a single molecule of RNA of some 6400 nucleotide and has a length of about 33,000 A. The RNA is wound in a helix about the central aqueous core of the particle with 49 nucleotides per turn (inner diameter, 34 A; outer diameter, 80 A; pitch, 23 A). The RNA is held in the arrangement by an outer sheath of 2, 130 identical protein units.

The T-even and lambda phages of bacteria are DNA viruses in which the DNA is a single, linear, double-stranded molecule. The lambda phage DNA is about 17 μ long, and the T2 and

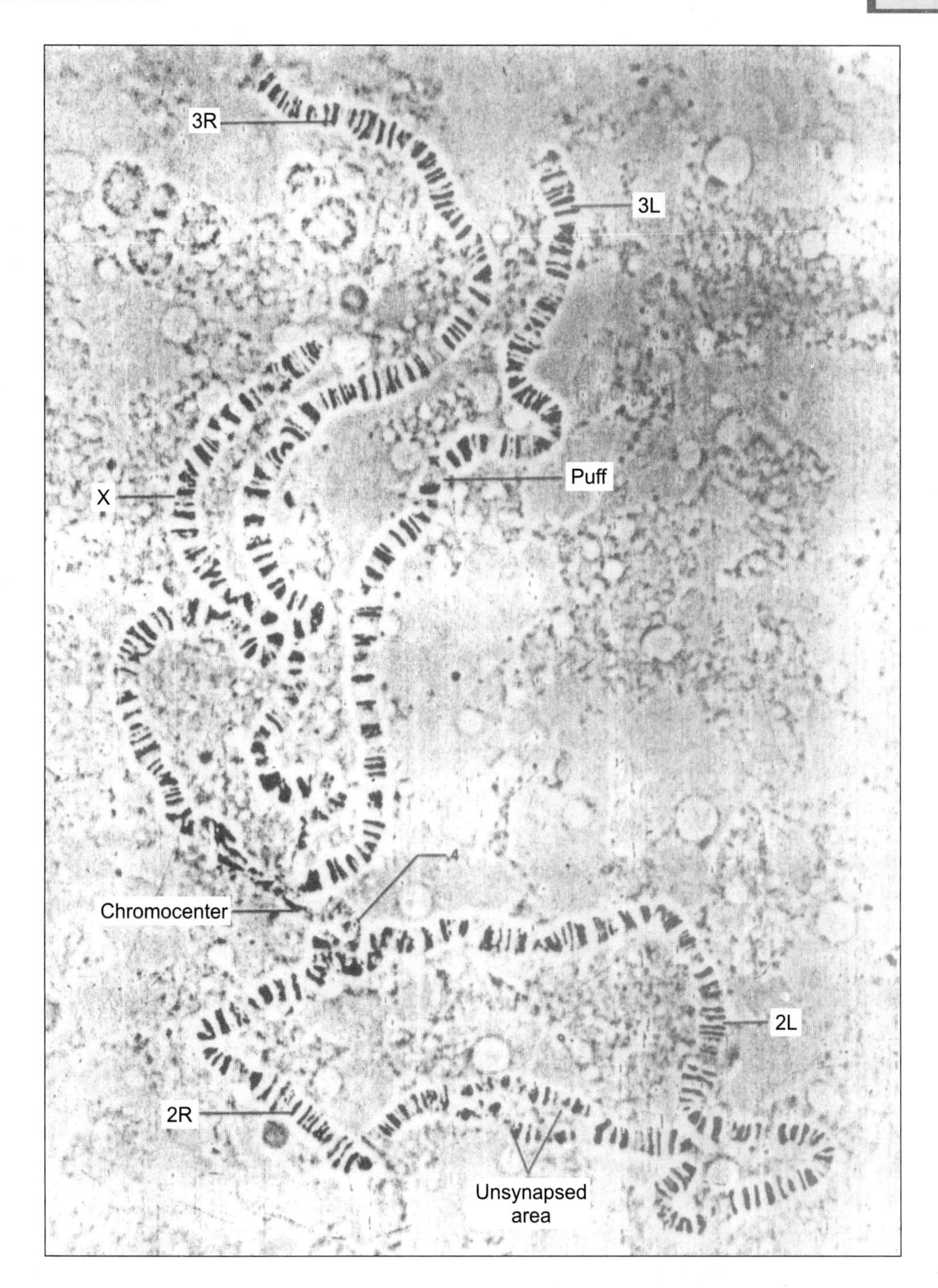

Fig. 8.7 Glant polytene chromosomes from a diploid salivary, gland cell of *Drosophlla melanogaster*. The chromosome arc labeled (X, 2R, 2L, 3R, 3L and 4). A puff is shown on 3L and an unsynapsed region of 2R reveals the two chromosomes (Courtesy of Dr. Lefevre).

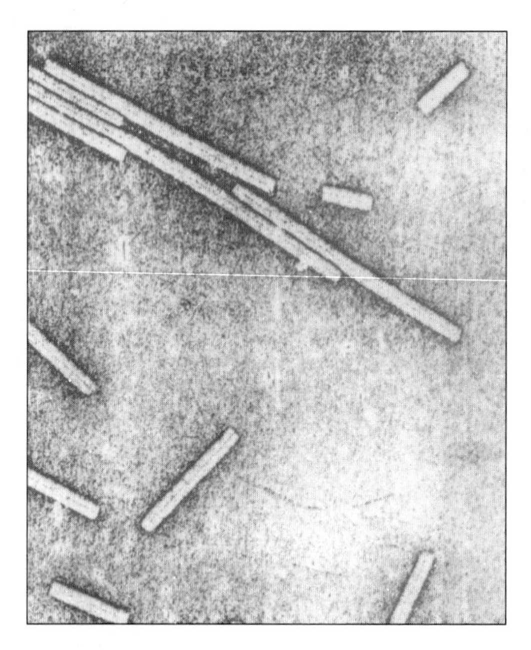

Fig. 8.8 Tobacco mosaic virus (photomicrograph courtesy of Dr. F. A. Eiserling).

T4 phage DNAs are about 100 u long. The lambda phage DNA assumes a circular shape after injection into the host bacterium. Numerous other viruses have been studies (Table 8.2).

THE VIRUS LIFE CYCLE

There are 6 basic steps in the life cycle of a virus:

1. **Virus adsorption.** The virus is adsorbed to the surface of the host cell. The specificity between the virus and the host is achieved through the recognition by specific viral adsorption proteins called pilot proteins, for specific host cell-surface receptors. The pilot protein not only attaches the virus to the host but may also orient the viral DNA (or RNA in the case of the RNA viruses) for insertion into the most appropriate part of the cell.
2. **DNA (or RNA) insertion.** Penetration of the viral DNA through the host cell envelop and cell membrane and into the interior of the cell may be aided by the pilot proteins, which maybe carried into the cell with the DNA.
3. **Expression of viral nucleic acid.** Once in the host cell, the viral DNA is expressed in same manner. In the case of the small single-stranded viruses like X 174 and M 13, a second strand complementary to the single strand is synthesized, yielding a duplex form. For the double-stranded DNA viruses, certain segments necessary for virus replication are transcribed and translated. While for the RNA viruses, certain portions are directly translated.

Table 8.2 Sizes and other characteristics of certain viruses.

Virus	Related viruses	Particle shape	DNA			Comments
			Mol. Wt. x 10^{-6}	Number of base pairs, in kb	Shape	
SV 40	Polyoma	Polyhedron	3.4	5.1	Duplex, circular	Animal cell host
$\phi \times 174$	S13	Polyhedron	1.8	5.4	Single strand, circular	Duplex replicative form
M13	fd, f1	Filament	1.9	5.7	Single strand, circular	Duplex relicative form
T7	T3	Head, short tail	23	35.4	Duplex, linear	Terminal redundancy
λ	ϕ80, 434 P2, 186	Head, tail	32	49	Duplex, linear	Cohesive ends form replicative circles, lysogenic
T5		Head tail	76	115	Duplex, linear	Nicked
T4	T2, T6	Head tail	120	180	Duplex linear	Terminal redundancy; permuted
R17	MS2, f2					
	Qb	Polyhedron	1.0	3.0	Single strand, linear	RNA, not DNA

Source: From A. Kornberg, *DNA Synthesis*, W.H. Freeman and Co., San Francisco, Copyring © 1974.

4. **Replication of virus nucleic acid.** The mechanism of DNA replication varies with type of virus, but ultimately many complete double (or single) strands are produced for final "packing".
5. **Virus assembly.** The mature DNA strands are assembled within a shell or coat of proteins.
6. **Release.** Finally, the assembled viruses are released from the host cell. The stages are summarized in Fig. 20-22, in some cases, viruses emerge from the host cell in buds and do not cause serious disruption of the cell surface; in other cases, the host cell ruptures (lyses), spilling the viruses into the surrounding medium.

SUMMARY

Eucaryotic nuclei are delimited by a pair of enveloping membranes that are continuous with each other around the margins of nuclear pores. Within the nucleus are the **chromosomes, nucleoli, nucleoplasm, ribosome-like** particles, and a variety of irregularly shaped particles. The chromosomes during mitosis and meiosis assume a distinctive shape as result of the condensation of their constituent DNA and proteins. Metaphase chromosomes are characterized by a primary constriction or centromere and a varying number of secondary constrictions. The arms of the chromosome contain a framework of a double helix of DNA about which are packed a number of proteins. Each strand of DNA, called a chromatin fiber, is supercoiled. The amount of coiling is responsible for the observation of two types of chromatin fibers, the narrower type A and the broader type B. The packing ratio, or length of

DNA divided by the chromatin fiber length, is about 56-1 for interphase chromosomes and 100: 1 for metaphase chromosomes.

During interphase the chromatin fibers show varying degrees of condensation. The more condensed portions of interphase chromatin are called heterochromatin, and the less condensed regions, euchromatin. Euchromatin appears to contain most of the structural genes, whereas heterochromatin may be more concerned with the control, of gene expression. The chromatin fibers have a repeating pattern of subunits called nucleosomes, which are specific associations of histones with DNA. DNA replication begins during the S phase of the cell cycle at random sites in euchromatin and later shifts to the heterochromatin.

The chromosomes of bacteria and viruses are dingle nucleic acid molecules. They may be single-stranded DNA or RNA or may consist of a double helix of DNA. These chromosomes are not contained with in a nuclear membrane but in bacteria may be attached to the cell membrane infolding of the cell membrane called a mesosome.

Replication is a semiconservative process, each strand of the double helix of DNA serving a template formation of new strands. Replication begins at one point on the chromosomes proceeds along both strands, adding nucleotides of both in the 5' → 3' direction. A swivel mechanism, consisting of enzymes for backing the polynucleotide strands (nickases) and resealing the strands (ligass), prevents twisting of the DNA during replication.

Much of the pioneering work on replication was done with prokaryotic, especially circular DNA molecule of *E. coli*. Technology has reached the stage where recombination (the exchanges of genetic information between chromosomes resulting in new combinations of genes in successive generations), can be accomplished experimentally using chromosomes of different organisms.

CHAPTER 9

DNA Structure

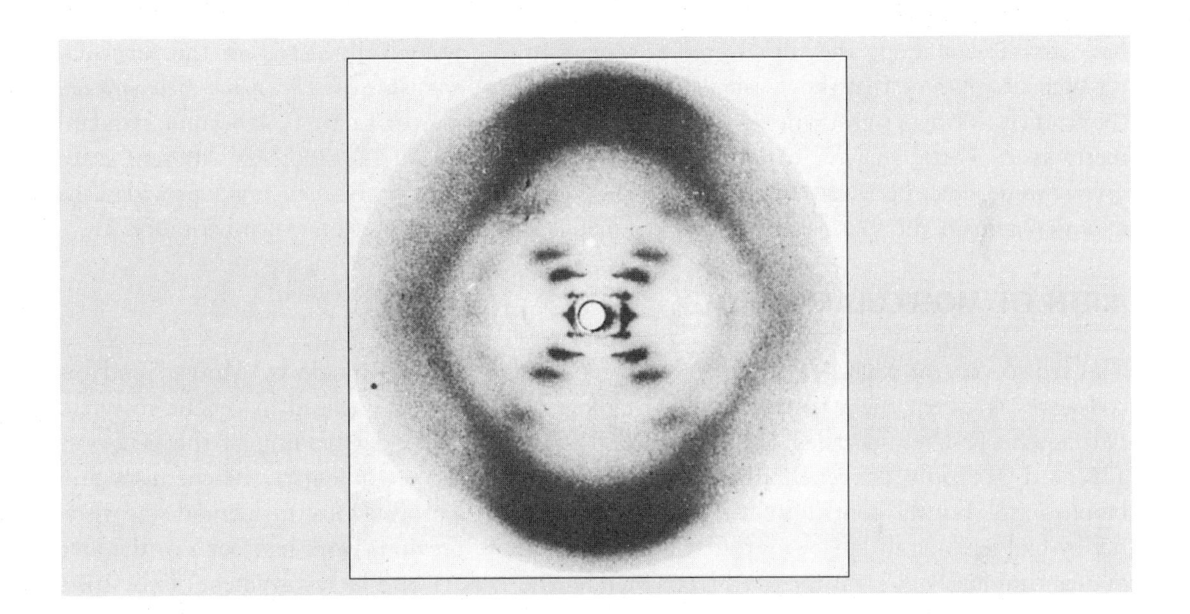

INTRODUCTION TO MOLECULAR BIOLOGY

Geneticists before 1930 were mainly interested in how genes are transmitted from parents to their offspring during reproduction and how different genes act together to control variable traits such as height and eye colour. A change in emphasis occurred during the 1930s when it was recognised that genes are physical entities, then, like other cell components, they must be made of molecules and it should, therefore, be possible to study them directly by bio-physical and biochemical methods. This led to a new branch of genetics, called *Molecular Biology*, which had as one of its initial aims the identification of the chemical nature of the gene. This new approach led to new concepts, and soon biologists ceased to regard individual genes simply as units of inheritance and instead began to look on them as units of biological information. The aim of geneticists and molecular biologists over the last 45 years has been to understand the way in which biological information is stored in genes and how that information is made available to the living cell.

FROM PEAS TO *DROSOPHILA*

Once the experimental basis to heredity was established and the chromosome theory accepted through studies on the inheritance of peas, the way was opened for an advancement in the understanding of genetics. This advancement occurred rapidly mainly, due to the intuition and imagination of Thomas Hunt Morgan and his research group, notably Calvin Bridges, Arthur Sturtevant and Hermann Muller. Morgan and his colleagues achieved something that many biologists dream about: they discovered an organism that was ideally suited for the particular research programme that they wished to carry out. This organism was *Drosophila melanogaster*, the fruit fly. Morgan's group developed many of the techniques that have now become standard methods in genetic analysis, including those that map the relative positions of different genes on a chromosome. Between 1911 and 1929 the 'fly-room' at Columbia University provided the data that remain the foundation to our knowledge of the gene as a unit of inheritance.

BIRTH OF MOLECULAR BIOLOGY

The rediscovery of Mendel's work and the remarkable advances made by Morgan and his colleagues attracted the attention of many biologists not actively engaged in genetic research. During the first few decades of the twentieth century, our understanding of the gene was advanced not only by geneticists but also by cytologists, physiologists, biochemists and biophysicists. But the contribution that in the long run has proved most influential was made not by biologists at all but by a group of physicists whose previous work had been in the area of quantum mechanics, far removed from Mendel and fruit flies. The involvement of physicists in biology was heralded by a lecture called "Light and Life" presented by Niels Bohr to an International Congress in Copenhagen in August 1932, and subsequently reached its culmination with the publication of the book "What is Life" by Erwin Schrodinger in 1944. Bohr and Schrodinger both attempted to interpret the central question in biology-the nature of life itself-in physical terms, but both were frustrated by the apparent limitations of trying to explain life by orthodox physical principle.

The lectures of Bohr and Schrodinger influenced many young physicists, like Max Delbruck, a German by birth who was forced by the political climate in Europe to emigrate to the United States. Delbruck learnt from Bohr about bacteriophages (viruses that attack bacteria) and decided that their infection cycle could be used as an experimental system with which to tackle the question of what genes are and how they work. He brought into existence in 1940 the "Phage Group" an informal association of physicists, biologists and chemists, all working indifferent laboratories but all with a common interest in the gene. The formation of this group stimulated the development of molecular biology as a discipline in its own right and led within twenty years to an understanding of the chemical nature of the gene.

BEGINNINGS OF EXPERIMENTAL PROOF OF DNA (Deoxyribose Nucleic Acid)

The necessity for an experimental identification action of the genetic material became pressing during the late 1930s, when it gradually became apparent that DNA, rather than being a simple molecule unsuitable as the genetic material, was in fact a long polymer and, like protein, could exist in an almost infinite number of variable forms. If both protein and DNA satisfy the fundamental requirement of the genetic material, and both are present in chromosomes, then which of the two are genes made of? Two critical experiments, radically different in their design, eventually proved that the genetic material is DNA and not protein. The first of these experiments was the identification by Oswald Avery, Colin MacLeod and Maclyn McCarty in the early 1940s of the chemical nature of the transforming principle, a substance that can change the bacterium *Diplococcus pneumoniae* from one form to another.

THE TRANSFORMING PRINCIPLE

Rough and Smooth Forms of *Diplococcus pneumoniae*

Variation is particularly important in bacteria that cause diseases, as often a single species may include both virulent and avirulent forms. The study of such forms in *Diplococcus pneumoniae,* the causative agent of one-type of pneumonia, led to the discovery of bacterial transformation by a London medical officer, Frederick Griffith, in 1928.

For some time three virulent serotypes of *D. pneumoniae.* called Types I, II and III, had been known. Bacteria of each serotype are distinguished by the nature of their capsule, a slimy coating that surrounds each cell. This capsule is made of a polysaccharide secreted by the bacterium with each serotype producing a slightly different polysaccharide and, therefore, slightly different capsule. The capsule gives colonies of *D. pnemoniae* a smooth, shiny appearance; hence the designation 'S' (for 'smooth') form.

In 1923 Griffith discovered that each serotype can also exist in a harmless, avirulent form, distinguished from the virulent by the absence of the capsule and a 'rough' appearance to the colonies. The avirulent types are, therefore, called 'R' forms. This discovery was not in itself particularly exciting since R and S forms had already been found in *Streptococcus* and, during the next few years, were to be discovered in several other species of bacteria. What interested Griffith about the R and S forms of *D. pneumoniae* was the appeared ability of one form to transform into another. At first his evidence was circumstantial, patients recovering from pneumonia often carried in their saliva the avirulent R form of the serotype responsible for their infection, possibly, Griffith suggested, because of attenuation of the bacteria during recovery from the illness. However, in 1928, Griffith published the results of a startling series of experiments that demonstrated that transformation from one form to another was caused by a *genetic* and *not a physiological change.*

Griffith carried out four particularly important experiments. These were as follows: (Fig. 9.1)

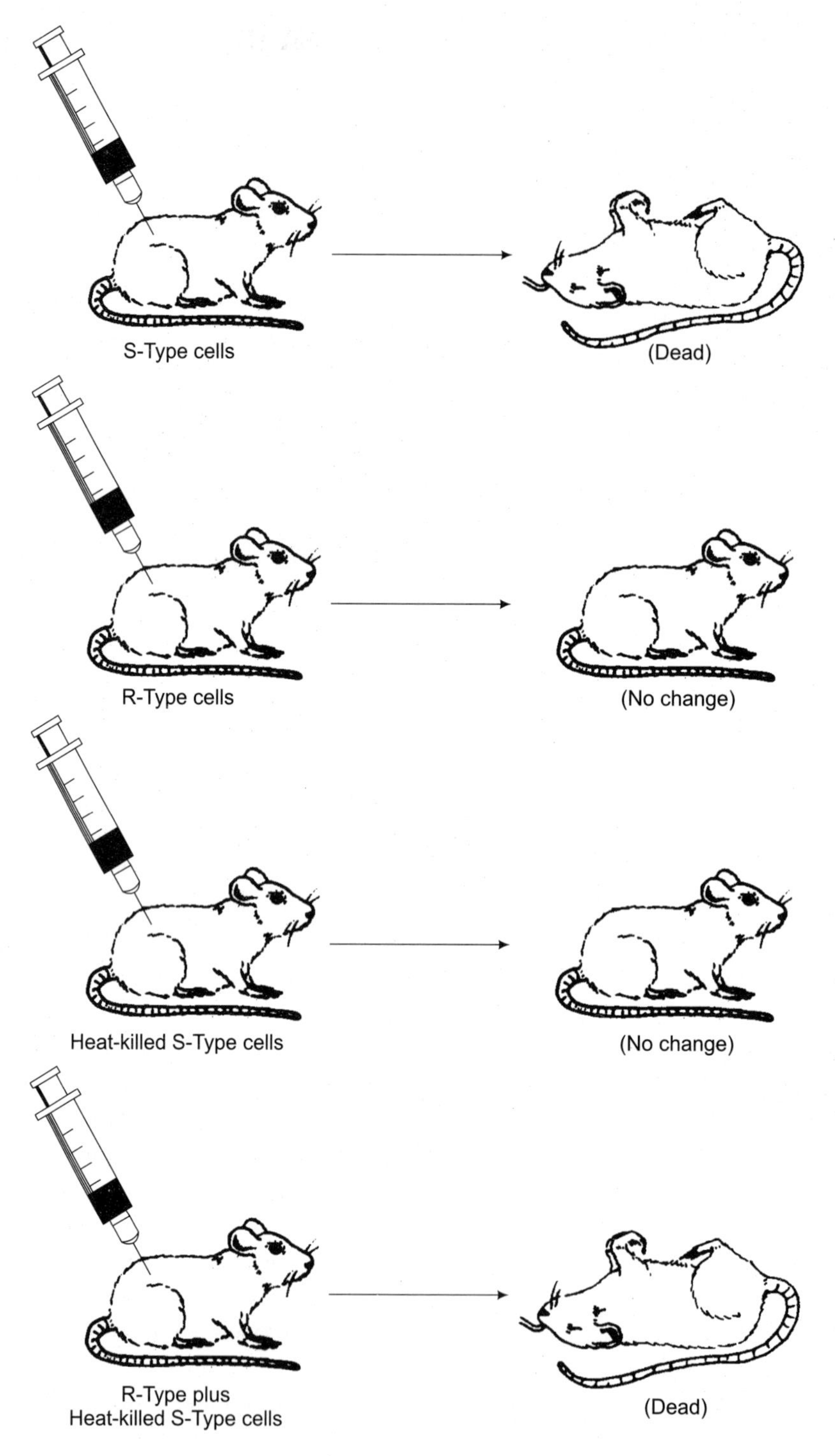

Fig. 9.1 Griffith's experiment with *Diplococcus pneumoniae.*

1. A mouse was injected with a sample of virulent, smooth bacteria; as expected, the mouse contracted pneumonia.
2. A second mouse was injected with avirulent, rough bacteria; again the expected result was obtained and the mouse remained healthy.
3. Next a sample of virulent, smooth bacteria were killed by heating and then injected into the mouse; the result was that the animal remained healthy because only live bacteria can cause the disease.
4. Finally, a sample of the heat-killed S bacteria, incapable of causing the disease, were mixed with live avirulent rough bacteria and the mixture injected into a mouse; this time a wholly unexpected result was obtained. Rather than remaining healthy, the mouse contracted pneumonia. In addition, from this animal large numbers of living smooth bacteria were eventually isolated, even though the only living bacteria in the original inoculum were avirulent R forms.

THE TRANSFORMING PRINCIPLE IS GENETIC MATERIAL

The conclusion to be drawn from Griffith's experiments is that a *component of the heat-killed S bacteria was able to enter the R cells and transform them into the virulent smooth form.* These new S bacteria then caused pneumonia after injection into the mouse. As the only difference between the R and S forms is the absence or presence of the capsule, the component taken up by the R bacteria must be able to induce the cells to start synthesising the polysaccharides that make up the capsule. The component could not be the capsular polysaccharide itself, because the heat-killed S cells would not yield enough material to coat all the living cells subsequently isolated from the mouse. The component taken up by the R bacteria, the transforming principle, must therefore be genetic material. Introduction of the transforming principle: into the R cells results in acquisition by these bacteria of a new heritable characteristic—the ability to synthesise the capsular polysaccharide.

THE TRANSFORMING PRINCIPLE IS DNA

Avery, MacLeod and McCarty, working at the Rockefeller Institute in New York, set out to identify the transforming principle. Their strategy was to prepare a filtrate from heat-killed S cells, containing the transforming principle, and then to digest the individual components of this filtrate with specific degradative enzymes. For instance, a protease could be used to degrade all the protein in the filtrate, and a ribonuclease to destroy the ribonucleic acid (RNA), the second type of nucleic acid present in living cells (though, unlike DNA, not a major constituent of chromosomes). After enzymatic digestion of one or more components, the filtrate was tested for retention of its transforming ability. Surprisingly (as genes were still thought to be made of protein), treatment with protease had no effect on the ability of the filtrate to transform R cells into the S form. Similarly, ribonuclease treatment had no effect on the transforming activity.

However, the digestion of the DNA with deoxyribonuclease totally destroyed the transforming ability so that the filtrate was no longer able to convert one cell type into the other.

Gradually it became clear to Avery and his colleagues that *the transforming principle must be DNA*. This idea was supported not only by the result of the enzymatic digestion experiments but also by analysis of the purified transforming principle by electrophoresis, ultracentrifugation and ultraviolet spectroscopy.

BACTERIOPHAGE GENES ARE MADE OF DNA

The second experiment directed at chemical characterisation of the genetic material was not carried out until 1951-52, seven years after Avery's work was published. By then experimental techniques had progressed and new approaches to problems in genetics were available. In particular, the use of bacteriophages as experimental tools for studying molecular genetics had become fully established.

BACTERIOPHAGES ARE VIRUSES THAT INFECT BACTERIA

Bacteriophages, or phages as they are commonly known, are viruses that specifically infect bacteria. Phage T2, for example, is one of several types of phage that are specific for the bacterium *Escherichia coli*. If T2 bacteriophages are introduced into a culture of *E. coli*, the cells will become infected and will produce large numbers of new phage particles. Details of how the infection cycle proceeds were not worked out until several years later but it was hypothesised that, as phages are not able to replicate on their own, the new phage particles must be synthesised by the bacteria. To do this, the bacteria would need to make use of the information carried by the phage genes and probably these genes would have to enter the bacterial cells, in order for their information to be used.

PHAGE PROTEIN AND DNA CAN BE LABELLED WITH RADIOACTIVE MARKERS

Identifying the substance injected by phages into bacterial cells is simplified by the fact that phage particles contain only protein and DNA. In 1951 a new method that would allow protein and DNA to be distinguished beyond doubt was developed. This procedure is called *radiolabelling* and involves attachment of a radioactive atom (or 'marker') to the molecule in question. Protein and DNA can be distinguished because protein can be labelled with ^{35}S, a radioactive isotope of sulphur, which will be incorporated into the sulphur-containing amino acids cysteine and methionine. DNA contains no sulphur and will not be labelled with ^{35}S. Conversely DNA can be labelled with ^{32}P, which will not be incorporated into proteins because proteins do not contain phosphorus. Once labelled, samples of DNA and protein can be distinguished from each other because the two radioisotopes, ^{32}P and ^{35}S, emit radiation of different characteristic energies.

THE HERSHEY-CHASE EXPERIMENT (Fig. 9.2)

A critical experiment was carried out by Alfred Hershey and Martha Chase at the Cold Spring Harbor Laboratory on Long Island, New York. They prepared a radioactive sample of T2 phage, one in which the protein was labelled with ^{35}S and the DNA with ^{32}P, from a T2-infected culture of *E. coli* that had been grown with ^{35}S- and ^{32}P-labelled nutrients. The labelled phages produced by this culture were used to infect a new, unradioactive culture of *E. coli*. However, this time infection was not allowed to proceed to completion, as a few minutes after inoculation the cells were agitated in a Waring blender (hence the popular name of the '*Waring blender experiment*'). These few minutes were long enough for the phage genes to enter the bacteria, but not enough time for new bacteriophages to be synthesised and the bacteria killed. Hershey and Chase believed that agitation would remove any phage material attached to the outside of the cell so that the only component retained by the bacteria would be the injected substance, the phage genes. The culture was then centrifuged so that the relatively heavy bacterial cells,

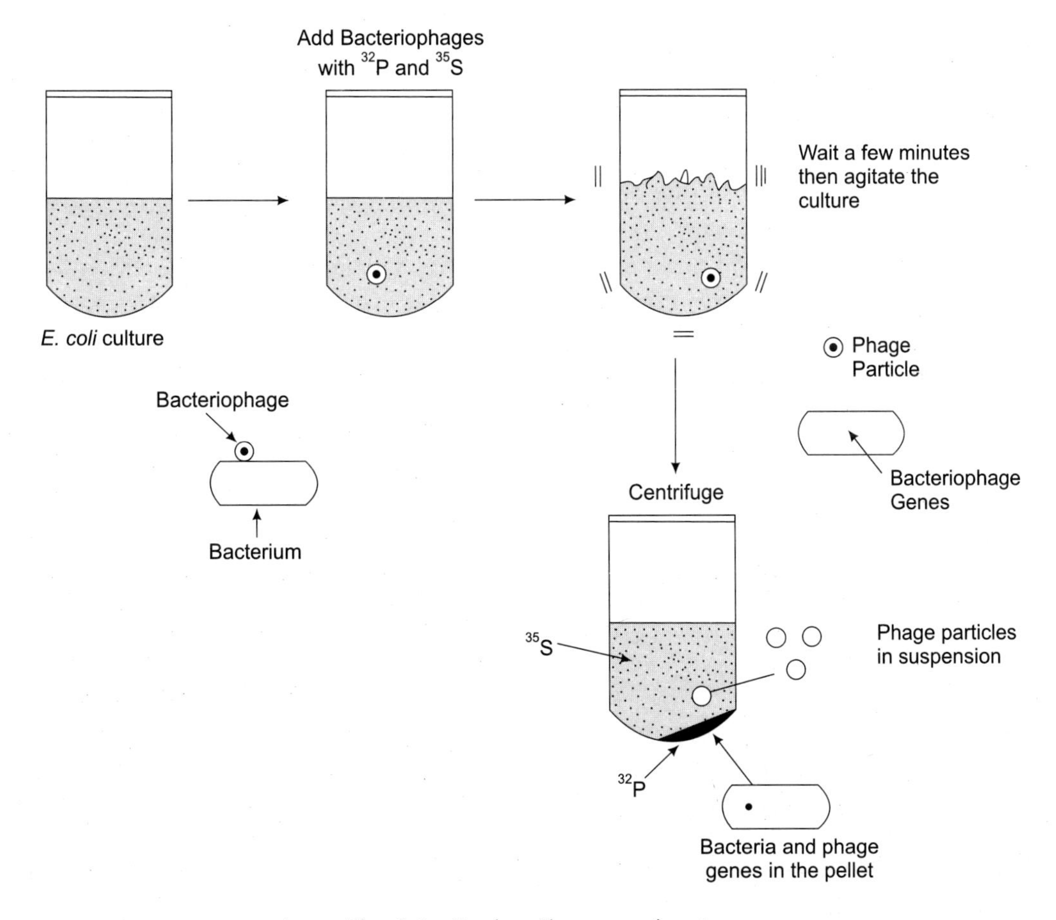

Fig. 9.2 Hershey Chase experiment.

containing the phage genes, collected at the bottom of the tube, leaving the empty phage particles in suspension. To identify what genes were made of, all that was needed was to determine which label, ^{35}S or ^{32}P, was present in the bacterial pellet. Hershey and Chase discovered that less than 196 of the ^{35}S, but over 80% of the ^{32}P, was present in the pellet. This experiment clearly proves that the genetic material was DNA and not protein.

ACCEPTANCE OF DNA AS THE GENETIC MATERIAL

Although very few doubts could be raised about the interpretation and validity of the Hershey-Chase experiment, several biologists still remained unconvinced that DNA was the genetic material in all organisms. *Strictly speaking, all that the Hershey-Chase experiment demonstrated was that genes of bacteriophages are made of DNA.* Phages are very unusual organisms, if indeed they can be called organisms at all, so perhaps the results of experiments with phages cannot be extrapolated to real organisms such as man.

Nowadays we do accept that the genetic material is DNA in all organisms except a few types of phage and virus, in which the second type of nucleic acid, RNA, replaces DNA. However, direct experimental evidence that genes are made of DNA in higher organisms has come only surprisingly recently, with the advent of recombinant DNA technology and the artificial introduction of pure DNA into bacteria, yeast, fungi, insects, plants and animals. These techniques became available during the 1970s showing the experimental evidence that the genetic material is DNA in higher organisms.

It was accepted from 1953 onwards that the genetic material is DNA. The reason for the belief in DNA was the discovery of structure of the double helix by Watson and Crick, and the realisation that the structure was absolutely compatible with the requirements of the genetic material.

THE STRUCTURE OF DNA

Having established that genes are made of DNA, the next step is to consider exactly what the structure of DNA is and how this relates to the properties and requirements of the genetic material. The answer to Schrodinger's conundrum 'What is Life?' lies in the DNA molecules.

DNA IS A POLYMER

The fact that DNA is a long polymeric molecule was first clearly understood in the late 1930s and led to the realisation that DNA, as well as protein, has the potential variability required of the genetic material. A polymer is a long chain-like molecule comprising numerous individual units called monomers, linked together in a series. Many important biological molecules including not only proteins and nucleic acids but also polysaccharides and lipids are polymers of one type or another.

NUCLEOTIDES—THE MONOMERS IN DNA

The basic unit of the DNA molecule is the *nucleotide.* Nucleotides are found in the cell either as components of nucleic acids or as individual molecules, in which form they may play several different roles. For instance, the nucleotides that make up RNA molecules are also important in the cell as carriers of energy used to power many enzymatic reactions.

The nucleotide is itself a complex molecule being made up of three distinct components: a sugar, a nitrogenous base and phosphoric acid (Fig. 9.3).

THE SUGAR COMPONENT

In DNA, the sugar component of the nucleotide is a pentose (containing five carbon atoms) called 2′-deoxyribose. Pentose sugars can exist in two forms, the straight chain or Fischer structure and the ring or Haworth structure. It is the ring form of 2′-deoxyribose that occurs in the nucleotide.

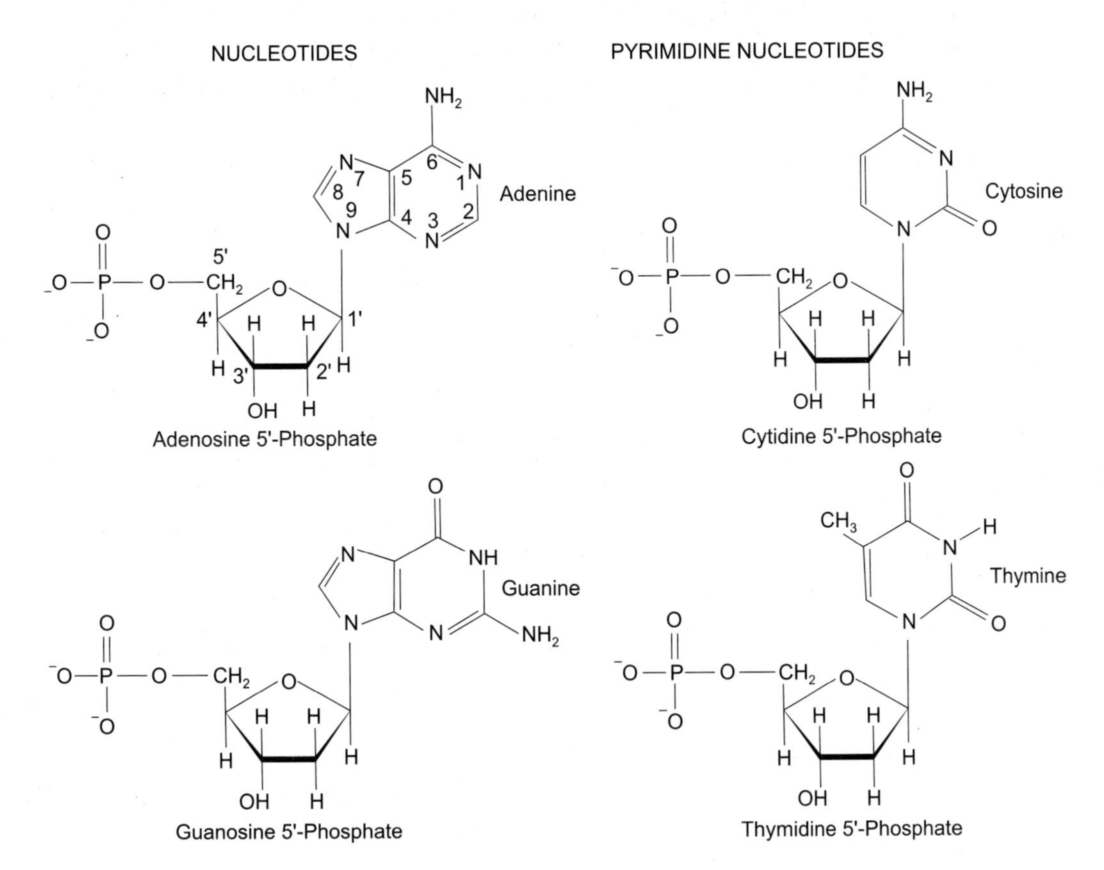

Fig. 9.3 Structure of the four Deoxyribose Nucleotides.

The name 2′-deoxyribose indicates that the standard ribose structure has been altered by replacement of the hydroxyl group (–OH) attached to carbon atom number 2′ with a hydrogen group (–H). The carbon atoms are always numbered in the same way, with the carbon of the carbonyl group (–C=O), occurring at one end of the chain form, numbered 1′. It is important to remember the numbering of the carbons because it is used to indicate at which positions on the sugar, other components of the nucleotide are attached. Note that the number is not just 1, 2, 3, 4, 5 and 6, but 1′, 2′, 3′, 4′, 5′ and 6′ (the dash is called the 'prime', and the numbers are called 'one-prime', 'two-prime', and so on). The prime is used to distinguish the carbon atoms in the sugar from the carbon and nitrogen atoms in the nitrogenous base. The latter are numbered 1, 2, 3, and so on, and it is important not to confuse these with the atoms on the sugar.

NITROGENOUS BASES

These are rather complex single- or double-ring structures that are attached to the 1′-carbon of the sugar. In DNA any one of four different nitrogenous bases can be attached at this position. These are called adenine and guanine, which are double-ring purines, and thymine and cytosine, which are single-ring pyrimidines. Their structures are shown in Fig. 9.4.

THE PHOSPHORIC ACID COMPONENT

A molecule comprised of the sugar joined to a base is called a nucleoside. This is converted into a nucleotide by attachment of a phosphoric acid group to the 5′-carbon of the sugar. Up to three individual phosphate groups may be attached in series giving a nucleoside monophosphate (NMP), nucleoside diphosphate (NDP) or nucleoside triphosphate (NTP). The individual phosphate groups are designated α, β and γ; with the α-phosphate being the one attached directly to the sugar.

NOMENCLATURE OF NUCLEOTIDES

The four different nucleotides that polymerize to form DNA are shown in Fig. 9.4. Their full names are:

2′-deoxyadenosine 5′-triphosphate

2′-deoxyguanosine 5′-triphosphate

2′-deoxycytidine 5′-triphosphate

2′-deoxythymidine 5′-triphosphate

Normally, however, these are abbreviated to dATP, dGTP, dCTP and dTTP, or even to just A, G, C and T, especially when writing out the sequence of nucleotides found in a particular DNA molecule.

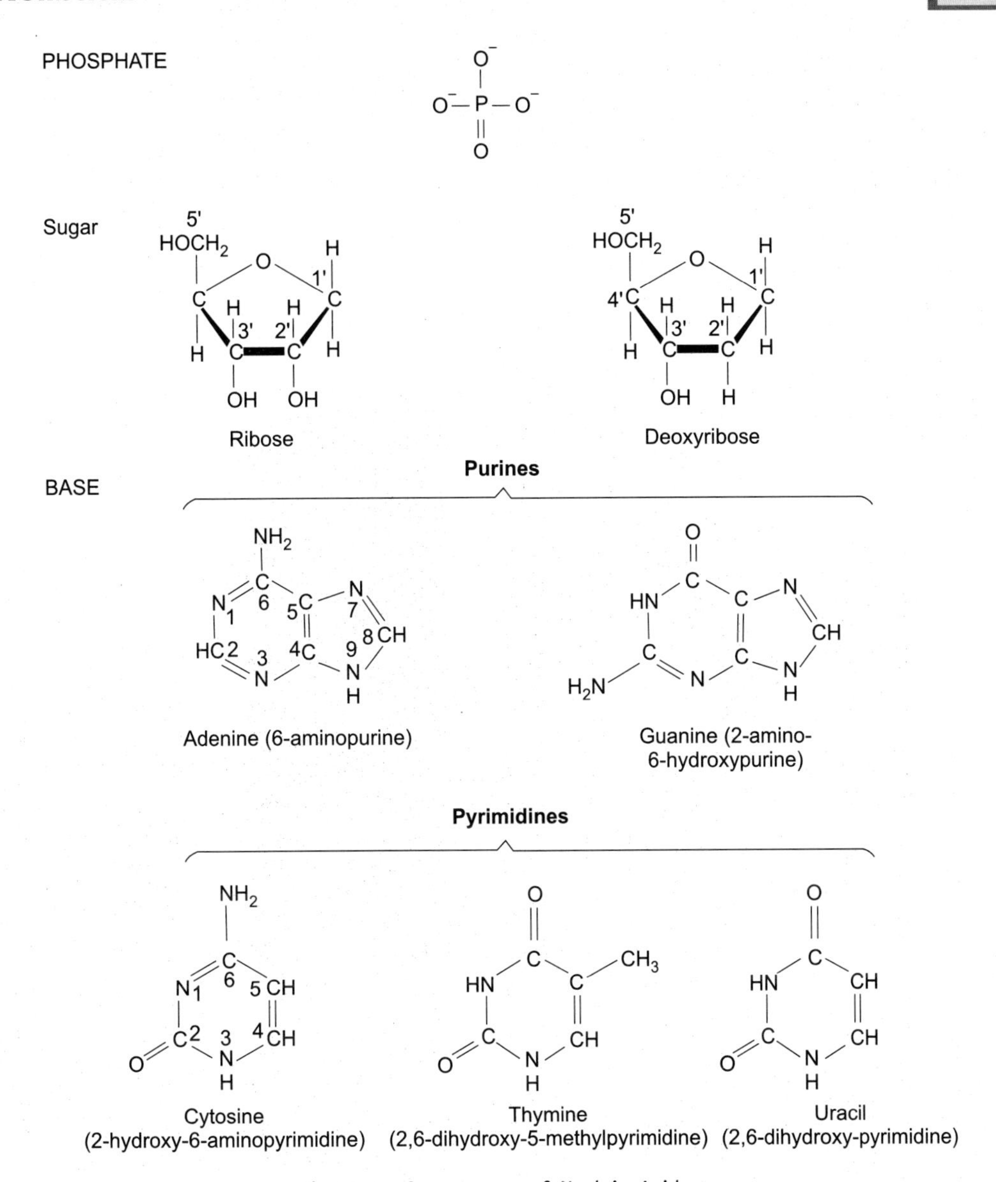

Fig. 9.4 Components of Nucleic Acids.

POLYNUCLEOTIDES

The next stage in building up the structure of a DNA molecule is to link the individual nucleotides together to form a polymer. This polymer is called a polynucleotide and is formed by attaching one nucleotide to another by way of the phosphate groups.

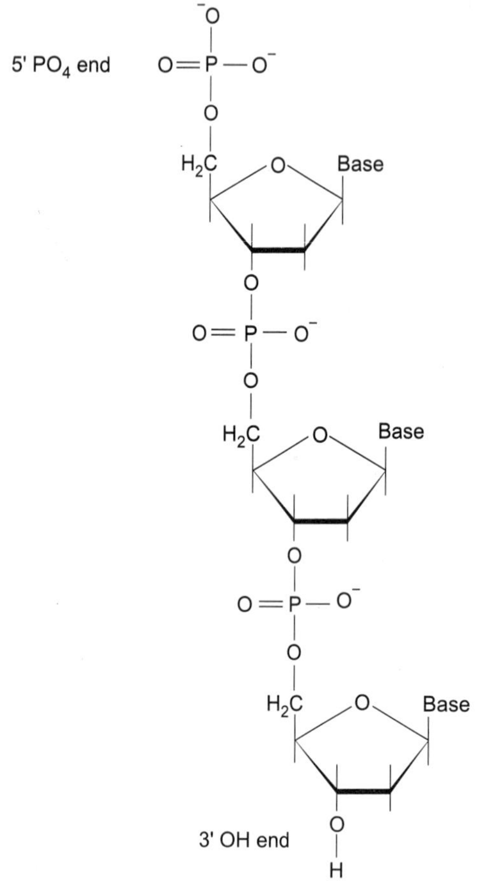

NUCLEOTIDES ARE JOINED BY PHOSPHODIESTER BONDS

The structure of trinucleotide, a short DNA molecule comprising three individual nucleotides is shown in Fig. 9.5. The nucleotide monomers are linked together by joining the α-phosphate group, attached to the 5′-carbon of one nucleotide, to the 3′-carbon of the next nucleotide in the chain. Normally, a polynucleotide is built up from nucleoside triphosphate subunits, so during polymerisation the β- and γ-phosphates are cleaved off. The hydroxyl group attached to the 3′-carbon of the second nucleotide is also lost.

The linkage between the nucleotides in a polynucleotide is called a phosphodiester bond: 'phospho' indicating the presence of a phosphorus atom, and 'diester' referring to the two ester (-C-O-P-) bonds in each linkage. To be precise we should call this a 3′-5′ phosphodiester bond so that there is no confusion about which carbon atoms in the sugar participate in the bond.

Fig. 9.5 Polymerisation of adjacent nucleotides to a sugar-phosphate strand.

POLYNUCLEOTIDES HAVE CHEMICALLY DISTINCT ENDS

An important feature of the polynucleotide is that the two ends of the molecule are not the same. This is clear from an examination of Fig. 9.5. The top of this polynucleotide ends with a nucleotide in which the triphosphate group attached to the 5′-carbon has not participated in a phosphodiester bond and the β- and γ-phosphates are still in place. This end is called the 5′ or 5′-P terminus. At the other end of the molecule, the unreacted group is not the phosphate but the 3′-hydroxyl. This end is called the 3′ or 3′-OH terminus. The chemical distinction between the two ends means that polynucleotides have a direction, which can be looked on as 5'–3′ (down) or 3′–5′ (up). The direction of the polynucleotide is very important in molecular genetics.

POLYNUCLEOTIDES CAN BE OF ANY LENGTH AND HAVE ANY SEQUENCE

There is apparently no limitation to the number of nucleotides that can be joined together to form an individual DNA polynucleotide. Molecules containing several thousand nucleotides are frequently handled in the laboratory and the DNA molecules in chromosomes are much longer, possibly several million nucleotides in length. In addition, there are no chemical restrictions on the order in which the nucleotides can join together. At any point in the chain the nucleotide could be A, G, C or T. Consider a polynucleotide just ten nucleotides in length: it could have any one of 4^{10} = 1,048,576 different sequences. Now imagine the number of different sequences possible for a polynucleotide one thousand nucleotides in length, or one million. This is the variability of DNA that enables the genetic material to exist in an almost infinite number of forms.

As the chain of nucleotides increases, the number of different possible messages increases exponentially (4^n, where n is the number of nucleotides in the molecule), and a DNA chain of 10 nucleotides could be constructed in more than a million (4^{10}) different ways. Thus, if a biological product were determined by a sequence of 10 nucleotides, and a total of 10,000 different biological products are needed by an organism, this could easily be accomplished by a total length of 10 × 10,000 or 100,000 nucleotides.

In reality, many more than 10 nucleotides are needed to code a biological product, since these are usually proteins of considerable length and complexity. At the same time, however, the number of nucleotides in any organism is undoubtedly greater than 100,000 (Table 9.1). In man, for instance, the haploid set of chromosomes contains DNA molecules with a total length of approximately two and three-quarter billion nucleotides. That amount of nucleotides could produce 10,000 different biological products even if each product is of such length and complexity as to be determined by more than 20,000 nucleotides. Since it is likely that organisms probably produce less than 10,000 biological products, and that each product is coded in a length much less than 200,000 nucleotides, we can consider that the amount of DNA is more than sufficient for the "coding" purposes needed. Table 9.1 gives the amount of DNA present in certain organisms listed.

The Double Helix

Following the establishment of a nucleotide structure for DNA (primary structure), the next difficult question concerned the relationship between the individual nucleotide chains. Were these chains present in only single condition; were they randomly placed alongside each other in groups of two, three, or more; or was there a regular relationship between them? In the 1940's, a number of findings already indicated that the DNA molecule was regularly organised. Chargaff and others had shown that, as a rule, the amount of 6-amino bases (A + C) equalled the amount of 6-keto bases (T + G) in any particular species (Table 9.2), signifying, perhaps, a paired relationship between these two types of bases.

RNA IS ALSO A POLYNUCLEOTIDE

DNA is not the only type of nucleic acid found in living cells. The closely related substance RNA (ribose nucleic acid) is also a nucleic acid and, like DNA, plays a vital role in molecular genetics. RNA is also a polynucleotide but with two important differences from DNA.

Table 9.1 Estimated amount of DNA and number of nucleotide pairs in haploid complement of chromosomes in different organisms.

Organism	*Weight of DNA (10^{-12} gram)*	*Number of nucleotide pairs*
man	3.2	2.87×10^9
cattle	2.8	2.51×10^9
chicken	1.3	1.17×10^9
turtle	2.5	2.24×10^9
toad	3.7	3.32×10^9
carp	1.6	1.43×10^9
snail	.67	6.01×10^8
echinoderm (sea urchin, *Lytechinus*)	.90	8.07×10^8
yeast (*Saccharomyces*)	.07	6.28×10^7
Escherichia coli	.0047	4.22×10^6
T2 bacteriophage	.0002	1.80×10^5
$\phi \times 174$ bacteriophage	.0000026	5.50×10^3

Table 9.2 Base composition of DNA in various organisms.

	Purines		*Pyrimidines*			
Organism	*Adenine*	*Guanine*	*Cytosine*	*Thymine*	*Uracil*	*A + G/C + T(U)*
man (sperm)	31.0	19.1	18.4	31.5		1.00
cattle (sperm)	28.7	22.2	22.00	27.2		1.03
rat	28.6	21.4	21.7	28.4		1.00
salmon	29.7	20.8	20.4	29.1		1.02
sea urchin	32.8	17.7	17.4	32.1		1.02
wheat germ	27.3	22.7	22.8	27.1		1.00
yeast	31.3	18.7	17.1	32.9		1.00
Escherichia coli	26.0	24.9	25.2	23.9		1.04
Mycobacterium tuberculosis	15.1	34.9	35.4	14.6		1.00
bacteriophage T2	32.6	18.2	16.6*	32.6		1.03
bacteriophage $\phi \times 174$	24.7	24.1	18.5	32.7		.95
tobacco mosaic virus (RNA)	29.3	25.8	18.1		26.8	1.23
polio virus (RNA)	30.4	25.4	19.5		24.7	1.26
influenza virus (RNA)	23.0	20.0	24.5		32.5	.75

*5-Hydroxymethylcytosine.

1. *Sugar in RNA is ribose* In RNA, the sugar component of the nucleotide is not 2'-deoxyribose, but ribose itself.
2. *RNA contains uracil instead of thymine.* Three of the nitrogenous bases-adenine, guanine and cytosine—are found in both DNA and RNA. However, the fourth base is different. In DNA it is thymine but in RNA it is another pyrimidine, *uracil* (Fig. 9.4).

The names of four nucleotides that polymerise to make RNA are: adenosine 5′-triphosphate; guanosine 5′-triphosphate; cytidine 5′-triphosphate; uridine 5′-triphosphate, which are abbreviated to ATP, GTP, CTP and UTP, or A, G, C and U, respectively. The polynucleotide structure of RNA is exactly the same as that of DNA, with 3′–5′ phosphodiester bonds linking together the individual nucleotides in the molecule. As with DNA there are no restrictions on the sequence of nucleotides within an RNA molecule. There is, however, one very important difference between RNA and DNA at this higher level of structure. RNA in the cell usually exists as a single polynucleotide whereas DNA is almost invariably in the form of two polynucleotides wrapped round one another. This is the famous double helix, the discovery of which convinced molecular geneticists that genes are indeed made of DNA.

The Double Helix

The discovery of the double helix by James Watson and Francis Crick at Cambridge in 1953 was one of the great deductive triumphs in the history of science and has influenced every aspect of molecular genetics, indeed of biology as a whole.

Watson and Crick made use of the results of several lines of investigation concerning the structure of DNA in living cells. They complemented these experimental results by building and assessing accurate scale models of various possible DNA structures. Only the double helix was compatible with all the data. Before examining the double helix itself we must, like Watson and Crick, consider the experimental foundation upon which the model was built.

CHARGAFF'S BASE RATIOS PAVED THE WAY FOR CORRECT STRUCTURE

The discovery by Avery, MacLeod and McCarty that the transforming principle is DNA-influenced another New York research worker, Erwin Chargaff of Columbia University to carry out a comprehensive analysis of the chemical composition of DNA. In particular, Chargaff, together with his colleagues, E. Vischer and S. Zamenhof, used new and sensitive paper chromatographic techniques to determine the exact amounts of each of the four nitrogenous bases in samples of DNA purified from different tissues and different organisms. Their results were quite startling; they revealed a simple mathematical relationship between the proportions of the bases in any one sample of DNA. The relationships that the number of adenine residues equals the number of thymines, and the number of guanines equals the number of cytosines, that is A–T and G–C. Implicit in this is that the total purines (A + G) will equal the total pyrimidines (T + C). This rule is illustrated by Table 9.3 which shows some of the original results published by Chargaff's group.

X-RAY DIFFRACTION ANALYSIS INDICATES THAT DNA IS A HELICAL MOLECULE

The second piece of evidence available to Watson and Crick was the X-ray diffraction pattern obtained when a crystallized DNA fibre is bombarded with X-rays. The theory behind X-ray diffraction analysis is very complex but is based on the fact that the angles at which X-rays are

Table 9.3 Summary of some of Chargaff's results indicating a consistent pattern to the ratios of bases in DNA from different organisms.

Organism	*Ratio of bases in DNA samples*	
	Adenine : Thymine	*Guanine : Cytosine*
Cow	1.04	1.00
Man	1.00	1.00
Salmon	1.02	1.02
Escherichia coli	1.09	0.99

deflected on passages through a crystal will be determined by the three-dimensional structure of the molecules in the crystal. The deflections can be recorded by allowing the X-ray beam, after passage through the crystal to expose a photographic film. The result is a pattern of spots, the positions and intensities of which may allow the structure of the molecule to be deduced.

X-ray diffraction analysis began in the late nineteenth century and has been successfully applied to crystals of many compounds. Its greatest achievement up to 1953 had been in providing the structure of several natural proteins, an example being keratin, a component of wool and hair. Since 1953, it has been used with increasing sophistication to help elucidate the complex three-dimensional structures of proteins such as haemoglobin. But the most famous X-ray diffraction pictures were taken with crystals of DNA by Rosalind Franklin at King's College, London during May 1952, using techniques previously developed by Maurice Wilkins (Fig. 9.6). These pictures show that DNA is a helix with two regular periodicities of 3.4 Å and 34 Å along the axis of the molecule. But how does this relate to Chargaff's base ratios and to the actual structure of DNA?

WATSON AND CRICK SOLVE THE STRUCTURE

The story of how Watson and Crick deduced that DNA exists as a double helix has been told many times, most entertainingly by Watson himself in his book *The Double Helix.* Watson and Crick put together all the experimental data concerning DNA and decided that the only structure that fitted all the facts was the double helix shown in Fig. 9.7. There are seven important features of this helix.

1. *The Double Helix Comprises Two Polynucleotides* The number of polynucleotides in the helix was a problematical question for some time. Several pieces of experimental evidence indicated that the number was two or three, and in fact the famous American Scientist Linus Pauling proposed an incorrect triple helix model for DNA in the months leading up to the Watson-Crick structure. Eventually, the measurement of the density of DNA was refined so that a double-stranded molecule became most likely.
2. *Nitrogenous Bases are Stacked on the Inside of the Helix* The experimental evidence also indicated that the sugar-phosphate 'backbone' of the molecule is on the outside, with the bases inside the helix. In fact, the bases are stacked on top of each other rather like a pile of plates.

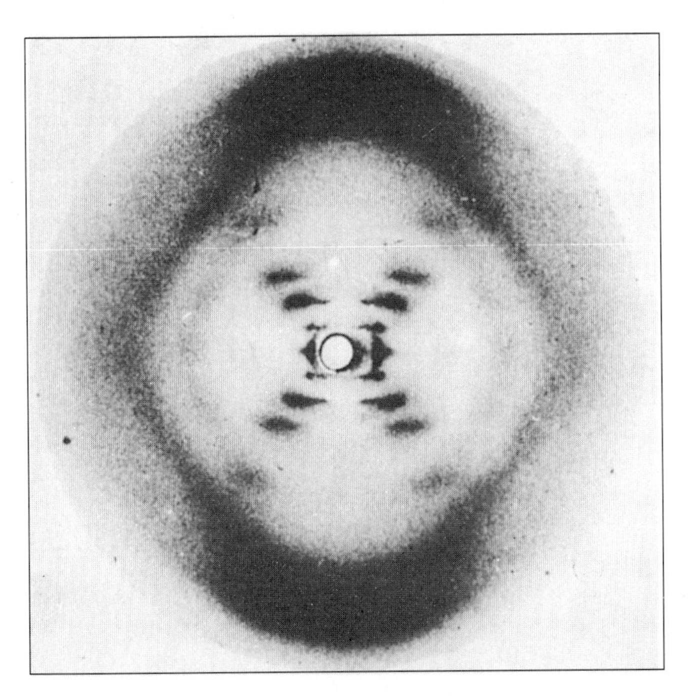

Fig. 9.6 X-ray diffraction picture of B-form of the paracrystalline form of DNA. The molecules are less regularly arranged.

3. *Bases of the Two Polynucleotides Interact by Hydrogen Bonding.* This is the explanation of Chargaff's base ratios. An adenine residue in one of the polynucleotides is always adjacent to a thymine in the other strand; similarly guanine is always adjacent to cytosine. These two pairs of bases, and no other combinations, are able to form hydrogen bonds between each other, two bonds between A and T, and three between G and C (Fig. 9.9). These hydrogen bonds are the only attractive forces between the two polynucleotides of the double helix and serve to hold the structure together.
4. *Ten Base Pairs Occur per Turn of the Helix* (Fig. 9.8) The double helix executes a turn every ten base pairs (abbreviated as 10 bp). The pitch of the helix is 34 Å; meaning that the spacing between adjacent base pairs is 3.4 Å and each is turned 36° from the preceding one, so that a complete turn of 360° involves 10 stairs or 10 box pairs, and is 34 Å long (1 Å = 0.0001 micron). These are the periodicities apparent from the X-ray diffraction pattern. The helix is 20 Å in diameter.
5. *Two Strands of the Double Helix are Antiparallel* One polynucleotide runs in the 5′–3′ direction, the other in the 3′–5′ direction. Only antiparallel polynucleotides will form a stable helix.
6. *The Double Helix has Two Different Grooves* The helix is not absolutely regular, a major and a minor groove can be distinguished. This feature is important in the interaction between the double helix and the proteins involved in DNA replication and in the expression of genetic information.

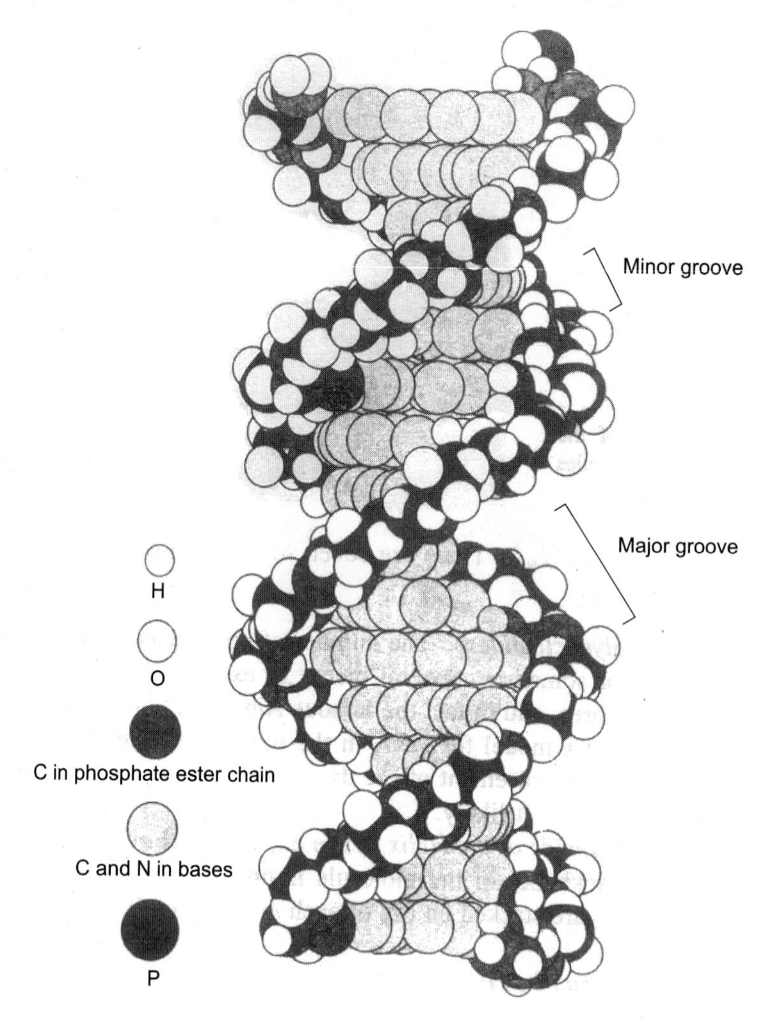

Fig. 9.7 Molecular model of DNA double helix.

7. *The Double Helix is Right-handed* This means that if the double helix was a spiral staircase that you were climbing up, the banister (= the sugar-phosphate backbone) would be on your right-hand side.

COMPLEMENTARY BASE PAIRING IS THE FUNDAMENTAL FACT OF MOLECULAR GENETICS (Fig. 9.9)

Of these various features, it is the base pairing that is most important. The rule is that A base pairs with T, and G base pairs with C. Other base pairs are not normally permitted because they will either be too large to fit in the helix (as the case with a purine-purine base pair), or too small for the helix (pyrimidine-pyrimidine), or will not align in a manner allowing hydrogen-

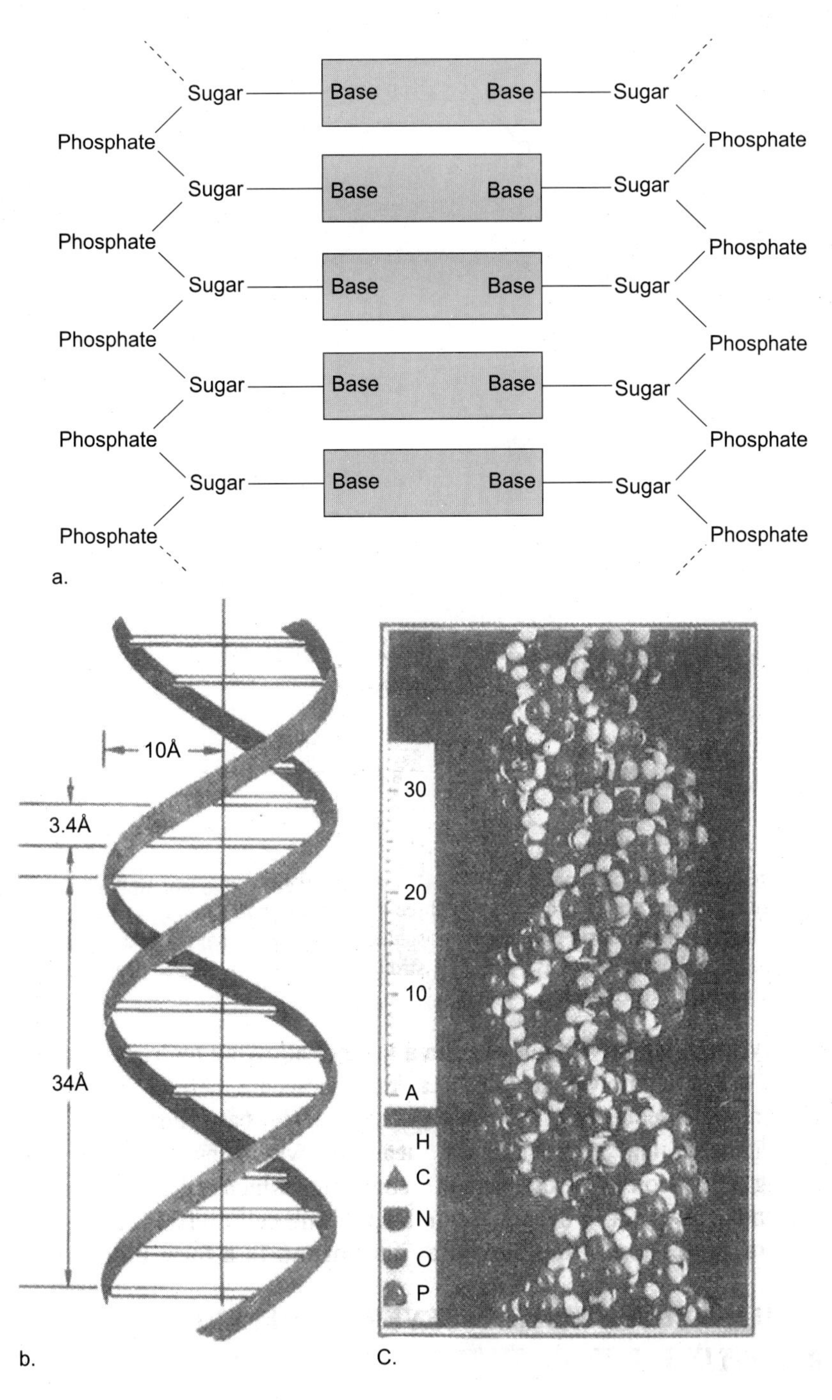

Fig. 9.8 Double Helix Structure of DNA: (a) Component parts; (b) Line drawing; (c) Space-filling model.

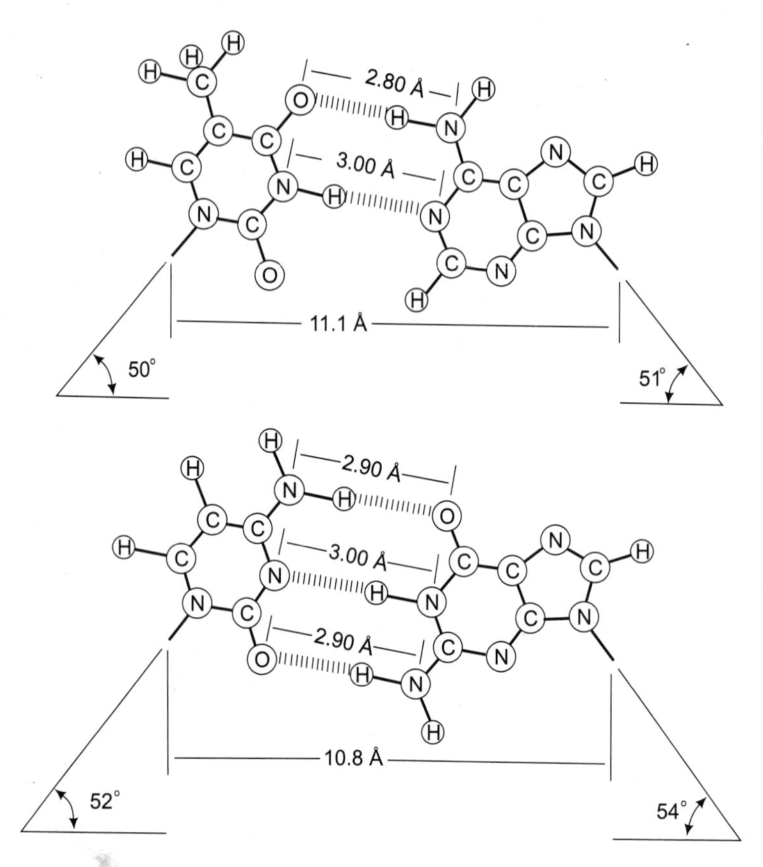

Fig. 9.9 Hydrogen-bonded base pairs in DNA. Adenine is always attached to thymine by two hydrogen bonds, whereas guanine bonds to cytosine by three hydrogen bonds. Obligatory pairing of the smaller pyrimidine with larger purine allows two sugar-phosphate backbones to have identical helical configurations. All the hydrogen bonds in both base pairs are strong, since each hydrogen atom points directly at its acceptor atom (nitrogen or oxygen).

bonding to occur (as with A–C, and G–T). The two polynucleotides in a double helix are, therefore, complementary, the sequence of one determining the sequence of the other. Complementary base pairing is of fundamental importance in molecular genetics and it provides the means by which the sequence of a DNA molecule is retained during replication of the double helix, something which is crucial if the biological information contained in the gene is not to become scrambled during cell division. Complementary base pairing is also vital for expression of the biological information in a form utilisable by the cell.

THE DOUBLE HELIX EXISTS IN SEVERAL DIFFERENT FORMS

By the time that Watson and Crick discovered the double helix it had already been shown by Rosalind Franklin that there are two distinct crystalline forms of DNA, the *A-form* and *B-form*.

The form taken up depends on the amount of water present in the DNA solution from which the crystals form. The Watson-Crick structure refers to the B-form, of which the classic X-ray diffraction pictures were taken. The B-form is, in fact, the structure that DNA takes up in the cell, at least under most conditions.

The A-form differs from the B-form in relatively minor though important ways. A-DNA is still a double helix, but more compact than B-DNA, with 11 bp per turn of the helix and a diameter of 23 A. Since 1953, four other forms of DNA have been found C-, D-, E- and Z-DNA, each with a slightly different conformation to the double helix. Z-DNA is the most strikingly different, as in this structure the helix is left-handed, not right-handed as it is with A-to E-DNA. These alternative forms of DNA have received increasing attention in recent years as it has been shown that the nucleotide sequence is one of the factors that influence the form taken by a segment of the double helix. Regions of DNA that are predisposed, by virus of their nucleotide sequences, to take up non-standard double helices (e.g. *Z-DNA regions*) have recently been discovered in chromosomes.

Note: Having known about DNA, its molecular structure and its scope, before actually going into a detailed discussion of the *definition of a gene*, in molecular terms, a summary at a glance is given here after the '*Molecular Genetics in a Nutshell*'. This will give the reader an idea about the extent to which Molecular Genetics can be discussed and understood. When once the structure of DNA is elucidated, then its function and role in the cell dynamics has to be understood in detail. This is the molecular aspect of the 'Gene' or Genetics. The ultimate goal of all genetics is to define in the clear cut terms, the 'gene'. It is a very complicated issue to define a gene in the modern sense, when compared to classical picture of gene presented in the section on Mendelian Genetics. The hypothesis of *one-gene-one enzyme* relationship is 'replaced by *one-gene-one polypeptide* hypothesis from a molecular genetics aspect at the present status of the gene. In future, it may still be broken up into pieces up to the atomistic precision with refined techniques in molecular genetics. From *E. coli* to man, the distance is very long to traverse on gene structure and function. The definition of gene is thus very complicated as can be seen in the few chapters, presented in this section of Molecular Genetics. At the same time, it is refined up to the nucleotide level that the present day understanding of the gene is so simple and crystal clear. This point will be borne out in the chapters on protein synthesis, genetic code, mutations, and recombination DNA technology, etc.

MOLECULAR GENETICS: AN INTRODUCTION

(A) Storage of Genetic Information

1. The genetic information of a cell is stored within its DNA macromolecules.
2. Structure of DNA:
   (a) The DNA is composed of nucleotides that are linked together.
   (b) A nucleotide consists of a nucleic acid base, a deoxyribose sugar and a phosphate group.

(c) Four nucleic acid bases occur in DNA: cytosine, guanine, adenine and thymine.

(d) The deoxyribose sugars are linked by phosphate diester bonds. The linkage establishes a directional orientation to the DNA. Nucleotides are linked by 3′–5′ phosphate diester linkages. At the ends of the DNA strand, there are no linkages and free hydroxyl groups are present. One end has a free hydroxyl group at the carbon 3 of the monosaccharide (3′-hydroxyl free end); the other end of strand has free hydroxyl group at the carbon 5 monosaccharide (5′-hydroxyl free end).

(e) DNA is a double-helix molecule composed of two primary polynucleotide chains. The chains are held together by hydrogen bonding between complementary nucleotide bases.

(f) Complementary base pairs are adenine and thymine, which are held together by two hydrogen bonds and guanine and cytosine which are held together by three hydrogen bonds.

(g) Complementary strands are antiparallel; where one strand has the 3'-hydroxyl free end, the complementary strand will have the 5'-hydroxyl free end.

3. A genome is a single copy of the genetic information of the cell.
4. The coded information within the genome is used to control the synthesis of proteins.
5. The sequence of bases determines the sequence of amino acids in proteins.
6. A gene is a segment of a genome.
7. Alleles are different forms of the same gene.
8. Prokaryotic cells have a single genome and, therefore, are haploid.
9. Eukaryotic cells generally have pairs of matching chromosomes, making them diploid.
10. Plasmids are small extrachromosomal genetic elements that permit microorganisms to store additional genetic information.
11. Plasmids contain genetic information for specialised features rather than for essential metabolic activities like F (fertility) plasmid codes for mating behaviour in *E. coli* and R (resistance) plasmids carry genes that code for antibiotic resistance.

(B) DNA Replication

1. Transmission of hereditary information necessitates the faithful replication of DNA.
2. DNA is replicated by a semiconservative process; after replication, half the double helix (that is, one strand) has been retained (conserved) from the original DNA, and the other strand (called the complementary strand) has been synthesised.
3. In DNA replication, one strand acts as a template for the complementary strand (which contains the complements of the bases on the template); adenine is aligned opposite thymine and cytosine is aligned opposite guanine.
4. The site of replication is called the replication fork.
5. Replication proceeds slightly differently on each of the parent strands; one strand of the DNA is called the leading (continuous) strand because its formation is in the direction of

the movement of the replication fork; the other strand is called the lagging or discontinuous strand because DNA polymerase synthesises small segments of DNA; ligases attach the short segments of DNA.

6. In prokaryotes, DNA replication starts at only one site; in eukaryotes DNA replication is initiated at multiple sites.

(C) Recombination

1. Recombination involves a reshuffling of genes.
2. In some recombination processes there is a transposition in the locations of genes.
3. In many recombination processes there is an exchange of genetic information from different sources that results in new combinations of genetic information.

(D) Genetic Exchange

1. Genetic exchange in prokaryotes occurs by transformation, transduction and conjugation.
2. In transformation, a free DNA molecule is transferred from donor to recipient bacterium.
3. In transduction, DNA is transferred from donor to recipient cells by a viral carrier; the virus acquires a portion of the genome of the host (donor) cell and transfers it to the recipient bacterium.
4. Conjugation (mating) requires physical contact between donor and recipient bacteria; there are different mating pairs that differ in the extent of successful genetic exchange (recombination) and whether the progeny are donor or recipient strains.
5. In eukaryotic microorganisms, genetic exchange normally occurs through sexual reproduction.
6. The vegetative cells of eukaryotic microorganisms typically are diploid, and sexual reproduction normally involves formation of specialised reproduction gametes or spores that are haploid; conversion of diploid to haploid states occurs during meiosis (reductive division); haploid nuclei of reproductive cells later fuse with nuclei of reproductive cells of an appropriate mating type (syngamy), re-establishing the diploid state.

(E) Genetic Engineering

1. Recombinant DNA technology can be used to engineer the creation of new microorganisms.
2. Genetic engineering often uses plasmids as carriers of unrelated DNA; it employs enzymes involved in normal recombination and replication of DNA for splicing foreign DNA into plasmid carriers.
3. A typical procedure for genetically engineering a new organism is to isolate plasmids or another vector; to cut open the plasmid DNA by using site-specific endonuclease (restriction enzyme) to create an insertion site for foreign donor DNA; to use

endonucleases to form a segment of donor DNA; to pair the plasmid and donor DNA; to use ligases to seal the circular plasmid; and to add the plasmid to a culture of recipient bacterium, which incorporates it as part of the bacterial genome; to clone the new organism.

4. Many economically beneficial products can be formed by genetically engineered organisms.
5. Genetic engineering permits the exchange of genetic information across species lines, including the intermixing of microbial, plants and animal genes; bacteria have been created by recombinant DNA technology that produce human proteins, and attempts are underway to create plants containing bacterial genes.

(F) Expression of Genetic Information

1. The genetic information that is contained within the genome represents the genotype, not all of the genotype is expressed at all times.
2. The genetic information that is expressed appears at the phenotype, the discernible characteristics of an organism.
3. Protein synthesis occurs in two stages: transcription and translation.
4. Transcription:
   (a) In transcription, the information in the DNA is transferred to RNA.
   (b) During transcription DNA serves as a template that determines the order of the bases in the RNA.
   (c) The RNA that is formed by transcription is complementary to the DNA.
   (d) In prokaryotes, transcription produces mRNA that is not extensively modified after synthesis; the mRNA is colinear with the bases of the DNA.
   (e) In eukaryotes transcription produces mRNA which is extensively modified to form the mature mRNA; the mRNA is not co-linear with the DNA because of excision of introns and other modifications during post-transcriptional processing.
5. Translation:
   (a) In translation, the mRNA is used to establish a sequence of amino acids that make up the protein.
   (b) Translation occurs at the ribosomes.
   (c) The genetic code has 64 possible codons; each codon is a triplet containing three nucleotides.
   (d) There is more than one codon for each amino acid; therefore, the genetic code is said to be degenerate because different codons can specify the same amino acid.
   (e) Nonsense codons are those for which there are no amino acids; the nonsense codons signal termination of synthesis of polypeptide chains.
   (f) The ribosome moves along the mRNA exposing one codon at a time.

(g) As each triplet is exposed by the ribosome, a tRNA brings the specified amino acid to the ribosome; the tRNA has an anticodon region that is complementary to the codon and is responsible for bringing the correct amino acid specified by the codon.

(h) The ribosome moves to the next triplet and the process is repeated.

(i) The amino acid of the second tRNA bonds to the amino acid carried by the first tRNA.

(j) As soon as the bonding between amino acids takes place, the bonds between the first tRNA and its amino acid and between it and the mRNA are broken and it leaves the ribosome.

(G) Mutations

1. A mutation is a change in the sequence of bases in the DNA.
2. Types of mutations:
   - (a) A lethal mutation results in the death of a microorganism or in its inability to reproduce; a conditionally lethal mutation exerts an effect only under certain environmental conditions.
   - (b) Temperature-sensitive mutations alter the range of temperatures over which the microorganisms may grow.
   - (c) Nutritional mutations alter the nutritional requirement for the progeny; nutritional mutants (auxotrophs) require growth factors not needed by the parental (prototrophic) strain.
3. Factors influencing rates of mutation:
   - (a) Some chemicals, called mutagens, increase the rate of mutation.
   - (b) Ames test procedure utilises *Salmonella typhimurium* as test organism for determining chemical mutagenicity and potential carcinogenicity.
   - (c) High-energy ionising radiation causes mutation and can be used for sterilising objects.
   - (d) Ultraviolet light can cause mutations.

(H) Regulation of Genetic Expression

1. The expression of genetic information can be regulated at the level of transcription.
2. Constitutive enzymes are continuously synthesised at a constant rate and are not regulated.
3. Inducible enzymes are made only at appropriate times, for example, when synthesis is induced by appropriate factors.
4. Operon model of gene control explains the basis of control of transcription.
5. An operon consists of *structural* genes that contain the code for making proteins; an *operator region*, which is the site where repressor protein binds and prevents RNA

transcription; a *promoter* region, which is the site where RNA polymerase binds; and a *regulator* gene which codes for the repressor protein.

6. The lac operon regulates the utilisation of lactose.
   (a) The regulator *r* gene codes for repressor protein.
   (b) The repressor protein binds to operator region and prevents structural genes from transcribing the mRNA needed to synthesise the enzymes for lactose catabolism.
   (c) In the presence of lactose, an inducer binds to the repressor protein preventing it from binding to the operator region of the operon; this results in depression of lac operon and structural genes needed for the utilisation of lactose are transcribed until the lactose has been broken down.
   (d) When lactose is used up, no inducer is present and the repressor protein is free to bind to the operator region; when this happens, the lac operon is repressed and the structural genes are no longer transcribed.
7. Catabolite repression is a generalised type expression.
   (a) Catabolite repression supersedes the control exerted by the operator region.
   (b) Catabolite repression acts via a promoter region of DNA by blocking the normal attachment of RNA polymerase; a catabolite activator protein is needed to bind RNA polymerases to the promoter region and cyclic AMP is required for efficient binding to occur.
   (c) In the presence. of glucose, the amount of cyclic AMP is reduced; therefore, the catabolite activator protein cannot bind to the promoter and transcription is unable to occur.

GENOME

Synopsis

Chromosome are the carriers of genetic information. A number of early observations led biologists to consider the genetic role of chromosomes. These included observations of the precision with which chromosomes are divided between the daughter cells during cell division; the realization that the chromosomes of a species remained constant in shape and number from one of division to the next; the findng that embryonic development required a particular complement of chromosomes; and the observations that the chromosome number is divided in half prior to the formation of the gametes and is doubled following the union of a sperm and egg at fertilization. Mendel's findings provided biologists with a new set of criteria for identifying the carriers of the genes. Sutton's studies of gamete formation in the grasshopper revealed the existence of homologous chromosomes, the association of homologues during the cell divisions that preceded gamete formation, and the separation of the homologues during the first of these meiotic divisions.

If genes are packaged together on chromosomes that are passed form parents to offspring, then genes on the same chromosome should be linked to one another to form a linkage group. The existence of linkage groups was confirmed in various systems, particularly in fruit flies where dozens of mutations were found to assort into four linkage groups that correspond in size and number to the chromosomes present in the cells of these insects. At the same time, it was discovered that linkage was incomplete; that is, the alleles originally present on a given chromosome did not necessarily remain together during the formation of the gametes, but could be reshuffled between maternal and paternal homologues. This phenomenon, which Morgan called crossing over, was found to occur as the result of breakage and reunion of segments of homologous chromosomes that occurred while the homologues were physically associated during the first meiotic division. Analyses of offspring from matings between adults carrying a variety of mutations on the same chromosome indicated that the frequency of recombination between two genes provided a measure of the relative distance that separates those two genes. Thus, recombination frequencies could be used to prepare detailed maps of the serial order of genes along each of the chromosomes of a species. Genetic maps of the fruit fly based on recombination frequencies were verified independently by examination of the locations of various bands in the giant polytene chromosomes found in certain larval tissues of these insects.

Experiments discussed in the Experimental Pathway provided conclusive evidence that DNA was the genetic material. DNA is a helical molecule consisting of two chains of nucleotides running in opposite directions with their backbones on the outside and the nitrogenous bases facing inward. Adenine-containing nucleotides on one strand always pair with thymine-containing nucleotides on the other strand, likewise for guanine- and cytosine-containing nucleotides. As a result, the two strands of a DNA molecule are complementary to one another. Genetic information is encoded in the specific linear sequence of nucleotides that make up the strands. The Watson-Crick model of DNA structure suggested a mechanism of replication that included strand separation and the use of each strand as a template that directed the order of nucleotide assembly during construction of the complementary strand. The mechanism by which DNA governed the assembly of a specific protein remained a total mystery. The model of DNA depicted in Fig. 10.9 is that of B-DNA, which is one of several right-handed helical forms. B-DNA contracts most markedly with Z-DNA, which takes the form of a left-handed helix in which the backbone assumes a zigzag conformation. The DNA molecule depicted in Fig. 10.9 is in a relaxed state having 10 base pairs per turn of the helix. DNA found whitin a cell tends to be underwound (contains a greater number of base pairs per turn) and is said to be negatively supercoiled, a condition that tends to facilitate the separation of strands that occurs during replication and transcription. The supercoiled state of DNA is altered by topoisomerases, enzymes that are able to cut, rearrange, and reseal DNA strands.

All of the genetic information present on a single haploid set of chromosomes of an organism constitutes on organism's genome. The variety of DNA sequences that make up the genome and the numbers of copies of these various sequences describe the complexity of genome. Understanding the complexity of a genome has grown out of early studies showing that the two strands that make up a DNA molecule can be separated by heat; when the temperature of the solution is lowered, complementary single strands are capable of reassociating to form stable double-stranded DNA molecules. Analysis of the kinetics of this reassociation process provides a measure of the concentration of complementary sequences, which in turn provides a measure of the variety of sequences that are present within a given quantity of DNA. The greater the number of copies of a particular sequence in the genome, the greater is its concentration and the faster it reanneals.

When DNA fragements from eukaryotic cells are allowed to reanneal, the curve typically shows three rather distinct steps, which correspond to the reannealing of three different classes of DNA sequences. The highly repeated fraction consists of short DNA sequences that are repeated in great number; these include satellite DNAs situated at the centromeres of the chromosomes, minisatellite DNAs, and microsatellite DNAs. The latter group tend to be highly variable, causing certain inherited diseases and forming the basis of DNA fingerprint techniques. The moderately repeated fraction includes DNA sequences that encode ribosomal and transfer RNAs and histone proteins, as well as various sequences with noncoding functions. The nonrepeated fraction contains protein-coding genes, which are present in one copy per haploid set of chromosomes.

The sequence organization of the genome is capable of change, either slowly over the course of evolution of rapidely as the result of transposition. The genes encoding eukaryotic proteins are often members of multigenes families whose members show evidence that they have evolved from a common ancestral gene. The first step in this process is thought to be the duplication of a gene which probably occurs primarily by unequal crossing over. Once duplication has occurred, nucleotide substitutions would be expected to modify various members in different ways, producing a family of repeated sequences of similar but not identical structure. The globin genes, for example, consist of clusters of genes located on two different chromosomes. Each cluster contains a number of related genes that code for globin polypeptides that are used at different stages in the life of an animal. The clusters also contain pseudogenes, which are homologous to the globin genes, but are nonfunctional. The duplication of genes over the course of generations can be followed in the laboratory by exposing cultured cells to drugs that select for cells that have increased copies of a particular protein.

Certain DNA sequences are capable of moving rapidly from place to place in the genome by transposition. These transposable elements are called transposon, and they are capable of integrating themselves randomly throughout the genome. The best-studied transposons occur in bacteria. They are characterized inverted repeats at their termini, an internal segment that codes for a transposase required for their integration, and the formation of short repeated sequences in the host DNA that flank the element at the site of integration. Bacterial transposons often contain genes for antibiotic resistance, which has contributed to the spread of resistance strains of bacteria. Eukaryotic transposable elements are capable of moving by several mechanisms. Some are replicated, and the copy is inserted into a target site, leaving the donor site unchanged. In other cases, the element is excised from the donor site and inserted at a target site. In most cases, the element is transcribed into an RNA, which is copied by a reverse transcriptase encoded by the element, and the DNA copy is integrated into the target site. The moderately repeated fraction of human DNA contains two large families of transposable elements, the *Alu* and L1 families.

A given sample of DNA can be cleaved into a defined set of fragments by treatment with a bacterial restriction endonuclease that recognizes a specific sequence of four to eight nucleotides and cuts both strands of the DNA duplex. Different enzymes cleave the same preparation of DNA inot different sets of restriction fragments. The sites within the genome that are cleaved by various enzymes can be identified and ordered into a restriction map. When DNA from different people (even nonidentical siblilngs) are treated with a restriction enzyme, the lengths of the fragments are similar, but not identical, due to differences in nucleotide sequences from one person to the next. Such differences are called restriction fragment length polymorphisism (RFLPs). RFLPs can be used to identify individuals by preparation of a DNA fingerprint or to track the inheritance of a particular gene.

THE CHEMICAL NATURE OF THE GENE

Three years after Gregor Mendel presented the results of his work on inheritance in pea plants, Friedrich Miescher graduated from a Swiss medical school and traveled to Tubingen, Germany, to spend a year studying under Ernst Hoppe-Seyler, one of the foremost chemists (and possibly the first biochemist) of the period. Miescer was interested in the chemical contents of the cell nucleus. To isolate material from cell nuclei with a minimum of contamination from cytoplasmic components, Miescher needed cells that had large nuclei and were easy to obtaine in quantity. He chose white blood cells, which he obtained from the pus in surgical bandages that were discarded by a local clinic. Miescher treated the cells with dilute hydrochloric acid to which he added an extract from pig's stomach that removed protein (the stomach extract contained the proteolytic enzyme pepsin). The residue from this treatment was composed primarily of isolated cell nuclei that settled to the bottom of the vessel. Miescher then extracted the nuclei with dilute alkali. The alkali-soluble material was further purified by precipitation with dilute acid and reextraction with dilute alkali. Miescher found that the alkaline extract contained a substance that had properties unlike any previously discovered: the molecule was very large, acidic, and rich in phosphorus. He called the material "nuclein." His year up, Miescher returned home to Switzerland while Hopper-Seyler, who was cautious about the findings, repeated the work. After the results were confirmed, the paper was published in 1871.

Back in Switzerland, Miescher continued his studies of the chemistry of the cell nucleus. living near the Rhine River, Miescher had ready access to salmon that had swum upstream and were ripe with eggs or sperm. Sperm were ideal cells in which to study nuclei. Like white blood cells, they could be obtained in large quantiy, and 90 percent of their volume was occupied by nuclei. Miescher's nuclein preparations from sperm cells contained a higher percentage of phosphours (almost 10 percent by weight) than those from white blood cells, indicating they had less contaminating protein. In fact, they were the first preparations of relatively pure DNA. The term "nucleic acid" was coined in 1889 by Richard Altmann, one of Miescher's students, who worked out the methods of purifying protein-free DNA from various animal tissues and yeast.

During the last two decades of the nineteenth century, numerous biologists focused on the chromosomes, describing their behaviors during and between cell division. One way to observe chromosomes was to stain these cellular structures with dyes. A botanist named E. Zacharias discovered that the same stains that made the chromosomes visible also stained a preparation of nuclein that had been extracted using Miescher's procedure of digestion with pepsin in an HCl medium. Furthermore, when the pepsin/HCl-extracted cells were subsequently extracted with dilute alkali, a procedure that was known to remove nuclein, the cell residue (which included the chromosomes) no longer contained stainable material. These and other results pointed strongly to nuclein as a componet of the chromosomes. In one remarkably far-sighted proposal, Otto Hertwig, who had been studying the behavior of the chromosomes during fertilization, stated in 1884, "I believe that I have at least made it highly probable that nuclein is the substance that is responsible not only for fertilization but also for the transmission of hereditary characteristics." Ironically, as more was learned about the properties of nuclein, the less it was considered as a candidate to be the genetic material.

During the fifty years that followed Miescher's discovery of DNA, the chemistry of the molecule and the nature of its components were described. Some of the most important contributions in this pursuit were made by Phoebus Aaron Levene, who immigrated from Russia to the United States in 1891 and eventually settled in a position at the Rockefeller Institute in New York. It was Levene who finally solved on of the most resistant problems of DNA chemistry when he determined in 1929 that the sugar of the nucleotides was 2-deoxyribose. To isolate the sugar, Levene and E.S. London placed the DNA into the stomach of a dog through a surgical opening and then collected the sample from the animal's intestine. As it passed through the stomach and intestine, various enzymes of the animal's digestive tract acted on the DNA, carving the nucleotides into their component parts, which could then be isolated and analyzed. Levene summarized the state of knowledge on nucleic acids in a monograph published in 1931.

Although Levene is credited for his work in determining the structure of the building blocks of DNA, he is also credited as having been the major stumbling block in the search for the genetic material. Through this period, it became increasingly evident that proteins were very complext and exhibited great specificity in catalyzing a remarkable variety of chemical reactions. DNA, on the other hand, was thought to be composed of a monotonous repeat of its four nucleotide building blocks. The major proponent of his view of DNA, which was called the tetranucleotide theory, was Phoebus Levene. Since chromosomes consisted of only two componets–DNA and protein–there seemed little doubt that protein was the genetic material.

Meanwhile, as the structure of DNA was being worked out, a seemingly independent line of research was being carried out in the field of bacteriology. During the early 1920s, it was found that a number of species of pathogenic bacteria could be grown in the laboratory in two different forms. Virulent bacteria, that is, bacterial cells capable of causing disease, formed colonies that were smooth, dome shaped, and regular. In contrast, nonvirulent bacterial cells grew into colonies that were rough, flat, and irregular. The British microbiologist J.A. Arkwright introduced the terms smooth (S) and rough (R) to describe these two types. Under the microscpe, cells that formed S colonies were seen to be surrounded by a gelatinous capsule, whereas the capsule was absent from the cells in the R. colonies. The bacterial capsule helps protect a bacterium from its host's defenses, which explain why the R cells, which lacked these structures, were unable to cause infections in laboratory animals.

Because of its widespread impact on human hleath, the bacterium responsible for causing pneumonia (*Streptoroccus pneumoniae*, or simply pneumonoccus) has long been a focus of attention among microbiologists. In 1923, Frederick Griffith, a medical officer at the British Ministr of Health, demonstrated that pneumococcus also grew as either S or R colonies and, furthermore, that the two forms were interconvertible; that is, on occasion and R bacterium could revert to an S form, or vice versa. Griffith found, for example, if he injected exceptionally large numbers of R bacteria into a mouse, the animal frequently developed pneumonia and produced bacteria that formed colonies of the S form.

It had been shonw earlier that pneumococcus occurred as several distinct types (types I, II and III) that could be distinguished from one another immunologically. In other works, antibodies could be obtained from infected animals that would react with only one of the three types. Moreover, a bacterium of one type never gave rise to cells of another type. Each of the three types of pneumococcus could occur as either the R or S form.

In 1928, Griffith made a surprising discovery when injecting various bacterial preparations into mice. Injections of large numbers of heat-killed S bacteria or small numbers of living R bacteria, by themselves, were harmless to the mouse. However, if he injected both of these preparations into the same mouse, it contracted pneumonia and died. Virulent bacteria could be isolated from the mouse and cultured. To extend the findings, he injected combinations of bacteria of different types. Initially, eight mice were injected with heat-killed type I S bacteria together with a small inoculum of a live type II R strain. Two of the eight animals contracted pneumonia, and Griffith was able to isolate and culture virulent type I S bacterial cells from the infected mice. Since there was no possibility that the heat-killed bacteria had been brought back to life, Griffith concluded that the dead type I cells had provided something to the live nonencapsulated type II cell s that *transformed* them inot an encapsulated type I form. The transformed bacteria continued to produce type I cells when grown in culture, thus the change was stable and permanent. Griffith went on to show that type I R strains could be permanently transformed into either type II or type III S, and vice versa. In all cases, transformation appeared to be type specific, predicatble, and inheritable.

Griffith's finding of transformation was rapidly confirmed by several laboratories around the world, including that of Oswlad Avery, an immunologist at the Rockefeller Institute, the same institution where Levene was working. Avery had initially been skeptical of the idea that a substance released by a dead cell could alter the appearance of a living cell, but he was convinced when Martin Dawson, a young associate in his lab, confirmed the same results. Dawson went on to show that transformation need not occur within a living animal host. A crude extract of the dead S bacteria, when mixed in bacterial culture with a small number of the nonvirulent (R) cells in the presence of anit-R serum, was capable of converting of the R cells into the virulent S form. The transformed cells were always of the type (I, II or III) characteristic of the dead S cells.

The next major step was taken by J. Lionel Alloway, another member of Avery's lab, who was able to solubilize the transforming agent. This was accomplished by rapidly freezing and thawing the killed donor cells, then heating the disrupted cells, centrifuging the suspension, and forcing the supernatant through a porcelain filter whose pores blocked the passage of bacteria. The soluble, filtered extract possessed the same transforming capacity as the original heat-killed cells.

For the next decade Avery and his colleagues focused their attention on purifying the substance responsible for transformation and determining its identity. As remakrable as it may seem today, no other laboraroty in the world was pursuing the identity of the "transforming principle," as Avery called it. Progress on the problem was slow. Eventually, Avery and his co-workers, Colin MacLeod and Maclyn McCarty, succeeded in isolating a substance from the soluble extract that was active in causing transformation when present at only 1 part per 600 million. All the evidence suggested the active substace was DNA: (1) it exhibited a host of chemical properties characterisitc of DNA; (2) no other type of material could be detected in the preparation; and (3) tests of various enzymes indicated that only those enzymes that were capable of digesting DNA were able to inactivate the transforming principle.

The paper published in 1944 was written with scrupulous caution and made no dramatic statements that genes were made of DNA rather than protein. The paper drew remarkably little attention. Maclyn McCarty, one of the three authors, describes an incident in 1949 when he was asked to speak at Johns Hopkins University along with Leslie Gay, who had been testing the effects of the new drug Dramamine for the treatment of sea sickness. The large hall was packed with people, and "after a short period of questions and discussion following {Gay's} paper, the president of the Society got up to introduce me as the second speaker. Very little that he said could be heard because of the noise created by people streaming out the hall. When the exodus was complete, after I had given the first few minutes of my talk, I counted approximately thirty-five hardy souls who remained in the audiene because they wanted to hear about pneumococcal transformation or because they felt they had to remain out of courtesy." But Avery's awareness of the potential of his discovery was revealed in a letter he wrote in 1943 to his brother Roy, also a bacteriolgist:

> If we are right, & of course that's not yet proven, then it means that nucleic acids are not merely structurally important but functionally active substances in determining the biochemical activities and specific characterisitcs of cells—& that by means of a known chemical substance it is possbile to induce predictable and hereditary changes in cells. This is something that has long been the dream of geneticists.... Sounds like a viurs—may be a gene. But with mechanisms I am not now concerned—one step at a time.... Of course the problem bristles with implications... .It touches genetics, enzyme chemistry, cell metabolism & carbohydrate synthesis—etc. But today it takes a lot of well documented evidence to convince anyone that the sodium salt of deoxyribose nucleic acid, protein free, could possibly

be endowed with such biologically active and specific properties and that evidence we are now trying to get. It's lots of fun to blow bubbles – but it's wiser to prick them yourself before someone else tries to.

Many articles and passages in books have dealt with the reasons why Avery's findings were not met with greater acclaim. Part of the reason may be due to the subdued manner in which the paper was written and the fact that Avery was a bacteriologist, not a geneticist. Some biologists were persuaded that Avery's preparation must have been contaminated with minuscule amounts of protein and that the contaminant, not the DNA, was the active transforming agent. Others questioned whether studies on bacteria published in a medical journal had any relevance to the field of genetics, and they viewed the phenomenon of transforamtion as a bacterial peculiarity.

During the years following the publication of Avery's paper, the climate in genetics changed in an important way. The existence of the bacterial chromosome was recognized, and a number of prominent geneticists turned their altenation to these prokaryotes. These scientists believed that knowledge gained from the study of the simplest cellular organisms would shed light on the mechanisms that operate in the most complex plants and animlas. In addition, the work of Ervin Chargaff and his colleagues on the base composition of DNA shattered the notion that DNA was a molecule consisting of a simple repetitive series of nucleoties. This finding awakened researchers to the possibility that DNA might have the properties necessary to fulfill a role in information storage.

Seven years after the publication of Avery's paper on bacterial transformation, Alfred Hershey and Martha Chase of the Cold Spring Harbour Laboratories in New York turned their attention to an even simpler system–bacteriophages, or viruses that infect bacterial cells. By 1950, researchers recognized that even viruses had a genetic program. The genetic material was injected inot the host cell where it directed the formation of new virus particles inside the infected cell. Within a matter of minutes, the infected cell broke open, releasing new bacteriophage particles, which infected neighboring host cells.

It was clear that the genetic material directing the formation of viral progeny had to be either DNA or protein because these were the only two molecules the virs contained. Electron microscopic observations had shown that, during the infection, the bulk of the bacteriophage remains outside the cell, attached to the cell surface by tail fibers Hersehey and Chase reasoned that the virus's genetic material must possess two properts. First, if the material were to direct the development of new bacteriophages during infection, it must pass into the infected cell. Second, it must be passed on to the next generation of bacteriophage. Hershey and Chase prepared two batches of bacteriophages to use for infection. One batch contained radioactively labeled DNA (^{32}P-DNA); the other batch contained radioactively labeled protein (^{32}S-protein). Since DNA lacks sulfur (S) atoms, and protein usually lacks phosphours (P) atoms, these two radioisotopes provided distinguishing labels for the two types of macromolecules. Their experimental plan was to allow one or the other type of bacteriophage to infect a population of bacterial cells, wait a few minutes, and then strip the empty viruses from the surfaces of the

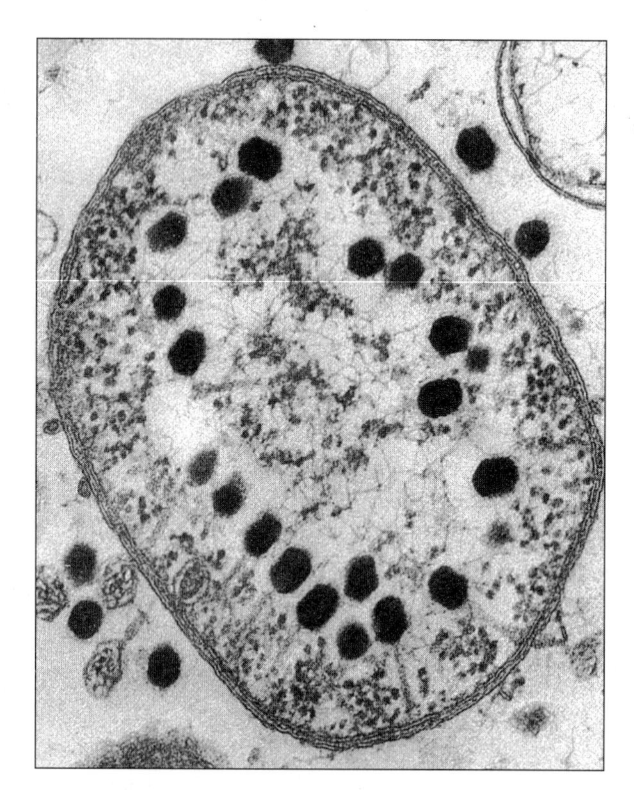

Fig. 9.10 Electron micrograph of a bacterial cell infected by T4 baceriophages. A phage is seen attached by its tail fibers to the outer surface of the bacterial cell, while new phage heads are being assembled in the hols cell's cytoplasm (Courtesy of Jonathan King and Erika Hartwing).

cells. After trying several methods to separate the bacteria from the attached phage coats, they found this was best accomplished by subjecting the infected suspension to the spinning blades of a Waring blender; Once the virus particles had been detached from the cells, the bacteria could be centrifuged to the bottom of the tube, leaving the empty viruses in suspension.

By following this procedure, Heershey and Chase determined the amount of radioactivity that entered the cells verus that which remained behind in the empty coats. They found that when cells were infected with protein-labeled bacteriophage, the bulk of the radioactivity remained in the empty coats. In contrast, when cells were infected with DNA-labeled bacteriophage, the bulk of the radioactivity passed inside the host cell. When they monitored the radioactivity passed onto the next generation, they found that less than 1 percent of the labeled protein could be detected in the progeny, whereas approximately 30 percent of the labeled DNA could be accounted for in the next generation.

The publication of the Hersehy-Chase experiments in 1952, together with the abandonment of the tetranucleotide theory, removed any remaining obstacles to the acceptance of DNA as the genetic material. Suddenly, tremendous new interest was generated in a molecule that had largely been ignored. The stage was set for the discovery of the double helix.

CHAPTER 10

DNA Replication

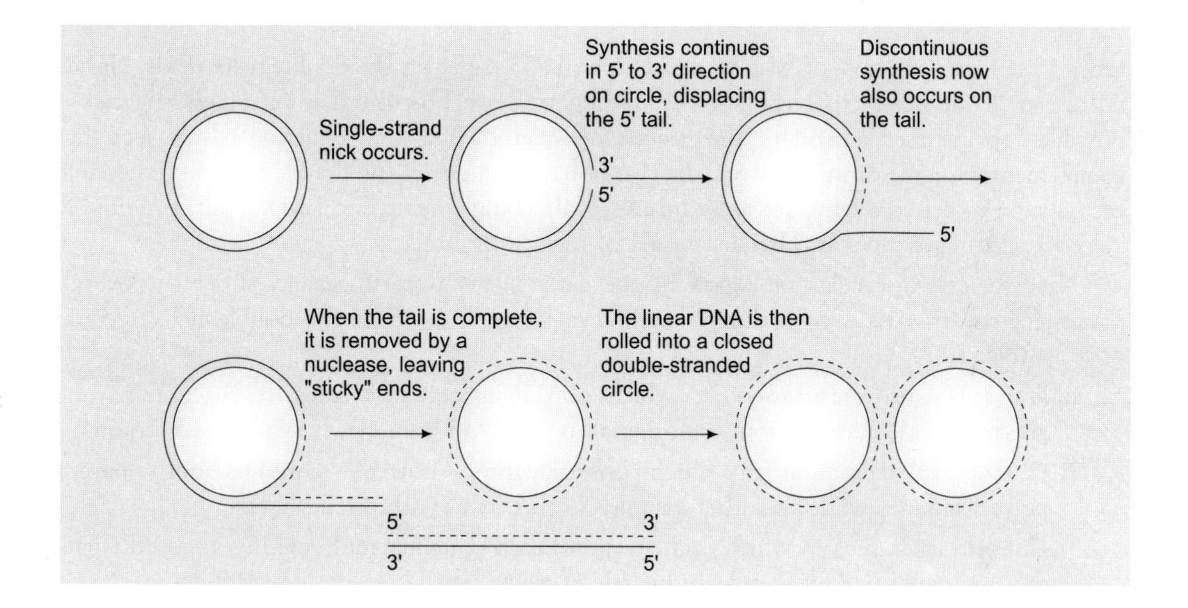

REPLICATION OF DNA MOLECULES

Every time a cell divides it must make a complete copy of all its genes. This is of course essential if the products of division, the two daughter cells, are each to receive a full complement of the biological information possessed by the parent. A dividing cell, therefore, has to carry out extensive DNA replication.

A considerable degree of accuracy must be achieved and maintained during DNA replication. Even an error rate of 0.01% (one mistake per 10,000 nucleotides) will cause a significant accumulation of alterations in the genes of a rapidly dividing organism, such as a bacterium, quickly leading to the vital DNA sequences becoming meaningless.

Three aspects of DNA replication must be considered by a molecular geneticist. The first concerns the overall pattern of replication and the question of how each of the strands of the parent double helix acts as a template for synthesis of new polynucleotides. This problem was solved by a single elegant experiment carried out in 1958. The second aspect is the biochemistry and enzymology of the process: which proteins are involved and what reactions do they participate in during DNA replication? This problem is now fairly well understood, although certain lacunae exist, especially surrounding the correction of errors, are still the subject of much research. The final aspect concerns the precise way in which DNA replication is handled in different organisms, as replication of a circular bacterial DNA molecule presents a completely different set of problems to replication of a eukaryotic chromosomes.

OVERALL PATTERN OF DNA REPLICATION

One of the final sentences of Watson and Crick's 1953 paper on the double helix reads, "It has not escaped our notice that the specific pairing we have postulated immediately suggests a possible copying mechanism for the genetic material." They were referring to the fact that complementary base pairing between the two polynucleotides of the double helix would enable each strand to act as a template for synthesis of its complement (Fig. 10.1). Thus, the double helix structure possesses an obvious means of replicating itself.

Nevertheless, molecular biologists in the early 1950s were uncertain about the overall pattern of the process. Three different schemes for replication of the double helix seemed possible (Fig. 10.2):

1. *Semi-conservative replication* in which the daughter molecules each contain one polynucleotide derived from the original molecule and one newly synthesised strand.
2. *Conservative replication* in which one daughter molecule contains both parent polynucleotides while the other contains both newly synthesised strands.
3. *Dispersive replication* in which each strand of each daughter molecule is composed partly of the original polynucleotide and partly of newly synthesised polynucleotide.

Although the semi-conservative scheme seems most likely purely on intuitive grounds, and in fact, was favoured by Watson and Crick, it was necessary to devise an experiment which would distinguish between these three modes of replication and confirm which scheme actually operates. The experiment that settled this question was carried out in 1958 at the California Institute of Technology by Matthew Meselson and Franklin Stahl.

THE MESELSON AND STAHL EXPERIMENT (Figs. 10.3 and 10.4)

Like many advances in biochemistry and molecular biology, the Meselson-Stahl experiment depended on the use of chemical isotopes. Nitrogen exists in several isotopic forms as well as the normal isotope, ^{14}N, which predominates in the environment and has an atomic weight of 14.008. There are a number of other isotopes that occur in much smaller amounts. These include ^{15}N, which because of its greater atomic weight is called 'heavy nitrogen'. Different isotopes can be distinguished from one another by a variety of means, so if a molecule

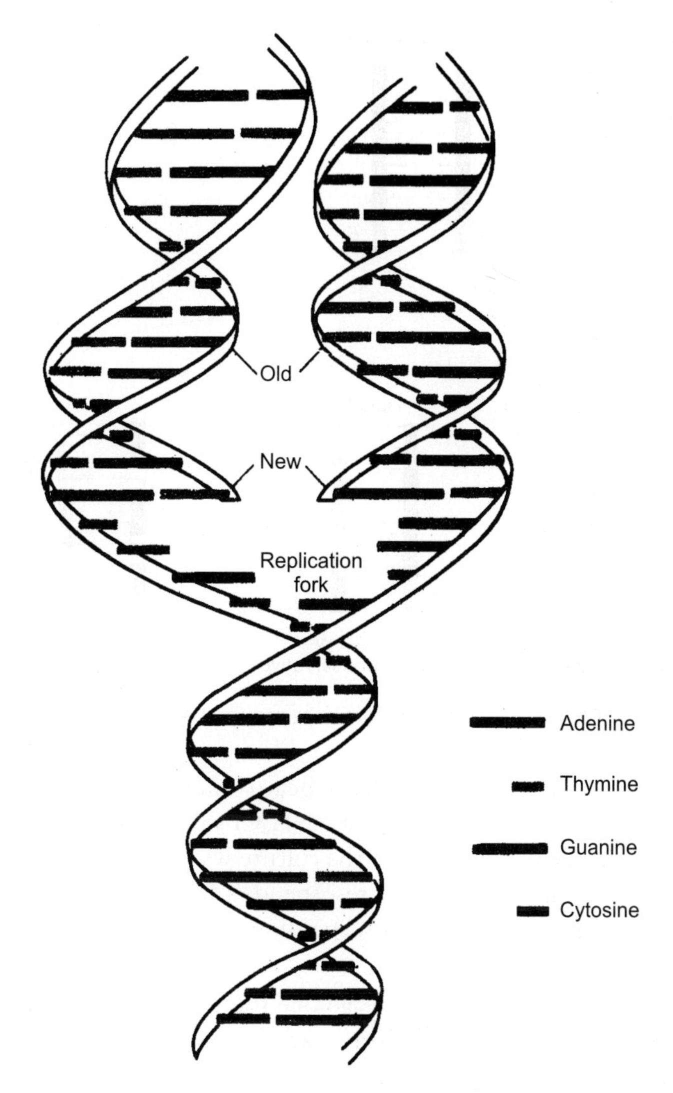

Fig. 10.1 The replication of DNA according to Watson and Crick.

containing an uncommon isotope is provided to a cell as a nutrient, the fate of the molecule (that is, the biochemical reactions that it undergoes in the cell) can be followed by identifying all the cellular metabolites that become labelled with the isotope. Of relevance to the Meselson-Stahl experiment is the fact that *E. coli* cells provided with heavy nitrogen in the form of $^{15}NH_4Cl$ will incorporate the labelled nitrogen into their DNA molecules.

Heavy nitrogen is a stable isotope, so does not emit radiation. However, DNA molecules containing heavy nitrogen can be distinguished from DNA molecules containing the normal isotope by density gradient centrifugation. This is because ^{15}N-DNA has a greater buoyant density than ^{14}N-DNA and will, therefore, form a band at a lower position in a CsCl density gradient.

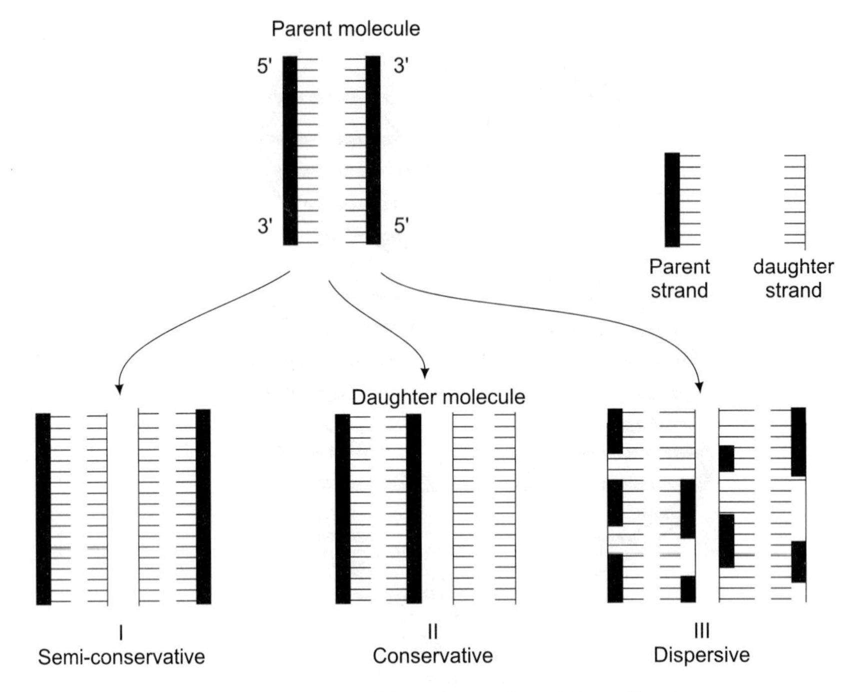

Fig. 10.2 Three possible schemes for DNA replication.

The Experiment: Meselson and Stahl utilised heavy nitrogen in the following way. First a culture of *E. coli* was grown in the presence of $^{15}NH_4Cl$ so that the DNA molecules in the cells became labelled with heavy nitrogen. Then the culture was spun in a low-speed centrifuge, the heavy medium discarded, and the bacteria resuspended in medium containing only $^{14}NH_4Cl$. New polynucleotides synthesised after resuspension, therefore, contained only the normal isotope of nitrogen.

The bacteria were then allowed to undergo one round of cell division, which takes roughly twenty minutes for *E. coli*, during which time each DNA molecule replicates just once. Some cells were then taken from the culture, their DNA purified and a sample analysed by density gradient centrifugation. The result, was a single band of DNA at a position corresponding to a buoyant density intermediated between the values expected for ^{15}N-DNA and ^{14}N-DNA, showing that after one round of replication each DNA double helix contained roughly equal amounts of ^{15}N-polynucleotide and ^{14}N-polynucleotide. If we examine the three schemes for DNA replication, we see that this result is compatible with both semi-conservative replication (each double helix comprises one ^{15}N-polynucleotide and one ^{14}N-polynucleotide) and dispersive replication (both strands are made up of a mixture of ^{15}N-polynucleotide and ^{14}N-polynucleotide). However, conservative replication can be ruled out at this stage, because this scheme would give two populations of DNA molecules, some entirely ^{15}N and some entirely ^{14}N: no hybrids would be seen.

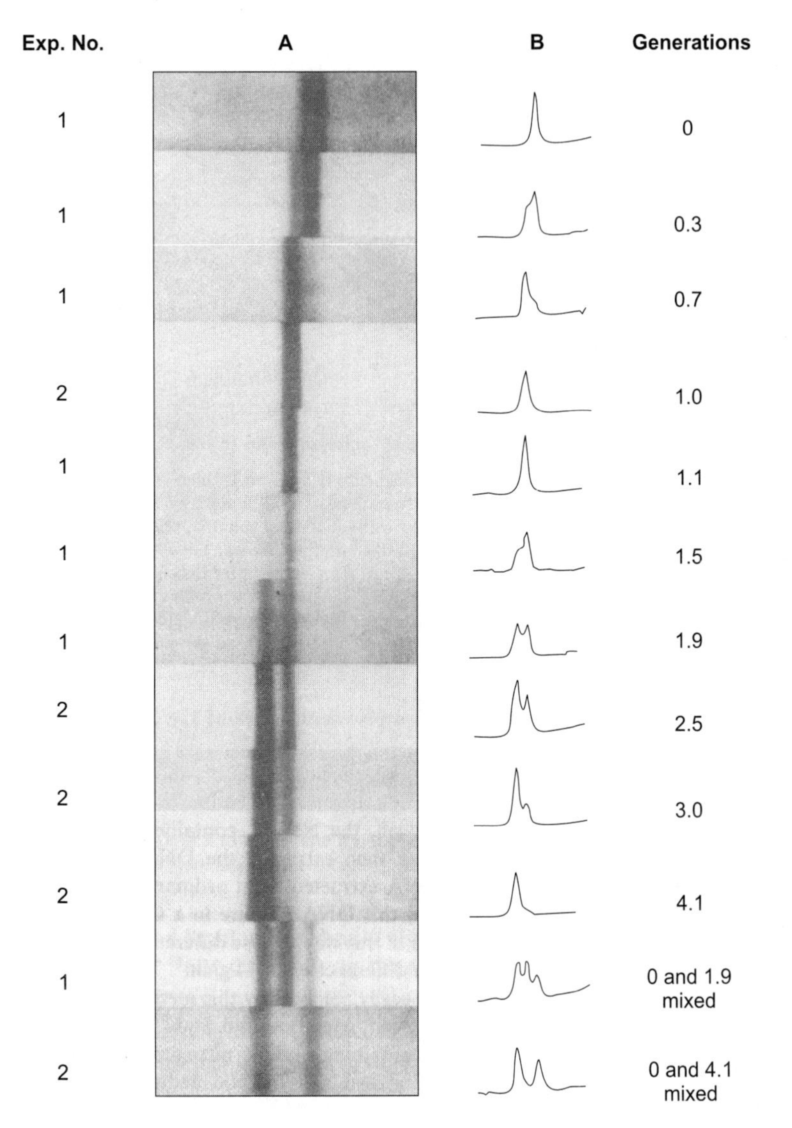

Fig. 10.3 Meselson-Stahl experiment to determine the mode of replication of DNA. Ultraviolet absorption photographs of bands resulting from density gradient centrifugation of bacterial DNA (a) and densitometer tracings of the same bands (b).

To distinguish between semi-conservative and dispersive replication, Meselson and Stahl allowed the *E. coli* culture to undergo a second round of cell division. Again, cells were removed and their DNA analysed in a density gradient. Two bands appeared: one representing the same hybrid molecules as before, but with an additional band corresponding to wholly ^{14}N-DNA. This result is entirely in agreement with the semi-conservative mode of replication, because according to this scheme there will now be some granddaughter molecules that are made up

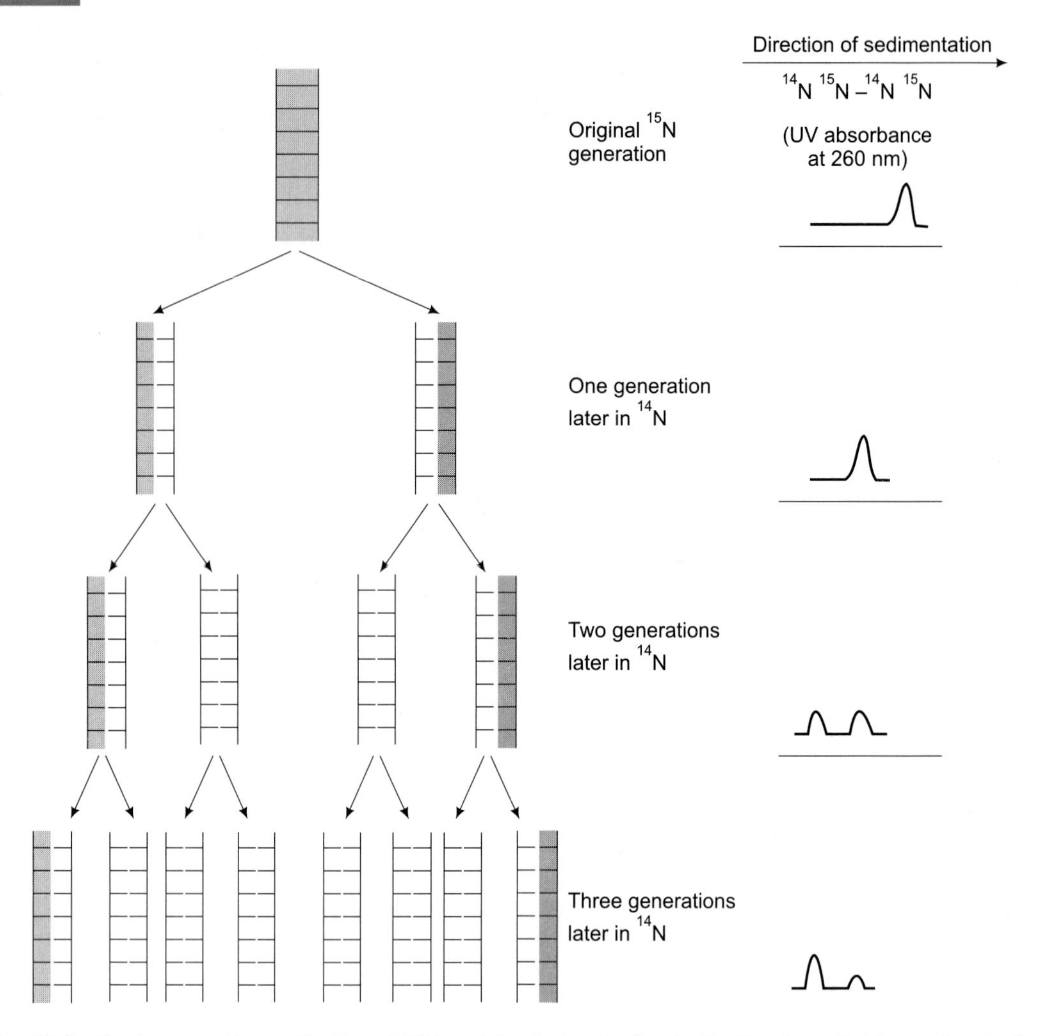

Fig. 10.4 Semiconservative replication of DNA and its demonstration in the Meselson-Stahl experiment. Gray colour indicates "old" (^{15}N-labelled) DNA while black indicates "new" (^{14}N- labelled) DNA.

entirely of ^{14}N-polynucleotides. In contrast, the dispersive mode can be discounted because that method would still produce only hybrid molecules, and in fact, would continue to do so for a very large number of cell generations.

The Meselson-Stahl experiment is rightly considered a classic example of scientific technique. This masterpiece enabled the attention of biochemists and molecular biologists to move directly on to the second aspect of DNA replication, the question of which enzymes are involved and how the process actually occurs.

MECHANISM OF DNA REPLICATION IN *E. Coli* (Fig. 10.5)

When a DNA molecule is being replicated only a limited region is ever in a non-base-paired form. The breakage of base pairing starts at a distinct position called the replication origin and gradually progresses along the molecule, possibly in both directions, with synthesis of the new polynucleotides occurring as the double helix unzips. The important region at which the base pairs of the parent molecule are broken and the new polynucleotides are synthesised is called a *replication fork.*

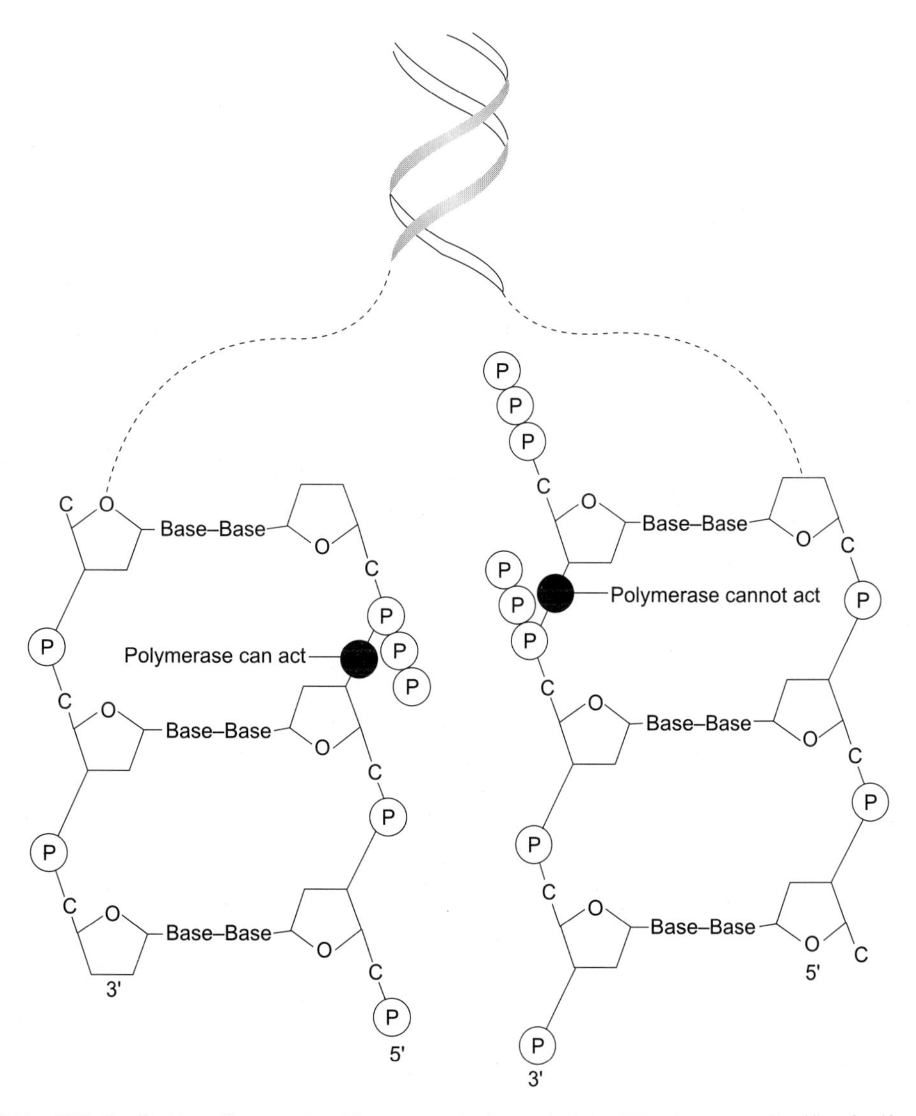

Fig. 10.5 *DNA Replication.* New nucleotides can only be added to DNA during replication in the 5′-3′ direction.

DNA POLYMERASE

An enzyme able to synthesise a new DNA polynucleotide is called a *DNA polymerase.* The chemical reaction catalysed by a DNA polymerase is very similar to that of RNA polymerase except, of course, that the new polynucleotide that is assembled is built up of deoxyribonucleotide subunits rather than ribonucleotides. In general terms, the reactions are the same: the sequence of the new polynucleotide is dependent on the sequence of the template and is determined by complementary base pairing and, as with RNA synthesis, DNA polymerisation can occur only in the 5′ to 3′ direction.

DNA POLYMERASE I AND DNA POLYMERASE III

In 1957, Arthur Kornberg and his colleagues isolated from *E. coli* an enzyme capable of DNA synthesis on a polynucleotide template. It was believed that this enzyme was the DNA polymerase responsible for DNA replication in the bacterial cell. Unfortunately, as the enzyme was studied in greater details, facts inconsistent with the role began to emerge. The most problematic of these arose with the discovery in 1969 of *E. coli* cells in which the gene coding for the enzyme, now called DNA polymerase I, was inactive; these mutant cells were still able to replicate their DNA. Eventually it was determined that although DNA polymerase I is involved in DNA replication, it is not the main replicating enzyme. The enzyme now believed to be primarily responsible for DNA replication in *E. coli* was isolated in 1972 and is called *DNA polymerase III.* The complexity of its role in the cell is reflected by the fact that it is a very large enzyme consisting of as yet undetermined number of subunits and with a molecular mass in excess of 2,50,000 daltons.

EVENTS AT THE REPLICATION FORK

Our knowledge of exactly how DNA is replicated has been built up over several years through the efforts of numerous molecular biologists and biochemists in laboratories all over the world. The consensus that has been reached is of a complex process involving a number of enzymes and other proteins.

BREAKAGE OF THE PARENT DOUBLE HELIX

During DNA replication there is a continual need for breakage of the base pairing between the two strands of the parent DNA molecule. This is carried out by two related enzymes called *helixases,* which work in conjunction with single-strand binding proteins (SSBs). The latter attach to the single-stranded DNA that results from helicase action and prevent the two strands from immediately reannealing. The result is the replication fork that provides the templates on which DNA polymerase III can work.

LEADING AND LAGGING STRANDS (Fig. 10.6)

The main complication in DNA replication is that the two disassociated strands of the parent molecule cannot be treated in the same way. This is because DNA polymerase III can synthesise DNA only in the 5′ to 3′ direction, which means that the template strands must be read in the 3′ to 5′ direction. For one strand of the parent, called the leading strand, this is no problem because the new polynucleotide can be synthesised continuously. However, the second or lagging strand cannot be copied in a continuous fashion because this would necessitate 3′ to 5′ DNA synthesis. Instead, the lagging strand has to be replicated in sections; a portion of the parent helix is disassociated and a short stretch of the lagging strand replicated, a bit more of the helix is disassociated and another segment of the lagging strand replicated, and so on.

At first, this process was just a hypothesis but the isolation by Reiji Okazaki in 1968 of short fragments of DNA between 100 and 1000 nucleotides in length, associated with DNA replication, confirmed that the suggestion is correct.

THE PRIMING PROBLEM AND JOINING UP THE OKAZAKI FRAGMENTS

Unlike RNA polymerase, DNA polymerase III cannot initiate DNA synthesis unless there is already a short double-stranded region to act as a primer. How can this occur during DNA replication? The answer appears to be that the very first few nucleotides attached to either the leading or the lagging strand are not deoxyribonucleotides but ribonucleotides that are put in place by a RNA polymerase enzyme. Once a few ribonucleotides (up to about sixty) have been polymerised onto the template, the RNA polymerase detaches, and polymerisation, now of DNA, is continued by DNA polymerase III. The RNA polymerase is called the *primase* and in *E. coli* it is a single polypeptide with a molecular mass of about 60,000 daltons. It acts in conjunction with six or more additional polypeptides which make up a structure referred to as the *primosome.*

On the lagging strand, DNA polymerase III can synthesise DNA only for a certain distance before it reaches the RNA primer at the 5′ end of the next *Okazaki fragment.* Here, DNA polymerase III stops and DNA polymerase I comes into action, continuing DNA synthesis and carrying out a further important function by removing the ribonucleotides of the primer of the adjacent Okazaki fragment and replacing them with deoxyribonucleotides. When all the ribonucleotides are replaced, DNA polymerase I either stops or possibly carries on a short distance into the DNA region of the Okazaki fragment, containing to replace nucleotides before it dissociates from the new double helix.

The final reaction needed to complete replication of the lagging strand is to join up adjacent *Okazaki fragments,* which are now separated by just a single gap between neighbouring nucleotides. All that is needed is to synthesise a phosphodiester bond at this position, a reaction catalysed by another new enzyme called *DNA ligase.*

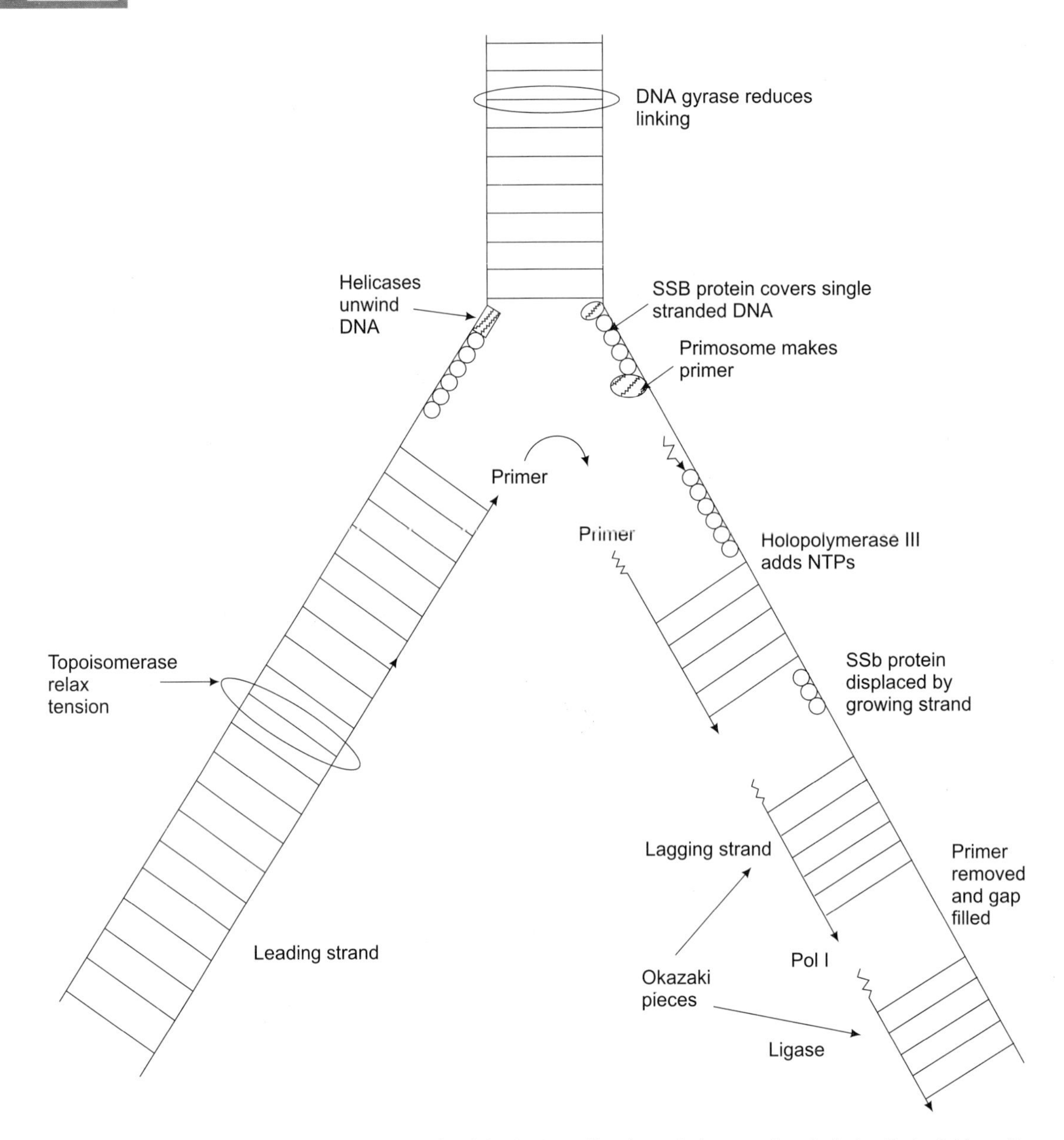

Fig. 10.6 Site of action of *enzymes* involved in DNA replication of the growing fork in *Escherichia coli.*

THE TOPOLOGICAL PROBLEM

The system just described brings about replication of the leading and lagging strands of the parent molecule and produces two daughter double helices. However, there is one outstanding problem to consider. The two polynucleotides of the parent helix are, of course, would round one another; this means that progression of the replication fork along the parent molecule requires the double helix not just to be unzipped but also to be unwound. This is in fact a significant problem if we consider that the *E. coli* DNA molecule is 4000 kb, or 4,00,000 turns of the helix, in length and must be replicated in twenty minutes. The implication is that the double helix is rotating at a rate of 6500 rpm.

This is so inconceivable that for many years molecular biologists sought solutions that avoided unwinding the helix. At one stage around 1979 things became so desperate that some scientists even suggested that the double helix was incorrect and that in fact the polynucleotides in a double-stranded DNA molecule are laid side by side and not wound round each other. Fortunately, a group of enzymes that solved the topological problem was eventually discovered.

DNA TOPOISOMERASES FUNCTION IN UNEXPECTED WAYS

These face-saving enzymes are the *DNA topoisomerases* and they fall into two classes, *Type I* and *Type II*. Both unwind a DNA molecule without actually rotating the double helix; they achieve this feat by causing transient breakages in the polynucleotide backbone.

Type I topoisomerase breaks just one of the polynucleotides and pass the other strand through the gap before reforming the back bone. Type II enzymes, which include the well, characterised DNA gyrase of *E. coli*, carry out the same sort of reaction but break both polynucleotides at adjacent positions whilst doing so. Both types of topoiosomerases produce the same result; portions of the double helix just in advance of the replication fork are unwound allowing the fork to progress unhindered along the parent molecule. DNA topoisomerases can also carry out the reverse reaction and introduce extra turns into a DNA molecule, this results in supercoiling as opposed to unwinding.

REPLICATION OF MOLECULES

From the foregoing, discussing the problem of how real DNA molecules in real cells are replicated may seem solved. This is not the case and difficulties arise even in the replication of a linear DNA molecule. Although bidirectional replication from one or more origins will give rise to copies of the bulk of a linear molecule, a problem arises when it comes to replicating the ends. This is because the extreme 5′ region of each newly synthesised polynucleotide will comprise an RNA primer which, according to our present knowledge, will not be removed. Although some ingenious solutions to this problem have evolved in different organisms, the basic question of how the terminal RNA primers are converted into DNA is still unanswered.

REPLICATION OF CIRCULAR DNA MOLECULES (Figs. 10.7, 10.8 and 10.9)

Circular molecules are, in fact, much easier to replicate and it is not surprising that the most rapidly dividing genomes, such as those of viruses and bacteria, are circular. There are two main strategies. Most bacterial genomes and viral DNA molecules replicate by bidirectional progress

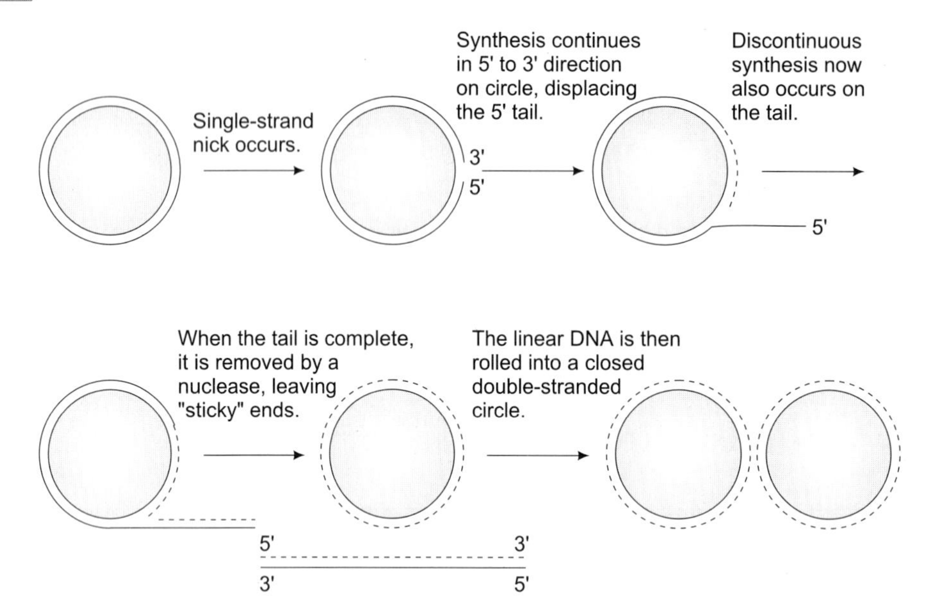

Fig. 10.7 Rolling-circle model of DNA replication in bacteria.

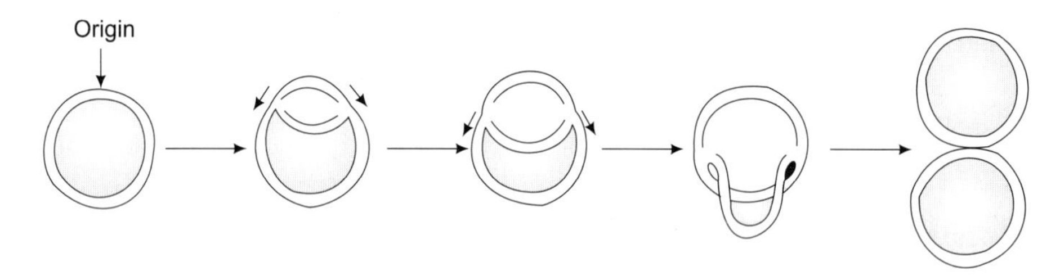

Fig. 10.8 Observable stages in the DNA replication of a circular chromosome, assuming bidirectional DNA synthesis. The intermediate figures are called theta structures.

of two forks around the circle from a single origin, producing an intermediate 8-form before the forks join up. The second system, employed by several types of phage is called rolling circle replication and begins with a nick in one strand of the parent molecule and extension of the free 3′-OH end by DNA polymerase. The original parent strand is, therefore, displaced and rolled off the molecule. DNA synthesis can stop after one revolution, resulting in two daughter molecules as shown in the figure, or can continue round the circle for several revolutions, very rapidly producing a series of *concatamers* (single-stranded copies of the genome linked end to end). In either case, discontinuous replication of the displaced strand will produce a long double-stranded molecule that, if necessary, can then be cut into individual genomes. Ligation of the ends of these will produce new circular molecules.

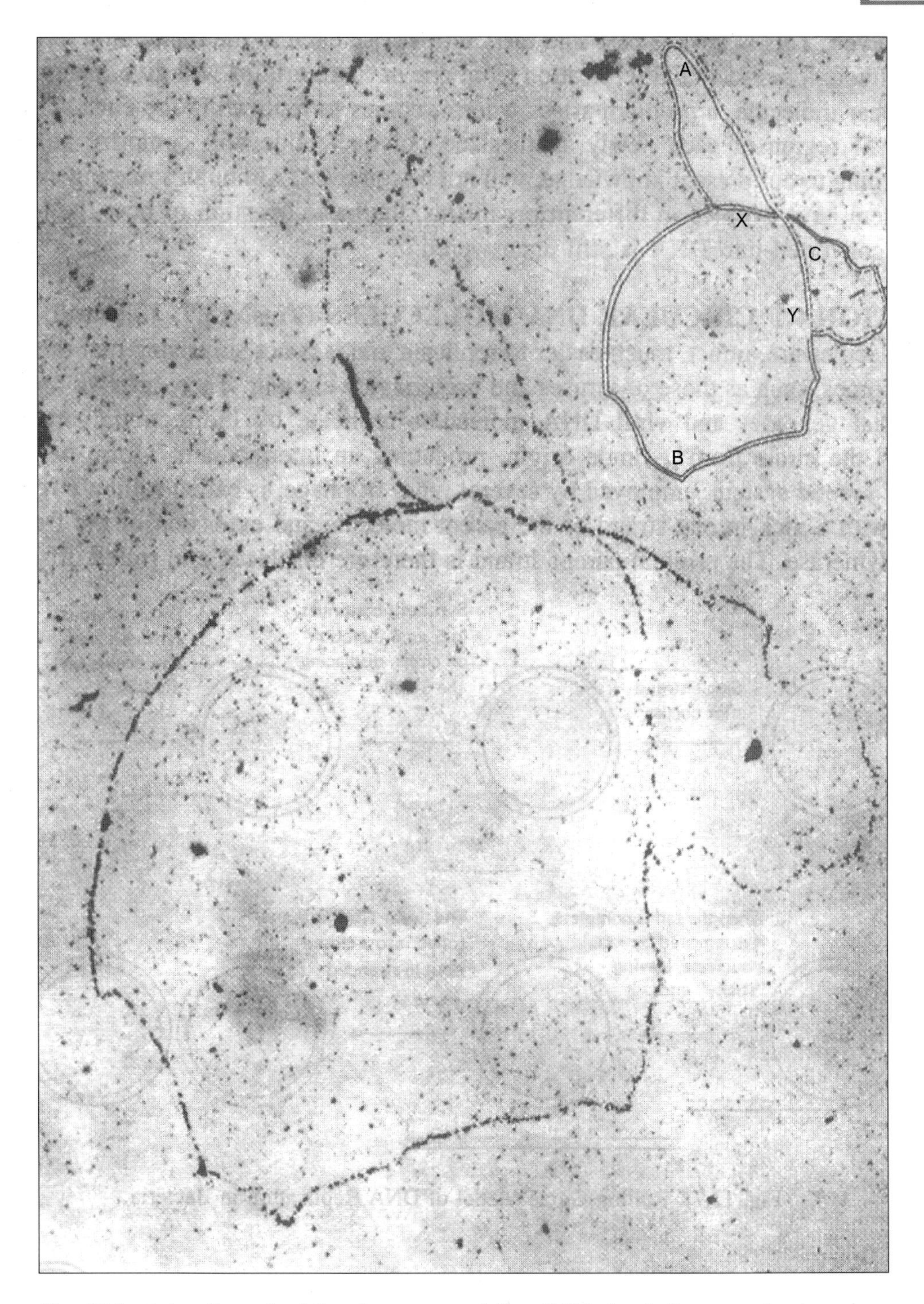

Fig. 10.9 Autoradiograph of the chromosome of *E. coli* K12. Inset, the same structure shown diagrammatically and divided into three sections (A, B and C) that arise at the two forks (X and Y).

SYNOPSIS

DNA replication is semiconservative, which indicates that one-half of the parent duplex is transmitted to each of the daughter cells during cell division. This mechanism of replication was first suggested by Watson and Crick as part of their model of DNA structure. They suggested that replication occurs by gradual separation of the strands by means of hydrogen bond breakage, so that each strand can serve as a template for the formation of a complementary strand. This model was soon confirmed in both bacterial and eukaryotic cells by showing that cells transferred to labeled media for one generation produce daughter cells whose DNA has one labeled strand and one unlabeled strand.

The mechanism of replication is best understood in bacterial cells. Replication begins at a single origin on the circular bacterial chromosome and proceeds outward in both directions as a pair of replication forks. Replication forks are sites where the double helix is unwoiund and nucleotides are incorporated into both newly synthesized strands.

DNA synthesis is catalyzed by a family of DNA polymerases. The first of these enzymes to be characterized was DNA polymerase I of *E. coli*. To catalyze the polymerization reaction, the enzyme requires all four deoxyribonucleoside triphosphates, a template strand to copy, and a primer containing a free 3′OH to which nucleotides can be added. The prime is required because the enzyme is unable to initiate the formation of a DNA strand. Rather, it is only capable of adding nucleotides to the 3′ hydroxyl terminus of an existing strand. Another unexpected characteristic of DNA polymerase I is that it is ony capable of polymerizing a strand in a $5' \rightarrow 3'$ direction. It had been presumed that the two new strands would be synthesized in opposite directions by polymerases moving in opposite directions along the two parental template strands. This finding was explained when it was shown that the two strands are synthesized quite differently.

One of the newly synthesized strands (the leading strnad) grows toward the replication fork and is synthesized continuously. The other newly synthesized strand (the lagging strand) grows away from the fork and is synthesized discontinuously. In bacterial cells, the lagging stand is synthesized as fragments approximately 1000 to 2000 nucletotides long, called Okazaki fragments, that are covalently joined to one another by a DNA ligase. In contrast, the leading strand is synthesized as a single continuous strand. Neither the continuous strand nor any of the Okazaki fragments can be initiated by the DNA polymerase, but instead begin as a short RNA primer that is synthesized by a type of RNA polymerase. After the RNA primer is assembled, the DNA polymerase continues to synthesize the strand or fragment as DNA. The RNA is subsequently degraded and the gap is filled in as DNA.

Events occurring at the replication fork require a variety of different types of proteins having specialized functions. These included a DNA gyrase, which is a type II topoisomerase required to relieve the tension that builds up as a result of DNA unwinding; a DNA helicase that unwinds the DNA by separating the strands; single-stranded binding proteins that bind selectively to single-stranded DNA and prevent their reassocation; an RNA primase, which is the enzyme that synthesizes the RNA primers at the beginning of each Okazaki fragment; and a DNA ligase that seals the fragments of the lagging strand into a continuous polynucleotide. DNA polymerase III is the primary DNA synthesizing enzyme that adds nucleotides to each RNA primer, whereas DNA polymerase I is responsible for removing the RNA primers and replacing them with DNA. Two molecules of DNA polymerase III are thought to move together as a complex in opposite directions along their respective template strands. This is accomplished by having the lagging strand template looped back on itself. Initiation of replication requires additional proteins that bind to a repetitive nucleotide sequence within the origin, triggering strand separation and the recruitment of proteins required for replication.

DNA polymerases possess separate catalytic sites for polymerization and degradation of nucleic acid strands. Most DNA polymerases possess both $5' \rightarrow 3'$ and $3' \rightarrow 5'$ exonucleases. The first of these two nucleases is used to degrade the RNA primers at the beginning of Okazaki fragments, and the second is used to remove an inappropriate nucleotide following its mistaken incorporation, thus contributing to the fidelity of replication. It is estimated that approximately one in 10^9 nucleotides are incorporated incorrectly during replication in *E. coli*. This high degree of fidelity is maintained by three separate checkpoints. (1) The DNA polymerase is able to measure the geometry of the base pair formed by an incoming nucleotide before that nucleotide is covalently linked to the end of the primer; (2) if an incorrect nucleotide is incorporated, it is usually removed by the $3' \rightarrow 5'$ exonucleas activity of the polymerase, which acts as a proofreading mechanism; (3) and incorrectly incorporated nucleotide can be removed following replication by a process of mismatch repair.

Replication in eukaryotic cells follows a similar mechanism and employs similar proteins to those of prokaryotes. To date, five different DNA polymerases have been isolated from eukaryotic cells, all of which elongate DNA strands in the $5' \rightarrow 3'$ direction. None of them is able to initiate the synthesis of a chain without a primer. Most possess a $3' \rightarrow 5'$ exonuclease activity, ensuring that replication occurs with high fidelity. Unlike prokaryots, replication in eukaryotes is initated simulataneoulsy at many sites along a chromosome with replication forks proceeding outward in both directions from the site of initiation. Studies on yeast indicate that origins of replication are sites that contain a specific binding site for an essential multiprotein complex.

Replication in eukaryotic cells is initmately associated with nuclear structures. Evidence indicates that much of the machinery required for replication is associated with the nuclear matrix. In addition, replication forks that are active at any given time are apparently localized within about 50 to 250 sites called replication foci, which also contain the enzymes responsible for methylating DNA after it is synthesized. Newly synthesized DNA is rapidly associated with nucleosomes. Histone octamers present prior to replication remain intact and are passed on to the daughter duplexes, while newly synthesized histones assemble into new core particles. The new and old core particles are thought to be distributed randomly between the two daughter complexes.

DNA is subject to damage by many environmental influences, including ionizing radiation, common chemiclas, and ultraviolet radiation. Cells possess a variety of systems to recognize and repair the resulting damage. It is estimated that less than one base change in a thousand escapes a cells's repair systems. In some cases, the damage can be repaired directly, but in most cases, the damaged section must be excised and replaced by newly synthesized DNA. Three major types of DNA repair systems are known. Nucleotide excision repair (NER) systems operate by removing a small section of a DNA strand containing a bulky lesion, such as a peyrimidine dimer. During NER, paired incisions are made by an endonuclease, the damaged oligonucleotide is stripped away by a helicase, the gap is filled by a DNA polymerase, and the strand is sealed by a DNA ligase. The template strands of genes that are actively transcribed are preferentially reparied by NER. Base excision repair acts to remove a variety of altered nucleotides that produce minor distortions in the DNA helix. Cells possess a variety of glycosylases that are capable of recognizing and removing various types of altered bases. Once the base is removed, the remaining portion of the nucleotide is removed by an endonuclease, the gap is enlarged by a phospohdiesterase, and the gap is filled and sealed by a polymerase and ligase. Mismatch repair is responsible for removing incorrect nucleotides incorporated during replication that escaped the proofreading activity of the polymerase. The newly synthesized strand is selected for repair by virtue of its lack of methyl groups compared to the parental strand.

CHAPTER 11

Ribosomes and The Synthesis of Proteins

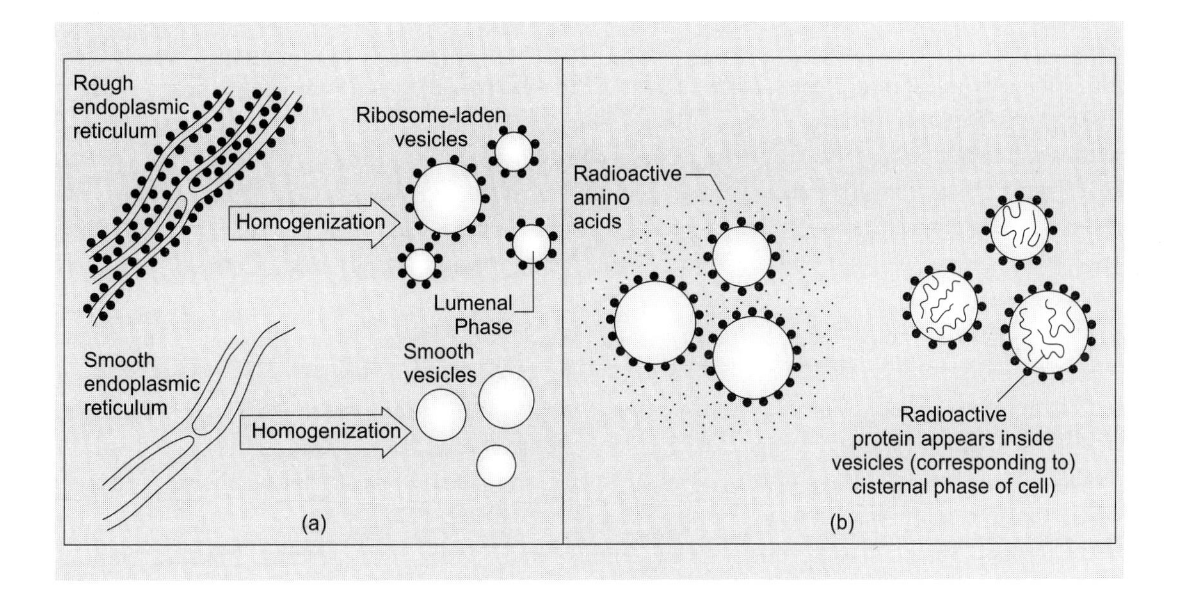

Until the 1930s, it was the prevailing view that DNA was found only in animal cells and RNA only in plant cells. This view was dispelled by a number of findings in the 1930s that definitively established that both DNA and RNA are present in animal and plant cells. Moreover, J. Brachet and T. Caspersson showed that the bulk of the RNA was present in the cytoplasm and that cells actively engaged in protein synthesis (such as pancreas cells and the silk-gland cells of silk worms) contain greater amounts of RNA then cells that do not actively produce protein. Albert Claude showed in the 1940s that the cytoplasmic RNA was included in tiny particles of ribonucleo protein later to be called "ribosomes".

PROTEIN TURNOVER IN CELLS

The rate of breakdown and replacement of protein in cells was badly misunderstood prior to 1939. In growing animals generally and in secretory tissues in particular (e.g. the liver, pancreas, and endocrine glands), active synthesis of protein was known. However, the amount of protein synthesis taking place in other tissues of the adult was believed to be very low and confined to that necessary to replace protein lost through damaged or dying cells. These small protein losses, together with the catabolism of dietary amino acids, were believed to be responsible for the urea and ammonia measurable in urine. Proteins were thus thus regarded as highly stable constituents lasting virtually the entire life-time of the cell.

The first serious challenge to the "wear and tear" view of protein turnover came as a result of the work of R. Schoenheimer in 1938. Schoenheimer synthesized a number of amino acids in which the ^{15}N content of the alpha-amino nitrogen was considerably increased over the natural amount of this isotope. Schoenheimer then injected ^{15}N-containing glycine and leucine into rats and noted that these labeled amino acids were incorporated into the proteins of many tissues very rapidly. Although ^{15}N is not a radioactive isotope of nitrogen. It may nonetheless be distinguished chemically from the more common ^{14}N form and is called a "heavy" isotope of nitrogen. Then results clearly indicated that protein synthesis in adult animals is not restricted to growing and secretory tissues but occurs in nearly all cells and that tissue proteins are in a continuous state of metabolic flux, being broken down and replaced by newly synthesized molecules.

Although the radioactive isotope of carbon ^{14}C, was produced in the Berkeley cyclotron in 1940, it was not until 1947 that ^{14}C-labeled amino acids became available. The availability of radioactive amino acids was followed by a series of classical tracer experiments by H. Borsook, T. Hultin. P. Zamecnik, and P. Siekevityz which verified the findings of Schoenheimer that most tissues readily incorporate amino acids into protein and also added crucial details to the newly emerging view of protein synthesis and metabolic turnover.

The first attempts to determine the subcellular site of amino acid incorporation into protein were carried out in 1950 by Borsook. Minutes after injecting ^{14}C-labeled amino acids into the bloodstreams of guinea pigs, Borsook removed the animals livers and using the technique of differential centrifugation. prepared subcellular fractions of the tissue. Borsook showed that it was the microsomal fraction that contained the highest degree of radioactivity and suggested that the microsomes were the repository of the cell's protein-synthesizing apparatus. In the same year, Hultin demonstrated that it was the microsomal fraction of chick liver tissue homogenates that incorporated intravenously injected ^{15}N-glycine into protein.

By 1952, Siekevitz and Zamecnik had been able to demonstrate the in vitro incorporation of ^{14}C- labeled amino acids into liver cell proteins by both tissue slices and tissue homogenates. By measuring and comparing protein synthetic activity in cell-free whole homogenates, individual cell subfractions, and various combinations of subfractions. Siekevitz showed that the incorporation of amino acids into proteins by microsomes was dependent on an energy source provided by the mitochondrial subfraction and required enzymes and other factors present in the cytosol. The denonstration that amino acid incorporation into protein required metabolic energy laid to rest a view popular in the 1940s that polypeptide synthesis might be brought about by the reversal of protein hydrolysis. It is especially intesesting to note that Siekevitz demonstrated the existence in the cytosol of a $MgCl_2$-precipitable factor required for protein synthesis. Since $MgCl_2$ was known to precipitate RNA Siekevitz suggested that RNA might somehow be involved in protein synthesis, a fact not fully recognized until many years later.

The studies described above established the general cytological and chemical basis of protein biosynthesis, Exhaustive research since the 1950s by dozens of groups of investigators has revealed the step-by-step, reaction by-reaction details of the process and has given us an astounding insight into molecular organization of the cell protein-syntesizing apparatus. Although much of this chapter is devoted to the examination of ribosome structure and to the chemical events that accompany protein synthesis, much of what is to be presented is better understood if *first* placed in perspective with a brief, preliminary on review of the subject; this will set the foundation for the more comprehensive study that follows. For simplicity, this synopsis will be concerned only with *cytoplasmic* protein synthesis in eucaryotic cells.

A PRELIMINARY OVERVIEW OF PROTEIN BIOSYNTHESIS

The variety and specific amino acid composition of the proteins synthesized by a cell are ultimately governed by the cellular DNA. This DNA is enzymatically transcribed in the cell nucleus to produce a host of RNAs, including ribosomal RNA, (rRNA), messenger RNA (mRNA), and transfer RNA (tRNA). The base sequences of these RNAs are *complementary* to the base sequences of the DNA molecules transcribed. rRNA is ultimately incorporated into the cytoplasmic ribosomes, which may be *free* in the cytosol or attached to the surface of the intracellular membrane network that faces the cytosol. Each ribosome consists of two parts or **subunits**—a small subunit and a large subunit. The small subunit binds mRNA entering the cytosol from the nucleus, and the functional complex is completed with the subsequent addition of the large subunit. Attached ribosomes are linked to the endoplasmic reticulum via the large subunit.

The nucleotides of mRNA are arranged as linear sequence of **codons**, each codon consisting of three successive nitrogeneous bases (also known as *triplet*). The codon sequence of each mRNA molecule contains all the information necessary to (1) properly initiate polypeptide synthesis on the ribosome, (2) designate the specific sequence of amino acids to be incorporated (i.e., the primary structure of the polypeptide), and (3) terminate polypeptide synthesis and release the completed polypeptide. Table 11.1 shows the various codons of mRNA and their meanings in protein synthesis. This is called the "genetic code". The code is said to be degenerate because in certain instances, a single amino acid may be coded for by more than one codon.

Moleules of tRNA entering the cytosol from the nucleus combine with amino acids; this is a molecule-specific association in that each amino acid species is enzymatically combined with a particular type (or species) of tRNA. The products, called **aminoacyl-tRNA**, represent the

Table 11.1 The genetic code.

First Base	Second base U	C	A	G	Third Base
U	phe	ser	tyr	cys	U
	phe	ser	tyr	cys	C
	leu	ser	"stop" (ochre)	"stop" (opal)	A
	leu	ser	"stop" (amber)	try	G
C	leu	pro	his	arg	U
	leu	pro	his	arg	C
	leu	pro	gln	arg	A
	leu	pro	gln	arg	G
A	ile	thr	asn	ser	U
	ile	thr	asn	ser	C
	ile	thr	lys	arg	A
	met ("start")	thr	lys	arg	G
G	val	ala	asp	gly	U
	val	ala	asp	gly	C
	val	ala	glu	gly	A
	val ("start")	ala	glu	gly	C

form in which amino acids are incorporated into newly synthesized protein. Each species of tRNA contains, among other functional groups, an **anticodon** (a sequence of three bases) that is reognized by a corresponding (probably complementary) codon of mRNA and ensures that the correct amino acid will be incorporated into its proper position in the primary structure of the polypeptide being synthesized.

Once the mRNA-ribosome complex has been formed amino acids bound to their specific tRNA molecules are sequentially brought to the ribosome and incroporated into the growing polypeptide chain. This process called transaction is believed to take place by an orderly and linear movement of the mRNA along the ribosome (or vice versa) so that each codon is translated in sequence. The elongation of the polypeptide chain takes places by a series of enzyme-catalyzed reactions occurring on two adjacent sites of the ribosome; these are the **amino acid** (or **acceptor**) site and the **peptide** (or **donor**) site. To understand the process of elongation, consider an intermediate stage in the synthesis of a polypeptide. At this time, the growing polyeptide chain is attached to the peptide side of the ribosome by a molecule of tRNA and in termed peptidyl-tRNA. The mRNA codon located in the vacant amino acid site specifies the, form of aminoacyl-tRNA that can be bound there. With a new aminoaccyl-tRNA in position in the amino acid site, the bond linking the growing polypeptide to its tRNA is broken and replaced by a peptidc bond with the amimo acid of aminoacyl-tRNA. This leaves the peptidyl-tRNA (which is now one amino acid longer) temporarily in the amino acid site. The tRNA molecule released in the process reenters the cytosol where it may combine with another amino acid to be used in protein synthesis. Formation of the peptide bond is followed by a shift of the peptidyl-tRNA to the peptide site, once again leaving the amino acid site vacant. This shift is accompanied by the movement of the ribosome and or mRNA so that the next codon is in position in the amino acid site and may now be translated. Thus tRNA molecules employed in bringing amino acids to the ribosome are transiently bound first to the amimo acid site and then to the peptide site before returning to the cytosol.

Amino acids are sequentially added to the growing polypeptide until its primary structure is complete. Once the end of the message coded in the strand of mRNA is reached, the completed protein is released from the ribosome The ribosome separates from the mRNA and dissociates into its two subunits; these may be used again in another round of protein synthesis.

Many mRNA molecules are large enough to be simultaneously translated by a number of ribosomes. These ribosomes move in a series along the mRNA, translating its coded message into a, number of *identical* proteins. The release of one ribosome at the end of the message is accompanied by the attachment of a new ribosome at the beginning of the message. Such strings of ribosome are called **polysomes**, and most protein synthesis that takes place in cells occurs on these structures, Although each mRNA molecule may be attached to several ribosome, each ribosome synthesizes but a single protein chain before dissociating into its subunits. The mechanies of protein syhnthesis described briefly here for perspective only is treated in detail in later sections of this chapter.

STRUCTURE, COMPOSITION AND ASSEMBLY OF RIBOSOMES

In this, we are concerned with the organization composition, and assembly of the cytoplasmic ribosome of procaryotic and eucaryotic cells. Organellar ribosome (e.g., chloroplast and mitochondrial ribosomes) will be considered separately later in the chapter. Although functionally analogous many differences exist between the ribosomes of procaryotic and eucaryotic cells (Table 11.2). Considerably more is known about the structure and composition of bacterial ribosomes than ribosomes of eucaryotic cells, as will become evident during the discussion that follows. Most of the work on procaryotic ribosomes has been carried out using *E. coli*. Although some variations are observed among the procaryotes, findings using *E. coli* are generally representative.

Ribosomes in the cytoplasm of eucaryotic cells have a sedimentation coefficient of about 80 S (M.W., about 4.5×10^6) and are composed of 40 S and 60 S subunits. It procaryotic cells, ribosomes are typically about 70 S (M.W., about 2.7×0^6) and are formed from 30 and 50 S subunits. The complete ribosome formed by combination of the subunits is also referred to as a monomer. Although ribosomes from both procaryotic and eucaryotic sources are about 30 to 45% protein (by weight), with the remairder being ribonucleic acid, the specific protein and RNA components of these two major classes of ribosomes differ (Table 11.2 and Fig. 11.1): carbohydrate and lipid are virtually absent. Magnesium ions (and perhaps other cations) play an important role in maintaining the structure of the ribosomes. Dissociation into subunits occurs when Mg^{++} is removed. The precise role (or roles) of Mg remains uncertain, although interaction with ionized phosphate of subunit RNA is presumed.

Table 11.2 Properties and composition of eucaryotic and procaryotic ribosomes.

	Eucaryotes	*Procaryotes*
Monomers		
Sedimentation coefficient	80 *S*	70 *S*
Molecular weight	4.5×10^6	2.7×10
Number of RNAs	4	3
Number of Proteins	70	55
Small subunit		
Sedimentation coefficient	40 *S*	30 S
Molecular weight	1.5×10^6	0.9×10^6
RNAs present	18 *S* (M.W., 0.7×10^6) (2110 nucleotides)	16 *S* (M.W., 0.6×10^6) (1600 nucleotides)
Number of proteins	30 (total M.W., 0.78×10^6)	21 (total M.W., 0.3×10^6)
Large subunit		
Sedimentation coefficient	60 *S*	50 *S*
Molecular weight	3×10^6	1.7×10^6
RNAs present	5 *S* (M.W., 3.2×10^4) (120 nucleotides) 5.8 S (M.W., 5×10^4) (150 nucleotides) 28 *S* (M.W., 1.7×10^4) (5000 nucleotides)	5 *S* (M.W., 3.2×10^4) (120 nucleotides) 23 S (M.W. 1.1×10^4) (3200 nucleotides)
Number of proteins	40 (total M.W., 1.37×10^6)	(total M.W., 0.5×10^6)

Procaryotic Ribosomes

RNA Content. The small subunit of procaryote ribosomes contains one molecule of RNA called 16 S RNA (M.W., 0.6×10^6) while the large subunit contains two RNA molecules, a 23 S RNA (M.W., 1.1×10^6) and as 5 S RNA (M.W., 3.2×10^4) (see Table 11.2). All three rRNAs are products of closely linked genes transcribed in the sequence 16 S → 23 S → 5 S. This assures an equal proportion of each unit. A polynucleotide containing the 16 S *transcript* is enzymatically cleaved (by an *endoribonuclease)* from the growing RNA strand once transcription has entered the 23 S region of the DNA (referred to as rDNA). A second polynucleotide containing the 23 S transcripts is similarly released once the 5 S region is reached. A final product contains the 5 S transcript. The initial transcription products are successively "trimmed" to form the 16 S, 23 S, and 5 S RNAs finally incorporated into the ribosomal subunits. Fig. 11.2 presents the scheme of maturation of the procaryotic rRNAs. For clarity, the incorporation of the ribosomal proteins is not shown. Ribosomal proteins combine. with the tRNAs at various stages of subunit assembly: some are incorporated during "pre-rRNA"

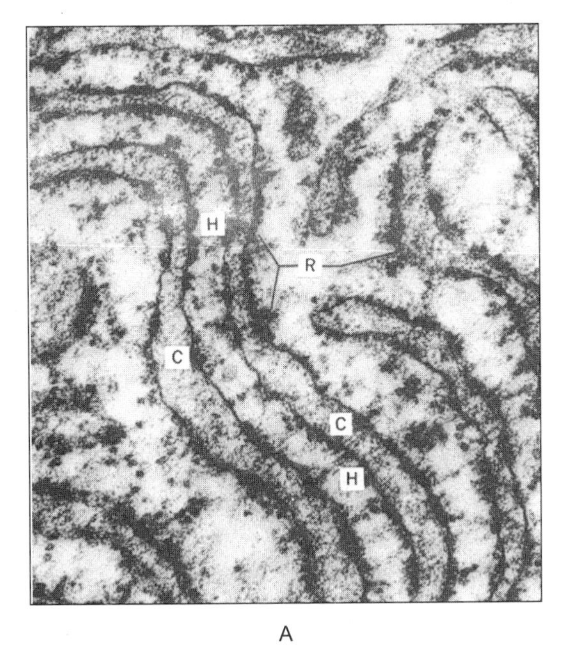

A

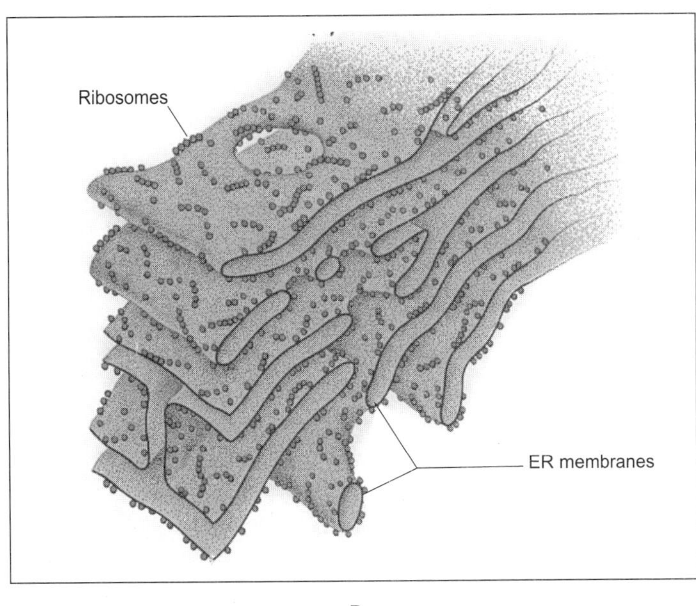

B

Fig. 11.1 (a) Electron micrograph of rough endoplasmic reticulum (RER) with ribosomes (arrow) attached to membrane. (b) Smooth endoplasmic reticulum (SER).

transcription, others following pre-rRNA cleavage from the growing polynucleotide and during trimming, and still others' once the mature rRNA products are formed. Certain proteins bind to the rRNAs only transiently and are not found in the fully assembled subunits.

Table 11.3 Reiteration of rRNA genes in various cells.

Cell or Tissue	*Number of genes per genome*
Liver	750
HeLa cells	1100
Xenopus (toad)	900
Drosophila melanolagasler	260
Tobacco leaves	1500
Saccharomyces cerevisiae	140
E. coli	5-10
Bacillus subtilis	9-10
B. megalerium	35-45

Multiple copies of the rRNA genes in the genomes of procaryotic (and eucaryotic) cells (Table 11.3); this is known as **reiteration.** In *E. coli* the number of rRNA genes is estimated to be between 5 and 10 and accounts for about 0.4% of the cell's total DNA. The primary structures of the three procaryotic rRNAs have been extensively studied. 5 S RNA was RNAs (about 120 nucleotides), was sequenced first (in 1967). Fig. 11.2 compares the prmary structures of several 5 S RNAs. All the sequences are compatible with the existence of a "stem" formed by base-pairing of the 3′ and 5′ ends of the molecules.

The sequencing of 16 S RNA (1600 nucleotides) and 23 S RNA (3200 nucleotides) is rapidly approaching completion by studies being carried out in a number of research laboratories. The known sequence for 16 S RNA is shown in Fig. 11.3. methylation of certain bases in the sequence of 16 S RNA (and also in the sequence of 23 S RNA) occurs while transcription is taking place. No methylation of 5 S RNA nucleotides occurs. Unlike 5 S RNA in which duplication of certain sequences occurs, no repeated sequences are found in 16 S and 23 S RNA. Although the rRNAs are single linear polynucleotides, they contain a number of double-helical regions that form "hairpins" stabilized by conventional, complementary base-pairing. Several **palindromes** (base sequences reading the same from either the 5′ or 3′ ends) exist in 16 S RNA, and these may play a role in restricting the formation of the double-helical regions. In 16 S RNA, a 7-nucleotide segment of the chain at the 3′ end is believed to interact with mRNA, leading to its binding during the initiation of translation. The 5 S and 23 S RNAs interact with one another in the large subunit, and both appear to be involved in amino acyl-tRNA and peptidyl-tRNA binding during polypeptide chain elon gation. Since several proteins of the small subunit interact with 23 S RNA, the latter may also have a role in subunit association.

It is generally believed that ribosomal RNA transcripts are not translated into protein (i.e., rRNA cannot serve as messengers); however, ribosomal proteins are the products of a typical transcription-translation process.

Protein Content. Nomura, Kurland, and others have established that the small procaryotic ribosomal subunit contains 21 proteins molecules (identified as S1, S2, S3 . . . S21) and the large subunit 34 proteins (LI, L2, L3 . .. 1.34). All the ribosomal proteins have been isolated

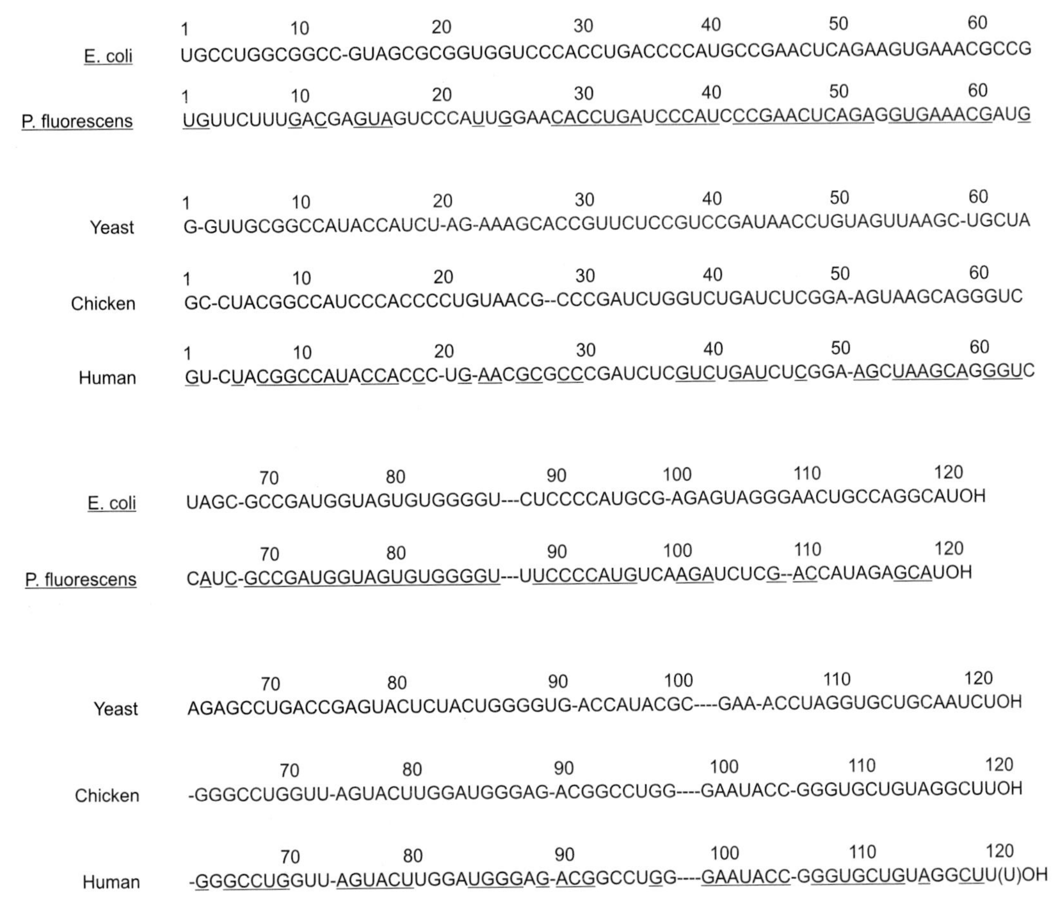

Fig. 11.2 Comparison of the primary structures of *E. coli, Pseudomonas flourescens*. Yeast chicken, and human ribosomeal 5 S RNA. Homologous sequences within the procaryote and within the eucaryote RNAs are underlined.

and characterized. The small subunit protein range in molecular weight from 10,900 (SI7) to 65,000 (SI); the large subunit proteins vary in molecular weight from 9600 (L34) to 31,500 (L2). Most of the ribosomal proteins are basic in nature, being rich in basic amino acid and having isoelectric points around pH 10 or higher. About 33 of the 55 ribosomal proteins have been fully sequenced (13 from the small and 20 from the large subunit). This together with the RNA observations described earlier and the discussion of protein synthesis later in the chapter, suggests that the procaryotic ribosome may well be the first organelle completely understood in term of molecular structure and function. An exhaustive analysis of the primary structures of procaryotic ribosomal protein in order to evaluate their degree of *homology*, indicates their these proteins did *not* have a common evalution ary ancestor. Homologies among them do not occur more often than would be expected on a random basic.

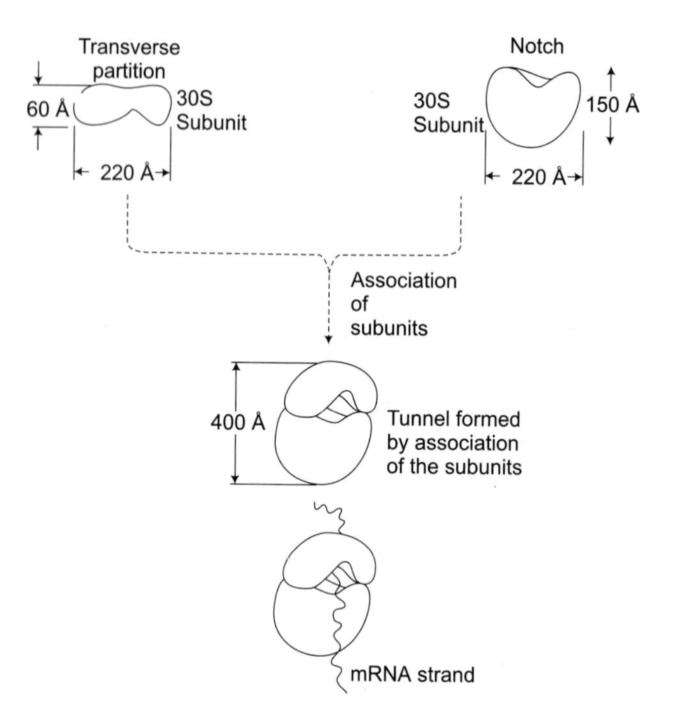

Fig. 11.3 Model of the procaryote ribosome.

Whittman, Traut, Stuffier, Kurland, Nomura, and others have studied the relationship between the three rRNA and the ribosomal proteins are shown that some 28 proteins bind specifically and directly to the rRNA (i.e., the **primary** binding proteins); 14 bind to 16 S RNA, 3 bind to 5 S RNA, and 11 bind to 23 S RNA. Those proteins that do not bind directly to rRNA (i.e., the **secondary** binding proteins) presumably intract with the primary binding proteins in the assembled ribosome.

Best understood is the RNA-protein interaction in the 30 S subunit. The approximate regions within the primary structure with which the primary binding proteins associate; the relative positions of the secondary binding proteins are also indicated. In addition to RNA-proteins interaction, there is considerable protein-protein interaction, including interactions among the primary binding proteins and with the secondary binders. This information, together with the known sizes of the proteins and other parameters, allows the construction of a scheme depicting the spatial relationship among the proteins that make up the small subunit. For simplicity, the 16 S RNA molecule is omitted from the scheme, but it is believed that the backbone of the polynucleocleotide winds its way among the proteins with interactions occurring between hairpin turns of the RNA and surface residues of the protein molecules.

The association of the RNA and protein complements of the 50 S ribosomal subunits is not so completely understood as the 30 S subunits; however, the situation appears to be analogous in that certain proteins bind to specific regions of the 23S and 5 S RNAs and protein-protein interactions are numerous.

Each ribosomal subunit contains *no more than one copy* of each of the S and L proteins, and not all ribosomes in a population contain all the proteins. The interface between the large and small subunits of the ribosome is an important functional area. Proteins S9, S11, S12, S15, S20, L26, and L27 are located in the interface; moreover, it appears that S20 and L26 are identical proteins and can associate with either subunit.

Assembly of Procaryotic Ribosomes

Since all of the proteins and RNAs of the procaryotic ribosome subunits may be isolated, it is possible through recombination studies to examine the assembly process.

Nomura and others have shown that the assembly of individual subunits and their association to form functional ribosomes (i.e., ribosomes capable of translating mRNA into protein) occurs spontaneously in vitro when all the individual rRNAs and protein components are available. Thus the ribosome is capable of *self-assembly*, and this is believed to be the mechanism in situ. The assembly is promoted by the unique and complementary structures of the ribosomal protein and RNA molecules and proceeds through the formation of hydrogen bonds and hydropholic interactions, There is *order* to the assembly in that certain proteins combine with the rRNAs prior to the addition of others. Cooperativity also exists, since addition of certain proteins to growing subunit facilitates addition and binding of others.

No self-assembly takes place when L proteins are added to 16 S RNA or when S proteins are added to 5 S and 23 S RNA. However, it is interesting to note that RNA from the 30 S subunit of one procaryotic species will combine with the S proteins of another procaryote to form functional subunits. The same is true for 50 S subunit proteins and RNAs from different procaryotes. Assembly of hybrid *subunits* and formation of functional monomers from these occur in spite of the fact that ribosomal proteins and RNAs from different procayotes have different primary structures. It is clear that their secondary and tertiary structures, which are very similar, are more important in guiding rRNA-protein interactions. Although some proteins from yeast, reticulocyte, and rat liver cell ribosomes can be replaced by *E. coli* ribosomal proteins, *hybrid monomers* formed from these procaryotic/eucaryotic subunits will not function in protein synthesis.

In spite of all that is known about the composition of ribosomes and the interaction of its molecular components, it is still difficult to propose a viable model of ribosome structure. Although electron microscopy has been immensely helpful in working out the gross structure and organization of other of other organelles, ribosomes are small enough to elude detailed analysis. Moreover, most techniques used to isolate and then prepare ribosomes for electron microscopy unavoidably alter the ribosome's native shape and organization.

Notwithstanding these limitations, several reasonable proposals can be made about the structure of the ribosome monomer and its subunits based upon the available electron-microscopic data, results of small-angle x-ray analysis and of course, chemical studies. The 30 S subunit approximates an oblate ellipsoid of revolution having dimensions of 60 Å × 200 Å × 200 A. A transverse partition or groove encircles the long axis of the subunit, dividing it into segments of one-third and two-thirds (Fig. 11.4). The 50 S subunit is somewhat more spherical,

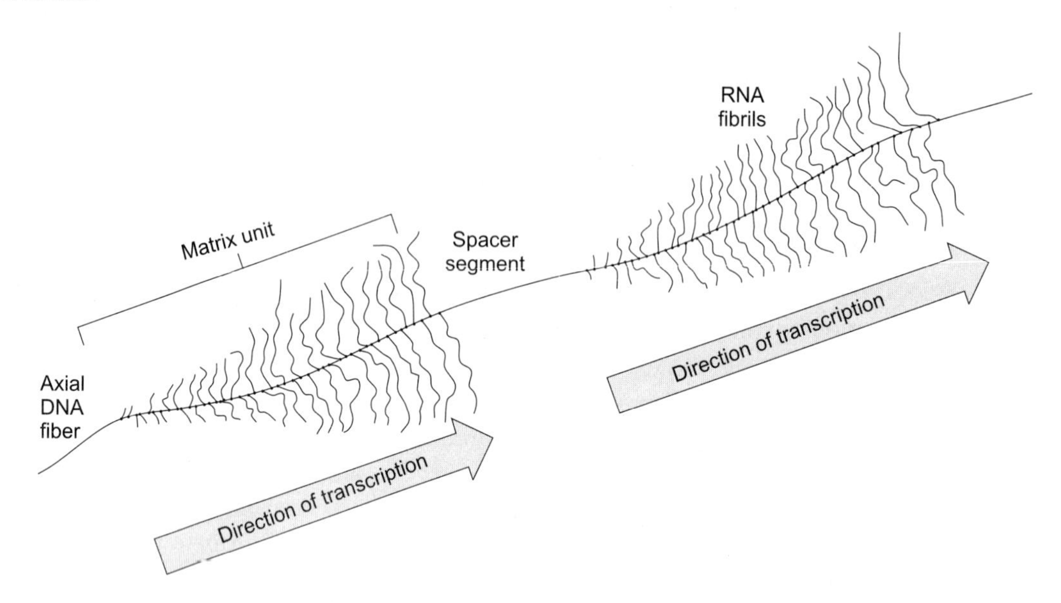

Fig. 11.4 Visualization of transcription of rDNA genes. The thin axial DNA fiber is being transcribed simultaneosly by number of RNA polymerase enzymes (small black dots; see also fig. 21.9). The transcripts (riboucleoprotien in complexes) appear as fine fibrlls extending radially away from the DNA axis. Magnification, 18,000 X (Eleatron photomicrograph courtesy of Dr. O.L. Millier, from O.L. Miller, and B.R. Beatty Science 164, 956,1969 Copyright 1969 by the American Association for the Advancement of Science.

having dimensions of 150 Å × 200 Å × 200 Å, and possesses a flattened or notched region on one surface (Fig. 11.4). Association of the subunits to from 70 S monomer accompanied by a deformation of the 30 S subunit a at transverse partition. The subunits are thereby joined in two regions on either side of a tunnel formed by the 50 subunits notch and the 30 S subunits groove. The 70 S monomer has maximum diameter of about 400A. There is considered morphological and biochemical evidence supporting the idea that the tunnel in the monomer accommodates minnesinger RNA and the aminoacyl-tRNA during protein synthesis. For example, (1) in many electron photomicrographs of polyribosomes, the thin mRNA strand seems to "disappear" into the ribosome; (2) in vitro experiments have shown that when the synthetic messenger polyU is associated with the 70 S monomer, the polynucleotide is protected from ribonuclease attack over a length of about 70 to 120 nucleotides; and (3) transfer RNA is protected from cleavage by nucleases when associated was the ribosome. The observation that nascent (i. e., growing) polypeptides are also protected from proteolysis suggest that they, too, are located within the ribosome-in either the same or perhaps a separate channel (see later).

Genes for Ribosomal RNA and Protein

The genome of *E. coli* and other procaryotes consists of a single, long, circular DNA molecule tightly packed into the nuclear, region of the cell. The *E. coli* chromosome is about 1300 μ long and appears to contain at least three separate regions coding for rRNA. Each region contains closely linked 5 S, 23 S, and 16 S rDNA genes, Since some 5 to 10 copies of each gene occur

Fig. 11.5 One of the riboosomal protein operons of the *E. coli* genome, Note that genes for both S and L proteins occur in the same operon, Promotor genes.

in the genome, more than one copy of each gene is likely present in each rDNA region. Genes coding for ribosomal proteins are present in at least two separate regions of the *E. coli* chromosome. The same regions appear also to contain genes for RNA polymerases, some transfer of RNAs, and the elongation factors required for protein biosynthesis (see later). The genes are distributed among at least four operons, each operon containing genes for, a dozen or more proteins (Fig. 11.5).

Eucaryotic Ribosomes

The cytoplasmic ribosomes of eucaryotic cells differ from those of procaryotes in both size and chemical composition (Table 11.2, Fig. 11.1). The monomer has a sedimentation coefficient of 80 S and is formed from 40 S and 60 S subunits. In addition, ribosomes occur in two states in the cytoplasm. They may be associated with cellular membranes such as those of the endoplasmic reticulum (.i.e.. "attached" ribosomes) and engaged in the synthesis of secretary or vesicle proteins, or they may be freely distributed in the cytosol and synthesize proteins retained within the cell, The functional differences between attached and free ribosomes will be pursued later, but let us turn first to a consideration of the chemical and morphological characteristics of eucaryotic ribosomes.

RNA Constant. The small subunit of the eucaryotic ribosome contains one molecule of 18 S RNA (M.W. 0.7×10^6), while the large subunit contains 28 S (M.W. 1.7×10^6), 5 S (M.W. 3.2×10^4) and 5.8 (M.W. 5×10^4) RNAs. Hence in addition to molecular weight or size differences a major distinction between the RNA complements of procaryotic and eucaryotic ribosomes is the precsence of an additional molecule of RNA in the large subunit of eucaryotes. Of the four rRNA, the 5.8 S molecule has only recently been discovered and characterized (5.8 S RNA has variously been referred to previously as IRNA, 7 S RNA and 5.5 S RNA). The 5.8 S RNA eluded earlier identification because of its intimate association with 28 S RNA in the ribosome.

18 S, 5.8 S, and 28 S rRNAs are the transcription products of closely linked genes in the chromosomes of the nucleolar organizing region (NOR) of the cell nucleus. Considerable redundancy exists since hundreds, perhaps even thousands, of copies of these rRNA genes are believed to be present (see Table 11.3). The genes for 5 S RNA are *not* present in the NOR but occur elsewhere in the nucleus. Consequently, unlike prokaryotes in which the 5 S RNA genes of eucaryotes occur separately in the nucleus. Thus difference, together with other observations to be noted later, support a contention that the 5 S rRNAs of procaryotic and eucaryotic ribosomes are not analogous; instead, it is the eucaryotic 5.8 S RNA that is the "counterpart" of procaryotic 5 S RNA.

The transcription and post-transcriptional modification of eucaryotic rRNAs. It should be noted that 5 S RNA is a primary transcription product and is not the product of post-transcriptional trimming (another distinction from procaryotic 5 S RNA). Where the precursors of the procaryotic rRNAs are sequentially cleaved from the growing transcript a single, high-molecular-weight transcript, 45 S RNA, containing the precursors of 18 S, 5.8 S and 28 S rRNAs is produced in eucaryotes. About half of the 45 S RNA molecule is represented by spacer sequences that are trimmed during final processing. The first processing step divides the 45 S RNA into two parts; the larger of these (41 S RNA) eventually gives rise to 5.8 S and 28 S RNA, while 18 S RNA is derived from the smaller product.

It is natural when comparing procaryotic and eucaryotic cells to look for structures or molecules of similar or even identical function. With regard to ribosome structure and composition, the analogy of 16 S RNA (of procaryotes) and 8 S RNA (of eucaryotes) is obvious, since both are parts of the small subunits of ribosomes and also have other features in common. Similarly, an analogy exists between 23 S RNA (of procaryotes) and 28 S RNA (of eucaryotes). But, what about procaryotic 5 S RNA and eucaryotic 5 S and 5.8 S RNA? Eucaryotic 5 S RNA is similar in, size (about 120 nucleotides) to procaryotic 5 S RNA and also lacks modified nucleotides. In contrast, eucaryotic 5.8 S RNA is larger (about 150 nucleotides) and contains small numbers of modified nucleotides. Notwithstanding these differences, there is significant albeit, not yet conclusive evidence for the contention that eucaryotic 5.8 S (not 5 S!) RNA is analogous to procaryotic 5 S RNA. For example, (1) the additional nucleotides of 5.8 S RNA occur for the most part in two sections at the 5′ and 3′ ends of the polynucleotide chain; the central portion reveals primary nucleotide sequences more closely related to procaryotic 5 S RNA than to eucaryotic 5 S RNA; (2) as noted earlier; procaryotic 5 S and eucaryotic 5.8 S RNAs are transcription products of closely linked rRNA genes and undergo post. transcriptional processing; and (3) there is evidence to support the proposal that 5.8 S RNA, like procaryotic 5 S RNA, interacts at the. A site of the ribosome, whereas eucaryotic 5 S RNA interacts with tRNA during the initiation phase of protein synthesis.

Protein Content. Various studies have established that the small subunits of eucaryotic ribosomes contain 30 proteins (S1, S2, S3 etc.), and the large subunits, 40 proteins (L1, L2, L3, etc.). The proteins of eucaryotic riboosomes are not only mare numerous but also have greater average molecular weights (Table 11.4). From a chemical standpoint, eucaryotic ribosomal proteins have similar general properties as those in procaryotes (e.g., rich in basic or amino acids, high isoelectric point, etc.). Certain eucaryotic and procaryotic ribosomal proteins reveal homologous regions; and these homologous proteins appear also to be functionally similar.

Nucleolar Organizing Region. Eucaryotic cells contain several hundred copies of the genes encoding for rRNA. These genes are arranged in a tandem fashion on one or more

Table 11.4 Average molecular weights of procaryotic and eucaryotic ribosome proteins.

	Procaryote	*Eucaryote*
Small subunit	18,900	25,300
Large subunit	16,400	28,100

chromosomes of the nucleus, The DNA sequences between successive rRNA genes cannot be transcribed and represent spacer DNA. The rRNA genes and the spacer segments are usually looped off the main axis of the chromosome and are referred to as the **nucleolar organizing region** (NOR). It is here that most of the rRNA is synthesized. The NOR coalesces with nuclear proteins and forms the visible bodies known as **nucleoli.** Most eucaryotic cells contain one or a few nucleoli, but certain egg cells are a striking exception. The oocytes of amphibians (e,. g., the clawed toad, *Xenopus laevis*) are extremely large cells and are engaged in the synthesis of especially large quantities of cellular protein. These cells produce large numbers of ribosomes in order to provide the means to sustain such quantitative protein synthesis. Accordingly, it is not unusual to find hundreds or thousands of nucleoli (and NORs) in the nuclei of these cells. Such large numbers of nucleoli are the result of gene amplification-the differential replication of the rRNA genes of the genome. The ribosomes produced in the oocyte serve its needs for protein synthesis from the period prior to fertilization through the first few weeks of embryonic development.

By gently dispersing nuclear fractions isolated from oocytcs of the amphibian. *Triturus viridescens* and "spreading" the material on grids, O.L. Miller and B.R. Beatty in 1969 were able to obtain photomicrographs of transcription in progress. Since then, the same approach has been extended by a number of other invesitigators to mammalian oocytes and to spermatocytes and embryo cells from various organisms. The visualization of transcriptional activity is achieved most easily with spread nucleoli because of the high degree of rDNA gene amplification. The tandem rDNA genes are serially trascribed by RNA polymerases to produce 45 S rRNA. The rRNA (apparently complexed with protein) appears as a series of fibrils of varying length extending radially from an axial, linear DNA fiber (Fig. 11.6 & 11.6A). These feather-shaped or "Christmas tree" suggesting that replication may not require dissociation of nocleosomes or that nucleosomes are almost immediately reformed. Transcriptional activity can be identified within a replicon, indicating that the newly synthesized DNA is almost immedialtely available for transcription. The growing RNA fibrils are seen in homologous regions of *both* choromatid arms of the replicon.

Assembly of Eucaryotic Ribosomes. The assembly of eucryotic ribosomes is more complex than that of procaryotes. Transcribed 45 S RNA combines with proteins in the nuncleolus to from ribonucleoprotein complexes (RNP). However, not all the protein molecules of the complex become a part of the completed ribosomal subunit.

In high magnification views (Fig. 11.6) even the RNA polymerase enzyme molecules carrying out the transcription of the DNA are visible along the axial DNA fiber.

Success in visualizing transcription has been not restricted to nucleolar genes. Almost identieal results have been obtained with nonnucleolar chromatin. Here, however, the RNA transcripts represent messenger RNA.

Dispersed and spread nuclear fractions contain nontranscribing DNA as well as matrix units (Fig. 11.6). The succession of nucleosomes reveals itself as a series of beadlike structures along the DNA fiber. Regions in which DNA is undergoing replication (called replicon) can also

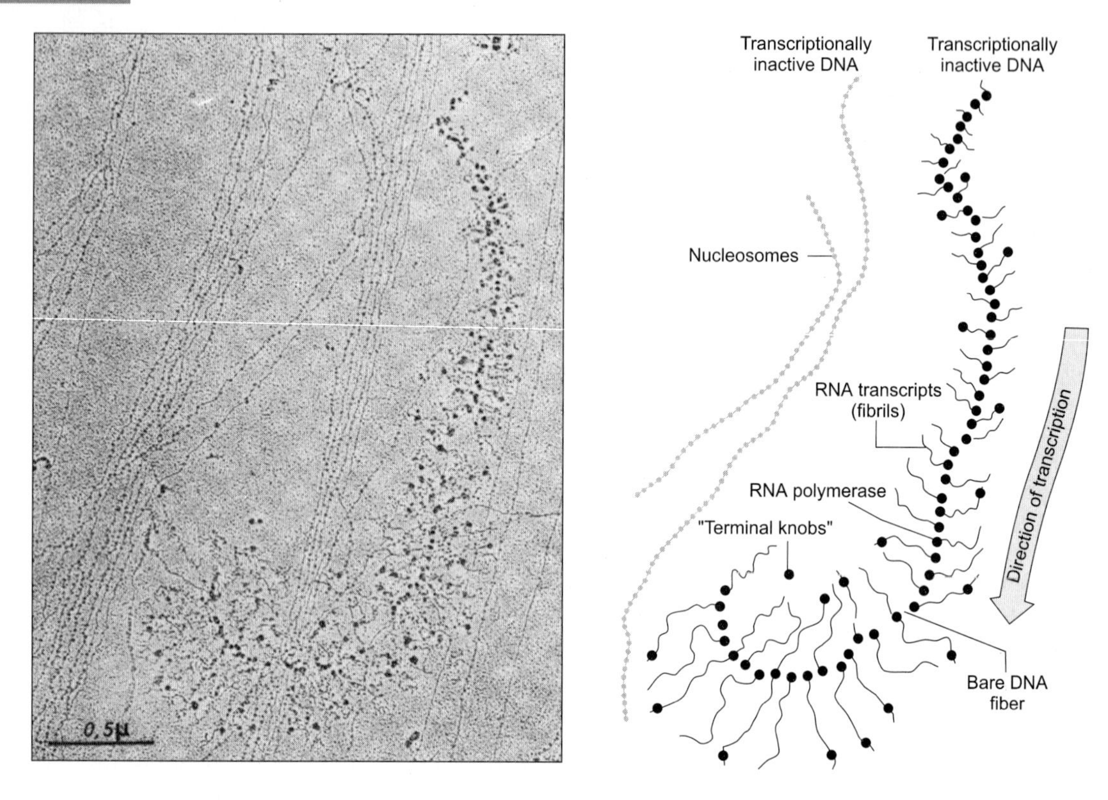

Fig. 11.6 High magnification electron photomicrograph of matrix until and neighboring non-transcribed DNA fibers. See legend of Fig. 11.5 and text for explanation (Photo courtesy of Dr. F. Puvion-Dutilleul).

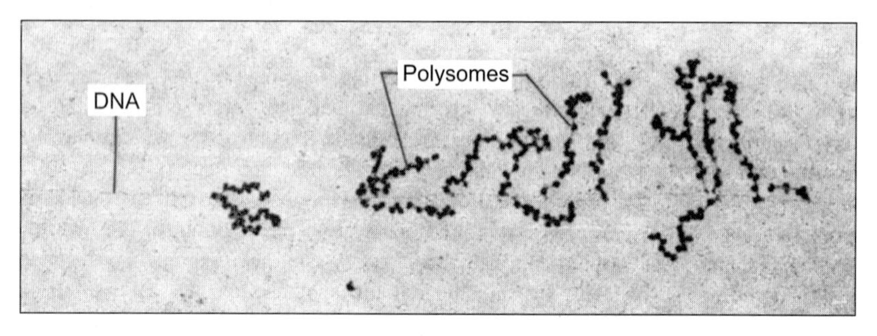

Fig. 11.6A Visualization of the simuitaneous transcription and translation of *E. coli* chromosomal DNA. Magnification, 43.000 × (Courtesy of Drs. O.L. Miller and B.A. Hamkalo, from O.L. Miller, B.A. Hamkalo, and C.A. Thomas, Science 169, 394 1970. Copyright 1970 by the American Association for the Advancement of Science).

be seen. S.L. McKnight and O.L. Miller have shown that DNA of homologous "daughter" fibers of the replicon also occurs as chains of nunleosomes.

Instead, certain proteins are released as RNA processing ensues; these "nucleolar proteins" return to a nocleolar pool and are reutilized. Those proteins that are ratained during processing and become part of the completed subunits are, of course, legitimately called "ribosomal proteins." Enzymatic cleavage of the RNP complex during processing produces three classes of fragments. One fragment contains spacer RNA and nucleolar proteins. (It should be noted that the spacer RNA is produced by transcription of rDNA and *not* the spacer DNA between genes.) The spacer RNA is hydrolyzed, and the free nucleolar proteins return to the pool. A second RNP fragment contains a complex of 18 S RNA certain ribosomal proteins that give rise to 40 S ribosome subunits in the cytoplasm. The third RNP fragment, which contains 28 S and 5.8 S RNA and ribosomal proteins, combines with 5 S RNA transcribed from extranuleolar rRNA genes, and the complex exits the nucleus to give rise to 60 S subunits in the cytoplasm. Like the genes for 45 S RNA, the extranucleolar 5 S RNA genes occur in multiple tandem copies. Among the various proteins synthesized in the cytopalsm using ribosome subunits derived from the nucleus are the ribosomal proteins themselves. These apparently reenter the nucleus for incorporation into new RNP complexes.

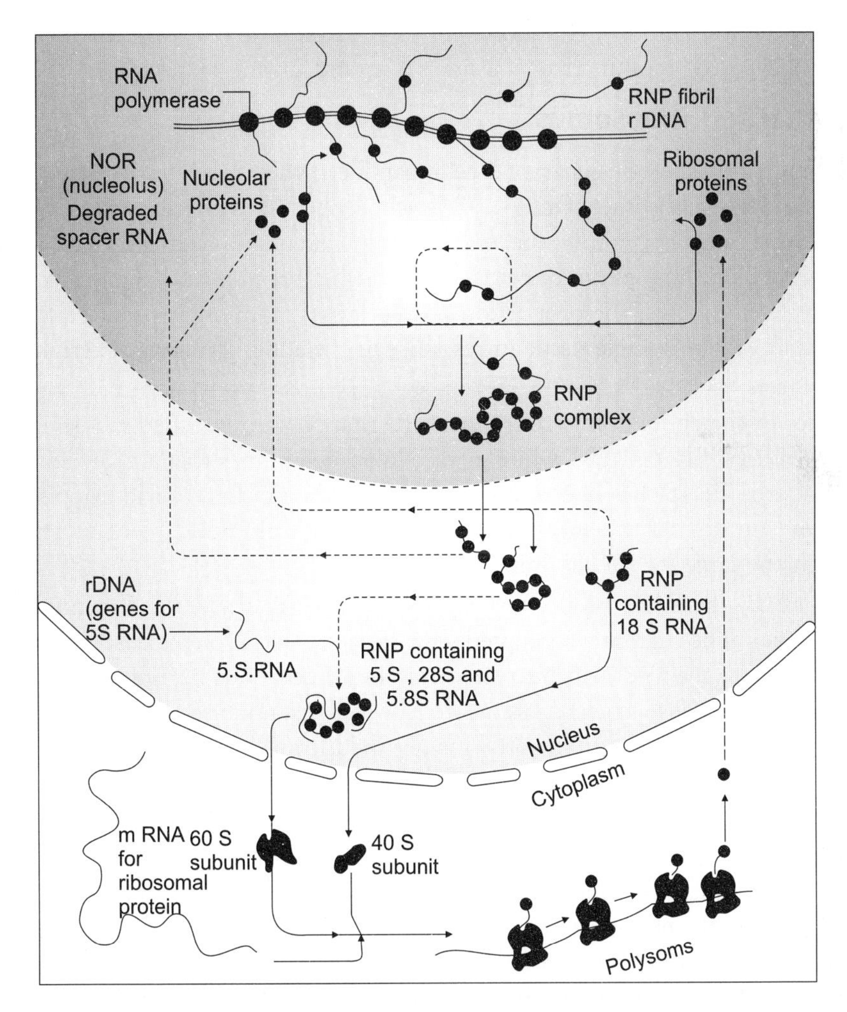

Fig. 11.7 Synthesis and assembly of the components of eucaryotic ribosomes.

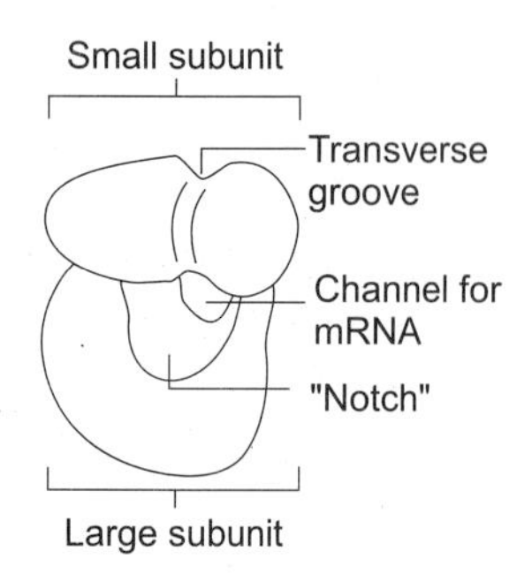

Fig. 11.8 Model of the eucaryotic cytoplasmic ribosome.

Model of Eucaryotic Ribosomes

In spite of the difference in overall sizes (as manifested in the greater molecular weights, sedimentation constants, sizes, and numbers of rRNA and proteins), the ribosomes of eucaryotes are remarkably similar in morphology to those of procarytes.

The 40 S subunit approximates a slightly flattened ellipsoid of revolution, having dimensions of 115 Å × 140 Å × 230 Å. As in 30 S subunits of procaryote ribosomes, the 40 S eucaryote subunit is divided into segments of one-third and two-third by a transverse groove (Fig. 11.8). The 60 S subunits is generally rounder in shape, having a diameter of about 200 A. One side of the large subunit is somewhat flattened, with a notch that becomes confluent with the formation of the monomer (Fig. 11.8). The resulting channel through the ribosome is believed to accommodate the mRNA strand during translation.

Free and Attached Ribosomes

The cytoplasmic ribosomes of eucaryotic cells can be divided into two classes (1) **attached** ribosomes and (2) **free** ribosomes (Fig. 11.9). Attached ribosomes are ribosomes associated with intracellular membranes, primarily the endoplasmic reticulum, whereas free ribosomes are distributed through the hyaloplasm or cytosol. Although all animal and plant cells contain both attached and free ribosomes, the proportion of each varies from one tissue to another and can be caused to shift within a single tissue in response to the administration of certain substances, notably hormones and growth factors.

Membranes of the endoplasmic reticulum (ER) that contain attached ribosomes constitute what is called "rough" ER (or RER), white intracellular membranes that are devoid of ribosomes are called "smooth" ER (SER). The ribosomes of RER are attached to the hylopisna surface of membranes (as opposed to the lumenal or external surface). Attachment to the membrane occurs through the large (60 S) subunit.

For many years, there has been considerable controversy about the functions of attached and free ribosomes. The currently acceppted view suggests that proteins destined to be secreted from the cell or to be incorporated into subintracellular bodies as lysosomes and peroxisomes (which) may or not release their contents to the cell exenor are synthesized on attached ribosomes, whereas most (but not all) proteins destined for the cytosol are synthesized on free ribosomes. For example, many of the proteins circulating in the blood plasma are derived via secretion by the liver, and these plasma proteins are known to be synthesized exclusively by the attached ribosomes of the liver cells.

Thyroglobulin, which is secreted by the thyroid gland, is also synthesized by attached ribosomes. So too are the milk proteins produced by mammary gland cells. There is also a good deal of evidence indicating that some of the proteins that make up the membranes of intercellular organelles may be synthesized on ribosomes attached to the endoplasmic reticulum.

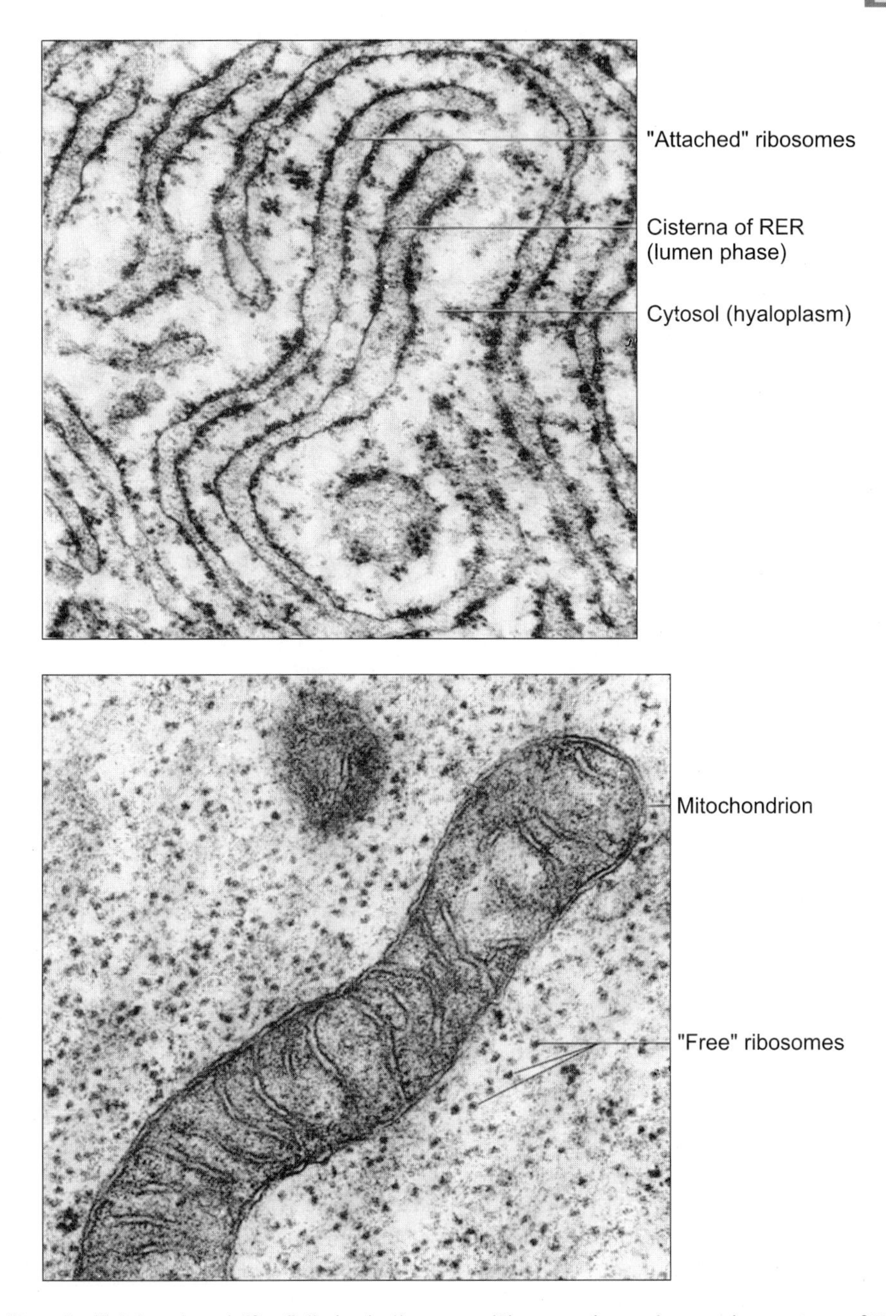

Fig. 11.9 "Attached" (above) and "free" (below) ribosomes (Electron photomicrographs courtesy of R. Chao).

Included in this category would be integral, and extrinsic proteins asymmetrically distributted in the exterior half of these membranes.

When cells are disrupted, the sheets of endoplasmic reticulum are broken into small vesicular fragments (Fig. 11.10a). Which may be isolated by centrifugation with the microsomal

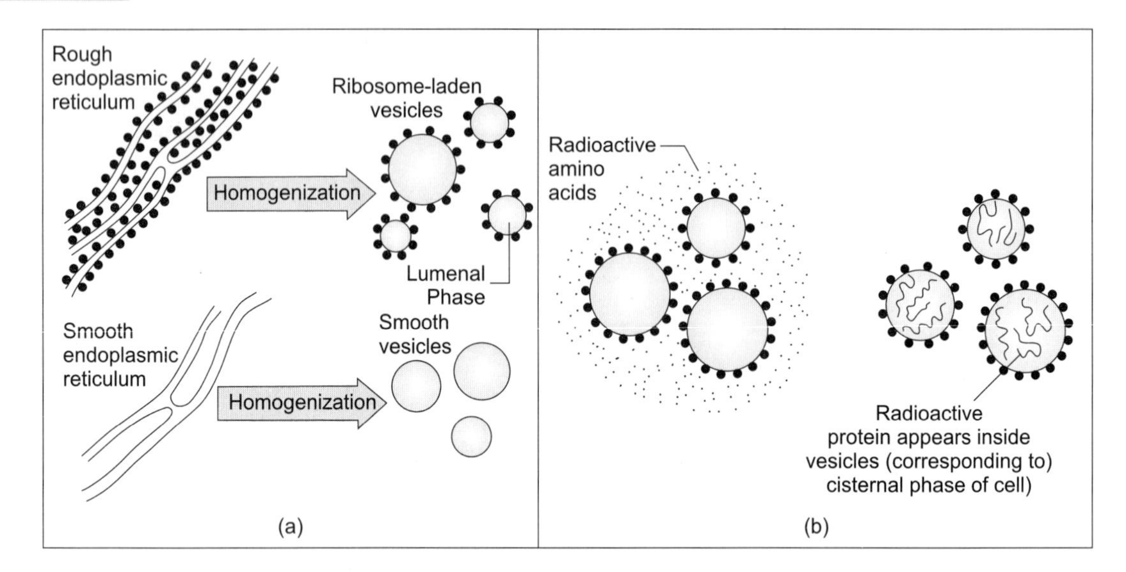

Fig. 11.10 (a) Production of microsomal vesicles during homogenization of endoplasmic reticulum. (b) Vectorial synthesis of protein by ribosome- laden vesicles.

phasee. As noted in Chapter12 this phase is quite heterogeneous and contains a variety of small particles in addition to fragmented endoplasmic reticulum. Fragmentation of the endoplasmic reticulum produces two kinds of vesicles; fragments of RER from vesicles whose either surface is studded with ribosomes, while SER vesicle: are free of ribosomes. The volume within the vesicle corresponds to the lumenal or cisternal phase of the cell.

In a series of elegant experiments, D. Sabatini and C.M. Redman examined protein synthesis in vitro by ribosome-laden vesicles. They were able to demonstrate that radioactive amino acids incorporated into protein by the ribosomes did not appear in the suspending medium put were recovered instead within the vesicles. These results argue strongly in favor of the proposal that attached ribosomes synthesize proteins for secretion, since the interior of the vesicles corresponds of the cisternal phase of the cell. Sabatini and Redman called this directional synthesis of proteins **vectorial synthesis**, Once the protein is released is released into the cisternae, it is transported to the Golgi apparatus for packaging.

Although it is widely accepted that proteins destined to be secreted from the cell are synthesized on membrane-bound ribosomes, there are several pieces of evidence tht indicate that membrane-bond ribosomes may have other functions as well. For example, J.J. Tata and others, working with muscle and nerve tissue, have observed the synthesis by RER ribosomes of small quantities of proteins for intracellular utilization. The rapid proliferation of RER during periods of active cell growth also suggest a nonsecretory function for attached ribosomes. The enzyme serine dehydrogenase has been shown to be specifically synthesized by attached ribosomes in live cells, and yet this is an intracellular enzyme. Preliminary evidence suggests that some mitochondrial enzymes may be differentailly synthesized by RER ribosomes.

Several independent lines of investigation support the idea that attached and free ribosomes may be structurally different. The **reticulocyte** (a developmental form produced during the differentiation of the mature red blood cell see later) is primarily engaged in the synthesis of hemoglobin molecules, but small amounts of other cellular proteins are also produced. The globin chains for hemoglobin are syntehsized on free ribosomes. (i. e., there is no endoplasmin reticulum in reticulocytes), while other cell proteins appear to be synthesized on ribosomes attached to the internal surface of the plasma membrane. These two ribosome populations have been isolated and separately studied ; discontinous electrophoretic analysis of dissociated ribosomes reveal several inportant differences in their constituent proteins.

Working with liver tissue, Subatini and his colleagues have shown that ER membrane to which ribosomes are attached contain two proteins absent in ribosomes-free ER membranes. Biochemical studies coupled with freze-fracture electron microscopy suggest that these membrane proteins from are interconnecting network of binding sites for ribosomes. The network is characterisic of RER but absent in SER. Fig. 11.11 depicts three contemorary models for the translation of mRNA by attached ribosoms. According to one of these models (Fig. 11.11a), the attachment of the ricosome to the membrane *precedes* the initiation of polypeptide chain synthesis. An alternative mechanism, for which there is growing experimental evidence involves synthesis of a "signal" region of the polypeptide before the ribosomes attaches to the membrane. The signal amino acid sequence interacts with membrane proteins directing the association of the large subunit with receptor sites in the membrane. In both Fig. 11.16a and b the protein is discharged into the cisterna of the ER. Fig. 11.11c depicts the model that accounts for the synthesis of cytoplamic proteins by attached ribosomes. In th latter case, the protein is not discharged through a channel in the large subunit. In all the models of Fig. 11.11. mRNA is shown attached to the membrane at its 3 end.

RIBOSOMES OF ORGANELLES

The mitochondria and chloroplasts of eucaryotic cells contain their own DNA and protein-synthesizing apparatus. Although reports of the presence of DNA in mitochondria and cholroplasts appeared periodically in the scientific literature since that 1920s, such an astounding proposition was not generally accepted until the 1960s, when H. Ris and M. Nass independently verified the existence of DNA fibrils in animal and plant cells using electoron-microscopic methods originally developed to visualize DNA in prcaryotic cells (Fig. 11.12).

Little attention was given to the possibility that mitochondria and chloroplasts might contain ribosomes and other elements involved in protein synthesis until the presence of DNA in these organelles was established. With such an inpetus provided, it did not take long before ribosomes were indeed indentified in and isolated from chloroplasts and mitochondria.

Chloroplast Ribosomes

Chloroplast ribosomes have a sedimentation coeffcient of 70 S and consist of 50 S and 33 S subunits. In this respect, chloroplast ribosomes are similar to those of procaryotic cells but

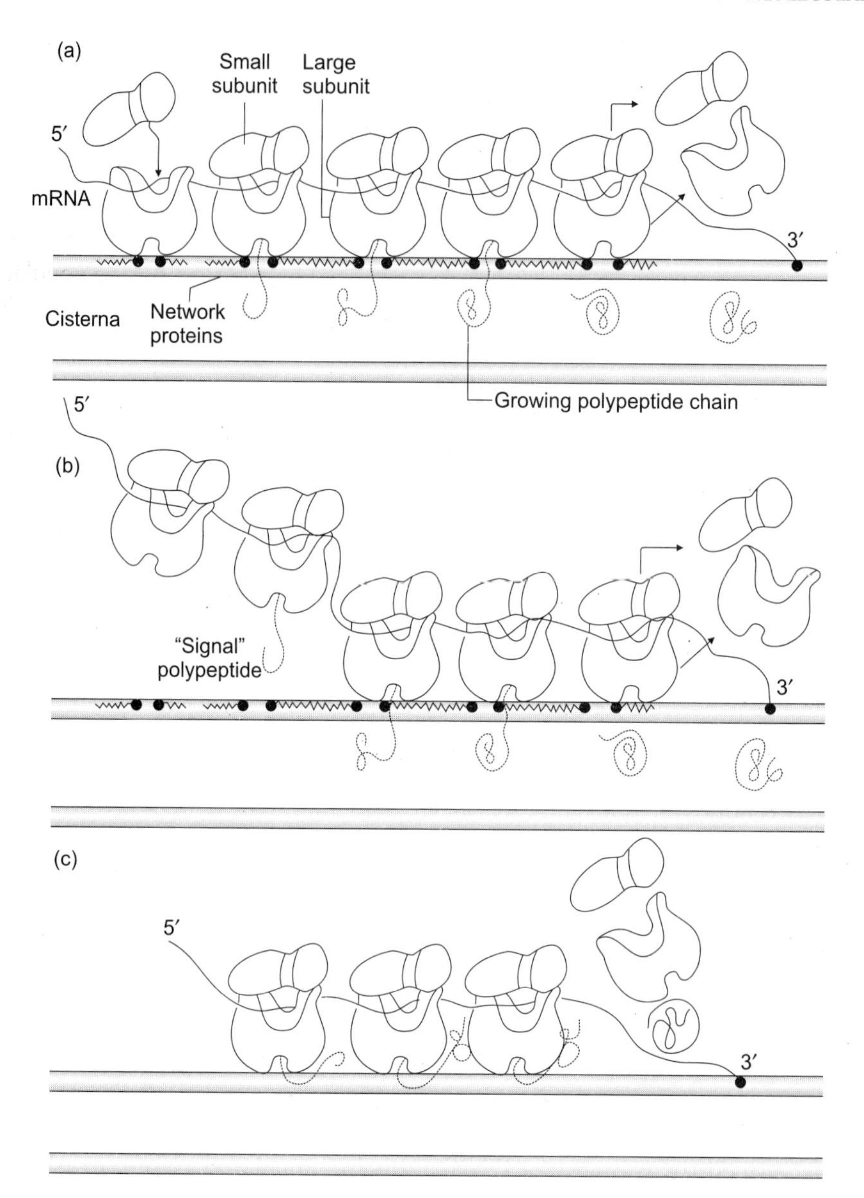

Fig. 11.11 Models for the synthesis of protein by attached ribosomes (See text for details).

distinct from eucarotic cytosol ribosome. The large subunit of the chloroplast ribosomes contains 5 S and 23 RNAs and the small subunit 16 S RNA (Table 11.5).

Mitochondrial Ribosomes

Unlike cholroplast, ribosomes which are similar in all groups of organisms studied, mitochondrial ribosomes are quite heterogeneous. With respect to their disposition within the mitochondrion, the ribosomes occur either free in the *matrix* of the organelle of are associated with the cristeal membranes. (Ribosomes attached to the cytosol side of the outer mitochondrial membrane are cytoplasmic ribosomes engaged in vectorial synthesis of certain intramitochondrial proteins; see below).

Although mitochondrial ribosomes from all sources studied consist of two subunit, there is cnosiderable variation in the sizes of the subunits and the monomers formed from them (Table 11.5). Mitochondrial ribosomes of yeast, fungi, protists, and higher are characterized by a sedimentation coefficient of 70 to 80 S, whereas those of animal cells have a sedimentation coefficient of 50 to 60 S and are therefore unusually small. Unlike the ribosomes of chloraplast, procryotic cells, and eucaryotic hyaloplasm, mitochondrial ribosmes contain only *two* species of rRNA.

Protein Synthesis in Chloroplasts and Mitochondria

The amounts of DNA present in chloroplasts and mitochondrial is only about 10 to 15% of that necessary to encode for the hundreds of different proteins present in these organelles. Therefore, most of the products of genetic information in the cell nucleus. Experimentally, ribosomal RNAs of chloroplasts and mitochondrial can be shown to hybridize with their organellar DNA, and it is generally agrreed that these RNAs are synthesized in the organelles. The origins of only a small number of chloroplast and mitochondrial proteins have been established to date, but the results are interesting and perhaps also surprising. Most if not all, of the proteins that make up the

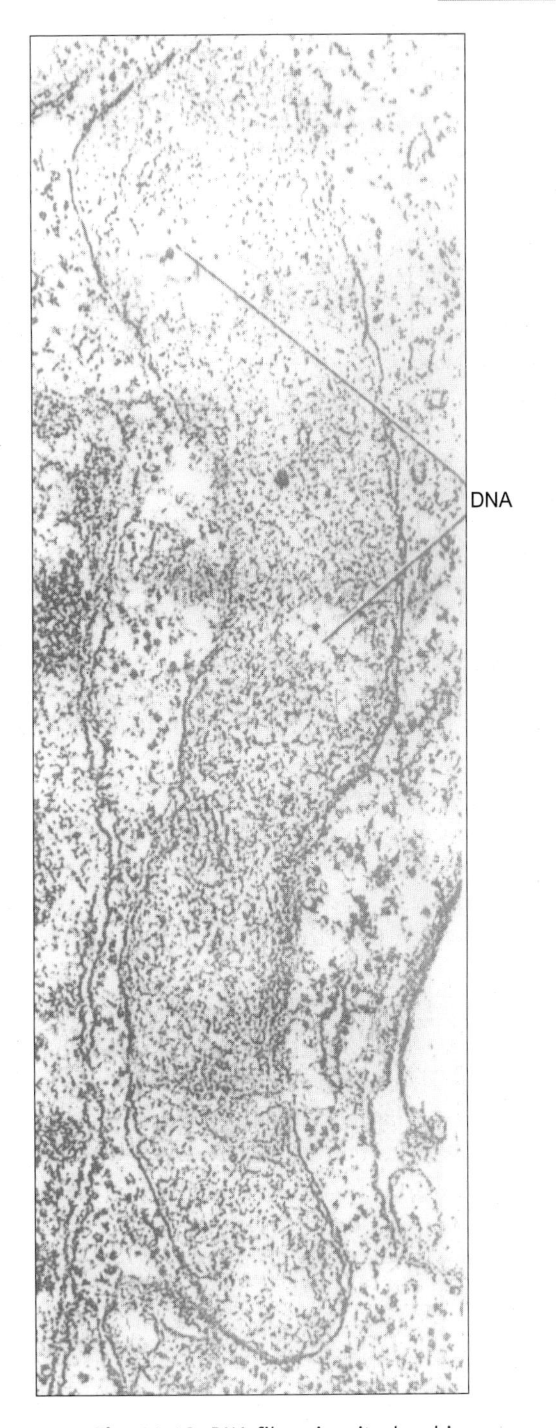

Fig. 11.12 DNA fibers in mitochondria (Electron photomicrograph courtesy of Dr. E.G. Pollock).

Table 11.5 Poperties of chloroplast and mitochondrial ribosomes.

	Sedimentation Coefficient	
	Particle	RNAs
Chloroplast ribosomes		
Monomer	70 *S*	
Large subunit	50 *S*	5 *S* and 23 *S*
Small subunit	33 *S*	16 *S*
Mitochondrial ribosomes		
Animal cells		
Monomer	50-60 *S*	
Large Subunit	40-45 *S*	16-18 *S*
Small Subunit	30-35 *S*	12-13 *S*
Yeast, fungi, and protists		
Monomer	70-80 *S*	
Large subunit	50-55 *S*	21-24 *S*
Small subunit	32-38 *S*	14-16 *S*
Higher plant cells		
Monomer	70-80 *S*	
Large subunit	50-60 *S*	> 23 *S*
Small subunit	40-44 *S*	>16 *S*

organelles' ribosomes are the synthesized in the organelle for assemble (along with rRNA) intc ribosomes. Entry into the organelle may be via vectorial discharge. The synthesis of several organelle enzymes, including *ribulosediphosphate* carboxylase (of chloroplasts) and *cytochrome* oxidase, and an ATPase (of mitochondrial) appears to be a "joint operation" of the organelle and the cytoplasm. For example, of the seven plypeptides that make up cytochrome oxidase, four are synthesized on cytosol ribosomes and three on mitochondrial ribosomes. There is some evidence that a few of the chloroplast membrane proteins are synthesized on chloroplast ribosomes.

Much remains to be learned about the total function of organelle ribosomes; however, it is clear that while most chloroplast and mitochondrial proteins are the products of the nucleocytoplasmic protein-synthesizing machinery of the cell, a small number of proteins and portions of certain multisubunit enzymes are the products of genetic material intrinsic to the organelle and the result of translation on organelle ribosomes.

CHAPTER 12

RNA and Protein Synthesis

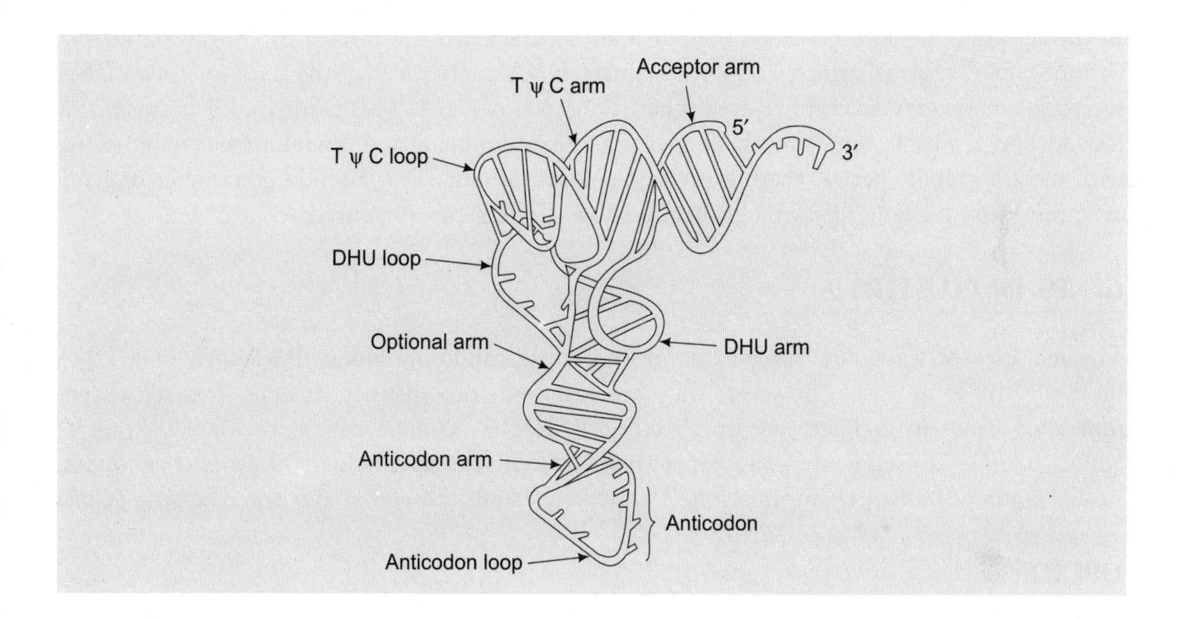

INTRODUCTION

A gene is simply a segment of a DNA molecule. This segment may be any length from about 75 nucleotides to over 40 kb (kilobases).

NUCLEOTIDE SEQUENCE—The Crucial Feature of Gene

The biological information carried by a gene is contained in its nucleotide sequence. This information is in essence a set of instructions for synthesis of an RNA molecule that may subsequently direct the synthesis of an enzyme or other protein molecule. This two-step process is called *gene expression.*

The biological information of a gene is, in fact, carried by just one of the two polynucleotides of the double helix. This polynucleotide is called the *coding strand* to distinguish it from the non-coding complement. The latter should not be looked on as unimportant even though it does not contain biological information, it plays a vital role in processes such as DNA replication. Two genes on the same DNA molecule do not necessarily have the same polynucleotide as their coding strand: a polynucleotide that is the coding strand for one gene may be the non-coding strand for a second gene.

ORGANISATION OF GENES ON DNA MOLECULES

In the cells of a higher organisms, such as man, all the genes are carried by a small number of chromosomes, each of which contains a single DNA molecule. Clearly each of these DNA molecules must carry several hundred if not thousands of genes. This holds for lower organisms too. Although a bacterium is much simpler than a human being and, therefore, has fewer genes; with most bacterial species, there is only one 'chromosome', so again a large number of genes are present on a single DNA molecule. How are these genes organised?

GENES IN CLUSTERS

The majority of genes are spaced out more or less randomly along the length of a DNA molecule. In some cases, however, they are grouped into distinct clusters. Sometimes the individual genes in a cluster are unrelated and there is no apparent reason or advantage of having them organised in this way. More frequently clusters are made up of genes that contain related units of biological information. Two examples are the *operon* and the *multigene family.*

OPERONS

Operons are fairly common features of the organisation of genes in bacteria. *An operon is a cluster of genes coding for a series of enzymes that work in a concerted and in an integrated biochemical pathway.* The first operon to be discovered was the *lactose* or *lac operon* of *E. coli*, a cluster of three genes each coding for one of the three enzymes involved in conversion of the disaccharide lactose into its monosaccharide units glucose and galactose. These enzymes will not be required all the time, but only when lactose is available to the bacterium.

MULTIGENE FAMILIES

Operons have no direct counterparts in organisms other than bacteria. In contrast, multigene families are found in many organisms. *A multigene family is also a cluster of related genes, but in multigene families the individual genes have identical or nearly identical nucleotide sequences.* Genes related in this way are referred to as homologous genes, and usually have a figure, such as 90%, attached to them to indicate the degree of sequence homology between members of the family.

Examples of well-studied multigene families are provided by the globin polypeptides of vertebrates.

We now have information on a whole range of multigene families in a variety of organisms, including some families with which the individual genes are not in fact clustered but scattered at different positions on a single or on more than one chromosome.

DISCONTINUOUS GENES

One of the most startling findings of recent years was made in 1977 when several investigators discovered that the biological information carried by some genes is split into several distinct units separated by regions of non-coding DNA (Fig. 12.1). The sections containing biological information are now called *exons* with the intervening non-coding sequences referred to as *introns.* Discontinuous genes (also called *split* or *mosaic genes*) are now known to be common in higher organisms and in many types of viruses, but are thought to be absent in bacteria, though even this idea has been challenged.

Why some genes should be split into exons and introns has been the subject of intense speculation. An interesting, though not necessarily correct, theory has been put forward by

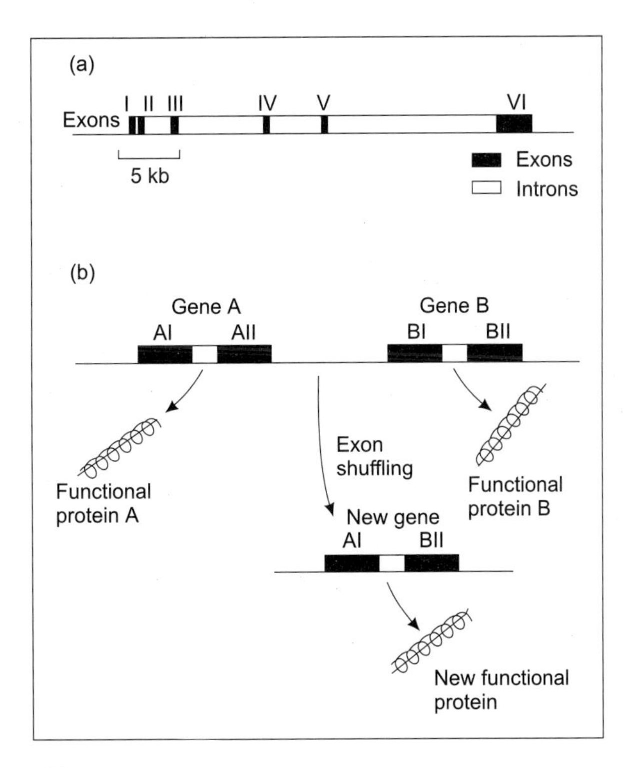

Fig. 12.1 Discontinuous genes. (a) The mouse gene for dihydrofolate reductase is made up of six exons separated by five introns. (b) the hypothesis of 'Exon Shuffling' suggests that new functional genes could be formed by rearranging exons of existing genes.

Walter Gilbert and associates, who suggest that each exon of discontinuous gene contains a different subcomponent of the biological information carried by the gene as a whole. Although these subcomponents are not sufficient on their own to code for the complete protein, the information they contain is meaningful in that it specifies a recognisable portion of the protein's function, for instance enabling the protein to bind a particular substrate or attach to a specific site in the cell. Gilbert's hypothesis is that, during evolution, exons from different discontinuous genes can be 'shuffled', creating new combinations of biological information. This kind of process would be more likely to produce new functional proteins than an entirely random rearrangement of existing genes.

THE CENTRAL DOGMA

The process that we now call gene expression was first put forward by Francis Crick in 1958 in a lecture called 'On Protein Synthesis' presented to the Society for Experimental Biology, Crick postulated that the biological information contained in the DNA of the gene is transferred first to RNA and then to protein (Fig. 12.2). We now accept that this is, in its simplest form, what happens. Crick also stated that the information flow is unidirectional, that proteins cannot themselves direct synthesis of RNA and that RNA cannot direct synthesis of DNA. This part of the *Central Dogma* was shaken in 1970 when Howard Temin and David Baltimore independently discovered that certain viruses do transfer biological information from RNA to DNA.

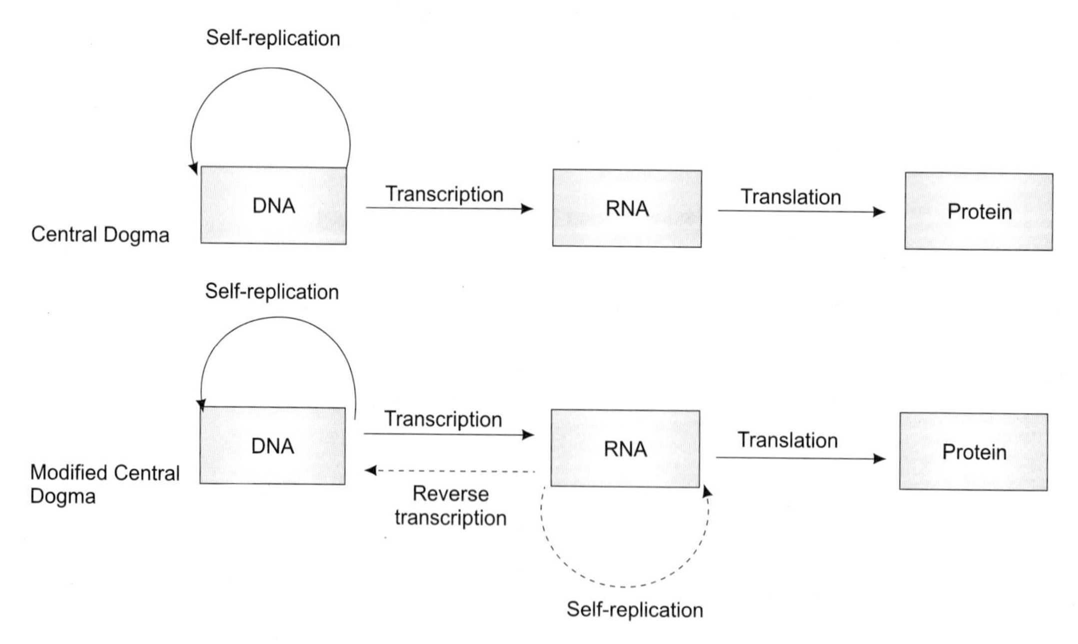

Fig. 12.2 Central dogma for protein synthesis.

However, the Central Dogma still remains today as one of the underlying concepts of molecular genetics.

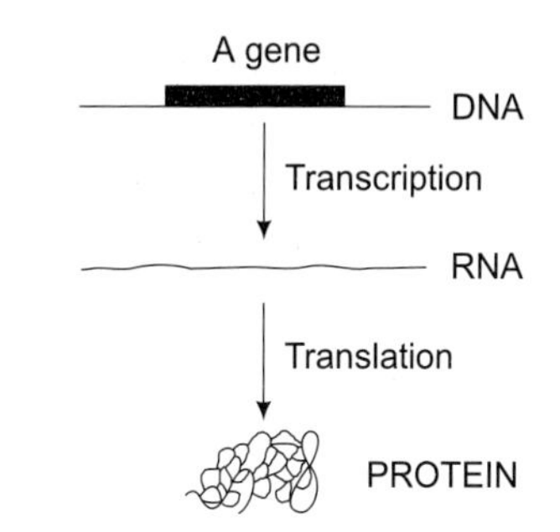

Fig. 12.3 The two stages of gene expression.

TRANSCRIPTION—The First Stage of Gene Expression

All genes undergo the first stage of gene expression which is called *transcription* (Fig. 12.3). During transcription, the coding strand of the gene acts as a template for synthesis of an RNA molecule. The sequence of this RNA molecule is determined by complementary base pairing with the template, so the RNA transcript is in fact the complement of the coding strand.

TRANSLATION—The Second Stage of Gene Expression

For some genes, the RNA transcript is itself the end product of gene expression. For others, the transcript undergoes the second stage of gene expression called *translation*. During translation, the RNA molecule (called a *messenger* or *mRNA*) directs synthesis of a polypeptide, the amino acid sequence of which is determined by the nucleotide sequence of the mRNA (which is of course itself derived from the nucleotide sequence of the gene). Each triplet of adjacent ribonucleotides specifies a single amino acid of the polypeptide, the identity of the amino acid corresponding to each triplet being set by the genetic code.

PROTEIN SYNTHESIS—The Key to Expression of Biological Information

For all genes, the end-point of gene expression is synthesis of either an RNA molecule or a polypeptide. How can this simple process bring about the utilisation of the biological information contained in genes and thereby enable construction of a living organism?

The answer lies in the functional flexibility of protein molecules. This is because polypeptides of different amino acid sequences can have quite different chemical properties and can, therefore, play quite different roles in the cell. Consider a few of the roles played by different proteins:

1. *Structural proteins* form parts of the framework of organisms. Examples are *collagen* which is associated with tendons, bone and cartilage in vertebrates, *sclerotin* in the exoskeleton of insects and virus coat proteins.
2. *Contractive proteins* enable organisms to move. Examples are *actin* and *myosin* in muscles, and *dynein* in clilia and flagella.
3. *Enzymes catalyse* the multitude of biochemical reactions that bring about the release and storage of energy (catabolism) and the synthesis of new compounds (anabolism).

Examples are: *hexokinase*, the enzyme that catalyses the first step in glycolysis, the pathway by which energy is released from glucose; *RNA polymerase*, which is responsible for transcription of DNA into RNA during gene expression; *tryptophan synthetase*, which catalyses the last step in the lengthy biosynthetic pathway that converts phosphoenolpyruvate, itself derived from glucose, into the amino acid tryptophan.

4. *Transport proteins* carry important molecules around the body. Examples are *haemoglobin*, which carries oxygen in the blood-stream of vertebrates; *haemocyanin*, which performs the same function in some invertebrates; and *serum albumin*, which transports fatty acids in the bloods.
5. *Regulatory proteins* control and coordinate biochemical reactions in the cell and in the organism as a whole. Examples are *catabolite activator protein*, which regulates expression of the genes involved in sugar metabolism in *E. coli*, and hormones such as *insulin*, which controls glucose metabolism in vertebrates.
6. *Protective proteins* have evolved to greatest sophistication in vertebrates and protect against infectious agents and injury. Examples are *immunoglobulins* and other *antibodies* which form complexes with foreign proteins, and *thrombin* and other components of the blood clotting mechanism.
7. *Storage proteins* which store compounds and molecules for future use by the organism. Examples are *ovalbumin*, which stores amino acids in egg white, and *ferritin* which stores iron in the liver.

The development and functioning of a living organism can be looked on as nothing more than the coordinated activity of a vast range of different protein molecules. The biological information contained in the genes, the blueprint for life itself, is simply the instructions for synthesising these proteins at the correct time and in the correct place.

TRANSCRIPTION

During the first stage of gene expression, the coding strands of the gene act as a template for synthesis of a complementary strand of RNA. This process is called *transcription* and the RNA molecule that is synthesised is called the *transcript*.

EUKARYOTES AND PROKARYOTES

Eukaryotes and prokaryotes are distinguished by their fundamentally different cellular organisations. The typical eukaryotic cell is usually larger and more complex than a prokaryotic cell, with a membrane-bound nucleus containing the chromosomes, and with other distinctive membranous organelles such as mitochondria, vesicles and Golgi bodies. Prokaryotes, in contrast, lack an extensive cellular architecture; membranous organelles are about and the genetic material is not enclosed in a distinct structure.

The distinction between eukaryotes and prokaryotes is important because although the basic features of genes were laid down before the two diverged, there are nevertheless important differences between them in gene structure and expression. We have already noted that some eukaryotic genes are discontinuous whereas prokaryotic ones are not. For reasons to do mainly with ease of handling in experiments, most of the pioneering work in molecular genetics has been carried out with a prokaryote, *E. coli.*

The equivalent processes in eukaryotes are generally less well understood but appear to be similar in outline though different in detail.

NUCLEOTIDE SEQUENCES

From this point on, increasing use will be made of nucleotide sequences to illustrate aspects of gene structure and expression. Usually the sequence presented will be that of the non-coding strand of the gene in the 5′ to 3′ direction. This may at first seem illogical as we know that the biological information is carried by the coding strand and not by its complement. However, as will become clear in later chapters, it is the sequence of the RNA transcript (which is the same as that of the non-coding strand) that the molecular biologist is really interested in, because it is from this sequence that the important conversion to amino acids is made by way of the genetic code. For this reason, when detailing a gene sequence, the convention is to describe the non-coding strand.

RNA SYNTHESIS

The underlying chemical reaction in transcription is synthesis of an RNA molecule. This occurs by Polymerisation of ribonucleotide subunits and can be summarised as:

$$n(NTP) \rightarrow (NTP)_n + n(PP_i)$$

During polymerisation the 3′-OH group of one ribonucleotide reacts with the 5′-P group of the second to form a Phosphodiester bond. This results in loss of a pyrophosphate molecule (PP_i) for each bond formed. In transcription, the chemical reaction is modulated by the presence of the DNA template, which directs the order in which the individual ribonucleotides are polymerised into RNA, and is catalysed by the enzyme DNA-dependent RNA polymerase (usually referred to simply as RNA polymerase).

RNA POLYMERASE

The enzyme that catalyses RNA synthesis during transcription in *E. coli* was first discovered in 1958 and has since been subjected to detailed study. Like many enzymes involved in molecular genetic processes, RNA polymerase has to perform several tasks, as will be seen when the events involved in transcription are described later in the chapter. It is not surprising, therefore, that RNA polymerase is a large protein made up of several different polypeptide subunits.

The RNA polymerase of E. coli comprises five subunits. Each *E. coli* cell contains about 7000 RNA polymerase molecules, between 2000 and 5000 of which may be actively involved in transcription at any one time. The structure of the enzyme is described as $\alpha_2\beta\beta^1\sigma$, meaning that each molecule is made up of two α polypeptides plus one each of β, the related β^1, and σ (Fig. 12.4). This version of the enzyme is called the holoenzyme (molecular mass 4,50,000 daltons) and is distinct from a second form, the core enzyme (molecular mass 3,80,000 daltons), which lacks the α subunit and is just $\alpha_2\beta\beta^1$. The two versions of the enzyme have different roles during transcription, as will be described later.

RNA POLYMERASE SYNTHESISES RNA IN THE 5′ TO 3′ DIRECTION

An important aspect of the chemical reaction catalysed by RNA polymerase is that the RNA molecule is synthesised in an ordered fashion. Rather than polymerising a large number of ribonucleotides at once, RNA polymerase starts by joining two ribonucleotides and then progresses in one direction, sequentially adding new ribonucleotides onto the free 3′ end of the existing polymer (Fig. 12.5). The RNA molecule is, therefore, synthesised in the 5′ to 3′ direction. Remember that in order to base pair, complementary polynucleotides must be antiparallel; this means that the template strand of the gene must be read in the 3′ to 5′ direction.

EUKARYOTES POSSESS MORE COMPLEX RNA POLYMERASES

All *E. coli* genes are transcribed by the same type of RNA polymerase enzyme. In contrast, eukaryotes possess three different RNA polymerases (called RNA polymerase I, II and III), each of which transcribes a different set of genes. The eukaryotic enzymes are each larger than the *E. coli* version with molecular masses in excess of 5,00,000 daltons and have more complex structures. Each is comprised of two large subunits and up to ten smaller ones. In addition, a group of polypeptides called transcription factors, which are not integral components of the RNA polymerase enzymes, play ancillary roles in transcription in eukaryotes. Large and complex enzymes are difficult to purify in active form and it is not surprising that study of the eukaryotic RNA polymerases has lagged behind comparable work with *E. coli*.

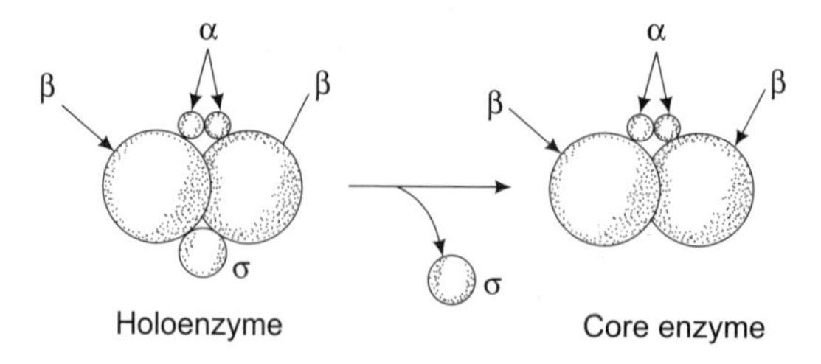

Fig. 12.4 The relationship between the holoenzyme and core enzyme versions of *E. coli*. RNA polymerase.

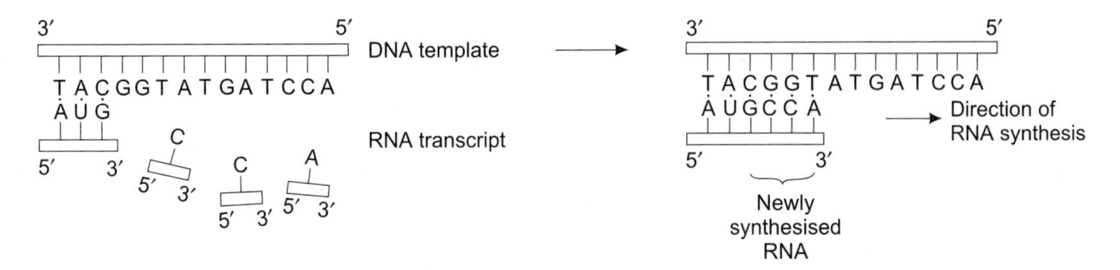

Fig. 12.5 RNA polymerase Synthesises RNA in the 5′ to 3′ direction.

EVENTS IN TRANSCRIPTION

The actual process by which a gene is transcribed is conveniently divided into three phases, viz. *initiation*, *elongation* and *termination*.

INITIATION

The key point about the initiation of transcription is that an RNA polymerase enzyme must transcribe genes rather than random pieces of DNA. This means that the initial binding of an RNA polymerase enzyme to a DNA molecule must occur at a specific position, just in front ('*upstream*') of the gene to be transcribed. Specific attachment points are essential because in all organisms, especially eukaryotes, a large proportion (possibly over 99%) of the total DNA is non-genic and efficient transcription of genes will not occur purely by chance.

THE TRANSCRIPTION INITIATION SITE IS SIGNALLED BY THE PROMOTER

These attachment points are called promoters. A promoter is a short nucleotide sequence that is recognised by an RNA polymerase enzyme as a point at which to bind to DNA in order to begin transcription. Promoters occur just upstream of genes and nowhere else.

PRIBNOW BOX

Clearly, all promoters must have the same or very similar sequences if each is to be recognised by the same enzyme. In *E. coli*, the promoter sequence has been dissected and found to be made up of two distinct components, the – 35 box and the – 1 0 box (Fig. 12.6). The latter is also called the *Pribnow box* (Fig. 12.7) after the scientist who first characterised it. The names refer to the location of the boxes on the DNA molecule relative to the position at which transcription starts. The sequences of the boxes are:

-35 box 5′-TTGACA-3′

-10 box 5′-TATAAT-3′

(Remember that these are the sequences found on the non-coding polynucleotide).

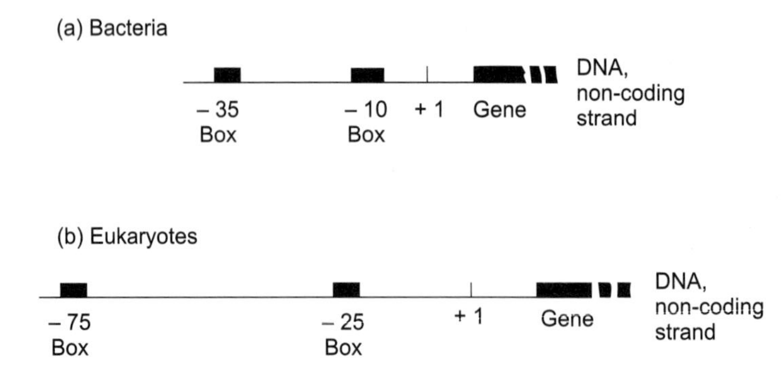

Fig. 12.6 The position of promoter sequences in (a) Bacteria and (b) Eukaryotes. Transcription begins at the point marked as '+ 1'. Note that this point is a short distance upstream of the start of the gene itself.

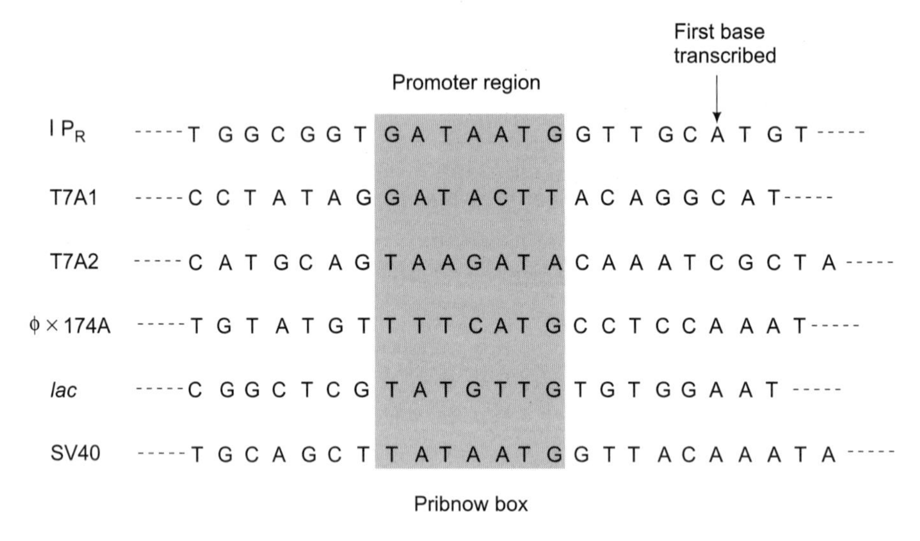

Fig. 12.7 Nucleotide sequences of the *Promoter* region and the first base transcribed on several different DNA sequences. Lambda (λ), T7, and ϕ × 174 are phages. Lac is an *E. coli* gene and SV40 is an animal virus.

In fact, the actual sequences of the components of the promoter vary from gene to gene, but all are related to and recognisable with the consensus sequences shown above. Whether there is any importance in the slightly different genes is not known. It has been suggested that the actual sequence may affect the efficiency with which RNA polymerase locates and binds to the promoter, thereby influencing the extent to which the gene is expressed. These proposals are complicated by the observation that certain minor alterations in the promoter sequence can completely block recognition by RNA polymerase.

EUKARYOTIC PROMOTERS ARE SLIGHTLY DIFFERENT

Eukaryotic genes are also preceded by promoters but of a slightly different construction (Fig. 12.6). For instance, genes transcribed by RNA polymerase II are often preceded by:

-75 box 5′-GGNNCAATCT -3′

-25 box (or Hogness box) 5′-TATAAAT-3′

Again these are consensus sequences, with N indicating a position that may be occupied by A, T, G or C.

The -25 box is found upstream of most vertebrate and insect genes transcribed by RNA polymerase II; the -75 box on the other hand is a less universal feature and is often absent.

THE SUBUNIT OF RNA POLYMERASE RECOGNISES THE PROMOTER

In *E. coli* the holoenzyme version of RNA polymerase ($\alpha^2\beta\beta^1\sigma$) is required for initiation; in fact, it is the subunit of the enzyme that is responsible for promoter site recognition. If the subunit is absent, the core enzyme will still be able to bind to DNA, but in a more random fashion at a variety of sites and not specifically to promoters.

The initial structure formed between RNA polymerase and DNA is called the closed *promoter complex.* In this complex, the enzyme covers or 'protects' about 60 bp of the double helix, from just upstream of the -35 box to just downstream of the -10 box. Experimental evidence suggests that the RNA polymerase holoenzyme specifically recognises the -35 box as the DNA binding site, although these conclusions are still controversial. However, it is clear that the -10 box is the region where breakage of base pairing (calling '*melting*') and unwinding of the DNA double helix first occurs, forming a second structure, the open promoter complex. Melting is, of course, an essential prerequisite for transcription because the bases of the coding strand must be exposed in order to act as the template for transcript synthesis. The fact that the -10 box is made up entirely of A–T base pairs, which comprise just two hydrogen bonds each, compared with three for a G-C pair, is presumed to make melting of the helix easier in this region.

When the open promoter complex has formed, the subunit dissociates and the holoenzyme is converted into the *core enzymes.* At the same time the first two ribonucleotides can be base paired to the coding polynucleotide at positions +1 and +2, and the first phosphodiester bond of the RNA molecule synthesised.

ELONGATION

The elongation stage of transcription is relatively straightforward. The RNA polymerase core enzyme migrates along the DNA molecule, melting and unwinding the double helix as it progresses, while sequentially attaching ribonucleotides to the 3′ end of the growing RNA molecule; base pairing to the code polynucleotide of the gene determines the identity of the ribonucleotide added at each position (Figs. 12.8 and 12.9). Just three points need to be noted.

1. The Transcript is Longer than the Gene: Position +1

Where synthesis of the RNA transcript actually begins is rarely the start of the gene itself. Almost invariably the RNA polymerase enzyme transcribes a leader segment before reaching the

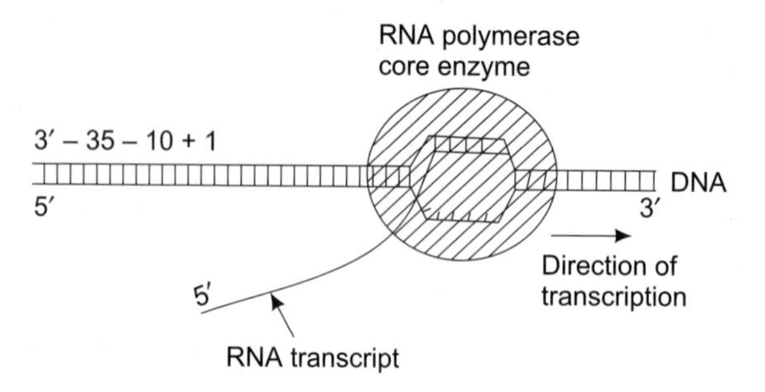

Fig. 12.8 Elongation step of transcription.

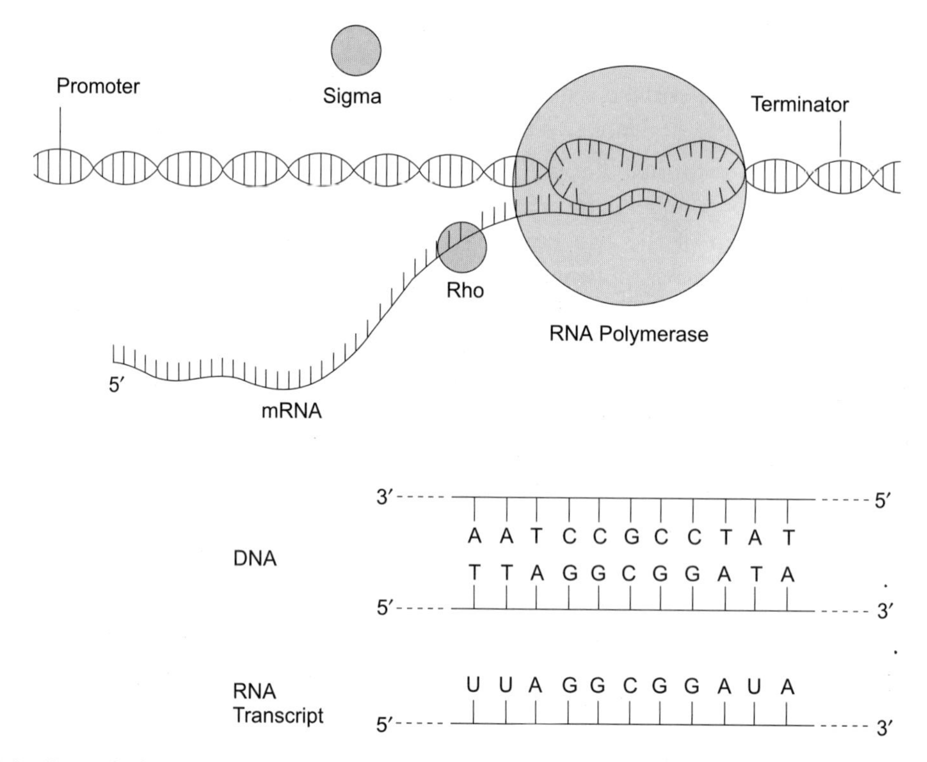

Fig. 12.9 Transcription and RNA polymerase molecules. RNA polymerase is transcribing a gene. The *sigma* factor is seen dissociated near the promoter. The *rho* factor is shown on the newly formed RNA.

gene (Fig. 12.10). The length of this leader varies from gene to gene in *E. coli* it may be as short as 20 nucleotides or longer than 600 nucleotides. Similarly, when the end of the gene has been reached the enzyme continues to transcribe a trailer segment before termination occurs.

2. Only a Small Region of the Double Helix is unwound at any One Time

Once elongation has begun, the polymerised portion of the RNA molecule can gradually dissociate from the coding strand, allowing the double helix to return to its original state. This

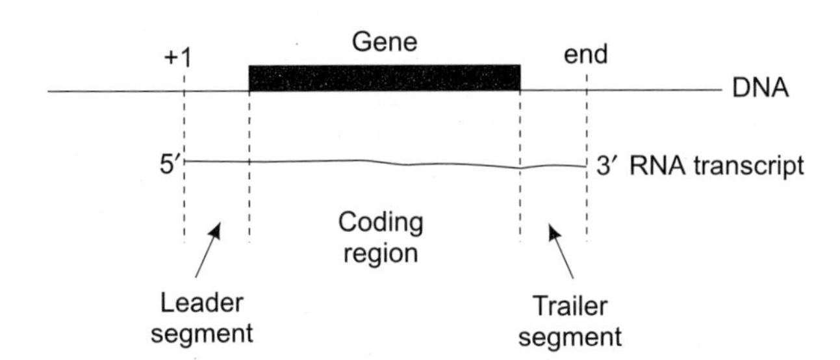

Fig. 12.10 Transcripts are usually longer than region genes as they include leader and trailer segments.

means that only a limited region of the DNA molecule is melted at any one time (see Fig. 112.8). The open region or transcription 'bubble' contains between 12 and 17 RNA-DNA base pairs and so is only a very short stretch of the gene as whole. This is important because unwinding a portion of the double helix necessitates overwinding the adjacent areas and there is only limited molecular freedom for this to occur.

3. The Rate of Elongation is not Constant

A feature of transcription that is attracting increasing interest from molecular biologists is the variation in transcription rate that can occur as RNA polymerase moves along a DNA molecule. Occasionally the enzyme will slow down and then reaccelerate, or even go so far so as to reverse over a short stretch, removing ribonucleotides from the end of the newly synthesised transcript before resuming its forward course.

TERMINATION

As with initiation, termination of transcription is not a random process but must occur only at suitable positions shortly after the ends of genes. Surprisingly, perhaps, termination is not mediated by a specific sequence analogous to the promoter, but by a more complex signal.

Terminators are Complementary Palindromes

Termination signals for *E. coli* genes vary immensely in their actual nucleotide sequences, but all possess a common feature; they are complementary palindromes. This means that base pairing can occur not only between the two strands of the double helix, but also within each strand and also within the RNA transcript (Fig. 12.11). The result of intrastrand base pairing is a cruciform structure in double-stranded DNA or a stem-loop structure in single-stranded RNA.

A cruciform structure is unlikely to form in double-stranded DNA because the interstrand base paired double helix is more energetically stable due to the much larger number of potential base pairs that can form. The complementary palindrome is more likely to exert its influence on transcription by enabling a stem-loop to form in the newly synthesised RNA molecule (Fig. 12.12). Unfortunately, exactly how this results in termination is not known.

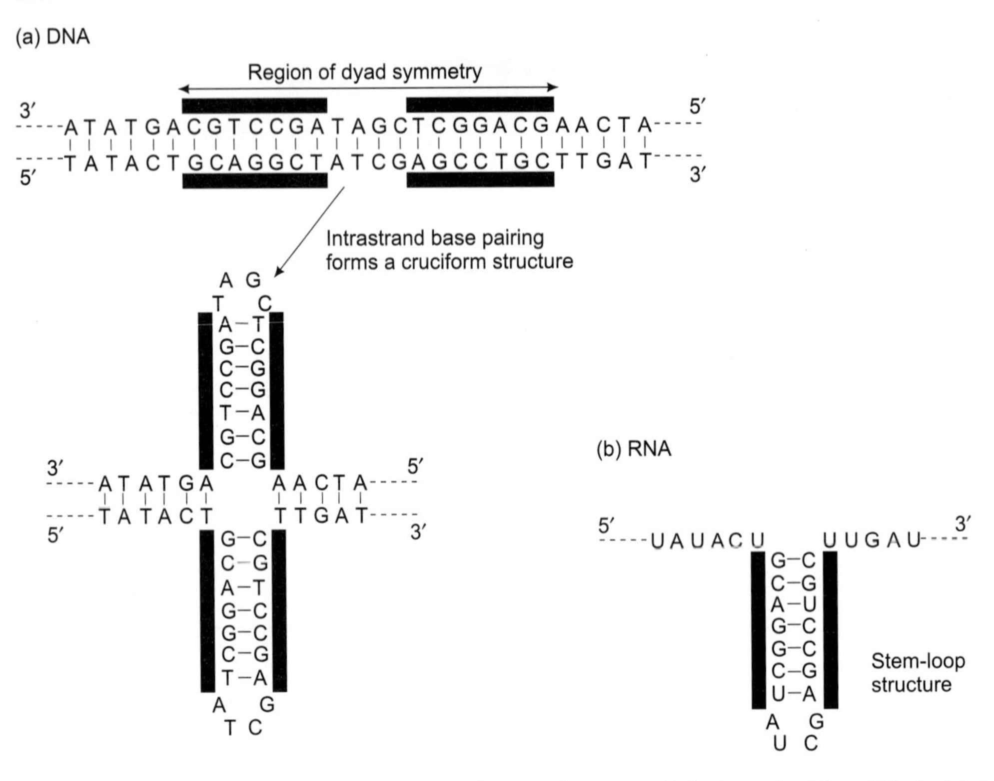

Fig. 12.11 Possible consequences of the presence of a complementary Palindrome in (a) a DNA double helix and (b) the RNA molecule transcribed from it. Base pairing between nucleotides in the same strand can result in a cruciform structure in a double-stranded DNA molecule, or a stem-loop structure in the single-stranded RNA.

Termination may require an Additional Polypeptide Component of RNA Polymerase

Some termination signals are recognised by the *E. coli* RNA polymerase core enzyme alone, while others require the presence of an additional polypeptide, called the ρ (rho) factor (molecular mass 55000 daltons) which forms a temporary association with the core enzyme. Little is known about the role of the ρ factor, although it is clear that the precise structures of ρ-dependent and ρ-independent terminators are different, since the latter usually have a series of A-T base pairs immediately following the complementary palindrome. These A-T base pairs give rise to a series of 5 to 10 Us at the 3' end of a transcript formed by ρ-independent termination.

COMPLETION OF TRANSCRIPTION

The final event in transcription is dissociation of the RNA polymerase enzyme from the DNA and release of the RNA transcript. The core enzyme is now able to reassociate with a sigma

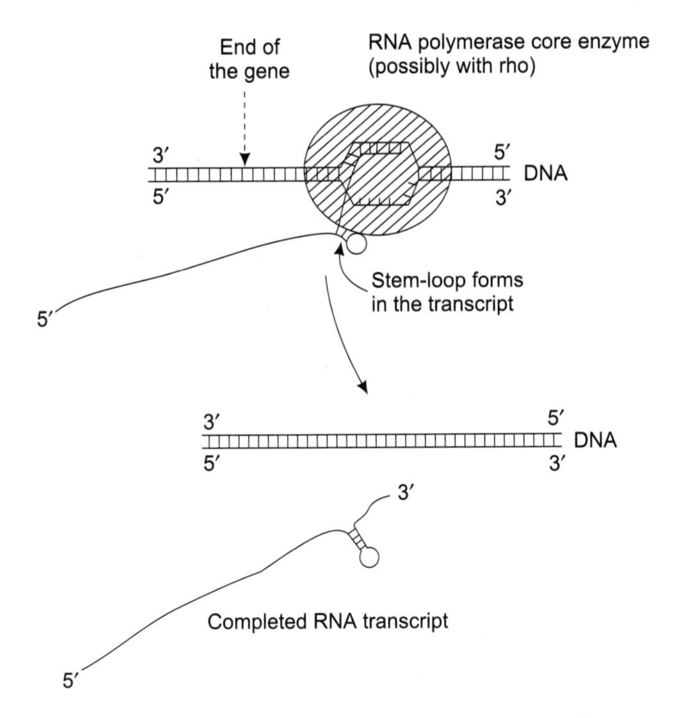

Fig. 12.12 Termination of transcription *in E. coli.*

subunit and begin a new round of transcription on the same or a different gene. The RNA transcript is ready to play its role either as the end product of gene expression or as the message for synthesis of polypeptide by translation.

TYPES OF RNA MOLECULE: rRNA AND tRNA

The RNA molecules produced by transcription can be grouped into three major classes according to function. These are *ribosomal* or rRNA, *transfer* or tRNA, and *messenger* or mRNA. Ribosomal and transfer RNAs are the end products of gene expression and perform their roles in the cell as RNA molecules. Messenger RNA, on the other hand, undergoes the second stage of gene expression, translation, and has no function beyond acting as the intermediate between a gene and its final expression product, a polypeptide.

Ribosomal and transfer RNAs are often referred to as stable RNA, indicating that these molecules are long-lived in the cell, in contrast to mRNA which has a relatively rapid turnover rate, possibly as short as a minute for some bacterial mRNA. Both synthesise even though they are not themselves translated.

RIBOSOMAL RNA (rRNA)

Ribosomal RNA molecules are components of ribosomes, the large multimolecular structures that act as factories for protein synthesis. During translation, ribosomes attach to mRNA molecules and migrate along them, synthesising polypeptides as they go, analogous in a way to the role of RNA polymerase in transcription.

Ribosomes are made up of rRNA molecules and proteins, and are extremely numerous in most cells. An actively growing bacterium may contain over 20,000 ribosomes, comprising about 80% of the total cell RNA and 10% of the total protein.

THE STRUCTURE OF RIBOSOMES

The prokaryotic ribosome has a total molecular mass of 2,520,000 daltons and is very roughly ovoid with approximate dimensions of 29 nm × 21 nm (1 nm = 10 A). Eukaryotic ribosomes are larger: 4,220,000 daltons and 32 nm × 22 nm.

Ribosome Sizes are determined by Sedimentation Analysis

Ribosomes, like many macromolecules and multimolecular assemblies, are so large that estimates of their molecular weights are very difficult to obtain. Instead the sizes of these structures are determined by velocity sedimentation analysis, a procedure that measures the rate at which a molecule or particle sediments through a dense solution often of sucrose while subjected to a very high centrifugal force (700,000g or more). The sedimentation coefficient is expressed as an S value (S = Svedberg unit), after the Swede. The Svedberg, who built the first ultracentrifuge in the early 1920s and is dependent on several factors, notably molecular mass and the shape of the macromolecule or macromolecular assembly. Prokaryotic ribosomes have a sedimentation coefficient of 70S, somewhat smaller than eukaryotic ribosomes, which are variable but average about 80S.

Ribosomes are made of Two Subunits

In *E.coli* these subunits have sedimentation coefficients of 50S and 30S. The large subunit contains two rRNA molecules, of 23S and 5S, together with 31 different polypeptides. The smaller subunit has just a single 16S rRNA plus 21 polypeptides.

Eukaryotic ribosomes are also made of two subunits but in this case the sizes are 60S and 40S. The large subunit has three rRNAs (28S, 5.8S and 5S) and 49 polypeptides; the small subunit has a single 18S rRNA and 33 polypeptides. The additional rRNA of the eukaryotic large subunit is the 5.8S molecule which in *E. coli* is present as an integral part of the 23S rRNA.

Molecular Structure of the Ribosome

The traditional view of ribosomal structure is that rRNA molecules act as scaffolding onto which proteins, which provide the functional activity of the ribosome, are attached. To fulfil this role the rRNA molecules must be able to take up a stable three-dimensional structure. This is achieved inter-and intra-molecular base pairs, with different rRNAs of a subunit base pairing in an ordered fashion with each other, and also more importantly with different parts of themselves.

The Three-Dimensional Structure

The base paired structures are two-dimensional representations and tell us little about the real three-demensional structure of the ribosome. In fact, this is the problem that is being pursued most actively at present. Ribosomes can be examined with the electron microscope, albeit towards the limits of resolution, and approximate three-dimensional reconstructions have been built up by analysing the images obtained (Fig. 12.13). These reconstruction have been refined by a variety of techniques that have, located the positions of individual ribosomal proteins on the three-dimensional image. But, the most significant advance has been the application of X-ray diffraction analysis to the ribosome. Diffraction patterns have been obtained and approximate three-dimensional structures were derived from them. The complexity of the diffraction patterns makes precise interpretation difficult, but rapid advances are being made in this area, particularly with the help of computer-aided analysis.

rRNA MOLECULES MAY HAVE ENZYMATIC ROLES DURING PROTEIN SYNTHESIS

The traditional view of the rRNA as the scaffolding and the proteins as attachments that provide the real biological activity of the ribosome is now being challenged by the latest exciting ideas about the potential functions of RNA. For many years it has been believed that enzymatic catalysis is uniquely a feature of proteins and that RNA molecules cannot act as enzymes. Accordingly, any catalytic activity that the ribosomes display during protein synthesis must be a property of the ribosomal proteins. Recently, however, it has been discovered that certain RNA molecules have enzymatic activity. This was first demonstrated in 1982 with the self-splicing intron of *Tetrahymena* and has subsequently been proven or implied for other RNA

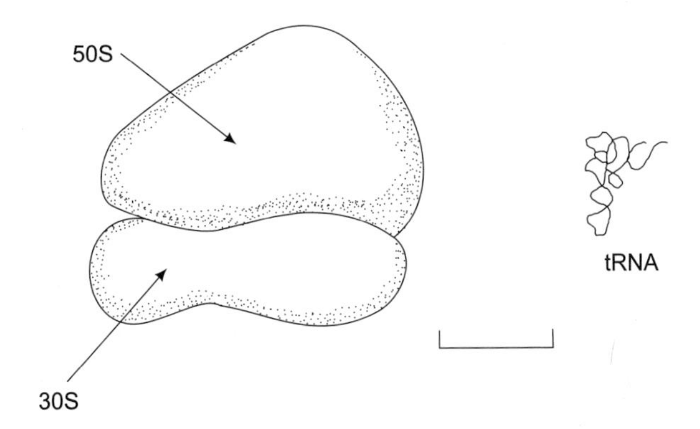

Fig. 12.13 The three-dimensional structure of *E. coli* ribosome.

molecules also. As yet there is no proof that rRNA has an enzymatic function in ribosomes, but a discovery along these lines could well be just around the corner.

SYNTHESIS OF rRNAs

Each ribosome contains one copy of each of the different rRNA molecules, three rRNAs for the prokaryotic ribosome or four for the eukaryotic version. The most efficient system would be for the cell to synthesise equal numbers of each of these molecules. Of course, the cell could make different amounts of each one but this would be wasteful because some copies would be left over when the least abundant rRNA was all used up.

THE rRNA TRANSCRIPTION UNIT

Synthesis of equal numbers of each rRNA molecule is assured by having an entire complement of rRNA molecules transcribed together as a single unit. The product of transcription, the primary transcript, is therefore a long RNA precursor, the pre-rRNA, containing each rRNA separated by short spacers. The spacers are removed by processing events that release the mature rRNAs.

A similar series of events brings about the synthesis of eukaryotic rRNAs, with the exception that only the 28S, 18S and 5.3S genes are transcribed together. The 5S genes occur elsewhere on the eukaryotic chromosomes and are transcribed independently of the main unit.

THERE ARE MULTIPLE COPIES OF THE rRNA TRANSCRIPTION UNIT

The actively growing *E. coli* cell that contains 20,000 ribosomes will divide once every twenty minutes or so. Therefore, every twenty minutes it needs to synthesise 20,000 new ribosomes, an entire complement for one of the two daughter cells. This necessitates a considerable amount of rRNA transcription, to such an extent that a single transcription unit would not be able to cope with the demand. In fact the *E. coli* chromosome contains seven copies of the rRNA transcription unit. In eukaryotes, there can be an even greater demand for rRNA synthesis and 50 to 5000 identical copies of the rRNA transcription unit may be present depending on species. In eukaryotes these units are usually arranged into *multigene families*, with large numbers of copies following one after the other, separated by non-transcribed spacers (Fig. 12.14). Even this may not satisfy the demand for rRNA synthesis under certain circumstances, and in some eukaryotic cells (e.g., amphibian oocytes) an additional strategy known as *gene amplification* may be called upon. This involves replication of rRNA genes into multiple DNA copies which subsequently exist as independent molecules not attached to the chromosomes (Fig. 12.14); transcription of the amplified copies produces additional rRNA molecules. Gene amplification is not restricted to rRNA genes and occurs with a few other genes whose transcription is required at a greatly enhanced rate in certain situations.

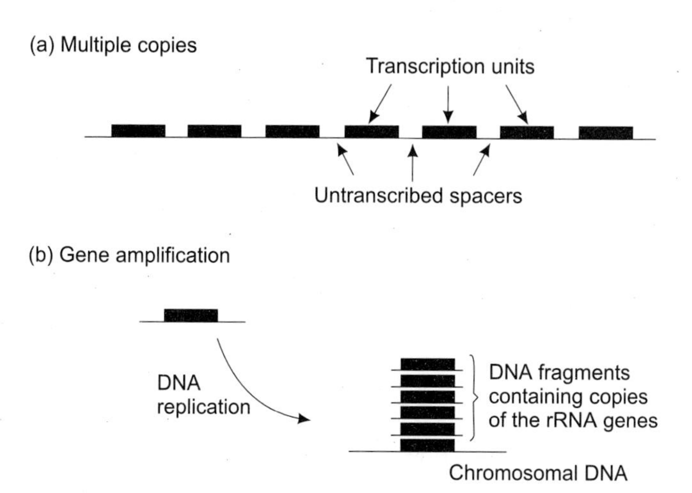

Fig. 12.14 Strategies for increasing a cell's capacity for rRNA synthesis. (a) Multiple copies of the rRNA transcription unit, as seen in most eukaryotes. (b) Amplified copies of the transcription unit, a feature of oocytes in amphibians.

TRANSFER RNA

Transfer RNA molecules are also involved in protein synthesis but the part they play is completely different from that of rRNA. Transfer RNAs are in fact the adaptor molecules that read the nucleotide sequence of the mRNA transcript and convert it into a sequence of amino acids. The existence of tRNA was postulated by Crick in the 1950s and individual tRNA molecules were first isolated in 1959 by Robert Holley, who six years later presented the complete nucleotide sequence of one of these molecules.

The exact function of tRNA in protein synthesis will be covered in Chapters on Genetic code and Protein synthesis. Here we will look at the structure of tRNA molecules, the manner in which they are transcribed, and the processing and modification events that they undergo after transcription.

STRUCTURE OF tRNA

Transfer RNA molecules are relatively small, mostly between 74 and 95 nucleotides for different molecules in different species. Each organism synthesises a number of different tRNAs, each in multiple copies. However, virtually all tRNAs take up the same structure after synthesis.

THE tRNA CLOVERLEAF MODEL (Figs. 12.15, 12.16 and 12.17)

Almost every tRNA molecule in every organism can be folded into a base paired structure referred to as the cloverleaf (Figs. 12.15 and 12.16). This structure is made up of the following components:

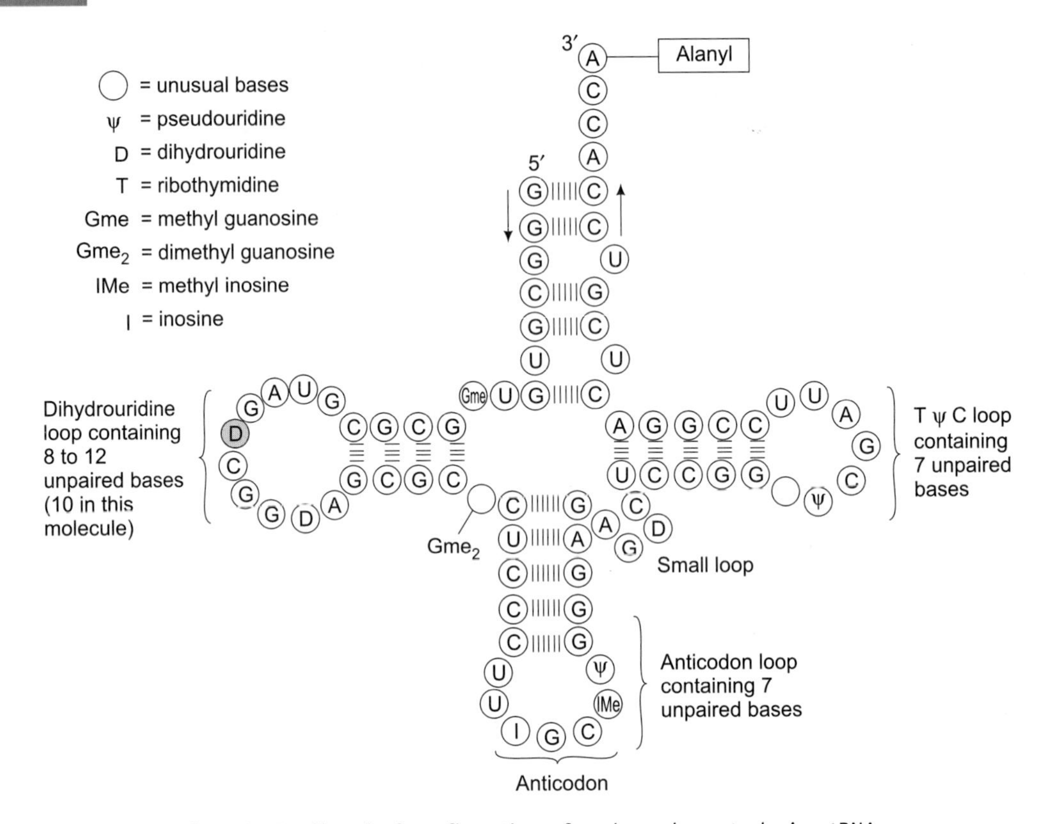

Fig. 12.15 Cloverleaf configuration of a charged yeast alanine tRNA.

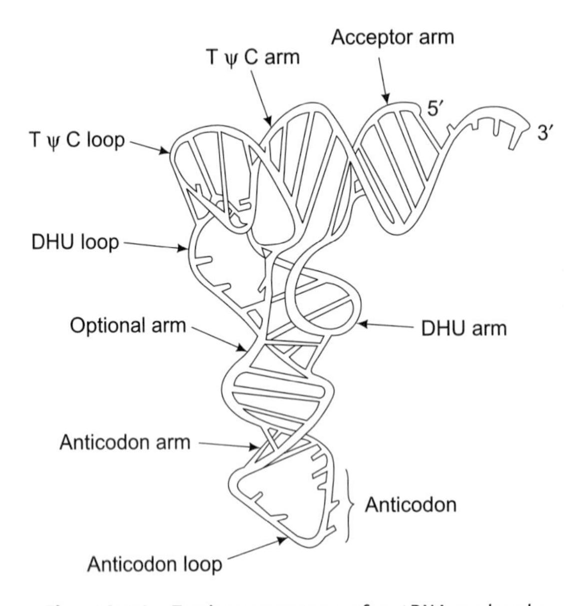

Fig. 12.16 Tertiary structure of a tRNA molecule.

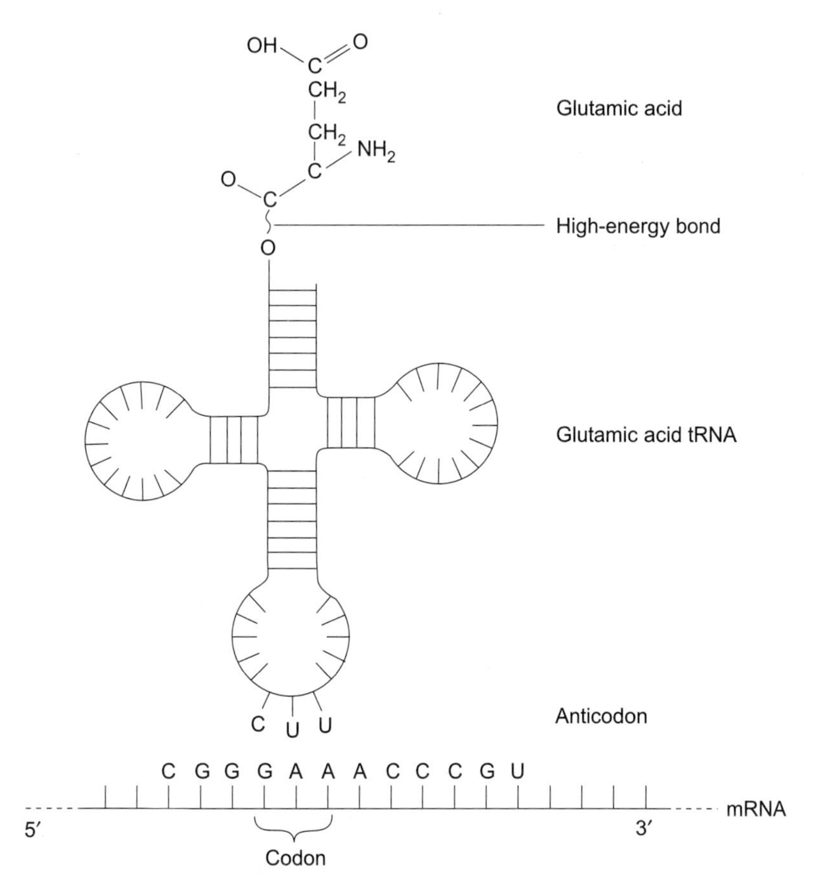

Fig. 12.17 Specificity of the genetic code resides in the tRNA where a particular anticodon is associated with a particular amino acid.

1. The acceptor arm, formed by a series of usually seven base pairs between nucleotides at the 5′ and 3′ ends of the molecule. During protein synthesis an amino acid is attached to the acceptor arm of the tRNA.
2. The D or DHv arm, so named because the loop at its end almost invariably contains the unusual pyrimidine dihydrouracil.
3. The anticodon arm which plays the central role in decoding the biological information carried by the mRNA.
4. The extra, optional or variable arm, which may be a loop of just 3 to 5 nucleotides (Class I tRNAs, about 75% of all tRNAs) or a much larger stem-loop of 13 to 21 nucleotides with up to 5 base pairs in the stem (Class II tRNAs).
5. The TC arm, which named after the nucleotide sequence, C (= a nucleotide containing pseudouracil, another unusual pyrimidine base) which its loop virtually always contains.

In addition to having a common secondary structure, different tRNAs also display a certain amount of nucleotide sequence conservation. Some positions are invariant: in all tRNAs such a position will be occupied by the same nucleotide. Others are semi-invariant and will always contain the same type of nucleotide, either one of the pyrimidine nucleotides or one of the purine nucleotides.

THE TERTIARY STRUCTURE (Fig. 12.16)

It should be appreciated that although the cloverleaf is a convenient way to represent the structure of a tRNA, the molecule itself will have a much less recognisable three-dimensional tertiary structure in the cell. This tertiary structure has been determined by X-ray diffraction analysis and is shown in Fig. 12.16. The base pairings in the stems of the cloverleaf are maintained in the tertiary structure, but several additional base pairings form between nucleotides that appear widely separated in the cloverleaf. This folds the molecule into a compact conformation. Many of the nucleotides involved in the tertiary base pairing are the invariant or semi-invariant ones that are conserved in different tRNAs. The tertiary conformation places the acceptor arm and anticodon loop at opposite ends of the molecule.

PROCESSING AND MODIFICATION OF tRNA TRANSCRIPTS

In both prokaryotes and eukaryotes, tRNAs are transcribed initially as precursor-tRNA, which is subsequently processed to release the mature molecules. In *E. coli* there are several separate tRNA transcription units, some containing just one tRNA gene and some with as many as seven different tRNA genes in a cluster. A pre-tRNA molecule is processed by a combination of different ribonucleases that make specific cleavages at the 5′ (RNase P) and 3′ (RNase D) ends of the mature tRNA sequence. RNase P is particularly interesting because the enzyme is a complex between a small protein (molecular mass 20,000 daltons) and a 377 nucleotide RNA molecule. Recently it has been shown that the RNA molecule in RNase P possesses at least part of the enzymatic activity of the complex and is hence an example of an RNA enzyme or 'ribozyme'.

Eukaryotic tRNA genes are also clustered and furthermore occur in multiple copies, a reflection of the huge demand for tRNA synthesis in the eukaryotic cell. The events involved in processing of eukaryotic pre-tRNA have not been characterised to any great extent as yet but enzymes similar in activity to RNase P and RNase D are implicated.

THREE NUCLEOTIDES AT THE 3′ END MAY BE ADDED AFTER TRANSCRIPTION

All tRNAs have at their 3′ end the trinucleotide sequence 5′-CCA-3′. When the genes coding for tRNAs in eukaryotes are examined, it is found that in most cases the CCA sequence does not occur at the expected position in the DNA molecule. Instead, the 3′-terminal CCA is added after transcription by another processing enzyme, this one called tRNA nucleotidyl transferase.

In prokaryotes the final CCA is more frequently coded by the tRNA gene and is, therefore, transcribed in the normal manner. Nevertheless, it appears that occasionally this sequence, or part of it, is removed by RNase D during processing of the pre-tRNA and has to be replaced by a prokaryotic nucleotidyl transferase enzyme. This CCA is crucial because it is the point at which the amino acid is attached to the tRNA during protein synthesis.

CERTAIN NUCLEOTIDES UNDERGO CHEMICAL MODIFICATION (Fig. 12.18)

We have already noted the existence of unusual nucleotides in certain stem-loops of the tRNA cloverleaf. In fact, a number of different nucleotides in a single tRNA will undergo chemical modification after transcription; each will always be modified into the same unusual nucleotide, though a number of different modifications are known. The most common types are:

1. Methylation–addition of one or more methyl (CH_3) groups to the base or sugar component of the nucleotide. Examples: guanosine to 1-methylguanosine (Fig. 12.18a) and uridine to ribothymidine (Fig. 12.18b).
2. Base rearrangements–interchanging the positions of atoms in the purine or pyrimidine ring. Example: uridine to pseudouridine (Fig. 12.18c).
3. Double-bond saturation–converting a double bond in the base to a single bond. Example: uridine to dihydrouridine (Fig. 12.18d).
4. Deamination–removal of an amino group (-NH_2). Example: guanosine to inosine (Fig. 4.18e).
5. Sulphur substitution–for the oxygen atom of guanosine or uridine. Example: uridine to 4-thiouridine (Fig. 12.18f).
6. Addition of more complex groups. Example: guanosine to queosine (Fig. 12.18g).

Over 50 types of chemical modification have been discovered so far with tRNA nucleotides, each catalysed by a different tRNA modifying enzyme. The reasons for most of these modifications are unknown, although roles have been assigned to some specific cases, in particular those involving a nucleotide within the anticodon.

TYPES OF RNA MOLECULES: mRNA

Messenger RNA acts as the intermediate between the gene and the polypeptide translation product. Its existence was, like many of the major features of gene expression, postulated by Crick and the associates during the 1950s. Among the circumstantial evidence for mRNA at that time was the knowledge that in eukaryotic cells, the genes reside on the chromosomes in the nucleus, whereas protein synthesis occurs in ribosomes to the cytoplasm. The physical separation of genes and ribosomes means that some sort of messenger molecule must carry the biological information from nucleus to cytoplasm. In bacteria the physical separation of DNA and ribosomes is less distinct but nevertheless a messenger molecule is still necessary.

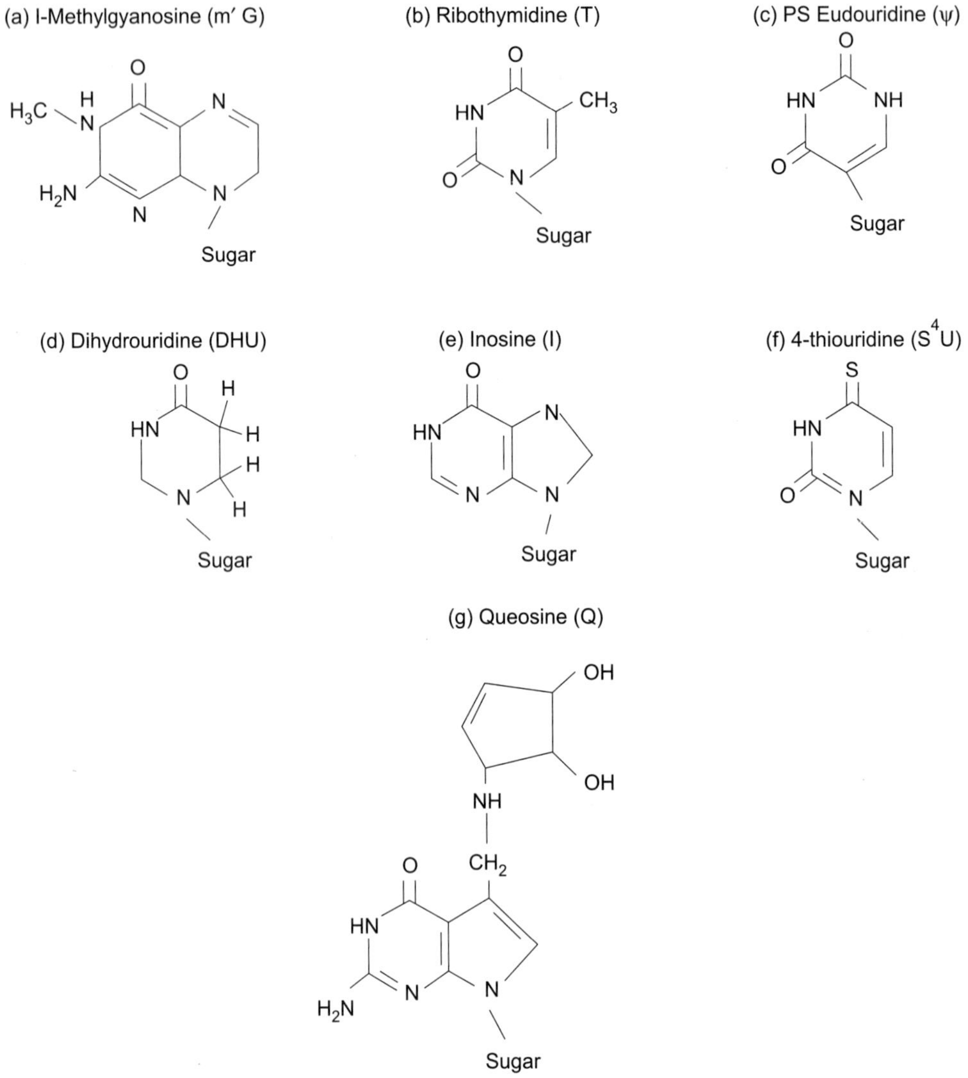

(a) I-Methylgyanosine (m′ G)

(b) Ribothymidine (T)

(c) PS Eudouridine (ψ)

(d) Dihydrouridine (DHU)

(e) Inosine (I)

(f) 4-thiouridine (S^4U)

(g) Queosine (Q)

Fig. 12.18 Some of the modified nucleotides (unusual) found in tRNA.

DISCOVERY OF mRNA

Evidence that this messenger is in fact RNA came first from Elliot Volkin and Lazarus Astrachan of the Oak Ridge National Laboratory in 1956, but more convincingly from Sol Spiegelman and Benjamin D. Hall of Illinois University five years later. Both groups demonstrated that after infection with a bacteriophage, the new RNA that is synthesised is related in sequence to the phage DNA, suggesting that the phage genes are copied into RNA before synthesis of the phage proteins occurs. Shortly afterwards two independent groups–

Sydney Brenner, Francois Jacob and Matthew Meselson, who worked at different universities but did the crucial experiment at the California Institute of Technology, and James Watson's group at Harvard-directly identified mRNA molecules in *E. coli* cells.

MOST mRNA MOLECULES ARE UNSTABLE

Messenger RNA molecules are, in general, not long-lived in the cell. Most bacterial mRNA have a half-life of only a few minutes and in eukaryotic cells mRNA is turned over at a comparable rate. Rapid turnover of mRNA is important because it means that the absolute amount of a particular mRNA in the cell can be controlled by adjusting the rate at which the relevant gene is transcribed. If the transcription rate for the gene decreases, then the level of the mRNA in the cell will also decrease until a new steady state is reached. Regulation of gene expression primarily involves control over transcription, a system that would not work if mRNA molecules were long-lived.

There are a few exceptions and examples of long-lived mRNAs are known. For instance, globin mRNA in reticulocytes is almost fully stable. In these cells regulation of globin gene expression is not important because the maximum rate of globin synthesis is required virtually all the time. Other long-lived mRNAs are being discovered as different types of specialised cells are studied.

MODIFICATION AND PROCESSING OF mRNA

In bacteria, the mRNA molecules that are translated are direct copies of the genes. In eukaryotes, the situation is different and most mRNAs undergo a fairly complicated series of modification and processing events before translation occurs.

The difference between prokaryotic and eukaryotic mRNA possibly reflects the higher structural complexity of the eukaryotic cell. In prokaryotes there is only one cell compartment and transcription and translation can occur together; indeed, the two processes may be coupled so that translation of an mRNA starts before transcription of the molecule has been completed. This cannot happen in eukaryotes because the transcribed mRNA must be transported to the cytoplasm before translation can occur. Although the functions of the modification and processing events are not understood, they may aid transport of the mRNA molecule to the correct site in the cytoplasm.

ALL EUKARYOTIC mRNAS ARE CAPPED

An RNA molecule synthesised by transcription and subjected to no additional modification will have at its 5′ end the chemical structure pppNpN..., where *N* is the sugar-base component of the nucleotide and *p* represents a phosphate group. Note that the 5′ terminus will be a triphosphate. With mature eukaryotic mRNA this is not the case as the 5′ terminus has a more complex chemical structure described as 7mGpppNpN, where 7mG is the nucleotide carrying the modified base 7-methyl-guanine. The 7mG nucleotide is added to the mRNA molecule after transcription by a two-step process, with methylation occuring only after a standard G has been added. As can be seen from the phosphate linkage between the 7mG and the first nucleotide of the transcripts, it's an unusual one rather than the normal 5′-3′ bond found in a polynucleotide, the bond in the cap structure is between the two 5′ carbons of the adjacent nucleotides.

As well as this basic cap structure, some eukaryotic mRNAs may undergo further modification of the 5′ end with the addition of further methyl groups to one or both of the next two nucleotides of the molecule. No role has yet been assigned to capping but it appears to be a prerequisite for translation of the mRNA.

MOST EUKARYOTIC mRNAS ARE POLYADENYLATED

A second modification of most eukaryotic RNAs is the addition to the end of the molecule of a long stretch of up to 200 A residues, producing a poly(A) tail:

$$...pNpNpA\ (pA)_n pA$$

Polyadenylation does not occur at the 3′ end of the primary transcript. Instead, the final few nucleotides are removed by a cleavage event that occurs between 10 and 30 nucleotides downstream of a specific polyadenylation signal (consensus sequence 5′-AAUAAA-3′; to produce on intermediate 3′ end to which the poly(A) tail is subsequently added by the enzyme poly(A) polymerase.

The reason for polyadenylation is not known although several hypotheses have been built round the idea that the length of the poly(A) tail determines the time the mRNA survives in the cytoplasm before being degraded. All theories must take into account the fact that certain eukaryotic mRNAs, notably those coding for histones, are not polyadenylated.

INTRONS MUST BE REMOVED FROM THE PRIMARY TRANSCRIPT

Transcription produces a faithful copy of the coding strand of the gene. Therefore, if the gene contains introns then the primary transcript will include copies of these. However, the introns must be removed and the exon regions of the transcript attached to one another, before translation can occur. This process is called *splicing* and it is very complex.

SPLICING OCCURS IN THE NUCLEUS

Primary transcripts containing introns are much longer than the mature mRNA molecules. These primary transcripts remain in the nucleus and form the RNA fraction that was once called heterogeneous nuclear RNA (hnRNA) but is now more usually referred to as pre-mRNA. Splicing takes place in the nucleus and the spliced, intronless mRNA is then transported to the cytoplasm.

For some time only two clear facts were known about splicing of nuclear pre-mRNA. The first was that all the introns in nuclear pre-mRNA genes possess the same dinucleotide sequences at the 5′ and 3′ ends. These sequences are GT for the first two nucleotides of the intron (GT is of course the sequence in the DNA, it is GU in the pre-RNA) and AG for the last two nucleotides. In fact the GT-AG rule expresses only the central positions of what are actually slightly longer consensus sequences found at the 5′ and 3′ splice junctions of all pre-mRNA introns.

The second established fact about pre-mRNA splicing was that members of a distinct class of RNA molecules called small nuclear RNAs (snRNAs) are involved in splicing. Six different snRNAs are known in vertebrates; these are called U1 to U6 and vary in size from 100 nucleotides for U6 to 215 nucleotides for U3.

There is also an additional snRNA called U7, but this one is probably involved in polyadenylation of mRNA. The sequences of each analogous molecules are also present in lower eukaryotes such as yeast. Small nuclear RNAs are in fact another class of stable RNA and are just as important in eukaryotes as tRNA and rRNA. In the cell snRNAs exist within particles called small nuclear ribonucleoproteins (snRNPs), each of which contains one snRNA (except for U4 and U6 which are present in the same snRNP) and several proteins. snRNPs can be purified from nuclear extracts as particles with sedimentation coefficients of 10S to 12S.

SPLICING PATHWAY FOR NUCLEAR PRE-mRNA (Fig. 12.19)

Our understanding of the splicing process itself has advanced immensely during recent years, mainly thanks to the development of experimental procedures for purifying pre-mRNA molecules and subjecting them to splicing under controlled conditions in the test tube. The splicing reaction appears to involve the following three steps (Fig. 12.19):

1. Cleavage occurs at the 5′ splice site.
2. The resulting free 5′ end is attached to an internal site within the intron to form a lariat structure. This requires an unusual 5′-2′ phosphodiester bond to be formed.
3. The 3′ splice site is then cleaved and the two exons joined together.

The main task at the moment in understanding pre-mRNA splicing is to assign roles to the snRNPs. The ones that have been clearly implicated in splicing (those containing U1, U2, U5 and U4/U6) work in conjunction with additional proteins within a complex RNA-protein structure called the *spliceosome.* However, this is probably not a stable structure like the ribosome because its constitution appears to change as the splicing reaction proceeds.

OTHER TYPES OF INTRONS

The scheme just described refers specifically to introns present in nuclear pre-mRNA genes, in other words those that obey the GT-AG rule. These are the most numerous introns in eukaryotic nuclear genes but they are not the only type. There are two other distinct classes which occur primarily in tRNA and rRNA genes:

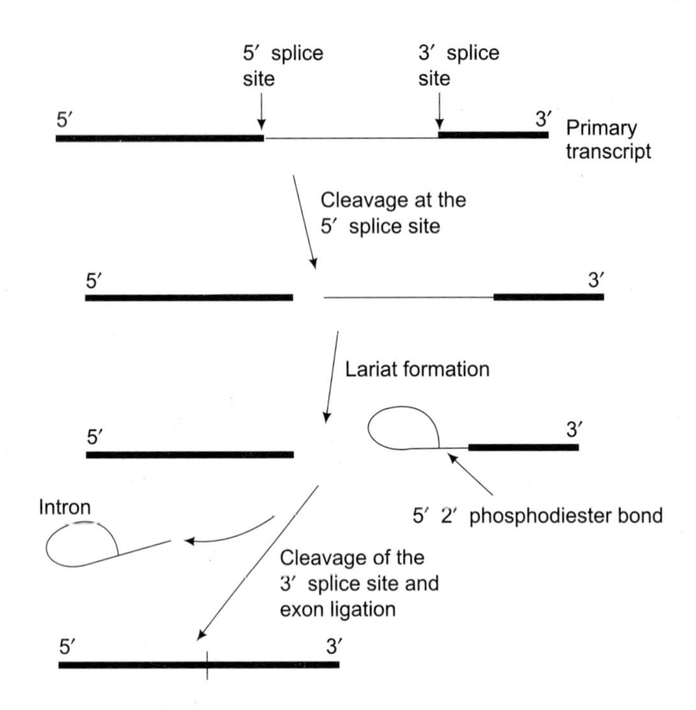

Fig. 12.19 The splicing pathway for introns in nuclear pre-mRNA.

1. *tRNA introns*: Many eukaryotic tRNA genes contain short introns in the anticodon loop. The splicing system for these introns seems to be much more straightforward than that for nuclear pre-mRNA, possibly because the cloverleaf structure of the tRNA forms in the unspliced precursor, bringing the splice junctions into close contact and aiding the overall process.
2. *The self-splicing intron*: The rRNA genes of certain protozoa, notably the ciliate *Tetrahymena*, contain introns that have a unique splicing mechanism. The intron folds up by intramolecular base pairing into a complex tertiary structure that possesses catalytic activity. This was the first RNA enzyme to be discovered and it is now clear that the ribozyme catalyses its own splicing process without the involvement of any protein molecules. This remarkable scheme may also hold for introns in some mitochondrial genes in yeast and filamentous fungi, as these introns also can take up the characteristic ribozyme base paired structure, and some are able to self-splice in the test tube. In the mitochondrion, however, splicing is probably aided by additional proteins.

CHAPTER 13

The Genetic Code and Protein Synthesis

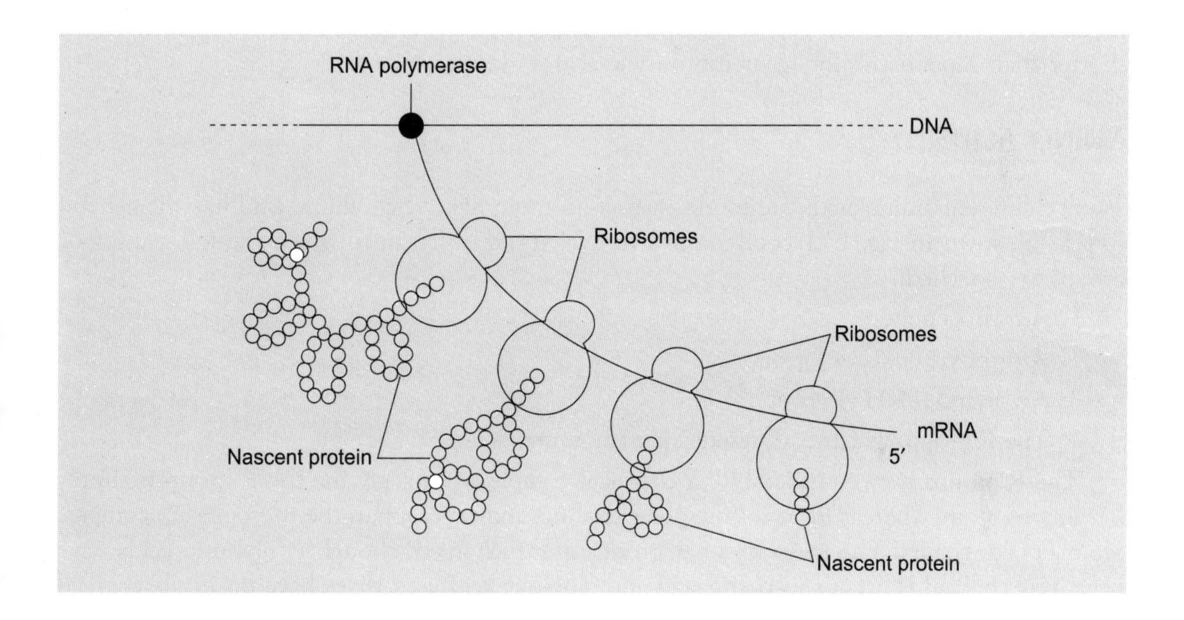

INTRODUCTION

The three major types of RNA molecules that are produced by transcription—messenger (mRNA), ribosomal (rRNA) and transfer RNA (tRNA)—work together to synthesise proteins by the process of *translation*. The central fact about translation is that the sequence of amino acids in the polypeptide being synthesised is specified by the sequence of nucleotides in the mRNA molecule being translated. The rules that determine which sequence of nucleotides specifies which sequence of amino acids are embodied in the *genetic code*.

The genetic code and protein synthesis are interlinked. This view goes back to the 1950s when the concept of the genetic code and the experiments designed to examine it were very much the province of molecular biologists, whereas unravelling the intricacies of how proteins are made was a problem for biochemists. But molecular biology and biochemistry are overlapping and are separable. First, genetic code will be dealt with and then the mechanism of protein synthesis. Before doing so one should know the basic facts concerning protein structure.

POLYPEPTIDES ARE POLYMERS

At the most fundamental level the structures of polypeptides and nucleic acids are the same: both are polymers made up of a series of linked monomers. In a polypeptide the monomers are called amino acids and the polymeric chain is generally less than 1000 units in length, much shorter than most naturally occurring nucleic acid molecules.

AMINO ACIDS

Twenty different amino acids are found in protein molecules. Each amino acid has the general structure shown in Fig. 13.1 comprising a central α-carbon atom to which the following four groups are attached:

1. A hydrogen atom.
2. A carboxyl ($-COO^-$) group.
3. An amino ($-NH_3^+$) group.
4. The R group, which is different for each amino acid.

The R groups vary considerably in chemical complexity: for glycine the R group is simply a hydrogen atom, whereas for tyrosine, phenylalanine and tryptophan the R groups are complex aromatic side-chains. The majority of R groups are uncharged, though two amino acids have negatively charged R groups (aspartic acid and glutamic acid) and three have positively charged R groups (lysine, arginine and histidine). Some R groups are polar (for example, glycine, serine and threonine), others are non-polar (e.g., alanine, valine and leucine). These differences mean that although all amino acids are closely related, each has its own specific chemical properties.

AMINO ACIDS ARE LINKED BY PEPTIDE BONDS

The polymeric structure of a polypeptide is built by linking a series of amino acids by peptide bonds, formed by condensation between the carboxyl group of one amino acid and the amino group of a second amino acid (Fig. 13.1). The structure of a polypeptide comprising eight amino acids, is shown in Fig. 13.2. Note that, as with polynucleotides, the two ends of a polypeptide are chemically distinct: one has a free amino group and is called amino–, NH_2–, or N-terminus, the other has a free carboxyl group and is called the carboxyl-, COOH-, or C-terminus.

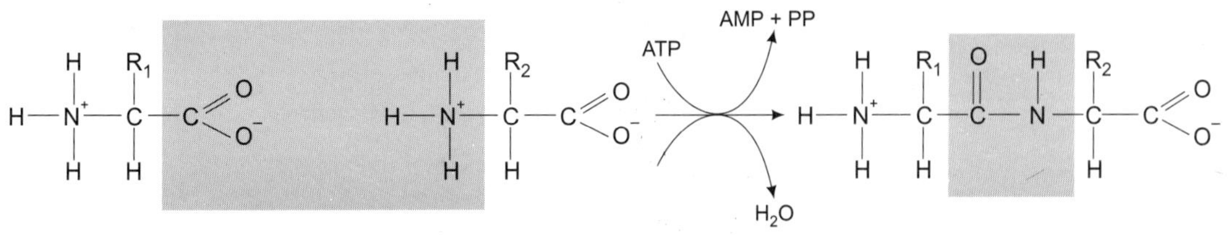

Fig. 13.1 Formation of a peptide bond between two amino acids.

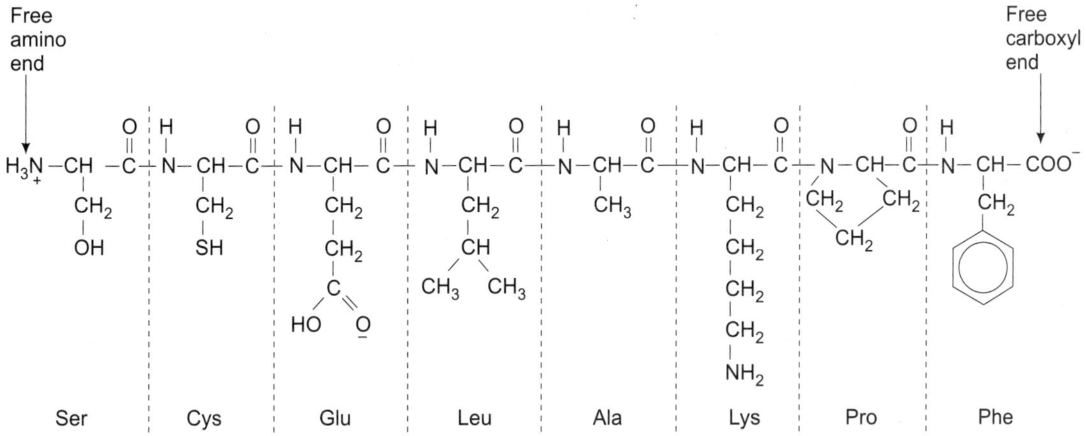

Fig. 13.2 A polypeptide has a free amino end and a free carboxyl end.

DIFFERENT LEVELS OF PROTEIN STRUCTURE (Fig. 13.3)

Four levels of structures are recognised in protein molecules:

1. The primary structure which is the amino acid sequence itself.
2. The secondary structure which is the regular or repeating configuration taken up by the amino acid chain. The two standard secondary structures are the α-helix and β-sheet. Both are stabilised primarily by hydrogen bonding between the carboxyl and amino groups of different amino acids. Usually different regions of a polypeptide will take up different secondary structures, so that the whole is made up of a number of α-helices and β-sheets, together with less-organised regions.
3. The tertiary structure, which is the three-dimensional conformation formed by folding together the secondary structural components of the polypeptide. The tertiary structure is held together by a variety of interactions between different amino acids, including hydrogen bonds, covalent linkages called disulphydryl bridges which can form between two cysteine residues and the natural tendency of the polypeptide chain to fold up so that non-polar R groups are shielded from water.
4. The quaternary structure, which refers to the way in which two or more polypeptides are oriented together to form a multisubunit structure. The quaternary structure may involve

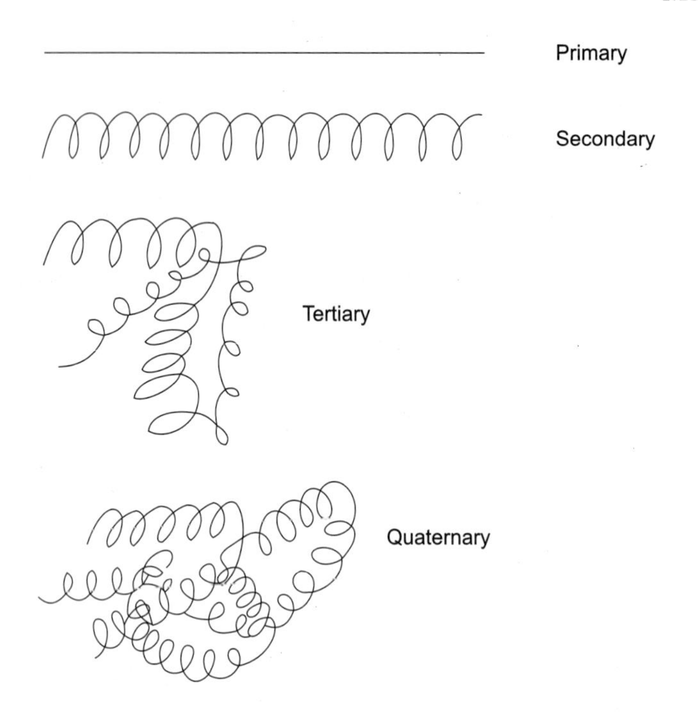

Fig. 13.3 Proteins have primary, secondary, ternary and quaternary structures.

several molecules of the same polypeptide or may comprise different polypeptides, as with RNA polymerase. In some cases, the quaternary structure is built up from a very large number of polypeptide subunits to give a complex array; the best examples are the protein coats of viruses such as that of tobacco mosaic virus, which is made up of 2150 identical protein subunits. The commonly found 20 amino acids are shown in Fig. 13.4.

AMINO ACID SEQUENCE IS THE KEY TO PROTEIN STRUCTURE AND FUNCTION

Each of the higher levels of protein structure–secondary, tertiary and quaternary–is specified by the primary structure, the amino acid sequence itself. This is most clearly understood at the secondary level, where it is recognised that certain amino acids, because of the chemical and physical properties of their R groups, stimulate the formation of an α-helix, whereas others promote formation of β-sheet. Conversely, certain amino acids more frequently occur outside regular structures and may act to determine the end-point of a helix or sheet. These factors are now so well understood that rules to predict the secondary structures taken up by amino acid sequences have been developed.

Although not very clear, the tertiary and quaternary structures of a protein depend on the amino acid sequence. The interactions between individual amino acids at these levels are so complex that predictive rules, though attempted, are still unreliable. However, it has been established for some years that if a protein is denatured, for instance by heating, so that it loses

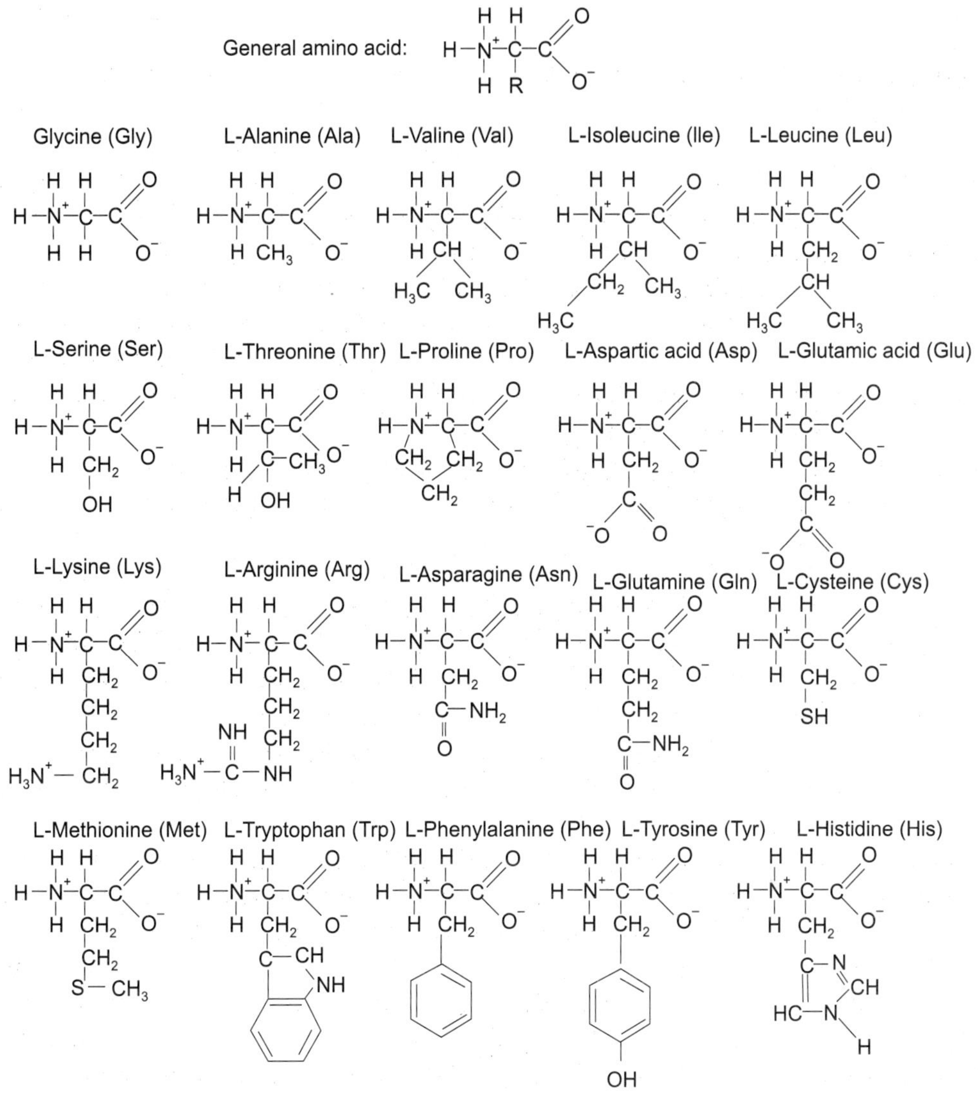

General amino acid:

Glycine (Gly) L-Alanine (Ala) L-Valine (Val) L-Isoleucine (Ile) L-Leucine (Leu)

L-Serine (Ser) L-Threonine (Thr) L-Proline (Pro) L-Aspartic acid (Asp) L-Glutamic acid (Glu)

L-Lysine (Lys) L-Arginine (Arg) L-Asparagine (Asn) L-Glutamine (Gln) L-Cysteine (Cys)

L-Methionine (Met) L-Tryptophan (Trp) L-Phenylalanine (Phe) L-Tyrosine (Tyr) L-Histidine (His)

Fig. 13.4 Structural formulas of 20 common amino acids. Each is an L-α-amino acid. The structures differ in other constituents.

its higher levels of structure and takes up a non-organised conformation. It still retains the innate ability upon renaturation (by cooling down again, for example) to reform spontaneously the correct tertiary structure. Once the tertiary structure has formed, subunit assembly into a multimeric protein again occurs spontaneously. This shows that the instructions for the tertiary and quaternary structures must reside in the amino acid sequence.

Function Depends on Amino Acid Sequence

From these considerations it is only a small step to appreciate that the function of a protein is also determined by its amino acid sequence. As an illustration we will consider those proteins that must attach themselves to a DNA molecule in order to perform their function in the cell. These DNA binding proteins form a large and diverse group that includes, for instance, RNA polymerase and a number of important regulatory proteins which modulate, and at times block, transcription of individual genes. Although their overall functions are different, several of these DNA-binding proteins possess common features. They are usually dimers made of two identical polypeptides, each of which has an α-helical region comprised of about eight amino acids. In the active form of the protein, the two helices are exactly 34 Å apart and therefore fit into two adjacent sections of the major groove of a DNA molecule. If the helices are absent, or oriented incorrectly, the DNA-binding ability of the protein will be lost. DNA binding depends, therefore, on the protein assuming the correct secondary, tertiary and quaternary structures, which is, in turn, dependent on the amino acid sequence.

IMPORTANCE OF AMINO ACID SEQUENCE

This is a crucial fact in molecular genetics: the function of a protein depends on its amino acid sequence which, in turn, is specified by the sequence of nucleotides in the mRNA, which is itself a copy of the gene. The biological information carried by the gene, codes in essence, for a protein function.

THE GENETIC CODE

The problem of exactly how the genetic code works was the major intellectual preoccupation of molecular biologists from 1953, when the double helix was postulated, to 1966 when the last remaining detail of the genetic code was finally worked out. Several of the more outlandish speculations made in the early days seem incredibly farfetched today, but it must be appreciated that until 1960 the available techniques made it very difficult to tackle directly the question of the code.

Code from First Principles

During the 1950s the most enlightened molecular biologists, centred as usual around Crick, strove to prevent the ideas about the code becoming too complex. Gradually the simplest system that was compatible with the established facts emerged as a working hypothesis. This system was based on two assumptions, both of which were subsequently confirmed by experimental analysis.

COLINEARITY BETWEEN GENE AND PROTEIN

It was more or less assumed right from 1953 that genes are colinear with the proteins they code for, meaning that the order of nucleotides in the gene correlates directly with the order of amino acids in the corresponding polypeptide. This is clearly the most straightforward way in which genes could code for proteins but was not in fact proven experimentally until 1964.

The type of experiment needed was fairly obvious: alter the nucleotide sequence of a gene at a specific point and determine whether the resulting change in the amino acid sequence of the corresponding polypeptide occurs at the same relative position or elsewhere. The problem was that although at that time alterations could be introduced into genes without too much trouble determining exactly where in the gene an alteration had occurred was much more difficult. Eventually, Charles Yanofsky and his colleagues managed to carry out the necessary analysis, with the *E. coli* gene coding for the enzyme called *tryptophan synthase* (which is responsible for catalysing the final step in the biosynthetic pathway that results in synthesis of the amino acid tryptophan). The result was a clear demonstration of *colinearity* between *gene and protein.* Very shortly afterwards, Sydney Brenner's group completed a similar analysis of a gene and its protein from the phage called T4, confirming Yanofsky's result. Later it was shown that the amino-terminus of the protein corresponds to the 5′ end of the gene.

EACH CODEWORD IS A TRIPLET OF NUCLEOTIDES

The second fundamental assumption about the genetic code concerned the size of the codeword, or *codon,* the group of nucleotides that code for a single amino acid. Clearly codons cannot be just single nucleotides (A, T, G or C) because that would allow only four different codewords when twenty are required, one for each of the twenty amino acids found in proteins. Similarly, a doublet code (codons such as AT, TA, TT, GC, etc.) can be ruled out as this would contain only $4^2 = 16$ different codons. However, the next stage up, a triplet code (codons AAA, AA T, TAT, GCA, etc.), would be feasible as this would yield $4^3 = 64$ codewords, which would be more than enough.

As with colinearity, experiments to prove the triplet nature of the code were straightforward in concept but difficult in practice. It was known that a group of chemicals called the acridine dyes of which *proflavin* is an example, cause single base pair deletions or additions in double-stranded DNA molecules. It was also known that certain proteins contain segments where the amino acid sequence can be changed without altering the function of the protein although other regions of the same protein will not tolerate such changes. What if a series of insertions and or deletions were introduced into the region of a gene coding for a tolerant segment of a protein? If the code is triplet, then a single insertion or deletion would give rise to a non-functional protein, because all the code words downstream of the mutation would be altered, including those in the non-tolerant segment following the tolerant regions (Figs. 13.9 and 13.9A). Two insertions or deletions (though not one of each) would have the same effect. But three insertions or deletions would restore the correct reading frame in the nontolerant region and would be predicted to have no effect on protein function. An elegant experiment of this kind was eventually carried out successfully by Crick, Brenner and others, using the gene called *rIIb* from T4 phage as the target, and determining the effect of proflavin-induced alterations by observing whether or not the treated phage were able to infect *E. coli.*

This work established the triplet nature of the code; however, by the time it was published in 1961 the first direct experimental attack on the code itself had begun.

ELUCIDATION OF THE CODE

Gradually, as the 1950s progressed, the necessary technical advances were made that would eventually allow the question of which triplet codon specifies which amino acid to be answered. There were two major technical developments.

1. *Synthesis of artificial RNA molecules.* If translation could be directed by an RNA molecule of known nucleotide sequence, then it would be possible to assign individual codons by looking at the amino acid sequence of the protein that is synthesised. To do this it would be necessary to make artificial RNA molecules of known or predictable sequence. This became possible with the discovery by Severo Ochoa in 1955 of polynucleotide phosphorylase, an enzyme which in the cell degrades RNA but which, in the test tube, can be 'forced' to catalyse the reverse reaction and synthesise RNA. This reaction does not require a DNA template and is unrelated to transcription.
2. *Cell-free protein synthesis.* To use the artificial RNA molecules as a message for translation, a cell extract able to synthesise proteins is needed. Such a cell-free, or *in vitro,* system must contain all the cellular components necessary for protein synthesis (for example, ribosomes, tRNAs, amino acids, and so on) but must lack endogenous mRNA so that protein synthesis occurs only when the artificial message is added.
3. *Cell-free protein synthesis with homopolymers.* Marshall Nirenberg and Heinrich Matthaei at the US National Institute of Health laboratories eventually perfected a cell-free system, prepared from *E. coli*, able to synthesise polypeptides specific to an added artificial RNA message. At about 6 am in the morning of Saturday 27th May 1961 Matthaei discovered that when a homopolymer made up entirely of uridine nucleotides poly (U) was added to the cell-free system, polyphenylalanine was synthesised. The first codon assignation could be made: 5′-UUU-3′ codes for phenylalanine.

Poly(U) was used in this first experiment because homopolymeric RNAs are relatively simple to synthesise. If the reaction mixture from which polynucleotide phosphorylase builds up, the RNA molecule contains only uridine mononucleotides then only poly(U) will be made. The three other RNA homopolymers were each synthesised, added individually to samples of the cell-free system and the assignations AAA-lysine and CCC-proline were made. For reasons that have never been entirely clear poly(G) would not work and so the amino acid coded by GGG could not at this stage be identified.

Random heteropolymers The next step was to construct heteropolymers, artificial RNA molecules containing more than just one nucleotide. This can be achieved by polymerising mixtures of more than one nucleotide with polynucleotide phosphorylase. There is a problem, however, as the random nature of the polymerisation means that the absolute sequence of the resulting RNA molecule will not be known. For instance, a heteropolymer of A and C will contain eight different codons (AAA, AAC, ACA, CAA, ACC, CCA, CAC, CCC) in a random order and will code for a polypeptide containing six different amino acids (proline, histidine, threonine, asparagine, glutamine and lysine). Clearly, these six amino acids are close by the eight possible codons, but which is coded by which?

An answer is at least partially provided by using a known relative proportion of each nucleotide in the reaction mixture for RNA synthesis. For example, if the ratio of C to A is 5 : 1 then the probability of a CCC codon occurring is much higher than the probability of an AAA. In fact the frequencies of each of the eight possible codons can be worked out and compared with the amounts of each amino acid in the resulting polypeptide. The codon allocations provided by this method are not definite but are statistically probable and can be cross-checked by changing the composition of the heteropolymer. These techniques allowed Nirenberg, Mathaei and Ochoa to propose meanings for most of the 64 codons of the genetic code (Figs. 13.6, 13.7, 13.8 and 13.9).

Completion of the Code Although homopolymers and random heteropolymers allowed most of the genetic code to be worked out, unambiguous determination of each codon required two additional types of experiment. The genetic code was mainly deciphered by Hargobind Khorana, Marshall Nirenberg and Robert Holley, for which they were awarded the Nobel Prize in 1968.

1. Ordered heteropolymers were synthesised by Hargobind Khorana. These are made by polymerising not nonnucleotides, but known dinucleotides such as AC (producing ACACACAC, which contains two codons, ACA and CAC) or trinucleotides such as UGU (producing UGUUGUUGU, codons UGU, GUU, UUG). The simpler polynucleotides produced by these messages allowed the meaning of several difficult codons to be determined.
2. The triplet binding assay was developed by Nirenberg and Philip Leder in 1964 as a modification of the cell-free protein synthesising system. It was discovered that purified ribosomes will attach to an mRNA molecule of only three nucleotides (a single codon, in fact) and bind the correct amino acid-tRNA molecule. As triplets of known sequence could be constructed in the laboratory, the binding assay allowed the previously assigned codons to be checked and virtually all the remaining triplets to be identified.

COMPLETION OF GENETIC CODE (Figs. 13.5 AND 13.6)

Finally, 1966, five years after Matthaei's first experiments, the meaning of the final codon was confirmed and the genetic code was completed. In the Proceedings of Cold Spring Harbor Symposium on Quantitative Biology: 1966: *Genetic code*, all papers regarding Genetic code were published.

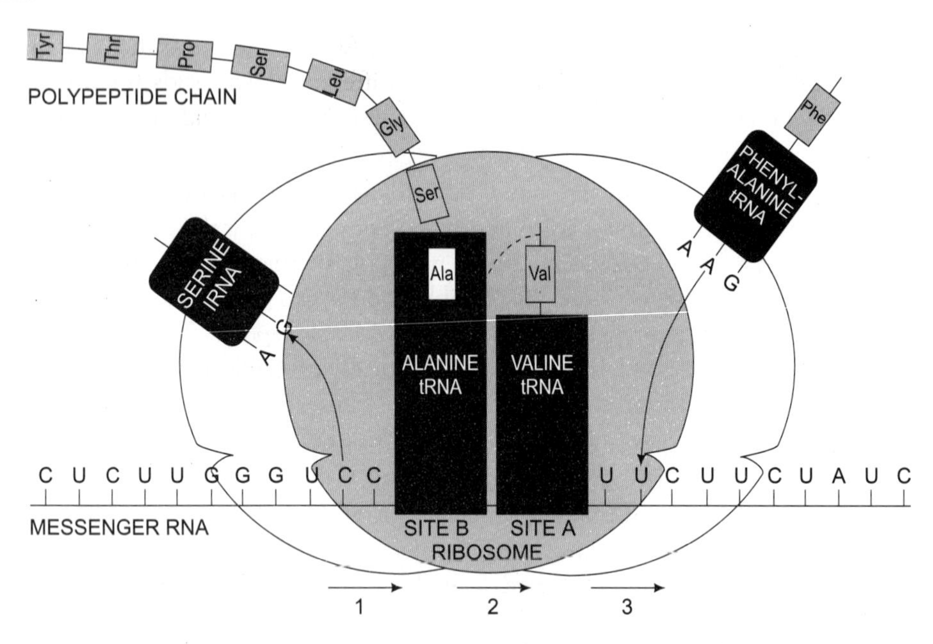

Fig. 13.5 *Synthesis of Protein Molecules.* Protein synthesis is accomplished by the intracellular particles called ribosomes. The coded instructions for making the protein molecule are carried to the ribosome by a form of ribonucleic acid (RNA) known as 'messenger' RNA. The RNA code 'letters' are four bases: uracil (U), cytosine (C), adenine (A) and guanine (G). A sequence of three bases, called a codon, is required to specify each of the 20 kinds of amino acids, identified by their abbreviations. When linked end to end, these amino acids form the polypeptide chains of which proteins are composed. Each type of amino acid is transported to the ribosome by a particular form of 'transfer' RNA (tRNA), which carries an anticodon that can form a temporary bond with one of the codons in messenger RNA. Here the ribosome is shown moving along the chain of messenger RNA, "reading off" the codons in sequence. It appears that the ribosome has two binding sites for molecules of tRNA: one site (A) for positioning a newly arrived tRNA molecule and another (B) for holding the growing polypeptide chain. (After Crick, F.H.C. 1966).

FEATURES OF THE CODE

The code is shown in Fig. 13.6. It has three important features. Important definition of terms used in Genetic Code are listed in Table 13.1.

1. *The code is degenerate* All amino acids except methionine and tryptophan have more than one codon, so that all the possible triplets have a meaning, despite there being 64 triplets and only 20 amino acids. Most synonymous codons are grouped into families (GGA, GGU, GGG and GGC, for example, all code for glycine), a fact that is relevant to the way the code is deciphered during protein synthesis.
2. *The code contains punctuation codons* Three codons, UAA, UGA and UAG, do not code for an amino acid and if present in the middle of a heteropolymer cause protein synthesis to stop, resulting in shorter polypeptides than expected. These are called termination codons and one of the three always occurs at the end of a gene at the point where translation must stop.

First Position (5′ end)	Second Position: U	C	A	G	Third Position (3′ end)
U	Phe	Ser	Tyr	Cys	U
	Phe	Ser	Tyr	Cys	C
	Leu	Ser	*stop*	*stop*	A
	Leu	Ser	*stop*	Trp	G
C	Leu	Pro	His	Arg	U
	Leu	Pro	His	Arg	C
	Leu	Pro	Gln	Arg	A
	Leu	Pro	Gln	Arg	G
A	Ile	Thr	Asn	Ser	U
	Ile	Thr	Asn	Ser	C
	Ile	Thr	Lys	Arg	A
	Met *(start)*	Thr	Lys	Arg	G
G	Val	Ala	Asp	Gly	U
	Val	Ala	Asp	Gly	C
	Val	Ala	Glu	Gly	A
	Val	Ala	Glu	Gly	G

Fig. 13.6 The genetic code.

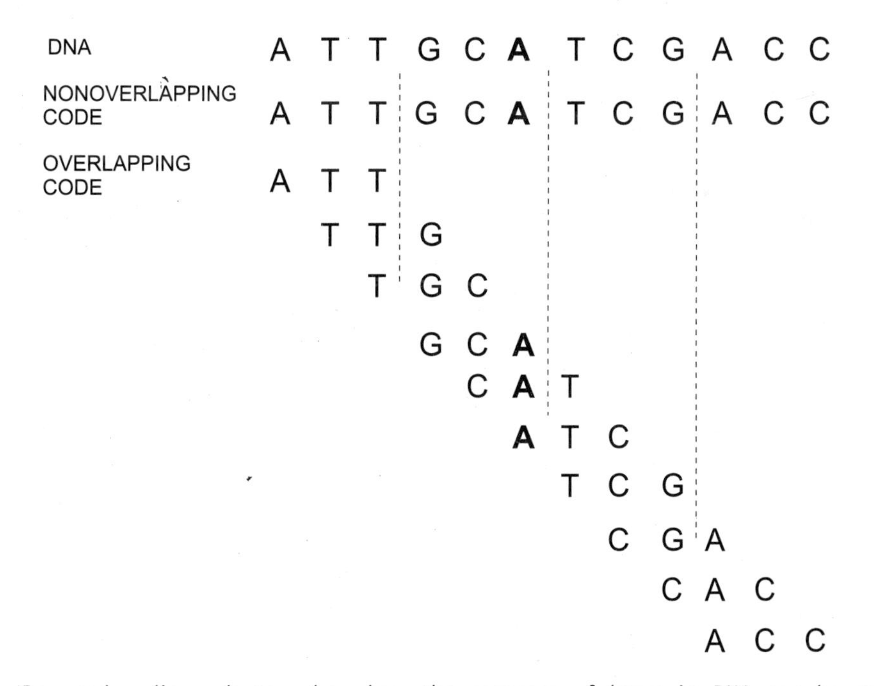

Fig. 13.7 Proposed coding schemes show how the sequence of bases in DNA can be read. In a nonoverlapping code, groups are read in simple sequence. In one type of overlapping code, each base appears in three successive groups.

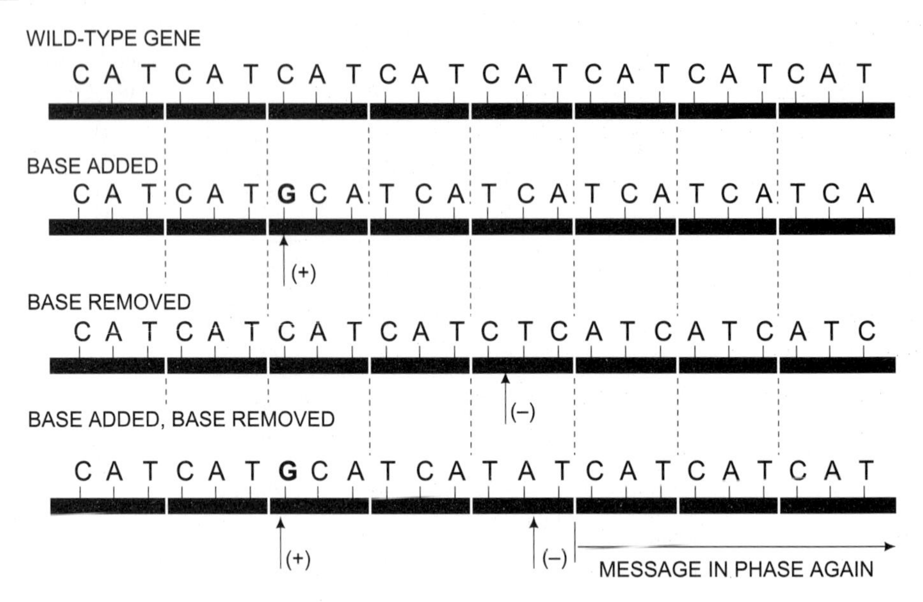

Fig. 13.8 Effect of mutations that add or remove a base is to shift the reading of the genetic message, assuming that the reading begins at the left-hand end of the gene. The hypothetical message in the wild-type gene is CAT, CAT... Adding a base, shifts the reading to TCA, TCA... Removing a base makes it ATC, ATC... Addition and removal of a base puts the message in phase again.

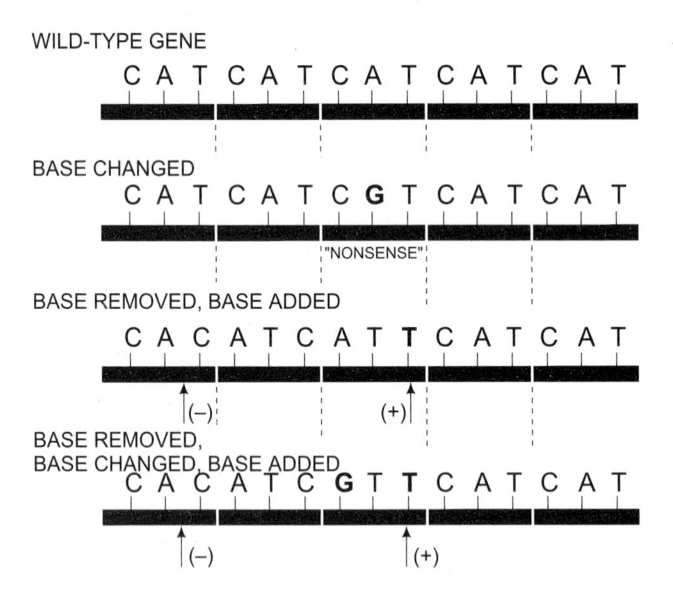

Fig. 13.9 Nonsense mutation is one creating a code group that evidently does not represent any of the 20 amino acids found in proteins. Thus it makes the gene inactive. In this hypothetical case a nonsense triplet, CCT, results when an A in the wild-type gene is changed to G. The nonsense triplet can be eliminated if the reading is shifted to put the G in a different triplet. This is done by recombining the inactive gene with one containing a minus-with-plus combination. In spite of three mutations, the resulting gene is active.

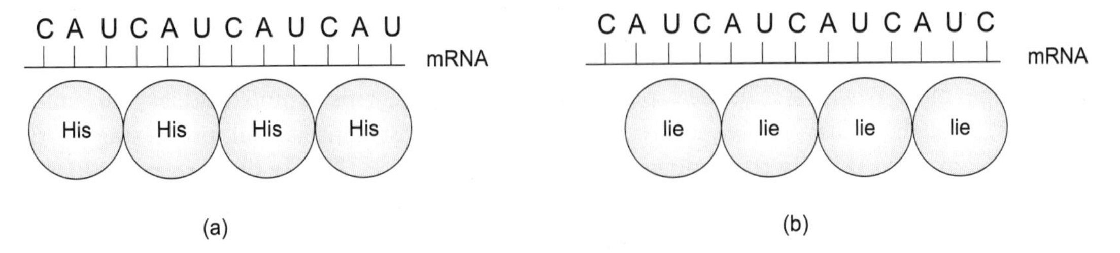

Fig. 13.9A (a) Normal reading of the messenger RNA, (b) Shift in the reading frame of the messenger RNA.

Table 13.1 Definition of some common terms used in describing the genetic code.

Terms	*Meaning*
Code letter	Nucleotide, e.g., AUGC (in mRNA) or ATGC (in DNA)
Code word, or codon	Sequence of nucleotides specifying an amino acid, e.g., UUU = phenylalanine
Anticodon	Sequence of nucleotides on tRNA that complements the codon, e.g., AAA = anticodon for phenylalanine
Genetic code or Coding dictionary	A table of all the code words or codons that specify amino acids
Word size	The number of letters in a code word, e.g., three letters in a triplet code (this is the same as *coding ratio* in a nonoverlapping code)
Nonoverlapping code	When only as many amino acids are coded as there are code words in end-to-end sequence, e.g., (triplet code), UUUCCC = phenylalanine (UUU) + proline (CCC)
Overlapping code	When more amino acids are coded for than there are code words in end-to-end sequence, e.g., UUUCCC = phenylalanine (UUU) + phenylalanine (UUC) + serine (UCC) + proline (CCC)
Degenerate code	When there is more than one codon for a particular amino acid, e.g., UUU, UUC = phenylalanine
Ambiguous code	When one codon can code for more than one amino acid, e.g., GGA = glycine, glutamic acid
Commaless code	When there are no intermediatary nucleotides (spacers) between words, e.g., UUUCCC = two amino acids in triplet nonoverlapping code
Reading frame	The particular nucleotide sequence that starts at a specific point and is then partitioned into codons. The reading frame may be "shifted" by removing or adding nucleotides, thereby causing a new sequence of codons to be "read"
Sense word	A codon that specifies an amino acid normally present at that position in a protein.
Missense mutation	A change in nucleotide sequence, either by deletion, insertion, or substitution, resulting in the appearance of a codon that produces a different amino acid in a particular protein, e.g., UUU (phenylalanine) $\rightarrow$ UGU (cysteine)
Nonsense mutation	A codon that does not produce an amino acid, e.g., UAG
Universality	Utilisation of the same genetic code in all organisms, e.g., UUU = phenylalanine in bacteria, mouse, man and tobacco

Similarly, AUG virtually always occurs at the start of a gene and marks the position where translation should begin. AUG is, therefore, the initiation codon and because it codes for methionine most newly synthesised polypeptides will have this amino acid at the amino-terminus, though it may subsequently be removed by post-translational processing of the protein. Note that AUG is the only codon for methionine, so AUGs that are not initiation codons may be found in the internal region of a gene.

3. *The code is not universal* When the genetic code was completed in 1966, it was assumed to be universal. It was postulated that the genetic code must be frozen and unable to evolve because a change in a codon assignation would cause almost every protein in the cell to be altered. The cell-free system used to elucidate the code was prepared from *E. coli* but it was soon shown that the mammalian code is the same and the assumption of universality became accepted dogma. It was a shock, therefore, when Frederick Sanger's group in Cambridge discovered in 1979 that the special genetic system in human mitochondria uses a slightly different code. Soon afterwards unusual codons were confirmed for mitochondrial genes of other organisms. More recently, slightly modified codes have been discovered for the nuclear genes of protozoans such a *Tetrahymena*, and the idea of universality has had to be revised. But the most significant discovery so far is that the code is not inviolable even for mammalian nuclear genes and that at least one mouse gene, coding for the enzyme glutathione peroxidase, uses the termination codon UGA (which is indeed a stop codon for many mouse genes) to specify the unusual amino acid selenocysteine.

TRANSLATION AND PROTEIN SYNTHESIS

The genetic code is utilised during the final stage of gene expression, when mRNA molecules are translated into polypeptides. Translation is very complex and involves many different components of the cell. However, compliance with the genetic code is central to the entire process, and is ensured by tRNA molecules which act as adaptors forming a physical and informational link between the nucleotide sequence of the mRNA and the amino acid sequence of the polypeptide. Later in this chapter we will look at the detailed mechanism by which proteins are synthesised in the cell. First, however, we must tackle the more fundamental question of exactly how tRNAs are able to ensure that translation proceeds in accordance with the rules of the genetic code.

ROLE OF tRNA IN TRANSLATION

Each cell contains a number of different types of tRNA molecules, distinguished from one another by their different sequences, though retaining the invariant and semi-invariant nucleotides. Each tRNA is distinguished in functional terms by its specificity for one of the twenty amino acids involved in protein synthesis: for instance, the tRNA molecule designated $tRNA^{tyr}$ is specific for tyrosine, $tRNA^{gly}$ is specific for glycine. A tRNA molecule forms a covalent linkage with its amino acids (and no other) and recognise and attach to a codon specifying that amino acid. There may be more than one type of tRNA molecule for a single amino acid, reflecting the fact that the genetic code is degenerate and that most amino acids are coded by more than one codon. Two tRNAs that bind the same amino acid are called isoacceptors.

AMINOACYLATION OF tRNA (Figs. 13.10 and 13.11)

Each tRNA molecule is able to form a covalent linkage with its specific amino acid by a process called aminoacylation, or charging. The amino acid becomes attached to the end of the acceptor stem of the tRNA cloverleaf. The linkage forms between the carboxyl group of the amino acid and the 3′-OH group of the terminal nucleotide of the tRNA. Remember that this terminal nucleotide is always adenosine because all tRNAs have the sequence 5′-CCA-3′ at their 3′ ends.

AMINOACYL-tRNA SYNTHETASES CONTROL CHARGING

Amino acylation is catalysed by a group of enzymes called the aminoacyl—tRNA synthetases. In most cells there is a single aminoacyl-tRNA synthetase for each amino acid, meaning that one enzyme can charge each member of a series of isoaccepting tRNAs.

Although aminoacyl-tRNA synthetases form a fairly heterogeneous group of enzymes, each catalyses the same basic reaction. The first step in aminoacylation involves synthesis of an activated amino acid intermediate in which the carboxyl group has formed a link with adenylic acid derived from ATP. This intermediate remains bound to the enzyme until the adenylic acid is replaced in the second stage of the reaction by the tRNA molecule, producing aminoacyl—tRNA and AMP.

The specificity of charging (ensuring that the correct amino acid is attached to the correct tRNA) is a function of the aminoacyl-tRNA synthetase, which is able to recognise both the correct amino acid, primarily on the basis of its unique P group, and the equivalent tRNA. How the tRNA is recognised is not understood in detail, though it is clear that certain of the variant nucleotides of the cloverleaf are characteristic for an individual tRNA and it is presumably these that are distinguished by the aminoacyl-tRNA synthetase.

CODON RECOGNITION (Figs. 13.12 and 13.13)

Once the correct amino acid has been attached to the acceptor stem of the tRNA, the aminoacylated molecule must complete the link between mRNA and polypeptide by recognising and attaching to the correct codon: that is, one coding for the amino acid that it carries. Codon recognition is a function of the anticodon loop of the tRNA, specifically of the trinucleotide called the anticodon. This trinucleotide is complementary to the codon and can, therefore, attach to it by base pairing. The specificity of the genetic code is, therefore, ensured because the

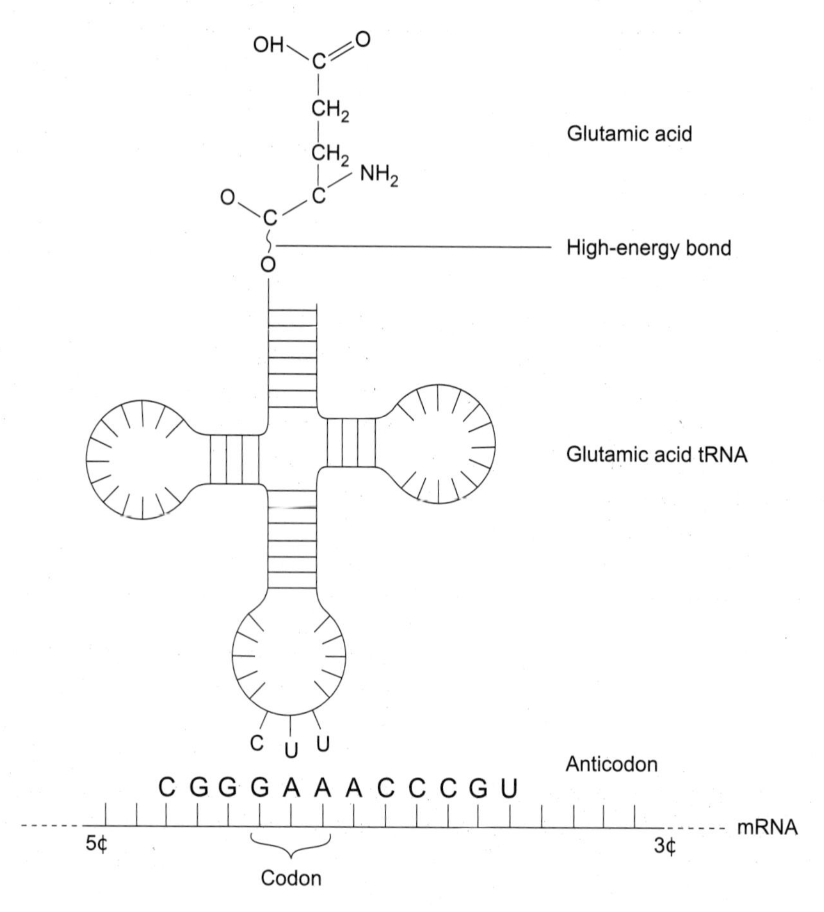

Fig. 13.10 Specificity of the Genetic Code resides in the tRNA where a particular anticodon is associated with a particular amino acid.

anticodon present on a particular tRNA is one that is complementary to a codon for the amino acid with which the tRNA is charged.

Codons and anticodons can wobble From what has been said so far it might be imagined that there are 61 different types of tRNA molecules in each cell, one for each of the codons that specifiy an amino acid. In fact it has been known since early 1960s that there are substantially fewer than 61 different tRNA molecules, usually between 31 and 40 depending on the organism which is explained by the Wobble hypothesis first expounded by Crick in 1966. (Figs. 13.14 & 13.15).

This hypothesis is based on the fact that the anticodon loop is just that, a loop, and the anticodon itself is not a perfectly linear trinucleotide. The short double helix formed by base pairing between the codon and anticodon does not, therefore, have the precise configuration of a standard RNA helix instead its dimensions are slightly altered. As a result, non-standard base pairs can form at the wobble position (between the first nucleotide of the codon). So a

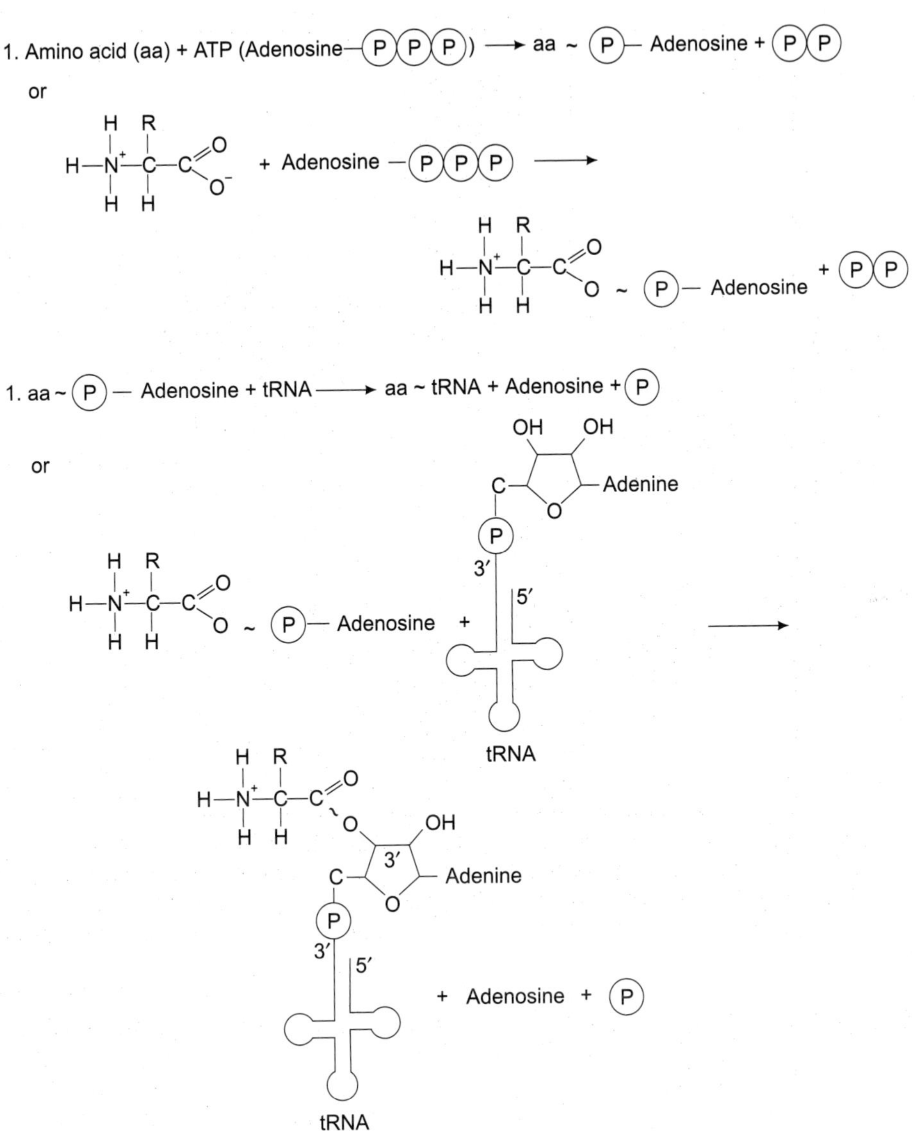

1. Amino acid (aa) + ATP (Adenosine—(P)(P)(P)) ⟶ aa ~ (P)— Adenosine + (P)(P)

or

H R / H—N⁺—C—C(=O)O⁻ / H H + Adenosine —(P)(P)(P) ⟶

H R / H—N⁺—C—C(=O)O ~ (P)— Adenosine / H H + (P)(P)

1. aa ~ (P) — Adenosine + tRNA ⟶ aa ~ tRNA + Adenosine + (P)

or

H R / H—N⁺—C—C(=O)O ~ (P)— Adenosine / H H + [tRNA: 3′ … (P)—C—ribose(OH, OH)—Adenine; 5′] ⟶

H R / H—N⁺—C—C(=O)—O—(3′ of ribose, OH)—Adenine / H H, C—(P), 3′, 5′, tRNA + Adenosine + (P)

Fig. 13.11 Aminoacylation of tRNA.

single anticodon may be able to base pair with more than one codon and single tRNA may decode more than one member of a codon family. However, the base pairing rules do not become totally flexible at the wobble position, and only a few types of unusual base pairs are allowed. Common examples are:

1. *G–U base pairs* A non-standard G–U base pair is allowable at the wobble position. Two possible consequences of this are shown in the Figs. 13.14 and 13.15.

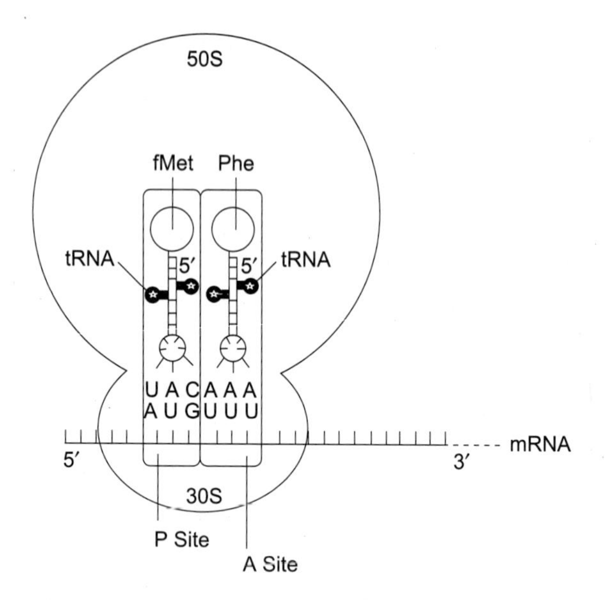

Fig. 13.12 Ribosome with two tRNAs attached in this case the second codon is for the amino acid phenylalanine.

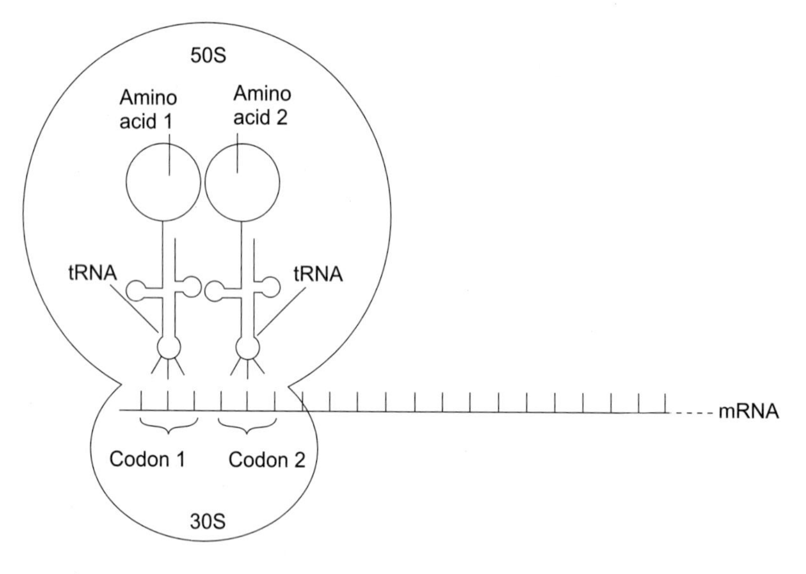

Fig. 13.13 Overview of the translation process at the ribosome.

2. *Inosine can base pair with* C, A *and* U In some tRNAs the wobble nucleotide of the anticodon is inosine (I), a deaminated form of G. I can base pair not only with C but also with A and U.

Wobble, therefore, decreases the number of tRNAs needed to decode the genetic code so that a single cell can get by with as few as 31 different tRNAs. Nevertheless, the rules of the

Anticodon	Codon
U	A
	G
C	G
A	U
G	U
	U
I	U
	C
	A

Fig. 13.14 "Wobble" Hypothesis of F.H.C. Crick.

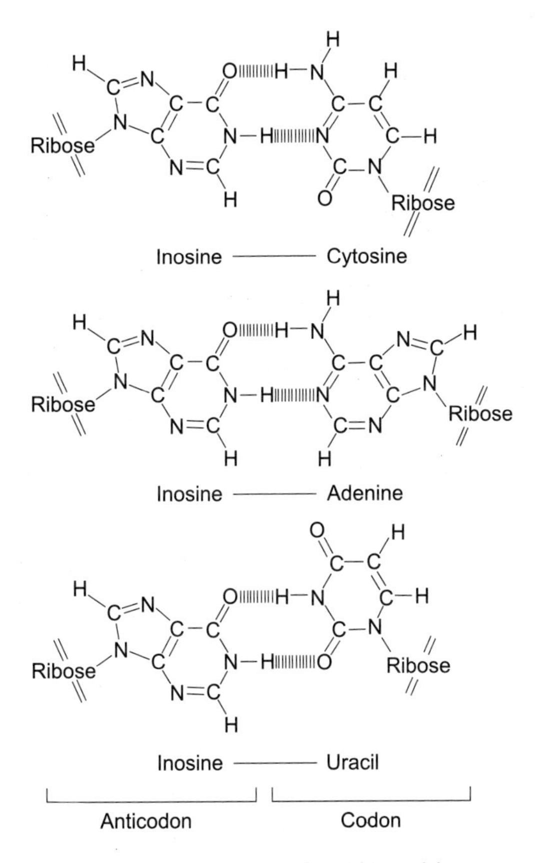

Fig. 13.15 Examples of wobble pairing.

genetic code still remain inviolate, and a polypeptide synthesised during translation is polymerised strictly in accordance with the nucleotide sequence of the relevant mRNA. We will now turn our attention to the mechanics of this process.

MECHANICS OF PROTEIN SYNTHESIS

Once the role of tRNA has been appreciated the exact mechanism by which translation occurs is much easier to follow. Traditionally the process is split into three *stages–initiation, elongation* and *termination*–although in practice the events are continuous.

Translation is very well understood in *E. coli* and what follows is a description of this process. The equivalent events in higher organisms are essential the same though some of the details differ. The most important of these will be mentioned here.

INITIATION OF TRANSLATION (Figs. 13.15, 13.16 and 13.16A)

The very first step in translation is attachment of the small (30S) subunit of a ribosome to an mRNA molecule. When not actually involved in protein synthesis, ribosomes dissociate into their component subunits so that the ribsome 'pool' in the cell is made up of large numbers of individual 30S and 50S subunits.

RIBOSOME BINDING SITES ENSURE THAT TRANSLATION STARTS AT THE CORRECT POSITION

During the transcription process, it is important to ensure that the RNA polymerase enzyme starts transcription at the correct point so that genes rather than random pieces of DNA are transcribed. An analogous situation exists with translation: the 30S ribosomal subunits must attach to the mRNA molecule to be translated not at random, but at a specific point just upstream of the initiation codon of the gene. The correct attachment point is signalled by the ribosome binding site which in *E. coli* has the consensus sequence:

5′-AGGAGGU-3′

This sequence, known as the Shine-Dalgarno sequence, is believed to form a transient base paired attachment to a part of the 16S rRNA polynucleotide, allowing the 30S subunit to bind to the mRNA.

Once the ribosome binding site has been located, the 30S subunit migrates downstream until it encounters an AUG codon, which it will usually find within 10 nucleotides of the binding site. This AUG triplet will be the initiation codon of the gene and will mark the position at which translation must begin.

FORMATION OF THE INITIATION COMPLEX (Figs. 13.17, 13.18 and 13.19)

The translation process itself starts when an aminoacylated tRNA base pairs with the initiation codon that has been located by the 30S subunit of the ribosome. This initiator tRNA is charged with methionine, because methionine is the amino acid coded by AUG. However, in bacteria (though not eukaryotes), this methionine is modified after charging by substitution of a formyl

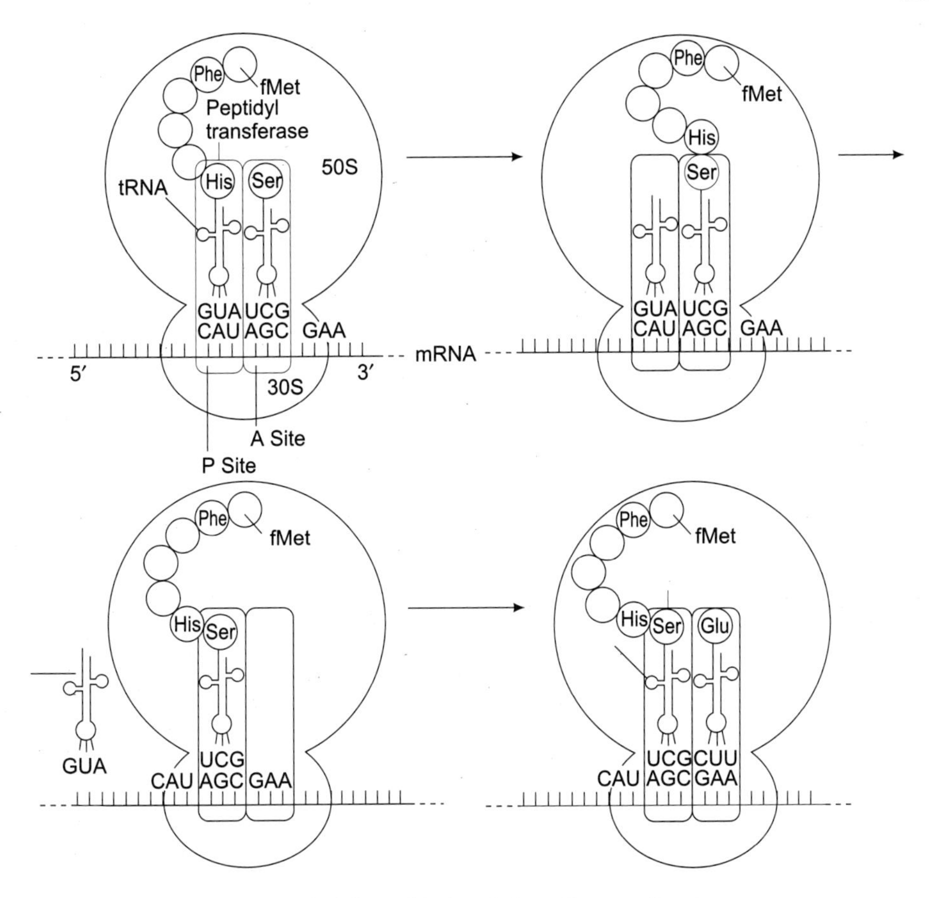

Fig. 13.16 Cycle of peptide bond formation on the ribosome.

group (–COH) for one of the hydrogen atoms of the amino group, to produce N-formylmethionine or *fmet*. This substitution blocks the amino group so that it cannot participate in the formation of a peptide bond; polymerisation of the polypeptide, therefore, can occur only in the amino to carboxyl direction.

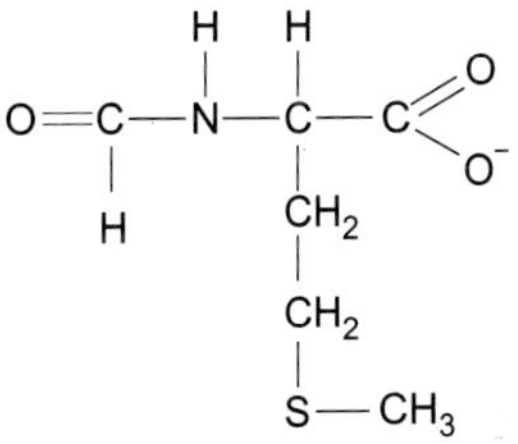

Fig. 13.17 Structure of N-formyl methionine.

The resulting structure, comprising primarily the mRNA, 30S subunit and aminoacylated-$tRNA^{fmet}$ is called the initiation complex and its formation marks the end of the initiation stage of translation.

Initiation Factors

The main outstanding area of translation that is not fully understood is the role of various non-ribosomal protein factors. For example, initiation in *E. coli* requires three proteins called

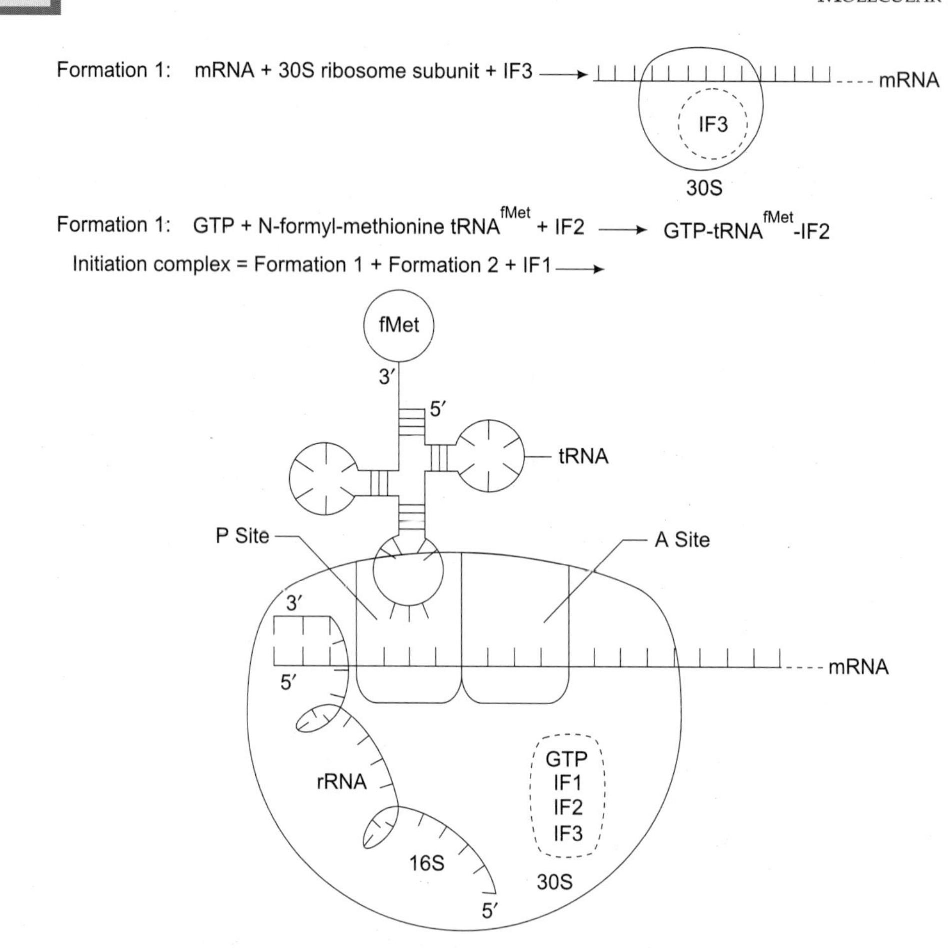

Fig. 13.18 Formation of the initiation complex.

initiation factors. *IF1* and *IF3* appear to be responsible for dissociation of the ribosome into 30S and 50S subunits, with IF3 possibly being involved in recognition of the ribosome binding site. IF2 participates in the attachment of the charged initiator-tRNA and also mediates binding to the initiation complex of a molecule of GTP, which provides the energy for the next step in translation. In eukaryotes, the situation is even more complex. At least nine initiation factors are involved in formation of the initiation complex in mammals.

ELONGATION OF THE POLYPEPTIDE CHAIN

Once the initiation complex has been formed the large subunit of the ribosome can attach. This requires hydrolysis of the GTP molecule associated with the initiation complex and results in two separate and distinct sites to which tRNAs can bind. The first to these, called the peptidyl- or P-site is at the moment occupied by the aminoacylated $tRNA^{fmet}$, which is still base paired

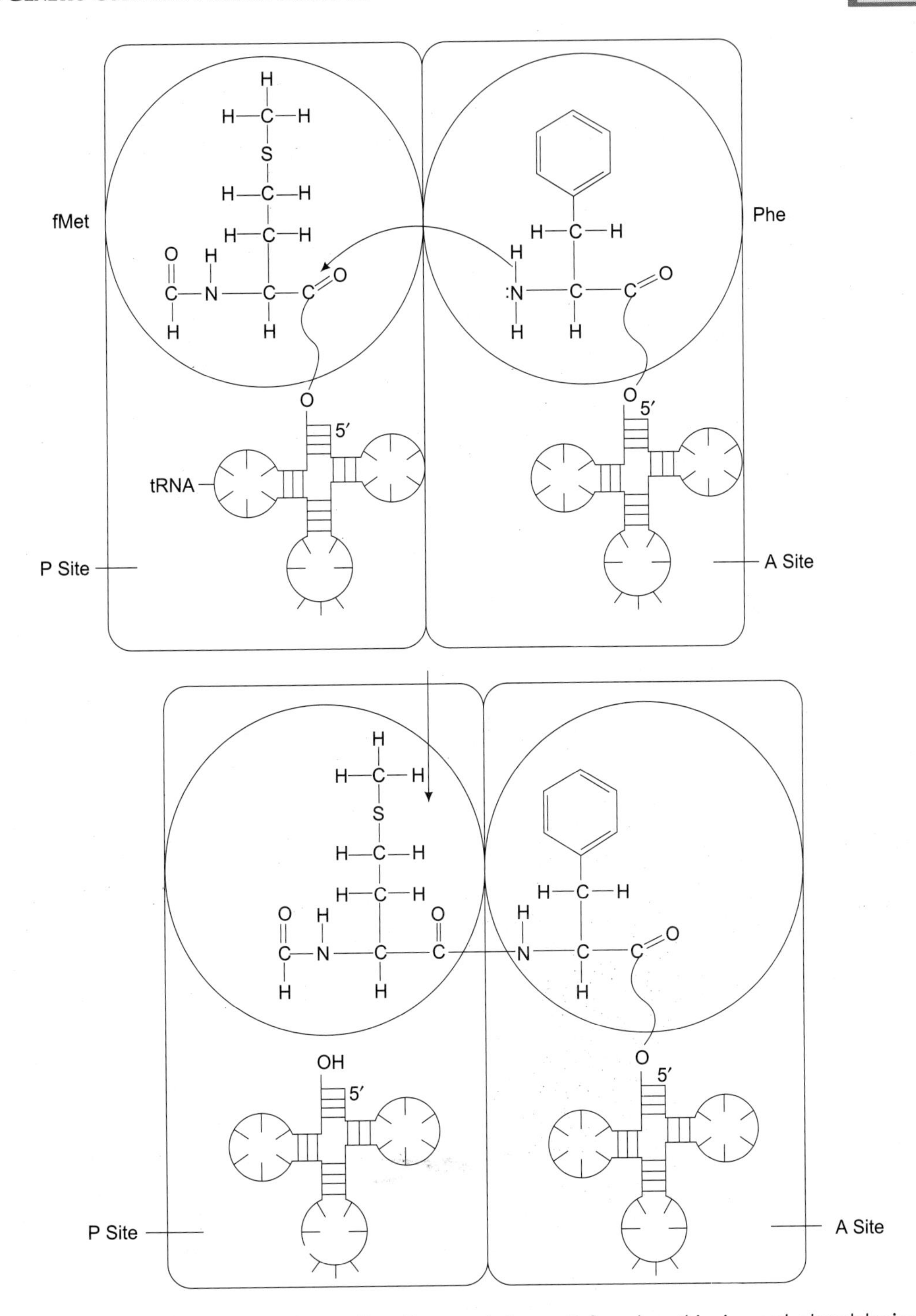

Fig. 13.19 Peptide bond formation at the ribosome between N-formyl methionine and phenylalanine.

with the initiation codon. The aminoacyl- or A-site is positioned over the second codon of the gene and at first is empty. Elongation begins when the correct aminoacylated tRNA enters the A-site and base pairs with the second codon. This requires two elongation factors. Ef–Tu and Ef–Ts, although only the former appears to be directly involved. GTP is needed to provide energy.

PEPTIDE BOND FORMATION AND TRANSLOCATION (Fig. 13.20)

Now both sites of the ribosome are occupied by aminoacylated tRNAs and the two amino acids are in close contact. The next step is formation of a peptide bond between the carboxyl group of the *fmet* and the amino group of the second amino acid. The reaction is catalysed by a complicated enzyme called *peptidyl transferase,* which is possibly a combination of several different ribosomal proteins. Peptidyl transferase acts in conjunction with a second ribosomal enzyme, tRNA deacylase, which breaks the *fmet*-tRNA link. The result is a dipeptide attached to the tRNA present in the A. site.

Translocation now occurs. The ribosome slips along the mRNA by three nucleotides so that the aa-aa-tRNA enters the P-site, expelling the now uncharged $tRNA^{fmet}$ and making the A-site vacant again. The third aminoacylated tRNA now enters the A-site and the elongation cycle is repeated. Each cycle requires hydrolysis of a molecule of GTP and is controlled by a third elongation factor, EF–G.

EACH mRNA CAN BE TRANSLATED BY SEVERAL RIBOSOMES AT ONCE

After several cycles of elongation, the start of the mRNA molecule is no longer associated with the ribosome, and a second round of translation can begin. A second 30S subunit can attach to the ribosome binding site and form a new initiation complex. The end result is a polysome, an mRNA that is being translated by several ribosomes at once. Polysomes have been seen in electron microscopic images of both prokaryotic and eukaryotic cells.

CHAIN TERMINATION (Figs. 13.21, 13.22 and 13.23)

Termination of translation occurs when a termination codon (UAA, UAG or UGA) enters the A-site. There are no tRNA molecules with anticodons able to base pair with any of the termination codons; instead one of three release factors enters the A-site and, in conjunction with hydrolysis of a molecule of GTP, cleaves the now completed polypeptide from the final tRNA. The ribosome dissociates into 30S and 50S subunits which enter the cellular pool before becoming involved in a new round of translation. The polypeptide may be processed by cleavage of a portion of its amino acid sequence, and possibly modified by attachment of chemical groups (sugar molecules, for example) to specific amino acids. In conjunction with these events, the polypeptide chain folds up into its tertiary structure and begins its functional life within the cell.

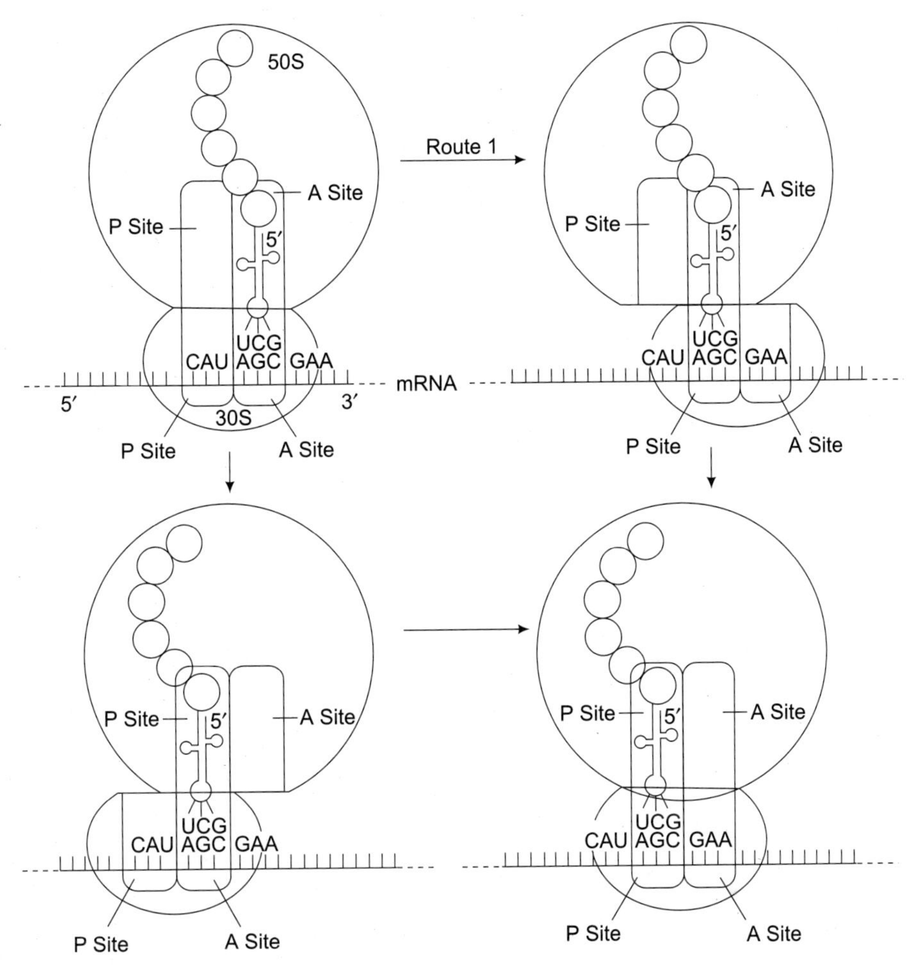

Fig. 13.20 Hybrid site model for translation of the ribosome.

ANTIBIOTICS

Antibiotics are substances that are produced by living organisms and that are toxic to other organisms. Antibiotics are of interest to us for two reasons. First, they have been extremely important in fighting human diseases and the diseases of farm animals. Second, they are a very useful tool for analysing many of the steps of protein synthesis. Antibiotics impede the process of protein synthesis in a variety of ways and selectively poison bacteria. The effectiveness of antibiotics usually derives from the metabolic differences between bacteria and eukaryotes. For example, an antibiotic that will block a bacterial ribosome will be an excellent antibiotic.

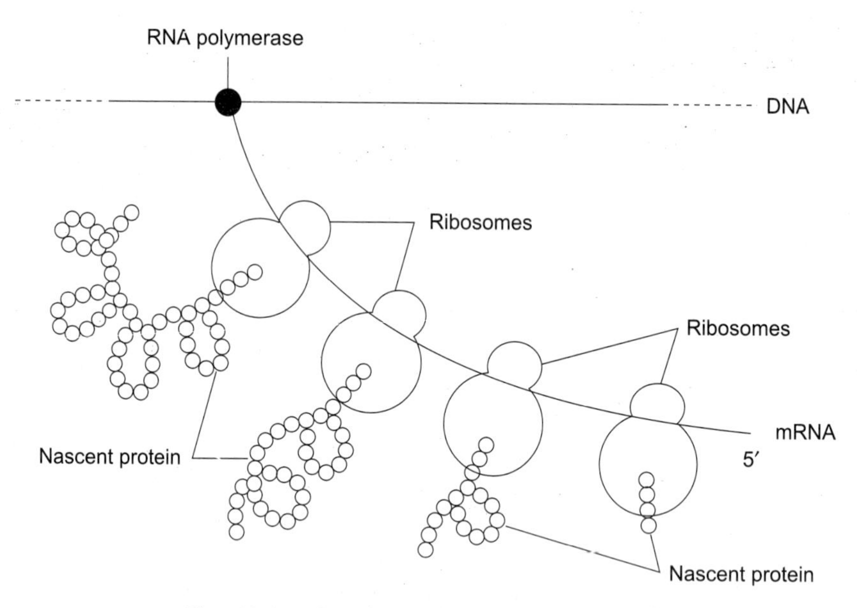

Fig. 13.21 Protein synthesis at a polysome.

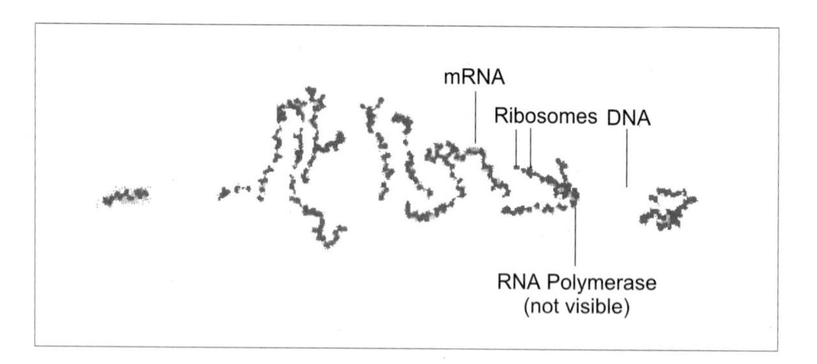

Fig. 13.22 Electron micrograph of mRNA transcription in *E. coli;* ribosomes are associated with the mRNA.

PUROMYCIN

Puromycin works by resembling the 3′ end of an aminoacyl-tRNA (Fig. 13.24). It is bound to the A site of the bacterial ribosome, where peptidyl transferase will transfer a bond from the existing peptide attached to the tRNA in the P site to puromycin. The peptide chain in then prematurely released and protein synthesis at the ribosome is terminated.

TETRACYCLINE

Tetracycline blocks protein synthesis by preventing the aminoacyl-tRNA from binding to the A site on the ribosome (Fig. 13.25A).

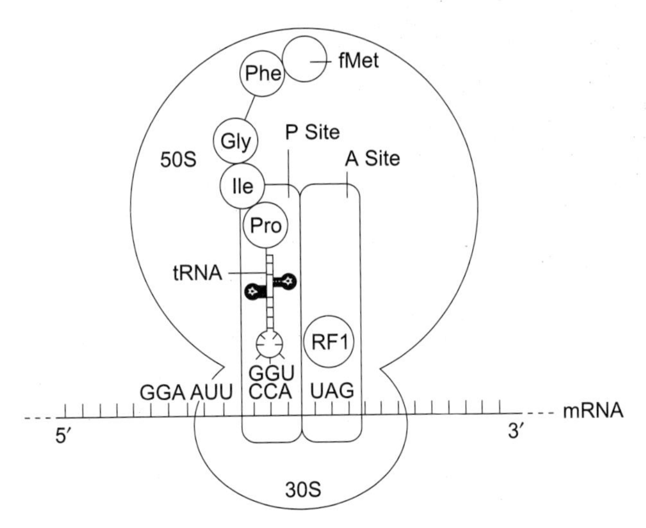

Fig. 13.23 A nonsense codon in the A site is recognised by a release factor (RF1).

CHLORAMPHENICOL

Chloramphenicol blocks protein synthesis by binding to the 50S subunit of the prokaryotic ribosome, where it blocks the peptidyl transfer reaction (Fig. 13.25C). Chloramphenicol does not affect the eukaryotic ribosome, but it does affect the mitochondrial ribosomes, since there are similarities between mitochondrial ribosomes and prokaryotic ribosomes.

STREPTOMYCIN

Streptomycin binds to one of the proteins (protein S12) of the subunit of the prokaryotic ribosome and inhibits initiation of protein synthesis (Fig. 13.25B). It also causes misreading of codons if chain initiation

CHAPTER 14

Control of Gene Expression

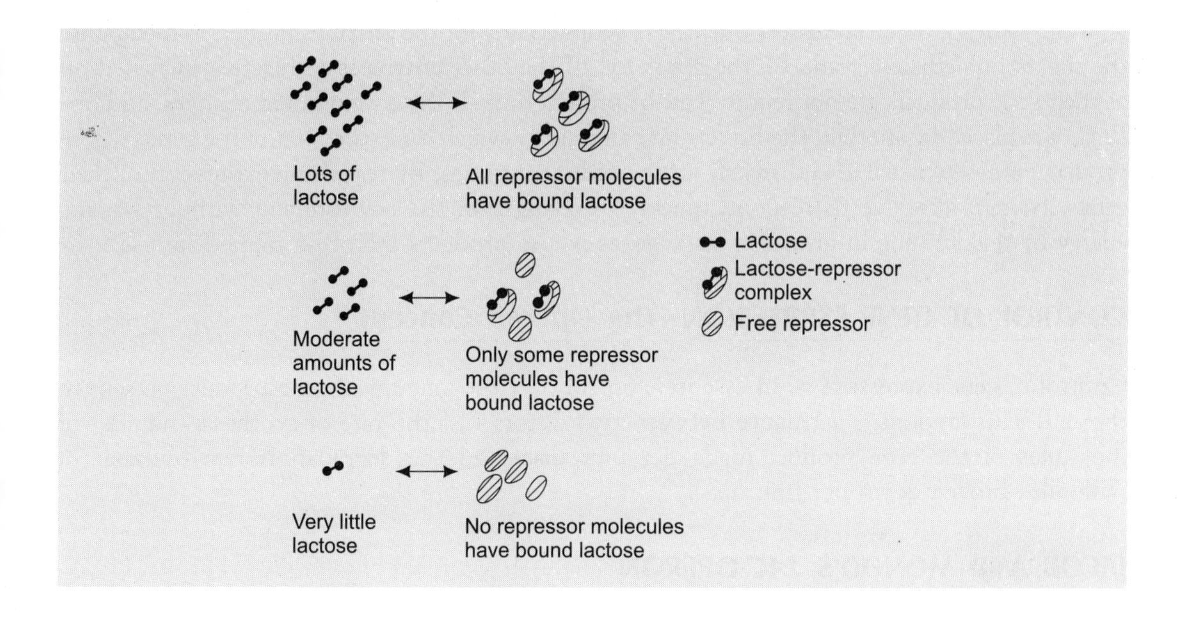

Chapters 12 and 13 have outlined the process of gene expression and described how the biological information contained in the DNA sequence of a gene is released and utilised by the cell. In this chapter the regulation of gene action is dealt with by what are called *operons*.

WHY CONTROL GENE EXPRESSION?

The entire complement of genes in a single cell represents a staggering amount of biological information. Some of this information is needed by the cell at all times: for instance, most cells continually synthesise ribosomes and so have continuous reqirement for transcription of the rRNA and ribosomal protein genes. Similarly, genes coding for enzymes such as RNA polymerase or those involved in the basic metabolic pathways will be active in all cells all of the time. These genes are sometimes called *housekeeping genes,* reflecting the role in the cell of the biological information they carry.

On the other hand, many genes have a more specialised role and their biological information is needed by the cell only in certain circumstances. This is true of genes in all organisms from the simplest bacteria to complex multicellular forms such as man. However, as an example, we can consider those bacterial genes involved in utilisation of different sugar compounds as sources of carbon atoms and energy. Most bacteria possess a number of genes coding for enzymes that enable the cell to take up and metabolise any one of several different sugars and other carbon compounds. Exactly which enzymes are actually required by an individual bacterium depends on the nature of the sugar molecules present in the environment. Of course, the bacterium could continuously express all its genes and so have molecules of each enzyme present all the time, but this would waste energy, something all living organisms attempt to avoid. Instead, bacteria express only those genes coding for the enzymes needed to metabolise the sugars immediately available; the genes for all the other enzymes are inactive, switched off as their gene products are not required at the present time. If the environment changes, and one sugar is replaced by another, the bacterium can quickly switch on expression of the genes whose products are now needed, and switch off the redundant ones. By regulating expression of their genes, bacteria are able to respond quickly to changes in the environment without wasting energy by maintaining in an active stage genes whose products are of no immediate use.

CONTROL OF GENE EXPRESSION—The Operon Concept

Control of gene expression is, in essence, control over the amount of gene product present in the cell. This amount is a balance between two factors *viz.*, the rate of synthesis (number of molecules of the gene product made per unit *time*) and the degradation rate (number of molecules broken down per unit *time*).

JACOB AND MONOD'S *LAC* OPERON

Francois Jacob and Jacques Monod in 1961 published a classic paper on how bacteria regulate expression of their genes, based on their genetic analysis of lactose utilisation by *E. coli.* These investigations were later on confirmed by DNA sequencing and analysis of DNA-protein interaction.

Important Terminology Used in Operon Concept

Operon: A cluster of genes that are transcribed into a single RNA molecule.

Promoter: The DNA binding site for RNA polymerase, to which the enzyme attaches to begin.

Repressor: A protein that is able to prevent transcription of a gene or protein.

Operator: The DNA binding site to which the repressor must attach in order to prevent transcription of gene or operon.

Inducer: A molecule that binds to a repressor and prevents the repressor from attaching to its operator. An inducer therefore switches transcription on.

Corepressor: A molecule that must bind to repressor before the repressor can attach to its operator. A corepressor therefore switches transcription off.

1. *Transcription:* If the number of transcripts synthesised per unit time decreases, the amount of the gene product in the cell will also decrease.
2. *mRNA turnover:* If mRNA molecules are degraded before translation can occur, synthesis of the gene product will be limited.
3. *mRNA processing:* Events such as capping, polyadenylation and splicing of eukaryotic mRNAs are in most cases prerequisites for translation. If these processing events are slowed, product synthesis will fall.
4. *Translation:* Control could be exerted over the number of ribosomes that can attach to a single mRNA, or over the rate at which individual ribosomes translate a message.

It is now becoming clear to molecular biologists that control of gene expression is a highly complex process that probably involves each of the above strategies. However, the best understood mechanisms, in both prokaryotes and eukaryotes, depend solely on the first possibility: regulatory control over transcription.

REGULATION OF LACTOSE UTILISATION (Fig. 14.1)

Lactose is a disaccharide composed of a single glucose unit attached to a single galactose unit. In order to utilise lactose as a carbon and energy source, an *E. coli* cell must first transport lactose molecules from the extracellular environment into the cell and then split the molecules, by hydrolysis, into glucose and galactose.

These reactions are catalysed by three enzymes, each of which has no function beyond lactose utilisation: *lactose permease,* which transports lactose into the cell; *β-galactosidase* which is responsible for the splitting reaction; and *β-galactoside transacetylase,* which plays a secondary role in the hydrolysis of lactose. In the absence of lactose only a small number of molecules of each enzyme are present in the *E. coli* cell, probably less than five each. When the bacterium encounters lactose, enzyme synthesis is rapidly induced (lactose is said to be the inducer) and within a few minutes, levels of up to 5000 molecules of each enzyme per cell are reached. Induction of the three enzymes is coordinate, meaning that each is induced at the same time and to the same extent. This provides a clue to the arrangement of the relevant genes on the *E. coli* DNA molecule.

THE LACTOSE UTILISATION GENES FORM AN OPERON (Fig. 14.2)

The three genes involved in lactose utilisation by *E. coli* are called *lacZ* (β-galactosidase), *lacY* (permease) and *lacA* (transacetylase). These genes lie in a cluster with only a very short distance between the end of one gene and the start of the next. In fact the three genes form an operon, each is transcribed onto the same mRNA molecule, delineated by a single promoter upstream of *lacZ* and a single terminator downstream of *lacA*.

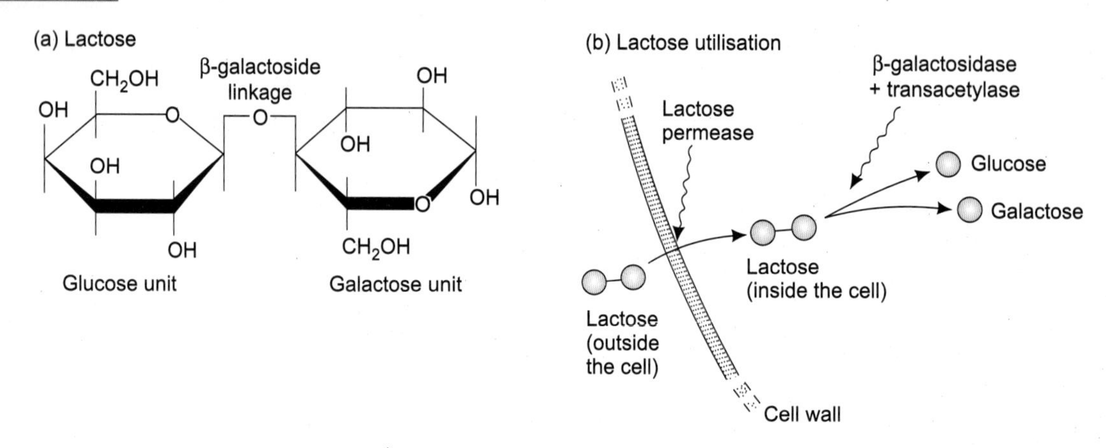

Fig. 14.1 Regulation of Lactose utilisation by E. *coli.*

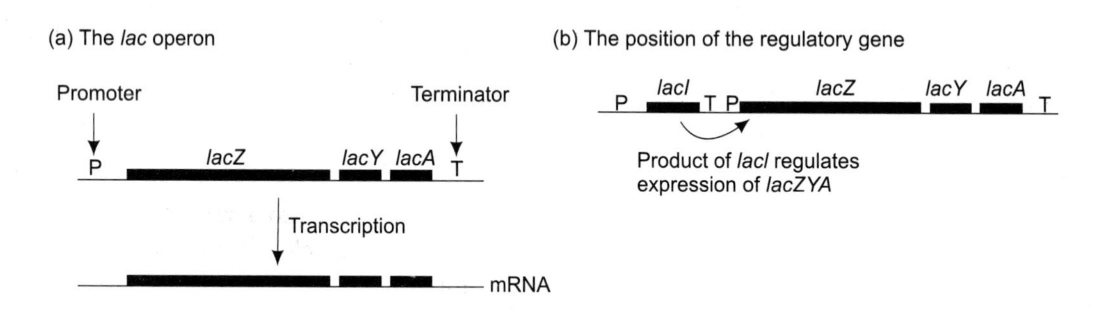

Fig. 14.2 The *lac* operon and its regulatory gene *lacI.*

Table 14.1 Regulatory genes and their products.

Gene	*Function*	*Gene product*
lacZ	LACtose utilisation	β-Galactosidase
trpA	TRyPtophan biosynthesis	Tryptophan synthase
rplA	Ribosomal Protein, Large subunit	Ribosomal protein L1
polA	DNA POLymerase	DNA polymerase 1
leuA	LEUcine biosynthesis	β-Isopropylmalate synthase
pyrG	PYRimidine biosynthesis	CTP synthetase

Jacob and Monod used genetic analysis techniques to identify *lacZ*, *lacY* and *lacA*, and to determine their relative positions on the *E. coli* chromosome. They also discovered an additional gene, which they designated *lacI.* This gene lies just upstream of the *lac* gene cluster but is not itself a part of the operon, because it is transcribed from its own promoter and has its own terminator. The gene product of *lacI* is intimately involved in lactose utilisation but is not an enzyme directly required for the uptake or hydrolysis of the sugar. Instead the *lacI* product regulates the expression of the other three genes; if *lacI* is inactivated by a mutation the *lac* operon becomes switched on continuously, even in the absence of lactose. The terminology used by Jacob and Monod is still important *lacZ*, *lac* Y and *lacA* are called structural genes because their products play enzymatic or structural roles in the cell; *lacI* is a regulatory gene, because its function is to control the expression of other genes; and the situation where *lacI* is inactivated and the *lac* operon is expressed continuously is called *constitutive expression.*

THE LACTOSE REPRESSOR (Fig. 14.3)

The gene product of *lacI* is a protein that Jacob and Monod called the *lactose repressor.* This protein is able to attach to the *E. coli* DNA molecule at a site between the promoter for the *lac* operon and the start of *lacZ*, the first gene in the cluster. This attachment site is called the operator, and was also located by Jacob and Monod by genetic means (and called *laco).* The operator, in fact, overlaps the promoter so that when the repressor is bound, access to the promoter is blocked so that RNA polymerase cannot attach to the DNA and transcription of the *lac* operon cannot occur. This is what happens if lactose is not available to the cell: if there is no lactose, transcription of the *lac* operon does not occur because the promoter is blocked by the repressor.

LACTOSE INDUCES TRANSCRIPTION (Fig. 14.4)

The *lac* repressor not only binds to the operator but also has a second binding site, at which it can attach a molecule of lactose. When lactose is present it binds to the repressor, causing

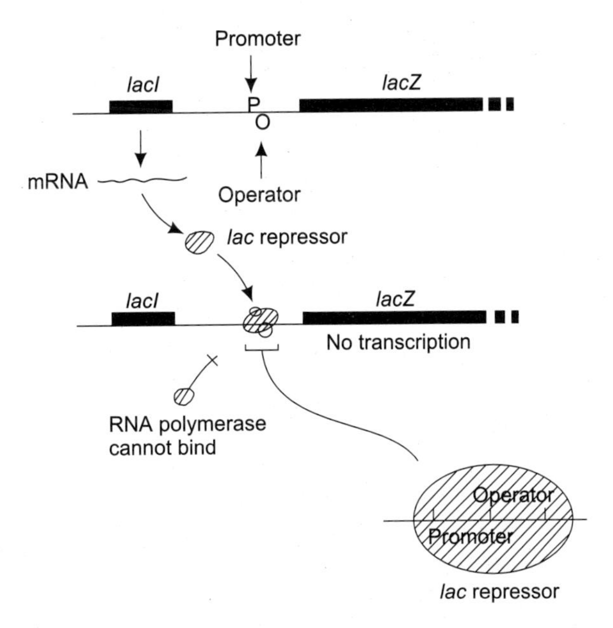

Fig. 14.3 In the absence of lactose, the *lac* repressor attaches to the operator site. RNA polymerase cannot bind to the promoter, so transcription does not occur.

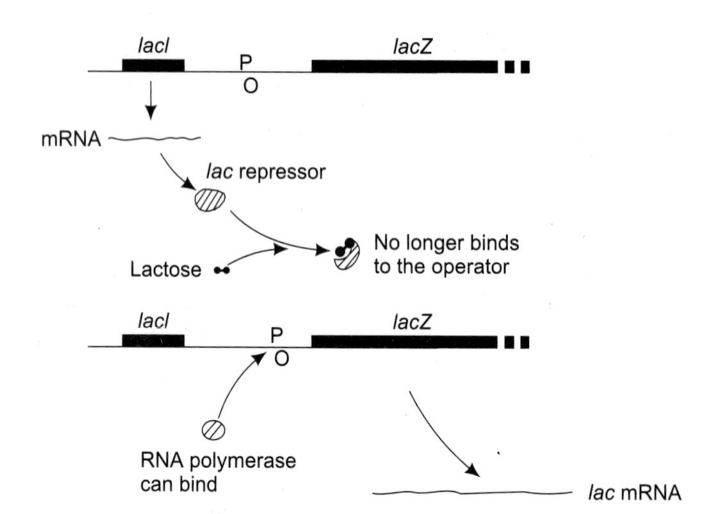

Fig. 14.4 In the p®resence of lactose, the *lac* repressor binds to a molecule of lactose. This causes a change in conformation of the repressor, so that it is no longer able to bind to the operator, and transcription occurs.

a change in the conformation of the protein in such a way that the repressor is no longer able to attach to the operator. The repressor-inducer complex dissociates from the DNA molecules and RNA polymerase can locate the promoter and begin transcription of the lactose operon. If lactose is present, transcription occurs as the repressor-inducer complex does not bind to the operator.

Eventually the lactose utilisation enzymes will exhaust the available supply of lactose. Repressor- inducer binding is an equilibrium event (Fig. 14.5), so when the free lactose concentration decreases, the number of repressor-inducer complexes will also decrease and free repressor molecules will start to predominate. These free repressors will have regained their original conformation and so can attach once again to the operator. When the lactose supply is used up, the lactose operon is switched off as the repressor reattaches to the operator.

REGULATION OF THE TRYPTOPHAN OPERON

Now a second type of regulatory control system, which is *repressible* rather than *inducible* will be considered. Tryptophan is synthesised in *E. coli* by a lengthy pathway, the last five steps of which are catalysed by a series of enzymes coded by five genes that form another operon (Fig. 14.6). Once again there is a repressor protein, the tryptophan repressor (a totally different protein from the *lac* repressor), which is able to bind to an operator adjacent to the promotor for the *trp* operon. Again, transcription is blocked when the repressor is bound. However, the *trp* operon has one important distinction from the lactose system. The *trp* repressor binds to the operator only in the presence of tryptophan. When tryptophan is required by the cell the *trp* operon must be switched on. Under these circumstances—that is, when the concentration of free tryptophan is low—the *trp* repressor does not bind to the operator. This is because the

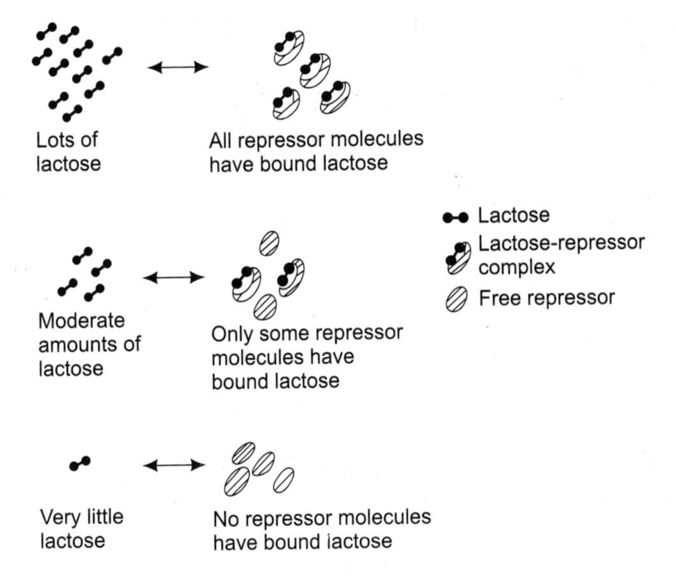

Fig. 14.5 Lactose operon repressor-Inducer binding is an equilibrium event.

repressor can take up the correct conformation for attaching to the operator only when it is also bound to tryptophan, which acts as the corepressor in this system (Fig. 14.6). In summary:

No tryptophan—repressor is not bound to the operator—operon is transcribed.

Tryptophan present-repressor-corepressor complex binds to the operator—operon is not transcribed.

The differences between an inducible and a repressible control system lies simply with whether the repressor on its own, or a repressor-corepressor complex, binds to the operator.

COMPLICATIONS WITH *lac* AND *trp* OPERONS

The systems that have been described are not over-simplified but they do ignore important aspects of the *lac* and *trp* operons. These concern the ways in which expression of the operons can be modulated by complementary through distinct control systems.

GLUCOSE REPRESSES *lac* OPERON

The presence or absence of lactose is not the only factor that influences expression of the *lac* operon. If the cell has a sufficient source of glucose (one of the breakdown products of lactose) for its energy needs it will not need to metabolise lactose even if lactose is also available in the environment. A system is, therefore, needed whereby glucose can override the lactose system and keep the operon switched off if need be.

This process is called *catabolite repression* and involves a second regulatory protein, the *catabolite activator protein* (CAP) and a second upstream binding locality, the CAP site. CAP

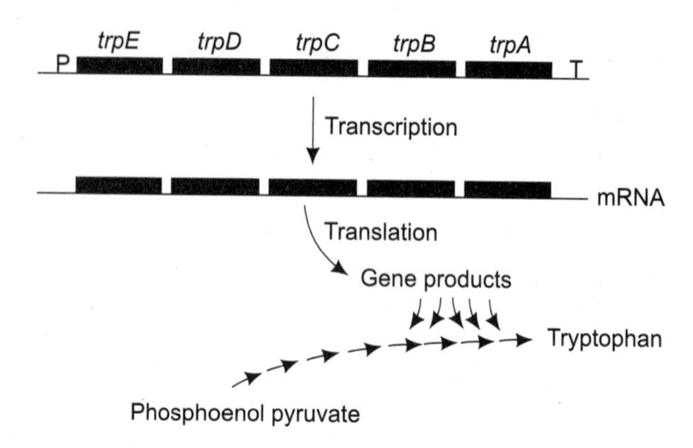

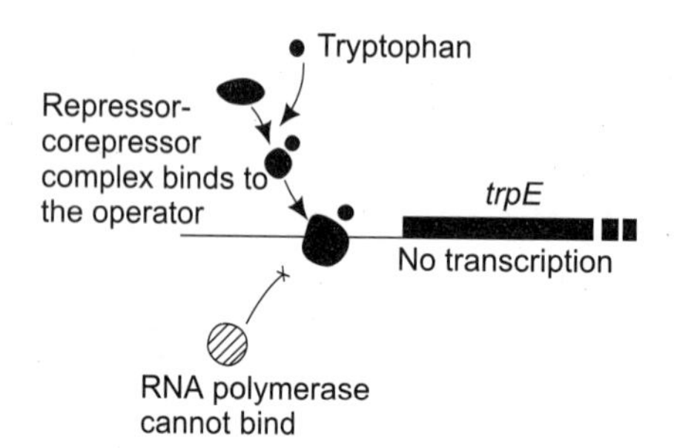

Fig. 14.6 Regulation of the *Tryptophan operon.* (a) The structure of the operon. (b) Tryptophan acts as the corepressor. When tryptophan is present, the repressor-corepressor complex binds to the operator and blocks transcription.

attaches to the CAP site only in the presence of cyclic AMP (cAMP), a modified nucleotide derived from ATP by a reaction catalysed by adenyl cyclase. The presence of the CAP-cAMP complex attached to the CAP site stimulates binding of RNA polymerase to the promotor and, therefore, stimulates transcription of the *lac* operon. In fact the *lac* operon cannot be transcribed efficiently unless CAP-cAMP is bound to the CAP site; if the CAP site is vacant,

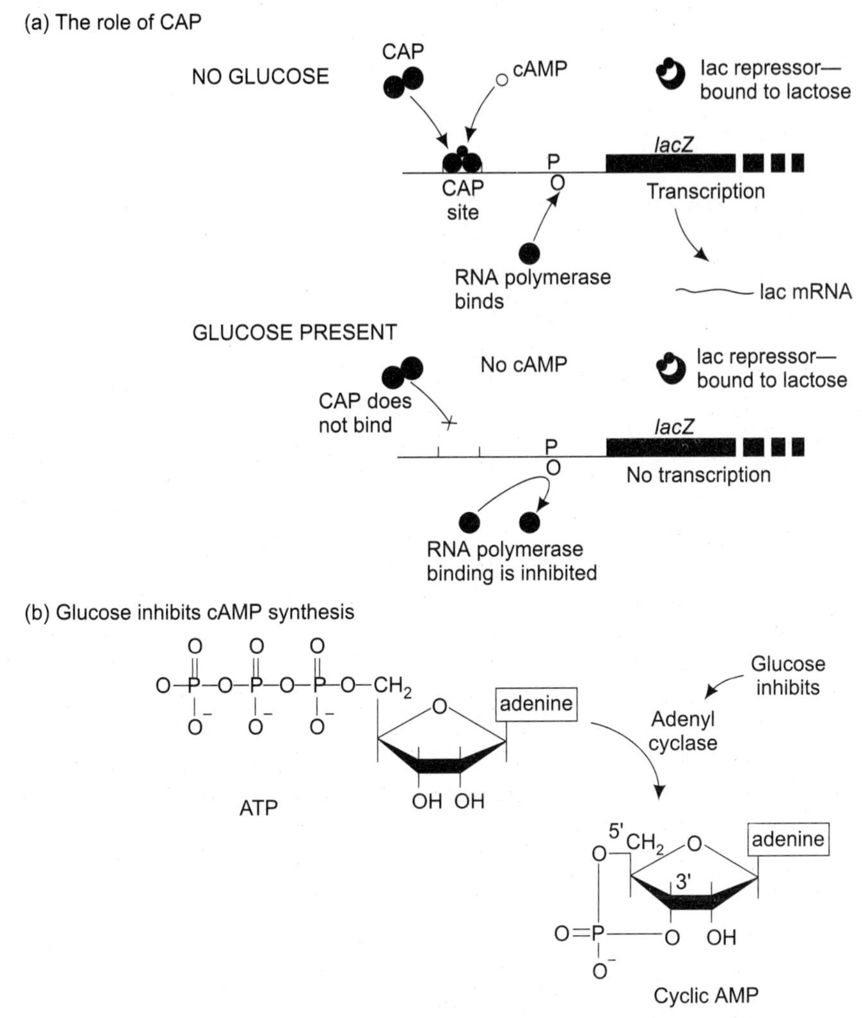

Fig. 14.7 Catabolite repression in *E. coli.* (a) In the presence of lactose, when the operator site is empty, the operon is transcribed efficiently only if the CAP site is filled by CAP-cAMP. (b) Glucose influences events by inhibiting synthesis of cAMP by adenyl cyclase.

the operon is transcribed at only a low rate, even when lactose is present and the *lac* repressor is not attached to the operator (Fig. 14.7). This is true , also of other sugar utilisation operons, as several different ones possess CAP-binding sites; the system is, therefore, of general importance and relevant not only to lactose utilisation.

Glucose influences these events by directly inhibiting the action of adenyl cyclase (Fig. 14.7). This is a straightforward inhibitory effect and does not itself involve transcriptional control. The system works as follows:

Glucose present–adenyl cyclase inhibited–cAMP levels low–CAP site vacant–*lac* operon is transcribed at only a low rate.

Glucose absent–adenyl cyclase active–cAMP levels high–CAP–cAMP attaches to the CAP site–*lac* operon is transcribed at a high rate.

Together with the *lac* repressor, CAP ensures that the extent of transcription of the *lac* operon is precisely in accordance with the requirements of the bacterium.

ATTENUATION OF *trp* OPERON (Fig. 14.8)

The final method for control of bacterial gene expression that we will look at is attenuation, which serves to finely tune the expression of several amino acid biosynthetic operons. Attenuation was first postulated quite recently and is still only partially understood, though one of the best-studied examples concerns the *trp* operon.

Attenuation depends on the fact that transcription and translation are linked in bacteria so that an mRNA molecule still being transcribed may already be undergoing translation by one or more ribosomes. The hypothesis is that the presence of a ribosome on an mRNA still being transcribed may influence transcription by preventing formation of a termination structure in the mRNA molecule. This stem-loop, if formed, would cause transcription to stop before the *trp* genes are copied into mRNA.

There are three components to an attenuation system (Fig. 14.8).

1. A short open reading frame (ORF, a short series of codons) immediately upstream of the first structural gene of the operon. This ORF codes for a peptide that contains several tryptophans, the amino acid for which the operon as a whole is responsible.
2. A stem-loop structure that can form in the transcript between the ORF and the first gene of the operon, and which acts as a terminator of transcription, causing premature termination so that the operon is not transcribed.
3. A larger stem-loop structure, upstream of the terminator. This large stem-loop does not itself stop transcription. The relative positions of the two stem-loops prevent both from forming at once; it must be either one or the other.

The central feature of the hypothesis of the role of the ribosome in influencing which of the two stem-loop structures is formed and hence whether premature transcription termination does or does not occur. In summary, the system with respect to the *trp* operon works as follows:

- Tryptophan is present in adequate amount—the ribosome rapidly translates the short ORF and moves into the stem-loop region–the ribosome prevents formation of the large, upstream stem-loop structure thereby allowing the terminator to form–transcription stops before the *trp* genes are copied into *mRNA-no tryptophan is synthesised.*
- Tryptophan is present in only low amounts–the ribosome stalls at the ORF as there is a lack of tryptophan to insert into the peptide. By being delayed, the ribosome cannot prevent formation of the large, upstream stem-loop structure (which is energetically more favourable and so forms instead of the terminator when the ribosome is absent–the terminator does not form–transcription continues–*tryptophan is synthesised.*

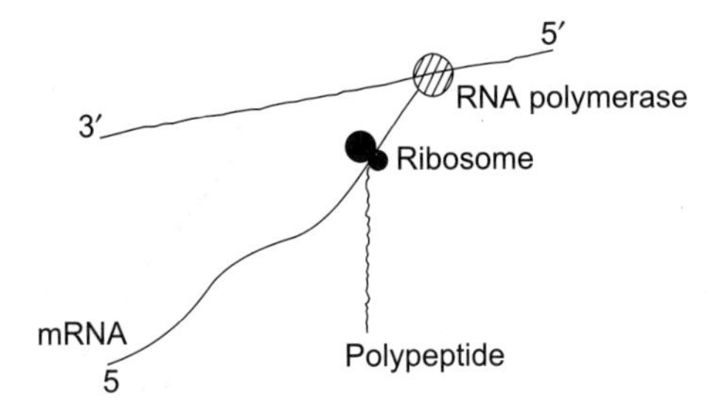

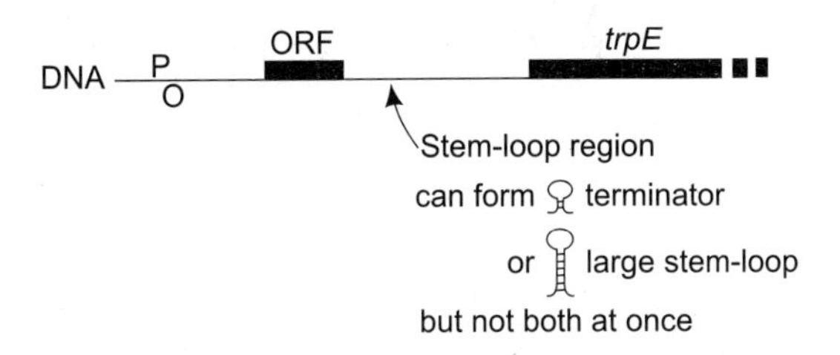

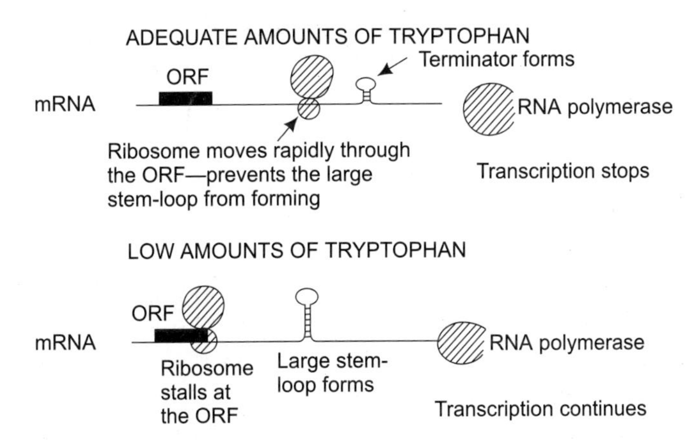

Fig. 14.8 Attenuation at the *trp* operon. See the text for details.

The important point is that the speed at which the ribosome translates the ORF will not be exactly the same for each transcript. When tryptophan is present in medium amounts (not abundant but not totally absent either), then some transcripts will terminate, but others will not. *Attenuation therefore modulates the synthesis of tryptophan.* On its own it probably cannot totally switch off a derepressed operon, but it can modify the amounts of the gene products that are synthesised and help maintain the tryptophan level in the cell at the optimum concentration.

CONTROL OF GENE EXPRESSION IN EUKARYOTES

In general terms, the more complex the organism, the more complicated are the systems that have evolved to control expression of its genes, and the less we understand about how those systems work. For simple eukaryotes, such as yeast, we are beginning to comprehend how the expression of individual genes is regulated, and the most recent research is directed at how gene expression in general is coordinated. In higher organisms, such as *Drosophila* and man, even the more straightforward answers are elusive.

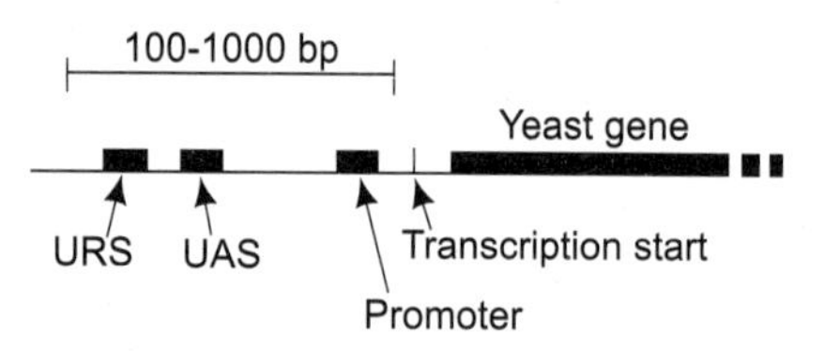

Fig. 14.9 Control elements for yeast genes. UAS—uPstream activating sequence. URS—uPstream repressing sequence.

CONTROL ELEMENTS FOR YEAST GENES

Saccharomyces cerevisiae genes do not occur in operons. Although related genes may be clustered, there are no known examplcs where yeast genes are contranscribed: each yeast gene must, therefore, have its own set of control elements. The most important of these are (Fig. 14.9).

1. The promoter, to which RNA polymerase binds. Molecular biologists working with yeast are moving towards a terminology, which includes all the control elements in the 'promoter'.
2. The upstream activating sequence (UAS), which has a positive efffect on transcription (in other words, switches the *gene on*).
3. The upstream repressing sequence (URS), which has a negative effect on transcription (it switches the *gene off*).

There are major differences between CAP sites in bacteria and yeast UASs, or between operators and URS sites. For one thing, UAS and URS sites can occur several hundred base pairs upstream of the gene they control, rather than in close relative positions as they do for the bacterial control sequences.

Not all yeast genes have UAS and URS sites and exactly how the sites control expression of their target genes is not clear. In some cases, a regulatory protein is known to bind to either a UAS or URS (or sometimes both?) causing a change in expression of the target gene. Models have been proposed for DNA binding, allowing the distant UAS/URS site to interact with the promoter, perhaps making the promoter more or less accessible for RNA polymerase binding. Presumably this interaction is mediated by the regulatory protein, but no clear descriptions have been produced and we can only guess at the precise nature of the system.

ENHANCERS FOR MAMMALIAN GENES

A striking, though probably misleading, similarity exists between yeast UAS sites and the mammalian control elements called *enhancers.* Enhancers also reside some distance away from their target genes and are able to modulate gene expression, probably through the binding of regulatory proteins. Enhancers differ from UASs, however, in that their positions relative to their target genes are highly variable: an enhancer may be several kilobases upstream of the gene it controls or may indeed be downstream of the target gene.

The relevance of enhancers to current ideas on the control of eukaryotic gene expression is that they may mediate tissue-specific gene expression. In multicellular organisms individual cells usually undergo specialisation, becoming differentiated into, for example, liver cells or brain cells. Each type of specialised cell requires a different set of proteins, so whole groups of genes must be switched on or switched off. This kind of change in gene expression is permanent: liver cells, for instance, cannot redifferentiate into brain cells. Enhancers have been implicated in the control of differential gene expression but exactly what their involvement is, has not yet been discovered.

CONTROL OF GENE EXPRESSION DURING DEVELOPMENT

Equally mysterious is how gene expression is regulated and coordinated during the development of an organism from fertilised egg cell to adult. The additional complication here is that the activities of a multitude of different cells must be coordinated, so we are not dealing with control of gene expression in a single cell, but with the coordinated regulation of gene activity in a number of different cells, perhaps in different parts of the organism.

Although this may appear an impossible process to decipher, considerable progress has been made, in particular with the fruit-fly *Drosophila melanogaster.* A major breakthrough has been the production by mutation of quite drastic changes in the body plan of the fly. An example is the mutation called *antennapaedia,* which causes the fly to grow legs instead of antennae (Fig. 14.10).

Genes that are altered by these developmental mutations have been studied and, as might be expected, are extremely complex both in terms of structure and expression. However, one intriguing feature of several of these genes has emerged: the *homoeobox,* a short segment of 180 bp coding for a 60 amino acid portion of the gene product, which is similar in at least ten of the relevant *Drosophila* genes. The fact that these genes have a structural relationship (that is, the biological information that each carries is similar in some respects) suggests that possibly the gene products are related in terms of function. But the biggest surprise to molecular biologists has been the discovery that homoeoboxes also exist in genes from frog, mouse and man. These vertebrate genes are totally different from the *Drosophila* genes except for the presence of the homoeobox and the fact that the vertebrate genes are also believed to be involved in development. Are the genetic instructions for developmental processes universal? Thomas Hunt Morgan and his colleagues showed that the study of a simple organism like *Drosophila* could provide information of general relevance to genetics. It is exciting to think that *Drosophila* could also provide the key to understanding the complex events that underlie development in man.

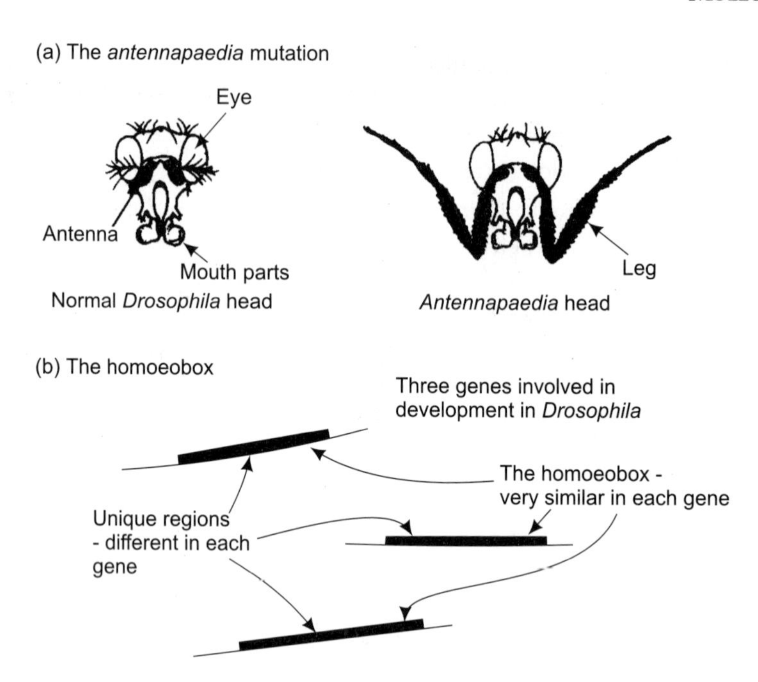

Fig. 14.10 Developmental genetics of *Drosophila*. (a) The *antennapaedia* mutation, (b) Several different genes involved in development contain homoeoboxes.

ROLE OF Z-DNA IN GENE REGULATION

The Watson-Crick B form of DNA is a *right handed* double-helix, whereas in the newly discovered Z-DNA, the DNA is *left handed* because of its zigzagged paths of the sugar-phosphate backbones of the molecule (Fig. 14.11). The Z-form of DNA occurs normally at high salt concentrations. There is evidence that Z-DNA exists in the interband regions of the salivary gland chromosomes of *Drosophila melanogaster* and in the transcriptionally active macromolecules of the ciliated Protozoan, *Stylonichia mytilus*. A. Rich and his coworkers have shown that the Z-DNA specific antibodies bind to the interband regions of the polytene chromosomes of *D. melanogaster*. It was supposed that Z-DNA participates in the regulating gene expression since the structures of certain regulatory proteins may bind in the major groove of left handed helices. Experiments have been designed to test the possibilities: (1) that B-form to Z-form transitions in DNA are involved in the regulation of gene expression and (2) that regulatory proteins may act by binding to and stabilising one or the other of these confirmations may lead to some novel developments in future.

HORMONAL CONTROL OF GENE EXPRESSION (Fig. 14.12)

Signals originating in various glands and/or secretory cells somehow stimulate *target cells* or *target tissues* to undergo dramatic changes in their metabolic patterns. Peptide hormones, such

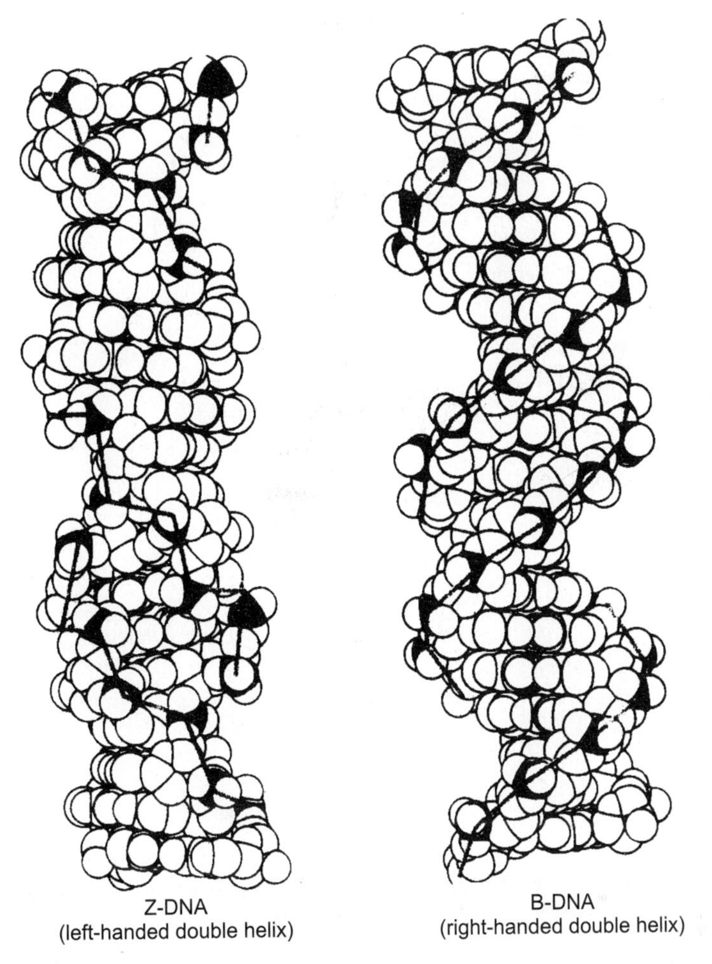

Fig. 14.11 Comparison of the structures of DNA in the well known B form (right) and the Z form (left). The heavy lines show the paths of the sugar-phosphate backbones in Z-DNA.

as insulin and steroid hormones, such as estrogen, and testosterone, represent two types of signal systems utilised in *intercellular communications.*

The steroid hormones are small molecules and readily enter cells through plasma membranes. The steroid hormones become tightly bound to specific receptor proteins. These receptor protein complexes rapidly accumulate in the nuclei of target cells. G. Tomkins and B.W. Omally have shown in mice and chickens that the hormone receptor protein complexes *activate* the transcription of *specific genes or sets of genes.* These hormone-receptor protein complexes would function as positive regulators (or 'activators') of transcription, much like that of CAP-cAMP complexes in prokaryotes.

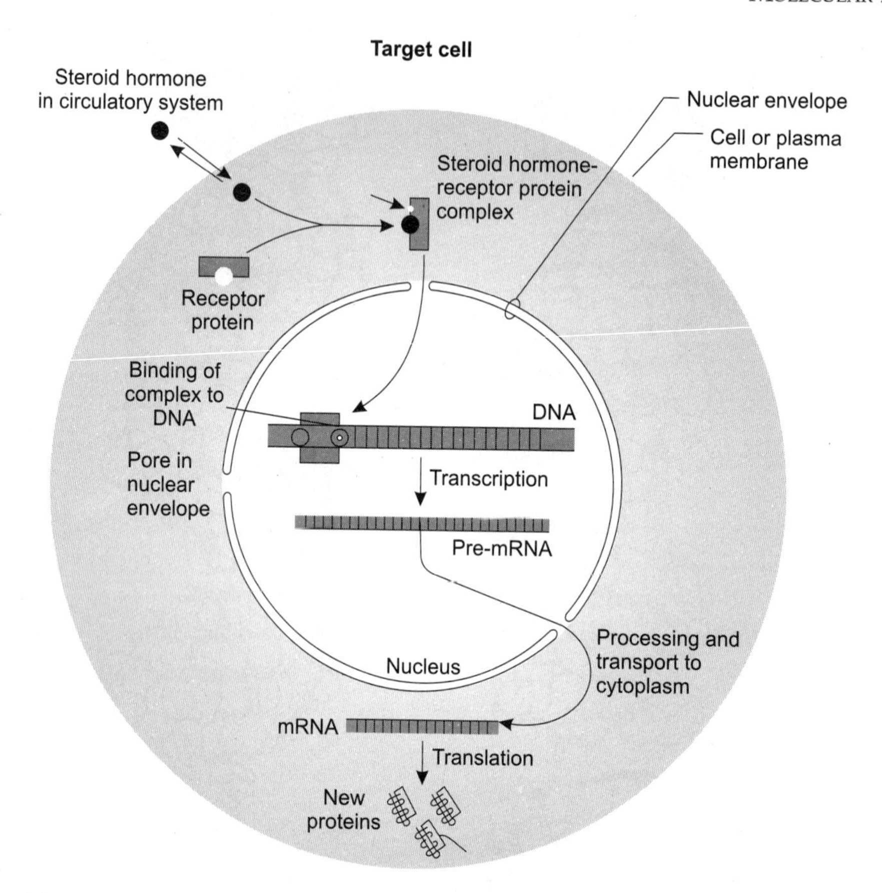

Fig. 14.12 Diagram illustrating the effects of steroid hormones on gene expression. Hormones are synthesised in specialised secretory cells and are distributed to the various tissues of the organism through the circulatory system. Their small size permits them readily to pass into cells through plasma membranes. ***The target cells*** (the cells responding to the presence of the specific steroid hormone) contain ***receptor proteins*** that specifically bind the hormone molecules. The steroid hormone-receptor protein complexes then pass through the pores in the nuclear envelope and accumulate in target cell nuclei. The hormone-receptor protein complexes then bind (probably as dimers) to a specific DNA sequence in an enhancer or promoter region of each gene that is activated by the hormone. Once bound, the hormone-receptor complex somehow stimulates transcription of the structural genes. Presumably, the bound complex enhances the binding of RNA polymerase to the promoter region; in any case, it functions as an activator of transcription of the regulated gene(s).

CAENORHABDITIS ELEGANS: A Model System for the Study of Development

Only *D. melanogaster* rivals *Caenorhabditis elegans* as a model system or the study of animal development. It is a small free living nematode worm, first chosen by S. Brenner in 1965. Its major advantages in the rapid progress in dissecting and understanding the morphogenesis lies in the following points: (1) Small size (1 mm in length), (2) short generation time (3-day life cycle under optimal conditions), (3) the existence of hermaphrodites and males, allows geneticists to produce worms homozygous or newly induced mutations very easily by self-fertilisation, like Pea plants, (4) large numbers of progeny (300 eggs during its life), (5) small genome (about 8×10^7 nucleotide pairs) greatly facilitates molecular analyses such as cloning, sequencing, and physical mapping of the genome, (6) no skeleton, transparent bodies—all cells of body can be observed under microscope, (6) precise and invariant developmental program—each adult hermaphrodite is composed of exactly 945 cells and these are produced from the zygote by precise pathways of cell division, cell migration, cell growth, and cell death; these are called *invariant cell lineages.*

CATABOLITE REPRESSION (Fig. 14.13)

An interesting property of the *lac* operon and other operons that catabolise certain sugars–for example, *arabinose* and *galactose*–is that they are all repressed by the presence of glucose in the medium. That is, glucose is catabolised in preference to other sugars; the mechanism (catabolite repression) involves cyclic AMP. In eukaryotes cyclic AMP acts as a "second messenger," an intracellular messenger regulated by certain extracellular hormones. Geneticists were surprised to discover cyclic AMP in *E. coli,* where it works in conjunction with another regulatory protein, the catabolite activator protein (CAP). First, cyclic AMP combines with CAP then, the CAP-cyclicAMP complex binds to a distal part of the promoter of the operon and thereby apparently enhances the affinity of RNA polymerase for the promoter. Without the binding of the CAP-cyclic-AMP complex to the promoter, the transcription rate is very low. The addition of glucose to *E. coli* cells depresses the quantity of cyclic AMP by some unknown mechanism and thus lowers the CAP-cyclic-AMP level. The transcription rate of CAP-cyclic-AMP-dependent operons will, therefore, be reduced. The same reduction of transcription rate is noticed in mutant strains of *E. coli* when this part of the distal end of the promoter is deleted.

his OPERON (REPRESSIBLE SYSTEM) (Fig. 14.14)

The inducible operons are induced when the metabolite that is to be catabolised enters the cell. Anabolic operons function in a reverse manner: They are "turned off" (repressed) when their end product accumulates in excess of the needs of the cell. Transcription of repressible operons appears to be controlled by two entirely different mechanisms. The first mechanism follows the basic scheme of inducible operons. The second mechanism involves secondary structure in mRNA, which is controlled by translation of an attenuator region of the operon.

HISTIDINE SYNTHESIS

One of the best studied repressible operons is the *histidine,* or *his,* operon in *Salmonella typhimurium,* a close relative of *E. coli.* The *his* operon in *Salmonella* contains the 9 genes necessary for the enzymes that transform phosphoribosyl pyrophosphate into histidine.

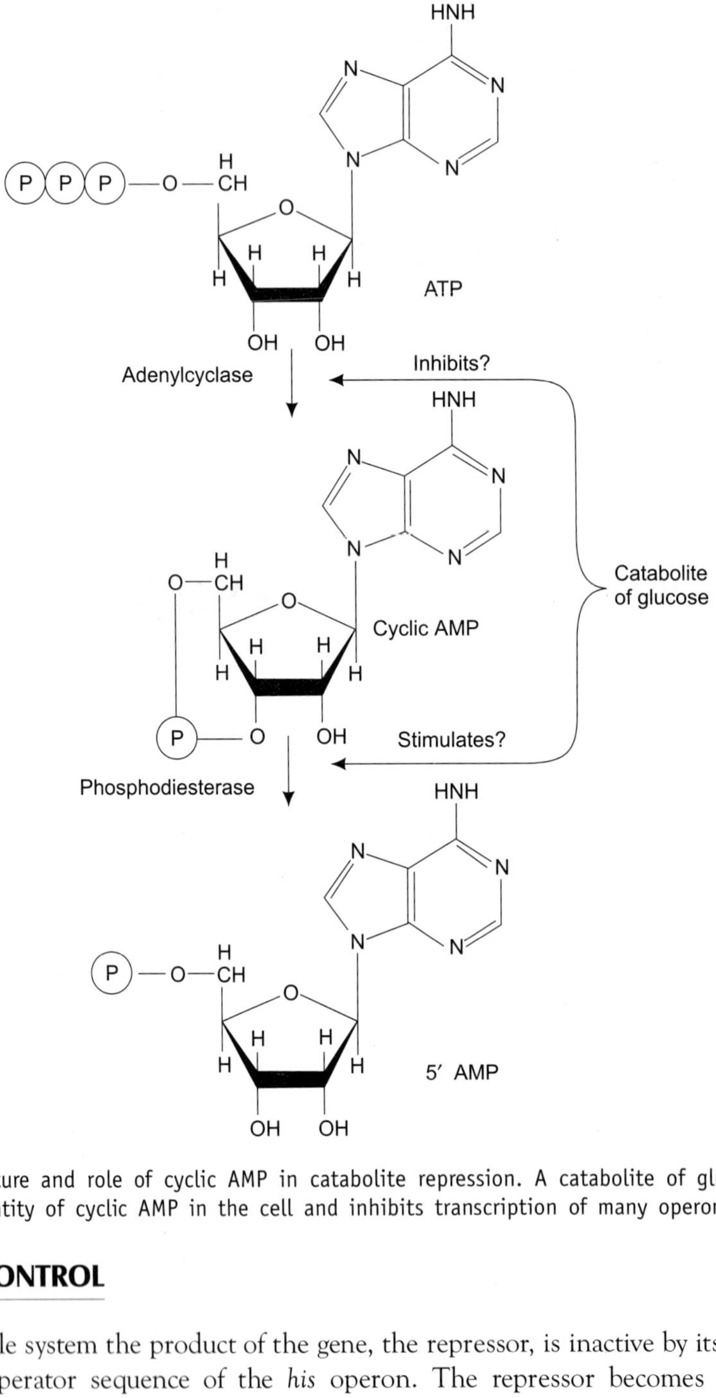

Fig. 14.13 Structure and role of cyclic AMP in catabolite repression. A catabolite of glucose lowers the quantity of cyclic AMP in the cell and inhibits transcription of many operons.

OPERATOR CONTROL

In this repressible system the product of the gene, the repressor, is inactive by itself. It does not recognise the operator sequence of the *his* operon. The repressor becomes active when it combines with the metabolite. The actual metabolite is not histidine itself but histidine bonded to its tRNA (histidinyl-tRNA). Thus, when there is an excess of histidinyl-tRNA, enough will be available to bind with and activate the repressor. The metabolite is referred to as the corepressor. The corepressor-repressor complex now recognises the operator, binds to it, and prevents transcription by RNA polymerase. After the available histidine in the cell is used up, there will be a paucity of histidinyl- tRNAs and eventually the last one will detach from the repressor. The repressor will then diffuse from the *his* operator. The transcription process will no longer be blocked and can proceed normally (it is now derepressed). Transcription will continue until enough of the various enzymes have been translated to produce a sufficient quantity of histidine. Then there will again be an excess of histidinyl-tRNA. Some will be available to bind to the repressor and make a functional complex. Thus the system will be again shut off and the process will repeat itself, assuring that histidine is being synthesised when it is needed.

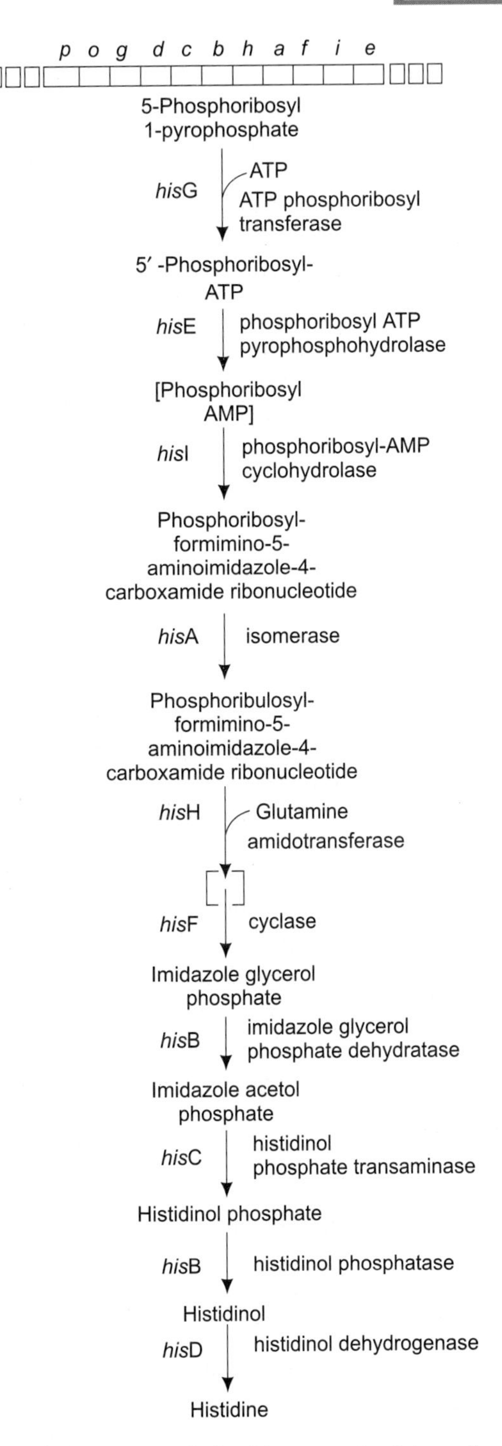

Fig. 14.14 Genes of the *his* operon in *Salmonella*.

trp OPERON (REPRESSIBLE SYSTEM) (Fig. 14.15)

Attenuator-Controlled Operons

Only recently have details of the second control mechanism of repressible operons been elucidated primarily by Yanofsky and his colleagues, who worked with the operon that transcribes the genes involved in the synthesis of tryptophan (*trp* operon) in *E. coli.* This type of operon control–that is, control by an attenuator region–has also been demonstrated for the leucine and histidine operons in *Salmonella.* These regulatory mechanisms may be the same for all operons involved in the synthesis of an amino acid.

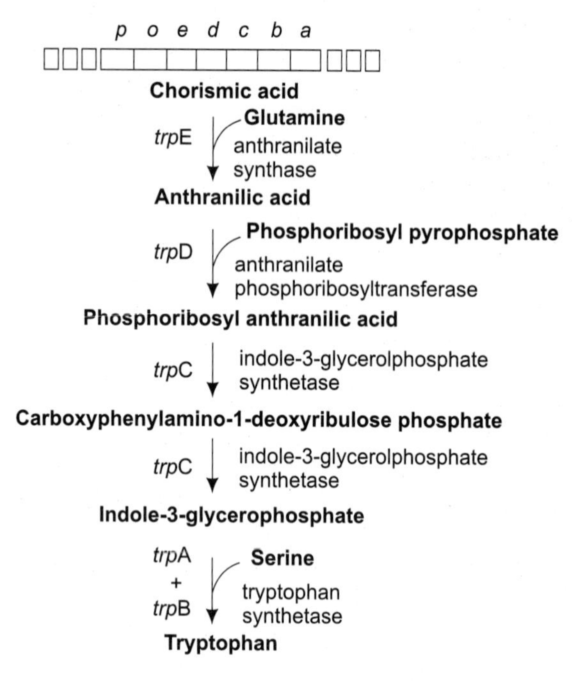

Fig. 14.15 Genes of the Tryptophan Operon in *E. coli.*

Leader Transcript

The *trp* operon in *E. coli* contains the 5 genes necessary for synthesis of the enzymes that transform chromic acid into tryptophan. In addition to the promoter and operator, there is an attenuator region in the *trp* operon between the operator and the first structural gene. The mRNA transcribed by the attenuator, region termed the leader transcript, has been sequenced, with two surprising and interesting facts emerging. First, the four sub-regions in (Fig. 14.17) are defined by the fact that they have base sequenced that are complementary to each other such that three different stem-loop structures can form in the RNA. Depending on circumstances, regions 1-2 and 3-4 can form two stem-loop structures or region 2-3 can form a stem and loop; the formation of other stem-loop structures is thus pre-empted and that the particular combination of stem-loop structures determines whether or not transcription will continue.

LEADER PEPTIDE GENE (Figs. 14.16 and 14.17)

The second interesting fact obtained by sequencing the leader transcript is that there is information for a small peptide from bases 27 to 68. The gene for this peptide is referred to as the leader peptide gene. It codes for 14 amino acids, of which two adjacent ones are tryptophan. These tryptophan codons are critically important in attenuator regulation. The presumed mechanism is as follows: *Excess Tryptophan*: Assuming that the operator site is not blocked, transcription of the leader RNA will begin. As soon as the 5' end of the leader peptide

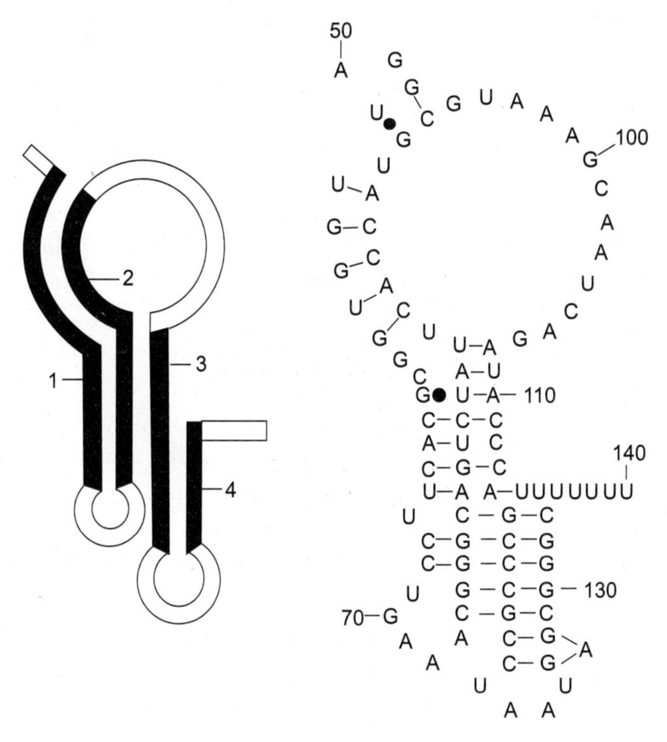

Fig. 14.16 Nucleotide sequence of part of the leader transcript of the *trp* attenuator region which includes bases 50 to 140. The proposed secondary structure (stem and loops 1-2, 2-3, 3-4) is shown.

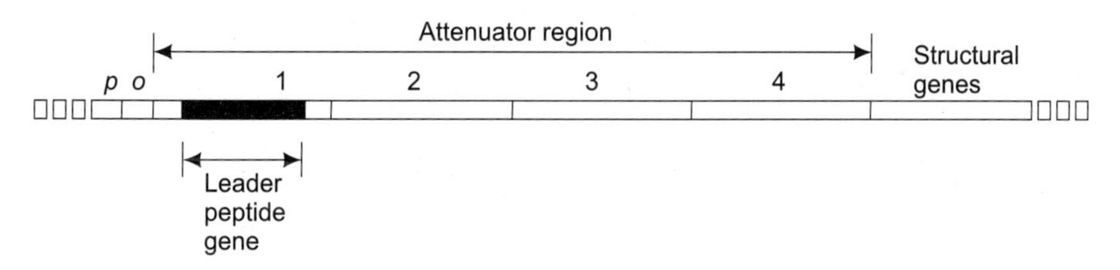

Fig. 14.17 Attenuator region of the *trp* Operon. This region is transcribed into a leader transcript, which contains a leader peptide gene.

gene has been transcribed, a ribosome will attach and begin the process of translation of this gene. Depending on the levels of amino acids in the cell, three different outcomes of this translation process can take place. If the concentration of tryptophan in the cell is such that tryptophanyl-tRNAs exist, translation will proceed down the leader peptide gene. The moving ribosome will overlap regions 1 and 2 of the transcript and allow the 3-4 stem endloop to form as shown in the configuration at the far left (Fig. 14.16). This loop will cause transcription to be terminated. (In other operons this stem-loop structure is variously referred to as the terminator or attenuator stem.) Hence, the existing quantities of tryptophan are adequate for translation of the leader peptide, and transcription is terminated.

DEVELOPMENTAL GENETICS

The earliest events in animal development are controlled by maternally synthesised factors. However, at some point, the genes in the embryo are selectively activated, and new materials are made. This process is referred to as zygotic gene expression, because it occurs after the egg has been fertilised. The initial wave of zygotic gene expression is a response to maternally synthesised factors. In *Drosophila*, for example, the maternally supplied *dorsal* transcription factor activates the zygotic genes *twist* and *snail*. As development proceeds, the activation of other zygotic genes leads to complex cascades of gene expression. We shall now examine how these zygotic genes carry the process of development forward.

Body Segmentation

In many invertebrates the body consists of an array of adjoining units called **segments.** An adult *Drosophila,* for example, has a head, three distinct thoracic segments, and eight abdominal segments. Within the thorax and abdomen, each segment can be identified by colouration, bristle pattern, and the kinds of appendages attached to it. These segments can also be identified in the embryo and the larva (Fig. 14.18). In vertebrates, a segmental pattern is not so evident in the adult, but it can be recognised in the embryo from the way that nerve fibers grow from the central nervous system, from the formation of branchial arches in the head, and from the organisation of muscle masses along the anterior-posterior axis. Later in development, these features are modified, and the original segmental pattern becomes obscured. Nonetheless, in body vertebrates and many invertebrates, segmentation is a key aspect of the overall body plan.

Homeotic Genes Interest in the genetic control of segmentation began with the discovery of mutations that transform one segment into another. The first such mutation was found in Drosophila in 1915 by Calvin Bridges. He named it bithorax (*bx*) because it affected two thoracic segments. In this mutant, the third thoracic segment was transformed, albeit weakly, into the second, creating a fly with a small pair of rudimentary wings in place of the halteres (Fig. 14.19). Later, other segment-transforming mutations were found in *Drosophila*, for example, *Antennapedia* (*Antp*), a mutant that partially transforms the antennae on the head into legs, which characteristically grow from the thorax. These mutations have come to be called **homeotic mutations** because they cause one body part to look like another. The word homeotic comes from William Bateson, who coined the term **homeosis** to refer to cases in which "something has been changed into the likeness of something else." Like so many other words Bateson, coined, this one has become a standard term in the modern genetics vocabulary.

The bithorax and Antennapedia phenotypes result from mutations in homeotic genes. Several such genes have now been identified in *Drosophila*, where they form two large clusters on one of the autosomes (Fig. 14.20). The **Bithorax Complex**, usually denoted **BX-C,** consists of three genes, *Ultrabithorax* (*Ubx*), *abdominal-A* (*abd-A*), and *abdominal-B* (*Abd-B*); the **Antennapedia Complex,** denoted **ANT-C,** consists of five genes, *labial* (*lab*), *proboscipedia* (*pb*), *Deformed* (*Dfd*), *Sex combs reduced* (*Scr*), and *Antennapedia* (*Antp*). Molecular analysis of these

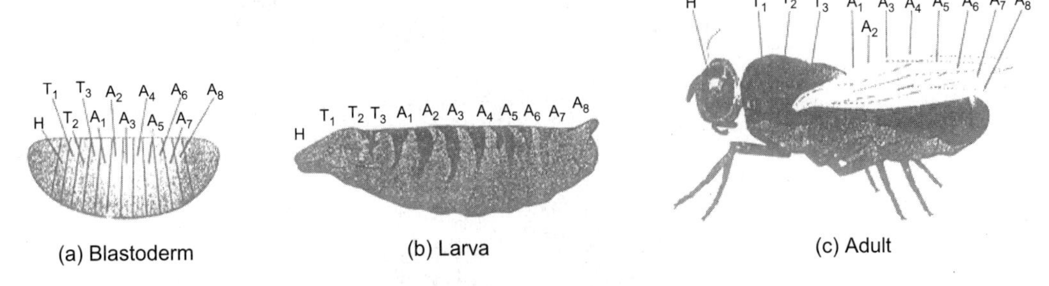

Fig. 14.18 Segmentation in *Drosophila* at the (a) blastoderm, (b) larval, and (c) adult stages of development. Although segments are not visible in the blastoderm, its cells are already committed to form segments as shown; H, head segment; T, thoracic segment; A, abdominal segment.

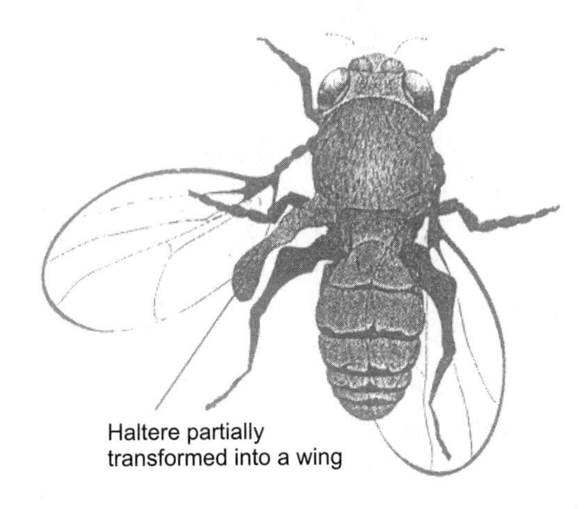

Fig. 14.19 The phenotype of a *bithorax* mutation in *Drosophila.*

genes has demonstrated that they all encode helix-turn-helix transcription factors with a conserved region of 60 amino acids. This region, called the homeodomain, is involved in DNA binding.

The BX-C was the first of the two homeotic gene complexed to be dissected genetically. Analysis of this complex began in the late 1940s with the work of Edward Lewis. By studying mutations in the BX-C, Lewis showed that the wild-type function of each part of the complex is restricted to a specific region in the developing animal. Molecular analyses later reinforced and refined this conclusion. Study of the ANT-C began in the 1970s, principally through the work of Thomas Kaufman, Matthew Scott, and their collaborators. Through a combination of genetic and molecular analyses, these investigators showed that the genes of the ANT-C are also expressed in a regionally specific fashion. However, the ANT-C genes are expressed more anteriorly than the BX-C genes. Curiously, the pattern of expression of the ANT-C and BX-C genes along the anterior-posterior axis corresponds exactly to the order of the genes along the chromosome; it is not yet clear why this is so. The analysis of genetic mosaics has shown

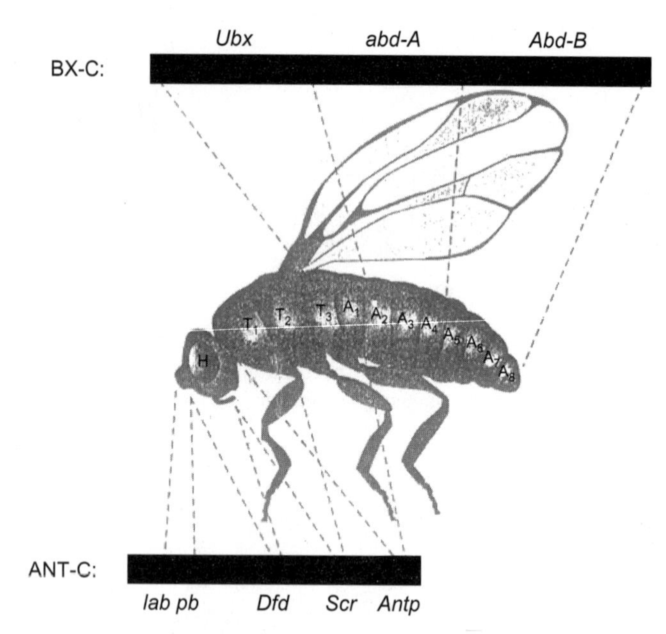

Fig. 14.20 The homeotic genes in the Bithorax Complex (BX-C) and Antennapedia Complex (ANT-C) of *Drosophila.* The body regions in which each gene is expressed are indicated.

that these homeotic genes control the identities of individual segments in a cell-autonomous way. In fact, the developmental pathway that each cell takes seems to depend simply on the set of homeotic genes that are expressed within it. Because the homeotic genes play such a key role in selecting the segmental identities of individual cells, they are often called **selector genes.**

The proteins encoded by the homeotic genes are homeodomain transcription factor. These proteins bind to regulatory sequences in the DNA, including some within the Bithorax and Antennapedia Complexes themselves. For example, the UBX and ANTP proteins bind to a sequence within the promoter of the *Ubx* gene–a suggestion that the homeotic genes can regulate themselves and each other. Other gene targets of the homeodomain transcription factors have been identified, including some that encode other types of transcription factors. The homeotic genes, therefore, seem to control a regulatory cascade of target genes, which in turn act to determine the segmental identities of individual cells. However, the homeotic genes do not stand at the top of this regulatory cascade. Their activities are controlled by another group of genes expressed earlier in development–the segmentation genes.

Segmentation Genes Most of the homeotic genes were identified by mutations that alter the phenotype of the adult fly. However, these same mutations also have phenotypic effects in the embryonic and larval stages. This finding suggested that other genes involved in segmentation might be discovered by screening for mutations that cause embryonic and larval defects. In the 1970s and 1980s, Christiane Nüsslein-Volhard and Eric Wieschaus carried out such screens. They found a whole new set of genes required for segmentation along the anterior-posterior axis. Nüsslein-Volhard and Wieschaus classified these **segmentation genes** into three groups based on embryonic mutant phenotypes.

1. ***Gap Genes*** These genes define segmental regions in the embryo. Mutations in the gap genes cause an entire set of contiguous body segments to be missing; that is, they create an anatomical gap along the anterior-posterior axis. Four gap genes have been well characterised: *Krüppel* (from the German for "cripple"), *giant, hunchback,* and *knirps* (from the German for "dwarf"). Each is expressed in characteristic regions in the early embryo under the control of the maternal effect genes *bicoid* and *nanos*. The gap genes encode transcription factors.
2. ***Pair-rule Genes*** These genes define a pattern of segments within the embryo. The pair-rule genes are regulated by the gap genes and are expressed in seven alternating bands, or stripes, along the anterior-posterior axis, in effect dividing the embryo into 14 distinct zones, or **parasegments**. Mutations in each of the several pair-rule genes produce embryos with only half as many parasegmets as wild-type. In each mutant, every other parasegment is missing, although the missing parasegments are not the same in different pair-rule mutants. Examples of pair-rule genes are *fushi tarazu* (from the Japanese for "something missing") and *even-skipped*. In *fushi tarazu* mutants, each of the odd-numbered parasegments is missing; in *even-skipped* mutants, each of the even-numbered parasegments is missing. Most of the pair-rule genes encode transcription factors.
3. ***Segment-Polarity Genes*** These genes define the anterior and posterior compartments of individual segments along the anterior-posterior axis. Mutations in segment-polarity genes cause part of each segment to be replaced by a mirror-image copy of an adjoining half-segment. For example, mutations in the segment-polarity gene *gooseberry* cause the posterior half of each segment to be replaced by a mirror-image copy of the adjacent anterior half-segment. Many of the segment-polarity genes are expressed in 14 narrow bands along the anterior-posterior axis. Thus, they refine the segmental pattern established by the pair-rule genes. Two of the best-studied segment-polarity genes are *engrailed* and *wingless*; *engrailed* encodes a transcription factor, and *wingless* encodes a signaling molecule.

These three groups of genes form a regulatory hierarchy (Fig. 14.21). The gap genes, which are regionally activated by the maternal-effect genes, regulate the expression of the pair-rule genes, which in turn, regulate the expression of the segment-polarity genes. Concurrent with this process, the homeotic genes are activated under the control of the gap and pair-rule genes to give unique identities to the segments that form along the anterior-posterior axis. Interactions among the products of all these genes then refine and stabilize the segmental boundaries. In this way, the *Drosophila* embryo is progressively subdivided into smaller and smaller developmental units.

Development in *Caenorhabditis elegans*

The orderly sequence of development has been worked out in many organisms by observing the movement pattern of cells in developing embryos. Often, simple observation is augmented by placing dyes that do not damage cells (vital dyes) on parts of the developing embryo and watching where these dyed regions or cells go. From these observations, it has been possible to

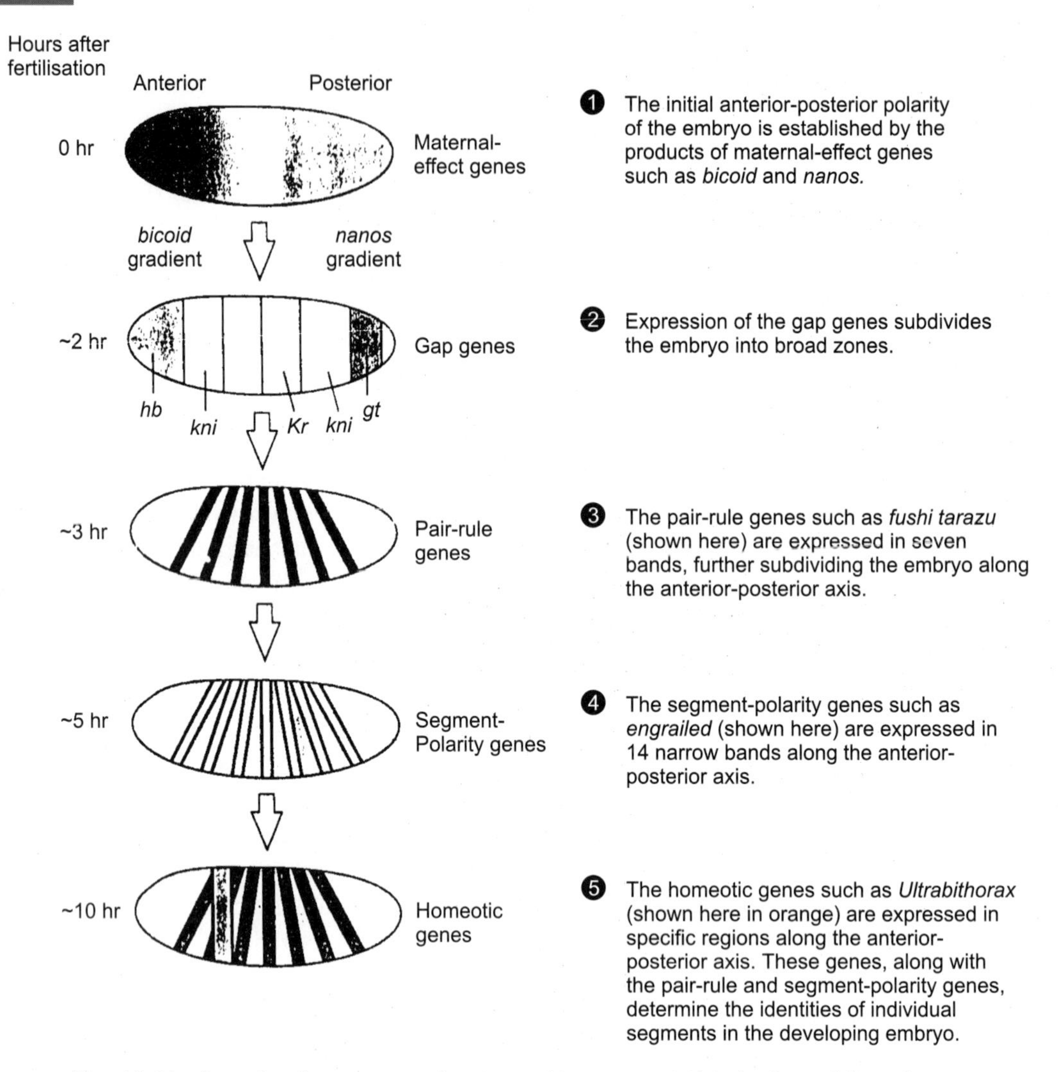

Fig. 14.21 Cascade of gene expression to produce segmentation in *Drosophila* embryos.

construct **fate maps,** which indicate the ultimate fate of various regions of a zygote of early embryo. The construction of fate maps is one part of the description of the gross aspects of development at the cellular level.

Historically, amphibians were the focus of developmental research because they have large eggs that are easily observed and manipulated. Then attention focused on *Drosophila* because, in addition to ease of manipulation, there is a tremendous amount of genetic information available on the fruit-fly. However, recently, the nematode, *Caenorhabditis elegans*, has emerged as another model organism for developmental studies because of its simplicity. The species consists of males and self-fertilising hermaphrodites (Fig. 14.22). Hermaphrodites have two sex chromosomes (XX), whereas males have only one (X0). Each individual consists of only about

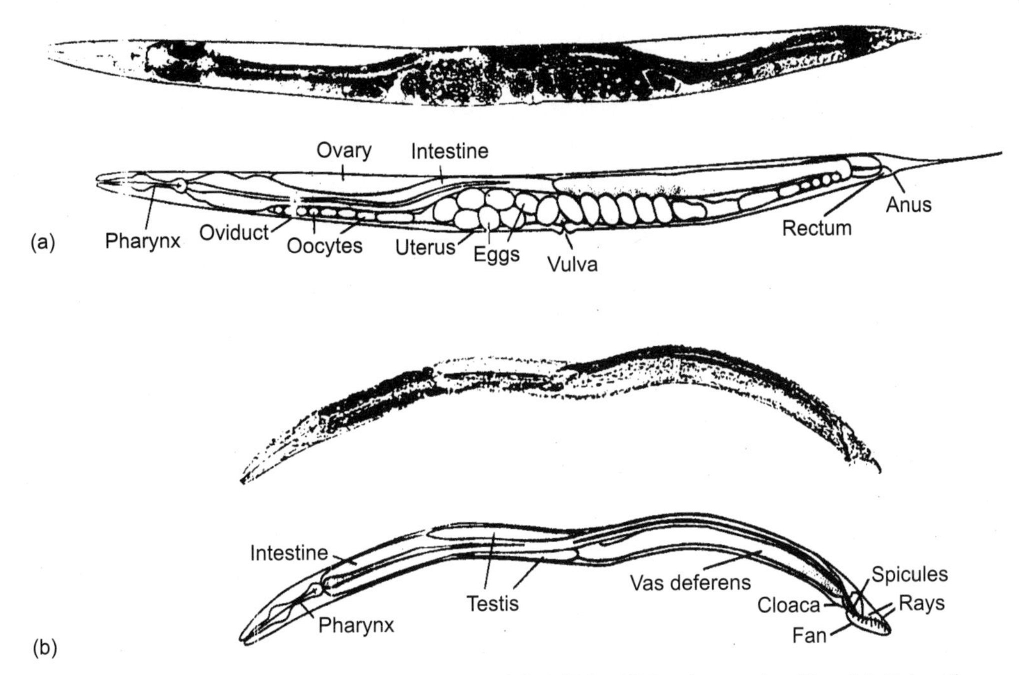

Fig. 14.22 The roundworm *Caenorhabditis* elegans. (a) Self-fertilising hermaphrodite. (b) Male. The worms are about 0.3 mm long.

one thousand cells; its life cycle lasts only 3.5 days, and with only 8×10^7 bp of DNA, it has the smallest genome of any multicellular organism. In 1963, S. Brenner proposed learning the lineage of every cell in the adult. With the efforts of numerous colleagues, that work has now been completed. From the fertilised egg to the adult, the division and fate of every cell of this nematode worm is known.

Cells were followed throughout development; as cell divisions took place and cell movement were noted. Starting with a few original cells, denoted by uppercase letters, observers kept track of daughter cells by appending a letter indicating relative position after each division. The designations were: a, anterior; p, posterior; l, left; r, right; d, dorsal; and v, ventral. Thus, in a newly hatched larva, one particular pharyngeal motor neuron cell is AB araapapaav. The process was found to be relatively invariant from one individual to the next.

Developmental steps

The first stage in our understanding of the genetics of development is now known for this species. How much has this description of development helped us understand the mechanism of developmental process? Several interesting concepts have emerged.

First has been the discovery that the fate of early embryonic layers is not absolute. In higher eukaryotes, early development gives rise to three embryonic tissue layers; the mesoderm, endoderm and ectoderm. These embryonic layers then give rise to all the adult tissues in a pattern that was believed to be invariant. All of the nervous system, for example, was believed to arise only from the ectoderm. In C. *elegans,* however, some neurons can arise from mesodermal cells. A progenitor cell can divide, producing one daughter that will become a neuron (ectoderm) and one that will become a muscle cell (mesoderm). This indicates that there is more flexibility in development than was previously recognised.

Second, cell death is often programmed into the developmental process. During development, one in six of all cells produced dies. These deaths are sex-specific and consistent. For example, there are cases in which a particular division leads to the death of a cell in one sex but not in the other.

Third, symmetry is not an accidental process. We might think that bilateral symmetry of an organism results from a mirror-image process of sister cells. In other words, we might predict that symmetry would be a natural outcome of cell divisions by daughter cells on opposite sides of the developing organism. This is not always the case. Sometimes, symmetry is the result of movement and division of cells that are not symmetrically related. Thus, bilateral symmetry is sometimes actively created.

Aberrant development

The next stage in understanding the development process is to observe aberrant development—development that is induced either by physical manipulation of the developing organism or by developmental mutants. In some of the experiments on developing C. *elegans,* specific cells were destroyed (ablated) by a laser microbeam. The conclusion from these studies is that cells are relatively fixed in their developmental pathways: If a precursor cell was ablated, its derivative cell types were missing in the adult. For example, when muscle cell precursors were ablated, the muscle cells that would have come from the destroyed cells never appeared; other cells did not adjust their developmental pathways to "fill in" for the missing tissue. The neighbouring cells seemed to be programmed for a specific developmental fate, rather than for responsiveness to the cellular milieu.

There were, however, several exceptions to this rule, such that an ablated cell was replaced by a cell that would normally have had a different fate. When AB.plapa was ablated, AB.plapaapa could not form. However, another cell, AB.prapaapa, took its place. Our conclusions from these experiments are that, for the most part, cells follow a relatively fixed path in development (the pattern is *hard-wired*), although some cells can respond to a changing environment by altering their developmental fates. The cells or higher eukaryotes may have more flexibility in determining their fates.

Homeotic Mutants

The study of mutants has also been valuable in understanding the control of developmental pathways. Over five hundred genes have been identified in C. *elegans* from studies of thousands of mutations. One group of mutants that is extremely interesting is that of **homeotic mutants,** in which one cell type follows the developmental pathway normally followed by other cell types. These mutants are important because they seem to indicate the presence of *master-switch* genes, genes whose function it is to regulate the development of whole groups of other genes. These genes presumably act as binary switches: They define whether development goes in one direction or another.

SYNOPSIS

Our understanding of the relationship between genes and proteins came as the result of a number of key observations. The first important observation was made by Garrod, who concluded that persons suffering from inherited metabolic diseases were missing specific enzymes. Later, Beadle and Tatum were able to induce mutations in the genes of *Neurospora* and identify the specific metabolic reactions taht were affected. These studies led to the concept of "one gene-one enzyme" and subsequently to the more refined version of "one gene-one polypeptide chain." The molecular consequence of a mutation was first described by Ingram, who showed that the inherited disease of sickel cell anaemia results from a single amino acid substitution in a globin chain.

The first step in gene expression is the transcription of a DNA template strand by an RNA polymerase. Polymerase molecules are directed to the proper site on the DNA by binding to a promoter region, which in almost every case lies just upstream from the site at which transcription is initiated. The polymerase moves in the 3′ to 5′ direction along the template DNA strand, assembling a complementary, antiparallel strand of RNA that grows from its 5′ terminus in a 3′ direction. At each step along the way, the enzyme catalyzes a reaction in which ribonucleoside triphosphates (NTPs) are hydrolyzed into nucleoside monophosphates as they are incorporated. The reaction is further driven by hydrolysis of the released pyrophosphate. Prokaryotes possess a single type of RNA polymerase that can associate with a variety of different sigma factors, which determine which genes are transcribed. The site at which transcription is initiated is determined by a nucleotide sequence located approximately 10 bases upstream from the initiation site.

Eukaryotic cells have three distinct RNA polymerase (I, II, and III), each responsible for the synthesis of a different group of RNAs. Approximately 80 percent of a cell's RNA consists of ribosomal RNA (rRNA). Ribosomal RNAs (with the exception of the 5S species) are synthesized by RNA polymerase I; transfer RNAs by RNA polymerase III: and mRNAs by RNA polymerase II. All three types of RNAs are derived from primary transcripts that are longer than the final RNA product. RNA processing requires a variety of small nuclear RNAs (snRNAs).

Three of the four eukaryotic rRNAs (the 5.8S, 18S, and 28S rRNAs) are synthesized from a single transcription unit, made from DNA (rDNA) that is localized within nucleolus, and are processed by a series of nucleolar reactions. Nucleoli from amphibian oocytes can be dispersed to reveal actively transcribed rDNA, which takes the form of a chain of "Christmas trees." Each of the trees is a transcription unit whose smaller branches represent shorter RNAs that are at an earlier stage of transcription, that is, closer to the site where RNA synthesis was initiated. Analysis of these complexes reveals the tandem arrangement of the rRNA genes, the presence of nontranscribed spacers separating the transcription units, and the presence of associated ribonucleoprotein (RNP) particles that are involved in processing the transcripts. The steps that occur in the processing of rRNA have been studied by exposing cultured mammalina cells to labeled precursors, such as ^{14}C-methionine, whose methyl groups are transferred to a number of nucleotides of the pre-rRNA. The presence of the methyl groups is thought to protect certain sites on the RNA from cleavage by nucleases involved in processing the transcript. Approximately half of the 45S primary transcript is removed during the course of formation of the three mature rRNA products. The nucleolus is also the site of assembly of the two ribosomal subunits.

The precursors of the 5S rRNA and the tRNAs are synhtesized by RNA polymerase III, whose promoter is located within the transcribed portion of the gene rather than in the upstream flanking region. Both types of RNA are synthesized on stretches of DNA in which the genes are organized in tandem array, with the transcribed portions alternating with nontranscribed spacers. Following processing, 5S rRNAs are transported to the nucleolus, where they are assembled with other components into the large ribosomal subunits. Transfer RNA genes are located in clusters that contain multiple copies of different genes.

Kinetic studies of rapidly labeled RNAs suggest that mRNAs are derived from much larger precursors. When eukaryotic cells are incubated for one to a few minutes in ^{3}H uridine or other labeled RNA precursors, most of the label is incorporated into a group of RNA molecules of very large molecular weight and diverse nucleotide sequence and are restricted to the nucleus. These RNAs are referred to as heterogeneous nuclear RNAs (or hnRNAs). When cells that have been incubated briefly with ^{3}H-uridine are chased with medium containing unlabled pre cursors for an hour or more, radioactivity appears in smaller cytoplasmic mRNAs. This and other evidence, such as the presence of 5′ caps and 3′ poly (A) tails on both hnRNAs and mRNAs, led to the conclusion that hnRNAs were the precursors of mRNAs.

Pre-mRNAs are synthesized by RNA polymerase II molecules in conjunction with a number of general transcription factors that allow the polymerase to recongnize the proper DNA sites and to initiate transcription at the proper nucleotide. In the majority of the genes studied, the promoter lies between 24 and 32 bases upstream from the site of initiation in a region containing the TATA box. The TATA box is recognized by the TATA-binding protein (TBP), whose binding to the DNA initiates the assembly of a preinitiation complex. Phosphorylation of a portion of the RNA polymerase leads to the disengagement of the polymerase and the intitiation of transcription.

One of the most important revisions in our concept of the gene came in the late 1970s with the discovery that the coding regions of a gene did not form a continuous sequence of nucleoties. The first observations in this regard were made on studies of transcrption of the adenovirus genome, in which it was found that the terminal portion of a number of different messenger RNAs are composed of the same sequence of nucleotides that are encoded by discontinuous segments in the DNA. The regions between the coding segments are called intervening sequences. A similar condition was soon found to exist for cellular genes, such as those that code for β-globin and ovalbumin. In each case, the regions of DNA that encode portions of the polypeptide (the exons) are separated from one another by noncoding regions (the introns). Subsequent analysis indicated that the entire gene is transcribed into a primary transcript. The regions corresponding to the introns are subsequently excised from the pre-mRNA, and adjacent codning segments are ligated together. This process of excision and ligation is called RNA splicing.

The major steps in the processing of primary transcripts into mRNA include the addition of a 5′ cap, formation of a 3′ end, addition of a 3′ poly (A) tail, and removal of intergening sequences. The formation of the 5′ cap occurs by stepwise reactions in which the terminal phosphate is removed, a GMP is added in an inverted orientation, and methyl groups are added to the added guanosine and the first nucleotide of the transcript itself. The final 3′ end of the mRNA is generated by cleavage of the primary transcript at a site just downstream from an AAUAAA recognition site and the addition of adenosine residues one at a time by poly (A) polymerase. Removal of the introns from the primary transcript depends on the presence of invariant residues at both the 5′ and 3′ splice sites on either side of each intron. Splicing is accomplished by a multicomponent spliceosome that contains a variety of proteins and ribonucleoprotein particles (snRNPs) hat assemble in a stepwise fashion at the site of intron removal. Insight into the evolution of splicing mechanisms has been gained from studies of the self-splicing reactions of group I and group II introns of RNAs in lower eukaryotes, mitochondria, and chloroplasts. These studies strongly suggest that the snRNAs of the splicesomes, not the proteins, are the catalytically active components of the snRNPs. One of the apparent benefits of split genes is the ease with which exons can be shuffled within the genome, generating new genes from portions of preexisting ones.

Information for the incorporation of amino acids into a polypeptide is encoded in the sequence of triplet codons in the mRNA. In addition to being triplet, the genetic code is nonoverlapping and degenerate. In a nonoverlapping code, each nucleotide is part of one, and only one, codon and the ribosome must move alnong the message three nucleotide at a time. To ensure that the proper triplets are read, the ribosome attaches to the mRNA at a precise site termed the initiation codon, AUG, which automatically puts the ribosome in the proper reading frame so that it correctly reads the entire message. A triplet code constructed of 4 different nucleotides can have 64 (4^3) different codons. The code is degenerate because many of the 20 different amino acids have more than 1 codon. Of the 64 possible codons, 61 specify an amino acid, while the other 3 are stop codons that cause the ribosome to terminate translation. The codon assignments are essentially universal, and their sequences are such that base substitutions in the mRNA teed to minimize the effect on the properties of the polypeptide.

Information in the nucleotide alphabet of DNA and RNA is decoded by transfer RNAs during the process of translation. Transfer RNAs are small RNAs (73 to 93 nucleotides long) that share a similar, L-shaped, three-dimensional structure and a number of invariant residues. One end of the tRNA bears the amino acid, and the other end contains a three-nucleotide anticodon sequence that is complementary to the triplet codon of the mRNA. The steric requirements of complementarity between codon and anticodn is relaxed at the third position of the codon to allow different codons that specify the same amino acide to use the same tRNA.

It is essential that each tRNA is linked to the proper (cognate) amino acid, that is, the amino acid specified by the mRNA codon to which the tRNA anticodon binds. Linkage of tRNAs to their cognate amino acids is guaranteed by a diverse group of enzymes called aminoacyl-tRNA synthetases. Each enzyme is specific for one of the 20 amino acids and is able to recognize all of the tRNAs to which that amino acid must be linked. The energy expended during formation of the aminoacyl-tRNAs is the primary energy-requiring step in the entire process of polypeptide assembly.

Protein synthesis is the most complex synthetic activity occurring in a cell, involving all the various tRNAs with their attached amino acids, ribosomes, messenger RNA, a number of proteins, cations, and GTP. The process is divided inot three distinct activities: initiation, elongation, and termination. The primary activities of initiation include the precise attachment of the small ribosomal subunit to the initiation codon of the mRNA, which sets the reading fram for the entire translational process; the entry into the ribosome of a special initiation $tRNA_i^{Met}$; and the assembly of the translational machinery. During elongation, a cycle of tRNA entry, peptide bond formation, and tRNA exit occurs that repeats itself with every amino acid incorporated. Aminoacyl-tRNAs enter via the A site, where they bind to the complementary codon of the mRNA. As each tRNA enters, the nascent polypeptide attached to the tRNA of the P site is transferred to the amino acid on the tRNA of the A site, forming a peptide bond. Peptide bond formation is catalyzed by a portion of the large rRNA acting as a ribozyme . The uncharged tRNA of the P site is then ejected and the ribosome translocates to the next codon of the mRNA, ready for the cycle to be repeated. Both initiation and elongation require GTP hydrolysis, which is thought to serve primarily to increase the accuracy of translation. Translation is terminated when the ribosome reaches one of the three stop codons. After a ribosome aseembles at the initiation codon and moves a short distance toward the 3′ end of the mRNA, another ribosome generally attaches to the initiation codon so that each mRNA is translated simultaneously by a number of ribosomes, which greatley increases the rate of protein synthesis within the cell. The complex of an mRNA and its associated ribosomes is a polyribosome.

CHAPTER 15

Cell Cycle, Mitosis and Meiosis

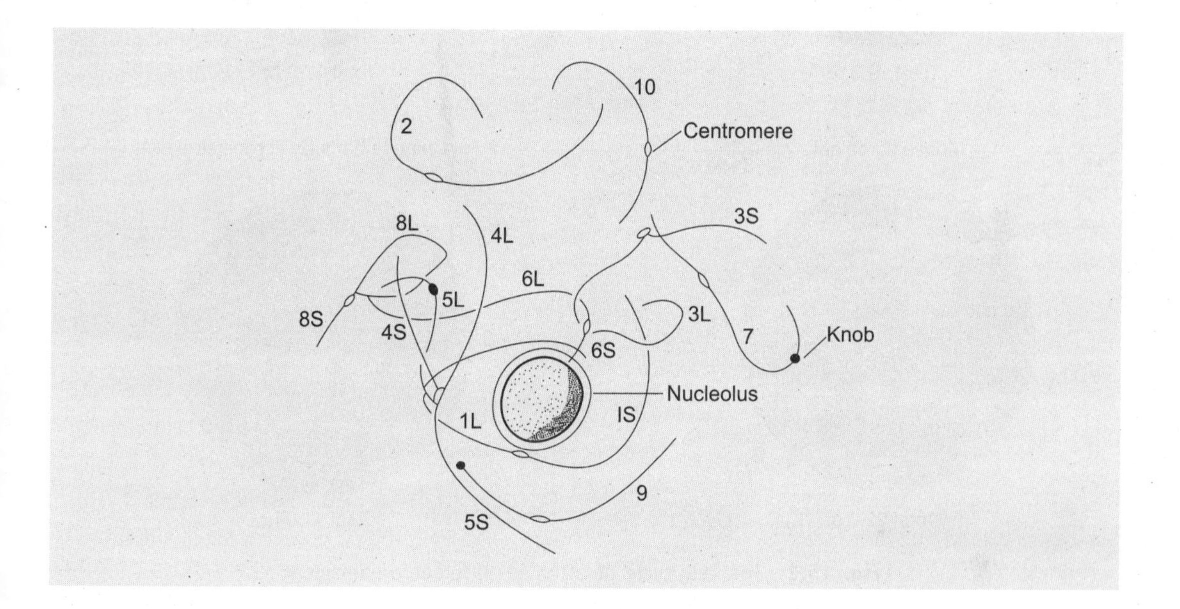

CELL CYCLE IN EUKARYOTES

In simple eukaryotes, such as yeast, the DNA synthesis leading to chromosome replication takes place throughout the interphase and ceases only during the brief period of mitotic, nuclear and cell division. In higher plants and animals, however, DNA replication occurs only during a discrete interval of the interphase known as the S *period* (S standing for synthesis). Before and after the S period, cells engage in growth and metabolic activity but not in chromosome replication and are said to be in the G_1 or G_2 (G standing for gap) periods of interphase. The G_2 period is ordinarily followed by another G_1, is known as the *Cell Cycle.* When nutrients become scarce the cells shift into a *stationary* or G_0 phase in which cellular metabolism essentially shifts into a holding pattern until nutrients are replenished.

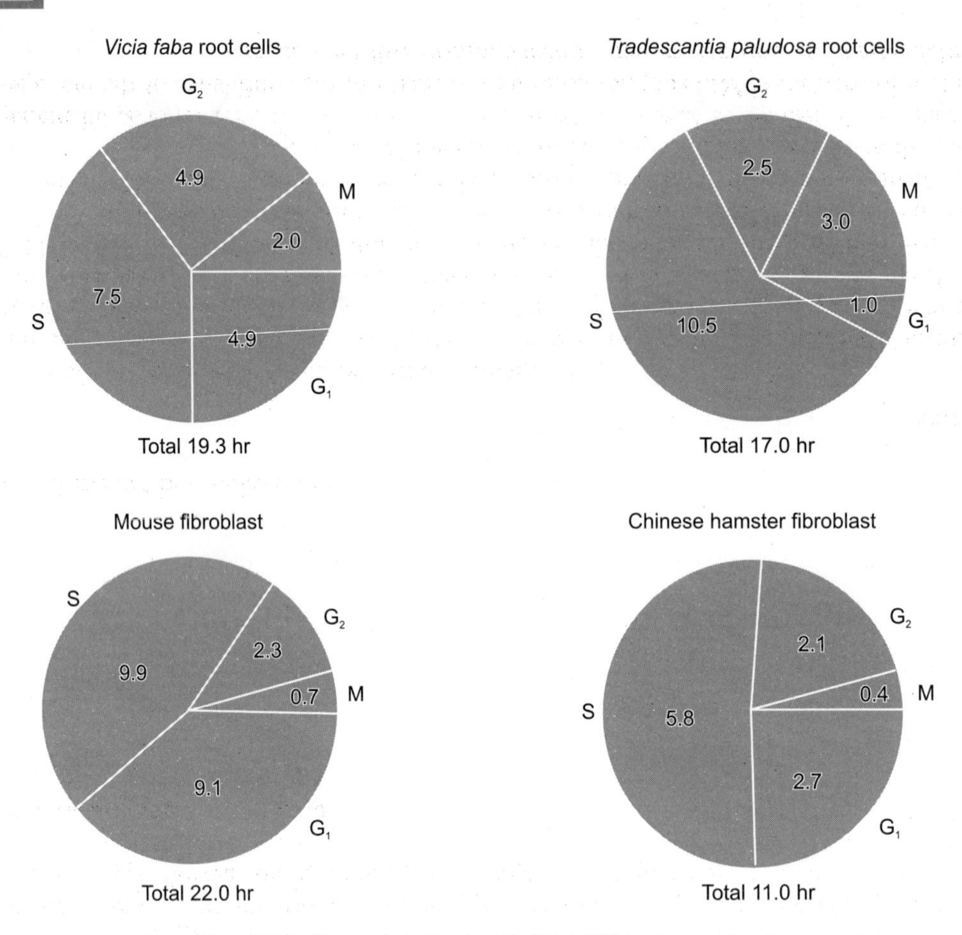

Fig. 15.1 The cell cycle duration in different organisms.

Different eukaryotic cells vary in the length of time (Fig. 15.1) they take to complete an entire cell cycle; they also differ in the relative proportions of time allotted to each of the four stages of the cycle. In general, the length of an S phase is a constant property of any one cell type, but the length of its G_1 phase may vary considerably with environmental fluctuations, such as nutrient supply. It is also obvious from the figure (Fig. 15.1) that the amount of time allotted to the actual mitotic segregation of daughter chromosomes is usually relatively brief. Exceptionally in grasshopper neuroblast, mitosis takes as long as eight hours after an interphase of 30 minutes only.

Considerable interest centres on characterising genes involved in the control of the cell cycle. About 50 temperature sensitive mutants of *Saccharomyces cerevisiae* have been isolated, each of which fails to pass start or cannot complete the cell cycle when the mutations are expressed at the non-permissive temperature. Some of these mutants have defects in enzymes important in DNA replication or cell division and probably do not represent genes actually involved in cell cycle control. However, one or two are more interesting. In particular, one of the genes which, when mutated, results in a yeast cell that cannot pass start has sequence similarities with homoeotic genes of *Drosophila* and other higher eukaryotes. Could it be that study of the cell cycle in yeast will prove directly relevant to developmental processes in higher organisms? Certain development depends on the coordinated division of different cells, so an understanding of cell division itself could provide valuable information.

Cell Cycle Genes

Some of the genes which, when mutated in yeast, result in a cell unable to pass start or unable to complete the cell cycle are listed below:

*Gene**	*Function of gene product*
cdc4	Initiation of DNA synthesis
cdc8	Thymidylate kinase-involved in DNA precursor synthesis
cdc9	DNA ligase
cdc16	Microtubule production—responsible for chromosome movement during cell division
cdc21	Thymidylate synthetase—involved in DNA precursor synthesis
tub2	-tubulin—a structural component of microtubules

*'cdc' stands for *cell division cycle.*

MITOSIS

Cell division occurs in all organisms in the growing tissues of plants and animals, but varies in a number of ways from one species to another, but the essential processes and consequences are basically similar in all organisms. Cell division is of two types, viz., *Mitosis* and *Meiosis*. *Mitosis* is for duplication of cells in different tissues and *Meiosis* takes place during the formation of *gametes*, the *sperms* and *eggs*. *Meiosis* is a phenomenon to reduce the chromosome number of the species (diploid, $2n$) to exactly half (haploid, n) and the original diploid chromosome number is restored after *fertilisation* (*Syngamy*) by the union of sperms and eggs. The product of union is called the *zygote* ($2n$).

Stages of Mitosis

For the sake of convenience, the process of cell division has been divided into five stages: *interphase, prophase, metaphase, anaphase and telophase* (Fig. 15.2). But it should be noted that mitosis is a dynamic process and also a continuous process, each step passing imperceptibly into the next stage. Metaphase and anaphase stand out as the most sharply discontinuous stages and as a result are the most easily defined. The entire process may be separated into *karyokinesis*, or nuclear division, and *cytokinesis*, or cytoplasmic division. Mitosis in an animal cell is shown in Fig. 15.4.

Interphase

Cells in interphase, or the *resting stage*, are characterised by a nucleus that shows little or no definable structure, except for the nucleoli and the prochromosomes. Both of these are seen as

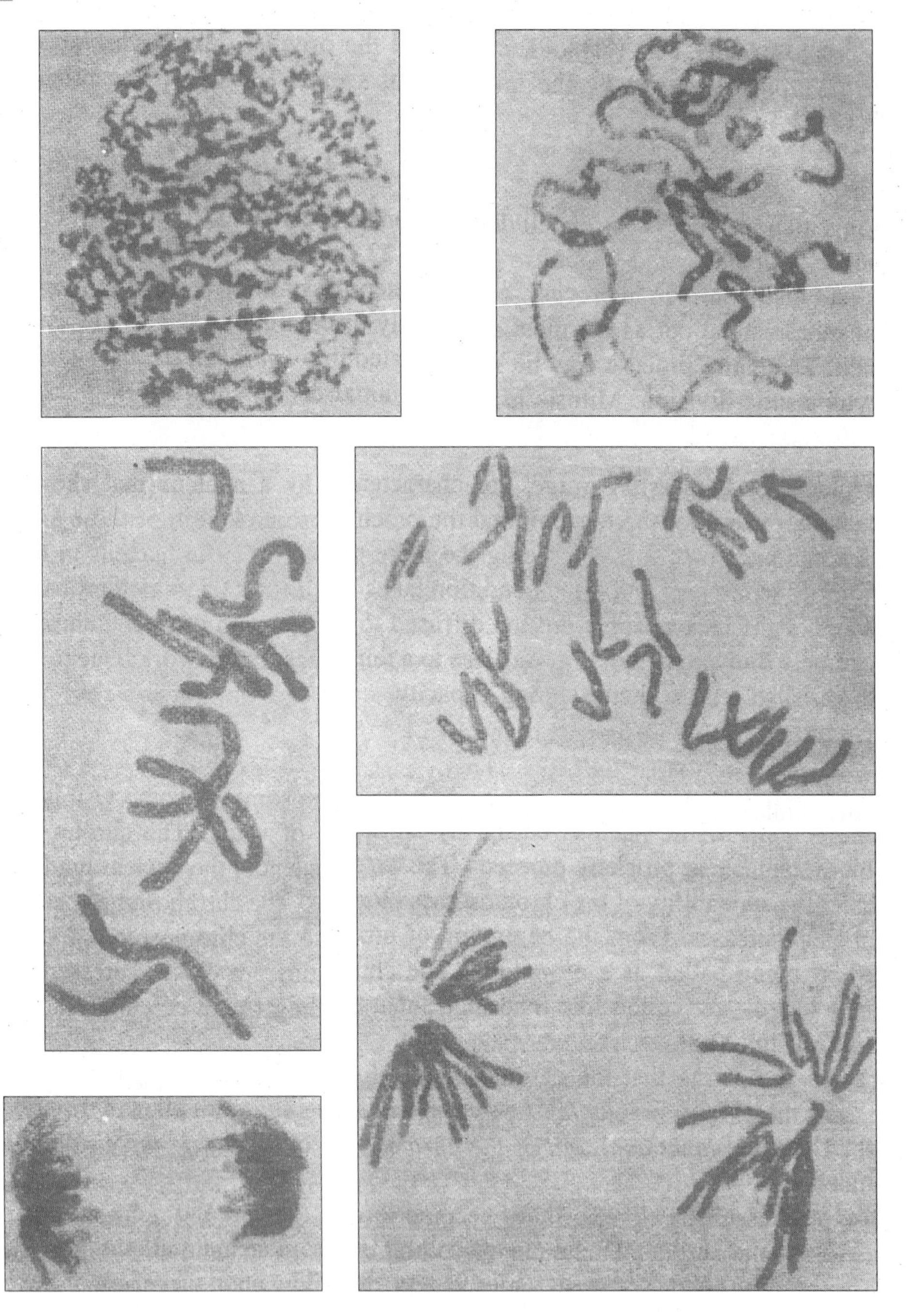

Fig. 15.2 Photomicrographs of mitosis in the peony *Fritillaria* ($2n = 20$). (A) Interphase nucleus; (B) early prophase; (C) late prophase; (D) metaphase; (E) early anaphase; (F) late anaphase; (G) telophase; (H) interphase daughter nuclei.

dark staining bodies, the nucleolus as a rule being the more pronounced. The nucleus in the living state is optically homogeneous, and in a fixed condition takes very little stain, which probably means that the nucleic acids of the chromosome are too diffused to absorb much dye. The chromosomes, as a result, are extremely thin and tenuous, giving rise to a faintly staining network. Due to high hydration the chromosomes also reduce their staining capacity.

Prophase

Prophase is said to be initiated at the moment when the chromosomes become visibly distinct. Prior to this, an enlargement of the nucleus occurs by an uptake of water. The chromosomes become increasingly more stainable as prophase proceeds. The water content of the nucleus gradually decreases and consequently the stainability of the chromosomes increases. The chromosomes become shortened and their diameter increases. From the beginning of prophase the chromosomes are longitudinally double, each half being called as a *chromatid.* The chromatids are closely pressed to each other throughout their length, are coiled like a spring. Such a coiling round of chromatids is known as *relational coiling* (Fig. 15.3).

Evidence exists indicating that the chromosome is further subdivided into *half-chromatids,* i.e., quadripartite, and these four strands are known as *chromonemata* (singular, *chromonema*). It has been established that *the functional unit of the chromosome in mitosis is the chromatid,* inspite of its multi-strandedness.

As prophase progresses the chromosomes become shorter, thicker and more distinct. There is an active contraction of the chromonemata due to coiling of chromonemata all along the length of the chromatids. The number of coils of *gyres* varies widely, depending upon the length of the chromosome and the diameter of the individual gyres. In passing from early to late prophase, the coils' number decreases as their diameter increases. Consequently, the chromosome undergoes a process of despiralisation. The initial period of coiling is called

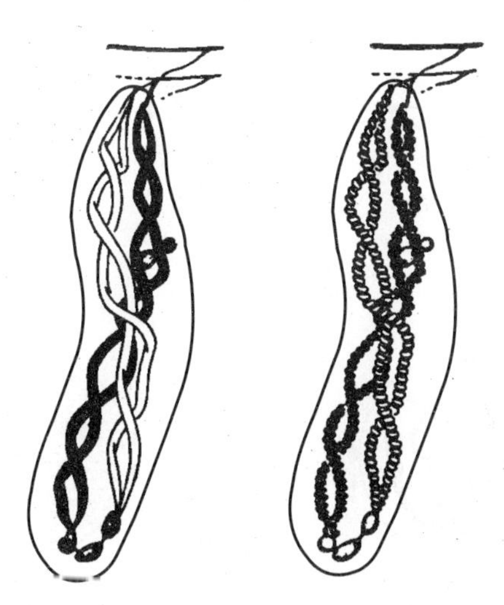

Fig. 15.3 Relational coiling of chromatids (with and without major coils showing) in the protozoan. *Holomastigotoides tusitula.*

spiralisation. The chromosomes during prophase are wound in a thread-like and apparently haphazard fashion throughout the nucleus. The whole prophase looks like a network made up of several threads which are loosely arranged and the threads never come in contact with each other. The nucleoli start disappearing and by the onset of metaphase they completely disappear.

In animal cells the centrosome undergoes a series of changes forming two centrioles, which migrate along the nuclear membrane for the formation of spindle at the metaphase stage (Fig. 15.4).

Metaphase

The nuclear membrane disappears at the onset of metaphase, followed by *spindle formation,* which brings the chromosomes onto the equatorial plate or the metaphase plate. Darlington (1937) recognised three stages in metaphase, viz., *congression*, *orientation*, and *distribution* on the metaphase plate. The first movement is apparently the result of an interaction between the poles of the spindle (or the centrioles) and the constricted area of the chromosome, which can be clearly seen in some metaphase chromosomes. This interaction brings the chromosomes to a

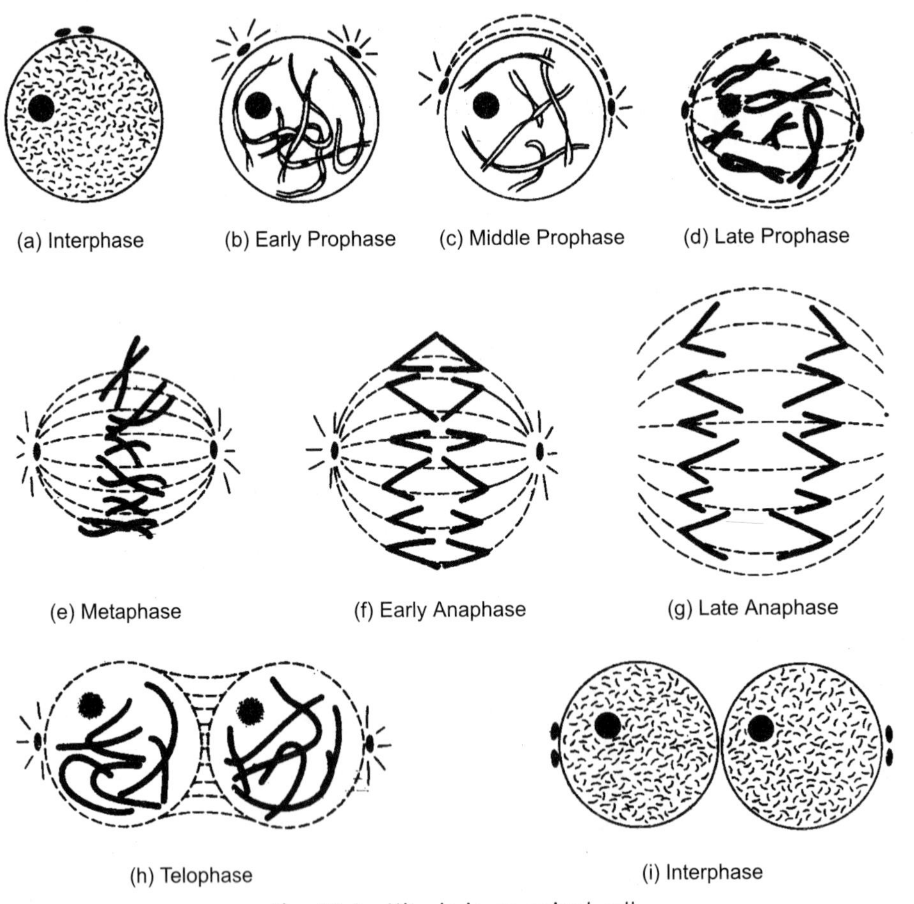

Fig. 15.4 Mitosis in an animal cell.

position of equilibrium midway between the poles. The term *centromere* or *kinetochore* is applied to the colourless, and unstained portion of the chromosome, which actually attaches the chromosomes to the spindle fibers.

The orientation of chromosomes on the plate is such as to arrange the centromeres in the longitudinal axis of the spindle, while the chromatid arms project at either side. Thus, the centromere is the only portion that is attached to the spindle.

The distribution of the chromosomes on the metaphase plate may be entirely random, or there may be a special arrangement, as in the case of certain animals like cells of a salamander tail, which exhibit a *hollow* or *central* spindle. The centromeres are fastened to the spindle's outer edge, and the arms of the chromosomes project into the cytoplasm, giving a hollow look at the centre. In animal cells the centrioles playa very important role in the organisation of the spindle and the arrangement of the chromosomes on the spindle fibers.

The centromeres are divided and lie parallel to each other on the spindle with the chromosome arms usually present in a horizontal position. Since the position of the centromere is always constant on the chromosome, during anaphase movement, they take different shapes, V, J, or rod shape depending upon whether the centromere is median, sub-median or sub-terminal, respectively.

Anaphase

Metaphase passes into anaphase at the time when the centromere becomes functionally double, and the chromatids begin to move toward the poles. The anaphase movement is complex and varies from organism to organism. During this process the chromatids are separated at the centromere, with the arms of the chromatids being passively dragged along. If a chromosome lacks a centromere, it will lag behind on the spindle fibers, and eventually is lost in the cell cycle. The *daughter chromosomes* eventually move to the polar regions where they stop.

Telophase

In *telophase* the chromosomes are regrouped into a nuclear structure within a nuclear membrane. Depending upon the genetic constitution of the organism, a telophase cell immediately goes into another mitosis, or if delayed, the chromosomes loosen their coils, lose their stainability, and finally take the appearance of an interphase nucleus.

Cytokinesis

During late anaphase itself the cytoplasm starts dividing and from this stage onwards there is considerable variation in plant and animal cells. In coenocytic algae and certain fungi, cytokinesis does not follow division of the nucleus, with the result that the thallus becomes multinucleate. *Karyokinesis* and *cytokinesis* have evolved independently in evolution and they are often shown one without the other. In plant tissues cytokinesis is accomplished by *cell plate* formation, while in animal cells it is by *furrowing.* During cell plate formation, the equatorial region of the spindle widens into the *phragmoplast* which looks barrel shaped in a side view, and fine droplets or granules accumulate cutting the cell into two daughter cells. Hence, the origin of cell plate is both from the spindle and cytoplasm. Furrowing takes place by the indentation of the outer membrane at the equatorial plate, and this indentation gradually moving inwards, divides the cell into two daughter cells.

MEIOSIS

Introduction

Meiosis is a special kind of cell division, and it is an antithesis of fertilisation, in that it halves the number of chromosomes. It consists of two divisions which follow each other in rapid sequence. One chromosome of each pair is of maternal, the other of paternal origin and their separation leads to the formation of haploid nuclei. The second division involves the longitudinal separation of chromatids in each of these two haploid nuclei, with the result that four haploid nuclei are produced. The entire process is complicated in detail, and varies widely from species to species.

Occurrence

In most animals, meiosis occurs just prior to fertilisation and results in the formation of sexual cells, the sperms and eggs. Their union in fertilisation gives a diploid zygote, and through cleavage, a diploid body. In some algae and fungi, meiosis immediately follows fertilisation, producing the gametophyte which is a haploid thallus. The gametes produced by mitosis of this gametophyte, unite producing a zygote (2n) which undergoes reduction. division immediately. Therefore, there is no diploid body. In higher plants, however, there is a stable diploid body called the sporophyte which produces the sex cells by meiosis. The sex cells, male and female, unite producing the zygote or the sporophyte. So, here, in higher organisms the diploid body is stable whereas the haploid body represented by the sex cells, is ephemeral or transitory in existence.

Meiosis has been defined as the occurrence of two divisions of a nucleus accompanied by one division of its chromosomes. The meiotic process has been extensively studied in the pollen mother cells of many angiosperms.

The process of meiosis has been divided broadly into two divisions: Division I and Division II. Again in these two divisions various stages have been recognised into Prophase, Metaphase, Anaphase and Telophase. Accordingly, the prophase of Division I is called as Prophase I and that of Metaphase as Metaphase I, and so on. Similarly, for Division II the various stages are numbered as Metaphase II, Anaphase II, and so on.

Division I (Fig. 15.5)

(A) LEPTOTENE The prophase of first division is the most extended stage of all the stages of meiosis. In the prophase I itself several sub-stages have been recognised as *Leptotene*, *Zygotene*, *Pachytene*, *Diplotene* and *Diakinesis*.

Leptotene

In *Leptotene* the chromosomes appear in the nucleus in their diploid number. They are single threads, not double as in mitosis. Each has an uneven granular structure which gives its

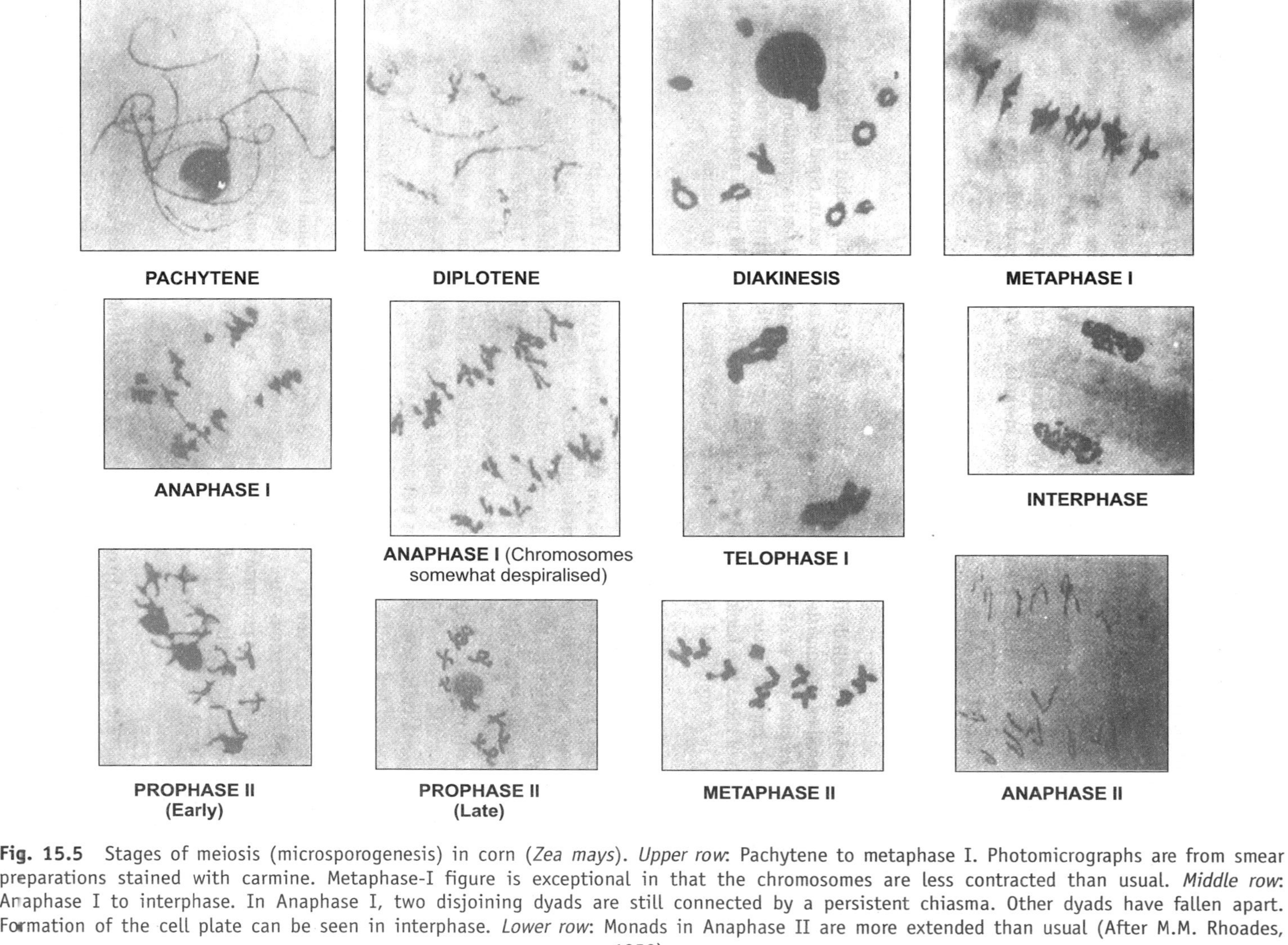

Fig. 15.5 Stages of meiosis (microsporogenesis) in corn (*Zea mays*). *Upper row*: Pachytene to metaphase I. Photomicrographs are from smear preparations stained with carmine. Metaphase-I figure is exceptional in that the chromosomes are less contracted than usual. *Middle row*: Anaphase I to interphase. In Anaphase I, two disjoining dyads are still connected by a persistent chiasma. Other dyads have fallen apart. Formation of the cell plate can be seen in interphase. *Lower row*: Monads in Anaphase II are more extended than usual (After M.M. Rhoades, 1950).

appearance of a string of unequal beads, unequally strung together. These beads are called *chromomeres.* Threads are disposed evenly in the nucleus. In some organisms these chromomeres can be counted and are constant for each chromosome, and also for a given species. Sometimes the chromosomes in some cells of certain species lie polarised to a side of the nucleus. This is called "*bouquet*" stage.

Zygotene

In *Zygotene* the pairing of chromosomes in intimate association begins. The chromosomes come together in pairs, corresponding chromomeres lying side by side. Pairing occurs between similar or homologous chromosomes lengthwise in a zipper-like manner and may begin at several places along the length of the chromosome. It may start at the ends and proceed towards the centromere or vice versa.

Pachytene

The *pachytene* chromosomes provide the most favourable material to characterise each chromosome by their relative length, arm ratio (short arm/long arm), and such topological features, such as heavily staining segments, prominent chromosomes or *knobs* and constrictions. The nucleolus-organising chromosome is identified by its association with the nucleolus. Cytogeneticists usually examine chromosomes at pachytene for evidence of structural aberrations and to establish both the site and extent of the aberration in specific chromosomes. Certain species, such as *maize, rice* and *tomato,* with superb pachytene chromosomes have been successfully exploited for detailed cytogenetic studies (Figs. 15.8 and 15.9).

Pachytene analysis in many plants like maize, tomato, rice, *Brassica, Physalis, Sorghum* and *Pennisetum* has been extensively studied to study *Karyotypic studies,* i.e., arranging the chromosomes based on their length measurements, arm ratio (short arm/long arm) centromere position from the longest chromosome to the shortest chromosome of the particular organism.

While the light microscope does not show a structure associated with the pachytene chromosomes, the electron microscope has revealed a highly organised structure of filaments, the *synaptinemal complex* (Figs. 15.6 and 15.7), between the paired chromosomes. The discovery of complex in crayfish spermatocytes by Moses (1956) has been confirmed in other animal and plant species. The complex appears as a triplet of parallel dense bands in a single plane, curving and twisting along the axis between the chromosomes. The association of the synaptinemal complex with pachytene but not mitotic chromosomes suggests some functional role in synapsis, crossing-over or chiasma formation. These speculations are supported by the observation that individual meiotic chromosomes (univalents) are not associated with a recognisable complex.

Synaptinemal Complex

In almost all eukaryotes the pairs of homologous chromosomes are locked together by a special structure, the synaptinemal complex (Moses, 1968). According to Catcheside (1978) the frequent spelling 'synaptonemal' is unsound. The synaptinemal complex is a ribbon-shaped axis consisting of two lateral components and a central component flanked by space which is less

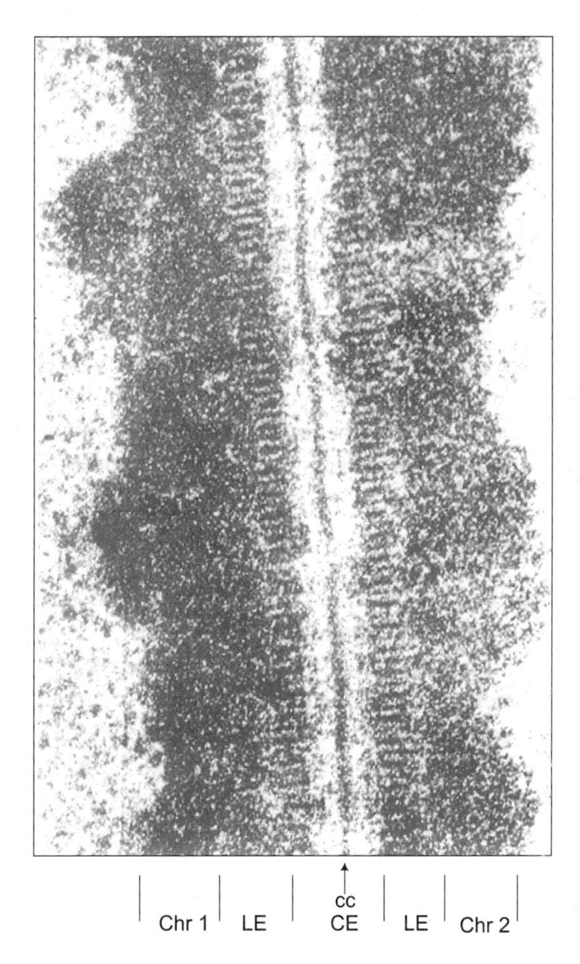

Fig. 15.6 Synaptinemal complex (SC) of the fungus *Neotiella* separating two homologous chromatid pairs (Chr 1 and Chr 2). The lateral elements (LE) are banded, which is not the case for many organisms. CC, central component; CE, central element.

electron dense. The space around the central component is traversed by thin filaments. The width of the central region is 90–120 nm with the central component 10–30 nm. Elements of the synaptinemal are composed of ribonucleo-proteins. All components are digested by trypsin and major portions of the lateral components are digested by RNase. However, DNase has no effect on the synaptinemal complex.

The homologous chromosomes, already two chromatids at zygotene, are paired at pachytene by the synaptinemal complex and held at a distance apart of about 100 nm, irrespective of chromosome size, morphology and DNA content. The union by the complex is over the whole length of the chromosomes including the centromeres. Chiasmata are not observed at pachytene by either light or electron microscopy.

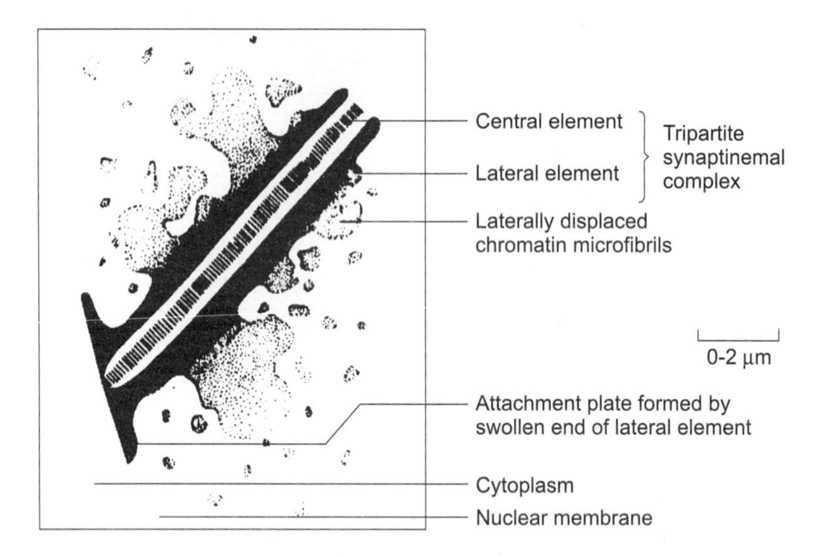

Fig. 15.7 Electronmicrograph of synaptinemal complex.

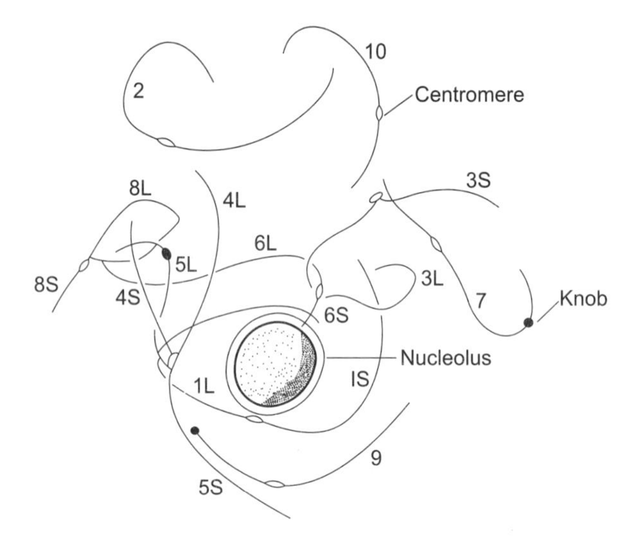

Fig. 15.8 Pachytene chromosomes of maize, $n = 10$; schematic diagram showing the site of the centromere in each chromosome.

In any event, synapsis, when once initiated, proceeds in a zipper-like fashion along the chromosome. At any place along the chromosome only *two chromosomes* pair, since in diploid organisms where three sets of similar or homologous chromosomes are present in a nucleus, pairing is two-by-two only. So the pairing is always two-by-two. The three homologous chromosomes are never associated together in a three-by-three arrangement in meiotic cells.

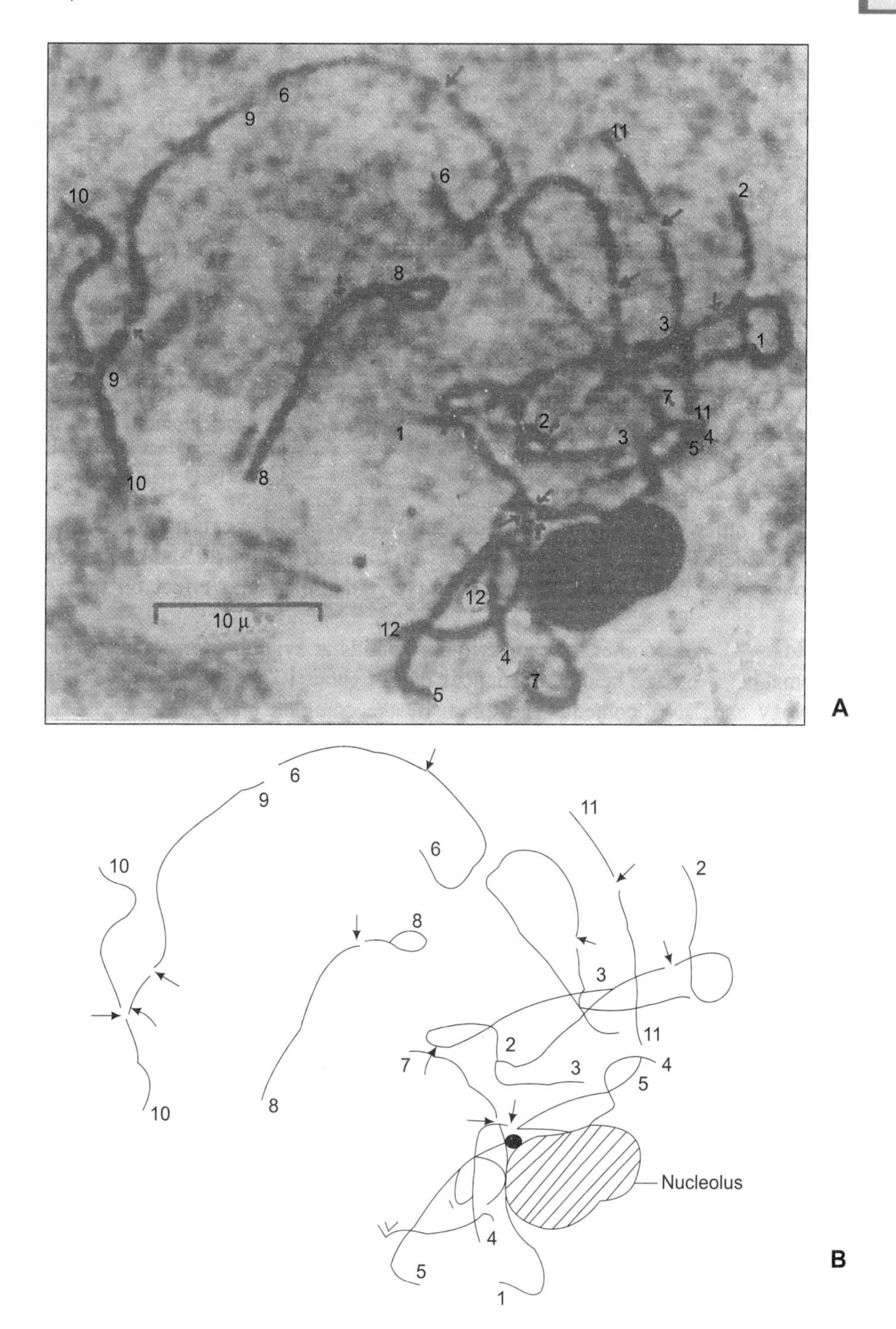

Fig. 15.9 Pachytene chromosomes in rice (*Oryza sativa* L) A—Photograph of pachytene, B—Line drawing of A. (Courtesy: A.V.S.S. Sambamurty).

Zygotene is the unstable stage of pairing while pachytene is the stable stage of pairing. The paired threads of each bivalent coil round one another. The chromosomes which are thicker than at zygotene are referred to as bivalents since they appear in two. The nucleolus is particularly evident here at this stage.

Diplotene

In *diplotene* longitudinal separation of paired chromosomes initiates. Here in this stage, each bivalent consists of four chromatids. Homologue move apart but are held together at one or more places all along the chromosomes. The bivalents take the appearance of a cross if there is only one contact in a chromosome or it may be loop-like, if there are two contacts or it may be a series of loops if there are more than three contacts. Each point of contact is a *chiasma(ta)*. The chiasmata cross each other in two of the chromatids. Chiasmata may be either interstitial or terminal according to their position on the chromosome as in the middle or at the ends, respectively. The number of chiasmata depend on the length of the chromosome, thus the longer the chromosome the greater the number of chiasmata, that occur.

During the diplotene stage the chromosomes start contracting or shortening, and they coil round each other. Recombination of characters results due to crossing over and chiasma formation, as at these places of contact chromatids exchange part of chromosomes.

Diakinesis

In *Diakinesis*, the nucleolus gradually starts disappearing from the nucleus and will be completely absent by metaphase stage. The chromosomes continue to shorten by coiling more tightly. The bivalents consequently assume a more rounded shape. The chiasmata which were formed at the diplotene stage gradually move towards the ends of the chromosomes. This process of movement of chiasmata along the length of the chromosomes to the ends is called *Terminalisation of Chiasmata.* Depending upon the chiasma terminalisation the individual bivalents assume different shapes by the time they reach the metaphase stage.

Metaphase I The nuclear membrane disappears and the spindle starts formation. The spindle is formed as in mitosis. The bivalents get oriented on the metaphase plate and the individual chromosomes are attached by their centromeres to the spindle fibers. In meiosis, the bivalent has two functionally individed centromeres. They lie in the long axis of the spindle. At this time, a strong repulsion appears to exist between homologous centromeres.

Anaphase I When the bivalent chromosomes have been arranged on the metaphase plate, so that their pairs of centromeres lie in the axis of the spindle they divide; the pairs of chromatids associated at the two centromeres move to opposite poles. Unlike mitosis, in which the centromere divides and sister chromatids pass to the opposite poles, the centromeres of each bivalent in meiosis are undivided as they move poleward with the result the whole chromosomes instead of chromatids segregate. Each anaphase group, therefore, is made up of a haploid number of chromosomes instead of a diploid number of chromatids. In this manner a reduction in chromosome number results from the first meiotic division.

Telophase I Telophase and Interphase are not necessarily components of meiotic cycle, since in some organisms the Anaphase I chromosomes, without a nuclear membrane formation, they directly pass to Metaphase II stage. The coiling of the chromosomes is retained and persists up to the end of the meiotic stage.

Division II

Metaphase II At an early stage the two chromatids are widely separated. They are then held together only at their centromeres. Later stages resemble mitosis, but metaphase is rapid.

Anaphase II The chromosomes are distributed as in an ordinary mitosis.

Telophase II Four daughter nuclei are formed, each receiving the haploid number of chromatids.

Three differences distinguish the second meiotic division from a mitotic division: (1) The chromosomes are present in haploid number; (2) Chromatids are widely separated and are not coiled relationally, and (3) each chromatid might be quite different genetically from its condition at the start of the meiosis since they recombined at diplotene in the process of chiasma formation by a process called crossing over.

Cytokinesis It occurs as in mitosis either by cell plate or by furrowing.

Significance of Meiosis Meiosis occurs as a rule in all diploid organisms during the process of reproduction. The occurrence of meiosis is a necessary pre-requisite for the constancy of chromosome number. Since, during fertilisation, sex cells, if they were not reduced in their chromosome number, will produce double the chromosome number. This newly formed organism in its turn produces a further doubling of chromosomes and this process goes on till an unimaginable number of chromosomes are realised in each cell. So to check this unusual and unimaginable doubling of chromosomes at every generation the chromosomes must be reduced to half before fertilisation or zygote formation so that at every generation only the same number of chromosomes are present.

I generation → $2x$ (diploid) = 14 chromosomes

↓ meiosis

x (haploid) = 7 chromosomes

Sex cells

male ($x = 7$) female ($x = 7$)

↓ Fertilisation

II generation → ($2x = 14$) diploid

The diploid chromosome number ($2x = 14$) is thus maintained at all generations.

During the process of meiosis at diplotene recombination of characters (or portions of chromosomes) occurs. This phenomenon is possible only in meiosis and such a similar mechanism is not possible in Mitosis. Recombination of characters or formation of new characters in the daughter cells or new organisms is an essential feature of evolution. If a species has to survive in diverse environmental conditions the newly born organisms must face those strange environmental fluctuations. For this reason, new gene combinations have got to be developed in the daughter organisms. This process of new gene combinations can be achieved only through meiosis.

It is in this process of recombination that the significance of meiosis lies most, apart from the constancy of chromosome number from one generation to other generation.

CHROMOSOME MOVEMENT IN MITOSIS AND MEIOSIS

When cells are viewed *in vivo*, by time exposure motion pictures, reveal the intricate and exacting manoeuvres performed by the chromosomes as they pass through the successive stages of mitosis and meiosis. Cell division presents a picture of constantly changing activity, an activity arising from an interplay of forces, that have their origin in the cytoplasm and within the chromosomes. Many hypotheses have been advanced to explain the various aspects of chromosome and cell dynamics, but until the systems themselves are understood in a physical and chemical sense the causal mechanisms involved can only be guessed at.

Chromosome Reproduction

The chromosome, presumably by an autocatalytic process, fashions a replica of itself out of the materials available in its immediate neighborhood. The nucleoproteins exert their profound effects, directly or indirectly, through enzymatic action. The genes and the chromosomes, with very rare exceptions, reproduce themselves exactly. It has been established beyond doubt that chromosome replication takes place in a semi-conservative way by molecular and biochemical studies.

Chromosome Contraction

A feature characteristic of chromosomes during cell division is the considerable contraction they progressively undergo as they pass from early prophase stages to metaphase. This shortening of the chromosome has a mechanical advantage to move on the spindle than a long and a tenuous one.

Preparations of metaphase cells of *Trillium* or *Tradescantia*, pretreated with ammonia vapours or with dilute solution of potassium cyanide (KCN), before fixation and staining, reveal the large chromosomes to be made up of a series of *coils* closely packed together in the manner of a wire spring. The frequency of coils is a function of temperature, genotype and probably nutrition. The prophase contraction of mitotic and meiotic chromosomes is largely a consequence of *coiling mechanism*. The contraction is greater in meiotic cells and not identical with those of somatic cells. The chromosomes at pachytene according to Sax and Sax are 7 to

11 times longer than they are at meiotic metaphase. In *Osmnuda regalis*, Manton (1950) has shown that leptotene chromosome are 50% longer than those in pachytene.

Coiling of Chromosomes

The individual turnings or *gyres*, make up the *somatic* or *standard* coil which is renewed at each division prophase. Chromatid is the unit of coiling. The loose spirals coming from Interphase, which undergo new somatic coiling cycle, are called *relic coils*. The loose wrapping of one chromatid about another constitutes the relation coils. In meiosis in addition to gyres or coils the chromonemata consist of *major* and *minor* coils, the former being big turns and the latter small turns of the chromonema.

Somatic coiling may *be plectonemic* or *paranemic*. The two chromatids in a chromosome may be relationally coiled about each other which do not separate laterally along their length without entanglement. This type of coiling is known as *plectonemic coiling* and in the other, the *paranemic coiling*, the two coiled chromatids are free to separate laterally without entanglement.

According to Darlington (1935) a molecular coil sets up an internal pace that determines the microscopically visible pattern of behaviour. This postulate assumes that a tightly coiled metaphase chromosome is under a state of tension. According to Kuwada (1939) and Nebel (1939) the matrix surrounding the chromosome determines the type of coiling. This hypothesis was further supported by Sax and Humphrey (1934), Huskins (1941), Coleman and Hillary (1941).

The process of coiling during the mitotic and meiotic Anaphase and Telophase stages is known as *spiralisation* or *gyre initiation*, while the loosening of the chromonemata or gyre elimination is known as *despiralisation*. During the early Prophase the number of gyres will be more and gradually decrease in number as the cell proceeds to Anaphase.

Synapsis

In meiotic cells homologous chromosome pairing is the rule and by definition, the process of *synapsis* begins in zygotene stage and reaches its full expression at pachytene, the stable stage of pairing. This is again lost in diplotene and an open bivalent results. The pairing is governed by homologies and the associations are two-by-two even in those cells where three or more homologous chromosomes exist simultaneously as in polyploids.

However, somatic synapsis is a rule in *Drosophila* species, which is not found in general, whether plants or animals. The homologous somatic chromosomes pair intimately and even crossing over takes place called *somatic crossing over*. The salivary gland chromosomes exhibit somatic pairing in *Drosophila*, *Chironomus* and *Sciara*.

Darlington (1937) contented that due to electrostatic forces which are present in the unsaturated or unsatisfied chromosomes, to become saturated they must pair homologously. When the chromosomes become double in late pachytene, the satisfied state is between sister chromatids instead of homologous chromosomes; the paired homologoues consequently fall apart and diplotene is initiated.

This theory has been greatly questioned by Faberge and Lamb (1942) who postulated that the initial contact is due to a hydrodynamic principle which involves long range forces. This theory assumes that all molecular systems are in a state of motion, each system having its own frequency of vibration. Attractions and repellings between the chromosomes depends upon the different systems of vibrational frequencies.

Delbruck (1941) hypothesised that 'pairing is an association of self-reproducing entities within the chromosome coupled with the chemical reduction of each paired pair of peptide bonds, so that between each pair the resonance bond can be formed.

Chiasma Terminalisation

Darlington proposed the term *chiasma terminalisation* for the movement of the chiasmata along the chromosome at diakinesis and Metaphase I of meiosis. Chiasmata are formed at diplotene. They tend to move to the ends as the chromosomes shorten continuously. The number of chiasmata per bivalent and per cell of an organism is usually constant and it reflects on the genetic variability of the organism. The mechanisms advanced to account for the process of terminalisation are: electrostatic forces (Darlington and Dark, 1932), tensions developed through the coiling of chromosomes (Swanson, 1942) and elastic chromosome repulsion (Ostergren, 1943).

Anaphase Movements

In addition to the above types of movements of chromosomes, a variety of other movements occur within the cell contributing to an orderly development of metaphase plate and anaphase movement. There are two forces: *polar* and *chromosomal.* The chromosomal force lies in the centromere responsive to the forces emanating from the poles. Polar and chromosomal forces reveal themselves through the responses exhibited by three structures: nucleus, chromosomes and spindle. In animals, a shortening of the spindle fibers which brings about chromosome movement has been observed by Belar (1929), Barber (1939) and Ris (1943).

SYNOPSIS

The stages through which a cell passes from one cell division to the next constitute the cell cycle. The cell cycle is divided into two major phases: M phase, which includes the processes of mitosis, in which duplicated chromosomes are separated into two nuclei, and cytokinesis, in which the entire cell is physically divided into two daughter cells; and interphase. Interphase is typically much longer then M phase and is subdivided into three distinct phases based on the time of replication, which is confined to a definite period within the cell cycle. G_1 is the period following mitosis and preceding replication; S is the period during which DNA synthesis (and histone synthesis) occurs; and G_2 is the period following replication and preceding the onset of mitosis. The length of the cell cycle, and the phases of which it is made up, vary greatly from one type of cell to another. Cell cycles can range from as short as approximately 30 minutes in rapidly dividing embryos to a period of months in very slowly growing tissues. Certain types of terminally differentiated cells, such as vertebrate skeletal muscle and nerve cells, have entirely lost the ability to divide. Most synthetic activities, other than that of DNA and histones, continue throughout interphase, but either drop markedly or cease entirely during mitosis.

The passage of a cell through the cycle is controlled by a variety of factors. The first indication that the cytoplasm contained factors that control the cell cycle came from cell fusion studies between cells in different stages of the cycle. These studies indicated that a replicating cell contained one or more factors that stimulate the initiation of DNA synthesis, and that a mitotic cell contains one or more factors that induce the condensation of the chromatin. Subsequent experiments showed that entry of a cell into M phase is triggered by the activation of a protein kinase called MPF (maturation promoting factor). MPF consists of two subunits: a catalytic subunit that transfers phosphate groups to specific serine and threonine residues of specific protein substrates, of proteins called cyclins. The catalyitc subunit is called a cyclin-dependent kinase (Cdk). When the cyclin concentration is low, the kinase lacks the cyclin subunit and is inactive. When the cyclin concentration reaches a sufficient level, the kinase is activated, triggering the entry of the cell into M phase.

The activities that control the cell cycle are focused primarily at two points: the transition between G_1 and S and the transition between G_2 and the entry into mitosis. Passage through each of these points requires the transient activation of a Cdk by a specific cyclin. In yeast, the same Cdk is active at both G_1-S and G_2-M, but it is stimulated by a different cyclin. The concentrations of the various cyclins rise and fall during the cell cycle as the result of changes in the rates of synthesis and destruction of the protein molecules. In addition to regulation by cyclins, Cdk activity is alos controlled by the state of phosphorylation of the catalytic subunit, which in turn is controlled in yeast by at least two kinases (CAK and weel) and a phosphatase (cdc25). In mammallian cells, at least eight different cyclins and a half-dozen different Cdks play a role in cell cycle regulation.

Cells possess feedback controls that monitor the status of cell cycle events such as replicaton and chromosome condensation, using the information to determine whether or not the cycle continues. The G_1-S and G_2-M transitions are points where primary control is exercise. Exogenous growth factors, for example, influence cells to enter S phase through the activity of cyclin-dependent kinases. If a cell is subjected to treatments that damage DNA, passage between G_1 and S is delayed until the damage is repaired. Arrest of a cell at one of the ckeckpoints of the cell cycle is accomplished by inhibitors whose synthesis is stimulated by events such as DNA damage. These inhibitors may have several targets including the Cdk-cyclins, the replication machinery, and various transcription factors that activate genes involved in cell growth. Once the cell has been stimulated by external agents to traverse the G_1-S checkpoint and initiate replication, the cell generally continues without further external stimulation through the subsequent mitosis.

Mitosis ensures that two daughter nuclei receive a complete and equivalent complement of genetic material. Mitosis is divided into prophase, prometaphase, metaphase, anaphase, and telophase. Prophase is characterized by the preparation of chromosomes for segregation and the assembly of the machinery required for chromosome movement. Whereas interphase chromosomes exist in a highly extended configuration, those of mitosis are highly condensed, rod-shaped structures. Each mitotic chromosome can be seen to be split longitudinally into two chromatids, which are duplicates of one another formed by replication during the previous S phase. The process by which extended interphase chromosomes are condensed into mitotic chromosomes is poorly understood. It is thought to be triggered by the activation of a cyclin-dependent kinase, which phosphorylates III histone molecules among other substrates. Chromosome condensation requires the presence of an active topoisomerase II, which is a major nonhistone protein of the mitotic chromosome scaffold. The primary constriction on the mitotoc chromosome markes the centromere, which houses a platelike structure, the kinetochore, to which the microtubules of the spindle attach. The first stage in spindle formation in a typical animal cell is the appearance of microtubules in a "sunburst" arrangement, or aster, around each of the two centrosomes during early prophase. Aster formation is followed by the movement of the centrosomes away from one another toward the poles. As this happens, the microtubules that stretch between them increase in number and elongate. Eventually, the two centrosomes reach opposite points in the cell, establishing the two poles. A number of types of cells including those of plants, assemble on mitotic spindle in the absence of centrosomes. The end of prophase is marked by the fragmentation of the nuclear envelope and the dispersal of its membranes in the form of small vesicels.

During prometaphase and metaphase, individual chromosomes are first attached to spindle microtubules emanating from both poles and then moved to a plane at the center of the spindle. At the beginning of prometaphase, microtubules from the forming spindle penetrate into the region that had been the nucleus and make attachments to the kinetochores of the condensed chromosomes. Initially, chromosomes can be seen to move actively along the lateral surfaces of microtubules, powered apparently by motor proteins located in the kinetochore. Soon, however, the kinetochores become stably associated with the plus ends of chromosomal microtubules from both poles of the spindle. Eventually, each chromosome is moved into position along a plane at the center of the spindle, a process that is accompained by the shortening of some microtubules due to loss of tubulin subunits and the lengthening of others due to addition of subunits. Once the chromosomes are stably aligned, the cell has reached metaphase. The mitotic spindle of a typical animal cell at metaphase consists of astral microtubules, which radiate out from the centrosome; chromosomal microtubules, which are attached to the kinetochores; and polar microtubules that extend from the centrosome past the chromosomes and form a structural basket that maintains the integrity of the spindle. The microtubules of the metaphase spindle may exhibit dynamic activity as demonstrated by the directed movement of fluorescently labeled subunits.

During anaphase and telophase, sister chromatids are separated from one another into separated regions of the dividing cell, and chromosomes are returned to their interphase condition. Anaphase begins as the sister chromatids suddenly split away from one another, a process that is probably triggered by the activity of a Cdk and requires a topoisomerase II. The separated chromosomes then move toward their respective poles accompanied by the shortening of the attached chromosomal microtubules, which resulsts primarily from the net loss of subunits at the kinetochore. The movement of the chromosomes, which is called anaphase A, is typically accompained by the elongation of the mitotic spindle and consequent separation of the poles, which is referred to as anaphase B. Telophase is characterized by the reformation of the nuclear envelope, the dispersal of the chromosomes, and reformation of membranous cytoplasmic netwroks. Microtubular motors, such as kinesin and dynein, are generally thought to be responsible for anaphase movements, but the role of microtubule depolymerization as a force-generating mechanism remains an alternate possibility.

Cytokineis, which is the division of the cytoplasm into two daughter cells, occurs by constriction in animal cells and by construction in plnat cells. Animal cells are constricted in two by an indentation or furrow that forms at the surface of the cell and moves inward. The advancing furrow contains a band of actin filaments, which are thought to slide over one another, driven by small, force-generating myosin II filaments. The site of cytokinesis is thought to be selected by a signal diffusing from the spindle poles. Plant cells undergo cytokinesis by building a cell membrane and cell wall in a plane lying between the two poles. The first sign of cell plate formation is the appearance of clusters of interdigitating microtubules and interspersed, electron-dense material. Small vesicle then move into the region and become aligned into a plane. The vesicles fuse with one another to form long chains in which the membrane of the vesicels becomes the plasma membranes of the adjoining cells and the enclosed material becomes part of the cell wall.

Meiosis is a process that includes two sequential nuclear divisions, producing haploid daughter nuclear that contain only one member of each pair of homologous chromosomes, thus reducing the number of chromosomes in half. Meiosis can occur at different stages in the life cycle, depending upon the type of organism. To ensure that each daughter nucleus has only one set of homologues, an elaborate process of chromosome pairing occurs during prophase I that has no counterpart in mitosis. The pairing of chromosomes is accompanied by the formation of a proteinaceous, ladderlike structure called the synaptonemal complex (SC). The chromatin of each homologue is intimately associated with one of the lateral bars of the SC. As they are paired, homologous chromosomes engage in genetic recombination that produces chromosomes with new combinations of maternal and paternal alleles. In subsequent stages, the chromosomes become further condensed, but the homologues remain attached to one another at specific points called chiasmata, which are thought to represent sites where recombination had occurred at a previous stage. In the oocytes and spermatocytes of many animal species, the chromosomes become dispersed during prophase I to form lampbrush chromosomes, which are sites of widespread gene expression. The paired homologues (called bivalents or tetrads) become oriented at the metaphase plate so that both chromatids of one chromosome face the same pole. During anaphase I, the homologous chromosomes separate from one another. Since there is no interaction between one tetrad and another, the maternal and paternal chromosomes of each tetrad segregate independently of one another. The cells, which are now haploid in chromosome contetn, progress into the second meiotic division during which the sister chromatids of each chromosome are separated into different daughter nuclei.

Genetic recombination during meiosis occurs as the result of the breakage and reunion of DNA strands from different homologues of a tetrad. Recombination is a poorly understood process in which homologous regions on different DNA strands are exchanged whithout the addition or loss of a single base pair. In an initial step, two duplexes that are about to recombine become aligned next to one another as the result of a homology search. Once they are aligned breaks occur in either one or both strands of one of the duplexes. In following steps, DNA strands from one duplexd invade the other duplexd, forming an interconnected structure. Subsequent steps may include the activity of nucleases and polymerases to created and fill gaps in the various strands similar to the way in which DNA repair occurs.

CHAPTER 16

The Golgi Apparatus

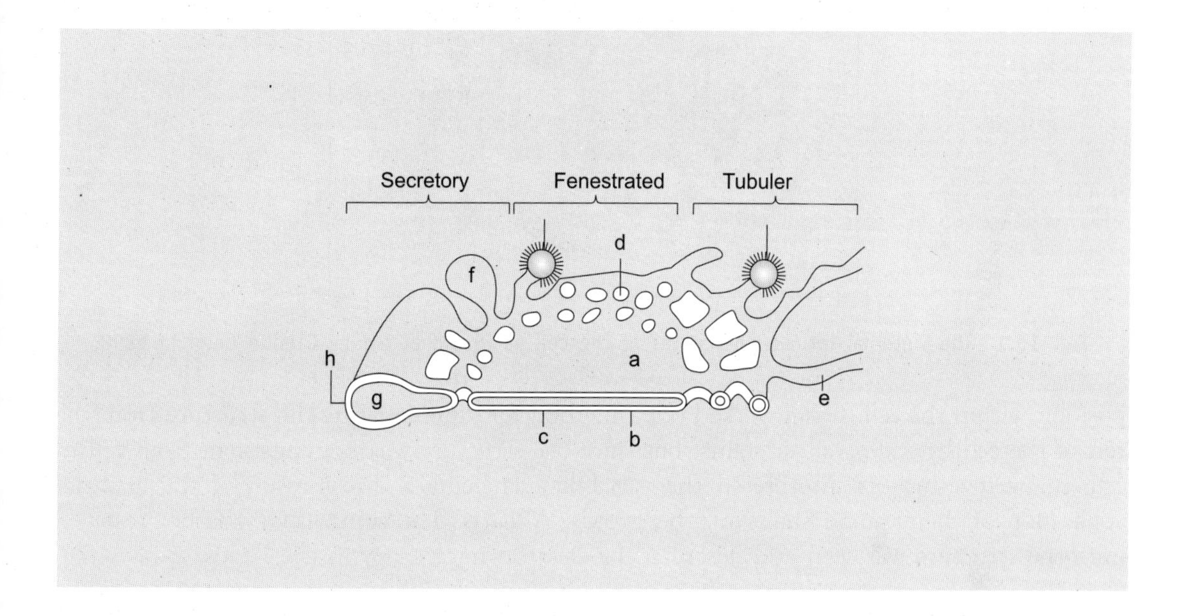

The Golgi apparatus (or Golgi body) is organelle along with the lysosomes, peroxysomes, and other vesicles that make up part of the endomembrane system of eucaryotic cells. The term "endomembrane system" is used to refer to the collection of cytoplasmic and vesicular membranes believed to arise from the same common source. The common source is either the outer nuclear envelope or the endoplasmic reticulum. The mitochondria and chloroplasts are not included in this class because they are known to arise de novo or by division of preexisting organelles of the same type.

The Golgi apparatus was first described by C. Golgi in 1898 (Fig. 16.1). Although he referred to the structures as the "internal reticular apparatus" they soon were identified with his name and become one of the more controversial cell structures. Part of the controversy arose from the fact that the Golgi apparatus is not easily seen and is variable in structure and

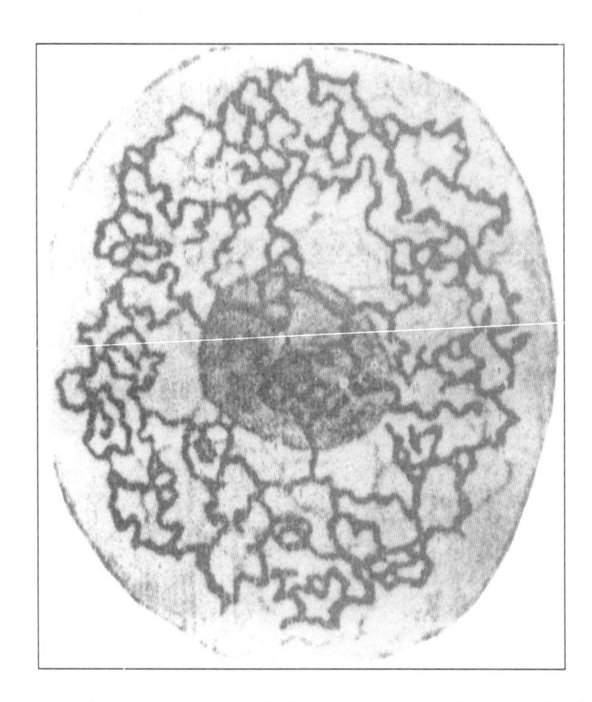

Fig. 16.1 The "internal reticular apparatus of the cell as first depicted by Camillo Golgi in 1898.

position within the cell. Golgi worked out a method for differentiating the structure from the rest ot the cell by adding silver stains, but since the structure was not consistent from cell to cell, many investigators interpreted the variability as being a direct result of the artificial application of chemicatand stains-in other words, *Artifacts.* The controversy was not resolved until the structure was seen and described by electron microscopy in the 1950s.

STRUCTURES OF THE GOLGI APPARATUS

The Golgi apparatus is complex of membrane-lined vesicles called cisternae. A cisterna is a fluid-filled sac or container, and the cisternae of the Golgi apparatus vary from large flattened vesicles to branching and anastomosing vesicles to individual spherical vesicles (Fig. 16.2). In most cells the Golgi apparatus is composed of layers of large flattened vesicles. These appear to be arranged in order, the small vesicles lying nearest the nuclear envelop or endoplasmic reticulum-the origin of the vesicles-and the large ones lying at the opposite of the structure. The stacks of vesicles nearest the nucleus comprise the "forming face" of the Golgi apparatus while the vesicles on the side closer to the plasma membrane constitute the "maturing face". The membranes bounding the vesicles are smooth, with no evidence of particles such as ribosomes. The different levels of organization of the Golgi structures may be described: the cisternae, the dictyosome, and the whole apparatus. Changes in the Golgi structure are interpreted from electron photomicrographs that show only static conditions. The sequence of changes is a judgment based upon changes in the activity of the cell.

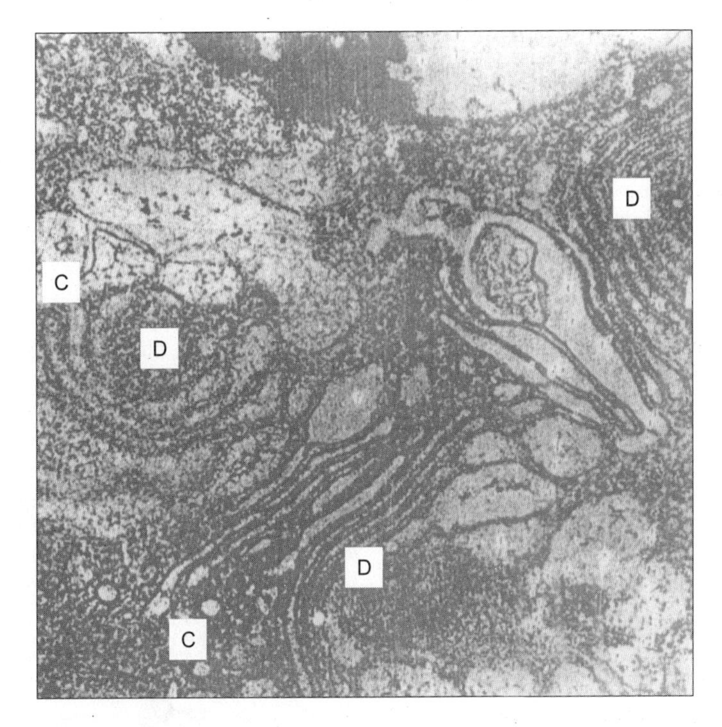

Fig. 16.2 Various shapes of cisternae (C) in a dictyosome (D) (Electron microphotograph).

The cisternae initially are relatively small spherical vesicles. Some of them subsequently become flattened sacs or tubular elements. Those that the become flattened may develop pores about the periphery, called **fenestrae,** and may form tubes that extend outward (Fig. 16.3). The ends of the tubular extensions may fuse with each other or enlarge. As these ends mature, they frequently detach and become secretary vesicles. In some cases, the mature edges of the cisternae become coated with electron-dense material.

The dictyosome is a collection of cisternae that forms a stack and acts as a unit (Fig. 16.4 and 16.5). The cisternae are formed at one end of the stack, progressively mature, and are released as vesicles on the opposite side of the stack. The size and number of dictyosomes vary from one type of cell to another and according to the metabolic activity of the cell. Some cells have been reported to have only a singe dictyosome; others may have hundreds. Since one of the functions of the Golgi apparatus is secretion, as one might expect, the size and occasionally the number of dictyosomes increase when the cells are actively secreting materials. In plant cells, the number of dictyosomes increases during cell division when these organelles secrete material for the cell plate, which then develops into the cell wall separating the two new cells. In goblet cells of intestinal epithelium there is a single dictyosome, but its size increase significantlt in periods of digestion.

The location of the dictyosomes in the cell is as variable as their number and size. Most often, the distribution appears to be at random; however, in some cases, the location is related

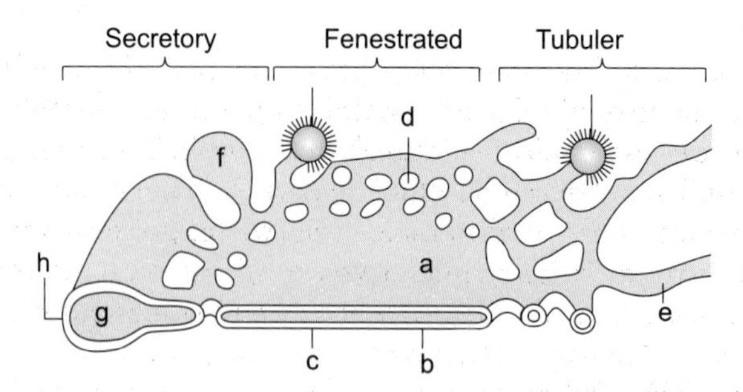

Fig. 16.3 Composite diagram of dictyosome cisternae: (a) Central plate-like region. (b) membrane; (c) lumen; (d) fenestrae (perforation); (e) peripheral trbules; (f) secretory vesicles; (g) vesicle lumen; (h) vesicle membrane; (i) coated vesicles (600to 750 A diameter).

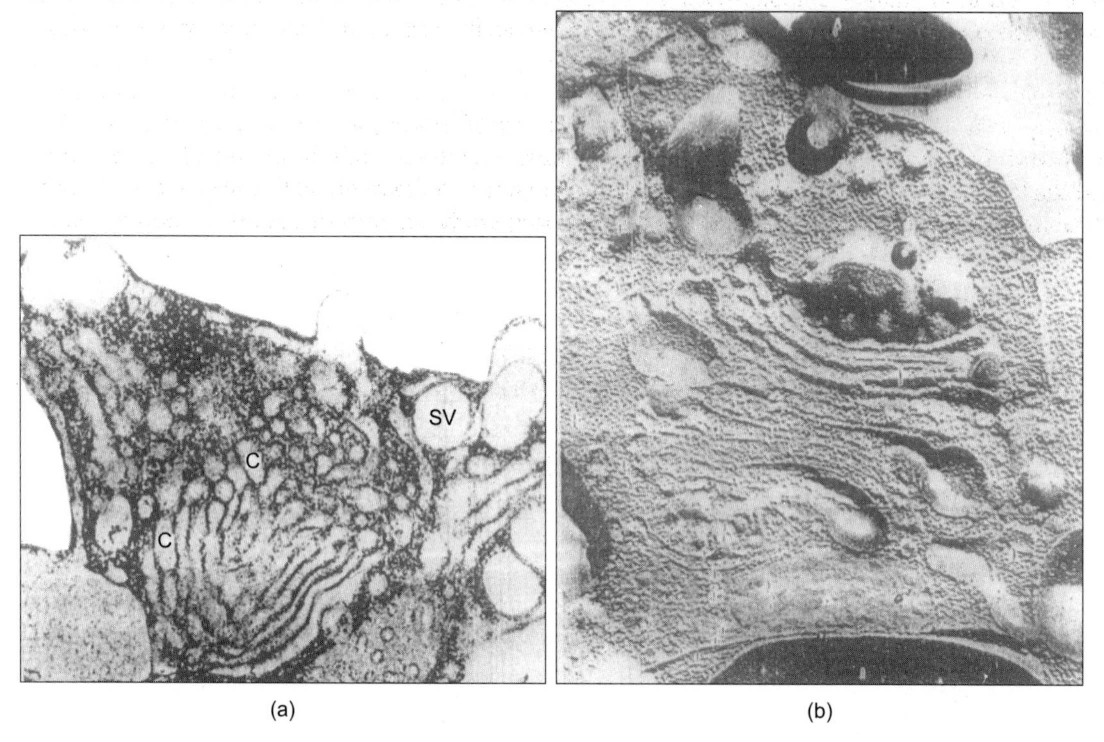

Fig. 16.4 Golgi apparatus. (a) Transmission electron photomicrograph showing large numbe of cisternae (c) and secetory vesicles (SV). (b) Freeze-fractured preparation showing arrangement of cisternae (Electron photomicrographs) courtesy of Dr. E.G. Pollock).

to a specific cell function. For example, the dictyosomes occur in greatest number near the site of cell plate formation in dividing plant cells; the dictyosomes in goblet cells of the intestinal epithelium are located next to the region of the cell where mucigen granules are stored prior to secretion (Fig. 16.6).

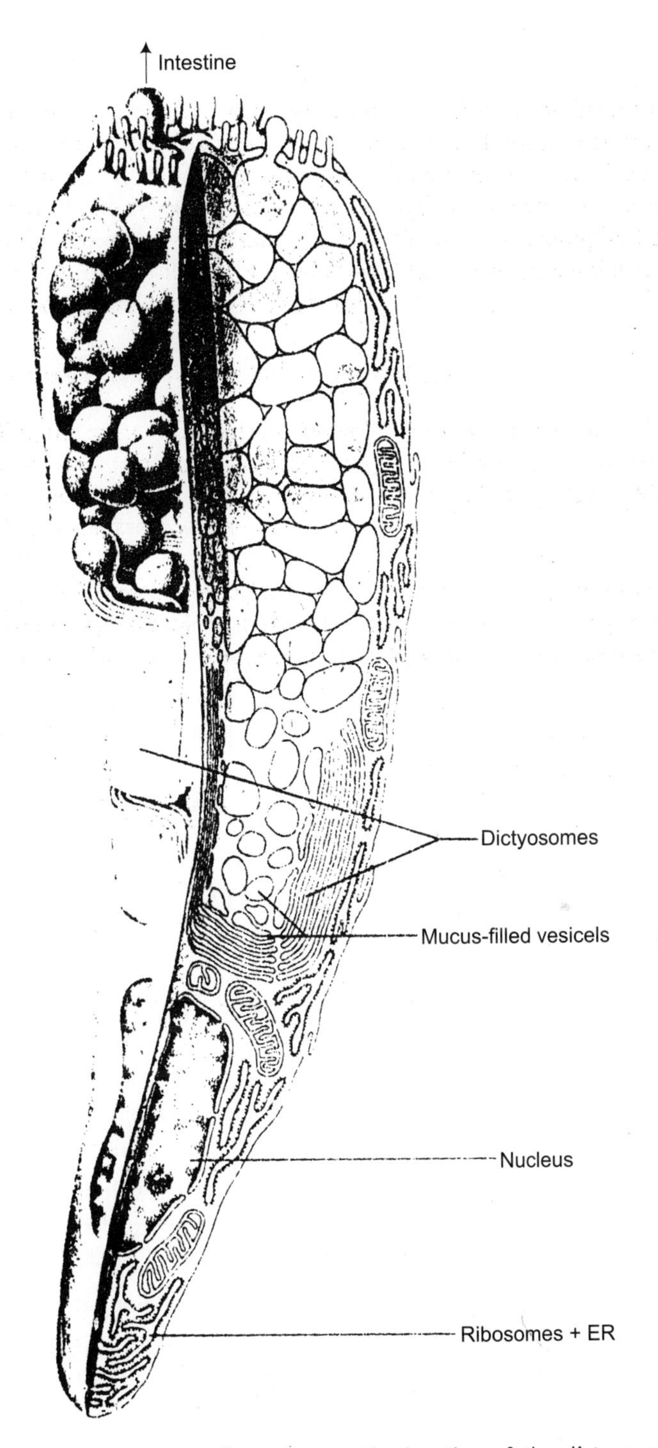

Fig. 16.5 Goblet cell the intestinal epithellum showing the location of the dictyosomes and mucigen granules.

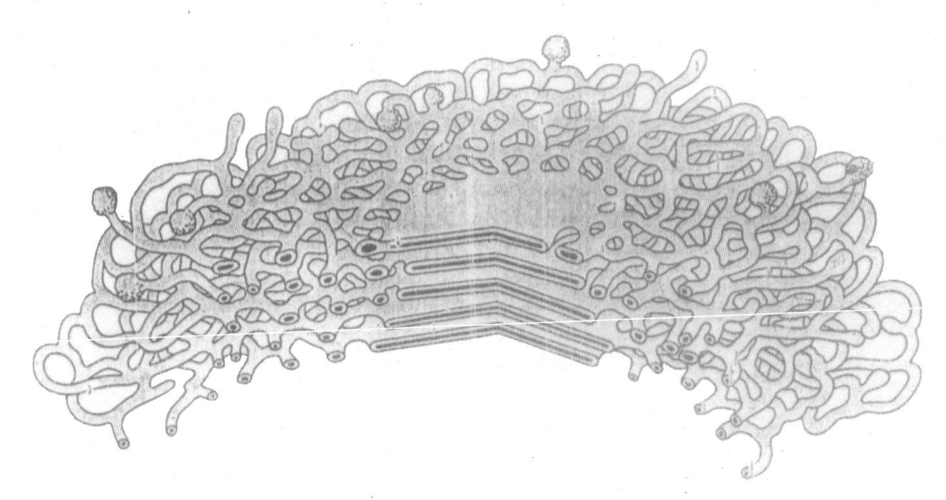

Fig. 16.6 Three-dimension drawing of dictyosome with highly fenestrated and tubular cisterne arranged in layers.

The Golgi apparatus, when used in the sense of hierarcny of structure, refers to the association of dictyosomes which together make up a cluster of distinct functional group. In some cases, the term "apparatus" refers to a single dictyosome together with the many surrounding small vesicles. In other cases, when there is evidence to suggest that they are associated in function, the term may apply to the entire collection of dictyosomes in the cell.

DEVELOPMENT OF THE GOLGI APPARATUS

Because it is not possible to observe living Golgi bodies clearly, it has not been possible to directly follow the developmental sequence of a single dictyosome. However, developmental sequences have been worked out by observing dictyosomes of cells at different stages of growth and then correlating differences in dictyosomes appearance with other development changes in the cell. During growth of the ciliate, *Tetrahymena pyriformis,* individual smooth surfaced cisternae are present about the oral region of the cell with small tubules between them. When the cells approach starvation (in the stationary phase of growth) stacks of lamellae form in the same region, apparently from the cisternae. During conjugation these cisternae are modified and assume the appearance of a typical Golgi apparatus. When fission occurs and the cells return to feeding and growth conditions, the Golgi apparatus disappears and isolated cisternae reappear.

The association of dictyosomes to form a complex Golgi apparatus could result either from aggregation of existing dictyosomes or from multiplication of dictyosomes in a confined area. Currently, the balance of evidence supports the first alternative. In spermatids, the dictyosomes fuse rapidly during development to form the Golgi apparatus of the acrosome. During zoospore formation in a fungal mold, the scattered dictyosomes aggregate in the cell. Observations of mammalian cells indicate that the dictyosomes frequently arise near the perinuclear region and migrate to other parts of the cell.

The differentiation of the dictyosome and Golgi apparatus often parallels the differentiation of the cell. In embryonic liver cells, the Golgi apparatus is made up of tubular cisternae. As the cells mature and differentiate, platelike cisternae are formed, and following this stage, secretory vesicles form at the outer edges of the cisternae. This sequence is illustrated in Fig. 16.7.

FUNCTIONS OF THE GOLGI APPARATUS

The Golgi apparatus is involved in many different processes in a variety of cells, but in general, the functions are associated with cell secretion or with membrane modifications. A third general function related to the first two is post-translational protein modification during the final assembly of glycoproteins.

Two sets of experiments bear on the role of the golgi apparatus in secretion. In 1964 L. Caro and G. Palade showed that the Golgi apparatus in the acinar cells of the pancreas is involved in the packaging of enzyme precursors into zymogen granules prior to secretion. Caro and Palade injected radioactive amino acids into rats and followed the movements of the "label" using autoradiography. This type of experiment is called a "pulse-chase" because the initial short-term application of labeled amino acids is immediately followed by the more prolonged application of unlabeled forms. Although amino acid metabolism and protein synthesis are not

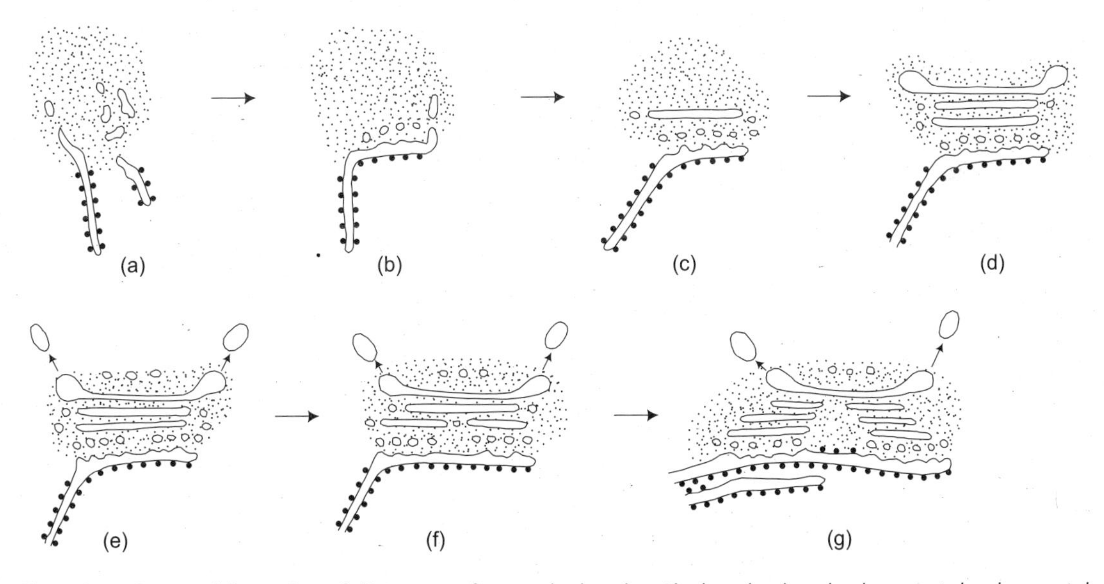

Fig. 16.7 Proposed formation of dictyosome from endoplasmic reticulum (a-c) and subsequent developmental stages: layered cisternae (a-c) formation of secretory vesicles (e), division of dictyosome (f-g). (Diagram courtesy of Dr. j. more et al., in Origin and Continuilty) of cell Organelles (Reinert and Ursprung, eds.), copyright 1971 Springer-Verlaq).

interrupted, the metabolic fate of the labeled amino acids can be traced through the cell with time as might be expected, after a 3-minute pulse the label appeared almost exclusively in the rough endoplasmic reticulum, since this is the region of protein synthesis. Following the 3-minute pulse, nonlabeled amino acids were added for 17 minutes (e.g., a total of 20 minutes from beginning of the pulse). Although some label was found in the rough endoplasmic reticulum as before, most of the label had shifted to the Golgi apparatus. When the chase was continued for an additional 100 minutes (120 total), almost all the label had left the endoplasmic reticulum and the Golgi apparatus and was now found in the zymogen granules and in the lumen outside of the cells (as a result of the release of the contents of the vesicles through plasma membrane). These experiments showed that the path of the amino acids is first into proteins in the rough ER and that there proteins are then transferred into the cisternae of the Golgi apparatus and then into the zymogen granules.

In 1966 using similar autoradiographic techniques M. Neutra and C.P. Leblond studied the secretion of mucous by the goblet cells of the intestinal epithelium. Mucose is a glycoprotein in which glucose and glucose derivatives are linked together forming polysaccharide side chains on the protein molecules (Fig. 16.8). Glucose labeled with tritium was used to follow the assembly and fate of the glycoproteins. Fifteen minutes after injection of the radioactive sugar, the label was most concentrated in the cisternae of the Goigi apparatus. This label did not enter or associate with the rough endoplasmic reticulum first. After a 20-minute chase the label appeared in the mucose vesicles, and after 4 hours, most had been released through the plasma membrane into the lumen. This experiment not only shows the path of glucose through the cell but also reveals that final stages of assembly of the glycoprotein occur in the Goigi apparatus. Using the goblet as an example, Fig. 16.9 depicts the central role of the Goigi apparatus in the packing of newly-synthesized proteins into vesicles for secretion. The assembly of large molecules in the Golgi is not unique to globlet cells. Cartilage cells assemble glycoproteins in the cisternae of their Goigi bodies and sulfate groups have been shown to be added as well. Pectins and cellulose are assembled in the Goigi bodies of plant cells prior to deposition onto the forming cell plate or cell wall.

The participation of the Golgi apparatus in the growth and modification of the cell membrane is well documented, but it is not clear at present whether this is a major or minor means of cell membrane synthesis. The plasma membrane is thicker and has a different composition of phospholipids and sterols than the membranes of the endoplasmic reticulum or the Goigi apparatus. The Goigi membranes seem to be intermediate in thickness between the thin ER membranes and the thick plasma membrane. When cells such as *Trichonympha* (a protozoan) are starved, the dictyosomes and many endoplasmic reticulum membranes disappear. Upon refeeding, the endoplasmic reticulum membranes reappear first, following which vesicles from these membranes appear to branch of and develop into cisternae. With time, the cisternae develop into dictyosomes, and one can observe what appears to be a migration of the outermost cisternae toward the plasma membrane and their fusion with the plasma membrane. It seems reasonable to speculate that the proteins for the membranes are synthesized on the endoplasmic reticulum and the initial membranes formed others as well. Most speculation identifies the

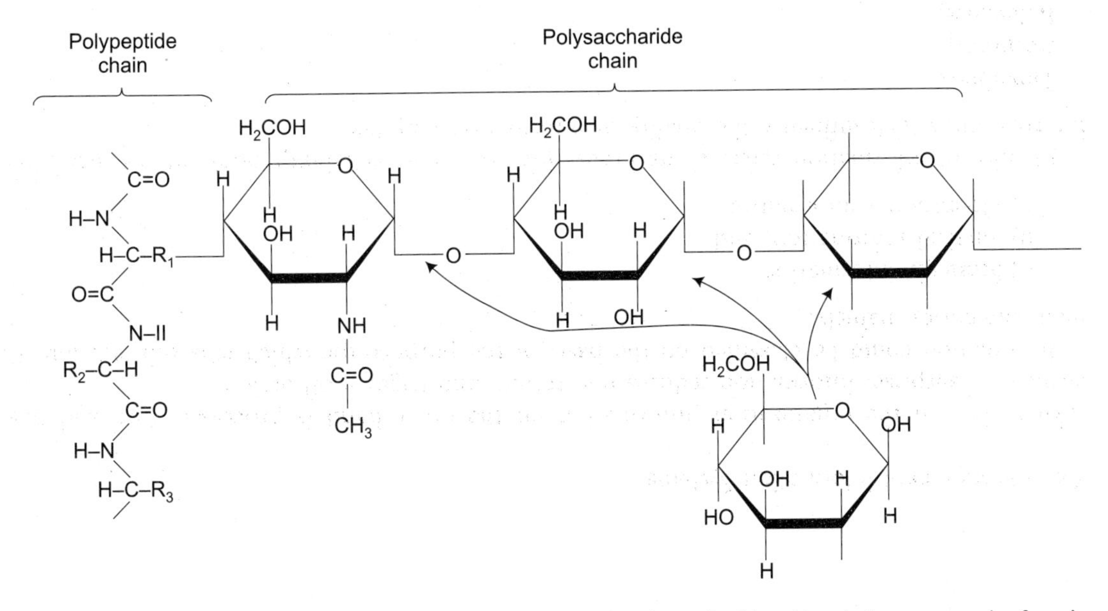

Fig. 16.8 Mucous glycoprotein showing attachment of polysaccharide side chain composed of various derivatives of glucose to the polypeptide chain.

smooth endoplasmic reticulum as the side of Goigi membrane assembly. The budding off of vesicles from the smooth ER to form cisternae and dictyosomes is probably accompanied by modification in the lipid components of the membrane. Further modification of both lipids and proteins probably occurs before fusion of dictyosome membranes with the plasma membrane.

CELL-SPECIFIC FUNCTIONS OF THE GOLGI APPARATUS

Formation of the Plant Cell Plate and Cell Wall

In plants, the cell plate and cell wall form during anaphase and telophase of mitosis and meiosis'I. During these final stages of nuclear division, the chromosomes have separated into two masses in the cell that will become nuclei. Between these two nuclei, pectin and hemiceullulose are deposited bit by bit, forming a plate in the center of the cell, which ultimately expands to the side walls, cutting and separating the protoplasts in two, thereby producing the two daughter cells. Prior to anaphase, the Golgi bodies are found outside the spindle. During anaphase, vesicles appear to be released from the Golgi apparatus, invade the center of the spindle, and aggregate about the spindle fibers. These vesicles provide the carbohydrate that forms the cell plate and eventually the wall. The nature of the carbohydrate secreted by the vessels is controversial. Some cell biologists believe that cellulose fibers are preformed and secreted, while others believe that the final stages of cellulose synthesis occur after secretion. In either case, the Golgi apparatus is clearly involved in the secretion of the carbohydrate that forms the wall between the two cell halves.

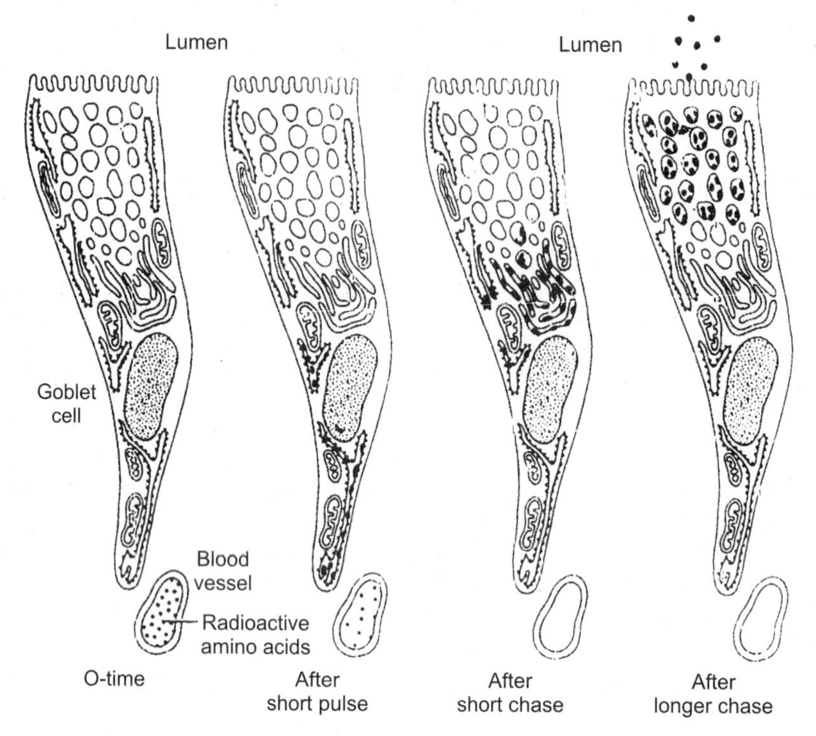

Fig. 16.9 Incorporation of amino' aclds into secretory proteins by the goblet cells of the intestine. Amino acids removed from the bloodstream are used in protein synthesis by rough endoplasmic reticulum. The proteins are conveyed to the Goigi apparatus for incorporation into secretory vesicles. Glycosylation of the proteins (to form glycoproteins) occurs within the Golgi apparatus. Vesicles detach from the maturing face of the Goigi apparatus and migrate to the plasma membranse whore they are discharged into the intestinal lumen.

The plasma membrane of plant cells does not pinch inward or grow inward during cell division as it does in animal cells. Instead, the membrane is formed on both sides of the developing cell plate and grows outward with it. Formation of the membrane results from the fusion of the vesicle membranes arising from the Golgi apparatus.

NEUROSECRETIONS

Nerve cells were the first cells described by Golgi in 1880 to contain the "internal reticular apparatus" (i.e.,Golgi apparatus). Since then, many studies have been conducted on the neurosecretion of the system, but the clarification that has been achieved with gland cells is still lacking. A great variety of substances are secreted by nerve tissue cells, including hormones (noradrenline, histamine, vasopressin, oxytocin, luteinizing hormone, follicle-stimulating hormone, etc) and other substances. Perhaps one the best known neurosecretion is acetylcholine, which is frequently described as a synaptic transmitter substance. It is this

Table 16.1 Specific functions of Golgi structures.

Cell	*Tissue or organ*	*Golgi function*
Exocrinc	Pancreas	Secretion of Zymogen (proteases, lipases, carbohydrases and nucleases)
Gland cell	Parotid gland	Secretion of zymogen
Goblet cell	Intestinal epithelium	Secretion of mucous and zymogens
Follicle cell	Thyroid gland	Prethyroglobulin
Plasma cells	Blood	Immunoglobulins
Myelocytes, sympathetic ganglia, Schwann cells	Nerous tissue	Sulfation reactions
Endothelial cells	Blood vessels	Sulfation reactions
Liver cells	Liver	Lipid secretion (lipid transformation?)
Alveolar epitheluim	Mammary gland	Secretion of milk proteins (and lactose?)
Paneth cells	Intestines	Secretion of proteins (chitnase?)
Brunner's gland cell	Intestines	Synthesis and secretion of mucopolysaccharides,enzymes, hormones
Connective tissue	Amblystoma limb	Synthesis (?) and secretion of collagen
Cornea	Avian eye	Secretion of collagen
Plant cells	Most	Secretion of pectin and cellulose

compound that is released by the end-bulb of many neurons (nerve cells) and the crosses the gap, called a synapse, between that neuron and the next acetylcholine is known to be present in vesicles (called synaptic vesicles) in the end-bulb. When a nerve impulse eaches the nerve endings, the vesicles all discharge through the plasma membrane and can be easily observed. When the acetylcholine reaches the plasma membrane of the neuron at the other side of the synapse, a new impulse is generated. The origin of the synaptic vesicles and the site of synthesis of the acetylcholine are not clear. Many different compounds are present in separate vesicles in the end-bulb; also present are many lysosomes. Whether the Golgi apparatus gives rise to each type of vesicle and perhaps also the lysosomes or whether the lysosomes themselves give rise to some to some of the vesicles remains unresolved. At the present time, most investigators favor the notion that the Golgi apparatus is the source of the synaptic vesicles. Accetylcholine released by synaptic vesicles into the synapse is broken down into acetate and choline by the enzyme acetylcholine esterase. These products are reused for acetylcholine synthesis. Whether the resynthesis occurs in the synapse, in the end-bulb cytoplasm after reabsorption, or in the Golgi apparatus is not known.

INTERRELATIONSHIP BETWEEN GOLGI, LYSOSOMES, AND VACUOLES

The membranes of lysosomes, vacuoles, and Golgi bodies are similar (although not idenitical) to those of smooth endoplasmic reticulum. For this reason, it has been suggested that they may be derived either directly or indirectly from the smooth ER. There is good evidence that dictyosomes accumulate hydrolytic enzymes in their outer more mature regions, and this has led to the proposal that lysosomes are derived directly from the Golgi apparatus. Some vacuoles in plants also have been found to contain mall amounts of hydrolytic enzymes, and thus their derivation from the Golgi apparatus has been prosposed.

ACROSOME DEVELOPMENT IN SPERM

The development of the acrosome of sperm cells is a good example of the involvement of both the membrane and contents of Golgi cisternae in the formation of another organelle. The acrosome is a membrane-bound structure at the anterior end of sperm cells of most animals. It is a part of the membrane of the acrosome that appears to contribute to the recognition and binding of the sperm to the egg cell in fertilization. The acrosome contain hydrolytic enzymes of which hyaluronidase is the most prevalent and which contributes to the breakdown of the protective surfaces of the egg. The singular large Golgi body buds off large coated vesicles that migrate to the forming acrosome. At the surface of the acrosome, the coated vesicle fuses, contributing its membrane to the growing membranes of the acrosome and the contents to the developing acrosomic granule.

Since the acrosomic granule is made up of hydrolytic enzymes, some have postulated that the acrosome is really a giant lysosome. As the acrosome expands, the Golgi becomes reduced in size, and in many, mature disappears entirely. The outer membrane of the acrosome merges with that of the plasma membrane. In mouse sperm, it has been shown that the area of the plasma membrane that has fused with the acrosomal membrane contains a large number of concanavalin. A binding sites. The increased number of carbohydrates (glycoproteins) in the membranes is attributed to the Golgi origin of the membrane.

SUMMARY

The membrane lined cisternae of the Golgi apparatus develop either from the outer portion of the nuclear envelope or from the endoplasmic reticulum. The cisternae may form a stack of vesicles that acts as a unit (called a dictysome), with vesicles joining the stack on some edge, maturing, and leaving the stack at the outer edge. A cluster of dictyosomes forms the functioning Golgi apparatus of the cell.

The Golgi apparatus functions in cell secretion and as a site for assembly of glycoproteins and other membraneous components. In the plant cells, Golgi bodies are associated; with the formation of the cell plate, and in nerve and gland cells, with various neurosecretions. The Golgi apparatus also gives rise to vacuoles, including those that become lysosomes and certain microbodies.

CHAPTER 17

The Mitochondrion

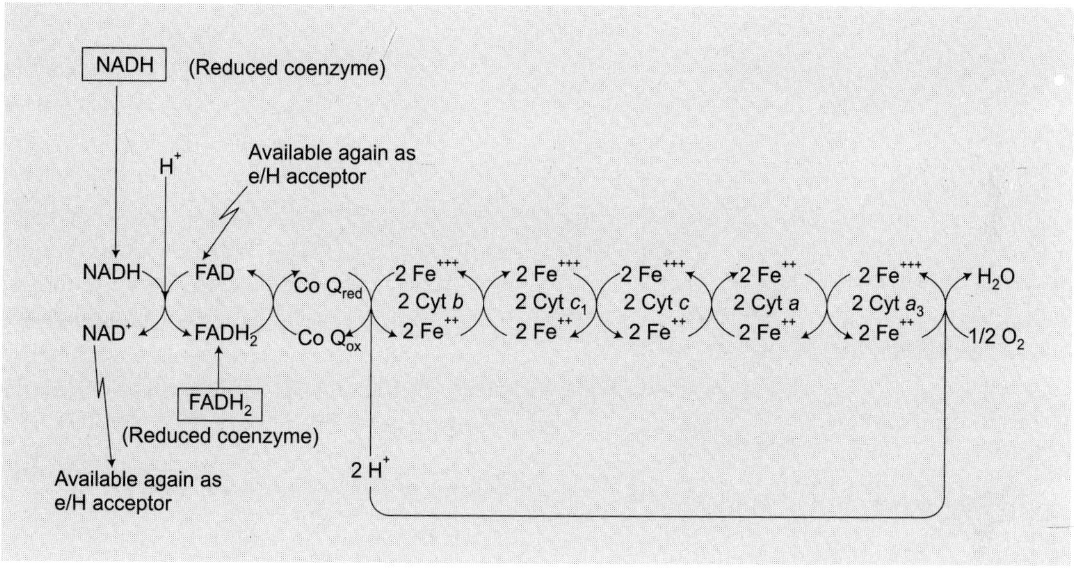

Most of the energy-requiring, or endergonic, reactions carried out in cells either directly or indirectly consume adenosine triphosphate (ATP). This "energy-rich" substance is converted in the process usually to adenosine diphosphate (ADP) and occasionally to adenosine monophosphate (AMP) (Fig. 17.1). Cells have evolved three major ways of producing this vital source of chemical energy: (1) ATP is produced in the cytosol during the chain of exergonic reactions called glycolysis in which sugars are catabolized; (2) ATP may be produced within the chloroplasts of certain plant cells, utilizing the energy of sunlight; and (3) ATP may be produced within the mitochondria present invirtually all plant and animal cells by the oxidation of a variety of elementary substrates. It is for this reason that mitochondria are often referred to as the cell's "powerhouse" their discussion of that subject is deferred until then. ATP synthesis in glycolysis occurs by substrate-level phosphorylations in which the phosphate is enzymatically transferred directly from the substrate to ADP to from ATP. Some ATP is generated in

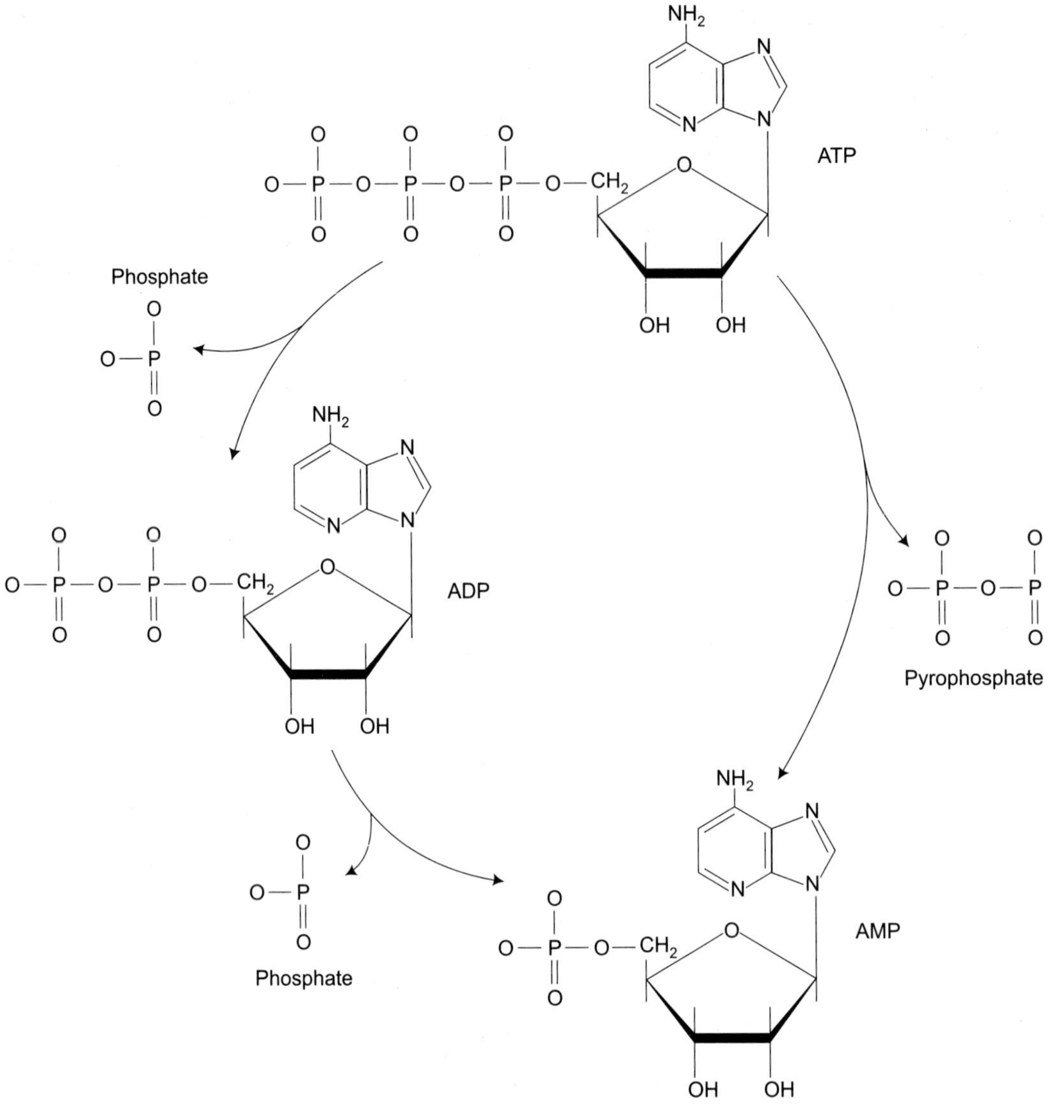

Fig. 17.1 ATP and its breakdown products ADP and AMP.

mitochondria by substrate-level phosphorylations but the greater amount is formed by special electron transport system (ETS) oxidation-reduction reactions. Although much of the present chapter will be devoted to the manner in which mitochondria function in ATP production it is to be noted that mitochondria also play several important roles; for example, in eucaryotic cells the mitochondria are also responsible for the oxidation of fatty acids and other lipids and are one of the sites for fatty acid chain lengthening. Some subunits of cytochromes are also synthesized in mitochondria, and final assembly occurs there.

In recent years, it has been clearly established that mitochondria contain their own genetic apparatus as well as the machinery for the synthesis of an array of enzymatic and structural mitochondrial proteins, and that mitochondria are capable of semiautonomous proliferation within the cell.

DISCOVERY OF MITOCHONDRIA

Mitochondria were first observed and isolated from cells about 100 years ago when Kollicker mechanically teased these organelles from insect striated muscle tissue and studied their osmotic behavior in various salt solutions. Kollicker, whose work extended over several decades beginning around 1850, concluded that these "granules" were independent structures not directly connected to the interior structure of the cell. In 1890, Altmann defined stains specific for these granules and named them "bioblasts", this term was superseded by Benda who introduced the expression "mitochondrion" (Greek: mito- = "thread" + chondrion = "granula") because of the threadlike appearance of these granules under the light microscope. In 1900, Michael introduced the use of the supervital dye Janus green B to specifically stain mitochondria to the exclusion of other cellular components and showed that oxidative reactions in the mitochondria caused color changes in the dye. Janus green B is still frequently employed as a cytological marker for mitochondria.

In 1910, Warburg showed that the "large granule" reaction isolated by low-speed centrifugation from tissues disrupted by grinding contained enzymes catalyzing oxidative cellular reactions, and Kingsbury in 1912 suggested that the mitochondria were the specific loci of the oxidation. Warburg's findings were confirmed in the 1930s by Clude, Bensley, and Hoerr, who employed more sophisticated methods to obtain a mitochondrial preparation essentially free of contaminating cell structures. They disrupted liver tissue using procedures almost identical to those currently used to prepare a conventional "homogenate" and isolated the mitochondrial by repeated differential centrifugation; the final mitochondrial preparation was then examined biochemically. At about the same time, Sir-Hans Krebs elucidated the various reactions of the tricarboxylic acid (TCA) cycle (or Kerbs cycle see later) and by 1950, Lehninger, Green. Kennedy, Hogeboom, and others had clearly shown that these reactions, as well as those of fatty acid oxidation and oxidative phosphorylation (i.e., ATP generation), were properties of mitochondria.

STRUCTURE OF THE MITOCHONDRION

The size, shape and structural organization of mitochondria, as well as the number of these organelles per cell and their intracellular location, vary considerably depending on the organism, tissue and physiological state of the cell examined.

Some cells, usually unicellular organisms, contain a single mitochondrion. Fig. 17.2 shows a photomicrograph of the single mitochondrion in the mobile swarm spore of *Blastocladiella emersonia*, a fungus, and a model of the single mitochondrion of *Polytomella agilies*, an alga. At

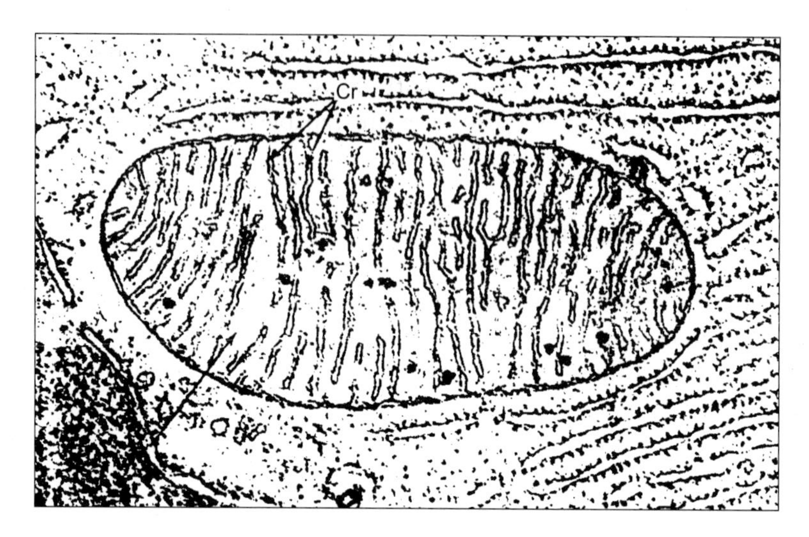

Fig. 17.2 Electron micrograph of a mitochondrion. Cristae (Cr) and matrix (M) are identified (Courtesy of K.R. Porter).

the other end of the scale are cells such as *Chaos Chaos*, an amoeba, which contains several hundred thousand mitochondria. Cells of higher animals also contain various numbers of mitochondria. Sperm cells have fewer than 100 mitochondria. Kidney cells generally contain less than 1000, while liver cells can contain several thousand. Procaryotic cells such as bacteria and blue-green algae do not contain mitochondria. The functions associated with mitochondria are carried out in the cytosol or associated with the cell (plasma) membrane in these-organisms.

The distribution of the mitochondria in the cell may vary, although in most cells the distribution appears to be at random. In *Blastocladiella*, the single mitochondrion is at the base of the flagellum. A concentration of mitochondria also appears in metabolically active areas of cells. In epithelial cells lining the lumen of the small intestine, the mitochondria occur in greater numbers near the surface of the cell adjacent to the lumen (where active absorption of digestive products is occurring). In general, when mitochondria are present in greater numbers is one part of the cell than another, it is usually near a site where significant ATP utilization is occurring. For example, is muscle tissue the mitochondria are aligned in rows parallel to the contractile fibrils. In many plant cells, cyclosis, the active streaming of the cytoplasm about the cells, tends to distribute the mitochondria in a uniform manner.

The number and distribution of mitochondria in a cell are closely related to the activity of the cell and its organelles. Cells that are actively growing, producing especially large amounts of some product such as digestive enzymes, actively transporting materials into the cell, or undergoing movement may actually increase the number of mitochondria during period of activity and reduce these numbers during period of quiescence. Yeast that are grown anaerobically produce cells in successive generations with fewer and fewer mitochondria per cell. However, cells that have been grown in this manner without oxygen will rapidly produce greater numbers of mitochondria per cell when oxygen and appropriate nutrients are added to the culture. An increased rate of cell growth and division is also observed as greater numbers of mitochondria are able to produce more ATP to facilitate absorption of nutrients and synthesis of cell constituents.

The size and shape of mitochondria, like the number in a cell, vary according to the organism's tissues and the physiological state of the cell. Most common mitochondria are between 0.5 and 1.0 μ in diameter and may be up to 7 μ in length. Usually, the smaller the number of mitochondria per cell, the larger the individual organelles are likely to be. The mitochondria may be spherical, ovoid, cylindrical, dumbbell-shaped, racket-shaped, or irregular with branches (Fig. 17.2). In contrast to the gross structure, the internal structure of the mitochondrion is unique.

The light microscope reveals little about the structure of mitochondria, since these organelles are so small. However, with the electron microscope, the mitochondrion is revealed as highly complex organelle. Regardless of source, all mitochondria exhibit features in common, and we may therefore describe "generalized" organelle (Fig. 17.3). The mitochondrion contains two distinct membranes, once inside the other, called the outer and inner membranes (Fig. 17.4). The inner membrane separates the internal volume into two phases: the matrix, which is a gel-like fluid volume enclosed by the inner membrane, and the fluid intermembrane space between the two membranes. The outer and inner membranes themselves, as well as the matrix and intermembrane space, contain a variety of enzymes (Table 17.1). The matrix contains a number of the enzymes of the Krebs cycle (tricarboxylic acid cycle, or TCA cycle) as well as salts and water. Suspended in the matrix are circular strands of DNA (Fig. 17.5) and ribosomes. A number of other inclusions have been described for mitochondria from diverse tissues. These include filaments and tubules, what appear to be crystalline proteins, and a number of small, nonribosomal granules. The granules vary in density, and little is known about their function except that some of them appear to strongly bind divalent cations.

The intermembrane space has been identified as containing a couple of enzymes (Table 17.1), but generally it appears to be devoid of inclusions. Several types of connections through this space that join the membranes have been reported.

The inner and outer membranes are distinctly different both in structure and function. Although determination of the thickness of the, membrane is difficult from electron photomicrographs because various fixatives cause different amounts of swelling, the inner membrane appears to be somewhat thicker (60-82 A) than the outer membrane (about 60 A). The inner membrane has a greater surface area because it is folded or extended into the matrix. These projections are called cristae and vary in number and shape. With distinct exceptions, it would appear that the cristae in higher animal cells are broad folds that sometimes almost bridge a matrix. Usually the cristae lie almost parallel to one another across the short axis of the mitochondrion, but they may run longitudinally in some organelles or form a branching network. In protozoa and many plants, the cristae form a set of tubes that project into the

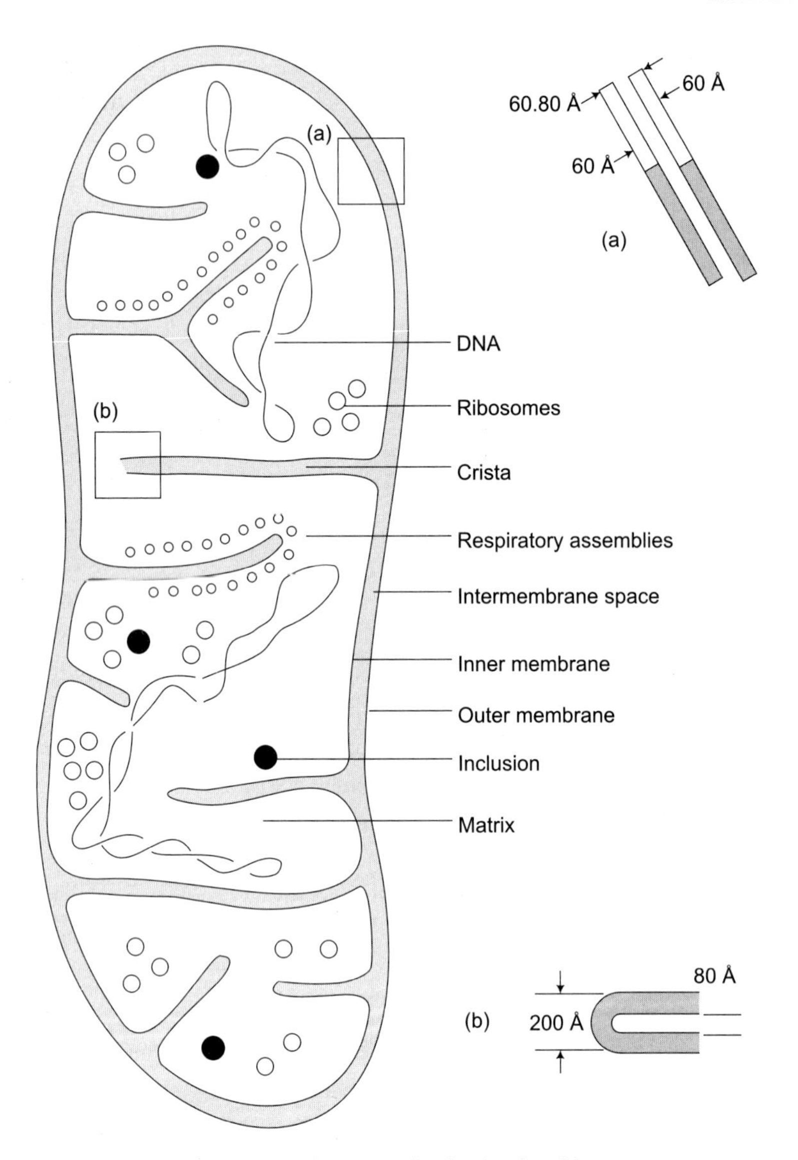

Fig. 17.3 The generalized mitochondrion.

matrix from all sides, sometimes twisting in different directions (Fig. 17.6). The number of cristae may increase or decrease depending upon aerobic activity. Active aerobic tissues producing great amounts of ATP generally have mitochondria with extensive cristae.

The difference in permeability between the two membranes is utilized to separate the outer membrane from the inner. Although several variations on the method exist, the basic technique is to disrupt the cell and isolate intact mitochondria by differential, density gradient or zonal centrifugation. The mitochondria are then placed in a hypotonic solution that causes them to swell. Usually, a phosphate buffer is used for this purpose. As the mitochondria swell, the outer

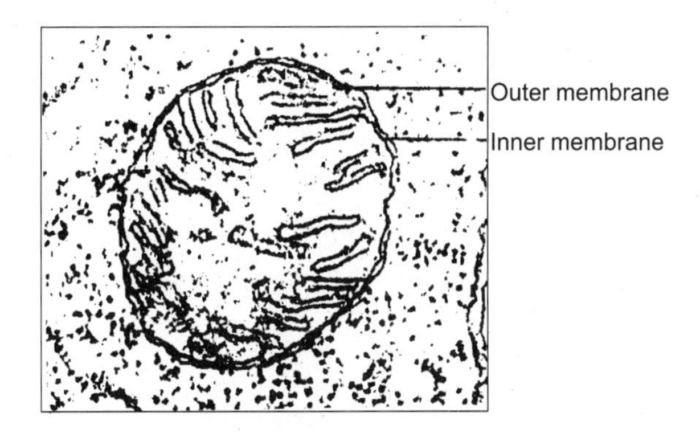

Fig. 17.4 Transmission EM of inner and outer membranes of the mitochrondrion.

Table 17.1 Location of some mtiochondrial enzymes.

Outer membrance:
Monoamine oxidase
Fatty acid thiokinases
Kynurenine hydroxylase
Rotenone-insensitive cytochrome *c* reductase
Space between the membranes:
Adenylate kinase
Nucleoside diphosphokinase
Inner membrane:
Respiratory chain enzymes
ATP-synthesizing enzymes
α-Keto acid dehydrogenases
Succinate dehydrogenases
D-β-Hydroxybuyrate dehydrogenase
Carnitine fatty acyl transferase
Matrix:
Citrate synthase
Isocitrate dehyrogenase
Fumarase
Malate dehdrogenase
Aconitase
Glutamate dehydrogenase
Fatty acid oxidation enzymes

Source: Courtesy A. Lehninger, *Biochemistry*, 2nd ed. Worth Publishing Co, New York, 1974, p. 512 (copyright Worth Publishing Co.).

membrane ruptures and fragments; the inner membrane also swells, causing the loss of cristae but normally does not break. The mitochondria are then placed in a hypertonic solution, causing the inner membrane matrix to shrink. The hypertonic solution is frequently a sucrose solution containing ATP and Mg^{++}. Since the inner membrane appears to be attached to the

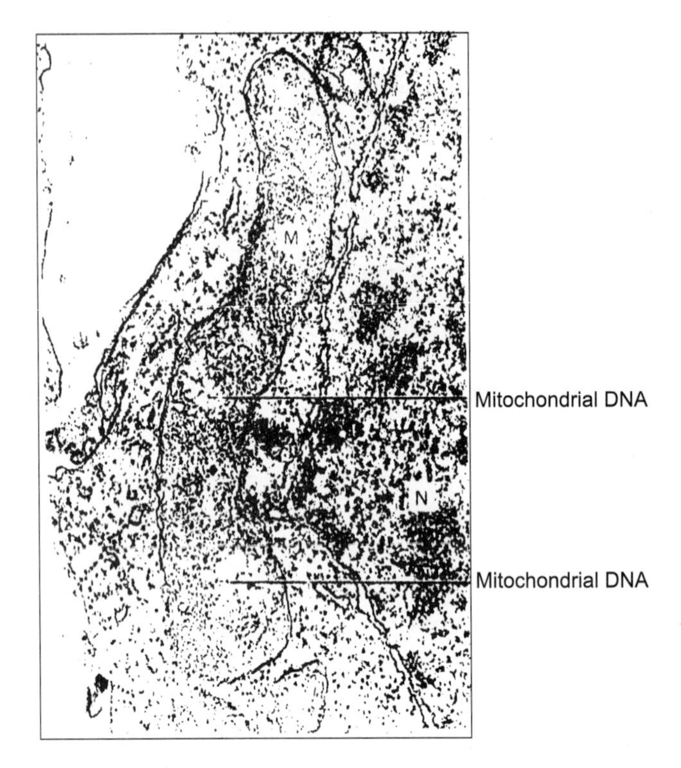

Fig. 17.5 DNA in the mitochondrion. M, mitochondrion; N, nucleus (Courtesy of Dr. E.G. Pollock).

outer membrane at several locations (probably by proteinaceous connecting strands), some workers use digitonin, EDTA, or sonication to aid the separation. Resuspension in an isotonic solution allows the inner membrane matrix to reestablish typical morphology. Separation of the membrane fractions can be done by differential centrifugations.

TCA CYCLE

The differences and relationship between the membranes are also apparent when the metabolic state of the mitochondria is altered. These reversible conformational states, illustrated in Fig. 17.9, were originally proposed by C. R. Hackenbrock. The orthodox conformation is the normal state and is transformed to the condensed conformation by binding of ADP to ADP-ATP translocase molecules on the inner membrane. The swollen and contracted conformations produced by ion shifts illustrate drastic osmotic changes.

The unique structure of the mitochondrion has prompted many investigators to try to correlate with function. As listed in Table-17.1, the location of many enzymes has been determined, and it is generally possible to assign specific metabolic functions to the interemembrane space, matrix, and inner membrane spheres. In some cases, the specific surface of the membrane can be assigned a function.

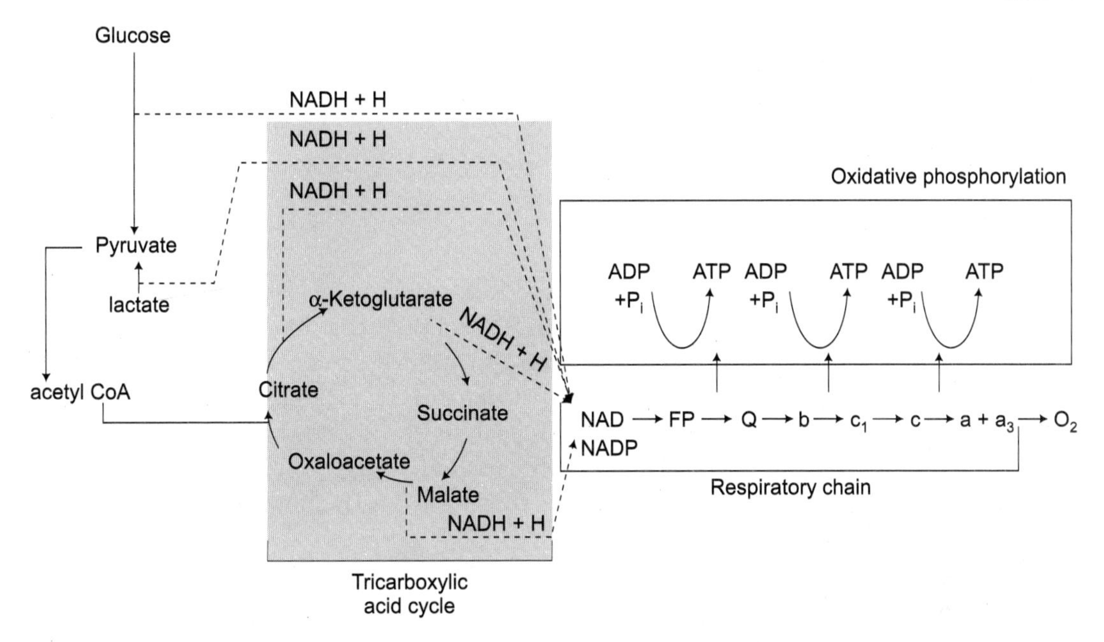

Fig. 17.6 Relationship between substrate oxidation, respirator chain oxidation reductions, and oxidative phosphorylations.

Among the most thoroughly studied processes unique to mitochondria are substrate oxidation, respiratory chain oxidation-reductions, and oxidative phosphorylation. In Fig. 17.6, these major events are diagrammed. Metabolic products of reactions in the cytosol (such as pyruvate formation during glycolysis) enter the mitochondrion to be oxidized by the Kerbs or tricarboxylic acid cycle enzymes. The enzymes for each of these reactions (except succinic dehydrogenase; see later are believed to be localized in the matrix or on or near the matrix facing surface of the inner membrane.

As a result of the Krebs cycle oxidations, CO_2 and water are formed as waste products, and a number of specialized oxidation-reduction compounds are reduced. These compounds (e.g., NADH act as the initial step in a sequence of oxidation and reduction reactions, called the respiratory chain that is specifically associated with the inner membrane of the mitochondrion. The result of this chain reaction is the reduction of O_2 form H_2O as a "waste product" and the induction of a third major process called oxidative phosphorylation. Oxidative phosphorylation, the net result of which is the production of ATP, is intimately associated with the inner membrane spheres.

In effect, pyruvate and other small molecules of cytosol metabolism must diffuse through the very permeable outer membrane and across the intermembrane space. It is upon entering the inner membrane that the three major reaction sequences (Kerbs cycle, respiratory chain oxidations-reductions, and oxidative phosphorylation) begin. During glycolysis in the cytosol, reduced pyridine nucleotides (e.g., NADH) are produced, as are others (e.g., NADPH) in the hexose monophosphate shunt and other metabolic pathways. These reduced compounds may also pass through the outer mitochondrial membrane or transfer their reductive capacity to the respiratory chain compounds and thereby innate the latter two major reaction sequences; respiratory chain oxidations-reductions and oxidative phosphorylation. Each of these three major sequences is described in more detail below.

TRICARBOXYLIC ACID REACTION

The tricarboxylic acid cycle is frequently called the Kerbs cycle because the major steps of the cyclic reactions were first proposed in 1937 by H. A. Kerbs. At that time radioactively labeled compounds were not available for experimentation, the cellular site of the oxidation reactions was not determined, and even the compounds that initiates the cycle (i.e., citric acid) was not known with certainty. Kerbs' experiments and his analysis of the data were outstanding. It was not until 1948, when E. P. Kenndy and A. L. Lehniger homogenized rat liver and fractionated and assayed the enzyme activity of its components that it was finally shown that the Kerbs cycle reactions were localized in the mitochondria.

The Kerbs cycle reactions are utilized by the cell to further metabolize a number of products of other reactions in the cytosol. These can be products as diverse as amino acids, fatty acids, and pyruvate. However, the greatest metabolic merit of the cycle is its oxidation of the pyruvate-the product of carbohydrate metabolism in the cytosol. The oxidation of this compound in the TCA cycle provides the reducing power to subsequently make significant amount of ATP. The reactions are summarized in Fig. 17.7 and are described below.

The oxidation of pyruvate is brought about by a complicated set of reactions catalyzed by the pyruvate dehydrogenase complex. This multienzyme complex is located in the matrix of the mitochondria. The first step is the decarboxylation of the pyruvate to yield CO and an A-hydroxyethyl unit attached through thiamine pyrophosphate (TPP) (a coenzyme) to that enzyme (E). That is, the first step of the Krebs cycle is a condensation reaction between acetyl CoA and oxaloacelate, a product of the cycle itself. The reaction is catalyzed by citrate synthetase, with citrate produced and CoA freed for reutilization in the prior reaction.

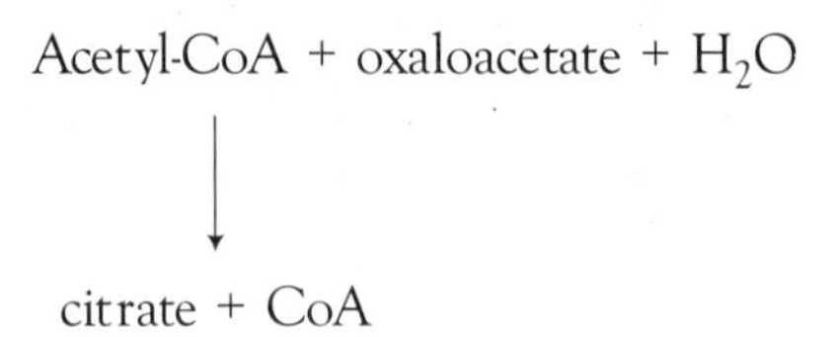

The G' is –7.7 kcal mole-1. This reaction is the "pace-maker" or primary rate-limiting reaction of the cycle. The rate is controlled by the availability of acetyl CoA and oxaloacetate. Succinyl CoA, the product of a later step in the cycle, is a competitive inhibitor of this reaction, since it competes with acetyl CoA for the active site on the enzyme.

The enzymatic conversion of citrate to isocitrate is a two-step conversion in which intermediate, cis-aconitate remains attached to the enzyme aconitase and therefore is frequently not shown in the Kerbs cycle. The equilibrium of this reaction lies toward citrate, G' is

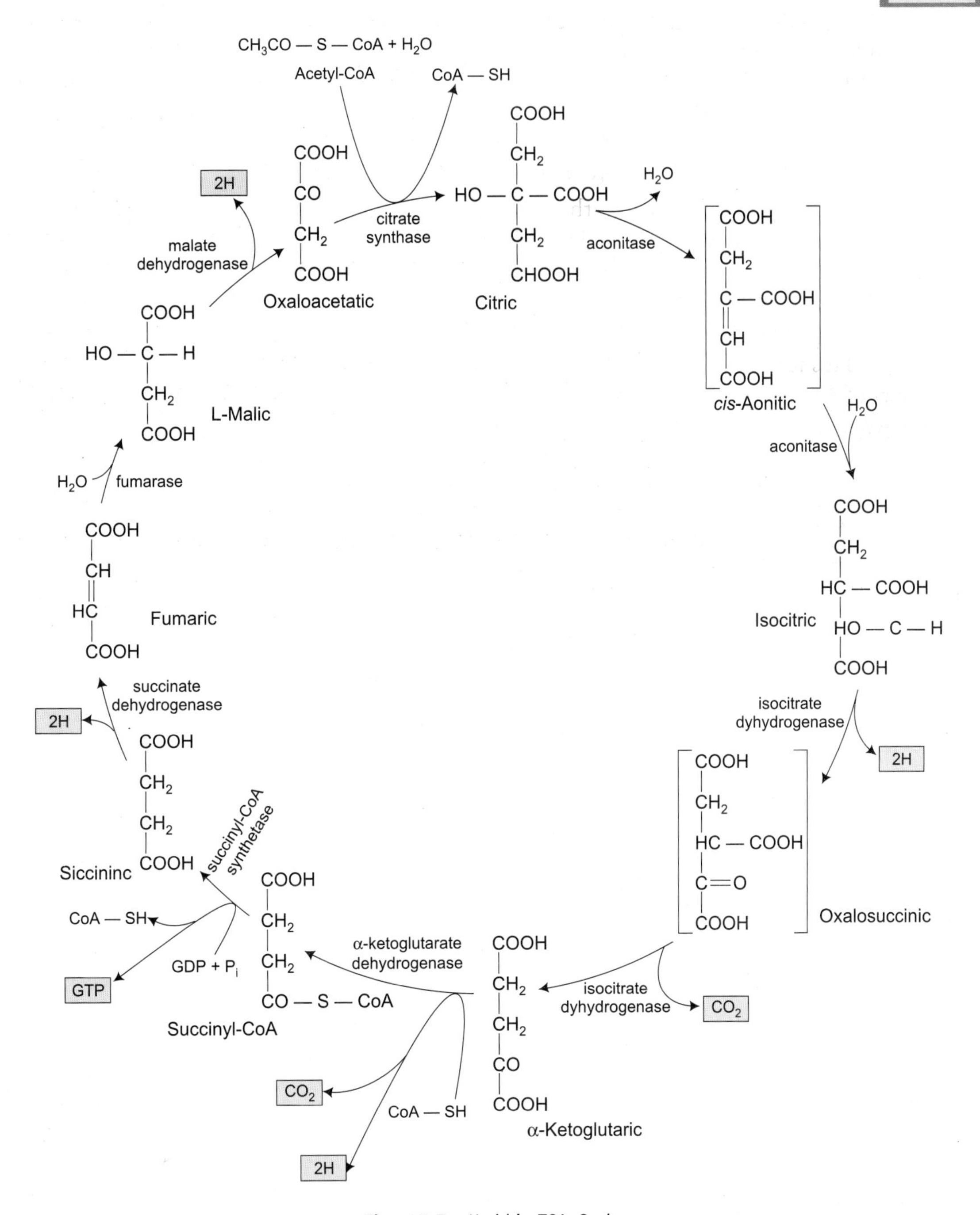

Fig. 17.7 Krebb's TCA Cycle.

1.59 kcal mole, but the isocitrate is rapidly oxidaized in the next step, thus shifting the direction of the reaction by removal of the product.

$$\text{Citrate} \xrightarrow{H_2O} \text{[cis-aconitic acid]} \xrightarrow{H_2O} \text{isocitrate} \tag{6}$$

The oxidation of isocitrate to A-ketoglutarate is also a two-step process, with the intermediate remaining attached to the enzyme isocitric dehydrogenase. The first part of the reaction is an oxidation with the two hydrogens being transferred to NAD+. Actually, there are two isocitric dehydrogenases in the mitochondrial matrix. One is linked to NAD+ and the other to NADP+. An NADP+ linked isocitric dehydrogenase is also found in the cytosol. However, it appears that the NAD+-isocitric dehydrogenase is the most active form in Krebs cycle reactions. This form of the enzyme is allosteric and specifically stimulated by ADP. When large amounts of ATP are consumed in the cell. ADP would be produced, acting as a stimulus for the Kerbs cycle. Alternately, when ATP accumulated there would be little ADP and the TCA cycle reactions would slow down owing to inactivity of ADP-dependent NAD+ -isocitric dehydrogenase. This point of rate control is considered secondary however to the citrate synthetase step.

The second part of the reaction catalyzed by isocitric dehydrogenase is the decarboxylation of the B-carboxyl group. The ΔG for the entire reaction is –5.0 kcal mole-l

Isocitrate

NAD

NADH + H^+

Oxalosuccinate

CO_2

α-Ketoglutarate (7)

The oxidation of A-Ketoglutarate to succinyl CoA is very similar in reaction sequences and enzyme complex to the conversion of pyruvate to acetyl CoA. The enzyme complex is called the α-ketoglutarate dehydrogenase complex. In this sequence of reactions, CO is removed by first complexing with thiamine pyrophosphate. During addition of CoA, two hydrogen atoms removed from lipoic acid and coneznyme A reduce Nad^+ to NADH + H^+. The G is –8.0 kcal mol.

$$\alpha\text{-ketogluarate} + NAD^+ + CoA \rightleftharpoons \text{succinyl-CoA} + Co + NADH + H^+ \tag{8}$$

The removal of the CoA is coupled with a substrate level phosphorylation of GDP catalyzed by succinyl CoA synthetase.

$$\text{Succinyl CoA} + P_1 + \text{GDP} \rightarrow \text{succinate} + \text{GTP} + \text{CoA} \quad (9)$$

The phosphate is actually attached to the enzyme-succinyl-CoA complex first and then transferred to GDP. In *E. coli*, the phosphate is transferred to ADP; in animals and most plant tissues GTP is formed and then the GTP donates the phosphate to ADP to Form ATP.

$$\text{GTP} + \text{ADP} \rightarrow \text{ATP} + \text{GDP} \quad (10)$$

The $\Delta G'$ is -0.7 kcal mole.

Succinic dehydrogenase catalyzes the oxidation of succinate to fumarate. This sis the one enzyme of TCA cycle reactions that has been shown to be firmly bound to the inner surface of the inner membrane and not associated with the matrix as are the other enzymes. This enzyme contains a flavin adenine dinucleotide (FAD) as a coenzyme, and it is this coenzyme that accepts the two hydrogens removed from the succinate during oxidation.

$$\text{Succinate} + \text{FAD} \rightarrow \text{fumarate} + \text{FADH} \quad (11)$$

The $\Delta G'$ is 0,

Fumarate is converted to malate by fumaqrase. The reaction has a $\Delta G' = -0.88$ kcal mole.

$$\text{Fumarate} + \text{HO} \rightarrow \text{malate} \quad (12)$$

Malate is oxidized by malate dehydrogenase, which is an NAD^+ -containing enzyme. Although the G' is 7.1 kcal mole and indicates that the isolated reaction.

$$\text{MALATE} + NAD^+ \rightarrow \text{oxaloacetate} + \text{NADH} + H^+ \quad (13)$$

Is endergonic, the products of the reaction are both readily removed in vivo and the reaction is therefore driven in the direction as written. Malic dehydrogenase is found not only is the matrix of mitochondria but also in the cytosol. The oxaloacetate produced by the reaction initiates the cycle again by combining with acetyl CoA form citrate.

SUMMARY OF THE TCA CYCLE

It is possible to account for all atoms entering the cycle. There are two acetyl carbons in acetyl CoA, and during the Krebs cycle, one is converted to Co at the isocitric dehydrogenase step (7) and the other at the A-ketoglutarate dehydrogenase step (8). Although these are not the same carbon that entered the cycle, a balance is achieved-two carbons enter the cycle and two leave the cycle.

The same accounting of carbons can be made when starting with glucose. This monosaccharide is broken down during glycolysis in the cytosol into two molecules of pyruvate. Each pyruvate molecules loses one carbon to CO as it enters the Kerbs cycle at the pyuvate dehydrogenase step (1, 2) and another at each of the two steps described in the Krebs cycle.

$$\underset{(6\text{ carbons})}{\text{Glucose}} \rightarrow \underset{(3\text{ carbons each})}{2\text{pyruvate}} \rightarrow 6\text{CO} \quad (14)$$

A balance can also be made of the hydrogen atoms during glycolysis and Krebs cycle reactions. One glucose ($C_6H_{12}O_6$) and 6 H_2O molecules contribute a total of 24 hydrogens. During glycolysis, 4 hydrogens are removed at the glyceraldehydes-3-phosphate dehydrogenase step (14) to form 2NADH + H^+ in the cytosol. Again, the two pyruvates from glucose each yield two more hydrogens (2NADH + H^+) at the pyruvate dehydrogenase step (1, 2). Four more are removed at the isocitric dehyrogenase step (7) to form NADH + H+, four more at the A-ketogluarate dehydrogenase step (16-8), four more at the succinate dehydrogenase step (11), and finally four more at the malate dehydrogenase step (13). This makes a total of 24 hydrogens. Again, although these are not exactly the same hydrogen atoms that were present in the original glucose and water molecules, a balance is achieved.

In a similar way, the oxygen also balances. Glucose contributes 6 oxygens, and the 6 waters added bring the total to 12 oxygens entering the system. Twelve oxygens are removed from the systems in the 6 CO_2 molecules produced. A summary of all of these balances is apparent in the overall equation for the respiration of glucose through the Krebs cycle reactions;

$$C_6H_{12}O_6 + 6H_2O + \{10NAD + 2FAD\}$$
$$\downarrow$$
$$6CO_2 + \{10NADH + H\}\{2FADH\} \quad (15)$$

the important part of the metabolism of the Krebs cycle reactions is the production of tremendous chemical reducing power in the matrix of the mitochondrion by the accumulation of NADH + H^+ and FADH. These compounds now enter the electron transport system reactions, and their potential drives oxidative phosphorylation thereby producing ATP.

ELECTRON TRANSPORT SYSTEM

The reduced NADH + H^+ and FADH that accumulate as the TCA cycle operates form an enormous potential energy pool. These compounds will be reoxidized and will transfer their reducing capabilities through a sequence of compounds to oxygen where the acceptance of electrons and H^+ will form water. The sequences of compounds through which the electrons and H^+ are passed is called the electron transport system. Each transfer of electrons (or H^+) from one compounds to the next results in the oxidation (electron loss) of the donor molecules and the reduction (electron acceptance of the acceptor molecules. Such a transfer occurred in reaction 16.3 in the matrix during pyruvate dehydrogenase action. The enzyme-linked FAD (E-FAD) accepted the electrons from the substrate and become reduced, that is, E-FADH. Subsequently, the E-FADH transferred hydrogens to NAD, which in turn became reduced while the FAD was reoxidized.

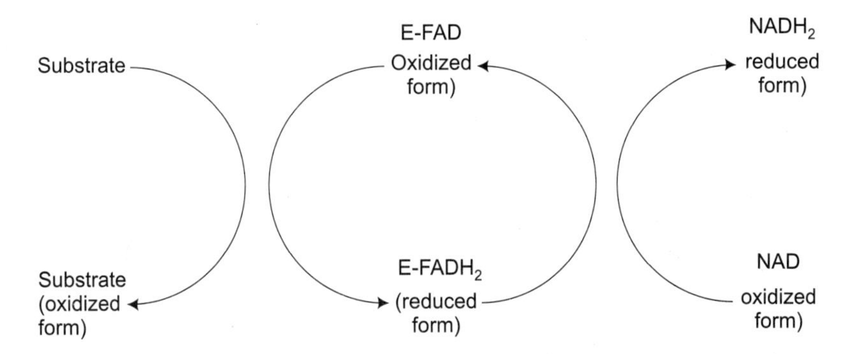

In the electron transport system the electrons are shuttled between about eight intermediate compounds before they eventually reach oxygen to form water. Although it is possible to reoxidize NADH + H+ directly by adding oxygen, in the mitochondria this is prevented. The enzymes of the electron transport system appear to be arranged physically in the inner membrane in such a manner that the transfer of the electrons must proceed through the specified series of compounds. With each transfer, there is an energy change. The energy released from electrons transfer is utilized to synthesize ATP—a process termed oxidative phosphorylation and described in the next sections.

OXIDATION-REDUCTION REACTIONS

Earlier the ability of acids and bases to lose or gain protons was described. In acid-base systems, one compound acts as the proton (H+) donor and the other as the proton acceptor. The proton donor is called the acid, and the proton acceptor is the base. The two form a pair that is called a conjugate acid-base pair. In a similar manner, oxidizing and reducing agents function as pairs. In this case, they are called redox pairs or redox couples. The member of the pair that donates the electron is called the reducing agents, or reductant, and the electron acceptor is called the oxidizing agent, or oxidant.

$$\text{Electron donor} \rightarrow e^- + \underset{\text{(oxidizing agent)}}{\text{electron acceptor}}$$

the terms donating and accepting fail to convey the forces or energy involved. In effect, the reducing agent has a certain ability (power) to hold electrons, as does the oxidizing agent. In a redox couple, one member attracts electrons more strongly than the other, and in effect the oxidizing agent can pull the electrons away from the reducing agent. This power to gain (or lose) electrons can be measured and is called the oxidation-reduction potential or redox potential.

It is frequently expressed volts. For convenience, the potential of most redox couples is standardized against a reference, which usually is a hydrogen electrode.

The mearurements are made using 1.0 M concentrations of oxidant and reductant at 25°C and at pH 7.0. The hydrogen electrode is equilibrated with H_2 gas as at 1 atmosphere and [i.e., $[H^+] = 10$ M], the redox potential of the reference hydrogen electrode indicates –0.42 volts. The

Table 17.2 Oxidation-reduction potentials of the electron transport system. (Values are based on two-electron transfers at ph = 7.0 and 25-30°C).

Electrode equation	E'_o V
$2H + 2e \rightleftharpoons H_2$	−0.421
$NAD^+ + 2H^+ + 2e^- \rightleftharpoons NAD + H^+$	−0.320
$NADP^+ + 2H^+ + 2e^- \rightleftharpoons$ a $NADPH + H^+$	−0.324
Ubiuinone + $2H^+ + 2e^- \rightleftharpoons$ ubiquinol	+0.10
2 cytochrome $b_{Kredh} + 2e^- \rightleftharpoons$ 2 cytochrome b_{klox}	+0.30
2 cytochrome $c_{red} + 2e^- \rightleftharpoons$ cytochrome c_{ox}	+0.254
2 cytochrome $a_{3red} + 2e^- \rightleftharpoons$ 2 cytochrome a_{3oxcy}	+0.385
$1/2O_2 + 2H^+ + 2e^- \rightleftharpoons H_2O$	+0.816

redox potentials of electron transport system compounds are given in Table 17.2. The greater the Eo' value, the ore strongly the oxidant binds electrons. However, an oxidant having a lower Eo' than another compound could become the reductant of electron donor to that compound.

Classes of Electron Transport System Compounds

There are live groups of compounds associated with the electron transport system (ETS). Three of these groups consist of enzymes whose coenzyme fraction or prosthetic group is known to be responsible for the transfer. They are (1) pyridine-linked dehydrogenases, which have either NAD+ or NADP+ as coenzymes; (2) flavin-linked dehydrogenases, which are linked to flavin adenine denucleotides (FAD) or flavin mononucleotides (FMN); and (3) the cytochromes, which contain iron-porphyrin prosthetic groups. Forming the fourth category is coenzyme Q or ubiquinone, a lipid-soluble coenzyme functioning in electron transport. The fifth group is composed of iron-sulfur proteins.

Pyridine-linked dehydrogenases require as their coenzyme either NAD^+ or $NADP^+$. As shown, both compounds can accept two electrons at a time; one as a hydride ion (II) and the other as in a hydrogen atom. There are about 200 dehydrogenases that employ NAD+ or NADP+ as coenzymes. Although the NAD+ and NADP+ dehydrogenases are found both in the cytosol and in the mitochondria and are known to transfer electrons between compounds in both places, it appears that only the NAD+ linked compounds are involved in the electron transport system.

Flavin-linked dehdrogenases require either FAD or FMN. Both are prosthetic groups and can accept two hydrogen atoms on the isoalloxazine ring. (Prosthetic groups are firmly bound to the protein and are not considered free carriers as are the coenzymes). Flavin-linked enzymes are involved in a number of enzyme systems, the more common of which are associated with fatty acid oxidation, amino acid oxidation, and Krebs cycle activity (pyruvate dehydrogenase and succinic dehydrogenase). It is not uncommon to have flavin prosthetic groups and NAD+ coenzymes linked to the same protein in dehydrogenases.

The cytochromes are proteins containing iron-porphyrin (or heme) groups. Most cytochromes are found in mitochondria, although some function in the endoplasmic reticulum and in chloroplasts. There are a large number of cytochromes in cells; in mitochondria there appear to be at least four different ones associated with the inner membrane. These are identified as cytochrome b, c1, a and a3. Some occur in two or more forms, but all transfer electrons by reversible valence changes of the iron atom ($Fe^{+++} \rightleftharpoons Fe^{++}$). In cytochromes b, c1, c and a, the manner of binding of iron in the porphyrin ring, and its association with the protein prevents the iron ligands from forming with oxygen and, therefore, these reduced cytochromes cannot be directly oxidized by molecular oxygen. Cytochrome a3, which is the terminal carrier in the electron transport system and which together with cytochrome a, forms the cytochrome oxidase complex, is an exception and can be directly oxidized by oxygen. In addition to cytochromes a and a3, cytochrome oxidase contains copper. Electrons received from cytochrome c are first picked up by cytochrome a and then transferred to a3. It is believed that their electrons are finally transferred from a3 to oxygen by a $Cu^{++} \rightarrow Cu^{+}$ intermediation.

Ubiquinones were so named because of their occurrence in so many different organisms and because of their structure (Fig. 17.8). They are found in several different forms and are related to the plastoquinones of chloroplasts. The forms present in mitochondria is often called coenzyme Q (CoQ) and will accept two hydrogen atoms at a time.

Electron Transport Pathway

The chain of electron transfers shown in Fig.17.9 has taken almost half a century to work out. During the early 1900s, workers discovered a number of dehydrogenases that remove hydrogen from substrates. Perhaps most notable of these early workers was T. Thunberg. In 1913,

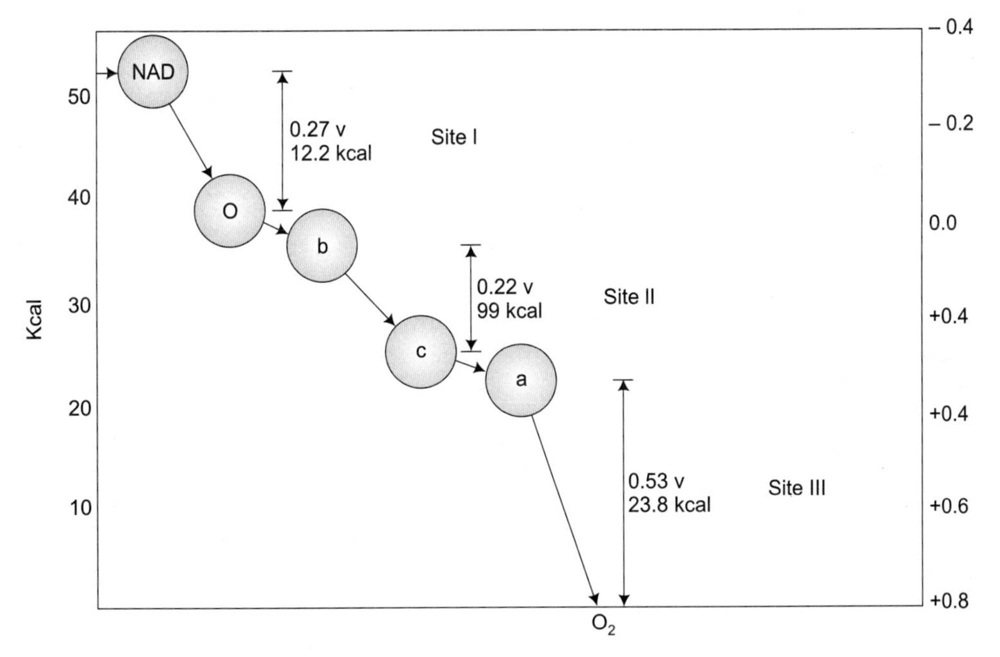

Fig. 17.8 Changes in free energy as electrons pass through the electron transport chain of enzymes, the three changes gengrating sufficient energy to drive the synthesis of ATP are indicated as sites I, II, and III.

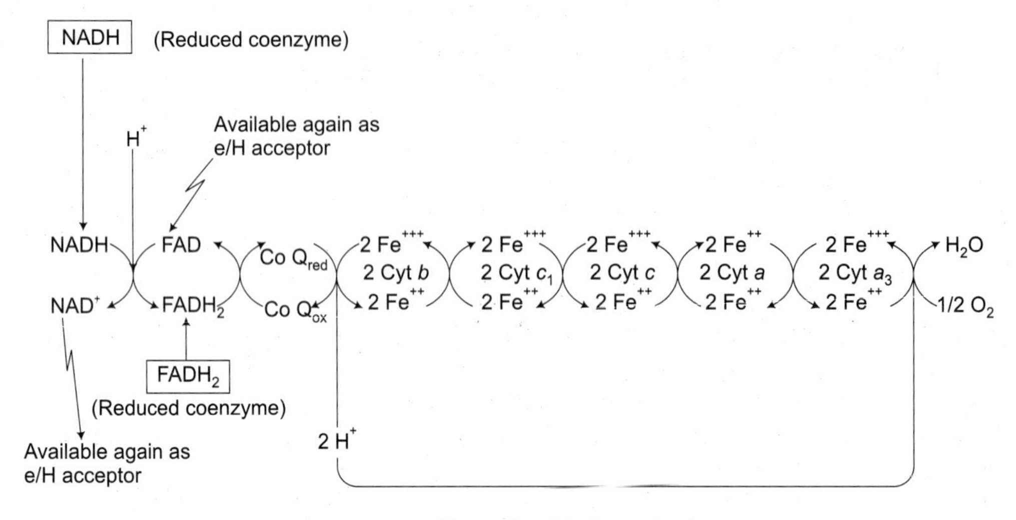

Fig. 17.9 Pathway of oxidation-reduction.

O. Warburg discovered that cyanide inhibits oxygen consumption but does not interfere with dehydrogenases. He proposed the existence of iron-containing "respiratory enzymes", now recognized as the cytochromes. The flavoproteins were identified by A.Szent-Gyorgyi as the intermediates between dehydrogenases and the respiratory enzymes. R.A.Morton added the information about ubiquinone and a number of workers, especially Keilin, Green, Okunuki, Singer, King, Chance, Williams, and Racker, added the details on compounds, structures, and sequences.

Today we visualize the electron transport chain as a sequence of compounds through which the hydrogens or electrons are passed in order, two at a time. The fact that the electrons do not jump or "short-circuit" to stronger oxidizing agents is probably the result of the physical positioning of the coenzymes in the inner membrane and the nonreactivity of some of the enzymes. However, hydrogen or electrons may enter the chain at various points.

Most electrons are removed from the metabolic substrates in the cytosol or matrix of the mitochondria by the NAD^+- (or $NADP^+$-) linked dehydrogenases. The reduced NADH + H^+ acts as the collection mechanism for carrying these electrons to the NAD+-flavoprotein-linked dehydrogenase in the inner membrane of the mitochondrion. As shown in Fig. 17.9, the reduced NADH + H^+ is then oxidized by FAD, which now being reduced is in turn reoxidized by CoQ, which is subsequently reoxidized by cytochrome b. The reduced iron in cytochrome b is then reoxidized by cytochrome c and finally the cytochrome oxidase accepts the electrons form cytochrome c.

Other Functions of Mitochondria

Mitochondria are generally described as the powerhouses of the cell, and as such, most interest is directed to the processes that evolve the most energy, namely the TCA cycle, electron transport, and oxidative phosphorylation. However, a great number of other reactions occur in mitochondria.

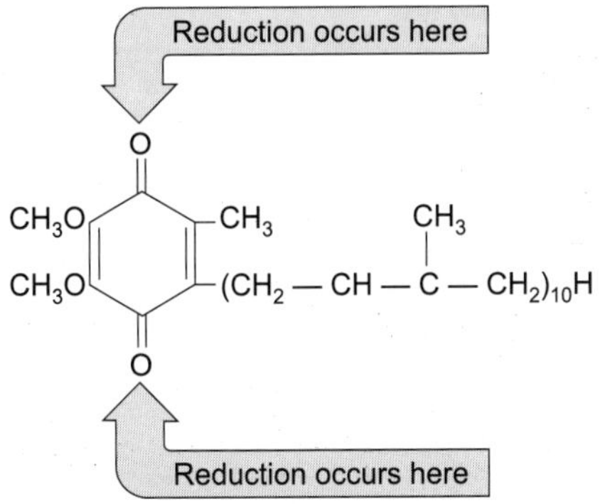

Reduction occurs here

Reduction occurs here

Cytochrome *c*

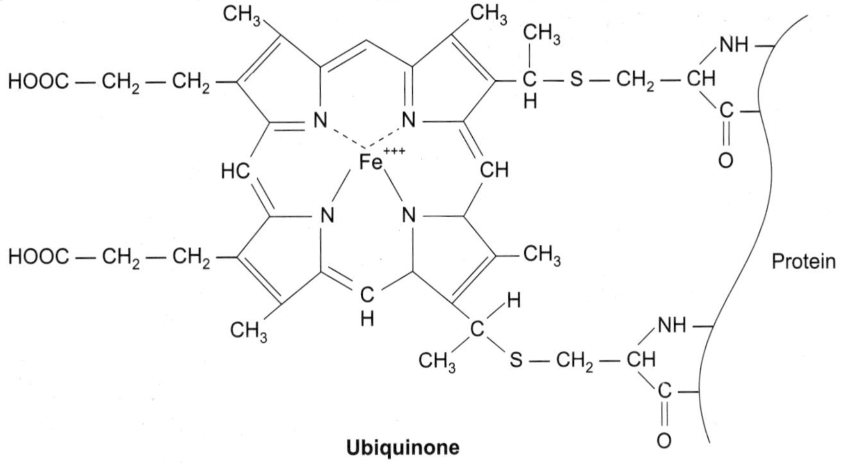

Protein

Ubiquinone

Fig. 17.10

The Glyoxylate Cycle

The enzymes are localized in the matrix, and some of them are the same as those of the TCA cycle. In effect, the glyoxylate cycle is a modified form of the Krebs cycle, but its function appears to be primarily associated with the conversion of acetate from fatty acid decomposition into oxaloacetate. Oxaloacetate is also an important intermediate in the conversion of fatty acids to carbohydrates. Animals lack certain enzymes of the glyoxylate cycle and are incapable of converting fatty acids to carbohydrates using this pathway. Both plants and microorganisms have functional glyoxylate systems. In higher plants, some of the enzymes of the system (e.g., isocitric lyase and malate synthetase) are localized in specific organelles called glyoxysomes as well as in mitochondria.

Fatty Acid Oxidation

Free fatty acids are rarely found in more than trace quantities in cells because they are highly toxic. The fatty acids associated with mono-, di-, and triglycerides and in phospholipids are

Table 17.3 TCA precursors of anabolic pathways.

TCA Cycle Intermedicite	*Pathway or anabolic reaction*
Citrate	Citrate + ATP + CoA → oxaloacctatc + ADP + P_i + acetyl CoA → fatty acid biosynthesis
Isocitrate	Isocitrate → glyoxylate → malate → carbohydrate synthesis
α-Ketoglutarate	α-Ketoglutarate + alanine → pyruvate + glutamate ↓
Succinyl CoA	Heme biosynthesis
Malate	Carbohydrate synthesis
Oxaloacetate	Oxaloacetate + alanine → pyruvate + asparate → protein synthesis

Table 17.4 Anapleurotlc reactions of the TCA cycle.

TCA cycle internediate produced	*Non-TCA generating the TCA cycle intermediate*
α-Ketoglutarate	Glutamate + pyruvate - a-ketoglutarate + alanine (from protein hreakdown)
Succinate	From glyoxylate cycle
Malate	Malic enzyme: pyruvatc + CO_2 + NADPH + H → malate + $NADP^+$
Oxalocetate	Pyruvate carboxyase: pyruvate + CO_2 + ATP + H_2O → oxaloacetate + ADP + P_i
Oxaloacetate	Aspartate + pyruvate → oxaloacetate + alaniae (from protein brcakdown)

Table 17.5 Mitochondrial membrane transport systems.

System	*Exchange*
Dicarboxylate carrier	Exchange on mole-for-mole basis of malate, succinate, fumarate, and phosphnte between matrix and cytosol.
Tricarboxylatc carrier	Exchange on mole-for-mole basis citrate and isocitrate, between matrix and cytosol.
	Exchange citrate or isocitrate for dicarboxylatr.
Aspartatc-glutamate carrier	Exchangc aspartate for glutamate across mcmhranc,
α-Ketogluturate-malate carrier	Specifically exchange α-ketoglutarute for malate across membrane.
ADP-ATP carrier	Exchange of ADP for ATP.

generally hydrolyzed from the glycerol in the cytosol and immediately activated for transport into the matrix of the mitochondrion where they are then oxidized.

Fatty Acid Chain Elongation

Fatty acids are generally sysnthesized on the smooth endoplasmic reticulum. However, there are a number of enzymes in mitochondria that catalyze the elongation of palmitic and other saturated fatty acids by successive additions of acetyl CoA to the carboxyl end. In smooth ER, both unsaturated and saturated fatty acids are elongated but by the addition of malonyl CoA rather than acetyl CoA.

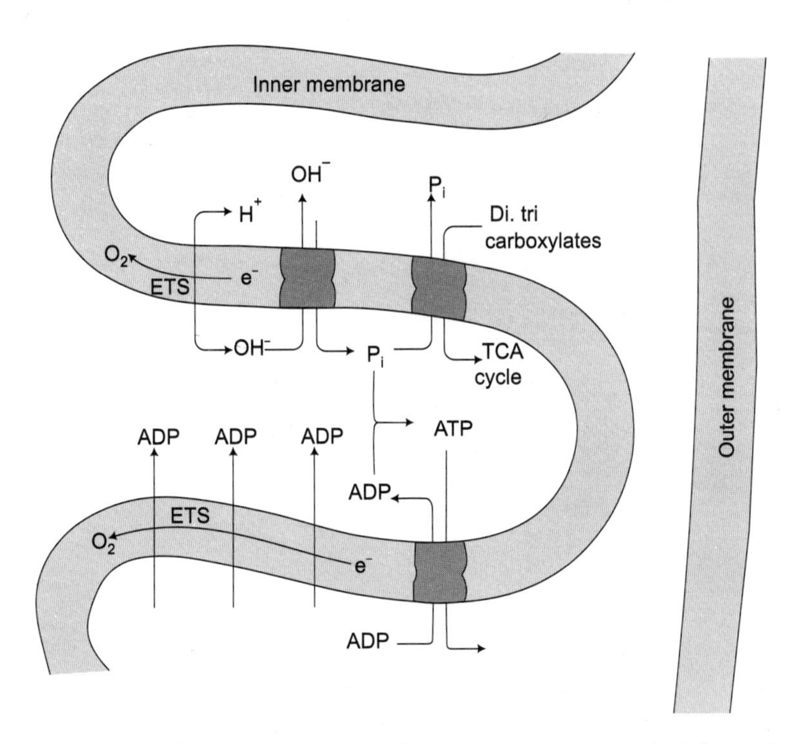

Fig. 17.11 The function of carrlers across the inner memorane, see text for explanation.

Superoxide Dismutase and Catalase

During electron transport, a number of toxic reductive products of oxygen are formed. The most common are superoxide (O_2^-) and hydrogen peroxide (H_2O_2). A protective enzyme has been identified in mitochondria called superoxide dismutase, when decomposes the O_2^- into hydrogen.

Total Energy Production from Catabolism of Glucose

The total energy produced from the complete oxidation of a molecule of glucose metabolized through glycolysis, the TCA-cycle, and electron transport is usually quoted as equivalent to a gross production of 40 ATP or a net production of 38 ATP. These ATP are produced by both substrate-level phosphorylations and electron transport coupled with oxidative phosphorylation.

SUMMARY

Mitochondria are the "powerhouses" of he cell and the primary sites of cell oxidations. Within the membranes of these organelles, elementary substrates produced by the breakdown of carbohydrates, lipids, or nitrogen macromolecules in other cell locations are oxidized to CO_2 and water. The energy released from the exergonic oxidations serves to form ATP.

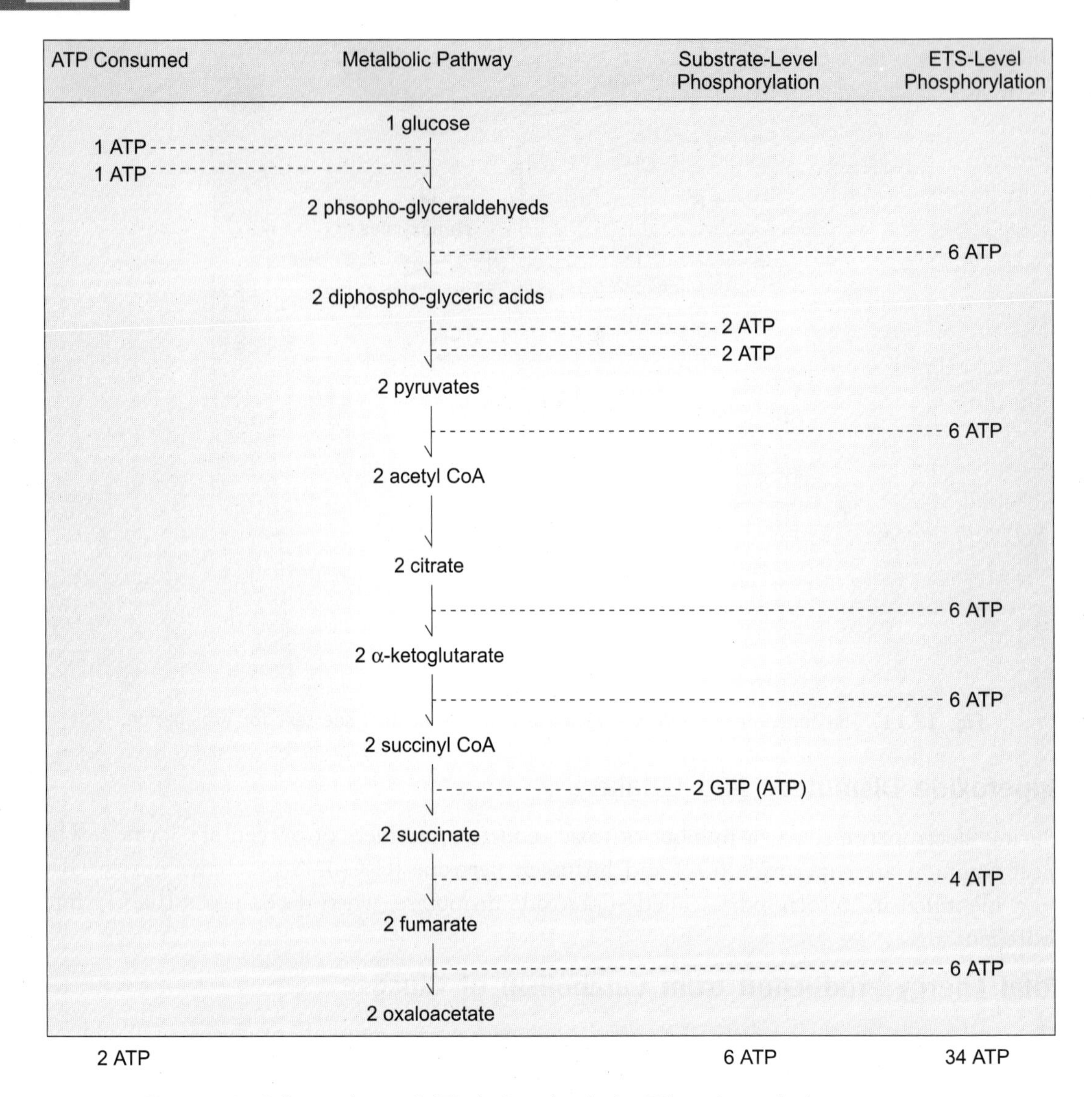

Fig. 17.12 Balance sheet of ATP during glycolysis, TCA cycle, and electron transport.

Mitochondria occur in almost every type of aerobic eucarylotic cell. Although their shape may vary, all mitochondria contain two structurally and functionally different membranes–an inner and an outer membrane. Between the two membranes is the fluid, intermembrane space. The inner membrane surrounds the innermost compartment or matrix. Projections of the inner membrane into the matrix are called cristae. New mitochondria form by pinching off from other mitochondria and may also arise from microbodies.

The oxidative and phosphorylation reactions occur in the inner membrane or in the matrix. The tricarboxylic acid or Krebs cycle reactions constitute the first phase of the oxidation of substrates such as acetate. In these reactions, the molecules are enzymatically degraded to CO_2 and water. Hydrogen and electrons from the substrates reduce NAD and FAD to NADH + H^+ and $FADH_2$. Some ATP is synthesized at a substrate level by these reactions. Most of the ATP generated in the mitochondrion is by the electron transport system (ETS) which oxidizes the NADH + H^+ and $FADH_2$ formed by the Krebs cycle. The ETS functions as a multistep series of oxidation-reduction reactions, transferring electrons through a set of intermediates associated with the inner membrane and ultimately reducing molecular oxygen to water. These intermediates include pyridine or pyrimidine-linked dehydrogenases, flavin-linked dehydrogenases, cytochromes, and ubiquinones.

Associated with the ETS is a mechanism for formation of ATP called oxidative phosphorylation. For each pair of electrons transferred from NADH + H^+, three molecules of inorganic phosphate are added to ADP to form ATP (two ATP for $FADH_2$). Oxidative phosphorylation is a function of the inner membrane. Three hypotheses for the mechanism of phosphorylation are the chemical-coupling hypothesis, the conformational-coupling hypothesis, and the chemiosmotic-coupling hypothesis.

The mitochondrion is also the site of the glyoxylate cycle reactions, fatty acid oxidation, fatty acid chain elongation, superoxide mutase, and catalase reactions and a number of amphibolic and anapleurotic reactions.

CHAPTER 18

The Chloroplast

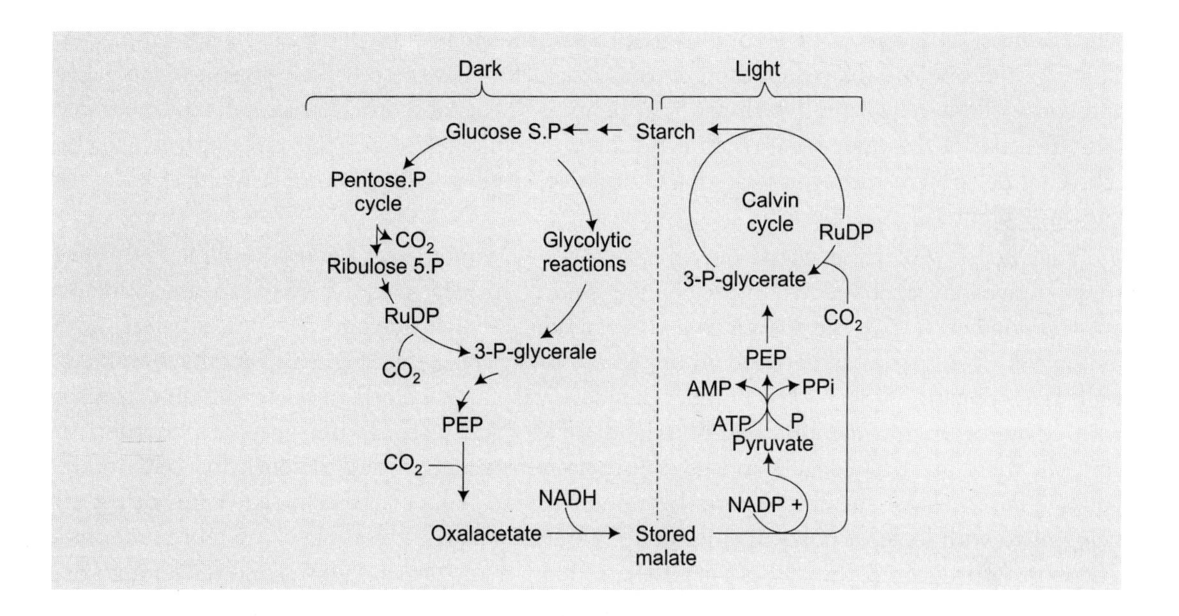

Chloroplasts are organelles in plant that are responsible for the absorption of light energy, the synthesis of carbohydrates, and the evolution of molecular oxygen. Light energy captured by the chloroplast is converted into potential chemical energy in the form of carbohydrate and in this state starts the "energy chain" in nature. The process is called **photosynthesis**. The energy is first made available to satisfy the needed of the cell and whole plant carrying out photosynthesis; the energy is then passed on to the consumers of the plant and their subsequent predators. The oxygen evolved during the capture of light energy becomes the ultimate oxidizing agent for cellular metabolic reactions, mitochondrial oxidations, as well as other oxidations in plant; animal and microbial cells depend on this primary source of oxygen. It is currently believed that the entire supply of oxygen in the atmosphere today was derived from and is presently maintained by phostosynthesis.

A chloroplast is any membrane-encased organelle containing chlorophyll that belongs to a group of related organelles in plant called plastids. The plastids have a variety of morphological forms, carry out diverse functions, and store many different compounds. For example, the amyloplast is the starch-storing plastid to potato tubers; the chromoplast is the lyeopene-containing plastid that gives the fruit to tomatoes its red color. Each of the diverse plastids is believed to arise from a common proplastid precursor. These organelles are found exclusively in the cells of green plants. All of the major groups of plants, with the exception of the fungi, contain chloroplasts. There may be a single chloroplast or dozens of chloroplasts in each cell. Most frequently, the simpler plants, such as the algae, contain single chloroplasts, while the higher plants, such as the cone-bearing and flowering plants, have many chloroplasts in each cell. For most laboratory studies of photosynthesis, the single-cells plant chlorella is employed. This alga has one cup-shaped chloroplast that practically fills the cell. The organism can easily and conveniently be cultured with artificial lighting in the laboratory in solutions of inorganic salts. Because of the one-to-one relationship between chloroplast and the ease with which the cells can be grown and enumerated, quantitative studies using chloroplast preparations of known content are possible.

For studies with chloroplasts of higher plants, leaves are generally used, and spinach and parsley leaves are probably the most popular source. In leaves, the chloroplasts are found in greatest numbers in two internal tissues, the palisade paraenchyma and the spongy parenchyma mesophyll. Both tissues lie between an upper and lower epidermis (Fig. 18.1) and have thin cell walls that are easily broken. The number of chloroplasts in each cell varies within an organism with changing environmental conditions and varies greatly from one species to another. In spinach, there are between 20 and 40 chloroplasts in each palisade parenchyma cell. In the palisade parenchyma, the chloroplasts lie along the side walls of the cell, the center of the ell being filled with large vacuoles. In the spongy parenchyma, the chloroplasts are more randomly distributed throughout the cytoplasm of the cell. In many genes, cytoplasmic streaming (i.e., cyclosis) moves, the chloroplasts about the cell, and in a few instances, an active amoeboid-type of movement of the chloroplasts has been observed.

Chloroplasts are routinely isolated from plant tissues by differential centrifugation following the disruption of the cells. Leaves are homogenized in an ice-cold isotonic saline solution such as 0.35 M NaCl buffered at pH 8.0. The disruption is generally carried out with short spins in a Waring blender. After a preliminary filtration through nylon gauze (20 μm pore size) to remove the larger particles of debris, cell nuclei, tissue fragments, and unbroken cells, the chloroplasts are separated by centrifugation at 200 g for 1 minute. The chloroplasts-rich pellet is then resuspended and centrifuged again at 2000 g for 45 seconds to resediment the chloroplasts. Chloroplast preparations obtained by this procedure are generally mixtures of intact and broken organelles. Since the chemical composition, rate of photosynthetic activity and other properties of intact chloroplasts differ significantly from those of damaged organelles, it is often desirable to separate the two populations. This may be accomplished by rate or isopycnic density gradient centrifugation of the chloroplast preparation using sucrose Ficoll, or Ludox gradients.

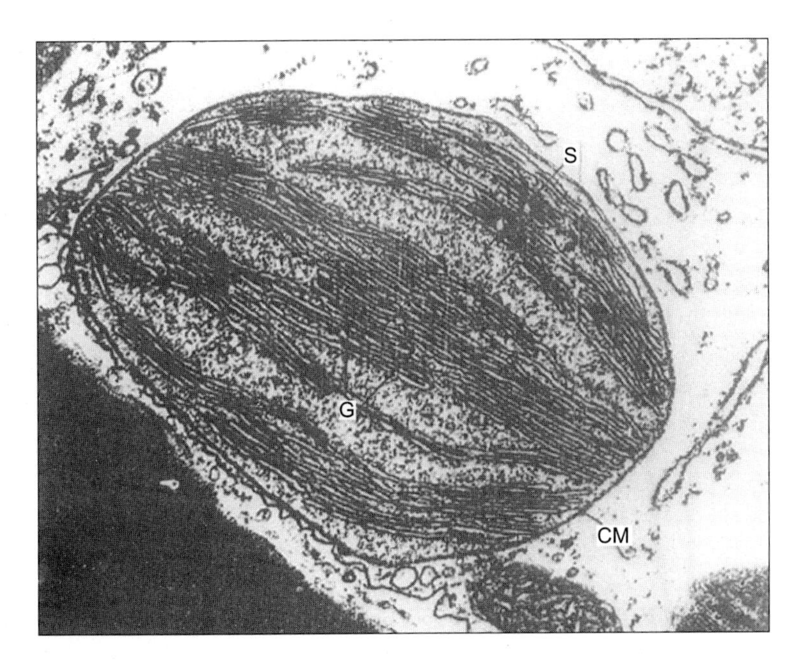

Fig. 18.1 Chloroplast of a higher plant cell, shown in cross section, contains grana (G) and matrix, or so-called stroma (S), with a chloroplast membrane (CM) surrounding the whole. The grana consist of stacks of internal membrane systems in the form of flattened sacs called *lamellae* or *thylakoids*. Grana are the site of photosynthetic phosphorylation (Courtesy of Dr. M.C. Ledbetter).

STRUCTURE OF THE CHLOROPLAST (Fig. 18.1 to 18.4)

Chloroplast size is quite variable. Although the average diameter of chloroplast in higher plant cells is between 4 and 6 um, the size may fluctuate according to the amount of available illumination. In sunlight, chlorophyll is more readily synthesized by the plant, and the chloroplasts increase in size; in the shade, chlorophyll synthesis declines, and there is a corresponding reduction in chloroplast size. Polyploidy cells contain larger chloroplasts than comparable diploid cells. Changes in chloroplast size and shape are also observed after short-term exposure of plants to light. Short-term light exposure produces a small but measurable decrease in chloroplast volume. Presumably, this due to a light-induced production of ATP, for the addition of ATP to chloroplast in the dark causes a reduction in volume.

The shape of most chloroplasts a higher pant is sphenoid, ovoid, or discoid (Fig. 18.1). Other irregular shapes sometimes occur but are more common in lower plants. For plasma and the interior of the organelle. The inner membrane parallels the outer membrane, but inward folds of this membrane are extensive. The inward growth of the inner membrane give rise to a series of internal parallel membranes called lamellae (Fig. 17.3b). The interior of the chloroplasts in which the lamellae are suspended is a granular fluid that appears somewhat electron-dense in electron micrographs. This matrix referred to as the stroma.

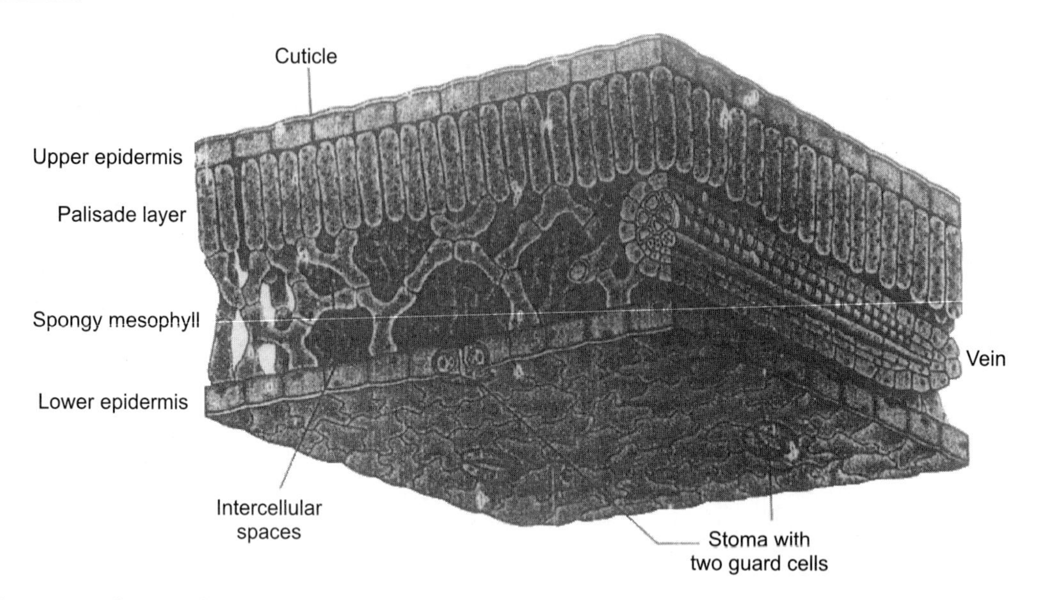

Fig. 18.2 Diagram of a cross section of a leaf showing the position of the chloroplast containing tissues, the palisade, and spongy parenchyma.

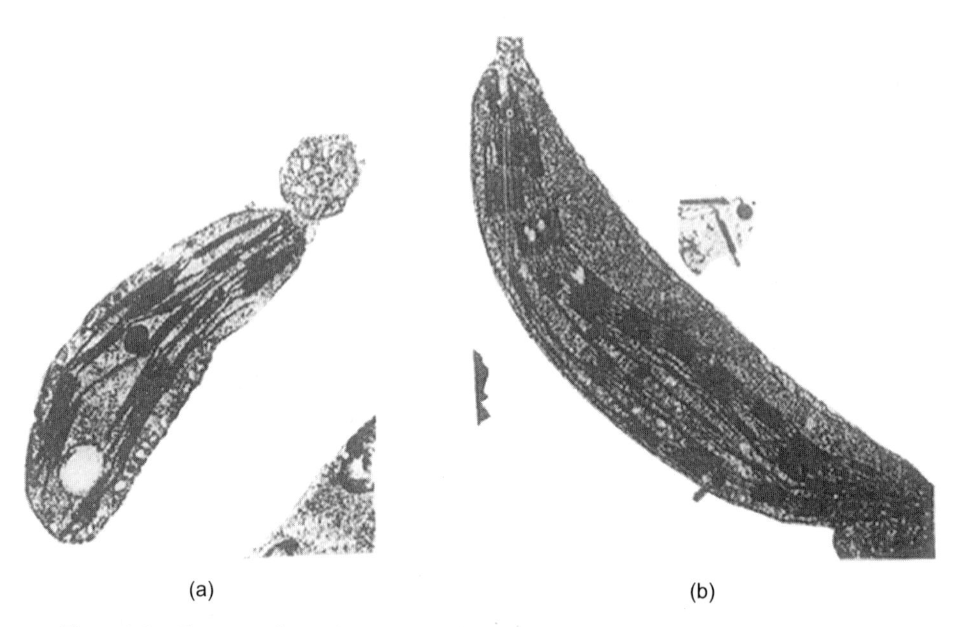

Fig. 18.3 Electron photomicrographs of chloroplasts of higher plants (a) tobacco.

Most of the lamellae are organized to form disk-shaped sacs called small thylakoids. The small thylakoids are often arranged in stacks called arena (sing. granum) having a diameter of about 3000 to 6000 A, Since these thylakoids are round, the grana appear much like a stack of coins (Figs. 18.5a and 18.5b). A typical chloroplast has between 40 examples, in algae, cup-

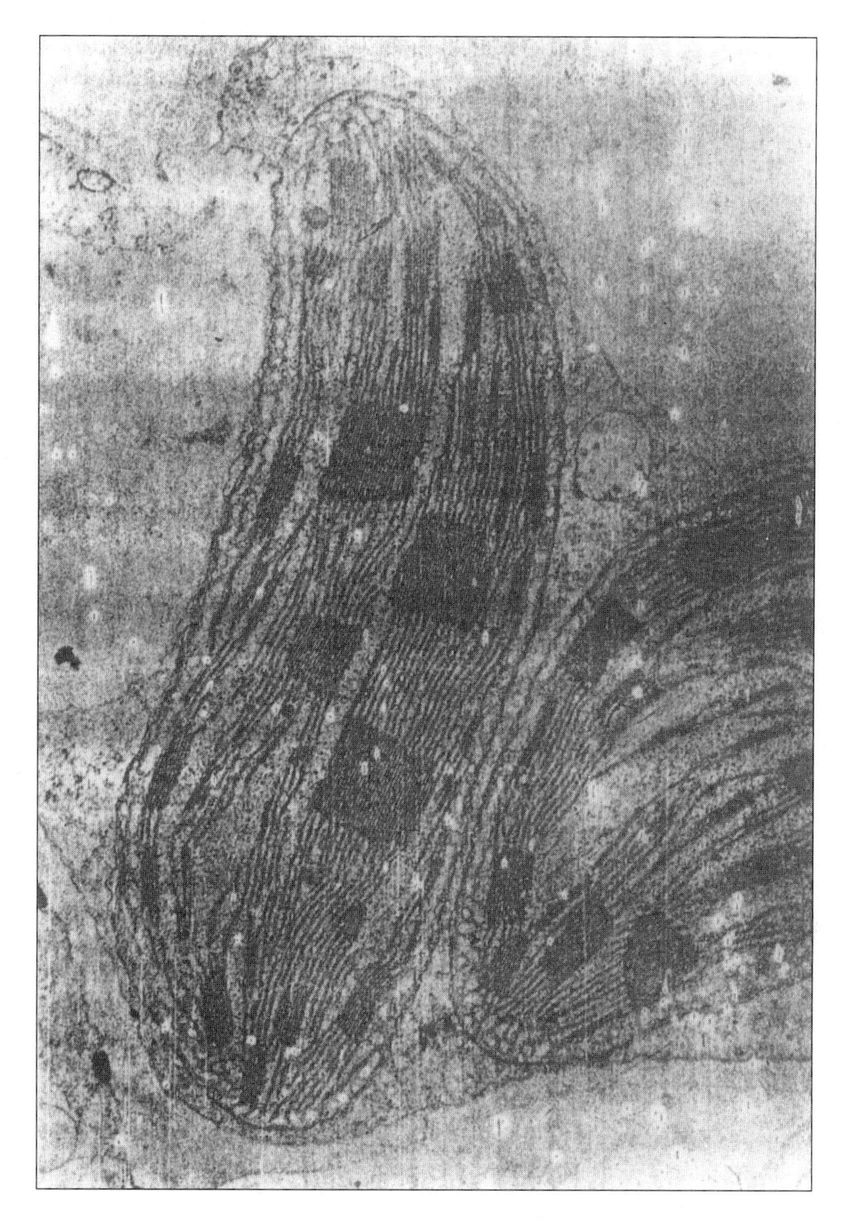

Fig. 18.4 Electron photomicrograph of chlorophasts of higher plants (c) grass (Courtesy of W. Laetsch).

shaped chloroplasts as well as spiral bands, star shapes, and digitate forms are observed. The shape and structure of chloroplasts can also be altered by the presence of starch granules. During periods of active photosynthesis, the sugars formed in the chloroplast are polymerized into starches that precipitate as small granules. The starch granules are usually ellipsoidal and up to 1.5 nm long.

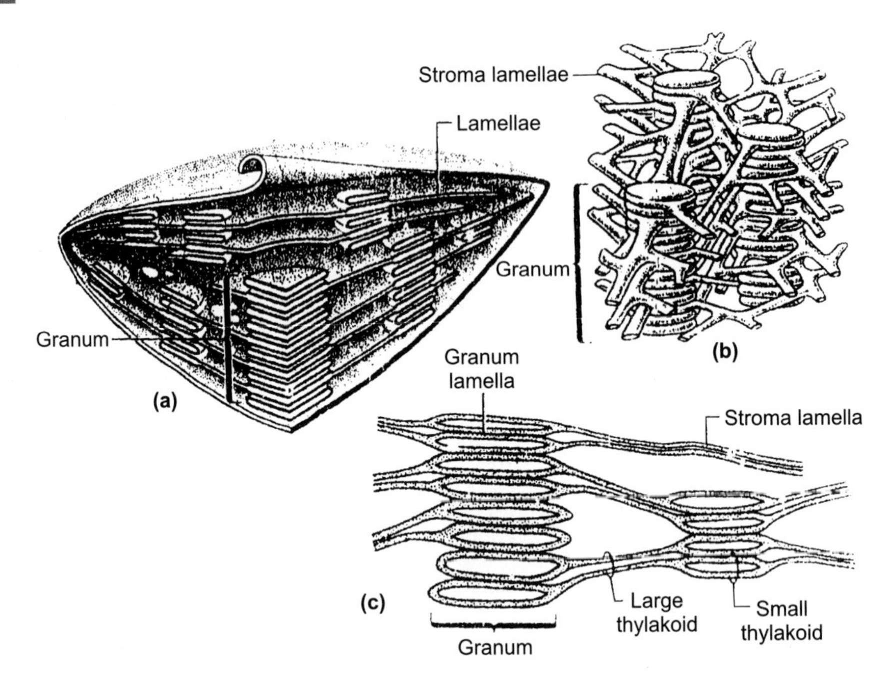

Fig. 18.5 Diagram of the membranes of the chloroplast. (a) Cross section showing the position of lamellae and grana within the chloroplast (b and c) Thylakoids of grana and stroma and relationship to lamellae.

FINE STRUCTURE OF THE CHLOROPLAST

The chloroplasts of higher plants are composed of two membrane layers similar to those of mitochondria. Each membrane is about 50 A thick. The outer membrane, which lacks folds or projections, serves to delimit the organelle and regulate the transport of materials between the cyto and 60 grana, and each may be composed of 2 to 1 00 small thylakoids. Frequently, a small portion of the thylakoid, extends radially into the stroma forming a branching tube, or large thylakoid that communicates with other small thylakoids and grana (Fig. 18.5c). Collectively, the branching and anastomosing network is called the stroma lamellae.

STRUCTURE OF THE THYLAKOID

The adjacent membranes of neighboring thylakoids within each of he grana form thick layers called grana lamellae. Electron photomicrographs of grana lamellae fixed with glutaraldhyde and stained with osmium reveal a five-layered arrangement consisting of three dark 40-A thick osmiophilic layers enclosing two 17-A thick light osmiophobic spaces. Freeze-fracture techniques indicate that these membranes contain numerous particles. The particles, which are undoubtedly protein in composition, appear to be of two basic sizes (Fig. 18.6). The stroma lamellae contain only the smaller size particles. However, the grana lamellae contain both large and small size particles.

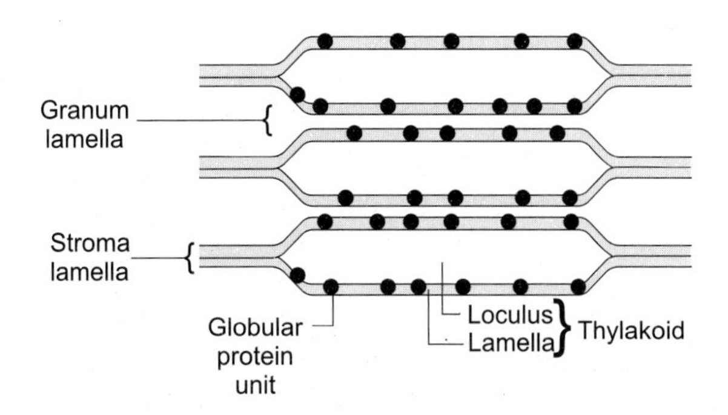

Fig. 18.6 Diagram of the stroma lamellae and grana lamellae showing layering and arrangement of large protein particles in membranes.

When the inner surface of the thylakoid of some plants such as spinach is examined by combined freeze-etch and shadowing techniques, a regular array of discrete internal units is observed. These units have been termed quantasomes. Each is about that protein and half lipid and measures 185 × 155 × 100 A. The quantasomes are occasionally arranged in a paracrystalline manner at the center of the thylakoid but are more random at the periphery. Each quantasome appears to be composed of four smaller subunits.

STROMA STRUCTURES

The granular stroma contains a variety of particles, the presence of starch granules was noted earlier. Electron micrographs also reveal a number of osmiophilic granules and groups of ellipsoidal structures called stromacenters. Of particular interest are the strands of DNA scattered through the stroma and ribosome like particles. Some of these structures are shown in Fig. 18.4, which also clearly depicts the relationship between the stroma and grana lamellae.

CHEMICAL COMPOSITION OF CHLOROPLASTS

The organic constituent present in greatest quantity in the chloroplast is protein. Up to 69% of the dry weight of the chloroplast may be protein. For leaf cells, 75% of the total gel nitrogen is found within the chloroplasts. Both structural and soluble proteins have been identified, but only a few of these have been extracted and purified. A peptide analysis by SDS gel electrophoresis shows compositional differences between stroma and grana lamellae, but the differences are primarily quantitative rather than qualitative. The relative compositions of stroma lamellae and grana lamellae are shown in Table 18.1.

Essentially all the pigments and cytochromes are located in the lamellae. The stroma lacks these compounds but contains DNA and RNA, which are not present in the lamellae. Most of the RNA is associated with the ribosomes of the stroma. The amount of DNA is low; estimates are 10 to 10 g per chloroplast or about 0.03% of its dry weight. However, this is enough to carry ample information for he synthesis of chloroplast proteins including many enzymes active in photosynthesis. The disposition of chloroplast DNA during chloroplast division is unclear.

Table 18.1 Chemical composition of spinach chloroplats.

Component	*Percentage of chloroplast dry weight*	
	Chloroplasts isolated in water	*Values corrected for loss of soluble protein*
Total protein	50	69
Water-insoluble protein	50	31
Water-soluble protein	0	38
Total lipid	34	21
Chlorophlyll	8	5
Carotenoids	1.1	0.7
Ribonucleic acids	–	1.0-7.5
Deoxyribonucleic acids	–	0.02-0.1
Carbohydrate (starch, etc.)	Variable	

Source: Modified from J.T.O. Kirk and R.A.E. Tilney-Bassett, The Plastids, W.H. Freeman and Co., San Francisco, 1967.

Table 18.2 Major components of strama and grand lamellae.

	Stroma lamellae	*Grana lamellae*
Total chlorophyll	278[a]	401
Chlorophyll a	238	281
Chlorophyll b	40	130
P_{700}	2.5	0.6
β-Carotene	21	17
Lutein	10	29
Violaxanthin	15	20
Neoxanthin	8	16
Phospholipid	76	66
Monogalactosyl diglyceride	231	214
Digalactosyl diglyceride	172	185
Sulfolipid	65	59
Cyt *b* (total)	1.0	3.4
Cyt *f*	0.5	0.7
Manganese	0.3	3.2

[a]Values in micromoles of component per gram of membrane protein.

Lipids and lipid-soluble pigments account for about 34% of the dry weight of the spinach chloroplast. An exceedingly large number of different lipid compounds have been identified. The more common lipids are the galactosyl diglycerides, phospholipids, quinines (including vitamin K), and sterols.

THE CHLOROPHYLLS

The green pigments of the chloroplast and the main sources of the color of green plants are the chlorophylls. Although a large number of chemically distinct plants, the structures of these

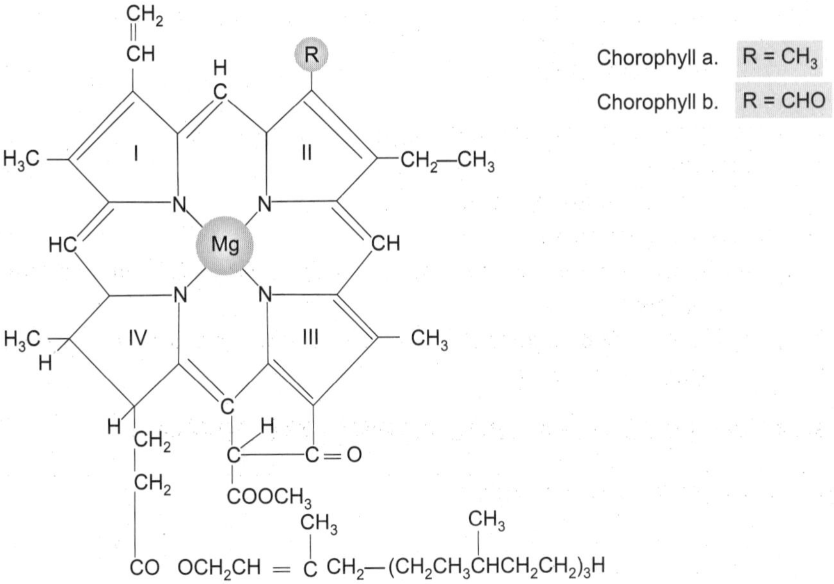

Chorophyll a. R = CH_3
Chorophyll b. R = CHO

Fig. 18.7 Structures of chlorophyll *a* and *b*.

chlorophylls are basically the same. It he structures of chlorophylls a and b are given in Fig. 18.7. It is customary to identify each chlorophyll by a different letter. All photosynthetic plants have been found to contain chlorophyll a, but the presence of the secondary chlorophylls b, c, or d depends on the type of plant. Higher plants usually have chlorophyll b. photosynthetic bacteria contain unique chlorophyll called bacteriochlorophyll. Together, chlorophylls a and b represent about 5% of the dry weight of the spinach chloroplast, with an a:b weight ratio of 2.05 to 3.52. In most plants, the a:b ratio varies according to the light intensity. For example, alpine plants have a ratio 0 5.5. the ratio is much lower in shade plants.

Each chlorophyll has a characteristic light absorption spectrum. The absorption spectra of chlorophylls a and b are shown in Fig. 18.8. Chlorophyll a has absorption maxima at 4.30 and 670 nm, whereas the absorption maxima of chlorophyll b occur at 455 and 640 nm. Absorption maxima of other plant and bacterial pigments are indicated in Table 18.3. The absorption spectrum and maxima of plant pigments vary according to the solvent used for extraction. Experiments conducted in vivo indicate that the native absorption maximum for chlorophyll a occurs at 677 nm. Absorption maxima determined in vivo or in extracts are the results of mixtures of chlorophyll molecules of the same type. However, each molecule exhibits its own characteristic maxima, and the can differ somewhat from the positions of the average peaks. One vary important from of chlorophyll a that is readily bleached by light has an absorption maximum at 700 nm. This form which represents only about 0.1% of the total chlorophyll a molecules present in a sample, is called P or chlorophyll a. The roles of the various chlorophylls in photosynthesis are discussed later in the chapter.

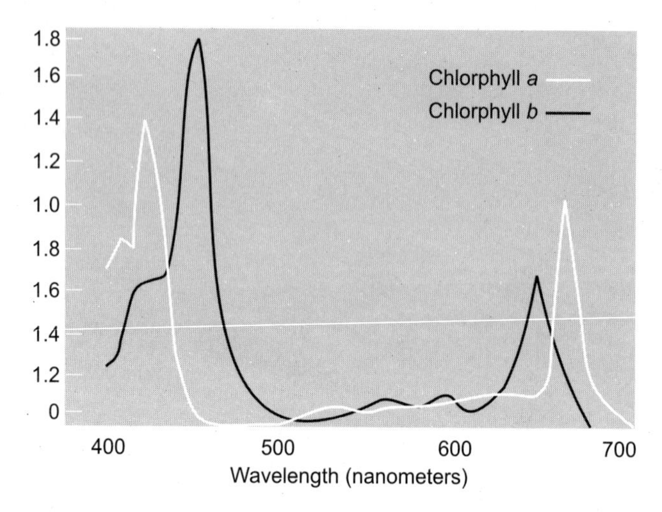

Fig. 18.8 Absorption spectra of chloraphylls *a* and *b* in an other solvent.

Table 18.3 Absorption maxima[a] of plant and bacterial pigments.

Pigment	*Wavelength (nm)*	*Occurrence*
Chlorophyll *a*	430, 670	All green plants
Chlorophyll *b*	455, 640	Higher plants; green algae
Chlorophyll *c*	445, 625	Diatoms; brown algae
Bacteriochlorophyll	365, 605, 770	Purple and green bacteria
α-Carotene	420, 440, 470	Leaves; some algae
β-Carotene	425, 450, 480	Some plants
γ-Carotene	440, 460, 495	Some plants
Luteol	425, 445, 475	Green leaves; red and brown algae
Violaxanthol	425, 450, 475	Some leaves
Fucoxathol	425 450 475	Diatoms; brown algae
Phycoerythrins	490, 546, 576	Red and blue-green algae
Phycoeryanins	618	Red and blue-green algae
Allophycoxanthin	654	Red and blue-green algae

[a]Absorption maxima vary according to the solvent in which the pigment is dissolved.

Source: From E. Rabinowitch and Govindjee, *Photosynthesis*, John Wiley & Sons, Inc., New York, 1969.

THE CAROTENOIDS

The carotenoids are all long-chain isoprenoid compounds having an alternating series of double bonds. Although these compounds are synthesized only in plant tissue and participate in photosynthesis, they also serve as precursors of vitamin A in animal tissues. Most carotenoids are yellow, orange, or red. The formulas of A-, B, and Y carotene are shown in Fig. 18.9, and their absorption maxima are given in Table 18.3. most of these pigments are located in the chloroplast lamellae and are believed to function as accessory pigments for light absorption during photosynthesis.

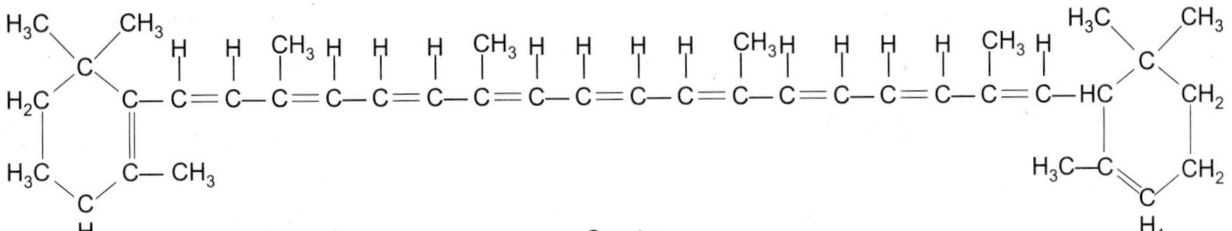

α—Carotene

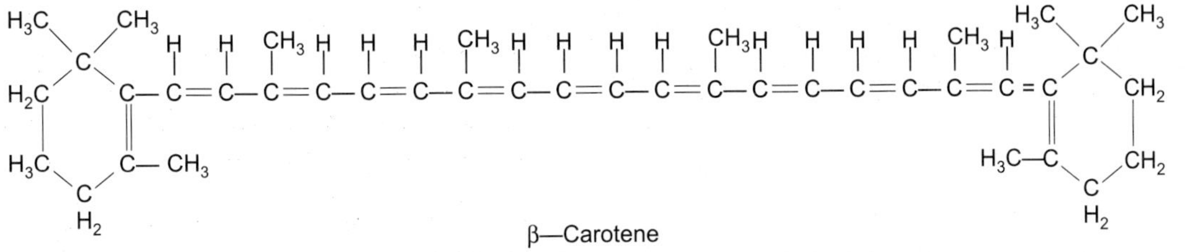

β—Carotene

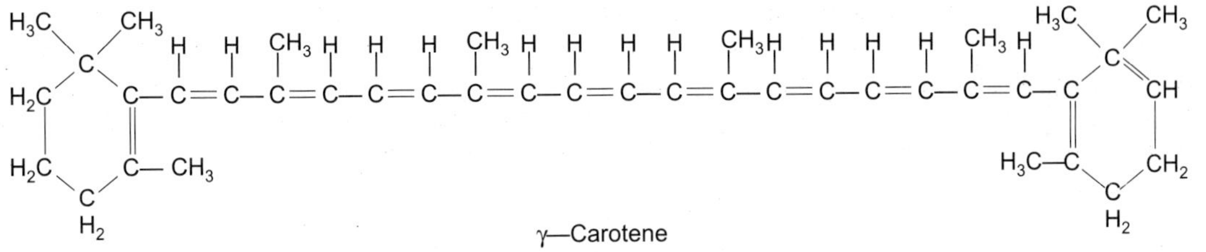

γ—Carotene

Fig. 18.9 Fomulae of alpha, beta, and gamma carotene.

LOCATION AND ARRANGEMENT OF THE PIGMENT

Both the chlorophylls and the carotnoids are located almost exclusively in the chloroplast lamellae. The lamellae are about 52% protein and 48% lipid, and the two pigments reside primarily in the lipid component. Some lipid is also found in the osmiophilic granules of the stroma, but these are not believed to contain chlorophyll. Because each chlorophyll molecules has a hydrophilic portion (the tetrapyrrole) and a lipophilic portion (the phytl chain), the chlorophyll molecules are thought t be aligned in a specific manner within the lamellae. The pyrrole groups form weak bonds with the lamellar protein, while the phytyl chains extend into the lamellar lipid. The carotenoids are dissolved in the lipid adjacent to the chlorophyll molecules.

It has been calculated that each quantasome contains about 230 chlorophyll molecules and 48 carotenoid molecules. Such a unit would also contain 1 molecule of P 700. On the basis of these calculations, as well as physiological studies, it has been proposed that the quantasome is the fundamental photosynthetic unit.

The chloroplast stroma contains many of the enzymes associated with photosynthesis. Chloroplast protein synthesis also takes place in the stroma. DNA strands about 150 μ long have been isolated from the chloroplast along with ribosomes and polyribosomes. Chloroplast ribosomes belong to the 70 S class and contain 23 S and 16 S RNA ; these ribosomes are smaller than those found in the cytoplasm of the plant cell.

Photosynthesis-Historical Background

The overall reactions of photosynthesis may be summarized by the equation.

$$6CO_2 + 12H_2O \xrightarrow[\text{Chlorophyll}]{\text{Light}} C6H_{12}O_6 + 6O_2 + 6HO_2$$

The studies of many individuals over the past 300 years have led to our present understanding of this process. One of the first studies was made in 1968 by Jan Baptista van Helmont (Dutch 1577-1640). Van Helmont planted a 5-lb willow shoot in a large pot containing 200 lb of soil. The plant was regularly watered over a 5-year period and was then carefully removed and weighed. While the willow tree was found to weigh almost 170 lb, the original soil weighed only a few ounces less than the original 200 lb. Van Helmont concluded that the increase in weight of the willow tree was due primarily to the addition of water, for he was not aware of the role played by gases in the air in the growth of the plant. Today, we also recognize that the small quantity of material removed from the soil included minerals, nitrogen, phosphate, and calcium salts.

Joseph Priestley (English, 1733-1804) was the first to show that plants exchange gases with the atmosphere. In particular, Priestley found that oxygen was evolved by plants during photosynthesis. Late in the eighteenth century, Jan Ingenhousz (Dutch, 1730-1799) showed that oxygen was produced by the green parts of plants when exposed to light. Furthermore, he observed that the amount of oxygen produced varied according to the amount of light to which the plant was exposed. In 1824, Nicholas de Saussure (Swiss, 1767-1845) found that the accumulation of carbon from carbon dioxide in the air. De Saussare also showed that leaves respire in darkness-that is, they take in oxygen and release carbon dioxide.

The studies of Julius Sachs in the late nineteenth century showed that chlorophyll was confined the chloroplasts and was not distributed throughout the plant cell. He also showed that sunlight caused chloroplasts to absorb carbon dioxide and that chlorophyll is formed in chloroplasts only in the presence of light. Sachs also noted that one of the products of photosynthesis was starch. It was not until 1918 that Wilstatter and Stoll isolated and characterized the green pigments chlorophyll a and b.

On the basis of these early findings, the reactions of photosynthesis were described by the equation.

$$6CO_2 + 12H_2O \xrightarrow[\text{Chlorophyll}]{\text{Sunlight}} C_6H_{12}O_6 + 6O_2$$

However, photosynthesis is not simply the fusing of carbon dioxide and water molecules through the use of light energy. Instead, two complex series of chemical reactions are involved. One set of reactions, called the photochemical or light reactions, occurs in the lamellae of the chloroplasts. In these reactions, light energy is absorbed and used to form ATP, and water molecules are split, releasing oxygen and hydrogen; the latter is used in the reduction NADP.

The second set of reactions, called the synthetic or dark reactions, occurs in the storma of the chloroplasts, and although these reactions do not require light, they do depend on the availability of ATP and NADPH2 from the photochemical reactions in order to reduce the carbon dioxide to form sugars.

PHOTOSYNTHESIS-PHOTOCHEMICAL (LIGHT) REACTIONS

The Absorption of Light by Chlorophyll

The absorption of electromagnetic energy by any atom or molecule often involves a shift of electrons from one atomic orbital to another. Each electron possesses energy, and the amount is determined by the location of the electron orbital in space and the speed at which the electron moves. When an atom absorbs light energy, an electron is either raised to an orbital of higher energy level or accelerated in its orbit. In either case, certain discrete quantities of energy are required, for when light protons have either too much or too little energy, they are not absorbed. Electrons may orbit in pairs within an orbital, the members of the same orbital spinning in opposite directions. Most atoms at their lowest energy level (i.e., ground state) have all their electrons paired in this fashion and are said to be in the singlet (i.e., S) state. When a photon of light is absorbed and an electron is thereby raised to a higher, unoccupied orbital, it may continue to spin in the direction opposite its former partner (in which case, the atom is still in the S state), or it may spin in the same direction as its former partner (in which case. the atom is said to be in the triplet or T state).

In a molecule, some electrons orbit exclusively about specific atomic nuclei; others may be shared between two nuclei forming a bond (called localized or H* electrons), or may orbit about several nuclei (called detocalized or H electrons). The absorption of a light photon may move a H electron to a H* position.

Chlorophyll is normally a singlet in its ground state. Although the absorption of light causes some H electrons to be raised to a H* orbital, chlorophyll remains in a singlet state. When red (680 nm) light is absorbed, an electron is raised to a higher. SH*, orbital. Blue (430 nm) light possesses more energy per photon than red light, and its absorption raises an electron to a higher (but still SH) energy state. These transitions are summarized in Fig. 18.10.

Once a molecule has absorbed light energy and is in an excited state, the ground state may be re-established in three different ways; (1) the energy may be reemitted in the form of radiation of longer wavelength (i.e., fluorescence); (2) the energy may be converted to heat; and (3) the excited molecule may transfer its excess energy to another molecule. The transfer of energy from one molecule to another often involves the exchange of a high-energy electron for one of lower energy. Referring to Fig. 18.10, the blue light absorbed by chlorophyll raises an electron to the Sb H* state. By losing energy in the form of heat, the electron could "drop" back to the Sa H* state. The electron could then drop back from the Sa H* state to the SH state (i.e., the ground state) by the immediate loss of energy as heat or fluorescence. Another possibility also exists, for the electron could drop from the Sa H* state to a triplet (i.e., TH*) state by heat loss and

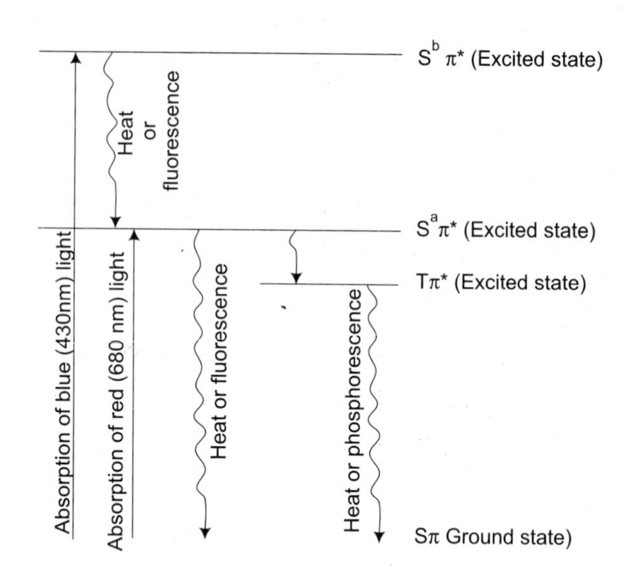

Fig. 18.10 Main electronic states in the excitement of chlorophyll by red and blue light.

then to the ground state by either phosphorescence (delayed fluorescence) or heat loss. Concentrated solutions of extracted chlorophyll will strongly fluoresce red when placed in a beam of sunlight or white light. Finally, it should also be noted that if the chlorophyll molecule is sufficiently excited, it may not return to the ground state by heat loss or reradiation; instead, the excited electron may be transferred to another molecule, leaving the chlorophyll in a temporary, oxidized state.

Primary Photochemical Events in Photosynthesis

As shown in Fig. 18.10, chlorophyll absorbs both blue and red light, and this raises electrons to Sb H* or Sa H* states. The return of an electron from the Sb H* state to the Sa H* state is extremely fast (about 10-12 seconds) and does not afford an opportunity for the energy to be lost by fluorescence or by transfer to another molecule. Consequently, the energy is lost as heat. The decay of an electron from the Sa H* state, whether brought there by red light absorption or by prior decay from the Sb H* state, does permit the transfer of energy to another molecule, and this is the event that initiates photosynthesis. Consequently, a photon of red light is just as effective in initiating photosynthesis as a photon of blue light, even though the former is much less energetic.

As we have already noted, the chloroplast contains many different pigment molecules (other chlorophylls, caroteinoids, phycobilins, etc) and the electrons of these molecules may be excited to various energy states by the absorption of light. As these excited accessory pigment molecules return to the ground state, the resulting energy is transferred to chlorophyll a molecules, causing their excitation. Since the chlorophyll a molecules present, in a quantasome vary in their absorption maxima, varying quantities of energy are required to raise their electrons to the Sa H* state. The molecule that requires the least energy is postulated to be a pigment absorbing long, red wavelengths-namely the P700 molecule. It seems reasonable that light energy captured by an accessory pigments and transferred to chlorophyll a is in turn transferred from the latter to P700. Since it appears that only one molecule of P700 is present in each quantasome, the quantasome may be the basic unit for the absorption of light energy. Each accessory pigment or chlorophyll a molecule can only pass its energy on to a pigment having an absorption maximum of longer wavelength because these require less energy to be activated to the Sa H* state. Since P700 has its absorption maximum at the longest wavelength, it serves as the final energy trap.

TWO PHOTOSYSTEMS

When some plants are exposed to light having a wavelength of 690 nm or longer, photosynthetic efficiency decreases. The effect is called the red drop (Fig. 18.11). Since the absorbed energy is funneled to P700, it would be expected that the absorption of light by the accessory pigments, chlorophyll a or even P700, should be equally efficient. The efficiency can increased through the addition of shorter wavelength radiations. This enhancement phenomenon can increase the photosynthetic rate 30 to 40% above the rate obtained by either the short wavelength or long wavelength alone. The synergistic effect of the two different wavelengths led early investigators to conclude that two distinct photochemical reactions exist.

The photoreaction sequence involving pigments absorbing light above 690 nm is called photosystem I. If the photosynthetic rate of a plant is enhanced by shorter wavelength of light, the second serried of photoreactions is called photosystem II. In higher plants, photosystem I is a unit containing about 200 molecules of chlorophyll of mostly type a, 50 carotenoids, one chlorophyll P_{700} molecule, one cytochrome f, one plastocyanin, two cytochrome b_{563}, and one

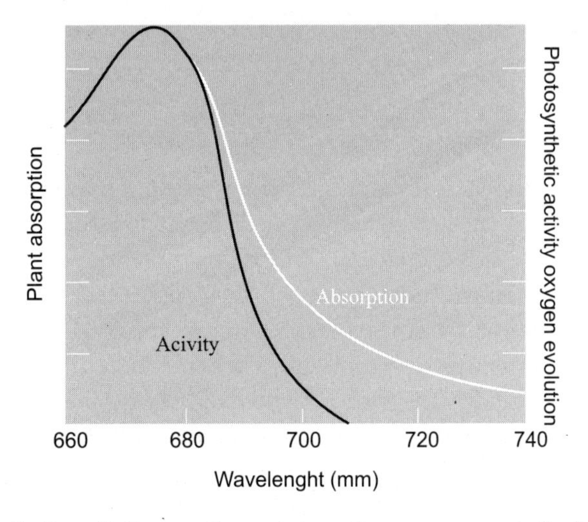

Fig. 18.11 The photosynthetic activity or action spectrum (i.e., oxygen evolution) and adsorption spectrum of the green alga Ulva. The difference between the two curves at wavelengths above 690 nm is called the *red drop.*

or two ferredoxin molecules. Photosystem II has about 200 molecules of chlorophyll of both a and b types, 50 carotenoids, a trapping chlorophyll (a primary electron do nor), a primary electron acceptor that is believed to be a quinine four plastoquinones, six Mn atoms, and two cytochrome b_{599} molecules.

SEQUENCE OF ENERGY (ELECTRON) FLOW

The absorption of light by chlorophyll alters the state of the orbiting electrons, and if the energy is not lost by reradiation or heat, the excited electrons can be transferred to another compound. Such an electron loss "bleaches" the chlorophyll and leaves it in an oxidized state. Oxidized P 700 may be reduced by absorbing an electron from photosystem II, while the reduction of the oxidized chlorophyll of photosystem II is brought about by the oxidation of water.

In 1938, Robin Hill demonstrated that isolated chloroplasts exposed to light could evolve oxygen and reduce a variety of compounds without consuming carbon dioxide. Three years later, S. Ruben and M. Kamen, using the isotope were able to show that the oxygen liberated during whole plant photosynthesis was derived from water molecules.

$$6CO_2 + 12H_2O^{18}O \xrightarrow{\text{Light}} C_6H_{12}O_6 + 6^{18}O_2 + 6H_2O$$

The method by which the water molecules are split is unknown, although four water molecules are required for the evolution of one oxygen molecule and for quanta of light is necessary:

$$4H_2O \xrightarrow{hv} O_2 + 4(H^+ + e^-) + 2H_2O$$

Protein-bound Mn and Cl may also be required. Some of these reactions are summarized in Fig. 18.12.

REDOX REACTIONS

Electrons are always transferred from activated chlorophyll, P700 and other molecules in the chloroplast to more positive or oxidized molecules. Each molecule has a different mount of electrical potential called redox potential. The redox potential can be measured and the molecules listed in order of the magnitude of their potential. During electron transfer, molecules may accept electrons from less positive (more negative) molecules and may donate electrons to more positive (less negative) molecules. When a molecule accepts electrons, it is said to be reduced, and when electrons are given up, the molecule is said to be oxidized. The transfer of electrons following the absorption of light energy by chlorophyll therefore involves a sequence of oxidation reduction reactions. This sequence is shown in Fig. 18.13, which also identifies the intermediate electron acceptors and their redox potentials.

Photosystem I is located in the membranes of the grana tylakoids and the stroma thylakoids. Photosystem II is found only in the membranes of the grana thylakoids. Therefore, in the stroma thylakoids, photosystem I functions independently but may function in conjunction

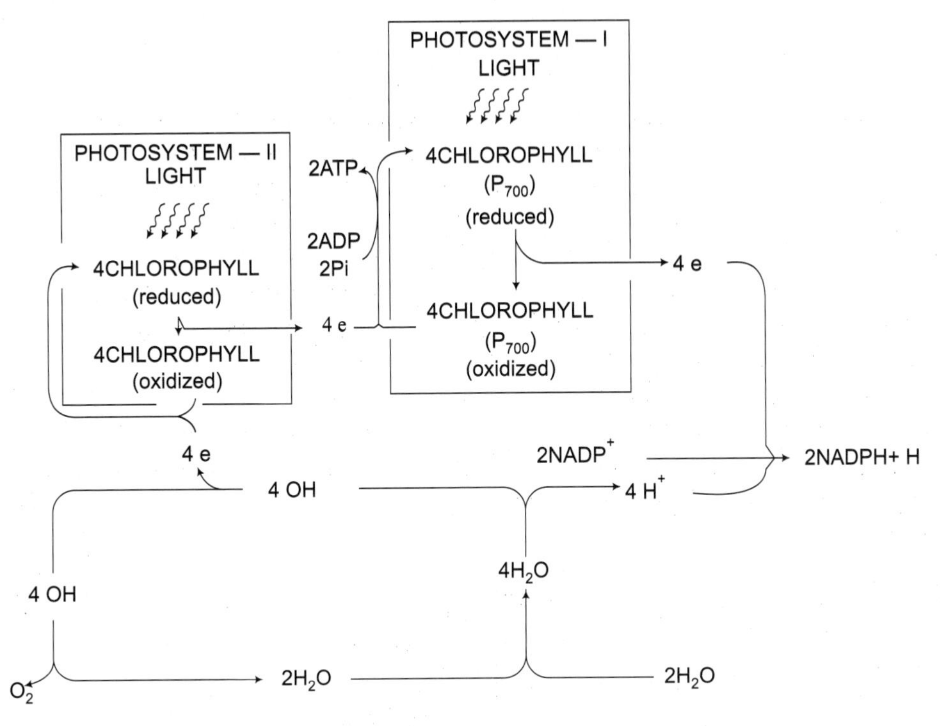

Fig. 18.12 Absorption of light energy transfer of electrons between photosystems, and the regeneration of reduced chlorophylls.

with photosystem II in the grana thylakoids. Absorption of light ultimately by P_{700} of photosystem I causes the ejection of electrons of ferredoxin (Fig. 18.13). Other molecules can be substituted as electron acceptors experimentally, but it would appear that the membrane-bound nature of ferredoxin and the positioning of P700 make this electron shift requisite in the chloroplast.

The ferredoxin appears to be reoxidized in the chloroplast by the transfer of electrons to either of two compounds, NADP or cytochrome b. The reduction of NADP requires a flavoprotein (ferredoxin-NADP reductase) and the absorption of H+ from the reduction pool created by the splitting of H_2O, that is,

$$NADP^+ + (H^+ + e^-) + H \rightarrow NADPH + H^+$$

The reduced NADPH + H^+ thus generated in the lamellae spills into the stroma, where it is reoxidized in the dark reactions (described later). The reduction of ferredoxin by cytochrome b initiates a sequence of redox reactions passing electron on to cytochrome, plastocyanin, and ultimately to P_{zoo} completing a cycle. The function of this cyclic set of redox reactions is coupled to the phosphorylation of ADP (cyclic photophosphorylation. described below).

In the grana lamellae the P_{700} oxidized by the absorption of light can be reduced by the cyclic return of he electrons from ferredoxin or by the transfer of electrons from photosystem II. In photosystem II (Fig. 18-13), light energy is transferred to an ultimate trap chlorophyll

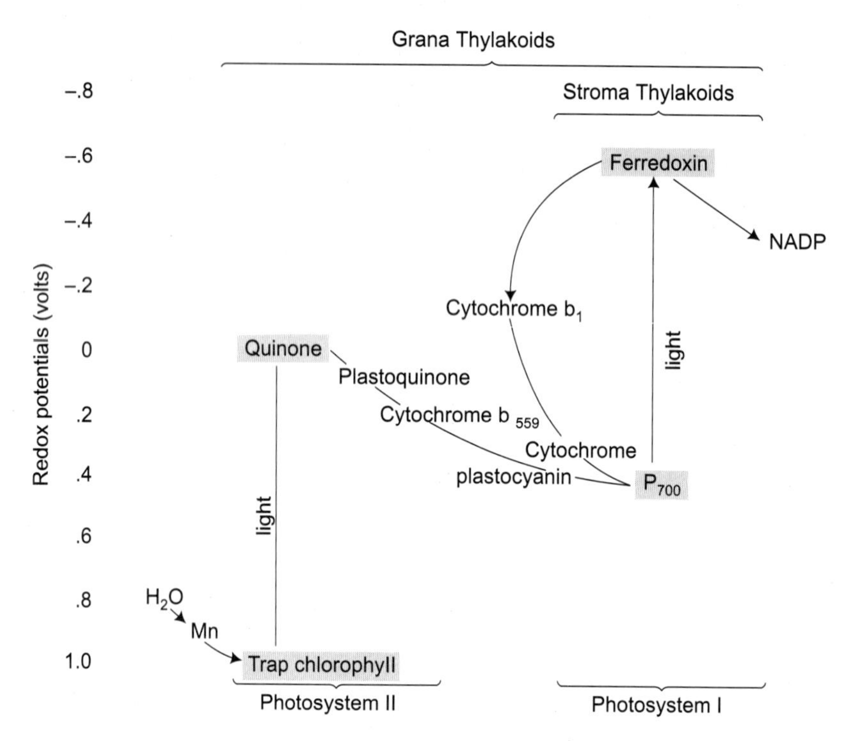

Fig. 18.13 Pathway of electron transport in photosynthesis.

molecule (or other ultimate electron donor) by the accessory pigments. The ejected electrons are absorbed by an electron acceptor, which is currently unidentified but could be quinine. This photoevent pass from the unidentified electron acceptor (quinine?) to plastoquinone to cytochrome b_{559} to the cytochrome f and plastocyanin of photosystem I and ultimately to P_{790}. The electrons in photosystem II do not cycle back to trap chlorophyll (or other electron donor), but their transfer through the redox reactions to P_{700} is coupled to phosphorylation of ADP (noncyclic photophosphorylation, described below).

The reduction of the oxidized trap chlorophyll (or other electron donor) is brought about by the oxidation of water via an enzyme system that is closely associated with the structure of the thylakoid and that also produces molecular oxygen. The enzyme system presumably processes four protons simultaneously to produce O.

$$2H_2O \rightarrow O_2 + 4H^- + 4e$$

Mn is a known cofactor in the system. The electrons produced act to reduce the light-oxidized trap chlorophyll, and the H+ forms a pool available for the reduction of NADP.

CYCLIC AND NONCYCLIC PHOTOPHOSPHORYLATION

Phosphorylation of ADP occurs in the chloroplast during the light reactions, as described briefly above. This photophosphorylation occurs in the lamellae of the stroma and grana thyakoids as

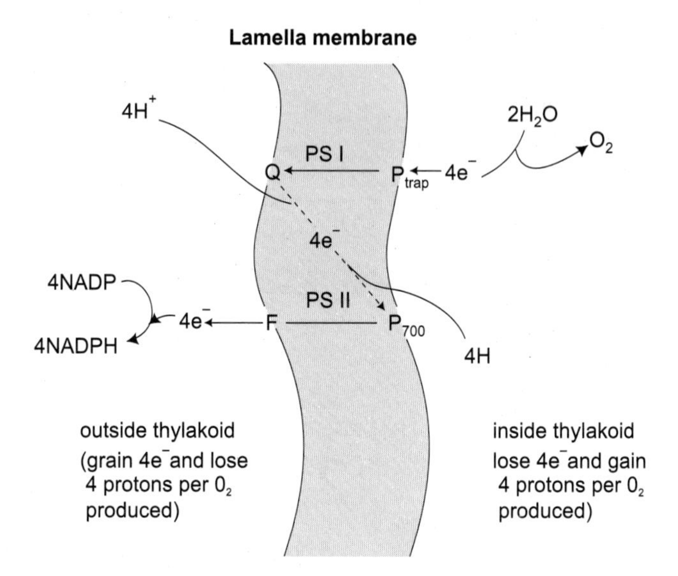

Fig. 18.14 Model illustrating accumulation of protons inside thylakoids during illumination. Q, electron acceptor of photosystem I (PS I); P_{trop} electron donor of photosystem II (PS II); F, electron acceptor of PS II.

a part photosystems I and II. There is a clear similarity between the mechanism of photophorylation in the chloroplast and electron transport system phosphorylation in the mitochondria. In both systems, the mechanism separated from the membrane and with a compartment separated from the rest of the organelle. In the chloroplast, the compartment is the loculus of the thylakoid; in the mitochondrion, it is the matrix within the inner membrane. In both systems, phosphorylation coccurs cupled to electron transport through a sequence of redox reactions. Several of the components of the redox reactions are vary similar, for example, the quinines, and cytochromes.

As described above, cyclic photophosphorylation occurs while electrons released from P by light are being shunted back to P700 through ferredoxin, cytochrome b, cytochromef, and plastocyanin. The energy from these exergonic redox reactions is coupled to phosphorylation. Noncyclic phosphorylation occurs while electrons released from trap chlorophyll of photosystem II are being shuttled via plastoquinone, cytochrome b, cytochrome f, and plastocyanin to P_{700} of photosystemI. The energy from these noncyclic exergonic redox reactions is coupled to phosphorylation.

At the present time, the mechanism of coupling is unknown, although evidence to support a proton gradient across the thylakoid membrane is growing medium quickly cause the medium to become alkaline, implying that protons are transported into the thylakoid sac. The system will reequilibrate in darkness. If the external medium is made alkaline in the dark, phosphorylation of ADP can be induced. Compounds that destroy the membrane, such as detergents, also cause leakage of the proton pool and prevent phosphorylation.

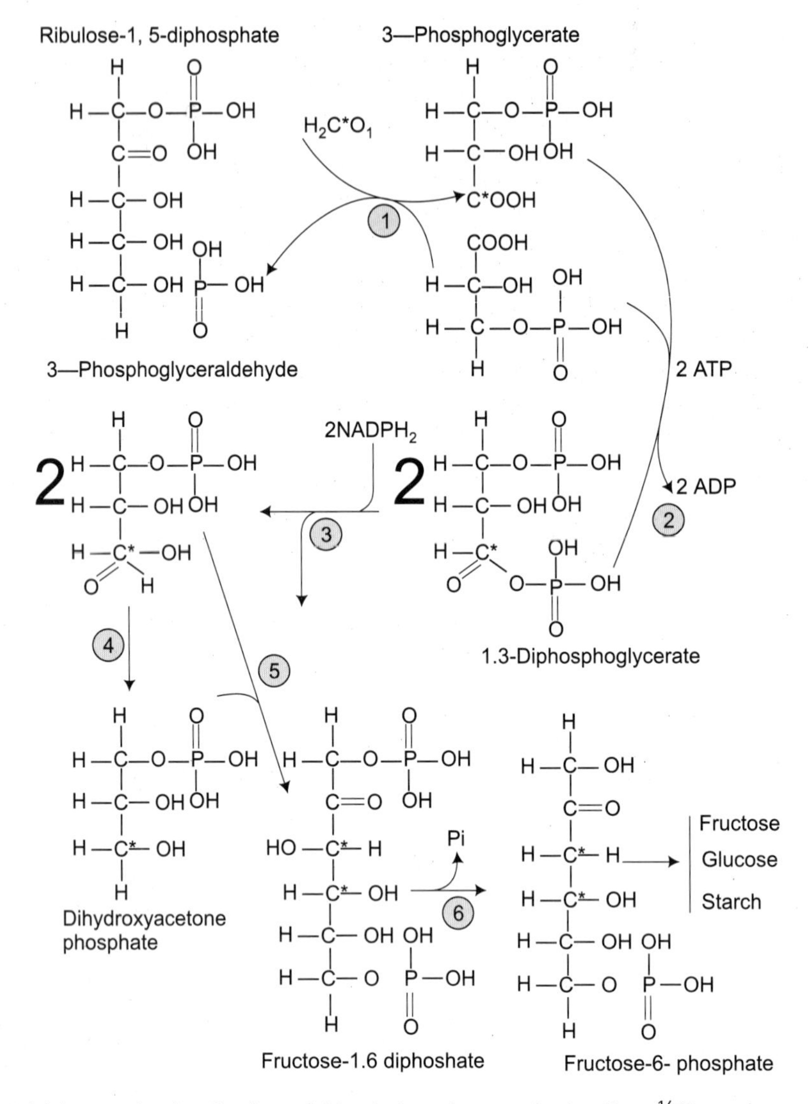

Fig. 18.15 Initial steps in the fixation of CO_2 during photosynthesis, C* = ${}^{14}_{6}CO_2$ and traces the path of carbon through the reaction sequence. Enzymes (1) Carboxydismutase; (2) phosphoglyceric acid kinase; (3) triose phosphate dehydrogenase; (4) isomerase; (5) fructose diphosphate aldolase; (6) fructose diphosphatase.

The standard free energy to generate ATP from ADP and P is about 0 kcal/mole, which is equivalent to a redox potential of about 0.43-0.61 volts. During the movement of a pair of electrons from photosystem I to II (noncyclic photophosphorylation), most measurements indicate that one ATP is produced. Recent measurement indicate that possibly more than one ATP is produced between O_2 evolution and NADP reduction, leading to strong speculation that the first ATP is produced between O_2 evolution (photolysis) and the second during electron transport between photosystems. For a proper balance between the light reactions and the dark reactions (discussed later), three ATP must be produced for each pair of electrons transported from H_2O to NADP (NADPH + H^+). The third ATP is probably generated by cyclic photophosphorylation.

SUMMARY OF THE LIGHT REACTIONS

Two photosystems function during the light reactions of photosynthesis. As each system absorbs four quanta of light energy, trap chlorophyll (system II) and P700 (System I) are activated, passing two pairs of electrons to acceptor molecules. Trap chlorophyll is then returned to its reduced state by the oxidation of four water molecules, and in the process one molecule of oxygen is evolved. P700 is returned to its reduced state by the flow of electrons from photosystem II through intermediate oxidation-reduction compounds. The transfer of two pairs of electrons from water through photosystem II to photosystem I causes the noncyclic photophosphorylation of two ADP molecules producing two ATP and two H2O. The two pairs of electrons released from photosystem I reduce two NADP to two NADPH + H^+.

Therefore, the net result is

$$2H_2O + 2NADP + (2ADP + 2P_1) \xrightarrow{\text{8 quanta}} O_2 + 2NADPH + H^+ + (2ATP + 2H_2O)$$

The two ATP and two NADPH + H^+ molecules produced by the light reactions occurring in the grana lamellae are used in the synthetic (dark) reactions that take place in the stroma. Te latter reactions fix CO_2 into sugars. For each CO_2 molecule fixed, two NADPH + H^+ and three ATP molecules are required (See below). Where the third ATP molecule is formed is not yet clear, it may be obtained from cyclic photophosphorylation.

PHOTOSYNTHESIS-SYNTHETIC (DARK) REACTIONS

The eludication of the sequence of chemical reactions that result in the incorporation of carbon dioxide into sugars and starches relied heavily on the use of radioactive isotopes. Using ${}^{14}_{6}C$-Labeled carbon dioxide, it was possible to add ${}^{14}_{6}CO_2$ at known times to an actively photosynthesizing system, halt the process a short time later, and then identify the compounds into which the carbon dioxide became incorporated. Identification of the ${}^{14}_{6}C$-containing intermediates was carried out using combined paper chromatography and autoradiography.

Ruben and co-workers first showed that the active form of CO_2 in the chloroplast was carbonic acid. Calvin and co-workers studied the sequence of reactions that follow the formation and entry of carbonic acid into the chloroplast. They added 146CO2 to cultures of the alga *Chlorella* and allowed the cells to photosynthesize for given periods of time (Usually between

2 and 60 seconds). *Chlorella* were then killed and the soluble cell components extracted and concentrated. The extracts containing radioactive carbon were chromatographed on paper, and the spots containing radioactivity were identified.

When photosynthesis in the presence of $^{14}_{6}CO_2$ was allowed to proceed for only 2 seconds, the major labeled compound identified was 3-phosphoglyceric acid (PGA). After 7 seconds, sugar phosphates and diphosphates were found in addition of PGA. A 60-second exposure to $^{14}_{6}$CO2 produced labeled phosphoenolpyruvic acid (PEP), carboxylic acids, and amino acids. Using many different time intervals, the entire sequence of reactions was uncovered, and it was found that many of the steps were the reverse of those in the glycolytc pathway.

In the chloroplast stroma, CO_2 in the form of carbonic acid reacts with the sugar ribulose diphosphate (RuDP) to form an unstable 6-carbon compound that immediately splits to form two molecules of 3-phosphoglyceric acid (PGA). The enzyme catalyzing this reaction is carboxdismutase and the radioactive carbon of 14/6CO_2 is incorporated into the carboxyl group of PGA.

PGA is then reduced to 3-phosphoglyceraldehyde (PGAL) in two steps. First, each PGA is phosphorylated by ATP and then reduced by NADPH + H_+. The ATP and NADPH + H^+ were produced by photochemical reactions in the grana lamellae. Thus, for each molecule of CO_2 fixed and converted to PGAL, two A TP and two NADPH + H_+ molecules are required. These reactions are catalyzed by a kinase and dehydrogenase.

Some of the PGAL is isomerized by triose phosphate isomerase to form dihydroxyacetone phosphate (DHAP). The enzyme aldolase then condenses PGAL and DHAP to produce fructose-I, 6-diphosphate (FDP). Fructose diphosphatase splits off the phosphate group of the first carbon atom, producing fructose-6-phosphate (F6P). F6P may then be converted to fructose, glucose, or starch.

F6P and PGAL are also used for the resynthesis of RuDP (Fig. 18.16). F6P and PGAL are converted to erythrose-4-phosphate (E4P) and xylulose-5-phosphate (X5P). An aldolase then catalyzes the condensation of E4P and DHAP to form sedoheptulose-I, 7-diphosphate (SDP) which is then converted to sedoheptulose-7-phophate (S7P). S7P and PGAL also react to form ribose-5-phosphate and X5P. The R5P is then isomerized to form ribulose-5-phosphate (Ru5P). Ru5P is also formed X5P finally ATP phosphorylates Ru5P to form RuDP.

For each CO fixed, one RuDP, two ATP, and two NADPH + H+ are consumed and one phosphate a sugar is produced. In actuality; all the RuDP is resynthesized Fixation of these molecules of CO results in the formation of six PGAL. The extra PGAL is the photosynthetic product and is used for sugar and starch synthesis. Fig. 18.17 summarizes all the steps and the requisite numbers of molecules participating in the dark reactions; the pathway is known as the Calvin cycle.

Since additional ATP is consumed in the formation of RuDP, a toal of inie ATP and six NADPH + H+ are required for the fixation of three CO molecules. Since six Co molecules are required to produce one 6-carbon sugar molecule, 18-ATP molecules are consumed in the fixation (or three ATP/CO).

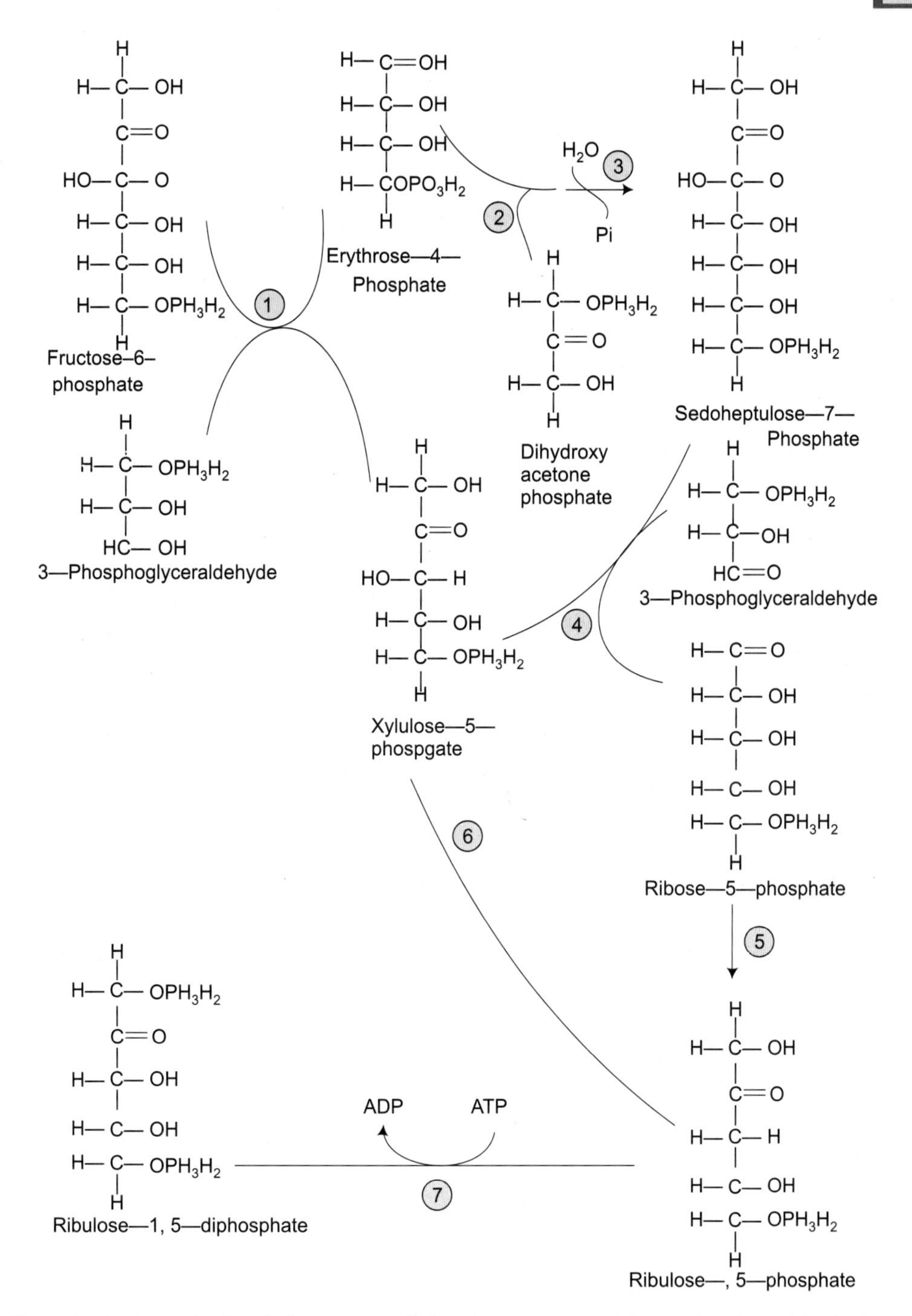

Fig. 18.16 Resynthesis of ribulose.1, 5-diphosphate. Enzmes; (1) transketolase; (2) aldoiase; (3) sedoheptulose.1 diphosphatase (4) transketolase; (5) isomerase; (6) epimerase; (7) kinase.

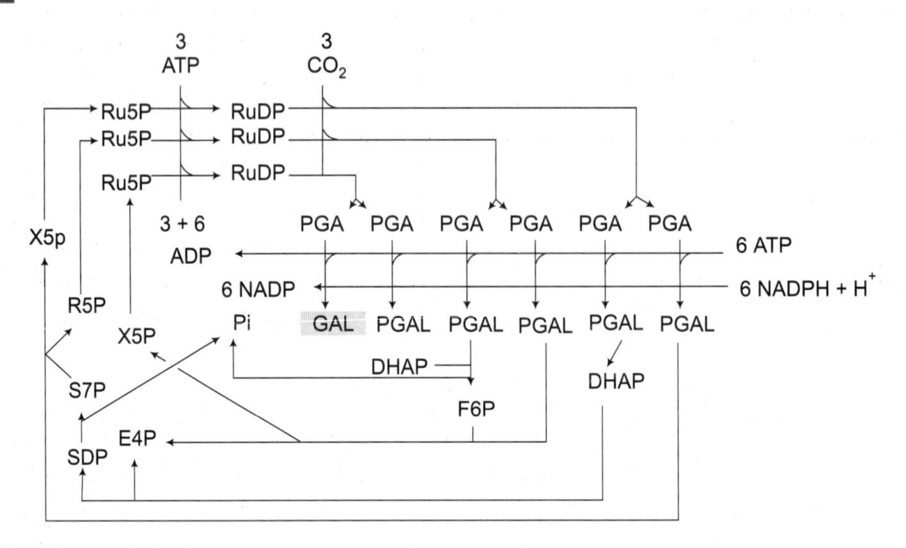

Fig. 18.17 The path of carbon in photosynthesis. The molecule of PGAL in the box is the photosynthetic product and is employed for sugar and starch synthesis. Ru5P, ribulose-5-phosphate; RuDP, ribulose-1,5-diphosphate; PGA, 3-phosphoglyceric acid; PGAL, 3-phosphoglyceraldehyde; DHAP, dehydroxyacetone phosphate; F6P, fructose-6-phosphate; SDP, sedoheptulose-1,7-diphosphate, S7P, sedoheptulose-7-phosphate; E4P, erythrose-4phosphate; X5P, xylulose-5 phosphate; R5P, ribose-5-phosphate; P inorganic phosphate.

It has been estimated that about 5 × 10 g of carbon are fixed annually by photosynthesis and this corresponds to storage of 4.8 × 10 kcal of energy. Since about 6.7 × 10 kcal of light energy falls on the earth each year, photosynthesis traps a mere 0.0072%.

OTHER CO-FIXATION PATHWAY

At the present time, three pathways for CO fixation during photosynthesis in plants are recognized the Calvin cycle or (C pathway) just described, the Hatch-Slack (or C pathway), and crassulacean acid metabolism (or CAM pathway). Species of higher plants can be characterized by the pathways they utilize. Although referred to as three "different" pathways, in effect, all these contain the Calvin cycle reactions, but two contain additional steps that provide CO in a more efficient manner.

The Hatch-Slack (C pathway) is common to corn, sugar cane, a number of other grasses, and several other plant species. How widespread this pathway is in the plant kingdom is not fully known, but it is believed to be substantial one of the characteristics of plants with this photosynthetic mechanism is a conservation of water, and therefore, plants that grow in relatively arid environments could be expected to have evolved such a mechanism. The primary characteristics of "C plants" are (1) high photosynthetic and growth rates, (2) low photorespiration rate, (3) an unusual leaf anatomy, (4) dimorphic chloroplasts, and (5) low rate of water loss.

Unlike the leaf anatomy described earlier in this chapter and typical for plants with Calvin cycle photosynthesis, c plants have an internal arrangement of chloroplast-containing cells that are oriented around the veins (vascular bundles). These cells are arranged in basically two layers (Fig. 17.18). These cells, which form a layer immediately about the vein, are called the bundle sheath cells. They contain chloroplasts, but the chloroplasts may lack grana (as in sugar cane) or the grana may be very reduced in size. These bundle sheath cells also accumulate starch in the chloroplasts during active photosynthesis. The bundle sheath cells also contain numerous microbodies and many large mitochondria.

Surrounding the bundle sheath cells is another layer of photosynthetic tissue called the mesophyll layer. The cells of this layer contain chloroplasts with extensive grana, have few mitochondria and microbodies, and do not accumulate starch. It is believed that the two tissue layers work together in photosynthesis. C plant vary as to the exact sequence of steps but in general the steps are as illustrated in Fig. 18.18. CO absorbed from the air spaces of the leaf into the mesophyll cells is fixed with phosphoenolpyruvate (PEP) in the chloroplast. Oxalacetate is the initial product that is subsequently converted to malate or aspartate. These dicarboxylic acids are then transported into the bundle sheath cells and decarboxylated releasing CO. The CO is then refixed by the Calvin cycle reactions in the chloroplasts, as described earlier. The 3-carbon compounds formed after decarboxylating malate or aspartate are transported back into the mesophyll where they are reconverted into PEP. The sugars that accumulate from the Calvin cycle reactions are temporarily converted to starch under active photosynthetic conditions. At night or during darkness or dim light the starches are converted back to the sugars and transported out of the leaf by the vascular tissue of the veins (vascular bundles).

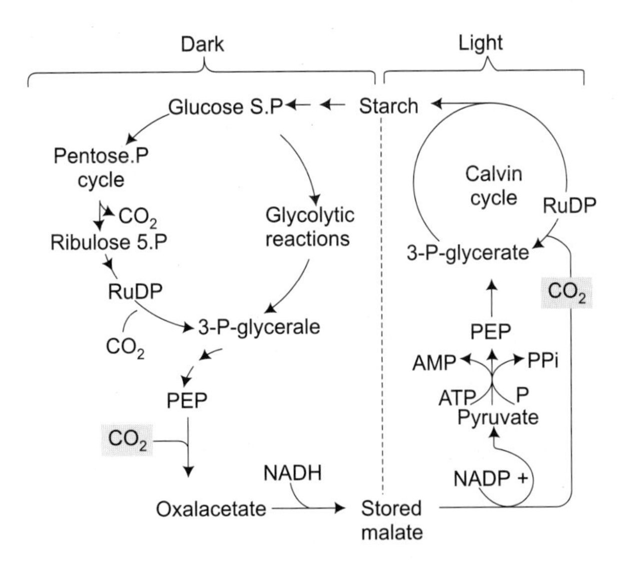

Fig. 18.18 Diagram of the major reaction steps of the light and dark phases of crassulacean aicd metabolism.

The presence of the C pathway seems like a needless addition to the Calvin cycle system. However, there is evidence indicating that the mesophyll is able to build up a high concentration of potential, CO by this method, which could provide evolutionary advantages. In addition, the rapid and efficient fixing and storing of CO decreases the leaf need to have a large number of stomates (openings in the leaf epidermis that allow CO and other gases to diffuse into the leaf). Open stomates, while allowing the passage of CO into the plant, also water to escape from the plant—a disadvantage to plants in arid climates.

Crassulacean acid metabolism (CAM) is a special form of metabolism associated with photosynthesis that is carried out by members of the plant family Crassulaceae (succulent herbs such as sedum). Plants carrying out this form of metabolism have closed stomates during the daylight hours and therefore cannot absorb sufficient CO for photosynthesis. But during the night (dark) hours, the stomates openand the leaf cells can fix CO in the dark by combining it with PEP to form oxalacetate. The oxalacetate is converted into malate for storage. During the following daylight hours, the malate is decarboxylated, and the CO is utilized by the Calvin cycle reactions. The 3-carbon pyruvate remaining after decarboxylation is converted first into PEP and then into phosphoglyceric acid is utilized by the Calvin cycle as well. The PEP required for the dark fixation of CO is derived from some of the starch produced from the Calvin cycle produced.

BACTERIAL PHOTOSYNTHESIS

Photosynthesis in prokaryotic organisms occurs in lamellar membrane system called chromatophores. The chromatophores contain the pigments for the photochemical reactions but none of the subsequent biosynthetic enzymes. The pigment system includes the chlorophylls. Carotenoids, and in some cases phycobilins. However, in bacteria, bacteriochlorophyll is the ultimate light-trapping molecule (not chlorophyll a).

The most important distinction between plant and bacterial photosynthesis is that water is not used as the reducing agent and oxygen is not an end product. The power to reduce CO may come from molecular hydrogen, H S or organic compounds. Two major groups of bacteria that carry out photosynthesis are the green and purple bacteria; these organisms utilize H S and produce sulfur and sulfate; that is

$$6CO + 12HS \xrightarrow[\text{bacteriochlorophyll}]{\text{Light energy}} CHO + 6HO + 12S$$

During photosynthesis, sulfur accumulates as granules of elemental sulfur and may be further metabolized later.

Nonsulfur purple bacteria use organic compounds such as acetic acid as electron donors. The acetic acid is anaerobically oxidized via the Krebs cycle reactions. Acetic acid can also be reduced to hydroxybutyric acid. Certain membranes of the sulfur and nonsulfur purple bacteria, can use molecular hydrogen to reduce either CO or acetic acid that is,

$$6CO + 12H \rightarrow CHO + 6HO$$

and

```
        O            OH        O
       /             |        /
2CH–C + H → CH–CH–CH–C + 2HO
       \                      \
        O                      OH
                         Hydroxybutyric acid
```

OTHER PLASTIDS

Chloroplast is only one of several different plastids found in plant cells. Other such as etioplasts, amyloplasts and chromoplasts have a different structure and function. They are all called plastids because they appear to develop from a common structure or from one another.

Proplastids are small, generally colcorless structures found in young or dividing cells. They have little internal structures but are delimited by a double membrane. Proplasmids give rise to other types of plastids.

Etioplasts are prevalent in the leaves of plants grown in the dark. Their ellipsoidal and sometimes irregular structures also bound by a double membrane. Internally, etioplasts contain a latticelike arrangement of tubules. Etioplasts are changed into chloroplasts upon exposure to light.

The outer membrane of the amyloplast encloses the stroma, containing one to eight starch granules. In certain plant tissues such as the potato tuber, the starch granules within the amyloplasts may become so large that they rupture the encasing membrane. Starch granules of amyloplasts are typically composed of concentric layers of starch.

Chromoplasts contain carotenoids and are responsible for imparting color (yellow, orange, red) to certain portions of plants such as flower petals, fruits and some roots. The chromoplasts of carrots contain large quantities of lipid that reduces their overall density to less than 1 g/ml; consequently, during centrifugation of root homogenates, the amyloplasts do not sediment but rise to the surface of the tube. Chromoplast structure is quite diverse; they may be round, ellipsoidal, or even needle-shaped, and the carotenoids that they contain may be localized in droplets or in crystalline structures. The function of the chromosomes is not clear but in may cases such as flowers and fruit, the color that they produce probably plays a role in attracting insects and other animals for pollination or seed dispersal.

A number of other, less frequently occurring plastids have been described, including the oil-filled elaioplasts, the protein-containing proteoplasts, and the sterol-rich sterino-chloroplasts.

SUMMARY

Photosynthesis, like mitochondrial reactions, is concerned in eukaryotes with the formation of energetically important ATP, involves hydrogen and electron transport in compounds like

NADPH + H and cytochromes, and occurs in or between membranes. The two process differ in that photosynthesis uses light rather than chemical substrates as the source of energy, CO_2 and water are consumed rather than produced, and O_2 and carbohydrate are produced rather than consumed. The overall reaction

$$6CO_2 + 12H_2O \rightarrow (C_6H_{12}O_6)_n + 6O_2 + 6H_2O$$

can be broken down into a light phase (photolysis and a dark phase (CO_2 fixation and dark reactions). In the light phase, visible light is absorbed by chlorophyll or a variety of other pigments located in the membranous thylakoids of the chloroplast. The light energy excites the molecules, inducing them to reemit light or heat or transfer the energy to a chlorophyll P_{700} molecule or trap chloroplyll. This activation of trap chlorophyll induces the reactions of photo system II, which generates ATP by the process called noncyclic photophosphorylation and terminates with the activation of P_{700} Activation of P_{700} by photosystem II or absorption of light energy directly initiates photosystem I.

This photosystem can also result in the formation of ATP (by the process of cyclic photophosphorylation) or the reduction of NADP to NADPH + H^+. The ATP and NADPH + H^+ transported into the stroma of the chloroplast are consumed in the dark reactions.

During the dark reactions, Co is fixed by binding to ribulose diphosphate and is subsequently reduced by NADPH + H^+. ATP, acts as the source of energy for these endergonic reactions. The final product is carbohydrate, usually in the form of a sugar.

Chloroplasts vary shape andare found only in ecaryotic plant cells. They form proplastids, which may be responsible for the formation of other plasmids, such as chromoplasts and leucoplasts. Plastids associated with C_3 photosynthesis have inner membranes arranged in layers and organized into grana. Plastids associated with C_4 photosynthesis may have a similar structure but frequently lack grana.

CHAPTER 19

Lysosomes and Microbodies

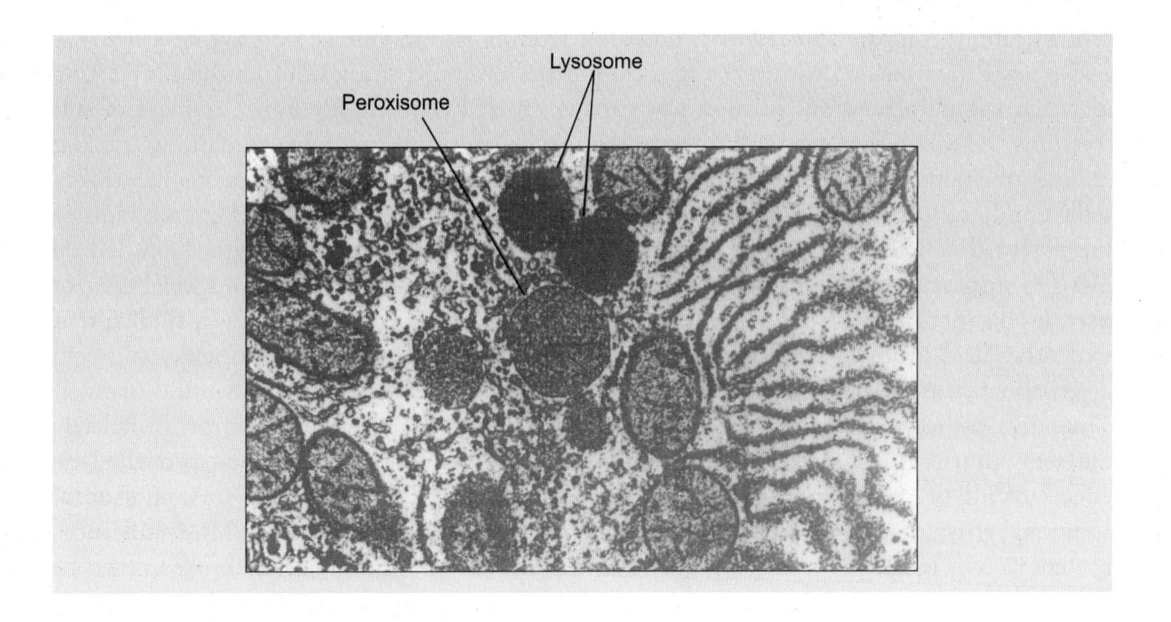

Our knowledge of the structure, composition, and function of lysosomes and micro bodies is considerably more recent than that of most their cell organelles. Although a variety of small oval bodies seen in plant and animal cells (including what are now termed lysosomes) had been variously called "microbodies" or "cytosomes" for many years, the diversity of their composition and action was net recognized until the 1950s. The "discovery" of lysosomes and micro bodies during the 1950s may be attributed to the growing sophistication of electron microscopy, the application of gentler procedures for dispersing tissue and cells, and the development of improved for separating, fractionating, and chemically characterizing the sub cellular complexes released from disrupted cells. At the present time, lysosomes are recognized as a separate category of organelles, whereas the microbodies include two major types' peroxisomes and glyoxysomes.

LYSOSOMES

The existence of lysosomes was subtly suggested for the first time in 1949 in the results of a series of experiments by c. de Duve and his co-workers that were designed to identify the cellular locus of the two enzymes glucose-6-phosphatase and acid phosphates. In these experiments, liver tissue homogentic were separated into nuclear, mictochdrial, microsomal, and soluble fractions by differential centrifugation and enzyme assays were performed on each of the collected fractions. Although at first, de Duve and his co-workers did not recognize that the acid phosphatase activity was associated with a distinct population of cellular particles. Instead, on the basis of the observed "latent" activity of the mitochondrial fraction obtained by differential centrifugation (compare values before and after freezing in Table 19.1), de Dave believed that acid phosphatase resided within the mitochondria: Continued studies during the early 1950s in which the mitochondrial reaction was further divided centrifugally into a number of sub fractions revealed that acid phosphatase was absent from fractions containing rapidly sedimenting mitochondria but was present in high concentrations in fractions containing slowly sedimenting mitochondria, this observation, together with a newly developed appreciation of the potential contamination of sediments occurring during differential centrifugation, led de Duve to suspect that the acid phosphatase might, in fact, be associated with a special class of particles distinct from the mitochondria. Added credence was given to this idea by finding that four other acid hydrolases, namely beta-glucuronidase, cathepsin, acid ribonuclease, and acid deoxyribonuclease, were distributed through the centrifugal fractions in an identical manner. Thus, five hydrolytic enzymes, each having an acid pH optimum and acting on completely different substrates, appeared to be present in the same cell particle. On the basis of the lytic effects of all of these enzymes, de Duve named the particles "lysosomes". A number of additional enzymes have subsequently been identified in lysosomes (Table l9.2) Most substances present in cells including proteins, polysaccharides, nucleic acid, and lipids are broken down by these enzymes.

It is interesting to note that he initial postulation of the existence of lysosomes was made by de Duve purely on biochemical grounds. However, in 1955, A Novikoff, working with de Duve, examined centrifugal fractions rich in acid provided the first morphological evidence supporting the existence of these particles. The lysosomes were identified as small, dense membrane-enclosed particles distinct from the mitochondria.

In recent years, sophisticated centrifugal methods have been devised for obtaining preparations rich in lysosomes. Nearly all preparations obtained by differential centrifugation are contaminated with quantities of mitochondria. Although the average sedimentation coefficient for mitochondria is greater than that of lysosomes, mitochondria are polydispersed with respect to size; so that the smaller mitochondria invariably sediment with the lysosomes. Moreover, in tissue containing peroxisomes (such as liver and kidney), the range of sedimentation coefficients for these organelles is almost identical to that of the lysosomes. Consequently, it is virtually impossible to obtain lysosomes preparations that do not also contain large numbers of peroxisomes. Somewhat greater success is obtained when isopyenic

Table 19.1 Distribution of acid phosphatase enzyme activity in liver tissue fractions prepared by differential centrifugation acid.

Fraction	*Acid phosphatase acitivity (µg phosphate released/20 min)*	
	Before freezing and storage	*After freezing and storage for 5 Days*
Whole homogenate	10	89
Nuclear fraction	2	10
Mitochondrial fraction	7	46
Microsomal fraction	6	10
Soluble fraction	6	9

Table 19.2 Some enzymes present in lysosomes.

Enzyme	*Substrate (S)*
Proleases and peptidases	
Cathepsin A, B, C, D and E	Various proteins and peptides
Collagenase	Collagen
Arylamidase	Amino acid arylamids
Peptidase	Peptides
Nucleases	
Acid ribonuclease	RNA
Acid deoxyribonuclease	DNA
Phsophatases	
Acid phosphatase	Phosphate monoesters
Phosphodiesterase	Oligonucleotides, phosphodiesters
Phosphatide aicd phosphatase	Phosphetidic acids
Enzymes acting on carbohydrated chains of glyco-proteins and glycolipids	
Beta-galactosidase	Beta-galactosides
Acetylhexosaminidase	Acetylhexosaminides, heparin sulfate
Beta-glucosidase	Beta-glucosides
Alpha-glucosidase	Glycogen
Alpha-mannosidase	Alpha-mannosides
Sialidase	Sialic acid derivatives
Enzymes acting on glycosaminoglycans	
Lysozymes	Mucopolysaccharides, bacterial cell walls
Hyaluronidase	Hyaluronic acid, chonodrotin sulfates
Beta-glucuronidase	Polysaccharides, mucopolysaccharides
Arylsulfatase A, B	Arylsulfates, cerebroside sulfates, chondroitin sulfate
Enzymes acting on lipids	
Phospolipase	Lecithin, phosphatidyl ethanolamine
Esterase	Fatty acid esters
Sphingomyclinase	Sphingomyclin

density gradient centrifugation is used in the last stages of the isolation procedure, for the equilibrium lenities of lysosomes (1.22 g/cm), mitochondira (1.19 g/cm) and peroxisomes (1.23-1.25g/cm) in sucrose density gradients are slightly different. Most density gradient procedures used to prepare lysosomes are modifications of the technique develop by W. C. Schneider and depicted in Fig. 19.1. Using this technique, most of the mitochondria are banded isopycnically at a density of about 1.19 g/cm, while most of the lysosomes form a separate zone at about 1.22 g/cm and can be recovered independently from the density gradient.

By far the greatest purity of lysosomes is obtained from tissues of animals previously treated with Triton WR-1339 (a polyethylene glycol derivative of polymerized p-tert-octylphenol), dextran (a polymer of glucose), and thorotrast (colloidal thorium hydroxide). These compounds are rapidly incorporated in large quantities by cell's lysosomes, significantly altering their density. For example, the incorporation of Triton WR-1339 reduces the average density of the lysosomes from 1.22 to about 1.10 g/cm. it is interesting to note that although the density of lysosomes in corporation Triton WR-1339 is significantly reduced, their size is increased; the result is that Triton WR-1339-loaded lysosomes have the same sedimentation coefficient as do normal lysosomes but have a lower density.

The latent enzymes effect originally noted by de Duve and his co-workers is still employed as a major criterion in evaluating the effectiveness of any lysosomes isolation. Accordingly, the lysosome preparation is incubated under the appropriate conditions with the hydrolase substrate before and after treatments known to disrupt the lysosome membranes. If the original preparation contains intact lysosomes, then no substrate is hydrolyzed before treatment (most substrate of the lysosomal hydrolases are unable to permeate the lysosome's membrane); however, distruption of the membrane (by sonification, repeated freezing and thawing addition of lytic agents such as bile salts, digitonin, Triton X-l00, etc.) and release of the lysosomal enzymes are quickly followed by hydrolysis of he added substrates.

Structure and Forms of Lysosomes (Fig. 19.2)

Lysosomes are a structurally heterogeneous group of organelles varying dramatically in size and morphology. As a result, it is difficult to identify lysosomes strictly on the basis of morphological criteria. When lysosome-rich fractions were initially isolated centrifugally by de Duve and Novikoff and examined with the electron microscope, it was found that the suspect lysosomes were generally smaller than mitochondria. Typically, they varied in diameter that from about 0.1 to 0.8 μ were bounded by a single membrane, and were usually somewhat electron-dense. Identification of lysosomes in sections of whole cells is considerably more difficult because other small, dense organelles are also bounded by a single membrane. The application of cytochemical procedures at he level of the electron microscope in which the lysosomes are identified on the basis of their enzyme content is much more reliable. Notable among such procedures is that introduced in 1952 by G. Gomori and which is routinely employed in variously modified forms for the identification of lysosomes on the basis of their high acid phosphatase content. In the Gomori method, the tissue to be examined is incubated at pH 5.0 in a medium containing beta-glycerophosphate (a substrate for acid phosphatase) and a lead salt (such as lead nitrate).

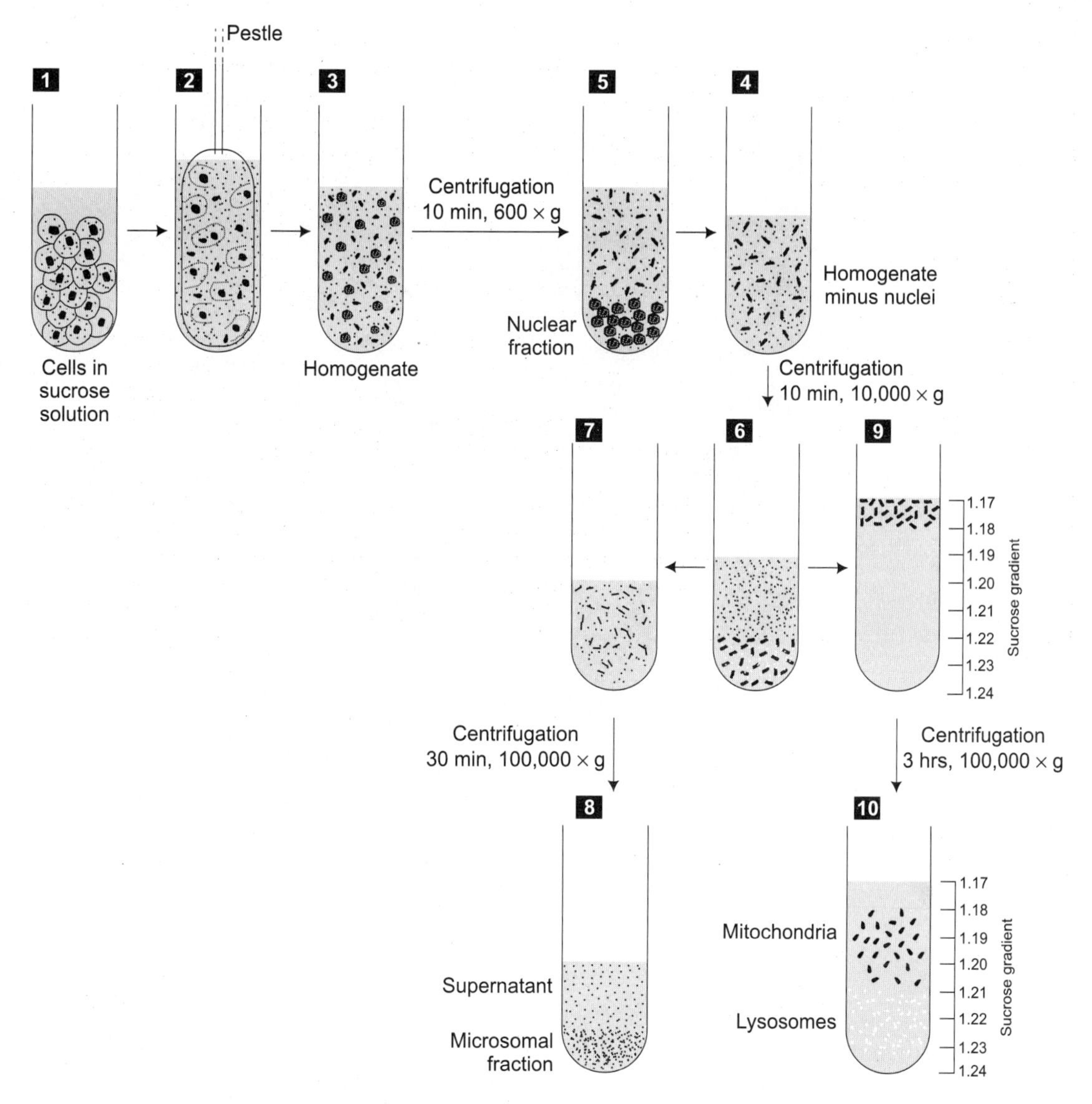

Fig. 19.1 Steps in the isolation of lysosomes by sucrose density gradient centrifugation.

Phosphate enzymatically cleaved from the substrate during incubation combines with the lead ions to from isoluble lead phosphate which precipitates at the locus of enzymes activity. Since the lead phosphate is electron-dense, lysosomes appear as particularly dark, granular organelles in the electron microscope (Fig. 19.2). For identification with the light microscope, ammonium sulfide may be employed to convert the lead phosphate produced to the black lead sulfide. The Gomori reaction may be carried out with fixed and sectioned material, as well as with fresh tissue, albeit with reduced efficiency as a result of some enzyme inactivation during and following fixation.

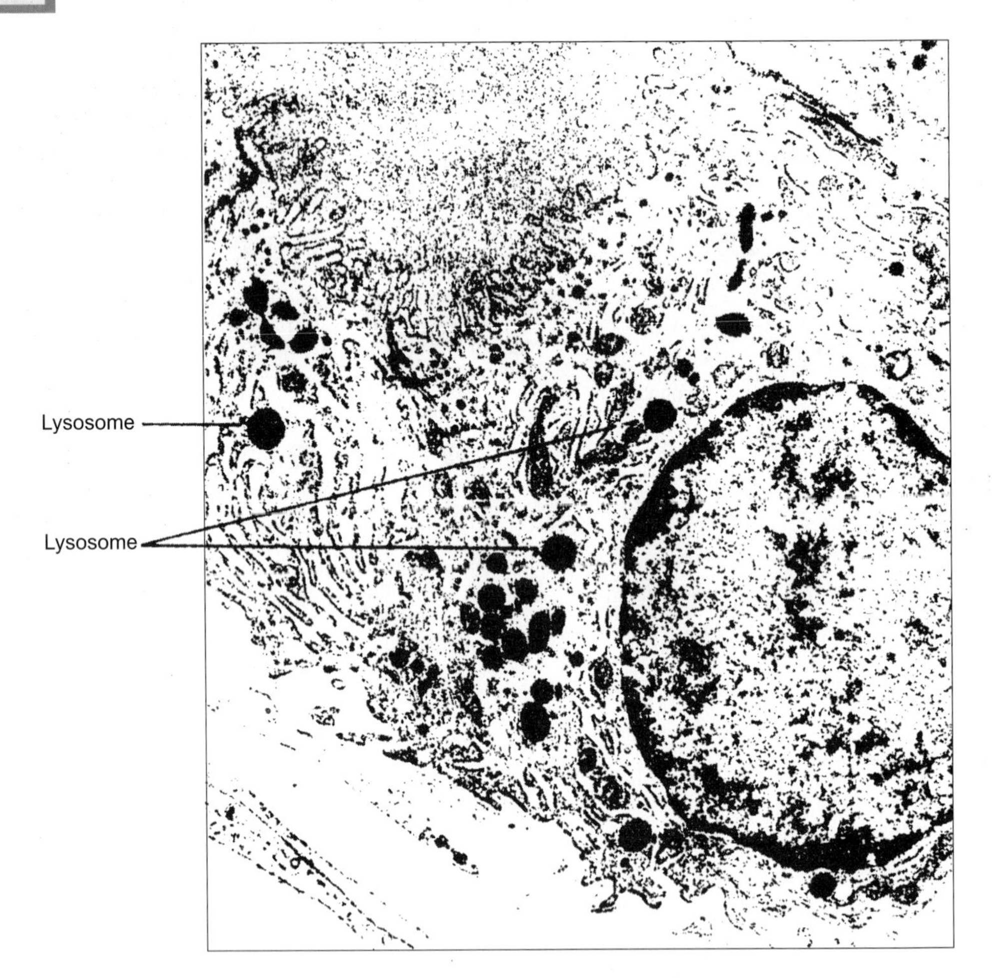

Fig. 19.2 TEM of lysosomes in thyroid follicle cell (Courtesy of R. Roberts, R.G. Kessel, and H. Tung, U. Iowa).

Several different lysosomal forms have been identified within individual cells including (1) primary lysosomes, (2) secondary lysosomes, and (3) residual bodies.

Heterophagy. Extracellular materials, brought into the cell by enocytosis are enclosed within vacuoles called phagosomes. These, materials may later be rejected unaltered by exocytosis, or the phagosomes may fuse with one or more primary lysosomes that empty their digestive hydrolases into the newly formed particle (now called a secondary lysosomes, Fig. 19.3. lysosomal digestion of endocytosed material is termed heterophagy.

The fusion of primary lysosomes with phagosomes has been demonstrated in vivo in a number of tissue using various exogenous markers introduced into the organism. These markers, which include horseradish peroxidase, ferritin, and hemoglobin, are engulfed by the tissue cells and are later detected within secondary lysosomes along with lysosomal hydrolases.

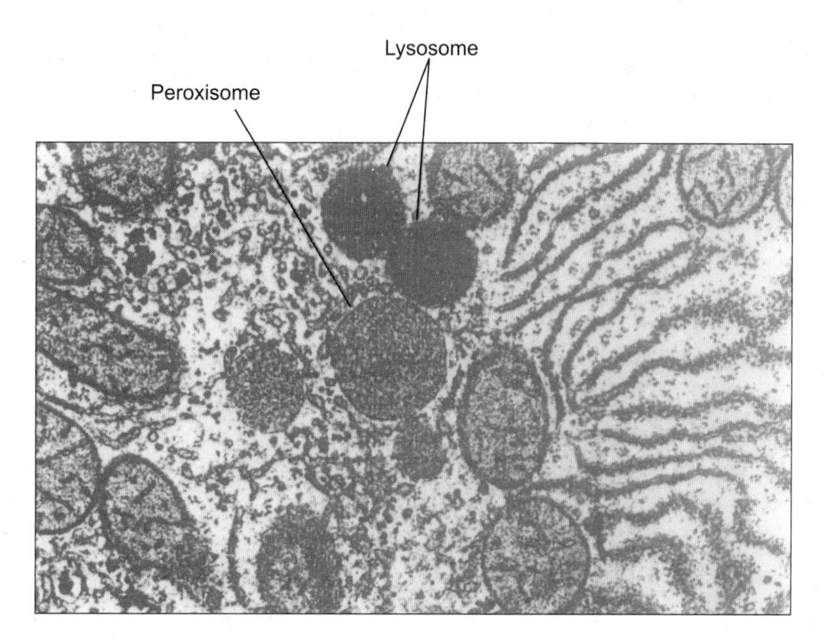

Fig. 19.3 TEM of peroxisomes in liver cell (Courtesy of R. Roberts, R.G. Kessel, and H. Tugn, U. lowa).

Autophagy. The isolation and igestion of portions of cells, own cytoplasmic constituents by its lysosomes occurs in normal cells and is termed autophagy (Fig. 19.4). The phenomenon is most dramatic in the tissue of organs undergoing regression (e.g., changes in the uterus following delivery, during metamorphosis in sects, etc.). Autophagic vacuoles containing partially degraded mitchondria, smooth and rough endoplasmic reticulum, microbodies, glycogen particles, or other cytoplasmic structures are frequently observed in tissue sections examined with the electron microscopy. Cellular autophagy results in a continuous turnover of mitochondria in liver tissue. The half-life of the liver mitochondrion is about 10 days and corresponds to the destruction of one mitochondrion per liver cell every 15 minutes.

Distribution of Lysosomes

Since their initial discovery in mammalian liver, lysosomes have been identified in many different cells and tissues; some of these are listed in Table-19.3. The greatest variety of tissues found to contain lysosomes occurs in animals. Although most studies have been carried out using mammalian tissue, lysosomes have been identified in insects, marine invertebrates, fish amhibians, reptiles, and birds. Lysosomes are particularly numerous in epithelial cells of obsorptive, secretory, and excretory organs (liver, kidneys, etc). they are also present in large numbers in the epithelial cells of the intestines, lungs, and uterus. Phagocytic cells and cells of the reticuloendothelial system (e.g. bone marrow, spleen, and liver have also been found to contain large numbers of lysosomes. Few lysosomes occur in muscle cells or in acinar cells of he pancreas. Lysosomes are produced by certain cells in tissue culture (He La cells, monocytes, lymphocytes etc.). Although it has a number of functions not shared by lysosomes of animal

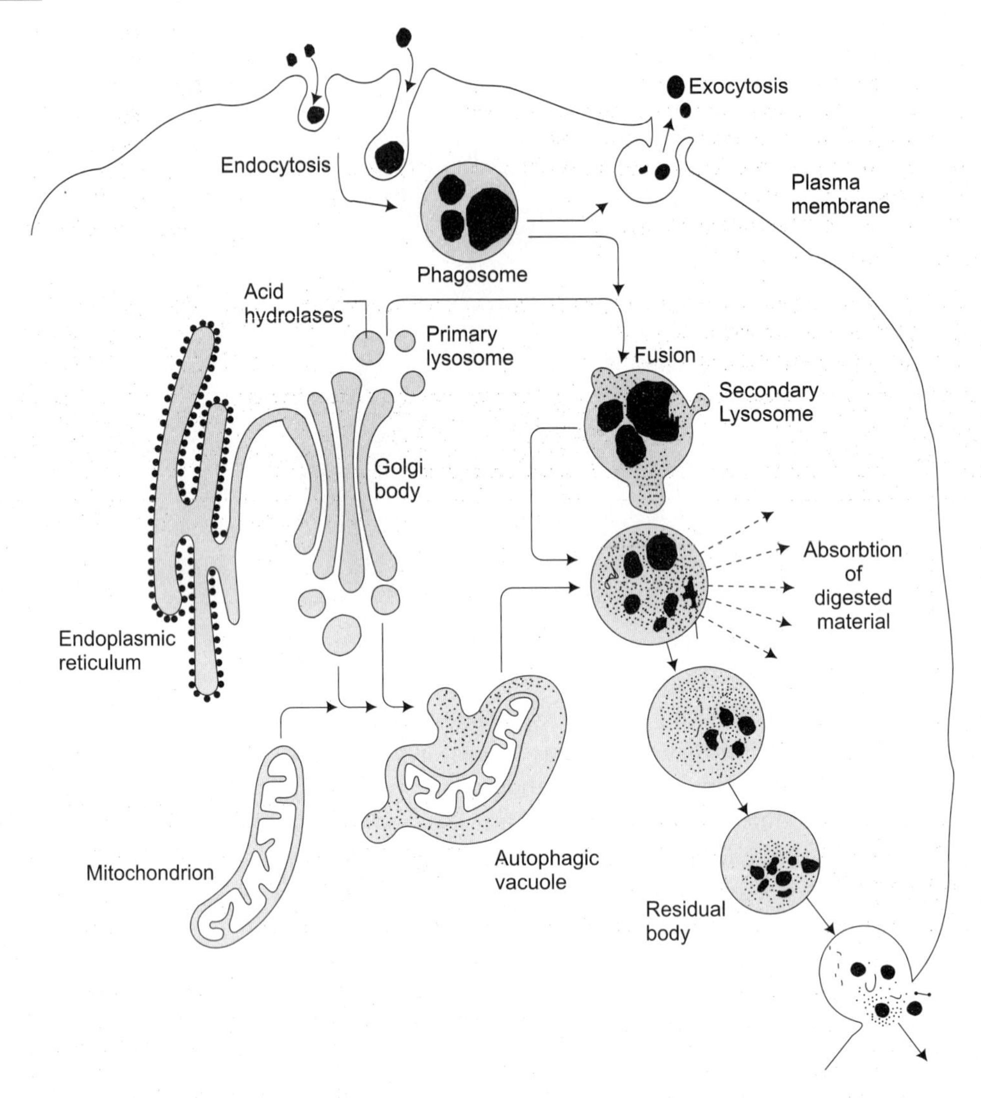

Fig. 19.4 Formation and function of lysosomes in cellular heterophagy and autophagy.

cells, the large vacuole of many plant cells is a modified lysosome. Some of the various roles played by the lysosomes are summarized in Table 19.4.

Leukocytes especially granulocytes, are a particularly rich source of lysosomes, and this is related to their physiological role as scavengers of microorganisms or other foreign particles in the blood. Following phagocytosis of a bacterium by a leukocyte, numerous lysosomes fuse with the endocytic vacuole containing the microorganism and initiate its digestion. The lysosomes of granular leukocytes are especially large and readily visible by light microscopy. Once the lysome content of the leukocyte is exhausted, the blood cells dies.

Table 19.3 Cells and tissues containing lysosome.

Protozoa	Nerve cells
Amoeba	Brain
Campanella	Intestinal epithelium
Tetrahymena	Lung epithelium
Paramecium	Uterine epithelium
Euglena	Macrophages of spleen, bone marrow liver and connective tissue
Plants	
Onion seeds	
Corn seedling	Thyroid gland
Tobacco seedling	Adrenal gland
Tissue culture cells	Bone
HeLa cells	Urinary bladder
Fibroblasts	Uterus
Monocytes	Ovaries
Macrophages	Blood (leukocytes and platelets)
Chick cells	
Lymphocytes	
Animal tissues	
Liver	
Kidney	

Table 19.4 Some functions of lysosmes.

1. Nutrition via a digestive role in protozoa and many metazoan cells
2. Nutrition via cellular autophagy during unfavorable environmental conditions
3. Lysis of organelles during cellular differentiation and metamorphosis
4. Scavenging of worn-out cell parts and denatured proteins
5. Destruction of aged red blood cells and dead cells
6. Defense against invading bacteria and viruses by circulating macrophages
7. Dissolution of blood clots and thrombi
8. Keratinization of skin
9. Secretion of hydrolases by sperm for egg penetration during fertilization
10. Yolk digestion during embryonic development
11. Bone resorption.
12. Reabsoption in kidney and urinary bladder

Lysosome Precursors in Bacteria

Although bacterial cells do not possess lysosomes, they do contain a variety of hydrolases that are believed to be localized in the space between the cell wall and the cell membrane. These hydrolases may be synthesized by ribosomes attached to the cell membrane and then dispatched through.

Disposition and Action of the Lysosomal Hydrolases

Many of the lysosome's enzymes are released into the surrounding environment when these organelles are physically or chemically disrupted. Those enzymes that are so readily solubilized are believed to be located in the interior of the organelle. Other lysosomal hydrolases cannot be solubilized or are extracted with great difficulty and are thought to be an integral part of the lysosome membrane together with other proteins and lipids. Some of the enzymes known to be present in the lysosomes are listed in Table 19.2; it is to be noted that while this list is extensive, it is by no means complete.

All the substrates of lysomal enzymes are either polymers or complex compounds and include proteins, DNA, RNA, polysaccharides, carbohydrate side chains of glycoproteins and glycolipids, lipids, and phosphates. The lysosomal breakdown of proteins into amino acids illustrates how these enzymes act in concert. The initial hydrolysis of protein is effected by cathepsins D and E and also by collagenase. These enzymes cleave peptide bonds and produce peptide fragments of varying length. These peptides, together with previously undigested proteins, are further hydrolyzed to individual amino acids by cathepsins A and B. cathepsin C, arylamidase, and the lysosomal dipeptidases act on specific peptides, producing additional amino acids.

The breakdown of DNA and RNA is initiated by the enzymes acid deoxyribonuclease and acid ribonuclease. The resulting oligonucleotides are then degraded first by phosphodiesterase and then by acid phosphates, producing nucleosides and inorganic phosphate. Lysosomes also possess all the enzymes necessary for hydrolysis of lipids and polysaccharides.

As noted earlier, some lysosomal enzymes are part of the membrane encasing the organelle. These are believed to form some sort of protective lining, for it is difficult to understand how the membranous portions of autophagocytosed mitochondria and endoplasmic reticulum are so readily hydrolyzed while the lysosome membrane remains impervious. In addition to having acid pH optima, he lysosomal enzymes are particularly resistant to autolysis. Among the enzymes found to be integral parts of the lysosome membrane are acetylglucosaminidase, glucosidase, and sialidase. Arylsulfatase, acid phosphatase, ribonuclease, and glucuronidase may also be bound to the membrane under certain conditions.

Enzymes freed from disrupted lysosomes exhibit a wide variation in stability. Some retain their activity for only a few hours following tissue disruption; others are stable for months when appropriately refrigerated. Of the few lysosomal enzymes isolated and characterized to date, several have been shown to be glycoproteins including cathepsin C, acid deoxyribonuclease, glucuronidase, and acetyl-glucosaminidase.

Microbodies

As noted at the beginning of the chapter, the term "microbody" has been used by cell biologists and cytologists for many years to describe a variety of different small cellular organelles. More recently, it has been restricted to organelles possessing oxidase, peroxidase, or catalase enzyme activity. Organelles possessing these activities are typically small ovoid structures having a diameter of about 0.5-1.5 μ and containing an amorphous granular matrix and, occasionally, crystalloid inclusions. The organelles vary somewhat in structure, appearance, and function from one tissue to another and from species to species Two distinct but related forms of microbodies are common in animal and plant cells and in microorganisms; these are peroxisomes and glyxysomes.

Peroxisomes (Fig. 19.3)

The modern usage of the term "microbody" dates back to 1954 and the work of J. Rhodin, who described the structure and proteins of these organelles in the mouse kidney. Since then, organelles of similar organization have been reported in many other tissues of both animals and plants.

In 1969, de Duve showed that microbodies of rat liver contained a number of oxidases that transfer hydrogen atoms to molecular oxygen, thereby forming hydrogen peroxide (Fig. 19.5). de Duve coined the term peroxisomes foe these organelles, although a true peroxidatic activity is generally demonstable only in vitro. In vivo, conditions favor the removal (or degradation) of hydrogen peroxide by catalase rather than by a peroxidase. However, since hydrogen peroxide is an intermediate in the reactions, the term "peroxisome" may be appropriate. The chemical and enzymatic relationships between an oxidase, peroxidase, and catalase are shown in Fig. 19.5.

A number of enzymes are characteristically present in peroxisomes including uric acid oxidase, D-amino acid oxidase, A-hydroxyacid oxidase, NADH-glyoxylate reductase, NADP-isocitrate dehydrogenase, and catalase. When uric acid oxidase is present in large amounts, it frequently takes the form of a paracrystalline "nucleoid" at the center of the organelle. The functions of peroxisomes in animal cells are not entirely clear. Peroxisomal catalase is thought to be involved in the degradation of H_2O; which is extremely toxic, the source of the peroxide being other peroxisomal reactions (e.g., those catalyzed by oxidases) or cytosol-produced H_2O. Uric acid oxidase may be important in degrading purines. The abundance of peroxisomes in cells engaged in lipid metabolism suggests that these organelles may be involved in gluconeogensis.

We saw earlier that one of the characteristic features of lysosomal enzymes activity is its latency. No latency is hibited by the peroxisomal enzymes, as relatively large molecues (including the peroxisomal enzyme substrates) readily permeate the peroxisome membrane.

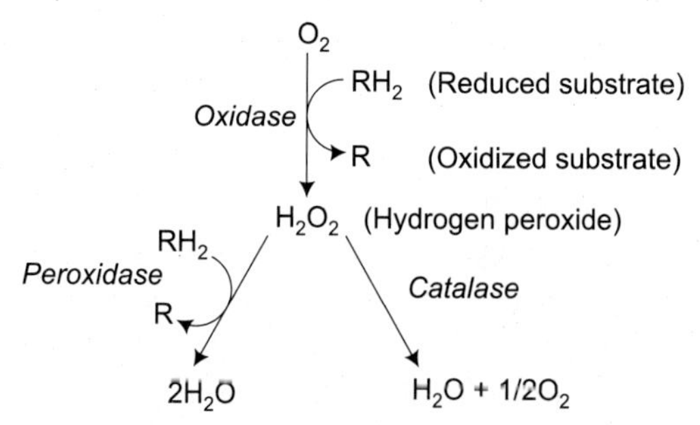

Fig. 19.5 Chemical interrelationship between oxidase, peroxidase, and catal...

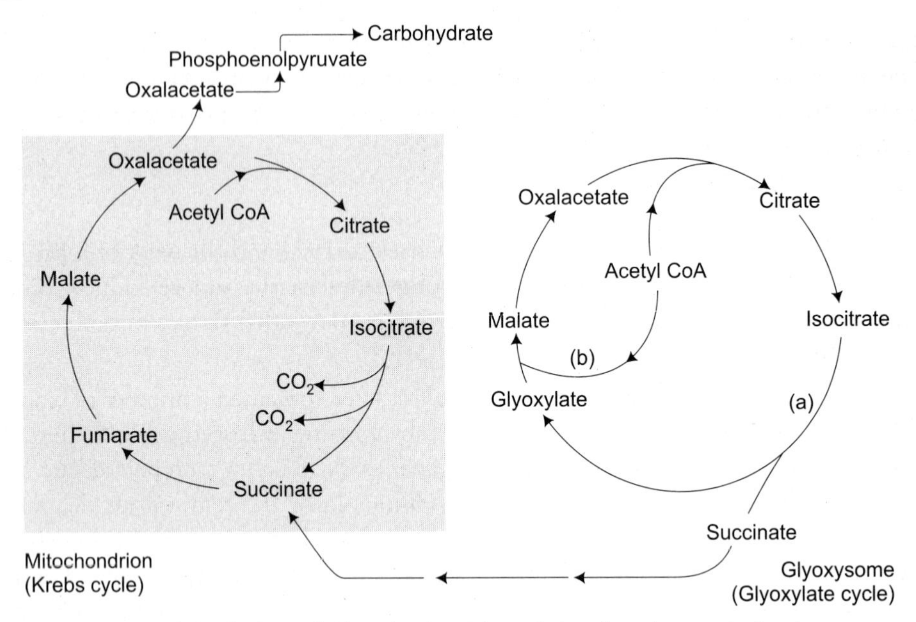

Fig. 19.6 Comparison of the krebs cycle in mitochondria and the glyoxylate cycle in glyoxysomes. Enzymes (a) and (b) are isocitrate lyase and malate synthetase. respectively.

Isolation of peroxisomes. The sedimentation coefficients and densities of peroxisomes are close to those of lysosomes and are not significantly different from mitochondria. Their similarities to lysosomes account for the fact that for some time peroxisomal enzyme activities were ascribed to lysosomes. Density gradient centrifugation of the "mitochondrial fraction" prepared by preliminary differential centrifugation is the method of choice for isolating peroxisomes. The greatest success in peroxisome purification is obtained if the lysosomes are first allowed to accumulate Triton WR-1339 (see earlier discussion). Triton-loaded lysosomes are considerably less dense than normal lysosomes and so are easily "floated" away from the peroxisomes during density gradient centrifugation.

Formation of peroxisomes. It is generally thought that peroxisomes arise as outgrowths of the endoplasmic reticulum and that the peroxisomal enzymes dispatched from attached ribosomes into the cisternae make their way to these out-growths prior to physical separation. Studies on peroxisomal catalase indicate that the subunits of the enzyme and the heme are assembled to ·om the functional molecule after entering the microbody.

'oxysomes

In 67, R. W. Briedenbach and H. Beevers discovered that microbodies of certain plant tissue con ed enzymes of the glyoxylate cycle in addition to peroxisomal enzymes. They used the term oxysomes for these particles. Glyoxysomes not only contain the glyoxylate bypass enzyme ocitrate lyase and malate synthetase but also contain several of the essential enzymes of the Ke s cycle, which therefore function simultaneously in both groups of organelles.

The relationship between the Kerbs cycle and the glyoxylate cycle is shown in Fig. 19.6. Both cycles employ the same reactions to produce isocitrate from acetyl Co A and oxalacetate, but beyond this point the pathways differ. In the Kerbs cycle, isocitrate is successively decarboxylated to produce succinate and two molecules of CO. in the glyoxylate cycle; isocitrate is converted to succinate and glyoxylate. Therefore, instead of being lost as two molecules of CO_2, the 2-carbon glyoxylate condenses with another acetyl Co A to from the 4-carbon dicarboxylic acid, malate. The four carbon atoms of the two acetyl CoA molecules are thus conserved as one 4-carbon compound which, after conversion to succinate and migration to the mitochondrion, may be converted to oxalacetate. The oxalacetate may then be utilized in gluconeogenesis.

Oxalacetate formed in mitochondria from glyoxysomal succinate is presumed to serve as a direct precursor phosphoenol pyruvate (PEP). The conversion of PEP to carbohydrate occurs by the reversal of the steps of glycolysis. Glyoxysome-containing tissues are thus able to convert simple 2-carbon sources such as acetate into carbohydrate. In some tissues, such as the fat-storage cells in seeds, the acetate is obtained through the degradation of a fatty acids (Chapter 10). The glyoxylate cycle is especially significant for cells growing exclusively on acetate or fatty acids (e.g., a number of microorganisms), where the cycle acts as a source of 4-carbon dicarboxylic acids.

Distribution and origin of glyoxysomes. Glyoxysomes are not as widespread as peroxisomes. They are primarily found in the endosperm, cotyledon, and aleurone tissues of plants. Microbodies possessing glyoxysomelike activites have also been reported to occur in a number of microorganisms including *Euglena*, *Chlorella*, *Neurospora*, and *Polytomella*. Reports have appeared in the literature from time to time indicating the DNA is present in glyoxysomes, raising the possibility that these organelles possess some degree of autonomy. However, this notion is not generally accepted. Instead, because of their regular intimate association with the endoplasmic reticulum, glyoxysomes, like peroxysomes are believed to be produced as outgrowths of the ER.

In concluding this discussion of microbodies, it is important to note that some organelles fitting the general microscopic description of microbodies do not clearly fit into either the peroxisome or glyoxysome category when evaluated interms of their enzymatic activities. It is entirely possible that microbodies may be associated with varying activities depending upon the specialization of the cell and the microbodies exist whose actions and cellular functions remain to be determined. It is clear that one cannot name particles solely on the basis of microscopic characteristics and expect that all will function identically.

SUMMARY

The realization that certain hydrolase activities are associated with a distinct class or organelles called lysosomes is quite recent. Previously, these activities were believed to be localized in mitochondria. Lysosomes function in the intracellular digestion of poorly functioning or superfluous organelles, as well as in endocytosed materials. Several forms of lysosomes may be identified including primary lysosomes, secondary lysosomes (e.g., heterophagic and autophagic vacuoles), and residual bodies. Primary lysosomes are formed at the peripheral edges of the Golgi complex, their hydrolase content derived through the cisternaeof he endoplasmic reticulum. Fusion of primary lysosomes with phagosomes forms the secondary lysosomes in which digestion occurs. Usable products of this digestive activity are transferred to the cytoplasm. Undigested or unabsorbed materials remain in the residual bodies, which may accumulate in the cell or fuse with the plasma membrane during exocytosis.

Two distinct but related classes of microbodies occur in cells; these are peroxisomes and glyoxysomes. Peroxisomes are probably formed as outgrowths of the endoplasmic reticulum and contain a number of oxidases that produce hydrogen peroxide during their degradative activity. The potentially harmful peroxides are further degraded by peroxisomal catalase. Glyoxysomes are found primarily in plant cells and contain enzymes of the Kerbs cycle and glyoxylate bypass, in addition to peroxisomal enzymes.

CHAPTER 20

Cilia, Flagella, Microtubules and Microfilaments

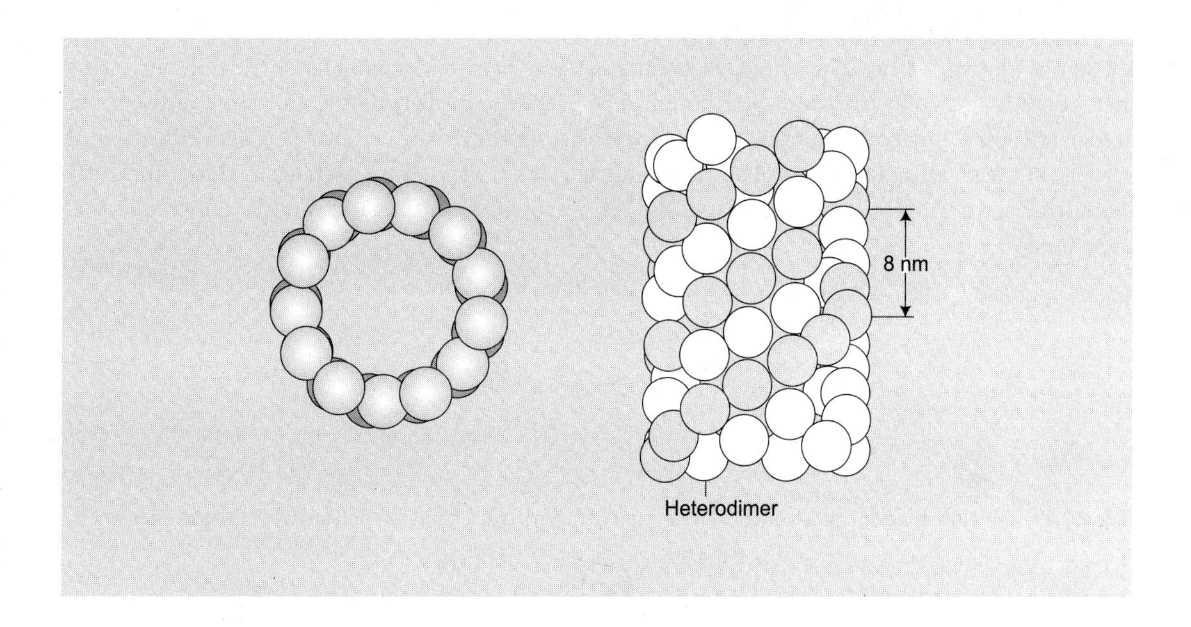

Microfilaments and microtubules are two distinctly different kinds of structures found in most cells. Although different, both are associated with cell movement and support and in some cases are so closely associated with the same functions in the cell that they collectively referred to as the microtubule-microfilament system. The presence of these two types of unbranched, elongated structures in the cells has been known since before the turn of the century, but only within the past decade have the techniques for proper fixation and staining been worked out so that good description could be obtained from electron photomicrographs. Microtubules in particular eluded description because their proteins depolymerize at low temperatures, and for many years, standard fixation of cells for electron microscopy was carried out at 0°C.

Microfilaments are elongated, unbranched, proteinaceous strands. The microfilaments usually consist of bundles or groups of proteins sometimes would in a helix. In most cells the microfilaments are 30-60 A in diameter but in other cells the microfilaments may be much thicker. Two types, thick and thin microfilaments, are recognized. The myosin filaments of striated muscle cells are about 100 A in diameters, while in amoebae and slime molds the thick microfilaments are 150-250 A wide and are tapered at the ends. Striated muscle cells contain two types of microfilaments (usually called myofilaments): myosin, the so-called thick filaments, and actin, the thin filaments. Actin or proteins closely related to actin, are found in most eucaryotic cells and are not restricted to muscle. In striated muscle cells, the thin F-actin filaments are composed of two strands of protein or protofilaments coiled around each other in a double helix (Fig. 20.1). Each protofilament is made up of monomers of G-actin which have been polymerized.

Microfilaments are disassociated by low concentrations of cytochalasin B (Fig. 20.2), a substance derived from the mold *Helminthosporium dematoiderum*. Celss treated with this metabolite lose certain functions attributed to the actions of microfilaments. Some of the more associated with microfilaments are phagocytosis, pinocytosis, ecocytosis, cytokinesis and cytoplasmic streaming (in plant cells). If the concentration of added cytochalasin B is sufficiently high, the actin microfilaments of muscle cells are dissociated and muscle contraction is prevented.

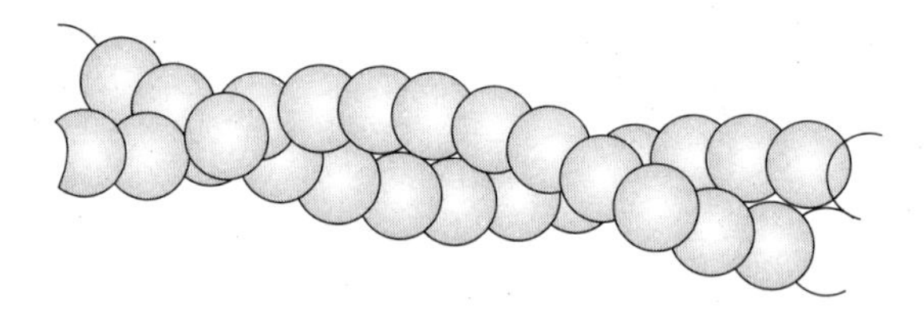

Fig. 20.1 The thin filament of striated muscle consisting of two chains of polymerized G-actin monomers.

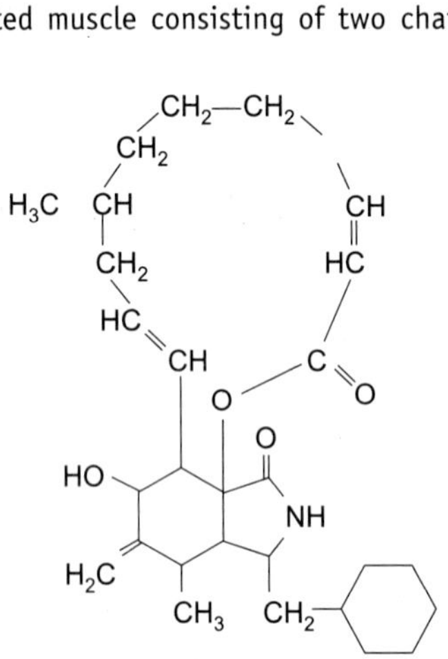

Fig. 20.2 Cytochalasin B structure.

Table 20.1 Comparison of the properties of most microfilaments and microtubules.

	Microfilament	*Microtubule*
Diameter	30-60 Å	100-250 Å
Structure	Double-helical protofilament	Hollow tube of 13 protofilaments
Protein	Actin or actinlike protein	Tubulin
Disassociating or inhibiting agent	Cytochalasin B	Colehieine vinblastine of vincristine
Subunit building agent	ATP	GTP

Microtubules differ from microfilaments in several fundamental respects (Table 20.1): (1) they are composed of different kinds of proteins, (2) they possess a different structure, and (3) they are dissociated by a different set of compounds. In microtubules, the protein monomer is tubulin, a heterodimer containing two subunits designated A and B. each subunit consists of a single polypeptide, each the same size (54,000 daltons) but different in composition. In electron photomicrographs, the chains of globular subunits appear to run parallel to each other as well as to the axis of the microtubule (Fig. 20.3). However, the A and B subunits form a heterodiamer and the arrangement of these paired subunits suggests a helical assembly of the heterodiamers about the hollow central core.

In the assembly of the subunits, the A and B polypeptides are each first activated by binding to GTP. The activated subunits assemble onto other subunits. As each subunits becomes bound to the growing tubule. GTP undergoes hydrolysis, with the resulting GDP and phosphate remaining tightly bound. Vincristine and vinblastine (Fig. 20.4) interfere with this assembly process, causing precipitation of the tubulin in polymeric, three-dimensional disarray. Colchicines (Fig. 20.4), another inhibitor of microtubule formation, acts by binding to the

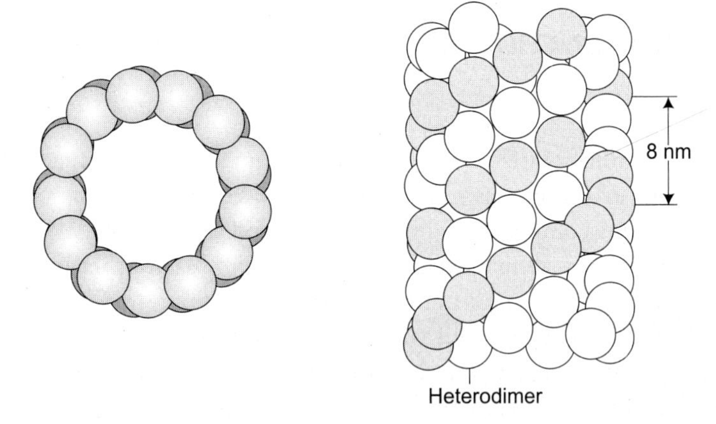

Fig. 20.3 Microtubule model seen in cross section and laterally. The circumference of the tubule usually has 13 subunits (see text for variations). Although subunits appear to be linear chains that are parallel to the axis of the tubule, the paired subunits (α + β heterodimers) join to form a helix about the central, hollow corc.

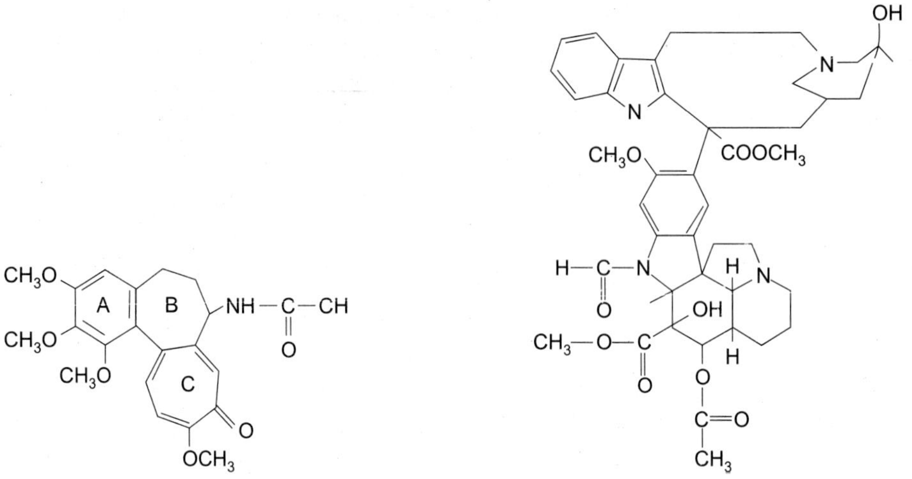

(a) (b)

Fig. 20.4 Microtubule inhibitors. (a) Colchlcine, (b) Vincristine (Vinblastine has a methyl group in place of the formyl group).

tubulin and preventing polymerization. A similar assembly process of protein units and ATP is seen in the formation of action microfilaments.

Microtubules are associated with a wide variety of functions in cells. In addition conferring shape to some cells, microtubules functions to mechanically extrude material (collagen from connective tissue) from cells, separate chromosomes during mitosis, and act as pat of the motion mechanism of flagella and cilia.

DISTRIBUTION AND FUNCTION OF MICROFILAMENTS

Muscle Cells

Striated, smooth, and cardiac muscle cells contain vast numbers of microfilaments that function during the contraction of these cells. There are two basic types of microfilaments in these muscle cells, thin filaments composed mainly of F-actin and thick filaments made up of myosin. In muscle cells, the two types of microfilaments interact with each other through cross-bridges that enable them to slide past one another the effect a shortening or contraction of the cell. In striated muscle cells (primarily those cells that make up the muscle that move the skeleton), the number and geometric arrangement of the two microfilaments are greatest. Equally spaced about each thick filament are six thin filaments, and equally spaced about each thin filament are three thick filaments (Fig. 20.5), units of several, hundred thick and thin filaments are grouped together to form a myofibril. Each striated muscle cell contains many myofibrils.

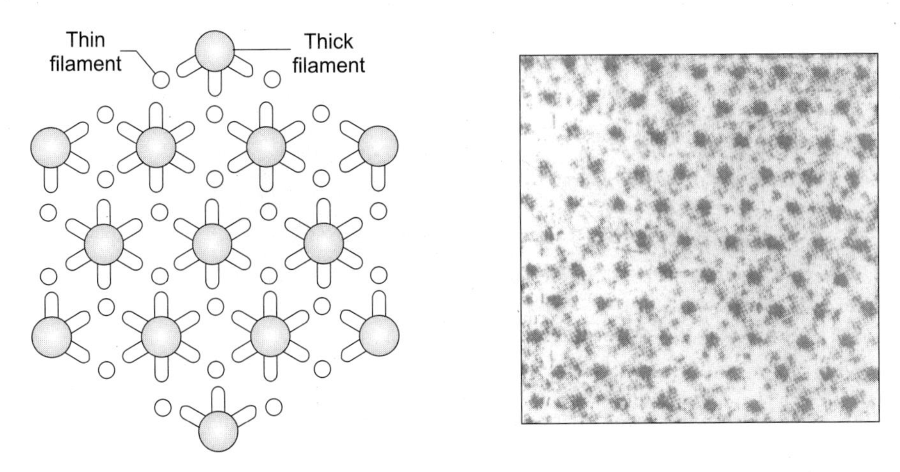

Fig. 20.5 Cross section of a striated muscle myofibril showing geometric arrangement of thick and thin fillaments (Courtesy of D. Fawcett. Copyrigt 1966 W.B. Saunders and Co., Atlas of fine structure p. 57).

CYTOKINESIS

Cell division in animal cells involves two separate mechanisms: mitosis and cell cleavage, or cytokinesis. The division and separation of daughter chromosomes is brought about in part by the action of the microtubules of the spindle. The separation of the chromosomes may also be aided by the constriction of the cell or in folding of the plasma membrane forming a girdle about the spindle. Cytokinesis usually begins toward the end of anaphase. In animal cells two events characterize the initiation of cytokinesis: (1) the cell begins to constrict about the midline of the spindle, and (2) dense material begins to collect about the peripheral spindle fibers also along the midline. Both processes continue as the plasma membrane moves inward, causing the cell to assume a "dumbbell" shape. The material collecting at the midline of the spindle becomes quite dense, forming a structure known as the midbody just before the in folding edges of the plasma membrane meet and fuse, the midbody fades and disappears (Fig. 20.6).

The furrowing or pinching-in of the plasma membrane is reminiscent of the action of a purse string or a rubber band tightening about a soft object. The origin of the force that causes the constriction is still being debated, but the involvement of microfilaments in the process is supported by their regular presence in the area of constriction and by the observation that cytochalasin B inhibits the process. The presence of specialized organelles just inside the plasma membrane in the area of constriction can be seen filaments (much like that known to occur during muscle cell contraction)

PLASMA MEMBRANE MOVEMENT

Intestinal epithelial cells have many small projections called microvilli that extend into the digestive cavity. The microvilli increase the surface area of the intestine, thus enhancing

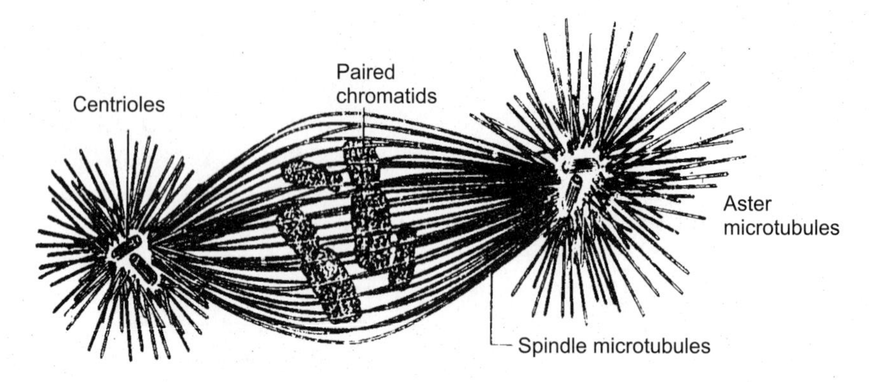

Fig. 20.6 Diagram of centriole, aster mictotubules, spindle microtubules and chromosomes during nuclear division.

absorption of the digest food. Microvilli also cyclically shorten and extend into the lumen of the intestine, a phenomenon that probably facilitates food absorption. Numerous actin microfilaments can be seen attached to the plasma membrane at he tip and side of each microvillus. M.S. Mooseke has shown that he actin microfilaments of intestinal epithelium are attached at one end to an A-actin like protein of the plasma membrane, just as the muscle thin filaments are attached to A-actin in of the Z-membrane. Mooseker also found that there is an association of the actin microfilaments with myosin filaments in the terminal web region of the cell. The movement of the microvilli may involve a mechanism such like that employed during myofibril contraction. It naturally follows that he movements of the microvilli may involve a mechanism similar to that which results in the contraction and relaxation of myofibrils.

The formation of pseudopodia and amoeboid; movement. Amoebae, slime, white blood cells, and a number of the other cells achieve movement by the formation of pseudopodia. A pseudopod is a portion of a cell that flows forward forming an extension into which the remaining cytoplasm of the cell subsequently follows. More than one pseudopod at a time can form, but reversal of the process in the nondominant pseudopodia. In pseudopod-forming cells, the outer portion of the cytoplasm is thick or gel- like cytoplasm is called ectoplasm; the more internal sol-like cytoplasm is called endoplasm. During cell movement, the endoplasm flows forward into the pseudopod, but as it reaches the anterior end of the pseudopod, it is transformed into gel, thereby forming part of the new ectoplasm. At the rear of the moving cell the ectoplasm solates, moves into the cell interior, and becomes endoplasm.

DISTRIBUTION AND FUNCTIONS OF MICROTUBULES

centrioles

Centrioles are an Assembly of Microtubules. These organelles are typically 150-250 nm in diameter and are composed of nine triple sets of parallel microtubules. Centrioles are usually associated with the nucleus and, although their function is generally described as related to nuclear division or, more specifically, to spindle fiber formation, the function of centrioles is more extensive than this. Indeed, their relationship of spindle fiber formation is questionable. Centrioles usually occur in pairs in a cell and are located in proximity to the nucleus (Fig. 20.7), as nuclear division begins, the centrole microtubules. Subsequently, as the chromosomes condense and the nuclear envelope disappears, spindle fibers appear, extending from the area of one centriole through the cell to the other centriole because of the close physical relationship between the spindle fibers and the centriole, it has been suggested that the centrioles may be associated with the spindle fiber microtubules. However, a number of observations argue against the idea. For example, the cells of cone-bearing and flowering plants do not have centrioles but do from spindles. Ferns have centrioles only in those cells developing flagella, and yet all fern cells undergoing mitosis have spindle fibers. The spindle fibers do not contact the centrioles, and no spindle fiber or aster microtubules are seen arising from a centriole.

Centrioles do play some role in the formation of microtubules found in flagella and cilia. In these locomotor or genelles, the centrioles become the basal bodies or kinetosomes (also blepharoplasts, basal granules, or basal corpuscles) which are structures located at the base of the flagellum or cilium. As described later in the chapter, basal bodies appear to act as organizing centers for tubulin and the assembly of flagellar microtubules.

Structure of the centriole. Centriole structure is basically the same in the cells of all species so far studied. The most notable characterstic, of centriole structure is the nine sets of microtutubles; each set contains three microtutules. The microtubules are arranged like vanes on a pinwheel seen in cross section (Fig. 20.8) and form the basic framework of the cylindrical structure of the centriole. Although there is no surrounding membrane, the nine triplets appear to be imbedded in an electron-dense material that does not extent into the center of the cylinder. While the diameter of the centriole varies only slightly (150- 250 nm), length is much more variable (160-800 nm). Centrioles usually occur in pairs, with their long axes oriented perpendicular to each other, but they do not touch each other, and there is no connecting material.

All triplets are identical. The innermost (or microtubule of each set is a complete, round tubule, but the middle (B) and outer (C) microtubules share the wall of the preceding tubule and are incomplete. Also, the C tubules may not run the length of the centriole as do the A and B tubules. The triplets, although generally parallel to each other, may be closer together at the proximal end of the centriole (that end when observed "end-on" which has the triplets tilted inward in a clockwise direction, as shown in Fig. 20.9. the triplets may also spiral somewhat about the axis of the centriole or individually twist. Less twisting of the microtubules is observed in the basal body form of the centriole. Dense material connecting A and C tubules has frequently been observed. Strands of material extend inward from each A tubule and join together at the central hub. These strands, when seen in cross section, give the centriole the appearance of a cartwheel (Fig. 20.8b). More dense material occurs in the center of the centrioles when it takes the from of a basal body.

Fig. 20.7 Final phase of cytokinesis in an animal cell. Only a few microtubules (m) of the spindle remain. Note midbody (mb) (Courtesy of B. Byers. Copyright 1968 springer-verlag, protoplasma 66, 423).

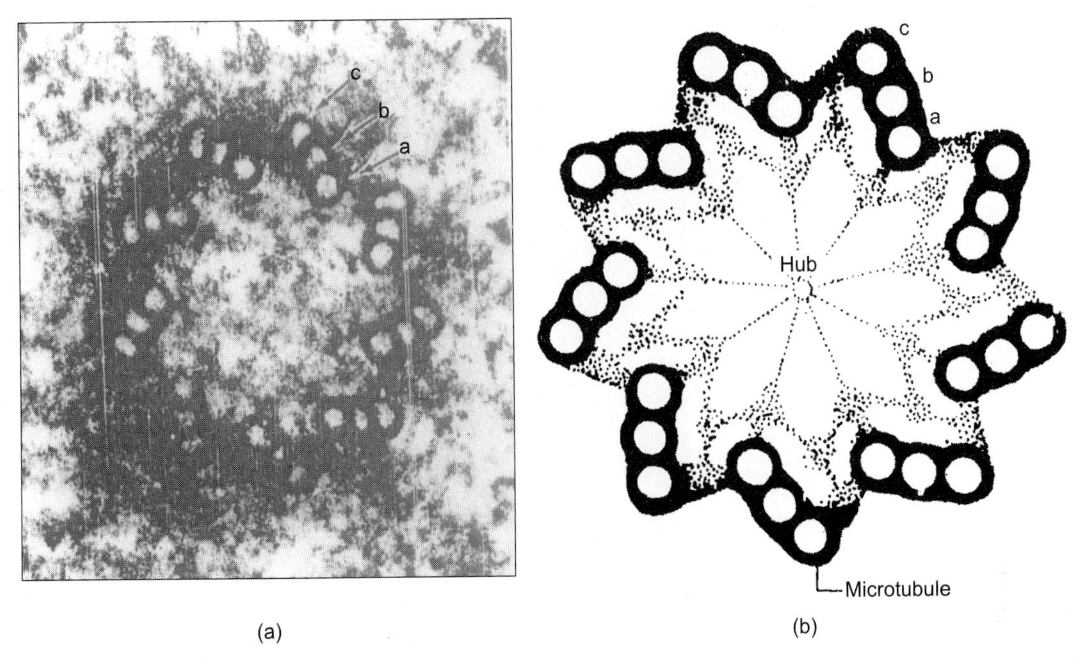

Fig. 20.8 (a) Cross section of a centriole as revealed by electron microscopy. The spokes of the "pinwheel" turn inward in a clockwise manner, indicating that this view is from the proximal end of the centriole (Courtesy D. Fawcett. Copyright 1966 W.B. Saunders and Co. Atlas fine structure, p. 185). (b) Diagram of a centricle showing the nine triplets that form the "pinwheel" and the electron-dense granular connectives.

CILIA AND FLAGELLA

Cilia and flagella are specialized extensions of the cell (Fig. 20.9), projecting from the cell surface into the surrounding medium, where they whip back and forth or create a corkscrew action. In some instances, ciliary or flagella movements propel the cells through their environment. The propulsion of the animal sperm cell by the whip-like movements of its "tail" is a classic example, although technically the sperm tail is not a flagellum but des have a similar basic structure. In other cases, the cell remains stationary (as in tissue), and the surrounding medium is moved past the cell by the beating of its cilia (as in the epithelial lining of the trachea or the collar cells lining the internal chambers of sponges).

Cilia are generally shorter than flagella (i.e., 5-10 nm vs. 150 nm or longer), but cilia are found in larger numbers per cell. Flagella may occur alone or in small groups, and infrequently they are present in large numbers, such as in a few protozoa and sperm of more advanced plants. The distinction is somewhat arbitrary, because other than differences in their lengths, the structure and action of cilia and flagella of eukaryotic cells are identical. (Bacterial flagella differ in structure; see later).

A cilium or flagellum is composed of three major parts: a central axoneme (a semirigid structure extending from the cell body through the long axis of the cilium), the surrounding

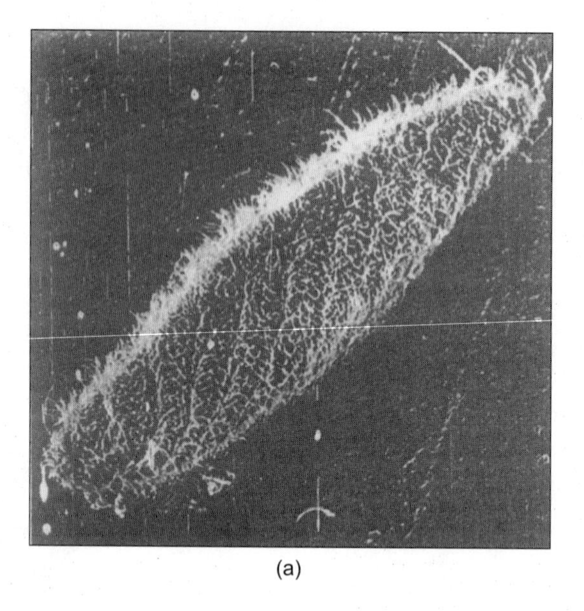

(a)

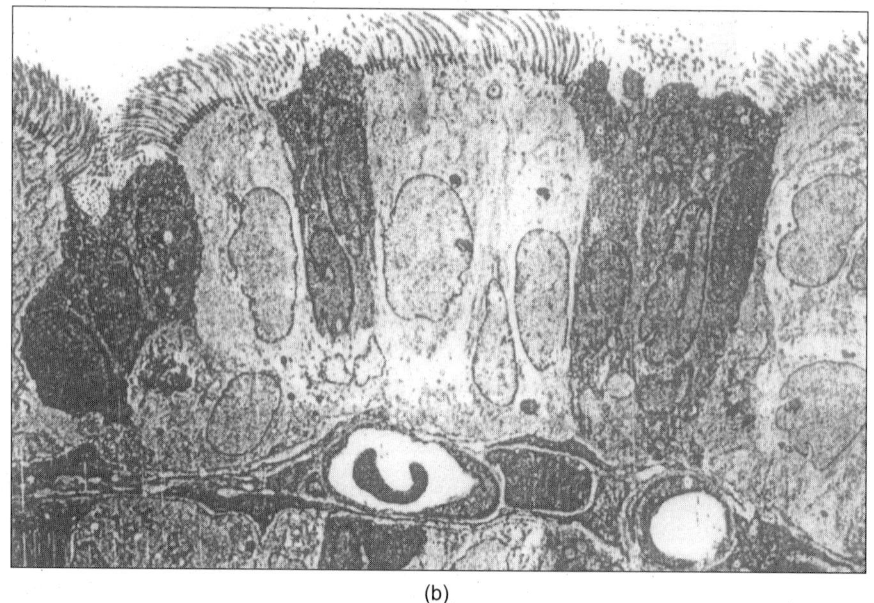

(b)

Fig. 20.9 (a) Scanning electron phtomicrograph of cilia on the dorsal surface of paramecium. The cilia beat rhythmically creating a wavelike appearance (a metachronal wave). (b) Cilia at the lumenal surfaces of epithelial cells lining the oviduct (Courtesy of J. Rhodin, in Paramecium, the beating cilia prople the cell whereas the cilia of the oviduct epithlium serve to push the egg released from the ovary toward the uterus).

plasma membrane, and some cytoplasm. The axonemal elements of nearly all cilia and flagella (as well as the tails of sperm cells) contain the same arrangement of microtubules. The microtubules are arranged in the now well-known "9+2" pattern. In the center of the axoneme

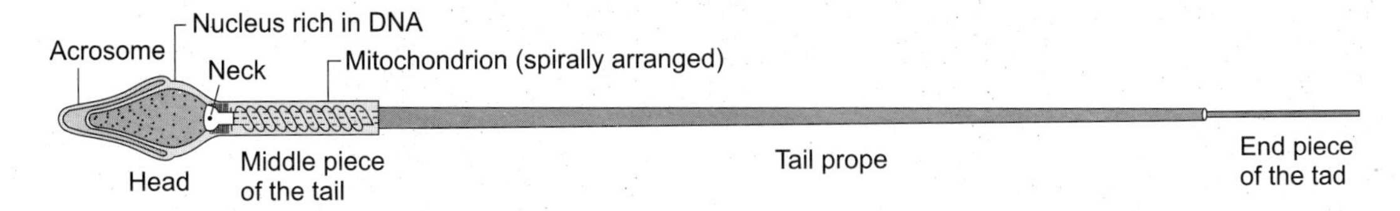

Fig. 20.10 A mature human sperm cell.

are two microtubules that run the length of the cilium and are joined together by a bridge (Fig. 20.11), cilium increase and beating cases. ATP is clearly the source of energy for movement and is produced by cellular respiration. In many cells, mitochondria are located adjacent to the basal body of the cilium or flagellum, and ATP diffuses toward the tip of the organelle. In sperm, large mitochondria are an integral pat of the tail (Fig. 20.10) and are wrapped in a spiral about the middle piece of the axoneme.

THE MITOTIC SPINDLE

The spindle fibers of the mitotic spindle are composed of both microtubules and microfilaments. Most of the studies of chromosomes movement during mitosis have been concerned with the microtubules and, at the present time, it is believed that these units alone can account for the movement. However, most workers agree that the microfilaments may also be involved, and several investigators support the idea that both microtubules and microfilaments must act to effect chromosome movement. The spindle fibers cause three distinct chromosomes movements during mitosis (chapter 20): (1) orientation of sister chromatids, (2) alignment of the kinetochores on the metaphase plate, and (3) division and separation of kinetocheres and movement of sister chromatids (segregation) to opposite poles of he spindle.

Three kinds of microtubules occur in the spindle: (1) the kinetochore microtubules, which terminate in a kinetochore; (2) the polar microtubules, which terminate at the poles; and (30 the free microtubules, which do not terminate in either a pole or a kinetochore. All three types can be dissociated into tubulin subunits by colchicine or cold temperatures. Also present in the spindle are microfilaments that are believed to be composed, at least in part, of actin. Immunofluorescene techniques indicate that the actin is present between the chromosomes and the poles of the spindle but is not present in the interzone between separating anaphase chromosomes.

SUMMARY

Microfilaments and microtubules frequently function together in cells to bring about movement or contribute to the cell's structural framework. The microfilaments are unbranched bundles of actin or actinlıne proteins would in a double helix These proteins are disassociated by low

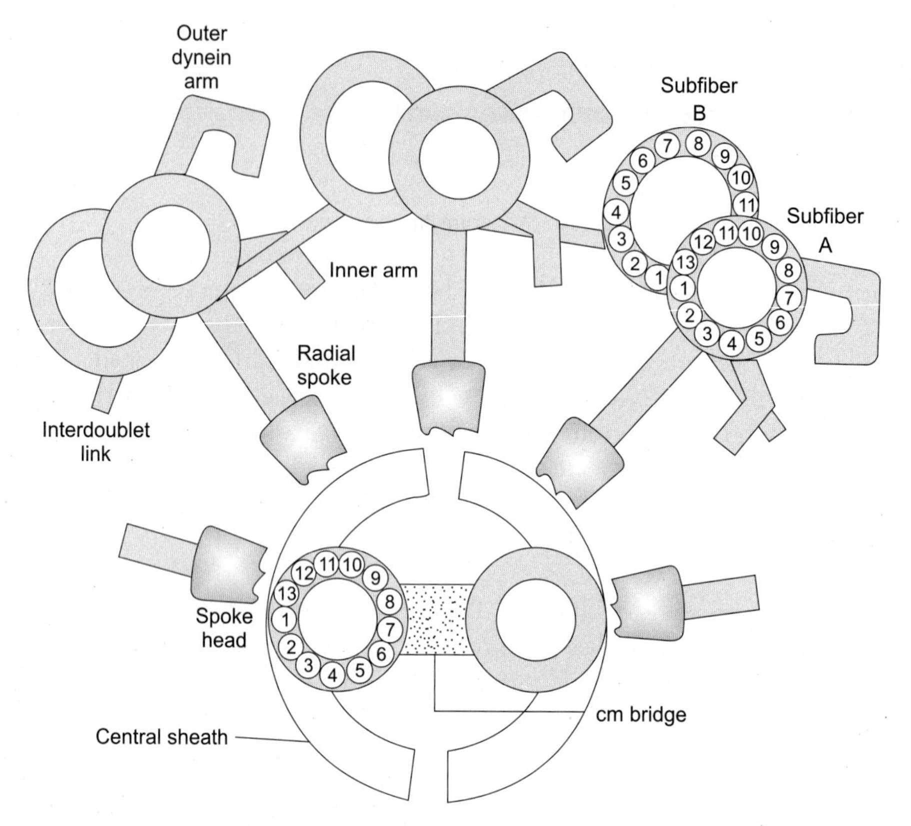

Fig. 20.11 Diagram of a cilium or flagellum seen in cross-section (See text for explanation) (Courtesy of F.D. Warner and P. Satir).

concentrations of cytochalasin. Microfilaments have been shown to play a role in phagocytosis, pincytosis, exocytosis, cytokinesis, and cytoplamic streaming.

Microtubules are cylindrical structures whose walls are composed to heterodimers of A and B tubulin. These globular proteins appear to be arranged in 13 chains that run parallel to each other and to the hollow axis of the tube. Microtubules are dissociated by colchicines, vinblastine, or vincristine. Microtubules are involved in the function of spindle fibers, flagella, and cilia and in mechanical extrusion of materials such as collagen from cells. Centrioles consist of nine triple sets of parallel microtubules.

CHAPTER 21

Cancer and Oncogenes

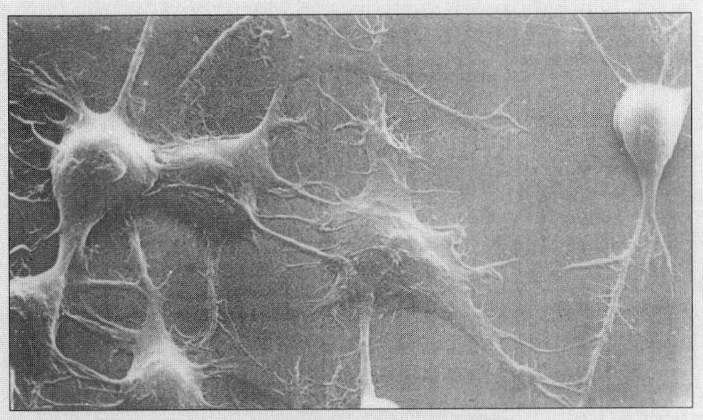

Phase-contrast photomicrograph of mouse neuroblastoma cells growing attached to a solid surface. Many elongated processes (neurites) extend from each cell. When such cells grow in suspension they have the rounded cancerous appearance characteristic of the cells which form tumors, but when they attach to a solid surface, they become flattened and begin to send out large numbers of neurites [Courtesy of Dr. Gunter Albrecht-Bühler, Cold Spring Harbor Laboratory].

In 1911, Peyton Rous of the Rockefeller Institute for Medical Research published a paper that was less then one page in length (it shared the page with a note on the treatment of syphilis) and had virtually no impact on the scientific community. Yet this paper reported one of the most farsighted observations in the field of cell and molecular biology. Rous had been working with a chicken sarcoma that could be propagated from one hen to another by inoculating a host of the same strian with pieces of tumor tissue. In the 1911 paper, Rous described a series of experiments that strongly suggested the tumor could be transmitted from one animal to another by a "filterable virus" which is a trem that had been coined a decade or so earlier to describe pathogenic agnet that were small enough to pass through filters that were impemeable to bacteria.

In his experiments, Rous removed the tumors from the breasts of hens, ground the cell in a mortar with sterile sand, centrifuged the particulate matereial into a pellet, removed the supernatant, and forced the supernatant fluid through filters of varying porosity including those small enough to prevent the passage of bacteria. He then injected the filtrate into the breast muscle of recipient hen and found that a significant percentage of the injected animals developed the tumor.

The virus discovered by Rous in 1911 is RNA-containing virus. By the end of the 1960s, similar viruses were found to be associated with mammary tumors and leukemias in rodents and cats. Certain strains of mice had been bred that developed specific tumors with very high frequency. RNA-containing viral particles could be seen within the tumor cells and also budding from the cell surface as. It was apparent that the tumors that developed in these inbred strains are transmitted vertically, that is, through the fertilized egg from mother to offspring so that the adults of each generation in variably develop the tumor. These studies provided evidence that the viral genome can be inherited through the gametes and subsequently transmitted from cell to cell via mitosis without having any obvious effect on the behavior of the cells. The presence of viral genomes is not a peculiarity of inbred laboratory strains, since it was shown that wild (feral) mice treated with chemical carcinogens develop tumors the often contain the same antigens characteristic of RNA tumor viruses and exhibit virus particles in the electron microscope.

One of the major questions concerning the vertical transmission of RNA tumor viruses was whether the viral genome is passed parents to progeny as free RNA molecules or is somehow integrated into the DNA of the host cell. Evidence indicated that infection and transformation by these viruses required the synthesis of DNA. Howard Temin of the University of Wisconsin suggested that the replication of RNA tumor viruses might occur by means of a DNA intermediate-a provirus-which could then serve as a template for the synthesis of viral RNA. But this model requires a unique enzyme-an RNA-dependent DNA polymerase-which had never been found in any type of cell. This situation changed in 1970 when and enzyme having this activity was discovered independently by David Baltimore of the Massachusetls Institute of Technology and by Temin and Satoshi Mizutani.

Baltimore examined the virions (the mature viral particles) from two RNA tumor viruses, Rauscher murine leukemia virus (R-MLV) and Rous sarcoma virus (RSV). A preparation of purified virus was incubated under conditions that would promote the activity of a DNA polymerase, including Mg^{2+} (or Mn^{2+}), NaCl, dithiothreitol (which prevents the –SH groups of the enzyme from becoming oxidized), and all four deoxyribonucleoside triphosphates, one of which (TTP) was labeled with H. It was found that the preparation incorporated the labeled DNA precursor into an acid-insoluble product that exhibited the properties of DNA. For example, the reaction product was rendered acid soluble (indicating that it had been converted to low-molecular-weight products) by treatment with pancreatic deoxyribonuclease or micrococcal nuclease, but it was unaffected by pancreatic ribonuclease or by alkaline hydrolysis (to which RNA is sensitive). The enzyme was found to cosediment with the mature virus particles, suggesting that it was part of the virion itself and not an enzyme donated by the host cell. Although the product was insensitive to treatment with pancreatic ribonuclease, the template was very sensitive to this enzyme particularly if the virions were pretreated with the ribonuclease prior to addition of the other components of the reaction mixture. These results strengthened the suggestion that the viral RNA was providing the template for synthesis of a DNA copy, which presumably served as a template for the synthesis of viral mRNAs required for infection that cellular transformation by RNA tumor viruses proceeds through a DNA intermediate, they overturned the long-standing concept originally proposed by Francis Crick and known as the Central Dogma, which stated that information in a cell *always* flows from DNA to RNA to protein. The RNA-dependent DNA polymerase became known as *reverse trancsriptase*.

During the 1970s, attention turned to the identification of the tumor virus genes that were responsible for transformation and the mechanism of action of the gene products. Evidence from genetic analyses indicted that mutant strains of viruses could be isolated that retained the ability to grow in host cells but are unable to transform the cell into one exhibiting malignant properties. Thus, the capacity ot transform a cell resides in restricted portion of the viral genome.

These findings set the stage for a series of papers by Harold Varmus, J. Michael Bishop, Dominique Stehelin, and their co-workers at the University of California, San Francisco. These researchers began by isolating mutant strains of the avian sarcoma virus (ASV) carrying deletions of 10 to 20 percent of the genome that render the virus unable to induce sarcomas in chickens or the transform fibroblasts in culture. The gene responsible for transformation, which is missing in these mutants, was referred to as *src* (for sarcom). To isolate the DNA corresponding to the deleted regions, which presumably carry the genes required for transformation, the following experimental strategy was used. RNA from the genomes of complete (oncogenic) virions was used as a template for the formation of a radioactively labeled, singlestranded complementary DNA (cDNA) using reverse transcriptase. The labeled cDNA (which is present as fragments) was then hybridized to RNA obtained from one of the deletion mutants. Those DNA fragment that failed to hybridize to the RNA represent portions of the genome that had been deleted from the transformation-defective mutant and thus were presumed to conhtain the gene required by the virus to cause transformation. The DNA fragments that did not hybridize to the RNA could be separated from those that are part of DNA-RNA hybrids by allowing the mixture to pass through a column containing a particular type of calcium phosphate salt called hydroxyapatite (which is extracted from bone tissue). Using this basic strategy, a DNA sequence referred to as cDNA*sarc* was isolated, which corresponds to approximately 16 percent of the viral genome (1,600 nucleotides out of a total genomic length of 10,000 nucleotides.

Once isolated, cDNA*sarc* proved to be a very useful probe. It was first shown that this labeled cDNA is capable of hybridizing to DNA extracted from a variety of avain species (chicken, turkey, quail, duck, and emu) indicating that the celullar genomes of these birds contain a DNA sequence that is closely related to *src*. In contrast, no homlogy between cDNA*sarc* and DNA from mammals, (mouse and calf) was detected in these early experiments. The kinetics of the reaction between the DNAs from these birds and the cDNA*sarc* indicated that the complementary sequence was present as part of the nonrepeated fraction of the genome and, thus, is present only once (or a very few times) per haploid set of DNA.

These findings provided the first strong evidence that a gene carried by a tumor virus that causes cell transformation is actually present in the DNA of normal (uninfected) cells and thus is presumably a part of the cells' normal genome.These results indicated that the transforming genes of the viral genome (the oncogenes) are not true viral genes, but rather are cellular genes that were picked up by RNA tumor viruses during a previous infection. Possession of this cell derived gene apparently endows the virus with the power to transform the very same cells in which this gene is normally found. The fact that the *src* sequence is present in all of the avain species tested suggests that the sequence has been conserved during avain evolution and, thus is presumed to govern a basic activity of normal cells.

In a subsequent study, hybridization between $cDNA_{sarc}$ and the cellular DNA from various animals was conducted under less stringent conditions, which allows DNA duplexes to form that contain a considreable percentage of mismatched base pairs. Under these conditions it was found that $cDNA_{sarc}$ is capable of binding classes, including mammals, but not to the DNA from sea urchins, fruit flies, or bacteria. One way to determine the degree to which such duplexes contain mismatched base pairs is to monitor the temperature at which the DNA-DNA complexes separate from one another (melt). The higher the melting temperature, the greater the percentage of proper (A-T and G-C) base pairs. The results of these experiment s are shown in Table 3. The reduction in melting temperature suggested about 3 to 4 percent mismatching of $cDNA_{sarc}$ with chicken DNA and 8 to 10 percent mismatching with the other vertebrate DNAs. Based on these results, it is possible to conclude that the *src* gene is not only present in the RNA of the ASV genome and the genome of the chicken cells it can infect, but a homologous gene is also present in the DNA of distantly related vertebrates, suggesting that it plays some critical function in the cells of all vertebrates.

These findings raised numerous questions, foremost among these were (1) what is the function of the *src* gene product and (2) how does the presence of the viral *src* gene (referred to as v-*src*) alter the behavior of a normal cell that already possesses a copy of the gene (referred to as c-*src*)?

The product of the *src* gene was initially identified by Ray Erikson and co-workers at the University of Colorado by two independent procedures: (1) the precipitation of the protein from extracts of transformed cells by antibodies prepared from RSV-infected animals, and (2) the synthesis of using the isolated viral gene as a template. Using these procedures, the *src* gene product was found to be a protein of 60,000 daltons, which they named pp60*src*. When the pp60 *src* protein was incubated with ^{32}P-ATP, radioactive phosphate groups were transferred to the heavy chains of the associated antibody (lgG) molecules used in the immunoprecipitation. This finding suggested that the *src* gene codes for an enzyme that possesses protein kinase activity. Confirming evidence was obtained on studies of the protein produced by a temperatur-sensitive ASV mutant that was able to transform cells when grown at 35°C but not at 41°C. Evidence had indicated that the *src* gene product encoded by these mutants is irreversibly denatured at the elevated temperature. To study the activity of the *src* mutant gene product, parallel cultures of chick cells were infected with either the mutant (designated NY 68) or the nondefective virus (designated nd), and the cells extracts were grown at either 41 °C or 35°C. Cell extracts were prepared from each of the four cultures, the $pp60^{src}$ was immunoprecipitated, and the precipitate was analyzed for protein kinase activity. It can be seen from Table 4 that, wherease the immunoprecipitated protein encoded by the nondefective virus exhibited elevated protein kinase activity at 41 °C, the same protein from the temperature-sensitive strain had greatly decreased activity. These data provided further evidence that the *src* product was the protein kinase. When cells infected with ASV were fixed, sectioned, and incubated with ferritin-labeled actibokies against $pp60^{src}$, the antibodies localized on the inner surface of the plasma membrane, suggesting a concentration of the *src* gene product in this part of the cell.

These were the first studies to elucidate the function of an oncogene. A protein kinase is the type of gene product that might be expected to have protential transforming activity because it can regulate the activities of numerous other proteins, each of which might serve a critical function in one or another activity related to cell growth. Further analysis of the role of the *src* gene product turned up an unexpected finding. Unlike all the other protein kinases whose function had been studied, $pp60^{src}$ transferred phosphate groups to tyrosine residues on the substrate protein rather than to serine or threonine residues. The existence of tyrosine residues had escaped previous detection because phosphorylated serine and threonine residues are approximately 3000 times more abundant in cells and because phosphotreonine and phosphotyrosine residues are difficult to separate from one another by traditional electrophoretic procedures. Not only did the product of the viral *src* gene (v-*src*) code for a tyrosine kinase, so too did c-*src*, the celullar version of the gene. However, the number of phosphorylated tyrosine residues in proteins of RSV-transformed cells was approximately eight times higher than that of control cells. This finding suggested that the viral version of the gene may induce transformation because it functions at a higher level of activity than the cellular version. The phosphorylation of additional tyrosine residues in cellular proteins was presumed to be essential to the malignant transformation of cells by RSV. Whether or not the v-*src* and c-*src* gene product are functionally equivalent remained to be determined.

The results from the study of RSV provided preliminary evidence that an increased activity of an oncogene product could be a key to converting a normal cell into a malignant cell. Evidence soon became available that the malignant phenotype could also be induced by an oncogene that contains an altered nucleotide sequence. A key initial study in this effort was conducted by Robert Weinberg and his colleagues at the Massachusetts Institute of Technology using the technique of DNA transfection.

Weinberg began the studies by obtaining 15 different malignant cell lines that were derived from mouse cells that had been treated with a carcinogenic chemical. Thus, these cells had been made malignant without exposing them to viruses. The DNA from each of these cell lines was extracted and used to transfect a type of nonmalignat mouse fibroblast celled and NIH3T3 cell. NIH3T3 cells were selected for these experiments because they take up exogenous DNA with high efficiency and they are readily transformed into malignant cells in culture. After transfection with the DNA from the tumor cells, the fibroblasts were grown in vitro and the cultures were screened for the formation of clumps (foci), which contained cells that had been transformed by the added DNA. Of the 15 cell lines tested, live of them yielded DNA that was capable of transforming the recipient NIH3T3 cells. DNA from normal cells lacked this capability. These results demonstrated that carcinogenic chemicals produced alterations in the nucleotide sequences of genes that gave the altered genes the ability to transform other cells. Thus, cellular gene could be converted into oncogenes in two different ways as the result of becoming incorporated into the genome of a virus or by becoming altered by carcinogenic chemicals.

Up to this point, virtually all of the studies on cancer causing genes had been conducted in mice, chickens, or other organisms whose cells were highly susceptible to transformation. In 1981, attention turned to human cancer, when it was shown that DNA isolated form human tumor cells was also capable of transforming mouse NIH3T3 cells following transfection. Of 26 different human tumors that were tested in this study, two provided DNA that was capable of transforming the recipient mouse fibroblasts. In both of these cases, the DNA had been extracted from cell lines taken from a bladder carcinoma (identified as EJ and J82). Extensive efforts were undertaken to determine if the genes had been derived from a tumor virus, but no evidence of viral DNA was detected in these cells. these results provided the first evidence that some human cancer cells contain an activated oncogene that can be transmited to other cells, causing their transformation.

The finding that cancer can be transmitted from one cell to another by DNA fragments provided a basis for determining which genes in a cell, when activated by mutation or some other mechanism, are responsible for causing the cell to become malignant. To make this determination, it was necessary to isolate the DNA that was taken up by cells, causing their transformation. Once the foreign DNA responsible for transformation was isolated, it could be analyzed for the presence of cancer-causing alleles. Within two months of one another, in 1982, three different laboratories reported the isolation and cloning of an unidentified gene from human bladder carcinoma cells that is capable of transforming mouse NIH3T3 fibroblasts. In one of these studies, the cloned DNA that is capable of transforming cells was used to examine normal human DNA for the presence of homologous sequences. It was found that DNA isolated from normal human fibroblasts contains a DNA fragment that hybridizes with the bladder carcinoma oncogene probe. This finding indicated that normal human DNA contains a sequence that, at the very least, is highly similar in sequence to the active oncogene, thus arguing against the possibility that the sequence entered the cell by viral infection. When the two DNAs–the transforming gene isolated from human bladder carcinoma cells and the homologous gene isolated from normal human fibroblasts–were compared by their susceptibility to restriction enzymes, no difference in the positions of restriction sites were revealed, indicating that if differences between the two versions of the gene do exist port for the proposal that cancer cells arise as the result of alterations in preexisting cellular genes (i.e., proto-oncogenes) that convert them into active, cancer-causing genes (oncogene).

By 1982, a number of different viral transforming genes had been identified, each of which was shown to have a ceilualr homologue. Once the transforming gene from human bladder cancer cells had been isolated and cloned, the next logical step was to determine if this gene bore any relationship to the oncogenes carried by RNA tumor viruses. Once again, within two months of one another, three papers appeared from different laboratories reporting similar results. All three showed that the oncogene from human bladder carcinomas that is able to transform NIH3T3 cells is the same oncogene (named *ras*) that is carried by the Harvey sarcoma virus, which is a rat RNA tumor virus. preliminary comparisons of the two versions of *ras*–the viral version and its cellular homologue–failed to show any diffenences, indicating that the two genes are either very similar or identical. These findings suggested that cancers that develop spontaneously in the human population are caused by a genetic alteration that is similar to the changes that occur in cells that have been virally transformed in the laboratory. It is important to note that the types of cancers induced by the. Harvey sarcoma virus (*sar*comas and erythroleukemias) are quite different from the bladder tumors, which have and epithelial origin. This was the first indication that alterations in the same gene-*ras*–can result in the development of a wide range of different tumors.

By the end of 1982, three more papers from different laboratories reported on the precise changes in the *ras* gene that lead to its activation as an oncogen. Comparison of DNA restriction fragments prepared from normal and malignant cells indicated that activation of *ras* does not involve any gross in th DNA. After pinning down the section of the large DNA fragment that is responsible for causing transformation, nucleotide sequence analysis indicated that the DNA from the malignent bladder cells was activated as a result of a single base substitution within the coding region of the gene. Remarkably, cells of both bladder carcinomas studied (identified as EJ and T24) contain DNA with precisely the same alteration a guanine-containing nucleotide at a specific site in the DNA of the proto-oncogene had been converted to a thymidine in the activated oncogene. This base substitution results in the replacement of a valine for a glycine as the twelfth amino acid residue of the polypeptide.

Determination of the nucleotide sequence of the v-*ras* gene carried by the Harvey sarcoma virus revealed alteration in base sequence that affects precisely the same codon found to be modified in the DNA of the human bladder carcinomas. The change in the viral gene substitutes an arginine for the normal glycine. It was apparent that this particular glycine residue plays a critical role in the structure and function of this protein. It is interesting to note that the *ras* gene is a proto-oncogene that, like *src*, can be activated by linkage to a viral promoter. Thus, *ras* can be activated to induce transformation by two totally different pathways: either by increasing its expression or by altering the amino acid sequence of encoded polypeptide.

The research described in this Experimental Pathways provided a great leap forward in our understanding of the genetic basis of malignant transformation. Much of the initial research on RNA tumor viruses stemmed from the belief that these agents might be important causal agents in the development of human cancer. The search for the viral cause of cancer led to the discovery of the oncogene, which led to the realization that the oncogene is a cellular sequence that is acquired by the virus, which ultimately led to the discovery that an oncogene can cause cancer without the involvement of a viral genome. Thus, tumor viruses, which are not themselves directly involved in most human cancers, have provided the necessary window through which we have been able to view our own genetic inheritance for the presence of information that can lead to our own undoing.

SYNOPSIS

Cancer is a disease involving heritable defects in cellular control mechanisms that result in the formation of invasive tumors capable of releasing cells that can spread the disease to distant sites in the body. Many, of the characteristics of tumor cells can be observed in culture. Whereas normal cells proliferate in culture until they form a single layer (monolayer) over the bottom of a dish, cancer cells continue to grow, piling on top of one another to form clumps. Other characteristics often revealed by cancer cells are an ability to grow while suspended in soft agar; tendency to exhibit an abnormal number of chromosomes; an ability to continue to divide indefinitely; a disorganized cytoskeleton; and a lack of responsiveness to neighboring cells, particularly with regard to locomotor activities.

Normal cells can be converted to cancer cells by treatment with a wide variety of chemicals, ionizing radiation, and variety of DNA- and RNA-containing viruses; all of these agents act by causing changes in the genome of the transformed cell. Analysis of the cells of a cancerous tumor almost always shows that the cells have arisen by growth of a single cell (the tumor is said to be monoclonal). The development of malignant tumor is a multistep process characterized by a progress of genetic alterations that makes the cells increasingly less responsive to the body's normal regulatory machinery and better able to invade normal tissues. The genes involved in carcinogenesis constitute a specific subset of the genome whose products are involved in such activities as control of the cell cycle, intercellular adhesion, and DNA repair. It is the combined effect of the various mutations rather than the particular sequence of mutation that is thought to be most important in the development of the malignant state. In addition to genetic alterations, growth of tumor cells is also influenced non genetic or epigenetic influences that allow the cell to express its malignant phenotype. Estrogen, for example, appears to promote the development of breast tumors.

Genes that have been implicated ill carcinogenesis are divided into two broad categories: tumor-suppressor genes and oncogenes. Tumor-suppressor genes encode proteins that restain cell growth and prevent cells from becoming malignant. Tumor-suppressor genes act recessively since both copies must be deleted or mutated before their protective function is lost. Oncogenes, in contrast, encode proteins that promote the loss of growth control and malignancy. Oncogenes arise from proto-oncogenes–genes that encode proteins having a role in a cell's normal activities. Mutations that alter either the protein or its expression cause the proto-oncogene to act abnormally and promote the formation of a tumor. Oncogenes act dominantly; that is, a single copy causes the cell to express altered phenotype. Most tumors contain alterations in both tumor-suppressor genes and oncogenes. As long as a cell retains at least one copy of all of its tumor-suppressor genes, it should be protected against the consequences of oncogene formation. Conversely, the loss of a tumor-suppressor function should not be sufficient, by itself, to cause the cell to become malignant.

The first tumor-suppressor gene to be identified was *RB*, which is responsible for a rare retinal tumor called retinoblastoma, which occurs with high frequency in certain familial form of the disease inherit one mutated copy of the gene. These individuals develop the cancer only after sporadic damage to the second allele in one of the retinal cells. *RB* encodes a protein called pRb, which is involved in regulating passage of a cell from G_1 to S in the cell cycle. The unphosphorylated form of pRb interacts with certain transcription factors, preventing them from binding the DNA and activating the genes required for certain S-phase activities. Once a pRB has been phosporylated, the protein releases its bound transcription factor, which can then activate gene expression, leading to the initiation of S phase.

The tumor-suppressor gene most often implicated in human cancer in *p53*, whose product (p53) may be able to suppress cancer formation by several different mechanisms. In one of its actions, p53 acts as a transcription factor that activates the expression of a protein (p21) that inhibits the cyclin-dependent kinase that moves a cell throught the cell cycle. Damage to DNA triggers the synhtesis of p53, leading to the arrest of the cell cycle until the damage can be repaired. p53 may also be capable of redirecting cells that are on the path toward malignancy onto an alternate path leading to programmed death, or apoptosis. p53 knockout mice begin to develop tumors several weeks after birth. Other tumor-suppressor genes include *APC*, which, when mutated, predisposes a person to developing colon cancer, and *BRCA1* and 2, which, when mutated, predispose a woman to developing breast cancer.

Most of the known oncogenes are derived from proto-oncogenes that play a role in the pathways that transmit growth signals from the extracellular environment to the cell interior, particularly the cell nucleus. Unlike tumor-suppressor genes, oncogenes have not been implicated in inherited forms of cancer. A number of oncogenes have been identified that encode growth factors or their receptors, including *sis*, the gene that encodes the platelet-derived growth factor (PDGF), and *erb*B, the gene that encodes the epidermal growth factor (EGF) receptor. Malignant cells may contain a much larger number of one of these growth factor receptors in their plasma membranes when compared to normal cells. The excess receptors make the cells sensitive to lower concentrations of the growth factor and, thus, stimulate them to divide under conditions that would not affect normal cells. A number of cytoplasmic protein kinases, including both serine-threonine and tyrosine kinases, are inclded on the list of oncogenes. These inclde raf, which encodes a protein kinase in the MPA kinase cascade. Mutations in *ras* are among the most common oncogones found in human cancers. As discussed in Chapter 15, *Ras* acts to activate the protein kinase activitiy of Raf. If Raf remains in the activated state, it continues to send signals along the MAP kinase pathway, leading to the continual stimulation of cell proliferation. A number of oncogenes, such as *myc*, encode proteins that act as transcription factors. Myc is normally one of the first proteins to appear when a cell is stimutated to reenter the ceel cycel from the quiescent G_0 phase. Overexpression of myc may cause cells to continue to proliferate by overriding the inhibitory influences exerted by the products of the tumor-suppressor genes. Another group of oncogenes, such as *bcl-2*, appear to encode proteins involved in apoptosis. Evidence suggests that overexpression of the *bcl-2*, gene leads to the suppression of apoptosis in lymphoid tissues, allowing abnormal cells to proliferate to form lymphoid tumors.

Genes that encode proteins involved in DNA repair have also been implicated in carcinogenesis. The genome of pateints with the most common hereditary form of colon cancer, called hereditary nonpoloposis colon cancer (HNPCC), contains microsatellite sequences of abnormal nucleotide number. Changes in the length of a microsatellite sequence arise as an error during replication, which is normally recognized by mismatch repair enzymes, suggesting that defects in such systems may be responsible for the cancer. This conclusion is supported by findngs that extracts from HNPCC tumor cells display DNA repair deficiencies. Cells that contain such deficiences would be expected to display a greatly elevated level of mutation in tumor-suppressor genes and oncogenes that would lead to a greatly increased risk of malignancy.

CHAPTER 22

Cell Differentation and Specialization

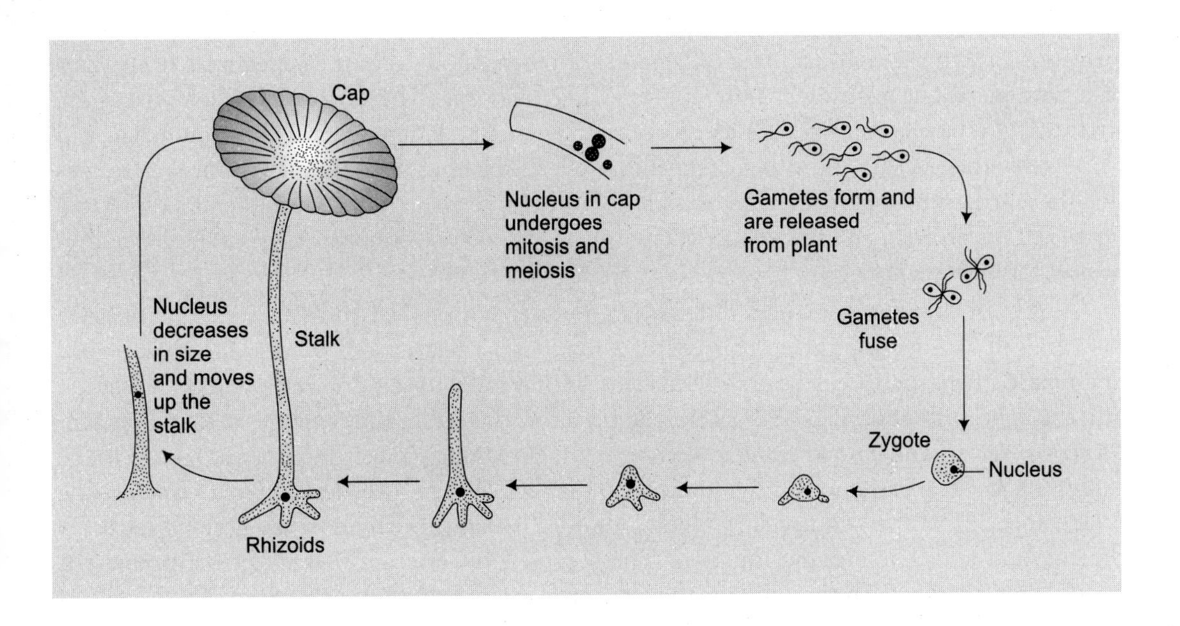

Differentiation is but one aspect of the more general field of developmental biology. Developmental biologists are concerned with the changes that organisms and their cells and molecules undergo in making transition from cells unspecialized in structure and /or function to forms having a permanent and specific structure and function. Four component processes characterize these changes: determination growth, differentiation, and morphogenesis.

Determination and growth are initial processes. New cells are usually not "committed" to a specific function. For example, a fertilized egg cell divides many times to produce a ball of cell called a morula. Normally, the morula develops into a single embryo, each cell or group of cell giving rise to specific organs or tissues. However, one cell separated from the others at the morula stage is capable of growing into a complete embryo. Even at later stages of embryo development, cells that normally become epidermal tissue (i.e., ectoderm) can be surgically transplanted to another part of the embryo and their develop into mesodermal tissue. At some early stage of organismal development, all cell of a given species have the potential to develop into the of a variety of different tissue and cell types of that species. This potential is based on the genetic composition of the chromosomes and ultimately on the specific sequence of DNA nucleotides in each gene. The generalization can also be made that at some point in development, cells become committed to a specific course of differentiation. The process that establishes the fate of a cell is called determination. During determination, some alternative modes of gene expression become permanently "turned off" while others are sequentially expressed, further and further restricting the course of differentiation of the cell. Prior to and during determination the cell may increase its biomass through growth.

During differentiation the cell acquires new properties. These can be structural (such as the formation of acting and myosin microfilaments) or biochemical (as in the appearance of enzymes of a new metabolic pathway). Differentiation may also take the form of loss of preexisting structures or biochemical processes. For example, in the differentiation of mammalian red blood cells, the nucleus and other organelles are lost, together with the biochemical processes that these structures provided. The gross result of the internal structural differentiation factors of the cell, its growth, and the effects of environmental factors is morphogenesis-the generation of new form or shape.

Above all, development and differentiation rests with DNA molecules in the nucleus. Before a cell can develop into a hair cell of a mammal, a feather cell of a bird, or a scale cell of a reptile, there must be genes in the nucleus of that cell whose transcription and translation into the appropriate enzymes and other proteins allow the cell to differentiate in that direction. Moreover, given the proper genetic complement, conditions must allow these genes to be expressed. Gene expression during development is regulated at three levels. First, there are the basic biochemical (i.e., molecular) regulatory mechanisms, such as mass action, feedback control, and allosteric enzyme function. The second level of control is effected through the interrelationship between the nucleus, the cytoplasm, and the cytoplasmic organelles. The third level of control involves the interactions between the cell and its environment. A variety of experiments have demonstrated the effects of the nucleus and cytoplasm on one another.

EXPERIMENTS WITH *ACETABULARIA*

Classic experiments with the unicellular alga *Acetabularia* demonstrated that the cytoplasm has a direct effect on gene expression. This unicellular plant contains a single nucleus and has the shape of a long tube with rootlike branches at the base. The nucleus is located in one of these branches. As the cell matures, an umbrella-shaped cap develops at the top of the cell. The nucleus ultimately divides (by meiosis and then mitosis) to produce many daughter nuclei that migrate into the cap and become encysted. Eventually, the cysts are released, break open, and the gametes swim out, conjugate, and start a new cycle (Fig. 22.1)

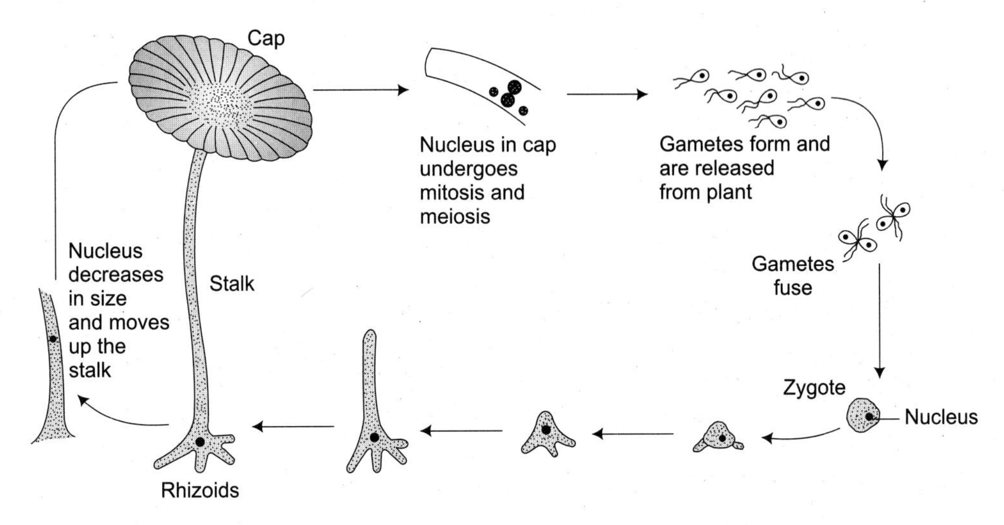

Fig. 22.1 Life cycle *Acetabularia.*

Acetabularia has good powers of regeneration, for if the cap is cut off, the cell will grow a new cap. This can be repeated many times with the same organism. If the cap is cut off and the nucleus then removed, a new cap will be regenerated once, but the regeneration will not occur if the experiment is repeated with the same anucleate organism. The regeneration of the cap in the absence of a nucleus indicates that nuclear information had previously been transferred to the cytoplasm and has persisted there, at least through the time required to form the new cap. The information in the cytoplasm is in the form of messenger RNA.

Different shape caps are formed by different species of *Acetabularia. A. mediterranea* has a complete cap, and A. *crenulata* has a fingerlike cap. The interactions of the cytoplasm and nucleus are strikingly illustrated by grafting experiments using these two species (Fig. 22.2). If the foot of A. *crenulata* (which contains the nucleus) is joined to an anucleate stalk of A. mediterranea, the two pieces fuse and a cap develops at the end of the stalk. The cap assumes characteristics of both species, presumably resulting from the mRNA cytoplasmic information originally present in the A. mediterranea stalk and the influence of perhaps both nucleus and some cytoplasmic mRNA of the A. *crenulata* foot. If the regenerated cap from the two grafted cell parts is removed and a second cap allowed to form, the new cap has the a. crenulated shape. Thus, any cap information (mRNA) in the original cytoplasm of A. *mediterranea* is "lost" after one regeneration, whereas the cap information (mRNA) of the A. *crenulata* nucleus continues to be formed and expressed.

ENVIRONMENTAL EFFECTS IN DIFFERENTIATION

A cell's neighbors and the surrounding environment have a direct effect upon differentiation. Since the fates of the late blastula-early gastrula stage cells of frog embryos (a well as embryos of a number of other species) have been determined, a map can be made of the types of tissues

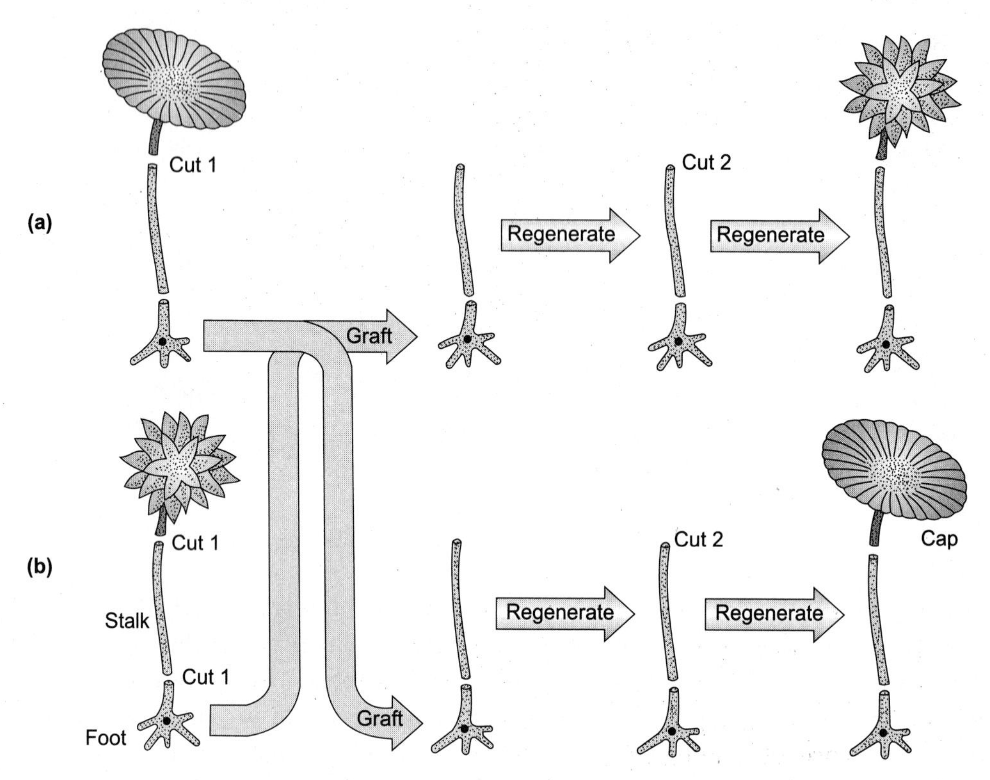

Fig. 22.2 Effects of grafting experiments with nucleate and anucleate cell segments of *Acetabularia crenulata* (b) and *A. mediterranea* (a) (See text for description and results of grafting).

to be formed by each group of cells (Fig. 22.5). If at the late blastula stage, cells that normally form the eye lens are surgically exchanged with cells that form the gut, embryonic development is still normal. The transplanted lens cells were influenced by their new position in the embryo and develop into gut tissue. Like wise, the transplanted presumptive gut cells develop into lens tissue.

RED BLOOD CELLS

The differentiation of mammalian red blood cells in bone marrow (a process called erythropoiesis) is one of the more striking examples of specialization occurring in nature. In adults, erythropoiesis begins with a pluripotent stem cell that gives rise in 7 to 10 days to mature, hemoglobin-filled erythrocytes. As noted earlier in the chapter; differentiation and specialization of the cells may be accompanied not only by the acquisition and development of specialized structures but also by the loss of internal structures of physiological properties. The latter is the case during erythrocyte diffentiation, for mature red blood cells lack nuclei, mitochondria, endoplasmic reticulum, golgi bodies, ribosomes, and most other typical cell organelles.

The mature erythrocyte is a simple cell, delimited at its periphery by a plasma membrane and containing internally a highly concentrated, para-crystalline array of hemoglobin molecules for oxygen transport. The apparent simplicity of the maturing erythrocyte is the principal reason for its selection over most other kinds of cells as the preferred object for studying plasma membrane structure and protein (i.e., hemoglobin) structure and biosynthesis.

ERYTHROPOIESIS

The development and differentiation of the mammalian red blood cell is depicted in Fig. 22.3. Development takes place in the extrasinusoidal Stroma of the bone marrow and begins with pluripotent stem cells capable of proliferating granulocytic leukocytes (white blood cells.), as well as erythrocytes. When primitive stem cells undergo division, one of the daughter cells remains undifferentiated and pluripotent, so that depletion of marrow stem cells does not normally take place. The erythopoietic activity of the bone marrow is under hormonal influence, increasing or decreasing according to the level of circulating erythropoietin (produced in the cortical region of the kidneys and secreted into the blood steam). The erythrocyte progenitors show an increasing sensitivity to erythropoietin through the pro-erythroblast stage and a parallel, ever decreasing pro-liferative potential (Fig. 22.3, as a result, by the pro-erythroblst stage, the cells are irreversibly committed to the maturation sequence leading to reythrocytes.

Because of its highly differentiated state, the mature erythrocyte is incapable of further proliferation. In humans, the average life span is 120 days. Since there are 5 billion red blood cells in each milliliter of blood, a few simple calculations, quickly reveal that in a typical adult

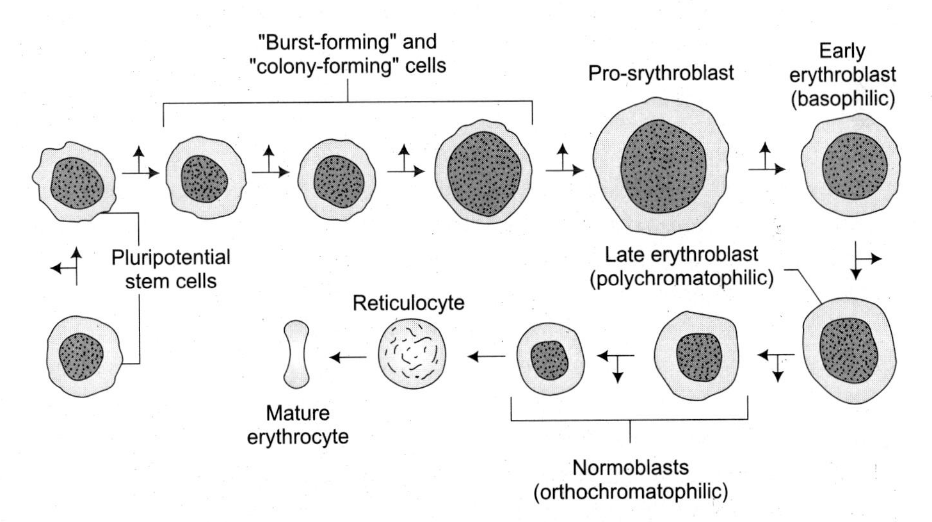

Fig. 22.3 Differentiation and maturation scheme of the mammalian erythrocyte. Differentiation begins with pluripotent stem cells in the bone marrow and ends 7 to 10 days later with the release of reticulocytes or mature erthrocytes into the bloodstream. Each double-headed arrow represents a mitotic division into two daughter cells only one of which is shown. Normoblasts becomes reticulocytes and then erythrocytes without division (the cell nucleus is lost during the normoblast stage).

the differentiation and maturation of 3 million erythrocytes is completed each second Obviously, an appreciable proportion of the body's energy and resources is continuously consumed to support erythropoietic activity. This is in stark contrast with other highly differentiated cells such as those of muscle and nerve, whose proliferation ceases shortly after birth.

GENIC AND MOLECULAR BASIS OF ERYTHROCYTE DIFFERENTIATION

Because of the intensity with which the red blood cell has, been studied by biochemists, cell biologists, molecular geneticists. physiologists, and others, the differentiation and specialization of this cell is better understood in molecular terms than any other; accordingly, erythrocyte differentiation may serve as a model for the genetic and molecular events that underlie cell differentiation generally.

MUSCLE CELLS

Vertebrates possess three types of muscle tissue: (1) smooth muscle, the contractile elements of most of the digestive system and most visceral organs. (2) cardiac muscle, found only in the heart, and (30 striated (or skeletal) muscle, responsible for most of the gross movements of the body. Each of these tissues is composed of cells called **muscle fibers**, which contain microfilaments of action, myosin, and other proteins responsible for the contractile nature of muscle. The microfilaments are arrangement is mast highly organized in striated muscle, and it is this type of muscle tissue that has been most extensively studied.

SUMMARY

Differentiation is the mechanism resulting in structural and biochemical change occurring in cells and is one stage in the development of organisms. The potential to differentiate depends upon the genetic composition of the DNA in the cell and the proper conditions for the expression of the genetic information. Not only must the genetic information be transcribed and translated into appropriate proteins and enzymes, but the conditions must also be proper for the metabolic regulatory mechanisms to functions. A third level of control involves the interactions between the cytoplasm and the nucleus, the **nucleocytoplasmic relationship**.

The cytoplasm of the cell accumulates a variety of products of metabolism, materils that have entered from the environment or adjacent cells, and products of nuclear activity such as mRNA. All of these materials can enter the nucleus and effect gene expression, initiating the transcription of some DNA segments and halting the transcription of other areas. Thus the cell changes or differentiates with time.

The development of red blood cells (erythrocytes) illustrates the sequential cause and effect mechanisms of differentiation. Stem cells in the bone marrow may give rise to red or white blood cells. Increasing levels of the hormone erythropoietin traveling in the blood stem to the marrow cause the erythrocyte progenator cells to cells to from mRNAs for the globin chains of hemoglobin. The cells become committed to be red blood cells, as 95% of all protein synthesis is directed to hemoglobin. The buildup of hemoglobin is then followed by the dissolution of cell structures such as the nucleus and residual nucleic acids. The cells are irreversibly differentiated into red blood cells.

Muscle cells are uniquely differentiated into a mechanism metrically differentiated parallel to the long axis of the cell and specially arranged about one another so that high, energy bonds can from between the filaments. When stimulation is applied or ATP and Mg^{++} are added, the bonds between action and myosin strands progressively break. Enzymes and proteins precisely located during differentiation react and contract, causing the filaments to slide past one another and bring about overall shortening of the cell.

Nerve cells are also examples of highly specialized differentiation. The cells are elongated and have developed membrane structure that can achieve a selected differential permeability. This enables the cell to maintain an electrical potential as well as to propagate an action potential along the length of the cell.

CHAPTER 23

Recombinant DNA Technology and Biotechnology

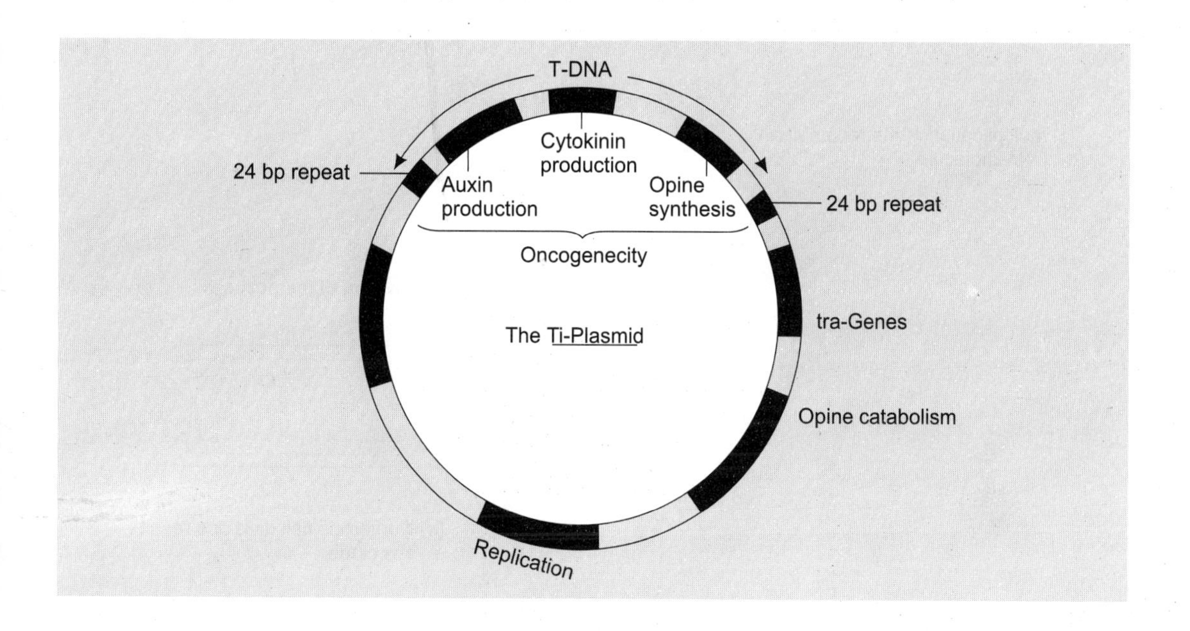

During the last two decades (1980s and 1990s), there have been great many advances in genetics and recombinant DNA technology. This technique has revolutionised the whole biological science both botany and zoology and virtually changed the whole scenario in the field of genetics. This chapter discusses the techniques and applications of recombinant DNA technology.

GENE CLONING

The basic events in a gene cloning experiment are (Fig. 23.1):

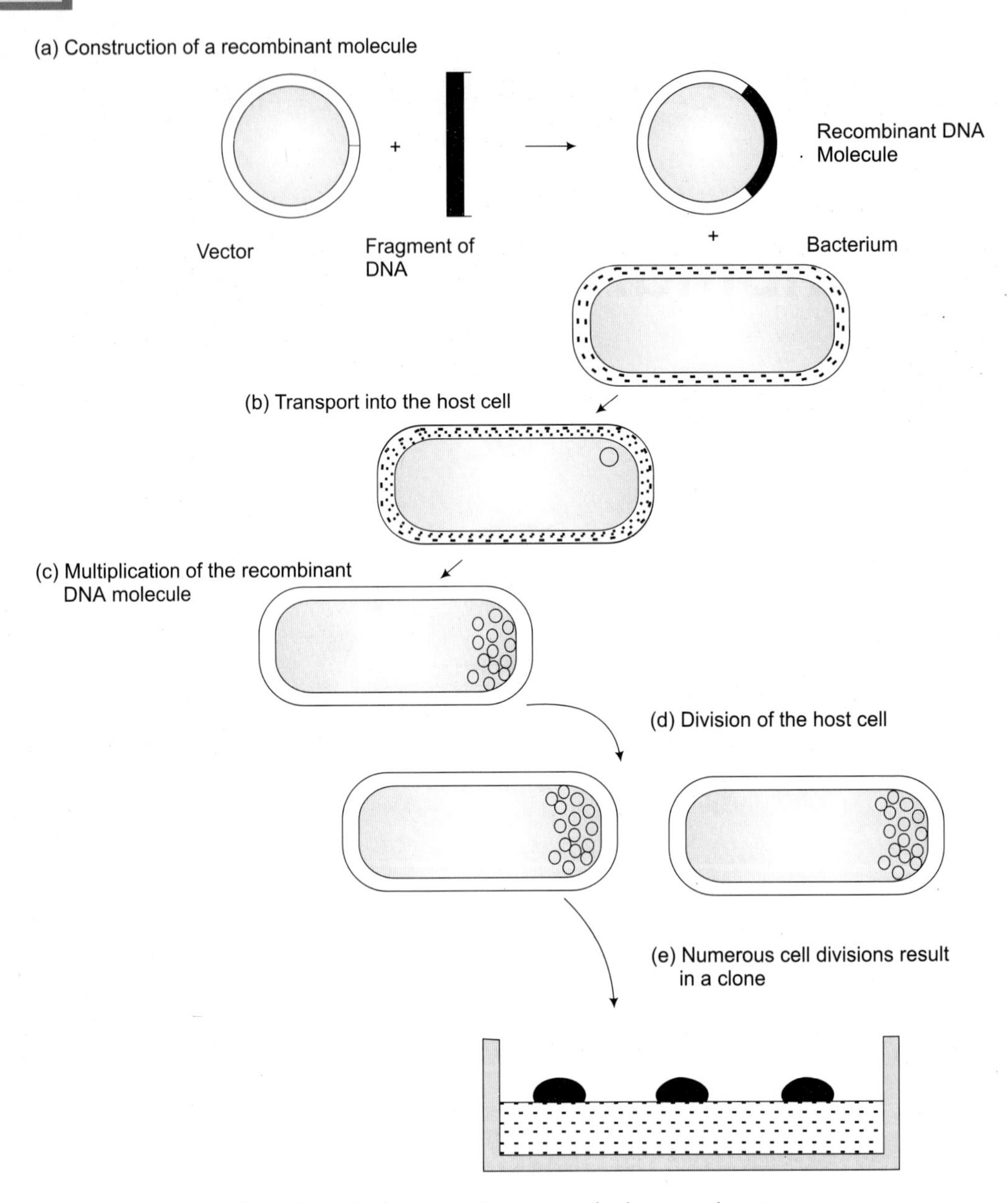

Fig. 23.1 Basic events in a gene cloning experiment.

1. A fragment of DNA containing the gene to be cloned is inserted into a second (usually circular) DNA molecule, called a cloning vector, to produce a recombinant DNA molecule.
2. The recombinant DNA molecule is introduced into a host cell (often but not always *E. coli*) by transformation or an equivalent procedure (Fig. 23.2).

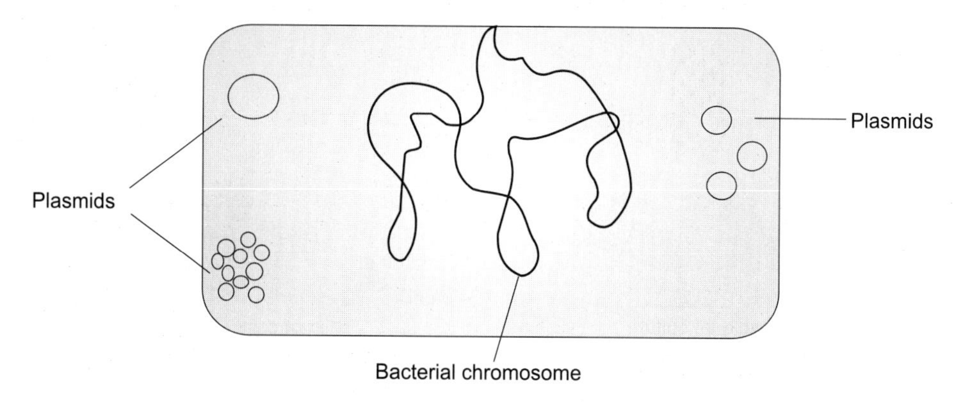

Fig. 23.2 Plasmids: Independent genetic elements found in bacterial cells.

3. Within the host cell the vector directs multiplication of the recombinant DNA molecule, producing a number of identical copies.
4. When the host cell divides, copies of the recombinant DNA molecule are passed to the progeny and further vector replication takes place.
5. A large number of cell divisions give rise to a clone, a colony of cells each containing multiple copies of the recombinant DNA molecule.

Gene cloning is, therefore, a relatively straightforward procedure; why then has it assumed such importance in biology? The answer is largely because cloning can provide a pure sample of an individual gene, separated from all the other genes that it normally shares the cell with (Figs. 23.3 and 23.4). To understand exactly how this arises, let us take a second look at a gene cloning experiment, drawn in a slightly different way. The DNA fragment to be cloned can be part of a mixture of many different fragments, each carrying different genes or parts of genes. This mixture could indeed be the entire genetic complement of a higher organism: human for instance. All these fragments will be inserted into different vector molecules to produce a family of recombinant DNA molecules, one of which carries the gene of interest. During transformation each host cell takes up just a single recombinant DNA molecule, so that although the final set of clones will contain many different recombinant DNA molecules, each individual clone will contain multiple copies of just one molecule. Once the clone containing the gene of interest has been identified, its recombinant DNA molecules can be purified from cell extracts and the gene recovered.

Before cloning techniques were developed it was impossible to obtain purified samples of individual genes. Information pertaining just to a single specified gene was very hard to obtain except by genetic analysis, and work on gene structure and expression could cope with genes only in general terms and could not concentrate on the particular features of individual genes. Cloning opened up new avenues of research.

Manipulating DNA and Constructing Recombinant DNA Molecules

The first step in a gene cloning experiment, i.e., construction of recombinant DNA molecules, involves cutting DNA molecules at specific points and joining them together again in a

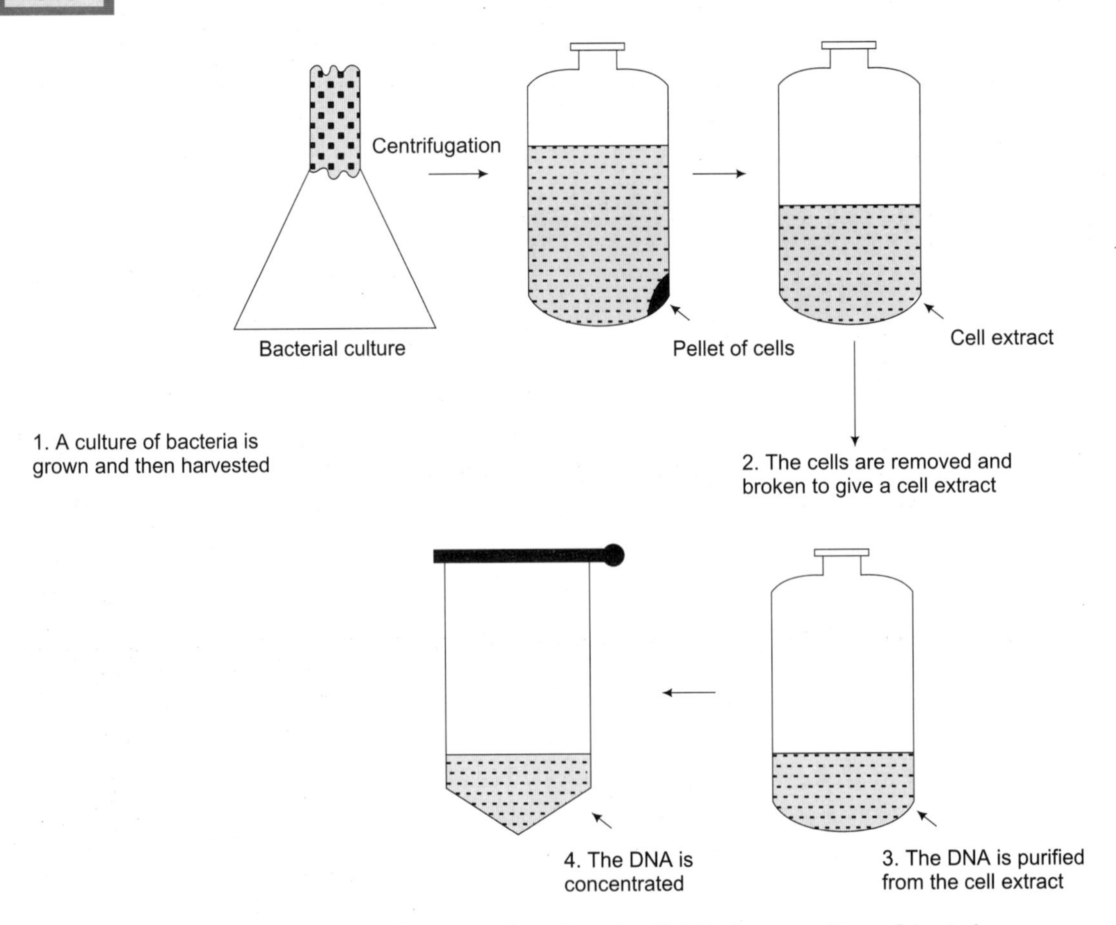

Fig. 23.3 Basic steps in the preparation of total cell DNA from a culture of bacteria.

controlled manner. DNA manipulative techniques of this type make use of purified enzymes that, in the cell, participate in processes such as DNA replication. The two main types of DNA manipulative enzymes used in gene cloning are restriction endonucleases and DNA ligases. (Fig. 23.5)

Enzymes for Cutting DNA: Restriction Endonucleases

During construction of a recombinant DNA molecule the circular vector must be cut at a single point into which the fragment of DNA to be cloned can be inserted. Not only must each vector be cut just once, but all the vector molecules must be cut at precisely the same position. Clearly, a very special type of nuclease is needed.

The relevant enzymes are called Type II restriction endonucleases. Each recognises specific nucleotide sequence and cuts a DNA molecule at this sequence and nowhere else. For instance, the restriction endonuclease called poul (isolated from *Proteus vulgaris*) cuts DNA only at hexanucleotide $\begin{matrix} 5' - \text{CGATCG} - 3' \\ 3' - \text{GCTAGC} - 5' \end{matrix}$. A second enzyme from the bacterium PvuII, cuts at a

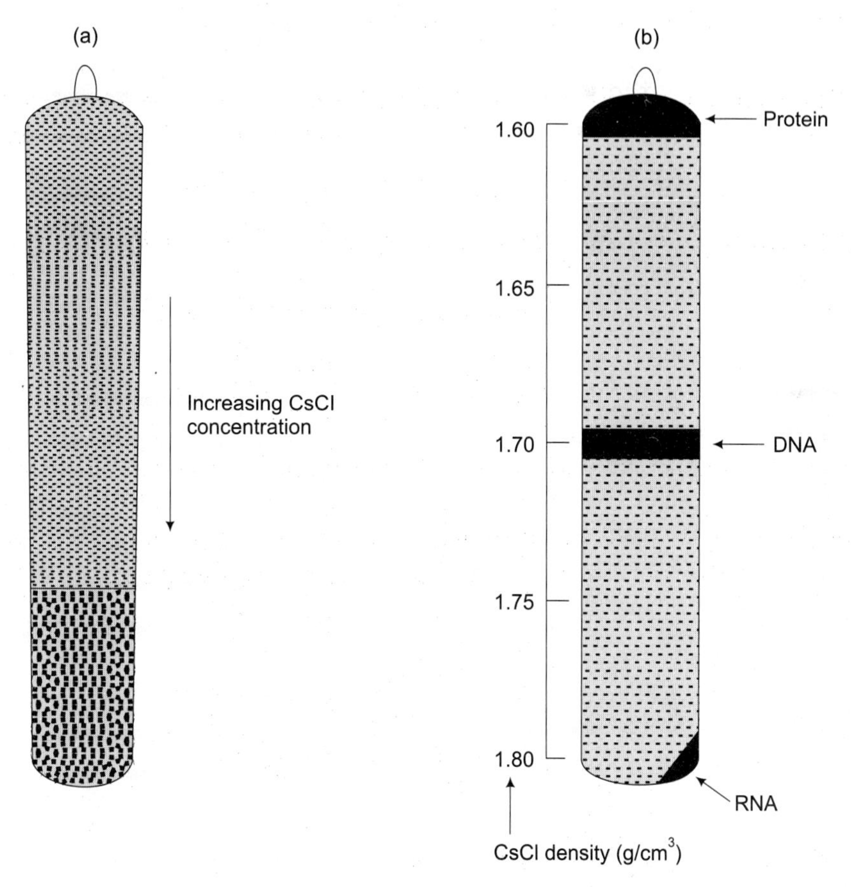

Fig. 23.4 CsCl density gradient centrifugation (a) A CsCl density gradient produced by high-speed centrifugation (b) Separation of protein, DNA and RNA in a density gradient.

different site $\begin{matrix} 5'-\text{CAGCTG}-3' \\ 3'-\text{GTCGAC}-5' \end{matrix}$. Many restriction endonucleases recognise hexanucleotides, but others cut at a sequence made up of four, five or eight nucleotides. There are also examples of restriction endonucleases with degenerate recognition sequences, which means that they cut DNA at any one of a family of related sites, HinfI (*Haemophilus influenzae*), for instance, recognises $\begin{matrix} 5'-\text{GANTC}-3' \\ 3'-\text{CTNAG-}5' \end{matrix}$ where N is any nucleotide.

The exact nature of the cut produced by a restriction endonuclease is very important in the design of a gene cloning experiment. Many restriction endonucleases make a simple double-stranded cut in the middle of the recognition sequence, resulting in a blunt end. Others cut the two strands in a staggered fashion, leaving short single-stranded overhangs at each end.

These are called sticky or cohesive ends because base pairing between them can stick the DNA molecule together again. Table 23.1 shows how two different restriction endonucleases may produce the same sticky ends.

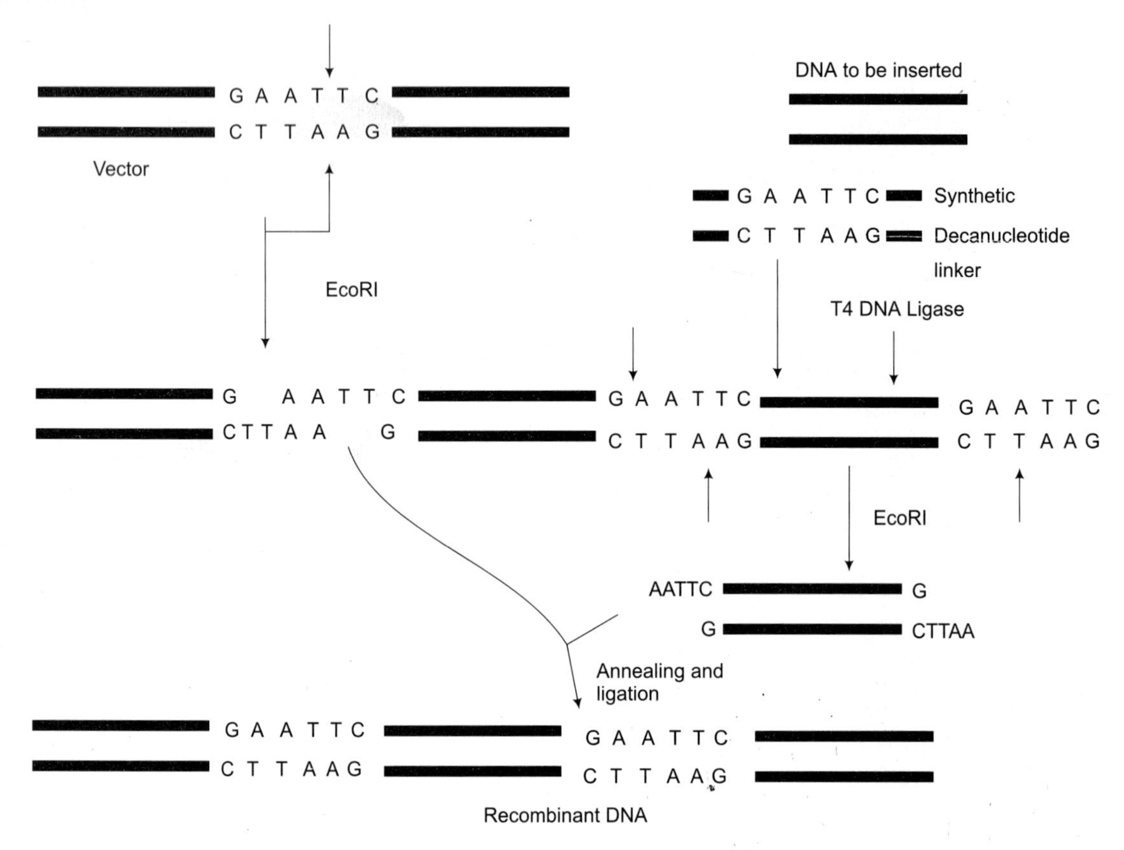

Fig. 23.5 Formation of recombinant DNA through the use of a LINKER. Short arrows indicate the sites of cutting by the EcoRI enzyme.

Ligases Join DNA Molecules Together (Fig. 23.6)

All living cells produce DNA ligases but the enzyme most frequently used in gene cloning is one involved in the replication of T4 phage, and is purified from infected *E. coli* bacteria. Within the cell the enzyme synthesises phosphodiester bonds between adjacent nucleotides during replication of the phage DNA molecules.

In the test tube purified DNA ligases will carry out exactly the same reaction and can join together restricted DNA fragments In either of the two ways:

1. By repairing the discontinuities in the base-paired structure formed between two sticky ends. This is relatively efficient because the ends of the molecule are held in place, at least transiently, by sticky-end base pairing.
2. By joining together two blunt-ended molecules. This is a less efficient process because DNA ligase cannot catch hold of the molecules to be ligated and has to wait for chance associations to bring the ends together.

Table 23.1 The recognition sequence of sequences of some commonly used restriction endonuclease.

Source	Enzyme		Sequence		
1. *Escherichia coli*	Eco R	|	5′. . . G A A T T C 3′. . . C T T A A G	. . . .	3′ 5′
2. *Serratia marcescens*	Sma	|	5′. . . C C C G G G 3′. . . G G G C C C	. . . .	3′ 5′
3. *Providedencia stuartii*	Pst		5′. . . C T G C A G 3′. . . G A C C T C	. . . .	3′ 5′
4. *Moraxelle bovis*	Mbo	|	5′. . . G A T C 3′. . . C T A G	. . . .	3′ 5′
5. *Escherichia coli*	ECORI	|	5′. . . C A A T T C 3′. . . G T T A A G	. . . .	3′ 5′
6. *Bacillus amyloliquefaciens*	Bam HI	|	5′. . . G G A T C C 3′. . . C C T A G G	. . . .	3′ 5′
7. *Proteus vulgaris*	Pvu I	|	5′. . . C G A T C G 3′. . . G C T A G C	. . . .	3′ 5′
8. *Proteus vulgaris*	Pvu II	|	5′. . . C A G C T G 3′. . . G T C G A C	. . . .	3′ 5′
9. *Haemophilus influenzae*	Hin II	|	5′. . . C A N T C 3′. . . G T N A G	. . . .	3′ 5′
10. *Staphylococcus aureus*	Sau3A	|	5′. . . C A T C 3′. . . G T A G	. . . .	3′ 5′
11. *Haemophilus aegyptus*	Hael II	|	5′. . . G G C C 3′. . . C C G G	. . . .	3′ 5′

Cloning Vectors and Their Way of Working

The vector is the central component of a gene cloning experiment. It provides the replicative function that allows the recombinant DNA molecule to multiply within the host cell. We have already encountered two types of naturally occurring DNA molecules that can replicate inside cells-plasmids and bacteriophage chromosomes and both are used extensively as cloning vectors.

Literally, hundreds of different cloning vectors are now available for use with different types of host cell. The largest number exists for *E. coli.* and the best known of these is the plasmid vector pBR322. The vector pBR322 is used as a typical example of a vector and is shown below how it works in a cloning experiment.

Useful Features of pBR 322 (Figs. 23.7 and 23.8)

The genetic and physical map of pBR322 gives an indication of why it has become such a popular cloning vector. Its first useful feature is its size, which, at 4363 bp, means that it and recombinant DNA molecules derived from it can be purified from *E. coli* cells with ease. Larger molecules (especially those of 50 kb and above) are much more difficult to purify intact.

A second useful feature of pBR322 is that it carries two antibiotic resistance genes, one coding for a β-*lactamase* (which modifies ampicillin into a form that is non-toxic to *E. coli*), and the second (actually a set of genes) coding for enzymes that detoxify tetracycline. The presence of these genes means that a bacterium transformed with pBR322 will be able to grow on a

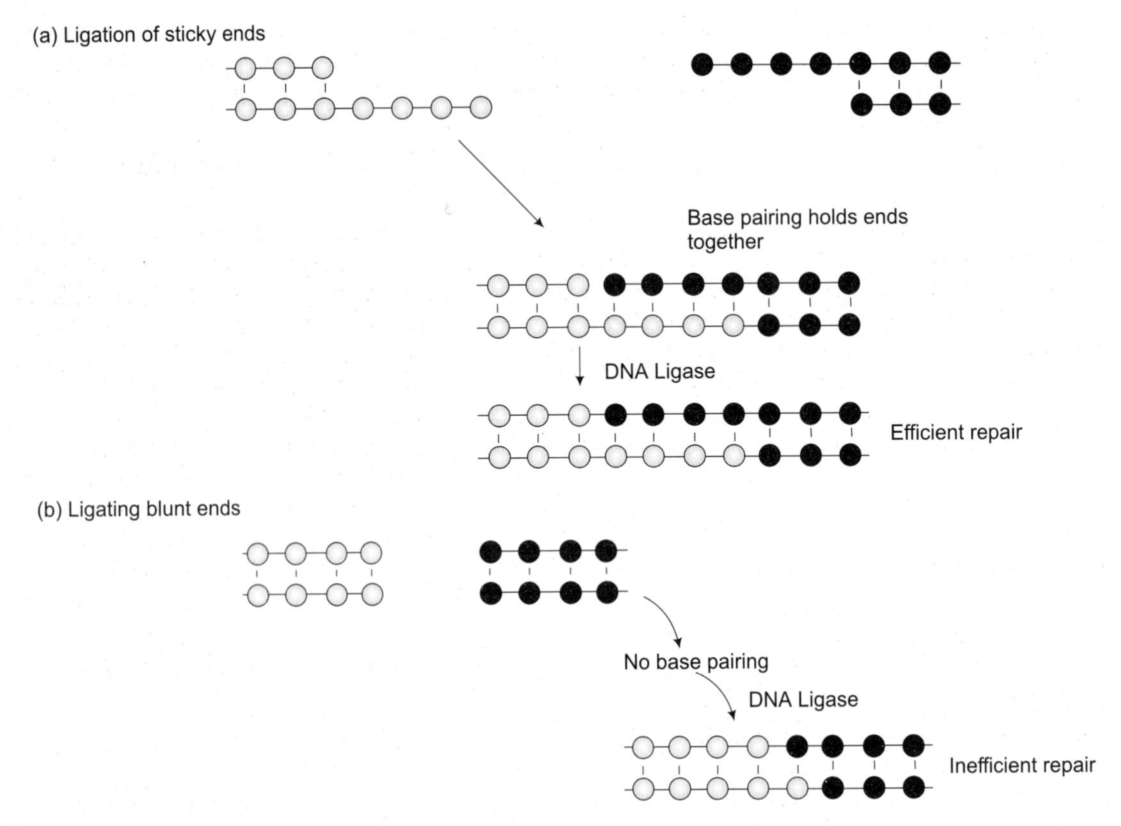

Fig. 23.6 Ligation of (a) Sticky ends and (b) Blunt ends.

medium containing *amplicillin* and/or *tetracycline*, whereas cells lacking the plasmid will not. Transformed bacteria can, therefore, be selected by checking for growth on the appropriate medium.

Finally, pBR322 is a relaxed plasmid with a copy number of about 15. This number can be further increased to as much as 1000 to 3000 by experimental means, so a cloning experiment using pBR322 will provide a good yield of recombinant DNA molecules.

Cloning with pBR322 (Fig. 23.9)

First, the vector molecule must be restricted to open the circle; pBR322 has several unique recognition sites for different restriction endonucleases and any one of these could be used for this purpose. In our example we have used BamHI, which will produce sticky ends at its cut site within the *ter*^{R} gene cluster.

The DNA to be cloned must also be cut with a restriction endonuclease, either the same one or another which will produce the same sticky ends. The DNA molecules are then joined together by ligation and mixed with competent *E. coli* cells, so that transformation will take place. The bacteria are then spread onto agar medium and colonies grown. The composition of the medium is designed with the following points in mind:

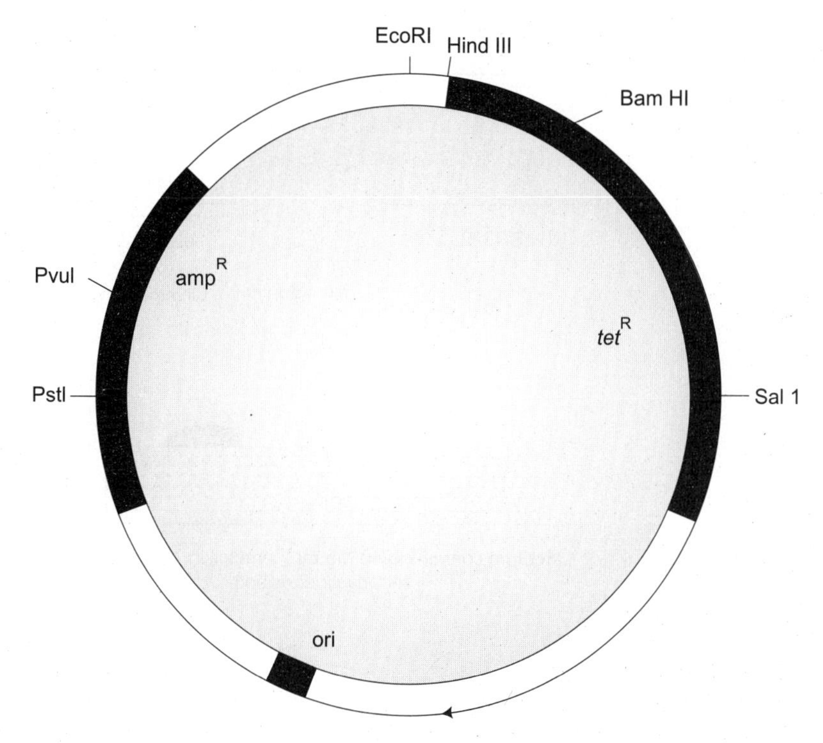

Fig. 23.7 The Cloning Vector pBR322.
(Note: *amp*^{R}: Ampicillin Resistance; *tet*^{R}: Tetracycline Resistance).

1. Cells that have not been transformed and so contain no plasmid molecules are *amp*^{S} *ter*^{S}.
2. Cells that have been transformed with a pBR322 molecule that has recircularised without the insertion of a DNA fragment (we call these cells transformants) will be *amp*^{R} *tet*^{R}.
3. Cells that contain a recombinant DNA molecule will have lost tetracycline resistance because of the presence of the inserted DNA fragment in the middle of the tetracycline resistance gene cluster. Recombinants are, therefore, *amp*^{R}*tet*^{S}.

By checking for growth on ampicillin agar we can exclude untransformed cells, but both transformants and recombinants will produce colonies. However, the recombinants can be identified by replica plating them on agar that contains both ampicillin and tetracycline. Colonies that do not grow are recombinants, and cells for DNA purification can be recovered from the ampicillin plate.

Other Types of Cloning Vector for *E. coli*

Although pBR322 is a typical example of an *E. coli* cloning vector, it is by no means the only one and many different types are now available, all designed for different roles. Of these the most important are those derived from phage λ. The main advantage of λ-based vectors is that the size of the DNA fragment that can be cloned is much larger than with plasmids such as

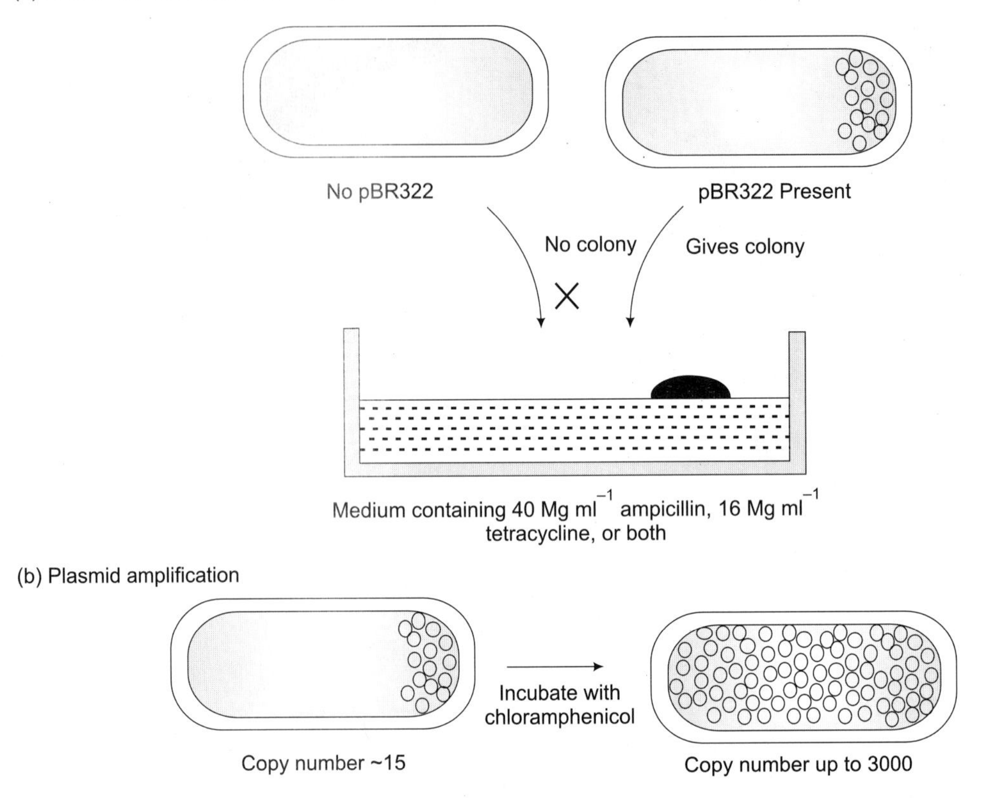

Fig. 23.8 Useful features of pBR322. (a) Cells containing pBR322 can be selected by plating on an agar medium that includes normally toxic amounts of ampicillin and/or tetracycline. (b) Copy number of pBR322 in the cell can be increased from about 15 to over 3000 by incubating the bacteria with chloramphenicol. This inhibits replication of chromosome but not replication of pBR322. Cell division, therefore, ceases but the plasmid continues to replicate until large numbers are present in each cell.

pBR322. The largest piece of DNA that can be cloned with the latter is about 5 kb, whereas fragments up to 25 kb can be handled by same λ vectors.

Libraries of Cloned Genes

Vectors, such as those based on, which allow relatively large pieces of DNA to be cloned, have made possible *the gene library, a collection of recombinant clones that together contain all the DNA present in an organism.* For instance, an *E. coli* gene library contains all the *E. coli* genes, so any desired one can be withdrawn from the library and examined.

Of course, gene libraries need to contain quite a large number of clones to be complete, or at least to have a high (95%) chance of containing a particular gene. For *E. coli* about 500 clones are needed, but for higher organisms the number is much greater. For *Drosophila* it is about 12000, and for man, several hundred thousand. These figures are by no means unreasonable in practical terms.

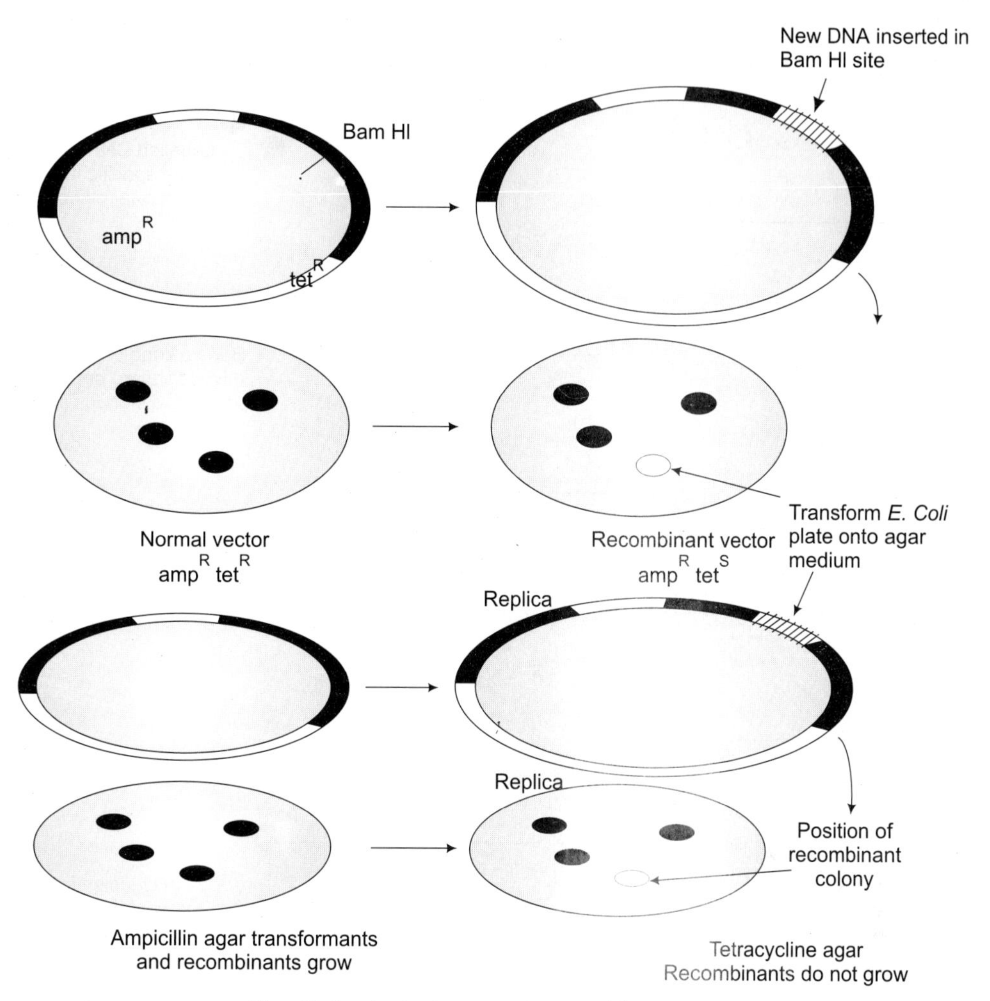

Fig. 23.9 A cloning experiment with pBR322.

Identification of a Gene from a Gene Library (Fig. 23.10)

Once a suitable library has been prepared, a number of procedures can be employed to identify a clone carrying a gene of interest. Many of these are based on hybridisation probing, during which DNA samples from different clones are immobilised on a solid support (usually a filter made of nitrocellulose or nylon); and then 'probed' with a second DNA molecule. This second molecule is radioactively labelled and is able to base pair with the desired gene. Base pairing is detected by placing a sheet of X-ray sensitive film over the filter, and the technique called autoradiography.

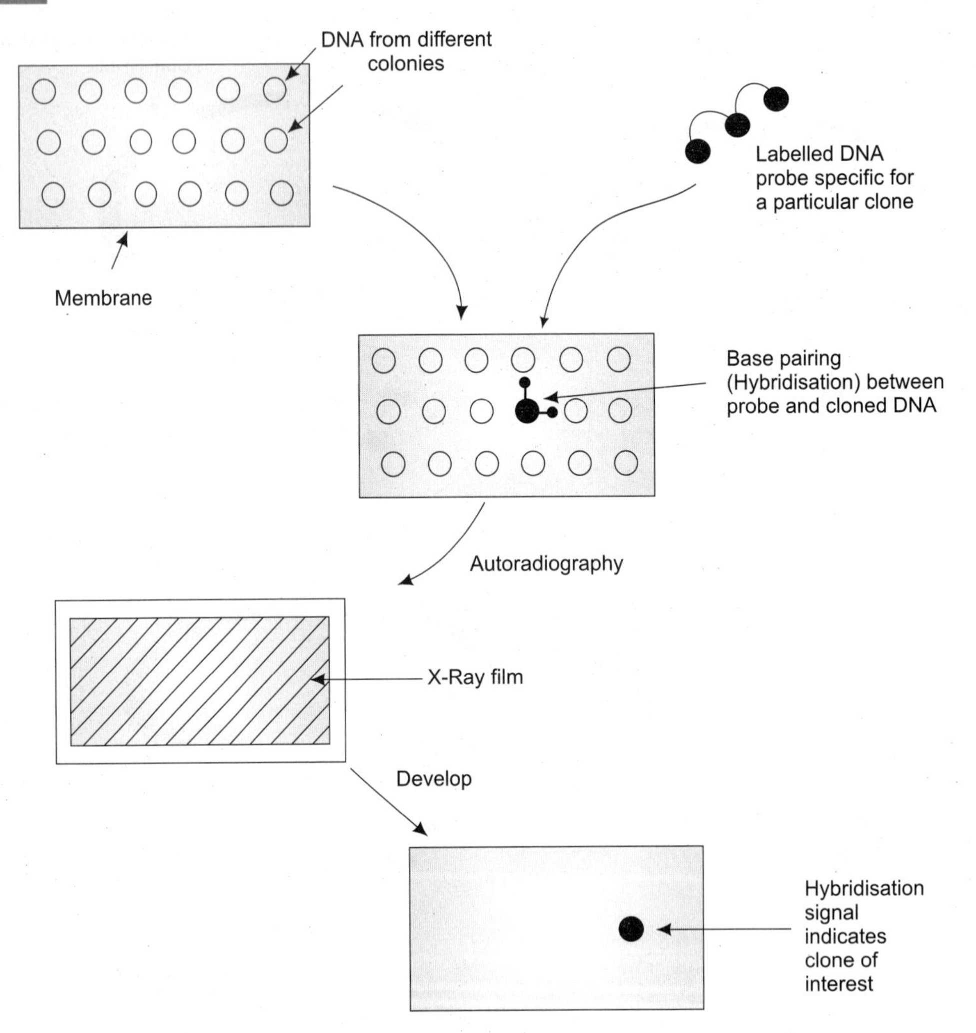

Fig. 23.10 Hybridisation probing enables a particular clone to be identified by virtue of base pairing between its DNA and a specific hybridisation probe.

DNA SEQUENCING

More advances in molecular genetics over past several years have come about through innovative technological developments. With the discovery of restriction endonuclease, which are DNA degrading enzymes that cleave DNA at specific points, it is possible to sequence DNA of a genome.

The basic concept of the sequencing method is that a small segment of DNA will be used as a primer for polymerisation of DNA up to a specific nucleotide ("*minus*") or for degradation of DNA down to a specific nucleotide ("*plus*"). This will generate very small segments of DNA of various sizes all ending with a known nucleotide. These segments can be separated from each other by gel electrophoresis, a technique for separating molecules of different sizes.

The first complete genome sequence was the single stranded DNA virus X174, which has 5375 nucleotides. DNA is isolated either from the virus itself or from an infected cell, either single stranded or double stranded DNA can be obtained.

Nucleic acid varying sizes will form bands such that the larger sequencing will move shorter distances. A step ladder is produced, with each ring representing a component *one less* nucleotide.

First DNA primer has to be produced by using restriction endonucleases which form uniform segments of varying sizes of DNA which are isolated by electrophoresis. A particular segment is placed with the single stranded viral DNA. It will form a double helix *at its point of complementarity* and thus form the *PRIMER.*

DNA polymerase is then added along with the four nucleotide triphosphates, at a very low temperature in order to slow the reaction.

Usually one of the nucleotide triphosphates has a radioactive label for future autoradiography. The system is sampled frequently. The goal is a mixture that contains primer plus extension products representing every single growth step, where *one step is the addition of* one more nucleotide. This mixture is called the *zero* sample. Unused nucleotides are washed out and eight sub-samples are separated; four of these are put into the '*plus*' system and four into the '*minus*' system. This technique is called "*plus and minus*" method of Sanger and Coulson. Upwards of 200 nucleotides can be read from a single gel. By *sequencing* each segment, the *whole genome can be read.*

Southern Blotting

Gel electrophoresis provides a powerful tool for the separation of macromolecules with different sizes and charges (Fig. 23.4).

These macromolecules can be further analysed. In 1975, E.M. Southern published new procedures that allowed investigators to identify the location of genes and other DNA sequences on restriction fragments separated by gel electrophoresis.

The essential feature of this technique is the transfer of DNA molecules separated by gel electrophoresis to a nitrocellulose or nylon membrane. The DNA is denatured either prior to or during transfer by placing the gel in the alkaline solution. After the transfer is complete, the DNA is immobilised on the membrane by drying or UV-induced cross-linking to the filter. A radioactive DNA ("*probe*") containing the sequence of interest is then hybridised or annealed with the immobilised DNA on the membrane (Fig. 23.10). The probe will anneal (from a double helix) only with DNA molecules on the membrane that contain nucleotide sequence complementary to the sequence of the probe. The non-annealed probe is then washed off the membrane and then the washed membrane is exposed to X-ray film, that detects the presence of the radioactivity in the bound probe. After the autoradiogram is developed, the dark bands show the position(s) of DNA sequences that have hybridised with the probe. Hence, this technique of identification of DNA nucleotides is called *Southern Blotting.*

Northern Blotting

When RNA is subjected to electrophoresis followed by the formation of RNA/DNA hybrids, then the technique is termed *Northern Blotting* as it is the mirror image or opposite of Southern blotting. The transfer of proteins from acrylamide gels to nitrocellulose membranes is called *Western Blotting* and is performed by using an electric current, so called "*electroblotting*".

These techniques require extensive purification of nucleic acid samples and the entire process is time consuming and expensive, although indispensable for structural analysis of nucleic acids. However, if only the detection and quantification of any given sequence is required, then it is possible to shorten the procedure to avoid purification stages, electrophoresis and blotting of the gel.

The major applications of this technique are: (1) the rapid detection of specific sequences, and (2) the determination of the relative amounts of any given species or sequence of RNA or DNA in a complex sample. The results obtained by quantification of any given nucleic acid sequence provide information on both the copy number or complexity of a gene or gene family, and on the developmental programme and regulation of specific gene activity.

DNA FINGERPRINTING

Amplification of DNAs by Polymerase Chain Reaction (PCR)

The *polymerase chain reaction* (PCR) is an extremely powerful new procedure that allows one to amplify a selected DNA sequence in a genome millionfold or even more. The PCR can be used to clone a given DNA sequence *in vitro without* using living cells during the cloning process. This PCR technique can be applied only when the nucleotide sequence of at least one short DNA segment on each side of the region of interest is known. The PCR procedure involves the following three steps: (1) The genomic DNA containing the sequence to be amplified is denatured by heating. (2) The denatured DNA is annealed to an excess of the synthetic oligonucleotide primers. (3) DNA polymerase is used to replicate the DNA segment between the sites complementary to the oligonucleotide primers.

Applications of PCR Technology

(1) It permits to obtain definitive structural data on genes and DNA sequences when very small amounts of DNA are available. (2) In diagnosing inherited human diseases in prenatal cases from foetal DNA. (3) In forensic cases, DNA fingerprinting is applied to detect criminals. By using PCR amplification, DNA sequences can be obtained from very small amounts of DNA isolated from small blood samples, sperm or even individual human hairs. These play a major role in legal cases involving disputed identity or paternity.

Restriction Fragment Length Polymorphisms (RFLPs) (Fig. 23.11)

Arabidopsis thaliana is a small, economically unimportant member of the family Brassicaceae (mustard family). It has been called as the '*Drosophila*' of plant kingdom judging from its

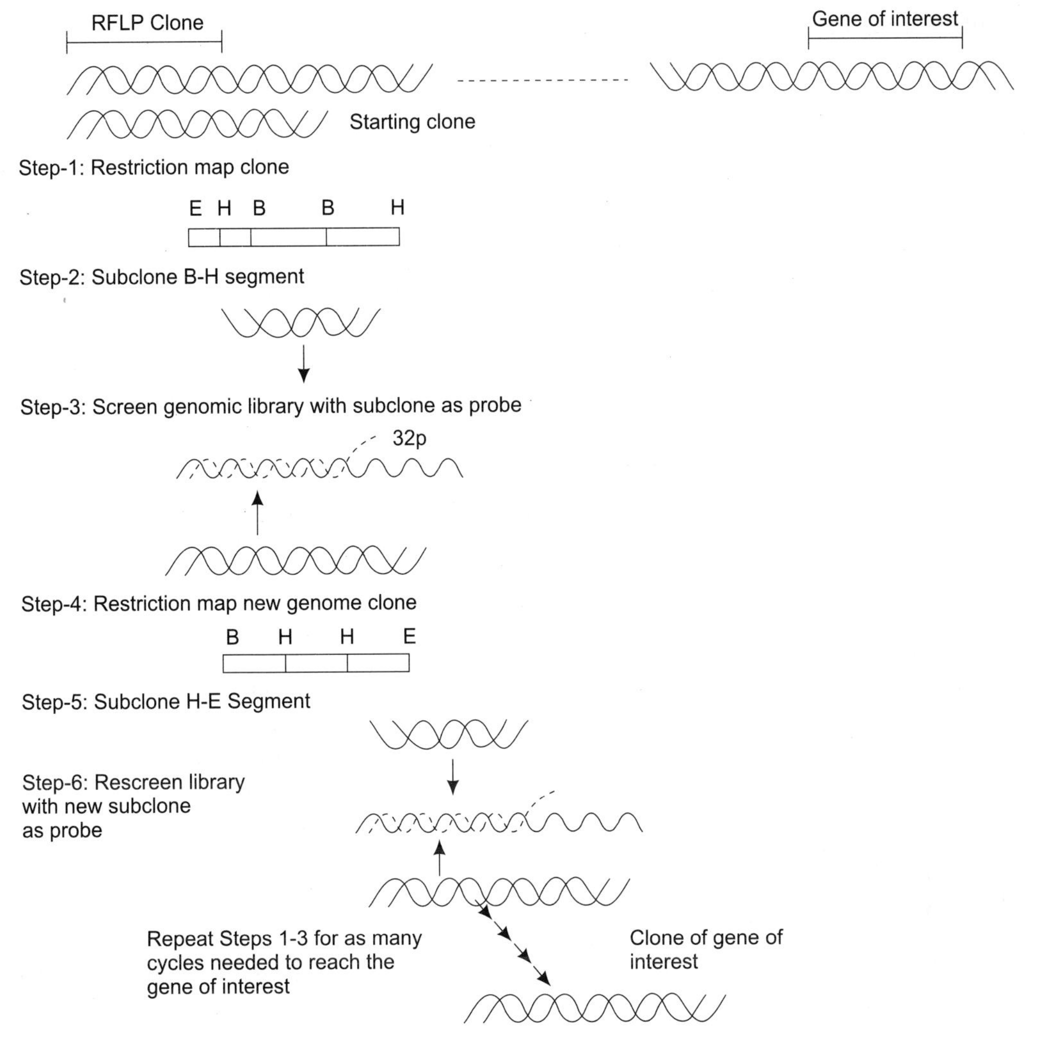

The '*work*' starts with the identification of the molecular marker closest to the gene that one wishes to isolate. In the example, the closest molecular marker is an RFLP marker clone. A restriction map is prepared for this initial RFLP clone, and the restriction fragment proximal to the gene of interest is subcloned. B, E and H identify cleavage sites for the enzymes BamHI, EcoRI, and Hind III, respectively. The subclone (a Bam III-Hind III restriction fragment is the example shown) is used as a radioactive hybridisation probe to screen the genomic library for overlapping clones. Restriction maps are then prepared for the new genomic clones. The restriction fragment proximal to the gene of interest is subcloned. This new subclone (a Hind III—EcoRI fragment) is, in turn, used as a probe to isolate a second set of overlapping genomic clones. Restriction maps of the new clones are proposed, and the process is repeated several times until the "walk" has covered the indicated distance.

Fig. 23.11 Chromosome "Walking" with RFLP markers.

advantages for genetic studies, viz., (1) Small size of the plant (1 foot ht.); (2) Short generation time (about 5 weeks); (3) High seed production (up to 40,000 seeds full plant); (4) Very small genome $- 7 \times 10^7$ nucleotide pairs (about 20 times larger than *E. coli* genome and less than 1/100th the size of maize) (5) Very little interspersed repetitive DNA—unique respective of DNA are on average 1,20,000 nucleotide pairs in length and (6) Natural self-pollination.

Chromosome Walking (Fig. 23.11)

Mutations are induced by EMS (ethyl methane sulfonate) to detect mutants in a particular gene. A major advantage of *Arabidopsis* is that once a mutation has been identified in a gene of interest, the mutant allele can be mapped and used as a tool in cloning the gene. Because of the small genome size and the rare occurrence of interspersed repetitive DNA sequences in *Arabidopsis* genes can be cloned by the procedure "Chromosome Walking". The technique involves mapping the mutant allele of interest relative to the standard molecular markers by standard genetic crosses, and the one "*walks*" along the chromosome from the closest marker by isolating overlapping (cross hybridising) genome clones until one reaches the gene of interest. Because of repeated sequences of DNA it is difficult to use "chromosome walking" in higher plants and one has to "walk" too far up to several thousand clones or steps. The procedure involves isolating large quantities of specific DNA fragments for subcloning and sequencing.

The molecular markers that are used to initiate chromosome "walks" may be cDNA or genomic clones of genes that have previously been mapped using mutant alleles.

Restriction Fragment Length Polymorphisms (RFLPs) are simply variations (of different ecotypes or geographical isolates) in the lengths of the DNA fragments *produced by cleaving DNA* molecules with specific restriction endonucleases (Fig 23.12). Since restriction enzymes cut DNA in a sequence–specific manner, every homologous DNA molecule from every cell of a totally homologous organism will be cut at exactly the same site. All genes and DNA sequences of the entire organism have to be cloned. In different ecotypes, many base pair substitutions, deletions and additions will take place which are called "*silent mutations*". These silent mutations occur in intergenic regions, in *introns* and *exon* positions that correspond to the 3′ bases of codons (the "degeneracy" or "wobble" positions) which have no perceptible phenotypic effect.

EXAMPLES OF GENETIC ENGINEERING TECHNIQUES

The joining of different DNA molecules together in vitro is termed recombinant DNA technology or *genetic engineering*. This technique has revolutionised the study of eukaryotes and have provided sources of particular proteins in quantities hitherto considered impossible to obtain.

Circular plasmid DNA is especially important in genetic engineering. The most important goal of an experiment in genetic engineering is usually to insert a particular fragment of chromosomal DNA into a plasmid or a viral DNA molecule. Hence, there are techniques developed for breaking DNA molecules at specific sites and for isolating particular DNA fragments.

1. Isolation and Characterisation of Particular DNA Fragments

Many nucleases have been identified from a variety of organisms and most produce breaks at random sites within a DNA sequence (*endonucleases*) or remove nucleotides only from the termini (*exonucleases*). Other class of nucleases called *restriction enzymes* recognise only one short

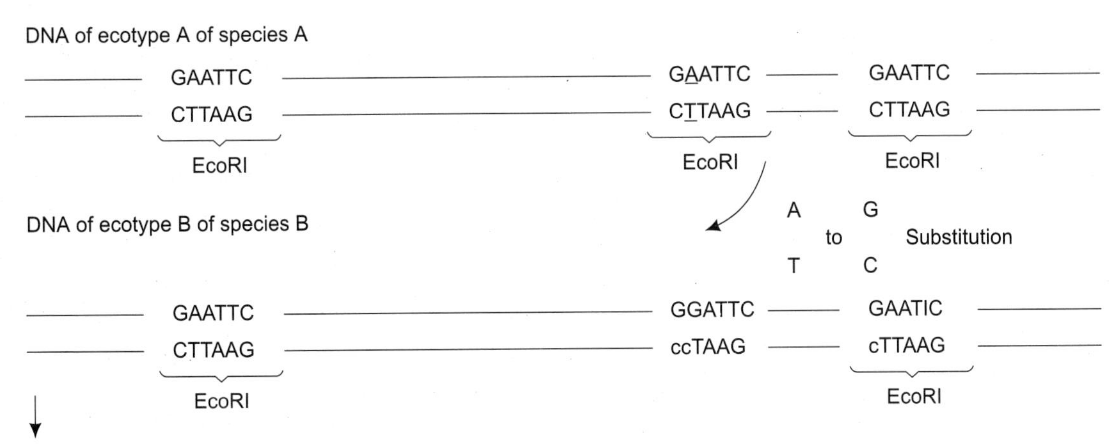

1. Isolate DNA from each ecotype
2. Cut with EcoRI restriction endonuclease
3. Separate fragments by agarose gel electrophoresis
4. Southern transfer to nylon membrane
5. Hybridise to radioactive gene A clone
6. Autoradiography

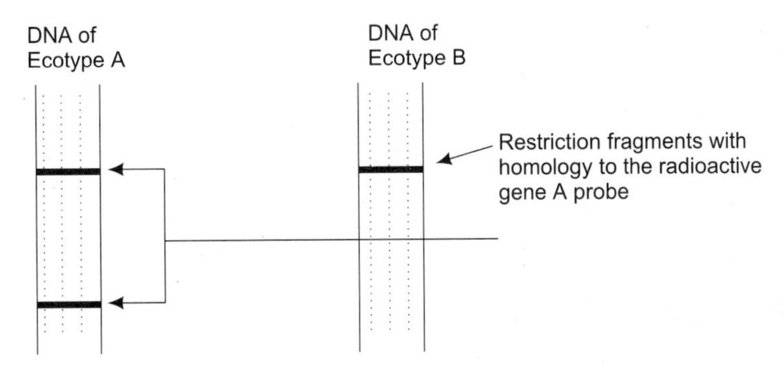

Fig. 23.12 Origin of RFLPs by DNA divergence in different ecotypes of a species. AT-GC pair substitution results in the loss of the central EcoRI recognition sequence present in gene A of ecotype A of species A.

base sequence in a DNA molecule and make two single strand breaks, one in each strand, generating 3′-OH and 5′-P groups each position. These recognised sequences are called *Palindromes,* i.e., the sequence has symmetry of the form:

$$\begin{array}{cccccc} A & B & C & C' & B' & A' \\ A' & B' & C' & C & B & A \end{array} \quad \text{or} \quad \begin{array}{ccccc} A & B & \times & B' & A' \\ A' & B' & \times & B & A \end{array}$$

$$\text{or} \quad \begin{array}{cccc} A & B & B' & A' \\ A' & B' & B & A \end{array}$$

in which the capital letters represent *bases,* A′ indicates a complementary base, × is any base and the vertical line is the axis of symmetry. These bases have 4–6 bases. These fragments spontaneously circularise and could be relinearised by heating. Circularisation will be permanent if they are treated; with *E. coli* DNA ligase, by joining 3′-OH and 5′-P groups. There are three properties exhibited by restriction enzymes namely, (1) restriction enzymes make breaks in symmetric sequences. (2) The breaks are usually not directly opposite one another. (3) The enzymes generate DNA fragments with complementary ends.

Circular plasmid DNA is very important in genetic engineering. Advances in enzymology, DNA sequencing techniques, hetero-duprex mapping, genetics and biophysics, coupled with the discovery of transposons (jumping genes) have resulted in the evolution of a new discipline called *biotechnology.* This field allows gene flow unrestricted among species and genera allowing wide crosses possible. Nuclear DNA libraries of some plant genomes have been constructed and genes for some storage proteins have been actively analysed.

VECTOR SYSTEMS

Use of Ti-plasmids (Figs. 23.13 and 23.14)

Tumor-inducing large plasmids (90 and 150 Kbs) of the virulent gram-negative soil bacterium, *Agrobacterium tumefaciens*, causes neoplastic disease (tumourous growth) around the root crowns of a wide range of dicotyledonous plants upon injury. Molecular hybridisation and DNA transformation experiments have conclusively shown that *Agrobacterium* introduces plasmids in the host (bacteria themselves do not enter): a discrete segment (23 kbp) of Ti-plasmid called T-DNA (transfer DNA) is stably incorporated with the nuclear DNA of the host at multiple sites.

The cloning of plant genes involves firstly, the splicing of foreign DNA within a site of T-DNA (Fig. 23.15).

The crown gall cells are inserted by opine synthetase gene of Ti-plasmids to produce nopaline and octopine (derivatives of arginine and lysine). The genetic information carried on T-DNA including newly introduced foreign genes will be transmitted through meiosis and expressed a transposon which confers resistance to streptomycin, Tn 904, has been transferred on a Ti-plasmid.

Cauliflower Mosaic Virus (CMV) (*as vector*). Cauliflower Mosaic Virus (a circular double stranded DNA Virus) and golden mosaic virus (a single stranded virus) can be used as vectors in gene cloning in plants (Cocking et al. 1981). These can be spliced into an *E. coli* plasmid, amplified in the bacterium, removed and used for infection of a range of plants. The virus vectors will multiply independently and therefore, transform the host without integration in the genome and the following steps are involved as shown in Fig. 23.16.

Hydrogen uptake gene (hup) confers the ability to utilise molecular hydrogen evolved in nitrogenase reaction during symbiotic nitrogen fixation (Fig. 23.17).

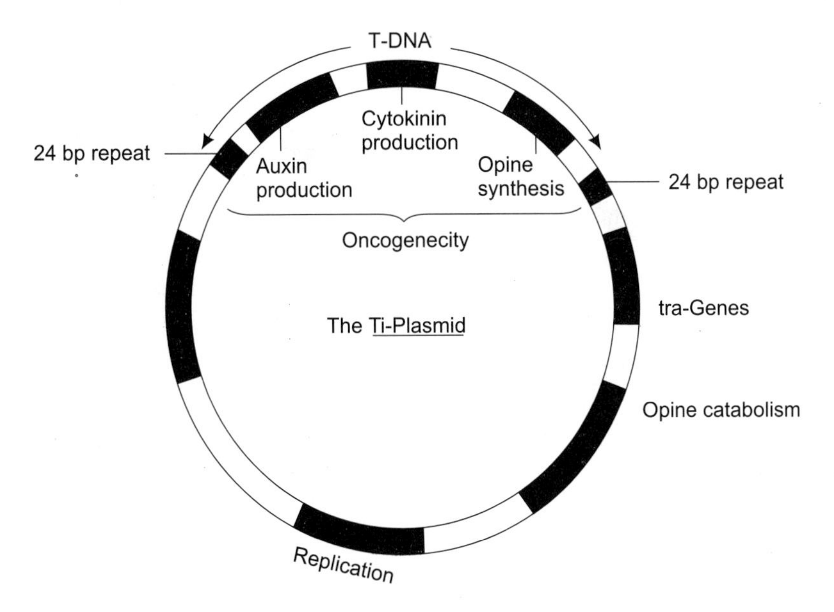

Fig. 23.13 Physical map showing the various regions of the Ti-plasmid with known function. The T-DNA is located between the two 24 base pair repeats. The *tra* genes are involved in conjugative transfer.

BIOTECHNOLOGY—GENETIC ENGINEERING INSTITUTES AND SCOPE

Universities and institutions currently engaged in research in genetic engineering in India.

1. Genetic Engineering–Indian Institute of Science, Bangalore.
2. Madurai Kamaraj University, Madurai.
3. Bose Institute–Kolkata
4. Jawaharlal Nehru University–New Delhi (JNU)
5. Animal Cell Culture and Virology–Poona University, Pune
6. Plant Tissue Culture, Photosynthesis and Plant Molecular Biology–Indian Agricultural Research Institute–New Delhi (IARI)
7. Oncogenes, reproduction Physiology, cell transformations, molic acid and Protein Sequences–Centre for Cellular and Molecular Biology (CCMB)–Hyderabad
8. Immunology—National Institute of Immunology (NII)–New Delhi
9. Enzyme Engineering–Institute of Microbial Technology (IMTECH)—Chandigarh

In addition;

Biotechnology Network (BTNET)

Creation of Infrastructural facilities:

Department of Biotechnology, New Delhi (DBT)

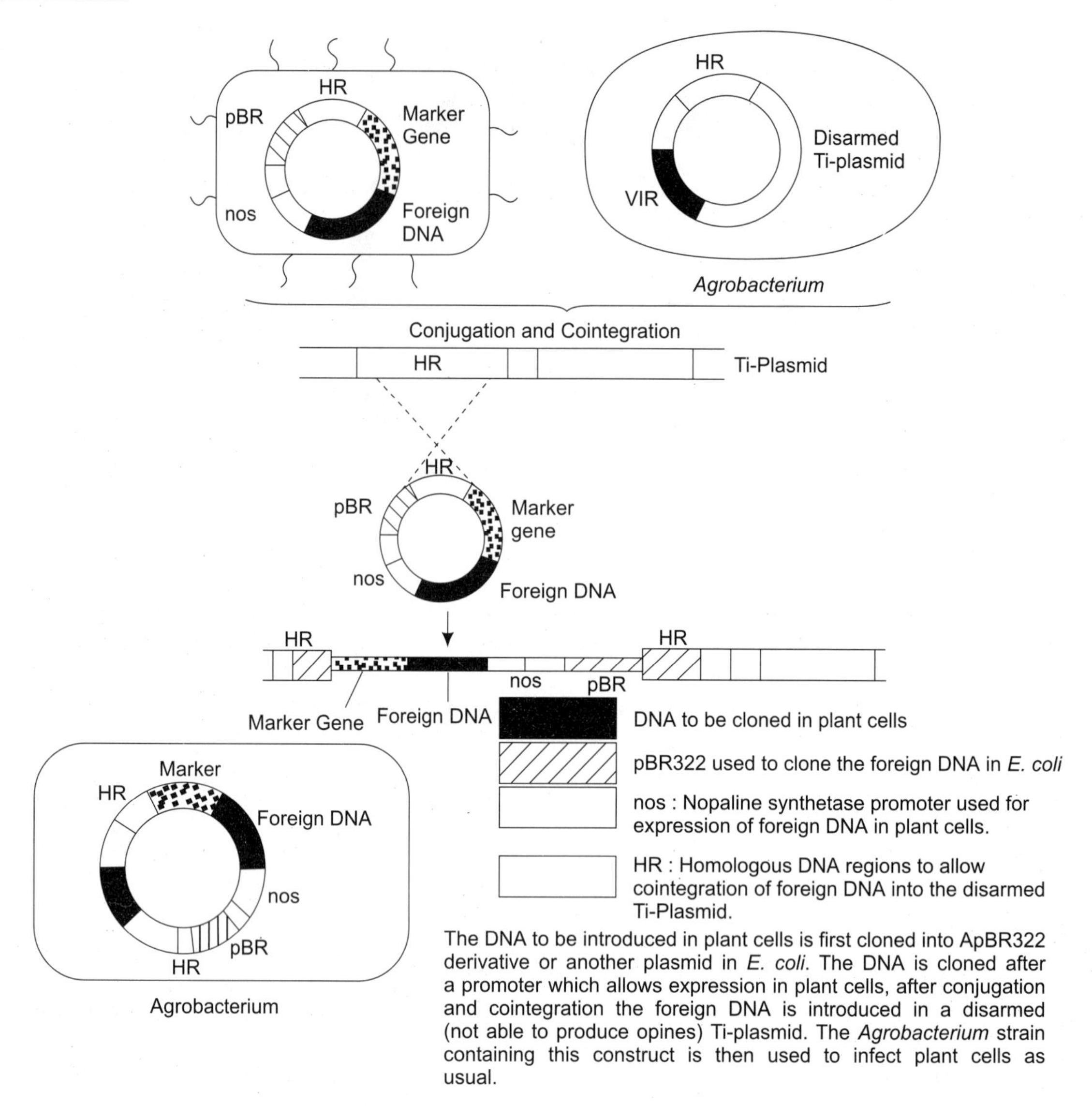

Fig. 23.14 The principle of cloning foreign DNA into the T-fragment using cointegration vectors and disarmed Ti-plasmids.

Well developed scientific and technological infrastructure, include the following:

(a) germ plasm banks for plants, animals;

(b) algae and microbes;

(c) animal houses;

(d) facilities for oligonucleotide synthesis;

(e) production, import and distribution of enzymes, reagents and radio-labelled compounds;

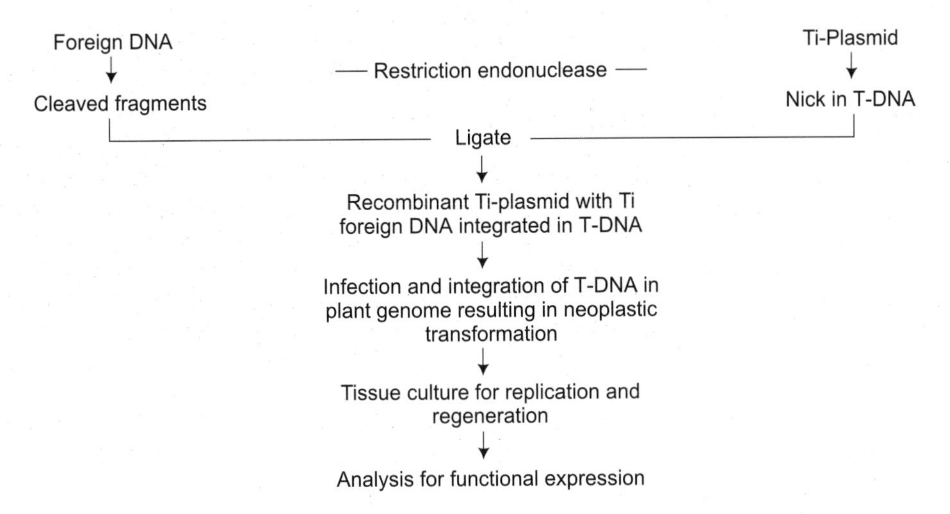

Fig. 23.15 Use of Ti-plasmid in cloning of plant genes.

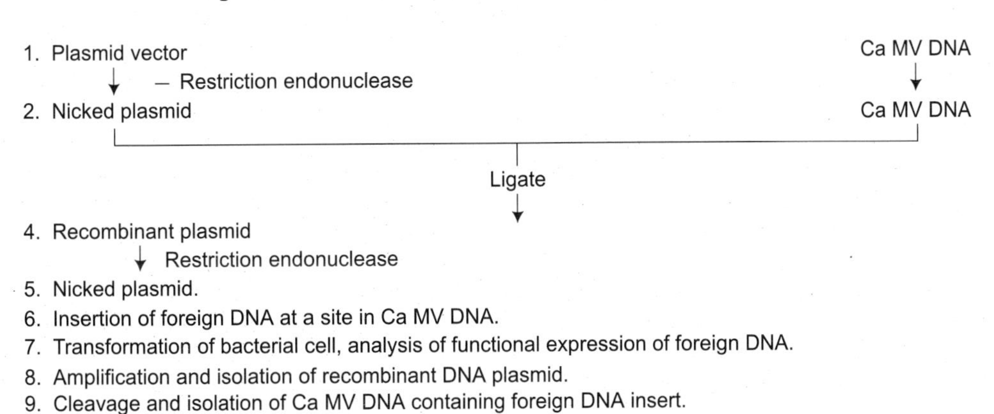

Fig. 23.16 Use of cauliflower mosaic virus (Ca MV) in cloning of plant genes.

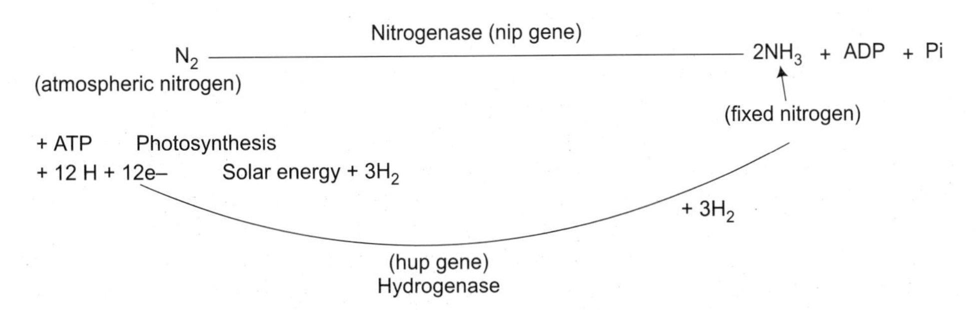

Fig. 23.17 Recycling of hydrogen gas by hydrogenase in some species of *Rhizobium* (*R. japonicum*).

(f) bioprocess optimisation and piles plants;
(g) genetic engineering R and D units;
(h) centre for reproduction biology and molecular endocrinology;
(i) facility for carbohydrate cell surface and cellular transport;
(j) facilities for protein and peptide sequencing;
(k) national NMR facility;
(l) marine cyanobacterial germ plasm collection, and
(m) antibiotic developments consortium.

Adequate trained/highly skilled manpower is essential for meeting expanding R & D (Research and Development) and manufacturing needs in the multidisciplinary areas of biotechnology.

R&D-cum-production Programmes in India

Intensive research is going on towards the possibilities of genetics of products, processes and sources for ultimate commercial exploitation like:

(a) Development of larval insecticide against mosquitoes (Biocide-S).
(b) Extraction of copper through bacterial leaching.
(c) Tissue culture propagation of *Bamboo.*
(d) Tissue culture propagation of *Cardamom.*
(e) Tissue culture propagation of coconut.
(f) Enhancement of alcohol production through the use of an improved yeast strain developed at IMTECH (Institute of Microbial Technology, Chandigarh) and better demonstration processing.
(g) Increase in production of antibiotics like *Penicillin* and *Streptomycin.*
(h) Production of sex hormones—follicle stimulating hormone (FSH), human chorionic gonadotropin (BCG).
(i) Production of *Insulin.*
(j) Hormone treatment for increased production of major crops and production of transgenic varieties.
(k) Animal birth control vaccine (*Talsur*) for reduction of scrab cattle and stray dogs.
(l) Establishment of two pilot plants, at Delhi and Pune, for raising elite varieties of species of *Eucalyptus, teak, sandal wood, bamboo* and *poplar* for forest regeneration, agroforestry and social forestry.
(m) Prawn aquaculture.

Health Programmes

Production of vaccines

Immuno diagnostics (fertility control, communicable diseases, etc.)

Agricultural Programmes

Tissue culture micropropagation of elite trees.

Bamboo—Biopepticides

Biofertiliser—Rhizobium-biofertiliser

Cardamom tissue culture

Oil Palm

Veternary biotechnology, aquaculture

Embryo transfer technology

Animal birth control injection

Recombinant-DNA Safety Guidelines

Advances in manipulation of genetic material have generated a sense of concern among scientists about the safety of work involving pathogenic microorganisms and genes encoding virulence factors.

On the basis of recommendation of the Recombinant-DNA Advisory Committee and available scientific information, DBT (Department of Biotechnology) has brought out a document on "Recombinant-*DNA safety guidelines*". The guidelines cover areas of research involving genetically engineered organisms, genetic transformation of green plants and animals, recombination-DNA. (r-DNA) technology in vaccine development and large scale production and handling of products obtained by r-DNA technology.

The monoclonal-antibody business is expected to touch two billion dollars (2 billion) per year as several countries are investing heavily, in terms of both manpower and financial resources, in the development of vaccines based on antigens produced through recombinant DNA techniques for virtually all the major infectious diseases.

Integrated Approach

Biochemistry

Molecular biology

Cell biology

Immunology

Genetics and viruology—many overlap each other.

Tagging of economic traits with *Restriction-Fragment-Length Poly-morphism* (RFLP) markers has opened up an entirely new approach for efficient selection of desired gene combination for a breeding population.

This technique is primarily based on the thesis that restriction enzymes, which can cut DNA at specific sites, can distinguish between plant strains even if they differ by a *single base* in the restriction enzyme site.

Thus, a changed pattern of DNA fragments is produced by cleavage with a restriction enzyme, when a target site is present in the genome of one individual but is absent in the other, owing to the loss of recognition by the enzyme due to the mutation. This polymorphism at the level of the DNA has been used successfully *for tagging the economically important traits, governed* by either single or many genes. For instance, in tomato, strains have been selected that carry high solid content, a character controlled by many genes. In India, RFLP studies have been initiated and are being used for molecular mapping and character-tagging studies in mustard (*Brassica sps*), Chickpea (*Cicer arietinum*) and rice, at the Biotechnology Centre, IARI, New Delhi, NCL, Pune, and the International Centre for Genetic Engineering and Biotechnology (ICGEB), New Delhi.

PROTOPLAST FUSION

Tissue culture methods. Creation of usable genetic variability through *somaclonal variation, e.g., Brassica* at IARI, New Delhi and sugarcane at Sugarcane Breeding Institute, Coimbatore for economically important characters through *Wide Hybridisation–embryo* rescue, *Somatic Hybridisation and Protoplast fusion.*

In vitro selection

Selection for disease resistance

Selection for tolerance of basic stresses.

PEG—Polyethylene glycol or Dextran induces drought conditions—salt tolerant varieties.

Micropropagation of Elite Genotypes

Mass multiplication of plants by tissue culture, e.g., orchids in Japan.

RECOMBINANT-DNA TECHNOLOGY

1. *Engineering plants for disease resistance* Spectacular success has been achieved with regard to viral diseases following the use of r-DNA technology. A major achievement has been the transfer and expression of coat-protein genes of Tobacco Mosaic Virus (TMV) and Alfalfa mosaic virus (AIMV) in tobacco, resulting in the protection against or delay of disease development in the *transgenic plants.*
2. *Engineering plants for resistance to pests* *Bacillus thuringiensis* has genes to encode insecticidal proteins. *B. thuringiensis* strains specifically toxic to Dipteran, Lepidopteran and Coliopteran insects have been identified and the insectidal protein gene cloned. Using Ti-plasmid vectors or *Agrobacterium tumifaciens*, the gene encoding the insecticidal protein has been transferred to tobacco, potato, rice, corn and other crop plants.

Such *transgenic plants* incorporate resistance to specific insects that feed on these crops. Another approach of producing insect-resistant transgenic plants is to introduce into plants, genes that inhibit enzymes responsible for the breakdown of protein (protease inhibitors). Transgenic plants with cow pea (*Vigna sinensis)* protease inhibition genes have demonstrated resistance to tobacco bud worm, corn earworm, armyworm and horn worm of tomato and tobacco etc.

Engineering Plants for Resistance to Herbicides

By using mutant EPSP (5-enolpyruvyl shikimate-3-phosphate) synthetase enzyme, crop plants with resistance to a particular herbicide have been engineered.

Engineering Plants for Nutritional Quality

From the mutational point of view, plant proteins *per se* suffer from amino acid imbalances. In situations such as prevailing in India, when the major source of dietary protein is of plant origin, distortions of amino acid *balance reduce* the nutritional efficiency of protein.

By genetic engineering, it may be possible to correct the *imbalance of amino acid profiles* in seed proteins. Storage protein genes are expressed during specifically limited periods of seed development and are, therefore, relatively easy to identify and isolate as in the case of *Zein genes* in *maize-* Lysine content improvement, *Lathyrus sativus*—aflatoxins which cause paralysis in animals and human beings. Engineering plants devoid of neuroprotein can be achieved by (1) mutating the gene of toxin synthesis so that the toxin is either not produced or produced in low quantities. (2) Developing transgenic plants harbouring the antisense gene so that the gene product is not formed. The latter strategy is expected to work via complementary base-pairing between the antisense m-RNA and the coding sense m-RNA, which will prevent translation of the latter into the gene product.

Basically, biotechnological approaches involve cellular intervention using tissue culture methods and r-DNA technology.

DNA Profiling and Its Applications

A structure of DNA that can detect a target sequence in the genome is called a *PROBE.*

Techniques for DNA Profiling

1. Isolation of DNA from the biological sample (such as blood, blood strains, semen strains, vaginal swabs or bone marrow).
2. Digestion of DNA with restriction enzymes.
3. Fractionation of the resultant DNA fragments on the basis of *size* by agarose gel electrophoresis.
4. Transfer of the fractionated DNA fragments onto *blotting membrane.*
5. Hybridisation of the DNA on the membrane with labelled *probe.*
6. Removal of non-hybridised probe by washing, and
7. Detection of the hybridised probe by autoradiography.

DNA is isolated by SDS-Proteinase digestion and phenol chloroform extraction. After digestion of the DNA with the appropriate *restriction enzyme* the samples are run on a 0.8 per cent agarose horizontal slab gel at 3-4V cm^{-1}. Some workers prefer vertical agarose gels. Field Inversion Gel Electrophoresis (FIGE) has been introduced to increase the resolution of the bands. Samples are transferred to blotting membranes by *vacuum blotting, Southern blotting* or *electrotransfer*. A variety of blotting membranes are available. These include nitrocellulose, charged or uncharged nylon and polyvinyl difluoride (PVDF) membranes. It was observed that the combination of vacuum blotting and nylon membranes result in quick and efficient transfer.

The DNA on the blotted membrane is hybridised with the labelled probe in the appropriate solution and at the right temperature. The probe can be labelled with a variety of methods, which include, nick translation, random hexamer–primed labelling, and single strand M13 labelling using a M13 probe primer. A variety of labelled nucleotides are available including 32p, biotin and digoxigenin-labelled nucleotides. Then, hybridise the membrane in solution containing per cent SDS in 0.5M phosphate buffer (pH 7.5) at 60°C. The hybridised blot is washed with a series of solutions that provides the desired level of stainability of hybridisation. The radioactivity labelled hybridised probe on the membrane is detected by autoradiography. Non-radioactive labelled probe is detected by appropriate enzymatic or chemiluminescent deletion method. Samples are transferred to blotting membranes, vacuum blotting, Southern blotting and electrotransfer.

Encoded in human DNA are about 1,00,000 genes. These represent only about 5 per cent of the total DNA in the chromosomes. The function of the remaining 95 per cent of the genome is not yet understood. One of the components of this "*extra*" DNA consists of sets of base sequences repeated numerous times and are called *mini satellites.* Botstein *et al.* suggested the restriction fragment length polymorphism (RFLP) technique of DNA analysis as an approach to mapping human genome. Wymann *et al.* discovered the first hyper-variable locus (HVR) in the human genome which was shown to be tandemly repeated.

APPLICATIONS OF BIOTECHNOLOGY

Every individual in the world can be identified at the molecular level on the basis of an extremely high level of polymorphism in the sequence of his or her DNA, which he or she inherits from his or her biological parents and is identical in every cell of the body.

DNA fingerprinting, as this technique of identification is called, can confirm with certainty the parentage of an individual. DNA probes derived from a DNA sequence first isolated from an Indian snake, *Bungarus* (banded krait) which are useful in forensic investigations by DNA fingerprinting. The following are the applications:

1. In pedigree analysis and establishing paternity–family relationships for immigration authorities.
2. In rape cases–hair roots, sample blood, bucaal smear, semen spots or skin tissue left behind by the criminals, vaginal swabs taken up to 20 hours after intercourse can be used to isolate DNA from sperm.
3. Identification of mutilated dead bodies with the help of DNA fingerprinting of close relatives.
4. In social security and record identification.
5. In solving murder cases by DNA fingerprint analysis of blood swabs taken from murder weapons or blood spots.

6. In sexing biological samples by *in situ* hybridisation with Y chromosome specific probes.
7. In animal breeding programmes, DNA fingerprinting can be used in providing assurance in livestock breeding.
8. In plant breeding programmes authentication of seed stocks and germ plasm.
9. In characterisation of cell cultures.
10. In detecting specific bands in close linkage with disease loci in large pedigrees.
11. In identification of post-transplant cell populations.

APPLICATIONS OF GENETIC ENGINEERING—SOME MORE EXAMPLES

Genetic Engineering or Recombinant DNA has several applications like construction of industrially important bacteria, studies of regulation by sub-cloning, genetic engineering of plants, production of synthetic drugs, vaccines, chemicals, site specific mutagenesis, production of single cell protein and restriction mapping etc.

1. Restriction Mapping

It is frequently of value to determine the location of restriction cuts in a particular DNA molecule. A map showing the positions is called *restriction map*. The most common way to obtain a restriction map is the *Double-digest procedure*, which uses two different restriction enzymes. The procedure is to make three samples of a particular DNA species, treat each of these with a separate enzyme and one sample with both enzymes (a double digest), and then compare the three sets of fragments.

2. Site Specific Mutagenesis (Figs. 23.18, 23.19 and 23.20)

Mutations are introduced into organisms in the laboratory for a variety of reasons. In modern molecular genetics to achieve a deep understanding of the molecular mechanisms underlying such phenomena as gene expression and gene regulation often requires mutations in particular base sequences (for example, binding sites). In general, the gene to be altered is carried on a plasmid. The plasmid is isolated, manipulated and then returned to a bacterium for replication and gene expression. Site specific mutagenesis is achieved by production of deletions, by producing point mutations at random sites in a particular region of DNA and by producing point mutations at a particular base pair.

3. Single-Cell Proteins or Production of Proteins from Cloned Genes

One of the goals of genetic engineering is the production of a large quantity of a particular protein that is otherwise difficult to obtain (for example, proteins for which there are normally only a few molecules per cell). In principle, the method is straightforward; the gene is inserted adjacent to a promoter and tested to determine whether it is oriented such that the coding strand will be transcribed. With a high copy number plasmid or an actively replicating phage, synthesis of a gene product may reach a concentration of about 1 to 5 per cent of the cellular protein. The term '*Single Cell Proteins*' (SCP) refers to the dried cells of microorganisms (algae,

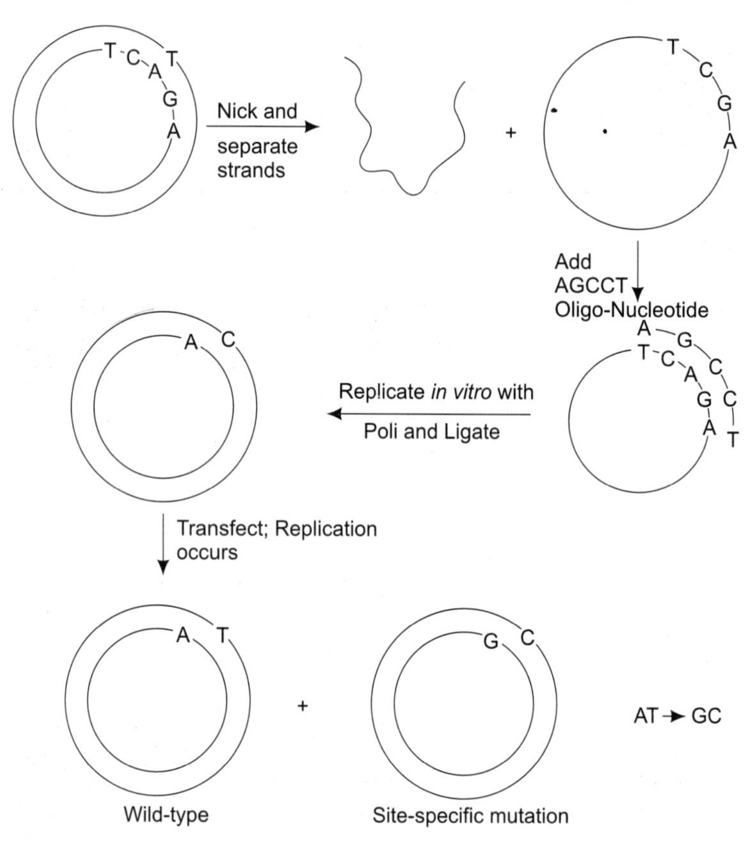

Fig. 23.18 A method for site-specific mutagenesis. The method generates point mutations.

bacteria, yeasts, etc). Genetic engineering now allows scientists to manipulate bacteria to construct new bacteria that *greatly overproduce a given protein.* This is done by joining the coding sequence of the gene to a strong promoter, inserting the promoter-coding sequence unit into a high copy number plasmid, and introducing the recombinant plasmid into an appropriate host.

4. Industrial Chemicals

Bacteria with novel phenotypes can be produced by genetic engineering, sometimes by combining the features of several other bacteria. For example, several genes from different bacteria have been inserted into a single plasmid that has been placed in a marine bacterium, yielding an organism capable of metabolising petroleum; this organism has been used to clean up oil spills in the oceans. In 1972, A. Chakrabarthy developed a technique in a strain of *Pseudomonas* that could degrade several compounds of crude petroleum (octanes, camphor, etc.). Furthermore, many biotechnology companies are at work designing bacteria that can synthesise industrially important chemicals. Enormous efforts are currently being made to create organisms that can convert biological waste to alcohol.

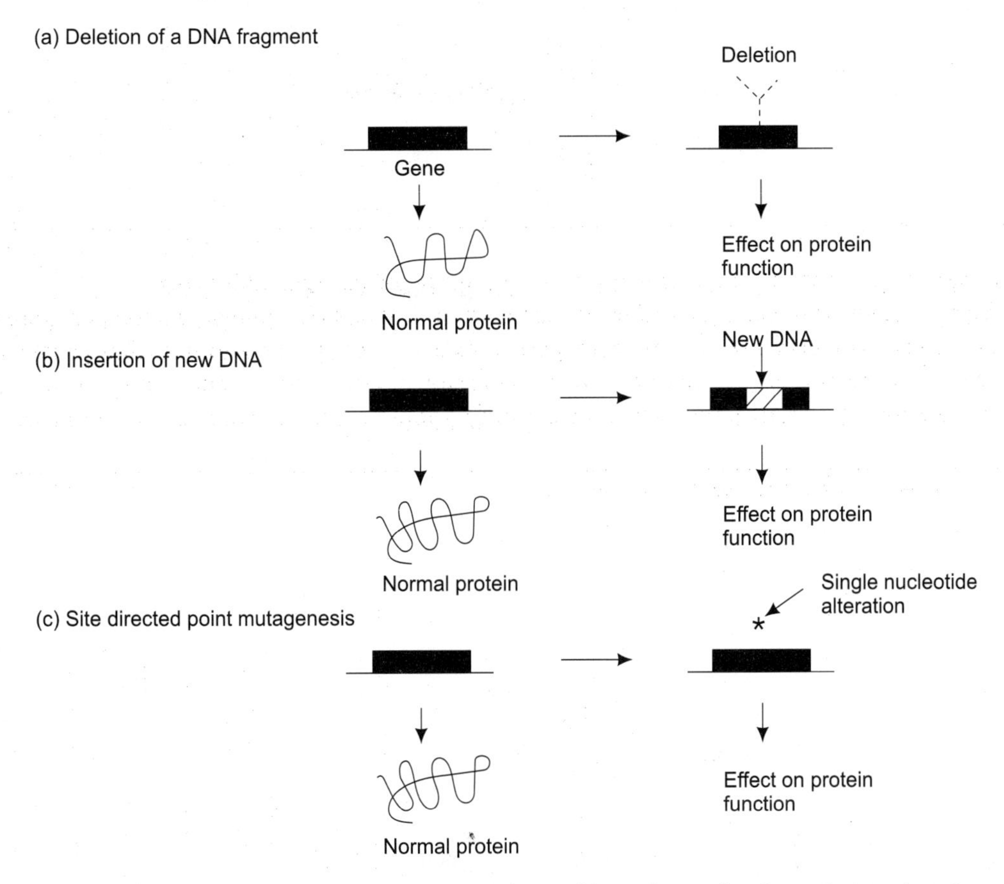

Fig. 23.19 Directed mutagenesis. *In vitro* mutagenesis can be used to make directed alterations in genes. Three techniques are illustrated. In each, the effect of the change on the nucleotide sequence of the gene would be known precisely. The codon, or group of codons, to be altered can be chosen and the exact nature of the alteration (Add new codons, take some away, change just one nucleotide) can be predetermined.

5. Proteins with Industrial Applications

Major tools of molecular biology today are themselves enzymes, viz., a vast array of restriction endonucleases, ligases, polymerases and reverse transcriptases, that are used to manipulate DNA molecules *in vitro*. These enzymes are now commonly produced by cloning the respective genes and expressing these genes on a high-copy number plasmids in bacteria.

6. Production of Drugs

Genetic engineering is having an impact on clinical medicine. The initial focus was on developing organisms that would overproduce antibiotics, thereby reducing production costs, and this has been accomplished for several antibiotics. *Human insulin* was the first commercial success of the new recombinant DNA technologies in the field of pharmaceuticals. Other human proteins that have been synthesised in bacteria and have medicinal roles are *blood-clotting*

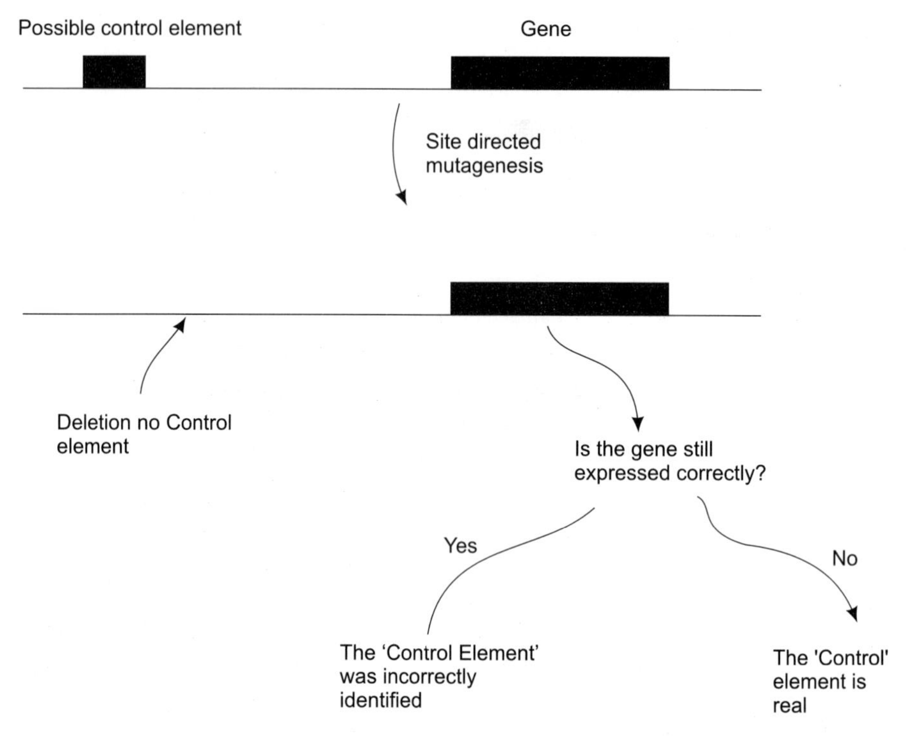

Fig. 23.20 The way in which *in vitro* mutagenesis has been used to identify control elements important in regulating gene expression.

factor VIII (lacking in individuals with one type of haemophilia), *plasminogen activator* (a protein that disperses blood clots), and *human growth hormone* (HGH) (a protein deficient in certain types of dwarfism); *α-interferon,* an antiviral agent which reduces the duration of viral infection. Synthetic vaccines are also produced by genetic engineering such as *anti-hepatitis B*, *influenza virus* and *vesicular stomatitis virus.*

7. Genetic Engineering of Plants

Altering the genotypes of plants is an important application of recombinant DNA technology. The bacterium *Agrobacterium tumefaciens* and its plasmid *Ti*, produces crown gall tumours in plants. These tumours result from integration of the plasmid DNA into the plant chromosome. It is possible by genetic engineering to introduce genes from one plant into this plasmid and then, by infecting a second plant with the bacterium, transfer the genes of the first plant to the second plant (actually genes are first cloned in *E. coli* plasmid and then recloned in *Ti*). Attempts are being made to perform plant breeding in this way. An example is the attempted alteration of the surface structure of the roots of wheat, by introducing certain genes from legumes (peas, beans, etc.) in order to give wheat plant the ability of the legumes to establish root nodules of nitrogen-fixing bacteria. If successful, this would eliminate the need for the addition of nitrogenous fertilisers to cereal growing soils.

The first engineered recombinant plant of commercial value was developed in 1985. An economically important herbicide (weed killer) is GLYPHOSATE, which inhibits a particular essential enzyme in many plants. The target gene of glyphosate is also present in the bacterium S. *typhimurium.* A resistant form of the gene was obtained by mutagenesis and *Salmonella* grows in the presence of glyphosate; the gene was cloned in *E. coli* and then recloned in *Agrobacterium.* Infection of plants with purified *Ti* containing glyphosate-resistance gene has yielded varieties of maize, cotton and tobacco that are resistant to glyphosate. Thus, fields of these crops can be sprayed with glyphosate at any stage of growth of the crop.

Agrobacterium tumefaciens

The most promising vector system for genetic engineering in plants nowadays is based on the *Ti*-plasmid from *Agrobacterium tumefaciens*, a soil borne, gram-negative rod form. It has a broad host range and causes crown gall on plants of more than 90 different families.

Once tumor formation has been induced, agrobacteria are no longer required for tumor proliferation, as the tumor inducing principle TIP factor is transferred from the bacteria to the host cell. This transfer gives rise to a stable transformation of the plant cells to crown gall tumor cells in a time period of about 36 hours. This TIP factor is the *Ti-plasmid* with a molecular weight in the range of 120-160 megadalton (md). DNA-DNA hybridisation techniques have demonstrated that during infection the Ti-plasmid is somehow transferred to the plant cells, where a small part of it, approximately, one-tenth, is integrated in the nuclear DNA of the transformed plant cells. This fragment of the Ti- plasmid which becomes integrated is known as the *T-DNA*.

As a general rule plant tumours resulting from infection with *Agrobacterium* synthesise a variety of unusual amino acid derivatives, called *Opines*. The genes encoding for the synthesis of opines are lying on the T-DNA fragment and are not expressed by the bacteria themselves. The Ti-plasmid allows the host bacteria to utilise the opines they induce as specific energy, nitrogen and carbon, sources. One type of Ti-plasmid encodes *Agropine*, the other *nopaline* and the third *octopine* (Fig 23.21).

The T-DNA Fragment

The mechanism by which the T-DNA fragment is transferred is not known: *Agrobacterium* itself is not thought to enter the plant cell, therefore, the entire Ti-plasmid or the T-DNA alone must be transported across the bacterial and plant cell membranes. Since T-DNA has been found inserted at several different locations in the plant genome, there are apparently no specific DNA borders (has a *cis*—acting function on the T-DNA sequence in the plant DNA) which serve as insertion sites. It was, however, found that a 25-base pair direct repeat sequence close to the T-DNA transfer from, nopaline Ti-plasmids.

Ti-Plasmids as Vectors for the Transformation of Plants

Before Ti-plasmids can be used as vectors for the transformation of plants, a number of problems inherent to the Ti-plasmid have to be overcome: 1. The size of the Ti-plasmids is so

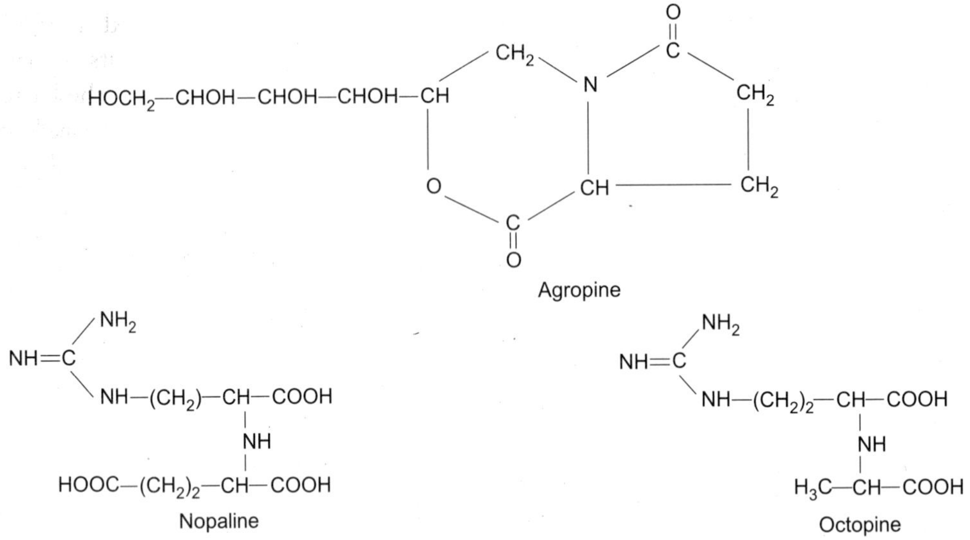

Agropine

Nopaline

Octopine

Fig. 23.21 The structures of the various opines synthesised in plant cells transformed by *Agrobacterium*.

big that it is impossible to find unique restriction sites within the T-region, in which foreign DNA fragments can be cloned (use intermediate vectors). 2. The T-DNA coding for plant hormone production makes it impossible to regenerate intact plants from transformed plant cells; since plant improvement will be the ultimate goal of most experiments in plant engineering, plant cells must be able to regenerate to intact plants (construct disarmed T-plasmids). 3. The proper expression of the genes cloned into the T-DNA fragment has to be assured by introducing promoters expressed in plants, linked to these genes.

Using various types of intermediate vectors, a number of foreign genes from bacteria, animals and plants have been introduced into tobacco and cauliflower plants.

Gene Therapy

Curing certain human diseases like diabetes, by DNA cloning is called *gene therapy.* Retroviruses are also being tested as vectors that might be used to alter the genotype of the animals. These viruses insert viral DNA into the chromosome of the infected cells. If the virus contained a cloned gene, that gene might be expressed in the cell. However, the use of retroviruses is not simple because many contain a gene that converts the recipient cell into a cancer cell. Thus, retroviral vectors usually have this gene removed, which also provides the space for a cloned gene.

In gene therapy some preliminary experiments have been done in human cells deficient in the synthesis of purines with Lesch–Nyhan syndrome and grown in culture. These cells have been converted to normal cells by transformation with recombinant retroviruses. Certain genetic defects can be corrected by gene therapy like for example in restoring the ability of a diabetic individual to make insulin or correcting immunological deficiencies. But retroviruses are not well understood and hence are potentially dangerous. Other difficulties in gene therapy are: there is no reliable way to ensure that a gene is inserted in the appropriate target cell or target tissue.

Genetic engineering has grown in leaps and bounds since 1980s and more than 600 biotechnology companies have been formed and numerous scientific journals and newspapers exclusively devoted to genetic engineering have been created. With legal consequences of developing new products, patent lawyers are even being required to take courses in biology. *Biotechnology is not new.* Cheese, bread and Yoghurt are products of biotechnology and have been known for centuries. However, the stock market excitement about biotechnology stems from the potential of gene manipulation, and its birth can be traced to the founding of the company Genetech. In 1976, Robert Swanson a capitalist and Herb Boyer a Californian Professor founded Genetech (Genetic Engineering Technology). Now in USA alone there are over 600 biotechnology companies since the founding of Genetech.

Genetech produced in 1977 first human protein (SOMATOSTATIN) in a microorganism; in 1978 Human insulin; in 1979 Human Growth Hormone; in 1982 first recombinant DNA drug (human insulin) marketed; in 1984 first laboratory production of Factor VII for therapy of Haemophilia; in 1985 Prototropin (human growth hormone) in children; in 1987 Activase (tissue plasminogen activator) for dissolving blood clots in heart attack patients; in 1990 Actimmune for treatment of chronic granulomatous disease; in 1990 Genetech and the Swiss Pharmaceutical company Roche completed a $ 2.1 billion merger.

RECENT ADVANCES IN PLANT GENETIC ENGINEERING (SUMMARY)

Plant Genetic Engineering

Restriction fragment length polymorphism (RFLP), random amplified polymorphic DNA (RAPD) and DNA fingerprinting techniques are widely used in gene mapping and identification of agronomically important crops. RFLPs have especially increased the efficiency of mapping the *quantitative trait loci (QTLs).* New technologies, such as yeast artificial chromosome ribozymes and gene tagging, are promising in gene cloning. *Polymerase Chain Reaction (PCR)* has become a widely applicable tool in biology, agriculture and medicine and may be replaced with a new method called *Ligase Chain Reaction (LCR).*

Gene Identification and Mapping

Restriction Fragment Length Polymorphism (RFLP)

The use of cloned fragment of chromosomal DNA as genetic markers is usually termed RFLP. This technique is dependent on natural variation in DNA base sequence and the digestion of DNA with restriction enzyme. Homologous restriction fragments of DNA which differ in size, or length, can be used as genetic markers to follow chromosome segments through genetic crosses. By this technique, RFLP linkage maps for many major crop plants such as tomato, potato, rye, maize, lettuce, wheat and rice have been useful to plant breeders.

Random Amplified Polymorphic DNA (RAPD)

The technical complexity of RFLP analysis, coupled with the widespread use of shortlived radioisotopes in the detection method, questions the suitability of routine use of RFLPs in large-scale crop improvement. Williams *et al.* described a new DNA polymorphism assay based on the polymerase chain reaction (PCR), amplified random DNA RAPD method is technically simple, quick to perform, requires small amounts of DNA and involves no radioactivity, and is well suited for use in large sample-throughout systems required for plant breeding, population genetics and studies of biodiversity.

DNA Fingerprint Analysis

In genomic DNA of various species, minisatellite and microsatellite DNA sequences have been found to detect several loci consisting of tandem repeats of short nucleotide sequences (10–60 base pairs). Probes that hybridise to fragments from several variable loci simultaneously have been found, producing an autoradiograph with a complex fragment pattern, called a DNA fingerprint. These fragments are inherited in a Mendelian fashion, and they therefore, provide a technique suitable for genetic analysis. DNA fingerprints are highly individual specific and are applicable in: genetic variation studies, forensic and ecological studies, breeding programmes, and population genetics, and in the analysis and characterisation of a plant genome, and cultivar identification.

Transposable Elements

Transposable elements (TE) were discovered by Barbara McClintock during the later 1940's and early 1950s in maize. Since then, genetic and molecular genetic understanding of these elements has provided information on how they operate within the genome of higher plants to affect the expression of a particular gene.

Transposable element is a DNA sequence that has the capability of moving from place to place in the genome through a process of excision and reintegration. If the site of integration is in or around a particular gene, the expression of that gene can be altered in a variety of ways depending on the proximity of the element to the regulatory or coding regions of the gene. A transposable element must first be cloned and characterised at the molecular level before it can be used in gene tagging experiments. The Ac-Ds, En/spm, and Mul systems have been cloned and, therefore, are amenable to gene tagging.

Transgenic Plants with Cloned Genes

Until recently, plant genes have been cloned, characterised and transferred into plant cells. These include: isopentyltransferase, fungal protection, nematode resistance, phaseolin gene, yeast ornithine decarboxylase, chalcone synthase, male sterility and cold resistance.

Transgenic Plants as 'Bioreactors'

Transgenic plants can be used as 'bioreactors' for obtaining virtually unlimited quantities of commercially useful proteins, biologically active peptides and antibodies for the large-scale production. This is mainly due to the fact that the techniques have been perfected to generate transgenic plants to the point where a foreign protein can be targeted to an organ of choice as well as to subcellular compartments. Hiatt et al. produced antibodies in tobacco transgenic plants by using cDNA derived from a mouse hybridoma mRNA.

Insect Resistance

Baculoviruses and *Bacillus thuringiensis* (Bt) provide an alternative to chemicals for controlling insect pests of agronomically important plant species. They can also be applied by spraying. Hammock et al. accomplished the production of genetically engineered baculovirus containing juvenile hormone esterase (JHE) gene which resulted in reducing insect-feeding habit.

Virus Resistance

Of all plant pathogens, viruses are most intimately associated with their hosts, their genomes are relatively small and well-characterised. Beachy *et al.* reported the engineering of resistance against tobacco mosaic virus (TMV) in tobacco by expression of TMV coat protein gene.

Herbicide Resistance

Herbicide resistance can be achieved with several different means such as (1) reduced herbicide uptake; (2) overproduction of the herbicide target site; (3) metabolism, modification, or conjugation of the herbicide; (4) mutational alterations in the herbicide; (5) tissue culture; (6) gene amplification; (7) mutation; and (8) genetic engineering. Transgenic plants have been obtained conferring herbicide resistance such as glyphosinate, gluphosinate and bromoxynil.

CONCLUSION

Genetic engineering can lead to 'molecular farming' by producing novel compounds in transgenic plants, which can be developed for use as 'bioreactors' for bulk production of commercial grade products with high commercial and economic potential. The biological control of insect pests and fungi would result in reducing the use of chemical pesticides and fungicides.

CHAPTER 24

Genomics

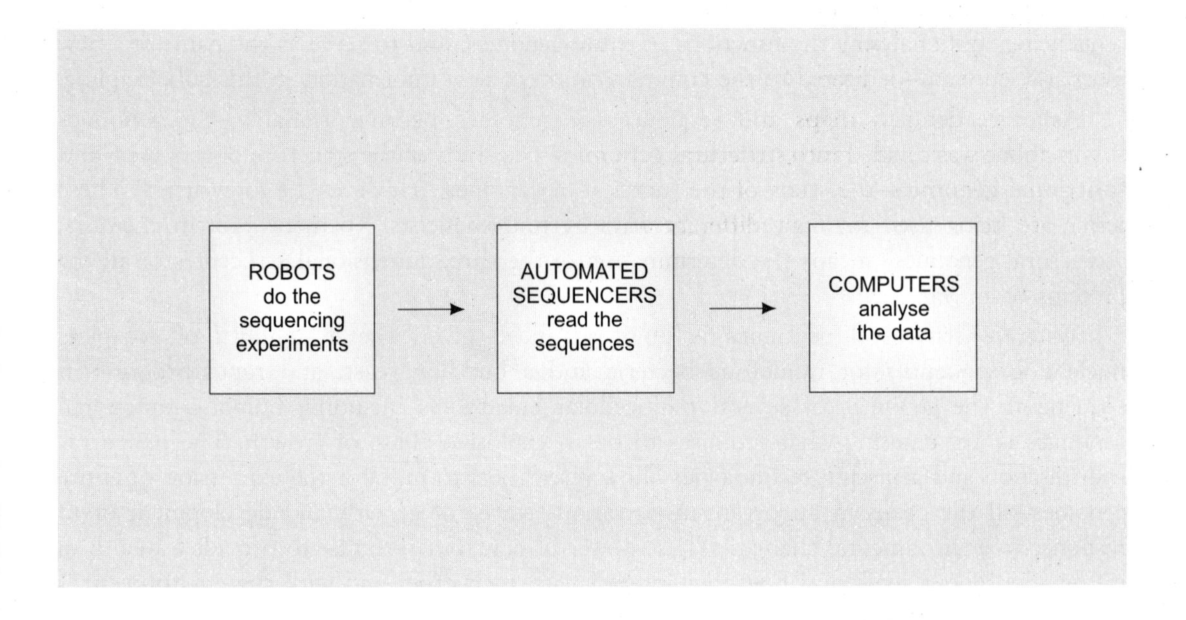

SECTION A

As of December 1998, complete nucleotide sequences are available for the genomes of eighteen species of bacteria, mitochondria from several species, chloroplasts from four species, and one eukaryote, the yeast *Saccharomyces cerevisiae*. The nucleotide sequence of the genome of the worm *Caenorhabditis elegans* is 99.9 per cent complete, and the genomic sequences of the fruit-fly *Drosophila melanogaster* and the plant *Arabidopsis thaliana* will soon follow. One of the original goals of the Human Genome Project was to determine the complete nucleotide sequence of the human genome by the year 2005. Now, the co-founders of a new company claim that they will sequence the human genome by the year 2001. Once the complete sequence of the genome of an organism is known, the new "DNA chip" technologies will allow scientists to study the expression of all the genes in the organism simultaneously.

In this chapter, we will examine some of the tools and techniques that are used to study the structure and function of genomes, especially the human genome, and we will look at what has been learned to date by the analysis of complete genomes.

GENOMICS: AN OVERVIEW

Geneticists have used the term *genome* for over seven decades to refer to one complete copy of the genetic information or one complete set of chromosomes (monoploid or haploid) of an organism. In contrast, genomics is a relatively new term. The word genomics appears to have been coined by Thomas Roderick in 1986 to refer to the genetics subdiscipline of mapping, sequencing, and analysing the functions of entire genomes, and to serve as the name of a new journal—Genomics—dedicated to the communication of new information in this subdiscipline.

As more detailed maps and sequences of genomes became available, the genomics subdiscipline was divided into **structural genomics**—the study of the structure of genomes—and **functional genomics**—the study of the function of genomes. (However, be forewarned! These terms are being used in many different ways by nongeneticists. To many protein chemists, "structural genomics" means the determination of the three-dimensional structures of all the proteins of an organism.)

Whereas structural genomics is quite advanced—given the availability of complete nucleotide sequences for numerous becteria and a budding yeast, and rapid progress in sequencing the genomes of several multicellular eukaryotes including humans—functional genomics is just entering what promises to be an explosive phase of growth. The new array hybridisation and gene-chip technologies allow researchers to monitor the expression of entire genomes—all the genes in an organism—at various stages of growth and development or in response to environmental changes. These powerful new tools promise to provide a wealth of information about genes and how they interact with each other and with the environment.

Key Point Genomics is the subdiscipline of genetics devoted to the mapping, sequencing, and functional analysis of genomes.

CORRELATED GENETIC, CYTOLOGICAL, AND PHYSICAL MAPS OF CHROMOSOMES

The ability of scientists to identify and isolate genes based on information about their location in the genome was one of the first major contributions of genomics research. In principle, this approach, called *positional cloning*, can be used to identify and clone any gene with a known phenotypic effect in any species. Positional cloning has been used extensively in many species, including humans.

Because the utility of positional cloning depends on the availability of detailed maps of the regions of the chromosomes where the genes of interest reside, major efforts have focused on the development of detailed maps of the human genome and the genomes of important model organisms such as *D. melanogaster*, *C. elegans*, and *A. thaliana*. The goal of this research is to construct correlated genetic and physical maps with markers distributed at relatively short intervals throughout the genome. In the case of the human and *Drosophila* genomes, the genetic and physical maps can also be correlated with cytological maps (banding patterns) of the chromosomes, we will discuss the construction of these maps in the following sections of this chapter.

Genetic maps (Fig. 24.1) are constructed from recombination frequencies, with 1 centi-Morgan (cM) equal to the distance that yields an average frequency of recombination of 1 per cent. Genetic maps with markers spaced at short intervals—high-density genetic maps—are often constructed by using molecular markers such as restriction fragments of different lengths (restriction fragment-length polymorphisms or RFLPs). Cytological maps are based on the banding patterns of chromosomes observed with the microscope after treatment with various stains. **Physical maps** (Fig. 24.1), such as the restriction maps Fig. 24.1, are based on the molecular distances—base pairs (bp), kilobases (kb, 1000 bp), and megabases (mb, 1 million bp)—separating sites on the giant DNA molecules present in chromosomes. Physical maps often contain the locations of overlapping genomic clones or *contigs* and unique nucleotide sequences called *sequence-tagged sites* or *STSs* (Fig. 24.1).

Physical maps of a chromosome can be correlated with the genetic and cytological maps in several ways. Genes that have been cloned can be positioned on the cytological map by *in situ* hybridisation. Correlations between the genetic and physical maps can be established by locating clones of genetically mapped genes or RFLPs on the physical map. Markers that are mapped both genetically and physically are called anchor markers; they anchor the physical map to the genetic map and vice versa. Physical maps of chromosomes can also be correlated with genetic and cytological maps by using (1) PCR to amplify short—usually 200 to 500 bp—unique genomic DNA sequences, (2) Southern blots to relate these sequences to overlapping clones on physical maps, and (3) *in situ* hybridisation to determine their chromosomal locations (cytological map positions). These short, unique anchor sequences are called **sequence-tagged sites (STSs).** Another approach uses short cDNA sequences (DNA copies of mRNAs) or **expressed-sequence tags (ESTs)** as hybridisation probes to anchor physical maps to RFLP maps (genetic maps) and cytological maps.

Physical distances do not correlate directly with genetic map distances because recombination frequencies are not always proportional to molecular distances. However, the two are often reasonably well correlated in euchromatic regions of chromosomes. In humans, 1 cM is equivalent, on average, to about 1 mb of DNA.

Restriction Fragment-length Polymorphism (RFLP) and Microsatellite Maps

When mutations change the nucleotide sequences in restriction enzyme cleavage sites, the enzymes no longer recognise them (Fig. 24.2 (a)). Other mutations may create new restriction sites. These mutations result in variations in the lengths of the DNA fragment produced by digestion with various restriction enzymes. Such restriction fragment-length polymorphisms, or RFLP, have proven invaluable in constructing detailed genetic maps for use in positional

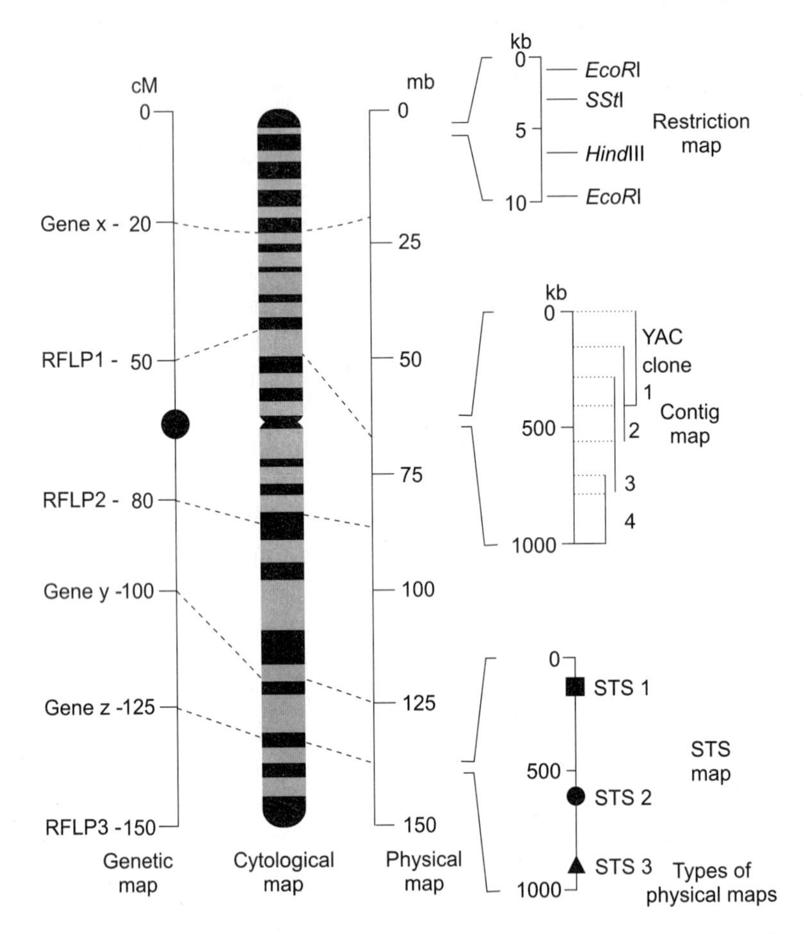

Fig. 24.1 Correlation of the genetic, cytological and physical maps of a chromosome. Genetic map distances are based on corossing over frequencies and are measured in percentage recombination or centiMorgans (cM), whereas physical distances are measured in kilobase pairs (kb) or megabase pairs (mb). Restriction maps, contig maps, and STS (sequence-tagged site) maps are described in the text.

cloning. The RFLP are mapped just like other genetic markers; they segregate in crosses as though they are codominant alleles.

The DNAs of different geographical isolates, different ecotypes (strains adapted to different environmental conditions), and different inbred lines of a species contain many RFLPs that can be used to construct detailed genetic maps. Indeed, the DNAs of different individuals–even relatives–often exhibit RFLPs. some RFLPs can be visualised directly when the fragments in DNA digests are separated by agarose gel electrophoresis, stained with ethidium bromide, and viewed under ultraviolet light. Other RFLPs can be detected only by using specific cDNA or genomic clones as radioactive hybridisation probes on genomic Southern blots (Fig. 24.2). The RFLPs themselves are the phenotypes used to classify the progeny of crosses as parental or recombinant. Fig. 24.3 illustrates the use of RFLPs as genetic markers in mapping experiments

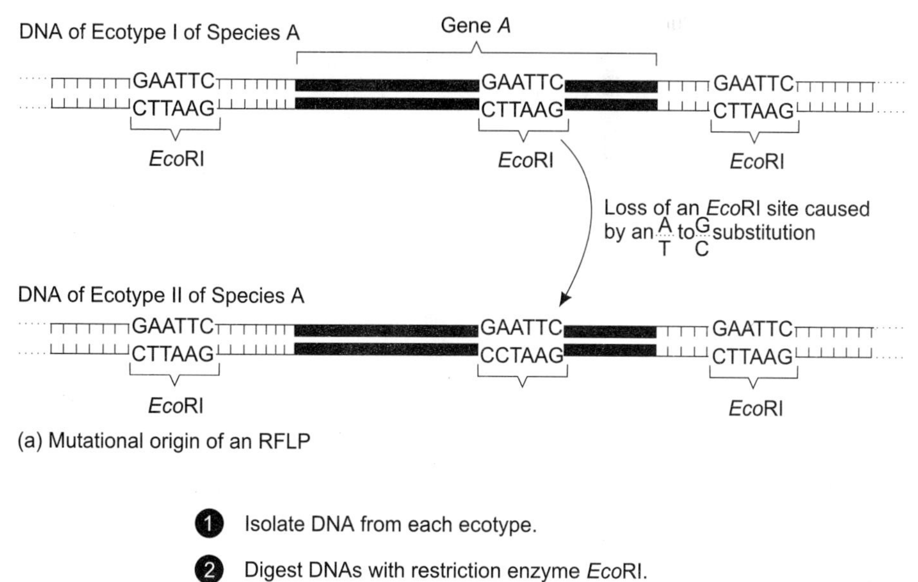

1. Isolate DNA from each ecotype.
2. Digest DNAs with restriction enzyme *Eco*RI.
3. Separate DNA restriction fragments by agarose gel electrophoresis.
4. Transfer DNA restriction fragments to nylon membrane.
5. Hybridise DNA fragments on Southern blot to radioactive gene *A* clone
6. Wash blot and expose it to X-ray film to produce autoradiogram.

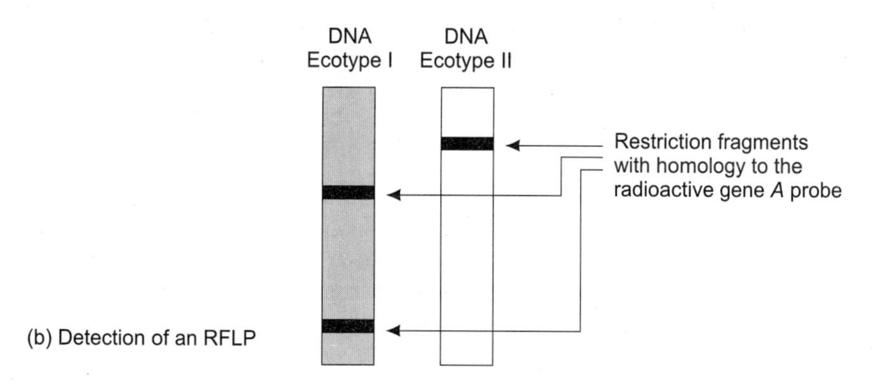

Fig. 24.2 The mutational origin (a) and detection (b) of RFLPs in different ecotypes of a species. In the example shown an A:T ® G:C base-pair substitution results in the loss of the central *Eco*RI recognition sequence present in gene A of the DNA of ecotype I. This mutation could have occurred in a common progenitor of the two ecotypes or in an ecotype II during the early stages of its divergence from ecotype I. In ecotype I, gene A sequences are present on two *Eco*RI restriction fragments, whereas in ecotype II, all gene A sequences are present on one large *Eco*RI restriction fragment.

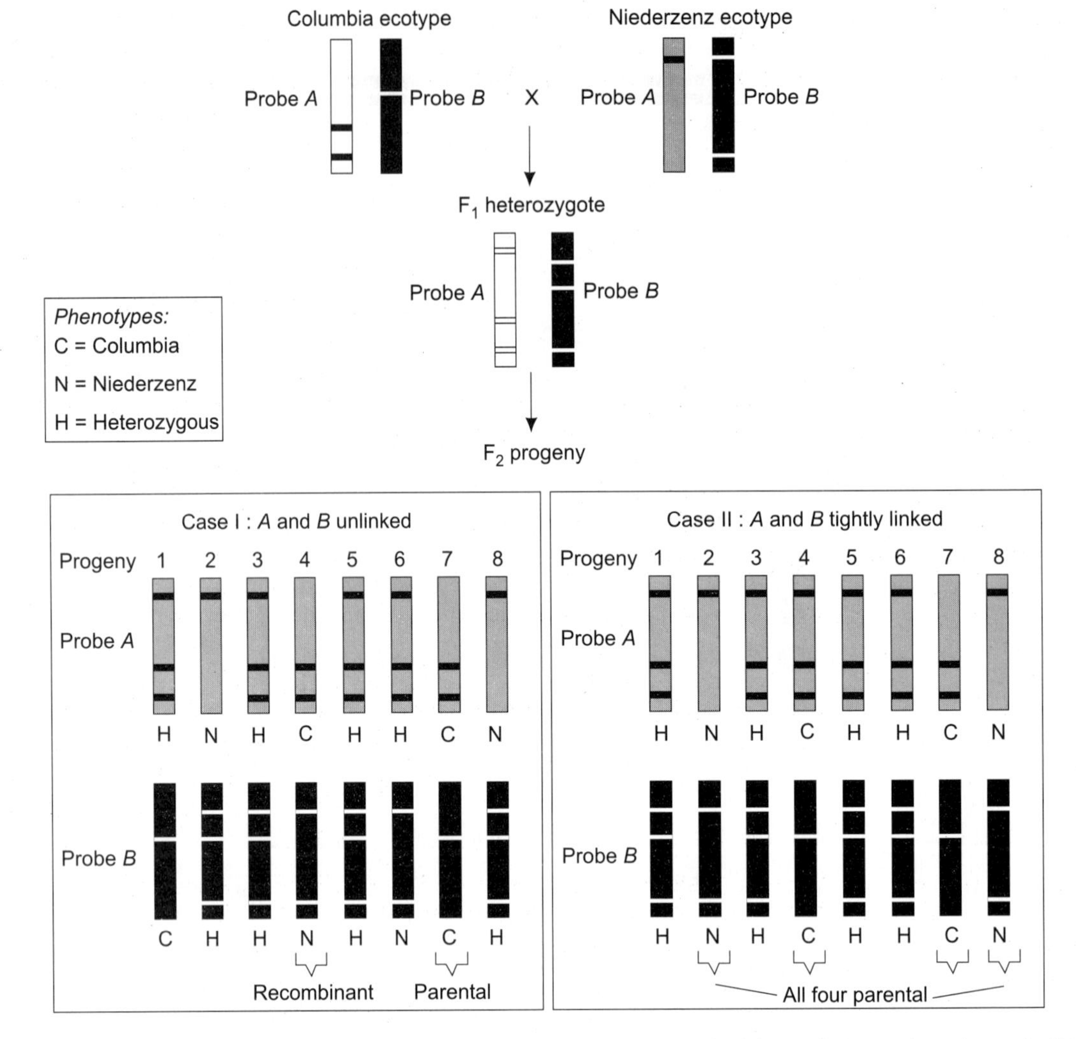

Fig. 24.3 Mapping RFLPs in *Arabidopsis.* All genomic DNAs are digested with restriction endonuclease *Eco*RI. The DNA fragments in all lanes (rectangular boxes) of the Southern blots are hybridised sequentially to probe *A* (blue) and probe *B* (green). The F_1 progeney are heterozygous, containing the restriction fragments of both parents as expected. When such F_1 plants are self-pollinated, a 1:2:1 (C homozygote:heterozygous:N homozygote) segregation ratio is expected for each RFLP. If the two RFLPs assort independently (*left*), parental and recombinant combinations should occur with equal frequency among the F_2 progeny. If they are tightly linked (*right*), recombinants should be rare. The frequency of recombinants can be used to calculate the map distance between RFLPs, just as for other genetic markers.

in the model plant *Arabidopsis*. An RFLP map of chromosome 1 of Arabidopsis is shown in Fig. 24.4. Because its genome is small (about 10^8 nucleotide pairs), *Arabidopsis* is an excellent system in which to clone genes based on their locations.

RFLP markers have proven especially valuable in mapping the chromosomes of humans, where researchers must rely on the segregation of spontaneously occurring mutant alleles in families to estimate map distances. Pedigree-based mapping of this type is done by comparing the probabilities that the genetic markers segregating in the pedigree are unlinked or linked by various map distances. In 1992, geneticists used this procedure to construct an early map of about 2000 RFLPs on the 24 human chromosomes. Fig. 24.5 shows an RFLP map of human chromosome 1 and its correlation with the cytological map of this chromosome.

In humans, the most useful RFLPs involve short sequences that are present as tandem repeats. The number of copies of each sequences present at a given site on a chromosome is highly variable. These sites, called **variable number tandem repeats,** or **VNTRs,** are therefore highly polymorphic. VNTRs vary in fragment length not because of differences in the positions of restriction enzyme cleavage sites, but because of differences in the number of copies of the repeated sequence between the restriction sites.

Microsatellites are another class of polymorphisms that have proven extremely valuable in constructing high-density maps of eukaryotic chromosomes. Microsatellites are polymorphic tandem repeats of sequences only two to five nucleotide pairs long. They are called microsatellites because they are a subset of the satellite sequences present in the highly repetitive DNA of eukaryotes. In humans, microsatellite sequences composed of polymorphic tandem repeats of the dinucleotide sequence AC/TG (AC in one strand; TG in the complementary strand) provide especially useful markers. In 1996, a group of French and Canadian researchers published a comprehensive map of 5264 AC/TG microsatellites in the human genome. These microsatellites defined 2335 sites with an average distance of 1.6 cM or about 1.6 mb between adjacent markers.

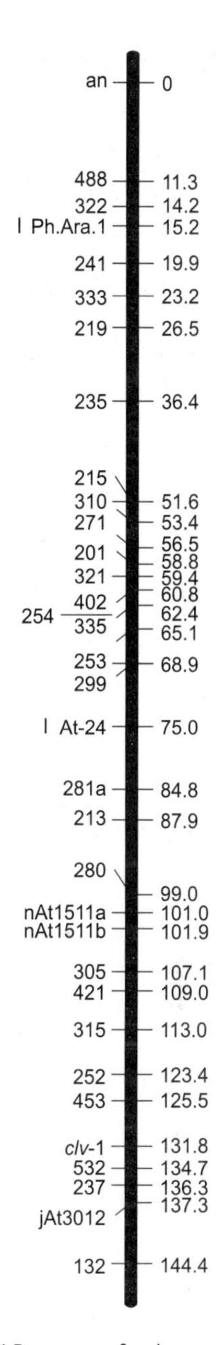

Fig. 24.4 RFLP map of chromosome 2 in *Arabidopsis*. Clone designations are shown on the left; map positions are given in centiMorgans on the right, with the uppermost marker arbitrarily assigned position 0. Clone nAT1511 hybridised to sequences at two sites, designated a and b on the map.

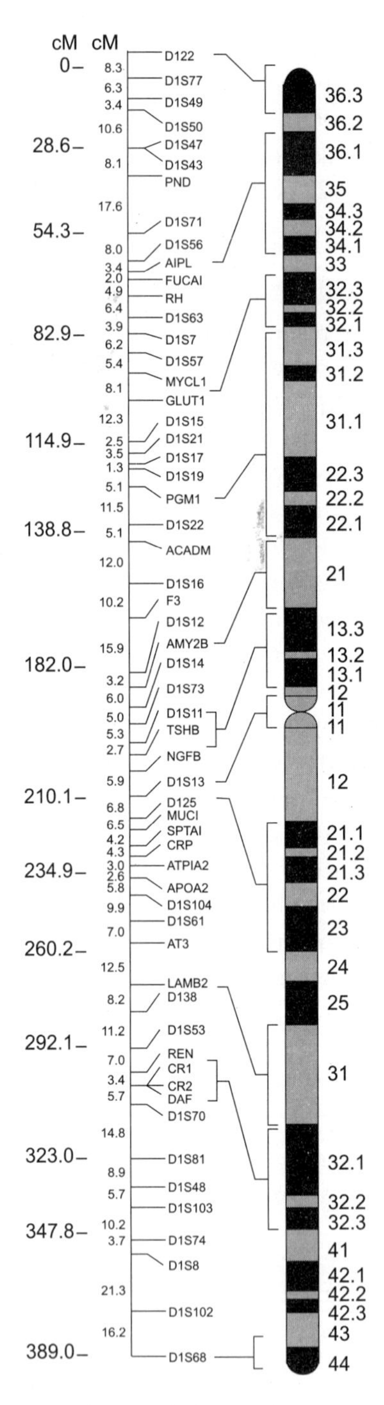

Fig. 24.5 Correlation of the RFLP map (left) and the cytological map (right) of human chromosome 1. Distances are in centiMorgans (cM), with the uppermost marker set at position 0 on the left and distances between adjacent markers shown in the center.

By 1997, a large international consortium had used RFLPs to map over 16,000 human genes (ESTs and cloned genes) and had integrated their map with the physical map of the human genome. In this collaborative study, over 20,000 STSs were mapped–many by researchers in two or more laboratories–to 16,354 distinct loci. These genetic maps composed primarily of RFLP markers have made it possible to identify and characterise mutant genes that are responsible for several human diseases.

Cytogenetic Maps

In some species, genes and clones can be positioned on the cytological maps of the chromosomes by *in situ* hybridisation. For example, in *Drosophila*, the banding patterns of the giant polytene chromosomes in the salivary glands provide high-resolution maps of the chromosomes. Thus, a clone of unknown genetic content can be positioned on the cytological map with considerable precision. In mammals, including humans, fluorescent *in situ* hybridisation (FISH; can be used to position clones on chromosomes stained by any of several chromosome banding protocols and the use of FISH to position clones on the cytological map of human chromosome 11. If RFLPs can be identified that overlap with these clones, they can be used as STS sites that anchor the genetic map of chromosome 11 to the cytological map producing a cytogenetic map. If the clones can be positioned on the physical map by Southern blot hybridisation experiments, they can also be used to tie the physical map to both the genetic and cytological maps of the chromosome. (Fig. 24.5).

Physical Maps and Clone Banks

The RFLP mapping procedure has been used to construct detailed genetic maps of chromosome, which, in turn, have made positional cloning feasible. These genetic maps have been supplemented with physical maps of chromosomes. By isolating and preparing restriction maps of large numbers of genomic clones, overlapping clones can be identified and used to construct physical maps of entire chromosomes and even entire genomes. In principle, this procedure is simple (Fig. 24.6). However, in practice, it is a formidable task, especially for large genomes. The restriction maps of large genomic clones in YAC or BAC vectors are analysed by computer and organised in overlapping sets of clones called contigs. As more data are added, adjacent contigs are joined; when the physical map of a genome is complete, each chromosome is represented by a single contig map.

The construction of physical maps of entire genomes require that vast amounts of data be searched for overlaps. Nevertheless, detailed physical maps are available for several genomes, including the human, *C. elegans*, and *A. thaliana* genomes. These physical maps are being used to prepare clone banks that contain catalogued clones collectively spanning entire chromosomes. Thus, if a researcher needs a clone of a particular gene or segment of a chromosome, that clone may already have been catalogued in the clone bank and be available on request. Obviously, the availability of such clone banks and the correlated physical maps of entire genomes are dramatically accelerating genetic research. Indeed, searching for a book in huge library with and without a computer index giving the locations of the books in the library.

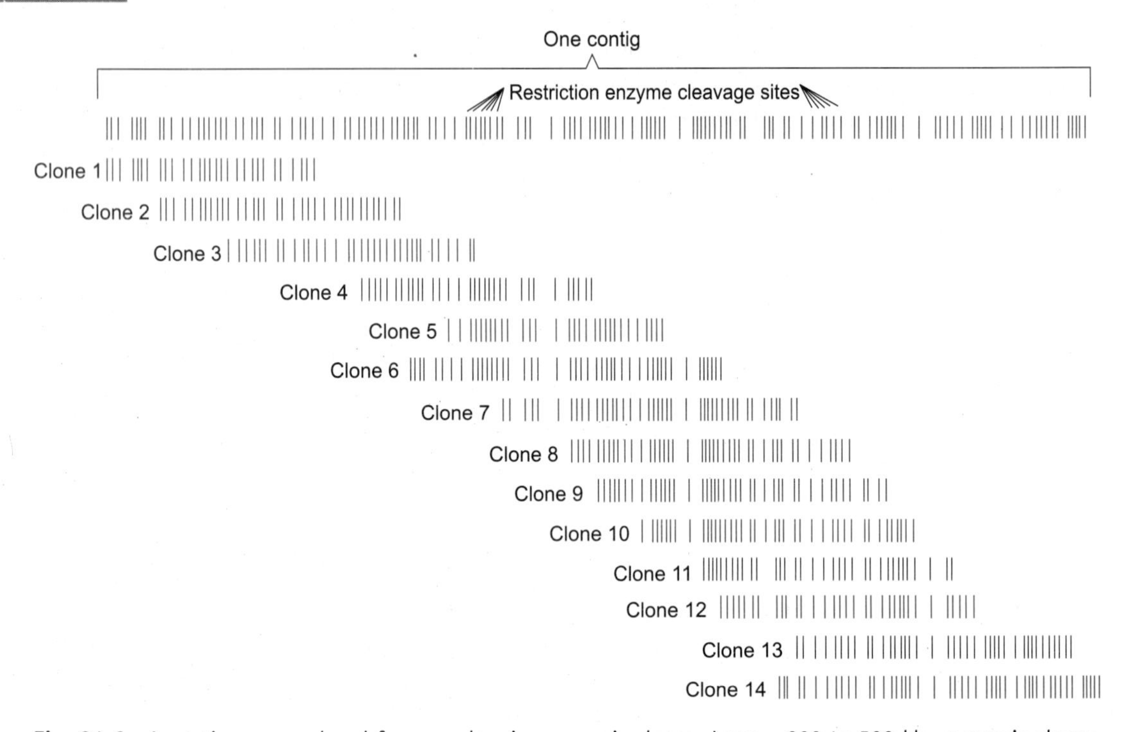

Fig. 24.6 A contig map produced from overlapping genomic clones. Large—200 to 500 kb—genomic clones, such as those present in YAC or BAC vectors (Chapter 20), are used to construct contig maps. Restriction maps of individual clones are prepared and searched by computer for overlaps. Overlapping clones are then organised into contig maps like the one shown here. When the physical map of a genome is complete, each chromosome will be represented by a single contig map.

Key Points Genetic maps of chromosomes are based on recombination frequencies between markers. Cytogenetic maps are based on the location of markers within cytological features such as chromosome banding patterns observed by microscopy. Physical maps of chromosomes are determined by the molecular distances in base pairs, kilobase pairs, or megabase pairs separating markers. High-density maps that integrate the genetic, cytological, and physical maps of chromosomes have been constructed for all of the human chromosomes and many chromosomes of other model genetic organisms.

MAP POSITION-BASED CLONING OF GENES

The first eukaryotic genes to be cloned were genes that are expressed at very high levels in specialised tissues or cells. For example, about 90 per cent of the protein synthesised in mammalian reticulocytes is haemoglobin. Thus, α-and β-globin mRNAs could be easily isolated from reticulocytes and used to prepare radioactive cDNA probes for genomic library screens. However, since most genes are not expressed at such high levels in specialised cells, how are genes that are expressed at moderate or low levels cloned? One important approach is to map the gene precisely and to search for a clone of the gene by using procedures that depend on its location in the genome. This approach, called **positional cloning**, can be used to identify any gene, given an adequate map of the region of the chromosome in which it is located.

The steps in positional cloning are illustrated in Fig. 24.7. The gene is first mapped to a specific region of a given chromosome by genetic crosses or, in the case of human, by pedigree analysis, which usually requires large families. The gene is next localised on the physical map of this region of the chromosome by analysis of the chromosome identified by physical mapping are then isolated from mutant and wild-type organisms or individuals and sequenced to identify mutations that would result in a loss of gene function. The human genes responsible for inherited disorders such as Huntington's disease and cystic fibrosis have been identified by using the positional cloning approach. In species where transformation is possible, copies of the wild-type alleles of candidate genes introduced into mutant organisms to determine whether the wild-type genes will restore the wild phenotype. Restoration of the wild phenotype to a mutant organism provides strong evidence that the introduced wild-type gene is the gene of interest.

Chromosome Walks

Positional cloning is accomplished by mapping the gene of interest, identifying an RFLP or other molecular marker near the gene, and then "walking" or "jumping" along the chromosome until the gene is reached. **Chromosome walking** is very difficult in species with large genomes (the walk is usually too far) and an abundance of dispersed repetitive DNA (each repeated sequence is a potential roadblock). Chromosome walking is easier in organisms such as A. *thaliana* and C. *elegans,* which have small genomes and little-dispersed repetitive DNA.

Chromosome walks are initiated by the selection of a molecular marker (RFLP or known gene clone) close to the gene of interest and the use of this clone as a hybridisation probe to screen a genomic library for overlapping sequences. Restriction maps are constructed for the overlapping clones identified in the library screen, and the restriction fragment farthest from the original probe is used to screen a second genomic library constructed by using a different restriction enzyme or to screen a second genomic library constructed by using a different restriction enzyme or to rescreen a library prepared from a partial digest of genomic DNA. Repeating this procedure several times and isolating a series of overlapping genomic clones allow a researcher to walk the required distance down the chromosome to the gene of interest (Fig. 24.8). Without information about the orientation of the starting clone on the linkage map, the initial walk will have to proceed in both directions until another RFLP is identified and it is determined whether the new RFLP is closer to or farther away from the gene of interest than is the starting RFLP.

Verification that a clone of the gene of interest has been isolated is accomplished in various ways. In experimental organisms such as *Drosophila* and *Arabidopsis*, verification is achieved by introducing the wild-type allele of the gene into a mutant organism and showing that it restores the wild-type phenotype. In Drosophila, this is done by transforming mutant embryos with transposable element vectors that carry the wild-type allele of the gene. In human, verification usually involved determining the nucleotide sequences of the wild-type gene and several mutant

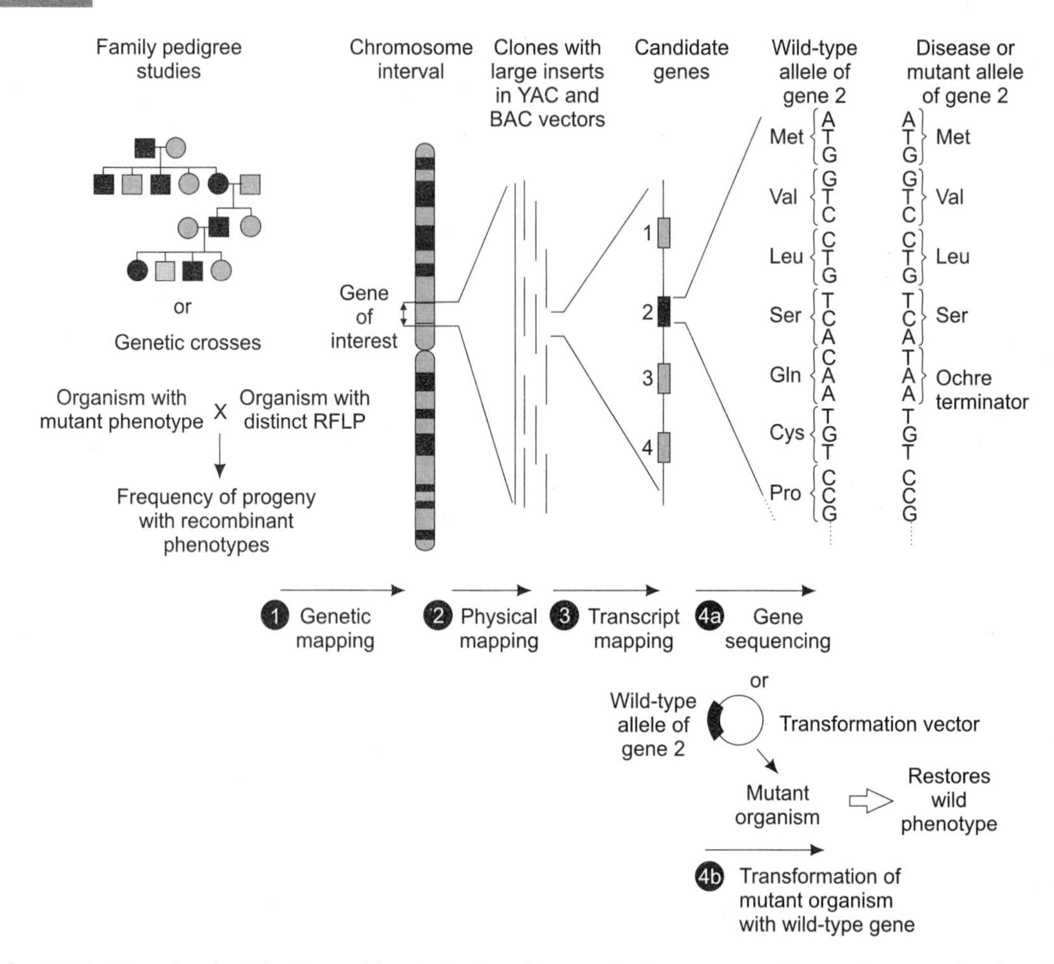

Fig. 24.7 Steps involved in the positional cloning of genes. In humans, genetic mapping must be done by pedigree analysis, and candidate genes must be screened by sequencing mutant and wild-type alleles (step 4a). In other species, the gene of interest is mapped by appropriate genetic crosses, and the candidate genes are screened by transforming the wild-type alleles into mutant organisms and determining whether or not they restore the wild-type phenotype (step 4b).

alleles and showing that the coding sequences of the mutant genes are defective and unable to produce functional gene products.

Chromosome Jumps

When the distance from the closest molecular marker to the gene of interest is large, a technique called **chromosome jumping** can be used to speed up an otherwise long walk. Each jump can cover a distance of 100 kb or more. The chromosome jumping procedure is illustrated in Fig. 24.9. Like a walk, a jump is initiated by using a molecular probe such as an RFLP as a starting point. However, with chromosome jumps, large DNA fragments are prepared by partial digestion of genomic DNA with a restriction endonuclease. The large genomic fragments are

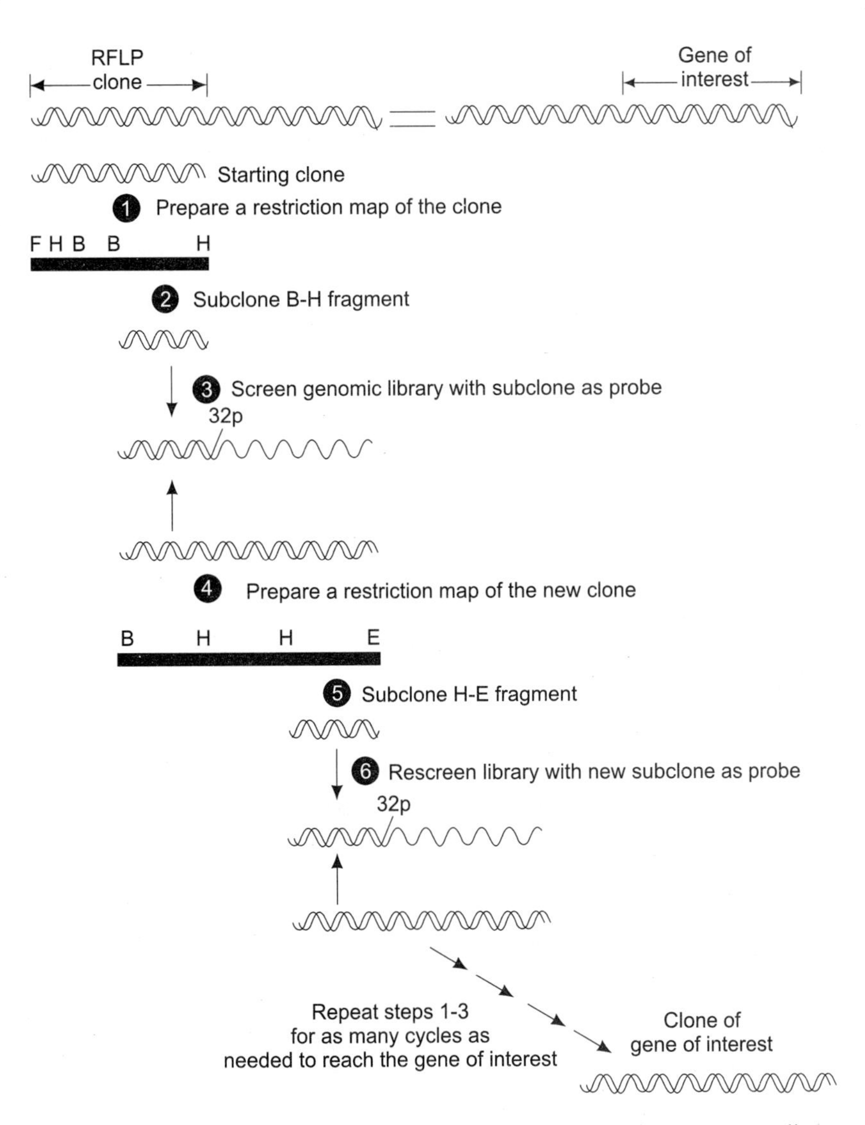

Fig. 24.8 Positional cloning of a gene by chromosome walking. A chromosome walk starts with the identification of a molecular marker—such as the RFLP shown (top)—close to the gene of interest and proceeds by repeating steps 1 through 3 as many times as is required to reach the gene of interest (bottom).

then circularised with DNA ligase. A second restriction endonuclease is used to excise the junction fragment from the circular molecule. This junction fragment will contain both ends of the long fragment; it can be identified by hybridising the DNA fragments on Southern blots to the initial molecular probe. A restriction map of the junction fragment is prepared, and a restriction fragment that corresponds to the distal end of the long genomic fragment is cloned and used to initiate a chromosome walk or a second chromosome jump. Chromosome jumping has proven especially useful in work with large genomes such as the human genome. Chromosome jumps played a key role in identifying the human cystic fibrosis gene.

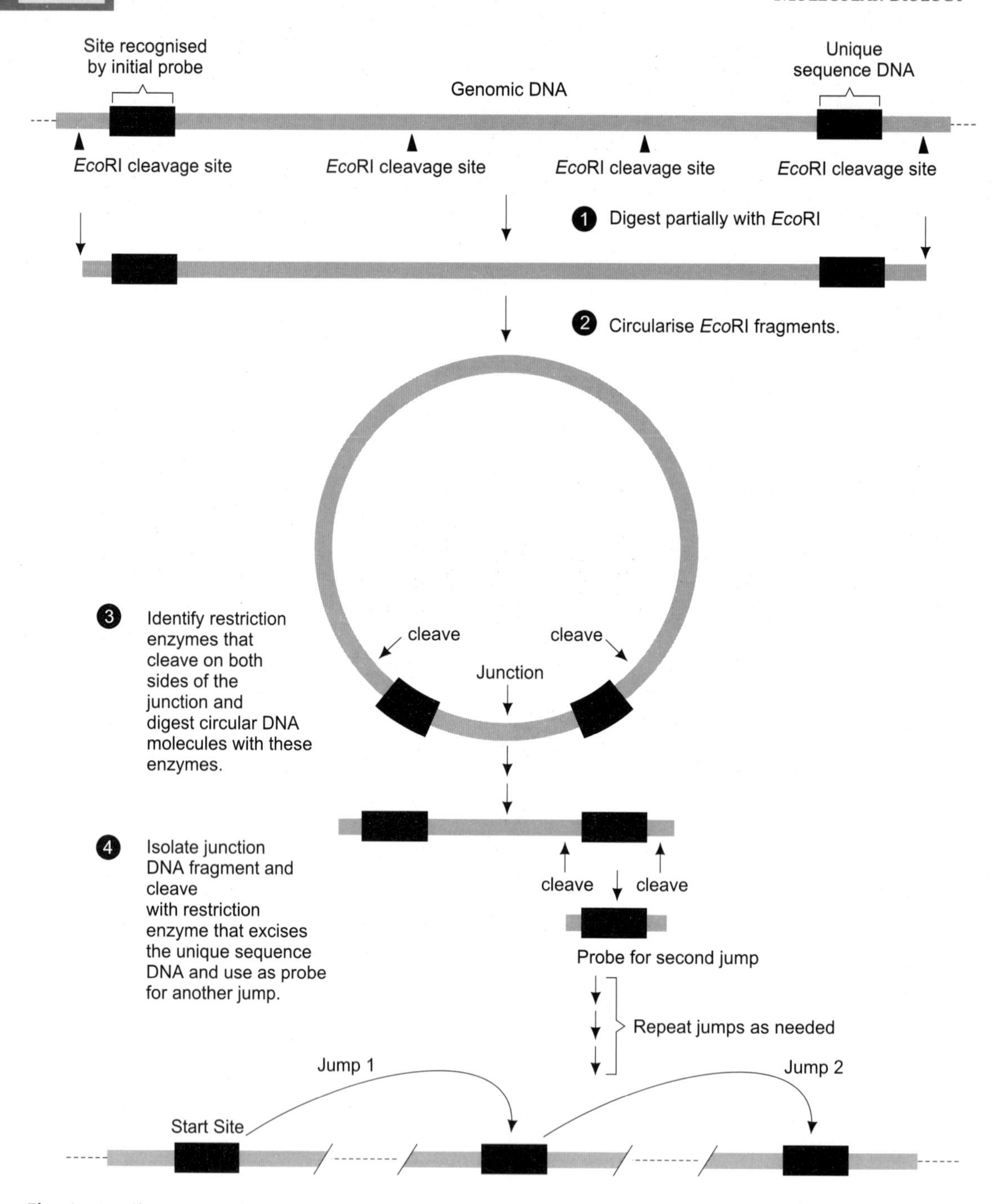

Fig. 24.9 Chromosome jumping as a short cut method for long chromosome walks. This procedure can also be used to jump over repetitive DNA sequences that block chromosome walks.

Key Points The availability of detailed genetic, cytogenetic, and physical maps of chromosomes permit researchers to clone genes based on their location in the genome. If a molecular marker such as a restriction fragment-length polymorphism (RFLP) maps close to a gene, that gene can usually be cloned by chromosome walking or chromosome jumping.

THE HUMAN GENOME PROJECT

As the recombinant DNA, gene cloning, and DNA sequencing technologies improved in the 1970s and early 1980s, scientists began discussing the possibility of sequencing all 3×10^9 nucleotide pairs in the human genome. These discussions led to the launching of the **Human Genome Project** in 1990. The initial goals of the Human Genome Project were (1) to map all of the 70,000 to 100,000 human genes, (2) to construct a detailed physical map of the entire human genome, and (3) to determine the nucleotide sequences of all 24 human chromosomes by the year 2005. Scientists soon realised that this huge undertaking should be a worldwide efforts. Therefore, an international **Human Genome Organisation (HUGO)** was organised to coordinate the efforts of human geneticists around the world.

James Watson, who with Francis Crick, discovered the double-helix structure of DNA, was the first director of this ambitious project, which was expected to take nearly two decades to complete and to cost in excess of $3 billion. In 1993, Francis Collins, who, with Lap-Chee Tsui, led the research teams that identified the cystic fibrosis gene, replaced Watson as director of the Human Genome Project. In addition to work on the human genome, the Human Genome Project has served as an umbrella for similar mapping and sequencing projects on the genomes of several other organisms, including the bacterium *E. coli*, the yeast *S. cerevisiae*, the fruit-fly *D. melanogaster*, the plant A. thaliana, and the worm *C. elegans*.

Bacterial Genomes

In 1995, *Haemophilus influenzae* became the first bacterium to have its genome sequenced in its entirety (Fig. 24.10). By late 1998, the complete sequences of 18 bacterial genomes were available in the public databases. The genomes range in size from 580,070 bp for *Mycoplasma genitalium*, which is thought to have the smallest genome of any self-replicating organism, to 4,411,529 bp for *Mycobacterium tuberculosis*, which causes more human deaths than any other infectious bacterium, to 4,639,221 bp for *Escherichia coli*, the best-known cellular microorganism. The genome of M. *genitalium* is of special interest because it may approximate the "minimal gene set" for a self-replicating organism–the smallest set of genes that will allow an organism to reproduce itself. Of course, the genome of M. tuberculosis is of great interest because of the pathogenicity of this organism and the hope that a complete understanding of its metabolism will suggest way to prevent tuberculosis in humans. The need for new ways to combat this pathogen has been enhanced by the recent evolution of antibiotic-resistant strains of M. *tuberculosis*.

Of the bacterial genomes sequenced to date, the sequence of the *E. coli* genome has undoubtedly caused the most excitement among geneticists. *E. coli* is the most studied and best understood cellular organism on our planet. Geneticists, biochemists, and molecular biologists

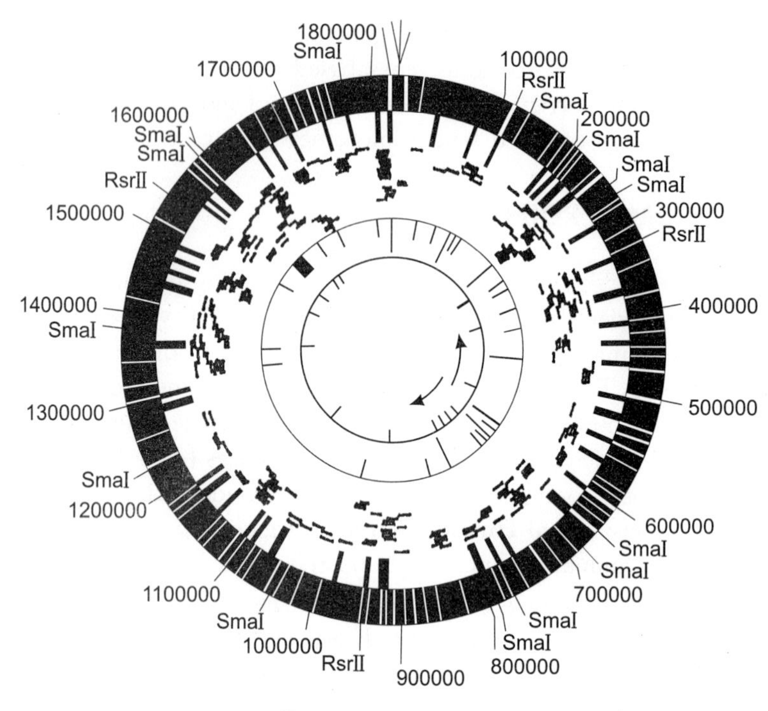

Fig. 24.10 Map of the chromosome of Haemophilus influenzae. This bacterium was the first free-living organism to have its entire genome—all 1,830,137 nucleotide pairs—sequenced. Distances in nucleotide pairs and the locations of NotI, RsrII and SmaI restriction sites are given outside the circular map. The outer concentric circle shows genes colour-coded according to function. For example, genes shown in green, yellow and pink encode products that are involved in energy metabolism, DNA replication and translation respectively.

have utilised *E. coli* as the preferred model organism for decades. Most of what is known about bacterial genetics was learned from research on *E. coli*. Thus, publication of the complete sequence of the *E. coli* genome in 1997 was a significant milestone in the history of genetics.

The *E. coli* genome contains 4288 putative protein-coding sequences or genes. Of these putative genes, about one-third are well-studies genes encoding known products, whereas 38 per cent are of unknown function. Putative protein-coding sequences that are not known to encode proteins are called open reading frames or ORFs. An ORF is nucleotide sequence that begins with a translation-initiation codon (usually AUG), continues with a sequence of triplet codons specifying amino acids, and ends with one of the three translation-termination codons.

The average distance between genes (size of intergenic regions) in the *E. coli* genome is 118 bp. Known and putative genes specifying proteins and stable RNAs make up 87.8 per cent and 0.8 per cent of the genome, respectively, and noncoding repetitive elements account for 0.7 per cent of the genome. Thus, 10.7 per cent of the genome must involve regulatory sequences, sequences with other unknown functions, and perhaps some nonfunctional nucleotide sequences. Six new tRNA genes were discovered during the analysis of the sequence, and similarity comparisons with sequenced regions of other genomes revealed two sets of genes encoding enzymes involved in the degradation of aromatic organic compounds and 12 new genes involved in the synthesis and assembly of flagella.

Once the complete sequence of a bacterial genome is available, computer searches can be performed to identify sequence similarities with other sequenced genomes. Such sequence comparisons can often be used to gain inferences about gene function. Because so much is known about gene function in *E. coli*, comparison of other sequented genomes with the E. coli genome are often very informative. For example, a comparison of the genomes of *Treponema pallidum*, the parasitic spirochete that causes syphilis, and *E. coli*, shows that *T. pallidum* contain the genes that encode proteins involved in DNA replication and repair, transcription, and translation, but carries few genes encoding biosynthetic enzymes and transport proteins. Thus, by comparing the coding potentials of entire genomes, geneticists can make important inferences about all aspects of metabolism, growth, and interactions with the environment in various bacteria. Such information should prove invaluable in developing strategies to protect human from bacterial pathogens.

The Human Genome

Progress in mapping the human genome has been quite spectacular. Complete physical maps of chromosomes Y and 21 and detailed RFLP maps of the X chromosome and all 22 autosomes were published in 1922. By 1995, the genetic map contained markers separated by, on average, 200 kb. A detailed microsatellite map of the human genome was published in 1996, and a comprehensive map of 16,354 distinct loci was released in 1997. All of these maps have proven invaluable to researchers cloning genes by their locations in the genome.

Unfortunately, the resolution of genetic mapping in humans is quite low–in the range of 1-10 mb. The resolution of fluorescent *in situ* hybridisation (FISH) is also approximately 1 mb. Higher resolution mapping (down to 50 kb) can be achieved by **radiation hybrid mapping**, which is a modification of the somatic-cell hybridisation mapping procedure. Standard somatic-cell hybridisation involves the fusion of human cells and rodent cells growing in culture and the correlation of human gene products with human chromosomes retained in the hybrid cells.

Radiation hybrid mapping is done by fragmenting the chromosomes of the human cells with heavy irradiation prior to cell fusion (Fig. 24.11). The irradiated human cells are then fused with Chinese hamster (or other rodent) cells growing in culture, usually in the presence of a chemical such as polyethylene glycol to increase the efficiency of cell fusion. The human Chinese hamster somatic-cell hybrids are then identified by growth in an appropriate selection medium.

Many of the human chromosome fragments become integrated into the Chinese hamster chromosomes during this process and are transmitted to progeny cells just like the normal genes in the Chinese hamster chromosomes. The polymerase chain reaction (PCR) is then used to screen a large panel of the selected hybrid cells for the presence of human genetic markers. Chromosome maps are constructed based on the assumption that the probability of an X-ray-

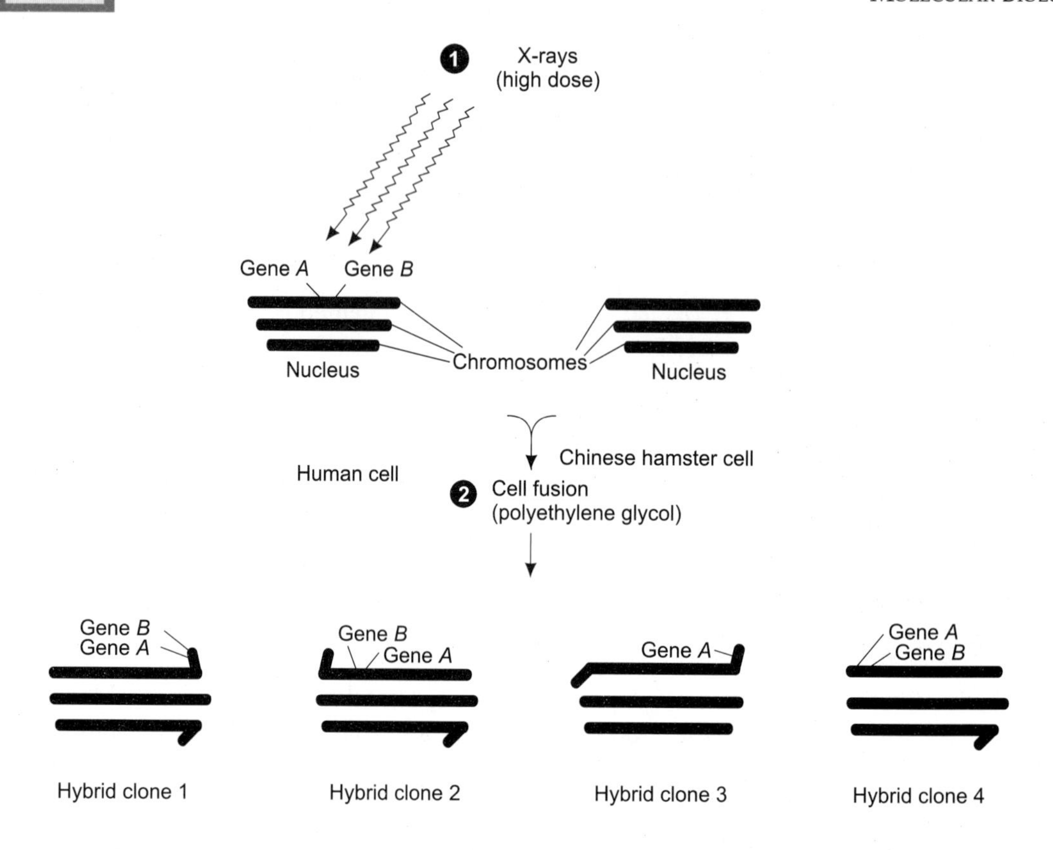

Fig. 24.11 The use of radiation somatic cell hybrids for high-resolution mapping of the human genome. The rationale behind radiation hybrid maps is that the probability of an X-ray-induced break between genes *A* and *B* is directly proportional to the distance between them on the DNA molecule. Note that genes *A* and *B* have remained together in hybrid clones 1, 2 and 4 but were separated by an X-ray-induced break during the formation of hybrid clone 3.

induced break between two markers is directly proportional to the distances separating them in chromosomal DNA.

Several groups have used the radiation hybrid mapping procedure to construct high-density maps of the human genome. In 1997, Elizabeth Stewart and coworkers published a map of 10,478 STSs based on radiation hybrid mapping; their map of human chromosome 1 is shown in Fig. 24.12.

Whereas the gene mapping work has advanced quickly, progress toward sequencing the human genome has lagged behind schedule. However, its status is expected to change rapidly. In July 1998, the National Human Genome Research Institute of the National Institutes of Health awarded grants totalling $60.5 million to seven sequencing in the United States with the combined goal of sequencing 117 million base pairs during the next year—twice the output of all the U.S. sequencing centres to date.

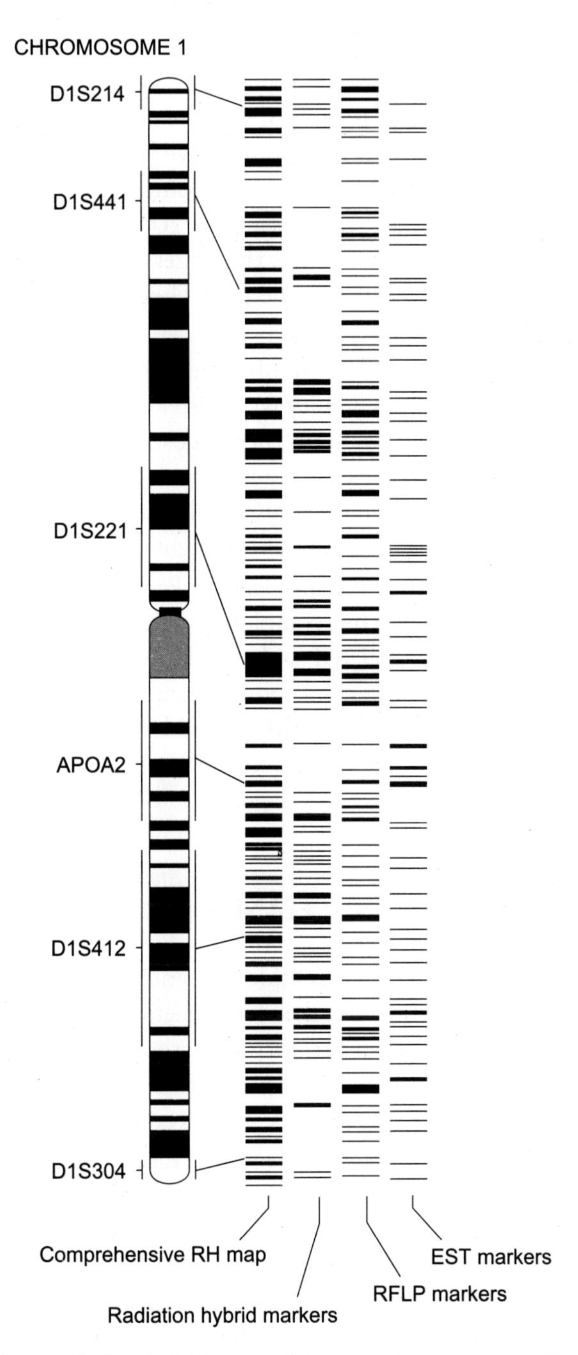

Fig. 24.12 A high-resolution radiation hybrid map of human chromosome 1. The cytogenetic map of chromosome 1 is shown on the left, along with the location of six anchor markers. To the right of the cytogenetic map are four genetic maps showing the locations of the ESTs (purple lines), RFLPs (green lines), high-confidence radiation hybrid markers (blue lines) and the comprehensive radiation hybrid map of all the markers (red lines).

However, the entire sequencing strategy of Human Genome Project laboratories has just been re-evaluated because of new breakthroughs in automated DNA sequencing and new private human genome sequencing initiatives. J. Craig Venter, past president of the Institute for Genomic Research where several bacterial genomes have been sequences, has joined forces with Perkin-Elmer Corporation and formed a private company, Celera Inc., with the goal of sequencing the human genome in just three years—by 2001. Perhaps at least in part because of the competition with Venter's private initiative, the leaders of the seven Human Genome Project sequencing laboratories have revised their schedule. They now plan to complete the sequence by 2003—two years earlier than originally proposed.

In any case, the complete sequence of the human genome will almost certainly be known within the next few years, and its availability will raise a whole new set of questions about the proper use of this new knowledge. Many of these questions focus on an individual's right to privacy. For example, if a mutation that causes a late-onset disorder such as Huntington's disease is discovered in a family, who should have access to this information? If such information were available to the public, widespread discrimination might occur.

Key Points Important goals of the Human Genome Project are to map all 70,000 to 100,000 human genes and to determine the complete nucleotide sequence of all 24 human chromosomes. Additional goals associated with the Human Genome Project are to map and sequence the genomes of several important model organisms. The complete sequences of the genomes of 18 species of bacteria, mitochondria and chloroplasts from several species, and the yeast *S. cerevisiae* are now available; 99.9 per cent of the sequence of the genome of the worm *C. elegans* is known; and the sequences of the genomes of the fruit-fly *D. melanogaster* and the plant *A. thaliana* will soon be available.

HUMAN GENE PROSPECTING IN ICELAND

The availability of detailed maps and nucleotide sequences of entire genomes of many species allows scientists to perform computer searches for sequences that encode enzymes with desired activities, to isolate these sequences, and to introduce them into the genomes of other species. Scientists have already engineered transgenic plants that are herbicide and insect resistant, as well as plant that synthesise antibodies, drugs, enhanced levels of vitamins, and even plastics.

Yes, the "genomics revolution" has begun, and the race to sequence the human genome and to identify genes that encode valuable products is underway. Although many scientists object to the patenting of living organisms and DNA sequences isolated from them, the courts have ruled that both genes and transgenic organisms are subject to patent protection. Indeed, patents that cover the nucleotide sequences of over a thousand genes and a large number of transgenic plant and animals have already been approved. The race to complete the sequence of the human genome and to patent important human genes is obviously on!

An interesting component of the effort to identity human genes with important pharmacological value is taking place in a somewhat unexpected location—the remote island of Iceland, located in the North Atlantic between Greenland and Scandinavia. Because of their history and geographical isolation, the 270,000 people of Iceland provide a unique resource for genetic studies. They are the genetically quite homogeneous descendants of Vikings who settled on this island more than 1100 years ago. This homogeneity has been enhanced by two "genetic bottlenecks" during which the population of Iceland was sharply reduced. During the fifteenth century, the population plummeted from about 70,000 to around 25,000 when bubonic plague ravished the island. During the 1700s, the population dropped below 50,000 on three occasions because of famine and disease caused in part by the eruption of the volcano Hekla. Thus, the human gene pool of Iceland is much more homogeneous than the gene pools of most other human populations. In addition, Iceland's national health service has kept superb family medical records since 1915.

In 1997, Kari Stefansson, a Harvard geneticist, recognised the uniqueness of Iceland's human gene pool and family records. He returned to his homeland to launch a private company, deCODE Genetics, with the goal of identifying human genes that would lead to the development of new pharmaceutical drugs and diagnostic tests. The company's first success was the identification of the *familial essential tremor* gene—a gene associated with shakiness in the elderly. In addition, deCODE scientists have made rapid progress in their studies of several other human disorders.

Based on these results, deCODE Genetics negotiated a $200 million contract with the Swiss pharmaceutical giant Hoffmann-LaRoche, which will give the Swiss firm exclusive rights to any drugs or diagnostic products resulting from the work of deCODE scientists. However, the contract specifies that Hoffmann-LaRoche must provide free of charge all drugs, diagnostic tests, and other products resulting from this research to the people of Iceland. Therefore, at least in this one case, the people who are providing the genetic date and the DNA samples for analysis will personally benefit from the results of research by a private company.

SECTION B

GENOMICS—HOW TO SEQUENCE A GENOME

A single chain termination sequencing experiment performed by hand produces about 400 nucleotides of sequence, and a single run in an automated sequencer gives about 750 bp. But the total size of a fairly typical bacterial genome is 4,000,000 bp and the human genome is 3200-000 000 bp (Table 24.1). Clearly a large number of sequencing experiments must be carried out in order to determine the sequence of an entire genome.

The situation is not hopeless but it requires the use of robotic systems to prepare the DNA for sequencing and to carry out the chain termination experiments, with the sequences read by automated sequencers that transfer the date directly into a computer (Fig. 24.13). In the factory-style laboratories that undertake these projects the main objective is to keep the automated sequencers operating at their full capacities. Each sequencer can run up to 96 experiments in parallel, generating 72,000 bp of sequence every 2 hours. The largest sequencing initiatives use up to 100 automated sequencers all working around the clock, representing a theoretical output

Table 24.1 Sizes of representative genomes.

Species	*Type of organism*	*Genome size (Mb)*
Mycoplasma genitalium	Bacterium	0.58
Haemophilus influenzae	Bacterium	1.83
Escherichia coli	Bacterium	4.64
Saccharomyces cerevisiae	Yeast	12.1
Caenorhabditis elegans	Nematode worm	97
Drosophila melanogaster	Insect	180
Arabidopsis thaliana	Plant	125
Homo sapiens	Mammal	3200
Triticum aestivum	Plant (wheat)	17000

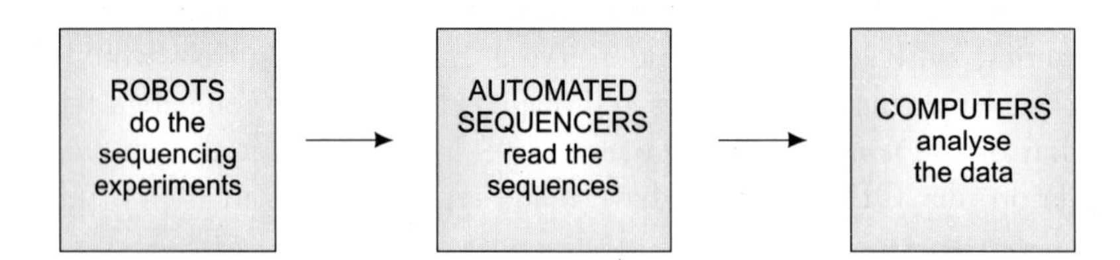

Fig. 24.13 The factory approach to large scale DNA sequencing.

of 50,000,000 bp per day. Looked at in these terms, genome sequences are not so daunting after all.

In practice, the generation of sufficient sequence data is one of the more routine aspects of a genome project. The first real problem that arises is the need to assemble the thousands or perhaps millions of individual 750 bp sequences into a contiguous genome sequence. Two different strategies have been developed for sequence assembly (Fig. 24.14).

(1) The **shotgun approach,** in which the genome is randomly broken into short fragments. The resulting sequences are examined for overlaps and these used to build up the contiguous genome sequence.

(2) The **clone contig approach,** which involves a presequencing phase during which a series of overlapping clones is identified. Each piece of cloned DNA is then sequenced, and this sequence place at its appropriate position on the contig map in order to gradually build up the overlapping genome sequence.

THE SHOTGUN APPROACH TO GENOME SEQUENCING

The key requirement of the shotgun approach is that it must be possible to identify overlaps between all the individual sequences that are generated, and this identification process must be accurate and unambiguous so that the correct genome sequence is obtained. An error in identifying a pair of overlapping sequence is obtained. An error in identifying a pair of overlapping sequences could lead to the genome sequence becoming scrambled, of parts missed

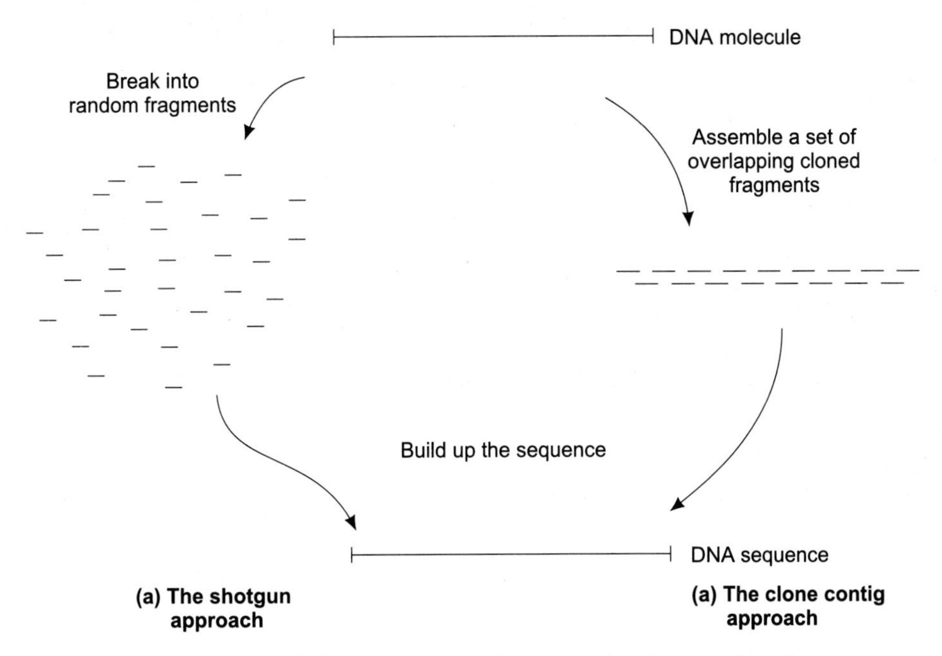

Fig. 24.14 (a) The shotgun, and (b) the clone contig approaches to assembly of a genome sequence.

out entirely. The probability of making mistakes increases with larger genome sizes, so the shotgun approach has been used mainly with the smaller bacterial genomes.

The *H. influenzae* Genome Sequencing Project

The shotgun approach was first used successfully with the bacterium *H. influenzae*, which was the first free-living organism whose genome was entirely sequenced, the results published in 1995. The first step was to break the 1830 kb genome of the bacterium into short fragments which would provide the templates for the sequencing experiments (Fig. 24.15). A restriction endonuclease could have been used but sonication was chosen because this technique is more random and hence reduces the possibility of gaps appearing in the genome sequence.

It was decided to concentrate on fragments of 1.6–2.0 kb because these could yield two DNA sequences, one from each end, reducing the amount of cloning and DNA preparation that was required. The sonicated DNA was, therefore, fractionated by agarose gel electrophoresis and fragments of the desired size purified from the gel. After cloning, 28 643 sequencing experiments were carried out with 19 687 of the clones. A few of these sequences - 4339 in all - were rejected because they were less than 400 bp in length. The remaining 24 304 sequences were entered into a computer which spent 30 hours analysing the data. The result was 140 contiguous sequences, each a different segment of the H. influenzae genome.

It might have been possible to continue sequencing more of the sonicated fragments in order eventually to close the gaps between the individual segments. However, 11, 631, 485 bp of sequence had already been generated–six times the length of the genome–suggesting that a

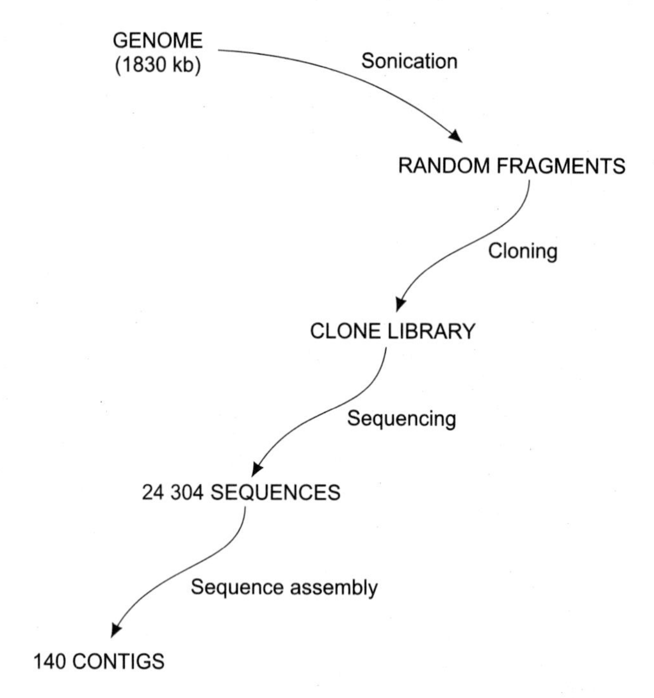

Fig. 24.15 A schematic of the key steps in the H. *influenzae* genome sequencing project.

large amount of additional work would be needed before the correct fragments were by chance, sequenced. At this stage of the project the most time-effective approach was to use a more directed strategy in order to close each of the gaps individually. Several approaches were used for gap closure, the most successful of these involving hybridisation analysis of a clone library prepared in a λ vector (Fig. 24.16). The library was probed in turn with a series of oligonucleotides whose sequences corresponded with the ends of each of the 140 segments. In some cases, two oligonucleotides hybridised to the same λ clone, indicating that the two segment ends represented by those oligonucleotides lay adjacent to one another in the genome. The gap between these two ends could then be closed by sequencing the appropriate part of the λ clone.

Problems with Shotgun Cloning

Shotgun cloning has been successful with many bacterial genomes. Not only are these genomes small, so the computational requirements for finding sequence overlaps are not too great, but they contain little or no repetitive DNA sequences. These are sequences, from a few base pairs to several kilobases, that are repeated at two or more places in a genome. They cause problems for the shotgun approach because when sequences are assembled those that lie partly or wholly within one repeat element might accidentally be assigned an overlap with the identical sequence present in a different repeat element (Fig. 24.17). This could lead to parts of the genome sequence being assembled in the incorrect position or left out entirely. For this reason, it has generally been thought that shotgun sequencing is inappropriate for eukaryotic genomes, as

(a) Prepare oligonucleotide probes

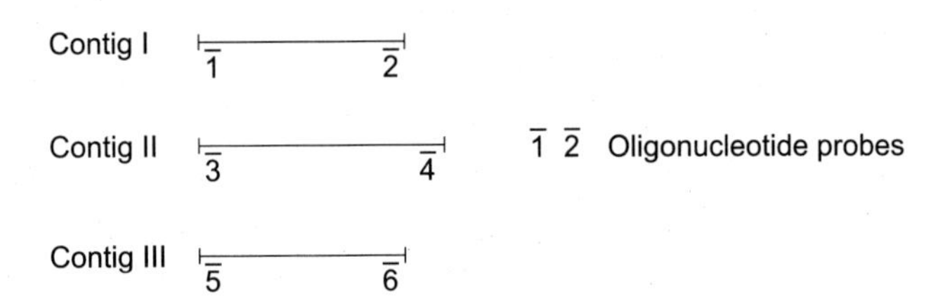

(b) Probe a genomic library

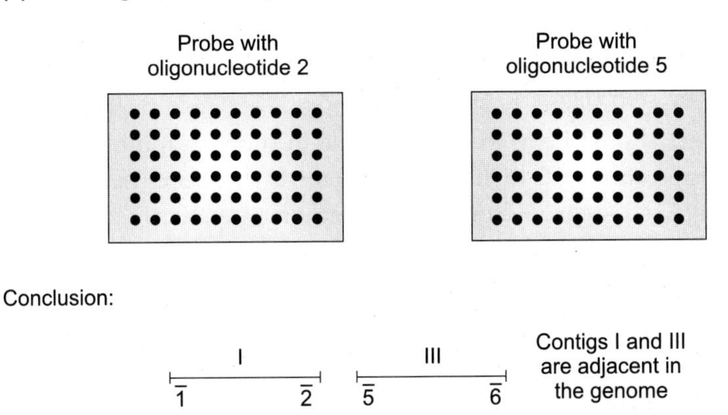

Fig. 24.16 Using oligonucleotide hybridisation to close gaps in the H. influenzae genome sequence. Oligonucleotides 2 and 5 both hybridise to the same λ clone, indicating that contigs I and III are adjacent. The gap between them can be closed by sequencing the appropriate part of the λ clone.

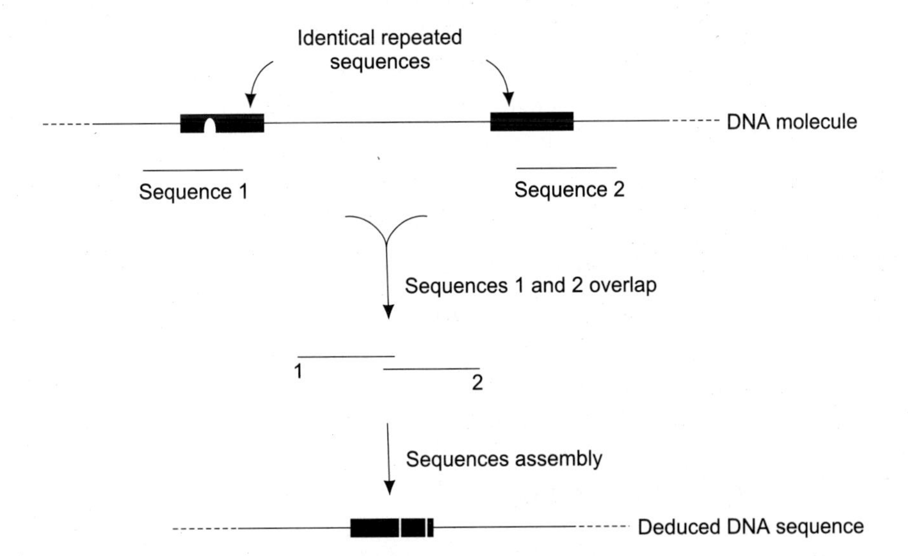

Fig. 24.17 One problem with the shotgun approach. An incorrect overlap is made between two sequences that both terminate within a repeated element. The result is that a segment of the DNA molecule is absent from the DNA sequence.

these have many repeat elements. Later in the chapter we will see how this limitation can be circumvented using a genome map to direct assemble of sequences obtained by the shotgun approach.

THE CLONE CONTIG APPROACH

The clone contig approach does not suffer from the limitations of shotgun sequencing and so can provide an accurate sequence of a large genome that contains repetitive DNA. Its drawback is that it involves much more work and so takes longer and costs more money. The additional time and effort is needed to construct the overlapping series of cloned DNA fragments. Once this has been done, each cloned fragment is sequenced by the shotgun method and the genome sequence built up step by step (Fig. 24. 14).

The cloned fragments should be as long as possible in order to minimise the total number needed to cover the entire genome. A high capacity vector is therefore used. the first eukaryotic chromosome to be sequenced - chromosome III of *Saccharomyces cerevisiae*–was initially cloned in a cosmid vector with the resulting contig comprising 29 clones. Chromosome III is relatively short, however, and the average size of the cloned fragments was just 10.8 kb. Sequencing of the much longer human genome required 300,000 bacterial artificial chromosome (BAC) clones. Assembling all of these into chromosome-specific contigs was a massive task.

Clone Contig Assembly by Chromosome Walking

One technique that can be used to assemble a clone contig is **chromosome walking**. To begin a chromosome walk a clone is selected at random from the library, labelled, and used as a hybridisation probe against all the other clones in the library (Fig. 24.18(a)). Those clones that give hybridisation signals are ones that overlap with the probe. One of these overlapping clones is now labelled and a second round of probing carried out. More hybridisation signals are seen, some of these indicating additional overlaps (Fig. 24.18(b). Gradually the clone contig is built up in a step-by-step fashion. But this is a laborious process and is only attempted when the contig is for a short chromosome and so involves relatively few clones, or when the aim is to close one or more small gaps between contigs that have been built up by more rapid methods.

Rapid Methods for Clone Contig Assembly

The weakness of chromosome walking is that it begins at a fixed starting point and builds up the clone contig step by step, and hence slowly, from that fixed point. The more rapid techniques for clone contig assembly do not use a fixed starting point and instead aim to identify pairs of overlapping clones: when enough overlapping pairs have been identified the contig is revealed (Fig. 24.19). The various techniques that can be used to identify overlaps are collectively known as **clone fingerprinting**.

Clone fingerprinting is based on the identification of sequence features that are shared by a pair of clones. The simplest approach is to digest each clone with one or more restriction endonucleases and to look for pairs of clones that share restriction fragments of the same size,

(a) Probe the library with clone A1

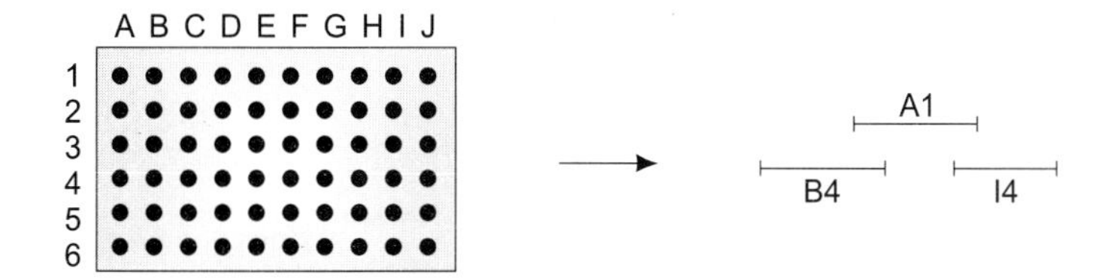

(b) Probe the library with clone 14

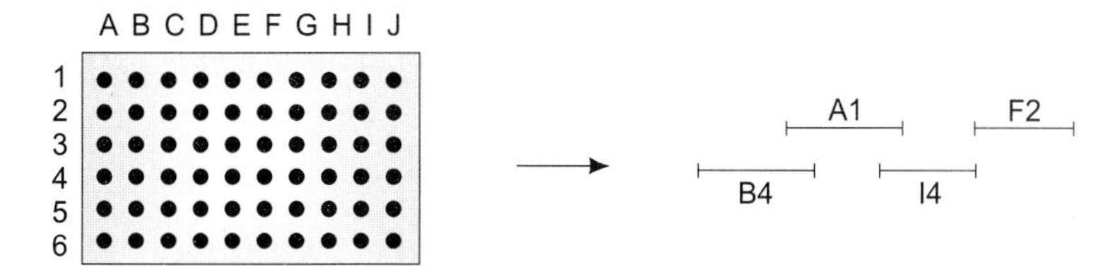

Fig. 24.18 Chromosome walking.

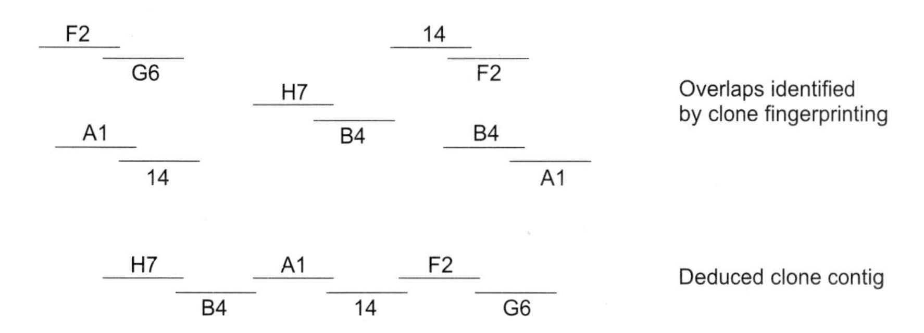

Fig. 24.19 Building up a clone contig by a clone fingerprinting technique.

excluding those fragments that derive from the vector rather than the inserted DNA. This technique might appear to be easy to carry out but in practice it takes a great deal of time to scan the resulting agarose gels for shared fragments. There is also a relatively high possibility that two clones that do not overlap will, by chance, share restriction fragments whose sizes are indistinguishable by agarose gel electrophoresis.

More accurate results can be obtained by **repetitive DNA PCR**, also known as **interspersed repeat element PCR (IRE-PCR)**. This type of PCR uses primers that are designed to anneal within repetitive DNA sequences and direct amplification of the DNA between adjacent repeats (Fig. 24.20). Repeats of a particular type are distributed fairly randomly in eucaryotic genomes, with varying distances between them, so a variety of product sizes are obtained when these primers are used with clones of eukaryotic DNA. If a pair of clones give PCR products of the same size, they must contain repeats that are identically spaced, possibly because the cloned DNA fragments overlap.

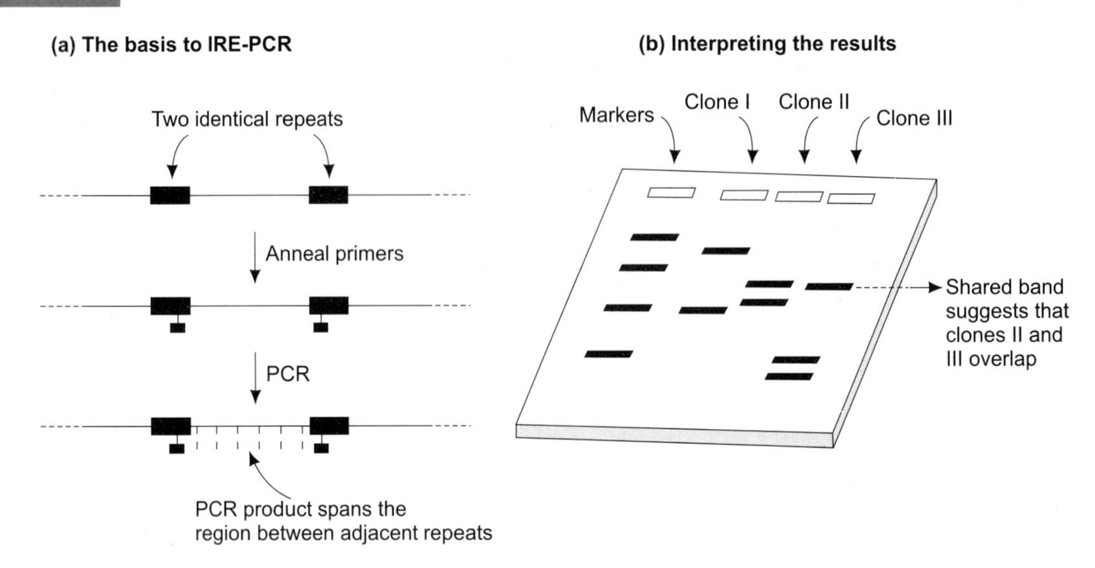

Fig. 24.20 Interspersed repeat element PCR (IRE-PCR).

Clone Contig Assemble by Sequence Tagged Site Content Analysis

A third way to assemble a clone contig is to search for pairs of clones that contain a specific DNA sequence of that occurs at just one position in the genome under study. If two clones contain this feature then clearly they must overlap (Fig. 24.21). A sequence of this type is called a **sequence tagged site (STS)**. Often an STS is a gene that has been sequenced in an earlier project. As the sequence is known, a pair of PCR primers can be designed that are specific for that gene and then used to identify which members of a clone library contain the gene. But the STS does not have to be a gene and can be any short piece of DNA sequence that has been obtained from the genome, providing it does not fall in a repetitive element.

USING A MAP TO AID SEQUENCE ASSEMBLY

Sequence tagged site content mapping is a particularly important method for clone contig assembly because often the positions of STSs within the genome will have been determined by **genetic mapping** or **physical mapping**. This means that the STS positions can be used to anchor the clone contig onto a genome map, enabling the position of the contig within a chromosome to be determined. We will now look at how these maps are obtained.

Genetic Maps

A genetic map is one that is obtained by genetic studies using Mendelian principles and involving directed breeding programmes for experimental organisms or **pedigree analysis** for humans. In most cases the loci that are studied are genes, whose inheritance patterns are followed by monitoring the phenotypes of the offspring produced after a cross between parents with contrasting characteristics (e.g., tall and short for the pea plants studied by Mendel). More

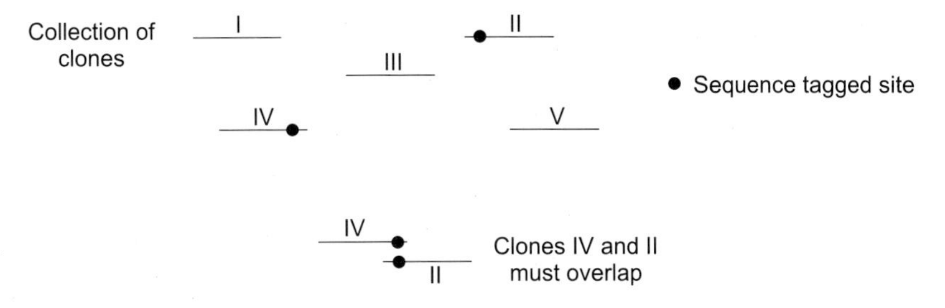

Fig. 24.21 The basis to STS content mapping.

recently, techniques have been devised for genetic mapping of DNA sequences that are not genes but which display variability in the human population. The most important of these DNA markers are:

(1) **Restriction fragment length polymorphisms (RFLPs)**, which are caused by a sequence variation that results in restriction site being changed. When digested with a restriction endonuclease the loss of the site is revealed because two fragments remain joined together. Originally, RFLPs were typed by Southern hybridisation of restricted genomic DNA, but this is a time-consuming process, so nowadays the presence or absence of the restriction site is usually determined by PCR (Fig. 24.22). There are approximately 100,000 RFLPs in the human genome.

(2) **Short tandem repeats (STRs)**, also called **microsatellites**, which are made up of short repetitive sequences, such as CACACA. The repeat units are 1-13 nucleotides in length and are usually repeated 5-20 times. The number of repeats at a locus can be determined by carrying out a PCR using primers that anneal either side of the STR, and examining the size of the resulting products by agarose or polyacrylamide gel electrophoresis (Fig. 24.23). There are at least 6,50,000 STRs in the human genome.

(3) **Single nucleotide polymorphisms (SNPs)**, which are positions in a genome where any one of two or more different nucleotides can occur (Fig. 24.24). These point mutations are typed by analysis with short oligonucleotide probes that hybridise to the alternative forms of the SNP. The number of SNP. The number of SNPs in the human genome is not yet known but is at least 1.4 million.

All of these DNA markers are variable and so exist in two or more allelic forms. The inheritance of the alternative alleles at a particular locus is followed by analysis of DNA prepared from the offspring of genetic crosses.

Physical Maps

A physical map is generated by methods that directly locate the positions of specific sequences on a chromosomal DNA molecule. As in genetic mapping, the loci that are studied can be genes or DNA markers. The latter might include **expressed sequence tags (ESTs)**, which are short sequences obtained from the ends of complementary DNAs (cDNAs). Expressed sequence tags are, therefore, partial gene sequences and when used in map construction they provide a quick

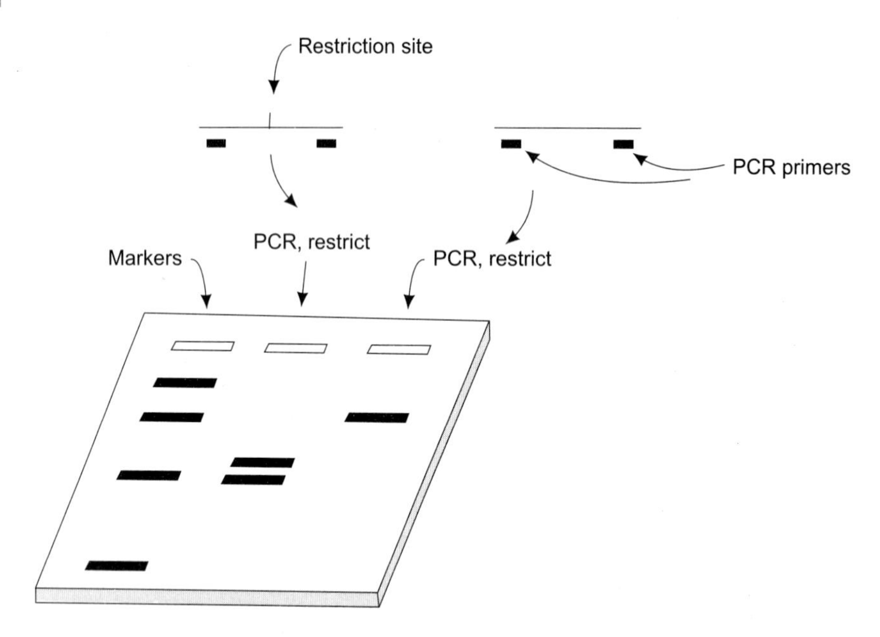

Fig. 24.22 Typing a restriction site polymorphism by PCR. In the middle lane the PCR product gives two bands because it is cut by treatment with the restriction enzyme. In the right-hand lane there is just one band because the template DNA lacks the restriction site.

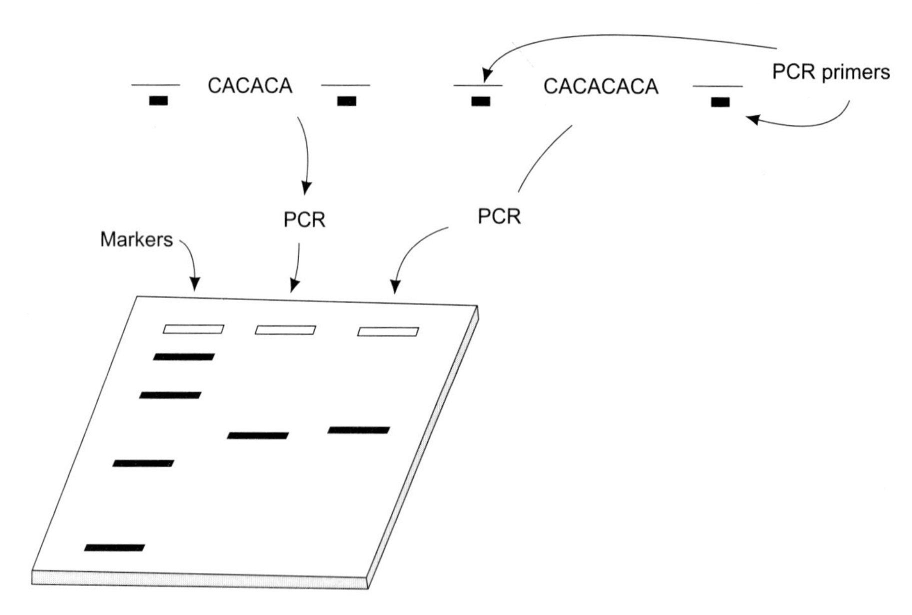

Fig. 24.23 Typing an STR by PCR. The PCR product in the right-hand lane is slightly longer than that in the middle lane, because the template DNA from which it is generated contains an additional CA unit.

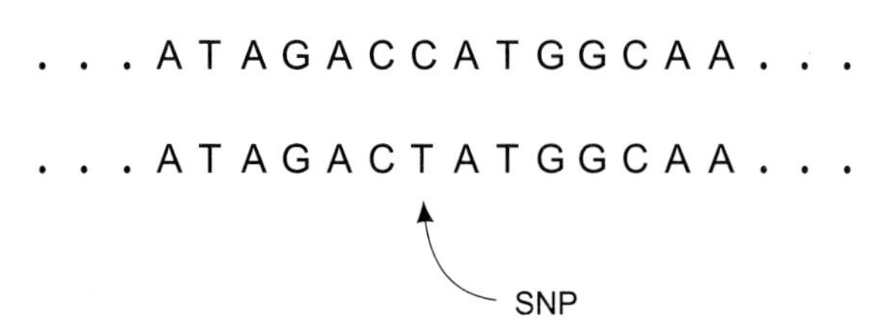

Fig. 24.24 Two versions of an SNP.

way of locating the positions of genes, even though the identity of the gene might not be apparent from the EST sequence. Two types of technique are used in physical mapping.

(1) Direct examination of chromosomal DNA molecules, for example by fluorescence *in situ* hybridisation (FISH). If FISH is carried out simultaneously with two DNA probes, each labelled with a different fluorochrome, the relevant positions on the chromosome of the two markers represented by the probe can be visualised. Special techniques for working with extended chromosomes, whose DNA molecules are stretched out rather than tightly coiled as in normal chromosomes, enable markers to be positioned with a high degree of accuracy.

(2) Physical mapping with a **mapping reagent**, which is a collection overlapping DNA fragments spanning the chromosome or genome that is being studied. Pairs of markers that lie within a single fragment must be located close to each other on the chromosome: how close can be determined by measuring the frequency with which the pair occurs together in different fragments in the mapping reagent (Fig. 24.25). **Radiation hybrids** are one type of mapping reagent that has been important in the Human Genome Project. These are hamster cell lines that contain fragments of human chromosomes, prepared by a treatment involving irradiation (hence their name). Mapping is carried out by hybridisation of marker probes to a panel of cell lines, each one containing a different part of the human genome.

The Importance of a Map is Sequence Assembly

It is possible to obtain a genome sequence without the use of a genetic or physical map. This is illustrated by the *H. influenzae* project that we followed, and many other bacterial genomes have been sequenced without the aid of a map. But a map is very important when a larger genome is being sequenced because it provides a guide that can be used to check that the

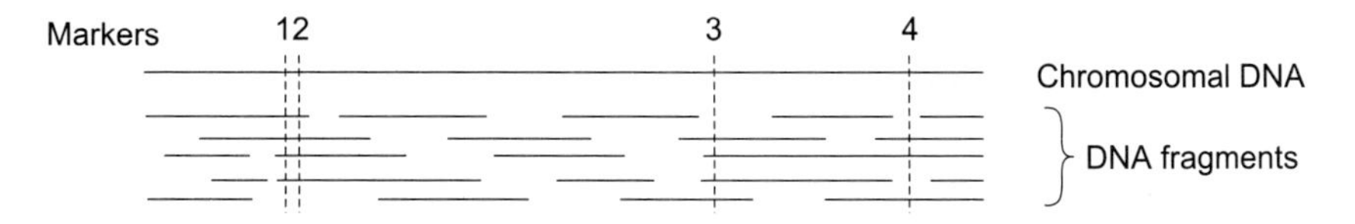

Fig. 24.25 The principle behind the use of a mapping reagent. It can be deduced that markers 1 and 2 are relatively close because they are present together on four DNA fragments. In contrast, markers 3 and 4 must be relatively far apart because they occur together on just one fragment.

genome sequence is being assembled correctly from the many short sequences that emerge from the automated sequences. If a marker that has been mapped by genetic and/or physical means appears in the genome sequence at an unexpected position then an error in sequence assembly is suspected.

Detailed genetic and/or physical maps have been important in the Human Genome Project, as well as those for yeast, fruit-fly, *C. elegans* and *A. thaliana*, all of which were based on the clone contig approach. Maps are also being used to direct sequence assembly in projects that use the shotgun approach. As earlier described the major problem when applying shotgun sequencing to a large genome is the presence of repeated sequences and the possibility that the assembled sequence 'jumps' between two repeats, so part of the genome is misplaced or left out (Fig. 24.17). These errors can be avoided if sequence assemble makes constant reference to a genome map. This 'directed shotgun approach' holds great promise as a rapid method of sequencing large genomes.

POST-GENOMICS—TRYING TO UNDERSTAND A GENOME SEQUENCE

Once a genome sequence has been completed the next step is to locate all the genes and determine their functions. This is far from trivial process, even with genomes that have been extensively studied by genetic analysis and gene cloning techniques prior to complete sequencing. For example, the sequence of *S. cerevisiae*, one of the best studied of all organisms, revealed that this genome contains about 6,000 genes. Of these, some 3,600 could be assigned a function either on the basis of previous studies that had been carried out with yeast or because the yeast gene had a similar sequence to a gene that had been studied in another organism. This left 2,400 genes whose functions were not known.

Despite a massive amount of work since the yeast genome was completed in 1996, the functions of the vast majority of these **orphans** have still not been determined. It is in this area that bioinformatics, sometimes referred to as molecular biology in silico, is proving of major value as an adjunct to convention experiments.

IDENTIFYING THE GENES IN A GENOME SEQUENCE

Locating a gene is easy if the amino acid sequence of the protein product is known, allowing the nucleotide sequence of the gene to be predicted, or if the corresponding cDNA or EST has been previously sequenced. But for many genes there is no prior information that enables the correct DNA sequence to be recognised. Under these circumstances, gene location might be difficult even if a map is available. Most maps have only a limited accuracy and can only delineate the approximate position of a gene, possibly leaving several tens or even hundreds of kilobases to search in order to find it. And many genes do not appear on maps because their existence is unsuspected. How can these genes be located in a genome sequence?

Searching for Open Reading Frames

The DNA sequence of a gene is an **open reading frame (ORF)**, a series of nucleotide triplets beginnig with an initiation codon (usually but not always AUG) and ending in a termination codon (TAA, TAG or TGA in most genomes). Searching a genome sequence for ORFs, by eye or more usually by computer, is therefore the first step in gene location. It is important to search all six **reading frames** because genes can run in either direction along the DNA double helix (Fig. 24.26(a)). With a bacterial genome the typical result of this search is identification of long ORFs that are almost certainly genes, with many shorter ORFs that are almost certainly genes, with many shorter ORFs partly or completely contained within the genes but lying in different reading frames (Fig. 24.22(b)). These short sequences are almost certainly combinations of nucleotides that by chance form an ORF but are not genes. If one of these short ORFs lies entirely between two genes there is a possibility that it might mistakenly be identified as a real gene, but in most bacterial genomes there is very little space between the genes so the problem arises only infrequently.

Gene location in eukaryotes is much more difficult: Eukaryotic genomes are not as densely packed as bacterial ones and there are much longer spacers between genes. This means that inspection of the sequence reveals many short ORFs that were placed in this 'questionable' category. Possibly some of these are real genes but probably most of them are not.

In humans and other higher eukaryotes, the search for genes is made even more complicated by the fact that many are split into exons and introns boundaries (Fig. 24.27) but these sequences are also found within exons and within introns. Working out which of these sequences mark true exon-intron boundaries can be very difficult.

(a) Every DNA sequence has six reading frames

```
1         GAC →
2       TGA →
3     ATG →
5' – ATGACCAATGACATGCAATAA – 3'
     |||||||||||||||||||||
3' – TACTGGTTACTGTACGTTATT – 5'
                      ← ATT   4
                    ← TAT     5
                  ← TTA       6
```

(b) The typical result of an ORF search with a bacterial genome

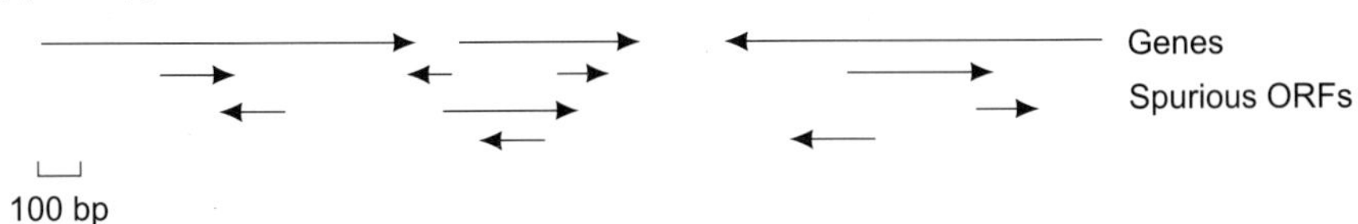

Fig. 24.26 Searching for open reading frames. (a) Every DNA sequence has six open reading frames, any one of which could contain a gene. (b) The typical result of a search for ORFs in a bacterial genome. The arrows indicate the directions in which the genes and spurious ORFs run.

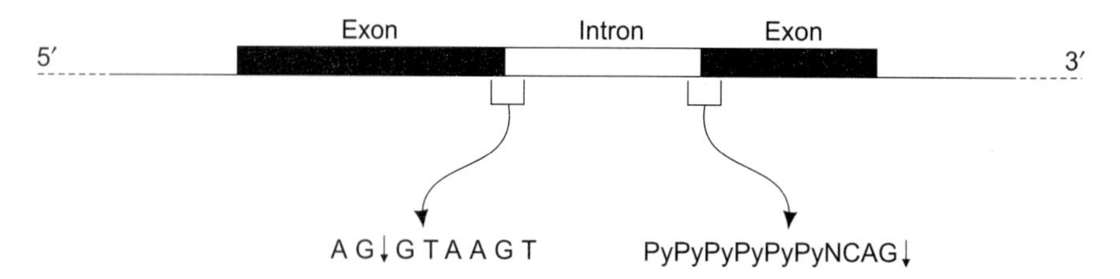

Fig. 24.27 The consensus sequences for the upstream and downstream exon-intron boundaries of vertebrate introns. Py = pyrimidine nucleotide (C or T), N = any nucleotide. The arrows are the boundary positions.

Distinguishing Real Genes from Chance ORFS

Some genomes provide helpful signposts that indicate the presence of a gene. The human and other vertebrate genomes are particularly helpful to the molecular biologist because 50–60% of the genes are accompanied by a **CpG island**, a distinctive GC-rich sequence whose position indicates the approximate start point for a gene. But features such as these are the exception rather than the rule and more general methods for identifying genes are needed.

With many genomes, **codon bias** provides a useful means of assigning a degree of certainty to a possible gene identification. All amino acids except methionine and tryptophan are specified by two or more codons. Alanine, for example, has four codons – GCA, GCC, GCG and GCT. In most genomes, not all members of a codon family are used with equal frequency. Humans are typical in this regard, displaying a distinct bias for certain codons: for example, within the alanine codon family, humans use GCC four times more frequently than GCG. If an ORF contains a high frequency of rare codons then it probably is not a gene. By taking account of the codon bias displayed by an ORF an informed guess can, therefore, be made as to whether the sequence is or is not a gene.

Tentative identification of a gene is usually followed by a **homology search**. This is an analysis, carried out by computer, in which the sequence of the gene is compared with all the gene sequences present in the international DNA databases, not just known genes of the organism under study but also genes from all other species. The rationale is that two genes from different organisms that have similar functions have similar sequences, reflecting their common evolutionary histories (Fig. 24.28).

To carry out a homology search the nucleotide sequence of the tentative gene is usually translated into an amino acid sequence, as this allows a more sensitive search. This is because there are 20 different amino acids but only four nucleotides, so there is less chance of two amino acid sequences appearing to be similar purely by chance.

The analysis is carried out through the internet, by logging on to the website of one of the DNA databases and using a search programme such as **BLAST** (Basic Local Alignment Search Tool). If the test sequence is over 200 amino acids in length and has 30% or greater identity with a sequence in the database (i.e., at 30 out of 100 positions the same amino acid occurs in both sequences), then the two are almost certainly homologous and the ORF under study can be confirmed as a real gene: Further confirmation, if needed, can be obtained by using transcript analysis to show that the gene is transcribed into (RNA).

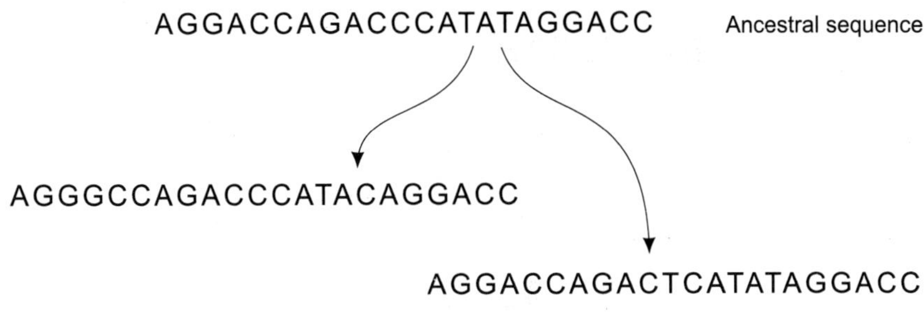

Fig. 24.28 Homology between two sequences that share a common ancestor. The two sequences have acquired mutations during their evolutionary histories but thier sequence similarities indicate that they are homologues.

DETERMINING THE FUNCTION OF AN UNKNOWN GENE

Homology searching serves two purposes. As well as testing the veracity of a tentative gene identification it can also give an indication of the function of the gene, presuming that the function of the homologous gene is known. Almost 2,000 of the genes in the yeast genome were assigned functions in this way. Frequently, however, the matches found by homology searching are to be other genes whose functions have yet to be determined. These unassigned genes are called orphans and working out their functions is one of the major challenges of post-genomics research.

In future years it will probably be possible to use bioinformatics to gain at least an insight into the function of an orphan gene. It is already possible to use the nucleotide sequence of a gene to predict the positions of α-helices and β-sheets in the encoded protein, albeit with limited accuracy, and the resulting structural information can sometimes be used to make inferences about the function of the protein. Proteins that attach to membranes can often be identified because they possess α-helical arrangements that span the membrane, and DNA binding motifs such as zinc finger can be recognised. A greater scope and accuracy to this aspect of bioinformatics will be possible when more information is obtained about the relationship between the structure of a protein and its function. In the meantime, functional analysis of orphans depends largely on conventional experiments.

Several techniques for studying the function of genes were described, and all of these can be applied to orphans. In gene knockout a deleted version of the gene is used to 'knock out' the functional version present in the organism's chromosomes. This is possible because recombination between the deleted gene, carried on a cloning vector, and the chromosomal copy can result in the former replacing the latter (Fig. 24.29). The effect of the gene knockout on the phenotype of the organism is then assessed in order to gain some insight into the function of the gene. The effect of a human gene knockout is inferred by studying a **knockout mouse**, which has a deleted version of the homologous mouse gene. Knockouts have helped to assign functions to a number of genes but problems can arise. In particular, some knockouts

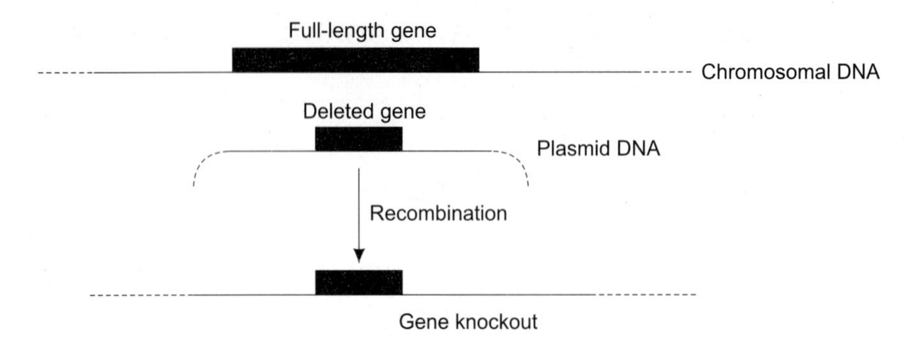

Fig. 24.29 Gene knockout by recombination between a chromosomal copy of a gene and a deleted version carried by a plasmid cloning vector.

have no obvious effect on the phenotype of the organism either because the gene is dispensable, meaning that after inactivation other genes can compensate for its absence, or because the phenotypic change is too subtle to be detected.

STUDIES OF THE TRANSCRIPTOME AND PROTEOME

So far we have considered those aspects of post-genomics research that are concerened with studies of individual genes. The change in emphasis from genes to the genome has prompted new types of analysis that are aimed at understanding the activity of the genome as a whole. This work has led to the invention of two new terms:

(1) The **transcriptome**, which is the messenger RNA (mRNA) content of a cell, and which reflects the overall pattern of gene expression in that cell.

(2) The **proteome**, which is the protein content of a cell and which reflects its biochemical capability.

STUDYING THE TRANSCRIPTOME

Techniques for studying the transcriptome were first developed as part of the yeast post-genomics project. In essence, these techniques involve a sophisticated type of hybridisation analysis. Every yeast gene – all 6000 of them—was obtained as an individual clone and samples spotted onto glass slides in arrays of 80 spots × 80 spots. This is called a microarray. To determine which genes are active in yeast cells grown under particular conditions, mRNA was extracted from the cells, converted to cDNA and the cDNA labelled and hybridised to the microarrays (Fig. 24.30). Fluorescent labels were used and hybridisation was detected by examining the microarrays by confocal microscopy. Those spots that gave a signal indicated genes that were active under the conditions being studied. Changes in gene expression when the yeast were transferred to different growth conditions (e.g., oxygen starvation) could be monitored by repeating the experiment with a second cDNA preparation.

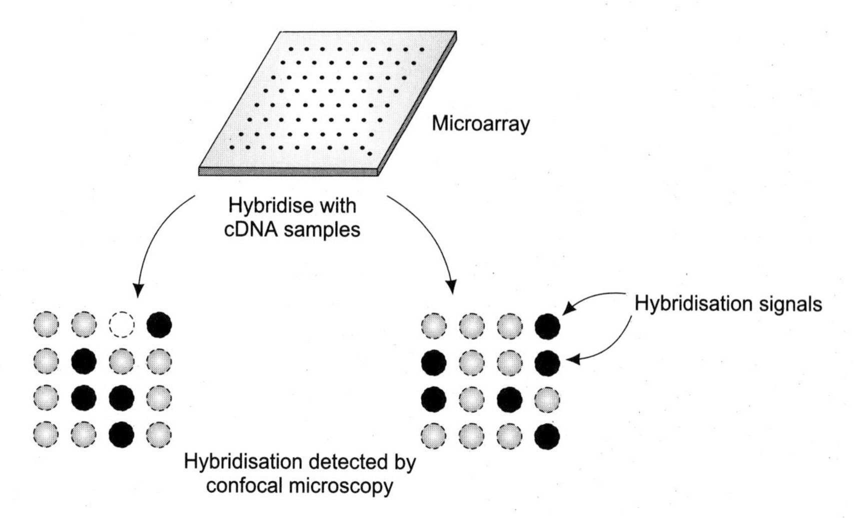

Fig. 24.30 Microarray analysis. The microarray shown here has been hybridised to two different cDNA preparations, each labelled with a fluorescent marker. The clones which hybridise with the cDNAs are identified by confocal microscopy.

Microarryas are now being used to monitor changes in the transcriptomes of many organisms. In some cases the strategy is the same as used with yeast, the microarray representing all the genes in the genome, but this is possible only for those organisms that have relatively few genes. A microarray for all the human genes could be carried by just 10 glass slides of 18 mm by 18mm, but preparing clones of every one of the 30,000-40,000 human genes would be a massive task. Fortunately this is not necessary. For example, to study changes in the transcriptome occurring as a result of cancer, a microarray could be prepared with a cDNA library from normal tissue. Hybridisation with labelled cDNA from the cancerous tissue would then reveal which genes are up- or down-regulated in response to the cancerous state.

An alternative to microarrays is provided by **DNA chips**, which are thin wafers of silicon that carry many different oligonucleotides (Fig. 24.31). These oligonucleotides are synthesised directly on the surface of the chip and can be prepared at a density of 1 million per cm^2, substantially higher than is possible with a conventional microarray. Hybridisation between an oligonucleotide and the probe is detected electronically. Because the oligonucleotides are synthesised *de novo*, using special automated procedures, a chip carrying oligonucleotides whose sequences are specific for every human gene is relatively easy to prepare.

STUDYING THE PROTEOME

The proteome is the entire collection of proteins in a cell. Proteome studies provide additional information that is not obtainable simply by examining the transcriptome, because a single mRNA (and hence gene) can give rise to more than one protein, because of post-translational processing (Fig. 24.32). In eukaryotes, most of the polypeptides that are synthesised by

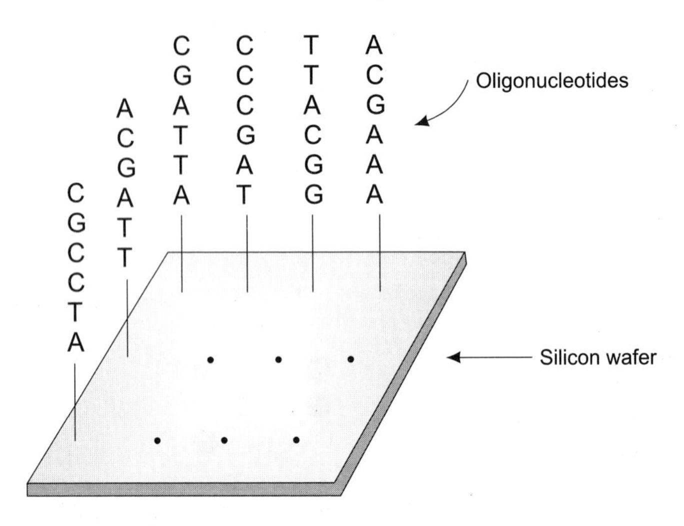

Fig. 24.31 A DNA chip. A real chip would carry many more oligonucleotides than those shown here, and each oligonucleotide would be 20-30 nucleotides in length.

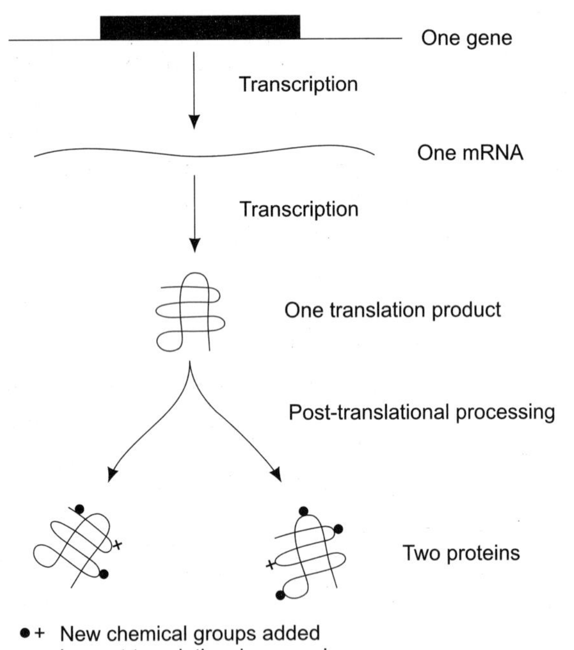

Fig. 24.32 A single gene can give rise to two proteins, with distinct functions, if the initial translation product can be modified in two different ways by post-translational processings.

translation are further processed by addition chemical groups. The particular additions that are made determine the precise function of the protein. Phosphorylation, for example, is an important modification used to activate some proteins.

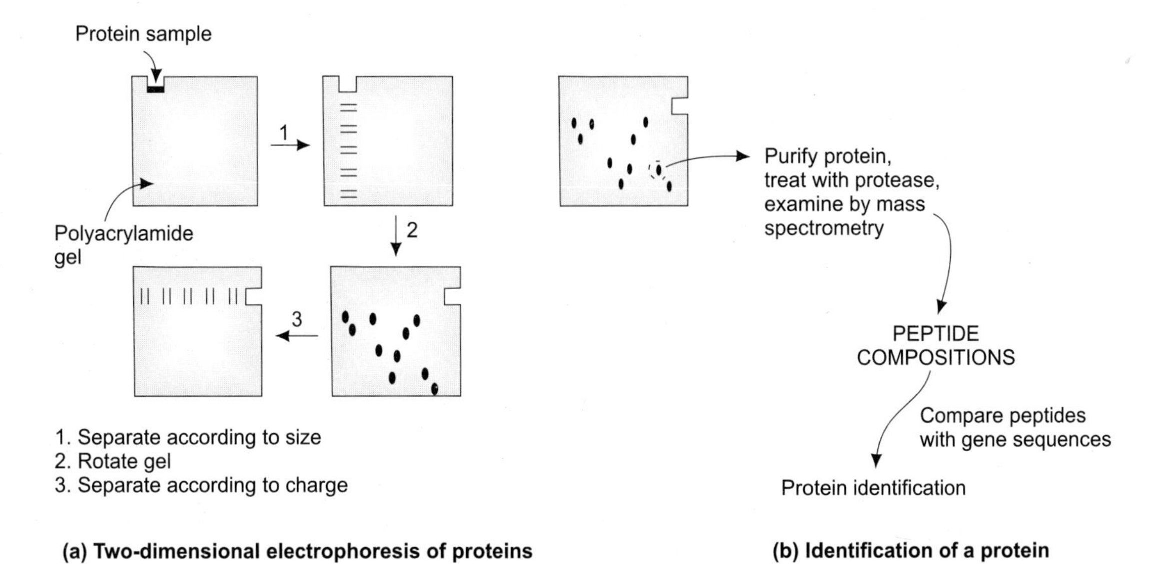

Fig. 24.33 Proteome analysis. (a) Two-dimensional polyacrylamide gel electrophoresis of proteins. (b) identification of the protein contained in a single spot by protease treatment followed by mass spectrometry of the resulting peptides.

To study the proteome the entire protein content of a cell or tissue is first separated by two-dimensional electrophoresis. In this technique, the proteins are loaded into a well on one side of a square of polyacrylamide gel and separated according to their molecular weights. The square is then rotated by 90° and a second electrophoresis performed, this time separating the proteins on the basis of their charges. The result is a two-dimensional pattern of spots, of different sizes, shapes and intensities, each representing a different protein or related group of proteins (Fig. 24.33(a)). Differences between two proteomes are apparent from differences in the pattern of spots when the two gels are compared. To identify the protein in a particular spot a sample is purified from the gel and treated with a protease that cuts the polypeptide at a specific amino acid sequence (similar in a way to the activity of a restriction endonuclease). The resulting peptides are then examined by mass spectrometry (Fig. 24.33(b)). The mass spectrometer determines the amino acid composition of each peptide. This information is usually sufficient to enable the gene coding for the protein to be identified from the genome sequence.

CHAPTER 25

Molecular Farming—Transgenic Plants

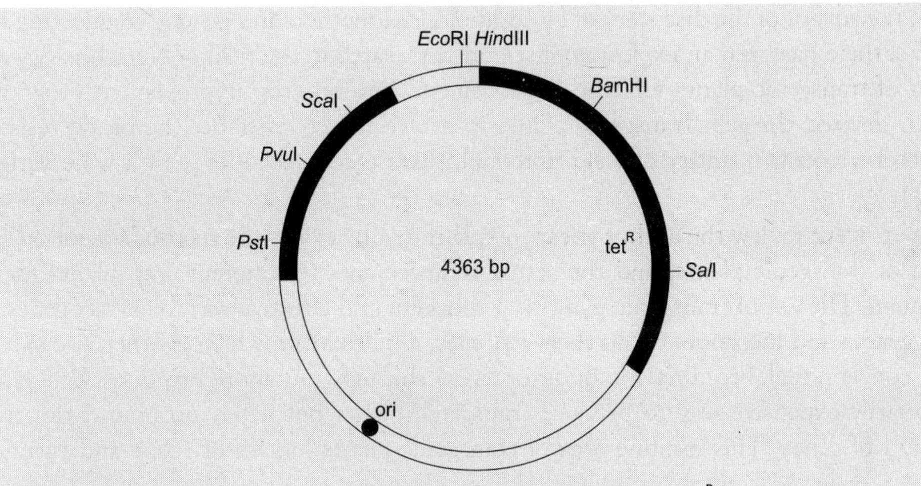

A map of pBR322 showing the positions of the ampicillin resistance (amp^R) and tetracycline resistance (tet^R) genes, the origin of replication (ori) and a selection of the most important restriction sites.

ABSTRACT

The use of transgenic plants is a short-cut and effective way to extract genes from different genera and incorporate into their genomes. Agrobacterium tumefaciens is an excellent system for introducing genes into plant cells because DNA can be introduced into whole plant tissue in relatively precise process. T-DNA gives consistent genetic maps and appropriate segregation ratio. Introduced traits have been found to be stable over at least five generations during cross breeding. Virulent strains of Agrobacteriums contain large Ti-(tumor inducing) plasmids which are responsible for the DNA transfer and subsequent disease symptoms. Ti-plasmids contain two sets of sequences necessary for gene transfer to plants, one or more tDNA (transferred DNA) regions that are transferred to the plant, and the Vir (Virulence) gene which are not themselves transposed during infection. Several selectable and non-selectable marker genes are easily available today for plant transformation and have been transferred in large number of plant species. Two basic approaches of cultivation of regenerating protoplasts and the leaf disc procedures are in use to obtain transgenic plants.

Keywords Production, Transgenic varieties, Recombinant technology, Tobacco, Sunflower, Potatoes, Anti-freeze protein.

INTRODUCTION

Crop plants have been subjected to rigorous improvement for their quality and quantity through conventional breeding and selection, since ancient time all over the world.

With the advent of the discovery of biotechnological method and genetic engineering in the late 1970's, there has been an explosion of scientific research in the field of biotechnology called evolution of transgenic plants for crop improvement. Several crop improvement programmes have been devised through transgenic plants to evolve insect resistance, herbicide resistance, quality improvement, quantity of yield potential, disease resistance, etc., in a wide variety of crop plants.

In the present review the author tries to explain in a brief way the methods involved in the evolution of transgenic plants, and the actual achievements in different crop plants for their improvement. The use of transgenic plants is a short cut and effective way to extract genes from different genera and incorporate into their genomes, a mechanism which is otherwise *impossible* through conventional breeding methods or even through mutation breeding. But it is an extraordinarily expensive way to produce transgenic plants but it has no boundaries for the exploitation of genes. This exciting area of transgenic plants has lot of scope and future.

How to Develop Transgenic Plants

Agrobacterium tumefaciens-remarkable genetic system

Agrobacterium tumefaciens, the gram negative, rod shaped aerobic bacterium caused little interest when Smith and Townsend (1907) reported that it is the causative agent of a widespread neoplastic plant disease crown gall, which is characterised by file overproduction of tissues at the infection site. This pathogen adsorbs on the surface of cell exposed at wound sites, most frequently just below the soil surface at the root crown and hence the name crown gall tumor. Later, a key discovery was made in 1974 by Zaenen *et al.* where they found that A. *tumefaciens* contains a large extrachromosomal element (more than 200 kb) harbouring genes involved in crown gall induction. This tumor inducing plasmid was given the name Ti plasmid by van Larebecke *et at.* (1975). However, this bacterium received extraordinary attention after the discovery that the bacteria is capable of modifying the genetic material of host cells, by transferring part of its Ti plasmid which gets incorporated into the host cell DNA and expresses itself with the normal host DNA. With researchers analysing and unraveling the molecular mechanism underlying crown gall induction, the *Agrobacterium* gene transfer system has now become one of the most significant discoveries triggering a series of existing developments in molecular biology and genetic engineering.

The first transgenic plants expressing engineered foreign genes were tobacco plants produced by the use of *Agrobacterium tumefaciens* vectors (Horsch *et al.*, 1984, DeBlock *et al.*, 1984). Transformation was confirmed by the presence of foreign DNA sequences in both primary transformants and their progeny and by an antibiotic resistance phenotype conferred by a chimeric neomycin phosphotransferase gene.

Agrobacterium constitutes an excellent system for introducing genes into plant cells, because of the following reasons:

1. DNA can be introduced into whole plant tissues, which bypass the need for protoplasts, and
2. The integration of T-DNA is a relatively precise process.

The region of DNA to be transferred is defined by the *border sequences* occasional rearrangements do occur, but in most cases an intact T-DNA region is inserted into the plant genome. This contrasts with free DNA delivery systems in which the plasmids routinely undergo rearrangements and concentration reactions before insertion and can lead to chromosomal rearrangements during insertion in both systems. Sequencing of insertion sites shows that only small duplications or other changes occur in flanking sequences during T-DNA integration. The stability of expression of most genes that are introduced by *Agrobacterium* appears to be excellent.

Survey of literature reveals that integrated T-DNAs give consistent genetic maps and appropriate segregation ratio. Introduced traits have been found to be stable over at least five generations during cross breeding and seed increase on in genetically engineered tomato and oilseed rape plants. This stability is critical to the commercialisation of transgenic plants.

Agrobacterium-mediated gene transfer

The utility of this bacterium as a gene transfer system was first recognised when it was demonstrated that the crown galls were actually produced as a result of the transfer and integration of genes from the bacterium into the genome of the plant cells. Virulent strains of *Agrobacterium* contain large Ti- (tumor inducing) plasmids, which are responsible for the DNA transfer and subsequent disease symptoms. Genetic and molecular analyses showed that Ti-plasmids contain two sets of sequences necessary for gene transfer to plants, one or more T-DNA (transferred DNA) regions that are transferred to the plant, and the *Vir* (Virulence) genes which are not, themselves, transferred during infection. The T-DNA regions are flanked by 25 base pair right and left border DNA sequences that were shown to be responsible for the T-DNA transfer to infected plant cell and that it can be deleted. Phytohormones synthesis interference with plant regeneration and replaced with marker gene or desirable gene.

Vectors for *Agrobacterium* mediated plant transformation

Plant transformation vectors based on *Agrobacterium* can generally be divided into two categories those that co-integrate into a resident Ti-plasmid (co-integrate vectors) and those that replicate autonomously (the binary vectors). Both the vectors have several common features imposed upon them by the requirements of *Agrobacterium*.

Co-integrating vectors

Co-integrating transformation vectors must include a region of homology between the vector plasmid and the Ti-plasmid. This requirement for homology means that the vector is capable of integrating into a limited number of Ti-plasmids. The first utilises the disarmed *Agrobacterium* Ti-plasmid PG 3850. In this plasmid the phytohormones gene of the C 58 plasmid have been excised and replaced by pBR 322 sequence. Any plasmid containing the pBR 322 sequence homology can be co-integrated into the disarmed Ti-plasmid. The border sequences as well as a nopaline synthase gene are part of the Ti-plasmid, and the co-integration places the new sequences between the T-DNA borders.

Phytohormones genes According to Fraley *et al.* (1985) when the right border and all the phytohormone genes are removed from the Ti-plasmid, a left border and a small part of the original T-DBA referred to as the Limited Internal Homology (LIH), remains intact. The vector to be introduced into *Agrobacterium* contains the LIH region for homologous recombination as well as a right border. The co-integrated DNA reconstructs a functional T-DNA with a right border and left border. This system has been used extensively for introduction of many genes into plants. Once the co-integrate has been formed, the plasmid is stable in *Agrobacterium* and is virtually impossible to lose.

Binary vectors, on the other hand, are not completely stable in *Agrobacterium* in the absence of drug selection. There is also evidence that a co-integrating vector can transform tomato at a higher frequency than a binary vector.

Binary vectors

These vectors are different from co-integrating vectors. Instead of a region of homology with the Ti-plasmid, they contain origins of replication from a broad host-range plasmid. These replication origins permit autonomous replication of the vector in *E. coli* and *Agrobacterium*. Since the plasmid does not need to form a cointegrate, these plasmids are considerably easier to introduce into *Agrobacterium*. A major advantage to binary vectors is their lack of dependence on a specific Ti-plasmid. The vector may be introduced into virtually any *Agrobacterium* host containing any Ti or Ri plasmid as long as the *vir* helper functions are provided. This may be important in the transformation of some plant species, since different *Agrobacterium* strains exhibit major differences in their abilities to infect different plant species.

Selectable markers for plant transformation

Several selectable and non-selectable marker genes, *viz.*, neomycin, phosphotransferase, luciferase, streptomycin phosphotransferase, chloramphenicol acetyl transferase, hygromycin phosphotransferase, β-glucuronidase, and dihydrofolate reductase are widely available today for plant transformation and have also been transferred in large number of plant species (Tables 25.1 to 25.5). Several requirements must be considered in the development of a truly useful selectable marker system.

Neomycin phosphotransferase type II (NPT-II) enzyme It is most widely used selectable marker. It was originally isolated from the prokaryotic transposon Tn 5. It detoxifies

Table 25.1 Transgenic field crops.

Crop	*Method and gene transferred*
Tobacco	At, NPT-II
Alfalfa	At, NPT-II
Cotton	At, NPT-II
Sunflower	At, NPT-II
Moth Bean	At, NPT-II
Soybean	At, GUS
Safflower	At, GUS
Green Bean	At, GUS
Cowpea	EL* GUS
Chickpea	At, NPT-II

At: *Agrobacterium tumefaciens,* NPT-II: phosphotransferase-II. GUS β-glucuronidase. EL*: Electroporation.

Table 25.2 Transgenic vegetable crops.

Crop	*Method and gene transferred*
Carrot	Ar, NPT-II
Tomato	At, NPT-II
Cucumber	Ar, NPT-II At, Ar, NPT-II
Potato	At, NPT-II
Oil seed rape	At, NPT-II
Lettuce	At, NPT-II
Cauliflower	Ar, Mannopine At, NPT-II
Celery	At, NPT-II
Brinjal	At, NPT-II
Watermelon	At, NPT-II
Sugarbeet	At, NPT-II
Muskmelon	At, NPT-II
Bell pepper	At, GUS
Peas	At, NPT-II
Sweet potato	MB* GUS
Cabbage	At, GUS

At: *Agrobacterium tumefaciens,* Ar: *Agrobacterium Rhizogenes,* NPT II: Neomycin phosphotransferase-II, GUS: β-glucuronidase, MB*: Micro-projectile bombardment.

Table 25.3 Transgenic fruit crops.

Crop	*Method and gene transferred*
Walnut	At, NPT-II
Apple	Ar, NPT-II
Peach	At, Octopine
Grapevine	At, NPT-II
Papaya	MB*, NPT-II, At, cp gene
Strawberry	At, GUS
Citrus spp.	At, NPT-II
Apricot	At, cp gene
Pecan	At, GUS

At: *Agrobacterium tumefaciens,* cp gene: Coat protein gene, NPT-II: neomycin phosphotransferase-II, GUS: β-glucuronidase, MB* Microprojectile bombardment.

Table 25.4 Transgenic forest plants.

Crop	*Method and gene transferred*
Populus hybrida (Poplar)	At, NPT-II, 'aroA'
Pseudotsuga menziesii (Douglas fir)	At, Octopine, NPT-II
Azadirachta indica (Neem)	At, Octopine, NPT-II
*Picea mariana** (Black spruce)	EI*, CAT
*Pinus banksiana** (Jack Pine)	EI*, CAT
Salix spp. (willows)	At, NPT-II
Catharanthus roseus (Madagascar)	Ar, Agropine
Digitalis purpurea (Foxglove)	At, GUS
Picea rubens, P. glauca (Red spruce, White spruce)	MB*, GUS, NPT-II
Larix X eurolepis (Larch)	MB*, GUS, NPT-II
Liquidambar styraciflua (Sweetgum)	At, NPT-II, Bt gene
Santalum album (Sandalwood)	At, NPT-II

At: *Agrobacterium tumefaciens,* aro A' gene: Glyphosate resistance, Bt gene: -endotoxin gene, NPT-II: neomycin phosphotransferase-II, GUS: β-glucuronidase, MB*: Microprojectile bombardment, CAT: Chloramphenicol acetytransferase, EL: Electroporation

aminoglycoside compounds such as kanamycin and G 418 by phosphorylation. This gene, fused to constitutive plant transcriptional promoters, has been used successfully to transform a large number of plant species and has been incorporated into numerous plant transformation on vectors.

Transformation of plants

There are two basic approaches which have been used to obtain transgenic plants:

1. Co-cultivation of regenerating protoplasts
2. The leaf disc procedure

The production of transgenic plants can be divided into four main steps:

1. Introduction of foreign genes into modified *Agrobacterium strain*

Table 25.5 A list of higher plants where transgenic plants have been produced using different methods.

I. Herbaceous Dicotyledons	
1. *Nicotiana tabacum* (tobacco)	
2. *N. plumbaginifolia* (wild tobacco)	
3. *Petunia hybrida* (petunia)	
4. *Lycopersicon esculentum* (tomato)	
5. *Solanum tuberosum* (potato)	
6. *Solanum melongena* (eggplant)	
7. *Arabidopsis thaliana*	
8. *Lactuca, sativa* (lettuce)	
9. *Apium graveolens* (celery)	
10. *Helianthus annus* (sunflower)	
11. *Linum usitatissimum* (flax)	
12. *Brassica napus* (oilseed rape: canola)	
13. *Brassica oleracea* (cauliflower)	
14. *Brassica oleracea var, capitata* (cabbage)	
15. *Brassica rapa* (syn. B. campestris)	
16. *Gossypium hirsutum* (cotton)	
17. *Beta vulgaris* (sugarbeet)	
18. *Glycine max* (soybean)	
19. *Pisum sativum* (pea)	
20. *Medicago sativa* (alfalfa)	
21. *M. varia*	
22. *Lotus corniculatum* (lotus)	
23. *Vigna aconitifolia*	
24. *Cucumis sativus* (cucumber)	
25. *Cucumis melo* (muskmelon)	
26. *Cichorium intybus* (chicory)	
27. *Daucus carota* (carrot)	
28. *Armoracia sp.* (horse radish)	
29. *Glycyrrhiza glabra* (licorice)	
30. *Digitalis purpurea* (morning glory)	
31. *Ipomoea batatas* (sweet potato)	
32. *Ipomoea purpurea* (morning glory)	
33. *Fragaria sp.* (strawberry)	
34. *Actinidia sp.* (Kiwi)	
35. *Carica papaya* (papaya)	
36. *Vitis vinifera* (grape)	
37. *Vaccinium macrocarpon* (cranberry)	
38. *Dianthus caryophyllus* (carnationa)	
39. *Chrysanthemum sp* (chrysanthemum)	
40. *Rosa sp.* (rose)	
II. Woody Dicotyledons	
41. *Populus sp.* (poplar)	
42. *Malus sylvestris* (apple)	
43. *Pyrus communis* (pear)	
44. *Azadirachta indica* (neem)	
45. *Juglans regia* (walnut)	
III. Monocotyledons	
46. *Asparagus sp* (asparagus)	
47. *Dactylis glomerata* (orchard grass)	
48. *Secale cereale* (rye)	
49. *Oryza sativa* (rice)	
50. *Triticum aestivum* (wheat)	
51. *Zea mays* (corn)	
52. *Avena sativa* (oats)	
53. *Festuca arundinacca* (tall fescure)	
IV. Gymmosperms (a conifer)	
54. *Picea glauca* (white spruce)	

2. Co-cultivation of *Agrobacterium* strains with protoplasts, plant cells or tissues
3. Selection and regeneration of transformants, and
4. Analysis and verification of gene expression in transformed plants.

Most transgenic plants produced to date, were created through the use of the *Agrobacterium*-mediated gene transfer system. Other methods that have the potential to influence the production of transgenic plants include, microinjection, electroporation, direct injection into reproductive organs, PEG mediated gene transfer, particle gun and liposome mediated gene transfer.

Progress in this exciting area of research for production of transgenic plants has been so spectacular that by the turn of the century, we hope to be growing crops which have been tailored to market specification by the addition, subtraction or modification of genes.

Transgenes will also be important in increasing the efficiency or crop production systems. For instance, transgenic plants resistant to herbicides insects, viruses and a host of other stresses have already been produced. Transgenic plants have also been produced, which are suitable for food processing (*e.g.*, bruise resistance and delayed ripening in tomato). Another exciting example is the production of male sterile (due to barnase gene) and fertility restorer (due to barstar gene) plants in *Brassica napus,* so that hybrid seed in future will be conveniently produced without manual emasculation and controlled pollination as practised in maize. This has also eliminated the need for a search of cytoplasmic male sterility (cms) and fertility restoration system in crops, where hybrids are intended to be produced for higher yields. Another major goal for production of speciality chemicals and pharmaceuticals is described as 'molecular farming or molecular pharming'. The transgenic plants have also been produced for identification of regulatory sequences for many genes, using gene constructs with overlapping deletions. These achievements and prospects of their future use will be briefly described below.

Transgenic Plants for Crop Improvement

Transgenic plants in dicotyledons

Despite the totipotency of plants cells, thus obviating the need of transforming specifically the germ line cells as required in animals, the production of large number of transformed plants was till recently (in 1980's) restricted mainly to tobacco, pentunia or tomato of the family Solanaceae. There are, however, recent reports of transgenic plants from other dicotyledonous families like Cruciferae. Leguminosae, etc., and from monocotyledonous families like Liliaceae and Gramineae. Some of these examples, where transgenes of economic value have been utilised are included in Table 25.5 and will be described in this section.

Herbicide resistance in transgenic plants Due to increasing concern about contamination of environment by herbicides, they are being developed that are safer and biodegradable. This has necessitated the development of resistance in crop plants against these new and safer herbicides. Herbicides, normally affect processes like photosynthesis or biosynthesis of essential amino acids (Table 25.6). Two approaches have, therefore, been used for the development of resistant plants: (i) in the first approach, we try that either the target molecules should become insensitive to herbicide or the target protein should be overproduced, (ii) in the second approach, a pathway is introduced that will detoxify the herbicide.

(a) *Modification of the target* The target has been modified for developing resistance against at least three herbicides (glyphosate, sulphonylureas and imidazolines). The transgenic petunia plants resistant to glyphosate (active ingredient of **Roundup** herbicide) were developed by transfer of a gene for **EPSPS (5-enol-pyruvyl-shikimat-3-phosphate synthase)**, that overproduces this enzyme. This overexpressing gene was isolated from plants selected for herbicide resistance. In some other cases a gene *ero*A was isolated from the bacteria *Salmonella typhimurium* or *E. coli* and was transferred to tomato and/or tobacco. Another class of herbicides includes sulphonylurea compounds (active ingredients of **Glean and Qust** herbicide) and imidazolinones that inhibit the enzyme **acetolactate synthase (ALS)**, which is involved in the biosynthesis of branched chain amino acids like leucine, isoleucine and valine. Transgenic

Table 25.6 Mechanisms of action of different herbicides and basis of achieving resistance against them in transgenic plants.

Active principle of herbicide	*Inhibited pathway*	*Target product*	*Use*	*Basis of resistance*
I. Amino acid biosynthesis inhibitors				
1. Glyphosate (Roundup) biosynthesis	Aromatic amino acid	EPSPS	Broad spectrum	Overexpression of EPSPS gene, bacterial *aroA* gene
2. Sulphonyl urea and Imidazolinones	Branched chain amino acids	ALS	Selected crops	Mutant ALS gene
3. Phosphinothricin (Basta)	Glutamine biosynthesis	GS	Broad spectrum	Gene amplification bar gene: detoxification
II. Photosynthesis inhibitors				
4. Atrazine (Lasso)	Photosystem H	Qn (32 kDa protein)	Selected crops	Mutant *PsbA* gene GST gene: detoxification
5. Bromoxynil (Buctril)	Photosynthesis	—	Selected crops	*bxn* gene: detoxification

tobacco plants expressing a mutant ALS gene from tobacco or *Arabidopsis*, were produced that were tolerant to sulphonylurea herbicides. Similarly, transgenic tobacco plants were produced by incorporation of genes for resistance against (1) **L-phosphinothricin (PPT)**, which is the active ingredient of herbicide 'Basta' and inhibits **glutamine synthase (GS)**, and (ii) atrazine which inhibits photosynthesis. These genes were isolated from *Medicago sativa* and *Amaranthus hybridus* respectively. The transgenic plants with this gene for protection against the herbicide 'Basta'.

(b) *Detoxification or degradation of herbicide* Detoxification or degradation of herbicides is the basis of selective use of herbicide, so that the latter will kill the weeds and not the crop. A number of detoxifying enzymes have been identified in plants as well as in microbes. Some of these enzymes include (i) glutathione-transferase or GST (in maize and other plants), which detoxifies the herbicide *atrazine*; (ii) nitrilase (coded by gene *bxn* in *Klebsiella pneumoniae*, which detoxifies the **herbicide bromoxynil**, and (iii) phosphinothricin acetyl transferase or PAT (coded by *bar* gene in *Streptomyces* spp.), which detoxifies the herbicide **PPT (l-phosphinothricin)**.

Transgenic tomato plants using the *bxn* gene from *Klebsiella* and bar gene from *Streptomyces* and transgenic plants in potato, oilseed rape (*Brassica napus*) and sugarbeet using bar gene from *Streptomyces* have been obtained and were found to be herbicide resistant. Other target crops for engineered herbicide tolerance include soybean, cotton and corn. Field trials with transgenic plants in some cases are either being conducted or will be conducted in the near future, so that the utility of transgenic herbicide resistant plants and the risk of growing them at the commercial scale will be known. It is estimated that the first transgenic plants will be commercially grown and that a number of other transgenic plants will be grown commercially before the turn of the century.

Insect resistance in transgenic plants (a) ***Genes for Bt toxins.*** The use of pesticides and insecticides is a common measure in plant protection programmes, since pests and insects cause appreciable damage to our crops. Most of these pesticides and insecticides are chemically synthesised. However, an exception is the **Bt toxins** produced by a bacterial species (*Bacillus thuringiensis*), so that a spore preparation of this bacterium has been used as biological insecticide during the last 20 years. Insecticidal activity of this species depends, on a protein (*delta endotoxins*) synthesised during sporulation. Since these toxins are very specific in their action, they are safe insecticides, but there use is limited due to high production cost and due to instability of crystal proteins when exposed in the field.

The above toxin gene (*bt2*) from *B. thuringiensis* has been isolated and used for *Agrobacterium* Ti-plasmid mediated transformation of tobacco, cotton and tomato plants. The transgenic plants were resistant to the *Manducta sexta*, a pest of tobacco. Experiments of feeding the leaves of these plants to larvae of M. *sexta*, showed 75 per cent - 100 per cent mortality of the larvae, while the control plants carrying no transgenes were severely damaged. The presence of the gene, *bt2* as well as that of the toxin protein synthesised under its control was also demonstrated by appropriated experiments. When inheritance of insect resistance was studied using crosses with normal control plants. F_1 showed resistance and F_2 generation exhibited expected segregation.

Field tests using transgenic insect resistant plants were also conducted and the results were excellent with tobacco and tomato. In view of this, one can expect that transgenic crop plants for this trait may be released for commercial cultivation in the near future. There were also reports that India may acquire technology from USA for introduction Bt toxin gene in cotton for the development or resistance against pests in this major cash crop of India.

(c) *Genes for protease inhibitors (e.g., gene for cowpea trypsin inhibitor or CpTI)* In cowpea (*Vigna unguiculata*), trypsin inhibitor (CpTI) level was shown to be responsible for its resistance to attack by the major storage pest of its seeds (*i.e.*, brucide beetle=*Callosobruchus macultus*). **CpTI** was later shown to be toxic to a variety of insects (Table 25.5) but cowpea seeds with high level of CpTI are not toxic to humans. It was therefore, considered desirable to transfer genes) for CpTI for production of transgenic insect resistant plants. A number of binary vectors were developed using Ti-plasmid, where CpTI gene was joined with CaMV 35S promoter, and one or more marker genes. The vector was mobilised into *Agrobacterium*, which was used to infect tobacco leaf discs, which led to the production of transgenic tobacco plants expressing high level of CpTi (as shown by western blotting) imparting resistance against a variety of insects (Table 25.6) The CpTI gene in transgenic plants is stably inherited and there is no serious 'yield penalty'. Thus like Bt toxin. CpTI can also be used as a protectant against insect attack in transgenic plants, However, extensive field trials will be necessary before these transgenic plants can be released to the farmers for cultivation.

(d) *Genes for other insecticidal secondary metabolites* Secondary metabolites produced by plants have also been implicated in the resistance to insect attack (Table 25.7 and 25.8). However, biosynthesis of each of these metabolites involves a series of steps (sometimes even more than one biosynthetic pathway), each controlled by a separate gene. Furthermore, these genes are tissue specific in expression. These features make the production of transgenic plants difficult in this case. CpTI discussed in the previous section is also a secondary metabolite, but its transfer is easier, since it is single gene controlled.

The production of transgenic plants by transfer of genes for entire **multi-enzyme biosynthetic pathway** (or for its augmentation) is though not yet achieved, but seems possible in future. For this purpose not only the transfer of genes is required, but the gene expression needs to be regulated, otherwise it leads to **'yield penalty'** and also to toxic effect when consumed by humans and livestock. Defined promoter sequences are now available, which will, allow in the transgenic plants, a control over the temporal and spatial expression of genes for biosynthetic pathways of secondary metabolites. Utilising these facilities, **transgenic plants with insect** resistance due to secondary metobolites will be available in the near future.

Resistance against viral infection. (a) *Cross protection* It has been shown that if susceptible strain of a crop is inoculated with a mild strain of a virus, then the susceptible strain develops resistance against more virulent strains. This phenomenon is known as cross protection and has been used to reduce yield losses in crops like tomatoes against tomato mosaic virus (TMV), in potato against **Potato spindle tuber viroid** and in citrus against citrus tristeza virus. In most cases of cross protection, the symptoms of infection are delayed, and even the replication of virus is suppressed, although eventually the server strain may be able to overcome the

Table 25.7 Insecticidal efficacy of CpTI in artificial diets and in transgenic plants (after Gatehouse *et al.*, 1991).

Insects killed by CpTI in artificial diets	*Insects, against whom resistance noticed in transgenic tobacco plants (with CpTI)*
I. Lepidoptera	
Heliothis virescens	*Helothis virescens*
H. zea	*H. zea*
Spodoptera littoralis	*Spodorera littoralis*
Chilo partellus	
II. Coleoptera	
Callosobruclus maculatus	
Anthononnus grandis	
Diabnotica underinpunclata	these insects do not attack control
Tribolium conjusum	tobacco plants
Costelytrazealardica	
III. Orthoptera	
Locusta migratoria	

Table 25.8 Examples of secondary compounds from legume seeds with demonstrated insecticidal properties (after Gatehouse *et al.*, 1991).

Compound	*Bruchid (insect)*	*Legume*
I. Non-protein antimetabolites		
1. Alkaloids		
(i) 2,5-dihydroxymenthyl-dihydroxypyrrolidine (DMDP)	3,4-*Callosobruchus maculatus*	*Lonchocurpus sp.*
(ii) Castanospermine	*Callosobruchus maculatus*	*Custanaspermum australe*
2. Non-protein amino acids		
p-Aminophenylalanine	*Callosobruchus maculatus* *Zabrotes subfasciatus*	*Vigna sp.*
3. Rotenoids	*Callosobruchus maculatus*	*Lonchocarpus salvadorensis*
4. Saponins	*Callosobruchus maculatus* *Callosobruchus maculatus*	*Phaseolus vulgaris* *Glycine max*
5. Polysaccharides		
(i) Pectosams	*Callosobruchus maculatus*	*Phaseolus vulgaris*
(ii) Heteropolysaccharide	*Callosobruchus maculatus*	*Phaseolus vulgaris*
	Acanthoscelides obtectus	*Phaseolus vulgaris*
(iii)*Polysaccharide*	*Callosobruchus maculatus*	*Vigna spp.*
II. Protein antimetabolites		
1. Lectins (phytohacmagglutinins)	*Callosobruchus maculatus*	*Phaseolus vulgaris*
2. x-amylase inhibitors	*Zabrotes subjasciatus*	*Phaseolus vulgaris*
	Callosobruchus maculatus	*Phaseolus vulgaris*
3. Protease inhibitors	*Callosobruchus maculatus*	*Vigna turganculata*
4. Arcelin/LLP	*Zabrotes subjasciatus*	*Phaseolus vulgaris*
	Abrotes subjasctams	*Phaseolus vulgaris*

protection. There are various disadvantages of this practice of cross protection such as (i) possibility of mutation in inducing mild virus strain (ii) possibility of synergism between inducing virus and another unrelated virus. (iii) possibility unnecessary spread of mild virus causing threat for future yield losses and (iv) possibility of some yield losses is transferred and transgenic plants are produced. Such transgenic plants have been produced in tobacco, tomato and potato using a broad spectrum of plant viruses. Following three kinds of genes have been used for this purpose.

(b) *Gene for virus coat or capsid protein (CP) from positive strand RNA viruses* Coat protein gene from tobacco mosaic virus (TMV), classified as a positive strand **RNA** virus, has been transferred to tobacco and in the transgenic plant, expression of coat protein (CP) was observed. Further, when inoculated with TMV, the infection in the transgenic plants was very low and delayed relative to control plants that were not transformed. This provides a new approach for producing virus resistant plants complementing the efforts of classical plant breeding in producing disease resistant crop varieties. However, in such transgenic plants, coat protein gene should constitutively be expressed and may thus have effects on the nutritional value of plants. Subsequently this approach has been applied to a range of crops (tomato, alfalfa, tobacco, potato, melons, rice) for developing resistance against a broad spectrum of positive RNA strand plant viruses (*e.g.*, alfalfa mosaic virus, potato virus X=PVX, potato virus Y =PVY and potato leaf roll virus). Potatoes have been produced which have coat protein genes and are tolerant in field tests to both PVX and PVY. DNA coding for a component of TMV replicase enzyme, was also transferred to tobacco plants, conferring resistance to TMV.

(c) *Gene for nucleocapsid (N) protein from tomato spotted with virus (TSWV)* In earlier reports of resistance against viruses in transgenic plants, a number of **positive (+) strand RNA** viruses were involved compared to no report of resistance against any negative (-) strand virus. An important **negative strand virus is tomato spotted wilt virus** (TSWV), which causes considerable yield losses in crops like tomato, tobacco, lettuce, groundnut, pepper and in ornamentals like *Impatiens, Ageratum* and *Chrysanthemum*. In this virus, genomic RNA is tightly associated with 'nucleocapsid (N) protein. This protein helps in (i) wrapping of viral RNA and also in (ii) regulation of transcription-to-replication switch during infection cycle.

For producing transgenic tobacco plants. N gene was associated with CaMV 35S promoter and a leader sequence (to enhance expression). The transgenic plants showed expression of N-gene and also significant resistance to TSWV. A correlation was also observed between the amount of N protein and the level of resistance. The mechanism involved in imparting resistance against negative strand RNA virus like TSWV, may differ from that involved in positive strand RNA virus, where coat protein (CP) gene has been used.

(d) *Satellite RNA and its use for transformation.* Satellite RNAs are species of RNA associated with specific strains of some plant RNA viruses, although it is not necessary for their replication. Replication of this satellite RNA depends on the virus, so that it gets packaged with it to cause infection elsewhere. Therefore, satellite RNA depends on virus for its replication and transmission, even though it is unrelated to viral genome. Presence of sat-RNA leads to reduction in severity of disease symptoms, and therefore, has been used for developing resistance against specific viruses.

Using the above principle, transgenic tobacco plants have been produced which carried a DNA fragment, which when transcribed gives a species of satellite RNA which can reduce the severity of disease. Following two such attempts were made: (1) In one case, transgenic tobacco plants using DNA for RNA satellite of cucumber mosaic virus (CMV) were produced and when satellite free aggressive CMV strains were used for infection, the presence of satellite RNA (transcribed from DNA) in transgenic plants led to reduction in disease symptoms. These transgenic plants are also protected against another related virus, i.e., **tomato aspermy virus.** (ii) In the other example involving tobacco ringspot virus (TobRV), transgenic tobacco plants could be produced, which carried DNA version of satellite RNA of tobacco ringspot virus (TobRV), transgenic tobacco ringspot virus (StobRV). These transgenic plants are protected against the deleterious effect of subsequent TobRV infection. The advantage of using satellite RNA strategy over coat protein, but is expressed only after virus infection.

Resistance against bacterial and fungal pathogens Several examples are now available, where transgenic plants with resistance against bacterial and fungal pathogens have been produced. Some examples are given below:

Some pathogens for which resistance has been transferred in some crop plants

	Pathogen	*Disease*	*Resistance*	*Source of gene*	*Transgenic crop*
1.	*Pseudomones syringae*	Wild fire	Acetyltransferase gene	—	Tobacco
2.	*Alternaria langipes*	Brownspot	Chitimase gene	*Serratia marcescens* (soil bacterium)	Tobacco
3.	*Rhconciama solum*		Chitimase gene	Bean	Tobacco
4.	*Phytopluhana infestans*	Late blight	Cosmotim gene	Potato	Potato

Altered fatty acid composition in *Brassica* seed oil In the past, specialised fatty acid composition in the seed oil, desired for edible and industrial purposes, has been achieved by conventional plant breeding or by mutagenesis programmes. Following are some examples: (i) removal of erucic acid from rapeseed oil to create 'canota'; (ii) reduction in linolenic acid content in flax; (iii) increase in stearate content (six times that in wild type) in safflower (up to 12 per cent stearate content (six times that in wild-type) in safflower (up to 12 per cent stearate) and soybean (up to 30% stearate).

Tissue specific antisense RNA expression has recently been used for reducing 'stearoyl-ACP desaturase' (ACP=acyl carrier protein) activity in seeds, leading to alteration in the ratio of saturated to unsaturated fatty acids. Due to tissue specific expression of anti-sense RNA, integrity of membrane lipids in leaf remained unaffected. 'Stearoyl-ACP desaturase' catalyzes the first desaturation step in seed oil biosynthesis, converting '*stearoyl-ACP*' to '*oleoyl-ACP*'. Seed specific anti-sense gene constructs of *B. rapa's* (say, 3. campestris) stearoyl-ACP desaturase were used for production of transgenic *B. rapa* and *B. napus* plants. This led to reduction in desaturase activity, resulting into a dramatic increase in the level of stearate in the seeds. This example demonstrated the potential to engineer the composition of fatty acids in seed oil.

Resistance against stress A number of genes responsible for providing resistance against stresses such as heat, cold, salt, heavy metals, phytohormones and nitrogen have been identified. Studies are also being conducted on metabolites like proline and betalnes, that are implicated in stress tolerance. With this background, transgenic plants resistant to a variety of stresses will be produced in future.

In a recent report (Nature 23; April, 1992) results were described, where resistance against chilling (1 °C for 10, days) was introduced into tobacco plants, by introducing a gene for **'glycerol-I-phosphate aryl transferase'** *enzyme from Arabidopsis* (*Arabidopsis* is resistant to chilling). The enzyme encoded by nuclear genome and later transported to chloroplast, determines the level of unsaturation of fatty acids in the phosphatidyl-glycerol of chloroplast membranes. The plants with high proportion of cis-unsaturated fatty acids (*e.g.*, spinach and *Arabidopsis*) are resistant to chilling, and so are the transgenic tobacco plants carrying the gene for the above enzyme.

Transgenic plants suitable for food processing A number of examples are available, where transgenic plants suitable for food processing have been developed. (i) Bruise resistant tomatoes were developed which express antisense RNA against polygalacturonase (PG), which attacks pectin in the cell walls of ripening fruit and thus softens the skin (ii) Tomatoes exhibiting delayed ripening were developed, either by using antisense RNA against enzymes involved in ethylene production (*e.g.*, Acc synthase), or by using gene for ACC deaminase, which degrades I aminocyclopropane-1 carboxylic acid (ACC), an immediate precursor to ethylene. This will increase the shelf life of tomato. These tomatoes can also stay on the plant longer, giving more time for accumulation of sugars and acids for improving flavour. Therefore, they are described as 'Flavr Savr. (iii) Tomatoes with elevated sucrose and reduced starch could also be produced using sucrose phosphate synthase gene. (iv) Starch content in potatoes could be increased by 20-40 per cent by using a bacterial **ADP glucose pyrophosphorylase gene (*ADP GPPase*).**

Male sterility and fertility restoration in transgenic plants (suitable for hybrid seed production) During 1990-1992, an exciting example producing transgenic plants with male sterility and fertility restoration genes has become available in *Brassica napus*. This should facilitate production of hybrid seed without manual emasculation and controlled pollination as often practised in maize.

In 1990, **C. Mariani** and others from Belgium, successfully used a gene construct having an **anther specific** promoter (from ***TA29*** gene at tobacco) and bacterial coding sequence for a **ribonuclease (*barnase*** gene from *Bacillus amyloliquefaciens*) for production of transgenic plants in *B. napus*. The results were spectacular in the sense that the transferred gene prevented normal pollen development leading to male sterility. The product of *barnase* gene is cytotoxic, killing the tapetal cells, thus preventing pollen development. Utilising this male sterility barnase gene construct (TA-29-RNase), it was possible to introduce male sterility in other crops also. These crops include tobacco, lettuce, cauliflower, cotton, tomato, corn, etc.

The same group of workers (Mariani *et al.*, 1992) used another gene construct later involving the same anther specific promoter, *i.e.*, **TA 29** and the **barstar** gene from *B. amyloliquefaciens*, for production of transgenic plants in *napus*. 'The product of *barstar* gene is a **ribonuclease-inhibitor.** It forms a complex with ribonuclease and neutralises its cytotoxic properties. In *Bacillus*, the ribonuclease is active extracellularly and the bacterium itself is protected by a ribonuclease inhibitor protein coded by *barstar* gene. When transgenic male sterile plants (with *barnase* gene) were crossed with transgenic male fertile plants (with *barstar* gene), the F_1 plants expressed both genes so that male fertility was restored due to suppression of cytotoxic ribonuclease complexes. This system of transgenic plants should facilitate hybrid seed production in crop plants in general.

Transgenic plants in monocotyledons

Production of transgenic plants in monocotyledons was initially not possible due to the following two limitations: (i) monocotyledons are ordinarily not infected by *Agrobacterium*, which was so widely used in dicotyledonous plants for carrying Ti-plasmids for transformation, and (ii) the regeneration of plants from protoplasts or single cells, which are commonly used for transformation, was not possible. Both these limitations have been overcome, since alternative methods for DNA uptake have now been developed and regeneration protocols for crops like rice and maize have been successful during the last few years.

Transgenic plants in rice In rice (both in *japonica* and *indica* varieties), there are reports (1998-1991) of successful production of transgenic plants. In one case, transgenic plants were produced, which carried a functional gene for **aminoglycoside phosphotransferase II(APT (3*) II)** along with the reporter gene for **neomycin phosphotransferase (NPT II).** In the other case, the transgenic rice plants possessed a bacterial *hph* gene, encoding **hygromycin B resistance** (Hm^1) along with the reporter gene encoding **β-glucuronidase (GUS).**

Transgenic plants in maize In maize, a reporter gene for neomycin phosphotransferase (NPT II) associated with 35S promoter region of the cauliflower mosaic virus (CMV) was used for production of transgenic plants. At the enzyme level, these plants exhibited I NPT II activity, and at the DNA level the presence of NPT II gene was demonstrated by Southern blot analysis.

In USA, the biotechnology company, **Monsanto** recently produced insect resistant transgenic maize plants, using the *Kurstaki* gene derived from *B. thuringiensis*. The gene was attached to CaMV 35S promoter, which enhanced the production of toxin, 1000 times. When each transgenic plant was inoculated with 100 larvae of European corn borer, complete control of insect was observed. Transgenic rice plants, resistant to a certain virus, were also produced using coat protein (P) gene from the virus.

Transgenic plants in wheat In wheat, transgenic plants were produced (in 1992) by a non resident Indian (NRI) scientist (Indra K. Vasil) and his coworkers. In these transgenic plants, resistance gene (*bar* gene) against the herbicide PPT (commercial name **'Basta'** =20% PPT) along with '*gus*' marker gene associated with CaNIV 35S promoter +Adh intron was shown to increase the activity of 35S promoter leading to enhanced expression of the associated gene) was introduced into wheat plants. 'The transgenic wheat plants showed expression of *gus* activity and also the resistance against the herbicide PPT. The development of these transgenic wheat plants has been considered a major breakthrough, since wheat is a major cereal crop and its improvement through transfer of foreign genes will be a great advantage to wheat breeders.

In the above successful reports for the production of transgenic plants in monocots, initially no useful genes were used for transformation and transformed plants were often sterile, although reports of fertile transgenic plants are now available. Therefore, they had mainly demonstrated that genes could be transferred into monocotyledons also. Efforts are being made now to produce transgenic monocotyledonous plants using unmodified desirable genes or suitably modified genes to meet the desired needs. Virus resistant rice plants using a coat protein gene have been produced. Similarly herbicide resistant wheat (also oats) plants have also been produced as discussed above. It is therefore apparent that transgenic cereals will be produced and field tested in future to be released for cultivation. However, before this is achieved, there is publicity against these transgenic plants questioning their safety for human consumptions. Therefore, fears of their safety will need to be dispelled before one can be successful in growing these transgenic plants at any large scale.

Transgenic plants for molecular farming

An additional major goal of biotechnology industry is also the use of transgenic plants as 'factories' for manufacturing **speciality chemicals and pharmaceuticals.** Sugars, fatty acids, starches, celluloses, rubber and wax are traditionally obtained from plants and genetic engineering can be used to increase their production. Following are some of the examples: (1) Transfer of a gene for **mannitol dehydrogenase** from *E. coli* to tobacco was achieved, which led to increase in the level of **mannitol** in transgenic tobacco plants. (2) Transfer of bacterial gene (from *Klebsiella*) for **cyclodextrin glucosyltransferasc (CGTase)** to potato was achieved successfully, leading to the production of a α and β **cyclodextrins (CDs)** in potato tubers. Cyclodextrins are cyclic oligosaccharides containing six, seven or eight glucose molecules in α, or γ CDs respectively. CDs can be obtained from starch by the action of **cyclodextrin glucosyl transferase (CGTase)** enzyme, and are useful for pharmaceutical delivery systems, flavour and odour enhancement, and for removal of undesired compounds (such as caffeine) from foods. The major use of CDs is due to their ability to form inclusion complexes with a wide variety of or organic compounds. Tissue specific expression or CGTase in potato tuber and its targeting to plastids or action on stored starch was achieved through the use of a chimeric gene construct consisting of the following: (i) a **patatin promoter** (for tuber specific expression of CGTase gene); (ii) a DNA sequence encoding small subunit of ribulose his phosphate carboxylase (**SSU**) **transitpeptide.** (iii) the CGTase structural gene from *Klebsiella,* and (iv) nopaline synthase (*nos*) 3* region. More transgenic plants with CGTase gene may be produced in view of the present high cost and world demand of CDs (Table 25.9) and due to availability of a number of CGTase genes (Table 25.10) (3) Transfer of genes for acetoacetyl CoA reductase (*phb B*) and polyhydroxybutyrate (**PHB**) synthase (*phb* C), which catalyze two steps in the production of

Table 25.9 Present and anticipated world market for cycle lextrins (Source: Consortium for electrochemische Industries GmbH).

Application	*Market (tonnes year^{-1})*	
	Present	*Expected for 1995*
Pharmaceuticals	50	2000
Food	700	2500
Cosmetics	50	500
Agriculture	10	100
Chemical industry (biotransformations, separation, catalysis)	30	300
Other purposes (*e.g.*, diagnostics)	10	200

Table 25.10 Bacteria which produce CGTases.

Organism	*CGTase type*	*Gene cloned*
Klebsiella oxytoca M5al	α	+
Bacillus macerans	α	+
Bacillus stearothermophilus	α	+
Bacillus circulans	β	+
Bacillus megaterium	β	–
Bacillus obbensis	β	–
Micrococcus sp.	β	–
Alkalophilic Bacillus 38.2	β	+
Alkalophilic Bacillus 17.1	β	+
Alkulphilic Bacillus 1011	β	+
Alkalophilic Bacillus 1-1	β	+
Bacillus subilis No. 313	γ	+
Alkalophilic Bacillus 290-3	γ	+

polyhydroxybutyrate or **PHB** (a biodegradable thermoplastic polymer). These genes were successfully transferred and their expression was demonstrated in transgenic plants of *Arabidopsis thaliana.* (4) Chimeric genes having CaMV promoter and encoding **human serum albumin (HSA)** were transferred successfully and transgenic potato plants obtained. The secretion of protein was achieved by using either the human preprosequence or the signal sequence from extracellular **PR-S** protein from tobacco. HAS was secreted in transgenic leaf tissue. (5) The production of pharmaceutically active compounds like **enkephalins** was achieved in transgenic oil seed rape.

CHAPTER 26

Eukaryotic Chromosome Structure

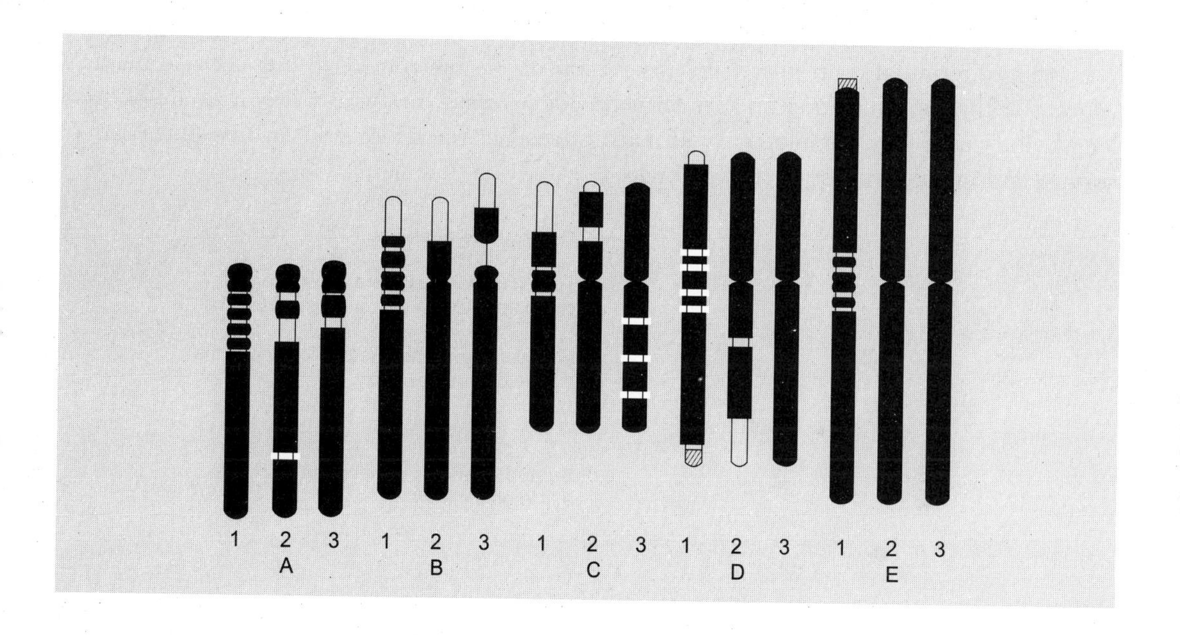

The role of chromosomes is unquestionably made clear in the heredity of the organisms whether plant or animal, as the carriers of genes and are faithfully replicated at each cell generation in the cell lineage of the organism. The morphology of chromosomes enables us to predict the genetic and cytogenetic consequences depending upon the length, shape, size, number and the distribution of the euchromatin and heterochromatin inside the chromosome of a particular species.

CHROMOSOME SHAPE (FIGS. 26.1 AND 26.2)

In somatic Metaphase or Anaphase the chromosomes reach their maximum contraction and attain a length that under ordinary environmental conditions remains constant from cell to cell. Shape characteristics are constant and is often used to identify the various chromosomes of a species.

The shape of the chromosomes is determined by the position of the *centromere.* The centromere can be seen as a constriction at metaphase and it may be *terminal, sub-terminal* or *median* position making the chromosomes look like a rod, J or V-shape as they move to the anaphase poles. In addition to these types, ring chromosomes have been found in maize and in *Drosophila* but their longevity is very unstable and are often lost in the cell cycle. The rod shaped chromosomes though possessing a terminal centromere, there is still a short arm exceedingly minute. The X chromosome of *Drosophila* was thought to be telocentric, i.e. possessing a terminal centromere. By misdivision of the centromere in a transverse manner, telocentric chromosomes arise and are termed *isochromosomes,* i.e., whose two arms are of equal length and genetically homologous with each other. But the telocentric chromosomes cannot survive and are always lost in the cell cycle.

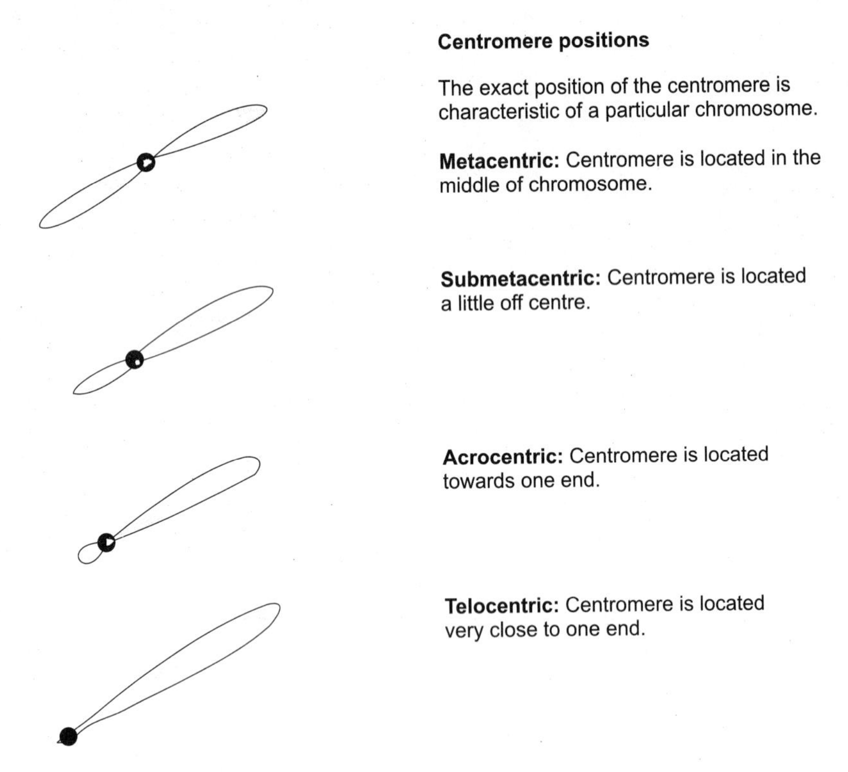

Fig. 26.1 Types of chromosomes.

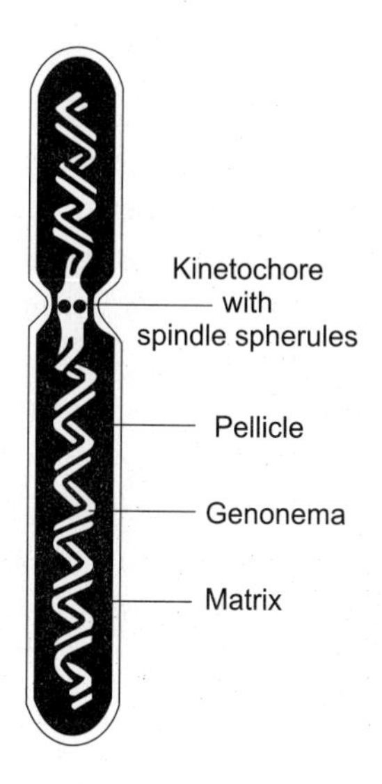

Fig. 26.2 Morphology of metaphase chromosome of mitosis.

Centromeres are often pin point structures sharply localised on the chromosome. However, sometimes a *diffuse centromere* occurs as in *Luzula*, a plant species and in insects of the order Hemiptera, where the spindles are attached all along the length of the chromosome and the entire chromatids are dragged apart at the Anaphase to the two poles in these cases.

In addition to the centromeric constrictions known as primary constrictions, some chromosomes possess *secondary constrictions* occurring as terminal bodies and are termed *trabants* or *satellites* and the chromosomes are referred to as SAT-chromosomes.

CHROMOSOME SIZE

Length measurements are made at the metaphase stage and are constant for a given species. The relative size of the chromosomes roughly correlates with the number of genes that it carries, but not a rule. Chromosome size is not an indication of gene content even in related genera or within a species. In the family Droseraceae the size ratio was of the order of 1000:1, by comparing the volume of the chromosomes in species complements. In *Lolium perenne* individual plants show marked size variations. Certain chemicals make the chromosomes contract very much and are environmentally controlled also like temperature.

The fungi generally have minute mitotic chromosomes. Among plants, monocots possess larger chromosomes than dicots but *Paeonia* is an exception with large chromosomes (fam: Ranunculaceae). *Trillium* has chromosomes up to 30 microns length; among animals, Orthopterans and Amphibians have large chromosomes. Human chromosomes measure from 4 to 6 microns on an average. *Drosophila* and maize average 3.5 and 8–10 microns at metaphase. There is no upper limit but chromosomes longer than the distance from pole of spindle to metaphase plate, would suffer loss of their terminal regions by the action of cell plate during cytokinesis.

CHROMOSOME NUMBER

The number of chromosomes in somatic cells are expressed as *zygotic*, *diploid* or *somatic* number while in the reduced chromosomes in the egg or sperm as *gametic* or *haploid* number. These are designated as 2*x* and *x* numbers, respectively. In polyploid series the primitive or the original number is known as *base number* and is called *n*. In wheat a series of species having 14, 28 or 42 chromosomes as their somatic (2*x*) number or 7, 14, or 21 as their gametic number, the base number (*n*) of the entire series is 7 and the 28- and 42-chromosome wheats are called *tetraploid* and *hexaploid* wheats.

Chromosome number varies from as low as two pairs in *Haplopappus gracilis* (fam. Compositae), three pairs in *Crepis capillaris* ranging to several hundred as in *Aulacantha*, a Radiolarian protozoan with 1600 chromosomes. *Ascaris megalocephala* var. *univalens* exhibits a single pair of chromosomes through the germ line.

Individuality of the chromosomes has been studied by obtaining *karyotypes* or *idiograms* (haploid complements) from a consideration of their shape and size measurements in several organisms.

DETAILED MORPHOLOGY OF THE CHROMOSOME

Chromonemata

The chromosome possesses a number of longitudinal sub divisions called chromonemata (singular chromoneme) which are carriers of genes. It is a matter of doubt always as to the exact number of chromonemata that a chromosome possessing in all organisms. However, the chromatid is the functional unit of chromosome in cell division and in gene segregation through crossing over.

A chromosome is divided into chromatids, chromatids into half-chromatids, and half-chromatids into several chromonemata. The chromonemata finally are composed of several DNA double helices. (Steffensen, 1959).

The bipartite chromosome of Prophase is sub divided into a quadripartite structure. The Prophase chromosome is also octipartite observed from treated chromosomes in certain coccid chromosomes. Whatever may be strandedness of the chromosome, it is the chromatid that is the functional unit in crossing over and coiling and not the smaller subdivisions.

Matrix. The chromonemata are found embedded in a mass of achromatic material called the matrix which is bounded by a sheath or pellicle which are non-genic materials. These structures have been served in maize, *Trillium* and *Tradescantia.* The matrix protects as a shield in keeping the chromonemata intact during the cell division. It also forms an insulating sheath during cell division.

Chromomeres. The chromosomes are particularly attenuated all along the length and are distinctly visible in pachytene chromosomes which are constant in size and position for a given chromosome which are known as chromomeres. These were first described by Balbiani in 1876. Stretching of the chromosomes does not disrupt the chromomeric pattern.

Centromeres. The centromere is the chief part of the chromosome forming a primary constriction. Chromosomes lacking a centromere fail to orient on the Metaphase plate, lag at Anaphase and are eventually lost. Since the position of the centromere is constant, the centromeres are responsible for the shapes of the chromosomes as they move poleward in anaphase.

Exceptionally in maize for chromosome-10 in a homozygous condition, Anaphase movement takes place prematurely (precocious) with distal ends of the chromosomes moving poleward instead of centric regions (neo-centric activity).

Centromere is functionally divided in the longitudinal axis of the chromosome at the beginning of the Anaphase.

The structure of the centromere is difficult to interpret. In somatic chromosomes it appears as a non-staining constriction with no morphological evidence of structure. In pachytene chromosomes of maize, it takes the form of a distinct ovoid body, non-stainable and structureless and somewhat larger in diameter than the rest of the chromosome. Sometimes the centromere is attached by thin threads to the rest of chromosome as in *Tradescantia.* Centromeres of *Secale* and *Agapanthus* show three distinct zones of differentiation. Two or three pairs of granules make up the chromomeric zone with centromere, and these are separated from each other by other fibrillae.

Non-localised centromeres are found in *Luzula purpurea* and *Ascaris megalocephela.*

Secondary Constrictions. The secondary constrictions arise as a result of nucleolar formation. The nucleolus diminishes by the advance of the stage from Prophase to Metaphase, At the pachytene stage when the nucleolus is large the nucleolar organising regions in chromosomes of maize and tomato are each associated with a large heteropycnotic region.

The Telomere. The chromosome at either end terminates in a structure called telomere. The meiotic chromosomes of tomato are all terminated by a telomere, whereas in rye, *Agapanthus* or *Sorghum* trail off into ghost like ends. The loss of telomere induces instability in the chromosome ends which is an unsaturated state and hence, it will unite with other broken ends of the chromosomes and "healing" takes place. The telomeres sometimes congress towards the nuclear membrane due to pronounced polarisations and forms what is known as the "bouquet stage".

HETEROCHROMATIN AND EUCHROMATIN (FIG. 26.3)

Chromosomes are made of two kinds of chromatin based on stainability *Euchromatin* and *Heterochromatin.* The former contains the Mendelising genes and the latter supposed to be inert genetically, but recent findings reveal genetic activity present in heterochromatin also. The euchromatin part of the chromosome stains lightly whereas the heterochromatin stains deeply. The difference in staining is attributed to differential condensation of the chromosomes in the two regions. Chromosomes which have clear distinction into euchromatin and heterochromatin are termed differentiated chromosomes and are characteristic of tomato, *Sorghum*, *Physalis*, etc.

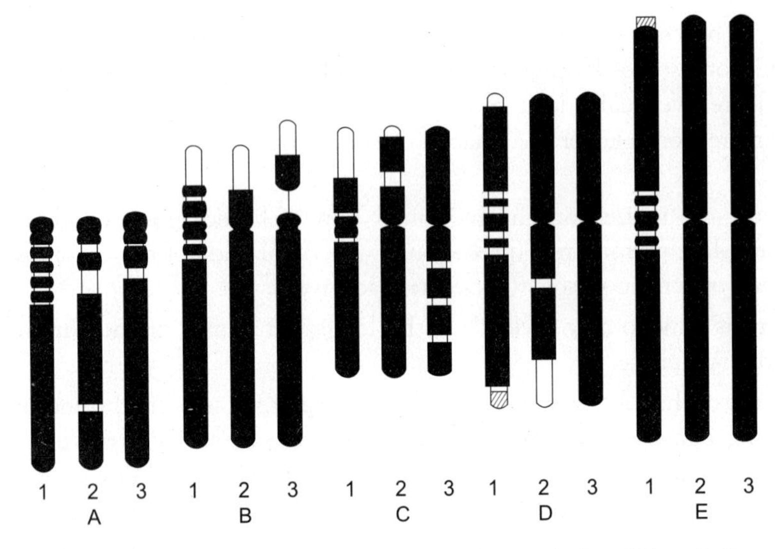

Fig. 26.3 Schematic representation of the size and position of the differentially staining regions (heterochromatin) in five haploid somatic chromosomes of *Trillium* when cell division takes place at low temperatures. 1. *T. erectum*; 2. *T. grandiflorum*; 3. *T. undulatum*. Letters indicate the different chromosomes of the haploid set. Cross-hatched regions appear only rarely even under conditions of low temperature.

Chromosomes which cannot be distinguished clearly into these two regions are termed undifferentiated chromosomes.

There are two general types of heterochromatin, called facultative and constitutive. Facultative heterochromatin may revert to the euchromatic state at certain times, in response to physiological and developmental conditions.

On the other hand constitutive heterochromatin is permanently condensed, genetically stable, late replicating material all of the time, e.g., centromere region of all chromosomes and mammalian X chromosome(s).

The heterochromatic X chromosome is visible which is a dense blob in interphase nucleus - Barr body which is used in the identification of the sex of embryo.

XY (♂) No barr body.

XXY (♀) One barr body (Klinefelter's syndrome).

XO (♀) No barr body (Turner's syndrome).

Only the euchromatic X is active and genetically functional.

X inactivation is random.

Lyonisation (Mary Lyon)

Lyon hypothesis in Mouse.

SPECIAL TYPES OF CHROMOSOMES

Lampbrush chromosomes (Fig. 26.4)

In majority of the organisms the chromosome shows a constant picture so far as its morphology is concerned. But some exceptions are there where special types of chromosomes are formed whose origin is unknown. These are the lampbrush chromosomes of vertebrate oocytes, salivary gland chromosomes of Diptera chromosomes of plants and animals.

In the developing oocytes of vertebrates like *Rana temporaria* (a frog), *Triturus viridescence* (a newt), the chromosomes are exceptionally large during the egg development reaching to an

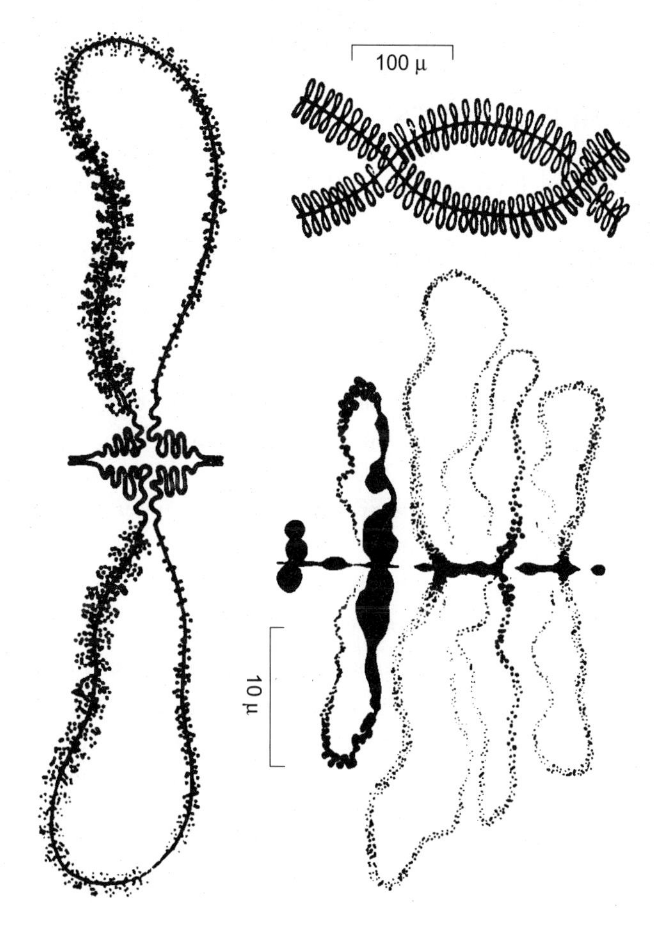

Fig. 26.4 Diagrammatic representation of the lampbrush chromosomes of the newt, *Triturs viridescens.* Top right, a bivalent with homologous chromosomes joined by two chiasmata, and showing the paired arrangement of loops, Bottom right, semi-diagrammatic view of a section of a single chromosome showing the variation which exists in loop and chromomere morphology. It will be noticed that the loop is always heavier in structure on one side than on the other. Left, postulated structure of a chromomere and its attached loops.

enormous length of 800 to 1000 microns. In *Triturus*, the 11 chromosomes reach a length of 5900 microns at the period of greatest extension.

The chromosomes resemble a single plastic cylinder in which, at specific loci, are embedded chromatic granules distinguished into "chromioles" and "chromomeres" which are ellipsoidal in appearance. These are in the form of loops which are 9.5 microns in length but in *Triturus* they measure 200 microns. A decrease in the number of loops occurs by disintegration rather than reabsorption back into the chromomere. Hence chromatin material is synthesised for the developing oocyte for eventual utilisation and not an integral part of the chromonemata in the form of a major coil. The chromosomes possess remarkable elasticity.

Nucleoli formation in lampbrush chromosomes follows an unusual pattern and as many as thousand nucleoli are found floating in nucleoplasm.

SALIVARY GLAND CHROMOSOMES

The giant chromosomes were discovered in 1881 by Balbiani in the salivary gland chromosomes of dipteran flies. Kostoff pointed out that the bands are similar to the linear arrangement of genes on the chromosome. Painter and Baner established the fact that the visible chromosomes consisted of homologous paired chromosomes intimately synapsed (Figs. 26.5, 26.6, 26.7 and 26.8).

Quite a large amount of work was done on the salivary gland chromosomes of *Drosophila melanogaster* and the band pattern was used to study the chromosomes genetically and cytologically. The salivary gland chromosomes are 100 times larger than the normal mitotic chromosomes and measure from 1180 to 2000 microns. In *Chironomus tentans* another dipteran genus, a pair of synapsed chromosomes is 20 microns in diameter and 270 microns in length. In *Rhyncosciara* also the chromosomes are much larger.

The characteristic features of the salivary gland chromosomes are:

(1) Giant size, when compared to mitotic chromosomes.

(2) Intimate synapsis of homologous chromosomes, like pachytene chromosomes of meiosis.

(3) Distinct transverse bands, which consist of alternating chromatic and achromatic regions. These bands differ in thickness, and they are so specific, that the entire chromosome can be mapped (Fig. 26.8). Small chromosomal abnormalities can be located like translocations, inversions, deletions and duplications.

(4) All chromosomes are united at one point called chromocentre.

(5) The chromosomes are called *polytene* chromosomes because, there are many strands present on a single transverse band. As many as 16 strands have been observed on a single bond. The polytene concept has been confirmed by observation of "Balbian rings" (chromosome "*Puffs*") (Fig. 26.7).

(6) Giant chromosomes are also found in the cells of Malpighian tubules, fat bodies, ovarian nurse cells and gut epithelia.

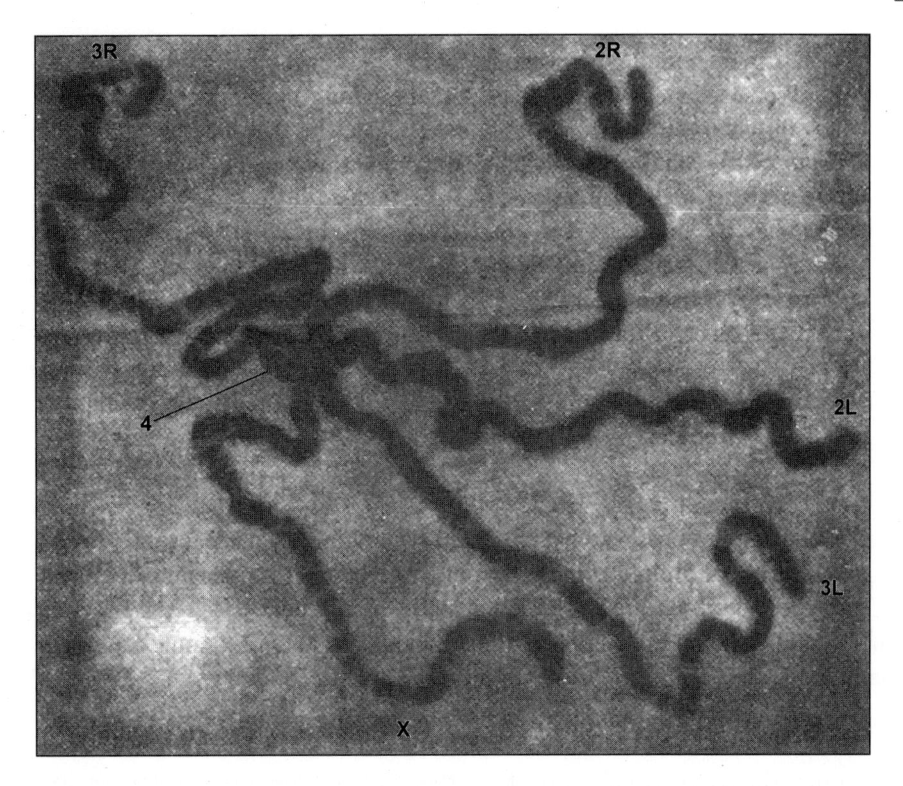

Fig. 26.5 Salivary gland chromosomes of a *Drosophila melanogaster* female showing the X chromosome, the arms of the two autosomes (2L, RR, 3L, and 3R), and the small 4 chromosome. The diploid number of chromosomes is present, but the homologues are in intimate synapsis.

Supernumerary Chromosomes

Many plant and animal species possess what are called supernumerary chromosomes or B-chromosomes, i.e., in addition to the normal diploid complement of chromosomes (A-chromosome), supernumerary chromosomes are present which are unstable and vary in number. These are genetically inert and produce little phenotypic expression, and are largely heterochromatic. Examples are: *Sciara* a dipteran fly; and hemipteran insect *Metapodius*; in plants like maize, *Sorghum*, *Pennisetum*, *Secale*, *Anthoxanthum*, *Poa*, etc.

ORGANISATION OF EUKARYOTIC GENOMES

It has been calculated that a human cell contains between 30,000 and 40,000 different genes, and that the average size of a gene is a little less than 5,000 bp. This suggests that the total coding region of the human genome takes up on more than 200, 000, 000 or 2×10^8 bp. However, the total length of the DNA molecules contained within the 23 human chromosome are in the region of 2×10^9 bp, which is about 10 times more than is needed for coding sequences. Even if we take into account promoters, enhancers, replication origins and other

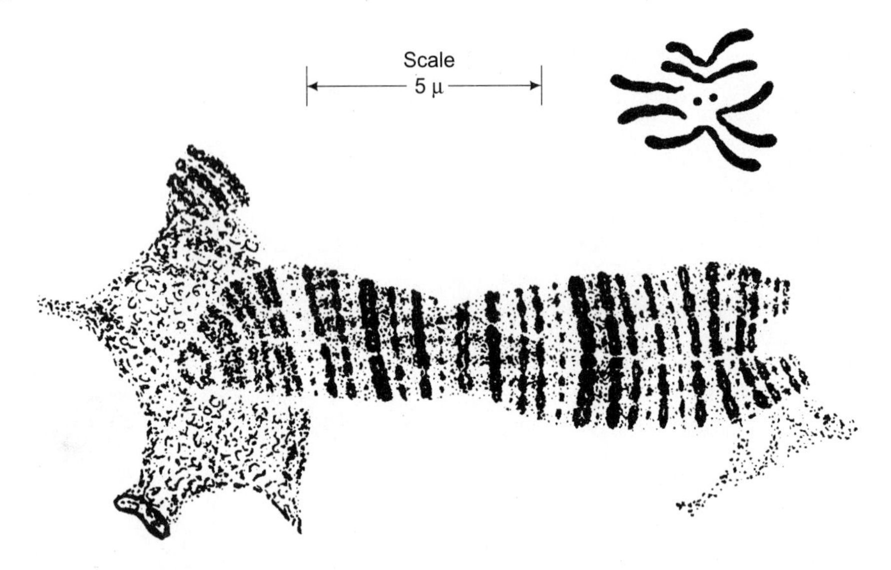

Fig. 26.6 The IV chromosome of *D. melanogaster,* as it appears in the cells of the salivary gland and in gonial cells (upper right, and indicated by arrow). Mass of material at the left end of the IV chromosome is the chromocentre to which the other chromosomes are also attached. Comparison of gonial and salivary gland chromosomes gives an indication of the wealth of detail which can be obtained from these giant structures.

essential elements we are still left with about 80% of the genome apparently unused. Similar calculations offer the same conclusions for other vertebrates; only when we look at lower eukaryotes, such as yeast and fungi, do we find a closer agreement between the expected and actual genome sizes.

MOST EUKARYOTIC GENOMES CONTAIN REPETITIVE DNA

The nature of the apparently redundant component of the eukaryotic genome was indicated first by biochemical study of purified DNA during the 1960s and 1970s, and more recently by direct sequence analysis during the 1980s. A substantial proportion of the eukaryotic genome is made up of repetitive DNA: individual sequence elements that are repeated many times over, either in tendem arrays or interspersed with the single copy DNA that comprises most of the coding regions. In the human genome repetitive DNA makes up about 50% of the total, and accounts for the bulk of the redundant portion.

Repetitive DNA is conventionally thought of as falling into two classes:

1. Highly repetitive DNA, made up of sequences that are repeated several hundred to several million times in the genome.
2. Moderately repetitive DNA, comprising sequences repeated less frequently, between 10 and several hundred times.

In fact, this distinction is rather artificial and it probably places misleading emphasis on the two extremes of what is more probably a continuous spectrum of repetitive DNA subclasses.

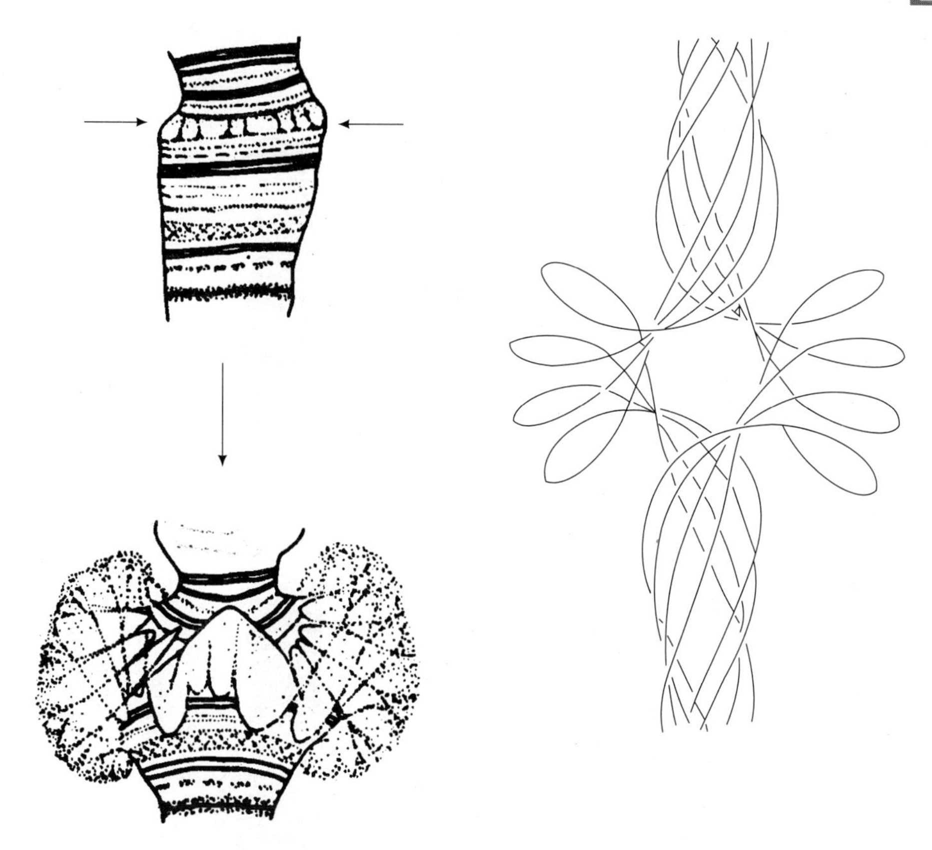

Fig. 26.7 Chromosomes of *Chironomus*. *Upper left*: normal banded appearance of a portion of a salivary gland chromosome which will expand locally (indicated by arrows) into a Balbiani ring (lower left). *Right*, Beermann's interpretation of the chromonematal structure of the Balbiani ring, but with only a few of the many strands indicated.

Each species has a number of different repeated elements in the genome, with the sizes of the elements anything between a few base pairs to several kb (kilo bases). It was found that sequences are interspersed in the genome, and the positions of repeated sequences are not random, as was thought at first. Whether this implies a function for the repetitive sequences, (possibly in packaging DNA into chromosomes), has not been determined.

The Alu Family of Mammalian Repetitive DNA

Experimental problems in analysing repetitive DNA have meant that very few repeated sequences have been fully characterised. An exception is the Alu family, a type of repetitive DNA found in several species of mammals, including man. In humans the basic Alu element is about 300 bp in length and occurs about 300,000 times in the genome. The individual repeat elements are not identical but share substantial sequence homology, usually greater than 80%, which means that on an average eight out of every ten nucleotide positions are the same. The equivalent family in mouse is called B1 and has 50,000 copies in the genome.

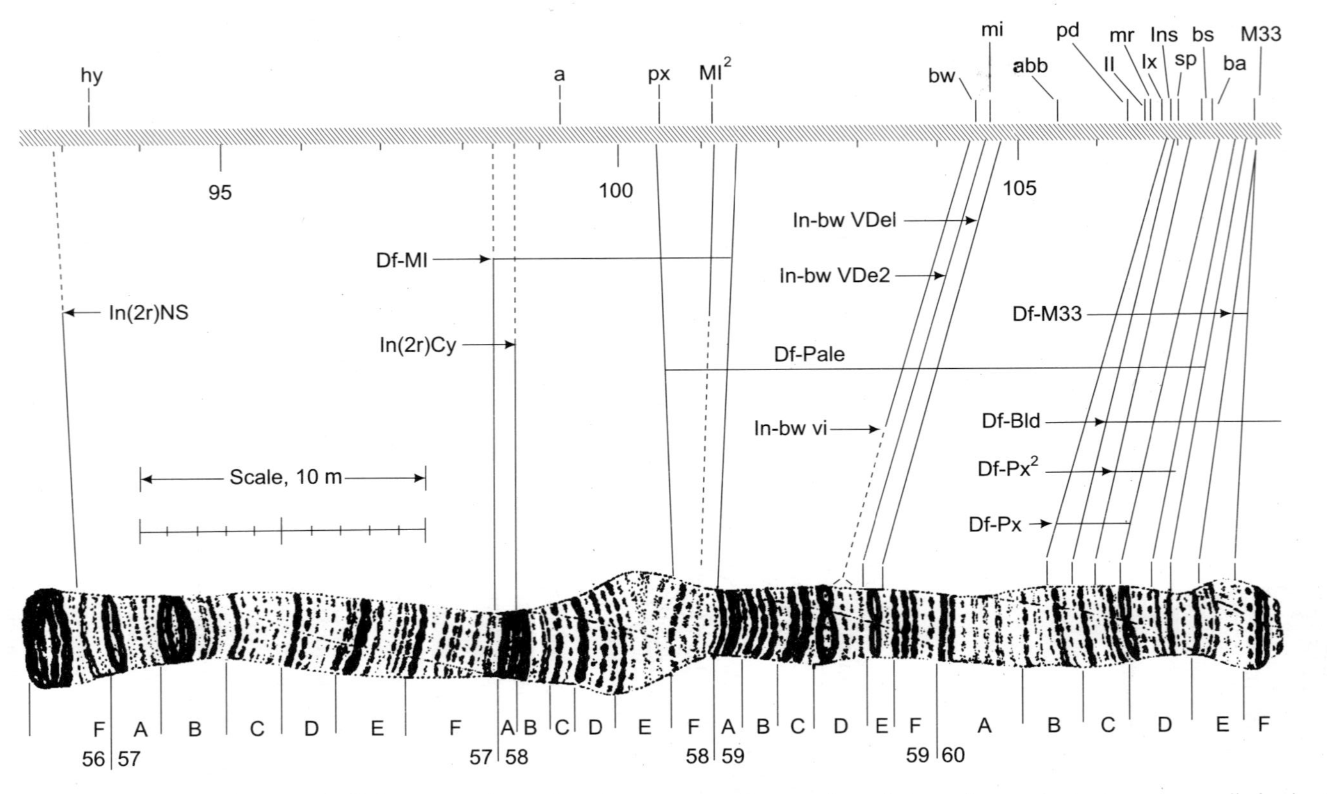

Fig. 26.8 A comparison of the genetic linkage map (above) and a corresponding section of the salivary-gland chromosome (below). Region is the end portion of the right limb of the second chromosome of *Drosophila melanogaster*.

MOLECULAR STRUCTURE AND ORGANISATION OF EUKARYOTIC CHROMOSOMES

Much of our information about the structure and mechanisms of DNA replication has come from studies of prokaryotes. The reason for this is that prokaryotes are less complex, both genetically and biochemically, than eukaryotes. Prokaryotes are monoploid (mono = one; equivalent to the haploid state of diploid organisms); they have only one set of genes, one copy of the genome. Most higher animals and many higher plants, by contrast, are diploid, having two complete sets of genes, one from each parent. Some higher plants are polyploid (poly = many), that is, carry several copies of the genome. Much of the genetic information of most viruses and prokaryotes is stored in a single chromosome, which in turn contains a single molecule of nucleic acid (either RNA or DNA).

The smallest known RNA viruses have only three genes. In fact, the complete nucleotide sequences of the genomes of several of these viruses are now known. (The phage MS2 genome is 3569 nucleotides long and contains four genes). The smallest known DNA viruses have only 9–11 genes. Again, the complete nucleotide sequences are known in a few cases. (Part of the nucleotide sequence of the phage X174 chromosome, which is 5387 nucleotides long and contains 10 genes.

The largest DNA viruses, like bacteriophage T2 and animal pox viruses, contain about 150 genes. Bacteria like *E. coli* have 3,000–4,000 genes, most of which are present in a single molecule of DNA.

GENOME COMPLEXITY

Eukaryotes have from 2 to 10 times as many genes as *E. coli* but have orders of magnitude more DNA. One of the most challenging problems being studied by geneticists today is the question of the function(s) of this "*excess*" DNA (DNA not required for structural genes that code for proteins).

Not only do most eukaryotes contain many times the amount of DNA of prokaryotes, but this DNA is packaged in several chromosomes, and each chromosome is present in two (diploids) or more (polyploids) copies. Recall that the chromosome of *E. coli* has a contour length of 1100 μm or about 1 mm. Now consider that the haploid chromosome complement, or genome, of the human contains about 100 mm of DNA (or about 2000 mm per diploid cell). This meter of DNA is, of course, subdivided among 23 chromosomes of variable size and shape, each chromosome containing from 15 to 85 mm of DNA. Until recently, we had little information as to how this DNA was arranged in the chromosomes. Is there one molecule of DNA per chromosome as in prokaryotes, or are there many molecules of DNA per chromosome? If many, how are they arranged relative to each other? How does the 85 mm of DNA in the largest human chromosome get condensed into a mitotic metaphase structure that is about 0.5 μm in diameter and 10 μm long? What are the structures of the metabolically active interphase chromosomes?

CHEMICAL COMPOSITION OF EUKARYOTIC CHROMOSOMES

Interphase chromosomes are not visible with the light microscope; moreover, electron microscopy of thin sections cut through eukaryotic nuclei has provided essentially no information about their structure. Recently, however, chemical analysis, electron microscopy, and X-ray diffraction studies on isolated chromatin, the complex of the DNA, chromosomal proteins, and other chromosome constituents isolated from nuclei have provided a solid framework for a rapidly emerging picture of chromosome structure in eukaryotes.

When chromatin is isolated from interphase nuclei, the individual chromosomes are not recognisable. Instead, one observes an irregular aggregate of nucleoprotein. Chemical analysis of isolated chromatin shows that it consists primarily of DNA and proteins with lesser amounts of RNA. Two major classes of protein are: (1) basic proteins (positively charged at neutral pH) called histones and (2) a heterogeneous, largely acidic (negatively charged at neutral pH) group of proteins collectively referred to as *nonhistone chromosomal proteins.*

Histones play a major structural role in chromatin. They are present in the chromatin of all higher eukaryotes in amounts equivalent to that of DNA. The histones of all higher plants and animals consist of five different major proteins. These five major histones, called *H1, H2a, H2b, H3,* and *H4,* are present in almost all cell types (a few exceptions exist, most notably some sperm, where the histones are replaced by another class of small basic proteins called protamines).

The five histones are present in molar ratios of approximately 1H1:2H2a:2H2b:2H3:2H4. They are specifically complexed with DNA to produce the basic structural subunits of chromatin, small (approximately 110 Å in diameter by 60 Å) ellipsoidal "*beads*" called *nucleosomes.*

The histones have been highly conserved during evolution, four of the five types of histones being very similar in all higher eukaryotes.

MOLECULAR STRUCTURE OF EUKARYOTIC CHROMOSOMES

Proteins, like nucleic acids, are large macromolecules composed of a large number of small subunits covalently linked together into long polymers. In the case of proteins, the subunits are called amino acids, of which 20 different species make up all proteins. Most of the amino acids are neutral in charge, i.e., have no charge at pH 7. However, a few are basic and a few are acidic. The histones are basic because they contain 20–30 per cent arginine and lysine, two positively charged amino acids. The exposed $-NH_3^+$ groups of arginine and lysine allow histones to act as polycations. This is important in their interaction with DNA, which is polyanionic because of the negatively charged phosphate groups.

The remarkable constancy of histones H2a, H2b, H3 and H4 in all cell types of an organism and even between widely divergent species is consistent with the idea that they are important in chromatin structure ("*DNA packaging*") and are only nonspecifically involved in the regulation of gene expression.

On the other hand, the nonhistone protein fraction of chromatin consists of a large number of heterogeneous proteins. Moreover, the composition of the nonhistone chromosomal protein fraction varies widely among different cell types of the same organism. Thus, the nonhistone chromosomal proteins are likely candidates for roles in the regulation of expression of specific genes or sets of genes.

ONE GIANT DNA MOLECULE PER CHROMOSOME

How is 1–120 cm (10^4 to 2×10^5 μm) of DNA, which is present in an average eukaryotic chromosome, arranged in the highly condensed mitotic and meiotic structures that are seen with the light microscope? Are there many DNA molecules that run parallel throughout the chromosome (the "*multineme*" or "*multistrand*" model), or is there just one DNA double helix extending from one end of the chromosome to the other end (the "*unineme* or "*single strand*" model)? (note that "strand" here refers to the DNA double helix, not the single strand of DNA.) Are there many DNA molecules joined end to end or arranged in some other fashion in the chromosome, or does one giant, continuous molecule of DNA extend from one end to the other in a highly coiled and folded form? The evidence supporting the unineme model of chromosome structure is now overwhelming. In addition, solid evidence presently supports the concept of chromosome size DNA molecules. That is, each chromosome appears to contain a single giant molecule of DNA that extends from one end through the centromere all the way to the other end of the chromosome.

"Lampbrush" Chromosomes in Vertebrate Oocytes

Some of the strongest evidence supporting the unineme model of chromosome structure has come from studies on the large, so called "lampbrush" chromosomes, which are present during prophase I of oogenesis in many vertebrates, particularly Amphibians. Lampbrush chromosomes are up to 800 μm long; they thus provide very favourable material for cytological studies. The homologous chromosomes are paired, and each has duplicated to produce two chromatids at the lampbrush stage. Each lampbrush chromosome contains a central axial region, where the two chromatids are highly condensed, and numerous pairs of lateral loops. The loops are transcriptionally active regions of single chromatids. The integrity of both the central axis and the lateral loops is dependent on DNA. Treatment with DNase fragments both the axis and the loops. Treatment with RNase or proteases removes surrounding matrix material, but does not destroy the continuity of either the axis or the loops. Electron microscopy of RNase and protease treated lampbrush chromosomes reveals a central filament of just over 20 Å in diameter in the lateral loops. Since each loop is a segment of one chromatid, and since the diameter of a DNA double helix is 20 Å, these lampbrush chromosomes must be unineme structures. This conclusion is also supported by studies on the kinetics of nuclease digestion of lampbrush chromosomes. That is, the kinetics observed are those expected if a single double helix of DNA is the central filament in the loops. The axial region then contains two DNA molecules, one from each of the two tightly paired chromatids.

Lampbrush chromosomes are meiotic or "germ line" chromosomes. Their structure is thus particularly relevant to an understanding of genetic phenomena. "Nongerm line" or somatic cell chromosomes may have different structures. Although most are unineme, some are known to be multineme structures.

VISCOELASTOMETRIC EVIDENCE FOR CHROMOSOME SIZE DNA MOLECULES

The question of whether the unineme chromosomes of eukaryotes contain a single large molecule of DNA or many smaller molecules linked end to end has proven very difficult to answer with rigorous experimental evidence. A centimeter-long molecule of DNA has a length to width ratio of 5 million to 1. Such a structure is very shear sensitive. If such a DNA molecule is in solution in a test tube, the slightest vibration will break the molecule into many fragments. For this and other reasons, accurate estimates of the molecular sizes of eukaryotic DNAs cannot be obtained by conventional biochemical methodology. Recently, biophysicists have used a technique called *viscoelastometry* (a procedure for analysing viscosity parameters of molecules in solution) to estimate the sizes of DNA molecules from eukaryotic chromosomes. When DNA molecules in solution, are exposed to a driving force (e.g., by rotating a cylinder in the solution) they are stretched into an extended conformation. When the driving force is removed (e.g., by stopping the rotation of the cylinder), the molecules will return to their lowest energy or relaxed state.

Recoil time is a function of the size of DNA molecules in the solution being analysed. The recoil decays exponentially over long time intervals, and the time constant, called the retardation time (T), is a sensitive function of the size of the largest DNA molecules in the solution. Thus, even if some of the DNA molecules are broken by shearing forces, the viscoelastometric procedure will permit one to estimate the size of the largest DNA molecules present. Moreover, the cells can be lysed right in the chamber of the viscoelastometer to minimise the chance of shearing the molecules when they are released from the cells.

The largest DNA molecules in *Drosophila melanogaster* cells were estimated to have a mass of 4.1×10^{10} daltons by viscoelastometry. (*A dalton is the unit of mass equal to the mass of a single hydrogen atom.* It is the most frequently used unit in dealing with the size of macromolecules like DNA.) The largest chromosomes of *Drosophila* have been shown to contain about 4.3×10^{10} daltons of DNA (total, whether one molecule or many) by direct microspectrophotometric analysis of metaphase chromosomes. Thus, the viscoelastometric estimate of the size of the largest DNA molecules in *Drosophila* nuclei correlates almost exactly with the total amount of DNA present in the largest chromosome.

Autoradiographic Evidence for Chromosome Size DNA Molecules

Researchers have also used the technique of autoradiography to attempt to detect chromosome size from DNA molecules of eukaryotic cells in the same way that Cairns was able to visualise the intact *E. coli* chromosome. Like viscoelastometry, autoradiography permits one to lyse cells very gently so as to minimise shearing forces. DNA molecules from the lysed cells are permitted to diffuse onto a membrane that is used to pick up the DNA molecules and expose them to the beta-particle-sensitive emulsion. Thus, breakage of molecules should not be a problem. The difficulty with autoradiographic examination of DNA molecules that are very large is that it is almost impossible to get all segments of the molecule sufficiently spread out on the membrane with no tangles or overlaps so that the entire length of the molecule is visible.

Nevertheless, autoradiographic analysis of DNA molecules from *Drosophila* have been successful, and the results of these studies also support the concept of chromosome size DNA molecules. The largest molecules observed have a contour length of 1.2cm. Although these are not as large as the viscoelastometric estimates of the largest molecules, such molecules correspond to a mass in the range of 2.4 to 3.2×10^{10} daltons, two-thirds to three-fourths the size of the largest chromosome size DNA molecules.

PACKAGING THE GIANT DNA MOLECULES INTO CHROMOSOMES

The largest chromosome in the human genome contains about 85 mm (85, 000 μm or 8.5×10^8 Å) of DNA that is believed to exist as one giant molecule. This DNA molecule somehow gets packaged into a metaphase structure that is about 0.5 μm in diameter and about 10 μm in length. *This represents a condensation of almost 10^4* fold in length from the naked DNA molecule to the metaphase chromosome. How does this condensation occur? What components of the chromosomes are involved in the packaging processes?

Are DNA molecules packaged in different chromosomes in different ways or is there a universal packaging scheme? Are there different levels of packaging? Clearly, meiotic and mitotic chromosomes are more extensively condensed than interphase chromosomes. What additional levels of condensation occur in these special structures, that are designed to assure the proper segregation of the genetic material during cell divisions? Are DNA sequences of genes that are being expressed packaged differently from those of genes that are not being expressed?

NUCLEOSOME STRUCTURE (Figs. 26.9, 26.10 and 26.11)

When isolated chromatin is examined by electron microscopy, it is found to consist of a series of ellipsoidal "*beads*" (about 10 Å in diameter and 60 Å high) joined by thin threads. Further evidence for a regular, periodic packaging of DNA has come from studies on the digestion of chromatin with various nucleases. These studies indicated that segments of DNA of 146 nucleotide pairs in length were somehow protected from degradation by certain nucleases. Moreover, the partial digestion of chromatin with these nucleases yielded fragments of DNA in a set of discrete sizes that were integral multiples of the smallest size fragment. These results are neatly explained if chromatin has a repeating structure, supposedly the "bead" seen by electron microscopy, within which the DNA is packaged in a nuclease resistant form. This "bead" or chromatin subunit is called the *nucleosome*. According to this model the "interbead" threads of DNA or linkers, are susceptible to nuclease attack.

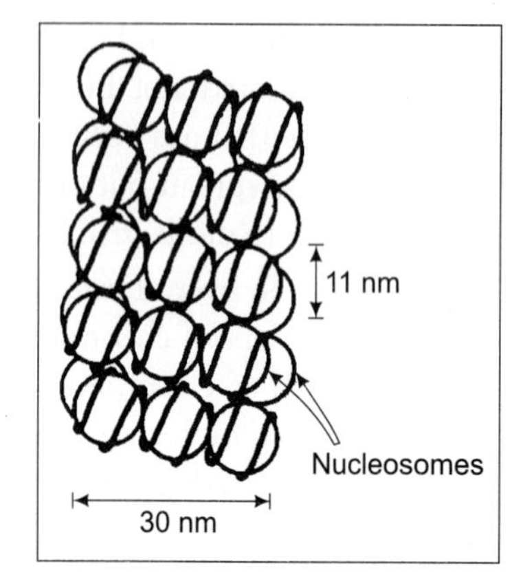

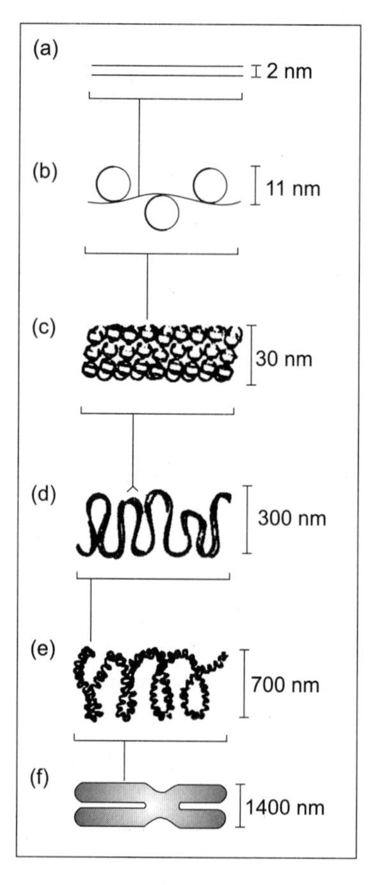

Fig. 26.9 The solenoid model for the 30 nm chromatin fibre. A summary of the different levels of DNA packaging in a chromosome: (a) naked DNA; (b) beads-on-a-string; (c) the 30 nm chromatin fibre; (d) the 30 nm fibre made up of looped 30 nm fibre made up of looped 30 nm chromatin; (e) 70 nm supercoiled structure that comprises each arm of a metaphase chromosome; (f) the metaphase chromosome.

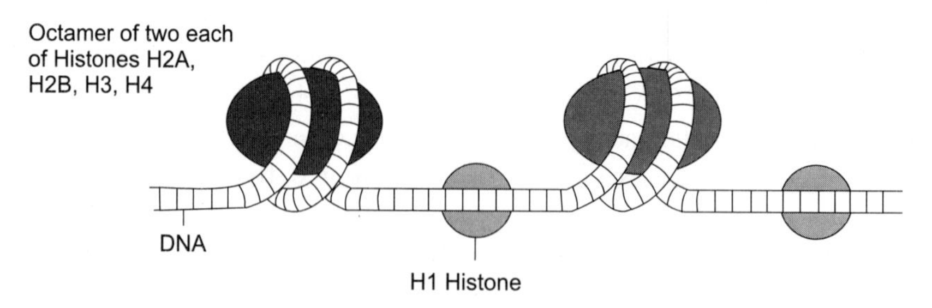

Fig. 26.10 Presumed structure of nucleosomes: Two loops of DNA interact with two copies each of histones H2A, H2B, H3 and H4. Histone H1 is associated with the DNA strands (linkers) between nucleosomes.

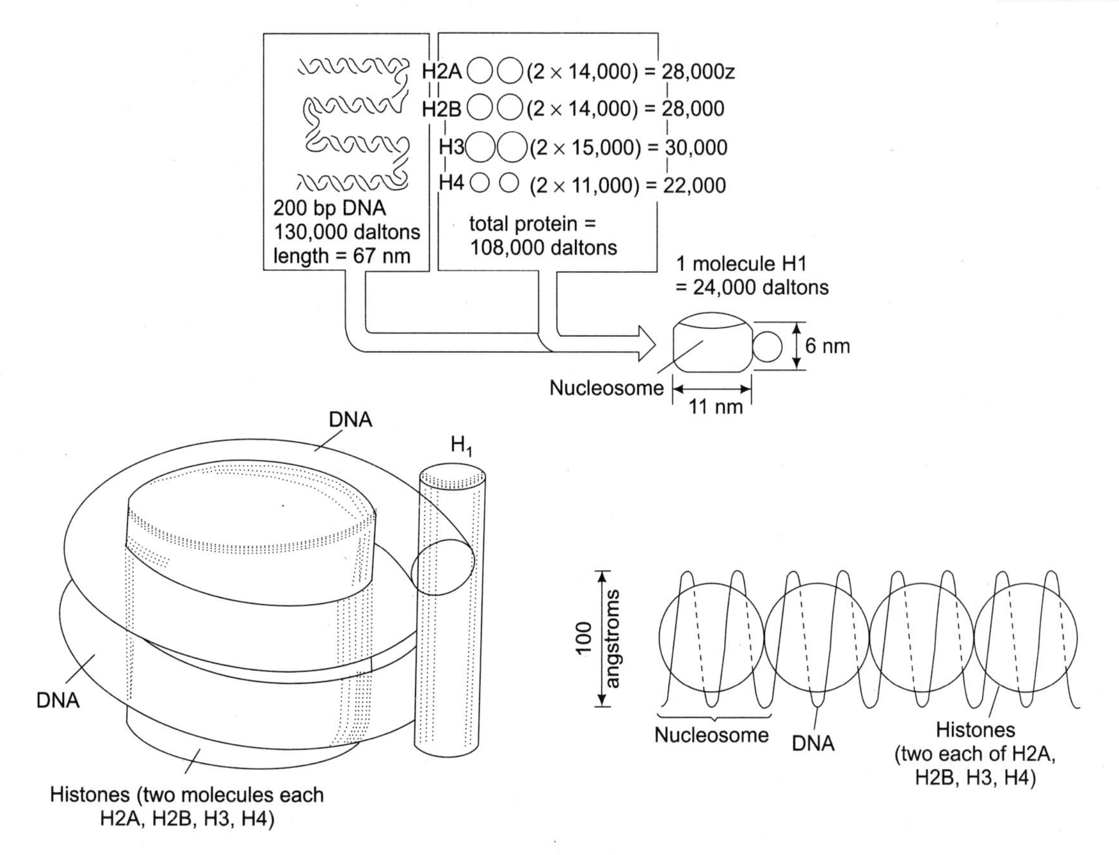

Fig. 26.11 Detailed structure of the nucleosome.

After partial nuclease digestion of chromatin, an approximately 200-nucleotide pair length of DNA is found associated with each nucleosome (produced by a cleavage in each linker region). After extensive nuclease digestion, a 146-nucleotide pair long segment of DNA remains present in each nucleosome. This nuclease resistant structure is called the *nucleosome core.* Its structure is essentially invariant in all eukaryotes, consisting of a 146-nucleotide pair length of DNA and two molecules each of histones H2a, H2b, H3, H4. Physical studies (X-ray diffraction and similar analyses) of nucleosome core crystals have shown that the DNA is wound as 1¾ turns of a superhelix around the outside of the histone octamer.

The complete chromatin subunit consists of the nucleosome core, the linker DNA, an average of one molecule of histone H1, and the associated nonhistone chromosomal proteins. However, note that it has not been firmly established that histone H1 is evenly distributed, one molecule per nucleosome or linker, in chromatin. The size of the linker DNA varies from species to species and from one cell type to another. Linkers as short as 8 nucleotide pairs and as long as 114 nucleotide pairs have been reported. Some evidence suggests that the "complete nucleosome" (as opposed to the nucleosome core) contains one molecule of histone H1, which stabilises two full turns of DNA superhelix (a 166 nucleotide pair length of DNA) on the surface of the histone octamer. Other evidence indicates that histone H1 is involved in the coiling or folding of the nucleosome fiber to form a 300-Å chromatin fiber and may be involved in other higher levels of organisation of chromatin.

Clearly, the basic structural component of eukaryotic chromatin is the nucleosome. But are the structures of all nucleosomes the same? How does the replication fork move past the nucleosome during replication? What role(s), if any, does nucleosome structure play in gene expression and the regulation of gene expression? The structure of nucleosomes in genetically active regions of chromatin is known to differ from that of nucleosomes in genetically inactive regions. But what are the details of this structure-function relationship? Present and future studies of the fine structure of nucleosomes will undoubtedly prove very informative with regard to these and other questions.

The 300-Å Chromatin Fibre

Electron micrographs of isolated metaphase chromosomes show masses of tightly coiled or folded lumpy fibers. These chromatin fibers have an average diameter of 300-Å. When the structures seen by light and electron microscopy during earlier stages of meiosis are compared, it becomes clear that the light microscope simply permits one to see those regions where these 300-Å fibers are tightly packed or condensed.

What is the substructure of the 300-Å fibre seen in mitotic and meiotic chromosomes? Although we do not have a firm answer to this question, we do know that the DNA is wound in a supercoil about a histone octamer to yield the roughly 100-Å in diameter nucleosome. *In vivo,* the nucleosomes are probably in direct juxtaposition with each other without detectable linker regions; if so, they will form a 100-Å nucleosome fiber. If this 100-Å fibre, in turn, is wound in higher order supercoil (a "*super-super coil*" or *solenoid*), a 300-Å fibre can easily be generated. Although we still do not understand all the details of the structure of this 300-Å chromatin fibre, there is good evidence that it represents some type of a solenoid like structure as that shown in Fig. 26.11.

"Scaffolds" Composed of Nonhistone Chromosomal Proteins

Metaphase chromosomes contain the maximum degree of condensation observed in normal eukaryotic chromosomes. Clearly, the role of these highly condensed chromosomes is to organise and package the giant DNA molecules of eukaryotic chromosomes into structures that will facilitate their segregation to daughter nuclei without the DNA molecules of different chromosomes becoming entangled and, as a result, being broken during the anaphase separation of the daughter chromosomes. The basic structural unit of the metaphase chromosome is the 300-Å chromatin fibre. However, the next obvious question is how are these 300-Å fibers further condensed into the observed metaphase structure? Unfortunately, there is still no clear answer to this question. There is evidence that the metaphase structure is not dependent on histones. Electron micrographs of isolated metaphase chromosomes from which the histones have been removed reveal a central core or "scaffold", which is surrounded by a huge pool or "halo" of DNA. Note the absence of any apparent ends of DNA molecules in the micrograph, again supporting the concept of one giant DNA molecule per chromosome.

SUMMARY

In summary, at least three levels of condensation are required to package the 10^3 to 10^5 μm of DNA in a eukaryotic chromosome into a metaphase structure a few microns long. The first level of condensation involves packaging DNA as a supercoil into nucleosomes. This produces 100-Å diameter interphase chromatin fibre. This clearly involves an octamer of histone molecules, two each of histones H2a, H2b, H3 and H4. The second level of condensation involves an addition folding and/or supercoiling of the 100-Å nucleosome fibre to produce the 300-Å chromatin fibre characteristic of mitotic and meiotic chromosomes. Histone H1 is involved in this supercoiling of the 100-Å nucleosome fibre to produce the 300-Å chromatin fibre. Lastly, nonhistone chromosomal proteins form a "scaffold" that is involved in condensing the 300-Å chromatin fibre into the tightly packed metaphase chromosomes. This third level of condensation appears to involve the segregation of segments of the giant DNA molecules present in eukaryotic chromosomes into independently supercoiled domains or loops. The mechanism by which this third level of condensation occurs is not known.

EUCHROMATIN AND HETEROCHROMATIN

When chromosomes are stained by various procedures, such as the Feulgen reaction, which is specific for DNA, and are examined by light microscopy, some regions of the chromosomes are observed to stain very darkly, whereas other regions stain only lightly. When examined by electron microscopy, the intensely staining chromatin, called heterochromatin, is seen to consist of densely packed chromatin fibers (300-Å in diameter). The lightly staining chromatin, called euchromatin, is composed of less tightly packed 300-Å fibers. Heterochromatin can often be shown to remain highly condensed throughout the cell cycle, whereas euchromatin is not visible with the light microscope during interphase.

Genetic analyses indicate that heterochromatin is largely genetically inactive. Most of the genes of eukaryotes that have been extensively characterised are located in euchromatic regions of the chromosomes. A structure function correlation is thus evident: the highly condensed chromatin tends to be genetically inactive; the less condensed chromatin to be genetically active. Furthermore, heterochromatin is often enriched in highly repetitive tandemly arranged DNA sequences.

REPETITIVE DNA AND SEQUENCE ORGANISATION

The chromosomes of prokaryotes contain DNA molecules with unique (nonrepeated) base pair sequences. That is, each gene (consider a gene to be a linear sequence of a few thousand base pairs) is present only once in the genome. (The genes for rRNA molecules are an exception.) If prokaryotic chromosomes are broken into many short fragments, each fragment will contain a different sequence of base pairs. The chromosomes of eukaryotes are much more complex in this respect. Certain base sequences are repeated many times in the haploid chromosome complement, sometimes as many as a million times. DNA containing such repeated sequences, called repetitive DNA, often represents a major component (20–50%) of the eukaryotic genome.

Satellite DNAs

The first evidence for *repetitive* DNA came from density gradient analysis of eukaryotic DNA. When the DNA of a prokaryote, such as *E. coli*, is isolated, fragmented, and centrifuged to in a cesium chloride (CsCl) density gradient, the DNA usually forms a single band in the gradient. For *E. coli*, this band will form at a position where the CsCl density is equal to the density of DNA containing about 50 per cent A-T and 50 per cent G-C base pairs. DNA density increases with increasing G-C content. The extra hydrogen bond in a G-C base pair is believed to result in a tighter association between the bases and thus a higher density than for A-T base pairs. On the other hand, CsCl density gradient analysis of DNA from eukaryotes usually reveals the presence of one large band of DNA (usually called "*mainband*" DNA) and one to several small bands. These small bands of DNA are called *satellite bands*, and the DNA in these bands is often referred to as *satellite DNA*. *Drosophila virilis* DNA, for example, contains three distinct satellite DNAs. When isolated and analysed, each satellite DNA was found to contain repeating sequence of seven base pairs. One satellite repeat sequence is 5'AGAAACT-3' (*one strand; the other strand will have the complementary sequence). A second satellite DNA has a* 5'ATAACT-3'. Thus, they differ from each other at only two positions. In three releated species of crabs, a satellite DNA is present that contains 97 per cent A-T base pairs. Some satellite DNAs in other eukaryotes have longer repetitive sequences.

The chromosomal locations of several satellite DNAs have been determined by a technique called *in situ hybridisation.* The complementary strands of DNA molecules are separated by heat or alkaline denaturation. Conditions can then be reversed by lowering the temperature or lowering the pH, and the separated strands will renature or reanneal to reform base paired double helices, a process called DNA renaturation. DNA renaturation is a specific type of nucleic acid hybridisation, the formation of hydrogen-bonded double. helices by single-stranded molecules containing complementary base sequences. Hybridisation will occur between complementary base sequences. Hybridisation will occur between complementary single strands regardless of their source. If both participating strands are DNA, the process is called DNA hybridisation. If one strand is DNA and the complementary strand is RNA, the process is called DNA-RNA hybridisation.

In situ hybridisation in the case of the satellite DNAs usually involves annealing single strands of isolated radioactive satellite DNA, or complementary RNA sequences synthesised using satellite DNA as template, directly to denatured DNA in chromosome squash preparations. After washing out the nonhybridised radioactive material, the locations of the satellite DNA sequences in chromosomes are determined by autoradiography. All satellite DNAs characterised to date are located in heterochromatic regions of chromosomes. In several cases, the satellite DNA sequences are located in heterochromatic telomeres (chromosome termini).

Function of Highly Repetitive DNA Sequences

The function(s) of highly repetitive DNA—most, if not all, of which is located in genetically inactive heterochromatic regions of chromosomes—is also completely unknown. Postulated functions for highly repetitive DNA include (1) structural or organisational roles in chromosome; (2) involvement in chromosome pairing during meiosis; (3) involvement in crossing over or recombination; (4) "protection" of important structural genes, like histone, rRNA, or ribosomal protein genes; (5) a repository of unessential DNA sequences for use in the future evolution of the species; and (6) no function at all—just "*junk*" *DNA* that is carried along by the processes of replication and segregation of chromosomes. The validity of any of these postulated roles remains to be established.

CHAPTER 27

Karyotype

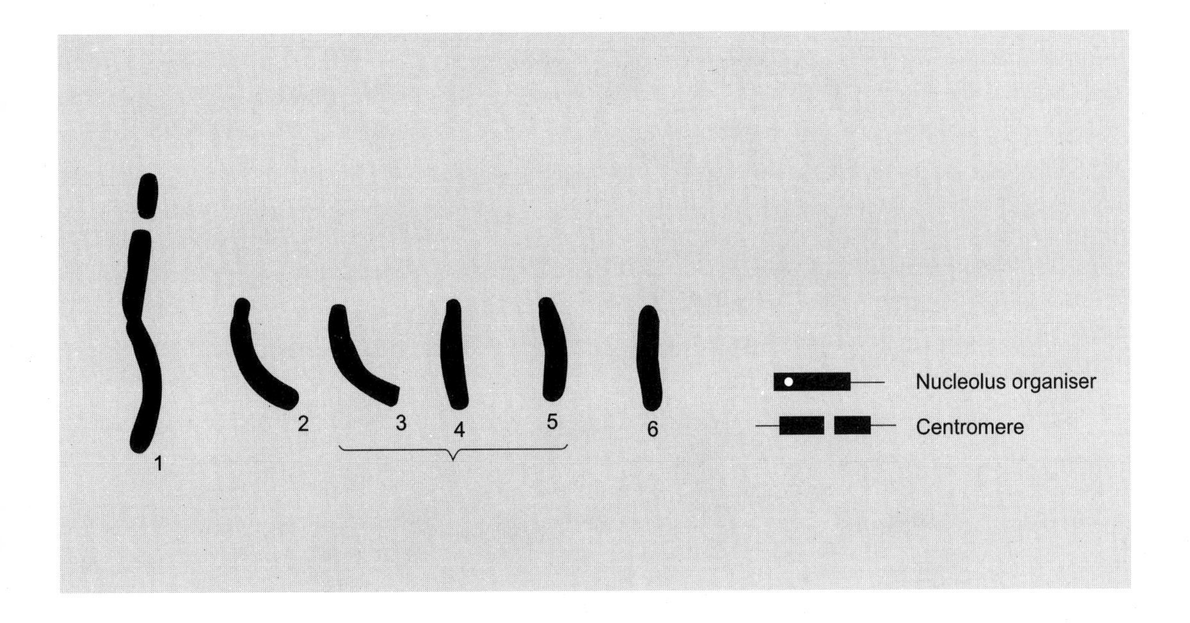

A complete description of all the chromatid pairs possessed by a given cell type or individual constitutes *KARYOTYPE*, and is usually prepared by cutting out individual chromatid pairs from a photograph and arranging them in series according to their size, centromere position, and arm ratio. In a human karyotype, the chromosomes are arranged as *autosomes* and *sex chromosomes* (1–22 and either XX or XY) (Fig. 27.1).

Staining procedures have been developed in the past two decades which reveal discrete bands along the lengths of sister chromatids. These readily permit the recognition of individual chromatid pairs and the matching of homologues having similar banding patterns. Of the many staining protocols that have been devised in various laboratories to produce banded chromosomes, the two most generally used are the Acid-Saline-Giemsa (ASG) technique, which reveals *G-bands* (for Giemsa stain), and the Quinacrine mustard technique, which produces

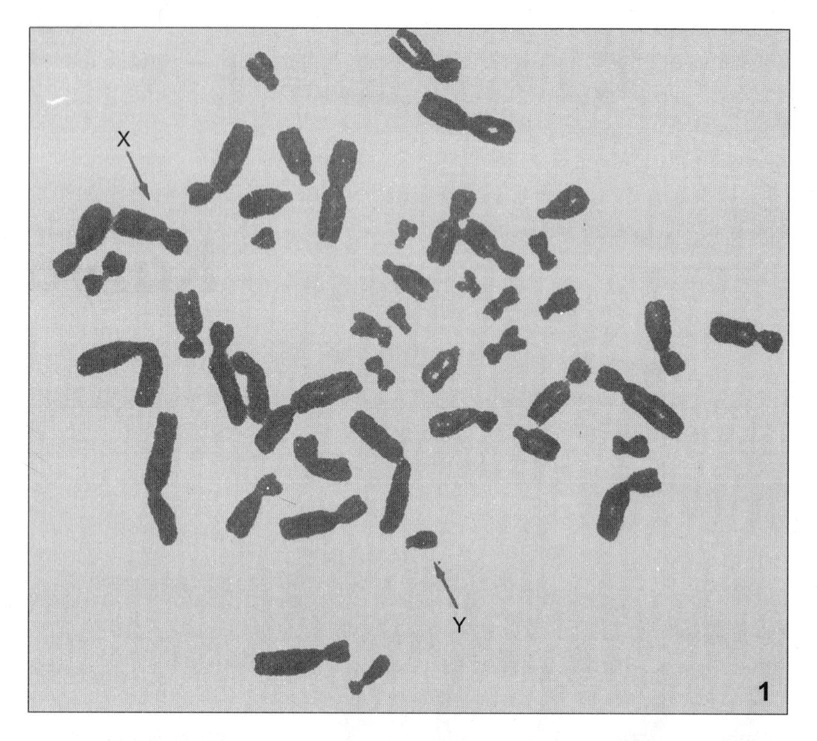

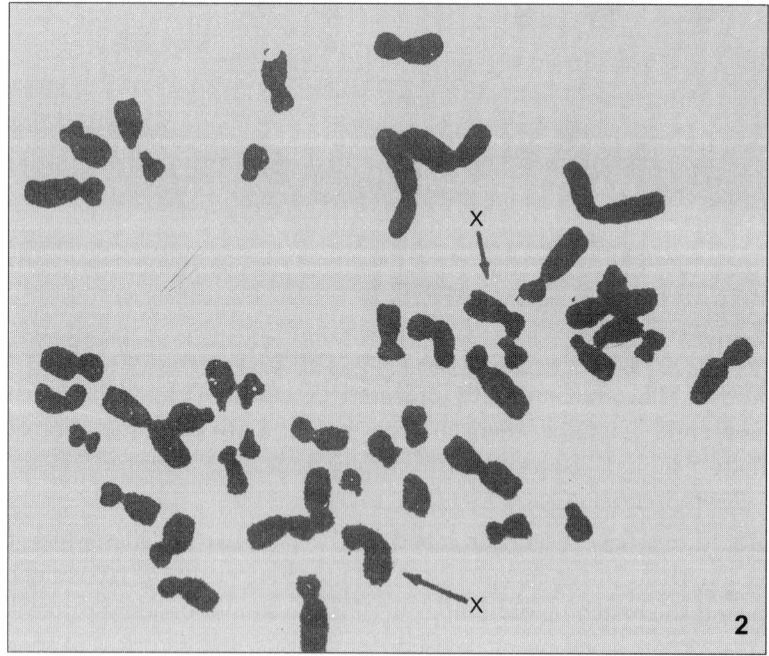

Fig. 27.1 Human Karyotype (1) Male (XY) shown by arrows; (2) Female (XX) (Courtesy: Dr. C.S.K. Raju).

fluorescent *Q bands;* G- and Q-bands have the same locations and are presumed to reveal the same underlying chromosome structures. These techniques have revolutionised the field of cytogenetics and many plant and animal species have since been studied for their karyotypic studies.

STAINING AND BANDING CHROMOSOMES (Figs. 27.2, 27.3 and 27.4)

1. *Feulgen Staining.* Cells are subject to a mild hydrolysis in 1 N HCl at 60°C, usually for about 10 minutes. This treatment produces a free aldehyde group in deoxyribose molecules. Aldehyde will then react with a chemical known as Schiff's reagent (basic fuchsin bleached with sulfurous acid) to give a deep pink colour. Ribose of RNA will not form an aldehyde under these conditions, and the reaction is thus specific for DNA.
2. *Q Banding.* The Q bands (from *quinacrine*) are the fluorescent bands observed after quinacrine mustard staining and observation with ultraviolet light. The distal ends of each chromatid are not stained by this technique. The Y chromosome becomes brightly fluorescent both in the interphase nucleus and in metaphase.
3. *R Banding.* The R bands (from *reverse*) are those located in the zones that do not fluoresce with quinacrine mustard, that is, they are between the Q bands and can be visualised as green, brightly fluorescent bands with acridine orange staining.
4. G *Banding.* The G bands (from *Giemsa*) have the same location as Q bands and do not require fluorescent microscopy. Many techniques are available, each involving some pretreatment of the chromosomes. In the acid-saline-Giemsa (ASG) technique, for

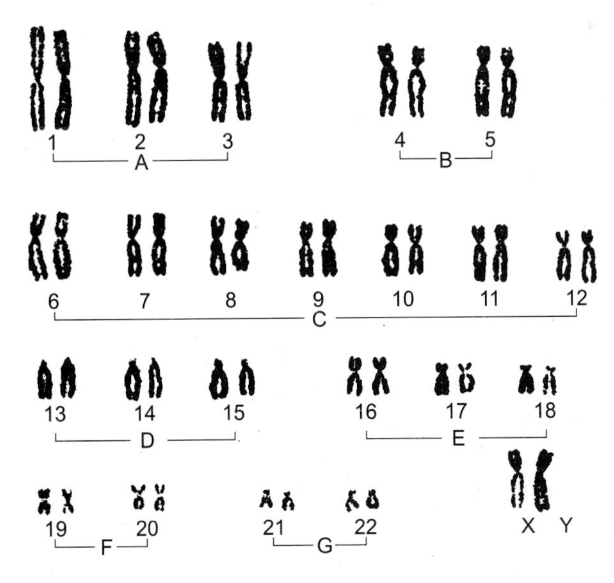

Fig. 27.2 Human karyotype conventionally stained and arranged according to the Denver system. Chromosomes are allocated to groups as many cannot be identified or paired with certainty.

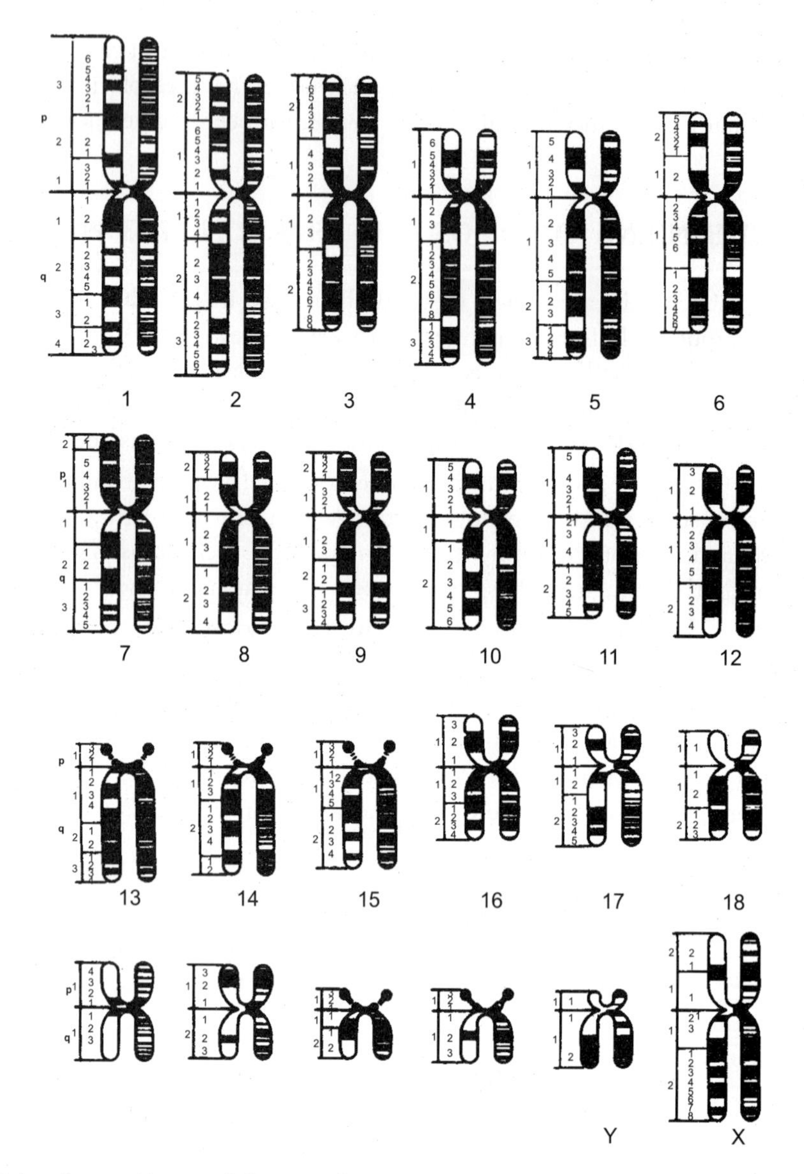

Fig. 27.3 G-banding patterns of human chromosomes at mid-metaphase (left chromatid of each chromosome) and at late prophase (right arm). Numbering system for the mid-metaphase bands is that established at the Paris Conference, 1971, with *p* and *q* being the short and long arms, respectively. The uppermost band of the first chromosome is designated 1p36, the next as 1p35, and so on.

example, cells are incubated in citric acid and NaCl for 1 hr at 60°C and are then treated with the Giemsa stain. Proteolytic enzyme treatment also reveals these bands.

5. C *Banding.* The C bands correspond to *constitutive heterochromatin.* The heterochromatic regions in a chromosome distinctly differ in their stainability from the

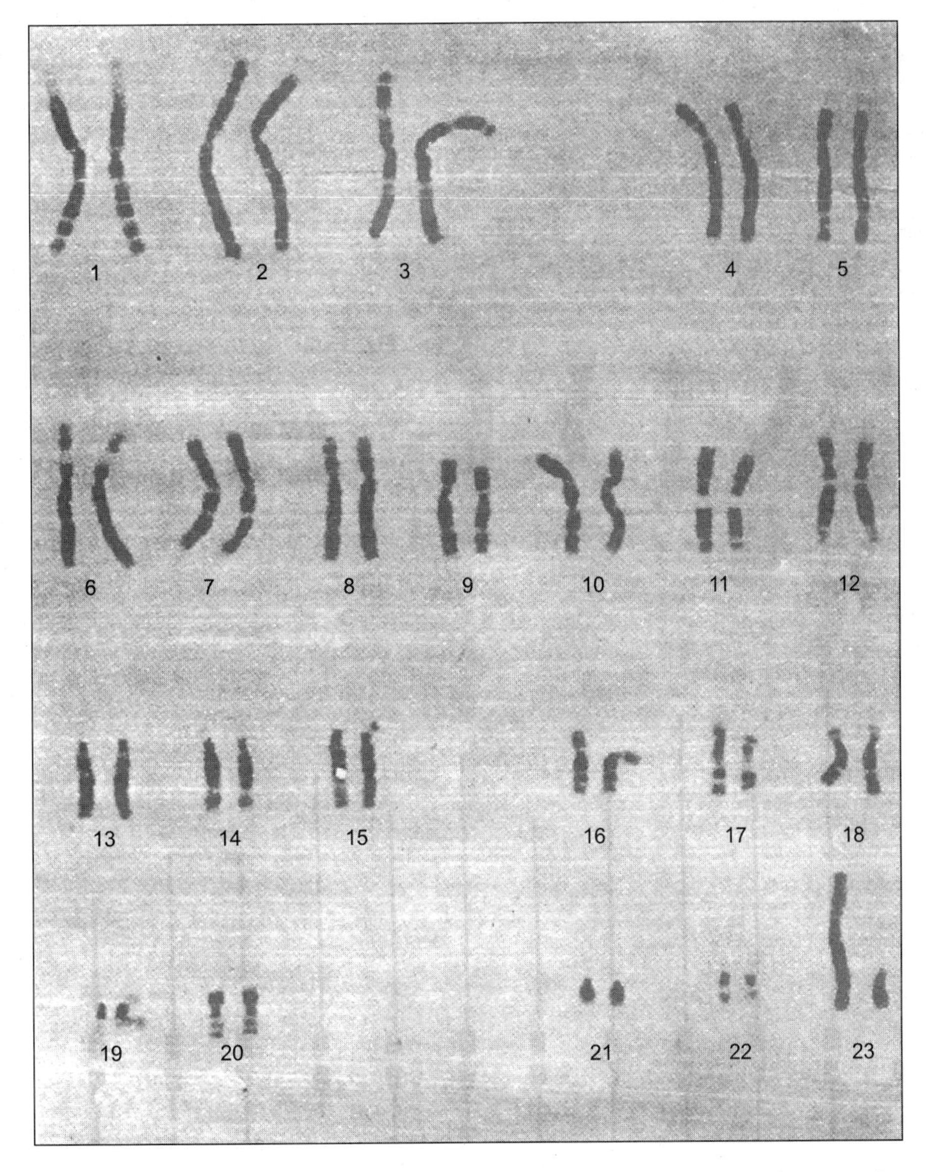

Fig. 27.4 G-banding pattern of metaphase chromosomes from a normal human male.

euchromatic regions, as in the differentiated chromosomes of tomato, *Physalis* and the knobs of maize chromosomes.

Karyotype is the organisation of the chromosomes and arranging the total complement of the genome in an order from the longest chromosome to the shortest chromosome. After the chromosomes have been arranged (by cutting the figures and arranging) as shown in different figures (Figs. 27.2 to 27.5). These length measurements will be graphically represented in the form of an *Idiogram* (Fig. 27.6) as shown for the pachytene chromosomes of rice

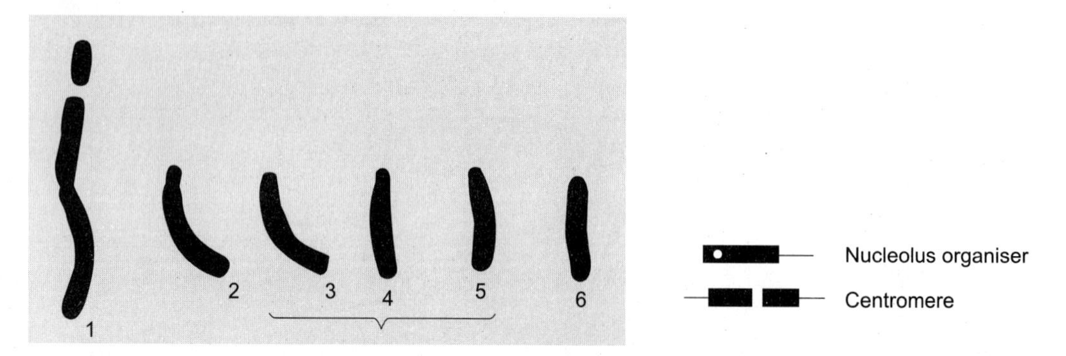

Fig. 27.5 C-mitosis and karyotype of *Vicia faba* ($2n = 12$). (Courtesy: A.T. Natarajan).

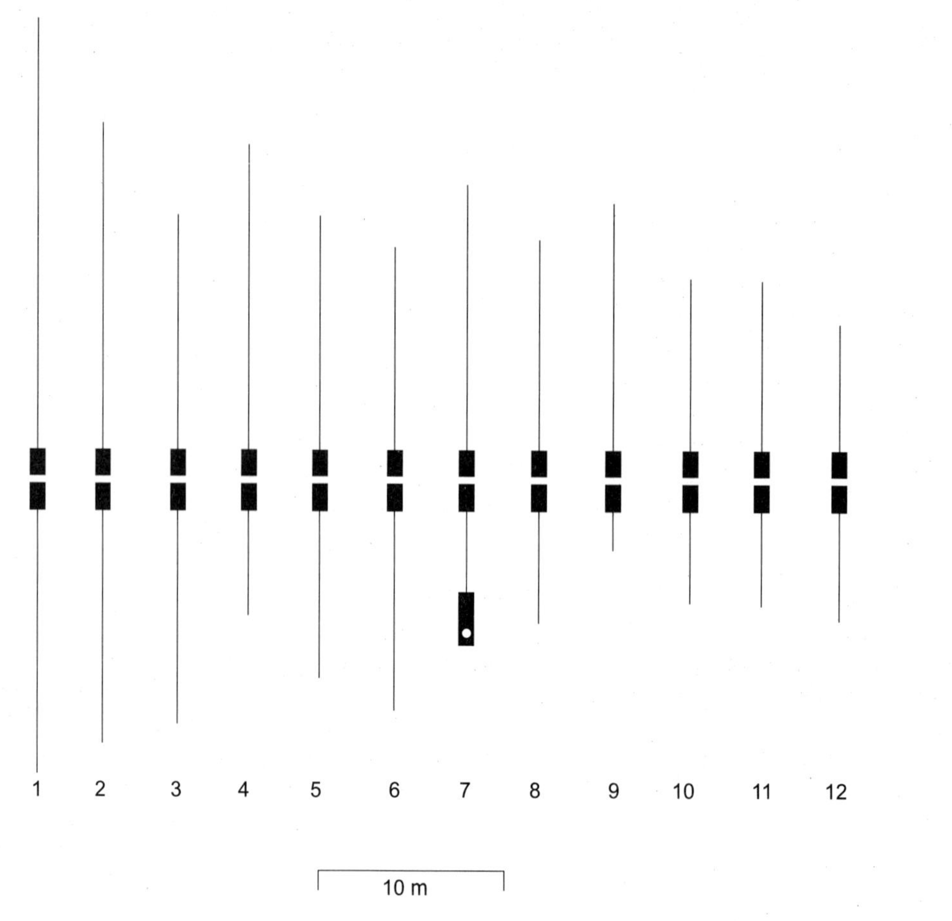

Fig. 27.6 Idiogram of rice *Oryza sativa* chromosomes pachytene (Courtesy: A.V.S.S. Sambamurty).

(*Oryza sativa L.*). All the twelve (2n = 24) haploid chromosomes are arranged by their length measurements from chromosome 1 to chromosome 12 which is the shortest. In plants the karyotypic studies are made either from C-mitosis (C for *Colchicine*) or from pachytene chromosomes. Pachytene analysis is a perfect system of identifying the chromosomes and also the morphology of the chromosomes from end to end like chromomere positions, centromere position, special structures like knobs as in maize, differentiated chromosomes like tomato and *Physalis,* where the chromosomes show distinct euchromatin and heterochromatin regions in the chromosome. Fig. 26.6. shows an *Idiogram* of two karyotypes of pachytene chromosomes of rice.

CHAPTER 28

Chromosomal Aberrations

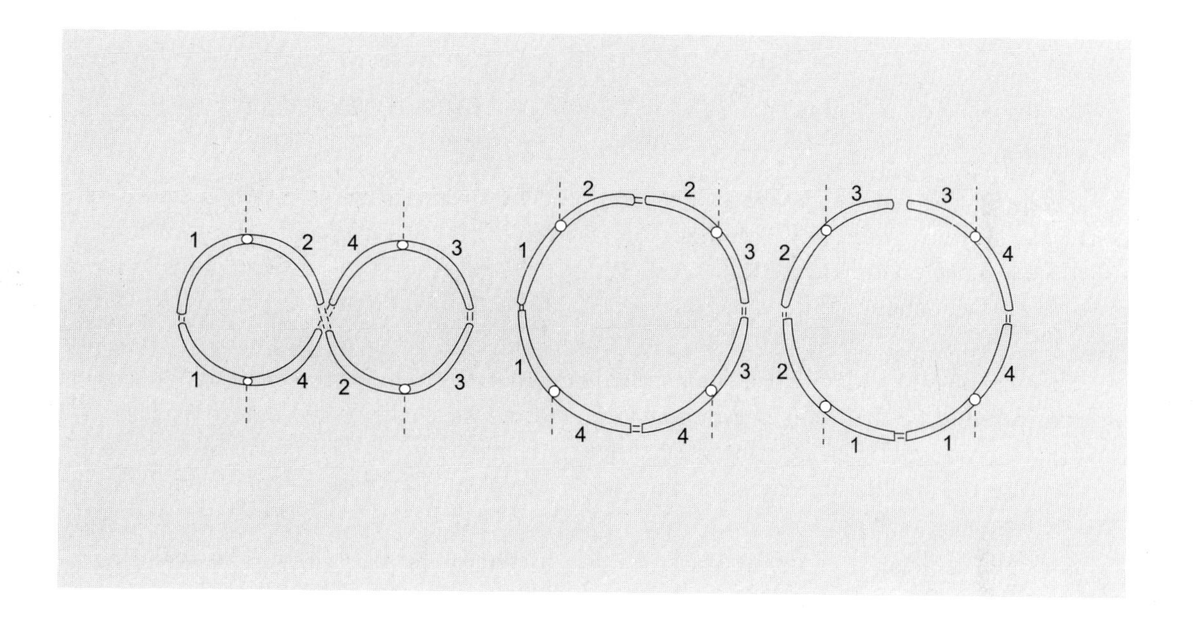

Chromosomal aberrations are either induced or spontaneously occurring mutations, i.e., mutations taking place on single chromosome. As opposed to gene mutations that are cryptic, chromosomal aberrations can be observed under the microscope. These are classified as: *translocations, inversions, duplications and deficiencies.* This chapter presents, cytogenetic discussion of chromosomal aberrations.

Mutations can be classified into the following main categories:

1. Genome mutations.
2. Chromosome mutations: (a) Structural rearrangements, and (b) Gene mutations.
3. Extranuclear mutations.

GENOME MUTATIONS: ALTERATIONS OF CHROMOSOME NUMBER (POLYPLOIDY, HAPLOIDY AND ANEUPLOIDY)

Polyploidy implies that one or more complete chromosome sets are added to the diploid chromosome number (in the case of triploids $2x + x = 3x$, tetraploids $2x + 2x = 4x$, where x denotes the basic number). Haploid (from diploidy) or polyhaploidy (from polyploidy) is the status of individuals with half the chromosome number ($2x \rightarrow x$, $4x \rightarrow 2x$). Aneuploidy implies that one or more extra chromosomes are added or subtracted from a complete diploid or polyploid number. For instance

trisomic, $2x + 1$		polyploid	$3x + 1$	(triplo-trisomic)
monosomic, $2x - 1$			$4x - 1$	(tetra-monosomic)
tetrasomic, $2x + 2$	or		$4x + 2$	(tetra-tetrasomic)
nullisomic, $2x - 2$				

Autopolyploid cells or tissues can be produced by treatment with certain chemicals like colchicine, nitrous oxide, Ethyl Methane Sulphonate (EMS) and also after the formation of unreduced gametes within a species (AA, AAA or AAAA). Allopolyploid or amphiploid forms arise after hybridisation of different species and the formation of unreduced gametes (AA × BB, AB, AB, AABB).

Spontaneous haploids occur infrequently. Some mutants induced by colchicine in *Sorghum* are considered to be the result of gene mutation and mitotic chromosome reduction. Haploids can also be produced by the use of irradiated pollen. Haploids are useful in plant breeding, since by doubling the haploid chromosome number, a complete homozygote ("pure line") can be directly formed.

Aneuploidy appears in many ways: (a) irradiation of meiosis (b) in the offspring of translocation and inversion heterozygotes, and ill crossings with triploids. Aneuploids are generally unbalanced and have decreased viability and fertility.

CHROMOSOME MUTATIONS OR CHROMOSOMAL ABERRATIONS

*(a) **Structural rearrangements of the chromosome***

Changes in the linear order of genes detectable cytologically or after special genetic procedures are commonly associated with chromosome breakage. They can be divided into four groups, viz., translocations, inversions, deficiencies and duplications.

Translocations (Figs. 28.1 to 28.10)

A translocation is formed when breaks occur in two chromosomes simultaneously in a nucleus and the broken chromosomes rejoin in a new manner. Chromosome breakage and exchange are easily induced by means of ionising radiations and radiomimetic chemicals. They also occur spontaneously.

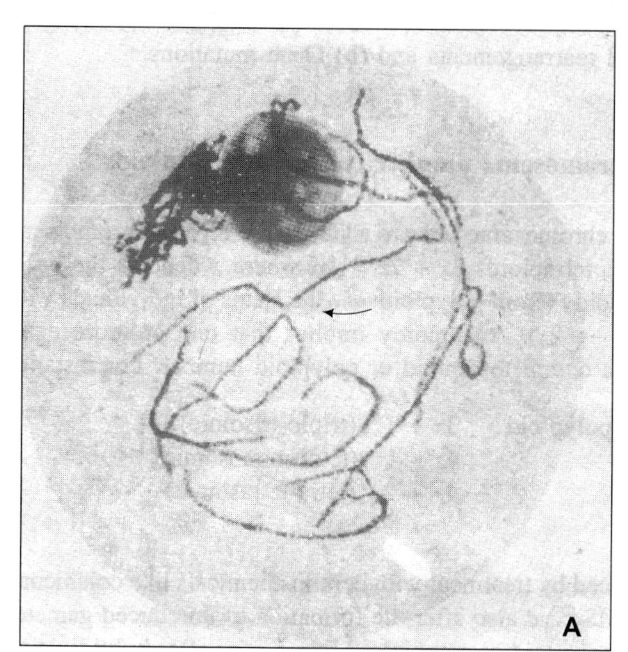

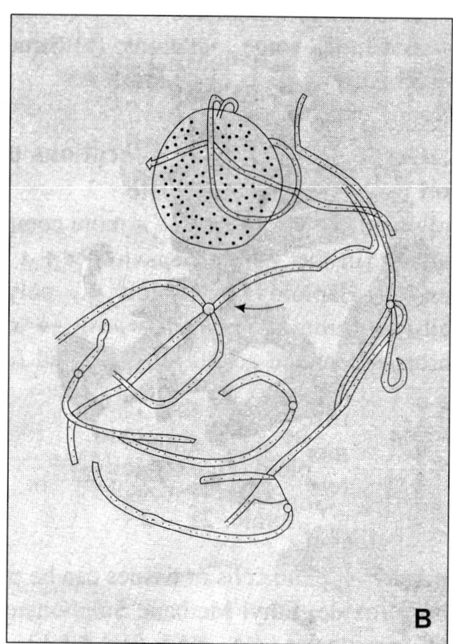

Fig. 28.1 Cross-shaped translocation (indicated by arrow) at pachytene in a *maize-Tripsacum* hybrid. A-Microscopic photograph, B-Line drawing through camera lucida (Courtesy: Dr. C.S.K. Raju, Harvard University, USA).

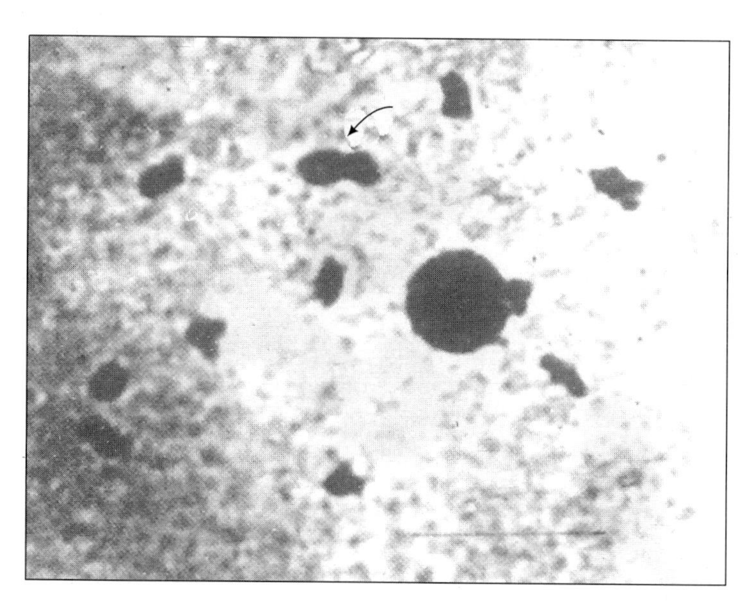

Fig. 28.2 Translocation in rice $10_{II} + 1_{IV}$ (indicated by arrow). (Courtesy: A.V.S.S Sambamurty).

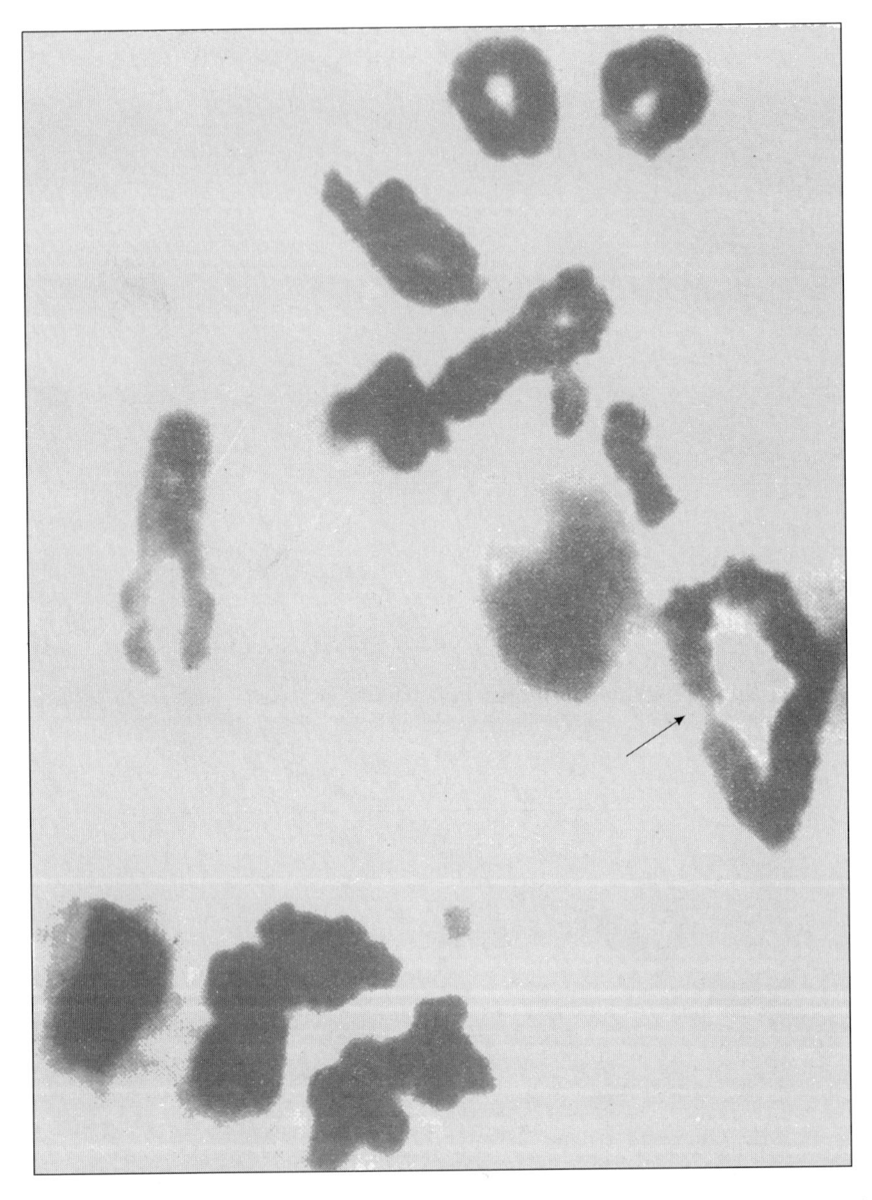

Fig. 28.3 Translocation in *Chlorophytum variegatum* 12II + 1IV (shown by arrow) fragments also are seen (Courtesy: A.V.S.S. Sambamurty).

Translocations are of *two types–simple* and *reciprocal.* In simple translocations only one exchange is realised and one part of a chromosome (a fragment) is lost; while in reciprocal no chromosome material is lost, it is only a mutual exchange of chromatin material giving rise to recombinations.

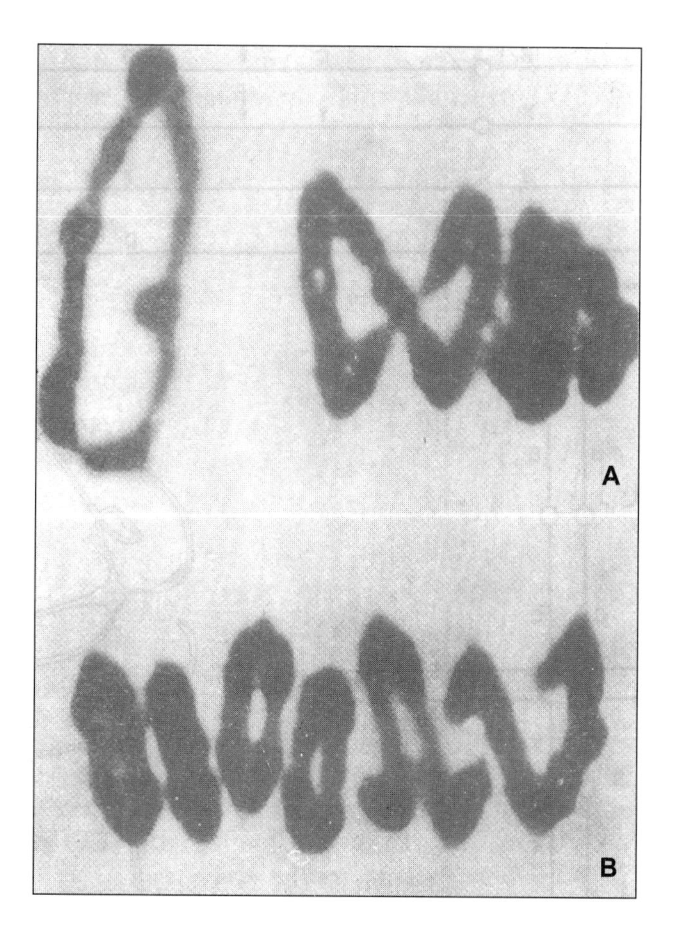

Fig. 28.4 Translocations in barley induced by X-rays: (A) 2 rings of 4 + 3 bivalents; (B) 1 ring of 4 + 4 bivalents (Courtesy: A. Hagberg).

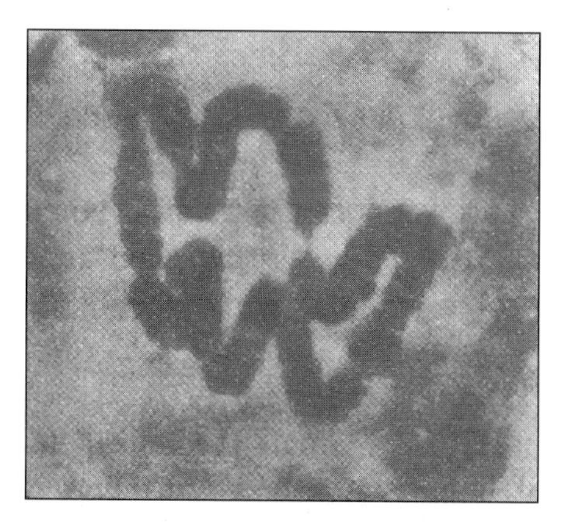

Fig. 28.5 X-ray induced translocation in Einkorn wheat—a ring of 14 chromosomes (Kihara, 1951).

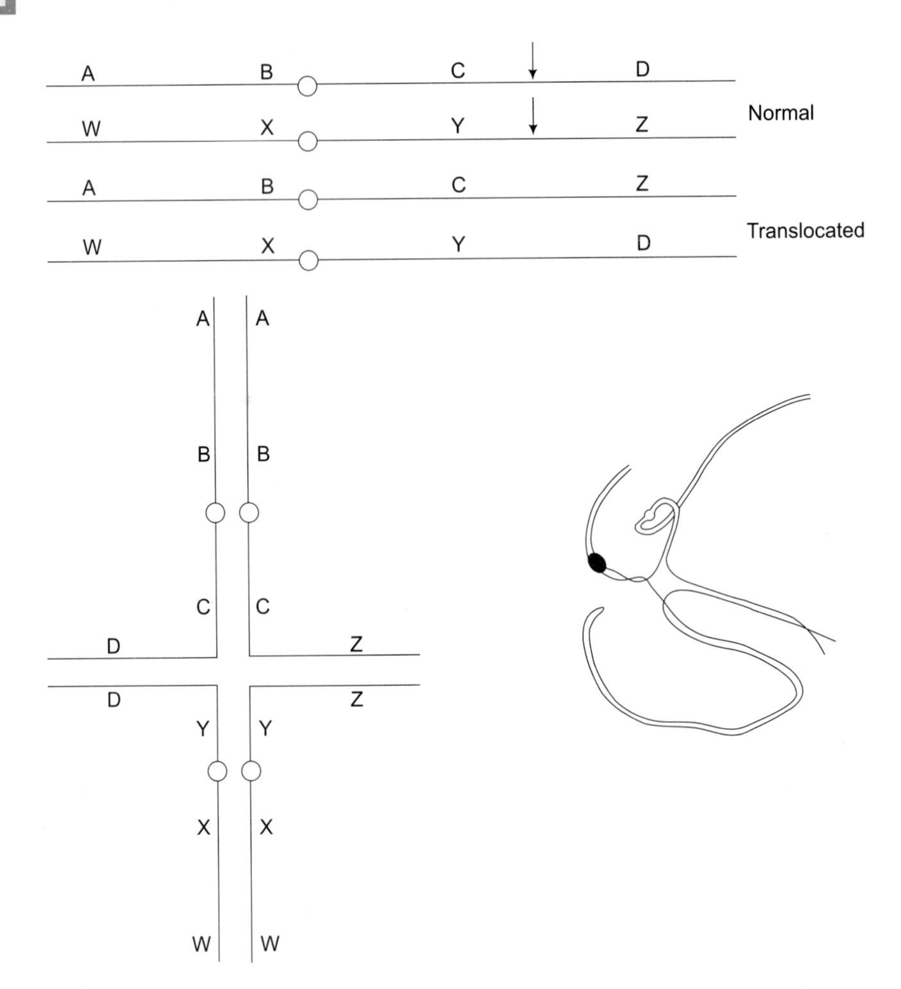

Fig. 28.6 Diagram of origin and the pattern of pachytene pairing in a translocation heterozygote. Lower right part shows a heterozygous translocation in the pachytene stage of maize, with the two pairs of centromeres being represented as open circles, and a knob on the long arm of chromosome-2 as dark body.

Multivalents are often formed during meiosis of translocation heterozygotes. In the case of one reciprocal exchange, one quadrivalent appears; in the case of several translocations, one or more multivalents occur. Translocated heterozygotes often give rise to unbalanced gametes, leading to *sterility*. Duplications and deficiencies result due to irregular chromosome separation. At pachytene stage, a cross-shaped configuration results in a translocation heterozygote. Spontaneous translocations have been found in numerous animals and plant species like *Datura*, *Oenothera*, *Godetia*, *Rhoeo*, *Paeonia*, *Pisum* and *Chlorophytum*.

Radiation induced translocations have been used to determine gene linkage and gene order in individual chromosomes.

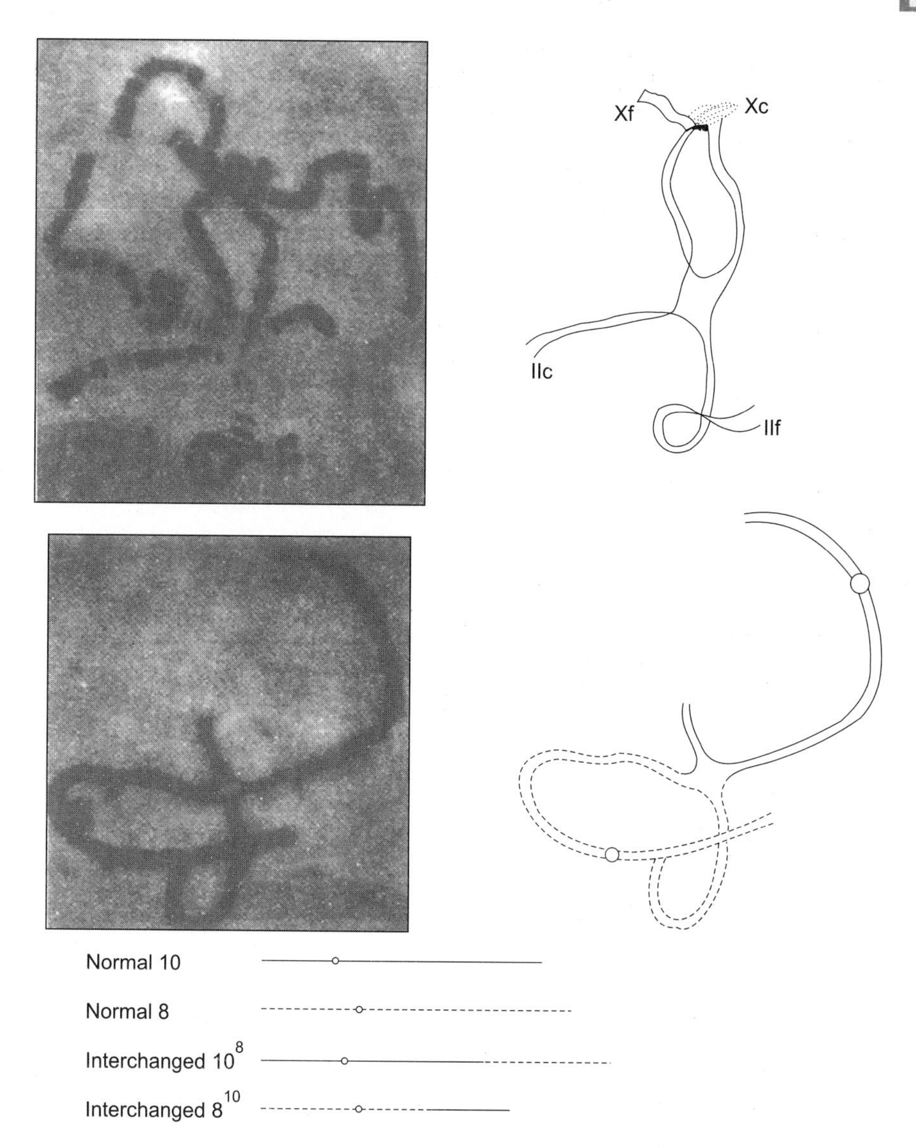

Fig. 28.7 Translocations in *Sciara* and maize. Above, an X–II translocation as seen in the salivary gland chromosomes of *Sciara coprophila*. *c* represents the centromeric ends of the two chromosomes, of the free ends. Below, an 8–10 translocation as seen in the pachytene nucleus of a microsporocyte of maize (left), with a diagram (right) to indicate the identity of chromosomes in the complex. Normal and translocated chromosomes are at the bottom.

When a translocation ring of four chromosomes reaches the metaphase plate, several arrangements are possible, and each, since the arrangements determine anaphase disjunction, has its own genetic consequences.

The three possible types are *alternate, adjacent*-1 and *adjacent*-2. In adjacent-1 and adjacent-2 types, the rings are so oriented that adjacent chromosomes go to the same anaphase pole, and

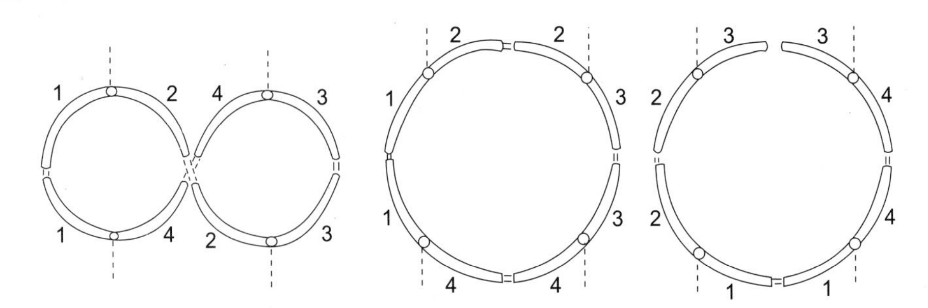

Fig. 28.8 Alternate (left), adjacent-1 (middle) and adjacent-2 (right) arrangement of a translocation ring of 4 chromosomes. Segregation from these positions should lead to viable gametes from the alternate position; duplication and deficiency from adjacent-1, with homologous centromeres going to the same pole; and duplication and deficiency from adjacent-2, with non-homologous centromeres going to the same pole.

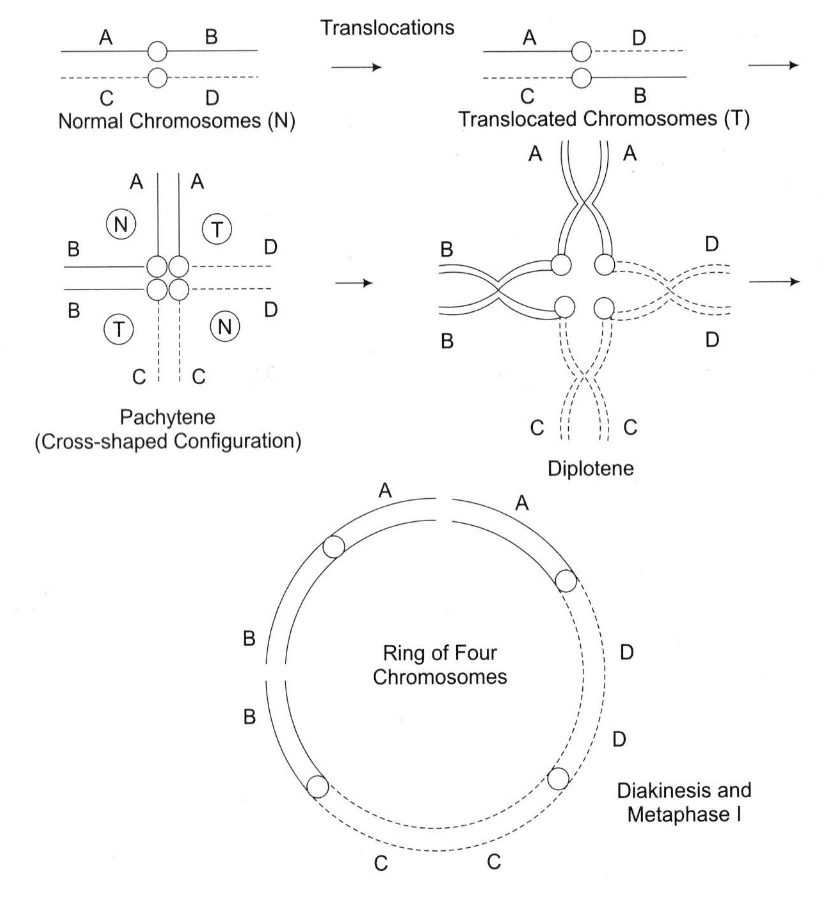

Fig. 28.9 Meiosis in a translocation heterozygote.

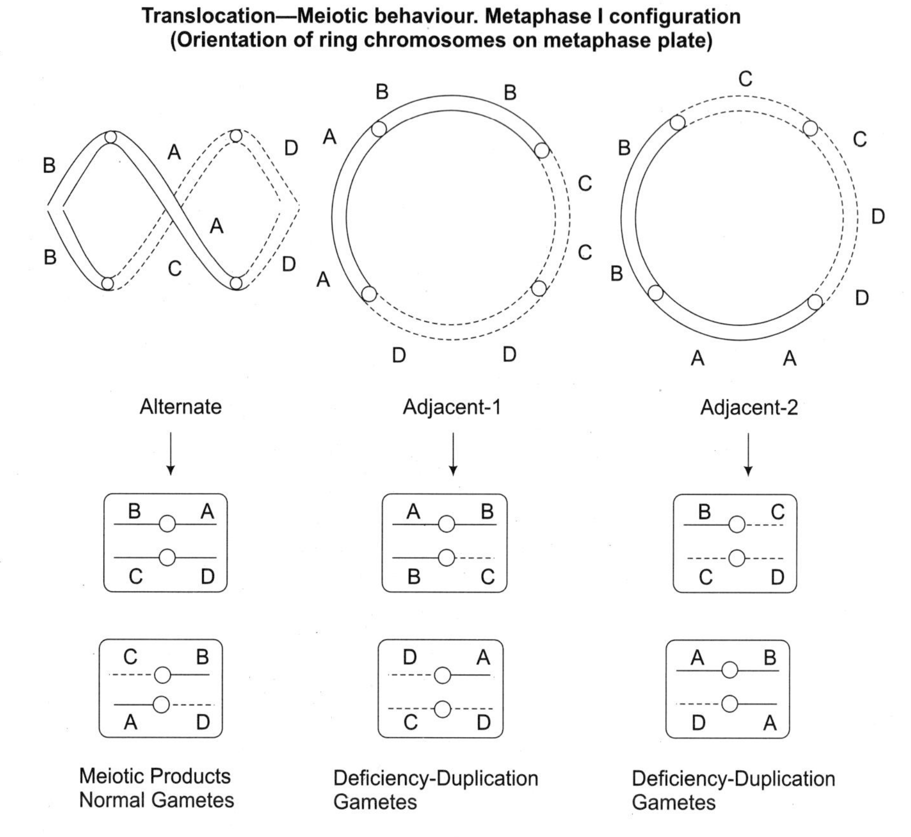

Fig. 28.10 Meiotic behaviour of a translocation heterozygote.

the gametes formed, while different from each other are both deficient and duplicated for certain regions of the chromosome. In alternate arrangement, normal chromosome (1.2; 3.4) go to one pole and the translocated chromosomes (1.4; 2.3) go to another resulting in viable gametes without duplications or deficiencies (Figs. 28.8 to 28.10).

In plants, especially in maize, the sterility in translocation heterozygotes has been extensively studied and about 50% defective pollen was always accounted.

If the area of one of the translocated chromosome be involved in a second interchange with a third non-homologous chromosome, a ring or chain of three would form at metaphase. The process can go on until the entire complement of chromosome is involved to produce a translocation complex as in *Rhoeo discolor* where a ring of 12 chromosomes occurs (Commelinaceae) and in *Paeonia californica*, etc. ...translocations play great part in evolution.

Inversions (Figs. 28.11 to 28.17)

Inversions arise as a result of two simultaneous breaks in one chromosome and their subsequent restitution after the broken piece has turned round. Inversions are *paracentric* if the centromere

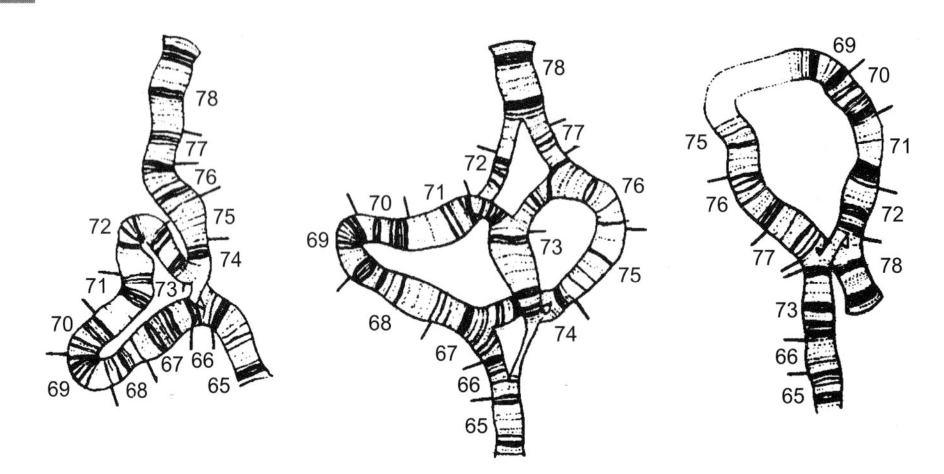

Fig. 28.11 Inversions in salivary chromosome B[1] of *Drosophila aztec* as revealed by loop formation. Left and right: two single inversions; centre a double inversion which is the combination of the two single ones.

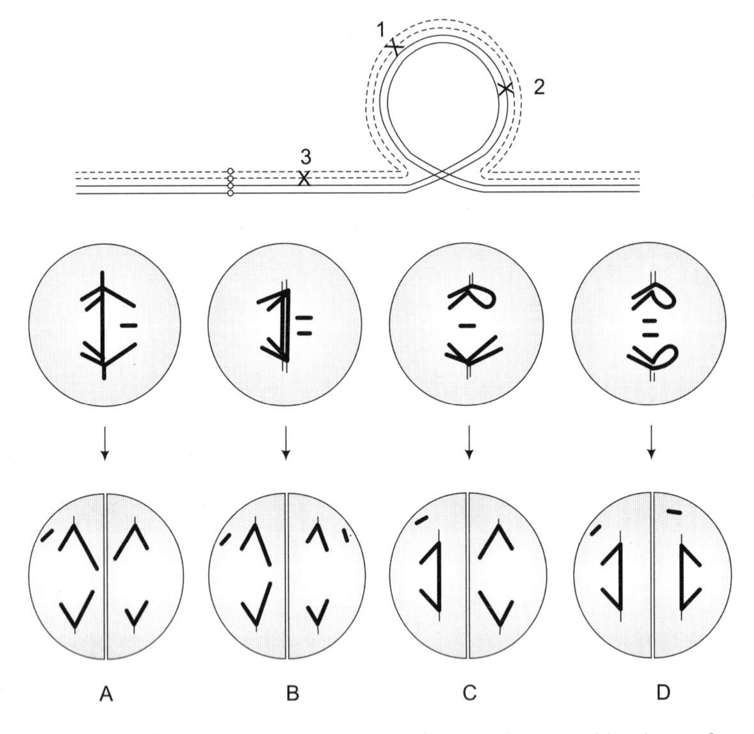

Fig. 28.12 Anaphase I and II configurations resulting from various combinations of crossovers within a paracentric inversion. A—from single crossovers at positions 1 or 2 within the inversion; B—from 4-strand double crossing over at positions 1 and 2; C—from two crossovers within and outside of the inversion at positions 2 and 3; D—from a triple crossover at positions 1, 2 and 3.

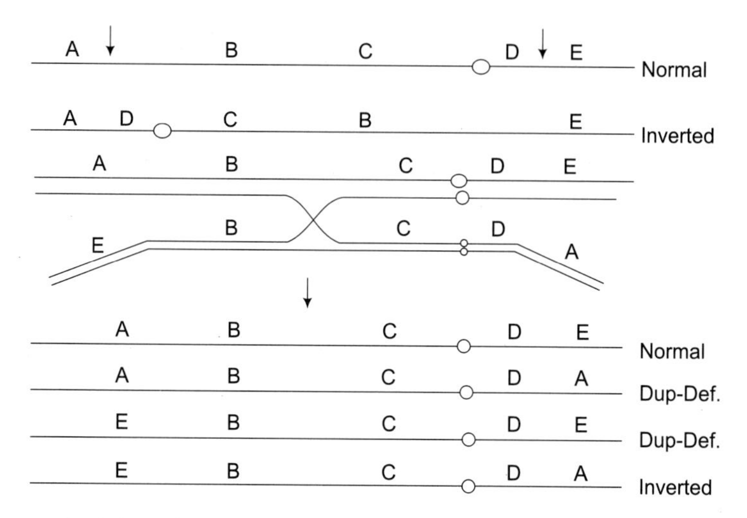

Fig. 28.13 Manner of origin and the consequences of crossing over within a pericentric inversion. Arrows indicate the break points. Duplication-deficient chromatids are indicated.

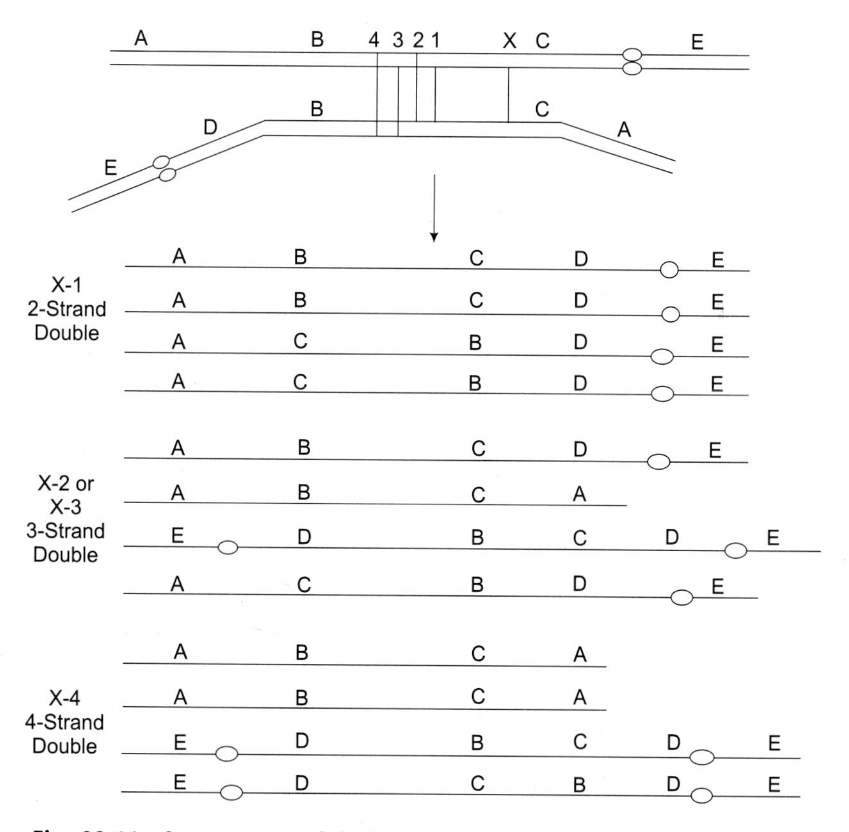

Fig. 28.14 Consequences of double crossovers within a paracentric inversion.

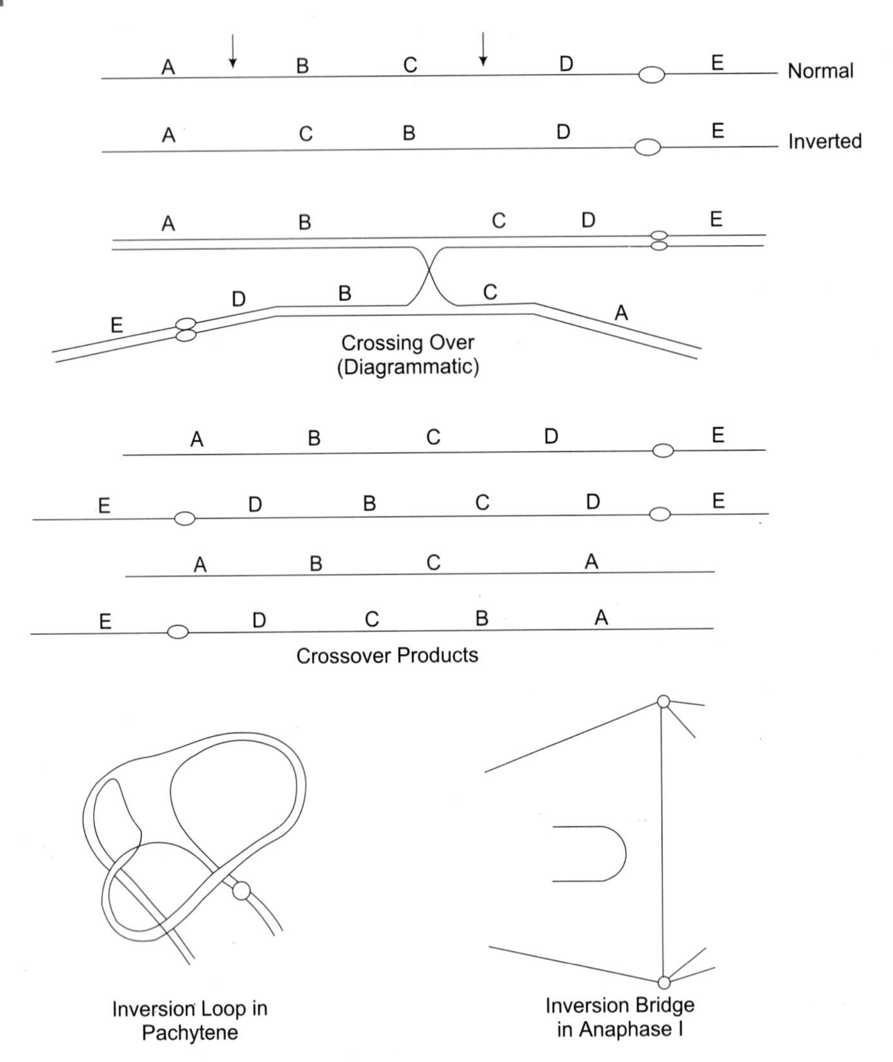

Fig. 28.15 Pairing and crossing over within a paracentric inversion, with the crossover products and anaphase I bridge formation. Pachytene configuration (lower left) in maize.

lies outside the inverted segment and *pericentric* if lies in the inverted segment. Pericentric inversions often give rise to quite *new Karyotypes*. Their heterozygotes, owing to irregular meiotic pairing, may form aneuploid gametes.

If for example, the sequence of arrangement of a normal chromosome is represented as ABCDEFGH, an inverted homologue might be *ABFEDCGH*, with the *CDEF* section being shifted in its relation to neighbouring loci.

In pachytene and salivary gland chromosomes, an inversion in a heterozygous condition can be recognised by the inversion "loop". At anaphase I an *inversion bridge* and an *acentric fragment* results when a single cross over takes place within the inverted segment (Fig. 28.15 and 28.16).

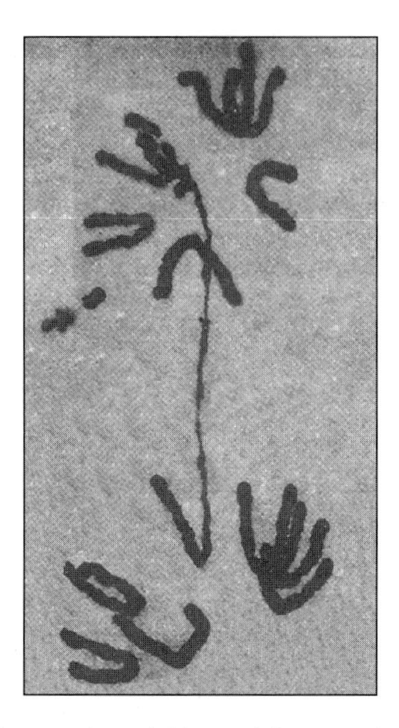

Fig. 28.16 Inversion showing anaphase bridge and fragment in *Vicia faba* induced by X-ray (Courtesy: S. Abraham).

Due to irregular breakage of inversion bridge at anaphase I and the loss of acentric fragment, inviability of gametes results.

Inversions often prevent crossing over in heterozygotes. Small inversions may be difficult to detect, since meiosis in such cases is nearly normal, inversion bridges and fragments are not formed and abnormal gametes do not arise.

Inversions occur spontaneously (maize, *Drosophila*) and have been induced and isolated by using radiations and chemical mutagenes. In general, seed sterility in inversion heterozygotes is less than in translocation heterozygotes. Translocations and inversions have greatly contributed to species differentiation.

Deficiencies (Figs. 28.18 to 28.20)

A deficiency involves the detachment and loss of a block of chromatin from the remainder of the chromosome. The detected portion of the chromosome without centomere gets lost since it cannot move to the poles during anaphase. The portion of the chromosome carrying the deficiency functions as a genetically deficient chromosome.

Deficiencies can be either *terminal* or *interstitial*. Terminal deletions arise by a single break in a chromosome followed by a "healing" of the broken end; and interstitial results from two breaks followed by the reunion of broken ends. If the deletions are large enough, they can be detected at the pachytene by a loop. Such structures have been observed in the salivary gland chromosomes of *Drosophila*.

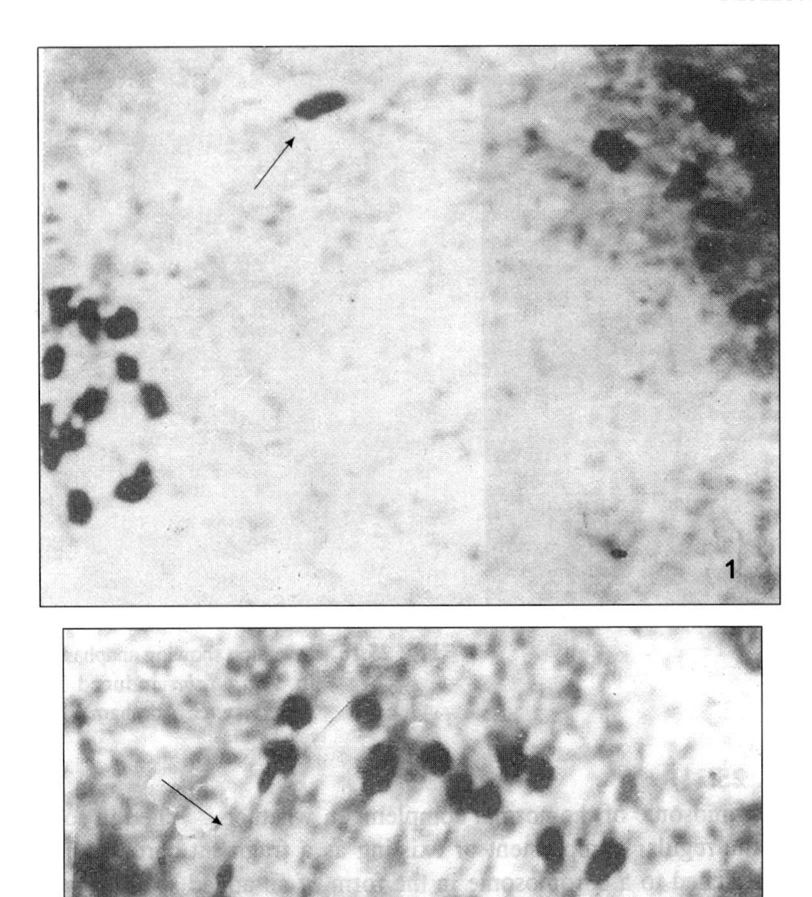

Fig. 28.17 Inversion in rice: (1) Laggard (2) Inversion bridge (Indicated by arrows) (Courtesy: Dr. A.V.S.S. Sambamurty).

In the terminal deletions their formation results in the loss of the normal end of the chromosome, or telomere, and the transformation of bipolar segment of chromosome into a unipolar segment results in the broken end and must heal for a stable structure and the chromatin ends fuse eventually.

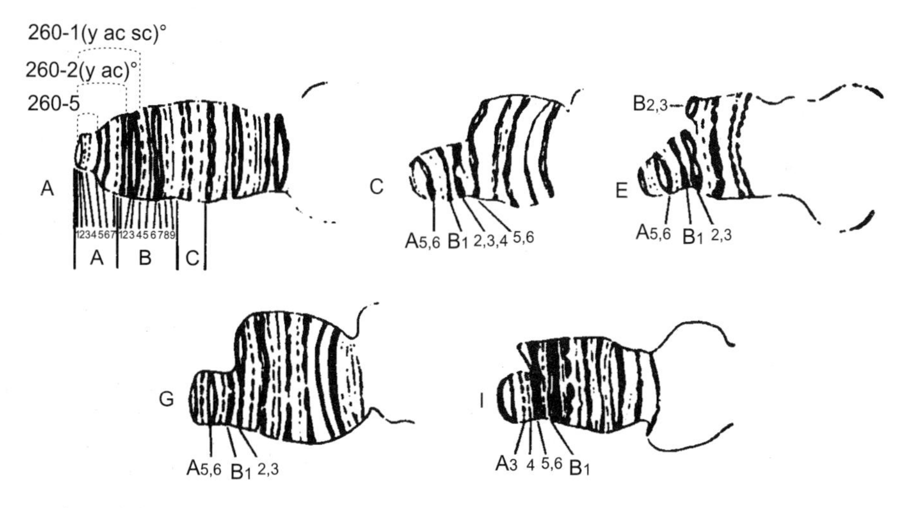

Fig. 28.18 Three deficiencies at the tip of X chromosome of *D. melanogaster.* A, normal tip; C, deficiency of 10 or 11 bands (260–1), which includes the genes *y, ac* and *sc,* and which is cell and organism lethal when homozygous; E and G, two drawings of the same 8-band deficiency (260–2), which includes y and *ac,* and which is organism but not cell lethal when homozygous; I, deficiency of 4 bands (260–5) which does not include these three genes, is not cell or organism lethal, and only reduces fertility in homozygous females.

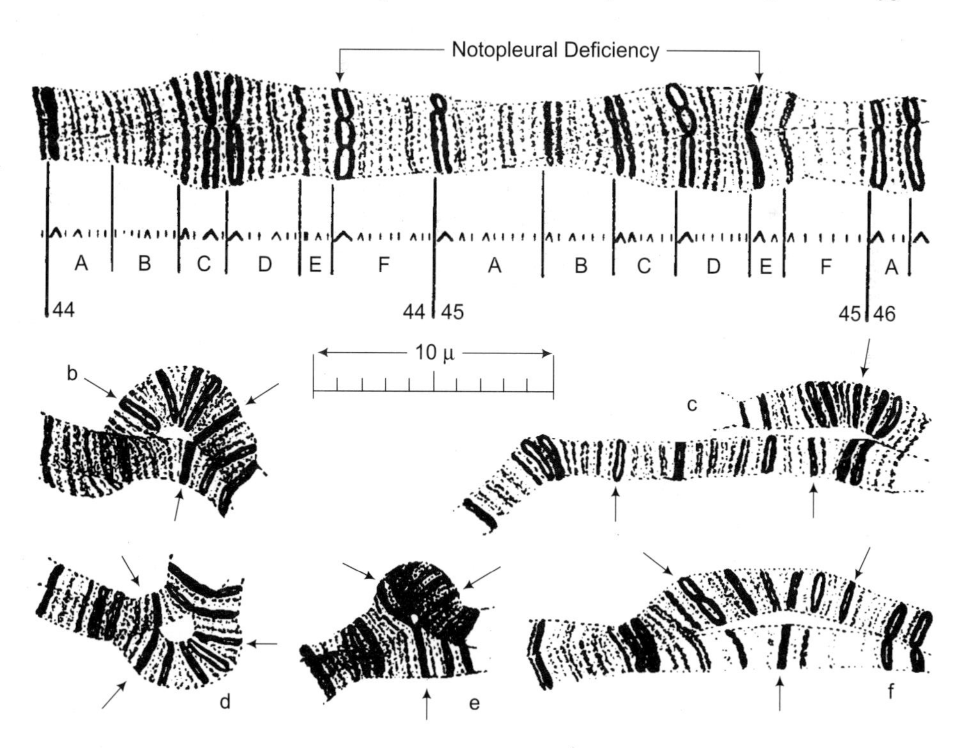

Fig. 28.19 Salivary gland chromosomes of *D. melanogaster* showing Notopleural deficiency in the right arm of chromosome 2. Above, normal chromosome indicating the limits of the deficiency; b–f, Notopleural heterozygotes; b, synapsed at the right limits of deficiency; c and f, non-synapsed strands; d, synapsed at the left limit; e, synapsed at both the limits.

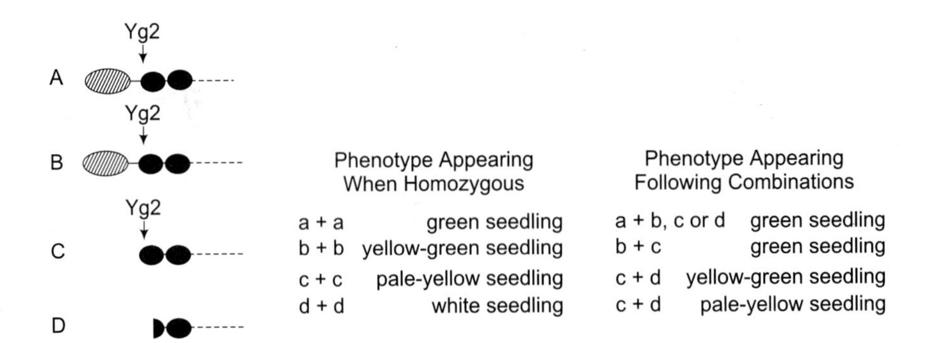

Fig. 28.20 *Deficiencies in Maize*: Schematic representation of chromosomes at the end of short arm of chromosome 9 of maize, together with the progressive deficiencies which have been detected. These give the equivalent of a pseudoallelic series, and the phenotypes of homozygous and heterozygous combinations are indicated (McClintock 1944). Crosses and their phenotypic effects are given in Table 28.1.

Depending upon the amount of genic material that is lost corresponding deleterious effects will be manifested on the organism in the form of mutations which are recognisable. In *Drosophila* the deficiencies induced the mutants like *Blond*, *Pale*, *Beaded*, *Carved*, *Plexate*, *Notch*, *Minute*, etc.

In man a number of congenital abnormalities have been traced to chromosomal deficiencies: chronic myeloid leukaemia is associated with the Philadelphia chromosome, identified as chromosome-21 minus substantial portion of its long arm.

The *Cri-du-chat* (*cat cry*) syndrome, so named because of the characteristic mewing cry of the affected child and mental retardation and other physical abnormalities, results from a loss of the short arm of chromosome 4–5 group (see Chapter 11 for details).

Deficiencies can be employed to locate genes on the salivary gland chromosomes of *Drosophila*. In some instances in *Drosophila* there is clear correlation between the phenotypic effect and the number of bands deleted.

McClintock has provided evidence that viable morphological variations can be produced in maize by homozygous tiny deficiences.

In Chromosome-9 of maize, the chromosome loss of chromatin produced mutant phenotypes when in a homozygous condition involving the terminal chromomeres (Fig. 28.20).

Table 28.1

Phenotype appearing when homozygous		*Phenotype appearing in combinations*	
a + a	green seedling	a + b, c or d	green seedling
b + b	yellow-green seedling	b + c	green seedling
c + c	pale-yellow seedling	b + d	yellow-green seedling
d + d	white seedling	c + d	pale-yellow seedling

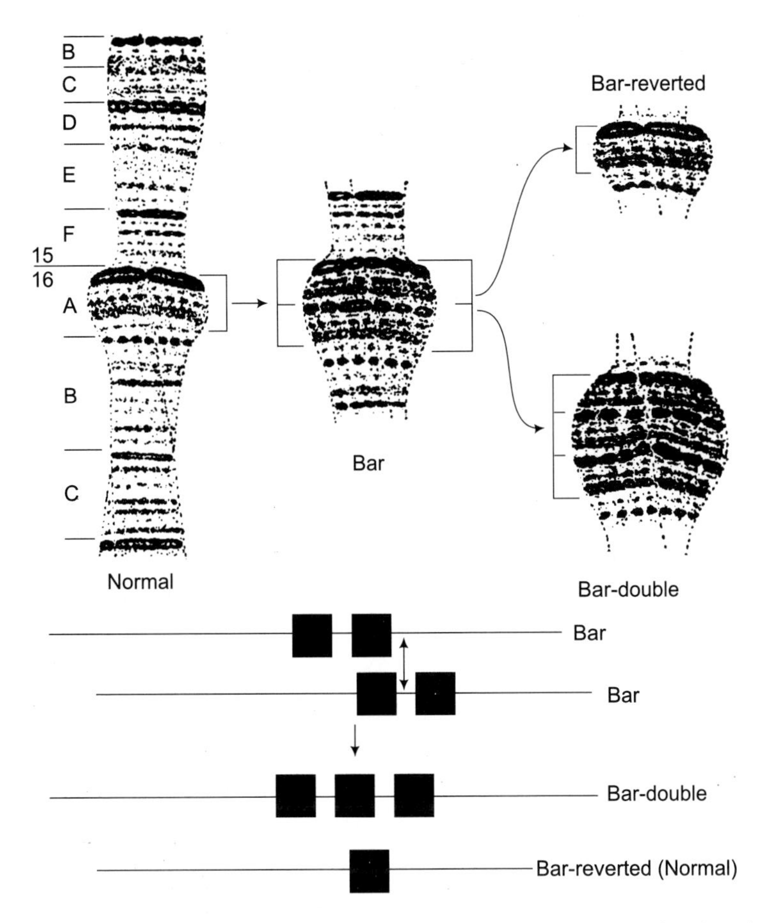

Fig. 28.21 *Bar-locus* in the salivary gland X chromosome of *D. melanogaster,* indicating that duplication of the 16A region is responsible for the *Bar* phenotype while triplication gives the *Bar-double* phenotype. Below, diagram to indicate the manner by which unequal crossing over causes *Bar* to give rise to normal (*Bar-reverted*) and *bar-double* chromosomes.

Duplications (Fig. 28.21)

An extra piece of chromosome of the normal complement, whether attached in some manner to one of the members of the regular complement or existing as a fragment chromosome, is known as a *duplication*. When attached to a chromosome in the form of an added section, the duplication may be in tandem, in reverse tandem, or as a displaced one. Thus if the duplicated piece is represented by the letters *def*, a tandem duplication would be, abc *defdef* ghi, a reverse tandem would be, abc *deffed* ghi, and a displaced duplication would be, rst *def* uvw or rst *fed* uvw.

Bar Phenotype in *Drosophila*

Four or five bands in the *16A* region are involved, and the salivary gland picture shows that the normal males have the *16A* region represented once, *Bar* females twice, and Bar double males three times.

As shown by Sturtevant (1925), Bar has a tendency to segregate altered phenotypes, and the frequencies of reversions of *Bar* to normal and to a more exaggerated *Bar-double* are similar. Their occurrence can be accounted for by the phenomenon of *Unequal crossing over*. Duplications are more lethal than the deficiencies. Should the mutated gene be present as a duplication along with the normally functioning gene, the possibilities of its retention and continued mutations, possibly in new directions, become considerably enhanced.

(b) Gene mutations

These are sub-microscopic changes within the fine structure of a gene locus ("Cistron"). They are characterised by : (1) No cytological irregularities are present, (2) Segregation of the heterozygote is normal; lethality phenomena of extreme kinds do not occur, and (3) the mutation is capable of reversion.

Reversions lead to unstable mutations and unstable genetic loci as are found widely in *Zea mays, Hordeum, Oryza, Antirrhinum, Delphinium, Neurospora, Drosophila,* etc., leading to Position Effect, Paramutation, Controlling elements and Unstable alleles.

EXTRANUCLEAR MUTATIONS

These are hereditary changes that are independent of mutations in genome and chromosomes. Heredity through cytoplasm and at least partially independent of the nucleus was shown in *Mirabilis, Pelargonium, Oenothera, Antirrhinum, Nicotiana, Epilobium, Zea, Humulus, Streptocarpus, Lymantaria, Drosophila* and *Paramaecium.*

Cytoplasmic inheritance can be divided into two groups: (1) Plasmon, and (2) Plastidom inheritance. Plastidom inheritance is related to the functions and properties of the plastids. Properties controlled plasmatic ally comprise male sterility, sexual differentiation, chlorophyll formation and also differences in height and vigour.

CHAPTER 29

Microbial Genetics: Viruses and Bacteria

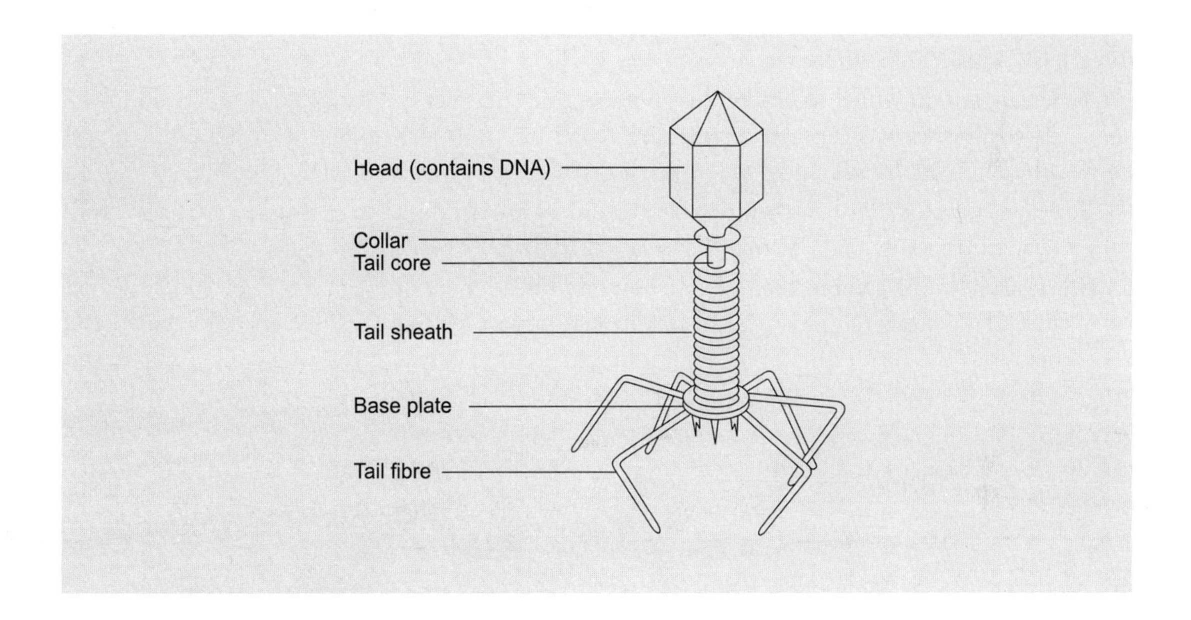

VIRUSES-The Simplest Form of Life

The viruses are the first and simplest form of life that exist on the Earth. In fact viruses are so simple in biological terms that the question has frequently been asked as to whether they can actually be considered to be living organisms. Doubts arise partly because viruses are constructed along lines different from all other forms of life–viruses are not cells–and partly because of the nature of the virus life cycle. Viruses are *obligate parasites* of the most extreme kind; they can live and reproduce *only within a host cell* which is living and in order to express and replicate their genes viruses must subvert at least part of the host's genetic machinery to their own ends. Some viruses possess genes coding for their own RNA polymerase and DNA polymerase enzyme but many depend on the host enzymes for transcription and DNA replication. All viruses make use of the host's ribosomes and translation apparatus for synthesis of the polypeptides that will make up the protein coats of their progeny. This means that virus genes must be matched to the host genetic system. Viruses are therefore quite specific for particular organisms and individual types cannot infect a broad spectrum of species. Viruses that infect bacteria are called *bacteriophages* and have been of the greatest interest to molecular biologists because they are excellent experimental tools for the analysis of basic genetic events in bacteria.

BACTERIOPHAGES (Figs. 29.1, 29.2, 29.3 and 29.4)

Bacteriophages are constructed from two basic components: protein and nucleic acid. The protein forms a coat or capsid within which the nucleic acid genome is contained. There are three basic capsid structures, viz.,

1. *Icosahedral,* in which the individual polypeptide subunits (protomers) are arranged into a three-dimensional geometric structure that surrounds the nucleic acid, e.g. MS2 phage which infects *E. coli,* and PM2 which infects *Pseudomonas aeruginosa.*
2. *Filamentous* or *helical,* in which the protomers are arranged in a helix producing a rod-shaped structure, e.g., *E. coli* phage called M13.
3. *Head-and-tail,* a combination of an icosahedral head, containing the nucleic acid, attached to a filamentous tail plus possibly additional structures that facilitate entry of the nucleic acid into the host cell. This is a common structure possessed by, for example, the *E. coli* phages T4 and A, and phage SP01 of *Bacillus subtilis.*

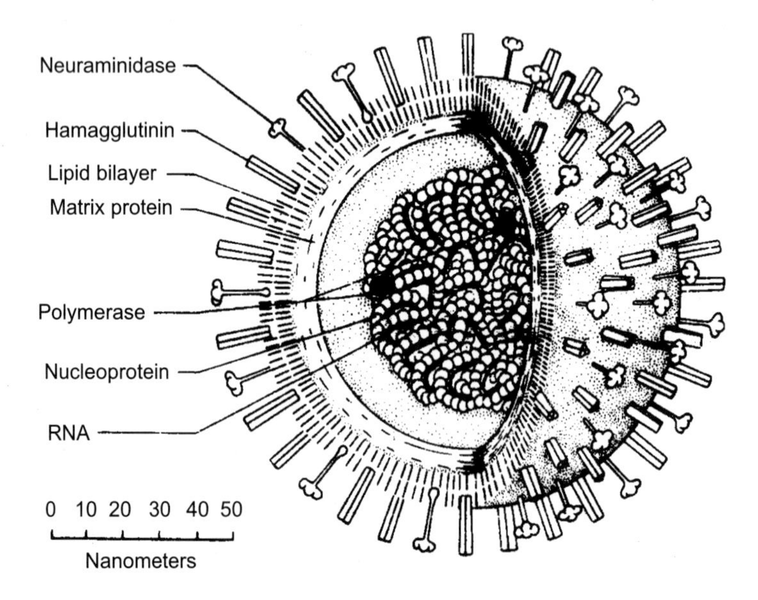

Fig. 29.1 Structure of the influenza virion.

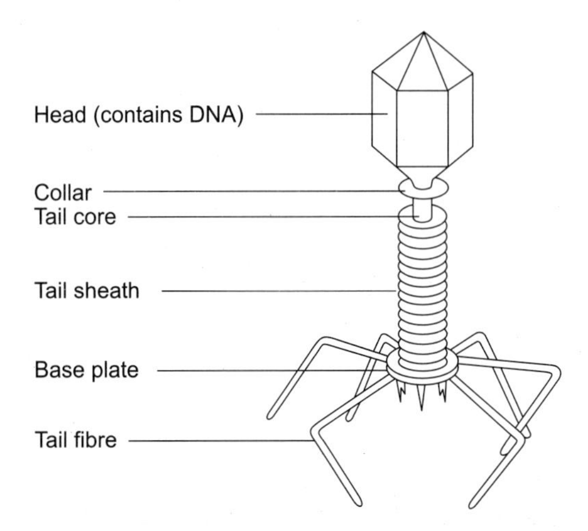

Fig. 29.2 Structure of T-2 bacteriophage.

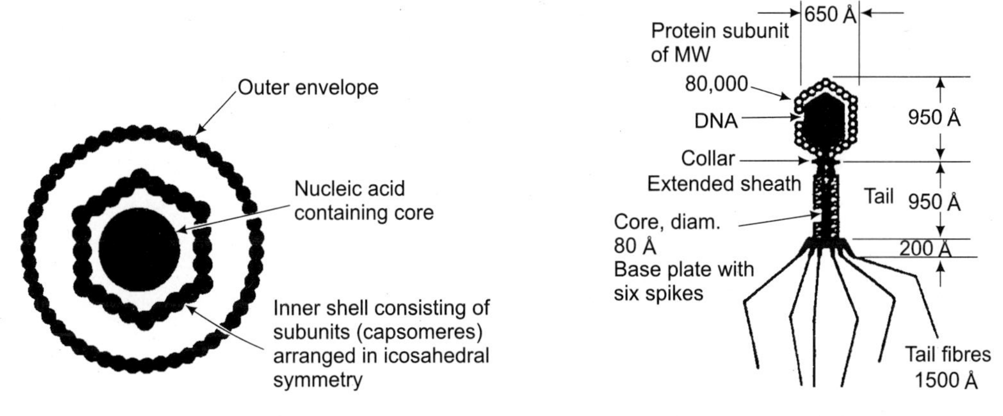

Fig. 29.3 The morphology of Herpes, a large DNA-containing virus which multiplies in animal cells.

Fig. 29.4 The structure of the T-even (2, 4 and 6) phage particle.

Bacteriophage Genomes can be DNA or RNA

The term 'nucleic acid' had been used when referring to phage genomes because in some cases these molecules are made of RNA. Viruses are the one form of 'life' that contradict the conclusion of Avery and his colleagues and of Hershey and Chase that the genetic material is DNA. Viruses, including phages, also break another rule: their genomes, whether of DNA or RNA, can be single-stranded as well as double-stranded.

With most types of phage there is a single DNA or RNA molecule that comprises the entire genome (Fig. 29.5). However, this is not always the case and a few RNA phages have segmented genomes, meaning that the entire complement of genes is shared between a number of different nucleic acid molecules. The size of the genomes vary enormously (Fig. 29.6), from about 1.6 kb

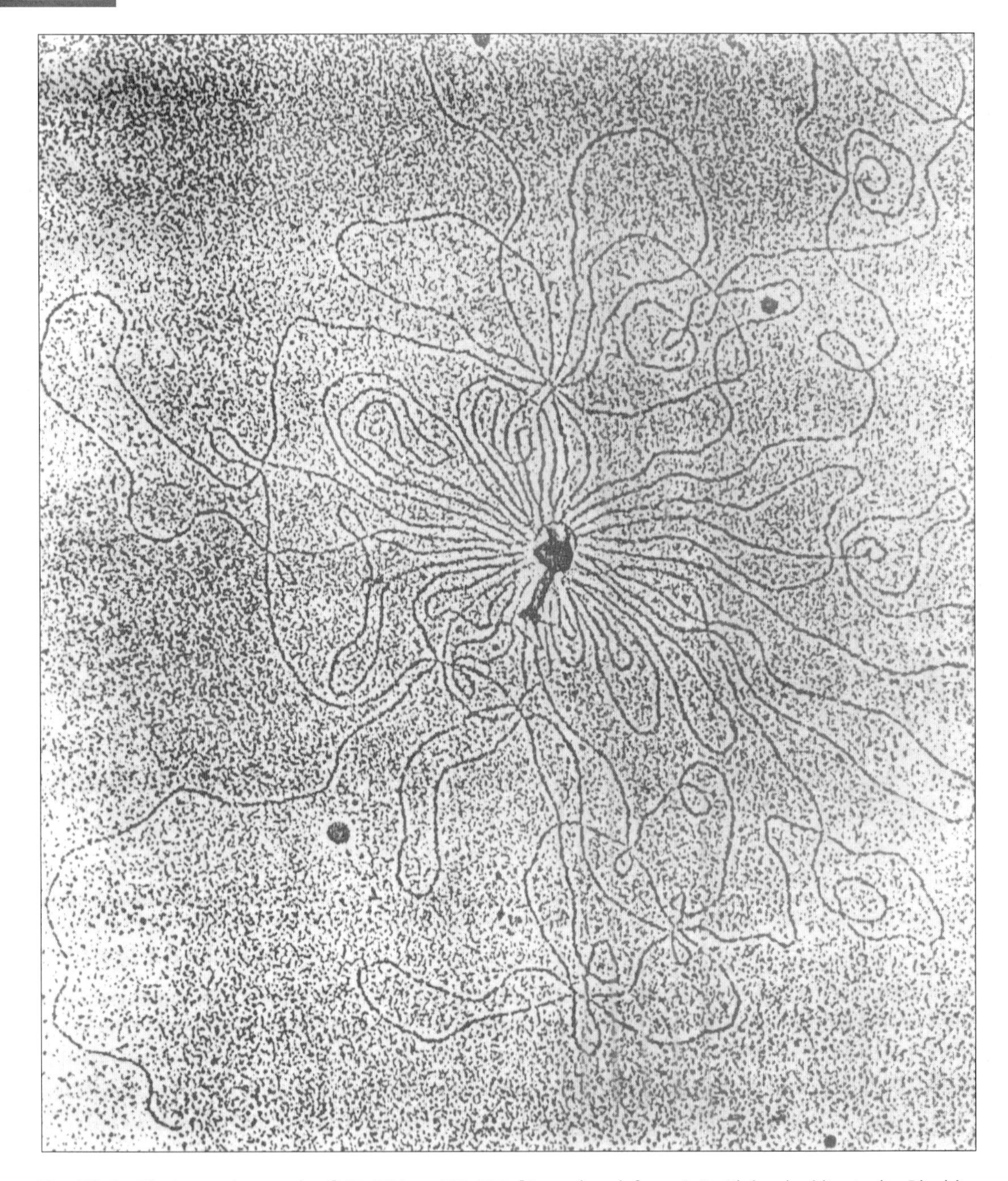

Fig. 29.5 Electron micrograph of T2 DNA × 100,000 [Reproduced from A.K. Kleinschmidt et al., Biochim. Biophys. Acta, 61, 857 (1962), with permission].

for small phages such as MS2 to over 150 kb for large ones such as the T-even phages (T2, T4, T6, etc.) and the number of genes carried also varies from less than 10 to over 200.

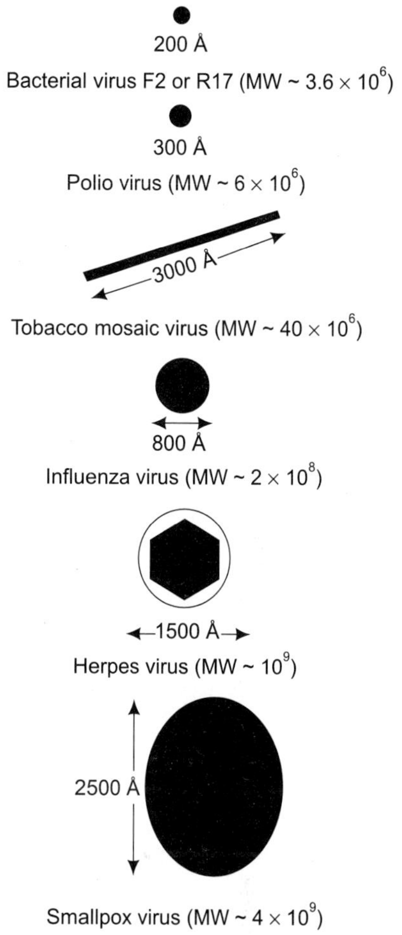

The Phage Life Cycle

Bacteriophages are classified into two broad groups according to their life cycle: *lytic* and *lysogenic*. The fundamental difference between these groups is that a lytic phage kills its host bacterium very soon after the initial infection, whereas a lysogenic phage can remain quiescent or dormant within its host for a substantial period of time, even throughout numerous generations of the host cell. These two life cycles are typified by two *E. coli* phages: the *lytic* (or *virulent)* T4 and the *lysogenic* (or temperate) λ.

Lytic Infection Cycle of T4 (Fig. 29.7)

The T series of *E. coli* phages (Tl, T2, T3, etc.) were isolated between 1939 and 1941 by M. Demerec, a Yugoslavian geneticist working in the United States. These were the first phages to become available to molecular geneticists and have been the subject of much study ever since. T2 for instance was used by Hershey and Chase to prove that genes are made of DNA. T4 is one of the best characterised viruses.

The initial event in infection of *E. coli* by T4 is the attachment of the phage particle to a receptor protein on the exterior of the bacterium. Different types of phage have different receptors: for T4 the receptor is a protein called *OmpC* ('*Omp*' stands for '*outer membrane protein*'). After attachment, the phage injects its DNA genome into the cell through its tail structure.

200 Å
Bacterial virus F2 or R17 (MW ~ 3.6×10^6)
300 Å
Polio virus (MW ~ 6×10^6)
3000 Å
Tobacco mosaic virus (MW ~ 40×10^6)
800 Å
Influenza virus (MW ~ 2×10^8)
1500 Å
Herpes virus (MW ~ 10^9)
2500 Å
Smallpox virus (MW ~ 4×10^9)

Fig. 29.6 Size variations in different viruses.

The next 22 minutes are called the eclipse period when, to the outside observer, nothing very much seems to happen. Inside the cell, however, all activity is directed at synthesis of new phage particles. Immediately after entry of the phage DNA, the synthesis of host DNA, RNA and protein stops and transcription of the phage genes begins. Within 5 minutes the bacterial DNA molecule has depolymerised and the resulting nucleotides are being utilised in the replication of the T4 genome. After 12 minutes new phage capsid proteins start to appear and the first complete phage particles are assembled. Finally, at the end of the eclipse period, the cell bursts and the new phage particles are released. A typical infection cycle produces 200 to 300 T4 phages per cell, all of which can go on to infect other bacteria.

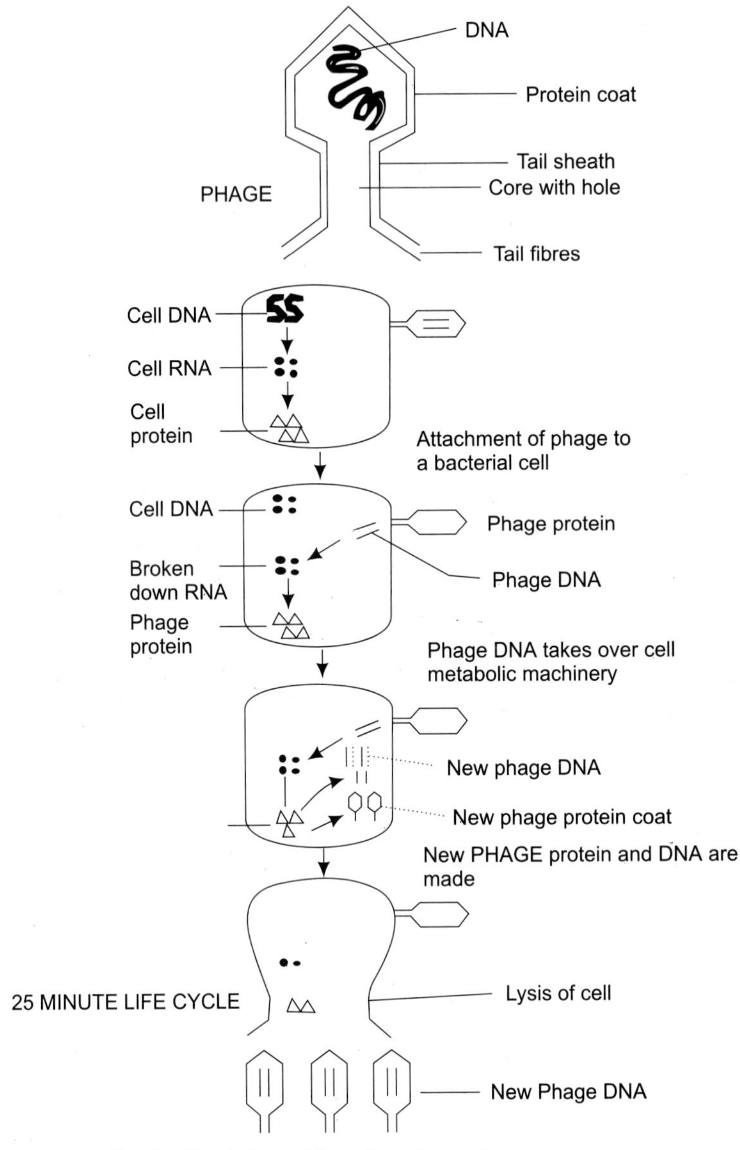

Fig. 29.7 Lysis in T4 bacteriophage.

Lysogenic Life Cycle of λ

During a lysogenic cycle the phage genome becomes integrated into the host DNA. This occurs immediately after the entry of the phage DNA into the cell, and results in a quiescent form of the bacteriophage, called the *prophage*. The DNA molecule and the *E. coli* genome share a short region of nucleotide homology, only 15 bp in length, but sufficient for the recombination event to be initiated. Note that this means that the genome always integrates at the *same position* on the *E. coli* DNA molecule.

Unusual Phage Life Cycles

Although lysis and lysogeny are the two most typical phage life cycles they are by no means the only ones. One or two other bacteriophages display *unusual infection cycles* that are neither truly lytic nor truly lysogenic. An example is provided by a third *E. coli* phage. M13, which has assumed importance in recent years in recombinant DNA technology as a source of single-stranded DNA versions of cloned genes. The M 13 phage particle contains a *single-stranded circular genome* which, after injection into the bacterium, is replicated by the synthesis of the complementary strand, producing a double-stranded, circular DNA molecule. This molecule then undergoes further replication until there are over 100 copies of it in the cell. At this stage the infection cycle takes on characteristics of both the lytic and lysogenic phages.

Viruses of Eukaryotes

All eukaryotes act as hosts for viruses of one kind or another. Indeed most eukaryotes are susceptible to infection by a broad range of virus types: think of the number of viral diseases that humans can catch. Because of this medical relevance, the viruses of man and animals have received most research attention, with plant viruses, capable of destroying crops, a rather distant second.

In many respects eukaryotic viruses resemble bacteriophages. We will not, therefore, go into great detail about their basic features, limiting ourselves to those that are distinctive. This will lead us into a brief review of the connection between certain animal viruses and cancer.

Structure of Eukaryotic Viruses

The capsids of eukaryotic viruses are either icosahedral or filamentous; the head-and-tail structure so common with bacteriophages is unknown with eukaryotic viruses. One distinct feature of eukaryotic viruses, especially those with animal hosts, is that the capsid may be surrounded by a lipid membrane, forming an additional component to the virus structure. This membrane is derived from the host when the new virus particle leaves the cell, but may be modified by insertion of virus-specific proteins.

Virus genomes may be made of DNA or RNA, and may be linear or circular, and single- or double-stranded. For reasons that no one has explained, the vast majority of plant viruses have RNA genomes. Genome sizes cover the same range as in phages.

Infection Cycles

Most eukaryotic viruses follow only a lytic infection cycle. However, these virulent viruses rarely take over the host cell's genetic machinery to the extent that a lytic phage will instead the virus and host cell coexist for some time and the host cell functions will cease only towards end of the infection cycle when virus progeny that have been stored in the cell will be released. Alternatively, new virus particles may be extruded continually throughout the infection period as membrane-bound units.

It should be remembered that eukaryotic cells are structurally more complex than bacteria and that the enzymes involved in DNA replication and transcription are confined to the nucleus after entry into the cell, and are therefore, able to make use of the host DNA and RNA polymerase enzymes for their own replication and gene transcription. Others, however, remain in the cytoplasm and must synthesise their own enzymes. Note that a cytoplasmic location means that the virus particle must contain a few molecules of RNA polymerase, which it must inject into the host cell along with its genome, so that viral gene expression can get started.

Animal Viruses and Cancer

A link between some types of animal virus and cancer has been suggested in the recent years. Renato Dulbecco in 1960, suggested that the lysogenic life cycle of a bacteriophage such as λ might be analogous to the ability of certain types of animal virus to cause cell transformation. This phenomenon (not to be confused with genetic transformation, as studied by Griffith) involves changes in cell morphology and physiology. In cell cultures, transformation results in a loss of control over growth, so that transformed cells grow as a disorganised mass, rather than as a monolayer. In whole animals, cell transformation is thought to underlie the development of tumours.

Two types of virus are thought to cause cancer: the *DNA tumor viruses,* which include a number of unrelated viruses, such as polyoma virus, SV40 and adenovirus, and a single group of RNA viruses called the *retroviruses.*

DNA Tumor Viruses

These viruses can follow a standard virulent infection cycle in some hosts, but in other organisms will cause cell transformation. "In these non-permissive hosts, virus replication cannot take place, as only the early viral genes are expressed: new virus particles are never produced. However, the viral genome can remain in the host as a provirus, integrated at possibly several positions in the host chromosomes. Integration is apparently sites on the *E. coli* DNA molecule.

Exactly why the provirus induces transformation of the host cell is not known. Expression of some virus genes continues after integration and the gene products can be detected in the transformed cell. However, even after several years of research, the link between the viral proteins and the transformed state is not understood.

Retroviruses

The retroviruses are a closely knit group of RNA viruses that include several, such as the AIDS virus, that have never been implicated in cancer. Even those retroviruses that do cause cell transformation do so only under special circumstances.

Integration of the viral genome into the host DNA is an essential stage in the infection cycle of all retroviruses. The fact that the retroviral genome is a single-stranded RNA molecule, between 6,000 and 9,000 nucleotides in length, whereas the integrated version is a double-stranded DNA, indicates that these viruses must contradict Crick's Central Dogma which states that information flows from DNA to RNA to protein. With retroviruses it is possible for information to flow from RNA to DNA. This fact was confirmed independently in 1970 by Howard Temin of the University of Wisconsin and David Baltimore of the Massachusetts Institute of Technology, both of whom isolated from infected cells retrovirus-specific enzymes capable of making a DNA copy of an RNA template. This enzyme is called *reverse transcriptase.* Note that as synthesis of the DNA copy of the genome occurs before viral transcription, it is necessary for molecules of reverse transcriptase to be carried within the retroviral particle and to be introduced into the cell with the RNA genome.

Integration of the viral genome into the host DNA is a prerequisite for expression of the *retrovirus genes.* There are three of these genes, called *gag, pol* and *env* coding for, amongst other things, the viral coat proteins and the reverse transcriptase. New virus particles are therefore, produced by cells that harbour the integrated provirus.

The captured gene (called a *v-onc,* with "*onc*" standing for oncogene) is the one that codes for a protein involved in cell proliferation. Normally the cellular version (*c-onc*) would be subject to strict regulation, and would be expressed only in limited quantities when needed. Possibly the *v-onc* is not subject to the same regulation, in which case its *uncontrolled expression* could be the basis to cell transformation causing cancerous tissues in animals.

INTRODUCTION TO BACTERIA AND VIRUSES

All organisms, including the viruses, have genes located sequentially in the genetic material, and all, with the possible exception of a small group of viruses, can have recombination between the homologous pieces of genetic material. Because such recombination does occur, it is possible to map the location and sequence of genes along the chromosomes of all organisms and viruses. The unique properties of the life cycles of bacteria and viruses have contributed to the study of genetics in the past four decades. It is through work with bacteria and viruses that we have entered the modern era of molecular genetics.

Bacteria, along with blue-green algae, make up the prokaryotes. The true bacteria can be classified according to shape. A spherical bacterium is called a *coccus;* a rod-shaped bacterium is called a *bacillus;* and a spiral bacterium is called a *spirillum.* Prokaryotes do not undergo mitosis or meiosis but simply divide in half after their chromosome, a circle of DNA, has replicated. Viruses do not even divide; they are mass-produced within the host cell. Several properties of bacteria and viruses have made them especially suitable for molecular genetic research.

Importance of Bacteria and Viruses in Genetic Research

First, bacteria and viruses have a very short generation time. One virus can become 100 in about an hour; an *E. coli* cell (*Escherchia coli,* the common intestinal bacterium, discovered by Theodor Escherich in 1885) doubles every 20 minutes. In contrast, there is a generation time of 14 days in fruit-flies, one year in corn, and 25 years in humans.

Second, bacteria and viruses have much less genetic material than eukaryotes, and the organisation of this material is much simpler. The term prokaryote arises from the fact that these organisms do not have true nuclei (*pro* means *before* and *karyon* means *kernel* or *nucleus*); they have no nuclear membranes and only a single "naked" chromosome (Fig. 29.8). Viruses are even simpler. They consist almost entirely of genetic material srrounded by a protein coat. Or, more precisely, the bacterial viruses in which we are interested, the bacteriophages or just phages (phage—one who eats)—are exclusively genetic material surrounded by a protein coat. Some animal viruses are more complex.

Viruses can be classified by their host preference (animal, plant, or bacteria) and by the nature of their genetic material RNA or DNA. Viruses are obligate parasites. Outside a host, they are inert molecules. Once their genetic material penetrates a host, they take over the metabolism of that host and construct multiple copies of themselves.

A third reason for the use of bacteria and viruses in genetic study is their ease of handling. Millions and millions of bacteria can be handled in a single culture with minimal amount of work as compared with the effort required to grow the same number of eukaryotic organisms, such as fruit-flies or corn. Some eukaryotes, such as yeast or *Neurospora*, can, of course, be handled using prokaryotic techniques.

Techniques of Cultivation of Bacteria

Since different groups of bacteria have different nutritional requirements, different media have been developed on which they are grown in the laboratory. All organisms need an energy source, a carbon source, nitrogen, sulphur, phosphorus, several metallic ions and water. Those that require an organic form of carbon are termed *heterotrophs*. Those that can utilise carbon as carbon dioxide are termed *autotrophs*. All bacteria obtain their energy either by photosynthesis or chemical oxidation. Bacteria are usually grown in or on a chemically defined or synthetic medium either in liquid or in test tubes or petri plates using an agar base for rigidity. Petri plates are circular dishes of glass or plastic about 2 cm deep with matching circular covers. When one

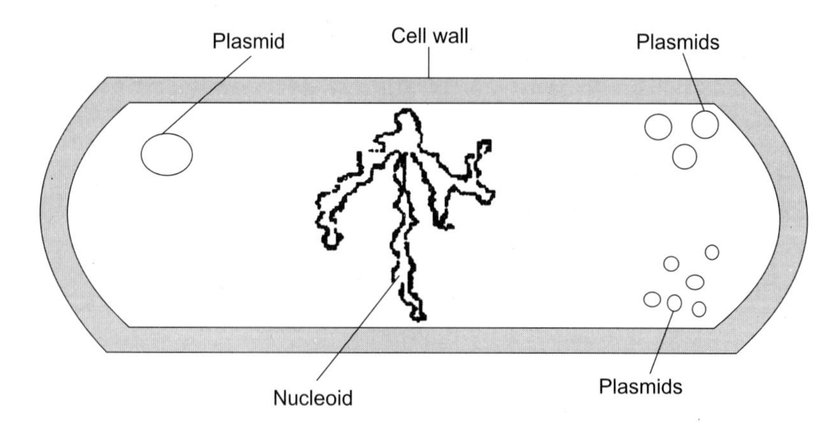

Fig. 29.8 *E. coli* with nucleoid and plasmids.

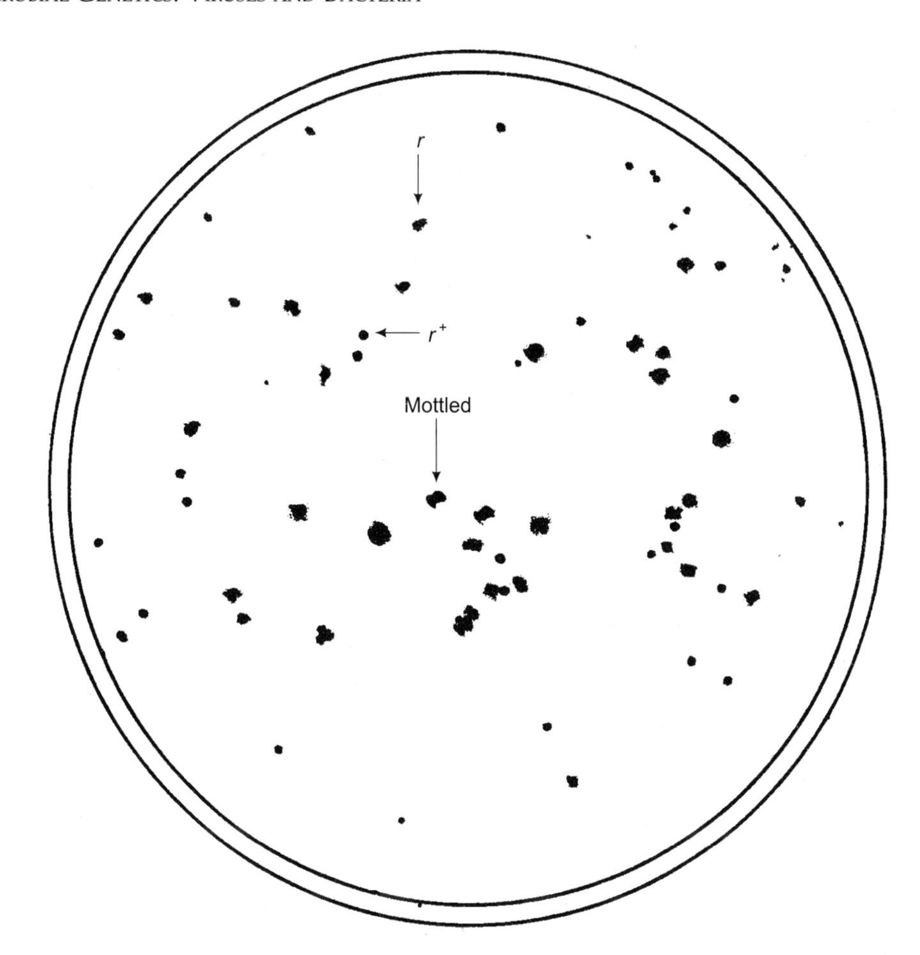

Fig. 29.9 Four types of plaques produced on a mixed lawn of *E. coli* by mixed phage T2.

cell is placed on the medium in the plate, it will begin to divide. After incubation, often overnight, a colony, or clone, will exist where there was previously only one cell. Overlapping colonies form a solid lawn of growth (Fig. 29.9). A culture medium that has only the bare minimal necessities required by the bacterial species being grown is referred to as a minimal medium. Table 29.1 shows a minimal medium for growing *E. coli.*

Alternatively, we could grow bacteria on a medium that supplies not only the minimal requirements but also all the requirements of a bacterium, including amino acids, vitamins, and so on. A medium of this kind will allow the growth of strains of bacteria that have specific nutritional requirements. (These strains are *auxotrophs* as opposed to the parent or wild types, which are *prototrophs.*) For example, a strain that has an enzyme defect in the pathway of the production of the amino acid histidine will not grow on a minimal medium because it has no way of obtaining histidine. If, however, histidine were provided in the medium, the organisms could grow. This histidine-requiring auxotrophic mutant could thus grow on an enriched, or

Table 29.1 Minimal synthetic medium for growing *E. coli,* a heterotroph.

Component	*Quantity*
$NH_4H_2PO_4$	1 g
Glucose	5 g
NaCl	5 g
$MgSO_4.7H_2O$	0.2 g
K_2HPO_4	1 g
H_2O	1000 ml

complete, medium, whereas the parent prototroph could grow on a minimal medium. Media are often enriched by adding complex mixtures of organic substances, such as blood, beef extract, yeast extract, peptone, a digestion product. Many media are made up of a minimal medium with the addition of only one other substance, such as an amino acid or a vitamin. These are called *Selective media.* In addition to minimal, complete and selective media, other media exist for purposes such as aiding in counting colonies, helping maintain cells in a non-growth phase, and so on.

VIRUS CULTURE

The experimental cultivation of viruses is somewhat different. Since viruses are obligate parasites, they can only grow in living cells. Thus, for the cultivation of phages, petri plates of appropriate media are inoculated with enough bacteria to form a continuous cover, or bacterial lawn. This bacterial culture serves as a medium for the growth of viruses added to the plate.

The Influenza Virion

The cause of animal influenza is a virion (virus particle) of extreme complexity compared to phage. The influenza virion is about 100 nanometres in diameter and covered with spikes of two types. The H spikes, so called because of their hemagglutining ability (cause red cells to clump), allow the virions to attach to the host cells. The other type of spike is an N spike, so called because it is the enzyme neuraminidase. Presumably, this enzyme allows the virions to get out of the host cells. Immunity, primarily to the H spikes, protects a person from being reinfected by the same strain of influenza.

The H and N spikes are embedded in a lipid bilayer which surrounds a protein matrix. Within the virion are eight segments of single-stranded RNA. Each segment is capable of directing the synthesis of one of the virion's proteins.

The exact sequence of events during an infection is not precisely known; however, much is understood, especially regarding the complex structure of the virions.

Since the virus attack eventually results in rupture, or lysis, of the bacterial cell, addition of the virus produces clear spots, known as *plaques,* on the petri plates (Figure 29.9). Different types of bacteria can be used to determine growth potentials of the various viral strains under study.

Bacterial Phenotypes

Bacterial phenotypes fall into three general classes: colony morphology, nutritional requirements and drug and infection resistance.

Colony Morphology

The first of these classes, colony morphology, relates simply to the form, colour and size of the colony that grows from a single cell. A bacterial cell growing on an agar slant or petri plate divides as frequently as once every 20 minutes. Thus, the number of cells doubles every 20 minutes. Each original cell will give rise to a colony, or clone, at the site of its original position. In a relatively short amount of time (overnight, for example), the colonies will consist of enough cells to be seen with the unaided eye. The different morphologies observed among the colonies are under genetic control.

Nutritional Requirements

The second basis for classifying bacteria on the basis of nutritional requirements reflects the failure of one or more enzymes in the biochemical pathways of the bacteria. If an auxotroph has a requirement for the amino acid cysteine that the parent strain (prototroph) does not have, then that auxotroph most likely has a nonfunctional enzyme in the pathway for the synthesis of cysteine. There are five steps in cysteine synthesis; it also shows that each step is controlled by a different enzyme. All enzymes are proteins, and the sequence in the strings of amino acids that make up those proteins are determined by information in one or more genes. A normal or wild-type allele controls a normal, functional enzyme. The alternate allele can control a non-functional enzyme. The one-gene-one-enzyme rule, although not strictly correct, is a useful rule or thumb at this point.

Screening Techniques: Replica plating, a technique devised by Lederberg and Lederberg, is a rapid screening technique that makes it possible to quickly determine if a given strain of bacteria is auxotrophic for a particular metabolite. In this technique a petri plate of complete medium is inoculated with bacteria. The resulting growth will have a certain configuration of colonies. This plate of colonies is pressed onto a piece of sterilised velvet. Then, any number of petri plates, each containing a medium that lacks some specific metabolite, can be pressed onto this velvet to pick up inocula in the same pattern as the growth on the original plate. If a colony grows on the complete medium but does not grow on a plate with a medium in which a metabolite is missing, the inference is that the colony growing in that location on the complete medium is an auxotroph that requires the metabolite absent from the second plate. This bacterial strain can be isolated from the complete medium for further study. Its nutritional requirement is its phenotype. The methionine-requiring auxotroph would be designated as Met$^-$ (methionine-minus or Met-minus).

Resistance and Sensitivity

The third class of phenotypes in bacteria involves resistance and sensitivity to drugs, phages, and other environmental conditions. For example, *Penicillin,* an antibiotic that prevents the final stage of cell-wall construction in bacteria, will kill growing bacterial cells. Nevertheless, we frequently find a small number of cells that do grow in the presence of penicillin. These colonies are resistant to the drug. This resistance is under simple genetic control. The phenotype is penicillin resistant or Pen^r as compared to penicillin sensitive (Pen^S), the normal condition. Many phenotypes of bacteria are resistant and sensitive to various antibacterial agents.

Screening Techniques–Drug sensitivity provides another rapid screening technique for isolating nutritional mutations. For example, if we were looking for mutants that lacked the ability to synthesise a particular amino acid (e.g., methionine), we could grow large quantities of bacteria (prototrophs) and then place them on a medium that lacked methionine but had penicillin. Here, any growing cells would be killed. But methionine auxotrophs would not grow and, therefore, they would not be killed. The penicillin could then be washed out and the cells reinoculated onto a complete medium. The only colonies that would result should be methionine auxotrophs (Met^-).

Screening for resistance to phage is similar to screening for drug resistance. When bacteria are placed in a medium containing phages, only those bacteria that are resistant to the phage will grow and produce colonies. They can thus be easily isolated and studied.

PLASMIDS

Plasmids are circular molecules of DNA that lead an independent existence in the bacterial cell (Fig. 29.8). Plasmids almost always carry one or more genes, and often these genes are responsible for a useful characteristic. For example, the ability to survive in normally toxic concentrations of antibiotics, such as chloramphenicol or ampicillin, is often due to the presence of a plasmid that carries antibiotic-resistance genes in the bacterium.

All plasmids possess at least one DNA sequence that can act as an origin of replication, so they are able to multiply in the cell independently of the main bacterial chromosome. The smaller plasmids make use of the host cell's own DNA replicative enzymes to make copies of themselves, whereas some of the larger ones carry genes that code for special enzymes that are specific for plasmid replication. A few types of plasmids are also able to replicate by inserting themselves into the bacterial chromosome. These integrative *plasmids* or *episomes* may be stably maintained in this form through numerous cell divisions, but will at some stage exist as independent elements.

Different Types of Plasmids

Virtually all species of bacteria harbour plasmids and a large number of different types are known. These are most usefully classified according to the genes that they carry and the characteristics that these genes confer on the host bacterium. According to this classification there are *five* main types of plasmids.

1. *Fertility* (F) *plasmids* are able to direct conjugation (see later) between different bacteria. An example is the F plasmid of *E. coli.*

2. *Resistance* (R) *plasmids* carry genes conferring on the host bacterium resistance to one or more antibacterial agents, such as chloramphenicol, ampicillin or mercury. R plasmids are very important in clinical microbiology because their spread through natural populations can have profound consequences for the treatment of bacterial infections. Examples are Rl00, found in *E. coli* and related bacteria, and RP4, commonly found in *Pseudomonas* but also occurring in other bacteria.
3. *Col plasmids* carry genes coding for colicins, proteins that kill other bacteria. An example is ColE1 of *E. coli.*
4. *Degradative plasmids* allow the host bacterium to metabolise unusual molecules, such as toluene and salicylic acid. Several examples occur in the *Pseudomonas* genus of bacteria, for instance TOL of *Pseudomonas putida.*
5. *Virulence plasmids* confer pathogenicity on the host bacterium. The best known example is the Ti plasmid of *Agrobacterium tumefaciens,* which induces crown gall disease in dicotyledonous plants.

The sizes of plasmids vary from about 1 kb for the smallest to over 250 kb (i.e. 6.25% of the size of the *E. coli* chromosome) for the larger ones. Some plasmids are restricted to just a few related species and are found in no other bacteria, others, such as RP4, have a broad host range and can exist in numerous species, although still displaying a preference for a few kinds of bacteria in which they will be more commonly found.

Plasmids and Bacterial Sex

Sex is always important to geneticists because it is the basis of analytical techniques that allow the relative positions of genes on chromosomes to be determined. The discovery that bacteria have distinct sexes was therefore, a major advance because it opened up a new avenue for genetic studies of prokaryotes.

The Discovery of Bacterial Conjugation

The breakthrough was announced at the Cold Spring Harbor Symposium of 1946 by Joshua Lederberg, a 21-year-old graduate student supervised by Edward Tatum at the University of Yale. They had discovered that individual cells of *E. coli* can exchange genetic material by a process that was subsequently called *conjugation,* and which has become the basis of genetic analysis in bacteria.

This quite unexpected discovery stimulated several geneticists and microbiologists around the world to examine conjugation. During the next ten years or so Joshua Lederberg and his wife Esther with amongst others, William Hayes of the Hammersmith Hospital, London, the Italian Luca Cavalli-Sforza, and Elie Wollman and Francois Jacob of the Pasteur Institute, Paris, demonstrated that transfer of genes is unidirectional from *donor* cells referred to as F^+ (F for fertility) to the *recipients* F^-, and that transfer involves a tube-like structure, called the *sex pilus,* which F^+ cells are able to develop. The exact role of the sex pilus is controversial. An P^+ cell forms a connection with an F^- cell via its sex pilus, and as the pilus is hollow it has been assumed that DNA transfer occurs through it. This has never been demonstrated, and it may be that the pilus acts only to draw the F^+ and F^- cells into close contact.

The F plasmid is responsible for bacterial sex. The first indication that a plasmid is involved in bacterial sex came during the 1950s when several biologists demonstrated that under certain conditions F^+ cells could lose their ability to set up conjugation, becoming indistinguishable from F^- cells. This led to the idea that fertility is controlled by a factor distinct from the main bacterial chromosome. Eventually the *F factor* was shown to be a *plasmid*, the F plasmid, which in *E. coli* is 95 kb in size. This plasmid carries approximately 30 genes (though additional genes may still remain to be discovered) including a large operon that contains the *tra* (for transfer) genes. These code for proteins involved in synthesis and assembly of the pilus and in the DNA transfer process itself. F^+ cells contain the F plasmid, whereas F^- cells do not.

CONJUGATION (Figs. 29.10, 29.11, 29.12 and 29.13)

During conjugation a copy of the F^+ plasmid passes into the F^- cell, presumably through the sex pilus, converting the F^- cell into F^+. A copy of the plasmid also remains in the donor bacterium so this cell remains F^+ also. However, bacterial conjugation is important to geneticists not because the F plasmid is transferred, but because chromosomal genes can also be passed from donor cell to recipient. How does this come about? There are several possible ways, the first being simply that a small random piece of the donor cell's chromosome is transferred along with the F plasmid. This is thought to account for the gene transfer phenomenon originally observed by Lederberg and Tatum, but probably occurs fairly infrequently. More important in genetics are the gene transfer properties of two additional types of donor cells, called Hfr and F′ ('F-prime'). In Hfr cells, the F plasmid has become integrated into the *E. coli* chromosome, by a recombination event analogous to that responsible for integration of the A genome into the *E. coli* DNA molecule. The integrated form of the F plasmid can still direct conjugal transfer, but in this case as well as transferring itself it will also carry into the F cell at least a portion of the *E. coli* chromosome to which it is attached. Consequently, a mating involving an *Hfr* cell will virtually always bring about transfer of some *E. coli* genes.

F′ cells, the second type of donor cell regularly associated with the transfer of chromosomal genes, occasionally arise from Hfr cells when the integrated F plasmid excises from the chromosomal DNA. Normally this event results in an F^+ cell, but sometimes excision of the F plasmid is not entirely accurate, and a small segment of the adjacent chrmosomal DNA is also snipped out. This will lead to an F′ plasmid that carries a small piece of chromosomal DNA, possibly including a few genes. Conjugation involving the F cell will always .result in the transfer of the plasmid associated chromosomal genes.

Conjugative Transfer in Other Species

A large number of bacterial species are able to conjugate in the same way as *E. coli* and in all other species the process is controlled by a plasmid analogous to the F plasmid. Fertility plasmids of this type are called self-transmissible, which means that they can set up conjugation

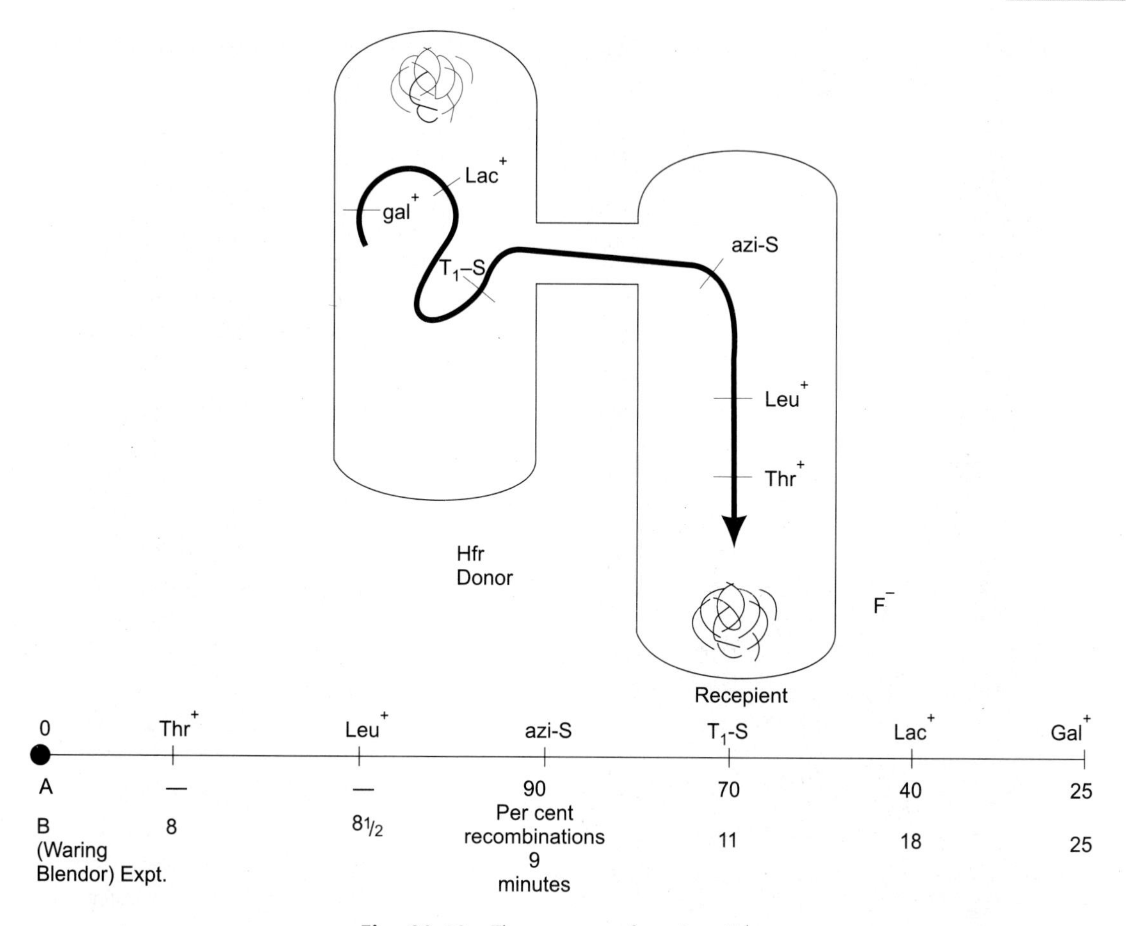

Fig. 29.10 The process of conjugation.

and mobilise themselves into the recipient cell. Conjugation (that is, setting up the initial contact) and mobilisation , (passage of plasmid DNA from donor Q recipient) are two distinct characteristics and not all plasmids are able to direct both. A few can set up only the conjugation contact between cells and not mobilise on their own; others can mobilise, but only if they are coexisting in the cell with a second plasmid that can set up the initial contact. Other plasmids are totally non-fertile and can neither conjugate nor mobilise.

Copy Number and Incompatibility

Two further features of plasmid biology must be mentioned. Firstly copy number, which refers to the number of the molecules of a plasmid that are present in a single cell and is a characteristic value for each plasmid type. Some plasmids, called stringent plasmids, have a low copy number of perhaps just one or two per cell; others, called relaxed plasmids, are present in multiple copies of 10 or more per cell. Secondly, incompatibility which refers to the fact that certain plasmid types cannot coexist in the same cell. An *E. coli* cell can contain up to seven

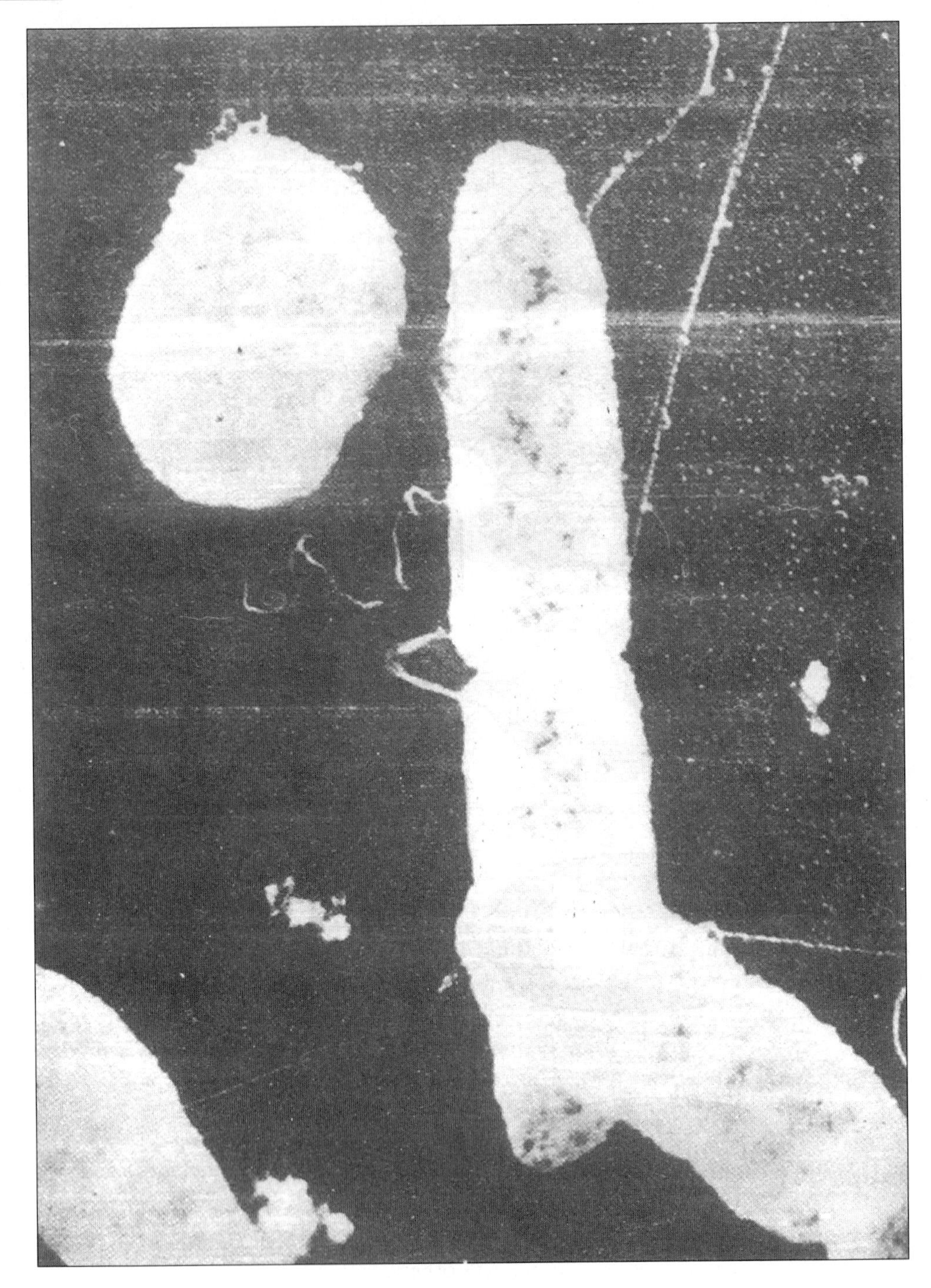

Fig. 29.11 An electron micrograph showing conjugation in *E. coli.* Long bacterium is *Hfr. Short, round bacterium* is F^-.

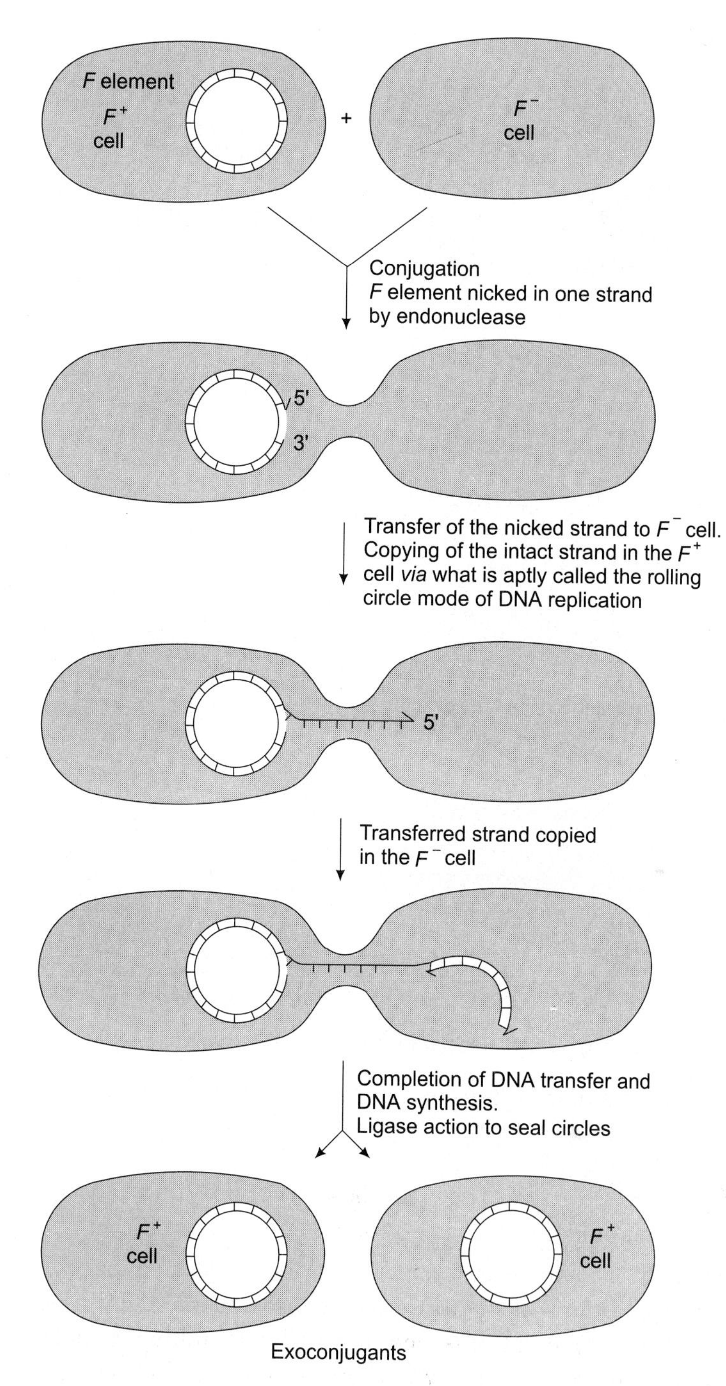

Fig. 29.12 Diagram illustrating transfer of the *F* element from an F^+ to an F^- cell during conjugation in *E. coli*.

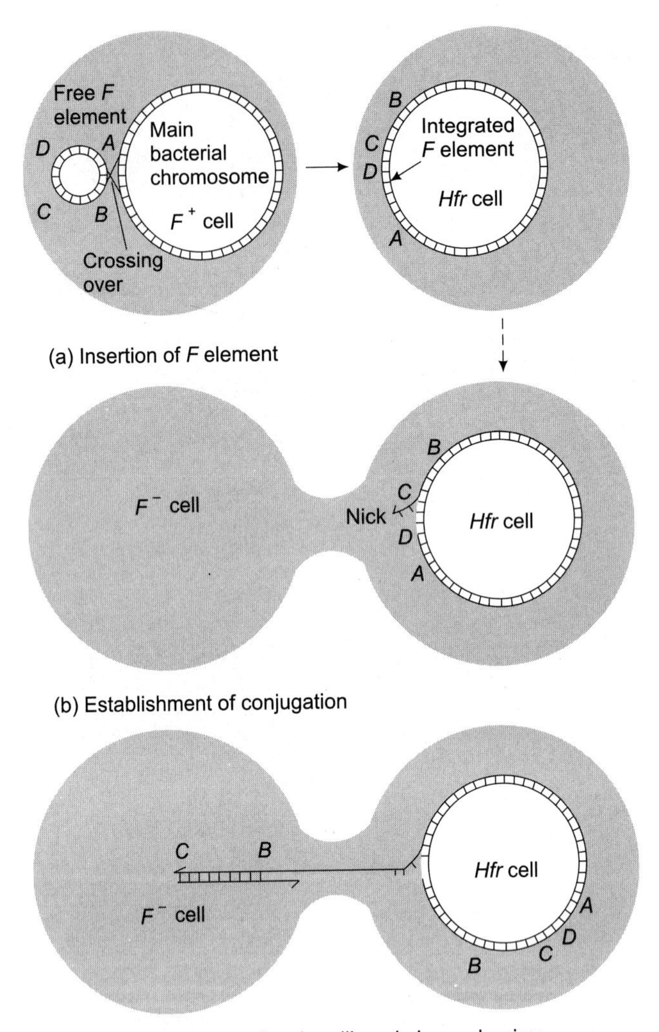

Fig. 29.13 Diagram illustrating the integration of the *F* element to form *Hfr* cell and transfer of *Hfr* chromosome to an F^- cell during conjugation in *E. coli.*

or more types of plasmid at the same time, but these must all belong to different incompatibility groups; two different plasmids of the same group cannot exist together in a single cell.

Copy number and incompatibility appear on the surface to be quite straightforward characteristics, but explanations of the mechanisms that underlie them have remained elusive. One theory holds that copy number is determined by an inhibitor molecule that prevents further replication of the plasmid once the characteristic value is reached. It has also been suggested that replication of incompatible plasmids is controlled by the same inhibitor molecule. A second theory suggests that plasmid replication requires attachment to a specific binding site on the cell membrane (recall that replication of the nucleoid involves attachment to the membrane) and that incompatible plasmids compete for the same attachment sites.

Bacterial Transposons

The next aspect of bacterial genes that we must examine concerns the genetic elements called *transposons.* During the 1960s geneticists were puzzled by a set of unusual mutations occasionally found in operons of *E. coli* and other bacteria. These are called polar mutations because they affect not only the gene in which the mutation occurs but also other genes downstream of the mutation. Several types of polar mutation are known and often they are the ones that introduce a transcription termination signal into an operon. However, the particular set of mutations that caused the problems in the 1960s are unusual in that unlike most other mutations they are almost always irreversible. A second very unexpected feature becomes apparent if a plasmid is introduced into a bacterium that carries a polar mutation of this type. Similar, irreversible polar mutations suddenly appear in genes carried by the plasmid almost as though the mutation is *able to jump from one DNA molecule to another.*

Insertion Sequences can move around the Genome (Fig. 29.14)

The first inkling of an explanation for this strange phenomenon appeared when the plasmids that had become mutated were studied. It was found that these plasmids had increased in size, generally by a kilo-base or more. When the DNA sequences of the mutated plasmids were examined it was discovered that the mutations were caused by the insertion into the target genes of new pieces of DNA, which were subsequently called insertion sequences or *ISs.* These insertion sequences possess the remarkable capability of being able to move around the genome from site to site by a process dependent on recombination and called *transposition.* When an IS transposes to a new site it leaves behind a pair of short direct repeat sequences, usually between 5 and 15 bp, which means that the gene which carried the IS probably remains mutated even after the element has departed.

Several different IS types are now known in *E. coli,* with sizes between 750 and 6000 bp, and copy numbers between one and about ten. The bulk of each IS is taken up by a single gene (except IS1 which has two) coding for a transposase enzyme that catalyses the transposition event. The element has at its ends a pair of inverted repeats between 9 and 41 bp in length depending on the IS type.

Insertion Sequences are Just One Type of Transposon (Fig. 29.15)

Insertion sequences were unique when they were first discovered in 1961. No similar type of genetic element was known at that time. Now we appreciate that IS elements are just one class of transposable elements and that others, similar in many respects to IS elements, also exist. There are *three main types* in addition to IS elements.

1. *Composite transposons* consist of two insertion sequences flanking an internal region which usually includes at least one gene, often coding for resistance to an antibiotic. Transposition occurs because of the transposases coded by the IS elements. There are

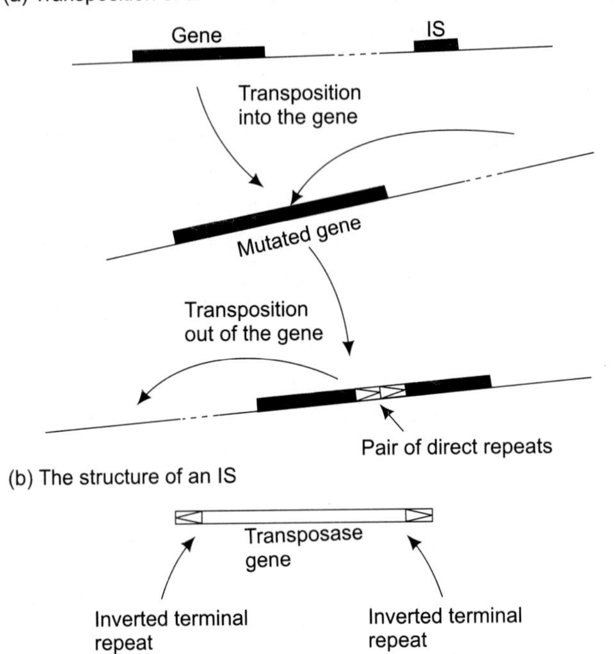

Fig. 29.14 Insertion Sequences.

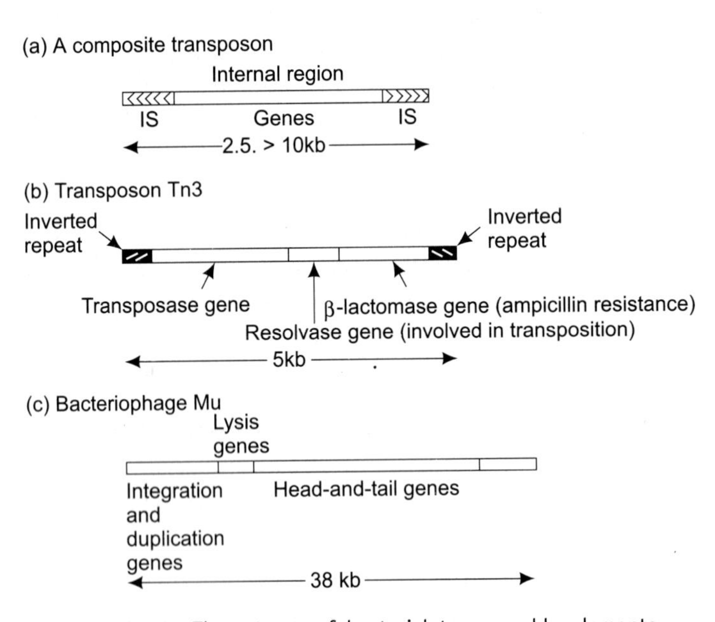

Fig. 29.15 Three types of bacterial transposable elements.

several different composite transposons, some over 10 kb in size. Often these types of elements are carried by plasmids.

2. *Tn3-type transposons,* which are not flanked by IS elements but are still able to transpose because they carry a different transposase.
3. *Transposable phages,* such as Mu and D108, which transpose as part of their normal infection cycle. This means that, as well as inserting themselves into the host DNA (like λ), the genomes of these phages are able to move about from place to place in the host DNA (unlike λ). Transposable phages may be transposons that have evolved genes for protective proteins that allow them to exist outside of cell, or they may be ordinary phages which have learned to transpose.

Why have transposable elements? We do not as yet understand the function of transposons. Indeed several molecular biologists have suggested that transposons have no function at all in fact are examples of selfish DNA, existing only to further their own existence and contributing nothing to the cell as a whole. However, it has been established that bacteria which carry a large number of transposons have a selective advantage over those with few transposons, possibly because transposons increase the mutation rate of the cell for greater adaptability.

MAPPING GENES IN BACTERIA

The person primarily responsible for placing the genetic analysis of bacteria on the same level as gene mapping with eukaryotes was Joshua Lederberg. Lederberg's first contribution was the demonstration in 1946 that genes can be exchanged between bacteria during conjugation. Over the next twenty years he played a part in most of the important advances in the area of bacterial gene mapping.

Three major techniques became available during the 1950s for mapping genes in bacteria. These are *conjugation mapping, transduction mapping* and *transformation mapping.* Each can provide valuable information, though with many bacterial species only one or two of the three procedures can be used.

Basic Features Common to Each Mapping Procedure

Each of the techniques used to map genes in bacteria is similar to eukaryotic organisms in that recombination has to occur between homologous pieces of DNA from two cells of different genetic constitutions. As with eukaryotic microorganisms, the genes are followed by means of genetic markers, which are biochemical characteristics generally obtained by mutagenesis, and easily distinguished by studying growth on agar media. The basic procedure, shared by each of the three different mapping techniques, is as follows:

1. Two strains of the bacterium are required, one to act as a *donor* and as a *recipient.* The recipient bacterium must possess mutated versions of the genes under study; the donor bacterium must carry unmutated wild-type copies of these genes.

2. Gene transfer is effected, and the DNA molecule (or more likely a small part of it) of the donor bacterium is transferred into the recipient bacterium. Exactly how this is achieved is the key distinction between the three mapping techniques.
3. Recombination occurs between the DNA from the donor bacterium and the intact, functioning DNA molecule of the recipient cell. Usually, to be of value for gene mapping, two individual recombination events must occur in such a way as to insert a piece of the donor cell DNA, containing one or more of the gene under study, into the recipient cell's chromosome. This double recombination event will result in the original copies of the relevant genes present in the recipient cell being replaced by the versions obtained from the donor cell.
4. The resulting change in genotype of the recipient bacterium is assessed by spreading samples onto a selective medium, on which only recipient cells containing the unmutated wild-type version(s) of the gene(s) obtained from the donor parent are able to survive. This will show whether the genes under investigation in the recipient cells have changed from mutant to wild-type.

The Gene Mapping Techniques

Although the three major gene mapping techniques for bacteria share the common features just described, each is different in detail, in particular in the way gene transfer between donor and recipient cells is achieved. The nature of the gene transfer is the main limitation on the amount of information available from each technique. We will look at each technique in turn and then briefly examine their limitations and strengths.

Mapping by Conjugation

The discovery of bacterial conjugation by Lederberg and Tatum in 1946 was the key breakthrough that enabled the first gene mapping experiments to be carried out with *E. coli*. The important observation was that transfer of some chromosomal genes could occur, along with the F plasmid, during certain types of mating. In particular, a mating involving an *Hfr* strain, as studied by Hayes, Cavalli-Sforza and others, almost always results in transfer of chromosomal genes from the *Hfr* cell to the F^- recipient. The transferred genes may then integrate into the recipient cell's DNA to produce recombinants. *Hfr* in fact stands for high frequency of recombination. Can recombination brought about in this way be used to map genes? The answer is yes, as was demonstrated first by Elie Wollman and Francois Jacob at the Pasteur Institute early in 1955, using the interrupted mating technique that is still the basis for conjugation mapping.

The interrupted mating experiment (Figs. 29.16 and 29.17): Wollman and Jacob's aim was to chart the transfer of chromosomal genes from an Hfr cell into an F^- cell. To do this they needed an F^- strain that carried several mutated genes and an Hfr strain that possessed correct copies of these genes. Transfer of each wild-type gene from the Hfr cell to the F^- cell would be signalled by a change in genotype of the recipient cell.

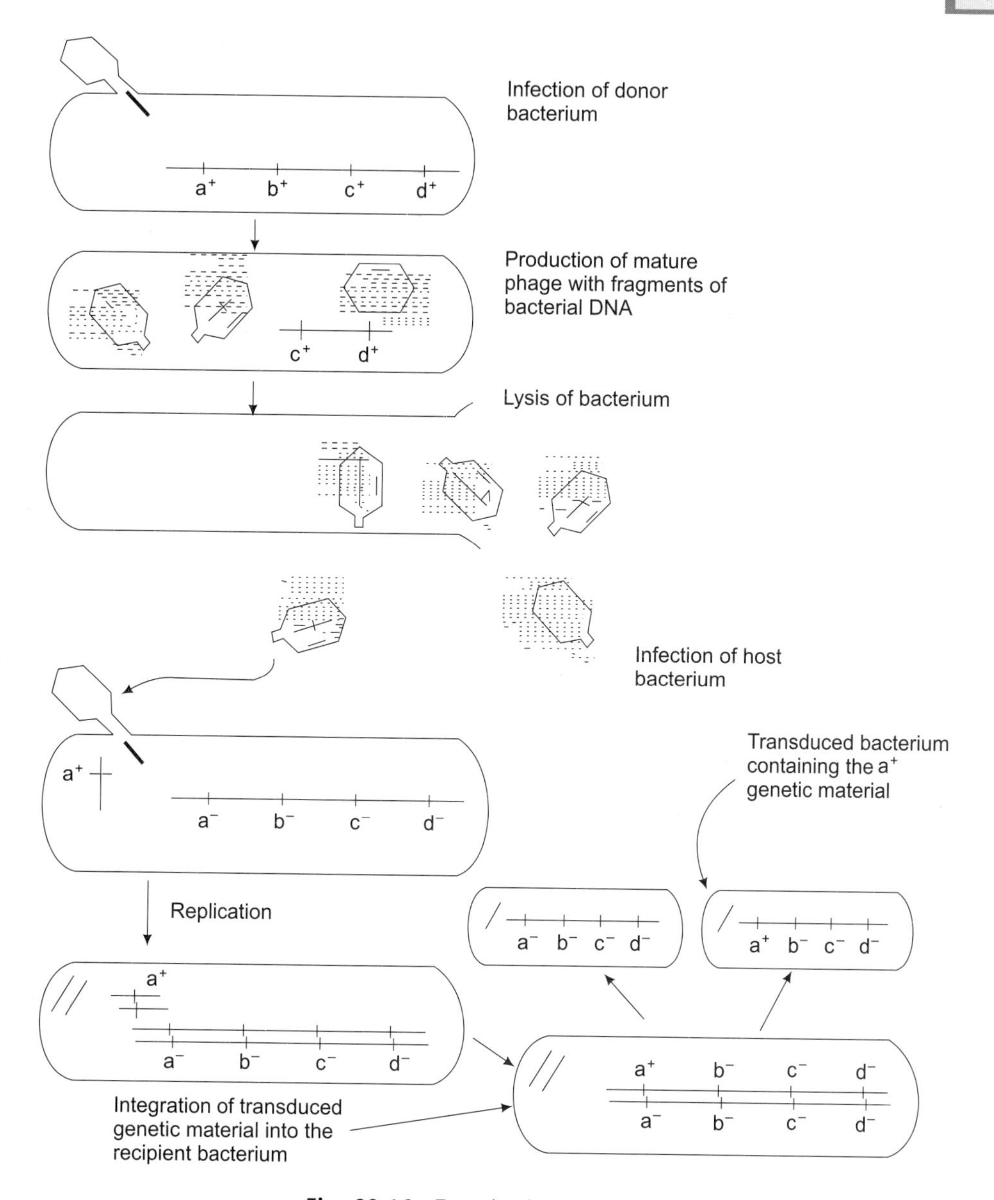

Fig. 29.16 Transduction in bacteria.

The series of markers studied by Wollman and Jacob are listed in Table 29.2. The cross could be described as:

$$\text{Hfr str}^{S} \times \text{F}^{-}\ \text{thr}^{-}\ \text{leu}^{-}\ \text{azi}^{R}\ \text{ton}^{A}\ \text{lac}^{-}\ \text{gal}^{-}\ \text{str}^{R}$$

The experiment was carried out by mixing the Hfr and F^- cells together and then removing samples from the culture at various time intervals. Each sample was immediately agitated in a Waring Blender to snap the conjugation tubes linking Hfr and F^- cells and so interrupt the

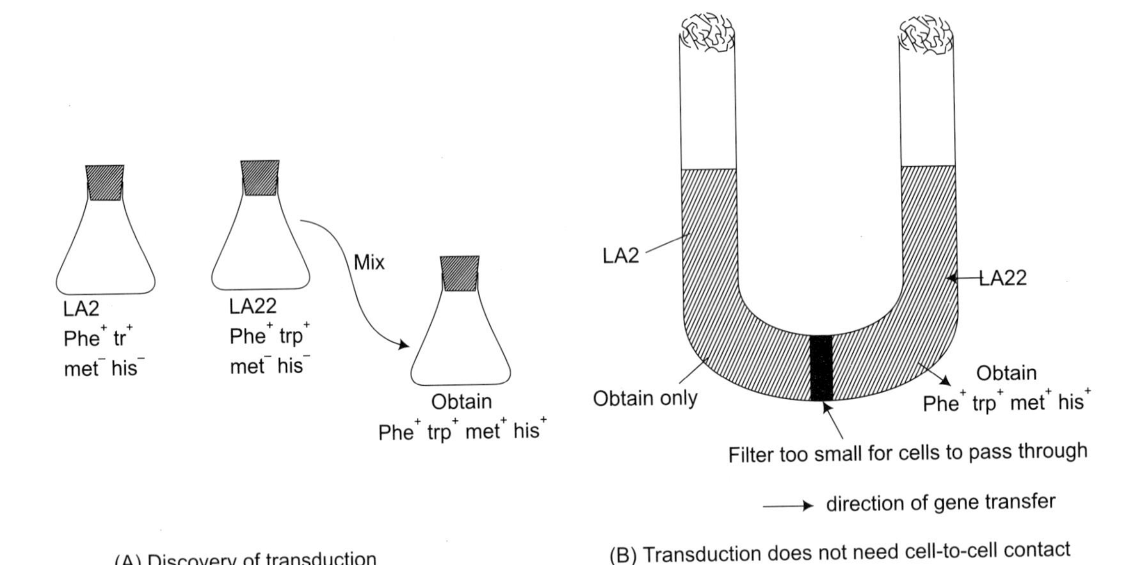

Fig. 29.17 Transduction.

Table 17.2 Series of *E. coli* genetic makers used by Wollman and Jacob in the first interrupted mating experiment.

Marker	*Phenotype*
thr'	Requires threonine
leu'	Requires leucine
azt^R	Resistant to azide
ton^A	Resistant to colcin and T1 phage
lac'	Unable to utilise lactose
gal'	Unable to utilise galactose

mating. The cells were: spread onto an agar medium containing streptomycin on which only the recipient cells would be able to grow (note the differences between the genotypes of the Hfr and F⁻ cells with respect to streptomycin resistance). The genotypes of the resulting colonies, which are derived only from recipient cells, were then determined by replica plating on to suitable media.

Gene Transfer Occurs Sequentially (Figs. 29.18 and 29.19)

Wollman and Jacob were not sure exactly what to expect from the interrupted mating experiment because, at that time, the precise details of *Hfr* conjugation had not been worked out. Although they had their own ideas about what might happen, the actual results came as a complete surprise. Rather than all the genes being transferred at once, *the genes were passed from the Hfr to F⁻ cells at different times.* First, after about eight minutes of mating, recipient cells that had lost their auxotrophic requirements for threonine and leucine began to appear. As the

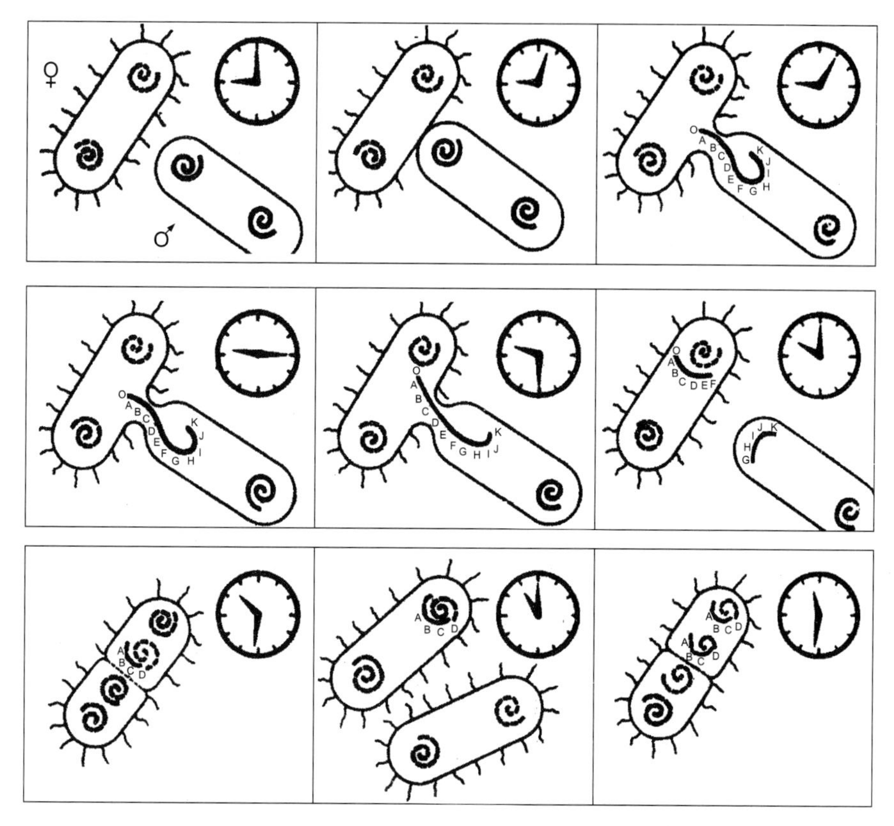

Fig. 29.18 Jacob and Wollman's interpretation of conjugation. Capital letters represent the position of various genes on the donor chromosome, with O representing the origin of transfer. Female is F^-, the male is Hfr.

mating continued, the proportion of cells displaying leucine and threonine prototrophy increased until, after a few more minutes, almost all the recipient cells were thr^+ leu^+. Next, after about ten minutes, azide sensitivity began to appear in the recipient cells, gradually increasing in frequency until about 80% of the population were azi^S. Eventually, each of the genes were transferred to the recipient cells. Wollman and Jacob represented the results in the form of a graph illustrating the kinetics of transfer of each gene.

Interpretation of the Interrupted Mating Experiment

The Wollman-Jacob experiment is easily explained, once we remember that *Hfr* cells transfer a portion of the host chromosome, possibly even the entire chromosome, as a linear molecule into F cells. During this transfer the F region is the first to enter the recipient cell, followed by the rest of the *E. coli* chromosome. If mating is interrupted, only those genes already transferred will be able to recombine with the recipient cell's DNA and give rise to recombinants. By measuring the times at which different genes appeared in the recipient cells, Wollman and Jacob were in effect mapping the relative positions of the genes on the *E. coli* chromosome. Indeed,

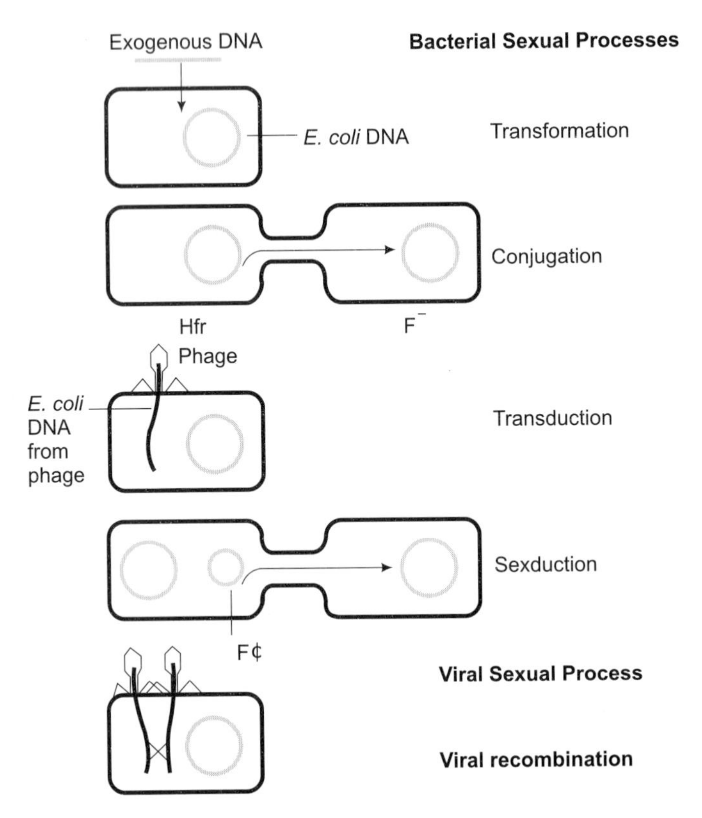

Fig. 29.19 Summary of bacterial and viral sexual processes.

map distances on the *E. coli* chromosome are now expressed in minutes, with one minute representing the length of DNA that takes one minute to transfer. There are 100 min. on the map because transfer of the entire chromosome takes about 100 min.

Transfer of the entire chromosome is not very likely, however, as many bacteria abandon mating naturally well before 100 min. have passed. To map the entire *E. coli* chromosome it has been necessary to carry out a series of interrupted mating experiments, with a variety of Hfr strains, each with the F plasmid inserted at a different position. The regions studied by the individual experiments overlap, enabling the complete map to be built up. It soon became apparent that although genes are transferred in a linear fashion, the *E. coli* genetic map is actually circular.

Conjugation mapping is a powerful technique for determining the relative positions of genes on the *E. coli* chromosome. However, it is not tremendously accurate, and genes that are less than two minutes apart on the genome cannot be separated by conjugation mapping. We know that two minutes corresponds to about 80 Kb of DNA, enough room for quite a few genes. Furthermore, conjugation mapping is only applicable to those species of bacteria that have F plasmids or their equivalent, many do not. We must, therefore, look at other methods for gene mapping in bacteria (Fig. 29.20 and Table 29.3).

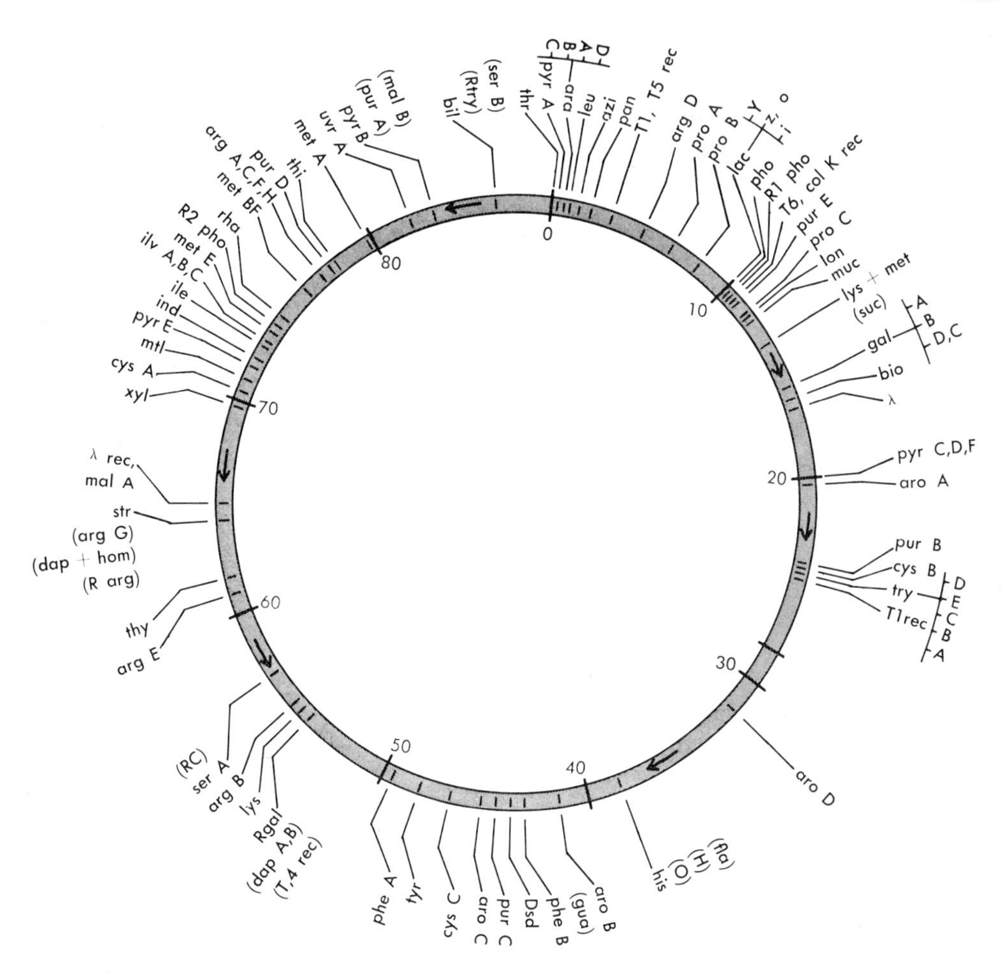

Fig. 29.20 Genetic map of *E. coli*. Symbols mark the locations of genes. Genes whose locations are approximately known are in parentheses. Numbers divide the map into time intervals corresponding to the time in minutes which it takes each male chromosomal segment to move into a female cell. Thus 89 minutes are now thought to be required for complete transfer. Arrows mark the points at which various Hfr chromosomes break prior to transfer into a female cell; direction of arrows indicate transfer direction.

Mapping by Transduction

Conjugation and transduction (as well as transformation) are based on totally different mechanisms for gene transfer from donor bacterium to recipient bacterium. Instead of physical contact between cells, with transduction, gene transfer is effected by a bacteriophage.

Discovery of Transduction

Transduction was discovered by Joshua Lederberg and Norton Zinder in 1952, not with *E. coli* but with the related enterobacterium *Salmonella typhimurium*. They mixed together two auxotrophic strains of this organism, one called LA22 (with the genotype phe$^-$ trp$^-$ met$^+$ his$^+$)

Table 29.3 Gene map of *E. coli*.

Genetic Symbols	*Mutant Character*	*Enzyme or Reaction Affected*
araD	Cannot use the sugar arabinose as a carbon source	L-Ribulose-5-phosphate-4-epimerase
araA		L-Arabinose Isomerase
araB		L-Ribulokinase
araC		
argB		N-Acetylglutamate synthetase
argC		N-Acetyl-γ-glutamokinase
argH		N-Acetylglutamic-γ-semialdehyde
argG	Requires the amino acid arginine for growth	Acetylornithine-*d*-transaminase
argA		Acetylornithinase
argD		Ornithine transcarbamylase
argE		Argininosuccinic acid synthetase
argF		Argininosuccinase
argR	Arginine operon regulator	
aroA. B, C	Requires several aromatic amino acids and vitamins for growth	Shikimic acid to 3-enolpyruvyl-shikimate-5-phosphate
aroD		Biosynthesis of shikimic acid
azi	Resistant to sodium azide	
bio	Requires the vitamin biotin for growth	
carA	Requires uracil and arginine	Carbamate kinase
carB		
chlA-E	Cannot reduce chlorate	Nitrate-chlorate reductase and hydrogen lysase
cysA	Requires the amino acid cysteine for	3-Phosphoadenosine-5-phosphosulfate to growth sulfide
cysB		
cysC		Sulfate to slfide; 4 known enzymes
dapA	Requires the cell-wall component diaminopimelic acid	Dihydrodipicolinic acid synthetase
dapB		N-Succinyl-diaminopimelic acid deacylase
dap + hom	Requires the amino acid precursor homoserine and the cell-wall component diaminopimelic acid for growth	Aspartic semialdehyde dehydrogenase
dnaA-Z	Mutation, DNA replication	DNA biosynthesis
Dsd	Cannot be the amino acid D-serine as a nitrogen source	D-Serine deaminase
fla	Flagella are absent	
galA	Cannot use the sugar galactose as a carbon source	Galactokinase
galB		Galactose-1-phosphate ridyl transferase
galD		ridine-diphosphogalactose-4-epimerase
glyA	Requires glycine	Serine hydroxymethyl transferase
gua	Requires the purine guanine for growth	
H	The H antigen is present	
his	Requires the amino acid histidine for growth	10 known enzymes[a]
hsdR	Host restriction	Endonuclease R

Table 29.3 Contd.

Table 29.3 Contd.

ile	Requires amino acid isoleucine for growth	Threonine deaminase
ilvA	Requires amino acids isoleucine	α-Hydroxy-β-keto acid rectoisomerase
ilvB		α, β-α, β-dihydroxyisovaleric dehydrase
ilvC		Transaminase B
ind (indole)	Cannot grow on tryptophan as a carbon source	Tryptophanase
λ (*att*λ)	Chromosomal location where prophage λ is normally inserted	
lacI	Lac operon regulator	
lacY	Unable to concentrate β-galactosides	Galactoside permease
lacZ	Cannot use sugar lactose as a carbon source	β-Galactosidase
lacO	Constitutive synthesis of lactose operon proteins	Defective operator
leu	Requires amino acid leucine for growth	3 known enzymes[a]
lip	Requires lipoate	
lon (*long form*)	Filament formation and radiation sensitivity are affected	
lys	Requires amino acid lysine for growth	Diaminopimelic acid decarboxylase
lys + met	Requires amino acids lysine and methionine for growth	
λ *rec, malT*	Resistant to phage λ and cannot use the sugar maltose	Regulator for 2 operons
malk	Cannot use sugar maltose as a carbon source	Amylomaltase (?)
man	Cannot use mannose sugar	Phosphomannose isomerase
melA	Cannot use melibiose sugar	Alpha-galactosidase
melA-M	Requires amino acid methionine for growth	10 or more genes
mtl	Cannot use the sugar mannitol as a carbon source	Mannitol dehydrogenase (?)
muc	Forms mucoid colonies	Regulation of capsular polysaccharide synthesis
nalA	Resistance to nalidixic acid	
O	The O antigen is present	
pan	Requires vitamin pantothenic acid for growth	
pabB	Requires p-aminobenzoate	
phe A, B	Requires amino acid phenylalanine for growth	
pho	Cannot use phosphate esters	Alkaline phosphatase
pil	Has filaments (pili) attached to the cell wall	
plasB	Deficient phospholipid synthesis	Glycerol 3-phosphate acyltransferase
polA	Repairs deficiencies	DNA polymerase I
proA	Requires amino acid proline for growth	
proB		
proC		
trpA	Requires amino acid tryptophan for growth	Tryptophan synthetase, A protein
trpB		Tryptophan synthetase, B protein
trpC		Indole-3-glycerolphosphate synthetase
trpD		Phosphoribosyl anthranilate transferase
trpE		Anthranilate synthetase

Table 29.3 Contd.

Table 29.3 Contd.

tyrA	Requires the amino acid tyrosine for growth	Chorismate mutase T-prephenate dehydrogenase
tyrR		Regulates 3 genes
uvrA-E	Resistant to ultraviolet radiation	Ultraviolet-induced lesions in DNA are reactivated
vals	Cannot charge Valyl-tRNA	Valyl-tRNA synthetase
xyl	Cannot use the sugar xylose as a carbon source	

[a]Denotes enzymes controlled by the homologous gene loci of *Salmonella typhimurium*.

and one called LA2 (phe^+ trp^+ met^- his^-), and obtained a small number of prototrophs (phe^+ trp^+ met^+ his^+), about one in every 1,00,000 cells. By now, one should appreciate that prototrophs can arise only by way of gene transfer between bacteria, followed by recombination, albeit in this case with a very low frequency.

At first it was thought that these results must be due to a conjugative process similar to that characterised in *E. coli,* but further examination suggested that this was not the case. In particular a modification of the basic experiment proved that cell-to-cell contact is not needed for gene transfer between LA2 and LA22: prototrophs were still obtained when individual cultures were placed in opposite arms of a U-tube, separated from each other by a sintered glass filter (Fig. 29.17). Media and small particles can pass through this filter, but not cells. Now it became apparent that gene transfer is unidirectional and that only strain LA22 was becoming prototrophic. These results suggested that this type of genetic exchange is mediated by a filterable agent that is produced by LA2 cells and transfers to LA22 cells.

Transduction Mediated by a Bacteriophage (Fig. 29.16)

The explanation of transduction turned out to be based on a familiar system, bacteriophage infection, but in an unexpected way. It was discovered that strain LA2 contains the prophage version of a phage called P22. Every now and then a P22 prophage is induced and enters into the lytic stage of its infection cycle. During the lead up to lysis the bacterial DNA molecule is broken into fragments, some of which by chance happen to be about the same length as the P22 phage DNA molecule. By mistake a portion of bacterial DNA can be packaged into a P22 phage coat, though this only happens very infrequently. The resulting phage particle is still infective, as infection is a function purely of the protein components of the phage coat, but the infection is abortive, since the bacterial DNA fragment injected into a new cell cannot itself direct synthesis of new phages.

The bacterial DNA fragment transferred into a new cell can, however, recombine with the new cell's DNA. This means that gene transfer can occur: the donor cell is the one that was lysed by the phage, and the recipient cell is the one that is infected by the transducing phage particle.

Mapping by Transduction

The amount of DNA carried by a transducing P22 phage particle is about 40 kb, or about 1% of the total *S. typhimurium* genome.

That is sufficient to carry a dozen or more genes. Contransduction, the transfer of two or more genes from donor to recipient, will occur if the genes are relatively close together on the S. *typhimurium* genome and hence can be contained in a single 40 kb fragment. A transduction mapping experiment, therefore, uses a wild-type donor and double mutant recipient and tests for contransduction of the two wild-type genes into the recipient cell. This type of experiment is very useful for establishing a relatively close proximity between two genes on a bacterial DNA molecule.

CONTRANSDUCTION (Fig. 29.21)

Transduction mapping can also be taken a stage further as the order of genes can be determined if three or more markers are used. For instance, genes *a, b* and c lie in that order in the same 40 kb region of the chromosome. Contransduction of the genes would be studied with a donor bacterium carrying wild-type copies (a^+ b^+ c^+) and a triple mutant recipient (a^- b^- c^-). Contransduction of all three genes would be signalled by the recipient becoming a^+ b^+ c^+. For this to occur, the two crossovers that result in recombination must occur outside the three genes. However, crossing-over could also occur between the genes, so the genes may be contransduced as pairs or even transduced individually. The crucial detail is that only adjacent genes (*ab* and *bc* in the example shown) will be contransduced as a pair with any great frequency. This is because contransduction of the adjacent genes requires just two crossover events, whereas contransduction of the outer genes alone (*ac* in this example) requires four crossovers. The contransduction frequencies for the different pairs of genes, therefore, allow the genes order to be determined.

Mapping by Transformation

Transformation is similar to transduction in the way it is used in gene mapping, but the basis of the genetic exchange is totally different. With transformation, 'naked' DNA molecules are taken up by the recipient cell (recall the transforming principle used to show that genes are made of DNA).

Competence and Uptake of DNA

Exactly how naked DNA enters a bacterial cell is not known. It is clear, however that some bacteria, such as *Bacillus* and *Haemophilus* species, have efficient mechanisms for DNA uptake whereas others, including *E. coli,* are transformed only with difficulty, although certain experimental treatments can improve the transformation efficiency by rendering the cells more competent.

To carry out a transformation mapping experiment a sample of DNA from the donor strain, wild-type for the genes under study, must be prepared. The DNA is mixed with competent recipient cells and uptake takes place followed by recombination. If two genes are close enough together on the chromosome, cotransformation may occur. Gene orders can be worked out by cotransformation, just as they can by contransduction.

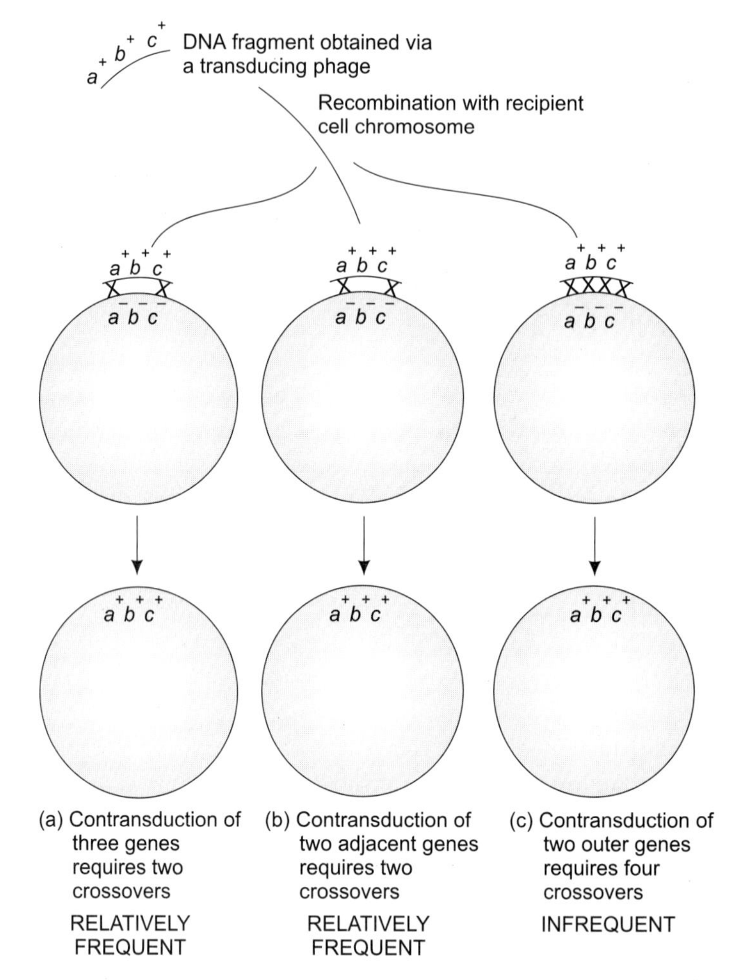

Fig. 29.21 Contransduction can be used to work out the order of the genes on a DNA fragment as adjacent genes will contransduce at a higher frequency than non-adjacent genes.

Specialised Transduction

The events described so far, are more properly called *generalised transduction* to distinguish the phenomenon from *specialised transduction,* a feature of several lysogenic phages. Specialised transduction arises when excision of an integrated prophage occurs slightly inaccurately, so that a short piece of bacterial DNA remains attached to the viral genome. This piece is then transferred via the phage particle to the next infected cell.

The critical difference between generalised and specialised transduction is that the former involves random pieces of bacterial DNA, whereas the latter results only in transfer of genes immediately adjacent to the prophage integration site.

Relative Merits of the Three Methods

The existence of three different mapping techniques has brought about great advances in understanding gene organisation in bacteria. To map a newly discovered gene in *E. coli.* a good approach would be first to perform an interrupted mating experiment, which will position the gene on the chromosome with an accuracy of about two minutes, equivalent to about 80 kb. The exact position of the gene can then be established with greater accuracy by testing for *cotransformation* or *contransduction* with genes known to lie in the same small area of the genome, followed by fine positioning by examining the gene order by contransduction and/or cotransformation. Comprehensive gene maps have now been determined for a variety of bacteria (Table 29.4).

Table 29.4 Bacterial species for which family extensive gene maps are now available.

Species	*Approximate number of genes mapped*
Escherichia coli	1000
Salmonella typhimurium	650
Bacillus subtilis	500
Pseudomonas aeruginosa	250
Caulobacter crescentus	95
Neisseria gonorrheae	85
Staphylococcus aureus	80
Pseudomonas putida	70
Proteus mirabilis	35

This situation does not hold for all bacteria. conjugation mapping is clearly limited to those species that possess F plasmids or equivalents. This means that it is not generally applicable to the Gram-positive group of bacteria, which includes several important genera (notably *Streptomyces,* from which 80% of all antibiotics are obtained). Transduction mapping is also limited in scope because transducing phages are known for only a few species. For many bacteria, transformation mapping is the only technique that can be used satisfactorily.

These limitations are not so important today because recombinant DNA techniques have provided new gene mapping procedures that do not depend on gene transfer in the conventional sense, and so are applicable to virtually all species, including higher organisms. *Recombinant DNA technology* also enables the researcher to go several stages beyond gene mapping and obtain detailed information on gene structure and expression. These powerful techniques are described in **Chapter** dealing with biotechnology.

RECOMBINATION

Recombination results in a rearrangement in the genetic material of the cell. It comes in several disguises and processes that may appear to be quite different and turn out to be based on the same type of recombination reaction.

Mechanism of Recombination

There are several models that attempt to explain how recombination occurs but all have the common features described here. Recombination generally requires two separate double-stranded DNA molecules that have a region of homology where the nucleotide sequences are the same or at least very similar.

These molecules line up adjacent to one another and interact by exchanging portions of polynucleotide. First, cleavages occur at the equivalent points of the identical polynucleotide by endonuclease action and strand displacement takes place so that a crossover branch IS formed, giving a *heteroduplex* in which the nicked polynucleotides are shared between the two double helices. The nicks are sealed by *DNA ligase* to stabilise the heteroduplex and strands are exchanged by branch migration, with the crossover point migrating along the two molecules. Remember that as the molecules are homologous the polynucleotides that are exchanged can base pair with the complementary strands of either DNA molecule. The sealed heteroduplex is called the Holliday structure, after Robin Holliday who first proposed its existence.

HOLLIDAY MODEL (Fig. 29.22, 29.23 and 29.24)

The main complication arises when the heteroduplex becomes cleaved to produce two separate molecules. The two-dimensional representation used so far is rather unhelpful at this stage because it obscures the true topological structure of the heteroduplex and makes the cleavage events difficult to follow. The heteroduplex shown is more correctly drawn in the chi form, an intermediate that has been confirmed by electron microscopic observation of recombining DNA molecules. Two different cleavage events can resolve the chi-form back into separate double-stranded DNA molecules. The first of these, is equivalent to cleavage directly across the crossover shown, so that after ligation of the cut strands two molecules that have exchanged a portion of polynucleotide result. As the exchanged strands are homologous the effect on the genetic constitution of each molecule will be relatively minor. More important is the second way of resolving the chi structure, the method that is not readily apparent from the representation. Now two radically different DNA molecules emerge, with the end of one molecule having been exchanged for the end of the other molecule, clearly a much more dramatic rearrangement.

Role of Recombination in Genetics

We now consider some of the general results of recombination. The first process to consider is one involving exchange of genetic material between individual homologous, linear, double-stranded DNA molecules. These events occur in eukaryotes during meiosis and bring about recombination of characteristics during sexual reproduction.

Consider, however, the effect if one or both of the DNA molecules involved in recombination are not linear, but circular. The result of resolution of the chi structure could now be insertion of one circle into the other. The two molecules need have only very limited regions of homology, perhaps just the few base pairs needed to set up a heteroduplex. The result of the

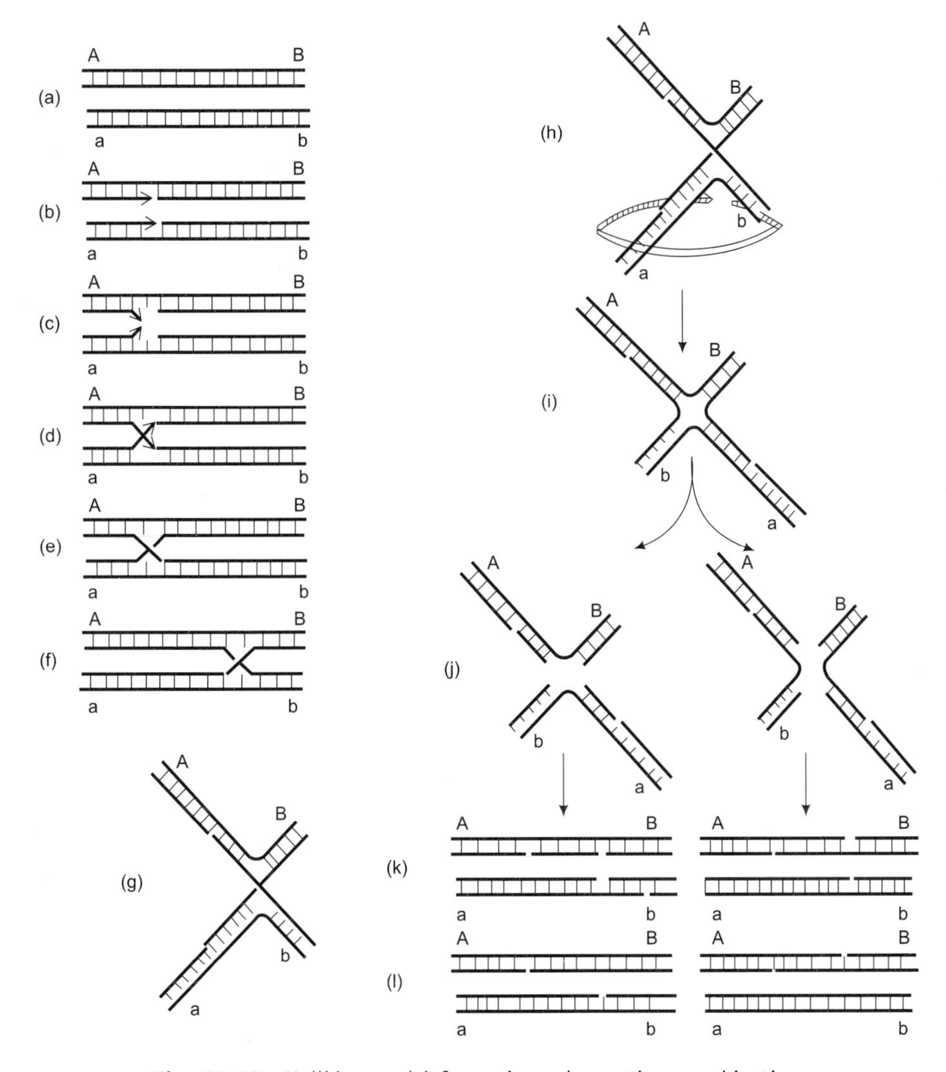

Fig. 29.22 Holliday model for reciprocal genetic recombination.

insertion will be that this homologous target sequence will become duplicated and will flank the inserted element. Intramolecular recombination between these two sequences could now bring about excision of the inserted region and recreation of the two original sequences.

Transposition. This is a special type of recombination which involves a short piece of DNA (usually between 750 bp and 10 kb) that is able to transpose from place to place on the same DNA molecule or between different molecules. When in place, a transposable element is flanked by short repeated nucleotide sequences, and transposition presumably involves recombination between these sequences, possibly with the formation of an intermediate circular molecule.

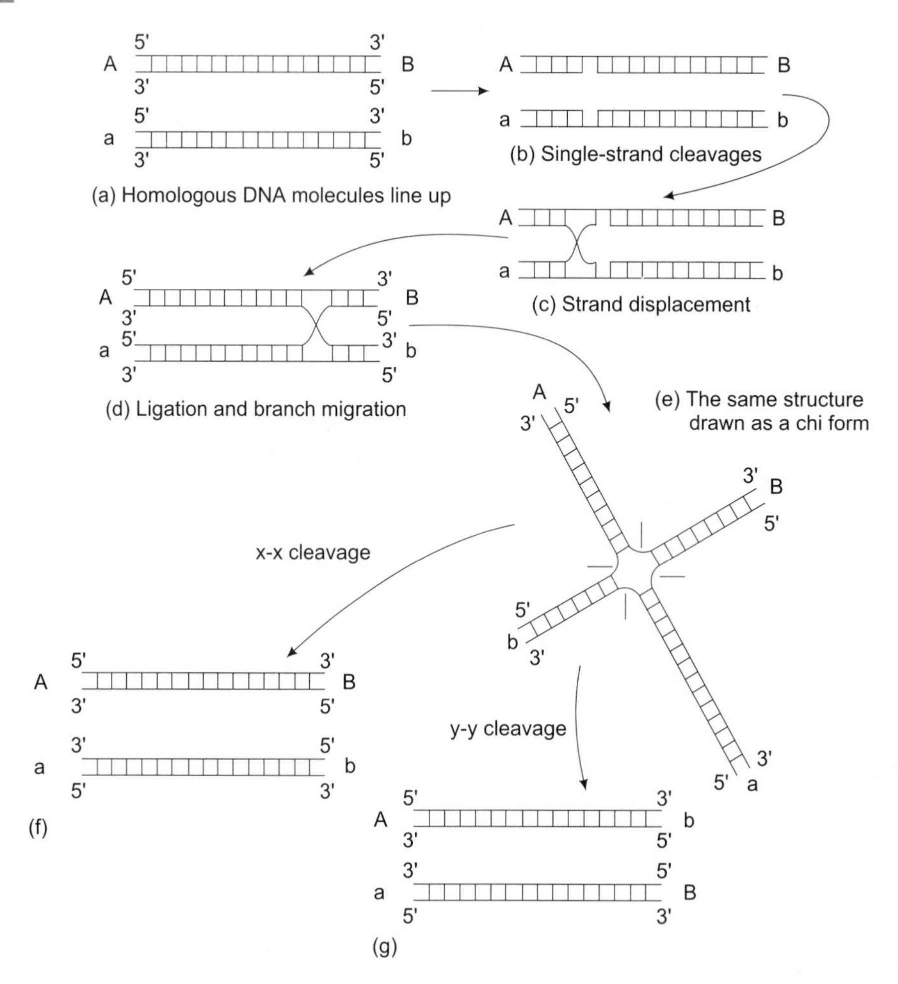

Fig. 29.23 Recombination between two homologous, linear, double-stranded DNA molecules.

Transposable elements are important in bacteria as they often carry genes for antibiotic resistance and may mediate spread of resistance in natural populations. In higher organisms the retroviruses which include some viruses implicated in cancer as well as the AIDS virus, resemble transposable elements. In the laboratory, transposable elements are used as mutagens because they can inactivate genes that they transpose into, or possibly modulate the expression of gene downstream from an insertion site in both cases causing a phenotypic change indistinguishable from a mutation.

Recombination between Non-Homologous Molecules

So far each of the recombination events we have described has involved molecules with at least a short region of nucleotide sequence homology. Although this is one of the basic requirements of most recombination processes it is now becoming clear that in some circumstances

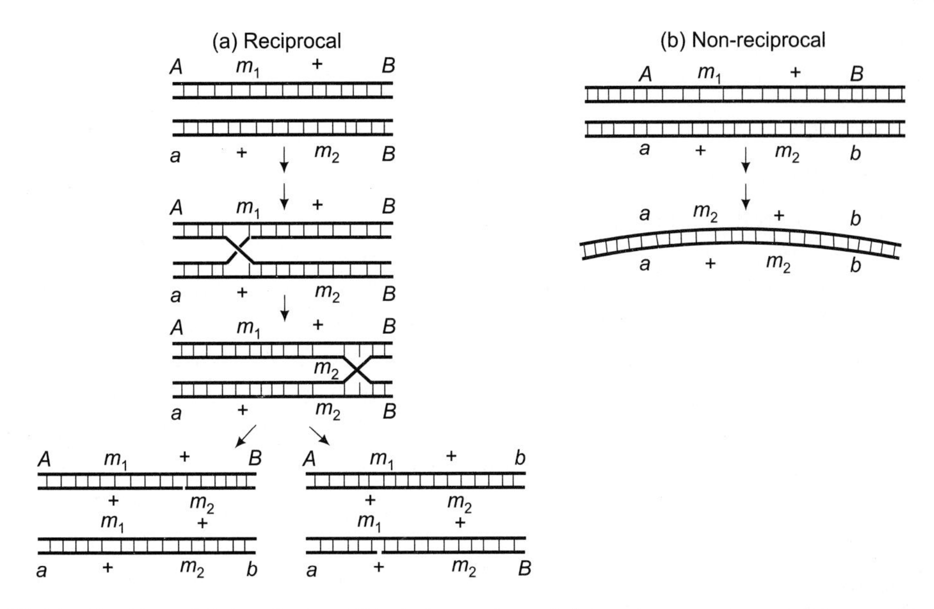

Fig. 29.24 Creation of heteroduplex DNA during reciprocal (a) and non-reciprocal (b) recombination.

recombination can occur between non-homologous DNA molecules which have no apparent regions of nucleotide sequence similarity. The resulting phenomenon, *heterologous recombination,* has been observed between chromosomal DNA and *small circular DNA* in higher organisms such as fungi and plants; in these instances it is not absolutely certain that an unsuspected region of nucleotide homology exists at the recombination sites but in at least a few cases this seems very unlikely. The mechanism by which heterologous recombination occurs is still a mystery.

GENES IN EUKARYOTIC ORGANELLES

Recent techniques in molecular genetics has revealed the presence of DNA in cytoplasmic organelles like mitochondria and chloroplasts other than the nucleus.

The idea that not all genes are located in the nucleus received additional support from Ruth Sager and her colleagues at the Rockfeller Institute, New York, who in 1954 identified in *Chlamydomonas* a mutation for streptomycin resistance which, although not as frequent as the petite phenotype, nevertheless displayed the same non-Mendelian mode of inheritance. By now biochemists had become alerted to the possible existence of extra-chromosomal genes, and DNA molecules that could be the locations for these genes were being sought. In 1951, DNA had been detected by staining in chloroplasts of a moss *Selaginella,* and subsequently DNA was also observed in the chloroplasts of a *Chlamydomonas* and in the mitochondria of the fungus *Neurospora crassa.* We now recognise that the mitochondria of all organisms, as well as the chloroplasts of all photosynthetic cells, contain DNA molecules that carry a limited number of genes. These code for RNA and protein molecules intimately involved in the functioning of the organelles.

Basic Features

Mitochondrial and *chloroplast DNA molecules* alone generally circular and double-stranded, though a few linear mitochondrial genomes are known. Copy numbers are not well understood: each human mitochondrion appears to contain about 10 identical molecules, which means that there are about 8,000 per cell. In yeast the figure is probably much smaller. The sizes of the organellar genomes depend on species with mitochondrial DNA molecules displaying a two hundred-fold variation, from 16459 bp in humans to over 2000 kb in some higher (that is, flowering) plants. Chloroplast genomes are less variable: those of all higher plants are between 120 and 180 kb and in other photosynthetic eukaryotes, such as algae, the sizes are not greatly different.

The results show that the yeast mitochondrial genome codes for eight proteins, seven of which are subunits of three complexes involved in the synthesis of ATP by oxidative phosphorylation. The eight proteins are:

1. The three largest subunits (I, II and III) of the cytochromes oxidase complex. Four additional subunits are coded by nuclear genes.
2. Three subunits (6, 8 and 9) of the mitochondrial ATPase. Five other subunits are coded by nuclear genes.
3. One subunit (apocytochrome b) of the cytochrome bc complex. In this case there are six additional subunits coded in the nucleus.
4. A protein associated in some way with the mitochondrial ribosome, coded by an unusual gene called *var?*

Mitochondrial Gene Products

(a) Common to all organisms

2 rRNAs (only ones in the mitochondrial ribosome) 20 to 35 tRNAs (all those needed for translation in the mitochondrion)

13 proteins:

Cytochrome cooxidase subunits I, II and III

ATPase subunits 6 and 8

Apocytochrome b

NADH dehydrogenase subunits 1 to 6, plus 4L

(b) In some organisms

ATPase subunit 9 (in yeast and filamentous fungi)

ATPase subunit B (in plants)

Ribosomal proteins (in yeast and filamentous fungi)

Maturases (in yeast and filamentous fungi)

Chloroplast Gene Products

Three or more rRNAs (all those in the chloroplast ribosome). About 30 tRNAs (all those needed for translation in the chloroplast). About 85 proteins, including:

ribulose bisphosphate carboxylase large subunit Photosystems I and II proteins. Cytochrome b/f complex proteins ATPase subunits

NADH dehydrogenase subunits Ribosomal proteins

RNA polymerase

Translation initiation factor

Various iron-sulphur proteins

Mitochondrial Genes (Fig. 29.25)

One of the most important offshoots of the recombinant DNA revolution was the emergence of DNA sequencing techniques that allow the precise order of nucleotides in a DNA molecule to be determined.

Human Mitochondrial Genome—Packed full of genes

The human mitochondrial genome is 16569 bp and carries genes for 13 proteins, 2 rRNAs and 22 tRNAs. These genes take up 15368 bp of the molecule, leaving only 1201 bp of non-coding sequences, part of which comprises the specific nucleotide sequence of the replication origin. Very little of the DNA molecule, only 87 bp or so, is genetically unimportant. This is much less than would be the case for a similarly sized portion of a chromosomal DNA molecule, which has extensive repetitive DNA regions.

In fact the human mitochondrial genome seems to have gone to great lengths to become as possible. The following features illustrate this point:

1. The rRNA genes are among the smallest known, coding for a large subunit rRNA with a sedimentation coefficient of only 165 and a small subunit RNA of 12S.
2. There are no intergenic regions, so the mRNAs are copies of just the gene: they have no leader or trailer sequences. This means that at the 5' ends of the mRNAs there are no obvious ribosome binding sites, so how translation is initiated is a mystery.
3. Several mRNAs are so truncated that the termination codons are incomplete: five mRNAs end with just a U or UA. The rest of the termination codon is provided by polyadenylation after transcription. This system is totally unique to metazoan mitochondria.
4. Genes are present for only 22 tRNAs, the absolute minimum needed to decode the mRNAs. In fact the standard codon-anti-codon pairing rules have to be modified by 'super-wobble', which allows some tRNAs to recognise all four codons of a single family two additions to the number of triplets that can act as initiation codons. We now know that non-stranded codes are general features of mitochondrial genomes, but that the differences are not the same in all organisms.

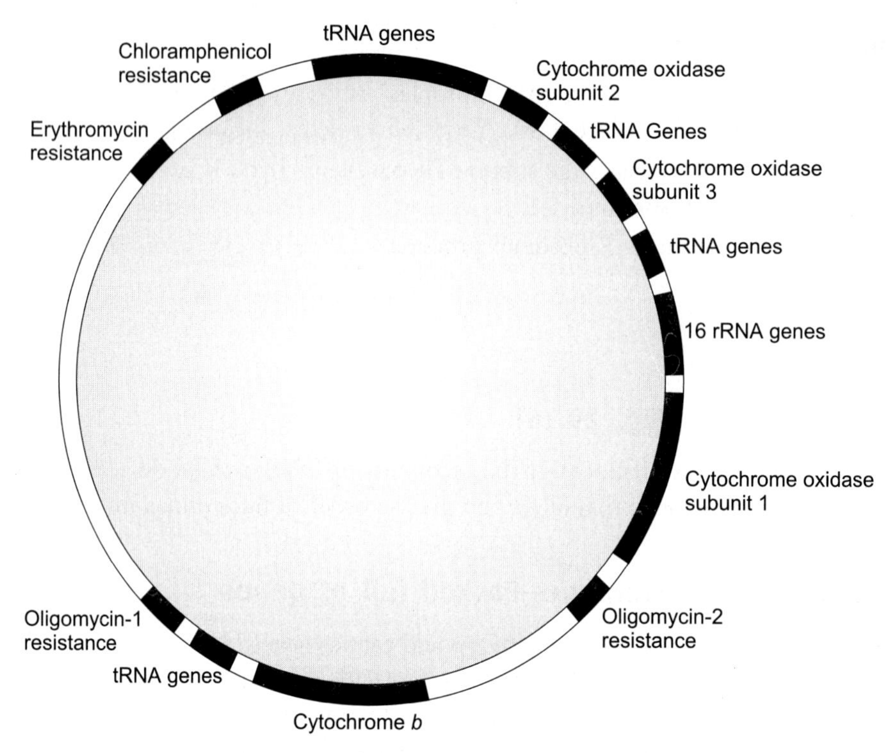

Fig. 29.25 Map of several loci on the mitochondrial DNA of yeast. The polarity gene is near chloramphenicol resistance.

Fungal Mitochondrial Genomes—Less compact having introns

DNA sequencing has now provided information about the organisation of mitochondrial genomes from a variety of organisms. All metazoan mitochondrial DNAs are organised along the same lines as the human genome, but different features appear when non-metazoan species are examined. In particular the fungal/yeast class of mitochondrial genome, of which *Aspergillus nidulans* provides a fairly typical example displays several interesting features. On the whole the genome is less compact and has larger rRNAs, a few more tRNAs (though still a non-standard genetic code) and spacers between the genes, so the mRNAs are not as unusual as the human ones.

However, the most striking feature of fungal mitochondrial genomes is the presence of introns, especially as these form a class of their own and are unrelated to the introns of eukaryotic nuclear genes. Two characteristics of mitochondrial introns emphasise their uniqueness. First, some *introns* contain a second gene that codes for a polypeptide completely different from the protein coded by the exon regions of the main gene. These *genes-within-genes,* first discovered in 1981 by a combination of genetic analysis and DNA sequencing carried out by Piotr Slonimski's group in Paris, are thought to code for maturases, enzymes involved in excision of the intron from the primary transcript of the gene. Their presence may aid the control of mitochondrial gene expression, thereby helping to ensure that synthesis of mitochondrial gene products is coordinated with events elesewhere in the mitochondrion and in the nucleus. Molecules can recombine to produce larger circles and, under some circumstances, the master genome containing all the DNA sequence can form. Plant mitochondria may, therefore, contain a bewildering array of different DNA molecules.

There is no adequate explanation for the large size of plant mitochondrial genomes. They do not contain any additional genes (except perhaps one or two) in comparison with the metazoan and fungal molecules, and although introns are present, their number is usually relatively small. The only indication we have of extra sequences come from the discovery that the mitochondrial genomes of at least a few plant species contain DNA segments obtained from the chloroplast. These sequences, called *promiscuous DNA*, may be several kb in length and may contain complete copies of some chloroplast genes. As yet we have no idea whether these genes are functional in the mitochondrion, or how the promiscuous DNA is transferred from one organelle to another.

Chloroplast Genes

The relatively large sizes of choloroplast genomes has made them less accessible to analysis by DNA sequencing and, as a result, investigations into chloroplast gene structure and expression have lagged behind equivalent work with mitochondria. However, thanks to the industry of two Japanese groups, one led by Kazou Shinozaki Nagoya, and the other by Kenji Ohyma at Kyoto, the complete sequences of two chloroplast genomes, for tobacco and the liverwort *Marchantia polymorpha,* now available. Overall, the genome organisations are very similar, and the same genes are present indicating that evolutionary change is acting relatively slowly on these molecules. We can expect more insights into the chloroplast genetic system as these sequences are examined in greater detail and used as the foundation for further research.

CHAPTER 30

Linkage and Crossing Over

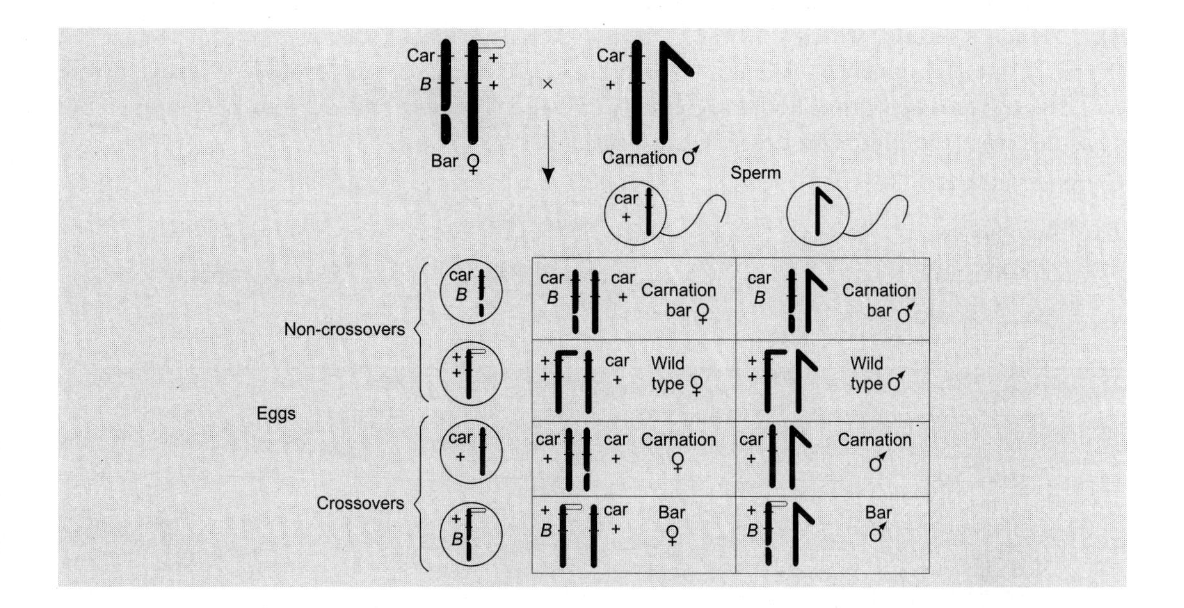

Mendel's Law of Independent Assortment holds good only when the genes are carried on different chromosomes, but if they are carried on the same chromosome, the genes do not assort independently. This was first discovered by Bateson and Punnett in 1906 in sweet peas who explained by two phenomena viz., *Coupling* and *Repulsion,* when crosses were made between two different alleles *AABB* × *aabb* and *AAbb* × *aaBB*. In the first cross the alleles tended to enter into the same gamete and in the second cross they entered into different gametes.

It was until T.H. Morgan in 1910 working with *Drosophila,* discovered similar situations and explained that the two phenomena are one and same called *Linkage.* He supposed that this tendency of linked genes to remain in their original combinations was due to their residence in the *same chromosome.* He also advocated that the degree of strength of linkage depends upon the distance between the linked genes in the same chromosome. This idea soon developed into the *theory of the linear arrangement of genes* in the chromosomes and later to the *construction of linkage maps* of chromosomes. This is the foundation of all construction of chromosome maps in all organisms three decades later.

LINKAGE IN MAIZE

Hutchison, crossed a variety of maize having seeds that were coloured and normally filled out (full): to one with colourless and shrunken seeds (Fig. 30.1 and 30.2). Colour gene C, is dominant over colourless c, and full grain S, is dominant over shrunken seed *s*. Accordingly, the parents were CCSS and *ccss,* and the F_1 was coloured, full with the *genotype* CS/*cs* (when the genes are linked they are represented together). If C and S assort independently, in accordance with Mendel's second principle, these F_1 plants should produce four types of gametes CS, C*s*, *c*S and *cs* in equal numbers. When a test cross is made (F_1 hybrid and double recessive parent *ccss*), the test cross progeny should segregate in a 1:1:1:1 ratio to all the four phenotypes. But in actual cross the following progeny was obtained:

Cross: CS/*cs* × *ccss*

Test Cross progeny:

Coloured, full	CS/*cs*	4,032	Parental combinations
Colourless, shrunken	*cs*/*cs*	4,035	- do -

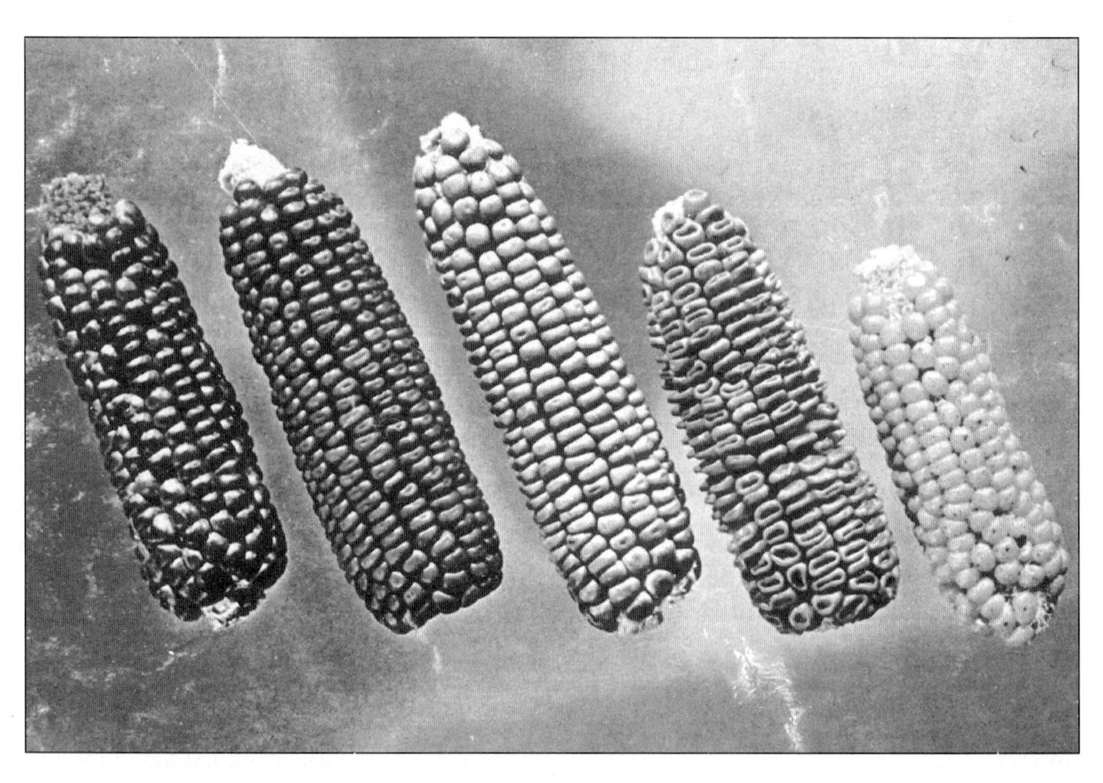

Fig. 30.1 Different varieties of maize for anthocyanin pigmentation in kernels.

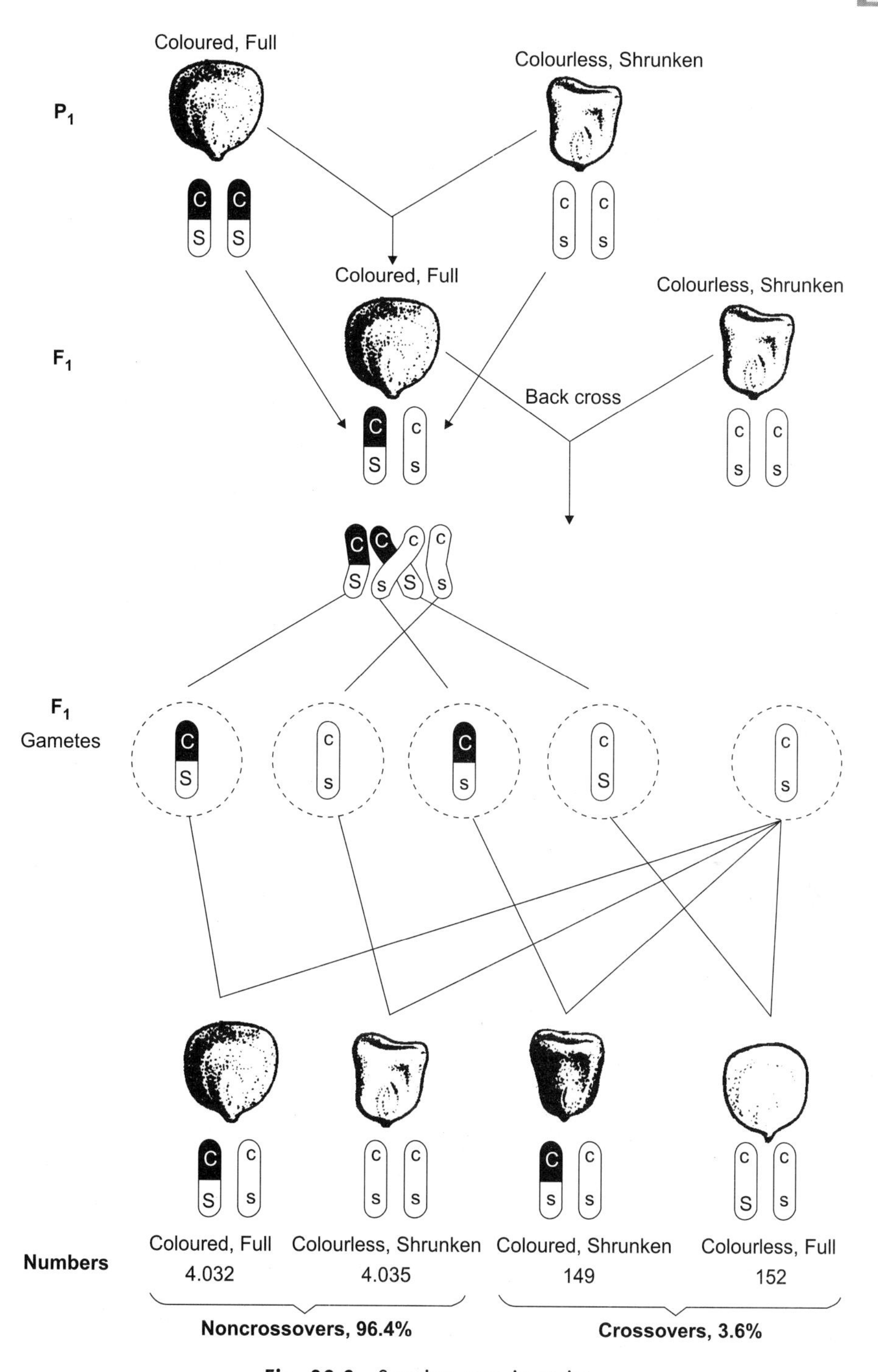

Fig. 30.2 Crossing over in maize.

Coloured, shrunken	*Cs/cs*	149	Recombinations
Colourless, full	*cS/cs*	152	- do -

The parental combinations are predominantly more (93.4 per cent) than the recombinations (3.6 per cent). Hence, the classical 1:1:1:1 test cross ratio was not obtained. It is obvious that the two pairs of genes C–*c* and S–*s* have not assorted independently and they remain combined or linked in 93.4 per cent and are recombined only in 3.6 per cent of the gametes.

The same situation was realised when a different cross was made, i.e., Colourless, full × Coloured, shrunken, the parental combinations were predominant than the recombinations. In this cross the parental combinations are just the reverse of the previous cross, since the parents are different. The results were as follows:

Test cross progeny:

Coloured, shrunken	*Cs/cs*	21,379	Parental combinations
Colourless, full	*cS/cs*	21,906	- do -
Coloured, full	*CS/cs*	639	Recombinations
Colourless, shrunken	*cs/cs*	672	- do -

In this cross, the parental combinations are in 97.06 per cent and the recombinations are 2.94 per cent. It is evident that whatever the parental combinations of two different pairs of linked genes may be, linkage tends to keep them together in about the same proportions.

These two experiments clearly prove that whenever a test cross is made between the F_1 hybrid and the double recessive parent, and in the test cross progeny if predominantly parental combinations occur (about 90-95 per cent) and very few of the recombinations (about 3–5 per cent), then it should be concluded that *Linkage* occurs between the genes involved in the cross.

CROSSING OVER

It was T.H. Morgan, who coined the term '*crossing over*' for the occurrence of recombination of linked genes to interchange of chromosome parts of homologous chromosomes. The linked genes separate due to crossing over between them. Thus, the genes for seed colour C, and for full or shrunken endosperm, S in maize remain associated in parental combinations in about 97 per cent of gametes but break apart in about 3 per cent as shown above. (Fig. 30.2). The diagram shows crossing over, leading to recombination of linked genes, due to interchange of sections of homologous chromosomes.

CHIASMATA FORMATION AT MEIOSIS

During *meiosis* the homologous (maternal and paternal) chromosomes come together and pair or synapse, during the prophase stage (zygotene and pachytene). The pairing is remarkably precise, similar sections of the chromosomes coming unfailingly together, like a zipper. Exchange of *chromatids* takes place at certain points on the chromosome at the *four strand stage* at diplotene stage. At each chiasma, two of the four chromatids have become broken and then rejoined, so that new chromatids are now formed along with original ones. Due to chiasma formation new chromosomes arise, or new gene combinations are formed which are called *recombinations.* Chiasma formation is a highly precise process, since the two chromatids at a chiasma exchange exactly equivalent segments, so that, with very rare exceptions, neither chromatid gains or loses any genes, but new gene combinations are formed.

TYPES OF CROSSING OVER

It has been well established from cytological, breeding and genetic experiments with various organisms, that crossing over takes place at the four strand stage of meiosis (prophase I). This is known as the *tetrad* stage of meiosis. At this stage, the homologous chromosomes have paired, each homologue, has split into two *chromatids*, held together by the *undivided centromere* and the chiasmata are observed in the tetrad (Fig. 30.3). At any point on the chromosome, only two chromatids take part in crossing over, whereas the other two chromatids remain in their original combinations. Depending upon the number of chiasmata, the number of chromatids involved in the crossing over, the types of crossing over are classified as: Two strand single, Two strand double, Three strand double type I and II, and Four strand double (Fig. 30.4).

In most animals and plants it is impossible to determine by observing the gene recombination in linkage experiments whether the crossing over that produced this recombination involved two, three, or all four strands of chromatids. The difficulty is that the products of each meiosis—the four haploid nuclei, each with one chromatid of the original four—cannot be exactly identified among the gametes, in the mass of which the cells coming from different meiotic divisions are randomly mixed. But, it has been found that in some lower organisms especially, certain Fungi, like *Neurospora* and *Aspergillus*, all the cells derived from a single meiotic division remain together and the individual gametes can be recognised, separated and tested independently.

TETRAD ANALYSIS IN *NEUROSPORA* (Fig. 30.3)

It is due to the extensive genetic studies of Dodge and Lindegren, that the bread mould, *Neurospora*, which was subjected to detailed studies of the meiotic products and traced them to specific gametes or ascospores. In *Neurospora*, each ascus, or fruiting body, contains eight haploid ascospores, which have arisen from a single diploid cell (zygote) through the two meiotic divisions followed by one mitotic division resulting in eight ascospores per ascus. These eight spores are arranged in a regular order in the ascus, which can be dissected out and grown separately, giving rise to haploid individuals that show the genetic constitution of each gamete. It was shown by Lindegren, that after crossing over in the linked genes, the four chromatids separate into *crossover* and *noncrossover* chromatids. The four meiotic chromatids undergo another mitotic division resulting in eight chromatids which form into *eight* ascospores.

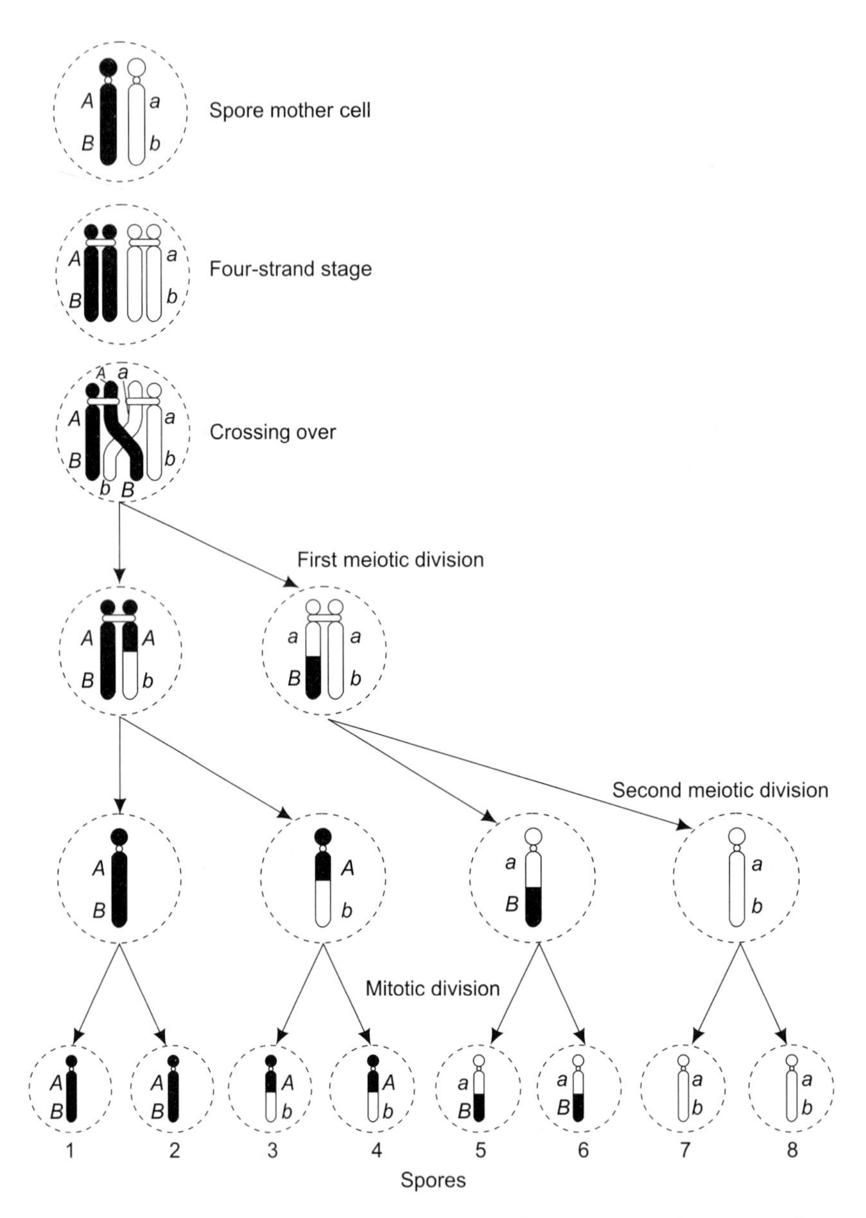

Fig. 30.3 Scheme of meiosis in *Neurospora,* illustrated by the behaviour of one pair of chromosomes.

The tetrads of *Neurospora* are accordingly ordered within the ascus. An analysis of segregation in this organism will reveal not only that a pair of alleles segregate from each other as predicted, but also the order of their segregation with respect to all four chromatids of a given bivalent. Wild-type *Neurospora* produces conidia that are pink in colour. A mutant form, known as *albino* (*al*), has white conidia. When a cross is made with these two strains, tetrad analysis

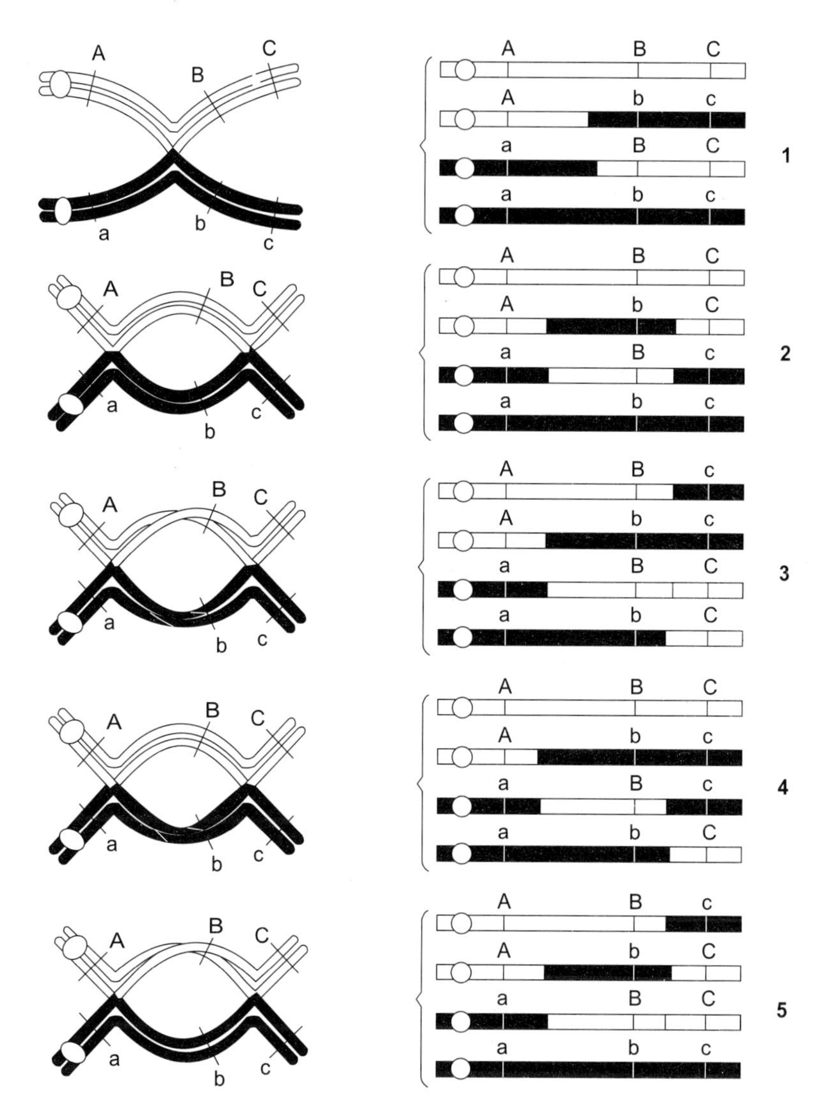

Fig. 30.4 Diagram showing chromosome bivalents with single chiasmata (above) and double chiasmata (below). 1. Single chiasma. 2. Two-strand double crossover. 3. Four-strand double crossover. 4 and 5. Three-strand double crossover.

reveals that there is a one-to-one segregation of wild type and *albino*. There are two patterns of segregation (Fig. 30.5) depending upon whether there was crossing over or not; in the first segregation it will be +, +, *al*, *al* and in the second pattern it will be +, *al*, +, *al* and accordingly, after mitosis, it will be +, +, +, +, *al*, *al*, *al*, *al* in one ascus (Fig. 30.5A) and in the other ascus it will be, +, +, *al*, *al*, +, +, *al*, *al* (Fig. 30.5B).

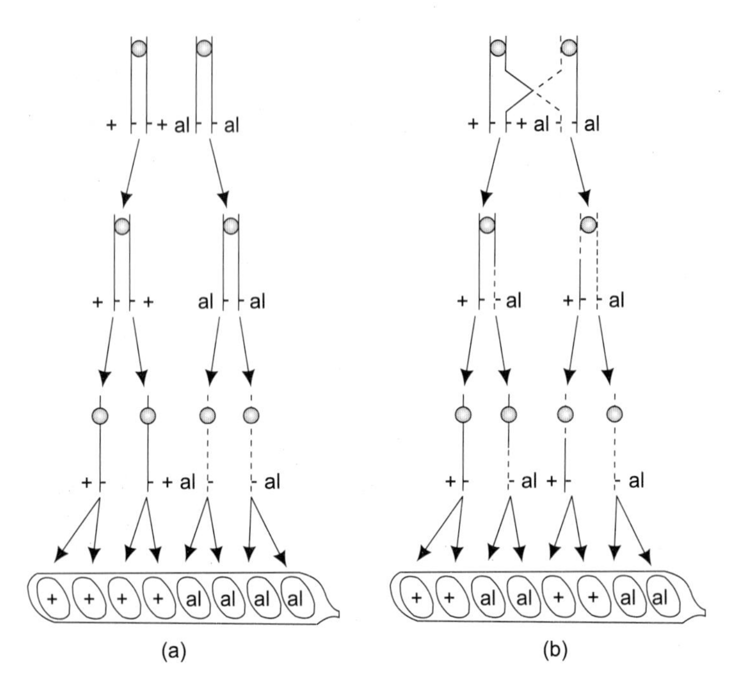

Fig. 30.5 First and second division segregation in *Neurospora.*

TETRAD ANALYSIS IN *CHLAMYDOMONAS*

A cross has been made between an *arginine-requiring* strain (*arg*) and *acetate-requiring* strain (*ac*) of *Chlamydomonas,* and the tetrads analysed after meiosis. The meiotic products are separated and grown to form colonies on medium containing both arginine and sodium acetate. The colonies are replica-plated to four different kinds of media: minimal medium, minimal medium plus arginine, minimal medium plus sodium acetate, and minimal medium plus arginine and sodium acetate.

The cross of the two mutant strains is symbolised as *arg* + × + *ac*, and can be made using either strain as the plus or minus mating type. Accordingly four kinds of progeny are expected from this cross: *arg* +, + *ac*, ++, and *arg ac.* They can be distinguished from each other by the media on which they will grow or not.

Tetrad analysis of the cross *arg* + × + *ac* are given in Tables 30.1 and 30.2. The first class consists of colonies derived from two sorts of meiotic products, *arg* + and + *ac*. Since two of the products have the genotypes of one parent and two the genotype of the other parent, this class of tetrad is called *parental ditype* (PD). Each pair of alleles is segregating in a one-to-one fashion. The second class of tetrad showed two new combinations, *arg ac* and + +. This tetrad, which consists of two types of products each genotypically different from either of the original parents, is called a *non-parental ditype* or NPD. In the third class four different genotypes are found, viz., parental, *arg* + and + *ac* and non-parental or recombinant *arg ac* and + +, types. A

Table 30.1 Types of progeny produced in the cross of *arg* + × + *ac* in *Chlamydomonas* and the media on which they will grow (Growth is indicated by a plus sign, absence of growth by a minus sign).

Type of progeny	*Minimal*	*Minimal + arginine*	*Minimal + acetate*	*Minimal + arginine and acetate*
arg ac	–	–	–	+
+ *ac*	–	–	+	+
arg +	–	+	–	+
+ +	+	+	+	+

Table 30.2 The three classes of tetrads produced in the cross *arg* + × + *ac* in *Chlamydomonas.*

Tetrad class	*arg* +	*arg ac*	*arg* +	
	arg +	*arg ac*	+ +	
	+ *ac*	+ +	*arg ac*	
	+ *ac*	+ +	+ *ac*	
Number	71	69	95	Total: 235

A PD:NPD ratio of one to one is indicative of the *independent* assortment of two pairs of alleles.

tetrad of this sort is called a *tetratype* or *T.* Each pair of alleles segregates one to one. (Figs. 30.6, 30.7 and 30.8).

FACTORS AFFECTING THE STRENGTH OF LINKAGE

The frequency of crossing over between genes, and hence the frequency of chiasma formation between their loci in the chromosome may be influenced by age, position of the chromosome, temperature, X-rays and the chemical composition of the food which are some of the agents that modify the chiasma frequency. Exceptionally crossing over can take place in somatic cells, Curt Stern called this phenomenon as '*somatic crossing over*'. Somatic crossing over was studied by Stern in *Drosophila* and in maize by Jones.

MEASUREMENT OF LINKAGE

The simplest technique for studying linkage is to obtain an F_1 hybrid by crossing two parents with linked genes and cross it with the double recessive parent (*Test cross*) (*AA/BB* × *aa/bb; AB/ab* or *Ab/aB* × *aa/bb*). In the test cross progeny, there will be parental combinations and recombinations, and the frequency of recombination is expressed in percentages of the total number of individuals examined in the experiments.

LINKAGE GROUPS AND CHROMOSOMES

If a certain gene, A, is linked to two other genes, B and C, then B and C are as well linked. The genes known to exist in a species may thus be divided into '*linkage groups*'. The members

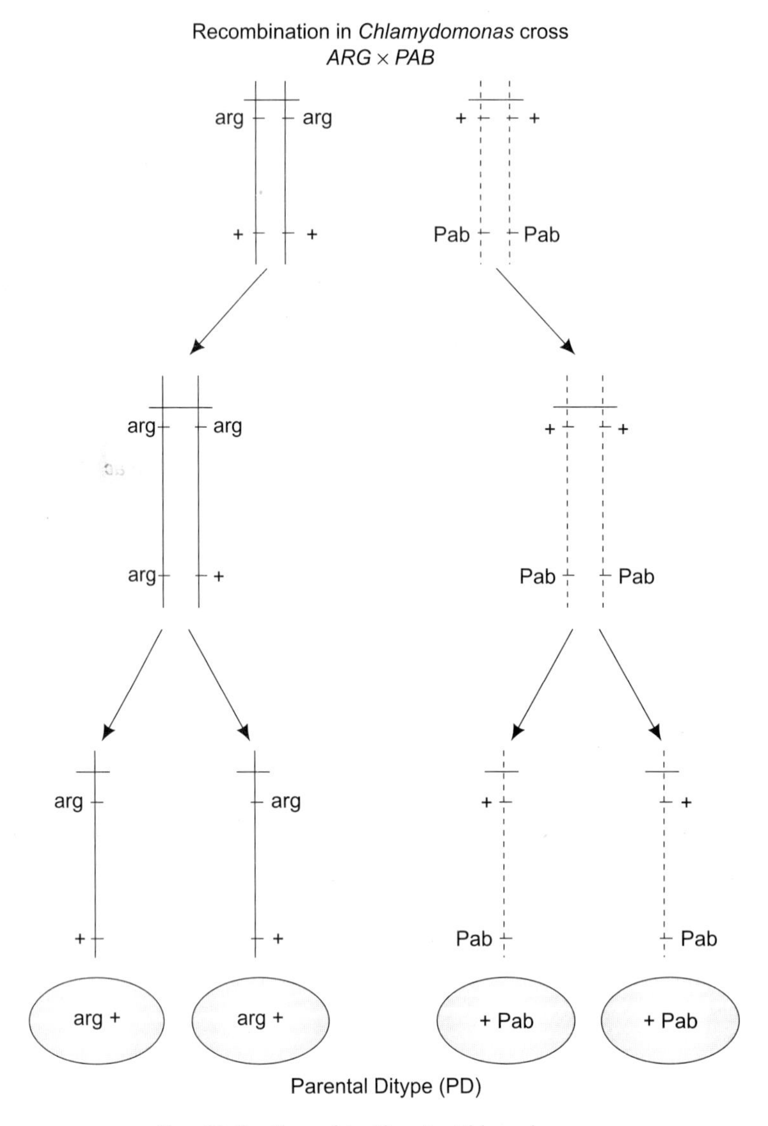

Fig. 30.6 Recombination in *Chlamydomonas.*

of a linkage group show linkage to each other. In genetically well studied species, the number of linkage groups is equal to the number of haploid chromosomes, which the species possess.

T.H. Morgan and his collaborators, especially Bridges and Brehme have subjected to detailed studies on linkage and crossing over, coupled with mutation studies of *Drosophila melanogaster*, the vinegar fly or the fruit fly, and constructed linkage maps. Some of the advantages of selecting *Drosophila* as the study material are as follows: 1. easy to breed under laboratory conditions, 2. their culture is simple and inexpensive, 3. life cycle is less than two weeks, 4. high fecundity producing thousands of offspring, 5. produce numerous mutations, 5.

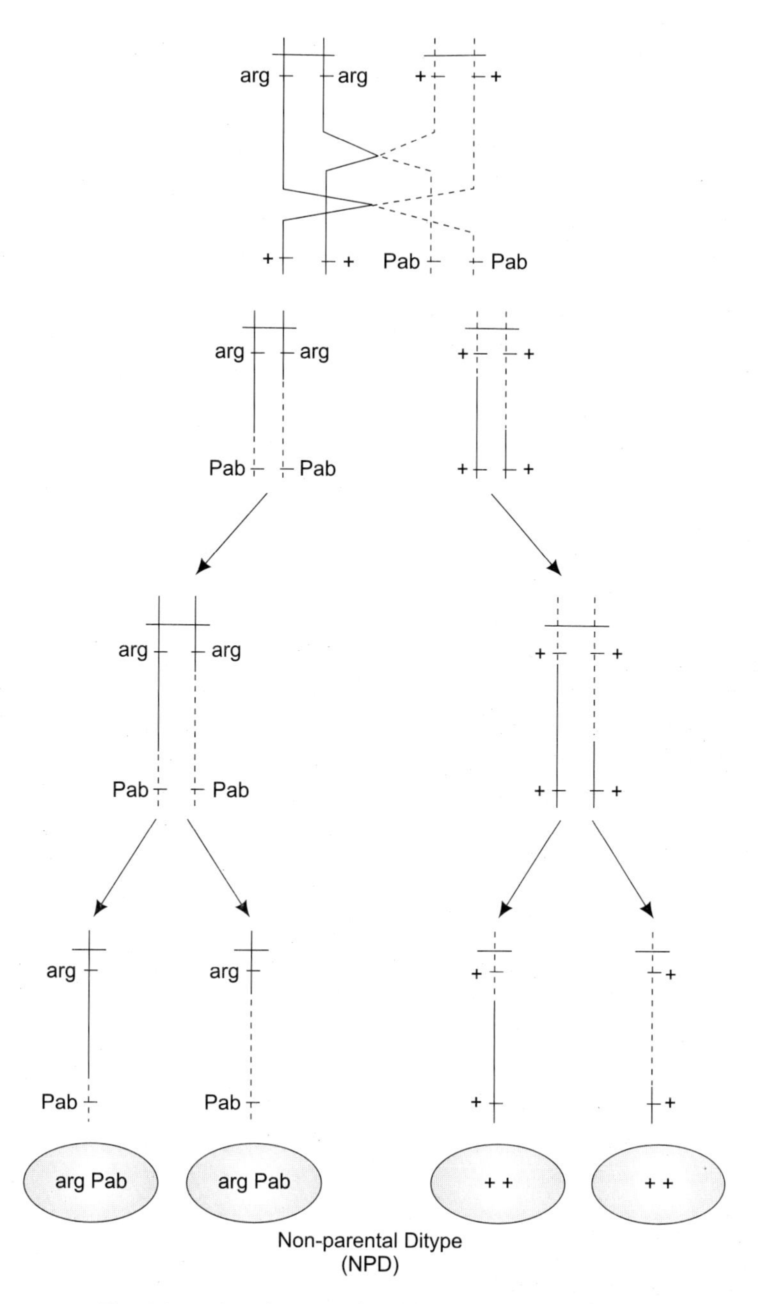

Fig. 30.7 Crossing over in *Chlamydomonas,* NPD type.

chromosomes are especially large called salivary gland chromosomes (Figs. 30.9 and 30.10). *Drosophila* has four linkage groups (Figs 30.11 and 30.12). In maize (*Zea mays*) there are 10 linkage groups (Fig. 30.13). In *Pisum sativum*, the pea plant there are 7 linkage groups, and in

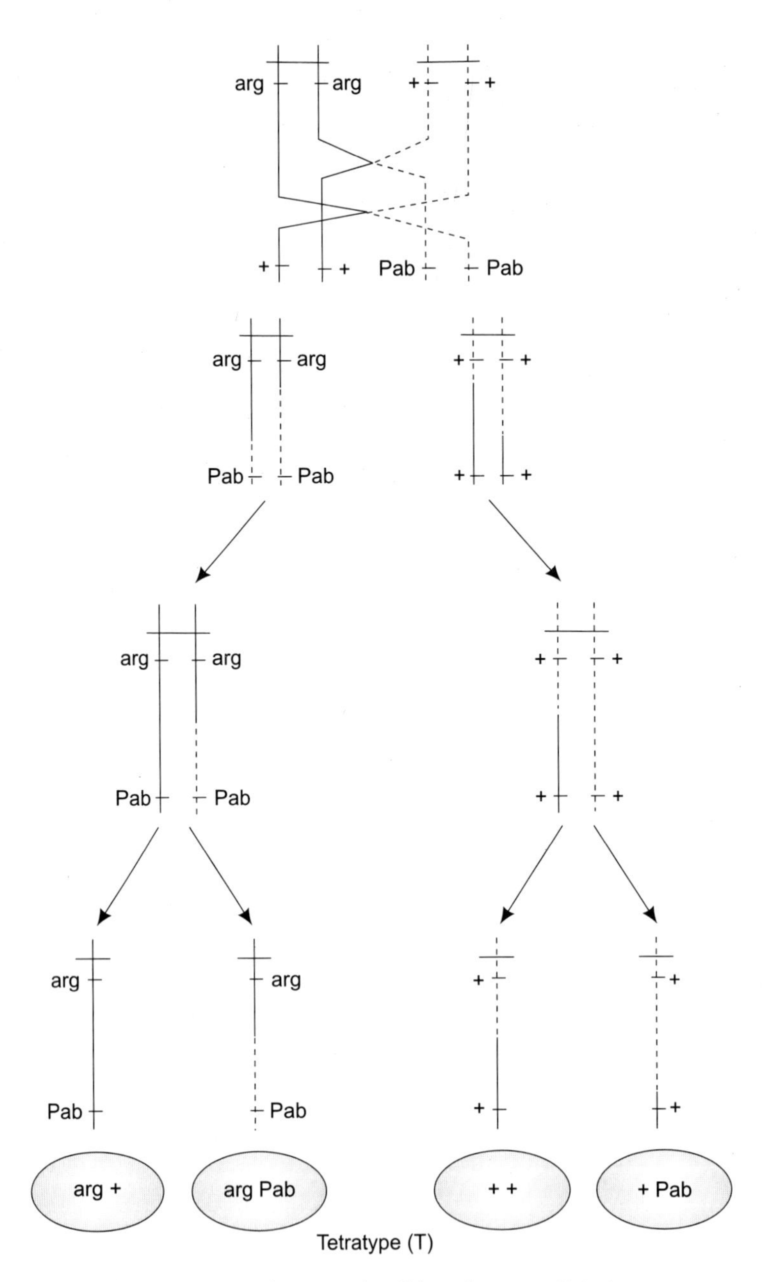

Fig. 30.8 Crossing over in *Chlamydomonas,* Tetratype.

Neurospora there are 7 linkage groups, and lastly there are 23 linkage groups in man, confirming the rule that every organism has as many linkage groups as there are haploid chromosomes in that organism.

Fig. 30.9 The vinegar fly, *Drosophila melanogaster,* male at left and female at right.

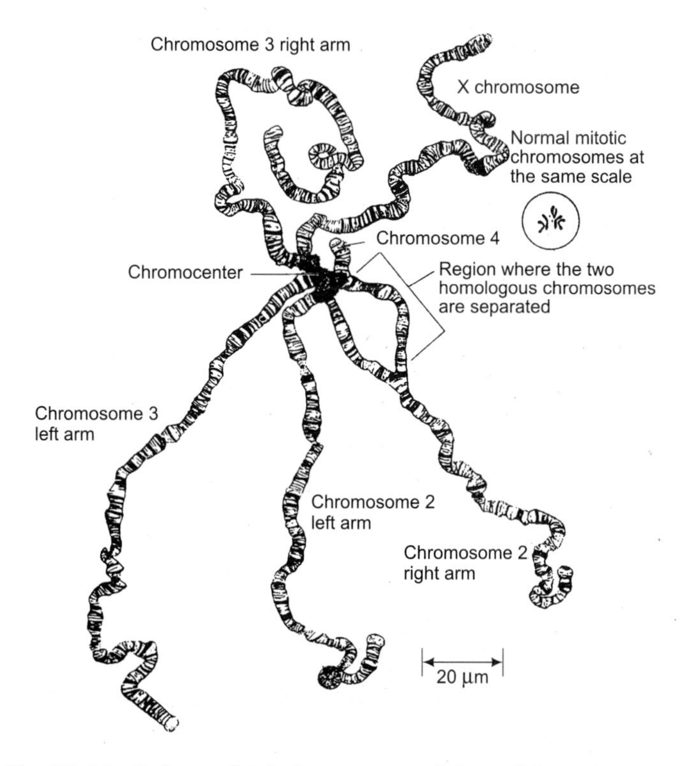

Fig. 30.10 Salivary gland chromosomes of *Drosophila melanogaster.*

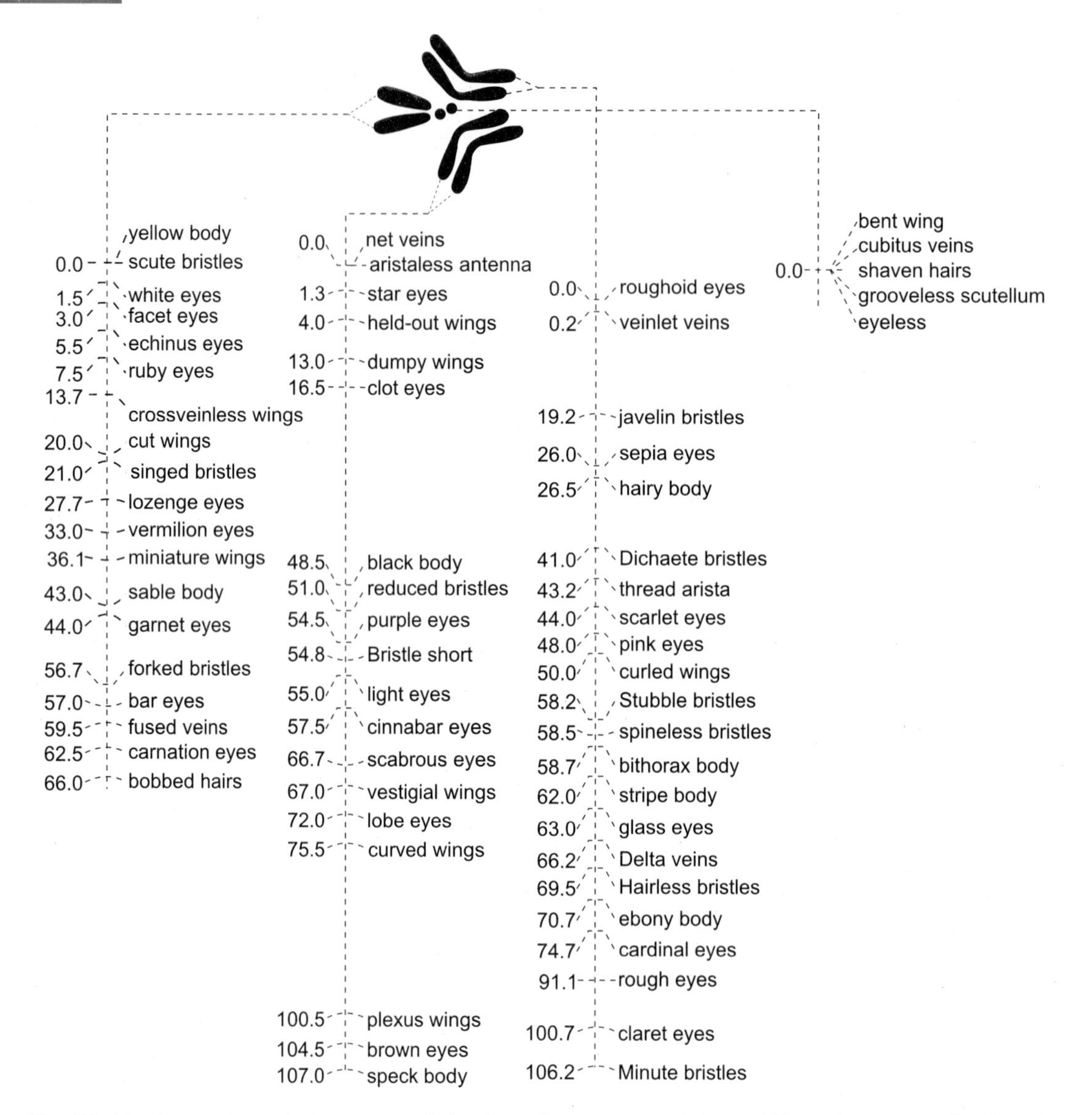

Fig. 30.11 A genetic or linkage map of the four chromosomes of *Drosophila melanogaster,* showing the relative positions of some of the more important genes. Figures refer to distances from the upper end of the chromosome as determined from the percentages of recombination observed in linkage experiments.

ABSENCE OF CROSSING OVER IN *DROSOPHILA* MALES

In genetically well studied species such as maize, peas, mice, poultry, man and others, recombination of linked genes takes place both in females and in males, except in the males of *Drosophila*.

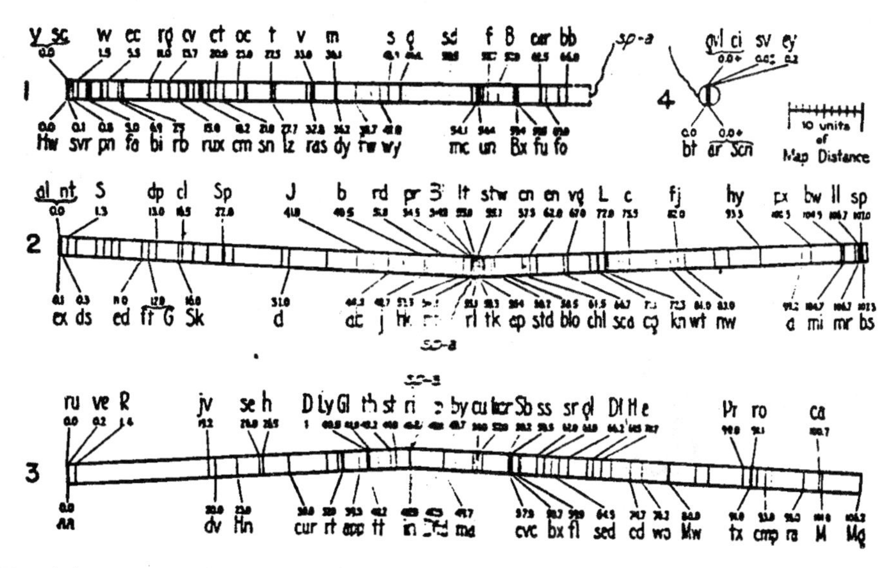

Fig. 30.12 Linkage maps of the 4 haploid chromosomes of D. *melanogaster.* The figures refer to map distances from the left end of each chromosome and as calculated from the percentage of recombination.

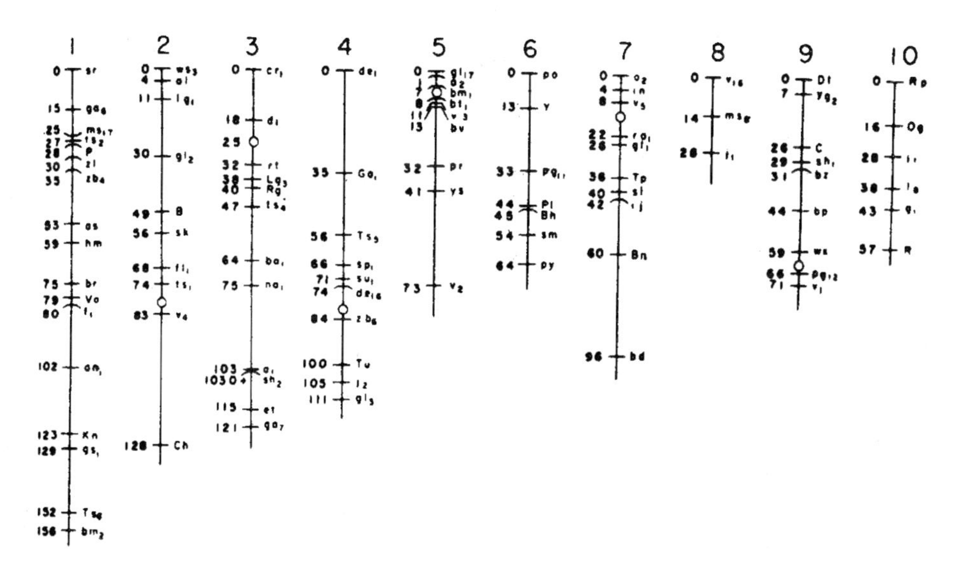

Fig. 30.13 Linkage maps of the 10 haploid chromosomes of maize. Centromeres are indicated by circles only when known with certainty; genes marked with asterisks are located approximately.

If a grey bodied vestigial winged fly is crossed to a black bodied long winged one, the F_1 generation consists of grey long winged normal flies (Figs. 30.14 and 30.15). Thus, the gene for grey body colour, *B*, is dominant over its allele, which causes black body colour, *b*; and the gene for long wings, *V*, is dominant over its vestigial allele, *v*. Now, if the F_1 male hybrids are crossed to double recessive females (black bodied vestigial winged females), only *two kinds* of offspring

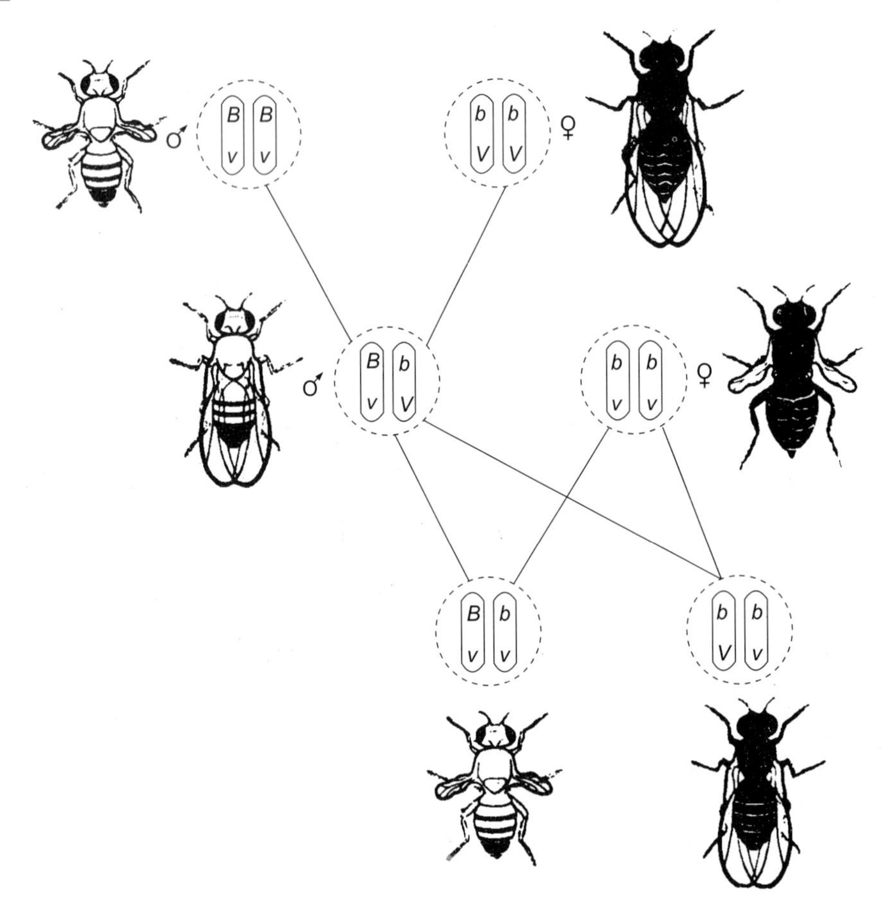

Fig. 30.14 Absence of linkage in the *Drosophila* male. A grey-bodied, vestigial-winged male crossed with a black-bodied, normal-winged female produces in the F_1 males and females with grey bodies and normal wings. Backcrossing such F_1 males with black, vestigial females produces only two types of offspring, which are like the original parents. There are no recombinations.

are produced: grey vestigial and black long (Fig. 30.14). The expected types of crossovers–grey long and black vestigial–do not appear at all. If, however, and F_1 female fly is crossed with a black vestigial male, the four expected types (Fig. 30.15) are produced in the following proportions:

Non-crossovers

Grey vestigial	–	41.5 per cent	
			83.0 per cent
Black long	–	41.5 per cent	

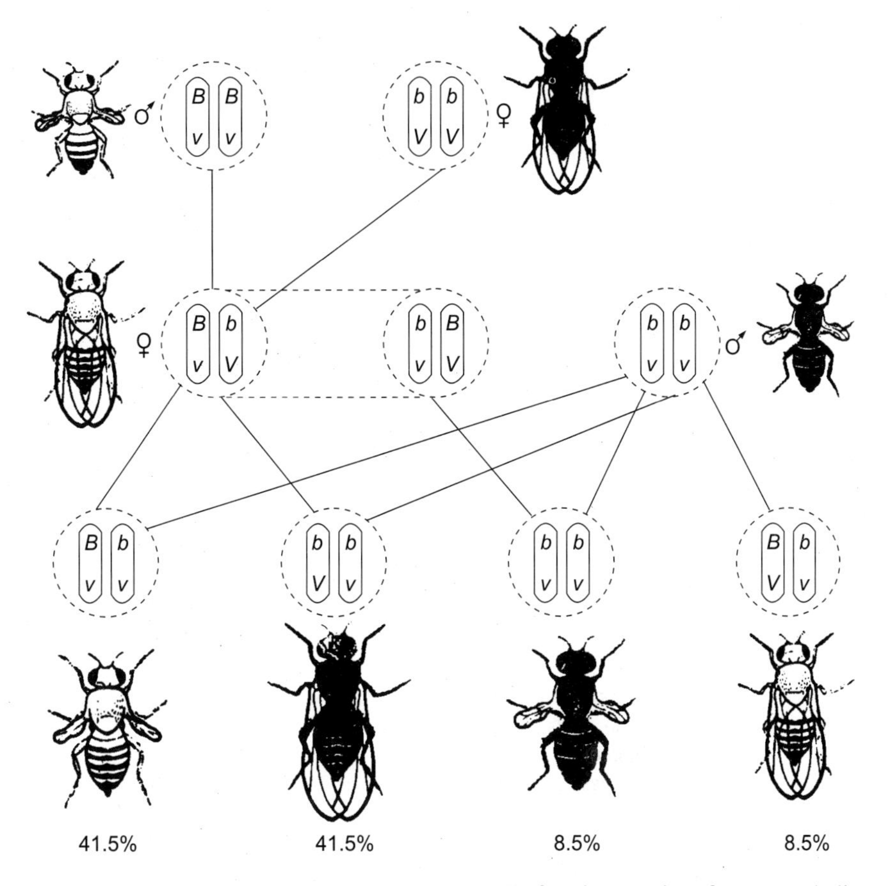

Fig. 30.15 Recombination of linked genes in the *Drosophila* female. Results of a cross similar to that in Figure 30.14 but with the F_1 females backcrossed with black, vestigial males (reciprocal cross).

Crossovers

Black vestigial	–	8.5 per cent	
			17.0 per cent
Grey long	–	8.5 per cent	

This experiment shows clearly that crossing over does not take place in the males of *Drosophila.*

CYTOLOGICAL PROOF OF CROSSING OVER

Curt Stern, working with D. *melanogaster*, and Creighton and McClintock, with maize, provided experimental evidence in 1931, for the cytological proof of crossing over.

Stern's Experiment The male parent in the experimental cross possessed a normal chromosomal set, with an X chromosome marked by the mutant eye-colour gene carnation (car) and the normal allele ($+^B$) of the genes Bar (*B*), which narrows the eye (Fig. 30.16).

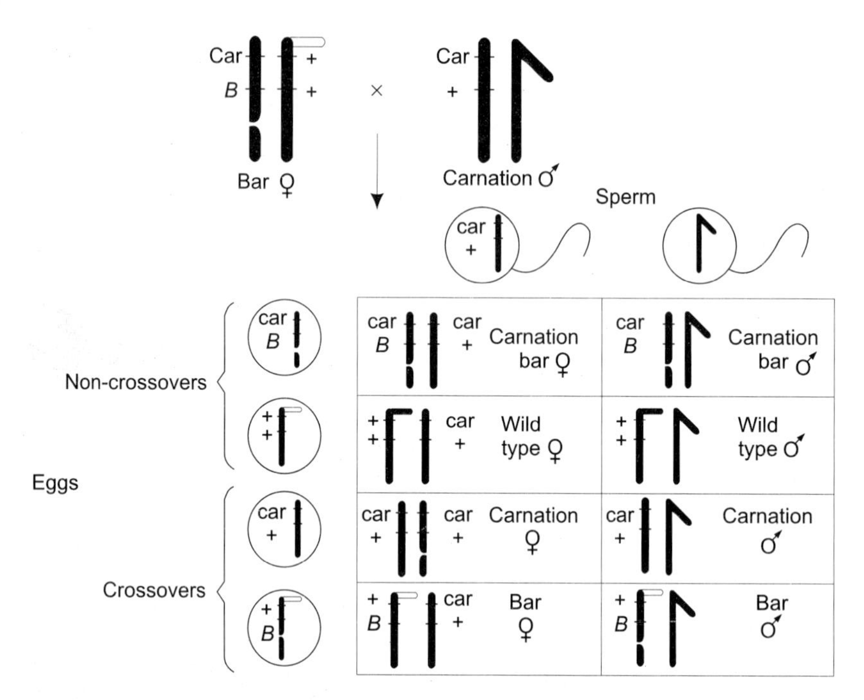

Fig. 30.16 Cytological proof of crossing over. Correlation between recombination of sex-linked genes and chromosomal exchange in *Drosophila melanogaster.* One X chromosome had a segment (*outlined*) of Y chromosome, and the other X chromosome was involved in a reciprocal translocation with chromosome IV. Progeny with eight or nine chromosomes and with a standard or the X-Y can be readily distinguished by a cytological examination of individuals with different phenotypes.

The female parent was derived from a cross between two strains, one of which had a large portion of the Y chromosome attached (translocated) to the end of X chromosome; this altered X chromosome carried the normal alleles of carnation ($+^{car}$) and Bar ($+ +^{B}$). In the other strain the X chromosome had been broken into two separate parts; the piece bearing the centromere carried the genes *car* and *B*, whereas the other fragment was translocated to the tiny IV chromosome.

Both of these strains were viable and when crossed, gave females that were heterozygous for the two genes in question. They also possessed two kinds of X chromosomes that could be morphologically distinguished from each other as well as from the normal X chromosome, which in the females, was derived from the father.

The heterozygous female parent can give crossover and non-crossover eggs, which, when fertilised by an X-bearing sperm containing the two genes *car,* and $+^{B}$, will give four types of female offspring (the males were discarded due to absence of crossing over). The two non-crossover types will be phenotypically *Car-Bar* and wild type $+^{car}$, $+^{B}$ respectively.

The former should, on cytological analysis, possess two X chromosome fragments, and the latter should possess the X chromosome bearing the piece of the Y chromosome. Both, of course, will have a normal X chromosome in addition.

The two crossover types will be carnations with normal shaped eyes, or *Bar* with red eyes (wild type).

The former should have two normal X chromosomes, and the latter should possess a normal X chromosome plus the two X chromosome fragments, one of which will bear a portion of the Y chromosome.

Stern made a cytological study of 364 crossover and non-crossover F_1 females, and these were in good agreement between genetic and cytological facts.

TETRAD ANALYSIS IN *SACCHAROMYCES CEREVISIAE*

Features of Genetic Cross *Saccharomyces Cerevisiae*

We will use the yeast *S. cerevisiae* to illustrate the fundamental principles of genetic analysis of eukaryotic microorganisms. The first point to appreciate is that the nature of the inherited characteristics studied with a microorganism will be different from those studied with *Drosophilia* and other multicellular organisms. Wing size, for example, is clearly inapplicable for yeast: rather than examining morphological traits, biochemical characteristics are generally used. A list of typical characteristics called genetic markers, for yeast is given in Table 30.3.

Table 30.3 Some typical genetic markers for S. *cerevisiae*. Each of these gives rise to a characteristic that is easily recognised by testing cells on a suitable agar medium. Markers are usually mutant versions of normal genes.

Marker	*Phenotype*	*Method by which cells carrying the marker are distinguished*
*ade*2	Requires adenine	Grows only when adenine is present in the medium
*can*1	Resistant to canavanine	Grows in the presence of canavanine
*Cup*1	Resistant to copper	Grows in the presence of copper
*cyn*1	Resistant to cycloheximide	Grows in the presence of cycloheximide
*leu*2	Requires leucine	Grows only when leucine is present in the medium
*suc*2	Able to ferment sucrose	Grows when sucrose is the only carbohydrate present
*ura*3	Requires uracil	Grows only when uracil is present in the medium

SETTING UP THE CROSS

Yeast has two mating types, equivalent to genders in higher organisms, called *a* and $\overline{a}$. Gametes of opposite mating type fuse during sexual reproduction. With a strain of yeast that has haploid vegetative cells, the gametes arise by differentiation (a subtle morphological and physiological change) of vegetative cells. All that is needed to carry out a genetic cross is to mix together cells of opposite mating types, which will then undergo meiosis, so long as the nutritional conditions are correct. Each zygote gives rise to four ascospores, each of which will divide to produce new vegetative cells.

As an example let us consider a cross between a *leu*2$^-$, mating type of strain of yeast and a wild-type, mating type of strain. The *leu*2$^-$ strain carries a mutation that inactivates the *leu*2 gene, which codes for β-isopropylmalate dehydrogenase, one of the enzymes involved in biosynthesis of leucine. The strain is therefore, a leucine auxotroph and can survive only if leucine is supplied in the growth medium. The wild-type strain will of course be a prototroph. We could write the cross as:

$$leuz^{-}\underline{\alpha} \times \underline{\alpha}\ leu2$$

Actually, when setting out the genetic analysis in this way, we assume that the strains are of opposite mating types. Also, the wild-type organism would be designated as '+'. So we would write the cross simply as:

$$leu\ 2^{-} \times +$$

ANALYSING THE RESULTS OF THE CROSS

Fusion of the leu2^{-} and wild-type gametes will produce diploid zygotes that will subsequently undergo meiosis, giving four haploid ascospores per zygote. From what Mendel has told us about the behaviour of genes during meiosis we can predict that the result will be equal numbers of *leu*2^{-} and wild-type ascospores

Parental gametes:	*leu*2^{-} x +
Ascospores:	*leu* 2^{-} *leu*2^{-} + +

In practice, one of two methods could be used to verify this result:

1. *Random spore analysis:* A sample of ascospores is spread on to an agar medium and the ability of each to survive without leucine is tested. This is done by first spreading on to a complete medium (containing leucine), on which all ascospores, whether *leu*2^{-} or wild-type, will be able to divide and produce colonies. The number of colonies tells you how many ascospores fare in the sample tested. Each colony is then transferred by replica plating on to minimal medium (containing no leucine), on which *leu*2^{-} cells will not be able to grow. The number of leucine auxotrophic ascospores in the original sample can then be worked out.
2. *Tetrad analysis:* This technique makes use of the fact that the four ascospores that result from meiosis of a single zygote are retained in a structure called an *ascus* or *tetrad.* Individual asci can be removed, broken open, and the four ascospores grown on agar medium. The number of leucine auxotrophic ascospores that arise from a single gamete fusion event can therefore, be determined, allowing the results of the cross to be examined in precise detail.

GENE MAPPING WITH *S. CEREVISIAE*

How are these techniques used to map genes in S. *cerevisae*? To illustrate the strategy we will look at a second cross, carried out the same way as the one just described, except that the parent yeast carries two makers, not one. We will use hypothetical markers, a^{-} and b^{-}, so our cross is:

$a^- b^-$ x + +

Ascospores arising from the cross could have any of four genotypes:

$a^- b^-$
parental combinations
+ +

a^- +
non-parental combinations
+ b^-

Are the genes linked? First, we will consider how to determine if the genes are on the same chromosome or on different ones. If the genes are unlinked, we would expect equal numbers of each ascospore. Random spore analysis would reveal a 1: 1 : 1 : 1 ratio for the four genotypes. If however, a^- and b^- are present on the same chromosome, then ascospores with the parental combinations (a^- b^- and + +) will be more frequent. If the genes are linked, any non-parental ascospores will have arisen from crossing-over and recombination between a^- and b^-.

What is the map distance between the genes? An approximate estimation of the map distance between a^- and b^- can be obtained from the number of recombinant ascospores detected by random spore analysis. By analogy with the strategy for gene mapping in higher eukaryotes, the recombination frequency (= the number of recombinant ascospores + the total number of ascospores) will correspond to the distance between the genes in centi-Morgans.

Unfortunately, a measurement of map distances by random spore analysis is only approximate. The method cannot take account of double crossover events, as two crossovers between a^- and b^- will not result in a recombinant ascospore. Because of this limitation, random spore analysis will not take account of all the crossover events occuring between a^- and b^-, and so will result in an apparent recombination frequency (and hence map distance) slightly less than the value.

Tetrad analysis is important because it can determine the number of double crossover events and so give a more correct map distance. Before explaining how, we should look at the three types of tetrads that are possible:

1. A parental ditype (PD), which contains two ascospores with the genotype of one parent, and two ascospores with the genotype of the other parents (a^-b^-, a^-b^-, + +, + +).
2. A non-parental ditype (NPD), which contains two of each of the non-parental combinations (a^- +, a^- +, +b^-, +b^-).
3. A tetratype (T), which contains four different ascospores, one of each possible combination (a^-b^- +, a^- +, +b^-, ++).

To understand how a map distance is arrived at, take careful stock of each of the following points:

1. If in a particular zygote no crossovers occur between $\bar{a}$ and b^-, then the tetrad that results will be a parental ditype, PD.

2. If a single crossover occurs between $\bar{a}$ and b^-, then a tetra-type, T, will result. Note that crossovers can occur between any two chromatids of a pair of homologous chromosomes.
3. (a) If two crossovers occur between $\bar{a}$ and b^-, then a PD, NPD or Tetrad will arise, depending on which chromatids are involved.
   (b) If two chromatids are involved then the tetrad will be PD. If three chromatids are involved then the tetrad will be T. If four chromatids are involved then the tetrad will be NPD.
   (c) Double crossovers will occur in the ratio 1 two-chromatid event: 2 three-chromatid events: 1 four-chromatid event.

Now remember that what we are trying to do is to work out the number of crossover events that have occurred between a^- and b^-. The number of single crossover events is determined as follows:

- All single crossovers give a T (see point 2), but some T's (equal to twice the number of NPD's) result from double crossovers (see point 3).
- Therefore, the number of single crossover events–T–2NPD. The number of double crossovers can also be calculated.
- The number of tetrads representing double crossovers—4NPD (see point 3).
- Therefore, the number of actual crossovers (two per each double crossover) = 8NPD.

The total number of crossovers occuring in the tetrads analysed is therefore:

$$T - 2NPD + 8NPD = T + 6NPD$$

The crossover frequency is therefore:

$$\frac{T + 6NPD}{2 \times \text{Total number of tetrads}} = \frac{T + 6NPD}{2(T + NPD + PD)}$$

Note that to obtain the crossover frequency we divide the number of crossovers by twice the total number of tetrads. This is because each ascospore contains four chromatids, but only two of these participate in an individual crossover event. Crossover frequency provides a direct, accurate measurement of map distance. The most recent S. *cerevisiae* map includes over 500 markers in 17 linkage groups, corresponding to each of the 17 yeast chromosomes.

GENE MAPPING IN YEAST—A WORKED EXAMPLE

Consider the cross:

$$a^-b^- \times ++$$

The following results are obtained by tetrad analysis:

PD tetrads	142
NPD tetrads	3

T tetrads	61
Total	206

The first conclusion is that these genes are linked. This is evident from the very low number of NPD tetrads that are seen. For unlinked genes the PD : NPD ratio should be about one.

The crossover frequency can be calculated from the formula:

$$CF = \frac{T + 6NPD}{2(T + NPD + PD)} = \frac{61 \times 18}{2 \times 206} = 0.192 = 19.2\%$$

This method for estimating map distances is accurate for genes up to about 40 map units apart. At greater distances the possibility of triple crossovers provides further complications.

CHAPTER 31

Extranuclear Inheritance

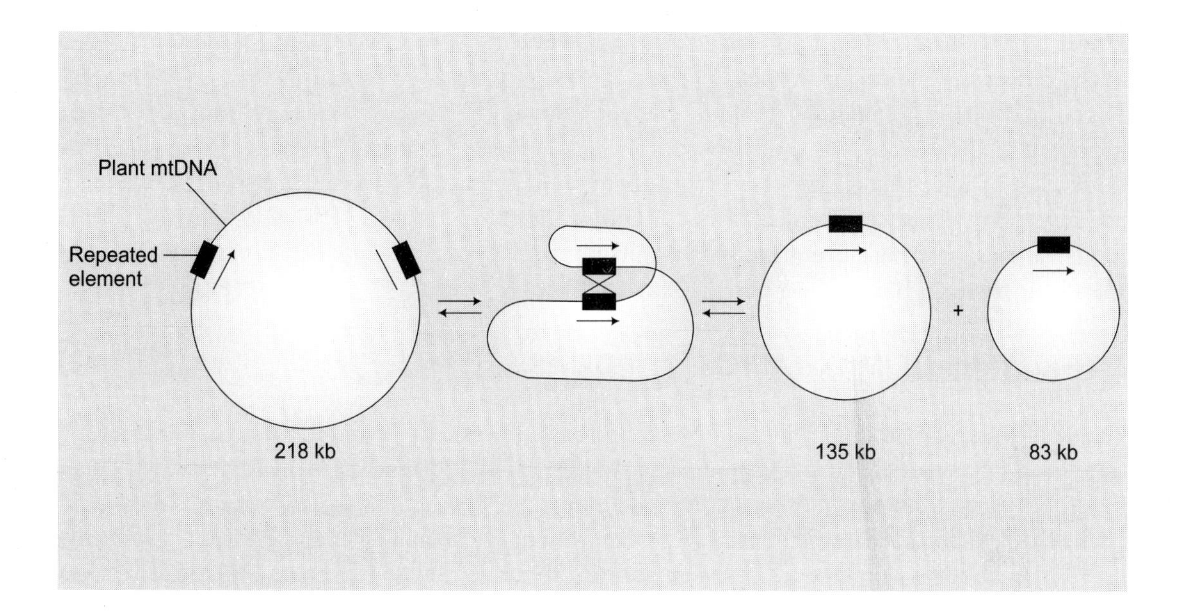

It has been proved beyond doubt that the expression of certain hereditary traits can be influenced by the *cytoplasm* in which the genetic material acts, or by the action of autonomous entities that reproduce and act in the cytoplasm.

MILK FACTOR IN MICE

Through careful and extensive inbreeding and selection, lines of mice have been developed that differ markedly in their incidence of *mammary cancer.* In certain lines, almost all of the female mice, generation after generation, die of this disease. In other lines, the mice die from other causes, rarely from this type of cancer. Once stabilised at a high or a low level, the mammary cancer incidence is transmitted as a characteristic of the line.

It is of interest to investigate the inheritance of this difference between lines. When reciprocal crosses are made, the outcome of a cross in the first and later generations seems to depend largely on the characteristic of the *female* parent. No matter which way the cross is made, the male makes little immediate contribution to the incidence of this cancer in his descendants. No chromosome or other known element in the nucleus follows this pattern of inheritance. This suggests that an inherited predisposition to mammary cancer in mice may depend on some *nonchromosomal,* or *extranuclear element* for its transmission. Furthermore, since the differences introduced by a female in a cross are transmitted for an indefinite number of generations, the responsible element seems to be capable of perpetuating itself, just as genes are.

Milk Factors and Placental Transmission. The explanation of the transmission of mammary cancer in mice came when baby mice from lines showing respectively high and low incidence of the cancer were taken from their mothers at birth and allowed to nurse on "foster mothers" of the opposite type. It soon became evident that mother mice from lines showing a high incidence of the disease transmit *through their milk,* an agent that later causes mammary cancer to develop in mice that have nursed from them. This agent, called a "milk factor", acts like an infective particle, and meets at least in some respects the criteria for a *filterable virus.* A female mouse that as a baby has been infected by nursing from an infected mother can in turn transmit the factor through her milk to her progeny.

INHERITANCE OF MILK FACTOR IN HORSES

One somewhat similar situation has been reported in horses. Breeders of horses have for a long time been bothered by the fact that an occasional stallion and mare seemed to develop a queer kind of incompatibility in reproduction. The mare, after having had three or more foals, would begin to have repeated difficulties rearing her young, and valuable foal after foal, apparently normal when born, would develop severe jaundice about 96 hours after birth, and die. Two facts have helped to define the situation. First, if the potentially jaundiced foal is prevented from nursing on its own mother and is given a foster mother, the foal does not get sick at all, but develops normally. And second, if the mare is bred to a different stallion, she commonly has no difficulty rearing his foals.

The basis of incompatibility in horses, which depends in large part on ordinary gene differences is shown in the diagram (Fig. 31.1). A foal inherits from his sire the ability to make a particular substance absent in his dam. This substance acts as a foreign material when it reaches the mother's circulation from the foetus, and she elaborates *antibodies* capable of reacting specifically with the substance. For the first three or more pregnancies, the antibodies, if they are present at all, are apparently too weak to cause any serious damage. In later pregnancies, however, the antibodies in the milk that the foal gets from his mother may react with the substance he makes by virtue of a gene inherited from his father, and this reaction is often fatal to the foal. The substance is part of his blood cells, and jaundice, a yellow colouring of tissues by the haemoglobin and derived pigments of the destroyed red cells, is one of the early symptoms of the disorder.

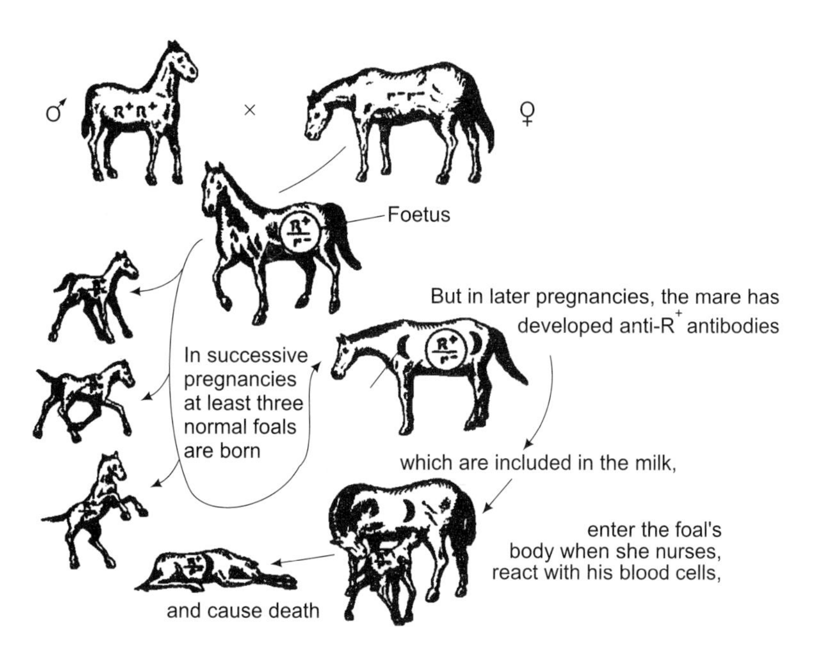

Fig. 31.1 Milk factor in horses. A foal sometimes dies because antibodies that react with his blood cells are contained in his mother's milk.

The severe reaction in the foal, then, depends on a two-component system. One of these components is a direct result of ordinary Mendelian inheritance: The foal inherits the ability to form an *antigenic* (antibody-inducing) substance from his sire. The other component of the system is an indirect result of the same Mendelian situation: When the mare lacks the antigenic substance present in the foal, she may elaborate antibodies against it. These antibodies, transmitted not through germ plasm but *through milk*, comprise the second component necessary for incompatibility between mare and foal.

Erythroblastosis Fetalis in Humans. The interpretation of this incompatibility in horses was relatively easy, because an essentially similar situation in human beings occurs widely known as *erythroblastosis fetalis* (haemolytic anaemia of the newborn). The condition is comparable to the severe jaundice of the foal described above, with one major exception: Instead of being transmitted primarily through milk, *Rh* antibodies are transmitted from mother to foetus primarily across the placenta (Fig. 31.2). A human foetus may inherit from its father a gene governing the ability to produce a substance absent in the mother. The mother may then elaborate antibodies against the foreign substance, and these antibodies, which in the human diffuse back across the placenta to the foetus, react with the developing foetal blood to cause its destruction.

Maternal Influence. In the examples previously discussed, the elements passively transferred from mother to child through milk or across the placenta, or even in the egg in the case of "chicken Rh", were not properties produced under the direct control of the mother's own genotype. They were more directly the result of influences external to the mother as an

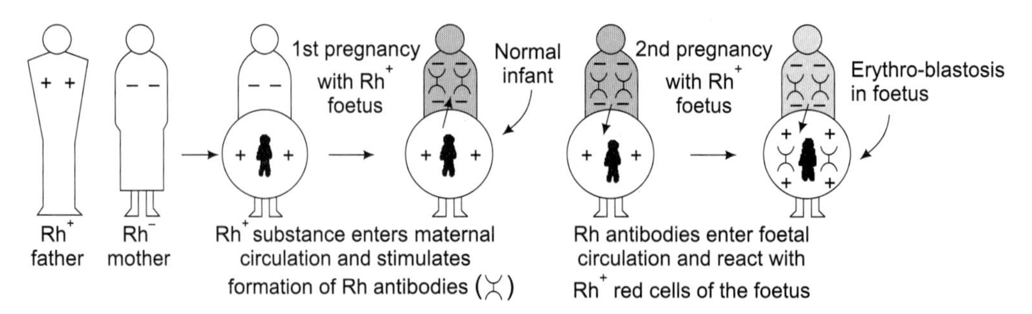

Fig. 31.2 Erythroblastosis fetalis in Man. Second or later Rh-positive babies of Rh-negative mothers are sometimes severely anaemic, because antibodies that react with the body's blood cells may be developed by the mother and passed across the placenta to the foetus.

individual—the virus-like agent in mouse mammary cancer, the foreign foetal antigens in the antibody-mediated incompatibilities. But a basic function of a mother provide, through the egg or across the placenta, materials of her own elaboration for use by the embryo. A few individual differences are known to depend on the extranuclear transmission of this kind of material.

INHERITANCE OF COILING IN SHELLS OF *LIMNAEA* (Fig. 31.3)

A presumably similar but less concrete example is provided by the direction of coiling in the shells of the snail *Limnaea.* Snail shells may coil in either of two directions, clockwise or counter-clockwise. These are commonly distinguished by the terms *dextral* and *sinistral*; if you hold a shell so that the opening through which the snail's body protrudes is facing you, this orifice may be either on your right (dextral) or on your left (sinistral). The difference is the same as that between a "right-handed" and a "left-handed" screw. Different species may be either dextral or sinistral, and within some species, races may differ in this regard. The race difference may be investigated through routine genetic techniques; the investigation has been facilitated by the fact that the particular species most studied (*Limnaea peregra*) is monoecious and can reproduce either by crossing or by self-fertilisation.

When reciprocal crosses are made, the F_1 progeny show the same direction of coiling as did their mothers. But F_2's produced by self-fertilisation of the F_1 all coil dextrally; and when the F_2's are in turn self-fertilised, each produces a uniform progeny, three producing dextral progenies to each one that produces sinistral progeny.

This rather puzzling situation is simply explained by postulating that the direction of coiling of the embryonic shell is impressed on the egg cytoplasm by the genotype of the diploid oocyte from which the egg came. The dextral-sinistral alternative depends on a pair of alleles in which the allele for dextral is dominant. The F_1 shells in either of the reciprocal crosses all agree with their homozygous mothers; the F_2 shells are all dextral, since their mothers are heterozygous; while the F_3 *progenies* reflect the expected 3:1 "phenotypic" ratio of the F_2 mothers. This explanation is the simplest that can be offered for the situation, although rare exceptions to the rule in the species most carefully investigated suggest that other conditions may also sometimes operate to modify the direction of coiling.

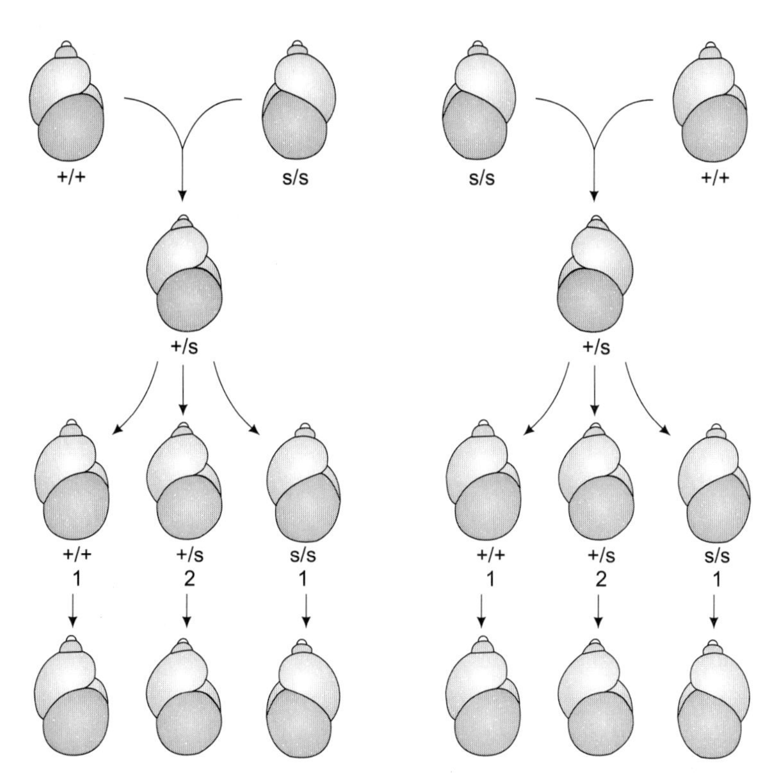

Fig. 31.3 Coiling in shells of *Limnaea*. The direction of coiling in the shells of certain snails depends on the mother's genotype, rather than on the genotype of the individual itself. Left, eggs produced by a homozygous dextral snail, fertilised by sperm from a homozygous sinistral individual, develop into dextral progeny. These heterozygous progeny, when self-fertilised, produce only dextral progeny in turn. But the three genotypes in this generation are reflected in the progenies produced in the next generation, again after self-fertilisation. Right, the reciprocal cross is different, in ways that confirm the maternal influence on direction of coiling.

In order for an extranuclear element to play a true and stable role in biological inheritance, it must have one characteristic, above all others, in common with genes. It must be capable of specific self-duplication from generation to generation. If it does not have this property, it is likely to be transient and of secondary consequence.

The Inheritance of Plastid Characteristics. Almost everyone is familiar with variegated plants that have areas of pale green or white in their otherwise normally green leaves. Sometimes the variegation is quite symmetrical and uniform; in other cases it is irregular, and whole branches may carry normal green leaves, while others are entirely pale or white, or mixtures of pale or white and dark-green areas.

In the four o'clock, *Mirabilis jalapa*, controlled pollination of flowers borne on these three kinds of branches of the same plant has given the provocative results, summarised in Table 31.1.

Table 31.1 Progeny of a variegated Four o'clock plant.

Pollen from Branch of Type	*Pollinated Flowers on Branch of Type*	*Progeny Grown from Seed*
Pale	Pale	Pale
	Green	Green
	Variegated	Pale, green and variegated
	Pale	Pale
Green	Green	Green
	Variegated	Pale, green and variegated
	Pale	Pale
Variegated	Green	Green
	Variegated	Pale, green and variegated

It will be noticed that the type of pollen used is not important; pollen grains from pale, green and variegated branches behave the same. The determining factor is the female's contribution. Seeds borne on pale branches produce only pale plants; those borne on green branches produce green plants, while those borne on variegated branches segregate in irregular ratios of pale, green and variegated.

The results can be interpreted on the basis of two different elements, transmitted only through the female and distributed in a variegated plant as follows:

Types of Branches	*Types of Elements Maternally Transmitted*
Pale	Pale
Green	Green
Variegated	Pale and Green

The cytoplasmic primordia of chloroplasts can mutate, and the mutant particles reproduce their own kind just as do mutated genes. In fact, the process of plastid mutation can be influenced by specific genes. In corn, for example, M.M. Rhoades has reported a gene (*iojap*) that when homozygous, greatly increases the mutation rate of the plastid primordia from normal to defective. The primordia do not appear to back-mutate under the influence of any allele at the *iojap* locus.

Killer Paramecia. Some strains of *Paramecia* produce a substance that kills other strains. In a brilliant series of investigations, T.M. Sonneborn and his co-workers at Indiana University have demonstrated that this characteristic involves a kind of extranuclear inheritance.

Paramecia that produce the lethal material are called *killers.* The individuals that are killed are called *sensitives.* Two different pairs of alternatives are concerned in the killer *versus* sensitiv system.

In the Nucleus	*In the Cytoplasm*
A gene *K* for killer	A self-duplicating cytoplasmic particle (kappa)

or and or

In allele *k* for nonkiller No kappa

Kappa, the cytoplasmic particle, is directly responsible for the production of the killing substance. But the continued reproduction of kappa in a *Paramecium* depends on the presence of the gene *K* in the nucleus of the cell; without this gene, kappa is soon lost. Even in the presence of the gene *K*, proper manipulation of the culture can cause the paramecia to reproduce more rapidly than does kappa. Once kappa is lost, it cannot be regained unless more kappa is introduced from another cell. In other words, gene *K* is important in the maintenance of kappa, but cannot initiate its production. In this way, a stock having the killer gene but free of kappa can be developed. When killers, *KK*, conjugate with sensitives, *kk*, the exconjugants are, of course, *Kk*, and should accordingly be killers. But if the conjugation is brief and little or no cytoplasm is exchanged, then a killer clone and a sensitive clone are produced (Fig. 31.4). A more prolonged conjugation results in exchange of cytoplasm, and then all descendant clones are killers and have kappa particles that can be shown by staining reaction to contain DNA. These particles are clearly vehicles of cytoplasmic transmission, but it has been shown that maintenance of the particles and the killer phenotype depends on the presence of the gene *K* in the nucleus. Thus a *Kk* individual may be either a killer or a sensitive depending on whether or not it has received kappa particles. However, *kk* individuals may inherit kappa particles in the cytoplasm but lose them after some fission generations. Certain killer clones, *KK* or *Kk*, can be converted to sensitives by making them undergo very rapid fissions during which the reproduction of the kappa particles fails to keep pace with the cell divisions, with the result that the particles are finally lost entirely.

Some very neat reasoning based on the number of generations required to lose kappa in this fashion resulted in an estimate of the number of kappa particles originally present. This figure was confirmed as an order of magnitude when it was found possible by means of a special technique to stain the kappa particles and to count them directly; There were as many as 1000 per animal in a strong killer stock.

Some biologists suppose that kappa is a parasite of paramecia and not a natural and inherent cytoplasmic particle at all. In some respects, the killer characteristic behaves like a disease, with which uninfected individuals can be infected under proper conditions. Kappa, then, could be regarded as a very small-microorganism that is pathogenic to some strains of paramecia but capable of duplication, indefinitely.

If kappa is indeed a parasite, it is a most interesting one; chronic infection with it confers a kind of immunity to its otherwise lethal product. It is transmitted through reproduction of its host and not by casual contact, and its multiplication in a cell depends on the presence of a particular allele in the nucleus of that cell. The *K* gene may be concerned with the production of a specific biochemical factor necessary to the growth and multiplication of this '*parasite*'.

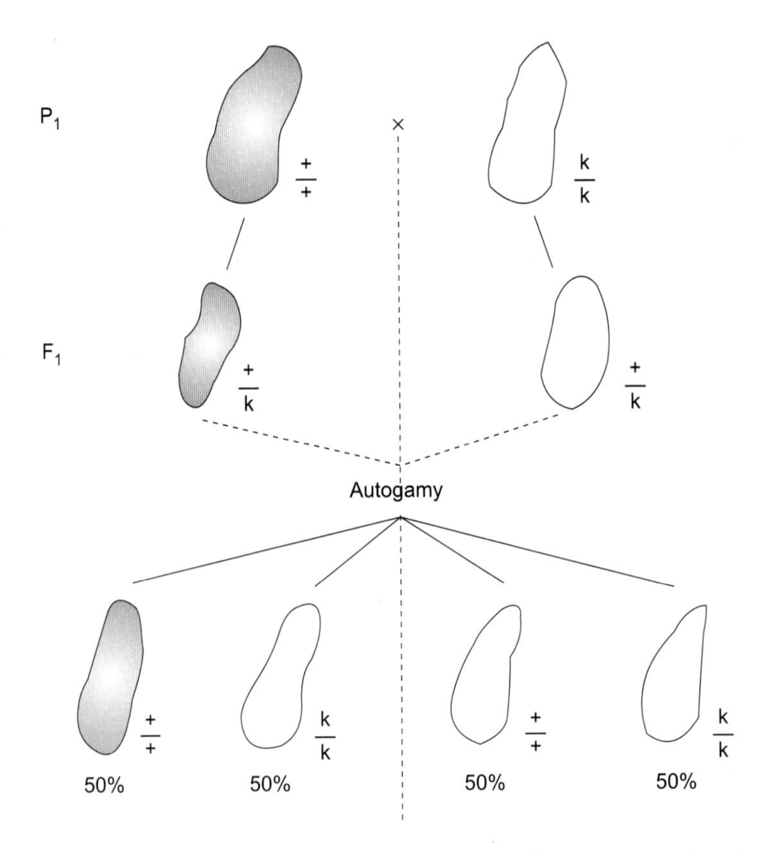

Fig. 31.4 Inheritance of the "killer" character in *Paramecium aurelia,* variety 4, when the killer gene *K* is transmitted but not the killer cytoplasm.

CO_2 SENSITIVITY IN *DROSOPHILA*

Among higher animals, perhaps the clearest example of cytoplasmic transmission is found in *Drosophila.* Normally, *Drosophila* can be anaesthetised by carbon dioxide; the flies tolerate high concentrations of this gas without apparent injury. A strain is known, however, that is killed by relatively low concentrations of CO_2. This characteristic, called *carbon dioxide sensitivity,* is normally transmitted almost exclusively through the egg and behaves in inheritance as though it were dependent on some specifically self-duplicating property of the cytoplasm. In certain respects, the controlling element acts like an infectious particle. For example, a fly can be rendered sensitive by, injecting with haemolymph from an affected individual. But the condition is not spread by a simple contact.

True cytoplasmic transmission from one sexual generation to another is as yet not certainly known among the vertebrates. There is, however, some reason to suspect that during development, the somatic cells of an individual may become different from each other as a result of the establishment of different stable, self-perpetuating cytoplasmic systems. No definite cytoplasmic particles have been identified with these differences, although they are in the cytoplasm of cells bodies that seem to enjoy a continuity and a degree of autonomy of their own.

MITOCHONDRIAL DNA

Mitochondrial DNA, or mtDNA as it is sometimes abbreviated, was discovered in the 1960s, initially through electron micrographs that revealed DNA-like fibers within the mitochondria. Later, these fibers were extracted and characterised by physical and chemical procedures. The advent of recombinant DNA techniques made it possible to analyse mtDNA in great detail. In fact, the complete nucleotide sequences of mtDNA molecules from several different species have now been determined.

Mitochondrial DNA molecules vary enormously in size, from about 16 to 17 kb in vertebrate animals to 2500 kb in some of the flowering plants. Each mitochondrion appears to contain several copies of the DNA, and because each cell usually has many mitochondria, the number of mtDNA molecules per cell can be very large. In a vertebrate oocyte, for example, it has been estimated that as many as 10^8 copies of the mtDNA are present. Somatic cells, however, have fewer copies, perhaps less than 1000.

Most mtDNA molecules are circular, but in some species, such as the alga *Chlamydomonas reinhardtii* and the ciliate *Paramecium aurelia*, they are linear. The circular mtDNA molecules, which have been studied most thoroughly, appear to be organised in many different ways. The simplest arrangement is that seen in the vertebrates, where 37 distinct genes are paced into a 16- to 17-kb circle leaving little or no space between genes are dispersed over a very large circular DNA molecule hundreds or thousands of kilobases in circumference. In these plants the mitochondrial genes may become separated onto different circular molecules by a process of intramolecular recombination (Fig. 31.5). This recombination is mediated by repetitive sequences located in the mtDNA. An exchange between two of the repetitive sequences can partition the "master" mtDNA circle into two smaller circles, a process that superficially resembles the excision of a lambda prophage from the *E. coli* chromosome. In some species, several DNA circles of different sizes are formed by recombination between pairs of repetitive sequences located at different positions around the master DNA circle. These molecules are difficult to study, and more research is needed to elucidate the mechanism that produces them.

The fine details of mtDNA organisation have been studied by DNA sequencing. Animal mtDNA is small and compact (Fig. 31.6). In human beings, for example, the mtDNA is 16, 659 base pairs long and contains 37 genes, including two that encode ribosomal RNAs, 22 that encode transfer RNAs, and 13 that encode polypeptides involvd in oxidative phosphorylation. In mice, cattle and frogs, the mtDNA is similar to that of human beings–an indication of a basic conservation of structure within the vertebrate subphylum. Invertebrate mtDNA is about the same size as vertebrate mtDNA, but it has a somewhat different genetic organisation. These differences seem to have been caused by structural rearrangements of the genes within the circular mtDNA molecule.

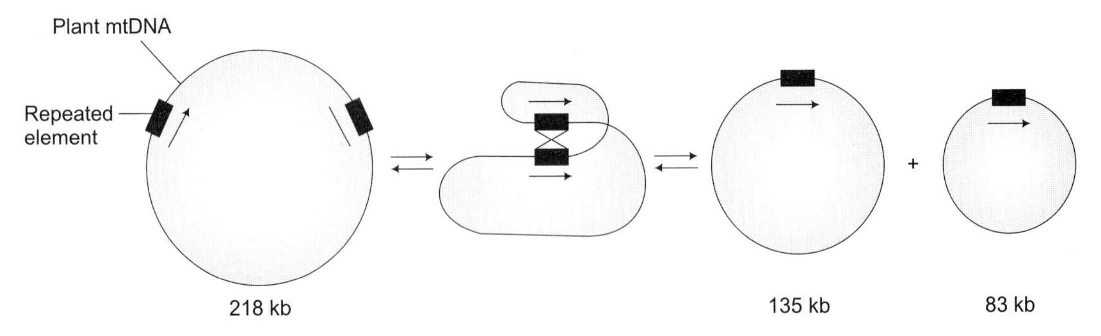

Fig. 31.5 Intramolecular recombination in the mtDNA of the Chinese cabbage, *Brassica campestris*. Recombination between the repeated elements in the large circular DNA molecule partitions this molecule into two smaller ones. Alternatively, the repeated elements in the two small molecules may recombine with each other, thereby joining the molecules into a single large molecule.

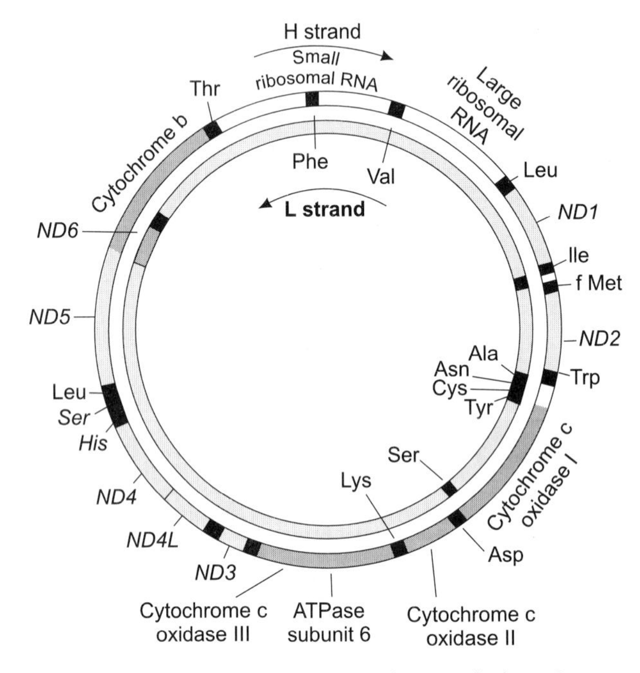

Fig. 31.6 Map of human mtDNA showing the pattern of transcription. Genes on the inner circle are transcribed from the L strand of the DNA, whereas genes on the outer circle are transcribed from the H strand of the DNA. Arrows show the direction of transcription. ND1–6 are genes encoding subunits of the enzyme NADH reductase; the tRNA genes in the mtDNA are indicated by abbreviations for the amino acids.

In fungi, the mtDNA is considerably larger than it is in animals. Yeast, for example, possesses circular mtDNA molecules 78 kb long. These molecules contain at least 33 genes, including two that encode ribosomal RNAs, 23 to 25 that encode transfer RNAs, one that encodes a ribosomal protein, and seven that encode different polypeptides involved in oxidative phosphorylation. The yeast mtDNA is larger than animal mtDNA because several of its genes contain introns and there are long noncoding sequences between plant mitochondrial transcripts into their constituent parts and also removes the introns, which are present in several plant mitochondrial genes. At present, the mechanisms of these events are poorly understood.

Another peculiarity of plant mitochondrial gene expression is that many of the mtRNA transcripts undergo editing; that is, some nucleotides are changed after the transcript has been synthesised. The most frequent change is C to U, but occasionally, U is changed to C. Thus, RNA editing alters the composition of codons in plant mitochondrial transcripts, including some that would otherwise signal the end of polypeptide synthesis. Editing corrects the information that is actually encoded in the mtDNA and allows functional polypeptides to be synthesised. Curiously, RNA editing is not found in the nonvascular plants (mosses and algae), even though all groups of higher plants (ferns, gymnosperms and angiosperms) seem to have it. Thus, the editing mechanism probably evolved sometime after plants had become established on the land. RNA editing also occurs in the mitochondria of protozoans, including the trypanosomes, where the mechanism has been studied in some detail. In these organisms, small RNA molecules that are partially complementary to the mtDNA transcripts serve as guides for the editing process. They are, therefore, called guide RNAs (gRNAs). A similar guiding mechanism may operate in plants, but the details are currently unknown.

Yet a third peculiarity of plant mitochondrial gene expression is that some mitochondrial messenger RNAs are formed by the process of ***trans*-splicing**. *Trans*-splicing occurs when segments of a gene are scattered over the mtDNA molecule. Each gene segment is transcribed independently, and the exons of the different transcripts are spliced together by interactions between the introns that flank them. For example, in wheat, the *nad1* gene, which encodes a subunit of NADH reductase, a protein of oxidative phosphorylation, is partitioned into four segments in the mtDNA. Each of these segments is separately transcribed, and the resulting transcripts are then spliced together to form the mRNA (Fig. 31.7). This process requires a single *cis*-splicing reaction and three *trans*-splicing reactions.

Interplay Between Mitochondrial and Nuclear Gene Products

Most—perhaps all—mitochondrial gene products function solely within the mitochondrion. However, they do not function alone. Many nuclear gene products are imported to augment or facilitate their function. Mitochondrial ribosomes, for example, are constructed with ribosomal RNA transcribed from mitochondrial gene and with ribosomal proteins encoded by nuclear genes. The ribosomal proteins are synthesised in the cytosol and imported into the mitochondria for assembly into ribosomes.

Many of the polypeptides needed for aerobic metabolism are also synthesised in the cytosol. These include subunits of several proteins involved in oxidative phosphorylation, for example, the ATPase that is responsible for binding the energy of aerobic metabolism into ATP. However, because some of the subunits of this protein are synthesised in the mitochondria, the complete protein is actually a mixture of nuclear and mitochondrial gene products. This dual composition suggests that the nuclear and mitochondrial genetic systems are coordinated in

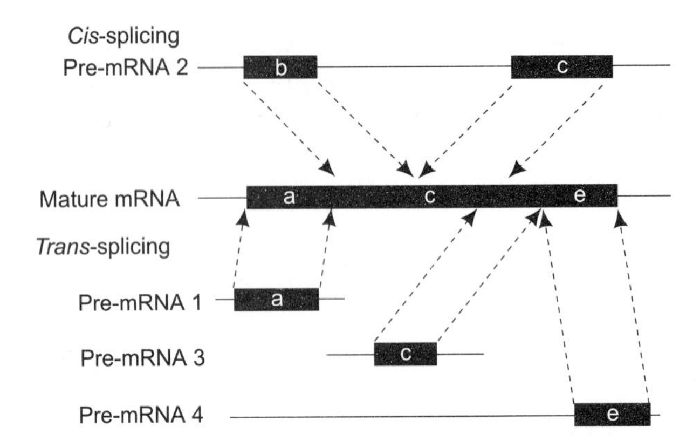

Fig. 31.7 *Trans*-splicing in wheat mitochondria. Four different RNAs contribute to the final mRNA encoding a polypeptide of the enzyme NADH reductase.

some way so that equivalent amounts of their products are made; possible molecular mechanism for this coordination are currently under investigation.

Key Points: Mitochondrial DNA (mtDNA) molecules range from 16 kb to 2500 kb in size, and most appear to be circular. These molecules contain genes for some of the ribosomal RNAs, transfer RNAs, and polypeptides used within the mitochondrion. The structure, organisation, and expression of these genes vary among species. In some groups of organisms, the transcripts of mitochondrial genes are edited after they are synthesised. Both mitochondrial and nuclear gene products are needed for proper mitochondrial function.

CHLOROPLAST DNA

In higher plants, cpDNAs typically range from 120 to 160 kb in size, and in algae, from 85 to 292 kb. In a few species of green algae in the genus *Acetabularia*, the cpDNA seems to be organised as a covalently closed circular molecule. However, in some species, especially those with large cpDNAs, a linear arrangement cannot be ruled out.

The number of cpDNA molecules in cell depends on two factors: the number of chloroplasts and the number of cpDNA molecules within each chloroplast. For example, in the unicellular alga *Chlamydomonas reinhardtii*, there is only one chloroplast per cell, and it contains about 100 copies of the cpDNA. In *Euglena gracilis*, another unicellular organism, there are about 15 chloroplasts per cell, and each contains about 40 copies of the cpDNA.

All cpDNA molecules carry basically the same set of genes, but in different species of plants these genes are arranged in different ways. The basic gene set includes genes for ribosomal RNAs, transfer RNAs, some ribosomal proteins, various polypeptide components of the photosystems that are involved in capturing solar energy, the catalytically active subunit of the enzyme ribulose 1,5-bisphosphate carboxylase, and four subunits of a chloroplast-specific RNA polymerase. A few cpDNA molecules have been sequenced in their entirety, including one from

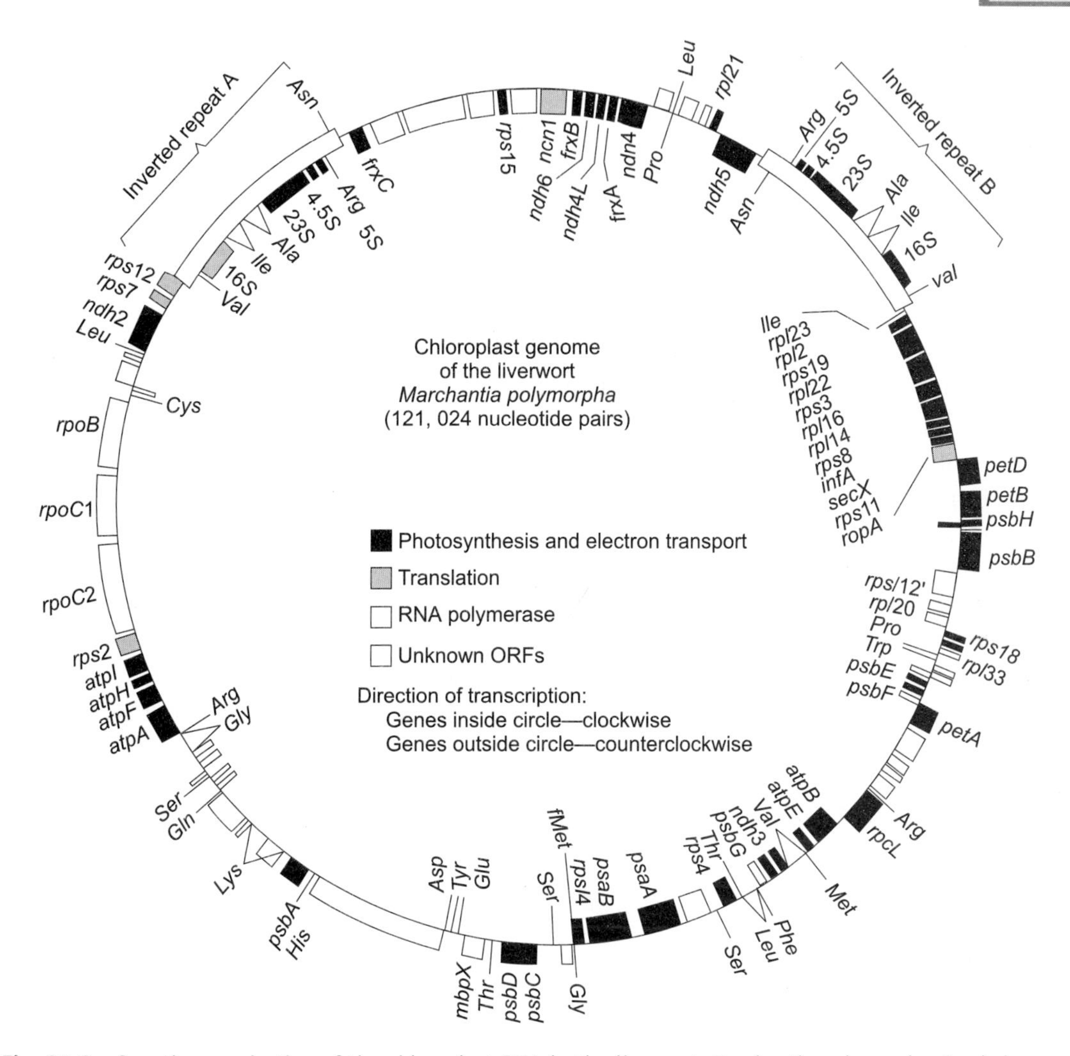

Fig. 31.8 Genetic organisation of the chloroplast DNA in the liverwort *Marchantia polymorpha*. Symbols: rpo, RNA polymerase; rps, ribosomal proteins of small subunit; rpl and secX, ribosomal proteins of large subunit; 4.5S, 5S, 16S, 23S, rRNAs of the indicated size; rbs, ribulose bisphosphate carboxylase; psa, photosystem I; psb, photosystem II; pet, cytochrome b/f complex; atp, ATP synthesis; infA, initiation factor A; frx, iron-sulfur proteins; ndh, putative NADH reductase; mpb, chloroplast permease (?); tRNA genes are indicated by abbreviations for the amino acids.

the liverwort, *Marchantia polymorpha* (Fig. 31.8), and another from the tobacco plant, *Nicotiana tobacum*. The tobacco cpDNA is larger (155,844 bp) and probably contains about 150 genes. The best estimate for the gene number in the liverwort cpDNA (121, 024 bp) is 136. Most cpDNA have a pair of large inverted repeats that contain the genes for the ribosomal RNAs. These repeats range anywhere from 10 to 76 kb in length and are variously located in different cpDNA molecules.

Chloroplast Biogenesis

As mentioned earlier, all plastids develop from pro-plastids. In the case of chloroplasts, this development is stimulated by light and involves the transcription of many genes, including some located in the nucleus. Illumination triggers a profound change in each proplastid. The organelle increase in size, and the inner of its two surrounding membranes expands to bud off vesicles that eventually arrange themselves in stacks forming structures called grana. All the proteins and chlorophyll pigments needed for photosynthesis are made and targeted to their appropriate locations within the emerging chloroplast. Some of these molecules are positioned in the thylakoid membranes that make up the grana, and others are localised in the stroma, which is the protoplasmic space surrounding the grana.

The formation of functional chloroplasts is a process referred to as biogenesis. Only some of the details are known. However, it is clear that light plays an important role. For example, the nuclear gene that encodes the small subunit of ribulose 1,5-bisphosphate carboxylase is vigorously transcribed when light is provided. A special class of pigmented proteins called phytochromes seems to mediate this and other responses to light. By absorbing light energy, these proteins acquire the ability to trigger other proteins to stimulate the transcription of genes involved in chloroplast biogenesis.

The formation of chloroplasts and the maintenance of their structure and function during the life of a plant depend on the coordinated expression of nuclear and chloroplast genes. Each genome produces a distinct set of mRNAs, which are translated in the cytosol and the chloroplast, respectively. Chloroplast biogenesis, therefore, involves considerable interplay between nuclear and chloroplast gene products.

CHAPTER 32

Human Genetics

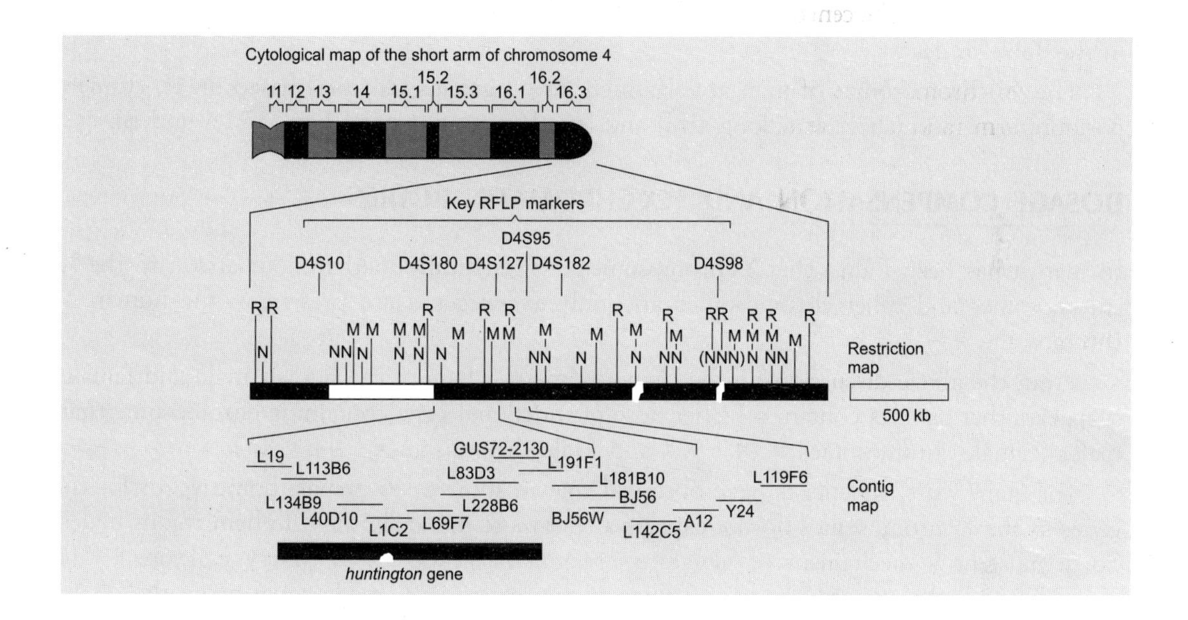

HUMAN CHROMOSOMES AND KARYOTYPE

Introduction

Chromosomes can be identified and classified by different banding techniques. Bands are defined as parts of the chromosomes that appear lighter or darker than adjacent regions when the chromosomes are treated with particular staining methods. When Quinacrine compounds are used as dye to stain chromosomes, then the chromosomes produce fluorescent Q-bands along the chromosomes. Similarly when Giemsa stain is used, chromosomes showed G-band, reverse staining methods produce R-bands and the heterochromatic regions and centromeric regions are stained C banding. The total number (2n) of human chromosome in a somatic cell is 46. Of these 46, 44 are autosomes and 2 are sex chromosomes or sex-determining chromosomes. Sex chromosomes are again of two types, X-chromosomes and Y-chromosomes. Every chromosome has 2 arms and the size of the arms are dependent on the position of centromere on the chromosome; p and q symbolise the two arms of the chromosome in which p represents the short arm and q the long arm. The extra chromosomes apart from the normal numbers, if present in human cells would be placed after the total number and sex chromosomes with a plus (+) or minus (–). For example, if the chromosome number is 47 then it should be written as 44XX or 44XY + extra chromosomes.

All the human chromosomes in normal numbers (46) can be arranged in groups based on size and positions of the centromere and the resulting classification of the chromosomes is given in the Table 32.1.

The 46 chromosomes of man are classified into 7 groups (A to G) based on centromere position, arm ratio (short arm/long arm) and length measurements (Figs. 32.1A and B).

DOSAGE COMPENSATION AND SEX-CHROMATIN BODIES

In man it has been found that Y-chromosomes are genetically inert in comparison to the X-chromosomes and other chromosomes and only a few genes are present in the human Y-chromosome.

From the above discussion on the chromosome numbers of the human male and female, it appears that females contain a higher dose of functional gene containing chromosome than males (Female chromosomes = 44 + XX and Male chromosomes = 44 + XY).

For many years, geneticists have observed that in some cases, female homozygous for the genes in the X-chromosomes do not express a trait more markedly than do hemizygous males. So, it must be a mechanism of "*dosage compensation*" (proposed by Marry F. Lyon, 1961), through which the effective dosage of genes of the two sexes is made equal or nearly so.

This mechanism of compensating the differential doses of functional sex chromosomes in male and female humans is effected by the inactivation of one X-chromosome in the normal female. The genetically inactive X-chromosome or condensed X-chromosome is called heteropycnotic X-chromosome or heterochromatin or sex-chromatin body or Barr body (according to the name of the geneticist M.L. Barr who first observed it) or Drum-stick (according to the shape of the inactive X-chromosome). Of the two X-chromosomes in females, which X-chromosome becomes inactive is a matter of chance: but it should be remembered that once an X-chromosome has become inactive, all cells arising from that cell will keep the same inactive X-chromosome.

In humans, inactive form of X-chromosome as a Barr-body have been observed by the sixteenth day of gestation, X-chromosome inactivation occurs in humans when two or more X-chromosomes are present.

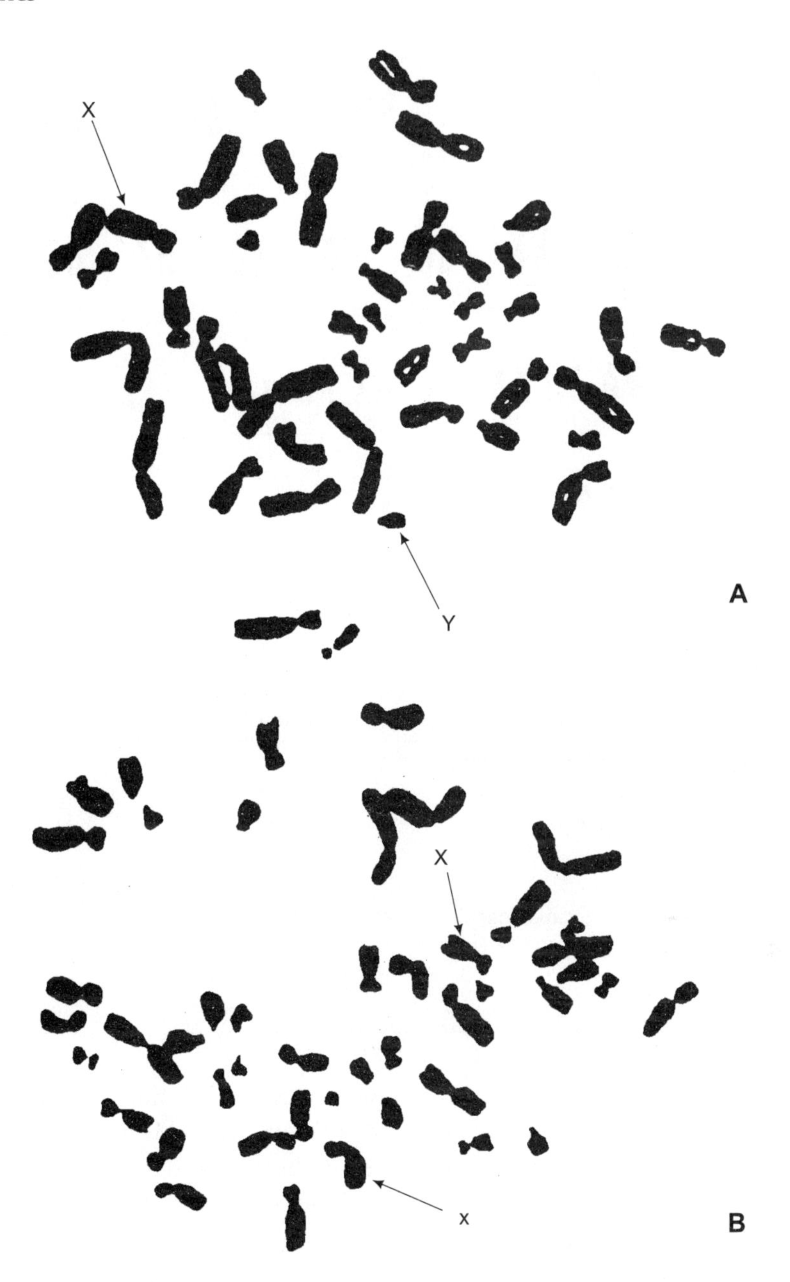

Fig. 32.1A Human karyotype. A: Male (XY); B: Female (XX).

CHROMOSOMAL ABERRATIONS AND VARIATION OF CHROMOSOME NUMBER IN HUMANS

It has been estimated that almost 20% of recognisable pregnancies miscarry and about 40% of early spontaneously aborted foetuses are chromosomally abnormal. About 5–7% of human

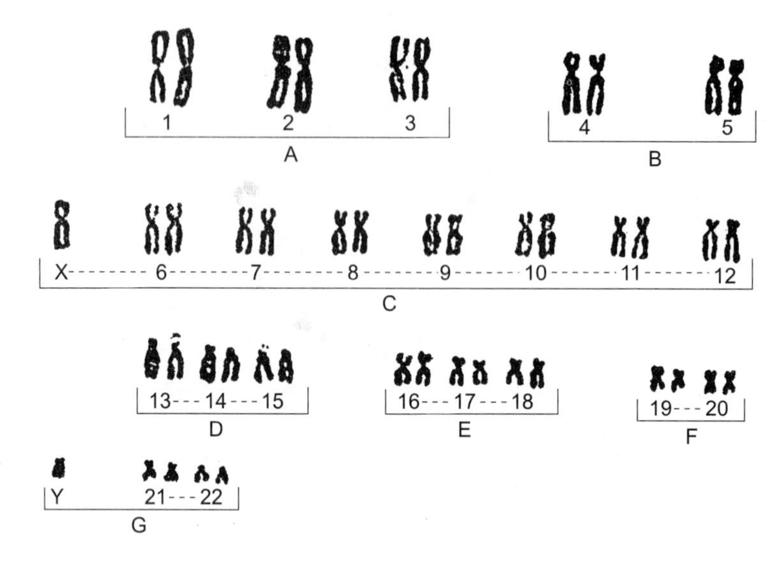

Fig. 32.1B Human karyotype classified from A to G groups.

Table 32.1 Classification of Human Karyotype.

Group	*Size and Centromere Position*	*Idiogram Number*	*Number in Diploid Cell*
A or I	Large; Metacentric/Submetacentric	1–3	6
B or II	Large; Submetacentric	4, 5	4
C or III	Medium; Submetacentric	6–12 and X	15 (male)
			16 (female)
D or IV	Medium; Acrocentric	13–15	6
E or V	Small; Metacentric/Submetacentric	16–18	6
F or VI	Small; Metacentric	19–20	4
G or VII	Smallest; Acrocentric	21, 22 and Y	5 (male)
		4 (female)	

pregnancies, the conceptus carries a chromosomal change that may be directly involved or responsible for abnormal development in some 90% of cases. Most of the developmentally abnormal conceptuses are eliminated as spontaneous abortion and a portion of the rest will die in perinatal period. The rest, i.e., almost 1 in 200 newborn survive as abnormal infants.

Mainly two types of anomalies occur in human karyotype and they are .numerical and structural chromosome aberrations. Aberrations of the human chromosome occur mainly as a result of non-disjunction of homologous chromosomes during one of the two meiotic divisions of gametogenesis.

The possibilities of a chromosomally aberrated child again depend upon the age of the mother, i.e., the older the mother the greater the risk of abnormal child. It is also well established that sometimes phenotypically normal persons are the carriers of balanced chromosomal anomalies and such a condition will result in the formation of imbalanced germ cells in gametogenesis in calculable proportion of all the gametes formed. Therefore, if a woman has two or more abortions, the karyotype of the aborted conceptus should be determined. If an anomaly is found or the karyotype cannot be established, both parents should have their chromosomes analysed, because the parents' chromosomal anomaly (particularly structural) may be responsible for the recurrent spontaneous abortions.

Any anomaly in the normal structure of a chromosome or any variation in the normal karyotype of humans causes an abnormal development of humans or resulting in zygotic loss, still births or infant deaths. The degree of the abnormalities in the human structure will ultimately depend upon the types of anomalies in the human normal karyotype. There are mainly two types of chromosomal anomalies and they are: *structural chromosomal anomalies* and *numerical chromosomal anomalies.*

STRUCTURAL CHROMOSOMAL ANOMALIES

This type of chromosomal anomalies arise mainly as a consequence of the breakage of chromosomes. If any segment of any chromosome is lost due to breakage, then it is known as *deletion,* but if it occurs at one end of any chromosome, it is known as *deficiency.* Similarly, a segment may become detached and then reattached–only the other way round–this is known as *inversion.* A segment may be included twice over and known as *duplication.* Similarly, chromosomes of too different pairs may exchange segments, known as *reciprocal translocation* or *interchange.* Likewise, special type of reciprocal, translocation may occur in between two acrocentric chromosomes (centromere is situated terminally, i.e., at the very near end of the chromosome), which can join in such a way that the two long arms are essentially preserved and this type is known as *centric fusion* or *Robertsonian translocation.* The basic difference between the reciprocal and Robertsonian translocation is that in case of the former, the total number of chromosomes remains unchanged whereas, in case of the latter, involving the possible loss of functionally irrelevant chromosome material and shows the chromosome complement is one short. Table 32.1A shows some of the chromosomally abnormal human beings and their symptomatic features.

CRI-DU-CHAT OR CAT'S CRY SYNDROME (Figs. 32.2 and 32.3D)

1. Introduction

This is due to a chromosomal deficiency. This condition was first described by Lefeune et al. (1963), associated with a deletion of the short arms of a member of the B4-5 group of chromosomes. The chromosome involved is almost certainly one of the member of the B5 pair.

2. Diagnostic Features

Some of the following features are not absolute in occurrence, and in that case, values in the parentheses are the frequency of occurrence.

Table 32.1A Variations in normal chromosome number of human and their phenotypic condition.

Chromosome picture	*Total number of chromosomes*	*Clinical syndrome*	*Estimated frequency at birth*	*Symptoms*
44 + XX or 44 + XY + 21	47	Down	1/700	Short broad hand, short stature, open mouth with large tongue, broad head.
Do + 13	47	Patau	1/20,000	Mental deficiency with deafness, cleft lip or palate, polydactyly, cardiac anomalies.
Do + 18	47	Edward	1/8,000	Malformed ears, small mouth and nose, mental deficiency.
44 + X	45	Turner	1/2,500	Female with retarded sexual development, usually sterile, short stature, hearing impairment.
44 + XXX	47	Triple 'X'	1/700	Female with usually normal genitalia and limited fertility, mental retardation.
44 + XYY	47	Born criminal	1/25,000 (male birth)	Rough attitude, coarse skin with hairs, facial toughness, normal IQ (male).
44 + XYY 44 + XXXY 44 + XXXY 44 + XXXXY 44 + XXXXXY	47 48 48 49 50	Klinefelter	1/500 (male birth)	Male subfertile with small testes, developed breast, feminine pitched voice, long limb, knock knees, rambling talkativeness.

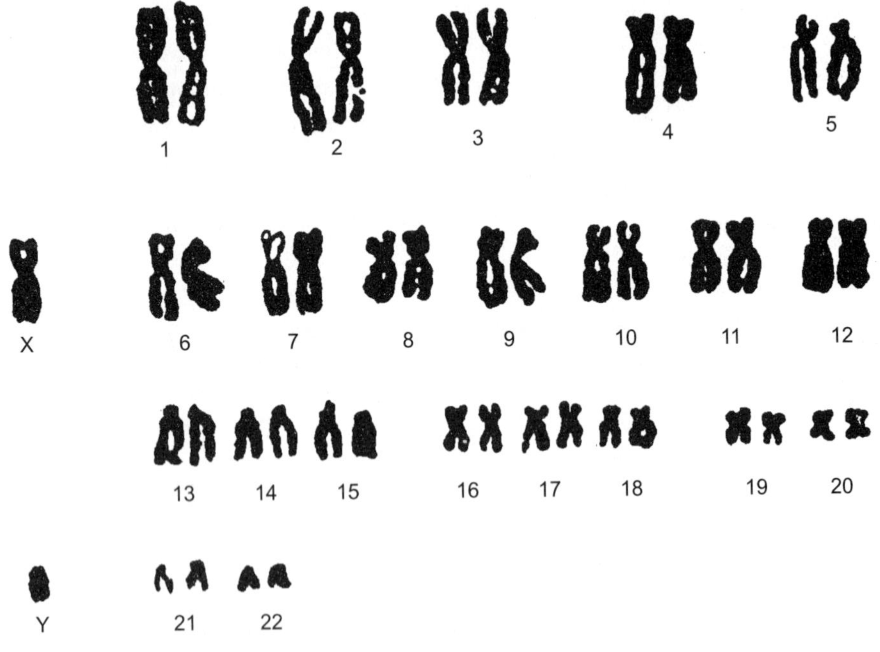

Fig. 32.2 Human karyotype showing deficiency (Cri-du-chat syndrome).

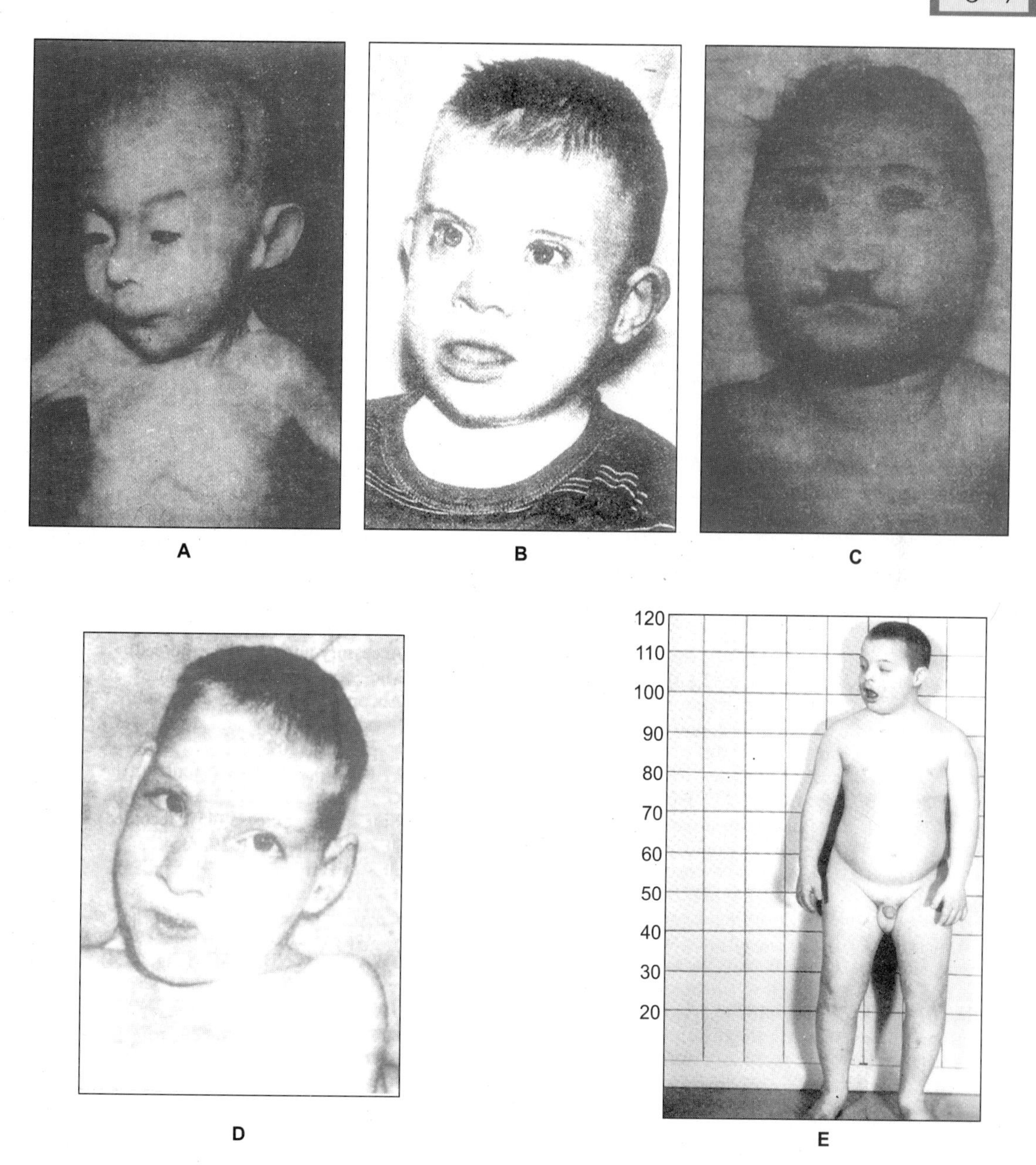

Fig. 32.3 A: Facial features associated with trisomy of No. 18 (47, XX18). Note elfin appearance, small palpebral fissures, small nose and mouth, and receding mandible and abnormal ears; B: Facial features of a child with Down's syndrome. Note slanting eyes, open mouth with large tongue, small protruding ears, and broad head; C: Child with trisomy for a member of the d group of chromosomes (47, XYD?) showing characteristic cleft lip; D: Patient with Cri-du-chat syndrome which is associated with a deletion to part of the short arm of chromosome No. 5 associated with microcephaly and cat-like cry; E: Patient with Cri-du-chat syndrome.

The name of this syndrome refers to the fact that babies exhibiting it have a peculiar face, plaintive viewing with cat like cry.

Mental, motor function and growth retardation.

Broad-headed microcephaly.

Epicanthis folds.

Occular hypertelorisus.

Downward and outwardly sloping palpebral fissures.

Broad face.

Saddle nose.

Micrognathia and outwardly sloping palpebral fissures.

Low set malformed ears.

Antimongoloid slant to eyes (64% recorded).

Rotated ears (11% recorded).

Accessory auricles (9% recorded).

Microcephaly (92% recorded).

Abnormal larynx (32% recorded).

Hypotonia (50% recorded).

Hypertonia (10% recorded).

Failure to urinate (88% recorded).

Oligophrenia (94 or 97% recorded).

Single palmar crease (45% recorded).

Distal + Triradius (62% recorded).

Absent C triradius (70% recorded).

Birth rates are low.

Gestation time is normal.

EDWARD SYNDROME (Fig. 32.4)

1. Introduction

Aneuploidy for the autosome in man often produces both physical and mental defects. One well-known condition of 18-trisomy or E-trisomy is a clinical syndrome and the child generally dies before birth.

2. History

In 1980, Edward et al. and Patau et al., simultaneously discovered a new syndrome associated with the presence of an extra group E chromosome, here considered to be a No. 18 chromosome.

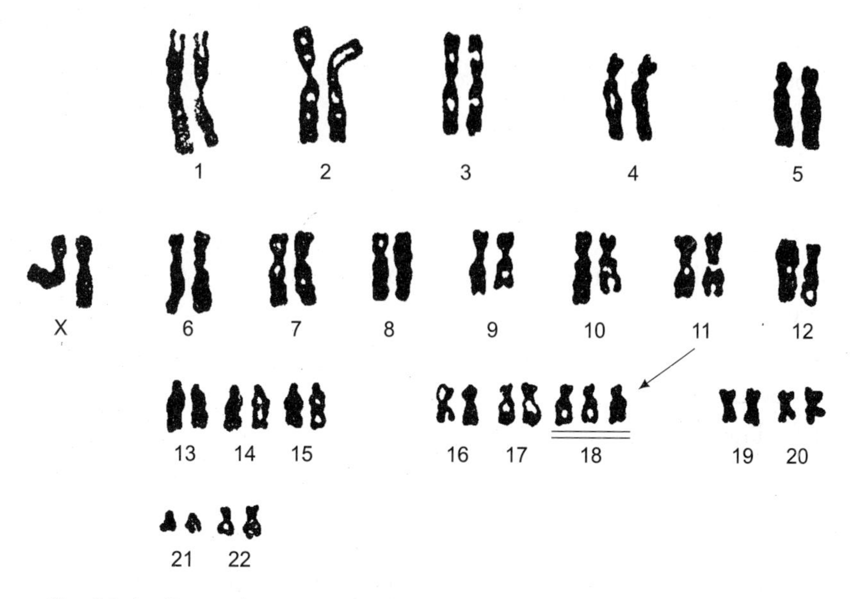

Fig. 32.4 Human karyotype showing Edward syndrome. 18-trisomy (arrow).

3. Origin

Nondisjunction during anaphase of mitotic or meiotic nondisjunction or nondisjunction in zygote lead to the mosaicism.

4. Longevity and Frequency

In 90% of cases, death occurs within the first six months (Weber et al. 1964). But Hook et al. (1965) recorded a mentally retarded female who was living at 15 years of age.

The ratio of affected females to male is 3 : 1 (Ferguson-Smith 1962, Weber 1967). This difference in sex ratio has been attributed to the survival advantages of the genetically more balanced XX females (Weber, 1967).

77% die during 3 months of post-mental condition.

5. Clinical Features

It is a syndrome characterised by multiple congenital malformation in which virtually every organ system is affected.

Birth weight is under six pounds (2.7 kg).

Multiple abnormalities, especially of bones characterised by short sternum, enlarged occiput, micrognathe, laterally compressed cranium with increased fronto-occipital length.

Hypertonicity

External auditory meatus slightly to extensively folded with certain auricular folding low set ears. Clenched fists with second digits overlapping the third or the fifth digit overlapping the fourth. Umbilical cord occasionally with single umbilical artery.

Male cryptorchidism.

Ventricular septal defect and latent ductus arteriosus.

Fingers are held permanently flexed but with the distal joint extended. The fingertips almost always have their dermal patterns in the forms of arches.

Horse shoe/or double kidney.

Evertration of diaphragm.

In a mentally retarded child having an isochromosome of the short arm of chromosome number 18 has recently been reported. (Balieck et al. 1976) Micrognathia.

Prominent occiput, failure to thrive.

Heterotropic pancreatic tissue.

Cytogenetical findings showed the following features:

(a) The additional E-chromosome was detected radio autographically (Yunis et al., 1964) as chromosome number 18 and later Quinacrine Mustard fluorescent microscopy (Caspirson et al. 1971), various modifications of Giemsa banding (G-banding techniques (Summer et al., 1971).

(b) Translocation E-trisomy individuals are known as translocation Mongols.

Remarks: Affected individuals have a short life-span and usually die in early infancy, although occasionally patients survive into early childhood. There is approximately a 3 : 1 female to male sex ratio.

MONGOLISM/DOWN'S SYNDROME/G-TRISOMY/21-TRISOMY (Figs. 32.5 and 32.3B)

The first autosomal abnormality described in man by John Langdon Down (1966) was known as Down's syndrome, more commonly used as mongolism. In this case, there is simple trisomy of chromosome 21, i.e., each cell containing three chromosomes of 21 rather than two. The term "Mongolism" has been applied due to their facial characteristics (round, full face with upper eyelids turned downwards) very similar to that of the Mongolian race. It is impossible to be certain whether the trisomy involves chromosome 21 or 22 since they are morphologically identical. For this, it is sometimes called G-trisomy.

Physical and Physiological Features

Mentally retarded child.

IQ–50.3.

Dull and happy looking.

Less sensitive to external stimuli.

Individuals having very low birth-weight (mean = 2. 83 kg).

Skull is brachycephalic (short from front to back).

A round, full face with epicanthic folds in their eyes. Here upper eyelid covers the lower eyelid. Brush filled sports, i.e., light speckles around the margin of the iris, are present.

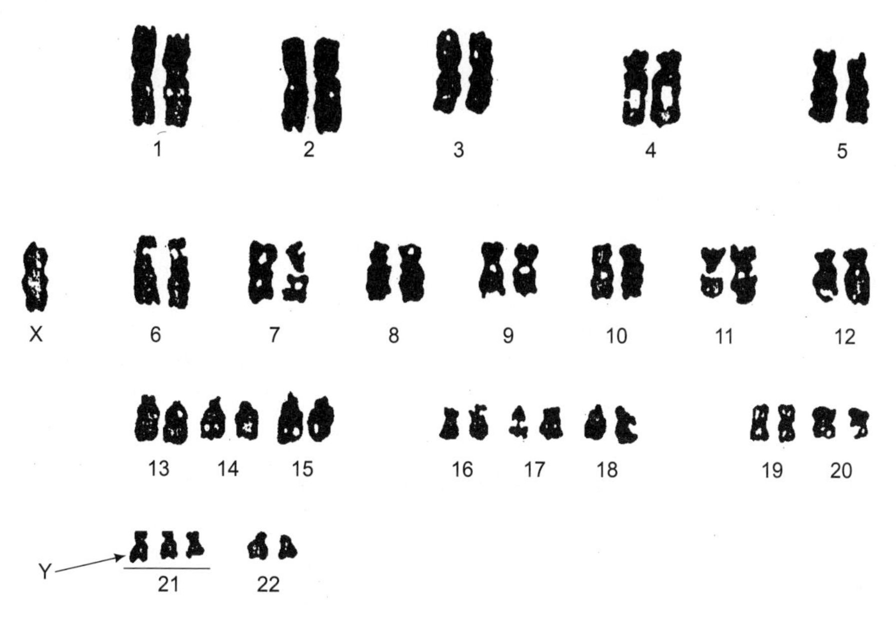

Fig. 32.5 Human karyotype showing Down's syndrome (21-trisomy) (arrow).

Nose from root to tip is short as well as flat and mouth with the lips in the shape of a *Cupid's bow*. Small, rotated ear.

A creased tongue.

Short stature, stubby hands and feet.

On the palm, the two normal more or less diagonal main creases may be replaced by a single transverse crease–"the so-called Simian crease". Dermatoglyphics increased ulnar loops on fingertips (except the index finger and thumb).

Usually loops in the fingertips.

Little finger short, flexion is usually absent; incurved.

Feet normal but with wide gap between the first and second toes. A few babies have serious intestinal malformations, such as duodenal atresia.

Congenital heart defects.

No sexual maturity.

In males, testicular degeneration and infertility seem to be the rule, with usually small, undescended testis.

In females the labia majora tend to be large and cushion-like and the labia minora are small or absent.

Chromosomal Findings in Down's Syndrome

There are four types of karyotypes associated with Down's syndrome. Richards (1967) collected data from 1103 cases. Majority are primary 21 trisomics with no great risk of recurrence, but the other 3 types may all be associated with familial Down's syndrome–in sibs and other relatives if there is an inherited translocation and in the offspring of the patient if he or she is a mosaic.

Incidence

Mongoloid idiocy occurs once in each 500 or 600 births, being more frequent when mothers are older than the average. So, the incidence of mongolism in a mongoloid family depends on the maternal age and not on paternal age.

Age of mother	Frequency of mongoloid child
20 years	1/30,000
40 years	1/40

KLINEFELTER'S SYNDROME (Figs. 32.6 and 32.7B)

The affected person of Klinefelter's syndrome is genotypically and phenotypically male but bears many female characteristics.

1. History

1. In 1942, Klinefelter, Refenstein and Albright found nine men with small testes, aspermia, elevated urinary gonadotrophins and gynecomastics. These persons are later designated as Klinefelter's syndromic.
2. Bradbury et al. (1956) gave that the person is chromatin positive having 2X chromosomes in its Karyotype.
3. Jacob and Strong (1956) described the presence of 2X chromosomes in addition to a Y chromosome in its Karyotype.
4. Court Brown et al. (1964) described that in more than 3/4th cases, Karyotype is 47/XXY.

But Klinefelter's syndrome will unusually be the result whenever more than one X occurs in the presence of a Y.

2. Symptoms

These individuals are normally tall, having sparse body hair, and knock knees; limbs tend to be long, hand and feet larger.

Mental retardation becomes more and more severe in addition with the increase of X chromosome number i.e., mental retardation is proportional to the no. of X chromosomes.

Immature personality leading to inadequacy and lack drive (Neilsen, 1965). Internally, exploratory laparotomy reveals that there is very little testes (2cm in greater diameter), occasionally with "Ghost" tubules. Microrchidism (small-sized testis) occasionally with cryptorchism (testis abdominal).

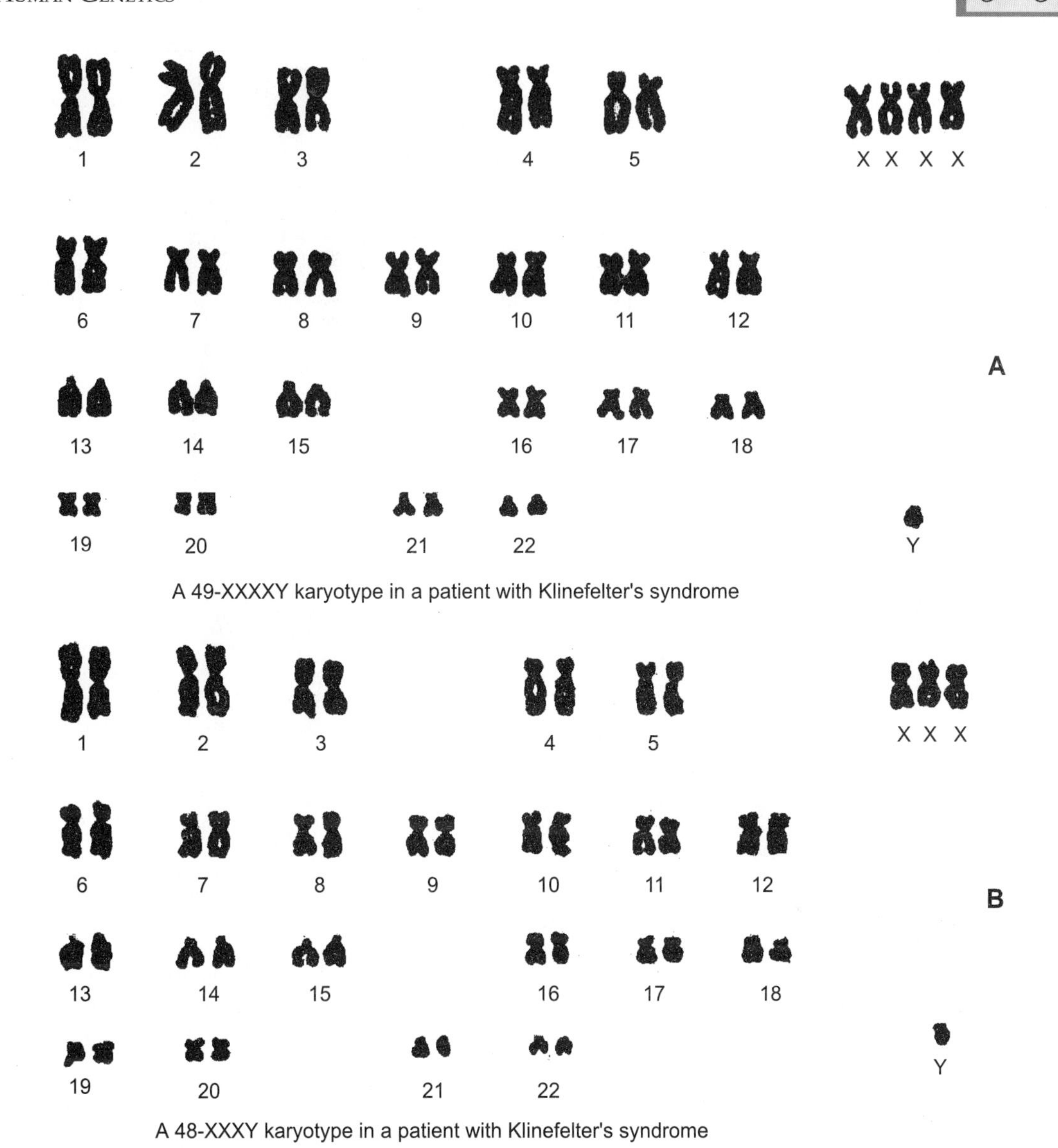

Fig. 32.6 Klinefelter's syndrome in man. A: 49-XXXXY; B: 48-XXXY (Courtesy: Dr S. Makino).

Secondary sexual characters, such as axillary hair, pubic hair, facial hair are little developed.

Complete aspermia, well-developed breasts, broad pelvis, horizontal limit of pubic hair, feminine voice, gynecomastia occur in about 80% of cases.

Depressed thyroid function.

Decreased amount of androgen in blood.

Temporal recession of the hair due to lack of androgen.

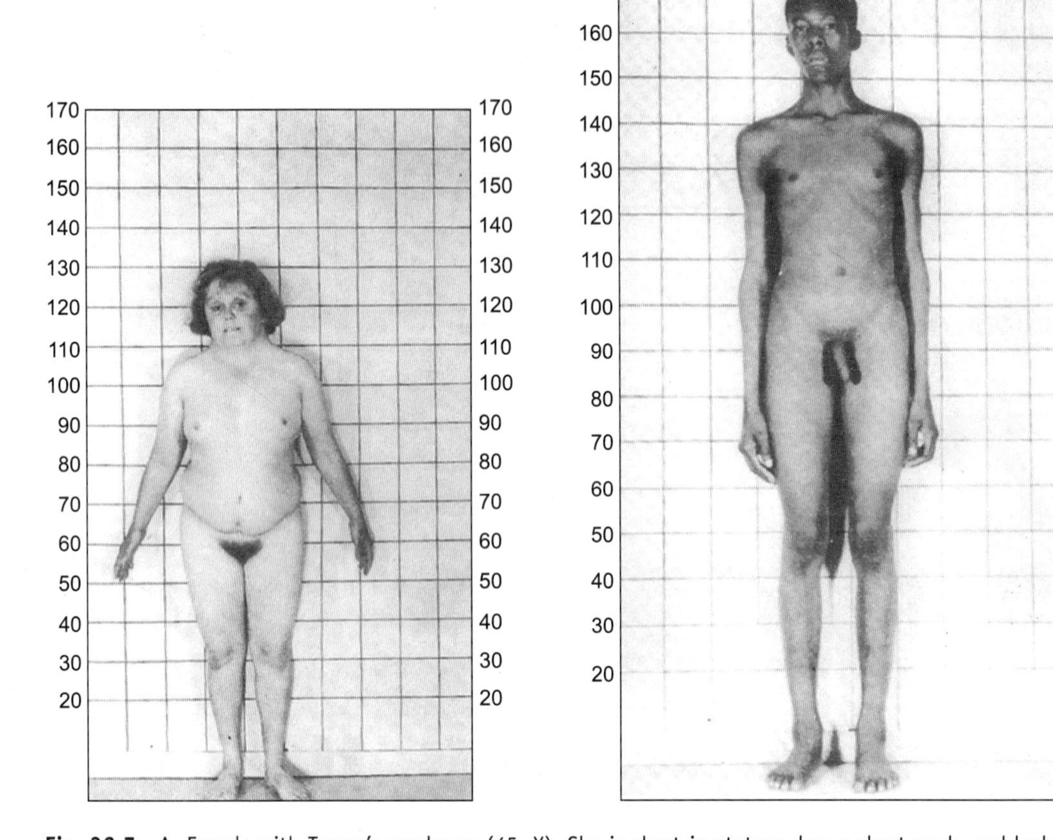

Fig. 32.7 A: Female with Turner's syndrome (45, X). She is short in stature, has a short neck, and lacks most female secondary sex characteristics. The breast development shown here has been induced by administering estrogen. Note the broad, shieldlike chest with widely spaced nipples, mild neck webbing and lowset ears. B: Male with Klinefelter's syndrome. The syndrome includes small testes and very little body hair. Affected males tend to be long-legged and thin. They often develop breasts (gynecomastia) like females and are sterile.

These individuals are sterile. The Klinefelter male is capable of excretion and intercourse with occasional aspermic ejaculation. Behaviourally Klinefelter person tends to be socially inadequate and of limited intelligence and ambition.

Chronic pulmonic condition, such as asthma, bronchiostasis, emphysema, etc., are found.

3. Ascertainment and Phenotype

The phenotype of Klinefelter's syndrome is somewhat variable and ascertainment is often done by X chromatic survey.

Probably Klinefelter individuals are of normal intelligence and leading normal lives except for subfertility problems and a proportion who are 46, XY/47, XXY mosaics may well be fertile. There is, however, a tendency toward lowering of I.Q.

4. Cytogenetic Findings

Sex chromatin, Drumstick, X-linked genes are present, so same phenomenon occurred as in normal females,

1. Majority of patients have a 47/XXY karyotype. A proportion are 46, XY/47, XXY mosaics.
2. A 46/XY karyotype, without apparent mosaicism, is of average intelligence.

5. Frequency

1. In XXY individuals—one within two male births.
2. In XXXY/XXXXY individuals—one within 360 male births.

TURNER'S SYNDROME (Figs. 32.7A and 32.8)

1. Introduction

Further clues to the nature of the sex-determining mechanism in human beings has come from studies of another sex abnormality, Turner's syndrome. In 1938, Turner first described this syndrome in post-pubertal females consisting of sexual infantilism, short stature, webbed neck, and cubitas valgus. These are phenotypically females, but at adolescence the adult characteristics do not develop normally and they never reach functional maturity. Wilkins and Fleischmann (1944) described "Streak" gonads devoid of ovarian follicles in such cases. Pollani et al. (1954) and Wilkins et al. (1954) demonstrated that most cases are chromatin negative whereas Ford et al. (1956) described an XO karyotype in a patient affected with Turner's syndrome.

2. Indications for Suspecting Turner's Syndrome

Infancy, childhood, adolescence, lymphedema, dwarfism, primary amenorrhoea, lax neck skin, retardation, webbed neck, skeletal anomalies and coarctation of aorta.

Also complete manifestation of X-linked traits in girls: e.g., colour blindness, haemophilia; Duchenne's muscular dystrophy, agamma globulinemia, glucose-6P-dehydrogenase deficiency.

3. Clinical Features

1. Lymphedema of dorsum of the hand and feet, and edcess of skin in the neck.
2. Short stature, never more than 5 and phenotypically females.
3. Nails are frequently small, narrow and set deep into the nailpit or are square with increased lateral curvature.
4. Webbed neck and a low posterior hairlines are seen.
5. Broad shield-like chest with widely spaced nipples and underdeveloped breasts.
6. Cubitas valgus—increased carrying angle of the arms is common.
7. Fourth metacarpal and metatarsal bones are short.
8. Congenital cardiac malformation is coarctation of the aorta.

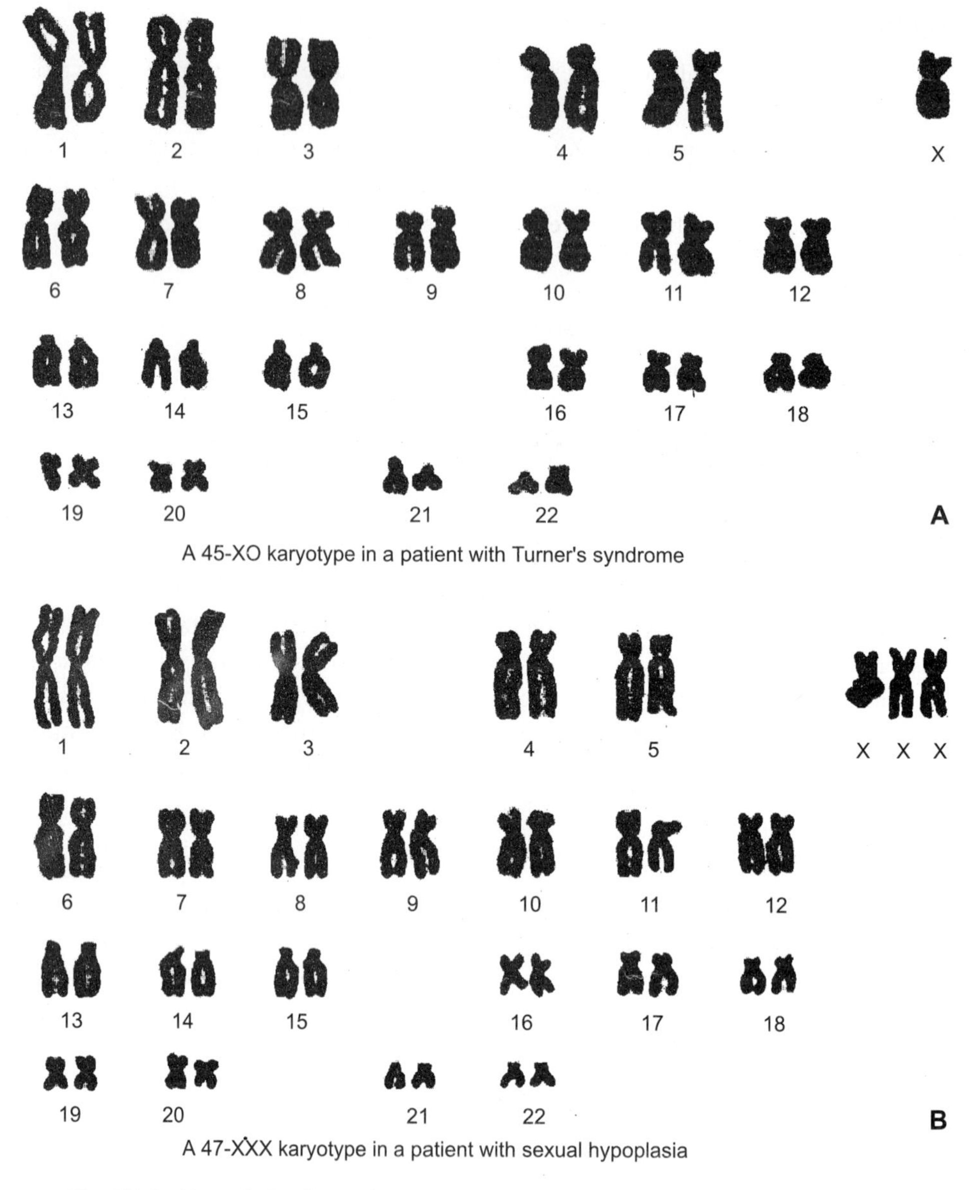

Fig. 32.8 Turner's Syndrome in man. A: 45-XO; B: 47-XXXO (Courtesy Dr. S. Makino).

9. Renal abnormalities such as horse-shoe kidney, double pelvis and ureter or minor rotational malformations may secondarily occur.
10. Intellectual impairment is occasionally present.
11. Sexual infantilism becomes evident at the time of expected puberty.
    (i) Primary amenorhoea.
    (ii) Usually underdeveloped breasts consisting mostly of fat.

(iii) Scanty growth of axillary and pubic hair.
(iv) External genitalia are infantile.
(v) Uterus may be represented by a hairstring in the middleline.
(vi) Dysgerminomas and gonadoblastomas are often found.
(vii) Patients are infertile generally.

12. Colour blindness is more common in patients than in normal woman.
13. Incidence of diabetes mellitus and congenital pyloric stenosis is higher than normal population.
14. Premature closure of epiphyses may occur.
15. Small uterus, ovary represented by fibrous streaks. Primordial follicles usually absent. Excretion of gonadotrophins is elevated after puberty.

4. Origin

The most common chromosomal finding in Turner's syndrome is 44 autosomes and only one X chromosome. These patients are chromatin negative. They have no barr body.

'XO'-constitution is caused by nondisjunction more commonly in spermatogenesis than in oogenesis.

On rare occasions, chromatin-negative patients with gonadal dysgenesis have a male sex chromosome constitution (46/XY). Another rare group of chromatin-negative patients have an XO/XY sex chromatin constitution. It has been described in women with streak gonads, in men with undescended testes and in intersexes.

5. Chromosome Patterns in Turner's Syndrome

(a) X-monosomy

Patients with an XO sex complement make the largest proportion of cases with Turner's Syndrome (Lindsten, 1963). The mechanism of formation of XO sex constitution is speculative. Use of colour blindness and Xga blood type as markers has shown that most of the cases which could be analysed has an X of maternal origin. Nevertheless, since mating that would reveal a paternally derived X chromosome are rare, the data are inconclusive (Lindsten, 1963).

(b) XO/XX mosaicism

Many patients with Turner's syndrome have two different cell populations, one has XO sex constitution and the other a normal XX sex complement. Height sometimes normal, spontaneous menstruation, lymphedema at birth, webbing of neck less frequent; sometimes normal ova are found in XO/XX patients. In XO/XX/XXX, buccal smears reveal nuclei with no barr body or with one or two. In XO/XXX, two or no barr bodies are seen.

(c) Isochromosome X

Sometimes a large metacentric chromosome resembling an extra No. 3 chromosome is found and one of the normal X-chromosomes lacks. The extra chromosome is assumed to be an isochromosome for the long arm of the "missing" X-chromosome. Lymphedema at birth, webbing of the neck and congenital heart disease are less frequent. Barr bodies are larger and more numerous than normal.

(d) Partial Loss of an X Chromosome

Partial deletion of the short arm of X chromosome and of the long arm has been reported in a few patients with Turner's Syndrome.

(e) Ring X Chromosome

This chromosome seems to be found by simple deletion of parts of both the short and long arms of the X chromosome with a subsequent fusion of chromatid ends.

(f) Minute Sex Chromosome

Probably deleted X and Y chromosomes. These are seen in association with gonadal dysgenesis.

XYY-MALES AND CRIMINAL BEHAVIOUR (Fig. 32.9)

The incidence of this type of abnormality (with XYY) in the new born males is about 0.2 per cent. It has been shown that XYY males are reproductively functional and more or less normal in outlook (in comparison to XY male). XYY males are different from the XY males in the following ways:

1. XYY males are more criminal in behaviour than XY males and XYY males are mainly involved in the crimes against property.
2. Mean age of the first criminal conviction is 13.1 yr in case of XYY males, whereas it is about 18 yr for XY males.
3. XYY males generally do not commit any crime against their siblings.
4. XYY males poorly respond to the corrective measures taken against their crime.

It has been suggested that the antisocial behaviour of XYY males is due to extra Y-chromosomes and the extra Y-chromosome have also some effect on the growth of the body.

AUTOSOME-LINKED DOMINANT GENETIC DISEASES

(a) The trait is passed on via one of the parents approximately 1/2 to their children. Thus a 50% inheritance probability for every child of a trait-bearer.

(b) The inheritance is totally independent of the sex of the bearer or the heir. Both sexes are equally affected and the defective gene may come from either father or mother.

(c) It also frequently happens with the diseases of the autosomal dominant variety that an individual inherits the defective gene and passes it on, while lie himself remains phenotypically unaffected. This phenomenon is known as *incomplete penetrance.*

(d) Though the diseases of autosomal-dominant mode of inheritance follow generation after generation, irrespective of sex, it may happen that physiological or other reasons make a disease far more dangerous for one sex than for the other. This is known as *sex limited effect of the gene.*

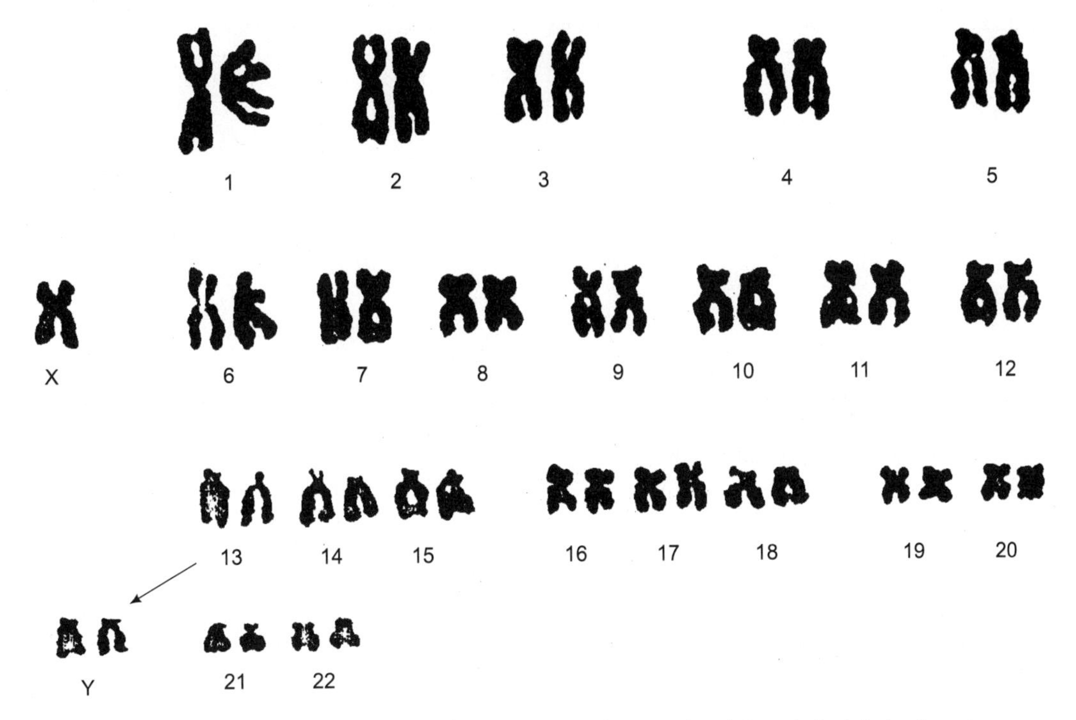

Fig. 32.9 Human karyotype showing XYY syndrome related to criminal behaviour.

SICKLE CELL ANAEMIA

Introduction

Sickle cell anaemia is an inherited abnormal disease caused by mutation of autosomal gene. The red blood cells of certain individuals have peculiar property of undergoing reversible alterations in shape when subjected to changes in the partial pressure of oxygen (Herrick, 1910; Hahn and Gillegpie, 1927). Sometimes RBC are sickle-shaped instead of being flat discs and are very inefficient oxygen carriers. The people with such red cells are called sickle cell trait but they are apparently healthy. The result is a progressive haemolytic anaemia, usually resulting in early death. This condition is caused by a mutant gene Hb_1^s.

SYMPTOMS OF SICKLE CELL ANAEMIA

The sickling leads to increased fragility of the red blood cell causing haemolytic anaemia and increased viscosity of the blood, causing the red cells to stagnate in small blood vessels and consequently form thrombi and infarcts. Regular phenomenon in sickle cell anaemia is autospleenectomy guided by repeated spleenic suspectibility of these patients to pneumococcal meningitis and other infections.

Table 32.2 Relatively frequent autosomal dominant and recessive genetic diseases in humans.

Name of the disease	*Genes located on chromosome number*
Dominant	
1. Familial hypercholesterolemia	19
2. Haemorrhagic telangietasis	(Not identified)
3. Marfan's Syndrome	15
4. Spherocytosis	11
5. Adult polycystic kidney disease	—
6. Huntington's	4
7. Acute porphyria	11
8. Osteogenesis imperfectarda	7
9. Von Willebrand's disease	12
10. Myotonic dystrophy	19
11. Idiopathic hypertropic subaortic stenosis	(Not identified)
12. Noonan's Syndrome	5
13. Neurofibromatosis	17
14. Tuberous sclerosis	—
Recessive	
1. Deafness	—
2. Albinism	11
3. Wo; spm's disease	13
4. Sickle cell anaemia	11
5. 3-Thalassemia	11
6. Cystic fibrosis	7
7. Phenylketonurea	12
8. Homocystinuria	21
9. Mediterranean fever	—

MOLECULAR BASIS OF SICKLE ANAEMIA

The molecular basis of the sickle cell character began to be understood when it was discovered that a special kind of haemoglobin (S) different from that of normal individual (A) is present in the blood cells of persons with the sickle trait or the sickle cell anaemia. The discovery that there are different molecular forms of haemoglobin stimulated research to determine the amino acid sequences of the different forms.

This work was greatly aided by a technique described by Ingram in 1956 and called "finger printing". When the beta chains of normal human haemoglobin was broken down piece by piece, the amino acid in the peptide chain designated as part 4 were found to be arranged in the following order:

Normal: Valine-histidine-leucine-threonine-proline-***glutamic acid***-glutamine-lysine.

Haemoglobin S of sickle cell patients were found to have all eight amino acids in same order except for number six glutamic acid which is replaced by *valine*.

Sickle cell: Valine-histidine-leucine-threonine-proline-***valine***-glutamine-lysine.

EXPLANATION OF SICKLE CELL ANAEMIA FROM GENETIC CODE

Data from studies of cell-free systems indicate that one of the DNA triplets that codes for glutamic acid is CTC. This triplet would code for the complimentary GAG in the mRNA transcript substitution of anA for a *T*(CAC) in DNA would result in GUG as in the following table and mRNA triplet codes valine. Substitution of a *T* for a C (TTC) would produce AAG, an mRNA triplet that yields lysine.

A single base change in the DNA sequence specifying an alternation of one amino acid at a specific point in a particular, protein molecule could account for the disease sickle cell anaemia.

Hb_1^A		Hb_1^S		Hb_1^C	
DNA	*mRNA*	*DNA*	*mRNA*	*DNA*	*mRNA*
C	G	C	G	T	A
T	A	A	U	T	A
C	G	C	G	C	G

ALKAPTONURIA (BLACK URINE DISEASE) (Fig. 32.10)

Alkaptonuria is a useful model for discussion of inborn errors of metabolism discovered by Archibald Garrod in 1909.

One of the end points of phenylalanine and tyrosine metabolism is the breakdown of homogentisic acid to CO_2 and H_2O. This reaction is accomplished under the influence of an enzyme that is present in the liver, named homogentisic acid oxidase. When the enzyme is defective, large amounts of homogentisic acid are excreted in the urine, which turns black upon exposure to the air. In addition, quantities of homogentisic acid accumulate in the body and become attached to the collagen of cartilage and other connective tissues. The metabolic pathways of phenylalanine and tyrosine are shown in Fig. 32.10.

Clinical Symptoms

The cartilage of the ears and the sclera, which is collagenous in nature are stained black. These manifestations are called ochronosis. In the joints such as those of the spine, the accumulation leads to arthritis. When alkaptonrics are fed increased quantities of phenylalanine or tyrosine, there is corresponding increased amount of homogentisic acid excreted in their urine.

Garrod's Interpretation

Garrod interpreted alkaptonuria as being caused by the congenital deficiency of a particular enzyme due to the presence in double dose of an abnormal Mendelian factor or gene. An important implication of the idea was that the normal allele of this gene must in some way be necessary for the formation of the enzyme in the normal organism. This was the first clue to the now well-established generalisation that genes exert their effects in the organism by directing

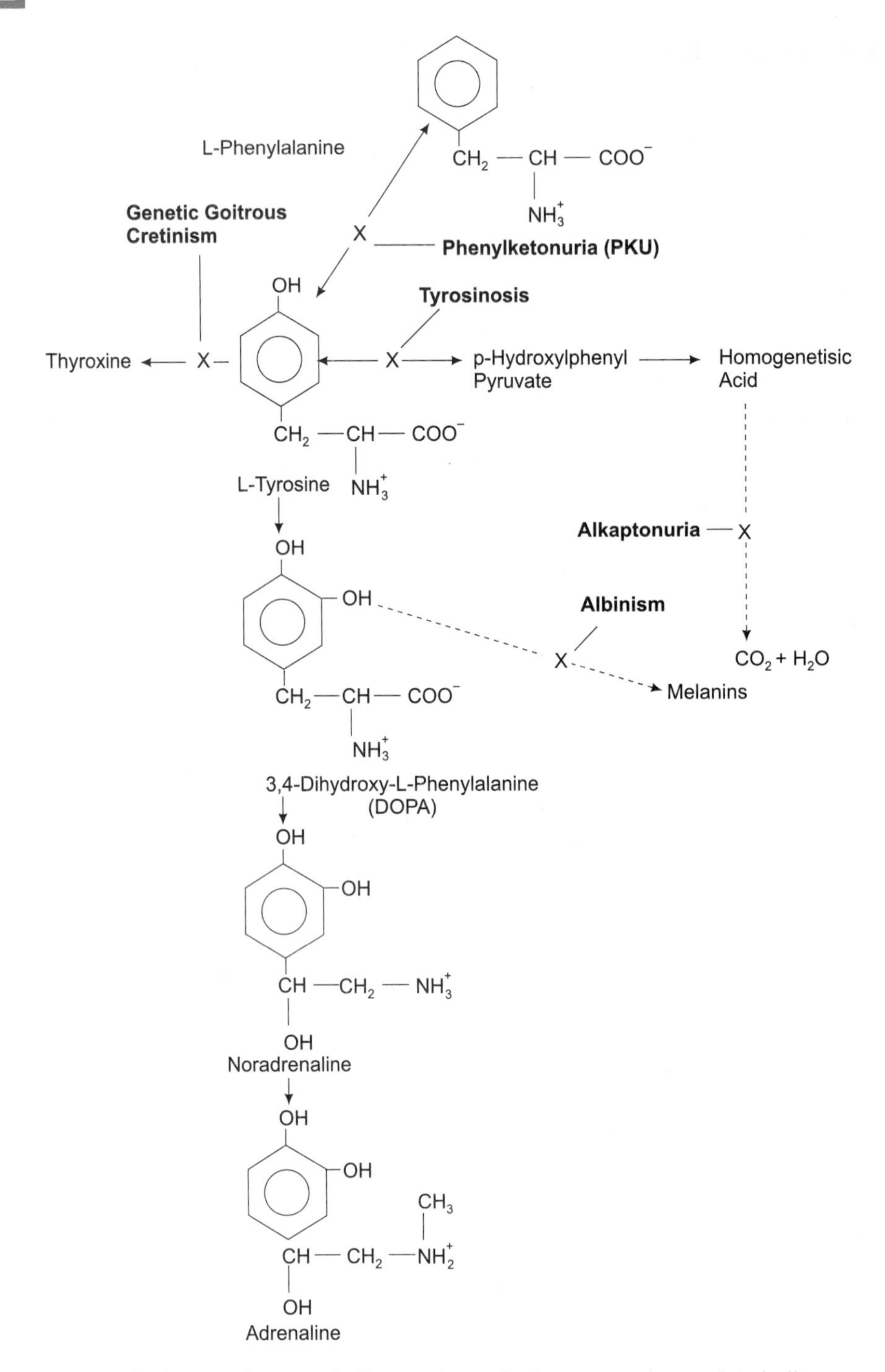

Fig. 32.10 Part of the tyrosine metabolism pathway in humans and associated diseases caused by homozygous recessive conditions. Broken arrow indicates that there is more than one step in the pathway.

the synthesis of enzymes and other proteins from which the concept of one-gene-one enzyme theory was developed later.

Inheritance Pattern

Alkaptonuria is inherited as an autosomal recessive trait and is an example of a genetic enzyme block in which the phenotypic features are due to accumulation of a substance just proximal to the block.

PHENYLKETONURIA (Fig. 32.10)

A mutation in the gene that determines a given enzyme may produce a disorder of the type, Garrod called, "inborn errors of metabolism". Much has been learned about the genetic control of enzymes and about intermediary metabolism by a study of mutant forms in the human species as well as in the micro-organism.

Phenylketonuria (PKU) is a such genetic defect in aromatic acid metabolism. The defect is in the enzyme involved in the conversion of phenylalanine to tyrosine. The deficient enzyme is phenylalanine-4 hydroxylase which in the normal individual occurs in the liver and catalyses para-hydroxylation of the amino acid phenylalanine to give tyrosine. The gene involved is an autosomal recessive and in homozygotes, the phenylalanine accumulates and is converted to phenylpyruvic acid. Phenylalanine is continuously being produced from the normal breakdown of tissue protein, and from the digestion of dietary protein. Its conversion to tyrosine in the liver is the first step in its catabolism and if this is blocked, it accumulates intracellularly and appears in high concentrations in the body fluids. The level of phenylalanine in blood serum in phenylketonuria is generally more than thirty times normal and there is increased excretion of the amino acid in urine.

Some of the Phenylpyruvic acid become concentrated in the cerebrospinal fluid while the rest is excreted in the urine. Excess accumulation of phenylalanine causes the production of several products such as phenylpyruvic acid, Phenyl lactic acid, Phenyl acetic acid in addition to phenyl acetyl glutamic acid, Phenyl-hydroxyphenyl acetic acid and several other products.

Phenylalanine ↔ Phenylpyruvic acid → Phenyl lactic acid → Phenyl acetyl glutamic acid.

These chemicals accumulate in blood due to low renal threshold excretion in urine. In plasma of normal individual phenylalanine level is one milligram per cent which increases to 20–60 mg per cent in PKU's but when detected early, this could be improved.

Frequency of Individuals with PKU

1 in 15,000 child birth.

Disease Symptoms

PKU individuals are also known as phenylpyruvic idiots and have very little IQ ($<$ 20) in 2/3rd and moderate IQ (between 21 and 50) in 1/3rd cases with severe mental retardation, certain physiological and physical disabilities. PKU individuals are feeble minded and have light pigmentation. The feeble mindedness is thought to be due to an impairment of the brain tissue by the phenylpyruvic acid in the cerebrospinal fluid. The light pigmentation is due to a decreased formation of tyrosine which is one of the precursors of melanin.

Genetic Aspects

PKU is an autosomal recessive condition. Patients, therefore, inherit an abnormal gene from each parent. Parents who are heterozygotes for the diseases although having a higher testing blood level of Phenylalanine, are clinically normal. Two heterozygotes have a 25% chance of having one normal child, a 50% chance of carriers and a 25% of children with PKU.

Advice for PKU'S Individuals

Removal of phenylalanine from diet and supplement of all the products of phenylalanine, so that the normal metabolism remains unaltered. Some also used activated charcoal for the phenylalanine removal. Lofenlac is the synthetic diet made for these babies. High proteins, e.g., fish, meat, eggs must be eliminated and more of fruits and vegetables should be introduced although little protein may interpretably be administered. All children necessitating early diagnosis and therapy will be treated by use of artificial enzymes.

SEX-LINKED GENETIC DISEASES AND THEIR MODE OF INHERITANCE

The known number of sex-linked mutations so far are nearly 170 and the most frequent combinations of the X-linked recessive mode of inheritance are as follows:

1. The mother is homozygous and normal (XX); the father is hemizygous and affected (X′Y). (Here X′ indicates the chromosome carrying the defective gene and X indicates the normal of chromosome). All the sons of this union will be normal and they will inherit the normal gene by way of the maternal X-chromosome. All daughters, however, are heterozygous (X′X); the defective gene is located on the paternal X′ chromosome. Half the sons of these daughters will inherit the defective gene.
2. The mother is a heterozygous carrier (X′X) herself phenotypically normal. The father is normal (XY). In this case, half of the sons will be affected (X′Y), whereas all the daughters will be normal. However, half the daughters will be heterozygous (X′X) carriers or conductors.
3. If an affected homozygous woman marries a normal man, all the sons will be affected, whereas all the daughters will be phenotypically normal and are carriers of the disease.

Therefore, the X-chromosomal recessive mode of inheritance is characterised by the fact that especially with rare diseases almost men are affected. The path of inheritance, however, runs only via the normal daughters of sick fathers and the half-normal sisters of sick men. This situation changes only if the brothers' anomaly can be traced to a new mutation, in which case the sisters will not be the conductors. The worst situation is that all the daughters of affected fathers will be the conductors.

The second sex-linked mode of inheritance is the X-chromosomal dominant one. The mode differs from the X-chromosomal recessive mode in that not only the hemizygotes (male) but also the heterozygotes (females) manifest the anomaly. Both men and women are affected evenly when the anomaly is a rare one. All the sons of the affected fathers are unaffected but all the daughters are trait-bearers as well as half of the sisters of the father. Among the children of the female patients there is a 1 : 1 segregation as in case of autosomal dominant mode of inheritance, i.e., independent of sex. So, in brief, male patients must have inherited the trait from the mother; among their siblings there is a 1 : 1 segregation independent of sex. Female patients may have inherited the defective gene from either father or mother.

So, the above facts alone serve to demonstrate that the insufficient data may make it terribly difficult to differentiate between the X-chromosomal dominant and the autosomal dominant modes of inheritance.

Examples (Table 32.3)

1. X-linked recessive diseases: Haemophilia A and B, colour blindness, Duchenne muscular dystrophy.
2. X-linked dominant diseases: Vitamin D resistant rickets with hypophosphatemia, some types of ectodermal anidrotic dysplasia, genetic defects of the enzyme glucose-6-phosphate dehydrogenase, or skin disease with additional symptoms such as missing teeth, oro-facio-digital syndrome (like cleft palate with other cleavages in the oral area), syndactyly, the defective enamel of the teeth, and one blood group Xg.

Y-LINKED GENES

The Y-chromosome has no corresponding locus in the X-chromosome and the mode of transmission of a Y-linked gene is very simple. The female has not Y-chromosome, so women cannot exhibit the trait. The normal male has only one Y-chromosome and so the gene is necessarily unpaired and, if present, it must be expressed. So, the question of dominant or recessiveness cannot arise. The gene simply follows the path of the chromosome and so is handed on by the affected male to all his sons.

The best known example of Y-linked human in gene is the growth of hair on the outer rim of the ear (*trichosis*) and *porcupine man*. The gene, present in Y-chromosome is also known as Holandric gene (Fig. 32.11).

A histocompatibility gene (H-Y), however, has recently been mapped on the short arm of the human Y-chromosome. However, one conclusion is that very few genes are located on the human Y-chromosome and most part of the Y-chromosome is permanently heterochromatic.

IS BALDNESS AN EXAMPLE OF SEX-LINKED GENES?

No, baldness is not an example of sex-linked genes. But it is an example of *sex-influenced genes.* It may occur in some cases, due to disease or other environmental factors, but is of generally

Table 32.3 Human sex-linked diseases.

Human Sex-Linked Traits

- Addison's disease (adrenal insufficiency) (one form)
- Agammaglobulinemia (immunoglobulin insufficiency)
- Albinism (various forms)
- Albright's osteodystrophy
- Aldrich syndrome (immunological disorder)
- Amelogenesis imperifecta (tooth malstructure)
- Anaemia, hypochronic
- Angiokeratoma (Fabry's disease) (kidney disfunction)
- Cerebellar ataxia
- Cerebral sclerosis, Scholz type
- Charcot-Marie-Tooth peroneal atrophy
- Choroideremia (progressive blindness)
- Cleft palate
- Colour blindness (complete and several partial types)
- Deafness
- Diabetes insipidusa (several types)
- Dyskeratosis
- Ectodermal dysplasia
- Ehlers-Danlos syndrome (bruising tendency)
- Endocardial fibroelastosis
- Faciogenital dysplasia
- Fibrin-stabilizing factor deficiency
- Focal dermal hypoplasia
- Glucose-6-phosphate dehydrogenase deficiency Glycogen storage disease
- Granulomatous disease
- Haemophilia A
- Haemophilia B (Christmas disease)
- Hydrocephalus
- Hypomagnesemic tetany
- Hypophosphatemia (vitamin D-resistant rickets)
- Hypoxanthine guanine phosphoribosyl transferase deficiency
- Ichthyosis (epidermal scaling)
- Incontinentia pigmentii
- Iris, hypoplasia of
- Keratosis follicularis spinulosa
- Lowe's oculocerebrorenal syndrome
- Macular dystrophy
- Megalocornea
- Menkes syndrome (kinky hair disease)
- Mental deficiency
- Microphthalmia
- Mucopolysaccharidosis II (Hunter syndrome)
- Muscular dystrophy, progressive Becker type (onset between age 30–40)
- Muscular dystrophy, Duchenne type (onset in childhood)
- Muscular dystrophy, tardive, Dreifuss type (not lethal)
- Night blindness
- Norrie's disease (blindness)
- Nystagmus (one type)
- Ophthalmoplegia (myopia)
- Oral-facial-digital syndrome
- Parkinsonism (one type)
- Pelizaeus-Merzbacher disease (cerebral sclerosis)
- Phosphoglycerate kinase deficiency
- Pituitary dwarfism
- Pseudohermaphroditism, male
- Reticuloendotheliosis
- Retinitis pigmentosa (blindness)
- Retinoschisis
- Spastic paraplegia
- Spinal and bulbar muscular atrophy
- Spinal ataxia
- Spondylo-epiphyseal dysplasia, late
- Testicular feminization syndrome
- Thrombocytopenia
- Thyroxine-binding globulin reduction
- Van den Bosch syndrome (mental deficiency)
- XG blood group system
- XM system (macroglobulin serum protein)

a hereditary character, which is more prevalent in men, than in women. Studies on the mode of inheritance of baldness have shown that it is not inherited in the same way as the recessive sex-linked genes. It cannot be due to a dominant sex-linked gene, because more women than men are not bald. It has been shown that baldness is due to peculiar genes, called sex-influenced genes. The character is dominant in men and recessive in women. A man is bald if he has only

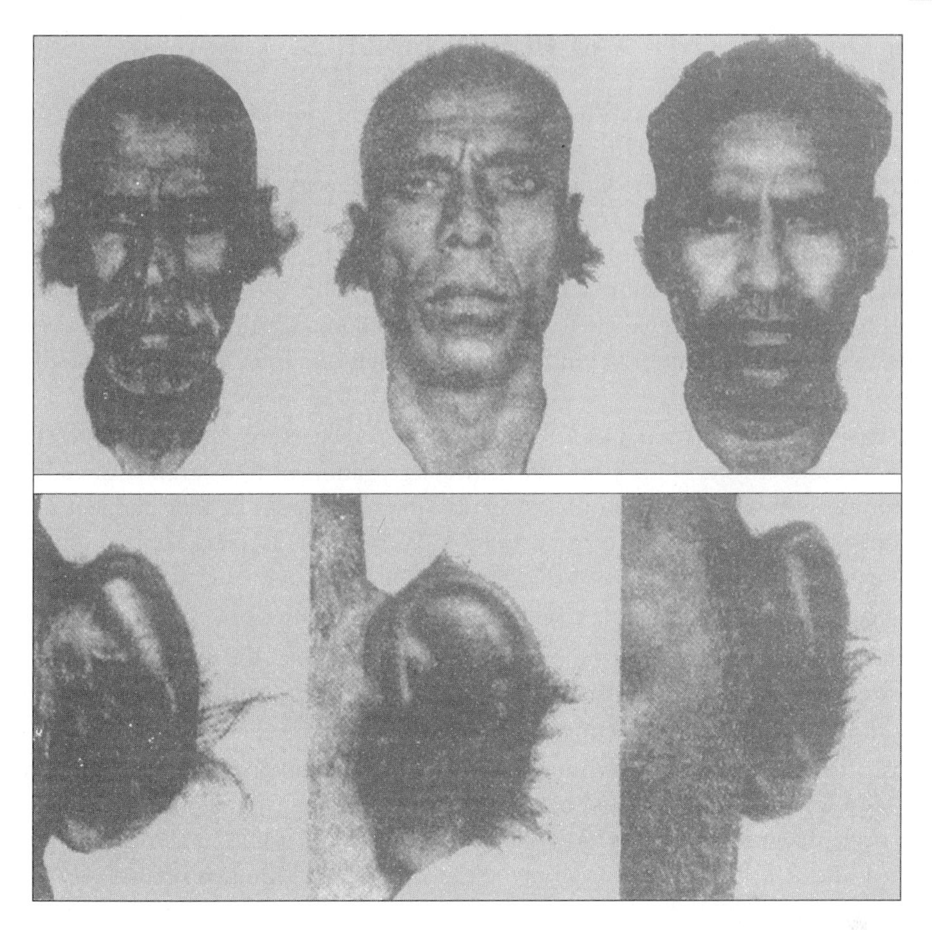

Fig. 32.11 "Hairy ears", an inherited trait common in parts of India.

one gene for baldness, but in a woman two genes are present. If 'B' presents a gene for baldness and 'b' for non-baldness, and the sex-influences is such that 'B' is dominant in man, and recessive in woman, the genotypes of various individuals will be:

Genotype	*Man*	*Woman*
BB	Bald	Bald
Bb	Bald	non bald
bb	non bald	non bald

However, baldness may be purely due to environmental factors, or due to certain diseases, e.g., syphilis, seborrhoea or thyroid diseases.

Horns of sheep and spotting in cattles are also the sex-influenced characters.

MEIOSIS, CROSSING OVER AND HUMAN GENETICS

The number of *kinds of gametes* is 2^n, where *n* signifies the number of allelic pairs. When 23 allelic pairs, on the 23 pairs of chromosomes, are involved, 2^{23} or 8,388,608 *kinds of gametes* could be produced. As large as the figure, 8, 388, 608 may seem, it is by no means large enough to give a true description of the number of possible combinations of maternal and paternal gametic elements in the gametes of a single individual. A chromosome pair contains not only one pair of loci, but a great many. Crossing over breaks up the original combinations and recombines maternal and paternal alleles in new chromosome strands. It thus increases greatly the number of hereditarily different kinds of gametes that one individual is potentially able to form.

Since crossing over in one pair of chromosomes take place largely irrespective of its occurrence in other chromosome pairs, and since segregation of four strands of each original pair of chromosomes is independent of segregation in other pairs, the total number of possible combinations is the *product* of the combinations in each of the 23 pairs. This product is $20 \times 20 \times 20 \ldots = 20^{23}$. The magnitude of this figure is beyond comprehension.

It is of course equivalent to $2^{23} \times 10^{23}$, or 8,388,608 followed by 23 zeros!

A woman during her reproductive years, produces only about 400 mature egg cells. Clearly, these 400 eggs are an infinitesimally small sample of the overwhelming variety of germ cells which she has the potential of forming.

A man's total production of germ cells is much larger than that of a woman; it has been estimated to be in the neighbourhood of 1,000,000,000,000 (1,000 billion). Even this immense number is negligible if compared to 20^{23}, the total number of possible combinations of maternal and paternal alleles in his sperm; it is only about one billion-billionth of 20^{23}.

Huntington's Disease

Huntington's disease (HD) is an insidious disorder caused by an autosomal dominant mutation, which occurs in about one of 10,000 individuals of European descent. Individuals with HD undergo progressive degeneration of the central nervous system, usually beginning at age 30 to 50 years and terminating in death 10 to 15 years later. To date, HD is untreatable. However, identification of the gene and the mutational defect responsible for HD has kindled hope for an efective treatment in the future. Because of the late age of onset of the disease, most HD patients already have children before the disease symptoms appear. Since the disorder is caused by a dominant mutation, each child of a heterozygous HD patient has a 50 per cent chance of being afflicted with the disease. These children observe the degeneration and death of their HD parent, knowing that they have a 50:50 chance of suffering the same fate. The gene for Huntington's disease has been identified to be present on the short arm of chromosome-4 in man (Fig. 32.12).

The *huntington* gene is expressed in many different cell types, producing a large 10- to 11-kb mRNA. The coding region of the *huntington* mRNA predicts a protein 3144 amino acids in length. Unfortunately, the predicted amino acid sequence of the *huntington* protein has provided

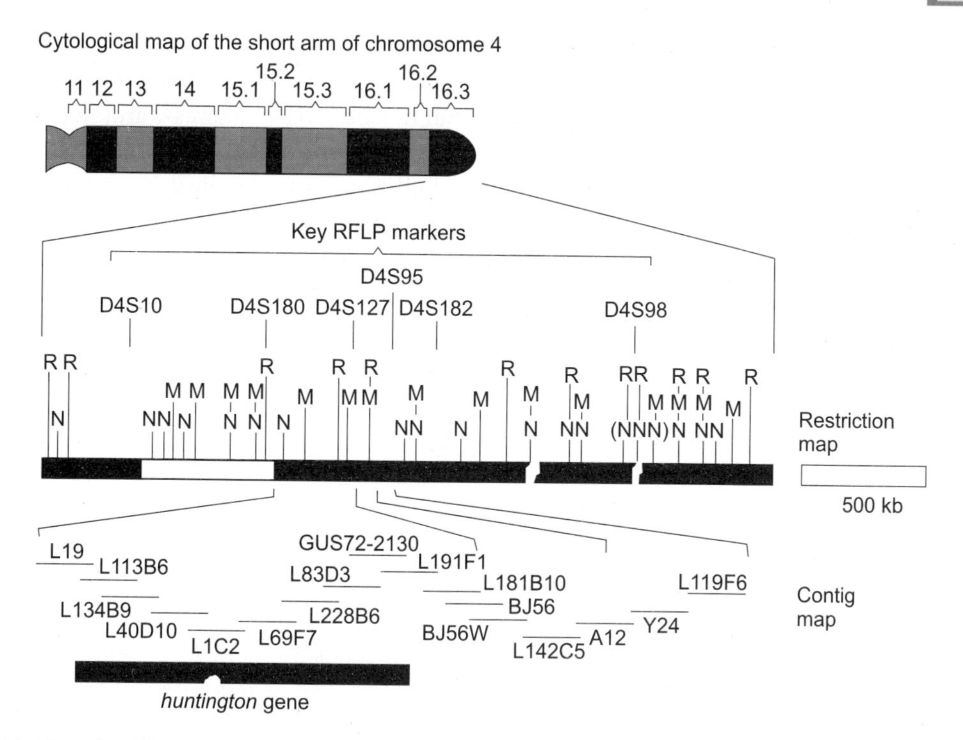

Fig. 32.12 Identification of the gene responsible for Huntington's disease by positional cloning. The cytological map of the short arm of chromosome 4 is shown at the top. The RFLP markers, restriction map, and contig map used to locate the *huntington* gene are shown below the cytological map. M, N and R represent *Mlu*I, *Not*I and *Nru*I restriction sites, respectively.

no hint as to its function. The amino acid sequence is not closely related to that of any known protein. The dominance of the HD mutation indicates that the mutant protein has some new function that causes the disease, but what this function might be is the focus of ongoing research.

HD was the fourth human disease to be associated with an unstable trinucleotide repeat. Fragile X syndrome (the most common form of mental retardation in human), as well as myotonic dystrophy and spinobulbar muscular atrophy (both diseases associated with loss of muscle control) had previously been shown to result from expanded trinucleotide repeats. More recently, two other neurodegenerative diseases, spinocerebellar ataxia type 1 and dentatorubro-pallidoluysian atrophy, have been shown to result from similar enlarged trinucleotide repeat regions. These results indicate that the expansion of such repeat regions may be a common mutational event in humans.

Cystic Fibrosis

Cystic fibrosis (CF) is one of the most common inherited diseases in humans, affecting 1 in 2000 newborns of northern European heritage. CF is inherited as an autosomal recessive mutation, and the frequency of heterozygotes is estimated to be about 1 in 25 in Caucasian populations. In the United States alone, over 30,000 people suffer from this devastating disease. One easily diagnosed symptom of CF is excessively salty sweat, a largely benign effect of the mutant gene. Other symptoms are anything but benign. The lungs, pancreas, and liver become clogged with a thick mucus, which results in chronic infections and the eventual malfunction of these vital organs. In addition, mucus often builds up in the digestive tract, causing individuals to be malnourished no matter how much they eat. Lung infections are recurrent, and patient often dies from pneumonia or related infections of the respiratory system. In 1940, the average life expectancy for a newborn with CF was less than two years. With improved methods of treatment, life expectancy has gradually increased. Today, the life expectancy for someone with CF is about 30 years, but the quality of life is poor.

Identification of the *CF* gene is one of the major successes of positional cloning. Biochemical analyses of cells from CF patients had failed to identify any specific metabolic defect or mutant gene product. Then, in 1989, Francis Collins and Lap-Chee Tsui and their co-workers identified the *CF* gene and characterised some of the mutations that cause this tragic disease. The cloning and sequencing of the *CF* gene quickly led to the identification of its product, which, in turn, has suggested approaches to clinical treatment of the disease and hope for successful gene therapy for *CF* patients in the future.

The *CF* gene was first mapped to the long arm of chromosome 7 by its co-segregation with specific RFLPs. Further RFLP mapping indicated that the gene was located within a 500-kb region of chromosome 7. The two RFLP markers closest to the *CF* gene were then used to initiate chromosome walks and jumps and to begin construction of a detailed physical map of the region.

Key Points: The human genes that cause Huntington's disease, cystic fibrosis, and several other disorders have been identified by positional cloning. The nucleotide sequences of these genes were used to predict the amino acid sequences of their polypeptide products and to obtain valuable information about the functions of the gene products.

Identification of these genes has led directly to new methods of treating the diseases, possible approaches to gene therapy, and DNA tests for the mutations that cause the diseases.

CHAPTER 33

Human Molecular Genetics

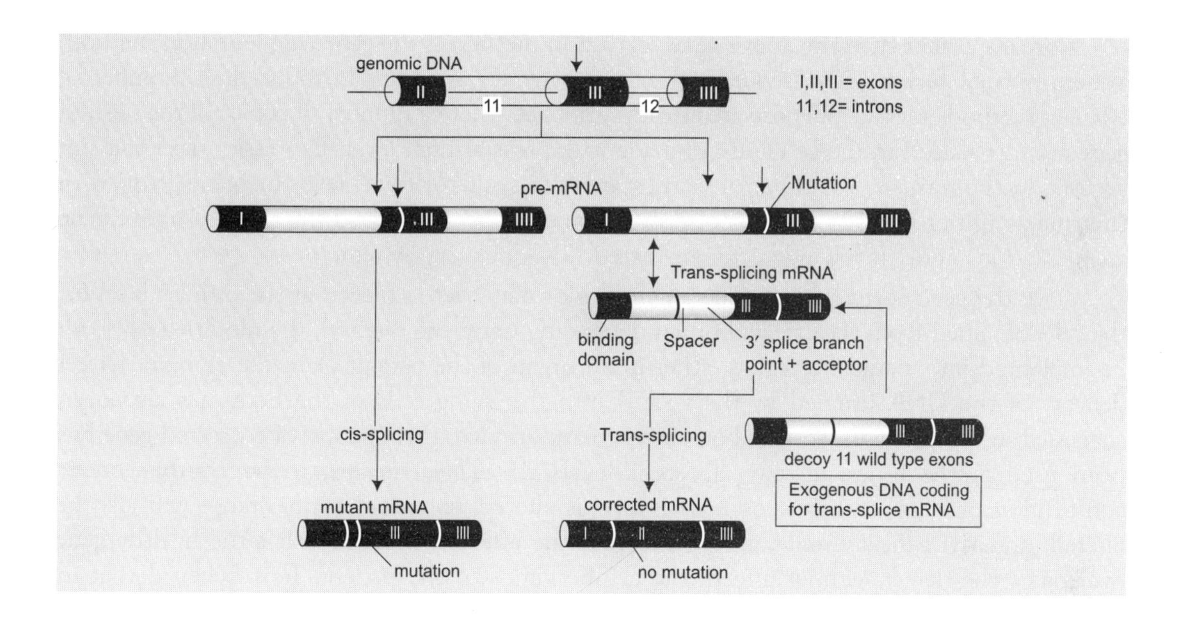

GENE THERAPY

During the last two decades of the 20th century (1980-2000), more than 200 genes responsible for a variety of genetic disorders in human beings were identified. The causes of disorders in the form of specific mutations in these genes have also been elucidated. Therefore, if the known defects in genes are rectified, or if the defective gene in each case replaced by a normal healthy gene, then the corresponding genetic disorder may be corrected. Such a treatment of human genetic disorders are described as gene therapy. However, gene therapy need not always be targeted at either the correction of the defective gene or suppression of its expression or even introduction of a normal healthy gene, but may also be targeted to peripheral or even epigenetic aspects of the pathogenesis pathway. For instance, while applying gene therapy to cancer, one may not like to correct the oncogene or introduce the cancer suppressor gene, but may only introduce either the genes having anti-tumor function in cells (e.g. tumor-infiltrating lymphocytes TILs) or introduce cytokinene genes into the tumor cells, thus indirectly suppressing tumor growth. In other words, gene therapy involves treatment of genetic disorders by insertion of a gene into the body of an organism, in order to correct the disorder. This sometimes involves replacement of the defective gene by the normal gene in a functional sense, but not in a physical sense. The healthy gene is introduced into the appropriate cells of the patient, a process that is technically described a gene delivery. Effective systems of gene delivery have been developed in many cases involving retroviruses and a variety of other viral and non-viral vector systems, More recently, nanotechnology has also been utilized for transfer of genes in particles that measure <500nm in diameter.

More important than the above gene correction method is the gene augmentation method. Where normal foreign gene sequences for the defective gene are introduced. A number of efficient methods for this purpose are already available, where a number of copies of the desired gene are introduced in the cell and are made to express at high level. Expression and transfer vectors, in the form of a number of viruses are now available to achieve this goal. In view of this, maximum progress in the area of gene therapy has been achieved through the above gene augmentation model involving vector mediated DNA delivery system.

Once the gene correction or gene augmentation has been achieved at the cellular level (in the cells obtained from the affected organ depending upon the disease), the modified cells can be implanted into a suitable region either in an organ of the patient or in the embryo. Direct delivery of the DNA (carried by the vector) into the living cells of the body has also been suggested in several cases, so that both in vitro and in vivo introduction of corrected gene has been used at the following two different levels (I) *embryo therapy*, in which the genetic constitution of embryo at the post-zygotic level is altered, so that the inheritance will also be altered, and (II) *patient therapy* gene overcomes the affected tissue, so that the healthy gene overcomes the defect without affecting the inheritance of the patient. It is believed that in future, gene therapy of both types will be possible.

Research in gene therapy, as a biomedical discipline, started in mid 1980s and the first gene therapy experiment was conducted in 1990 on a human patient (a four-year old child suffering with adenosine deaminase (ADA) deficiency, which had wiped out her immune system, crippling her ability to fight infectious diseases). During 1990-2000, hundreds of approved clinical trials, involving more than a thousand patients have been conducted in pilot phase I experiments, so that opportunist in gene therapy are attracting interest not only from research organizations in the public sector, but also from the small and large pharmaceutical companies. Different aspects of gene therapy including its possibilities and limitations will be discussed in this chapter.

HUMAN DISEASES TARGATED FOR GENE THERAPY

Human diseases, that are targeted for gene therapy, may be due to defects in single genes or may be acquired under a variety of circumstances due to genetic predisposition to these diseases (Table 33.1) . Diseases belonging to both these classes can be treated through gene therapy.

Table 33.1 Some human diseases targeted for gene therapy.

I. Diseases caused by single gene defects shown in parentheses (targets for gene therapy)

1. Cystic fibrosis (CFTR);
2. Diabetes mellitus;
3. Pituitary dwarfism (hGH);
4. Emphysema (α-I-antitrypsin);
5. Familial hypercholesterolemia (LDL receptor);
6. Haemoglobinopathies - thalassemia major and sickle cell anaemia - (β globin);
7. Haemophilia - A, B (Factor VIII, Factor IX);
8. Lysosomal disease;
9. Gaucher's disesase (glucocerebrosidase);
10. Glycogen strorage diseases;
11. Phenylketonuria (phenylalanine hydroxylase);
12. Severe combined immunodeficiency diseases or. SCID (ADA);
13. Duchenne muscular dystrophy (dystrophin);
14. CNS disorders - Leisch - Nyhan syndrome (hypoxanthine phosphoribosyl transferase) and Tay -Sachs diseases (hexosaminidase).
15. Immunodeficiency (adenosine dearminase);
16. Hyperammonaemia (ornithine transcarbamylase).

II. Diseases acquired due to genetic predisposition (amenable to gene therapy using gene shown in parentheses)

1. Cancer (interleukins, TK, tumor suppressor);
2. Cardiovascular diseases - Myocardial infarcts or MI (tPA); prevention of blood clots (tPA), hypercholesterolaemia (LDL receptor)
3. Neurodgenerative diseases - Alzheimer's disease (NGF), Parkinson's disease (TH);
4. Joint disorders - Rheumatoid arthritis (cytokine, IL-I antagonists);
5. Infectious disorders - AIDS (HIV antigens, cytokines, TK genes, cytomegalovirus or CMV infections).

CFTR = Cystic fibrosis transmembrane conductance regulator ; hGH = human growth hormone ; LDL = low density lipoprotein; ADA = adenine deaminase ; TK = thymidine kinase ; tPA = issue plasminogen activator; NGF = nerve growth factor ; TH = tyrosina hydroxylase ; IL -I = interleukin-I

Single gene controlled diseases are transferred from generation to generation. The genes for these diseases are generally carried in heterozygous condition in the normal individuals and the disease is caused by the recessive allele in homozygous condition. If the disease is caused by the dominant allele, gene therapy becomes difficult, because, not only a normal allele needs to be introduced, but the defective dominant allele needs to be removed. The heritable Mendelian diseases include phenylketonuria, haemophilia, cystic fibrosis (CF), etc. Acquired diseases include diseases, which are caused mainly due to somatic mutations in some genes, so that they are caused due to specific genes, but seldom transmitted to the germ line. Majority of cancers belong to this class, although heritable cancers are also known now. Acquired diseases are also caused due to genes, which predispose an individual to a disease, like cancer, cornary artery disease, Alzheimer's disease, etc. These diseases are sometimes Mendelian, but are more often caused by a combination of genetic predisposition and environmental factors, or even solely due to environmental factors. For instance, among heart diseases, a small proportion (e.g. type III hypercholesterolemia) is caused due to a deletion or mutation either in the gene encoding low density liporotein (LDL) receptor, which imports cholesterol into the cell, or in a gene encoding carrier protein apolipoprotein B. But most heart attacks are caused due to a mixture of genetic and environmental causes, which means that there is a predisposition to heart disease, which requires environmental triggers or abuse, to become apparent. Knowledge about such predisposition through genetic testing in such cases may enable the predisposed person to modify the environment. Somatic gene therapy can be used for all the above classes of diseases (Table 33.1).

VECTORS AND OTHER DELIVERY SYSTEMS FOR GENE THERAPY

Gene delivery for gene therapy, is achieved by one of the following two methods (Table 33.2 33.3, 33.4): (i) viral transduction, involving introduction of a modified virus carrying foreign gene and (ii) physical transfection involving the use of either 'chemical methods' like calcium phosphate, liposomes and molecular conjugates or 'physical methods' like electroporation and particle acceleration (biolistics). Since transduction gives stable integration and autonomous replication. Vector mediated gene transfer is preferred and a variety of vectors have been tried (Table 33.2). However one of the serious problems in the nonavailability of an ideal vector, which must achieve regulated and sustained expression of foreign genes in specific cells or tissues. This requires that a foreign gene be included in a gene construct, which either replicates autonomously or gets intergrated with an active region of the genome of the host. Furthermore, the whole process must be safe, efficient and selective. Among the available vectors, each vector satisfies some of these features but none fulfils all the requirements.

VIRUSES AS VECTORS

Retrovirus (RV) as a vector. Among the vectors, the most widely used vector is retrovirus, because it can introduce genes into a single active region of the chromatin, thus making it a permanent part of the genome for long term expression. However, there are a number of problems associated with the use of retroviruses as vectors (Table 33.5), but solutions to some of these problems have actually been made in retroviral victors through incorporation of synthetic and heterologous components (Table 33.3).

Adenovirus (AV) as a vector. Adenovirus is a common virus that infects the human respiratory tract and eyes. It can be easily grown in laboratory and infects both diving and nondividing cells. It expresses genes without inserting them into the host's genome, so that although it poses no risk of cancer, but it stimulates strong immune response, thus clearing the vector from the body and making it ineffective for long-term therapy. High doses may be required to transfer enough genes for mealth benefit and this may cause toxic effects.

Adeno-associated virus (AAV) as a vector. AA V is a parvovirus, which readily infects dividing and non-dividing cells. It needs a 'helper adenovirus to replicate, and intergrateds into host DNA at a safe location causing no risk for cancer. However, it is difficult to grow to high concentration (relative to adenovirus and has a small genome thus restricting the amount of

Table 33.2 Comparison of gene delivery systems for gene therapy.

Deliving system	*Integration*	*Transduction*	*Comments*
I. Viral vectors			
1. Retrovirus	+	–	Low titre
2. Adenovirus (AV)	+	–	
3. Adeno-associated virus (AAV)	+	–	Helper virus needed
4. Lentiviruses (Herpes simplex virus)	+	?	Pathogenic
II. Physical methods			
1. Receptor mediated	+	+	Enhanced by
2. Direct injection	–	+	Muscle
3. Lipofection	+	+	Gene gum

Table **33.3** Present day gene therapy vectors and their characteristic features.

Features	*Viral vectors*					*Non-viral vectors*		
	RV	*AV*	*HV*	*PV*	*AAV*	*Mol. conj*	*Transfection*	*Liopsomes*
Insert capacity (kb)	8	7-8	720	725	4.5	–	–	–
Integrates	Yes	No	No	No	No	No	No/rare	No
Oncogenic	Yes	Yes	?	?	?	?	?	?
Viral proteins expressed	No	Yes	Yes	Yes	Yes	No	No	No
In vivo delivery	Poor	Yes	Yes	Yes	Yes	?	Yes	Yes

RV = retrovirus; AV = adenovirus; HV = herpes virus; PV = AAV = adeno-associated virus

Table 33.4 Characteristics of different types of vectors used for delivery of genes.

Retrovirus	*Adenovirus*	*Other viral vectors*	*Non-viral vectors*
RNA genome (e.g.MMLV)	Linear double stranded DNA genome	DNA (adeno associated virus or AAV, HSVI)	DNA protein complex
Reverse transcribed	Does not integrate in host genome; slowly lost	It integrates	DNA entrapped in endosomes
Viral genome integration in host genome; is not lost	Human kidney cell line supplying EIA products- dividing cells not necessary		Cultured hepatocytes
Suitable host: proliferating tissue	EIA region of virus, needed for gene expression and replication, is deleted and replication by therapeutic gene		
Therapeutic gene is inserted in viral genome	Can infect non-dividing cells (major advantage)		

Table 33.5 Major problems with retroviral (MOMLV) vector and possible solutions.

Problem	Solution
1. Regeneration of replication competent retrovirus during vector production	1. Replace murine cells with human helper cells replace MLV vector with non-homologous vectors
2. Insert size limited to - 8 kb	2. Use synthetic particles or MAS vectors
3. Rapid inactivation in *vitro*	3. Modify helper cells, *gag* and *env* genes; use synthetic particles/conjugate/liposomes
4. Infect only dividing cells	4. Add HIV matrix protein gene to helper systems

therapeutic DNA, although techniques are being developed to increase AAV's gene carrying capacity. It has been used to transfer gene for clotting factor IX into patients suffering with hemophilis B.

Lentiviruses as a vector. HIV and Herpes Simplex Virus (HSV) are lentiviruses, which can introduce genes into dividing and non-dividing cells of nervous systems, which are resistant to other vectors.

Common Problems with Most Viral Vectors

Some of the common problems with majority of viral vectors include the following: (i) a limited DNA capacity, (ii) expression of viral genes, (iii) initiation of antiviral immune response, (iv) reversion to a replication competent state and (v) decreasing expression over time. In view of this, in the year 1995, national Institute of Health (NIH), also established three 'national gene vector laboratories', so that high quality gene transfer agents may become available for gene therapy. Each center will plan to focus on specific vectors gene delivery systems including the following (i) retroviruses, adeno-associated virus, (ii) liposomes, naked DNA and DNA-coated pellets and (iii) adenovirus and other viruses.

NON-VIRAL DNA DELIVERY SYSTEM

It is known that about 75% of recent clinical protocols involvinggene therapy make use of viral based vectors for DNA delivery. However, in clinical trials, none of these viral vectors could be shown to be safe. Therefore, non-viral DNA delivery systems have been developed both for gene therapy and DNA vaccination for treating and controlling diseases. Most of these available DNA delivery systems are though versatile and safe, they are less efficient than the virus-mediated DNA delivery systems discussed above. The low efficiency of non-viral DNA delivery is due to the following three major barriers to DNA delivery: (i) low uptake across plasma membrane, (ii) inadequate release of DNA molecules and (iii) lack of nuclear targeting. These three barriers have been addressed in the efforts to develop efficient DNA delivery systems, which can be broadly classified into (i) mechanical/electrical methods and (ii) chemical methods.

Mechanical /electrical methods. Direct injection of naked DNA into the cell is simple and therefore appealing, although there are problems associated with it, since only single cells can be manipulated, and the process is slow and laborious. The three methods that have been tried include: (i) low voltage electroporation or microinjection; (ii) particle bombardment and (iii) pressure-mediated DNA delivery. Electroporation and particle bombardment are commonly used with cultured cells except where limited local expression of delivered DNA is adequate to achieve immune response (e.g., DNA vaccination in epidermal or muscle cells). Pressure-mediated delivery gave 50% efficiency of delivery, when tried with cardio-vascular tissue. Hydrodynamic force has also been used for DNA delivery to hepatocytes with 40% efficiency. Ultrasonic nebulization has also been tried.

Chemical methods. The use of a chemical to enhance DNA uptake by the cell is considered easy and versatile. The techniques can be broadly classified into those based on ; (i) 2-diethylamino ether (DEAE)-dextran, (ii) calcium phosphate, (iii) artificial lipids, (iv) proteins, (v) dendrimers, etc. In all these techniques, DNA is complexed with a chemical and is deposited onto the cells, so that they are internalized by endocytosis and the DNA is released at the desired target site. The major limitation of these methods is toxicity upon systemic administration.

(a) **efficient uptake of DNA during the a pulse of DNA complexed with a chemical.** During 1965-75, DEAE- dextran-DNA and calcium phosphate-DNA complexes were used for internalization by the cells through endocytosis. These methods can be largely used for in vitro transfection and suffer with problems of unstable and variable transfection. During 1970-1990, a variety of artificial lipids-based DNA delivery systems were developed, which are the most commonly used methods today. Lipofectin-DNA complexes were the first chemical systems that could be used in animals and liposome-mediated DNA delivery has been used in humans. The cationic peptide poly-L lysine (PLL), polyamidoamine (PAMAM) dedrimers and polythyleneimine (PEI) are other chemicals, which vary in efficiency (20%-80%), but are toxic.

(b) **DNA complexed with polymers for controlled release on a long-term basis.** The above chemical methods involve exposure of cells to a pulse of DNA complexed with a chemical, so that a short period is available for uptake of DNA. Therefore, biocompatible polymers have been used, which allow controlled repeat administration. For instance, biodegradable poly (D) L-lactide-co-glycolide (PLG), microparticles/microspheres and poly ethylene-co vinyl acetate (EVAc) matrices have been used for long-term controlled release of naked DNA within the cell.

Synthesis Particles as Vectors

Synthetic paricles may combine useful attributes of different classes of vectors and avoid viral pathogenicity. These synthetic vectors include liposomes or molecularconjugate vectors or a combination of both. These synthetic vectors can accommodate genes, more than 8-kb long and can be engineered for entry into cell through receptor mediated endocytosis (RME). They are so engineered that after entery into the cell, they leave endosome. For this purpose, a receptor binding moiety (e.g. asialo-orosomucoid, which is a ligand for the asialoglycoprotein receptor of hepatocytes) and a DNA binding moiety are incorporated. Microtubule inhibitors may block uptake by lysosomes, and nuclear 1ocalization signals (NLS) may facilitate entry into the nucleus. In contrast to vector-mediated gene transfer, transfection is though safe and efficient but involves no mechanism for intergration or autonomous replication. It allows only transient expression. In view of this, efforts are also being made to combine transduction and transfection for efficient and safe gene transfer providing long-term expression.

Selectivity and Long-term Expression

The delivery system and the gene used for gene therapy should also provide selectivity at the following levels: (i) it should enter non-germline cells; (ii) it should selectively integrate at a specific site within a genome and (iii) it should be selectively expressed (over time and space). Even if this selectivity is achieved, long-term expression may still remain a problem. Stable intergration may partly solve the problem and may be achieved through the use of reteroviral reverse transcriptase, retroviral integrase, transposase and parvovirus 'rep' proteins. The enzymes may be used with the vector, or else DNA encoding these enzymes may be included in non-integrable form for transient expression to facilitate stable integration. Autonomous chromosomes (MACs) or autonomous replicating circular minichromosomes (found in some cancer cell lines) may also be used for achieving long-term expression.

Transient Expression for Gene Therapy

Depending on the disease being treated through gene therapy, expression of the introduced gene may be required throughout the body or its expression may be restricted to particular tissues. Also, although in majority of cases of gene therapy, sustained genetic modification is needed, in some cases, a transient expression of therapeutic gene may be desirable. Two such studies involving central nervous systems were conducted and published in September 2000 (nature Biotechnology). In which one of these studies, a viral vector was used, which carried a gene, whose transient expression was triggered by the neurological insult itself, In the other study, poliovirus replicon vector was rendered non. infectious and was used to carry the desired gene for transient expression in transgenic mice; the latter carried human poliovirus receptor (PVR). In the first study the gene expression could be modulated by pathophysiological variations due to the stroke itself, while in the second study, the vector itself limited the replication thus limiting the transgene expression (Fig. 33.1). This transiet expression of desired gene in the affected brain area provides neuroprotection from a stroke.

Splicesome Mediated Trans-RNA splicing (SMART)

Anovel approach for gene therapy, described as SMART, was reported in January 2002. In this approach, an exogenous DNA, coding for trans-splice mRNA (mRNA which will undergotrans splicing with native endogenous mRNA) is introduced as a part of gene therapy. The exogenous DNA is designed to contain the following sequences in 5' to 3' order ; (i) a binding domain (at 5'end) that hybridizes to the intron present towards 5'to the defective exon, to be corrected, (ii) a spacer and (iii) at its 3' end, a 3' splice branchpoint and acceptor, which lies 5' to the

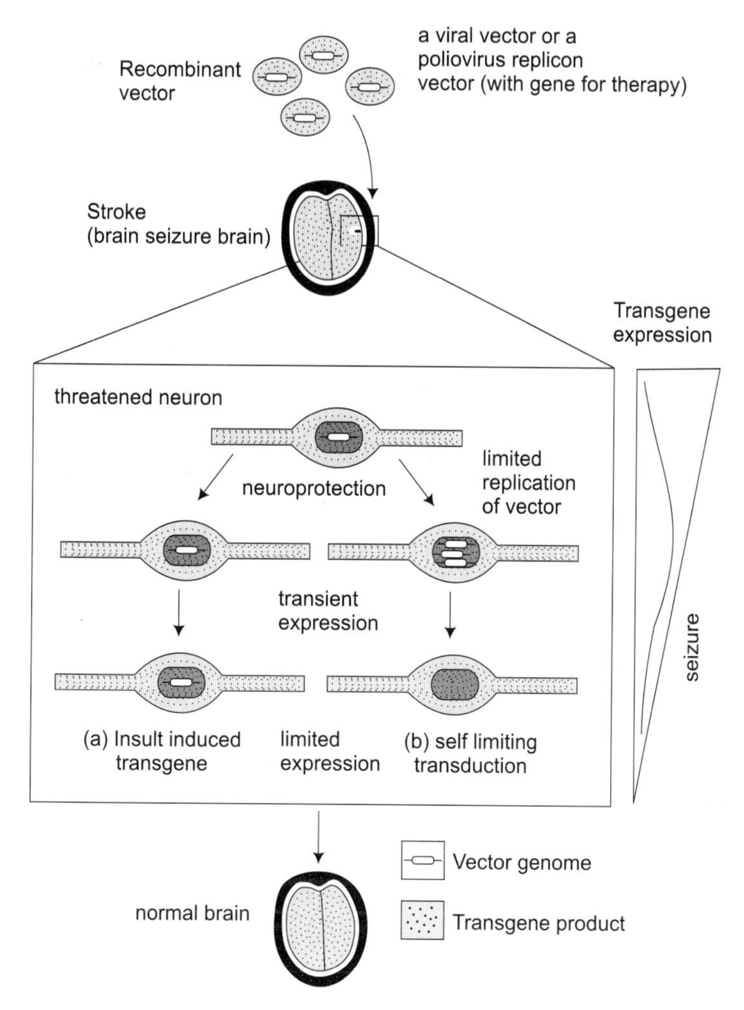

Fig. 33.1 Use of poliovirus replicon vector (it was modified, so that it could replicate, but is non-infectious) for transient gene expression of a therapeutic gene that can used during brain seizure (redrawn from Nature Biotechnology, Sepetember, 2000).

defective exon. The mRNA transcribed from this introduced exogenous DNA and the defective endogenous mRNA are spliced and rejoined in such a manner, that the native mRNA will be corrected for the mutation (Fig 33.2).

Ex Vivo vs In Gene Theory

In gene therapy, recombinant vector with the therapeutic gene is used for gene delivery to specified cells and tissues. Two different strategies are used for this gene delivery, i.e. ex vivo and in vivo approach the cells may be cultured and used for gene transfer, so that these transfected cells are then introduced in a targeted tissue. Alternatively in the in vivo approach, the gene may be delivered through a vector directly into the target cell or tissue. Ex vivo gene

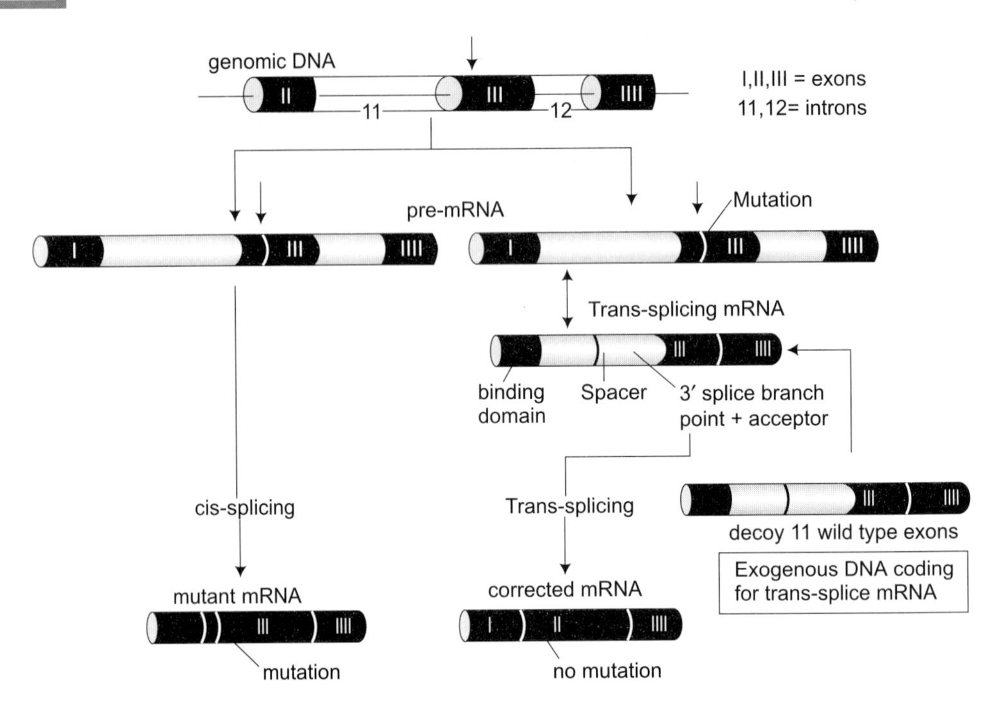

Fig. 33.2 Spliceosome- mediated trans-RNA splicing (SMART) to correct a hereditary disorder at the RNA level; notice that a mutation occurs in exon II, which gives a mutant mRNA due to cis-splicing (shown on the left), and gives a comected m RNA due to trans-splicing, if a DNA sequence carrying that wild tupe exons is introduced (shown on the right), (redrawn from Nature Biotechnology, January, 2002.

delivery is common and more certain, but at the same time more problematic, since; it requires (i) a mitotic cell population; (ii) a tissue culture method and (iii) a cell transplantation technology. There are also following advantages of an ex vivo approach, (i) gene transfer efficiency is generally high and retroviral vectors are particularly effective; (ii) the transduced cells can be enriched if the vector has a selectable marker gene and (iii) transduction efficiency can be assessed before reimplantation. Bone marrow cells, can be multiplied and modified and are the common candidates for ex vivo gene delivery, but for other types of cells, ex vivo modification may be difficult. However, the use of embryonic stem (ES) cells may be really useful. For this purpose, although earlier ES cells were available only from mouse, rat, pig, rabbit, etc. but human ES cells have also become available recently (1998-2002).

Despite some advantage of ex vivo gene delivery, in vivo gene delivery would be preferred, if feasible. However, one major precaution required during in vivo gene delivery is that the gene be delivered only to the targeted cells and no other cells, and in no case to germ line cells. Retrovirus delivers the gene only to dividing cells and therefore is safe for delivery of gene to cancer cells. But in other cases *receptor mediated endocytosis* (RME) may have to be used, so that the vector will have to carry a ligand for the receptor available on the target cells. Sometimes, instead of targeting specific cell types, selective expression of a transferred gene can also be achieved by the use of tissue specific enhancers/promoters. However, the disadvantages of in vivo gene delivery include the following (i) specificity and low efficiency of stable gene transfer, (ii) for clinically useful therapeutic application, repeated treatments may be needed raising the problem of a host immune response.

In view of the above advantages and disadvantages of ex vivo and in vivo approaches, the choice will depend on the relative ease of in vitro culture (ex vivo approach) and the gene transfer efficiency of the target tissue (in vivo approach).

TARGET TISSUE OF CHOICE FOR GENE-DELIVERY SYSTEM

Both in ex vivo and vivo approaches, there will be a target tissue in the body, where either the cells transfected in culture, will be transferred, or the vector carrying the therapeutic gene will be targeted, either for integration in the DNA of target cells (retrovirus) or for transient expression (adenovirus). Following are some of the target tissues.

BONE MARROW

Bone marrow is a hematopoietic system, which is a suitable target for gene therapy, since well developed procedures for bone-marrow transplantation are available. The pluripotent hematopoietic stem cells (HSC) constitute .01 to 0.1% of human bone marrow ceels which are therefore, considered as an ideal target tissue. Gene transfer to HSC is also considered to be the best example of ex vivo gene therapy, and transduction of even few HSCs is sufficient to give enough transduced cells by multiplications.

MUSCLE

Muscles consist of myoblasts, which are used as target cells, if muscles are the target tissues for gene therapy. Myoblasts infected with retovirus carrying a therapeutic gene, when injected. Exhibited sustained expression of gene for at least six months in animals models. The risk, however, is that myoblasts may migrate to other tissues and express there in the absence of tissue specific expression.

Adenovirus vector containing a reporter gene has show efficient transduction of muscle after direct injection in mouse. Myoblasts are also suitable for direct gene transfer using in vivo approach. Stable gene expression of injected DNA (present as episomes) was achieved in cardiac and skeletal muscles of mouse, although the efficiency was low in primates, when 'gene gun' tried.

LIVER

Liver contains hepatocytes, which can be used as target cells, using both ex vivo and in vivo approaches. However, they can not be extensively manipulated in culture, because they undergo few cell divisions. Further, the transduction efficiencies with retviral vectors are the low. Nevertheless, retovirus-mediated ex vivo transfer of disease related human genes has been successfully achieved hepatocytes, which are generally non-dividing cells, can be activated into cell division by hepatactomy, so that retovirus mediated in vivo gene transfer can be achieved. Adenovirus, adeno-associated virus and some non-viral vectors have also been used for in vivo targeting of DNA to the liver.

OTHER TISSUES (BRAIN, TRACHEA)

Cells like brain cells or neurons (neurological disorders) and those belonging to tracheal epithelium or lung (e.g. 'cystic fibrosis' diseases) are difficult target, because cells and not be easily harvested for ex vivo, approach and the tissues are generally inaccessible for in vivo approach. However, HSV-l vectors have been used to transducer neurons both ex vivo and in vivo. Similarly adenovirus mediated and liposome mediated gene delivery to respiratory epithelium has been achieved in some animal models (e.g. rat).

Cell types including keratinocytes, fibroblasts, myoblasts, and vascular endothelial cells/ smooth muscle cells are also used for delivering genes whose products are secreted into the circulating blood. In vivo gene transfer into a specific blood vessel and lymphocytes has been achieved.

IN UTERO GENE THERAPY OR IUGT (BEFORE SYMPTOMS APPEAR)

There are about 400 approved clinical gene therapy protocols available world-wide, but all of them treat pediatric and adult patients after birth. However, genetic diseases can also be treated by the gene transfer in the faetus and the technique is described as 'in utero gene therapy (IUGT)'. For a number of neurological genetic diseases (e.g. Tay Sachs, Lesh-Nyhan,etc.), where irreversiable damage may be caused during gestation before the birth, treatment may be needed before birth to allow the birth of a normal baby. However, decision about the target tissue and the availability of technique to safely and efficiently deliver the gene at the target tissue (e.g. brain) is the major limitations for treating these diseases involving an easily accessible cell type (e.g. haematopoietic stem or progenitor cells or HSCs) will be the choice for IUGT. These diseases may improve immuno-deficiency, thalassemia, osteopetrosis, etc.

Since IUGT should lead to correction for life-time, the gene should integrate in patient's genome, for which only the retroviral in patient's genome, for which only the retroviral vectors are suitable. For this purpose, HSCs may be genetically engineered ex vivo and transplanted back into the patient. Some success in this approach has been achieved in animal system suggesting that it is a viable approach. Several safetly and ethical issues need to be dealt with before clinical trails using IUGT will be permitted.

GENE THERAPY FOR GENETIC (HERITABLE) DISASES

Neurological and Other Metabolic Disorders

Central nervous system (CNS) is a difficult target for gene therapy, but efforts are being made to use gene therapy for a variety of CNS defects whose genetic basis is known now (Table 33.6). However the following aspect of neurological disorder make gene therapy difficult for diseases: (i) compart-mentalization of neural systems, (ii) complexity of neuromal interaction, (iii) poor accessibility of CNSdue to physiological blood-brain barrier, making introduction of gene-transfer vectors difficult. Despite these problems, at least some neurological disorders of central nervous systems (CNS) and peripheral nervous system(PNS) are becoming amenable to gene therapy. Both ex vivo and in vivo approaches are being tried for diseases like the following: (i) Huntington's disease (HD), (ii) brain tumors, (iii) familial Alzeimer's disease (FAD), (iv) familial amyotrophic lateral sclerosis (ALS), (v) Duchenne muscular dystrophy (MSD).

Ex vivo approaches to gene therapy for neurological disorders. Although success in treating neurological disorder through gene through gene therapy may take time, it seems likely that a graft of genetically modified cells (which can synthesize and secrete trophic factors or provide components of neurotransmitter pathway) might be effective atleast in some cases of neurological disorders, particularly those with focal dysfunction. **Parkinson's and Alzheimer's** diseases are candidates for this kind of gene therapy (the genetic basis of these diseases, if any, is not fully understood). Some success in this connection has been achieved using rat models. For instance, in Parkinson's diseases, the lack of ability to synthesize the neurotransmitter, dopamine is believe to be the main cause of the disease, so that the grafts of dopamine producing fetal cells or genetically modified cells capable of synthesizing dopamine have been shown to help in correcting the disease symptoms. Similarly, the supply of neurotrophic growth factor (NGF) from hippocampus plays an important role in survival and function of 'cholinergic neurons' which are essential for memory acquisition and retention .Degeneration of these cells in the absence of NGF causes Alzheimer's disease. It has been shown that the supply of NGF or genetically modified cells capable of synthesizing NGF will help correcting the symptoms of

Table 33.6 Gene therapy models for some neurologual disorders.

Disorder	*Transgene*	*Target tissue/organ*
I Genetic disorders		
1. Phenylketonuria	Phenylalanine hydroxylase	Hepatocytes or direct dilivery to liver
2. Leisch-Nyhan disease	HPRT	Surrogate cells or direct delivery to brain
3. Lysosomal storage disorder	Lysosomal exumed	Bone-marrow cells or delivery to brain
II Neurodegeneratine disorders		
4. Alzheimer's diseose	NGF or other trophic factors	Surrogate cells or direct delivery to hasal forebrain
5. Parkinson's disease	Tyrosine hydroxylase	Surrogate cells or direct delivery to basal ganglia or mid brain

Alzheimer's disease. Genetically modified bone marrow or the precursor CNS cells that are capable of differentiating into neurons or glial cells, can also be introduced to deal with some neurological disorders.

In vivo approaches to gene therapy for neurological disorders. DNA mediated gene transfer can be achieved by injection of naked plasmid DNA into the brain although the efficiency of transfer is low, so that more efficient viral vectors may need to be used in future.

Gene Therapy for Cystic Fibrosis (CF) Lung Disease

Cystic fibrosis involves failure of airway defense against bacterial infection and is attributed to loss of function of cystic fibrosis transmembrane conductance regulator (CFTR) protein due to a mutation in CFTR gene. CFTR regulates certain ion channels. The gene CFTR carried by a vector like adenovirus has been used both ex vivo and in vivo gene delivery. Liposomes have also been tried.

GENE THERAPY FOR ACQUIRED DISEASES

Gene Therapy for Infectious Disease

Two different strategies for the treatment of infectious diseases are being explored. Both strategies involve the use of genetically manipulated cells. (i) the first strategy termed 'interacellular immunization' is designed to render cells resistant to viral replication, thus limiting the spread of virus. (ii) The second strategy termed 'genetically engineered vaccines' or DNA vaccines' makes use of genetically modified cells expressing viral gene products to enhance antiviral cellular immune responses.

Intracellular Immunization (Rendering Cells Resistant to Viral Replication/ gene Expression)

In the year 1988, Baltimore coined the tenn 'interacellular immunization' for genetic modification of target cells (prone to viral infection), making them resistant to intracellular pathogens. However, 'intracellular resistance', 'intracellular interference' or 'intracellular inhibition' could be better terms AIDS is an example of the approach, because insertion of resistance genes into hematopoietic cells (that are prone to infection by HIV) or the use of gene based vaccines could reduce, if not eliminate, the spread of virus. The resistance may be achieved through (i) antisense RNA, (ii) ribozymes, (iii) RNA decoys (exploiting steps of HIV replication), (iv) use of altered HIV gene products having transdominant (TD) mutant phenotype (e.g. altered forms of Gog, Rev, Tat and Env.).

Genetically engineered vaccines (also called DNA vaccines). The approach involving genetically engineered vaccines can induce production of either the antibodies or the cytotoxic T Cell receptors. Introduction of genes that encode T Cell receptors has been considered a better alternative. The above resistance genes or the gene vaccines need to be delivered to hematopoiectic target cells of the patient and this may be achieved using either the retrovirus based gene transfer systems or the vectors based on adeno-associated virus (AAV). The advantages and disadvantages of using these vectors have been discussed elsewhere in this chapter

The efficacy and safety of the use of resistance genes or the genes encoding T Cell receptors for gene therapy have been tested in immortal cell lines and in animal models, but both these systems have their limitations that need to be dealt with, before clinical trials on human subjects can be conducted.

Gene Therapy for Cardiovascular Diseases

Several cardiovascular diseases are due to abnormalities in muscles of the hearts or blood vessels. In cases of arteries blockage, when arteries are opened using balloon angioplasty, reblockage (also called 'restenosis') occurs in 50% cases and can be treated by 'gene therapy'. Gene therapy has been attempted for several of these diseases. For instanes, blockage of arteries caused by proliferation of smooth muscles lining the artery walls, can be treated by gene encoding thymidine kinase makes the proliferating cells susceptible to a drug called genciclovir. Similarly, the gene. For *vegf* (vascular endothelial growth factor) has been used for growth of new blood vessels for the heart. The *vrgf* may be transferred through a virus or via liposomes.

Gene Therapy for Cancer (Somatic Mutations)

It has been shown that cancer in most cases has a genetic basis, but in majority of cases it is a diseases, which is acquired during the life span of an individual. Such an acquired cancer results due to somatic mutations, although it can also be inherited from the parents. In either case, gene therapy can be used to selectively target and destroy cells by using one of the following approaches: (i) Whenever, cancer is caused, either p53 gene or due to mutation in this gene, insertion of a copy of wild type p53 gene has been shown to results in the death of tumor cells. (ii) If cancer is caused by over expression of an oncogene like K-RAS, an antisense gene has been shown to disable over expression of this gene. (iii) Therapy using immunostimulatory genes, such as interleukin (IL-2) gene may also lead to immunological destruction of tumor cells. (iv) Insertion of an antisense gene that stops production of insulin like growth factor (IGF-I) in some tumors also leads to destruction of tumors since these tumors produce high level of IGF-I.

In all the above approaches the most important problems the need for a more efficient gene delivery systems. NO gene delivery system is 100% efficient, unless germ line therapy is used. However, non-integrating vector such as those found to be efficient. In contrast, if integrating vectors can be developed, advance delivery of genes in susceptible patients may prevent development of malignancies.

Clinical trials for untested technologies are also appropriate ethically for those cancers which are incurable, and when life expectancy is a few weeks/months, rather than years as in other genetic diseases. In such cases approval of trials is also easy.

Ex vivo gene transfer in tumor cells, fibroblasts or T lymphocytes. In this approach tumor cells from the patient are grown in culture, immunostimulatory genes are inserted in vitro and the genetically engineered tumor cells are reinjected into the patient. This not only destroys the tumors, but also vaccinates the patient against the recurrence of the tumor. In most such cases retroviral vectors are used for delivery of either the genes encoding cytokines (e.g. IL-2, IL-4, IL-6, IL-IB, TNF-, etc.), or the antisence genes blocking production of growth factors. In some cases, the gene is inserted into patient's fibroblasts, which are mixed with irradiated tumor cells from the patient and reinjected. This is preferred because culturing cells from a large number of individuals. Tumor infiltrating T lymphocytes (TILs) have also been used for genetic engineering to deliver cytokines directly onto the tumors, although their transformation is difficult and they down regulate expression of cytokine gene.

Sensitivity genes have also been used, which increase sensitivity of the tumor cells to certain drugs that will thus become toxic and kill tumor cells. One such gene is thymidine kinase gene of the herpes simplex virus (HSVtk). It imparts to the cancer cells sensitivity to anti-perpes drug 'ganiclovir (GCV)', which is phosphorylated due to HSVtk and inhibits DNA polymerase, causing cell death. The approach has been used in certain brain cancers and ovarian cancers.

Ex-vivo gene transfer in bone marrow stem cells. Gene therapy in cancer patients is also used for protecting hematopoietic stem cells from the toxic effects of chemotherapy. Gene conferring multidrug resistance (MDR-I) has been used for this purpose. Bone marrow cells with MDR-I, when injected have been shown to protect stem cells in vivo.

In situ genetic alteration of cancer cells. Both liposome mediated and retrovirus mediated gene transfer has been attempt to modify cells in situ. Genes for antigens like HLA-B7 and B2-microglobulin have been transferred through liposomes and expressed on cell surface inducing immune reaction against tumor cells. Fibroblasts, that produce retrovirus vector (also called retroviral vector producer cells or VPCs) have also been directly injected into growing brain tumors, from retrovirus mediated transfer of HSVtk gene. In animal models, and human patients, this led to complete tumor destruction in 80% of animals. In vivo retrovirus mediated transfer of p53 tumor suppressor gene, or antisense K-RAS gene or Y-interferon gene have also been tried for some kind of lung cancers in humans.

Nanotechnology for Drug Targeting and Gene Therapy

Selective delivery of drugs to specific areas in the body is necessary to maximize drug action and minimize side-effects. "Magic bullet concept" represents an early description of drug targeting paradigm. Such targeting can be achieved at the organ level, cell level or even at the intracellular level and the drugs may involve prodrugs (which release a drug at the targeted site), monoclonal antibodies and polymeric and colloidal carriers.

Nanotechnology involves the use of nano particles (< 1 μm, usually < 500 nm in diameter) used as particulate carriers systems for drug delivery. These particulates include not only solid structures in the form nanocapsules, but also includes liposomes and emulsions, that can reach specific tissues or even specific sites within a cell. They can be use for delivery of drugs, vaccine administration and for diagnostic imaging.

Designing of Targeting System, Keeping in View the Fate of Nanoparticles After Intravenous Administration

A targeting system need to be appropriately designed keeping in view the fate of nanoparticles after intravenous administration. After the particles are injected into the blood stream, they are conditioned or coated by elements like plasma proteins and glycoproteins available in the circulatory system. This process is called '**opsonization**', which renders the particles recognizable by the major defense system of the body. The macrophage cells (called **'Kupffer cells'**) of the liver and other macrophages remove these opsonized particles. Particles having hydrophobic surface are more efficiently coated, but those having hydrophilic surface resist this coating and are therefore cleared more slowly. Therefore, nanoparticles having hydrophilic surfaces should be preferred for gene therapy.

Passive targeting of nanoparticles. Although uptake of nanoparticles by Kupffer cells may be a hindrance in delivery of drug to the target, atleast in some cases, this feature can be exploited to help treatment. This is true in cases, where the macrophages themselves are directly involved in the disease process, as in case of **leishmaniasis, listeria** and **candasis.** In other cases, capture and degradation of nanoparticles by macrophages can also be exploited to achieve the controlled delivery of drugs to the blood circulation. This is because through macrophages, the particles reach endosomes and then to lysosomes, where they are degraded, so that the drug is released diffused into the circulatory system. This strategy can be applied in the delivery of anticancer agent **adriamycin,** and antifungal agent **amphotericin.**

In cases, where capture by Kupffer cells is a barrier to the targeting of drug, the particles may be disguised, so that they are no longer recognized by Kupffer cells and therefore escape capture by them. The nanoparicles can be disguised partly by making them hydrophilic and partly by stabilizing them using adsorbed polymers. Nondegradable polystyrene, polyoxyethylene or a block copolymer are some such adsorbed polymers, which do not allow thus prevent opsonization so that the nanoparticles persist within the circulation. Such 'stealth liposomes' are undergoing clinical trials for tumor targeting.

Active targeting of colloidal carriers. Targeting ligands (e.g. monoclonal antibodies, sugar residues, etc.) can also be attached to nanoparticles to achieve active targeting of those particles. The hepatocytes of liver can be an important target in case of hepatitis and also in other cases of gene therapy, when administered gene needs to be expressed in these cells of the liver. The nanoparticles having ligands for active targeting, should not only escape captive by kupffer cells of the liver. but also need to be small enough (>50nm or even >20nm in size) to be able to enter the hepatocytes with the help of specific receptors on these surfaces and through ferestrations that are 100-150nm in diameter.

Gene Therapy using Nanotechnology

Pilot phase I experiments involving gene therapy in human subjects have already been conducted for diseases like cystic fibrosis and muscular dystrophy. Some of the attractive targets for gene therapy include the following: (i) epithelial surfaces of the lungs and gastrointestinal tract; (ii) endothelial cells lining the blood vessels. muscle myoblasts and skin fibroblasts, and (iii) tumor cells. In all these cases. the plasmid DNA carrying the gene sequence should reach the nucleus of the target cell, overcoming all the hurdles (capture by endosome and lysosome) on its way to the nucleus. In the nucleus, it will stay as an episome and express itself.

Three main types of gene-delivery systems have been described: (i) viral vectors; (ii) non-viral vectors (particles and polymers), and (iii) 'gene guns' for direct injections of the genetic material into the target tissue. Since viral vectors pose some serious problems, the non-viral vectors are the gene-delivery systems of negatively charged plasmid DNA is condensed into a nonparticulate structure. 50-200nm in size. The use of cotionic lipids and cationic polymers gives a compact structure due to interaction between cationic material and anionic DNA. These compact structures also provide increased stability and uptake by the target cells. Some of the nonviral vectors for gene therapy based on nonoparticles are listed in Table 27.7. The targets for gene therapy using nonparticles include liver hepatocytes. endothelial cells. spleen and lymph nodes, where some success has already been achieved.

A marker gene in transferred to these cells (the marker gene is neomycin resistance gene= Neo R) using retroviral; the marked cell and unmarked cells are grown together and then transferred back to the patient. This is an example of a successful delivery of a marker gene. Thus deomonstrating the feasibility of gene delivery approach. (ii) **ADA gene therapy.** In 1900 the first trial of actual gene therapy was conducted in USA. A little (4 years old) girl suffering with adenosine deaminase deficiency (ADA), a lethal disorder, was transfected with lymphocytes bearing the ADA gene carried by retroviral vector. In 1991 another girl (9 years old) began treatment in USA and received more than a dozen transfusions by 1992. Both these patients were doing well. In Italy also. a five year old boy was given mixture of ADA gene-corrected T cells and bone marrow cells. In Netherelands also bone marrow gene therapy protocol of ADA deficiency was approved in 1992. In 1993. Donald Kohn in Los Angeles separated CD34 cells form umbilical cord blood and transformed them with ADA gene and reintroduced them in the 3 days old children (three). The experiment was partially successful. (iii) **Cancer gene therapy.** In 1991 two protocols for cancer gene therapy were initiated in USA. Tumor necrosis factor (1NF. an anticancer agent) agent of IL-2 (interleukin-2) gene ,was inserted in TIL tumors cells isolated from the patient and grown in culture. These gene-corrected cells were then injected into the body of patient. Other cancer gene therapy protocols were also approved.

In December 1991, a conference titled **'Human Gene Therapy'** was held at National Institute of Health in USA and concluded that gene therapy would soon become a potent force in medicine to deal with hear diseases. liver diseased. Diabetes, and a variety of cancers. It was hoped that gene therapy will eventually play a role in disease prevention by correcting deficiencies right at birth. The cellular vehicles for gene transfer also multiplied and included endothelial cells and myoblasts. besides lymphocytes. Efforts were also made at University of California. San Diego to treat the children with AIDS using gene therapy. Stem cells form umbilical cord blood were transformed with anti-HIV gene for an enzyme cleaving HIV (Science, Vol.268. 12 May, 1995).

Some Initial Success in Gene Therapy

In the past, for approved human genetic engineering (or gene therapy), experiments were conducted and partial success was achieved. These experiments include the following: (i) **Neo R/TIL gene marking.** In 1989, using Neo R/TIL protocol, the first gene-marked. immune cells (tumor-infiltrating lymphocytes=TIL) Were successfully transferred into patients with advanced cancer. The TILs are those lymphocytes that are isolated directly from tumor and run then grown to large numbers in tissue culture in the presence of cell growth factor (interleukin-2 =IL-2). In Neo R/TIL protocol, an aliqot of cells from TIL is taken.

Gene Therapy Deaths and the Future of Gene Therapy

After the first case of gene therapy attempted in 1990. gene therapy was considered to be a promising area of research for human healthcare. But after almost a decade. in the year 1999, a solitary case of death of an 18-years old patient (Jesse Gelsinger) at the University of Pennsylvania's Institute of Human Gene Therapy, set in motion a fresh debate on the desirability of conduction clinical trials on human subjects. The patient suffered with an heritable disease of liver and heavy dose of gene for **ornithine-transcarbamylase or OTC** (needed for removing excess of ammonia from blood) carried by a vector (crippled adenovirus) was introduced. The patient died on 17"' September, 1999, four days after the genetically altered virus was injected into his liver. The death 2was attributed to massive immuno-response caused by adenovirus and resulted in a major blow to the field of gene therapy in USA and elsewhere. All clinical trials at University of Pennsylvania's Institute of Human Gene Therapy were stopped and an investigation was ordered. Several other deaths of patients undergoing gene therapy were not reported and were sometimes attributed to caused other than gene therapy. FDA (Food and Drug Administration) of USA was therefore scrutinizing all gene therpay programmes across USA. As a consequence of the above, the use of adenovirus that was once considered to be the most efficient vector (capable of infecting both dividing and non-dividing cells and capable of expressing the gene within 24 hours, as against 3-6 weeks needed with other vectors), has been curtailed in several laboratories.

CHAPTER 34

Oncogenes

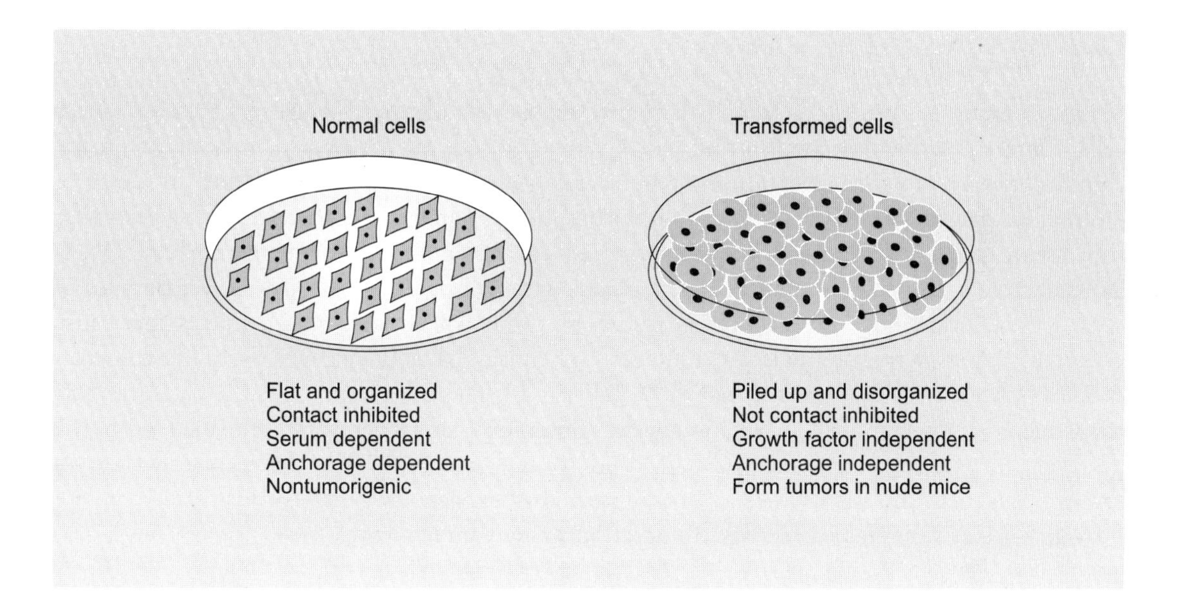

ANTI-ONCOGENES AND TUMOR VIRUSES

In no area of medicine has recombinant DNA had a greater impact than in our understanding of cancer. Although this disease has been inscrutable for decades, the seed for the recent explosive growth in our knowledge of cancer was planted in 1911. In the year, Peyton Rous discovered that filtered cell-free extracts from chicken tumors could cause new tumors when inoculated into healthy chickens. It was eventually recognized that what recognized that what Rous had discovered was a tumor virus. The realization that a tumor virus with an exceedingly small genome could trigger the entire panoply of traits that characterize a cancer cell was an intellectual watershed: the task of understanding how a cell's entire growth program is rewritten in cancer now seemed a feasible one. What was not anticipated was that tumor viruses would provide a window on our own genomes, a window opened wide by recombinant DNA technology.

What we have learned since the tools of recombinant DNA were unleashed on the cancer problem is that our own genome is littered with genes having the potential to cause cancer and the other genes having the power to block it. From the study of these genes, we have derived a greater understanding of the molecular events that occur in tumor cells. And as bonus, we have discovered many of the players that orchestrate normal cell growth and differentiation. The implications for cancer as disease are significant. Already recombinant DNA is a valuable tool for diagnosing cancer. Soon it may offer us new ways to treat the disease.

CANCER IS A GENETIC DISEASE

A fundamental. feature of cancer cells is that when they divide, both daughters are also cancer cells. Cancer is stably inherited during cell division. Indeed, many tumors can be shown to be clonal, derived from a single common progenitor cancer cell that divides incessantly to generate a tumor of identical sibling cells. That cells within a tumor are genentically related suggests that the disease phenotype is genetically determined that is, encoded within tumor cell DNA. Considerable circumstantial evidence for a genetic basis for cancer has built up over decades of study of chemical carcinogenesis. These studies showed that virtually all chemicals that cause cancer in experimental animals are mutagens, chemicals that damage DNA. This common property of carcinogens suggested that their ability to inflict genetic damage is the basis for their carcinogenic properties. The role for genetic alterations in cancer was confirmed with the discovery that normal cells can be converted to cancer cells by the direct introdudion of exogenous genetic information. As we will see. this exogenous genetic information can be introduced experimentally by DNA transfection or by tumor viruses.

TUMOR CELLS HAVE ABERRANT GROWTH PROPERTIES IN CELL CULTURE

When normal tissues are explanted from adult animals and placed into culture in the laboratory, they seldom grow. Tissues from fetal or newborn animals will often grow steadily for a few weeks and then cease. Rarely, cell lines arise from these cultures that are immortal. They grow indefinitely in culture but otherwise retain the growth properties of their normal counterparts. Cultures like this can be treated with carcinogenic agents such as chemicals. radiation and tumor viruses. and from these treated cultures. transformed cells with novel growth properties arise. These cells often have a distinctive shape, they are less dependent on specific factors in their growth medium. they continue to grow long after their neighbors have stopped-piling atop one another-and when injected into animals they form tumors (Fig. 34.1). These traits are stably passed from mother cell to daughter. Thus, these transformed cells behave very much like real tumor cells the ability to recapitulate the transformation process in cell culture has been the cornerstone for building the molecular view of cancer.

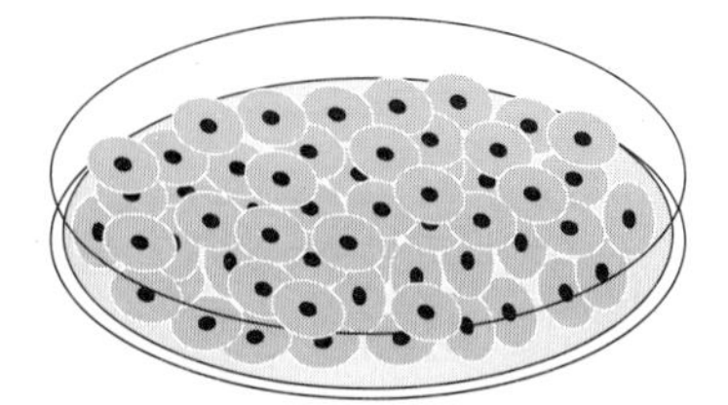

Fig. 34.1 Normal and transfermed cells, Normal nontumorigenic cells grow in a flat and organized pattern until they cover the surface of the culture dish. Then, when each cell is touching its neighbors they stop a phenomenon termed contact inbibition, In addition, cell growth requires a cocktail of growth factors, usually supplied in the form of serum (a cell-free preparation from animal blood), and the cells grow only when attached to a solid surface (*ancborage dependence*), In contrast, cancerous cells usually have a different shape. They disregard their neighbors and grow to high densities in disorganized piles, Their growth often does not require serum growth factors, and they can grow simply suspended in growth medium, without being attached to a surface, When transformed cells are injected into *nude* mice (mutant animals with a compromised immune system), they rapidly form tumors.

TUMOR VIRUSES OPENED THE STUDY OF CANCER TO MOLECULAR METHODS

The discovery that viruses, when inoculated into animals. elicited tumors considerably simplified our ideas about cancer. for somewhere within the tiny genomes of these viruses-nearly a millionth the size of the genome of an animal cell-lurked the genetic instructions to derail normal cell growth. So. theses viruses came under intense scrutiny. And while they did not give up their secrets easily. study of the tumor viruses uncovered many of the basic principles of carcinogenesis as well as those of eukaryotic gene organization and regulation.

Tumor viruses come in two types. depending on whether their genome is encoded in DNA or RNA. The genomes of DNA tumor viruses vary in size from 5 to 200kbp (Fig. 34.2). the smallest viruses encode only a few genes and depend heavily on host cell enzymes for transcriptions and replication of their genomes. In some cells, the viruses grow lytically: they enter multiply rapidly and kill the host cell by rupturing its cell membrane. Certain cells, however, resist lytic infection. and with a low frequency, become transformed into cancer cells. Invariably. transformation is associated with integration of the viral genome (or portions thereof) into host cell DNA and its stable passage to daughter cells. Thus, tumor viruses tumor viruses induce genetic alterations in transformed cells.

Tumor viruses carry discrete genetic elements called *oncogenes* that are responsible for their ability to transform cells. These genes include the T antigen genes of SV40 and polyoma virus, EIA and EIB of adenovirus. and E6 and E7 of papillomaviruses. Oncogenes encode proteins, often termed oncoproteins, that play a number of important roles in the viral life cycle, such

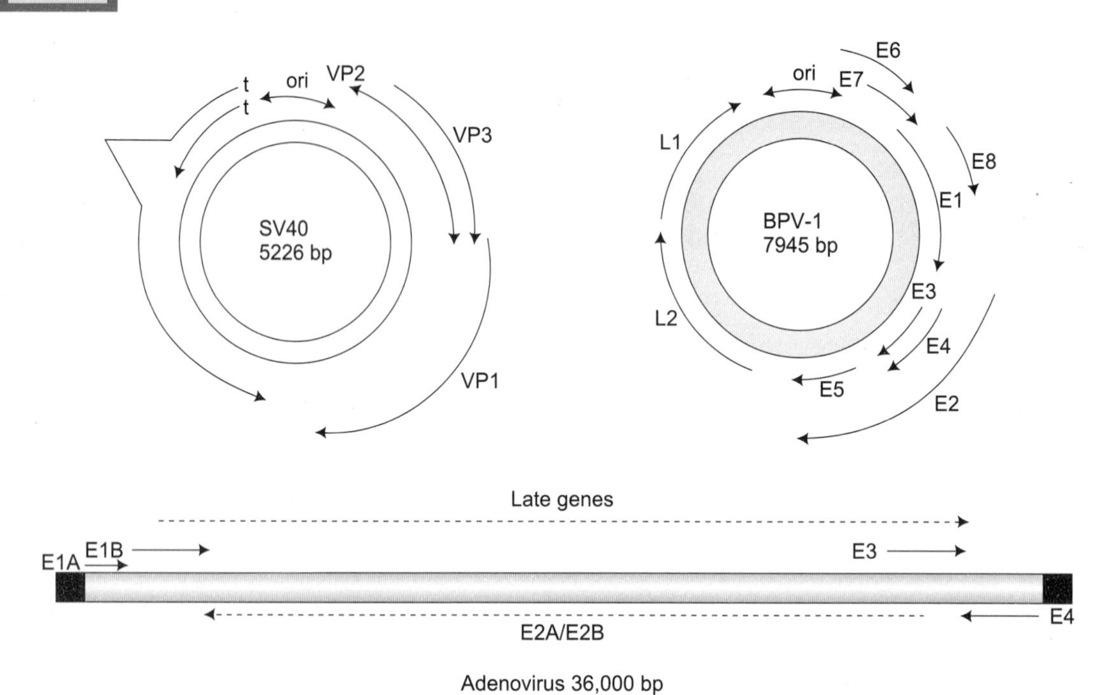

Fig. 34.2 Organization of three DNA tumor viruses SV40, a monkey virus and BPV-1 a bovine papillomavirus, have small circular double-stranded DNA genomes. Adenovirus, which causes colds and conjunctivitis in humans, has a longer linear double-stranded genome. Each virus contains genes responsible for their transforming activity (red). These genes presumably are important regulators of the viral life cycle. Other regulatory genes with no direct transforming properties are indicated in purple. These genes are generally expressed early in the viral infection. The late ganes (green) encode structural proteins involved in the assembly of new virus particles. The linear genome of adenovirus is replicated from the ends. The adenovirus gene map has been considerably simplified for clarity.

a as initiation of DNA replication and transcriptional control of viral genes, and the ability of oncogenes to transform cells appears to be a consequence of theses activities. Viruses normally infect non-growing cells, which represent the vast majority of cells in an animal, and because they require host enzymes to replicate their DNA, one of the critical things a virus has to do after infection a cell is to prod the cell into synthesizing the cellular DNA replication machinery. Activating dormant cells so viral DNA can be replicated is the chief role of the oncoproteins of the DNA tumor viruses. Thus, when the viral genome integrates and becomes a permanent part of the cell genome, the viral oncogene keeps the cell growing indefinitely and tumor forms.

RETROVIRAL ONCOGENES ARE CAPTURED FROM CELLULAR DNA

While researchers struggled with the refractory oncogenes of the DNA tumor viruses, those working with RNA tumor viruses, or retroviruses, made the astonishing discovery that their potent oncogenes originated from cellular DNA. During their evolution, these viruses captured cellular genes that gave them their dramatic growth-transforming properties. The demonstration that retroviral oncogenes are cellular in origin arose from studies of the virus originally discovered by Peyton Rous, now called Rous sarcoma virus (RSV). The isolation of mutant strains of RSV that were deficient in transformation provided the first evidence that the virus carried a specific genetic locus, termed sro, that encoded its ability to transform cells. One class of these transformation-defective (td) viruses carried extensive deletions in the viral genome. These deletions marked the site of the sro oncogene. By laboriously isolating the RNA molecules from large quantities of normal and transformation-defective viruses and converting the former to isotopically labeled cDNA, it was possible to remove from the normal viral cDNA the genetic information also present in the transformation-defective viruses. Left behind was isotopically labeled cDNA containing only sequence specific to the *sro* oncogene. When this *sro-specific* DNA preparation was annealed to DNA from uninfected cells, the cDNA annealed stably, suggesting that cellular DNA contained sequences very closely related to *sro*. These sro-related sequences were found not only in chicken, the natural host for RSV but in all vertebrates, including humans.

Thus, the *sro* oncogene carried by RSV is actually a derivative of a normal cellular gene, captured and stably incorporated into the viral genome by a poorly understood process called *transdution*. With the ability to generate molecular alones of retroviral genomes, it soon become apparent that each of the acutely transforming retroviruses, those that cause rapidly growing tumors within a few weeks of inoculation, carried oncogenes derived from cellular DNA, and these cellular genes were quickly isolated. In all, more than 25 distinct cellular genes, now referred to as proto-oncogenes, have been transduced and employed as oncogenes by retroviruses that infect birds, rodents, cats, and monkeys. By convention. the viral form of such oncogenes is labeled with the prefix "v" (e.g. v-*sro*), whereas the cellular gene is labeled with 'e" (e.g. c-sro). Although retroviruses of this class do not appear to be involved in hum:an cancer, study of these viruses has revealed a family of genes in the human genome that probably play a very direct role in the disease.

AN ACTIVATED HUMAN ONCOGENE IS CLONED

And then the race was on to clone the oncogenes from human tumors. With no knowledge of the structure or sequences of the oncogenes, more arduous methods were required to clone them. The method that eventually worked was the tagging of the oncogenes with markers that could be used to retrieve oncogene clone from a library.

In one case, a natural tag was exploited. Human DNA carries a widely dispersed family of repetitive DNA sequences termed *Alu* repeats. These repeats occur about every few thousand base pairs, conveniently placing a repeat within or near every human gene. Sequences closely related to the *Alu* repeats are not found in the mouse genome, so the presence of human DNA in transfected mouse cells can be monitored by hybridizing a southern blot of restriction enzyme-digested transfected cell DNA with an *Alu* probe (Fig. 34.3). Investigators at tempting to isolate the oncogene from a human bladder tumor cell line (EJ) passed the oncogene through

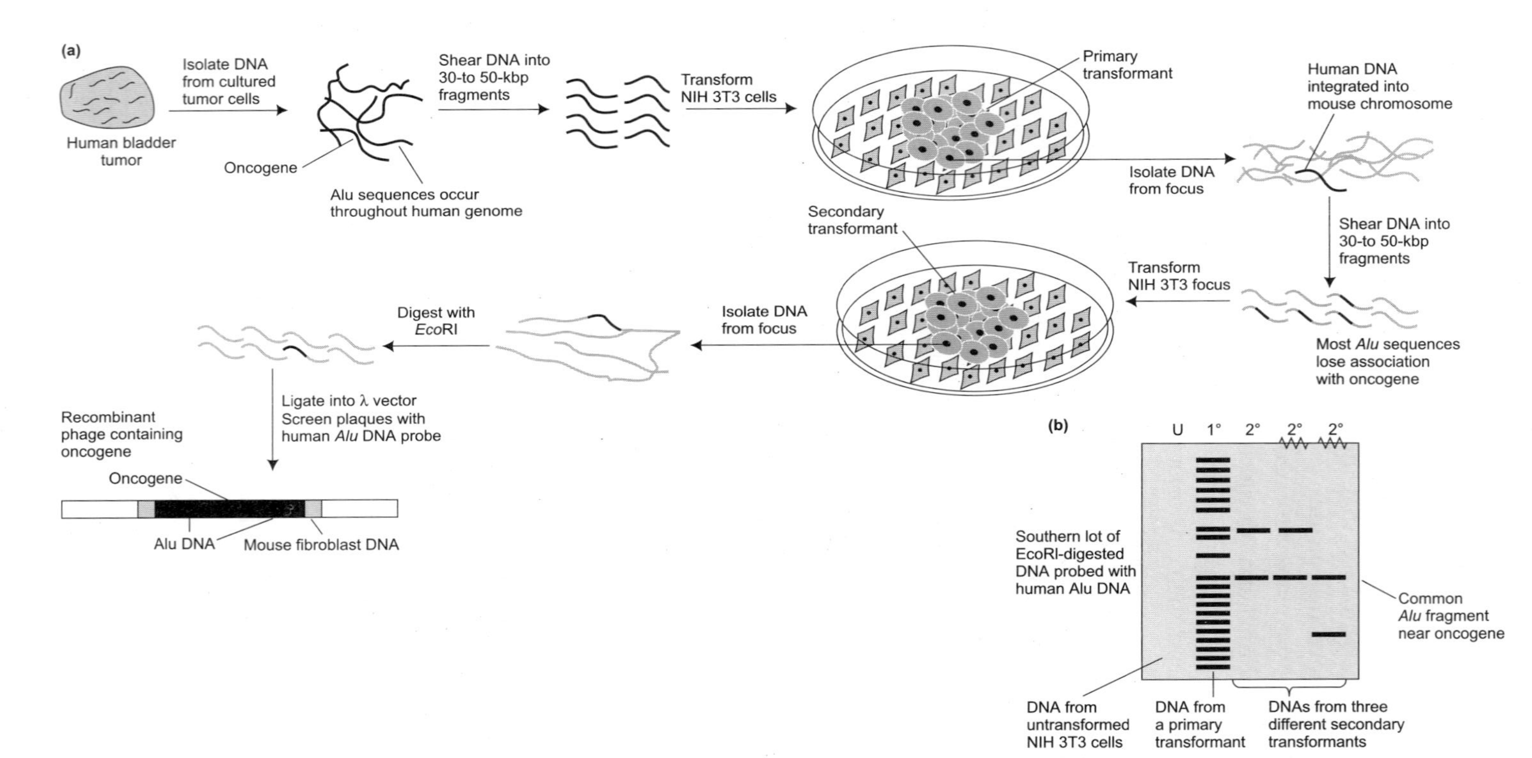

Fig. 34.3 Cloning of an activated human tumor oncogene. Transfer of DNA from a human bladder tumor cell line (EJ) to NIH 3T3 cells yielded foci of transformed cells. In the process the transformed cells acquired an activated human oncogene from the tumor cells plus approximately a million base pairs of irrelevant human DNA. Human DNA can be detected in mouse cells by using a probe for the *Alu* family of repetitive sequences, present every few thousand base pairs in the human genome. Therefore, an *Alu* probe hybridizes to many restriction fragments in a Southern blot of DNA from a primary transfectant (lane l° in b). After a second round of transfcction into NIH 3T3 cells (using the primary transformants as the source of the DNA), however, the human DNA is distributed among many recipient cells. Only a few nearby Alu sequences remain with the oncogene. Thus, in secondary transformants, only a few fragments hybridize in a Southern blot, and comparison of several independent foci showed that each carried a common Alu-reactive fragment (lanes 2° in b). This fragment could be isolated and cloned simply by preparing a library of genomic DNA from these cells and screening the library with an *Alu* probe. DNA from the phage isolated from this library carried the intact oncogene and very efficiently transformed NIH 3T3 cells.

two rounds of transfection into mouse fibroblasts, following the human DNA with an *Alu* probe. After the second round a single common *Alu*-reactive fragment remained in the transformed cells. Because the same fragment was found in several independently derived transformants, it must be closely linked to, or perhaps even carry, the active oncogene. It was then a relatively simple matter to make a library from transfected fibroblasts and screen the library with an Alu probe to find the clones carrying human DNA. In the case of the oncogene from the EJ bladder carcinoma cell line, DNA from the phage clone isolated in this manner yielded thousands of transformed foci when transfected into the mouse fibroblasts—a dramatic demonstration that a powerfully active human oncogene had been cloned.

THE HUMAN BLADDER CARCINOMA ONCOGENE IS AN ACTIVATED ras GENE

But what was this EJ oncogene? The frenzy of tumor virus cloning in the late 1970s and early 1980s enabled leading oncogene labs to develop libraries of DNA probes that reacted with the two dozen or so known retroviral oncogenes. Using this library of oncogene probes, researchers examined the DNA of recipient cultures transformed by human tumor DNA for the presence of new DNA fragments related to the oncogenes. When genomic DNA from normal mouse fibroblasts is hybridized to a given oncogene probe in a Southern blot, usually only a small number of DNA fragments of characteristic size hybridize. These fragments represent the endogenous proto-oncogene in mouse DNA. When DNA from normal cells was compared to DNA from fibroblasts transformed with DNA from the EJ cell line, the patterns of hybridization were the same for most oncogene probes. When examined with a probe derived from the *ras* oncogene of Harvey sarcoma virus (a mouse retrovirus), however, new fragment were detected in the transformed fibroblasts. Several fibroblast clones in dependently transformed by the human tumor DNA each carried new *ras*-related DNA fragments. Indeed, the transformed phenotype correlated perfectly with .the presence of new *ras* DNA in the cells. And the cloned bladder oncogene also hybridized to the ras probe. The conclusion: the oncogene in the human tumor passed from cell to cell by transfection is the cellular counterpart of the retroviral *ras* gene. At last there was a smoking gun. A cellular proto-oncogene, found in every cell, was responsible for the transformed phenotype being passed from the tumor to the fibroblasts.

ONCOGENES CAUSE CANCER IN TRANSGENIC MICE

Development of the ability of generates transgenic mice provided an opportunity to critically test the oncogene hypothesis in intact animals. And it quickly became apparent that mice born with an activated oncogene in their genome reproducibly developed cancer at a young age. This was a dramatic demonstration of the power of dominantly acting oncogenes in a real animal, not just in cells on a culture dish. In general, experiments with transgenic mice carrying active oncogenes strongly support the multiple-hit model for cancer. Although the oncogene is present in all cells, all cells expressing the oncogene were not transformed. Instead tumors arose sporadically in the affected tissue. And the tumors were clonal. That the tumors were sporadic and clonal means that additional, relatively rare events had to occur to allow oncogene containing cells to grow out into tumors.

With two lines of transgenic mice bearing different oncogenes, it was possible to breed animals with two hits. Mice expressing activated ras in their mammary tissues were bred to mice expressing deregulated myo in their mammary tissues. Mice in these individual lines developed clonal mammary tumors at roughly three months of age. What happens to mice expressing both *ras* and *myo* in their mammary cells? If two hits are sufficient for transformation the mammary tissue of these animals should be uniformly transformed. But this was not observed. Instead these mice still developed clonal mammary tumors, but much more quickly. These observations argued that each oncogene contributes significantly to tumor formation, but that still additional events are required. We will see how multiple events can be documented during development of a human cancer.

SUSCEPTIBILITY TO CANCER CAN BE INHERITED

Further evidences for the existence of tumor suppressor genes came from studies of rare cancers that run in families. Members of affected families appear to inherit susceptibility to cancer and develop certain kinds of tumors at rates much higher than the normal population. A dramatic example of inherited tumor susceptibility is retinoblastoma, a tumor of the eye that strikes young children. Retinoblastoma can occur sporadically, an isolated event in a family with no history of the disease. This form of the disease is the most common. But about one-third of retinoblastoma patients develop multiple tumors, usually in both eyes. And the children and siblings of such patients often develop the same disease, an observation suggesting that the susceptibility to retinoblastoma is inborn. The data are consistent with a mechanism involving a single responsible gene.

In 1971, Alfred Knudson suggested a prescient model to account for the sporadic and inherited forms of retinoblastoma. He supposed that the development of retinoblastoma required two rare mutations. In the sporadic form of the disease. both mutations would have to occur within a single retinoblast cell, an exceedingly infrequent circumstances that explains why these children never develop more than one tumor. In the inherited form of the disease, however, he suggested that one of the mutations is already. present (inherited) in all retinal cells. Therefore, only a single additional mutation is required for a full-blown tumor, and tumors are much more frequent in these individuals.

A decade later, analysis of the chromosome in tumor cells and normal tissues of retinoblastoma patients resoundingly confirmed Knudson's hypothesis. Many retinoblastoma patients carried deletions in chromosome 13. And the gene inherited by afflicted children, termed RB, was mapped to this same chromosome. Most important, while unaffected tissues in these children could be shown to carry one mutant RB allele and one normal one, tumor DNA carried only mutant RB alleles. Thus, development of retinoblastoma appeared to require that both copies of the RB gene be mutated; these are Knudson's two hits. Because in many cases the mutant RB alleles had large deletions, it was clear that the mutations eliminated the function of this gene. The RB gene must therefore be a tumor suppressor gene or anti oncogene that normally functions to arrest the growth of retinal cells. Even one copy is sufficient to keep growth in check. But loss of both copies of RB eliminates the block, and a tumor develops.

Genetically, mutations in tumor suppressor genes behave different from oncogene mutations. Whereas activating oncogene mutations are dominant to wild type-they emit their proliferative singles regardless of the presences of wild-type gene product-tumor suppressor mutations are recessive. Mutation in one gene copy usually has no effect, as long a reasonable amount of wild-type protein remains. Thus, tumor suppressor genes are sometimes called recessive oncogenes.

CANCER RESULTS FROM ACCUMULATION OF DOMINANT AND RECESSIVE MUTATIONS

With remarkable speed and steadily increasing clarity, the molecular defects in cancer have come into view. And the molecular picture provides a solid foundation for understanding decades of cancer research. Cancer is a genetic disease, and now we know many of the genes involved and, in some cases, what they do. Cancer often takes decades to develop, requiring multiple independent events to systematically dismantle the complex and redundant regulatory circuits that maintain normal growth in a cell. Now know that many of these events are.

Cancer results when a single cell sustains a mutation, to an oncogene or a tumor suppressor gene, the gives it a growth advantage over its neighbors. As the number of descendants of the original mutant in creases, so does the likelihood that one of them will sustain a second mutation that in turn allows its immediate decendants to grow even faster. And the terrible cycle continues as cells accumulate additional mutations that accelerate their growth further and allow them to invade surrounding tissue. Further events that about tumor growth may also arise from mutations. Some tumors secrete abundant quantities of oncogene factors, peptide hormones that direct the growth of blood vessels into the tumor (an adequate blood supply is critical for tumor growth). This change may be the result of a regulatory mutation. Additional mutations may account for metastasis, the deadly property of certain tumors that enables them to escape to seed new tumors elsewhere in the body. The individual events on the road to cancer are, but we live for a long time, and chance dictates that occasionally this lethal combination of events will unfold.

The increase in our understanding of cancer has been breathtaking, and finally this knowledge is beginning to reach the cancer clinic. Molecular analysis of the oncogenes and anti-oncogenes affected in tumors can predict the course of the disease and the suggest the appropriate treatment. The benefits of simply detecting cancer at an earlier stage should not be under-estimated. In coloertal cancer, mortality can be prevented if the disease is caught any tome prior to the last step; we know have four or five different molecular probes to do that. And many of the same genes will undoubtedly be involved in other cancers. In coming years, cancer diagnosis will be revolutionized by recombinant DNA.

But what about cancer treatment? Cancer is a complex disease, and our recent progress tells us that it is perhaps even more complex than we had appreciated.

But recombinant DNA technology is pushing toward new cancer treatment in several significant ways. First, it is uncovering new targets for cancer therapy, the genes and proteins that drive tumor growth or restore growth to normal. Second, molecular research into cancer has unerringly brought to our immediate attention the system that regulate cell growth, and it is an axiom of medicine that the deeper our understanding of the underlying biology of disease, the more effective our therapies. And third, our ability to produce critical cellular proteins at will and manipulate them to change their properties promises new therapeutic drugs and strategies unlike any available at the present time.

CHAPTER 35

Plasmids

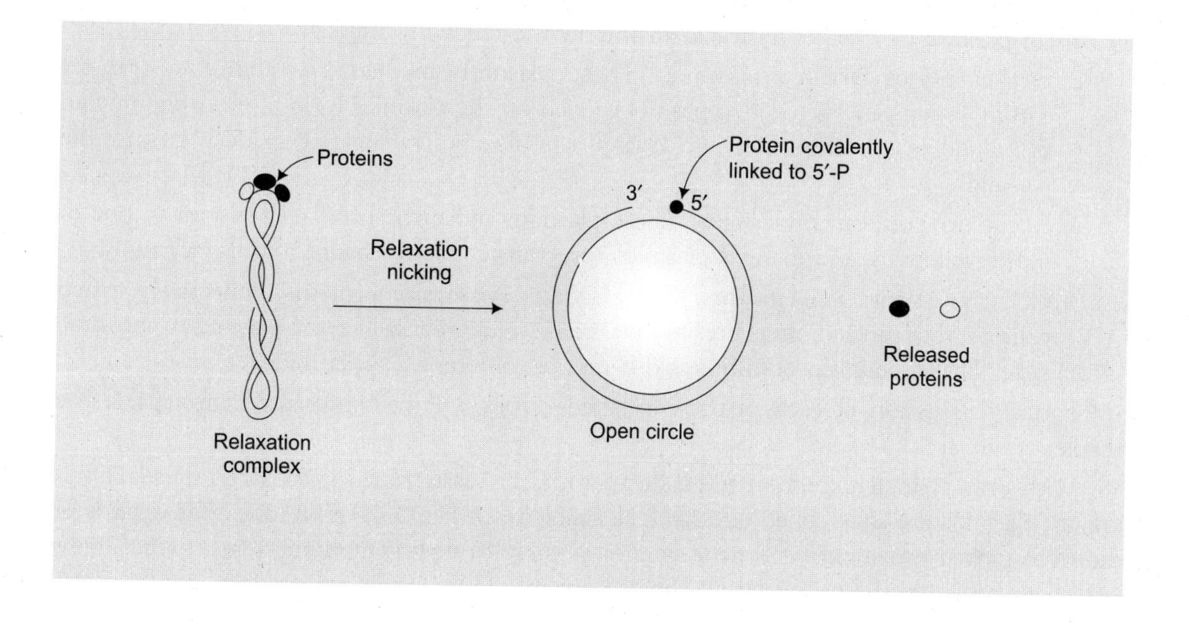

Plasmids are circular, supercoiled DNA molecules present in most species, but not all strains, of bacteria. They are small, ranging in size from about 0.2 to 4 that of the bacterial chromosome. Under most conditions of growth, plasmids are dispensable to their host cells. However, many plasmids contain genes that have value in particular environments, and often these genes are the main indication that a plasmid is present. For example, R plasmids render their host cells resistant to certain antibiotics, so in nature a cell containing such a plasmid can better survive in environments in which a fungal antibiotic is present.

In many bacterial species plasmids are responsible for a particular type of gene transfer between individual cells, a property that accounted for the initial interest in the 1950s. Furthermore, like phages, plasmids are heavily dependent on the metabolic functions of the host cell for their reproduction. They normally use most of the replication machinery of the host and hence have been useful models for understanding certain features of bacterial DNA replication. Finally, they have been exceedingly valuable to the microbial geneticist in constructing partial diploids and as a gene-cloning vehicle in genetic engineering.

TYPES OF PLASMIDS

This chapter will be concerned primarily with plasmids of *E. coli* except when otherwise noted. Many types of plasmids have been detected in variour *E. coli* strains, but the greatest amount of information has been obtained about three main types-the F, R, and Col plasmids-which share some properties but which are, for the most part, quite different. The presence of an F, R, or Col plasmid in a cell is indicated mainly by the following traits:

1. F, the sex, of fertility, plasmids. These plasmids mediate the ability to transfer chromosomal genes (that is, genes not carried on the plasmid) from a cell containing an F plasmid to one that does not. F itself can also be transferred to a cell lacking the plasmid.
2. R, the drug-resistance plasmids. These plasmids make the host cell resistant to one or more antibiotics, and many R plasmids can transfer the resistance to cells lacking R.
3. Col, the colicinogenic factors, Col plasmids synthesize proteins, collectively called colicins, that can kill closely related bacterial strains that lack a Col plasmid of the same type. The mechanism of killing is different for different Col plasmids.

Further discussion of each of these plasmids types will be represented throughout the chapter.

With only a single exception (the killer-plasmid of yeast, which is an RNA molecule) all known plasmids are supercoiled circular DNA molecules (Fig. 35.1) The molecular weights of the DNA range from about 10^3 for the smallest plasmid to slightly more than 10^6 for the largest one. Table 35.1 lists the molecular weights for several plasmids that are actively being studied.

DETECTION OF PLASMIDS

Plasmids can be detected by both genetic and physical experiments. The first plasmid that was discovered was F. An *E.coli* strain (A) with phenotype Met-Bio-Thr+Leu+ was mixed with a second strain (B) with phenotype *Met+Bio+Thr–Leu–*, and the mixture was plated on minimal agar. At a frequency of about 10-7, colonies formed on the minimal agar; these had the phenotype *Met+Bio+Thr+Leu+* and hence were recombinants. In a second experiment strain A was treated with streptomycin (and then washed free of the antibiotic) prior to mixing the cells; recombinant colonies still formed. However, if strain B was first treated with streptomycin, no recombinants were found. This experiment indicated that the recombinants were derived from strain B, and that mating somehow involved a third strain C, which could not transfer any genetic information to B. However, if A and C bacteria were mixed and allowed to grow together for a long time and then colonies of C were isolated, these colonies (call them C') could transfer genes to strain B. The explanation was that strain A contained a fertility element, called F, which mediated transfer of chromosomal genes from bacterium A to bacterium B, strain C lacked F, so no recombinants formed. However, when strains A and C were grown together, F was transferred to C, generating strain C', which could then (because it contained F) transfer genetic markers to B cells. Thus, the initial cross was

F+met−bio−thr+leu+X F−met+bio+thr−Ieu−

Because of the one-way nature of the transfer the cell containing F is said to be a donor or a male, and the cell without F is the recipient or female.

Variants of F, called F', are known that possess genetic markers. One that was studied in great detail carried the lac operon and was used in constructing the partial diploids. It is

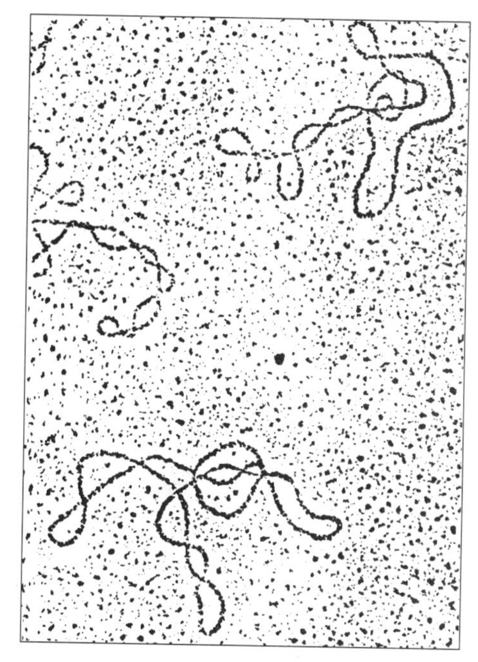

Fig. 35.1 Two supercoiled plasmid DNA molecules.

Table 35.1 Several plasmids and selected properties.

Plasmid	*Mass × 10^6*	*No. copies/ chromosome*	*Self-transmissible*	*Phenotypic features*
Col plasmids				
ColE1	4.2	10-15	No	Colicin E1 (membrane changes)
ColE2 (*Shigella*)	5.0	10-15	No	Colicin E2 (DNase)
ColE3	5.0	10-15	No	Colicin E3 (ribosomal RNase)
Sex plasmids				
F	62	1-2	Yes	F pilus
F'*lac*	95	1-2	Yes	F pilus; *lac* operon
Rplasmids				
R100	70	1-2	Yes	Cam-r Str-r Sul-r Tet-r
R64	78	(limited)	Yes	Tet-r Str-r
R6K	25	12	Yes	Amp-r Str-r
pSC 101	5.8	1-2	No	Tet-r
Phage plasmid				
λdv	4.2	≈50	No	λ genes *cro, cl, O, P*
Recombinant plasmids				
pBR322	2.9	≈20	No	High-copy-number
pBR345	0.7	≈20	No	Col E1-Type replication

designated F' lac. Transfer of an F' is recognized easily with color-indicator plates containing an antibiotic (for example, streptomycin) to which the donor cells are sensitive and the recipients are resistant. For example, in a cross, F'lac+/Iac-str-s X lac-str-r in which the donor cell carries the lac+ marker on the plasmid and lac- marker in the chromosome (note the use of the/to distinguish plasmid and chromosomal markers), plating the mixture on EMB-lactose agar containing streptomycin yields white colonies (*lac–str–r–*) and purple colonies (*F'lac+/lac–str–r* recombinants). If the cultures had not been mixed, plating the donor alone would yield no colonies (because all cells are Str-s) and plating the recipient alone would yield only white colonies (all would be *lac–*). Note that the function of the streptomycin marker is to prevent growth of donor cells; such a marker is termed the counterselection, or counterselective marker. Antibiotics are commonly used for counterselective purposes but other agents serve as well. For example, F' transfer can be detected in the cross *F'lac+/met–* X *lac–met+* with the mixture plated on minimal agar containing lactose. Donors will not form colonies because methionine is lacking in the agar (methionine is the counterselective agent), and females lacking a transferred *F'lac+* will not grow because they are Lac-and cannot use the lactose in the agar. In this case, the selected marker is *lac+*.

Since physical experiments are frequently carried out with bacterial extracts, a plasmid is often discovered incidentally as a nonchromosomal circular DNA molecule present in a DNA sample isolated from a culture. This has led to the development of the following screening technique by which the presence of a plasmid in the cells of a single bacterial colony can be detected electrophoretically. A single colony is taken from a plate, lysed by the lysozyme-detergent procedure (or sometimes lysed directly in the well), and then subjected to gel electrophoresis. The bacterial chromosome is very large and cannot penetrate the gel, but the plasmid DNA can do so. Since the rate of electrophoretic movement of DNA molecules through a gel increases with decreasing molecular weight, plasmid DNA, if present, will form a narrow band at a position in the gel characteristic of its molecular weight. The band can be visualized by the usual treatment of staining the gel with ethidium bromide, which binds tightly to the DNA and fluoresces upon irradiation with ultraviolet light. From the distance moved in a particular time interval relative to that for plasmids of known molecular weight, the molecular weight of the plasmid DNA can be calculated, as shown in Fig. 35.2.

PURIFICATION OF PLASMID DNA

Plasmid DNA can be isolated from bacteria in a simple way. Plasmid-containing bacteria are opened by the lysozyme-detergent treatment, and the resulting translucent solution, called a cell lysate, is centrifuged. The bacterial chromosome complex, which contains protein and RNA, is very large and compact and sediments to the bottom of the centrifuge tube; the smaller plasmid DNA remains in the clear supernatant, which is called a cleared lysate. Some chromosomal DNA is usually present in the cleared lysate, but since most of the plasmid DNA is covalently circular, this contaminating material can be easily removed by the following procedure. CsCl plus ethidium bromide is added to the cleared lysate, and the lysate is centrifuged to

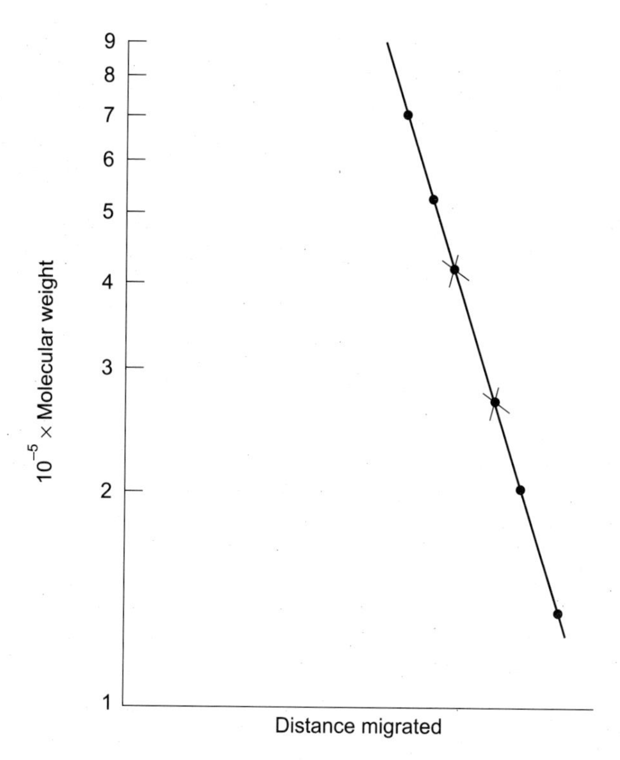

Fig. 35.2 Determination of the molecular weights of DNA molecules (× points) by gel electrophoresis with DNA molecules whose molecular weights are known (• point). The line is a plot of the • points; the molecular weights of the molecules weights of the molecules being studied are then determined from the positions of the ×; points.

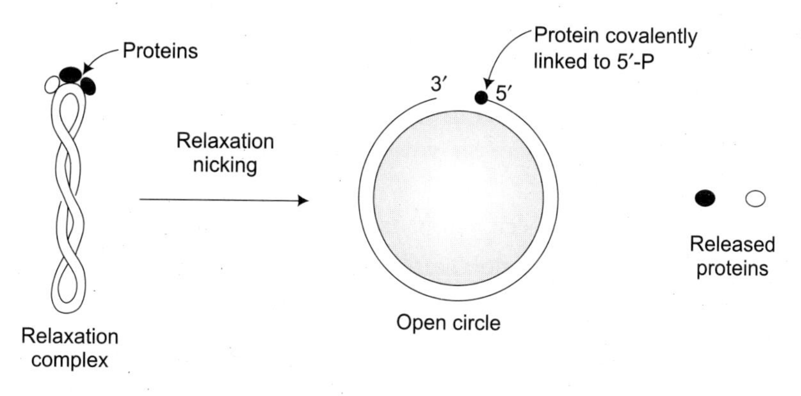

Fig. 35.3 Nicking of one strand of a supercoiled DNA relaxation complex.

equilibrium. When ethidium bromide is present, the covalently circular DNA has a higher density than the linear chromosomal fragments, so the plasmid DNA is purified. The only disadvantage of this technique, which is the most commonly used, is that both nicked circles (which often form accidentally during isolation of plasmid DNA) and nonsupercoiled replicating molecules are discarded, since they come to equilibrium within the chromosomal-DNA band.

The isolation of DNA usually entails a deproteinization step (for example, treatment with phenol). When the DNA molecules of some *E. coli* plasmids are isolated without such a step, about half of the supercoiled DNA molecules contain three tightly bound protein molecules. This DNA-protein complex is called a relaxation complex. If this complex is heated or treated with alkali, proteolytic enzymes, or detergents, one of these proteins, which is a nuclease, nicks one DNA strand, thereby "relaxing" the supercoil to the nicked circular form. The nick occurs in only one strand and at a unique site. During relaxation the two smallest proteins are released, but the largest protein becomes covalently linked to the 5'-P terminus of the nick. If, prior to relaxation, the supercoiled DNA is nicked by any of a number of laboratory techniques, the relaxation nuclease is unable to make its site-specific nick, indicating that the nuclease is active only on supercoiled DNA. This nicking, which presumably also occurs within cells, is an early step in plasmid transfer.

TRANSFER OF PLASMID DNA

In an earlier section we described the detection of F and F' plasmids by virtue of their ability to be transferred from a donor cells to a recipient. In this section we examine the transfer process in some detail.

Stages in Transfer Process

One of the earliest discoveries about recombination between donor and recipient cells is that it requires cell-to-cell contact. This was demonstrated by experiments in which F+ and F-cultures were separated either by a porous membrane or a porous plug. Recombinants were not produced in this arrangement, so recombination could not result from movement of genetic material through the growth medium. This led to the use of the term bacterial conjugation of the cell-mating event, and some years later, striking electron micrographs were obtained of conjugating bacteria. A variety of experiments have shown that plasmid transfer can be divided into four stages:

1. Formation of specific donor-recipient pairs (effective contact).
2. Preparation of DNA transfer (mobilization).
3. DNA transfer.
4. Formation of a replicative functional plasmid in the recipient.

When cells containing F and F' come in contact with a recipient, all four steps occur. However, many types of plasmids are genetically unable to carry out ail of these processes. Thus, the following terminology has been developed to describe different types of plasmids:

A conjugative plasmid is a plasmid carrying genes that determine the effective contact function.

A mobilizable plasmid can prepare its DNA for transfer.

A self-transmissible plasmid, such as F, is both conjugative and mobilizable.

All mating systems do not entail the use of pili. For example, some strains of the becterium *Streptococcus faecalis* carry a self-transmissible plasmid. Plasmid-free recipients produce a mating protein (analogous to the pheromones of female insects) that is not made in plasmid-containing (donor) cells. The pheromone causes the donor to synthesize a protein called adhesion that coats the donor cells and causes donor-recipient pairs to form. Once the plasmid is transferred, synthesis if the pheromone is inhibited.

Mobilization and Transfer

Mobilization begins when a plasmid-encoded protein, which is probably the nicking protein of the relaxation complex, makes a single-strand break in a unique base sequence called the transfer origin of in F, or T. (Relaxation complexes have not been detected for all transmissible plasmids but presumably there is a protein serving the function just described.) This nick initiates rolling circle replication and the linear branch of the rolling circle is transferred. It is thought that the nicking protein remains bound to the 5' terminus and that the replication mode is like the looped rolling-circle mechanism used by phage Q × 174. The sequence of events during transfer is shown schematically in Fig. 35.4.

Note the DNA synthesis occurs both in donor and recipient cells. The synthesis in the donor, called donor conjugal DNA synthesis, serves to replace the single strand that is transferred. Synthesis in the recipient cell (recipient conjugal DNA synthesis) converts the transferred single strand to double stranded DNA.

In the usual situation during transfer the transferred strand is simultaneously replaced by donor conjugal synthesis and converted to double-stranded DNA is the recipient. This would indicate that DNA synthesis and transfer are coupled were it not for the following observations: (1) Transfer occurs even if donor synthesis in inhibited by appropriate host mutations, such as a temperature-sensitive mutation in the polymerase III gene; (2) donor synthesis occurs even if transfer is prevented by appropriate plasmid mutations; (3) transfer occurs even though recipient conjugal synthesis is inhibited by mutations in the recipient cell. These observations raise the question of the identity of the motive force for transfer since clearly it is not DNA replication. This question has not yet been answered.

PLASMID REPLICATION

A plasmid can replicate only within a host cell, so we might expect that all plasmids native to the same host species would have the same mode of replication. However, as with phages, there is enormous variation in both the enzymology and mechanics of plasmid DNA replication, as will be seen in the following sections.

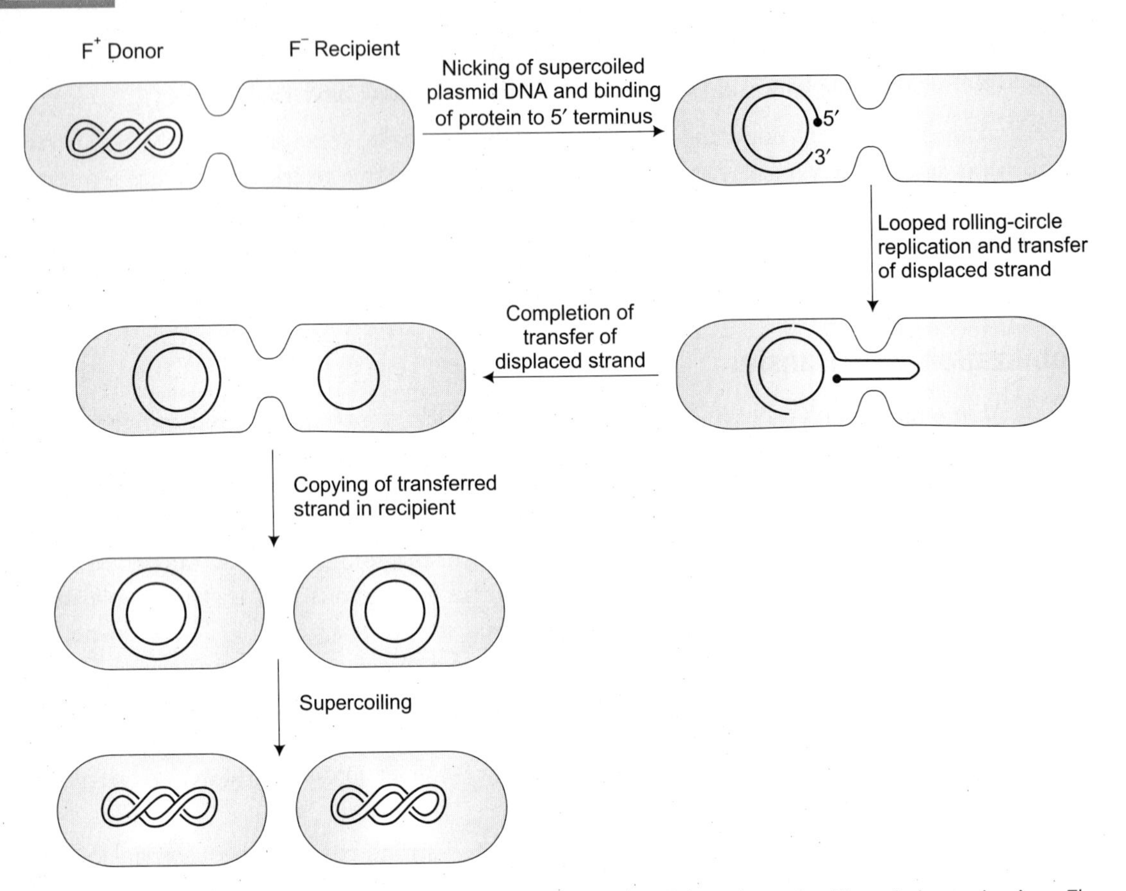

Fig. 35.4 A model for transfer of F plasmid DNA from an F^+ cell by a looped rolling-circle mechanism. The displaced single strand is transferred to the F^- recipient cell, where it is converted to double-stranded DNA. Chromosomal DNA and proteins of the relaxation complex is completed DNA molecules are omitted for clarity.

Variations in the Use of Host Proteins in Replication Mechanisms

Plasmids rely heavily on the host replication proteins for their replication. However, even though polymerase III is the major replication protein in *E. coli*, the particular DNA polymerase responsible for chain growth is not the same for all *E. coli* plasmids. For example, an F^+*polA*- culture, which has much reduced activity of DNA polymerase I, continues to produce F+ progeny whereas ColE1 fails to replicate unless the host possesses active polymerase I. This is because chain growth of F utilizes Pol III, but ColE1 uses Pol I for the initial stages of replication. Some plasmids use host gene products exclusively. For example, ColE1 can be replicated in vitro by adding purified ColE1 DNA to cell extract prepared from cells that do not contain ColE1 or any other plasmid; clearly, such an extract cannot contain any plasmid-encoded gene products. For instance, F' lac plasmids carrying a temperature-sensitive mutation in F have been isolated that fail to replicate at 42°C; that is, at this temperature, *F'(Ts)lac+/lac- cells* produce F-lac-daughter cells after several generations of growth at 42°C.

All plasmids examined to date replicate semiconservatively and maintain circularity throughout the replication cycle. However, there are significant differences in the replication pattern from one plasmid to the next. For example, some plasmid replicate unidirectionally and others bidirectionaaly. The plasmid RK6 replicates first in one direction and then later in the opposite direction from the same origin. The bidirectionally replicating plasmids terminate replication in two ways. In one type, termination occurs when the growing forks collide. Others have a fixed termination site that is sometimes reached by one growing fork before the other fork reaches it. In the most carefully studied plasmids, replication occurs by the so-called butterfly or rabbit's ears mode (Fig. 35.5) in which a partially replicated molecule contains untwisted replicated potions, as is usually the case in replication and a supercoiled unreplicated portion. When the replication cycle is completed, one of the circles must be cleaved in order to separate the daughters. The result after one round of replication is one nicked molecule and one supercoiled molecule. The nicked molecule is then sealed and, somewhat later, supercoiled. Whether this is a general mechanism for plasmid replication is not known.

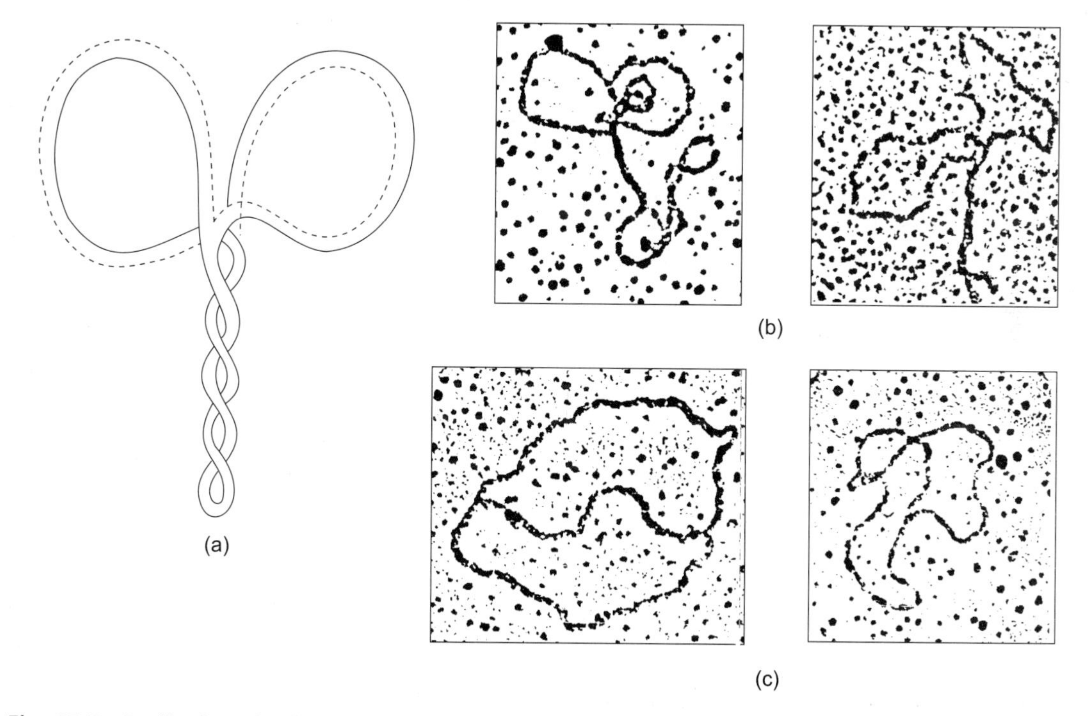

Fig. 35.5 Replication of ColE1 DNA. (a) Diagram of butterfly molecule. The newly synthesized strands are dotted. (b) Electron micrographs of butterfly molecules. (c) Nicked butterfly molecules showing that a nick converts a butterfly molecule to a q molecule [Courtesy of Donald Helinski].

PROPERTIES OF PLASMIDS

Bacterial Plasmids

In this section we examine certain plasmids with respect to plasmid-specific properties that have not yet been described.

F Plasmids

An important property of F is its ability to integrate into the bacterial chromosome to generate an Hfr cell. Integration is a reciprocal exchange much like that occurring when phage lysogenizes a bacterium. However, integration of F into the *E. coli* chromosome differs from prophage formation in that there are several possible exchange sites in F and many sites in the chromosome at which integration can occur. More than 20 major sites and nearly 100 minor sites in the chromosome are known. The affinity of F for each site is not the same.

Excision of F also occurs, though this is quite rare (at some locations, excision occurs more frequently than the average). Often excision is imperfect and one cut is made at one of the two termini within the integrated F and a second cut is in the adjacent chromosomal DNA. In such a case the new F plasmids contains chromosomal genes which may or may not be identifiable. This aberrant excision is the origin of the F' plasmids.

Integration of F into certain bacterial mutants that have defects in DNA replication gives rise to a phenomenon called integrative superession. F uses many *E. coli* replication proteins but nay not always use ,the bacterial dnaA-gene product, which is necessary to initiate chromosome DNA replication. As stated earlier, plasmids generally carry their own genes for replication initiation. Thus, if F as a free plasmid, is contained inan *E. coli* dnaA(Ts) mutant and the temperature is raised to 42°C (which inactivates the mutant dnaA protein), initiation of chromosomal replication is no longer possible; however, F can still replicate. In constrast, ina dnaA(Ts) mutant strain in which F is integrated (anHfr strain) chromosomal replication can occur indefinitely at the high temperature. However, replication is not initiated at the E.coli replication origin but instead is initiated at the oriV site for F replication. Thus, integration of F suppresses the DnaA(Ts) phenotype by providing a DnaA-independent replication origin.

Screening for integrative suppression is one way to isolate a strain in which F is integrated. For example, an *F'lac+/str-s* male can be mated with a *lac+dnaA*(Ts)str-r strain and *Lac+Str-r* cells can be selected by plating at 42°C on a color-indicator plate containing streptomycin. All surviving colonies should contain an integrated F' lac. As a test, cultures can be derived from a few colonies and grown in the presence of acridine organge. Lac-segregants should not appear because the plasmid is integrated.

The Drug-resistance (R) Plasmids

The drug-resistance, or R, plasmids were originally isolated from the bacterium *Shigella dysenteriae* during an outbreak of dysentery in Japan and have since been found in *E. coli* and other bacteria. Their defining characteristics are that they confer resistance on their host cell to a variety of fungal antibiotics and are usually self-transmissible. Most R plasimids consist of two contiguous segments of DNA. One of these segments is called **RTF (resistance transfer factor)**; it carries genes regulating DNA replication and copy number, the transfer genes, and sometimes the gene for tetracycline resistance (*tet*), and has a molecular weight of 11 × 10. The other segment, sometimes called the r determinant, is variable in .size (from a few million to more than 100 × 1 0 molecular weight units) and carries other genes for antibiotic resistance. Tesistance to the drugs ampicillin (Amp), chloramphenicol (Cam), streptomycin(Str), kanamycin (Kan), and sulfonamide (Sul), in combinations of one or more, appears commonly. Small drug resistance plasmids, lacking the ability to transfer but still containing the tet gene, are also known; one of these, pSC101, whose molecular weight is 5.8 × 10, is commonly used ingenetic engineering. The two component R plasmid are reminiscent of F' plasmids, but the drug-resistance gene are not acquired by integration of RTF followed by aberrant excision; instead, they result from acquisition of transposons that carry antibiotic-resistance genes.

R plasmids are of considerable medical interest since they can be transferred to bacteria that cause major epidemics, such as S. typhimurium and shigella dysenteriae, and to strains that cause infections in hospitals (various enterobacteria, *Pseudomonas aeruginosa*, and *Staphylococcus aureus*). In fact, it has become clear that since the beginning of the "antibiotic era", R plasmids have been on the increase in nature. For example, penicillin was introduced to general use in the early 1940s. By 1946 14 percent of strains of *S. aureus* isolated in hospitals were Pen-r. The fraction was 38 percent in 1947, 59 percent in 1969, and nearly 100 percent by the 1970s. the majority of these resistant strains either carry an R plasmid or a pen-r gene of the type found in R plasmids. A particular problem has been the transfer of R+ bacteria from animals to man. For example, animal carcasses, especially poultry, are frequently contaminated with R+ *E. coli*, which can colonize the human intestine, and *S. typhimurium*. Studies by the U.S. government have shown that handling of raw meat is the usual route for transmission to man; bacteria from the meat get on utensils and kitchen surface, and later to humans. Numerous drug-resistant epidemics of salmonellosis in farm animals occurred between 1960 and 1980 on Great Britain, and the causative organisms were quickly transmitted to man. The infection could not easily be combated because of resistance to standard antibiotics. The major cause of the infection of animals is apparently the extensive use of penicillin and tetracycline in animal feed; these substances lead to more rapid growth of the animals and hence are economically valuable. Nonetheless, in the early 1980s the United Kingdom Government Committee recommended that all antibiotics be eliminated from animal feed.

THE COLICINOGENIC OR COL PLASMIDS

Col plasmid are *E. coli* plasmids able to produce colicins, proteins that prevent growth of susceptible bacterial strains that do not contain a Col plasmid. They are one class of a general type of plasmid called a bacteriocinogenic plasmid, which produce bacteriocins in many bacterial species. Bacteriocins, of which colicins are one example, are proteins that can bind to the cell wall of a sensitive bacterium and inhibit one or more essential processes such as replication, transcription, translation, or energy metabolism. There are many types of colicins, each designated by a latter (e.g. colicin B) and each having a particular mode of inhibition of sensitive cells (Table 35.2) Colicin production is detected by an assay similar to that for detecting phage. A colin-producing cell is placed on a lawn of sensitive cells; the colicin inhibits growth of nearby bacteria, producing clear area, known as a lacuna, in the turbid of bacteria.

Colicin production is normally repressed but can be induced by a variety of agents. One that is commonly used is ultemviolet light. A small fraction of every population of Col+ cells also produces colicin constitutively. Thus, the presence of the Col plasmid has survival value for the population, which is thereby able to compete more effectively in the wild with colicin sensitive cells.

Colicins are probably of two-types true colicins and defective phage particles. The latter class is inferred from studies of many purified colicins. Only a few colicins are simple proteins; others look like phage tails when examined by electron microscopy and are thought to be gene products transcribed from remnants of ancient prophages. The hypothesis is that repeated mutation has resulted in loss of the genes for replication and head production, but genes encoding a repressor system, the lysis enzyme, and the tail proteins have survived intact. Presumably these phages share with phage T4 the property that adsorption without DNA injection causes an inhibition of macromolecular synthesis. True colicins also bind to specific cell functions. Table 35.2 shows that colicins act in a variety of ways. A surprising feature of colicin activity is that in some cases one colicin molecule is sufficient to kill a target cell.

In most cases colicins are inactive against a cell that contains a related Col plasmid. This is called immunity, Interstingly, immunity is conferred by an excreted small protein that binds to the larger colicin protein. The immunity protein not only confers immunity on Col+ cell but is also necessary for killing of Col-cell. In the case of the colicin clocin DF13 the role of the immunity protein is understood somewhat. The DF13 protein consists of three regions a receptor-binding region, an RNase and a segment that binds the immunity protein. The receptor-binding terminus is very hydrophobic and may be able to adsorb to the cell membrane. The immunity-binding segment has strong negative charge that is neutralized by the positively charged immunity protein. After binding to the receptor, the colicin is cleaved. The N-terminal region remains outside the cell and the RNase segment enters the cell, leaving the immunity protein on the cell surface. The RNase acts on the RNA in ribosomes and thereby kills the sensitive cells.

Table 35.2 Properties of several *E. coli* colicins.

Colicin	*Action of Colincin*
Colicin B Colicin 1b	Damages cytoplasmic membrance
Colicin E1 Colicin K	Uncouples energy-dependent processes by an unknown effect on the cell membrane
Colicin E2	Degrades DNA
Colicin E3	Cleaves 16S rRNA

The Col plasmids range in size from a molecular weight of a few million, for those that are not self-transmissible, to more than 60×10^6, for the self-transmissible plasmids. The best-studied Col plasmid is ColE1, Whose molecular weight is 4.4×10^6. It's complete sequence of 6646 base pairs is known. It is used extensively in recombinant DNA research (Chaper 20). Many of the large self-transmissible Col plasmids are hybrids between a small Col plasmid and F or F plasmids.

The *Agrobacterium* Plasmid Ti

A crown gall tumor found in many dicotyledonous plants is caused by the bacterium *Agrobacterium tumefaciens*. The tumar-causing ability resides in a plasmid called Ti. When a plant is infected, some of the bacteria enter and grow within the plant cells and lyse there, releasing their DNA in the cell. From this point on, the bacteria are no longer necessary for tumor formation. By an unknown mechanis a small fragment of the *Ti* plasmid (the T DNA), containing the genes for replication, becomes integrated fragment breaks down the hormonally regulated system that controls cell division and the cell is converted to tumor cell. This plasmid has recently become very important in plant breeding because specific genes can be inserted into the *Ti* plasmid by recombinant DNA techniques, and sometimes these genes can become integrated into the plant chromosome, thereby, permanently changing the genotype and phenotype of the plant. New plant varieties having desirable and economically valuable characteristics derived from unrelated species can be developed in this way.

Broad Host Range Plasmids

Most plasmids can exist in only a limited number of closely related bacteria; these are called narrow host range plasmids. However, the self-transmissible R plasmids of incompatibility group IncP of *E. coli* and of group IncP1 of *Pseudomonas aeruginosa* are notable in that they can be transferred to and maintained in bacteria of a large number of species. These are called broad host range plasmids. Why some plasmids have a narrow host range and others a broad host range is unknown. The broad host range plasmids have recently become exceedingly useful since they integrate (albeit at low frequency) into the host chromosome of numerous species and have thereby enabled the establishment of genetic systems in species in which mapping had not previously been possible. In this way, genetic maps have been obtained for several economically important bacteria; such mapping facilitates the construction of strains with desirable characteristics.

Most of these plasmids are able to mobilize the chromosome only at very low frequency (ca. 10^{-8} per cell). However, by various genetic and recombinant DNA techniques variants of these plasmids have been constructed in which chromosome mobilization is enhanced by a factor of 10°–10°.

Other Plasmids

Several plasmids render fairly innocuous plasmid-free cells pathogenic. For example, the Ent plasmids of *E. coli* synthesize enterotoxins that are responsible for travelers diarrhea. A plasmid called Hly (for hemolysis) has been found in *E. coli* strains isolated from pigs. Whereas the hemolysin destroys red blood cells in blood samples, the Hly plasmid does not seem to cause any pathogenicity. Certain plasmids residing in the human pathogen, staphylococcus aureus, enhance pathogenicity. A penicillinase (penicillin-destroying) S. aureus plasmid has been carefully studied.

Many species of *pseudomonas* can utilize several hundred organic compounds as carbon sources-in particular, such toxic substances as camphor, toluene, and octane. This metabolic ability sides in a set of plasmids collectively known as **degradation plasmids**. Each plasmid provides one or more metabolic pathways to degrade these compounds. Since many enzymes are needed, the plasmids are fairly large (molecular weight of 50 – 100 × 10*). These plasmids enable bacteria to degrade many synthetic compounds and hence are making an important contribution in the removal of environmental pollutants. For example, some strains can degrade persistent chlorinated hydrocarbons, herbicides, and various detergents. Many laboratories are attempting to use genetic engineering to construct "super plasmids", which can be used for pollution control and chemical syntheses.

Plasmids have also been isolated that confer resistance to toxic metal ions. Such plasmids, whose origin is unknown, are found in environments containing these ions, such as the sludge produced by industrial reprocessing to photographic film (resistance to the Ag^+ ion). In only one case has the biochemistry of resistance been elucidated: resistance to Hg^{2+} ions results from a plasmid-encoded reductase that converts Hg^{2+} to metallic mercury, which is sufficiently volatile that it evaporates away.

CHAPTER 36

Transposable Elements

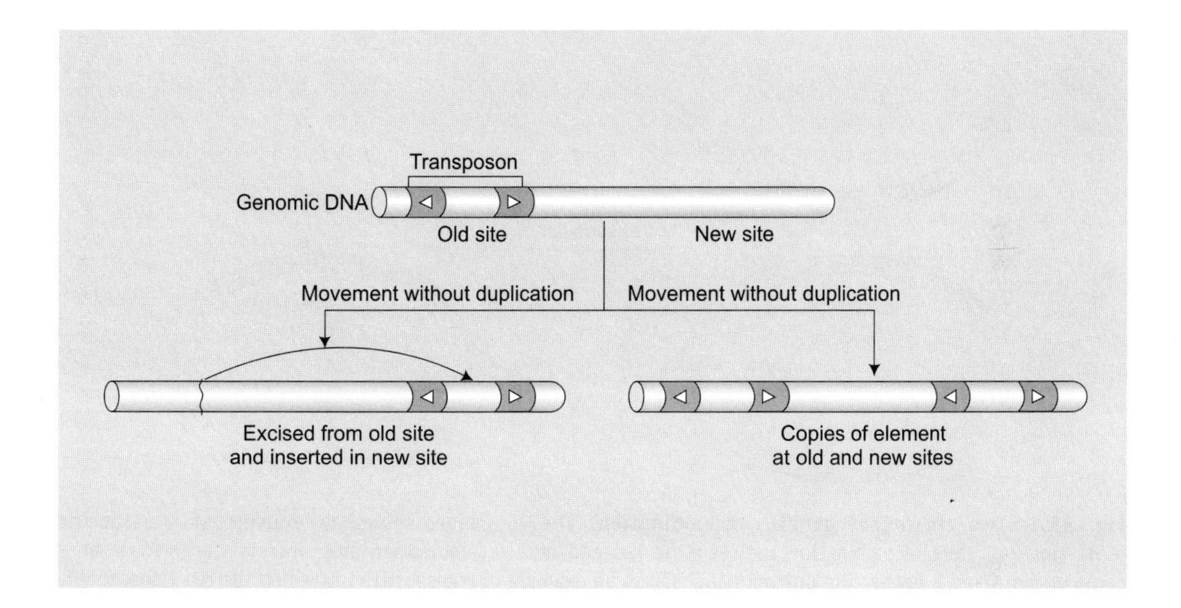

DNA is a very stable molecule. DNA replication, repair, and homologous recombination, as we have learned in the previous chapters, all occur with high fidelity. These processes serve to ensure that the genomes of an organism are nearly identical from one generation to the next. Importantly, however, there are also genetic processes that rearrange DNA sequences and thus lead to a more dynamic genome structure. These processes are the subject of this chapter.

Two classes of genetic recombination, conservative site-specific recombination (CSSR) and transpositional recombination (generally called transpostion), are responsible for many important DNA rearrangements. CSSR is recombination between two defined sequences elements (Fig. 36.1). Transposition, in contrast, is recombination between specific sequences and nonspecific DNA sites. The biological processes promoted by these recombination reactions include the insertion of viral genomes into the DNA of the host cell during infection,

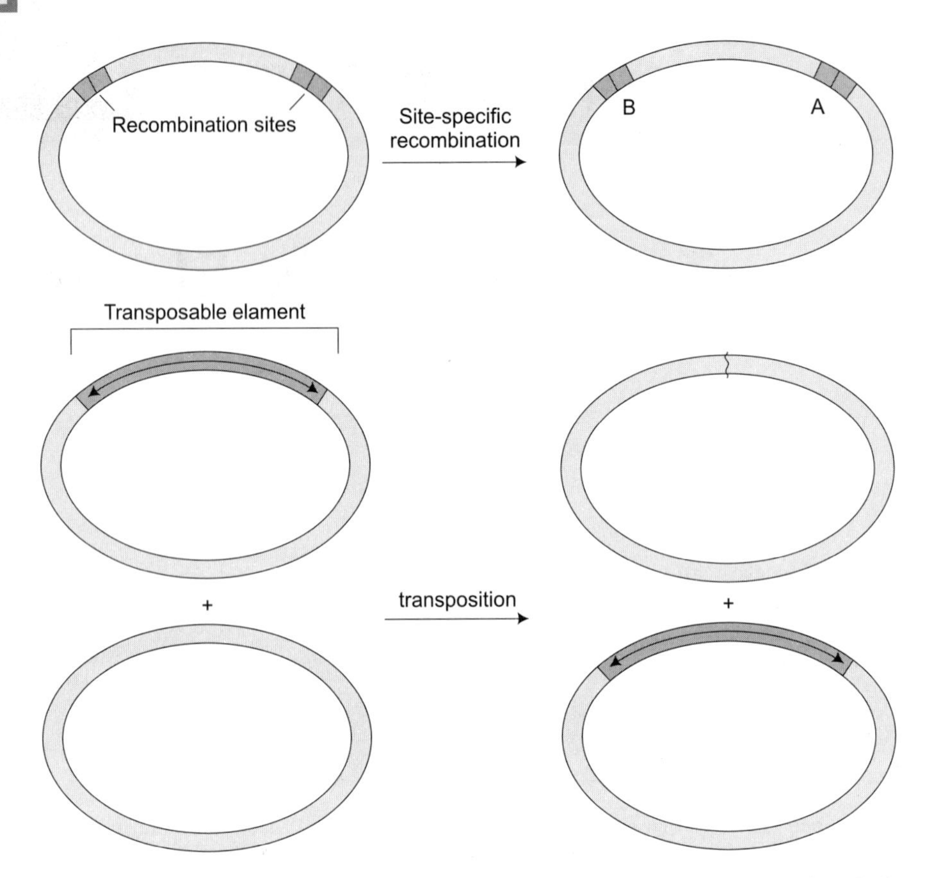

Fig. 36.1 Two classes of genetic recombination. The top panel shows an example of site-specific recombination. Here recombination between the red and blue recombination sites inverts the DNA segment carrying the A and B genes. The bottom panel shows an example of transposition in which the red transposable element excises from the gray DNA and inserts into an unrelated site in the blue DNA.

the inversion of DNA segments to alter gene structure, and the movement of transposable elements-often called "jumping" genes from one chromosomal site to another.

The impact of these DNA rearrangements on chromosome structure and function is profound. In many organisms, transpostion is the major source of spontaneous mutation and nearly half the human genome consists of sequences derived from transposable elements. Furthermore, as we will see, both viral infection and development of the vertebrate immune system depend critically on these specialized DNA rearrangements (Fig. 36.2).

Conservative site-specific recombination and transpostiton share key mechanistic features. Proteins known as recombinases recognize specific sequences where recombination will occur within DNA molecules. The recombinases bring these specific sites together to from a protein DNA complex bridging the DNA complex bridging the DNA sites, known as the synaptic complex. Within the synaptic complex, the recombinase catalyzes the cleavage and rejoining of

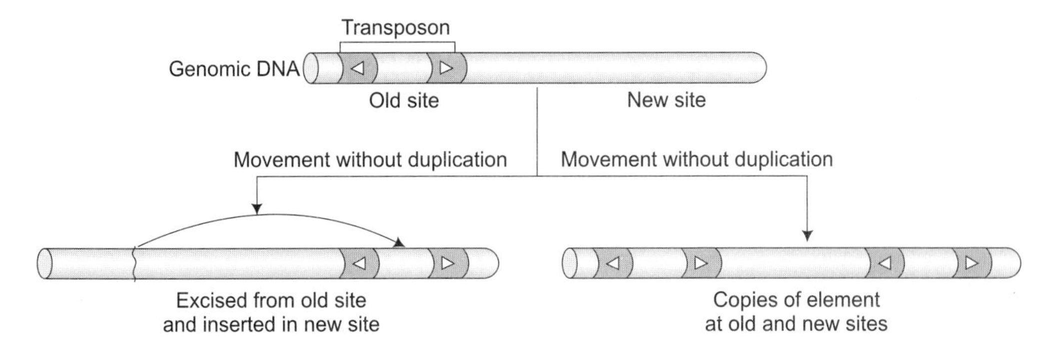

Fig. 36.2 Transposition of a mobile genetic element to a new site in the host DNA. Recombination, in some cases, involves excision of the transposon from the old DNA location (left). In other cases, one copy of the transposon stays at the old location and another copy is inserted into the new DNA site.

the DNA molecules either to invert a DNA segment or to move a segment to a new site, one recombinase protein is usually responsible for all these steps. Both types of recombination are also carefully controlled such that the danger to the cell of introducing breaks in the DNA, and rearranging DNA segments, is minimized. As we shall see, however the two types of recombination also have key mechanistic differences.

Transposable elements are present in the genomes of all life-forms. The comparative analysis of genome sequences reveals two fascinating observations. First, transposon-related sequence can make up huge fractions of the genome of an organism. For example, more than 50% of both the human and maize genomes are composed of transposon-related DNA sequence. This is in sharp contrast to the small percentage (< 2% in human) of the sequence that actually encodes cellular proteins. Second, the transposon content in different genomes is highly variable (Fig. 36.3). For example, compared to human or maize, the fly and yeast genomes are very "gene-rich" and "transposon-poor".

There are many different types of transposable elements. These elements can be divided into families that share common aspects of structure and recombination mechanism. In the following sections, we introduce the three major families of transposable elements and the recombination mechanism associated with each family. Some of the best-studied individual elements will then be described. In the description of individual elements, we focus on how transposition is regulated to balance the maintenance and propagation of these elements with their potential to disrupt or misregulate genes within the best host organism.

The genetic recombination mechanisms responsible for transposition are also used for functions other than the movement of transposons. For example, many viruses use a recombination pathway nearly identical to transposition to integrate into the genome of the host cell during infection. These Viral integration reactions will therefore be considered to gather with transposition. Likewise, some DNA rearrangements used by cells to alter expression patterns occur using a mechanism very similar to DNA transposition. V {D} J recombination,

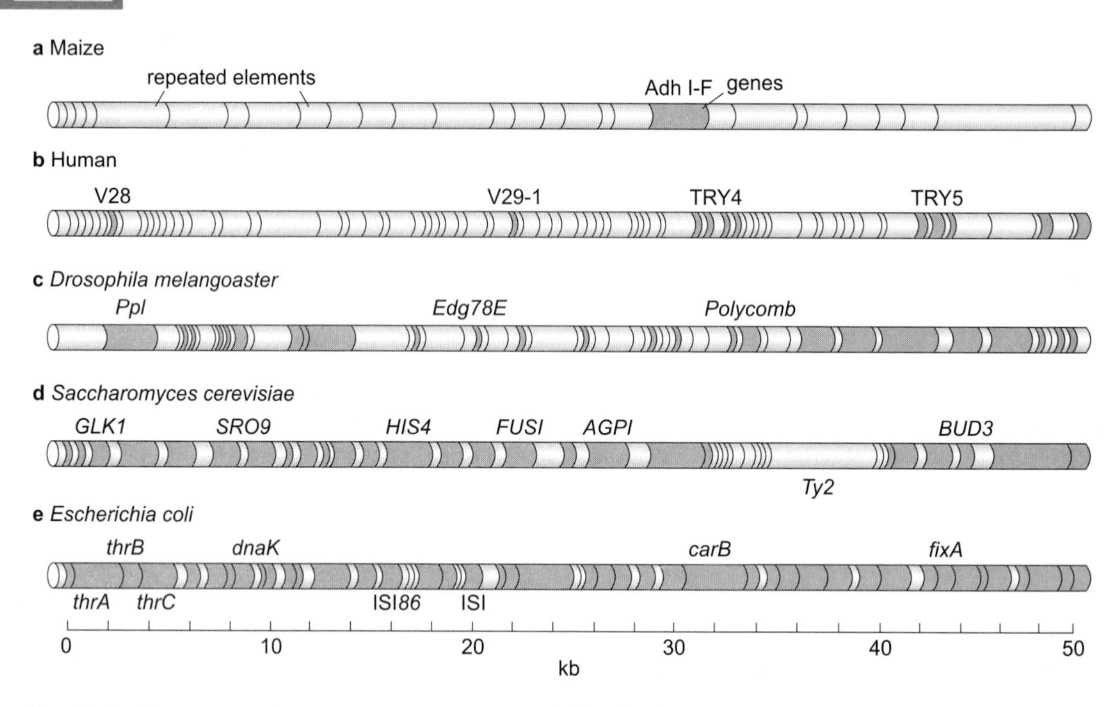

Fig. 36.3 Transposons in genomes: occurrence and distribution. Repeated elements, mostly composed of transposons or transposon-related sequences (such as truncated elements) are shown in green. Cellular genes are shown in blue. (a) Maize. (b) Human. (c) Drosophila. (d) Budding yeast. (e) *E. coli.* (Source: From Brown T.A. 2002. Genomes, 2nd edition, p. 34, Fig. 2.2 and references therein. Copyright © 2002).

a reaction required for development of a functional immune system in vertebrates, is well-understood example: V{D}J recombination is discussed at the end of this chapter.

THERE ARE THREE PRINCIPAL CLASSES OF TRANSPOSABLE ELEMENTS

Transposons can be divided into the following three families on the basis of their overall organization and mechanism of transposition:

1. DNA transposons.
2. Viral-like retrotransposons—this class includes the retroviruses. These elements are also called LTR retrotransposons.
3. Poly-A retrotransposons. These elements are also called nonviral retrotransposons.

Fig. 36.4 shows a schematic of the general genetic organization of each of theses element families. DNA transposons remain as DNA throughout a cycle of recombination. They move using mechanisms that involve the cleavage and rejoining of DNA strand, and in this way they are similar to elements that move by conservative site-specific recombination. Both types of retrotransposons move to a new DNA location using a transient RNA intemediate.

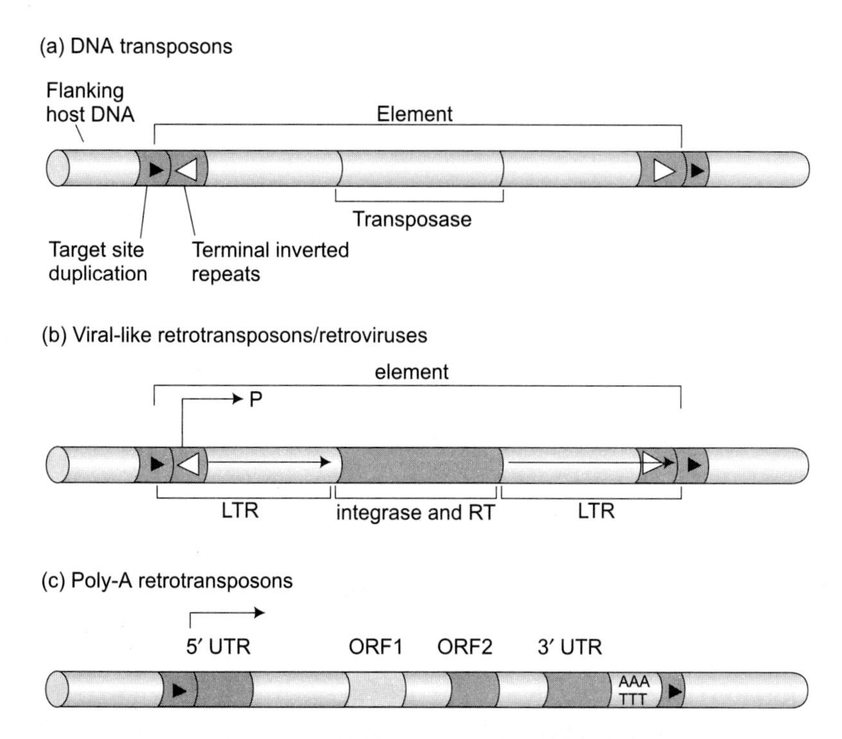

Fig. 36.4 Genetic organization of the three classes of transposable elements. (a) DNA transposons. The element includes the terminal inverted repeat sequences (green arrows) which are the recombination sites, and a gene encoding transposase. (b) Viral-like retrotransposons and retroviruses. The element includes two long terminal repeat (LTR) sequences that flank a region encoding two enzymes, integrase and reverse transciptase (RT). (c) Poly-A retrotransposons. The element terminates in the 5¢ and 3¢ UTR sequences and encodes two enzymes, an RNA-binding enzyme (ORF1) and an enzyme having both reverse transcriptase and endonuclease activities (ORF2).

DNA TRANSPOSONS CARRY A TRANSPOSASE GENE, FLANKED BY RECOMBINATION SITES

DNA transposons carry both DNA sequence that function as recombination sites and genes encoding proteins that participate in recombination (Fig. 36.4). The recombination sites are at the two ends of the element and are organized as inverted repeat sequences. These terminal inverted repeats vary in length from about 25 to a few hundred base pairs, are not exact sequences repeats, and carry the recombinase recognition sequences. The recombinases responsible for transposition are usually called transposases (or, sometimes, integrases).

DNA transposons carry a gene encoding their own transposase. They may carry a few additional genes, sometimes encoding proteins that regulate transposition or provide a function useful to the element or its host cell. For example, many bacterial DNA transposons carry genes encoding proteins that promote resistance to one or more antibiotic. The presence of the transposon, therefore, causes the host cell to be resistant to that antibiotic.

The DNA sequences immediately flanking the transposon have a short (2 to 20 bp) segment of duplicated sequence. These segments are organized as direct repeats, are called target site duplications, and are generated during the process of recombination, as we shall discuss below.

TRANSPOSONS EXIST AS BOTH AUTONOMOUS AND NONAUTONOMOUS ELEMENTS

DNA transposons that carry a pair of terminal inverted repeats and a transposase gene have everything they need to promote their own transposition. These elements are called autonomous transposons. However, genomes also certain many even simpler mobile DNA segments known as nonautonomous transposons. These elements carry only the terminal inverted repeats that is, the cis-acting sequences needed for transposition. In a cell that also carries an autonomous transposon, encoding a transposases that will recognize thee terminal inverted repeats, the nonautonomous elements will be able to transpose. However, in the absence of this "helper" transposon {to donate the transposase}, nonautonomous elements remain frozen, unable to move.

VIRAL-UKE RETROTRANSPOSONS AND RETROVIRUSES CARRY TERMINAL REPEAT SEQUENCES AND TWO GENES IMPORTANT FOR RECOMBINATION

Viral-like retrotransposons and retroviruses also carry inverted terminal repeat sequence that are the sites of recombinase binding and action (Fig. 36.4b). the terminal inverted repeats are embedded with in longer repeated sequences; these sequences are organized on two ends of the elements as direct repeats and are called long terminal repeats or LTRs. Viral-like retrotransposons encode two proteins needed for their mobility: integrase (the transposase) and reverse transcriptase.

Reverse transcriptase {RT} is a special type of DNA polymerase that can use an RNA template to synthesize DNA. This enzyme is needed for transposition because an RNA intermediate is required for the transposition reaction. Because these elements convert RNA into DNA, the reverse of the normal pathway of biological information flow (DNA to RNA), they are known as "retro" element. The distinction between viral-like retrotransposons and retroviruses is that the genome of a retrovirus is packaged into a viral particle, escapes its host cell, and infects, a new cell. In contrast, the retrotransposons can move only to new DNA sites with in a cell but never leave that cell. Like the DNA transposons, these elements are flanked by short target site duplications that are generated during recombination.

POLY-A RETROTRANSPOSONS LOOK LIKE GENES

The poly-A retrotransposons do not have the terminal inverted repeats present in the other transposon classes. Instead, the two ends of the elements have distinct sequences (Fig. 36.4c).

One end is called the 5'UTR (for untranslated region) whereas the other end has a region called the 3' UTR followed by a stretch of A- T base pairs called the poly-A sequence. These elements are also flanked by short target site duplications.

Retrotransposons carry two genes, know as ORF1 and ORF2. ORF1 encodes an RNA-binding protein. ORF2 encodes a protein with both reverse transcriptase activity and an endonuclease activity. This protein, although distinct from the transposases and integrases encoded by the other classes of elements plays essential roles during recombination. Like their DNA and viral-like transposon counterparts, poly-A tetrotransposons exist commonly in both autonomous and nonautonomous forms. Furthermore, genome sequence DNA analysis reveals that there are many truncated elements that do not have a complete 5'UTR sequence and have lost their ability to transpose.

A TRANSPOSITION BY A CUT-AND-PASTE MECHANISM

DNA transposons, viral-like retrotransposons, and retroviruses all use a similar mechanism of recombination to insert their DNA into a new site. First, let us consider the simplest transposition reaction: the movement of a DNA transposon by a nonreplicative mechanism. This recombination pathway involves the excision of the transposon from its initial location in the host DNA followed by integration of this excised transposon into a new DNA site. This mechanism is therefore called cut-and-paste transpostion (Fig. 36.5).

To initiate recombination, the transposase binds to the terminal inverted repeats at the end of the transposon. Once the transposase recognizes these sequences, it brings the two ends of the transposon DNA together to generate a stable protein-DNA complex. This complex is called the synaptic complex or transpososome. It contains a multimer of transposase-usually two or four subunits-and the two DNA ends (see below). This complex function to ensure that the DNA cleavage and joining reactions needed to move the transposon occur simultaneously on the two ends of the element's DNA. It also protects the DNA ends from cellular enzymes during recombination.

The next step is the excision of the transposon DNA from its original location in the genome. To achive this, the transposase subunits within the transposome first cleave one DNA strand at each of the transposon, exactly at the junction between the transposon DNA and the host sequence in which it is inserted (a region called the flanking host DNA). The transposase cleaves the DNA such that the transoposon sequence terminates with free 3′OH groups at each end of the element's DNA. To finish the excision reaction, the other DNA strand at each end of the elements must also be cleaved. Different transposons use different mechanisms to cleave these "second" DNA strands (those strands that terminate with 5′ ends at the transposon host DNA junction). These mechanisms are described in a following section.

After excision of the transposon, the 3′OH ends of the transposon DNA-the ends first liberated by the transposase-attack the DNA phosphodister bonds at the site of the new insertion. This DNA segment is called the target DNA. Recall that for most transposons, the

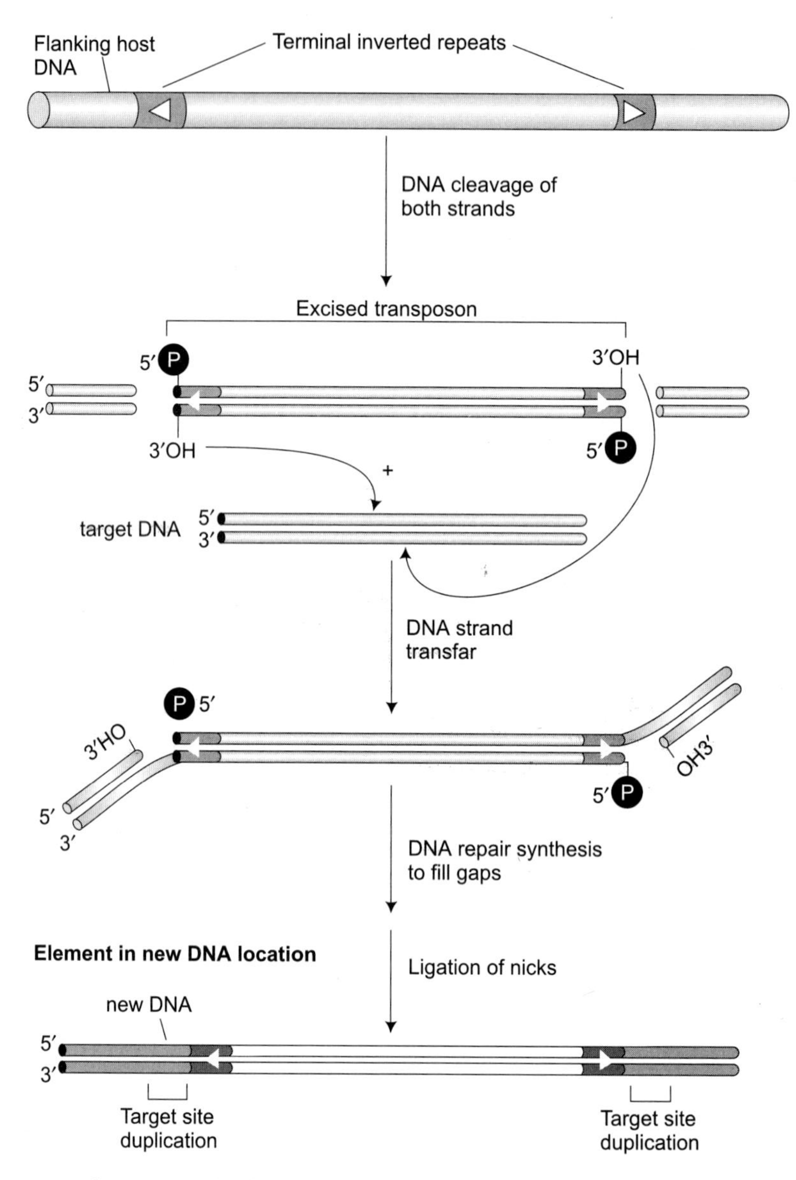

Fig. 36.5 The cut-and-paste mechanism of transposition. The Fig. shows movement of a transposon from a target site in the gray host DNA to a new site in the blue DNA. Note the staggered cleavage sites on the target DNA during the DNA strand transfer reaction that give rise to short repeated sequences at the new target site (the target site duplications) The DNA at the original insertion site (here in gray) will be left with a double stranded DNA break as a result of transposon excision. This break can be repaired by nonhomologous end joing or homologous recombination.

target DNA can have essentially any sequences as a result of this attack, the transposon DNA is covalently joined to the DNA at the target site. During each DNA joining reaction, a nick is also introduced into the target DNA (Fig. 36.5). This DNA joining reaction occurs.

EXAMPLES OF TRANSPOSABLE ELEMENTS AND THEIR REGULATION

Transposons have successfully invaded and colonized the genomes of all life-forms. Clearly they are very robust biological entities. Some of this success can be attributed to the fact that transposition is regulated in ways that help to establish a harmonious coexistence with the host cell. This coexistence is essential for the survival of the elements, as transposons cannot exist without a host organism. On the other hand, as introduced above transposons can wreak havoc in a cell, causing insertion mutations, altering gene expression and promoting large-scale DNA rearrangements. These disruptions are particularly noticeable in plants, a feature that led to the discovery of transposons in maize (Box 36.1: Maize Elements and the discovery of Transposons).

In the following sections we briefly describe some of the best-understood individual transposons and transposon families. (A larger list of transposons and some of their important features is summarized in Table 36.1). Each subsection provides a brief overview of a specific element and an example of regulation that is of particular importance to the element. As we will see two type of regulation appear as recurring themes:

- Transposons control the number of their copies present in a given cell. By regulating copy number, these elements limit their deleterious impact on the genome of the host cell.
- Transposons control target site choice. Two general types of target site regulation are observed. In the first, some elements preferentially insert into regions of the chromosome that tend not to be harmful to the host cell. These regions are called safe havens for transposons. In the second type of regulation, some transposons specifically avoid transposing into their own DNA. This phenomenon is called transposition target immunity.

IS4-F AMIL Y TRANSPOSONS ARE COMPACT ELEMENTS WITH MULTIPLE MECHANISMS FOR COPY NUMBER CONTROL

The bacterial transposn Tnl0 is a well-characterized representative of the IS4 family which also includes Tn5. Tn10 is a compact element of 9kb and encodes a gene for its own transposase and genes imparting resistance to the antibiotic tetracycline (Fig. 36.6)

Tnl0 transposes via the cut-and-paste mechanism (described above), using the DNA hairpin strategy to cleave the nontransferred strand (Fig.s 11.19 and 11.21). The Tnl0 sequence also has a site for IHF binding. IHF helps in the assembly of proper transposome complex needed for recombination as it does during phage integration (see above).

Box 36.1 Maize Elements and the Discovery of Transposons

Plant genomes are very rich in transposons. Furthermore, the ability of transposable elements to alter gene expression can often be readily observed as dramatic variation in the coloration of the plant (Box 36.1 Fig. 1). Thus, it is not surprising that transposable elements, and many of their salient features, were first discovered in plant.

Barbara McClintock discovered "controlling elements" in maize in the late 1940s. It was actually the ability of transoable elements to break chromosomes that first come to McClintock's attention. She found that some strains experienced broken chromosome very frequently, and she named the genetic element responsible for these chromosome breaks Ds (dissociator). Surprisingly, she observed that the sites of these "hotspots" for chromosome breaks were different strains, and could even be in different chromosomal locations in the descendents of an individual plant. This observation provided the first insight that genetic elements could move, that is "transpose" within chromosomes.

Ds, in fact, is a nonautonomous DNA transposon that moves by cut and paste transposition. Ds movement requires the Ac (activator) element-also discovered by McClintock-to be present in the same cell and provide the transposase protein. Ac is now recognized to be part of a large fly of DNA transposons called the hAT family named for the hobo elements from files, the Ac elements from maize, and the Tam elements from snapdragon.

Tn10 limits its copy number in any given cell by strategies that restrict its transposition frequency. One mechanism is the use of an antisense RNA to control the expression of the transposase gene. Near the end of ISIOR are two promoters that direct the synthesis of RNA by the host cell's RNA polymerase. The promoter that direct RNA synthesis inward (called Pin) is responsible for expression of the transposase gene. The promoter that direct transcription outward (P out), in contrast, serves to regulate transposase expression by making an antisense RNA, as follows. The RNAs synthesized from Pin and Pout overlap (by 36 base pairs) and therefore can pair by hydrogen bonding between these overlapping (complementary) regions. This pairing prevents binding of ribosomes to the Pin transcript, and thus synthesis of the transposase protein.

By this mechanism cells that carry more copies of Tn 10 will transcribe more of the antisense RNA, which in turn will limit expression of the transposase gene (Fig. 36.7), see legend for more details). The transposition frequency will, therefore be very low in such a strain. In contrast, if there is only one copy of TnlO in the cell, the level of antisense RNA will be low, synthesis of the transposable protein will be efficient, and transposition will occur at a higher frequency.

Fig. 1 Example of color variegation in snapdragon flowers due to Tam 3 transposition. The size of white patches is related to the frequency of transposition (Source: Chatterjee M. and Martin C. 1997. The Plant Journal 11:759–771, Fig. 2a, page 762).

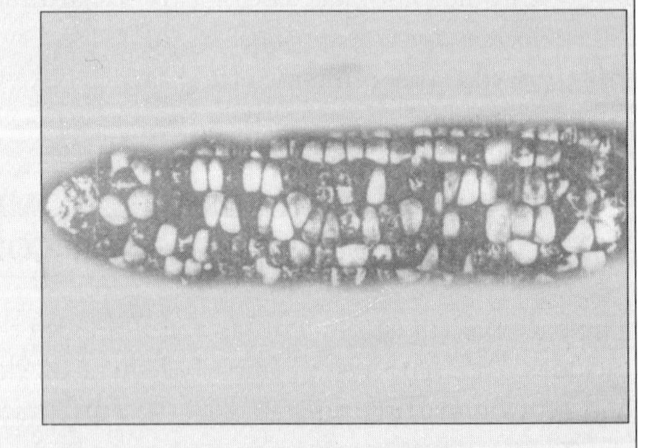

Fig. 2 Example of corn (maize) cob showing color variegation due to transposition (Source: Photograph taken by Barbara McClintock; image courtesy Cold Spring Harbor Laboratory Archives).

Table 36.1 Major Types of Transposable Elements.

Type	Structural Features	Mechanism of Movement	Examples
DNA-MEDIATED TRANSPOSITION			
Bacterial replicative transposons	Terminal inverted repeats that flank antibioctic-resistance and transposase genes	Copying of element DNA accompanying each round of insertion into a new target site	Tn3, γδ, phage Mu
Bacterial cut-and-paste transposons	Terminal inverted repeats that flank antibiotic-resistance and transposase genes	Excision of DNA from old target site and insertion into new site	Tn5, Tn*10*, Tn7, IS*911*, Tn*917*
Eukaryotic transposons	Inverted repeats that flank coding region with introns	Excision of DNA from old target site and insertion into new site	P elements (*Drosophila*) hAT family elements Tc 1/Mariner elements
RNA-MEDIATED TRANSPOSITION			
Viral-like retrotransposons	~250 to 600 bp direct terminal repeats (LTRs) flanking genes for reverse transcriotase, integrase, and retroviral-like Gag protein	Transcription into RNA from promoter in left LTR by RNA polymerase II followed by reverse transcription and insertion at target site	Ty elements (yeast) Copia elements (*Drosophila*)
Poly-A retrotransposons	3′ A- T-rich sequence and 5′ UTR flank genes encoding an RNA-binding protein and reverse transcript	Transcription into RNA from internal promoter; target–primed reverse transcription initialed by endonuclease cleavage	F and G elements (*Drosophila*) LINE and SINE elements (mammals) *Alu* sequences humans

Tnl0 is organized into three functional modules. This organization is relatively common, and elements that have it are called composite transposons. The two outermost modules, called IS 10L (left) and IS10R (right0, are actually mini transposons. "IS" stands for insertion sequence, IS10R encodes the gene for the transposase that recognizes the terminal inverted repeat sequence of ISIORt ISIOL, and Tnl0. IS10L, although very similar in sequence to IS 10R, does not encode a functional transposase. Thus, both ISIOR and Tn10 are autonomous, whereas ISIOL is a nonautonomous transposon. Both types of ISIO elements are found, as expected, unassociated with Tn10.

Tn10 TRANSPOSITION IS COUPLED TO CELLULAR DNA REPLICATION

Tnl0 also couples transposition to cellular DNA replication. Recall that bacteria such as *E. coli* (a common host for Tnl0) menthylate their DNA at GATC sites this methylation occurs after DNA replication, such that GATC sites are hemimethylated for the few minutes between passage of the replication fork and recognition of these sequence by the methylase enzyme.

It is during this brief period–when the Tnl0 DNA is hemimethylated that transposition is most likely to occur. This coupling of transcription to the methylation state is due to the

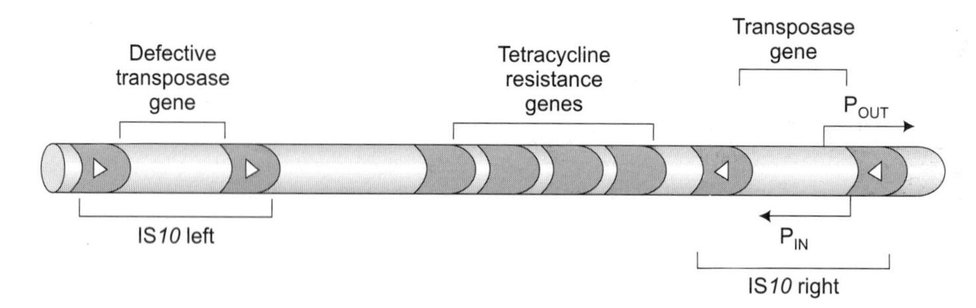

Fig. 36.6 Genetic organization of bacterial transposon Tn10. The map shows the functional elemens in the bacterial transposon Tn*10*. Tn*10*, like many bacterial transposons, actually caries two "mini-transposons" at its termini. For Tn*10*, these elements are called IS*10L* (left) and IS*10R* (right). Both types of IS*10* elements can transpose, and are found in DNA separately from Tn*10*. The white triangles show the inverted repeat sequences at the ends of the IS elements and Tn*10*. Although these four copies are not exactly the same in sequence, all are recognized by the Tn*10* transposase and are used as recombination sites.

presence of two critcal GATC sites in the transposon sequence. One of these sites is in the promoter for the transposase gene; the second is in the binding site for the transposase with in one of the inverted terminal repeats. Both RNA polymerase and transposase bind more tightly to the hemimethylated sequences than to their fully methylated versions. As a result, when the DNA is hemimethylated, the transposase gene is most efficiently expressed, and the transposase protein binds most efficient to the DNA. Therefore, transposition of tnl0 occurs at its highest frequency during this brief phase of the cell cycle just after its DNA has been replicated (Fig. 36.8).

Regulation of tnl0 transposition by DNA methylation serves to limit the overall frequency of transposition. It is also restricts transposition specifically to actively dividing cells. This timing ensure that there are two copies of the chromosomes present to "heal" the double-stranded DNA break left in the old target site as a result of transposon excision. These" empty target sites" are repaired via homologous recombination by the double-strand break repair pathway. This recombination reaction requires that two copies of the chromosomal region be present.

PHAGE Mu IS AN EXTREMELY ROBUST TRANSPOSON

Phage Mu, like phage, is a lysogenic bacteriophage, Mu is also a large DNA transposon, This phage uses transposition to insert its DNA into the genome of the host cell during infection and in this way is similar to the retroviruses (discussed above). Mu also uses multiple rounds of replicative treansposition to amplify its DNA during lytic growth. During the lytic cycle, Mu completes about 100 rounds of transposition per hour, making it the most efficient transposon known. Furthermore, even when present as a quiescent lysogen, the Mu genome transposes quite frequently, compared to traditional transposons such as Tn10. the name Mu is short for

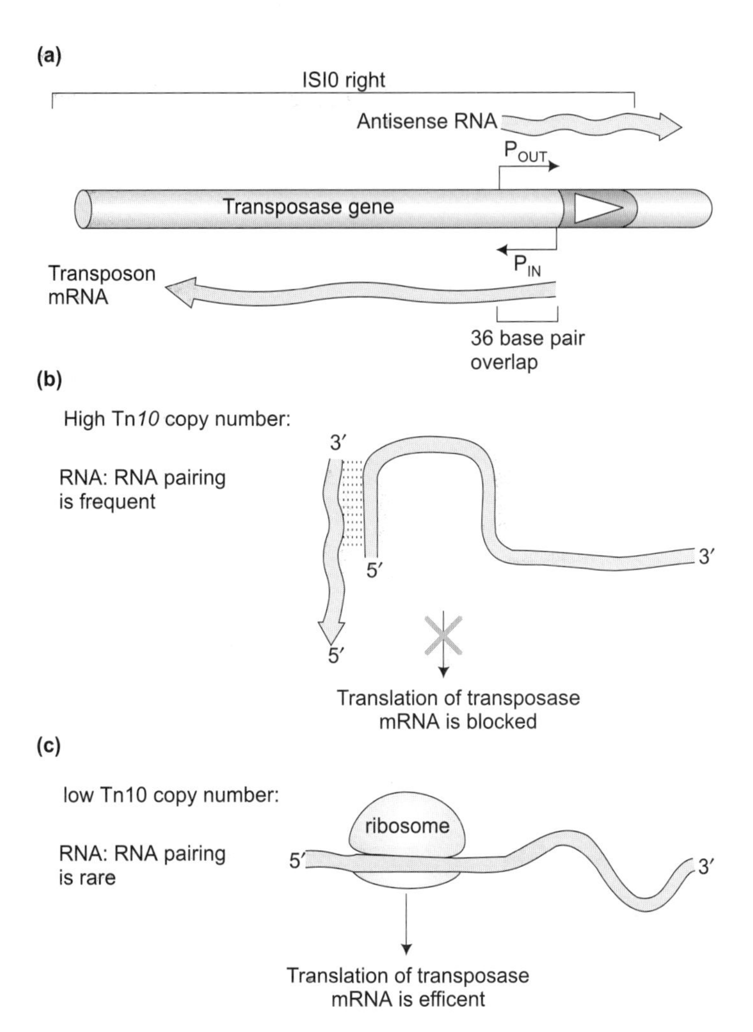

Fig. 36.7 Antisense regulation of Tn*10* expression. (a) A map of the overlapping promoter regions is shown. The leftward promoter (pIN) promotes expression of the transposase gene; the rightward promoter (pOUT), which lines 36 bases to the left of pIN promotes expression of an antisense RNA. The first 36 bases of each transcript are complementary to one another. Note that in cells the antisense transcript initiated at pOUT is longer lived than is the mRNA initiated at pIN. (b) In cells having a high copy number of Tn*10*, the RNA:RNA pairing occurs frequently and blocks translation of the transposase mRNA (thereby eventually reducing the copy number of the elements). (c) In cells having a low copy number of the transposon, RNA:RNA is rare; the translation of transposase mRNA is efficient and the copy number in the cell is increased.

carrying an inserted copy of the Mu DNA frequently accumulate new mutations due insertion of the phage DNA into cellular genes.

The Mu genome is about 40 kb and carries more than 35 genes, but only two encode proteins with dedicated roles in, transposition. These are the A and B genes, which encode the proteins MuA MuB. MuA is the transposase and is a member of the DDE protein superfamily

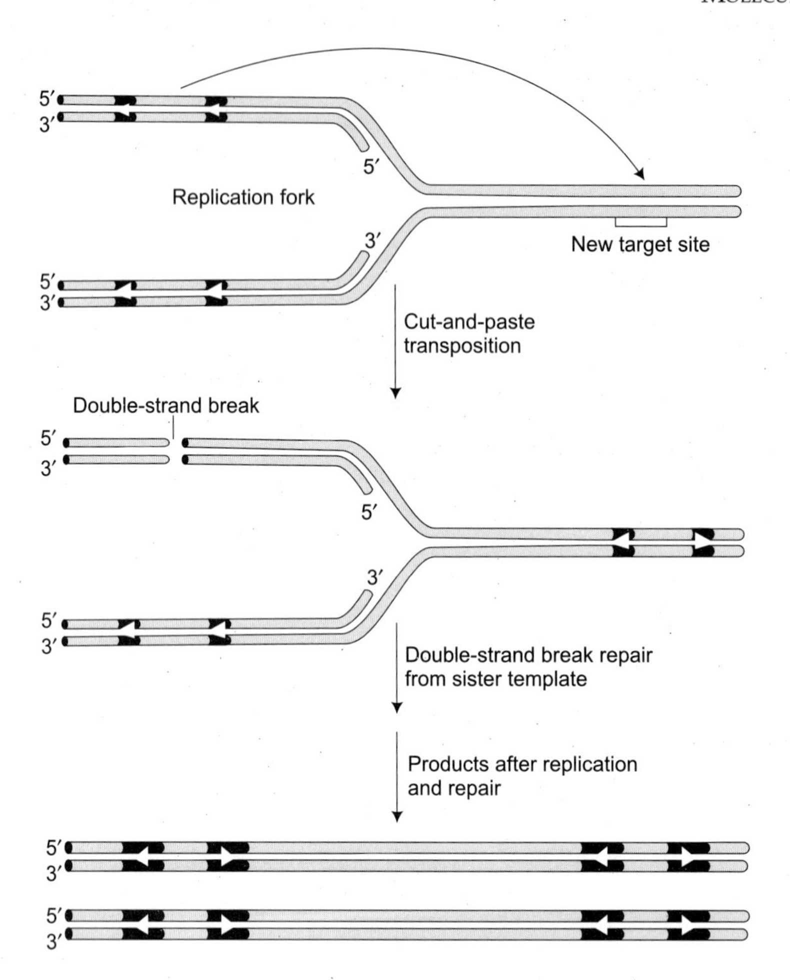

Fig. 36.8 Transposition Tn*10* after passage of a replication fronk. Transposition is activated by the hemimethylated DNA that exists just after DNA replication (methylation sites are not shown). During transposition, a double-stranded break is made in the chromosomal DNA where the element excised. This break can be repaired by the DSB-repair pathway (see Chapter 10), a process that regenerates a copy of Tn*10* at the site of excision. By this mechanism, transposition may appear to be "replicative" in nature, although the actual recombination process goes through the cut-and-paste (nonreplicative) pathway.

we discussed, MuB is an ATpase that stimulates MuA activity and controls the choice of the DNA target site.

CHAPTER 37

The Spliceosome Mechinery

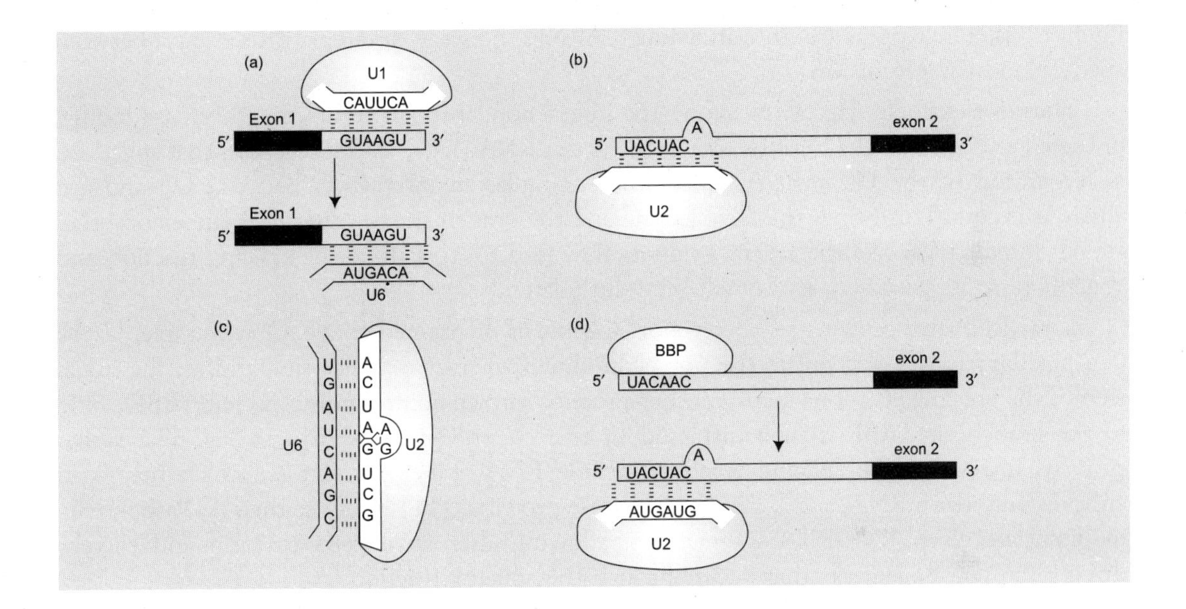

RNA SPLICING IS CARRIED OUT BY A LARGE COMPLEX CALLED THE SPLICEOSOME

A huge molecular "machine" called the spliceosome, comprises about 150 proteins and 5 RNAs and is similar in size to a ribosome. In carrying out even a single splicing reaction, the spliceosome hydrolyzes several molecules of ATP. Strikingly, it is believed that its RNA compontents rather than the proteins, again reminiscent of the ribosome, carries many of the functions of the spliceosome out. Thus, RNAs locate the sequence elements at the intron-exon borders and likely participate in catalysis of the splicing reaction itself.

The five RNAs (U1, U2, U4, U5, and U6) are collectively called small nuclear RNAs (snRNAs). Each of these RNA is between 100-300 nucleotides long and is complexed with several proteins. These RNA-protein complexes are called small nuclear ribonuclear proteins (snRNPs-pronounced "snurps"). The spliceosome is the large complex made up of these snRNAs, but the exact make up differs at different stages of the splicing reaction: different snRNPs come and go at different times, each carrying out particular functions in the reaction. There are also many proteins within the spliceosome that are not part of the snRNPs, and others besides that are only loosely bound to the spliceosome.

The snRNPs have three roles in splicing. They recognize the 5' splice site and the branch site; they bring those sites together as required; and they catalyze (or help to catalyze) the RNA cleavage and joining reactions. To perform these functions, RNA-RNA, RNA-protein, and protein-protein interactions are all important. We start by considering some of the RNA-RNA interactions. These operate within individual snRNPs, between different snRNPs, and between snRNPs and the pre-mRNA.

Thus, for example, Fig. 37.1a shows the interaction, through complementary base paring, of the Ul snRNA and the 5' splice site in the pre-m RNA. Later in the reaction, that splice site is recognized by the U6 snRNA. In another example, an interaction between U2 and U6 snRNAs occurs. This brings the 5' splice site and the branch site together. It is these and other similar interactions, and the rearrangements they lead to, that drive the splicing reaction and contribute to its precision, as we will see a little later.

Some RNA-free proteins are involved in splicing as mentioned above. One example, U2AF (U2 auxillary factor), recognizes the polypyrimidine (py) tract/3' splice site, and, in the initial step of the splicing reaction; helps another protein, branch-point binding protein (BBP), bind to the branch site. BBP is then displaced by the U2 snRNP, as shown in Fig. 37.1 Other proteins involve in the splicing reaction include RNA- annealing factors, which help load snRNPs onto the mRNA, and DEAD-box helicase proteins. The. latter use their ATPase activity to dissociate given RNA-RNA interactions, allowing alternative pairs to form and thereby driving the rearrangements that occur through the splicing reaction.

Finally, before turning to the spliceosome mediated splicing pathway it self, we look at one further interaction, Fig. 37.2 shows the crystal structure of a section of the Ul snRNA bound to one of the proteins of U1 snRNP.

SPLICING PATHWAYS

Assembly, Rearrangement, and Catalysis within the Spliceosome: The Splicing Pathway

The steps of the splicing pathway are shown in Fig. 37.3. Initially, the 5' splice site is recognized by the Ul snRNP (using base pairing between its snRNA and the pre-mRNA, shown in Fig. 37.3). One subunit of U2AF binds to the Py tract and the other to the 3'splice site. The former subunit interacts with BBP and helps that protein bind to the branch site. This arrangement of proteins and RNA is called the Early (E) complex.

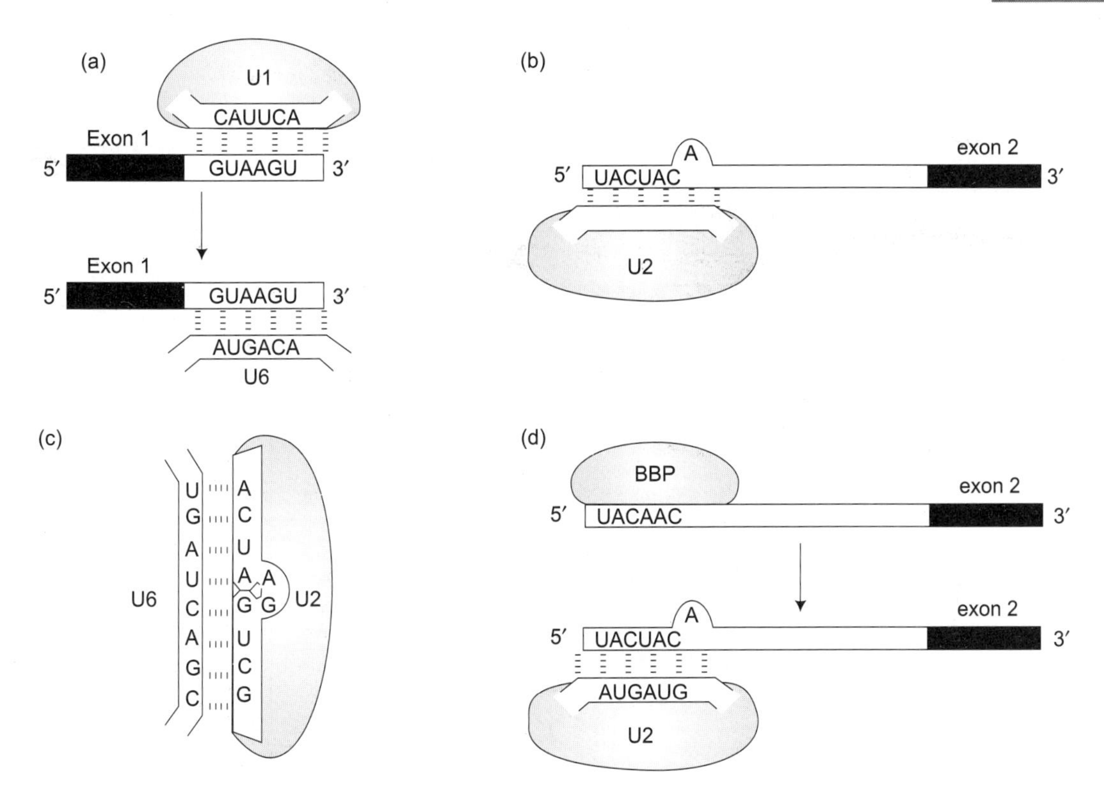

Fig. 37.1 Some RNA-RNA hybrids formed during the splicing reaction. In some cases, (a) different snRNPs recognize the same (or overlapping) sequences in the premRNA at different stages of the splicing reaction, as shown here for U1 and U6 recognizing the 5′ splice site. In (b) snRNP U2 is Shown recognizing the branch site. In (c) the RNA: RNA Pairing between the snRNPs U2 and U6 is shown. Finally, in (d), the same sequence within the pre-M RNA is recognized by a protein (not part of an snRNP) at one stage and stage and displaced by an snRNA) at another. Each of these changes accompanies the arrival or departure of components of the spliceosome and a structural rearrangement that is required for the splicing reaction to proceed.

U2 snRNP then binds to the branch site, aided by U2AF and displacing BBP. This arrangement is called the A complex. The base paring between the U2 sbRNA and the branch site is such that the branch site. A residue is extruded from the resulting stretch of double helical RNA as a single nucleotide bulge as) shown in Fig. 37.1b. This A residue is thus unpaired and available to react with the 5' splice site.

The next step is a rearrangement of the A complex to bring together all three splice sites. This is achived as follows: the U4 and U6 snRNPs, along with the U5 snRNP, join complex. Together these three snRNPs are called the tri-snRNP particle, within which the U4 and U6 snRNPs are held together by complementary base-pairing between their RNA components, and the U5 snRNP is more loosely associated through protein: protein interaction. With the entry of the tri-snRNP, the A complex is converted into the B complex.

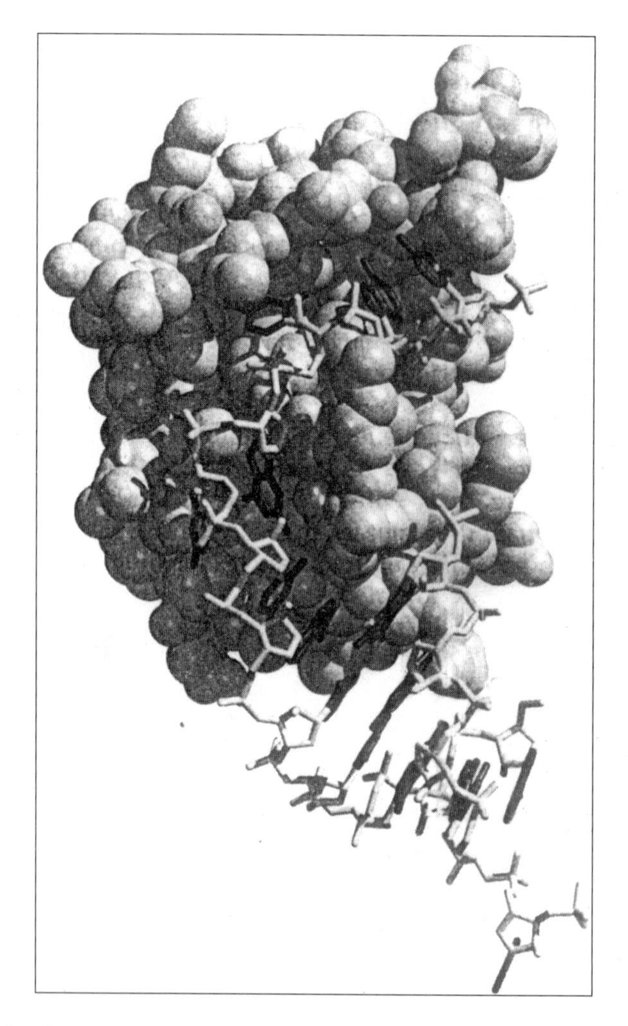

Fig. 37.2 Structure of spliceosomal protein-RNA complex: U1A binds hairpin II of U1 snRNA. (Oubridge C, Ito Evans P.R, Teo C.H, and Nagai K. 1994. *Nature* 372: 432).

In the next step, Ul leaves the complex, and U6 replaces it at the 5' splice site. This requires that the base-pairing between the Ul snRNA and the pre-mRNA be broken, allowing the U6 RNA to anneal with the same region (in fact, to an overlapping sequence, as shown in Fig. 37.1a).

Those steps complete the assembly pathway. The next rearrangement triggers catalysis, and occurs as follows; U4 is released from the complex, allowing U6 to interact with U2 (through the RNA: RNA basepairing shown in Fig. 37.1c). This arrangement, called the C complex, produces the active site. That is,the rearrangement brings together within the spliceosome those components believed to be solely regions of the U2 and U6 RNAs- that together form the active site. The same rearrangement also ensures the substrate RNA is properly positioned to be acted

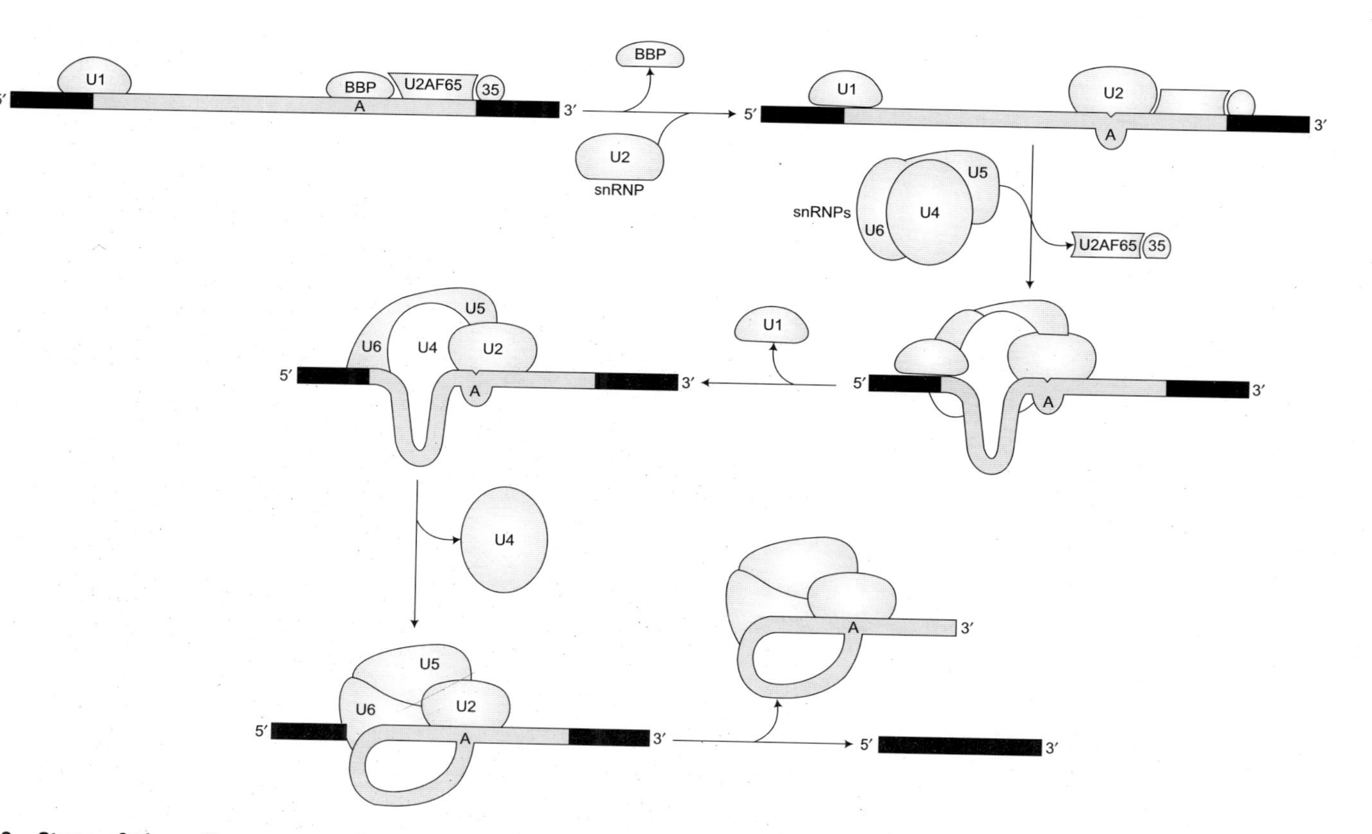

Fig. 37.3 Steps of the spliceosome-mediated splicing reaction. The assembly and action of the spliceosome are shown, and the details of each step are described in the text. Components of the splicing machinery arrive or leave the complex at each step, changes that are associated with structural rearrangements necessary for the splicing reaction to proceed. There is evidence to suggest that some of the components shown do not arrive or leave precisely when indicated in this Fig.; they may, for example, remain present but weaken their association with the complex rather then dissociating completely. It is also not possible to be sure of the order of some changes shown, particularly the two steps involving changes in U6 pairing: when it takes over from U1 at the 5′ splice site, compared to when it takes over from U4 in binding U2. Despite these uncertainties, the critical involvement of different stages of the splicing reaction, and the general dynamic nature of the spliceosome, are as shown.

upon. It is striking that, not only is the active site primarily formed of RNA, but also that it is only formed at this stage of spliceosome assembly. Presumably this strategy lessens the chance of aberrant splicing; linking the formation of the active site to the successful completion of earlier steps in spliceosome assembly makes it highly likely that the active site is available only at legitimate splice sites.

Formation of the active site juxtaposes the 5' splice site of the pre-mRNA and the branch site, facilitating the first transesterification reaction. The second reaction, between the 5' and 3' splice sites, is aided by the U5 snRNP, which helps to bring the two exons together. The final step involves release of the M RNA product and the snRNPs. The snRNPs are initially still bound to the lariat, but get recycled after rapaid degradation of that piece of RNA.

It might seem odd that the machinery and mechanism of splicing is so complicated. How did it evolve that way? Would it not have been simpler to fuse the exons in a single reaction, rather than undergo the two reactions just described? To consider this question, we turn to a group of introns that- unlike those we have considered thus far-can splice themselves out of pre-mRNA without the need for the spliceosome. They are called self-splicing introns.

SELF-SPLICING INTRONS REVEAL THAT RNA CAN CATALYZE RNA SPLICING

The three classes of splicing found in cells (not including tRNA processing, are shown in Table 37.1). thus far we have dealt only with nuclear pre-mRNA splicing, that mediated by the spliceosome found in all eukaryotes. Also shown in Table 37.1 are the so-called group I and group II self-splicing introns. By selfsplicing, we mean that the intron itself folds into a specific conformation within the precursor RNA and catalyzes the chemistry of its own release. In terms of a practical definition self-splicing means that these introns can remove themselves from RNAs in the test tube in the absence of any proteins or other RNA molecules. The self-splicing introns are grouped into two classes on the basis of their structure and splicing mechanism. Strictly speaking, self-splicing introns are not enzymes (catalysts) because they mediate only one round of RNA processing (Box 37.1).

A typical self-splicing intron is between 400 to 1,000 nucleotides long, and in contrast to introns removes by spliceosomes, much of the sequences requirement holds because the intron

Table 37.1 Three Classes of RNA Splicing

Class	*Abundance*	*Mechanism*	*Catalytic Machinery*
Nuclear pre-mRNA	Very common; used for most eukaryotic genes	Two transesterification reactions; branch site A	Major spliceosome
Group II introns	Rare; some eukaryotic genes from organelles and prokaryotes	Same as pre-mRNA	RNA enzyme encoded by intron (ribozyme)
Group I Introns	Rare; nuclear rRNA in some eukaryotes, organelle genes, and a few prokaryotic genes	Two transesterification reactions; branch site G	Same as group II

Box 37.1 Converting Group I introns into Ribozymes

Once a group I self-splicing intron has been spliced out, the active site it contains remains intact. So what prevents this splicing reaction from reversing itself? One thing is the high cellular concentration of G nucleotides-this strongly favors the forward reaction. But in addition, the intron undergoes a further reaction. that effectively prevents it from participating in the back reaction. Conveniently, at the extreme 3 end of the intron is a G, which can bind in the G-binding pocket. Meanwhile, the 5' end of the intron can bind along the internal guide sequence. Thus, a third transesterification reaction can occur to cyclize the intron. The new bond formed with the terminal G is labile and hydrolyzes spontaneously. As a consequence, the intron is relinearized, but it is truncated and so precluded from the back splicing reaction.

As explained earlier in the text, group I (and II) introns are not enzymes because they have a tunover number of only one. But they can be readily converted into enzymes (ribozymes) in the following way (Fig. 1): the relinearized intron described above retains its active site. If we provide it with free G and a substrate that includes a sequence complementary to the internal guide sequence, it will repeatedly catalze cleavage of substrate molecules. We will have converted a group I intron into a ribozyme, similar to the way that the self-cleaving harmmerhead could be converted to a ribozyme by separating the active site from the sequence of the internal guide sequence and thereby generate tailor-made ribonucleases that cleave RNA molecules of our choice.

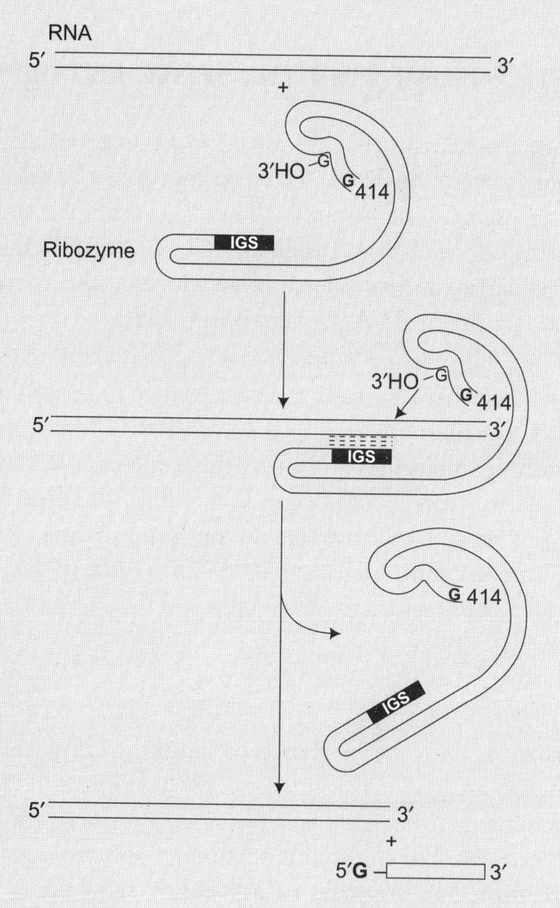

Fig. 1 Group I introns can be converted into true ribozymes.

must fold into a precise structure to perform the reaction chemistry. In addition, in vivo, the intron is complexed with number of proteins that help stabilize the correct structure-partly by shielding regions of the backbone from each other. Thus, the folding requires certain sections of the RNA backbone to be in close proximity to other sections, and the negative charges provided by the phosphates in those backbone regions would repel each other if not shielded. In vitro, high salt concentrations (and thus positive ions) compensate for the absence of these proteins. This is how we know that the proteins are not needed for the splicing reaction itself.

The similar chemistry seen in self- and spliceosome-mediated spliceing is believed to reflect an evolutionary relationship. Perhaps ancestral group II-like self-splicing introns were the staring point for the evolution of modern pre-m RNA splicing. The catalytic functions provided by the RNA were retained, but having the snRNAs relieved the requirement for extensive sequence specificity with in the intron itself and their associated proteins provide most of those functions in trans. In this way, introns had only to retain the minimum of sequences elements required to target splicing to the correct places. Thus, many more and varied sizes and sequences of introns were permitted.

HOW DOES THE SPLICEOSOME FIND THE SPLICE RELIABLY?

We have already seen one mechanism that guards against inappropriate splicing–the active site of the spliceosome is only formed on RNA sequence that pass the test of being recognized by multiple elements during spliceosome assembly. Thus, for example, the 5' splice site must be recognized initially by the UI snRNP and then by the U6 snRNP. It is unlikely both would recognize an incorrect sequence, and so selection is stringent. Yet, the problem of appropriate splice-site recognition in the pre-m RNA remains formidable.

Consider the following. The average human gene has eight or nine exons and can be spliced in three alternative forms. But there is one human gene with 363 exons and one *Drosophila* gene that can be spliced in 38,000 alternative ways. If the sn RNPs had to find the correct 5'and 3' splice sites one complete RNA molecules and bring them together in the correct pairs, unaided, it seems inevitable that many errors would occur. Remember, also that the average exon is only some 150 nucleotides long, whereas the average intron is approximately 3,000 nucleotides long (as we have seen, some introns can be as long as 800,000 nucleotides). Thus, the exons must be identified within a vast ocean of intronic sequences.

SUMMARY

Most genes encode proteins, and the sequence of amino acids within any given protein is determined by the sequences of "codons" in its gene. Each codon is made up of a group of three adjacent nucleotides. In almost all bacterial and phage genes, the open-reading frame is a single stretch of codons with no break. But the coding sequences of many eukaryotic genes is split into stretches of codons interrupted by stretches of noncoding sequence.

The coding stretches in these split genes are called exons (for "expressed sequences") and the noncoding stretches are called introns (for "intervening sequences"). The numbers and sizes of the introns and exons vary enormously from gene to gene. Thus, in yeast, only a relatively small proportion of genes have introns, and where they occur they tend to be short and few in number (one or occasionally two per gene). In multicellular organisms such as humans, the number of genes containing introns is much larger, as is the number of introns per gene (up to 362 in an extreme case). The sizes of exons do very but are often around 150 nucleotides; introns, on the other hand, vary from 61 to as much as a staggering 800 kb.

When a gene containing introns is .transcribed, the RNA initially retains those introns. These are then removed to produce the nature mRNA. **The process of intron removal is called splicing.**

Many intron-containing genes give rise to a unique m RNA species. That is, in each case, all the introns are removed from the original RNA, leaving an mRNA composed of all the exons. But in other cases, splicing can produce a number of different mRNAs from the same gene by splicing the original RNA in different patterns. Thus, for example, some genes contain alternative exons, only one of which ends up in a given m RNA. In other cases, a given exon might be removed (along with the introns) from some copies of the RNA-again producing alternative versions of mRNA from the same gene.

Sequences found at the boundary between introns and exons allow the cell to identify introns for removal. These splicing sequences are almost exclusively within the introns (where there are no restrictions imposed by the need to encode amino acids, as there are in exons). These sequences are called the 3' and 5' splice sites, denoting their relative locations at one or the other end of the intron. To splice out an intron also requires a sequences element, called the branch site, near the 3' end of the intron.

Intron removal proceeds via two transesterification reactions. In the first, an A in the branch site attacks a G in the 5' splice site. In the second, the liberated 5' exon attacks the 3' splice site. These reactions have two consequences. First and foremost, they fuse the two exons. Second, they release the intron in the form of a branched structure called a lariat.

Splicing of nucleosomal pre-RNSs requires a large complex of proteins and RNAs called the spliceosome. This is made up of so-called snRNPs, of which there are five-U1, U2, U4, U5, and V6 snRNPs. Each of these comprises an RNA molecule, called the U1 to V6 snRNA, respectively, and a number of proteins, the majority of which are different in each case.

The action of the spliceosome is particularly interesting in two regards. First, the RNA components have a central role in recognizing introns and catalyzing their removal. Second, the complex is very dynamic. That is, at different steps during the process of splicing, the spliceosome constitution alters different subunits of the machine join and leave the complex, each performing a particular function.

Thus, early on UI snRNP recognizes the 5' splice site, while the U2 snRNP recognizes the branch site. U4 and U6 then join together with U5 bringing the branch site and 5' splice site together and stimulating the first reaction concomitant with UI and U4 leaving. Finally, the 3' and 5' splice sites are brought together and exons are fused.

There are a few introns that can remove themselves from with in RNA molecules by a process known as self-splicing. Though not strictly an enzymatic reaction the RNA of the intron nevertheless mediates the chemistry of removal. These self-splicing introns come in two classes, one of which (group II) splice by the same chemical pathway as that mediated by the spliceosome. These introns probably represent the evolutionary origin of modern spliceosomal introns, and the two-step chemical pathway used by both reflects that evolutionary relationship (and perhaps explains why introns are not removed by a more direct single-step mechanism).

The splice sites described above are defined by rather short sequences with low levels of conservation. It thus represents a significant and splice only at correct sites. There are various mechanisms by which the spliceosome enhances accuracy. First it assembles on the sites soon after they have been synthesized. This ensures they are selected before downstream sites are available to compete. Second there are other proteins-SR proteins-that bind near legitimate splice sites and help recruit the splicing machinery to those sites. In this way, authentic sites effectively have a higher affinity for the machinery than do so called psuedo sites of similar sequences.

There are a large variety of SR proteins. Each binds RNA with one surface and with another interacts with components of the splicing machinery. Some SR proteins regulate splicing. That is, a given SR protein may be found only in one cell type and mediate a particular splicing event only in that cell type. Other SR proteins are only active in the presence of specific physiological signals, and so a given splicing event only occurs in response to that single. In this way, SR proteins resemble transcriptional activators, as we will see in later chapters. Also analogously with transcriptional regulation, there are repressors of splicing that exclude splicing of specific introns under certain circumstances.

Together with the other modifications splicing is required before mRNAs can be transported out of the nucleus through nuclear pores. This too can be regulated.

It is believed that a given exon typically encodes an independently folding (and functional) protein domain. Thus such an exon can readily function in combination with different exons. This suggests it bas been relatively easy, through evolution, to generate new proteins by shuffling existing exons between genes.

RNA editing is another mechanism that allows an RNA to be changed after transcription so as to encode a different protein from that encoded by the gene. Two mechanisms for editing are: enzymatic modification of bases (generating forms that alter how they are read by tRNAs) and the insertion or deletion of multiple U nucleotides within the message.

CHAPTER 38

Immunology and Immunity

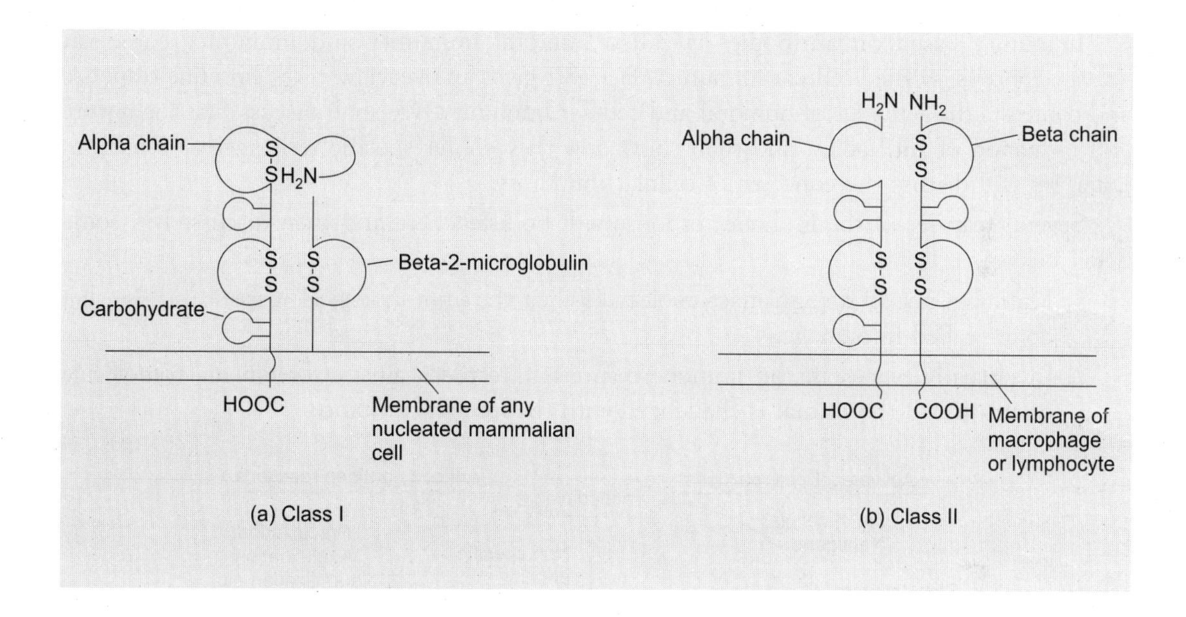

Higher animals posses a highly sophisticated mechanism, the immune response for developing resistance to specific microorganisms or viruses. The immunological response occurs because the body has a general system for the neutralization of foreign macromolecules and microbial cells. Foreign substances which elicit the immune response are called antigens. As a result of antigen stimulation, the immune system produces specific proteins called antibodies or immunoglobulins and specific cells called activated T cells. Since an invading microorganism contains a variety of macromolecules foreign to the host antibodies and activated T cells are generated against it and are able to recognize the foreign material and bring about its destruction.

The immune response shows three major characteristic: specificity, memory, and tolerance. (1) The specificity of the antigen antibody or antigen T cell interaction is unlike any of the host resistance mechanisms. Phagocytosis, inflammation and the other nonspecific host resistance mechanisms develop against virtually any invading microorganism, even those the host has never encountered before, whereas in the immune response, a specific interaction must occur for each new invaded. (2) Once the immune system produces a specific type of antibody or activated T cell, it is capable of producing more of the same antibody or T cell more rapidly and in larger amounts. This capacity for memory is of major importance is resistance of the host to subsequent reinfection or in the protection to the host provided by vaccination. (3) Tolerance exists because macromolecules on the surfaces of body cells are also potentially antigenic and would be damaged if antibodies and activated T cells to these body cells were produced during an immune response.

Immunity based on antibodies is called humoral immunity and immunity based on activated T cells is called cellular immunity. Fig. 38.1 gives an overview of the immune response and contrasts the activities of humoral and cellular immunity. We shall discuss first the nature and formation of antibodies, and then show how they confer specific resistance to infection. Later, we will discuss the concepts of cellular immunity.

Several features of the immune response will be listed here and then discussed in some detail below.

1. Many but not all foreign macromolecules elicit the immunological response; those that do are called Immunogens
2. In virtually every case, an immune response directed against a foreign macromolecule occurs only if the animal is challenged with the foreign substance

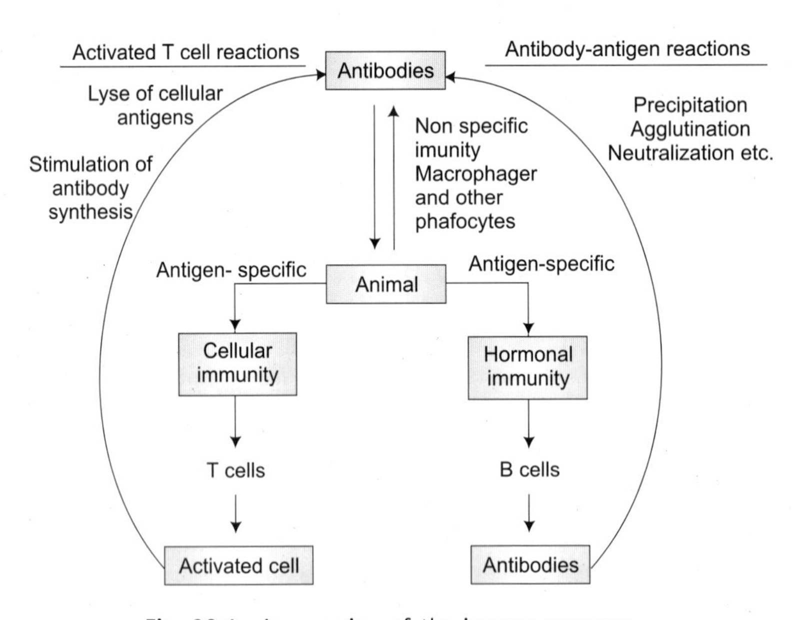

Fig. 38.1 An overview of the immune response.

3. There is a high specificity in the immune response; antibodies or activated t cells made against on antigen generally do not react against other antigens
4. Not all immune reactions are beneficial; some such as those involved in hypersensitivities and autoimmune reactions, are harmful
5. Antibodies are formed against a variety of foreign macromolecules, but ordinarily not against macromolecules of the animal's own tissues; thus the animal is able to distinguish between its own ("self") and foreign ("nonself") macromolecules
6. Microorganisms and viruses that invade the host contain large numbers of different macromolecules that can act as antigens. Thus the immunological response can be made the basis of specific immunization procedures for the prevention and control of specific diseases
7. The high specificity of antigen-antibody reactions makes them useful in many research and diagnostic procedures

KEY POINTS

The immune response is a specific reaction by the body to the presence of foreign material, generally macromolecules. The substance which induces the immune response is called an antigen or immunogen. As a result of antigenic stimulation, proteins called antibodies or immunoglobulins are produced, and specific cells called activated T cells are formed. Microorganisms and viruses capable of invading the body contain numerous antigens and the immune response participates in the prevention and control of infectious disease.

IMMUNOGENS AND ANTIGENS

Immunogens are substances that, when administered to an animal in the appropriate manner, induce an immune response. The immune response may involve either antibody production, the activation of specific immunologically-competent cells (called activated T cells), or both.Antigens are substances that react with either antibodies or activated T cells and most antigens are also immunogens. However, some substances are recognized by immune systems while not being true immunogens. For example, haptens are low-molecular weight substances that combine with specific antibody molecules but do not by themselves induce antibody formation. Haptens include such molecules as sugars, amino acids, and small polymers.

An enormous variety of macromolecules can. act as immunogens under appropriate conditions. These include virtually all proteins and lipoproteins, many polysaccharides, some nucleic acids, and certain of the teichoic acids. One important requirement is that the molecules must be of fairly high molecular weight, usually greater than 10,000. However, the antibody is directed not against the antigenic macromolecule as a whole, but only against distinct portions of the molecule that are called its antigenic determinants (Fig. 38.2). Chemically, antigenic determinants include sugars, amino acid chains, organic acids and bases, hydrocarbons, and

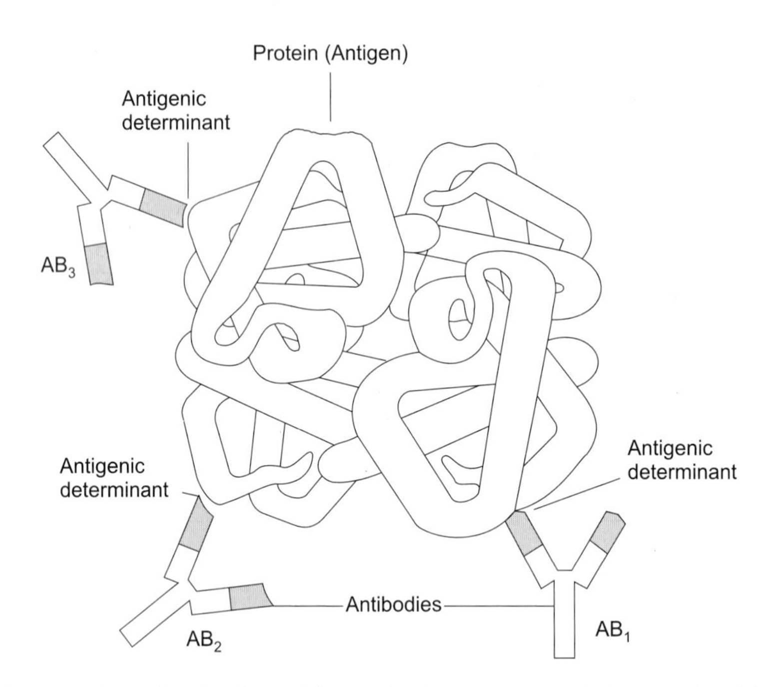

Fig. 38.2 Antigens and antigenic determinants. Antigens may contain several different antigenic determinants, each capable of reacting a specific antibody.

aromatic groups. Antibodies are formed most readily to determinants that project from the foreign molecule or to terminal residues of a polymer chain. In proteins, for example, the majority of antibodies are made to surface determinants, because the surface contains a continuum of antigenic sites. A region of as few as 4-5 amino acids can define an antigenic determinants. A cell or virus is a mosaic of proteins, polysaccharides, and other macromolecules, each of which is a potential antigen. Each antigen of the cell is also a mosaic of side chains and residues, each of which is a potential antigenic determinant. The immunological response to an invading microbe or virus is thus a complex phenomenon.

In general, the specificity of antibodies is comparable to that enzymes, which are able to distinguish between closely related substances. For instance, antibodies can distinguish between the sugars glucose and galactose, which differ only in the position of the hydroxyl group on carbon 4. However, specificity is not absolute, and an antibody will react at least to some extent with determinants related to the one that induced its formation. The antigen which induced the antibody is called the homologous antigen and others, if any that react with the antibody are called heterologous antigens.

An immunogen is a substance that induces an immune response; if the immunogen reacts with either antibody or activated T cell it is called an antigen. Although the reaction between antigen and antibody or activated T cell is highly specific, it is not directed at the antigen as a whole, but just against one or more restricted portions of the antigen called its antigenic determinants. Although the specificity of the immune response is comparable to the specificity of enzyme and substrate, antibodies or activated t cells can sometimes react with heterologous antigens.

IMMUNOGLOBULINES (ANTIBODIES)

Immunoglobulins (antibodies are protein molecules that are able to combine with antigenic determinants. They are found predominantly in the serum fraction of the blood, although they may also be found in other body fluids, as well as in milk .Serum is the fluid portion of the blood that is left when the blood cells and the materials responsible for clotting are removed. Serum containing antibody is often called antiserum. Immunoglobulins (abbreviated Ig) can be separated into five major classes on the basis of their physical, chemical, and immunological properties: IgC, IgM, IgD, and IgE (Table 38.1). Immunoglobulin IgC has been further resolved into four immunolobulines distict subclasses called IgG1, IgG2, IgG3, and IgG4. The basis of this separation will be discussed below. Antibody molecules specific for a given antigenic determinant can be found in each of the several classes, even in a single immunized individual. Upon initial immunization, the first immunoglobulin to appear is IgM, a pentameric immunoglobulin with a molecular weight of about 970,000; IgC appears later. In most individuals about 80 percent of the immunoglobulins are IgG proteins, and these have therefore been studied most extensively.

Immunoglobulin Structure (Fig. 38.3)

Immunoglobulin G is the most common circulationary antibody and thus we will discuss its structure in some detail. Immunoglobulin G has a molecular weight about 160,000 and is composed of four polypeptionary chains (Fig. 38.3). Both intrachain and interchain disulfide (S–S) bridges are present (Fig. 38.3). The two light (short) chains are identical in amino acid sequence, as are the two heavy (longer) chains. The molecule as a whole is thus symmetrical (Fig. 38.3). Each light chain consists of about 212 amino acids and each heavy chain consists of about 450 amino acids (fragment of antigen binding). The fragment containing the carboxy terminal half of both heavy chains, called Fc (fragment crystallizable), does not combine with antigen (Fig. 38.3c). Therefore, each antibody molecule of the IgC class contains two antigen combining sites (and is thus bivalent). When an IgG molecule is treated with a thiol-containing reducing agent and the proteolytic enzyme papain, it breaks into several fragments. The two fragments containing the complete light chain plus the amino terminal half of the heavy chain are the portions that combine with antigen and are called Fab fragments. The importance in understanding the manner in which some antigen-antibody reactions occur (see below). The antigen binding site is found in a small region of the amino terminal portion of both the heavy and the light chains (Fig. 38.3). Immunoglobulins also contain small amounts of complex carbohydrates consisting mainly of hexose and hexosamine, which are attached to potions of the heavy chain (Fig. 38.3); the carbohydrate is not involved in the antigen binding site.

Although the view of the IgG molecule shown in Fig. 38.3 is adequate for conveying the general structure of this molecule, since immunoglobulins are proteins they are likely to be

Table 38.1 Properties of human immunoglobulins.

Class designation	*Molecular weight*	*Proportion of total antibody (percent)*	*Concentration in serum (mg/ml)*	*Antigen binding sites*	*Properties*	*Distribution*
IgA	160,000 385,000 (secretory form)	13	3 0.05	2 4	Major secretory antibody	Secretions (saliva), colostrum, serum), cellular and blood fluids; exists as a monomer in serum and as a dimer in
IgG	146,000	8O	13	2	Major circulating antibody; four subclasses exist: IgG_1, IgG_2, IgG_3, IgG_4, binds complement weakly antitoxin	Extracellular fluid; blood and lymph; crosses placenta
IgM	970,000 (pentamer)	6	1.5	10	First antibody to appear after immunization; binds complement strongly	Blood and lymph; B lymphocyte surfaces (as monomer)
IgD	184,000	1	0.03	2	Minor circulating antibody; heat labile; high carbohydrate content	Blood and lymph; lymphocyte surfaces
1gE	188,000	0.002	0.00005	2	Involved in allergic reactions contains mast cell binding fragment	Blood and lymph only

highly twisted and folded in their final conformation and assume a complex three dimensional structure. We will see later how the higher order structure of the polypeptide chains that combine to form an intact Immunoglobulin expose a unique binding site on each antibody molecule and how this binding site is ultimately responsible for the specificity of antigen antibody reactions.

Light Chains of IgG

Each IgG light chain contains two amino acid regions, the variable region and the constant region. The sequence of amino acids in a major portion of the light chains of immunoglobulins of the class IgG are frequently identical, even in IgG's directed against completely different antigenic determinants. This is because the amino acid sequence in the carboxy terminal half of the light chain constitutes one of the two specific sequences, referred to as the lambda sequence or the kappa sequence. One IgG molecule will have either two lambda chains or two kappa chains but never one of each (Fig. 38.3a). By contrast, light chain variable regions always

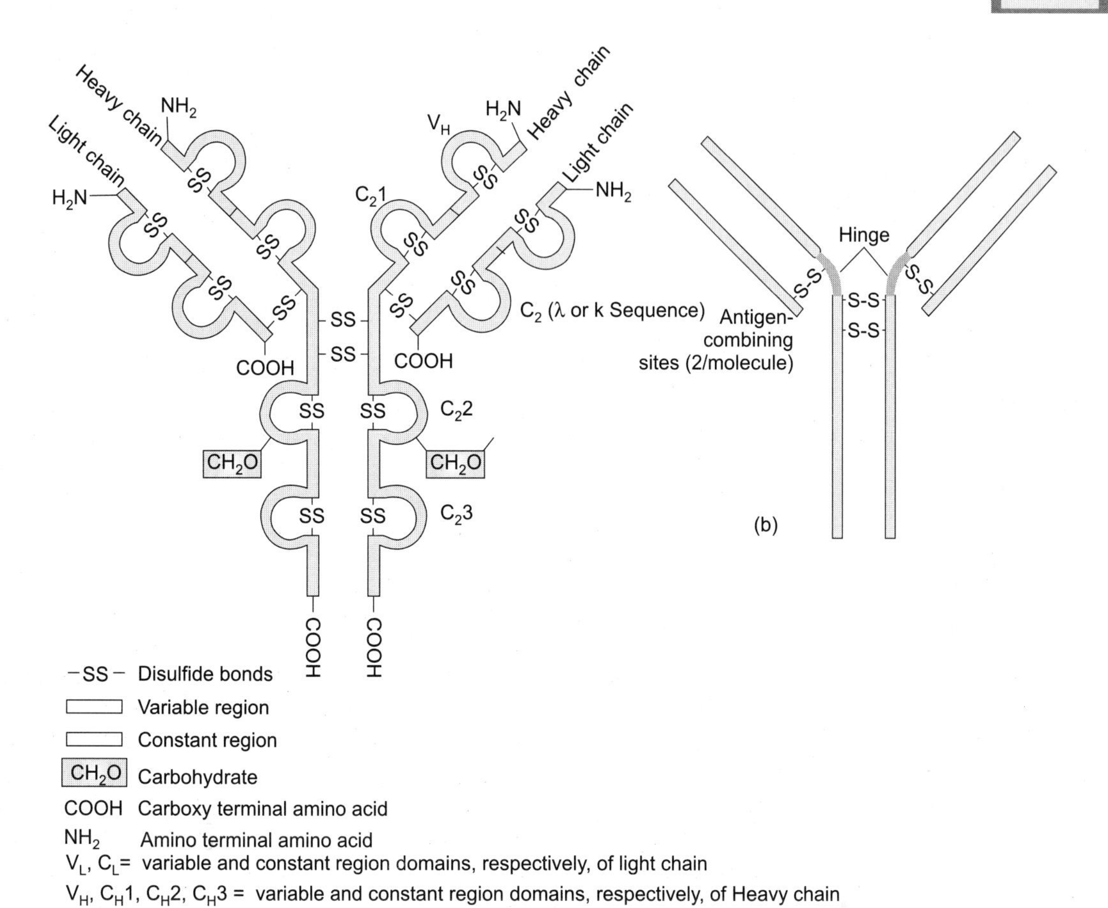

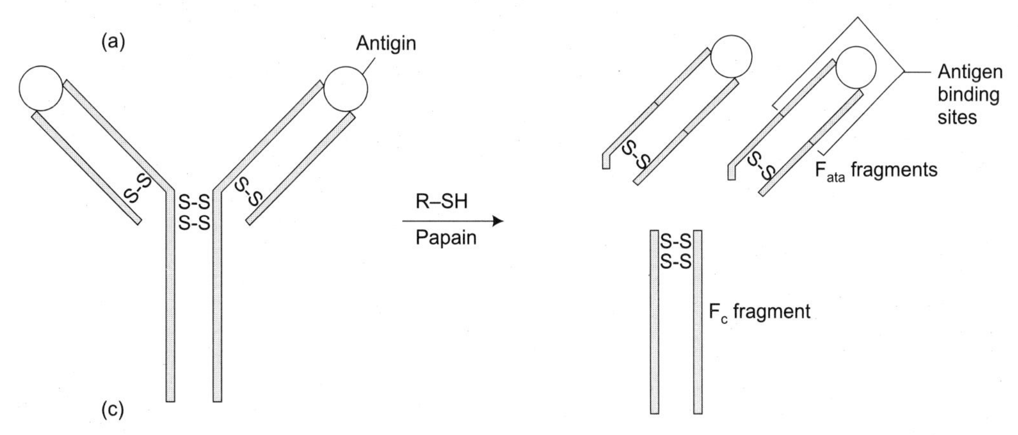

Fig. 38.3 Structure of immunoglobulin G (IgG). Structure showing disulfide linkages within and between chains. (b) Alternative structural diagram which de-letes the intrachain disulfide bonds and carbohydrates to simplify the diagram. (c) Effect of papain treatment on immunoglobulin structure. R—SH represents one of several organic thiols which will react with selected S—S bonds of the immunoglobulin molecule.

differ in amino acid sequence from one IgG molecule to the next unless both molecules bind the same antigenic determinant.

Heavy chains of IgG

Each IgC heavy chain contains four amino acid regions, one variable and three constant regions (referred to as variable and constant domains, respectively). Analogous to the situation that exists in the light chain, all Immunoglobulins of the IgG class have a portion of their heavy chain (the carboxy terminal region) in which the amino acid sequence is identical (Ch1, ch2, and ch3, see Fig. 38.3) from one IgG molecule to another. In addition, each heavy chain has a region in the amino terminal end (antigen binding site Vh) where considerable amino acid sequence variation occurs from one IgG to the next. The great specificity of a given antibody molecule for a particular antigen thus lies in the unique three-dimensional structure of the antigen binding site dictated by the amino acid sequence in the variable regions of the heavy and light chains.

Other classes of immunoglobulins

How do immunoglobulins of the other classes differ IgG? The heavy chain constant region of a given immunoglobulin molecule defines it as to class and can have one of five amino acid sequence: gamma, alpha, mu, delta, or epsilon. These sequences constitute the carboxy terminal three-fourths of the heavy chains of immunoglobulins of the class IgG, IgA, IgM, IgD, or IgE, respectively (Fig. 38.3a). Each antibody of the class IgM, for example, will contain a stretch of amino acids in its heavy chain which constitutes the mu sequence. If two immunoglobulins of different classes react with the same antigenic determinant, then the variable regions of their heavy and light chains would be identical, but their class-determining, sequences, specific to their heavy chains, would be different. It is not unusual in a typical immune response to observe the production of antibodies of two different classes to the same antigenic determinant.

The structure of immunoglobulin M (lgM) is shown in Fig. 38.4. It is usually found as an aggregate of five immunoglobulin molecules attached as shown in Fig. 38.4 by short peptides and accounts for 5-10 percent of the total serum immunoglobulins. Each heavy chain of IgM contains an extra constant region domain (CH_4), and IgM in general is very carbohydrate rich. IgM is the first class of immunoglobulin made in a typical immune response to a bacterial infection (see Fig. 38.3), but immunoglobulins of this class are generally of low affinity. The latter problem is compensated to some degree, however, by the high valency of the pentameric IgM molecule: 10 binding site are available for interaction with antigen (Table 38.1 and Fig. 38.4). The term avidity is used to describe the strength of multivalent antigen-binding molecules; thus, IgM is said to be of low affinity but high avidity.

Immunoglobulin A (IgA) is of interest because it is present in body secretions. 1Ig is the dominant antibody in all fluids bathing organs and systems in contact with the outside world: saliva tears, breast milk, and colostrums, gastrointestinal secretions, and mucus secretions of the respiratory and genitourinary tracts. IgA is also present in serum, but the IgA of secretions has

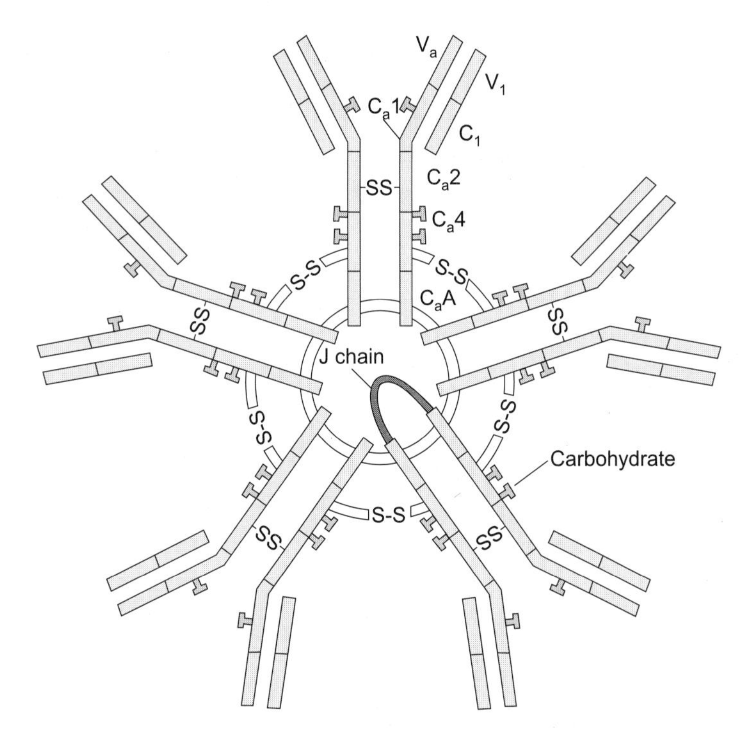

Fig. 38.4 Structure of lgM, a large immunoglobulin with five molecules (a pentamer). Note that each heavy chain has four rather than three constant regions and that the five molecules are themselves held together by disulfide bonds Also note 10 antigen binding sites are available.

an altered molecular structure, consisting of a dimeric immunoglobulin molecule together and possibly aid in the passage of IgA into secretions.

Immunoglobulin E (IgE) is found in serum in extremely small amounts (in an average human about 1 of every 50,000 serum immunoglobulinmolecules is IgE). Despite its low concentration it is important, since immediate-type hypersensitivities (allergies) are mediated by IgE. The molecular weight of an IgE molecule is significantly higher than that of other immunoglobulins (Table 38.1) because like IgM, it contains an additional constant region. This region is thought to function in binding IgE to mast cell surfaces, an important prerequisite for certain allergic reactions.

Imunoglobulin D(IgD) is also present in low concentrations and its function in the overall immune response is unclear. Experiments have shown that IgD is abundant on the surfaces of antibody- producing cells (B lymphocytes, see next section) and IgD may play a role along with monomeric IgM in binding antigen as a signal to the lymphocyte to begin antibody production.

Cells of the Immune System

The immune system enlists the activity of a number of organs and cell types which interact in various ways to elicit the observed immune response. Organs of the immune system are located throughout the body. A key cell involved in immune responses is a type of white blood cell called a lymphocyte, and any tissue that participates in the immune response is referred to as a lymphoid tissue.

Lymphocytes arise from undifferentiated stem cells in the bone marrow, the soft tissue in the hollow shafts of long bones. Stem cells differentiate into functionally distinct cell types during a maturation process which takes place in association with specific lymphoid tissues (Fig. 38.5). Although almost a dozen different cell types have been recognized, we limit our discussion her to lymphocytes and macrophages, because it is these cells which are primarily involved in specific immune responses. Because they are differentiated lymphocytic cells, we will discuss the nature of plasma cells and memory cells when we present the details of antibody production.

Lymphocytes

Lymphocytes are dispersed throughout the body and are one of the most prevalent mammalian cell types (the average adult human has about 10 12 lymphocytes). Two types of lymphocytes, B lymphocytes (B cells) and T lymphocytes (T cells), are involved in immune responses.

Immunoglobulins are made only by lymphocytes of the B type while T cells play a variety of alternative roles in the overall immune response. Both B and cells are derived from stem cells in the bone marrow and their subsequent differentiation is determined by the organ within which they become established (Fig. 38.6). B cells probably mature in the bone marrow and T cells in the thymus (hence the designation "T" Several subsets of T lymphocytes are known. T helped (Th) cells stimulate B cells to produce high levels immunoglobulin; in most cases little if any antibody is made without Th cells interaction Cytotoxic (Tc) cells secrete toxic substances

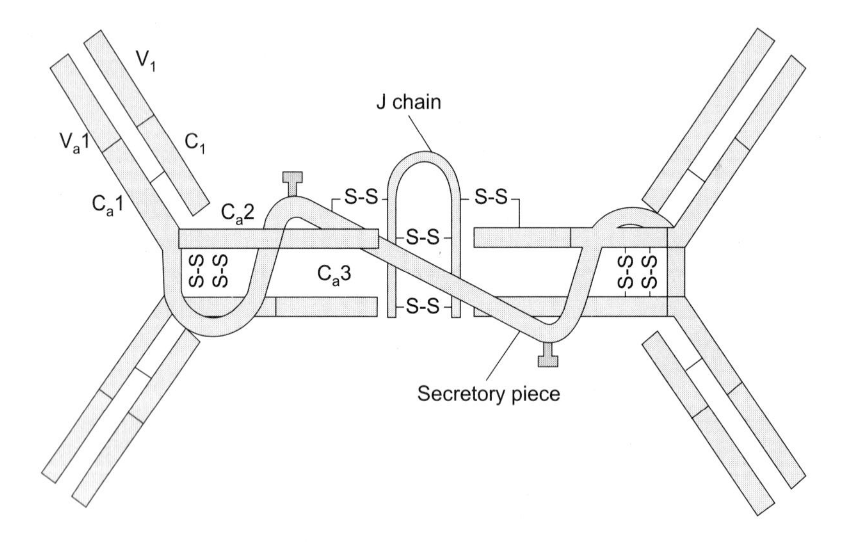

Fig. 38.5 Structure of human secretory immunoglobulin A (lgA). Although most lgA is monomeric, the secretory from of lgA is a dimeric molecule.

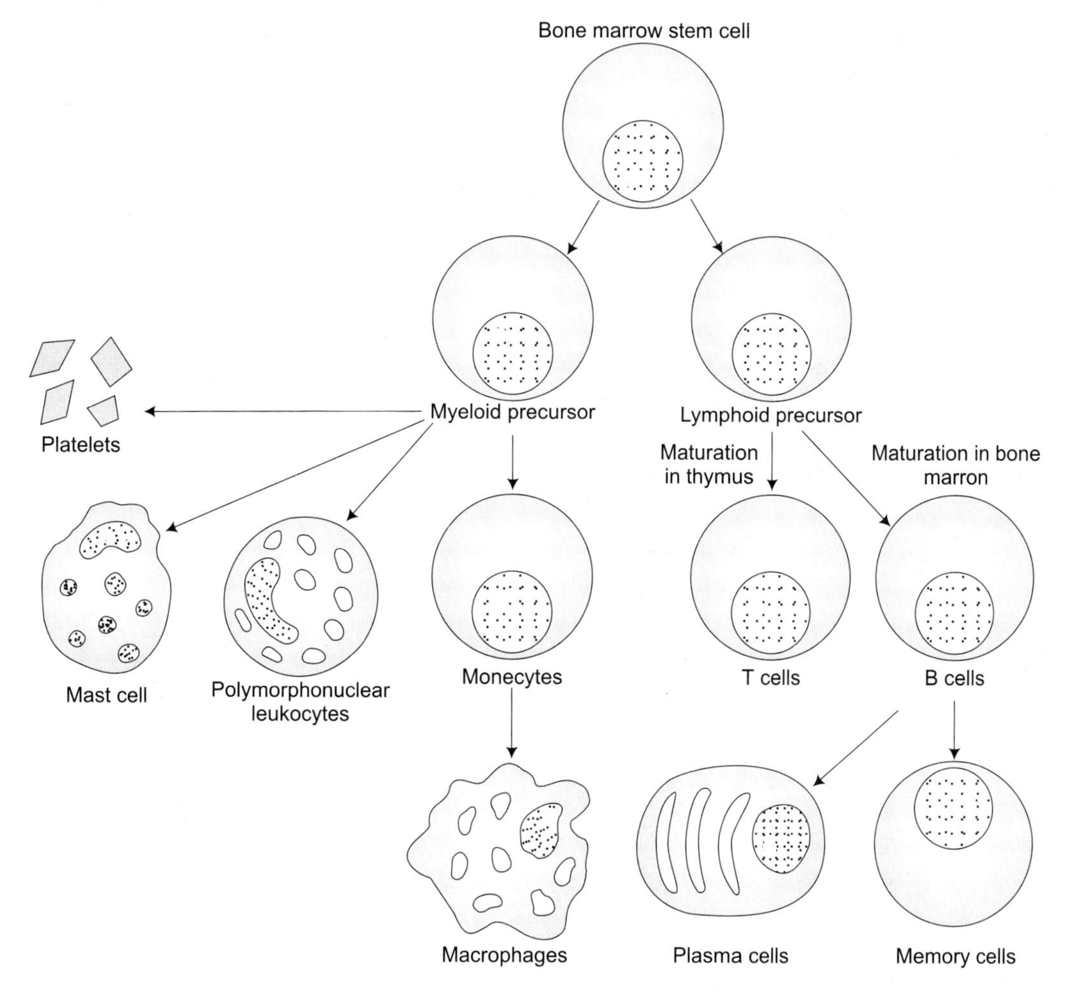

Fig. 38.6 Origin of major cells involved in the immune response. there are two major lines, one generating phagocytic cells, the other generating lymphocytic cells that participate directly in immune responses.

that can kill foreign cells, and together with T -delayed-typed hypersensitivity (Tdth) cells, play primary roles in the cellular immune responses (Table 38.2).

Macrophages

Macrophages are large phagocytic cells which are capable of ingesting antigens and destroying them as well as cooperating with lymphocytes in the production of specific antibody. Phagocytic cells are of two types, monocytes and polymorphonuclear granulocytes ("polymorphs"). Monocytes can differentiate to be come macrophages (Fig. 38.7). The term macrophage is generally used to describe phagocytes that are fixed to tissue surface, while the term monocyte is used to describe freely circulating phagocytes. Macrophages are abundant in lymphoid tissue and in the spleen, while monocytes are abundant in the blood and lymph.

Table 38.2 Comparison of B and T lymphocytes.

T cells	*B cells*
Origin: bone marrow	Origin: bone marrow
Maturation: thymus	Maturation: bone marrow; bursa in birds
Long-lived: months to years	Short-lived: days to weeks
Mobile	Relatively immobile (stationary)
No complement receptors	Have complement receptors
No immunoglobulins on surface	Immunoglobulins on surface
Restricted antigenic specificity	Restricted antigenic specificity
Proliferate upon antigenic stimulation	Proliferate upon antigenic stimulation into plasma cells and memory cells
Produce cell-mediated immunity (T_{dth} and T_c cells)	Synthesize immunoglobulin (antibody)
Show delayed hypersensitivity (T_{dth} cells)	
Produce lymphokines (T_{dth} cells)	
Participate in transplantation immunity (T_c, cells)	
Help in immunoglobulin production by B cells (T_h cells)	
Perform as killer T-lymphocytes in cell-mediated immunity (T_c cells)	

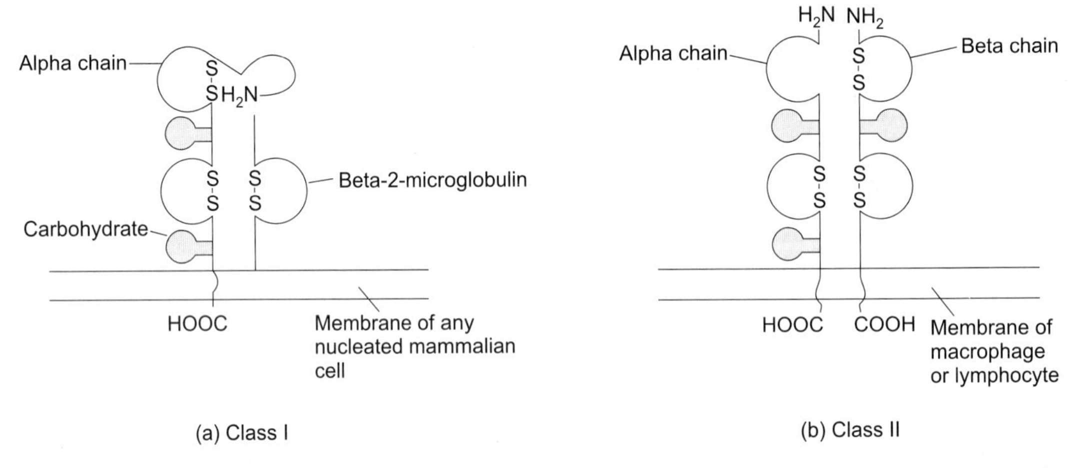

Fig. 38.7 Structure of major histocompatibility (MHC) antigens. (a) Class I type. (b) Class II type, Note that Class I molecules are present on the surfaces of all nucleated cells and that Class II molecules are only present on certain cells.

If an antigen penetrates an epithelial surface it will eventually come in contact with phagocytic cells. Phagoytic cells such as macrophages can engulf particles as large as bacterial cells and kill them by releasing lytic substances (proteases, nucleases, lipase, and lysozyme) from cytoplasmic vesicles within the cell. Lysis of a bacterial cells releases a variety of distinct bacterial antigens within the macrophage, and these, as well as antigens ingested directly, can be processed by the macrophage and used to initiate the early steps in antibody synthesis.

Macrophages serve as antigen-presenting cells because they imbed foreign antigens on their cell surfaces where they can come in contact with specific T cells and B cells; antigen recognition by T cells and B cells is the first step in antibody production.

Macrophages and monocytes are nonspecific cells. Unlike T cells and B cells, phagocytic cells cannot distinguish between antigens; any foreign substance is ingested, whether antigenic or not However, since many macromolecules are antigens and foreign cells contain numerous antigens, the majority of particles phagocytized by macrophages or monocytes will indeed be antigenic. The action of macrophages in antigen processing and antigen presentation is a key step in the overall process of antibody production, because the vast majority of antigens can only stimulated lymphocytes through macrophages acting as intermediaries.

SUMMARY

Cells of the lymphoid tissue involved in the immune response arise in the bone marrow and become differentiated into different classes of lymphocytes. Macrophages are lymphoid cells that engulf foreign particles and macromolecules and digest them. They are unspecific in their action and their role is to prepare the antigen for the specific and their immune response. Lymphocytes are involved specifically in the immune response, becoming differentiated into activated T cells and into B cells that are the forerunners of immunoglobulin-producing cells.

HISTOCOMPATIBILITY ANTIGENS AND T-CELLS RECEPTORS

Among the many molecules on the surface of animal cells there is a group of proteins that are highly specific for each animal species, or even each strain within a species. These proteins are referred to as major histocomatibility complex (MHC) antigens. These molecules can be thought of as species or strain" markers" molecular reference points by which cells from one species can be differentainted from those of another. MHC antigens are the major target molecules for transplantation rejection; if tissues from one animal are immunologically rejected when transplanted to a second, then their major histocompatibility antigens are likely to differ. The MHC of mice and human have been the best characterized and we focus here on the MHC of humans.

STRUCTURE OF THE HUMAN MAJOR HISTOCOMPATIBILITY COMPLEX

The major histocompatibility complex consists of group of proteins, known as class I, class II, and class III MHC antigens. Class I and class II molecules are cell surface proteins and intimately involved in immune recognition events. Class III molecules are soluble proteins important in complement activation.

Class I MHC antigens consist two polypeptides (Fig. 38.8), one of which is coded for by genes in the MHC region of human chromosome number 6. The other class I polypeptide,

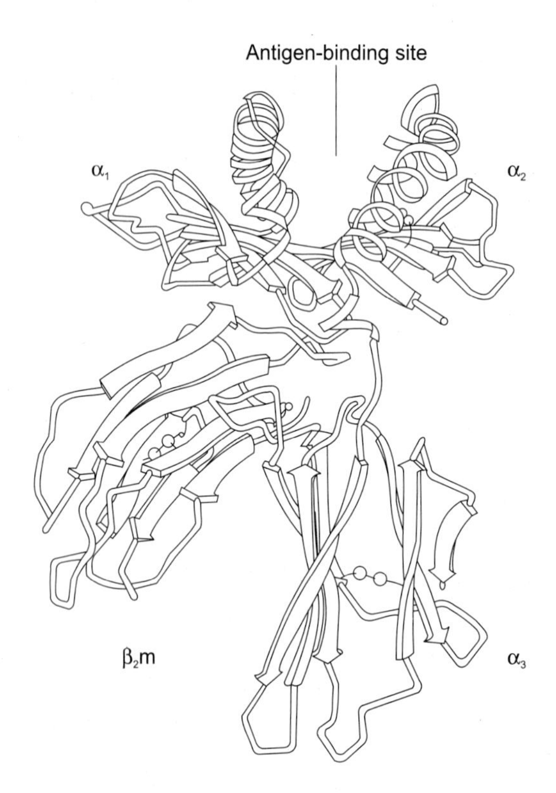

Fig. 38.8 Three dimensional structure of a MHC class I molecule Note the antigen-binding site which exists as a groove between the α_1 and α_2 alpha helices of the protein. β_2m, β_2 microglobulin. α_3 is the portion of the molecule that spans the plasma membrane.

called B-2 microglobulin, is coded for by non-MHC linked genes. The MHC- coded polypeptide is a glycoprotein firmly anchored in the cell membrane; B-2 microglobulin is a small polypeptide noncovalently linked to the MHC class I protein (Fig. 38.8). Class II MHC antigens consist of two noncovaliently-linked glycosylated polypeptides, called A and B. Like class I molecules, these polypeptides are imbedded in the plasma membrane and project outward as surface markers (Fig. 38.8).

Detailed studies of the crystal structure of class I MHC molecules have revealed a distinctive shape that suggest hoe these proteins interact with antigens and cells of the immune system. The alpha chain of a class I molecules forms three separate domains, called AI, A2, and A3, with the later domain spanning the plasma membrane to anchor the protein firmly on the cell surface (Fig. 38.9). Between the Al and A2 domains, a large groove exists and it is within this groove that the MHC molecule binds foreign antigens (Fig. 38.9). As we will see below, the embedding of a foreign antigen in the cleft of a class I molecules and the interaction of this bound antigen with T cells triggers the immune response.

MHC antigens are not structurally identical with in a given species. Instead, different individuals may show subtle differences in amino acid sequences of their MHC antigens. These

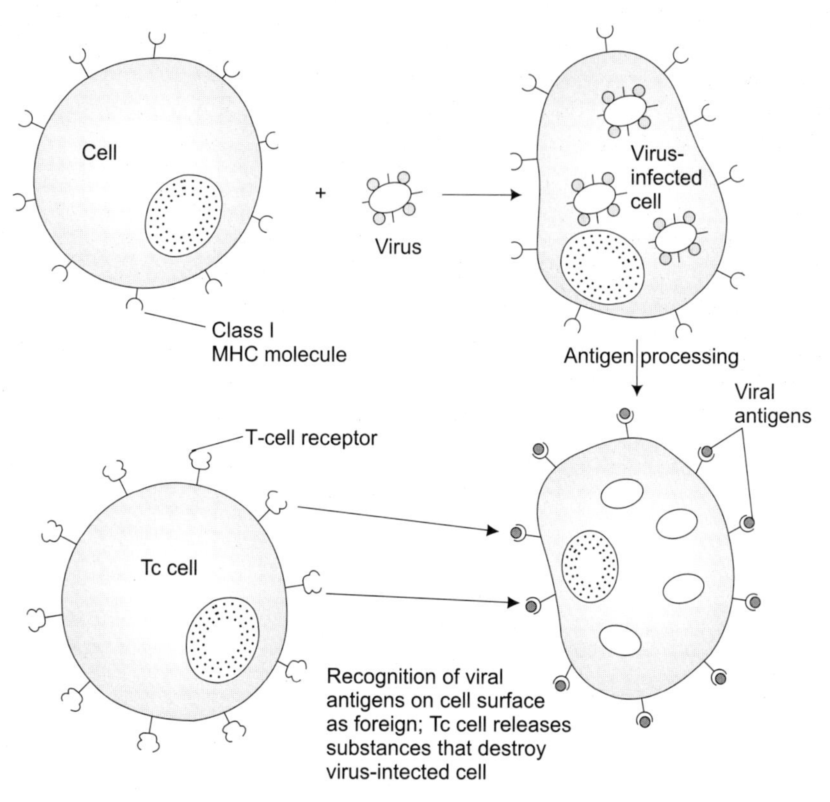

Fig. 38.9 Manner in which a T cell recognizes a foreign antigen. In this case, a virus infected cell is used as an example.

minor sequence variation, referred to as polymorphisms, occur in all classes of histocompatibility antigens. Structural changes in a given MHC antigen always exist between different animal species. These differences, both within tissues transplanted from one individual to another are frequently rejected.

Class I MHC antigens are found on the surfaces of all nucleated cells and on the surfaces of platelets. In mice, but in humans, class I proteins are also found on the surface of erythrocytes (red blood cells). Class II antigens are found only on the surface of B lymphocytes, certain types of T lymphocytes, and macrophages.

FUNCTIONS OF THE MAJOR HISTOCOMPATIBILITY COMPLEX

In the normal animal, T cells are constantly interacting with proteins or other potential antigens. It is critical that T cells be able to discriminate "self" from "non- self" antigens. The MHC proteins are thought to function as specific "self' reference points for the body's immune system. The T cell, through its T-cell receptor (see below), binds to MHC molecules and can then recognize foreign antigens; a T cell is "blind" to a foreign antigen until it is presented in the context of an MHC protein. How does this happen? When a foreign antigen is taken up by a host cell, the cell "processes" or degrades it. This processed antigen then becomes embedded in the MHC protein and the two are passed through the cell membrane and become attached to the surface of the cell.

In effect, the MHC molecule acts as a platform on which the foreign antigen is bound (see Fig. 38.9). For example, viral infection of an animal cell leads to the imbedding of viral antigens in Class molecules on the infected cell's surface (Fig. 38.10). Cytotoxic T cells (abbreviated T cells), surveying the overall cell population for those expressing foreign antigens, will recognize the viral antigen as long as it is bound by the Class I MHC molecules (Fig. 38.10). The T-cell receptor (TCR) on the surface of the T cells has both antigen-specific "(nonself") and MHC molecules- specific ("self") reactive sites. When contact between the TCR and the class I-boran foreign antigen is made, a group pf polypeptides on the surface of the T cells are activated (Fig. 38.11). This interaction induces the T cell to release cytotoxic proteins, killing the target (viral infected)cells.

In an analogous manner, T-helper (Th) cells recognize Class II MHC molecules and antigen molecules on the surface of the phagocytic cells and lymphocytes. In this case, foreign antigens digested by the phagocyte are presented by Class II MHC molecules, and when the Th cell is activated by contact with foreign antigen, it secretes molecules that stimulate antibody production by specific B cell clones.

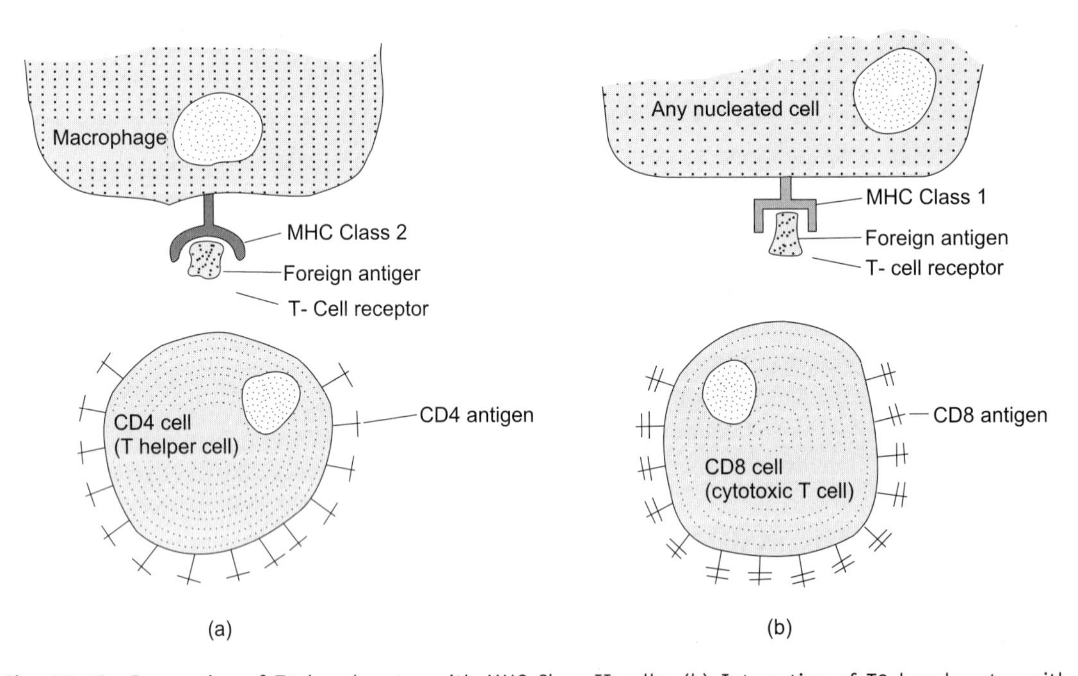

Fig. 38.10 Interaction of T4 lymphocytes with MHC Class II cells. (b) Interaction of T8 lymphocytes with MHC Class I cells. Note that the two T cells themselves are antigenically unique due to distinct T4 or T8 surface antigens.

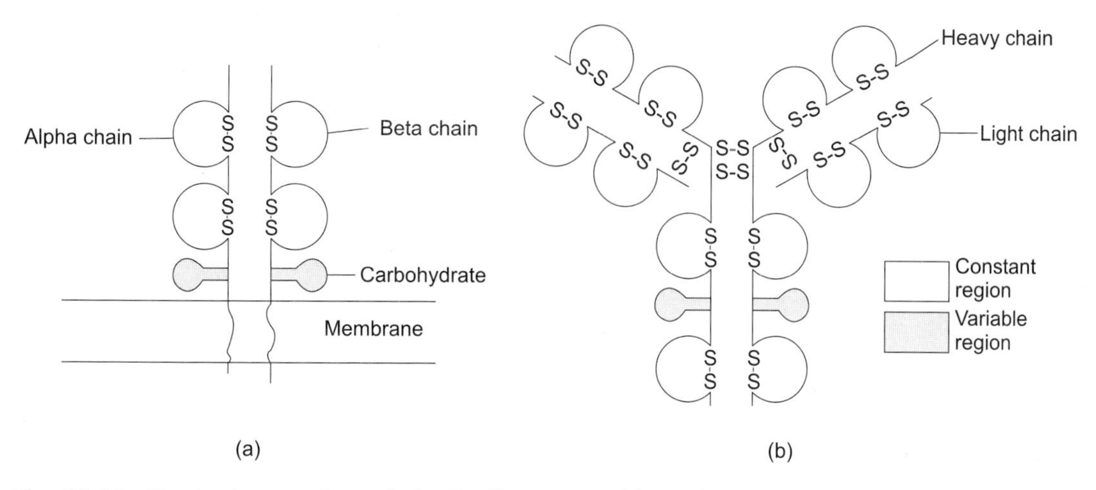

Fig. 38.11 Structural comparison of the T-cell receptor with an immunoglobulin. Note the presence of variable regions.

In summary, MHC molecules serve specific identifying markers for the host cells and also bind antigen such that T and T cells can "see" a foreign antigen in relation to a "self" reference molecule, the MHC antigen.

Proteins of the major histocompatibility complex (MHC) serve as specific markers, molecules reference points that permit the body to distinguish self from nonself.

These proteins are present on the surface of immune cell and serve as a platform to which the foreign antigen can bind. Under these conditions, the cell containing the foreign antigen is recognized as nonself and becomes the target for the immune response.

T-CELL RECEPTORS: TYPES AND DISTRIBUTION

T cells play a variety of complex roles in the overall immune response. Although T cells do not produce antibody, they do recognize antigen, and this recognition process is due to specific antigen receptor molecules located on T -cell surfaces called T-cell receptors. T -cell receptors have receptors have immunoglobulin –like specificity but are membrane integrated entities, like MHC molecules. Two major subpopulations of T cells have been identified: CD4 lymphocytes and CD8 lymphocytes. The two groups can be distinguished by whether they contain one or the other of two major identifying antigens on their surface. (The CD4 and CD8 antigens are present in addition to the MHC/foreign antigen receptor on the T -cell surface.

Lymphocytes of the CD4 type are classified as T -helper (Th)cells. Their major function is to recognize and bind foreign antigen and secrete substances that stimulate a complementary clone of B lymphocytes to divide and begin producing specific antibody CD8 lymphocytes are generally cytotoxic T cells (T) whose function is to identify foreign cells or cost cells displaying foreign antigens and kill them. However, besides specific identifying molecules and MHC receptors, T cells can recognize specific antigens by virtue of their T -cells receptors. How is this accomplished at the molecular level?

STRUCTURE OF T-CELL ANTIGEN RECEPTOR

Antigen receptors have to be highly specific-so specific in fact that their sepecificity must rival that of antibodies themselves. Although T -cell receptors are not antibody molecules, they resemble antibody molecules in many ways. Indeed, T-cell receptors and antibodies have much in common and in evolutionary terms are clearly related molecules.

The T-cell antigen receptor consist of two disulfide-linked polypeptides, called A and B. In analogy to the structure of antibodies, T-cell antigen receptors contain regions of constant and regions of highly variable amino acid sequence. Also as in antibody molecules, it is the variable region of the T -cell receptor which actually binds antigen. In contrast to antibodies, however, each polypeptide of the T-cell receptor contains only one constant region; antibody molecules contain one constant region on each light chain and three to four constant regions on each heavy chain. A comparison of the t-cell antigen receptor with a corresponding antibody molecules is shown in Fig. 38.12.

T-cell receptors are integral cell membrane proteins. Studies of polypeptide folding have shown that the secondary structure of T -cell receptor constant regions consist of relatively rigid B sheets, while the variable regions (antigen binding sites) are a-helices, with a much greater capacity to flex and fold (the transmembrane "anchors" are also a-helices). The a-helix form of secondary structure is ideal for generating highly specific binding sites. The constant and variable regions of antibody molecules also show B-sheet and a-helical folding, respectively.

A number of different types of T cells exist, each of which has specialized functions in the immune response. Activated T cells recognize antigen because of the presence on their surfaces of specific T-cell receptors which have immunoglobulin-like specificity. T-cell receptors are integral cell membrane proteins that are embedded in the membrane in such a way that the variable portion is exposed to the exterior, where it can combine with the antigen.

The Mechanism of Immunoglobulin Formation

How is it possible for animals to produce such a large variety of specific proteins, the immunoglobulins, in response to invasion by foreign macromolecules, viruses, and cells? The problem of antibody formation can be divided into two separate questions: (I) How do cells of the immune system interact to produce antibodies when stimulated by an antigen? (2) How is antibody diversity generated?

The production of immunoglobulins in response to stimulation by antigen is a complex process involving T cells, B cells, and antigen-presenting cells (for example, macrophages), and involves the intimate interaction of the various cell surface molecules discussed in the in the previous section. As we have described, macrophages act nonspecifically to ingest foreign antigens. Once the antigen is processed, the macrophage presents it to T cell: B cell pairs.

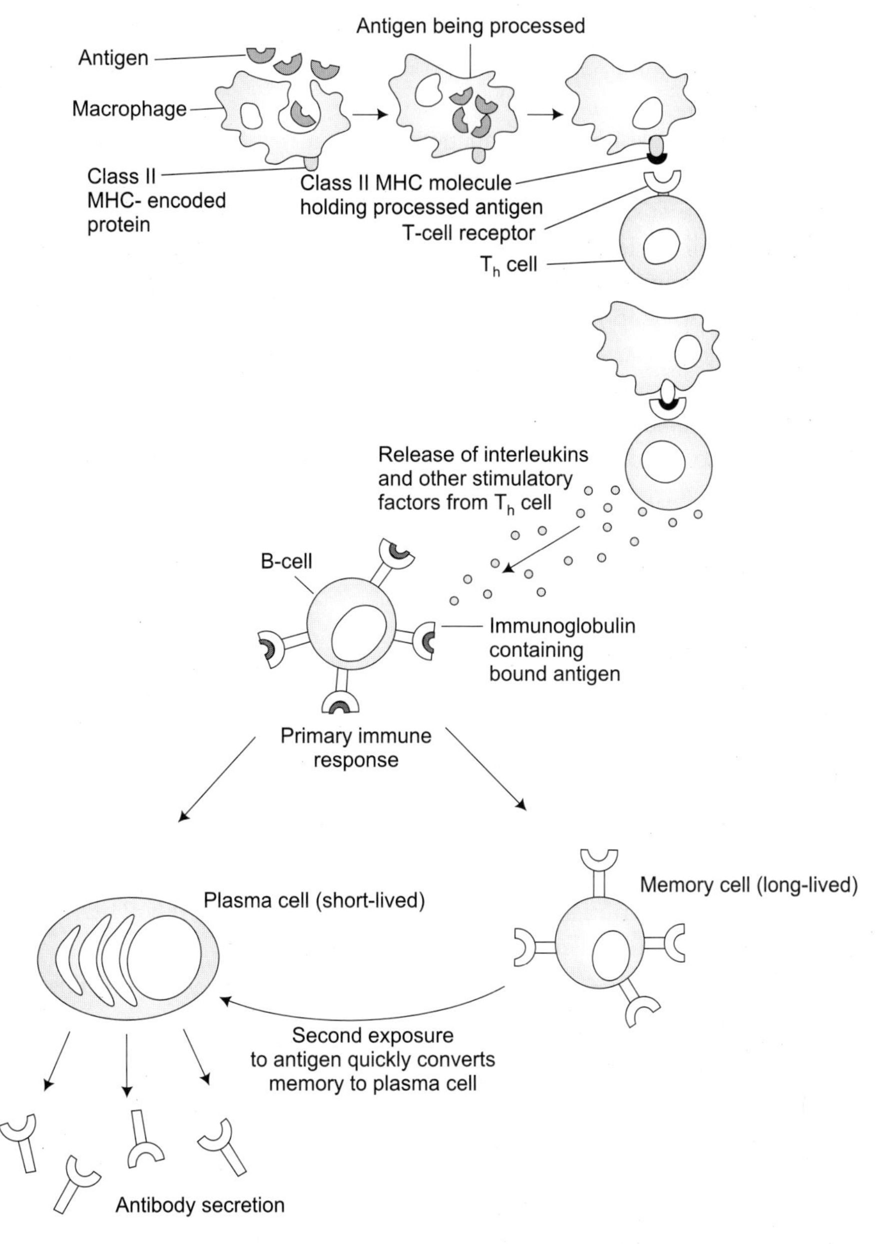

Fig. 38.12 Interaction between macrophages and B cells to produce plasma cells and memory cells. See text for details.

Following recognition, the T cell then releases substances that stimulate B cells to begin antibody synthesis. Although the initial interaction between antigen and macrophage is nonspecific, all further steps in the process of antibody production are highly specific; receptor molecules on the surfaces of both T cells (T-cell receptor) and B-cells (antibody molecules, see later) help ensure that antibodies made in response to an antigen will indeed be specific for that antigen.

The genetic control of antibody production is also a highly complex process. It has been estimated that an animal is capable of making over one billion structurally distinct antibody molecules. Although at first sight this seems like an enormous demand on an organism's genetic coding potential, we will see that only a relatively small number of genes is required to code for this immense antibody diversity because of a phenomenon known as gene rearrangement. During development of lymphocytic cells in the bone marrow, gene rearrangements and deletions occur in B cells to eventually yield two complete transcriptional units, one of which codes for the synthesis of a specific heavy chain and one for a light chain of the antibody molecule. The number of possible gene rearrangements, even of a relatively small number of genes, is sufficient to account for the diversity of immunoglobulin molecules. Similar rearrangements occur during T cell development to yield the diversity of T-cell receptors observed. We now detail the steps on antibody production, beginning with the injection of an antigen and ending with the production of a specific antibody that will react with that antigen.

Exposure to antigen

In considering the mechanism of antibody formation, we must first explore what happens to the antigen in the whole animal body. Antigens are carried to all parts of the body by the blood and lymph systems. The main sites of antigen localization in the body are the lymph, nodes, the spleen, and the liver. It has been well established that antibodies are formed in both the spleen and lymph nodes; the liver is not involved. If the antigen is injected intravenously, the spleen is the site of greatest antibody formation, whereas subcutaneous, intradermal, and intraperitoneal injections lead to antibody formation in lymph nodes. Fragments of lymph node or spleen form immunized animals can continue to produce antibody when placed in tissue culture or when injected into other, nonimmunized, animals.

Following the first introduction of an antigen, there is a lapse of time (latent period) before any antibody appears on the blood, followed by a gradual increase in titer (that is, concentration) and then a slow fall. This reaction to a single injection is called a primary antibody response (Fig. 38.13). When a second exposure to antigen is made some days or weeks later, the titer rises rapidly to a maximum 10 to 100 times above the level achieved following the first exposure. This large rise in antibody titer is referred to as the secondary antibody response (see Fig. 38.13). With time the titer slowly drops again, but later injections can bring it back up. The secondary response is the basis for the vaccination procedure known as a "booster shot" (for example, the. yearly rabies shot given to domestic animals) to maintain high levels of circulating antibody specific for a certain antigen.

Antibody production

How do B cells, T cells, and macrophages act together to produce immunoglobulin? Macrophages act nonspecifically, phagocytizing antigens, processing them, imbedding the

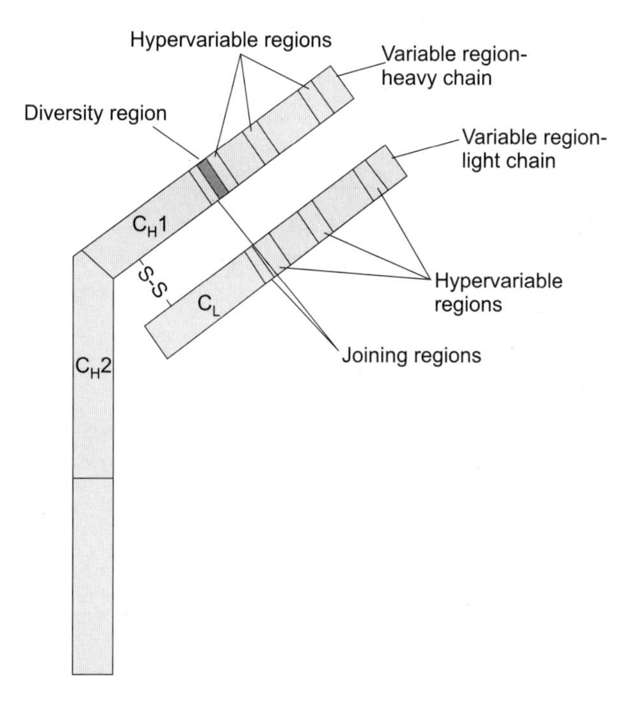

Fig. 38.13 Detailed structure of variable regions of immunoglobulin light and heavy chains. Only one-half of a typical immunoglobulin molecule is shown C_H, C_L, constant regions of heavy and light chains, respectively.

processed antigens in their surface MHC molecules) and presenting them to lymphocytic cells (Fig. 38.14). Macrophages are very sticky and attach well to surfaces; their stickiness probably also promotes the attachment of B and T cells. In this fashion, macrophages present the antigen to B and T lymphocytes, thus initiating the process of antibody production. Both B and T cells contain specific antigen receptors on their cell surfaces. We discussed the nature of the T -cell receptor in the previous section. The receptor on the B-cell surface are the immunoglobulin molecules that the B-cell is genetically programmed to produce. T -cell receptors on CD4 T cells recognize foreign antigen in the context of the MHC Class II molecules. Following binding of the T cell to the foreign antigen expressed on macrophage surface) the T cell releases interleukins and other soluble "helper factors" which stimulate B cell differentiation. It is thought that some T -cell helper factors are released T -cell receptors containing potions of bound antigen. This would account for how a specific T cell activates only a certain B cell clone and not others when it releases helper factors.

The T -cell helper factors stimulate growth and division of the B cell to form a clone of B cells, each capable of producing identical antibody molecules. Further differentiation of the activated B cell population then occurs, resulting in the formation of large antibody-secreting cells called plasma cells and a special form of B cell called a memory cell (see Fig. 38.14). Plasma cells are relatively short-lived (less than 1 week) but excrete large amounts of antibody during this period. Memory cells, on the other hand, are very long-lived cells, and upon second

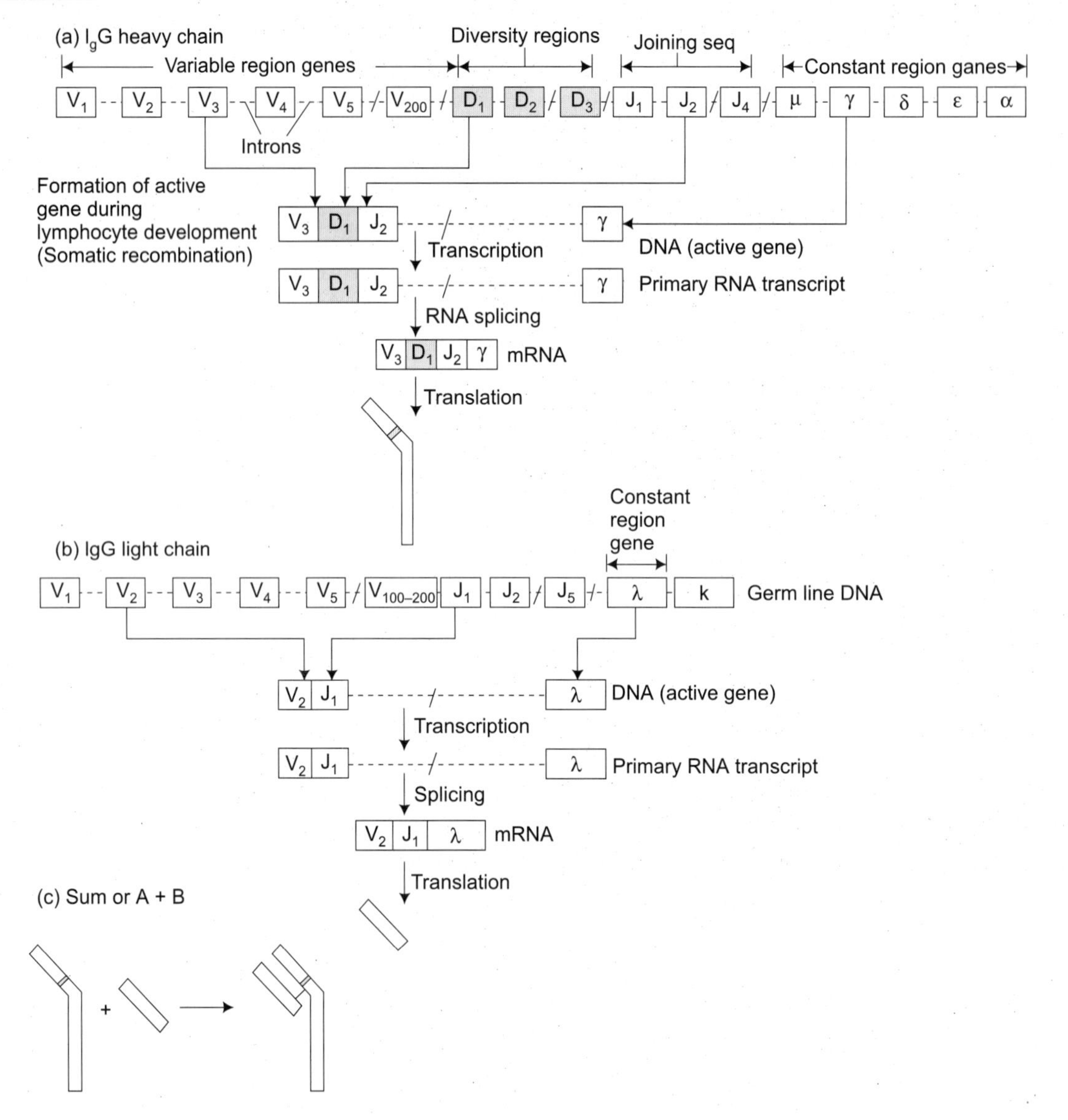

Fig. 38.14 Immunoglobulin gene arrangement in immature lymphocytes and the mechanism of active gene formation. (a) Heavy chain. (b) Light chain. (c) Formation of one-half an antibody molecule.

exposure to antigen, these cells are quickly transformed into plasma cells and begin secreting antibody. This account for the rapid and more abundant antibody response observed non antigenic stimulation the second time (secondary response; see Fig. 38.13).

One B lymphocyte may produce immunoglobulins of are than one class, but each will react, but will react with only one antigenic determinant (that is, their variable regions will identical). Surface receptor immunoglobulin are always the class IgM or IgD.

Certain antigens can stimulate low-level antibody production in the absence of previous T cell interactions (these are the so-called T-independent antigen). Most T-independent antigen are large polymeric molecules with repeating antigen determinants (for example, polysaccharides). The immunoglobulin produced to T-independent antigens are usually of the class IgM and are of low affinity. In addition, B cells which respond to T-independent antigen do not have immunologic memory.

The discussion above describes the general principles of antibody production. Each antigen (strictly speaking, each antigenic determinant) will catalyze, through the action of antigen-presenting cells and T-helper cells, the growth of a different B-cell line, the B-cell type in each case being genetically capable of producing antibodies that react specifically capable of producing antibodies that react specifically with that antigen. In this fashion it is thought that the normal animal can respond to perhaps as many as one billion distinct antigens by developing a specific B-cell clone in response to stimulation by antigen.

We now turn our attention to the genetic orchestration of these immune events and investigate the molecular diversity of antibodies and T-cell receptors at the DNA level.

The production of immunoglobulin is a complex process that involves the interaction of T-cells, B cells, and macrophages. The macrophage processes antigen and presents it to the T cell: B cell pair. The T cell: B cell pair is preconditioned to recognize the antigen, and following recognition the T cell releases substances that stimulate B cells to differentiate into immunoglobulin producing cells. The antigen serves a selective function, promoting the formation of a clone specific T and B cells. Eventually a population of immunoglobulin-secreting cells called plasma cells result.

GENETICS AND EVOLUTION OF IMMUNOGLOBULINS AND T-CELL RECEPTORS

Gene Organization

If one B lymphocyte produces immunoglobulins of a single specificity, it should theoretically require only one gene to code for the two identical light chains and one gene for the two identical heavy chains. However, this is not the case. A single light or heavy chain is actually coded for by several genes that undergo a complex series or regarrangements as B cells mature. How does this series of events occur?

To understand the genetics of immunoglobulin sysnthesis, it,is necessary to examine the structure of immunoglobulin in more detail. As illustrated in Fig. 38.3, variation in amino acid sequence exits in the "variable region" of different immunoglobulins. Further, amino acide variability is especially apparent in several so-called hypervariable regions (Fig. 38.15). It is at these hypervariable sites that combination with antigen actually occurs. Each variable region in the light and heavy chains has three hypervariable regions. A portion of the third hypervariable region on the heavy chain is coded for by a distinct gene called the D (for "diversity") gene, with the first two hypervariable sites being coded for by the variable region gene itself (Fig. 38.16). In addition, at the site of joining between the diversity region and the constant region genes,

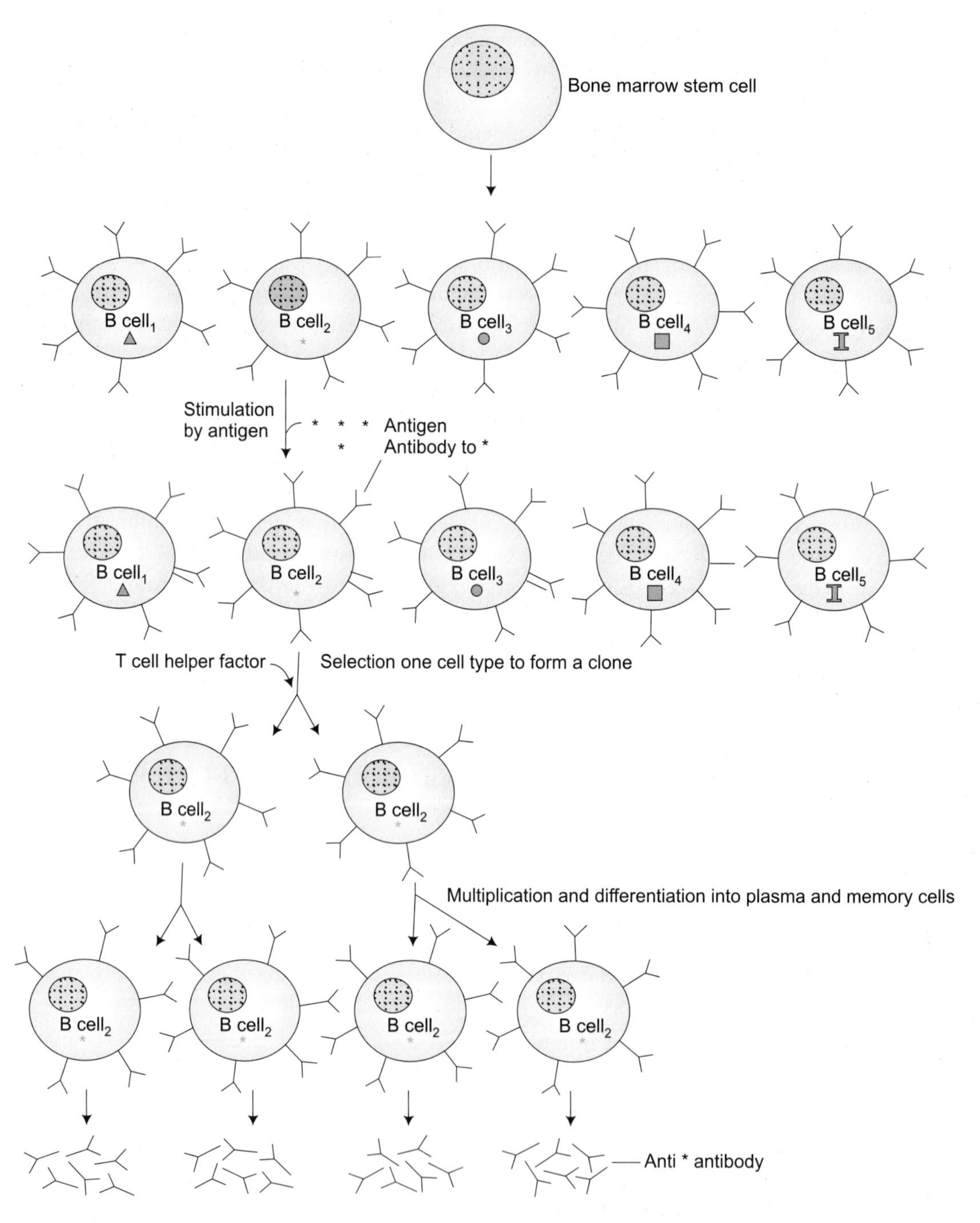

Fig. 38.15 Clonal selection. The expansion of a particular antibody-producing cell line after stimulation with antigen.

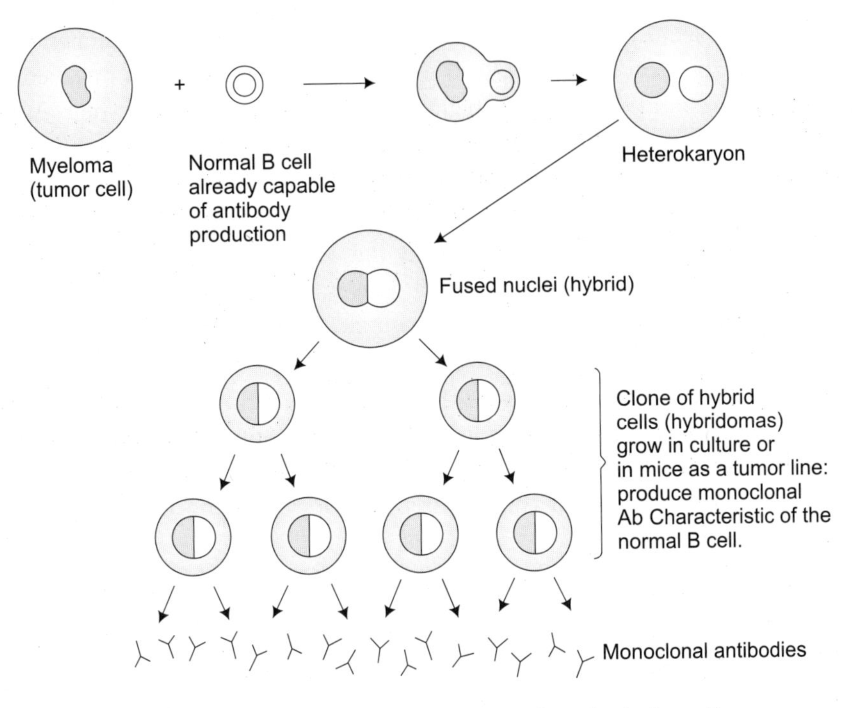

Fig. 38.16 Basic theory behind monoclonal antibody formation.

there is a stretch of nucleotides, about 40 bases in length, called the J (for "joiner") region that is coded for by a distinct gene (J gene). Finally, the class-defining constant region of the immunoglobulin molecule is coded for by its own gene, the C (for "constant") gene. Light chains are coded for by their own variable region genes, joining region genes, and constant region genes, but do not contain diversity regions as in the heavy chains (Fig. 38.16).

GENE REARRANGEMENTS

The origin of cells of the immune system was presented in Fig. 38.6. All the diversity required to make antibodies against perhaps as many as one billion different antigenic determinants is thought to exist in each lymphocyte as it is formed in the stem cells of the bone marrow. Each immature B cell is though to contain about 150 light-chain variable region. Genes and five distinct joining sequences, while 100-200 variable region genes, 4 joining sequences, and about 50 diversity region genes are thought to exist are the heavy chains (Fig. 38.16). In addition 5 heavy chain, constant-region genes and 2 light-chain, constant-region genes are present. These genes are not located adjacent to one another, but are separated by noncoding sequences (introns) typical of gene arrangements in eucaryotes. During matuation of B lymphocytes, genetic recombination occurs, resulting in the construction of an active heavy gene and an active light gene that are transcribed and translated to make the heavy and light chains of the immunologlobulin molecule, respectively Randomly selected V, J, and D segments are fused by enzymes that delete all intervening DNA. The active gene (containing an intervening sequence between the VDJ genes and the constant region gene) is transcribed, and the resulting primary RNA transcript spliced to yield the final mRNA (Fig. 38.16).

The Final light and heavy gene complement of a given B cell is a matter of chance rearrangement, and all possible gene combinations appear equally probable. The number of possible gene combinations is such that as many one billion different antibody molecules can be made. By the time each B cell has matured, however, gene rearrangements will yield a B cell capable of making immunoglobilins with only a single specificity. Put another way, each B cell makes only immunoglobulin with identical variable regions.

Additional antibody diversity occurs as the result of mutations arising during formation of the active heavy chain gene. The DNA splicing mechanism appears to be rather imprecise and frequently varies the site of VDJ fusions by a few nucleotides. This, of course, is sufficient to change an amino acid or two, and this apparently sloppy splicing mechanism leads to even greater anti body diversity.

It is truly impressive that such an enormous number of molecules can be coded for by a relatively small amount of DNA. The discovery of gene rearrangements in lymphoctic cells has had a marked impact on our understanding of eucaryotic genetics. It had always been assumed that every cell of a multicellular organism was genetically identical (barring any somatic cell mutation, of course) and that the differences between cell types were a function of differential gene expression. This clearly is not the case with B lymphocytes, because several million (if not billion) genetically distinct lymphocytes, a product of somatic cell recombination, exist in the mammalian body. It is not known whether lymphocytes constitute the only eucaryotic cell line in which gene rearrangements occur, but this simple yet elegant mechanism has shown the power of in vivo gene splicing for generation molecular diversity.

MONOCLONAL ANTIBODIES

In the whole animal, a typical immune response results in the production of a board of spectrum of immunoglobulin molecules of various classes, affinities, and specificities for the determinants present on the antigen. The immunoglobulin directed toward a particular determinant will represent only a portion of the total antibody pool. Each immunoglobulin would be produced by a single clone that had arisen as a result of antigen stimulation and an antiserum containing such a mixture of antibodies is thus referred to as a polyclonal antiserum. However, it is possible to produce antibodies of only a single specificity, derived from a single B-cell clone, A variety of such monospecific antibodies, called monoclonal antibodies, have been generated for use in research and clinical medicine. Table 38.3 compares the properties of antibodies prepared against an antigen in the usual way-by preparing a polyclonal antiserum-with monoclonal antibodies.

Table 38.3 Characteristics of monoclonal and polyclonal system.

Polyclonal	*Monoclonal*
Contains antibodies recognizing many determination on an Antigen	Contains antibody recognizing only a single determinant
Various classes of antibodies are present (IgG, IgM, etc.) (IgG only)	Single antibody class
To make a specific system, highly purified antigen is necessary	Can make a specific system using an impure antigen
Reproducibility and standardization difficult	Highly reproducible

Each B cell clone produces a monoclonal antibody-it is genetically programmed to do so.

Grow a single B cell clone for experimental production of monoclonal antibodies. Normal lymphoctes are not easy to grow and maintain in cell culture, hence this approach has not been found useful. However, B lymphocytes can be fused with myeloma (tumor) cells to from B cell lines will grow in culture, yet retain the ability to produce antibodies. This technique of B cell/ myeloma cell fusion is known as the hybridoma technique, and is summarized in Fig. 38.15.

How are hybridomas capable of producing monoclonal antibodies made? First, a mouse is immunized with the antigen of interest and a period of weeks allowed for B-cell clones to be selected and begin producing antibody by the normal sequence of events.

Then, spleen tissue, rich in B lymphocytes, is removed from the mouse and the B cell are fused with myeloma cell (Fig. 38.16). Although true fusion represent only a small fraction of the total cell population remaining in the mixture, addition of the compounds hypoxanthine, aminoptrin, and thymidine to the medium (the so-called HAT medium) strongly selects for fused hybrids. This occurs because unfused myeloma cells are unable to use the metabolites hypoxanthine and thymidine to bypass a metabolic block caused by aminpterin, a cell poison. By contrast, fused hybrid cells are able to use hypoxanthine and thymidine and thus grow normally in HAT medium. Unfused B cells die off in a week or two because they are unable to grow in culture for more than a few cell divisions. Following fusion, the most important job is to identify the clones of interest; that is, if individual fused cells are placed in the wells of microtiter plates and allowed to grow and produce antibody (Fig. 38.16), which ones are producing the antibodies of interest?

A variety of assay techniques can be used to identify clones producing monoclonal antibodies. In a typical experiment, several distant clones are isolated, each making a monoclonal antibody to a different determinate on the antigen. Once the clones of interest are identified they can be grown in cell culture, a rather laborious procedure that yields only relatively low antibody titers, or they can be injected into mice and be perpetuated as a mouse myeloma tumor. As a mouse tumor line, hybridomas secrete large amount of monoclonal antibody 9 in mice over 10 milligrams of pure monoclonal antibody can be obtained per milliliter of mouse peritoneal fluid).

Specific hybridomas of interest can also be stored frozen and later thawed and injected back into mice when more of a specific monoclonal antibody is needed.

Monoclonal antibodies are extremely useful reagents for research and medical science. For research purposes, monoclonal antibodies directed against specific cell markers, for example the CD4 and CD8 lymphocyte markers can be used to separate mixtures of these cells from. one another. By attaching a fluorescent dye to the monoclonal antibody, those cells to which the nonoclonal antibody binds will now fluoresce (Fig. 38.17). Fluorescing, cells can be separated from their nonfluorescing counterparts with a fluorescence activated cell sorter (FACS), an instrument employing a laser beam to place an electrical charge on fluorescing molecules (in this case fluorescing cells). Following laser exposure, the application of an electrical field to the cell mixture will orient fluorescing and nonfluorescing cells to opposite ends of the electrical field where they can be collected in separate receptacles. FACS machines have given the immunologist the ability to separate complex mixtures of immune cells into highly enriched populations of one type or the other for use in immunological research.

Other research applications of monoclonal antibodies include studies of the active site of enzymes or the combining site of antibodies themselves by observing the effect on an enzyme's activity or an antibody's binding ability when treated with monoclonal antibodies to specific determinates on the molecules. Monoclonal antibodies show great promise in the immunologocal typing of bacteria and in the identification of cell containing foreign surface antigens (for example, a virus infected cell). Monoclonal antibodies have also been used in genetic engineering for measuring levels of gene products not detectable by other methods.

Of considerable importance are the applications of monoclonal antibodies to clinical diagnostics and medical therapeutics. We discuss the use of monoclonal antibodies in clinical diagnostics later. In therapeutics, perhaps the boldest application of monoclonal antibodies may be in the detection and treatment of human malignancies. Malignant cells contain a variety of surface antigens not expressed on the surfaces of normal cells. These include differentiation antigens expressed during fetal development but not in the normal adult, and unique glycolipid cell surface antigens (the latter were originally detected by monoclonal antibodies prepared against random tumor cell surface antigens). Because monoclonal antibodies prepared against cancer-related antigenic determinants should be able to distinguish between normal and malignant cells, monoclonal antibodies may serve as vehicles for directing toxins or radioisotopes to malignant cells. Such specificity would greatly improve cancer chemotherapy because damage to normal cell by the toxic agents necessary to kill cancerous cells is one of the major problems with conventional cancer chemotherapy.

Monoclonal antibodies have also been prepared that can distinguish human transplantation antigens, thus improving the specificity of the tissue matching process critical for successful transplantations. Monoclonal antibodies also show great promise for increasing the specificity of a variety of conventional clinical tests including blood typing rheumatoid factor determination, and even. pregnancy determination; the later test employs monoclonal antibodies prepared against specific hormones associated with the pregnant state. Other

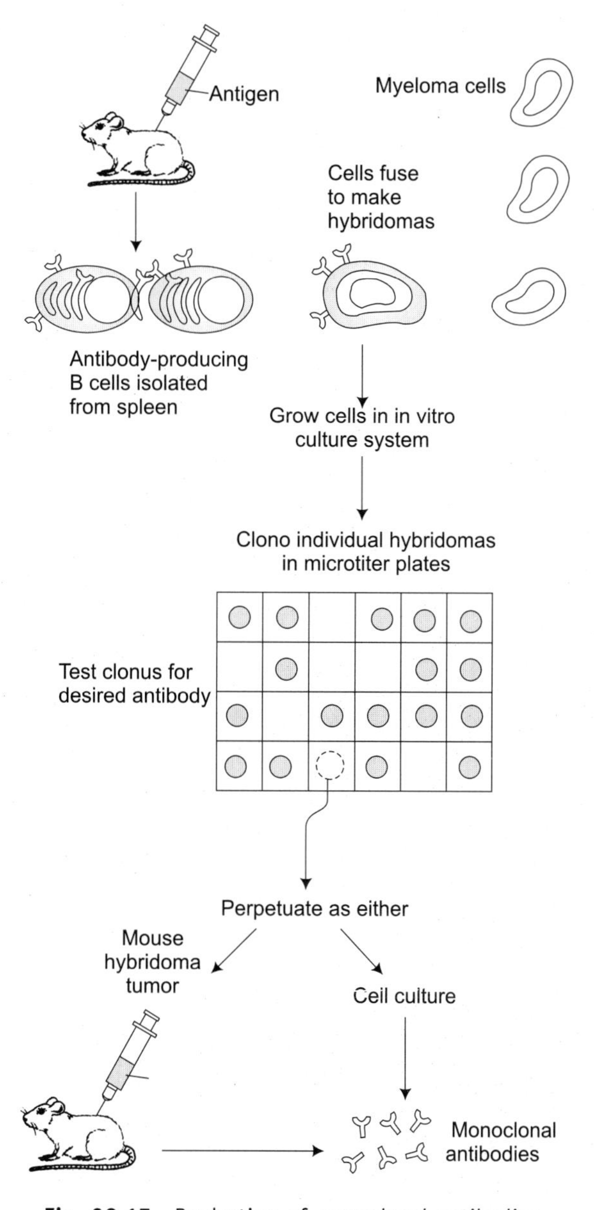

Fig. 38.17 Production of monoclonal antibodies.

monoclonals are being developed for in vivo diagnostics. For example, radioactively tagged monoclonal antibodies have been used experimentally to detect exposed myosin in heart muscle; myosin is a major protein in heart muscle cells that normally only becomes detectable following damages to the heart muscle. A myosin-specific monoclonal antibody would therefore be useful for diagnosing the extent of heart damage following a heart attack.

As with any clinical reagent, the greater the specificity the more useful it is in clincal diagnostics. Monoclonal antibodies now offer the physician and clinical microbiologist the ultimate in immunological specificity. (See Urine Testing for drug abuse box for a discussion of the use of monoclonal antibodies as one means of detecting illicit drugs).

The hybridoma technique permits the selection *by* cell cloning of large populations of cell all capable *of* producing immunoglobulin molecules active against a single antigen determinant. These monoclonal antibodies are useful immunological reagents and can also be used in certain types of disease therapy.

ELISA

The covalent attachment of enzymes to antibody molecules creates an immunological tool possessing both high specific and high sensitivity. The technique, called ELISA (for enzyme-linked immuno absorbent assay), makes use of antibodies to which enzymes have been covalently bound such that the enzyme's catalytic properties and the antibody's specificity are unaltered. Typical linked enzymes include peroxidase, alkaline phosphatase, and B-galactosidase, all of which catalyze reactions whose products are colored and can be measured in very low amount.

Two basic ELISA methodologies have been developed, one for detecting antigen (direct ELISA) and the other for antibodies (indirect ELISA). For detecting antigens I such as virus particles from a blood or fecal sample, the direct ELISA method is used. In this procedure the antigen is "trapped" between two layers of antibodies. The specimen is added to the wells of a microtiter late previously coated with antibodies specific for the antigen to be detected. If the antigen (virus particles) is present in the sample it will be trapped by the antigen binding sites on the antibodies. After washing unbound material away, a second antibody containing a conjugated enzyme is added. The second antibody is also specific for the antigen so it will bind to any remaining exposed determinants. Following a wash, the enzyme activity of the bound material in each microtiter well is determined by adding the substrate of the enzyme. The color formed is proportional to the amount of antigen originally present.

To detect antibodies in human serum, an indirect ELISA is used. The microtiter wells in this case are coated with antigen and a serum sample is added. If antibodies to the antigen are present in the serum, they will bind to the antigen in the wells. Following a wash, a second antibody as the "detecting" antibody. The second antibody is a rabbit or goat anti-human IgG antibody containing a conjugated enzyme. Following the addition of enzyme substrate a color is formed, and the amount of circulating human antibody to the specific antigen is quantitated from the intensity of the color reaction.

CHAPTER 39

Human Genome

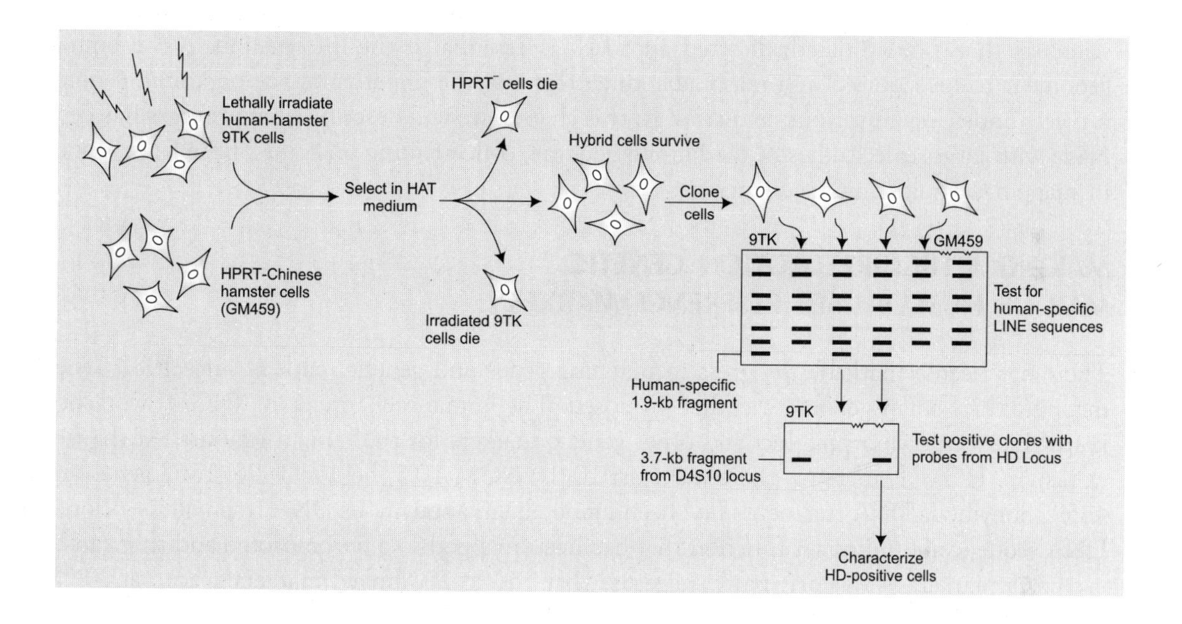

The various projects map and sequence the entire genomes of human beings and other organisms open up the most exciting research prospects ever in the biological sciences. Map and sequence data will generate a wealth of new knowledge about biological processes. We have seen that sequences data are invaluable in revealing functions for newly cloned genes and for generating insights into how organisms work. This approach is especially rewarding when sequence comparisons between different species point to evolutionarily conserved sequences that are likely to encode functionally important proteins. This is one of reasons that the genomes of species other than *Homo sapiens* are also the subject of intensive study. In addition to helping us to learn more about the basic mechanisms of life, we hope to learn more about the genetic basis of human diseases, especially about the genes that contribute to the development of polygenic disorders. It is these diseases-such as coronary heart disease and mental disorders—that are very common, and an understanding of the interaction between genes and the environmental factors involved could lead to significant improvement in preventative medicine. Detailed knowledge of disease genes will help us unravel their pathologies and may suggest new rationales for treatment.

Until recently, human gene mapping and sequencing were done by individual investigators working on particular parts of the genome, sometimes in collaboration, but more often in competition, with other scientists. But the human genome is immense. The 23 human chromosomes (Fig. 39.1) have a total genetic size of some 3300 cM, and the total number of base pairs is about 3 billion. To put this in context, even the whole sequences of *E. coli*, a mere 4 million bp, is not yet known, The formidable size of the human genome has persuaded many scientists that a coordinated, directed approach is essential if the information in the human genome is to be acquired in a reasonable time; this was the impetus for the organized genome projects under way in various countries. In this chapter we will describe what progress has been made with large-scale studies of the human genome, concertrating on some of the latest results in mapping, cloning, and sequencing.

MAKING A HIGH-RESOLUTION GENETIC MAP OF HUMAN USES REFERENCE MARKERS

There has been remarkable progress in mapping genes and genetic markers since 1980, when mapping the human genome was first proposed. The human gene mapping (HGM) workshop keeps track of polymorphic loci and other genetic markers for the human genome. At the first workshop, HGM 1, in 1973, there were just 10; by HGM 10.5, held in 1990, 1867 genes and 4859 anonymous DNA fragments had been mapped; (an anonymous DNA fragment is a cloned DNA sequence of unknown function that has been mapped to a chromosome and that can be used as a marker). It was originally suggested that 150 to 200 linked markers spaced at 20 cM intervals would be sufficient for mapping genes in the human genome, but it has become clear that markers need to be closer together than this; the current aim is to produce a map with markers on average 2 CM apart by 1996. Intermediate maps with reference markers unambiguously assigned to a locus and ordered on the chromosome are being prepared with a 10 cM resolution. The markers are being chosen for their high degree of polymorphism, so that they should be useful for linkage analyses in many families. Chromosome 1 provides a good example of the progress in producing these reference marker maps (Fig. 39.2)

Chromosome 1 is the largest human chromosome and makes up approximately 8.4 percent of the autosomal DNA. At HGM 10.5, 193 genes and 120 anonymous DNA polymorphisms were assigned to this chromosome. In a collaborative effort involving 11 laboratories, and 40 families form the Center du Polymorphisme Hmain collection, 58 loci could be accurately mapped on chromosome 1 so that there is less than a 1 in 1000 chance that the loci could be elsewhere on the chromosome. The loci are a mean of 6.7 cM apart, and there are only four gaps where the separation between loci is greater than 15 cM. As other chromosome 1 markers

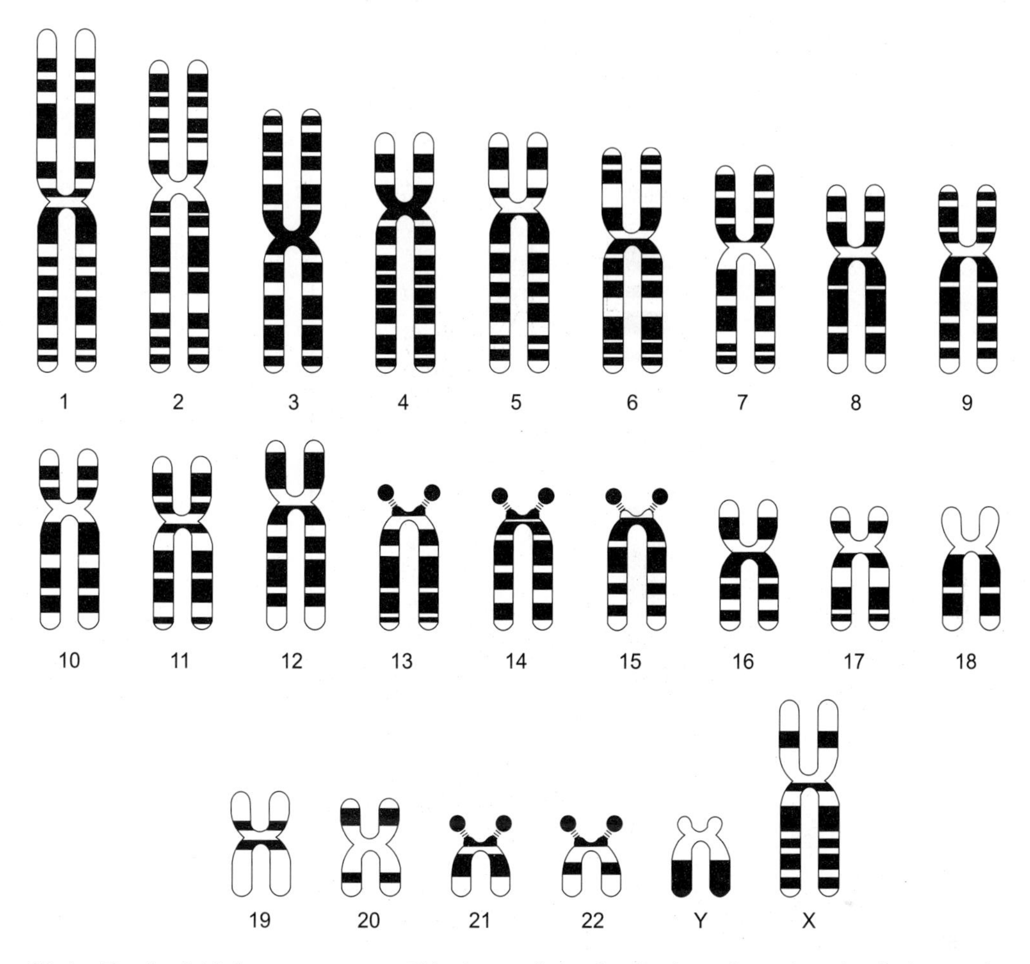

Fig. 39.1 The haploid human genome. This is a schematic drawing of 1 of each of the 23 human chromosomes, showing the pattern of staining seen with the Giemsa banding method. Chromosomes, are first treated with trypsin and then stained with Giemsa. The patterns of light and dark bands are characterisitc for each chromosome; and translocations, deletions, and other structural abnormalities can be identified. Typically 400 bands can be seen per haploid genome, and each band represents on average 7.5×10^6 bp, or twice as many base pairs as in the entire *E. coli* genome! Chromosome 1 constitutes 8.4 percent, and the Y chromosome about 2.0 percent, of the human genome. Taking the *E. coli* genome as a unit of genome size, a cytogenetic band is gnome units, and the Y chromosome is 15 genome units.

are cloned, they will be mapped on the chromosome using these reference markers. Because the reference markers are available to all investigators, they will provide a framework for integrating data from different laboratories to make a linkage map for the whole chromosome. This is good progress, but it does show the magnitude of the task ahead over the next 4 years if the whole genome is to be mapped with a resolution of 2 cM.

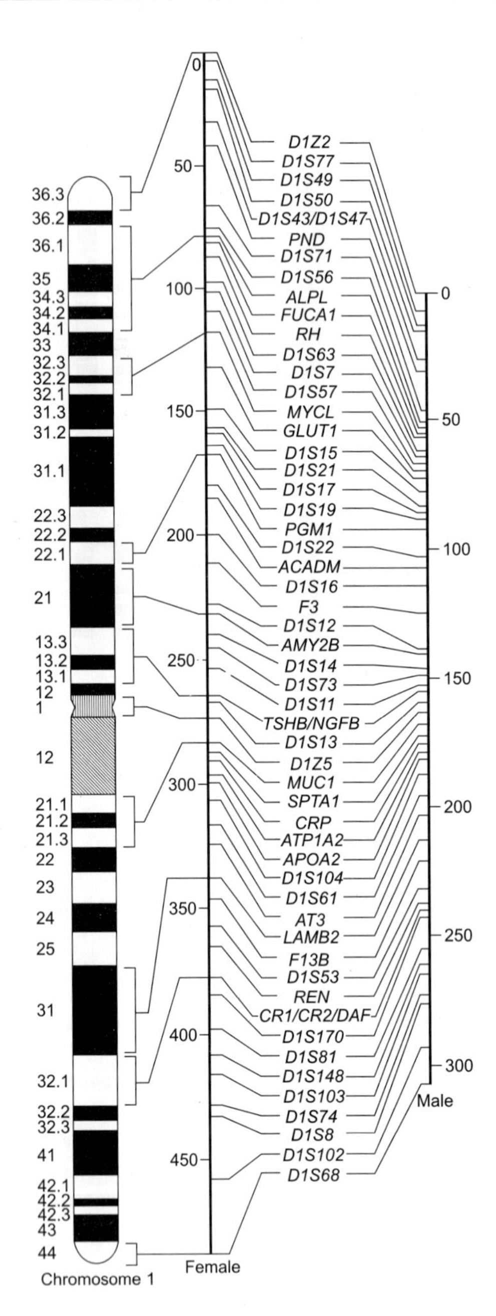

Fig. 39.2 High-resolution genetic linkage map of human chromosome 1, prepared by the CEPH Consortium. A drawing of the standard cytogenetic map is shown at the left, and the position of 58 markers are shown. There are two linkage maps, one for females and one for males. Recombination is higher in human females than in males, so the genetic map of chromosome 1 is longer in females (about 500 cM). This figure shows only the reference markers. The map is based on an analysis of 126 polymorphisms at 101 loci.

Human Chromosomes are Separated from Each Other Using Cell Sorting Machines

There have also been significant technical advances in cloning human genes. Because the human genome is so large, much effort has gone into isolating DNA form smaller portions of the genome. For example, it is much more efficient to search for an X-linked gene in a library prepared from just the X chromosome, rather than in a library containing clones from all 23 chromosomes. Chromosome- enriched libraries are made from chromosomes that are sorted by using fluorescence-activated cell sorters, of FACS (Fig. 39.3). These machines were originally designed as flow cytometers to measure the DNA content of single cells. They were adapted to sort and collect different cell types, usually different classes of white blood cells. For chromosome sorting, metaphase chromosomes are isolated form tissue culture cells that have been blocked in mitosis by treating the cultures with the drug colcemid. The mitotic cells attach loosely to the culture dish surface and can be collected by shaking the dishes. The cells are treated with a chromosome isolation buffer that helps maintain the structure of the chromosome through the mechanical trauma of passing through the FACS and that inactivates the cell nucleases that will destroy the DNA. The isolated chromosomes are then stained with DNA-specific dyes that stain AT-rich regions (for example, Hoechst 33258), and ones that stain GC-rich DNA (for example, chromomycin A). The stained chromosomes fluoresce when illuminated with light of the correct wavelength from lasers. The Hoescht dye fluoresces when exposed to UV light (wavelengths of 351 and 363 nm), while the chromomycin fluoresces with light of wavelength 458 nm. The amounts and proportions of the dyes taken up vary for each chromosome, and a computer recognizes the fluorescence pattern characteristic for each chromosome. In practice some chromosomes, particularly chromosome 9 through 12, are indistinguishable and cannot be separated. It is usual to sort as many as 1 × 10* chromosomes to make a genomic library. This requires several days with a standard FACS, although there are machines custom built specifically for chromosome sorting that can produce this number of chromosome in few hours. An early example of the usefulness of chromosome enriched libraries is the X-chromosome library that was the source of the probes that detected the first RFLPs linked to Duchenne muscular dystrophy. Now chromosome-enriched libraries technical developments will lead to genomic libraries prepared from increasingly pure chromosome isolates.

DNA for Cloning can be Microdissected from Human Chromosomes

While a FACS isolates single chromosomes, microdissection can be used to cut out DNA from specific regions of chromosomes. This technique was first used with those chromosomes that are easy to identify visually, without staining, using phase contrast light microscopy, for example, the giant polytene chromosomes of *Drosophila* and those of a mouse strain in which the X chromosome can be recognized because most of the chromosomes are fused in pairs. Similarly, clones could be prepared using DNA dissected from the short arm of human chromosome 2 because this also can be organized in unstained preparations were used because it was thought

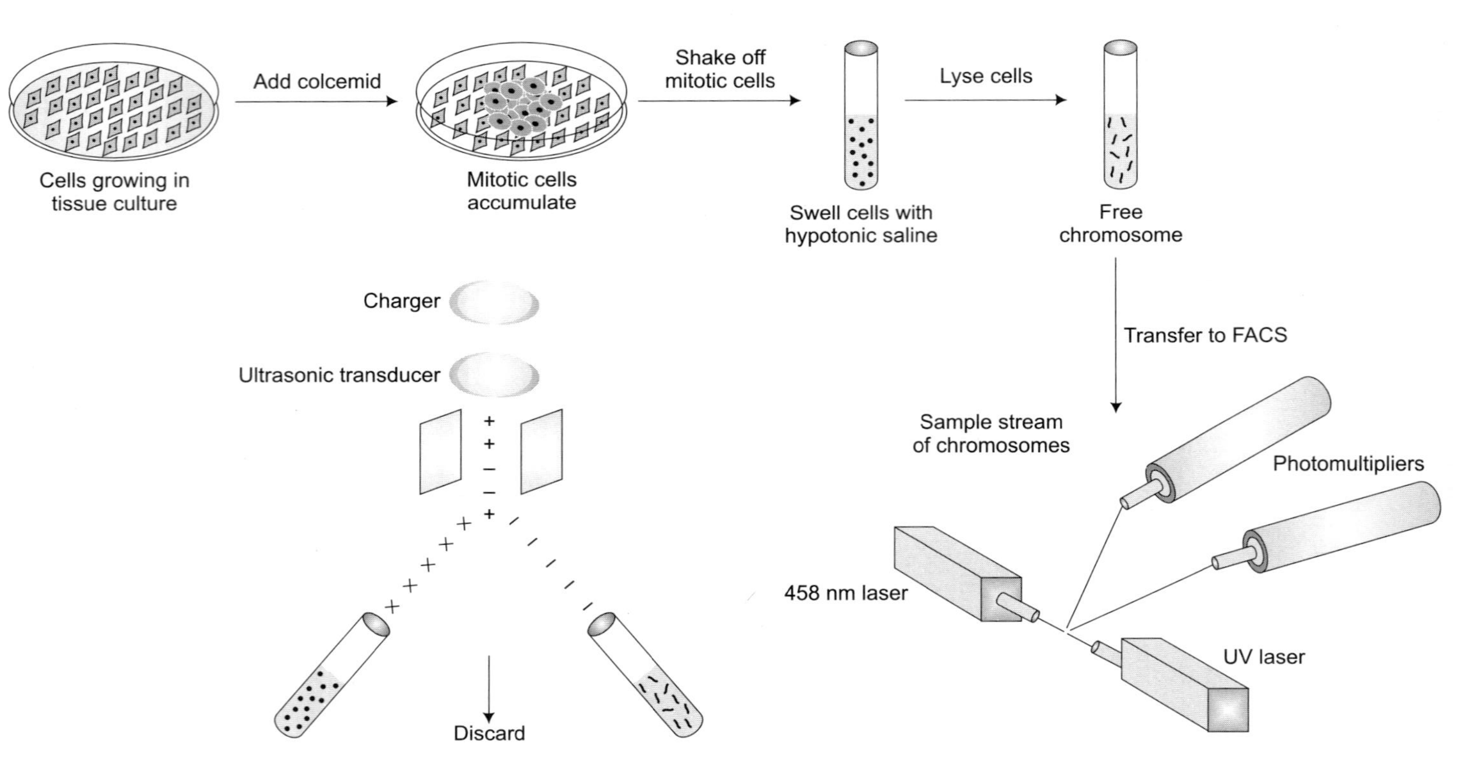

Fig. 39.3 Principles of using a fluorescence-activated cell sorter (FACS) for sorting chromosomes. Cultures of cells are treated with colcemid, a mitotic inhibitor that blocks cell division in metaphase. Mitotic cells are rounded and only loosely attached to the culture dish surface, and pure preparations (better than 90 percent) of mitotic cells are collected by shaking the dish to dislodge them. The cells are gently broken open and the chromosomes stained with DNA- specific dyes like Hoechst 33258 and chromomycin A. The chromosome preparation is transferred to a FACS, where a narrow stream of solution containing the chromosomes passes through two laser beams that excite the fluorescent dyes. The photomultipliers monitor the emissions from each dye, and a computer determines the fluorescence of each individual chromosome as it passes through the laser beam. Human chromosomes stained with these dyes have characteristic fluorescence emissions so that the computer can distinguish between them. The computer directs the stream to be electrically charged depending on which chromosomes are to he collected. The stream is then broken into droplets by an ultrasonic transducer so that each droplet contains a single chromosome. If the drop is not charged, it passes straight through a magnetic field and is discarded. If the drop is electrically charged, it will be deflected by a magnetic field and collected in one of two collection tubes, depending on the charge it carries.

that DNA subjected to the procedures required to reveal chromosome bands would be damaged or otherwise made unclonable. However, further research showed that cytogenetically stained chromosomes were a good source of DNA for cloning if PCR was used to amplify DNA from individual bands cut from the chromosomes. This meant that the technique could be applied to all human chromosomes. Spreads of human chromosomes are made and stained using the standard trypsin-Giemsa technique, and then chromosome fragments are cut from the appropriate bands by using ultrafine glass needles. DNA extraction (proctease digestion, phenol extraction) and cloning reactions (restriction enzyme digestion, phenol extraction, ligation) are carried out in minute volumes of solution put within oil drops to reduce evaporation. These restriction fragments are ligated into a plasmid, or synthetic olionucleotides ("linkers") are added to their ends. In either case, the fragments of unknown sequence are thus flanked by known sequences that are the targets for PCR primers. The amplified DNA is cloned into a suitable vector, and the chromosomal location of the DNA in each clone is subsequently determined by using in situ hybridization or panels of somatic cells hybrids. The amount of DNA dissected from 20 chromosomes in estimated to be as little as 300 femtograms (300×10^{-15} g), but this is sufficient DNA for PCR amplification. Once again, the PCR turned a technique that had limited applications into a very useful method.

Somatic Cell Hybrids Serve as Sources for Purified Human Chromosome DNA

The importance of somatic cell hybrids in gene mapping and as sources of DNA for cloning and cell lines made by fusing mouse or hamster cells with cells from patients with chromosomal translocations and deletions have proved especially valuable. Recently a new method has been devised that circumvents the need to find patients with chromosomal abnormalities; rearranged and deleted chromosomes can now be produced at will by using x-irradiation to break up the chromosomes in human cells. The irradiated cells are then fused with rodent cells, and a selectable marker is used to identify hybrid clones containing human DNA from the region of the marker. Hybrid cells containing DNA from the region of the genes implicated in a complex of disorders-of the genes implicated in a complex of disorders-the Wilms tumor, aniridia, genitourinary enomalies, and mental retardation (WAGR) syndrome-were isolated in this way. The cell surface antigen gene MICl is on the short arm of chromosome 11 at 11p 13, the same location as the WAGR syndrome. Hamster human hybrid cells containing the short arm of chromosome 11 were irradiated and fused with Chinese hamster cells, and so-called radiation hybrids expressing the MIC1 cell surface antigen were selected by "panning". The amount of human DNA in these hybrids was narrowed down even further by discarding clones that expressed a second cell surface antigen, MER2, located at 11 p 15. One cell line contained about 3 mb of DNA from the 11p13 region, representing about a 900-fold purification from the rest of the genome.

This method relies on a selectable marker in the region of the human chromosome that the investigator can use to select the radiation hybrid cells containing the irradiated human chromosome fragments. But what can be done if, a selectable marker is unavailable for the human DNA? A strategy that has been used to isolate DNA fragments from the region of the Huntington's disease (HD) gene offers a way out of this impasse (Fig. 39.4). The chromosomes of a human-hamster hybrid cell line containing human chromosome 4 were fragmented with a lethal dose of radiation. These cells were fused with hamster cells that were HPRT, and the cultures grown in HAT medium. The only cells that grow in this medium are radiation hybrid cells that contain an HPRT gene from the irradiated human- hamster hybrid cell line. (The lethal dose of radiation kills the latter unless they fuse with the hamster cells, and the hamster cells die because they are HPRT). The surviving cells were screened with a probe specific for human LINE sequences, and the positive cells were tested with probes form the HD region. One cell line contained a 10-megabase fragment from the region of chromosome 4 believed to contain the HD gene. This technique does not rely on the retention of a selectable human marker and could be used for isolating DNA from any region.

X-Irradiated Fragments of Human Chromosomes are Used for Gene Mapping

The x-irradiation hybrid method was originally devised for mapping genes, but it was of limited application because only a few markers were available at that time. This has changed dramatically with the cloning of many human DNA markers and with the availability of restriction fragment length and other polymorphisms. The power of this technique, named *radiation hybrid (RH) mapping,* has been demonstrated by the creation of a map spanning 40 megabases of human chromosome 21. Chinese hamster-human hybrid cells containing human chromosome 21 were x-irradiated. In situ hybridization with human genomic DNA showed that chromosome 21 was broken into an average of five pieces, each piece about 8 mb long. The irradiated cells were rescued by fusion with HPRT - cells, and the surviving hybrid cells were screened with 18 chromosome 21 DNA probes (Fig. 39.5). The Frequencies with which pairs of markers were retained in each clone were calculated. This is complicated because the cells may contain more than one DNA fragment, and statistical methods are used to determine the *breakage frequency* (O). This in turn is used to calculate the distance between pairs of markers, the distance being measured in *centirays,* analogous of the centimorgans of conventional mapping. The best order of the markers was determined by calculating the most probable of the 12 possible orders for sets of four markers taken together. This produces a probability score similar to the lod score of linkage maps. The accuracy of this RH map was determined by comparing it with a physical map of the region produced by using pulsed field gel electrophoresis. The two maps were similar, confirming the usefulness of RH mapping.

Cloned Human DNA Fragments must be Assembled into Megabase-Sized Contigs

Cloning the entire human genome is not technically difficult, and gene libraries, large enough to contain the human genome several times over, are available in phage or yeast artificial chromosomes. However, these libraries contain DNA cloned at random from the genome. To make sense, the DNA is these clones must be arranged in the correct order so that the genome,

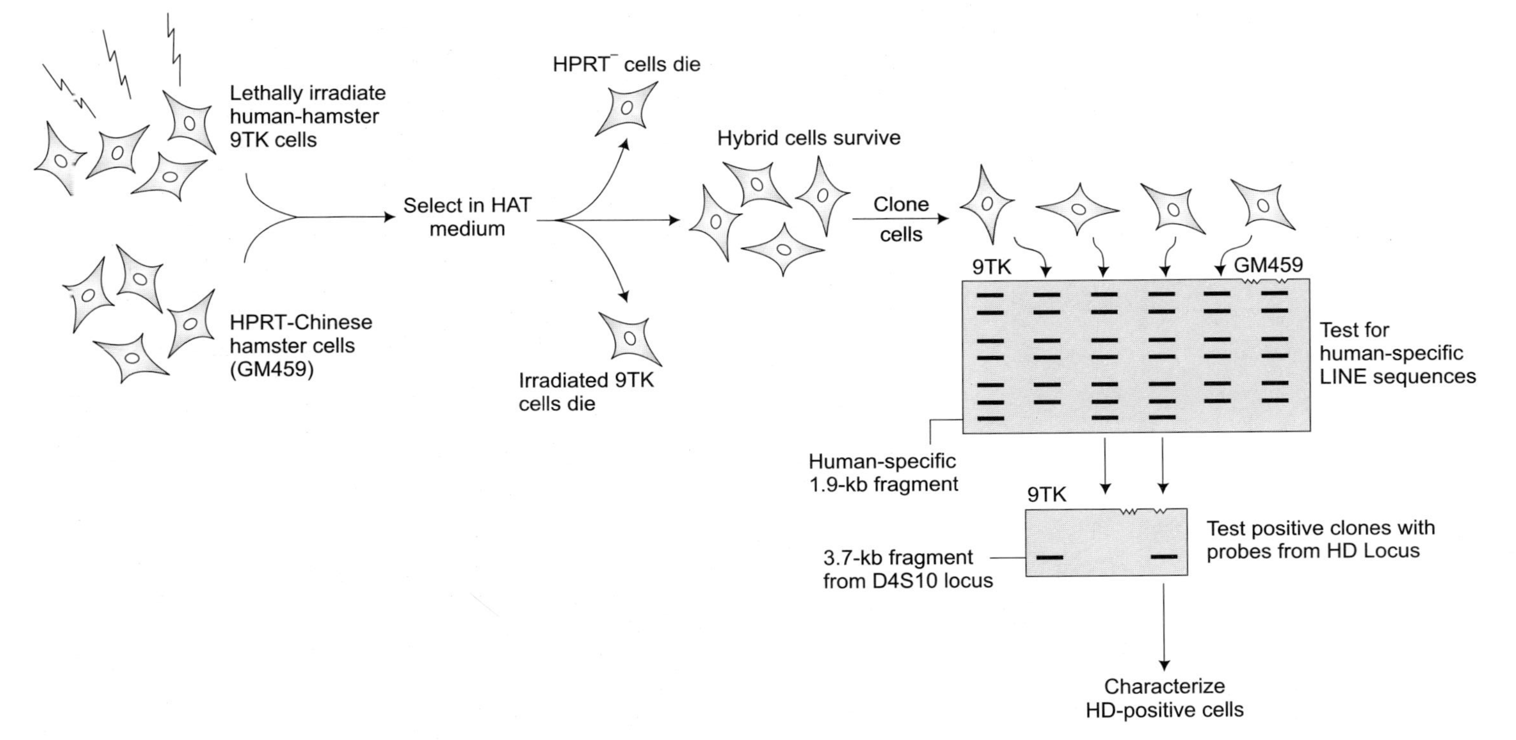

Fig. 39.4 Using x-irradiated hybrids for cloning. 9TK cells are a human-hamster hybrid cell line; the only human DNA present is chromosome 4. These cells are x-irradiated to fragment chromosome 4, and the dosage of radiation is sufficient to kill the cells. The GM459 Chinese hamster cell line is deficient in hypoxanthine phosphoribosyl transferase (HPRT), and these cells are killed when they are grown in medium containing hypoxanthine, aminopterin, and thymidine [HAT (Chapter 12)]. When the parental cells are fused and grown in HAT medium, the only cells that can survive are Chinese hamster cells that have acquired a hamster HPRT gene from the irradiated hamster-human cell line. Clones of surviving cells are first screened for the presence of human DNA using probes for *Alu* or LINE repeat sequences. Clones containing human DNA have a characteristic 1.9-kb fragment. Positive clones were screened with probes mapping close to the Huntington's disease (HD) locus at the rip of chromosome 4. For example, probe pHD2 from the *D4S10* locus detects a 3.7-kb band on Southern blots. In this way other DNA sequences mapping in the HD region were cloned.

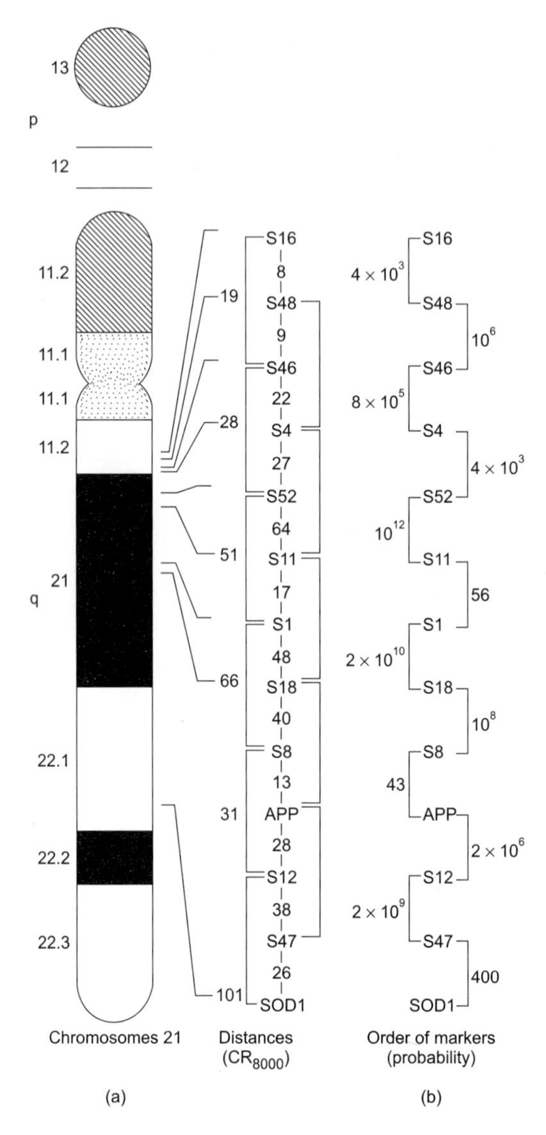

Fig. 39.5 Mapping genetic markers on human chromosome 21 using x-irradiated hybrids. Over 100 x-irradiation hybrid clones were made by the technique illustrated in Fig. 30-4, except that the human-hamster hybrid cell line contained chromosome 21 as its only human genetic material. Each cell clone was tested with DNA markers from the long arm of chromosome 21. An analysis of the patterns of 13 retained markers leads to the map shown in (a), where the distances between pairs of markers measured in centirays for the radiation dosage used(is determined by the frequency of breaks between markers, in a fashion similar to the way crossing over is used to establish distances (in centimorgans) in conventional mapping. In this case 8000 rads was used, so the unit is cR_{8000}. These can be used in turn to calculate the most likely order of the markers (b). For example, the probability that the order S8-APP-S12-S47 is correct compared with S8–S12–APP–S47 is 2 million to 1. The chromosome at p12 is composed of ribosomal genes that do not stain with conventional Giemsa staining, so that the DNA at the end of the chromosome appears separated from the main body of the chromosome.

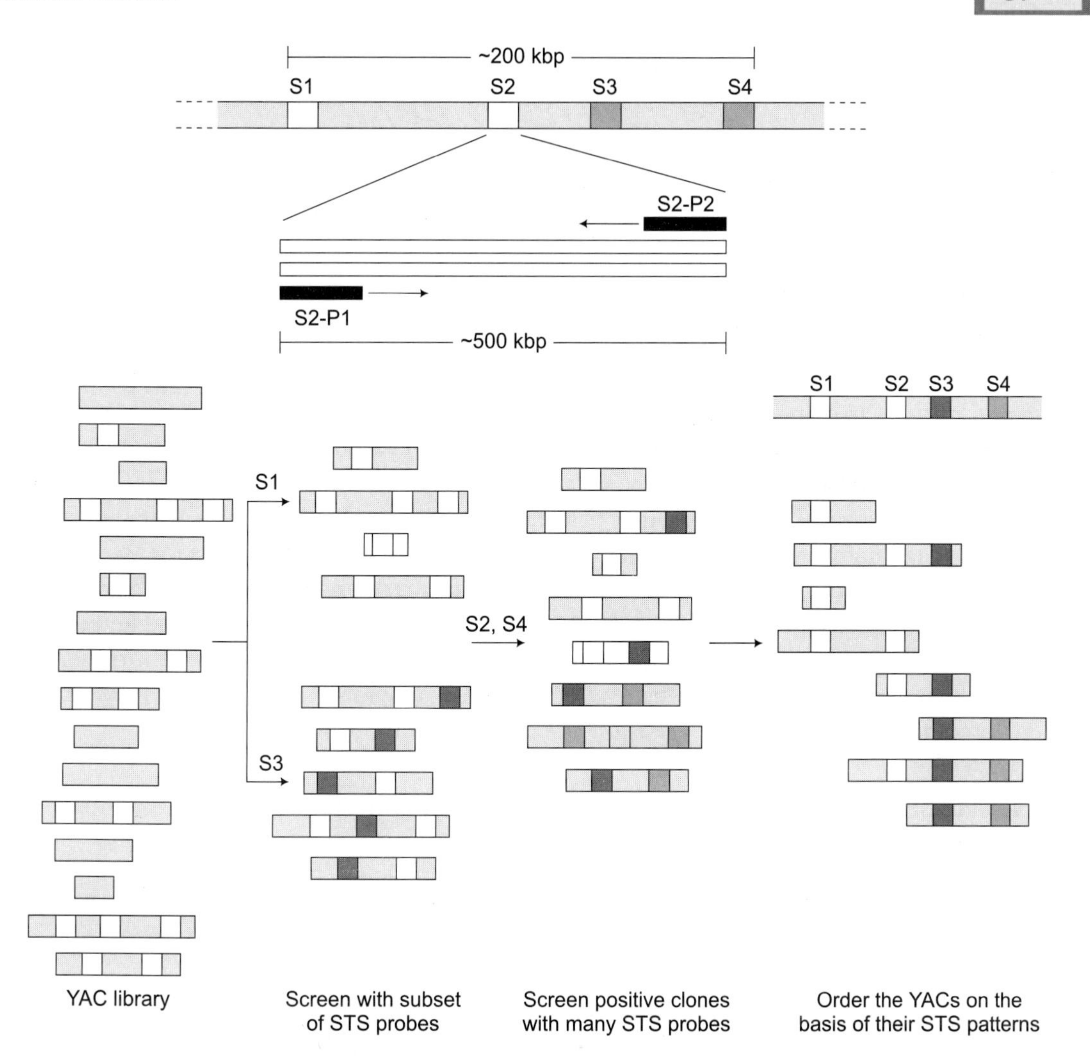

Fig. 39.6 Ordering YAC clones using sequence-tagged sites (STSs). A stretch of DNA some 200 kbp long has four STSs(S1 through S4). Each STS is about 500 bp long, and flanking sequences are known that can be used for making oligonucleotide primers for a polymerase chain reaction. S2 is shown in detail below (with primers designated P1 and P2). A YAC library is first screened by PCR using a subset of the STSs available (S1 and S3). Those clones that are positive are then screened with a large set of STSs, and the clones are ordered by the STSs that they contain.

of pieces sufficiently large to contain biologically interesting stretches of sequence, can be constructed. These reconstructions result in a series of overlapping clones in a computer database. Using some of the strategies, researchers have made progress in assembling some large sections of human chromosomes. Overlapping cosmid clones of chromosome 16 are being put together by a fingerprinting method which uses probes for the (dC- dA),,' (dG-dT),, repetitive sequence to identify restriction fragments common to two or more clones. The frequency of this repetitive sequence on chromosome 16 is such that there is a very high probability that clones with the same patterns of fragments hybridizing to a probe for (dC dA),, (dG-dT),, will have overlapping regions. Clones without (dC-dA),, (dG-dT),, repeats have been further fingerprinted using other repetitive sequences like Alu and Ll, 4000 clones have been fingerprinted in this way and assembled into 550 contigs, comprising about 60 percent of chromosome 16.

Sequence-Tagged Sites Identify Cloned DNA

Large-scale mapping and sequencing involve many laboratories, and the information generated must be collated. At present, this is difficult because different laboratories, are using different mapping and sequencing strategies. So-called sequence-tagged sites (STSs) may provide a common language to overcome these difficulties. An STS ia a unique sequence, from a known location, that can be amplified using the polymerase chain reaction. Each STS is 200 to 500 bp long, and the sequences for the unique primers that will amplify each STS will be listed in databases. A laboratory will be able to obtain information about the sequences for an STS from a database, synthesize the appropriate oligonuleotide primers, and "recover" the STS by PCR amplification of geneomic DNA. One disadvantage is that unique sequences suitable for use as. STSs have to be sought out. This requires knowledge, patience, and a degree of intuition; but interactive computer programs are being designed that may make the selection process faster and more efficient. STSs can be used for ordering clones by searching.

GENE MAPPING CAN BE FACILITATED BY COMPARING

We have given many examples of how molecular geneticists have exploited homologous relationship between genes for cloning and for guessing the probable functions of newly cloned genes. Genetic similarities between different species are not restricted to nucleotide sequences in homologous genes. The same set of genes grouped together on a chromosome in one species may be found grouped together on a chromosome in another species (Fig. 39.7). For example, there are 13 genes on mouse chromosome 11 that have homologues on human chromosome17. This is called conserved synteny. If the order of the genes (no markers) is the same in the two species, the genes demonstrate conserved linkage. Synteny and linkage conservation presumably reflect the preservation of ancient chromosome linkage groups in the mammalian ancestors, despite the rearrangements that occur over many generations. These preserved linkage group can be used as guides to the locations of genes. This may be especially useful in speeding the mapping of genes of agriculturally important animals like cattle, sheep, and goats, animals in which little genetic analysis has been done at the molecular level.

UNDERSTANDING OUR GENOME WILL BENEFIT HUMANITY

We believe that the benefits of the new genetic knowledge are undeniable- in research, where the richness of sequence data will keep scientists at work for generations to come, and in human

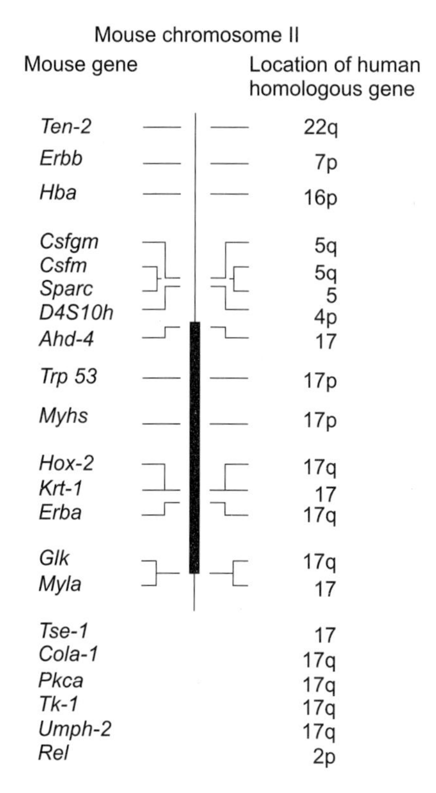

Fig. 39.7 Comparative gene mapping in mouse and human. Mouse chromosome 11 is shown with genes mapped to this chromosome listed in their correct locations. The locations of the human genes homologous to mouse chromosome 11 genes are listed opposite their mouse homologues. An extensive region of the mouse chromosome shows synteny with human chromosome 17. The genes listed below with their human homologues are known to be no mouse chromosome 11, but their positions on the chromosome map have not yet been determined.

genetics, where there are many families who have healthy children as a consequence of prenatal diagnosis. But these benefits are' tempered by other concerns; for example, DNA based prenatal diagnosis may be unacceptable to those who believe abortion is wrong, and DNA storage may alarm those who see in it a threat to individual privacy. It is encouraging that both the National Institutes of Health and the department of Energy human genome programs have set aside substantial funding to ensure that discussions of these issues take place, and that these debates involve the general public, not just experts. Many of the "big" projects of science the superconducting Super collider, the space station, "star Wars" are seen to be at best irrelevant and at worst potentially catastrophic for humanity. The various plans to analyze the human genome have as their ultimate goal improving the lives of many thousands of people. Is it too much to hope that we as a society will be able to use the new knowledge of our molecular genetics in such a way as to enhance our duties and respect for others?

CHAPTER 40

Plant Genomics

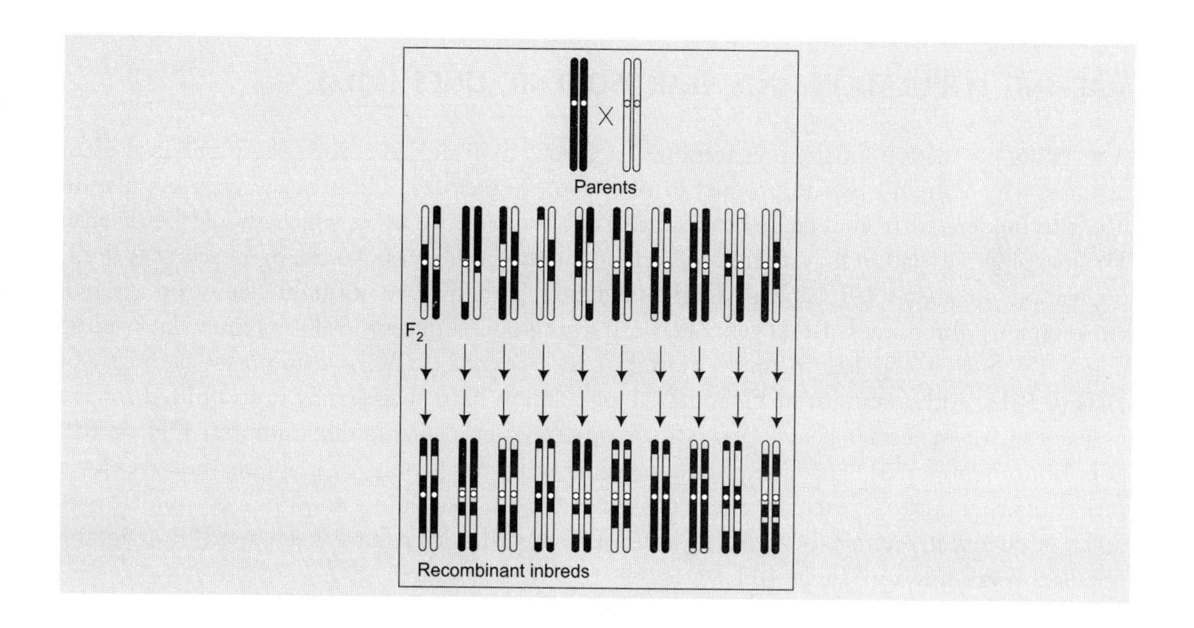

CONSTRUCTION OF MOLECULARS MAPS AND SYNTENY COLLINEARITY

The detection and exploitation of naturally occurring DNA sequence polymorphisms represent one of the most significant developments in molecular biology. The techniques of development of molecular markers like RFLP, RAPD, AFLP, SSR, STS (sequence tagged site) and EST (expressed sequence tag) were earlier described. The preparation of the molecular maps using these molecular markers are made for animal genomes. The preparation of similar molecular maps in crop plants has been a priority area of research in plant biotechnology also. The utility of these maps for crop improvement programmers has been emphasized during 1980s, so that the molecular maps have already been prepared for all major crop species. The techniques of the use of molecular markers have also proved useful for characterization of crop varieties and plant genetic resources through DNA fingerprinting. More recently YAC and BAC physical maps have also been prepared and used for "clone to clone" genome sequencing. These aspects of plant biotechnology will be briefly discussed in this chapter.

Preparation of Genetic Maps

In the preparation of molecular genetic maps, following steps are involved: (i) development of a mapping population, (ii) selection of one or more molecular marker systems, (iii) identification of adequate number of molecular markers showing polymorphisms in the mapping population, and genotyping of individual plants or lines of the mapping population, using each polymorphic molecular marker, and (iv) construction of molecular map using suitables computer software.

MAPPING POPULATION AND NEAR ISOGENIC LINES (NILs)

A mapping population for the preparation of a genetic map should consist of a number of plant or lines, which should be the product of one or more meioses. Therefore, such a population needs to be derived from a cross between two inbreds or pure lines, which should be diverse. The mapping population may, therefore, consist of either an F2 population, or a backcross (BCI) population, or a doubled haploid (DH) population each derived form FI plants or a set of recombinant inbred lines (RILs) generated using single seed decent (SSD) method of breeding (Fig. 40.1). Such a population can be mortal, as in case ofF2 or BCI, or immortal as in case of DHs or RILs. Although mortal F2 and BC1 population have successfully been utilized for the preparation of molecular genetic maps in a variety of crop plants, an immortal DH or RIL population is preferred, since it can be repeatedly utilized not only for mapping, but also for a variety of other purposes including gene tagging and QTL analysis for important economic traits (QTL = quantitative trait loci). A few sets of maize RILs that have been used as mapping population are listed in Table 40.1.

The backcross programme is also utilized for generation of near isogenic lines (NILs) for specific traits, where pair of lines (NILs) differs for small segment (S) for the concerned trait. In doing so, an initial cross is made between a recurrent parent (often a desirable genotype deficient for a Particular trait) and a donor parent (carrying the gene of interest, although otherwise agronomically poor). The F1 plants are backcrossed repeatedly. With the recurrent parent to reconstitute its genotype. While making back crosses, in each generation only selected plant carrying the gene of interest are used.

DIFFERENT MOLECULAR MARKER SYSTEMS

Different molecular marker systems including RFLP, RAPD, SSR, and AFLP. have been used for the preparation of molecular maps in several crops. However, the latest of these markers called single nucleotide polymorphisms (SNPs), which have already been used for the preparation of a molecular map of human genome, could not be used so far (till 2003) for the construction of

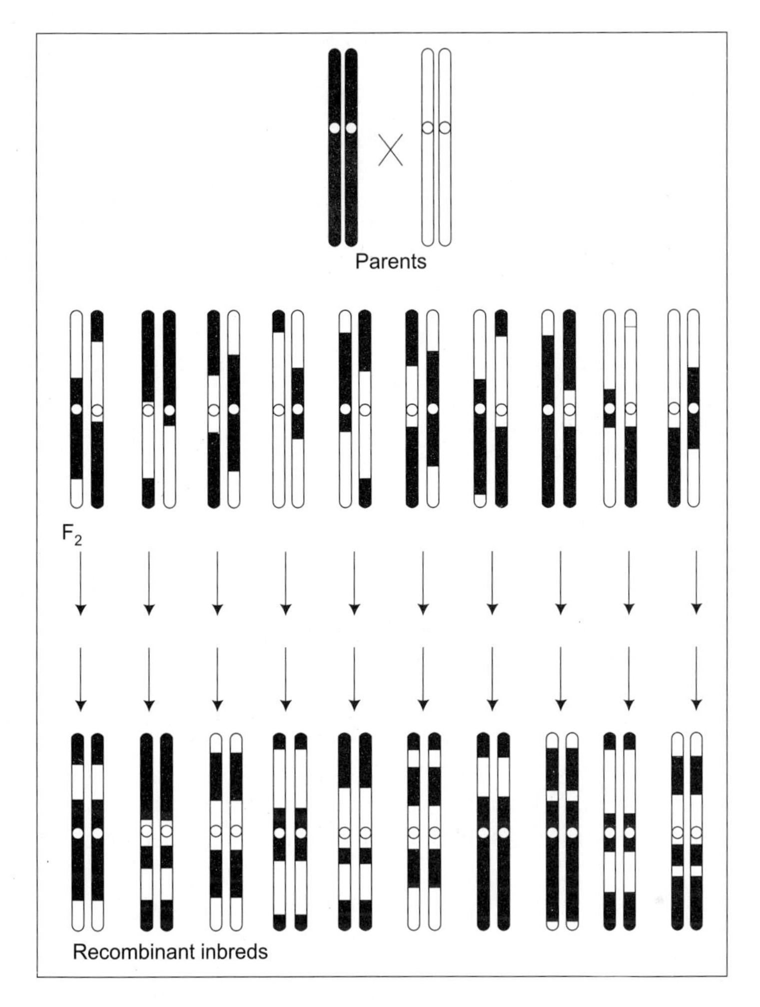

Fig. 40.1 The production of recombinant inbred lines. This figure depicts a hypothetical diploid organism with a single pair of chromosomes. The original parents are themselves inbred lines and homozygous at all loci. Individuals of a large F_2 population produced from a mating of these inbreds are self-pollinated, in case of maize for at least six generations to produce new recombinant inbred lines (RILs).

a map in any crop plant. But SNPs have already been used for the detection of polymorphism in crop plants suggesting that this new class of molecular markers will also be used in future for preparing molecular maps in crop plants.

POLYMORPHIC MARKERS AND GENTYPING THE MAPING POPULATLON

Before molecular markers are mapped, they need to be tested for detecting polymorphism between the two parents of the mapping population. For further mapping work, only those markers are used, which exhibit polymorphism between the parents mapping population, since

Table 40.1 List of available recombinant inbred populations in maize.

Parents	*No. of RI lines*	*Generation of inbreeding a*
T232 × CM337[b]	48	8
CO159 × Tx303[b]	41	8
CO159 × Tx303	160	5
B73 × Mol (IBM*)[b]	44	7
PA 326 × ND300	74	5
CK52 × A671	162	5
CG16 × A671	172	5
CH593 – 9 × CH606 – 11	101	5
CO220 × N28	173	5

[a]Generations of selfing beyond the F_2; *IBM=intermated BM
[b]Extensive mapping has been done in these families.

markers showing no polymorphism cannot be mapped genetically (although they can be mapped physically; see later for physical maps). The two polymorphic states of the marker recorded in two parents are given allelic symbols or I and (zero) states, and all individuals/lines (~ 100) of the mapping population are genotyped (assigned specific allele or 1/0 state). For co dominant markers, like RFLPs if mapping population belongs to F2 generation, there will be three genotypes (AA, a), which can be scored as 0, 1 and 2 to be used with the computer software. The data on genotyping the mapping population, involving molecular markers ranging in number from a few hundred to few thousand, is then used for the preparation of a genetic map, which is based on recombination frequencies that are automatically worked out by the software and utilized in a mapping function for determining genetic distances.

PREPARATION OF MAP USING THE SOFTWARE MAPMAKER

Keeping in view the amount of data (~100 individuals/lines, each genotyped with few thousand markers) to be handled for the preparation of a molecular genetic map, computer software are used. One such software in common use in MAPMAKER, which makes use of either Haldane's mapping function (no interference is assumed) or Kosambi's mapping function (interference is taken into consideration). Other softwares like LINKAGE were used earlier for the same purpose.

MOLECULAR GENETIC MAPS (LOW/MODERATE DENSITY)

During the last two decades (1980-2000) and in the early years of the 21st century, molecular genetic maps become available for all major crop and tree species (cereals, legumes, oil seeds, vegetables, trees, etc.). the mapping populations that were used for the construction of these maps included F2, Bc1, RIL and DH populations, and the molecular markers that have been used for this purpose mainly included RFLP, RAPD, SSR and AFLP, although other marker systems like STS, ISSR and DAR were also utilized. For constructing a mapping population, when sufficient variability was not available within the germplasm of the crop, a wild species had to be crossed with a cultivar to provide for the diversity, which is key to the success of constructing a map. For instance, in case of tomato, a wild species, *Lycopersicon hirsutum* or *L.pennellii* was often used; in case of barley, *Hordeum spontaneum* (wild barley) was used, and in case of bread wheat, a synthetic wheat (W7984) involving *Aegilops tauschii*, the D genome progenitor, was used for crosses leading to the construction of one or more mapping populations. In sugarcane, *Saccharum spontaneum* the wild relative of sugarcane was used either directly for the preparation of a molecular map or for crosses with cultivated species S. *officinarum* for generating a mapping population.

In each of the majority of crops used for molecular mapping, more than one molecular maps are often available. This is because, either separate maps were prepared using different marker systems, or more than one laboratories prepared independent maps, utilizing the same marker system, thus multiplying the number of maps for the same crop. This multiplicity of maps was sometimes necessary, because often in a single population, only a fraction of available markers segregated, so that several populations were needed to map all the available markers. In some crops, prepared, utilizing the information from all the available maps (see later). Some details about the available maps are presented in Table 40.2, where it can be seen that most of these maps are low to moderate density maps with variable resolution, suggesting that in many cases these maps still need to be saturated (see next section for high density maps). The present status of these molecular maps (at the beginning of the present century) in different crops will be briefly discussed in this section. For any details, readers should use the internet where more detailed information about most of these maps is readily available.

MAPS FOR CEREALS, MILLETS AND SUGARCANE (GRASS FAMILY)

Molecular maps have been prepared in all major cereals, millets and other graminaceous crops including maize, bread wheat, emmer wheat, oats, barley, rice, sorghum, pearl millet, sugarcane, etc. the most extensive maps among cereal genomes, however, are available in case of rice, which has become a model for all graminaceous crops and has therefore been the first food crop to be used for genomics research (due to its small genome size of 430 mb). As many as about 2300marker loci spanning a genetic distance of 1800 cM, have been mapped in this crop. Sorghum is another important crop that has been targeted for plant genomics research, because it is adapted to harsh environments, and is represented by diverse germplasm and a small genome (750 Mbp). In this crop about ten molecular genetic maps with moderate density of RFLP and SSR markers spreading over a genetic distance of up to ~ 1800cM were prepared during 1990s (for high density maps in rice and sorghum, see next section). In bread wheat, RFLP maps were initially prepared by joint efforts of different laboratories, which wee co-ordinated by the International Triticeae Mapping Initiative (ITMI). Subsequently, microsatellite (SSR) maps in bread wheat also prepared by several laboratories in different part of the world.

Table 40.2 Details or molecular maps in some crops.

Crop	Gametic chrom. no.	No. of marker loci	Map length (cM)	Genome size (Mbp)	Resolution bp/cM
I. Cereals					
Maize (*Zea mays*)	10	–2000	1000	2500	2.5×10^6
Rice (*Oryza sativa*)	12	>2300	1800	430	0.29×10^6
Bread wheat (*Triticum aestivum*)	21	–2000	5000	16,000	3.2×10^6
Emmer wheat (*T. dicoccoides*)	14	3000	3000	750	0.25×10^6
Barley (*Hordeum vulgare*)	7	>2000	1500	5200	3.5×10^6
II. Solanaceae					
Tomato (*Lycopersicon esculentum*)	12	>3000	1500	950	0.67×10^6
Potato (*Solanum tuberosum*)	24	1000	800	1800	2.25×10^6
Pepper (*Capsicum annuum*)	12	–350	2500	3200	1.28×10^6
III. Legumes					
Chickpea (*Cicer arietinum*)	8	350	–2000	–1000	0.5×10^6
Pea (*Pisum sativum*)	7	850	–1500	4200	2.8×10^6
Common bean (*Phaseolus vulgaris*)	11	1000	–1000	700	0.7×10^6
Soybean (*Glycine max*)	20	1000	1500	4200	2.8×10^6
Alfalfa (*Medicago sativa*)	16	–	500	1500	3.0×10^6
Lentil (*Lens atstivum*)	7	200	500	5000	10×10^6
Lotus japonicus	6	605	367	–450	1.23×10^6
Medicago truncatulus	8	550	1000	–500	0.50×10^6
IV. Crucifers (Brassicas)					
B. oleracea	9	550	900	700	0.78×10^6
B. napus	19	992	2500	1300	0.52×10^6
B. Juncea	18	>1000	1600	1000	0.6×10^6

Molecular maps with low to moderate density were also prepared in barley, rye, oats, pearl millet, sugarcane, etc. For instance, barley genetic maps were prepared independently in Europe and USA. An SSR map of barley was also prepared in the year 200, using 242 of more than 600SSR primers that were available in the public domain at that time. Molecular maps based on RFLPs and AFLPs have also been prepared for deploid and hexaploid oats. In sugarcane, both cultivated and wild species of Saccharum were used for preparation of molecular maps involving RFLPs and RAPDs. Detailed information on majority of cereal molecular maps is available on grain Gene and Gramene sites, although web sites are also available for individual crops (e.g. for maize, maize BD : <www. agron.missouri.edu> and for sorghum, httpt// sorghumgenome. tamu.edu>.

MAPS IN LEGUMES (SOYBEAN, PEA, COMMON BEAN, ALFALFA, CHICKPEA, LENTIL)

All major legumes have been used for constructing molecular maps, although a moderately dense map is available only in case of soybean, where more than one thousand loci involving all kinds markers (RFLP, RAPD ,SSR,AFLP) were mapped over a length of 1500 cM. In pea also, a fairly good map involving ~ 1000 marker loci spanning a genetic length of ~1500 cM became available toward the end of last century. Similarly in commen bean (*Phaseolus vulgaris*), a map with as many as 1000 RFLP and RAPD markers is available. In constrast to these moderate density maps in soybean, pea and common bean, the available maps inchikpea and lentil are sparse and low density maps, with about 500 mapped markers in chicpea and 200 mapped markers in lentil (Table 39.2). Maps have also been prepared in some other legumes including *Lotus japonicus* (2n = 12), alfalfa or *Medicago sativa* (2n = 32) and the model legume, M. truncatula (2n =16).

Maps in Brassicas

Molecular maps have been prepared in a number of *Brassica* species including both (i) the diploid *Brassics* liken *B. oleraceae* (2n=18) and *B. nigra* (2n=16), and (ii) the polyploidy Brassicas like *B. napus* (2n=38) and *B. juncea* (2n=36). In these crops, mainly RFLP, RAPD and AFLP markers were used, since SSR markers were not available in large numbers for mapping. For instance, in B. oleracea, 258 RFLP markers were mapped in 1990 and later an integrated map with 543 marker loci was published in 2001. A skeleton map with 67 RAPD loci was also prepared in *B.nigra* Similarly, in *B. napus*, a consensus map with 542 RFLP loci and 253 map in *B. juncea* (Indian mustard) with 1029 loci (996 AFLP and 33 RFLP) was published in 2002 from a group from Delhi University South Campus (UDSC), India.

Maps in cotton

A number of molecular genetic maps have been prepared in cotton most of them for tetraploid upland cotton (*Gossypium hirsutum*) with AADD genomic constitution, but some of them also for diploid cotton (G. *herbaceum*)with AA genomic constitution. The mapping populations used for mapping the genome of tetraploid cotton, generally consisted of F2 or F23 (bulk-sampled plots) or DH lines derived from interspecific cross G. *hirsutum* × G. *barbadense*. Some of these maps included the following: (i) In 1998, a detailed restriction fragment length polymorphism (RFLP) map, was prepared for tetraploid cotton, in which 705 RFLP loci, assembled into 41 linkage groups, covered a genetic distance of 4675 cM. The genomic origin (A vs D) of most, and chromosomal identity of 14 (of 26) linkage groups was shown. (ii) Anther molecular linkage map was constructed using 58 doubled haploid plants derived from the interspecific cross G. hirsutum x G. barbadense. Among a total of 624 marker loci were assembled into 43 linkage groups and covered 3,314.5 centi-Morgans (cM). (iii) A joinmap from four populations comprising 284 loci spanning a genetic distance of approximately 1,500 cM was also published in 2001. This was the first genetic linkage join map assembled in G. *hirsutum* with a core of

RFLP markers assayed on different genetic backgrounds of cotton population (Acala, Delta, and Texas plain). These maps taken also demonstrated that there are duplicate loci for many markers due to homoelogous relationship between the two genomes (A and B) that are present in tetraploid cotton.

MAPS FOR SUNLOWER

Several molecular maps have been prepared in sunflower. The molecular markers that have been used for this purpose included RFLPs, SSRs, DALPs and AFLPs. The maps, each consisted of 17 or 18 linkage group (n = 17) and differed in their genetic length from 1368 cM in the SSR map to 2,168 cM in DALP/AFLP map. The latest of these maps include an SSR map published in 2002. Nuerous duplicated loci were detected by many RFLP probes, suggesting that sunflower is a secondarily derived polyploidy.

MAPS FOR SOLANACEOUS CROPS

Solanceous crops that were used for molecular mapping, mainly included tomato (2n = 24), potato (2n = 48) and pepper (2n = 52, 104). Information on these maps is freely available on SolGenes database. A molecular map of tomato was perhaps the first molecular map that was prepared the first molecular map that was prepared in plants and should now contain more than 2000 mapped loci.

MAPS FOR FOREST TREES AND FRUIT TREES

Molecular maps have also been prepared in several tree species that are important for forestry. These tree species include species from both gymnosperms and angiosperms. The gymnosperm trees that have been used for genome mapping mainly include a number of pines {e.g.loblolly pine (*Pinus taeda*), radiatapine (*Pinus radiata*) slash pin (*Pinus eliottii*) longleaf pines (*Pinus palustris*), martime pine (*Pinus pinaster*), scots pine (*Pinus sylvestris*)}, at least two spruces {white sprue (Picea glauca) and Norway sprway spruce (*Pices abies*) and Doouglas fir (*Psedotsuga menziesii*). Similarly angiosperm trees, whose geneomes have been mapped include European larch (*Larix decidua*), white birch (*Betula alba*), poplar (*Populus* sp), and several species of *Eucalyptus* and *Quercus* (oaks). A variety of DNA-based molecular makers that have been used for preparing these maps include RFLPs, RAPDs, AP-PCR, DAF, STS, SSLPs, ESTPs, SCARs etc.

Among fruit tress, genomes of apple (*Pyrus malus*, syn *Malus sylvestris*), apricot (*Prunus armeniaca persica*), lemon (*Citrus* sp.), avocado (*PERSEA AMERICANA*), etc. have been subjected to molecular mapping. In some cases (particularly in case of apple), permanent mapping populations have been established for mapping. As in other cases, the molecular markers that have been used for mapping of fruit trees include RFLPs, RAPDs, SSR, and AFLPs. In both, forest tress and fruit tress, the mapped molecular markers are being used for gene tagging and QTL analysis, so that in forest and fruit tress also, molecular markers will be utilized in future for marker assisted selection and gene isolation.

Molecular Genetics MAPS High Density

Although molecular genetic maps are now available in majority of crops, most of these maps are inadequate for identification of candidate genes and for subsequent map based cloning of genes. Therefore, efforts have been underway to prepare high-resolution maps in several crops. One of the approaches followed in maize in this connection involved intermating among plants in F2and subsequent generations to provide for an opportunity for increased recombination. When such a population after five-generations of intermating was used in maize, it led to the production of a high resolution map, in which the distance increased from 1532 cM to 5917 cM. Intermated RILs were also used in maize to prepare a high density SSr map involving 978 SSrs. In sorghum also, high density map with-3000 marker loci spread over 3000 cM was later produced during 2000-2002. A high density genetic linkage map of tetraploid cotton was also developed, which had 2662 loci (RFLPa; %% of SSrs), which were distributed in 26 linkage groups and spanned a genetic distance of over 4700 cM. An integrated high density map in Aegilops tauschii genome, which represents the D genome of bread wheat and maize, efforts have also been made to identify gene-rich genomic regions (these regions are also shown to be the recombination hot–spots) in the maps. These gene rich regions are being subjected to high resolution mapping but also to genome-wide sequencing.

Integrated Genetic Maps

Integrated maps have also been prepared utilizing multiple maps involving either the same class of markers, or more than one marker classes. For instance, integrated maps have been prepared by integrating microsatellite maps into RFLP maps in several crops including rice, wheat, barley, maize, soyabean, potato, tomato, sugarbeet etc.

Resolution Gap Differs in Maps with Genomes of Different Sizes

In preparing genetic RFLPs maps, 1 cM(centiMorgan) is often considered to be the limit of resolution achievable without examining a very large population. At the molecular level, this limit of resolution mean 1 cM would be equal to different physical distances in organisms, with genomes of different sizes, as shown in Table 40.3. This gap, therefore, needs to be bridged, if a gene of interest is to be approached from a linked RFLP locus, that has been mapped. Bridging of this gap has become possible due to development of techniques like pulse field gel electrophoresis (PFGE) and cloning in yeast artificial chromosome (YAC) vectors, which will enable handling of large pieces of DNA. Similarly jumping and linking libraries can be prepared, so that chromosome walking need not be followed throughout the genome to identify, locate and isolate a gene of interest.

Table 40.3 The resolution gap as worked out in four different genomes (Shields, 1989).

Organism	Genome size (kbp)	Map length (cM)	kbp equal to 1 cM
Arabidopsis	7×10^4	501	139
Tomato	7.15×10^5	1,400	510
Maize	3×10^6	1,400	2.140
Human	3×10^6	2,700	1,108

USES OF MOLECULAR GENETIC MAPS

Mapping of Major Genes of Economic Value Using Molecular Markers (Tagging)

Molecular markers can be used to map positions of genes for simple traits as well as those for polygenic traits (alo described as quantitative trait loci or QTLs). Once these genes are mapped the molecular markers linked to specific genes or gene clusters (gene tagging) can be used in selection schemes for plant breeding (see later for details). Molecular markers mapped very close to important genes can also be used for isolation of these genes through map based cloning.

Use of near isogenic lines (NILs) for gene tagging

In recent years, near isogenic lines (NILs) have been utilized to identify linkages between molecular markers and conventional phenotypic markers. An NIL is produced, when a gene for conventional phenotypic marker is transferred from a donor payment (DP), into recurrent (RP) and the genotype of the recurrent parent is restored through 5-7 backcrosses, retaining the conventional phenotypic marker in each such backcross. Although an NIL, as the term near isogenic indicates, will contain alleles derived from DP at several loci on different chromosomes, due to selection of conventional marker, only about 50% of such alleles are expected to be present on the conventional marker introgressed into RP. It has been calculated that if DP and RP differ for 100 molecular markers, NIL will retain four of them in a species with n=20 (each chromosome with 50 cM length), and of these four, only two or three will be present on the marker chromosome. Therefore, allelic contrast between RP and its NIL and allelic equality in the corresponding DP and NIL for a molecular marker suggests linkage of molecular with the conventional phenotypic marker. This, however, needs to be confirmed through F2 cosegregation data generated from NIL X RP crosses. This will also give an estimate of the linkage distance between the conventional phenotypic marker and each of the associated molecular markers. This technique has been successfully employed for tagging genes of economic value in several crops including tomato, potato, maize, barley, rice and sugarbeet. Some examples of RFLP linkage mapping of genes of economic value that were initially detected are given in the Table 40.4. Hundreds of cases of such linkages of molecular markers with genes for economic traits in a variety of crop plants are now known.

Table 40.4 Some examples of initial RFLP linked mapping of major genes of economic value in some crops.

Crop	*Gene located on RFLP map*	*Lines used*	*Reference*
1. Tomato	(a) Tm-2a locus for resistance to TMV	Near isogenic lines (NILs)	Young et al., 1988
	(b) Gene 12 and II for resistance against *Fusarium axysporium*	Near isogenic lines (NILs) and BC population	Sarfatti et al., 1989, 1991
	(c) Gene Mi for resistance against root knot nematode *(Meloidogyne* spp.*)*	Near isogenic lines (NILs)	Klein-Lankhorst et al., 1991
	(d) Gene Sm for Stemphylium resistance	Near isogenic lines (NILs)	Bchare et al., 1991 resistance
2. Potato	(a) Gene for against cyst nematode *(Globodora rostochiensis)*	Two diploid clones	Barone et al., 1990
	(b) Two genes (Rxl. Rx2) for resistance against potato virus X (PVX)	Two F_1 populations	Ritter et al., 1991
3. Maize	Gene for resistance against maize dwarf mosaic virus (MDMV)	Near isogenic lines (NILs)	McMullen and Louie, 1989
4. Barley	*ml-o* locus for resistance against Erysiphe graminis f sp. hordei	Near isogonie lines (NILs) and two subspecies of *Hordeum vulgare*	Hinze at al., 1991
5. Sugarbeet	Gene for resistance against a nematode *(Heterodera schachtii)*	Addition of a *Beta procumbens* chromosome	Jung et al., 1990
6. Rice	Two genes for blast resistance, Pi.2 (t) and Pi-(t)	Near isogenic lines (NILs)	Yu et al., 1991

Identification and Mapping of Quantitative Trait Loci (QTLs) Using Molecular Markers

Many quantitative traits of economic value are under polygenic control and are selected for, directly. Such a selection is often ineffective, since the effect of each gene is small, which is also influenced by the environment. Therefore, one would welcome a procedure for indirect selection which is not influenced by the environment. For this purpose, linkage between genes for quantitative traits and marker loci should be known. Such markers involving morphological traits have been identified and mapped in *Drosophila* and in several crop plants. Isozyme marker loci linked to quantitative traits of economic value had also been earlier identified in tomato and maize. Initially. RFLPs were extensively used in a variety of crop plants to identifty and Map quantitative trait loci (QTLs). More recently (1995-2002). SSR and AFLPs became the markers of choice and were widely used. In future. SNPs will also be used in many crops.

For the study of marker-trait association. the experimental procedures differ for self pollinated and cross pollinated species. although with tile availability of inbred lines in cross pollinated species. similar methods can be used in the two cases. In self pollinated genotypes and among inbred lines (in cross pollinated species). parents are first selected which differ for a number of markers as well as for mean values of the quantitative traits. Once such parental lines are selected. they are crossed and F2 plants are derived. In the F2 population, a chromosomal segment representing linkage between a molecular marker and a QTL will be present in the, background of random genetic variation due to independent assortment and recombination. By growing Samples of F2 population in different environments. one may also study the linkage under different environments. In these cases if mean values of a particular quantitative trait are determined in two groups of plants representing altenative alleles of a molecular marker. a significant difference in means (of two groups) for quantitative trait will indicate marker-QTL linkage. The observed marker associated difference in quantitative trait value will be a function of allelic effects at the QTL. Map distance between the marker and a QTL can also be determined.

In Table 40.5 results of an initial experiment are shown where in tomato, linkage is evident between one of the two molecular markers (A1A2) and QTL for high soluble solids in the fruits. Two lines (cultivars VF 36 and its near isogenic line LA 1563) used in the corss had molecular markers (RFLPs) that identified L.esculentium (-5% soluble solids) and L.chmielewskii (-10% soluble solids). and also differed significantly for soluble solids (-5% and 7-8%). In F2 allelic 'c' of the marker Al showed association with the QTL for high mean value of soluble solids, while the allele 'e' of Al and both the alleles of A2 did not, suggesting that this QTL is found in the vicinity of AI. RFLPs , associated with the expression of complex traits like insect resistance and WUE, i.e, water use efficiency, were also identified and mapped in tomato. Subsequently, thousands of such marker-trait associations were reported in a number of crops. For instance, about once hundred such associations have been reported each in wheat and rice. In the experiments like the above, if parents differ for a sufficient number of molecular markers, which are spaced out throughout the genome, a single cross can allow identification and mapping of all those QTLs for which the parents differ.

Marker Aided Selection (MAS) in Plant Breeding

If a molecular marker locus is tightly linked with a gene of interest, indirect selection for the gene can be exercised by selecting for a particular market phenotype, which is linked with the desired phenotype for the economic trait. Such an indirect selection has several advantages

Table 40.5 The data on soluble solids content (mean per cent values) and two RFLPs in F_2. {F2 was derived from a cross between a tomato cultivar (–5% soluble solids) and its near isogenic line (NIL) carrying a QTL for high (7-8%) soluble solids derived from L. chmielewskii with 10% soluble solids; data from Osborn et al., 1987.

RFLP locus	*% Soluble solids*	
	RFLP allele e/e (L. esculentum)	*RFLP allele c/c (L. chmielewskii)*
A_1	5.82	6.24*
A_2	5.93	5.96

*significant difference at 5% level

including the following: (i) during backcrossing, selection for a recessive gene can be exercised even in a heterozygous plant, if co-dominant markers like RFLPs are used: this will eliminate the need of selfing after each backcross as required in conventional backcrossing; (ii) during backcrossing, indirect selection may also be desirable for dominant genes in some cases, because selection can be exercised at the seedling stage and the plants not carrying the desirable allele can be weeded out early, thus saving space and expense; (iii) during pyramiding of genes (when it is desired to incorporate more than one genes for the same trait) for traits like disease or insect resistance. it is not possible to select of the presence of an additional gene in the presence of an existing gene for the same trait; under such a situation, molecular markers associated with different genes will differ and can be conveniently used to follow any number of genes during the breeding programme; (iv) selection for mapped QTLs representing several chromosome segments contributing to the quantitative trait, can be exercised simultaneously with the help of individual molecular markets associated with each of these chromosome segments.

Identification of Breeding Lines and Varieties Using Markers

It has been demonstrated that even closely related individuals of some species (e.g., humans) can be distinguished using DNA fingerprinting. Molecular markers like SSRs and AFLPs. that are equivalent in resolution to minisatellites or VNTRs (see Chapter 26) of human beings have now become available in plants, so that genotypes specific RFLP/SSR/AFLP patterns have been detected and used for the identification of breeding lines and varieties and also for the characterization of plant genetic resources (PGRs).

Any character used for variety identification needs to fulfill the following basic criteria: (i) distinguishable intervarietal variation (ii) minimal intravariel variation (iii) environmental stability and (iv) experimental reproducibility (in cases of biochemical or molecular markers). These requirements are met in most of the cases where molecular markers have been identified for the detection of varieties. Hundreds of cases in dozens of crops were reported during 1990-2003, where molecular markers proved to be useful for identification of varieties and crop genetic resources (see next section). The genotype specific patterns have already been detected and their use for distinguishing varieties has been suggested in crops like maize, wheat, oats, barley, rice, oilseed. lettuce. soybean and potato. An exception, however, may be tomato where intravarietal differences were rare. In rice, cultivar specific DNA fingerprints were also obtained when a human minisatellite DNA probe was used.

However, the limitations for the practical use of molecular markers for variety identification include the following: (i) use of radioactive isotope is inconvenient, (ii) the technique is expensive, and (iii) the technique is labour intensive. As already discussed, nonradioactive labeling (biotin and digoxigenin) has already become possible for use of molecular probes. Similarly, automation (use of automatic quipments, reducing dependence on labour) and standardization of the technique will make it less expensive and less labour intensive, so that DNA fingerprinting using molecular markers will develop into a valuable and reliable varietal diagnostic tool.

Characterization of Genetic Resources and Estimation of Genetic Diversity Using Markers

Characterization of genetic resources becomes prohibitive due to limitation of time and resources, particularly when large collections are being evaluated. the collection may also have duplications, which should be eliminated, and a well characterized using biochemical and molecular markers in addition to morphological and agronomic traits. These biochemical and molecular markers may be used to assess genetic diversity within and among accessions and to monitor changes during maintenance and regeneration of genetic stocks. Work on DNA fingerprinting for characterization, conversation and maintenance of plant genetic resources is being undertaken in India at NFPTCR (National Facility for Plant Tissue Culture Repository), New Delhi and at CDFD (Centre for DNA Fingerprinting and Diagnostics), Hyderabad.

In the form of biochemical markers, isozymes have been used, but their use is limited due to lack of polymorphism and a small number of loci and alleles available for analysis. The use of molecular markers for characterization overcomes this limitation, since they provide substantial polymorphism and they can be unlimited in number. The only disadvantage of using molecular markers for characterization of genetic resources is the cost of material and labour which may gradually go down. Developments of more efficient PCR based markers will make the use of molecular marker more convenient and more economic.

Identification of Somatic Hybrids

One of the problems in production of somatic hybrids, often, is lack of appropriate methods for identification of these hybrids among developing regenerants. Some of the methods traditionally used, include the following: (i) Combining light beached protoplasts with green leaf mesophyll protoplasts, followed by manual isolation of heterokaryons; (ii) genetic complementation of auxotrophic mutants; (iii) Fusion of different antibiotic-and/or herbicide – resistant lines derived by plant transformation and (iv) use of hybrid vigour, isozymes, proteins, etc. These methods. however, are time consuming and costly. Molecular markers can be utilized for the same purpose. where a small DNA sample from developing regenerant can be used to determine hybridity. In case of PCR-based markers (RAPDs, SRs, etc.) primers may be used which give polymorphic PCR products from each of the parents involved in fusion, so that the hybrids can be identified by the presence of molecular profiles, which represented a combination of the profiles of the two parents. This has been achieved successfully in several cases including potato and the approach will certainly be utilized in future for identification of somatic hybrids in a variety of plant materials.

Molecular Cytogenetic Maps Using Cytogenetic Stocks

Through a study of linkage of molecular marker loci with morphological or biochemical (isozymes) markers having known chromosome location, it was possible to locate molecular marker loci not only on specific chromosomes, but also on specific regions of these chromosome. For this purpose. nullisomic-tetrasomic lines (in wheat), monosomics (in cotton.

maize), trisomic (in barley and rice) and various available translocations (in cotton) or deletion stocks (in wheat), were utilized. This resulted in the preparation of molecular chromosome maps for all maior crops.

The above technique also allowed mapping of duplicate loci, when more than one fragments homologous to a probe were detected on Southern blots or more than one PCR products obtained using same SSR printed pair. While in maize. 28.6% sequences detected duplicate loci, in tomato these duplicated loci were less that 2% of the loci tested. This gave new evidence about the presence of frequent duplicate chromosome segments in maize rather than in tomato. More recently, duplicate loci have also been detected using whole genome sequencing, as done in *Arabidopsis*.

Molecular chromosome maps are also being prepared for wheat, rye, barley, rice etc. and the availability of aneuploids has greatly facilitated this work. For instance in wheat the use of nullisomic–tetrasomic lines facilitated the location of marker loci on specific chromosomes and the use of telocentrics allowed their location on specific arms. This has thus allowed in wheat the identification of molecular markers for each of the chromosomes arms belonging to 21 different chromosomes. In this connection, under the name 'International Triticeae Mapping initiative (ITMI)' interanational collaboration was initiated to prepare maps for the tribe triticeae and each homoeologous group was assigned to one research group. Twelve such workshops to assess the progress made and to plan future work were organized during 1990–2002, Rapid progress in preparation of molecular maps of different species of the. Triticeae was made through these efforts.

Alien addition lines in wheat have also been utilized for assigning genes to specific chromosomes of the alien species involved. For instance wheat–barley addition lines have been successfully utilized for assigning genes on specific barley chromosomes and wheat-rye addition lines were used for a similar purpose for assigning genes on specific rye chromosomes. Oat-maize addition lines were similarly used for assigning markers to individual maize chromosomes. Genomic specific probes (isolated from DNA of *Aegilops squarrosa*) have also been isolated to identify the D genome chromosomes in wheat and several other polyploidy species carrying the D genome e.g. *Ae. Cylindrical*, CCDD: *Ae. Crassa*, DDMM; *Ae. ventricosa*, DDUU.

Molecular Physical Maps

The physical maps of DNA fragments, chromosome segments, whole chromosomes or whole genomes are based on physical distances. These physical distances are measured in terms of relative physical lengths or in terms of kilobase pairs between genes or molecular markers or in terms of relative distances of these genes/markers from the centromeres. For instance, the distance of a marker form the centromere may be expressed as a fraction of the length of the whole arm taken as unity, so that the distance of a marker from centromere will be indicated as 0.1, if its distance from the centromere measures one-tenth of the whole arm. Physical maps may also obtained in the form of restriction maps, where for one or more enzymes, the restriction sites are shown to be separated by distances expressed in terms of kilobase pairs.

Whole genome sequences are also being used for preparing physical maps in case of *Arabidopsis* and rice genomes. These physical maps were prepared in the past using deletions, restriction digestions of DNA fragment followed by electrohoretic fractionation, or by in situ hybridization with molecular probes as done in case of wheat.

With the availability of vectors like yeast artificial chromosomes (YACs), bacterial artificial chromosomes (BACs) or binary bacterial artificial chromosomes (BIBACs), physical maps are now being prepared with these clones, whose positions on the map are seprated physically by distances in terms of kilobase pairs.

Genome-wide Physical Maps Using Large-insert YACs, BACs and BIBACs

In all plants and animals, in order to facilitate genomics research, genome-wide physical maps are now being prepared using large-inserts in YAC, BAC and BIBAC vectors. These maps provide essential platforms for a variety of activities including the following: (i) large scale genome-sequencing, (ii) effective positional cloning, (iii) high throughout physical mapping of ESTs and (iv) target DNA-marker development. Several approaches that are available for the development of these large-insert physical maps include the following: (i) hybridization-based methods using interative hybridizations, (ii) restriction-based fingerprinting methods and (iii) methods for development of integrated maps involving BAC and sequencing, fingerprinting and whole-genome sequencing. Of these three methods, the hybridization method suffers with the problem of the occurrence of repeat sequences, while integrated-map approach, based on sequencing and fingerprinting, is expensive. Therefore, the method of choice during the year 2001 has been the restriction based fingerprinting. In this restriction fingerprinting approach, the restricted fragments are end labeled and fractioned on agarose or denaturing polyacrylamide gels. During 2001, automatic sequencers were also used for electrophoresis-based fingerprinting, leading to the preparation of large-insert BAC based physical maps in an indica rice, *Arabidopsis thaliana* and maize.

Integrated physical map of Arabidopsis thaliana During 1990-2000, several genome-wide physical maps of *Arabidopsis thaliana* genome were developed using large-insert YACs or BACs. These maps were also used for working out the whole genome sequences, which are published in December, 2000. The whole genome sequence, however, still had gaps in he heterochromatic and centomeric regions. In order to fill up these gaps, a whole-genome integrated physical map in Arabidopsis was prepared and published towards the end of the year 2001. For preparing this latest physical map, a new BIBAC library was prepared, which complemented the BAC libraries used earlier for mapping. As many as 10,752 BAC and BIBAC clones were used for restriction fingerprinting. Based on fingerprints, the BAC/BIBAC clones were assembled into 196 contigs, which included one contig belonging to chloroplast DNA, another contig belonging to mitochondrial DNA and the remaining 194 clones belonging to > 126Mb of the nuclear genome. The physical maps thus prepared was integrated with the available maps based on BACs and whole genome sequence. Approximately 95% of the new BAC/BIBAC contigs used for integrated physical maps were consistent with contigs of the existing physical and sequence maps, both in clone content and clone order. but 5% of contigs differed. The accuracy of the above integrated physical map was also checked using DNA markers. The large-insert clones earlier used for mapping were available in BAC vectors, which required subcloning before they could be used in transformation experiments for functional analysis. This problem was largely overcome through the use of a large number of BIBAC clones in the above integrated map, which being binary in nature. could be directly (without subcloning) used in transformation studies for functional. analysis.

Sequences analysis of Arabidopsis genome had earlier suggested that it contained 25,498 genes. However. the functions of >50% these genes remained to be characterized experimentally. a goal that is planned to be achieved by 2010 through the Arobidopsis "2010 project" (consult next chapter for details). This work will be greatly facilitated by the availability of the above integrated whole-geneome physical map.

GENOMIC-WIDE PHYSICAL MAP IN INDICA RICE

In the year 2001, in an indica rice, a total of 21.087 BACs covering about seven times the geneome wee used for the preparation of a rice genome-wide physical map. The BAC fingerprints were obtained due to restriction digestion and fractionation on an automatic sequencer. The resulting fingerprints were scanned into image files, edited, and created into FPC (fingerprint contig) database. Using FPC program, the overlapping clones were assembled into 585 contigs, each contig having 5-128 BACs (the contigs having less than 5 BACs were rejected). Two contigs were treated as overlapping, if the end clones of these contigs had 10 or more common bands. Using this criterion, extended contig's were developed and the number of contigs was reduced from 585 to 298, which covered 97% of the genome. The BACs of each contig were screened using DNA markers, and their correctness validated. In future, this physical map will certainly prove useful for genomics research in rice and other grass species.

PHYSICAL MAP OF MAIZE

Using inbred line B73, three BAC libraries (using three enzymes, HindIII, EcoRI and Mbol) were prepared. Together the three libraries provided 27 fold genome coverage of the genome. The BAC clones of these libraries are being screened using markers earlier mapped using IBM population. The BAC clones are fingerprinted by digesting BAC clones with HindIll, and fractionating the fragments on agarose gel. The gel image are digitized with IMAGE software (Sanger Center, UK). Contigs were assembled using fingerprint contig (FPC) software. In the first instances. 83,000 of the 450,000 BAC fingerprints were assembled into 13,000 contigs and 9,300 singletons (singletons are clones which have no overlap with any other BAC clone). A variety of markers including 90 RFLPs. ~1000 ESTs, a large number of AFLPs and mites from the maize genome are being used to screen BAC libraries. A number of cDNA clones from sorghum were also used to assign markers to individual BACs. It is also proposed to develop at least 2000 SNOs from ESTs. SNPs will also be developed through direct sequencing of BACs and BAC ends. The resulting information will provide important anchor points for assembling individual BACs into contigs.

Maize DB can also be accessed at http://www.agron.missour/edu, where maps, probes, primers, screening image and lab protocols are available. It also allows users to compare sequences to interest to all public maize sequences. For more information maize DB also provides database links including the following: (i) CUG 1 links to BAC contigs (ii) GenBank and Zm BD (Zea mays database at lowa State University) links for sequence and clone information.

PHYSICAL MAPS IN BARLEY

For construction of physical maps in barely, translocation break point in each of the seven chromosome involving 120 translocations were used as physical landmarks. Since each chromosome was involved in several translocations. it could be divided in sub-regions on the basis of breakpoints, in a manner analogous to that used in the deletion stock of bread wheat. STS primers from> 300 genetically mapped RFLP markers were prepared, and were used with, microdissected chromosome carrying translocated segments as templates for PCR, leading to the prepared Reactions. This allowed identification of markers flanking each translocation breakpoint leading to the preparation of physical maps.

Physical map in sorghum An integrated map of sorghum genome was also published in the year 2000. For this purpose BAC pools were screened with multilocus markers like AFLPs and MITES.

Physical Maps Using in situ Hybridization (ISH)

Utilizing molecular probes derived from cDNA or gDNA (even if they were not derived from the same species i.e. heterologous probes), in situ hybridization could be successfully utilized for mapping genes with multiple copies on specific chromosomes. For instance cDNA copies of wheat rDNA were used for in situ hybridization to mitotic metaphase chromosomes of bread wheat cv. Chinese Spring. It was found that 90% of the ribosomal genes were located on chromosomes, 1B and 6B and the remaining repeat units located in 5d. This had suggested that the diploid genomes like A genome, which had two chromosomal sites for rRNA genes (Gerlach et al., 1980) had undergone a change after incorporation into hexaploid (6x) bread wheat. Similarly, information was collected on physical location of the 5S rRNA loci in wheat. rye and barley. Using in the situ hybridization (ISH), it was found that 5S rRNA loci are located distal to the secondary constrictions on chromosome IB of wheat and 1R of rye and on non-nucleolar chromosome in barley. An additional 5s rRNA locus in rye was located on 5R.

In situ hybridization has also been used for physically locating the molecular markers earlier mapped in rice. It has been shown that distances on linkage maps (in cM) are not proportionate to physical distances on chromosomes.

Techniques have also been described, where each of two or three DNA sequences can be labelled with different non-radioactive haptens, each of which can be visualized with an independent detection system. It is, therefore, possible to detect two or three DNA sequences simultaneously on animals or plant chromosomes, in presence of each other, for physical mapping. For instance, in rye chromosomes, the location of pTa71 (rDNA sequence from wheat) and pSci l9.2 (a repeat sequence from rye) could be detected by labeling one with biotin and the other with digoxogenin and then using a mixture of the two for in situ hybridization.

Molecular Transcript (EST) Maps (Table 40.6)

With the emphasis on functional genomics research, and with accumulation of EST data for number of plant genomes, efforts were also underway at the beginning of the present century to prepare transcript maps in crop plants. These maps could be genetic, cytogenetic, or physical in nature. A composite map with 272 loci (spread over 1573 cM) of the expressed sequences

Table 40.6 Large genomic fragments sequenced in some cereal Species.

Crop and species	*Fragment length sequenced*	*Surrounding region of gene/locus*	*Reference*
1. Bread wheat			
(*Triticum aestivum*)	16 kb	Lrk10	Feuillet and Keller (1999)
	23 kb	*LrklO*	Stein et al. (2000)
2. Diploid wheat			
(*Triticum tauschii*)	16 kb	SBE-1	Rahman et al. (1997)
Triticum monococcum	211 kb		Wicker et al. (2001)
Triticum monococcum	180 kb	Vrn1	Dubcovsky et al. (2001)
Triticum monococcum	112 kb, 125 kb, 151 kb	X1	Li and Gill (2002)
Triticum monococcum	45 kb, 125 kb	Sh2	Li and Gill (2002)
3. Barley			
(*Hordeum vulgare*)	160 kb	HvLrk	Feuillet and Keller (1999)
	60 kb	mlo	Panstruga et al. (1998)
	204 kb	Mla	Wei et al. (1999)
	66 kb	rarl	Shirasu et al. (2000)
	90 kb	Vrnl	Dubcovsky et al. (2001)
4. Rice			
(*Oryza sativa*)	340 kb	Adh1-Adh2	Tarchini et al. (2000)
	50 kb	BAC 3615	Dubcovsky et al. (2001)
	150 kb	chrl/6	Rice Genome Project (http://www.staff.or .jp/ GenomeSeq html)
5. Sorghum			
(*Sorghum vulgare*)	78 kb	Adhl	Tikhonov et al. (1999)
6. Maize			
(*Zea mays*)	218 kb	*Adh1*	Tikhonov et al. (1999)
	140 kb	*A1/Sh2*	Civardi et al. (1994)
	78 kb	*Zein*	*Llaca* and Messing (1998)

and phenotypic traits in sunflower was published in 1999. The related species but also between more distantly related species. For instance, in adh region, collinearity is greater between closely related maize and sorghum than between either of these two species and the distantly related rice. In contrast to this, wheat shows greater collinearity with distantly related barley than with the closely related rye.

Construction of this map involved anonymous cDNA markers. A rice transcript map was published in 2002 and barley transcript map will be published in 2003. The rice transcript map contained as many as 6591 EST sites. It was shown that 40% of ESTs occupied only 21% of the genome, suggesting that genes are not randomly distributed and that there are gene rich and gene-poor regions in genomes of crop plants. A transcript map in barley is being constructed at IPK, Gatersleben (Germany) and should be published. Ten different laboratories in USA are also busy in preparing a transcript map in wheat involving 10,000 unique ESTs.

Genomics for Evolutionary Studies

DNA sequences have also been utilized for evolutionary studies, where phylogenetic trees could be prepared on the basis of similarities between sequences of individual genes like rRNA genes, 18S rDNA genes, rbcl and other genes. Already by 1980s, large data and sophisticated computer based analytical tools had become available for this purpose for redrawing the phylogenetic tree of life. However, in most of these earlier studies, sequences of one or two genes were utilized at a time, and the sequences of individual genes like 18S rDNA and rbcl gene did not find the most parsimonious trees. Therefore, in a stud conducted in 1999, DNA sequences of plastid genes rbcl and atpB and the nuclear gene for 18S rRNA of 560 species of angiosperms and seven non-flowering plants were utilized together for preparing a well-resolved phylogenetic tree. Similarly, a study of DNA sequences of five mitochondrial, plastid and nuclear genes (atpl, matR. atpB, rbcL, 18S rDNA) of 105 species belonging to 103 genera and 63 families suggested that Amborella. Nymphaeales and illicales-Trimeniaceae-Austrobaileya represent the first stage of Angiosperm evolution with *Amborella* being sister to all other angiosperms. Almost the same conclusions were also drawn through a study of duplicate phytochrome genes, PHYA and PHYC.

CHAPTER 41

Recombinant DNA Technology: Nucleic Acid Hybridization

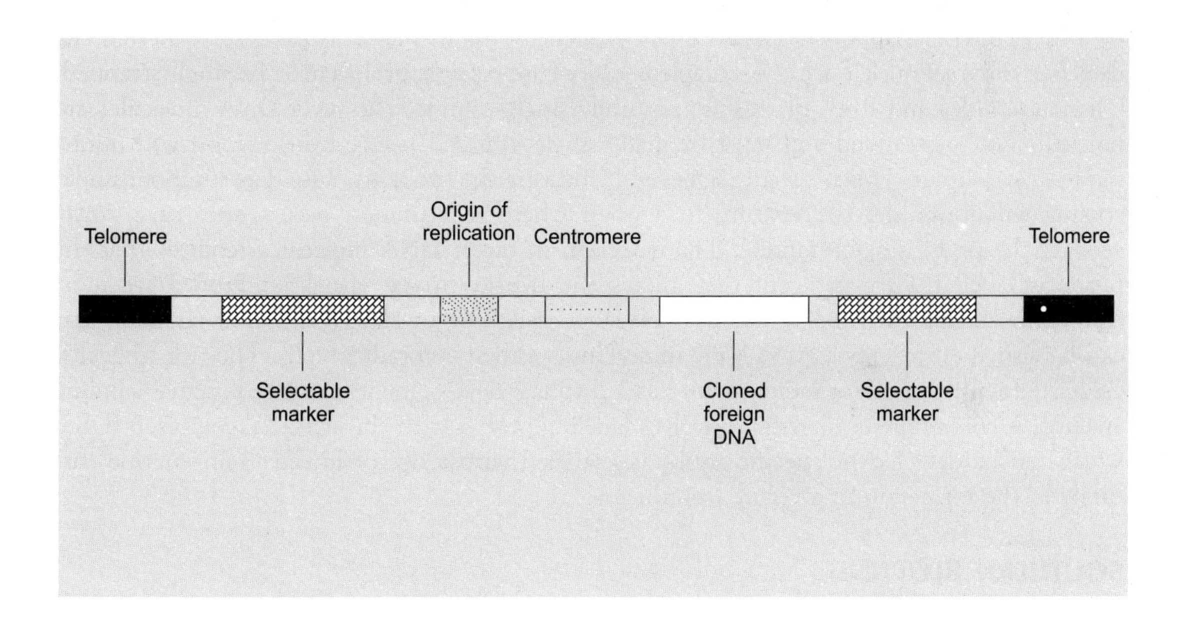

HYBRIDIZATION

When double-stranded DNA is heated the hydrogen bonds that stabilize the double helix are disrupted. The helix becomes unstable and the two strands of the DNA molecule separate. This process is called **denaturation**. If the temperature is then reduced the double helix reforms and the original double-stranded DNA molecule is recovered. This process is called **renaturation** or **reannealing**. In fact, any two single-stranded nucleic acid molecules are capable of forming a double-stranded molecule as long as their base sequences are mostly complementary. The double-stranded molecule is called a hybrid and its formation is called hybridization. Hybrids can form between two strands of DNA, between DNA and RNA and between two strands of RNA. This feature of nucleic acids is used in a series of analytical techniques in which specific DNA or RNA sequences in complex mixtures are detected by using a nucleic acid of complementary sequence.

PROBES

The detection of nucleic acid sequences using techniques based on hybridization requires the use of probes. A probe is simply a DNA or RNA molecule that can be used to detect nucleic acids of complementary sequence. Probes must be pure and free from other nucleic acids with different sequences. Typically, probes are cloned DNA sequences or DNA obtained by polymerase chain reaction (PCR) amplification. Synthetic oligonucleotides and RNA obtained by a in vitro transcription of cloned DNA sequences are also used as probes. To permit the detection of target nucleic acids by complementary base pairing, probes must be single-stranded. Oligonucleotides and RNA probes are naturally single-stranded, however DNA molecules are normally double-stranded and must be made single-stranded before hybridization with target sequence can occur. This is usually achieved by heating the probe to cause denaturation. Rapid cooling will lower the temperature to a point where renaturation occurs only very slowly keeping the probe single-stranded. The detection of target DNA molecules requires that the probe is labeled with an agent that allows the hybrids to be visualized Probes are most commonly labeled with radioactive isotopes such as P, S C, or H. Hybrids emit radiation and can be visualized by exposure to X-ray film. This is called autoradiography. The risk to health associated with the use of radioactivity has led to the development of nonradioactive laboling systems. A commonly used system involves labeling probes with the steroid digoxigenin (DIG) which can be detected by specific antibodies labeled with a dye or linked to an enzyme that catalyzes the formation of a colored product.

SOUTHERN BLOTTING

This was the first analytical technique based onnucleic acid hybridization to be developed and is named after its inventor, Ed Southern of the University of Oxford. Southern blotting is used to detect DNA molecules and is typically used to analyze the structure of genes (Fig. 41.1). for this application, the DNA for analysis is purified from cells and is obtained as large fragments of chromosomal DNA, typically 20000 bp or more in length. This first step in Southern blotting involves digestion of the DNA with a restriction enzyme. These enzyme are derived from bacteria and cut DNA molecules specifically at each point where a DNA sequence of 4-8 bases recognized by the enzyme occurs. Digestion of the chromosomal DNA produces thousands of DNA fragments ranging in size from just a few bases to several thousand bases.

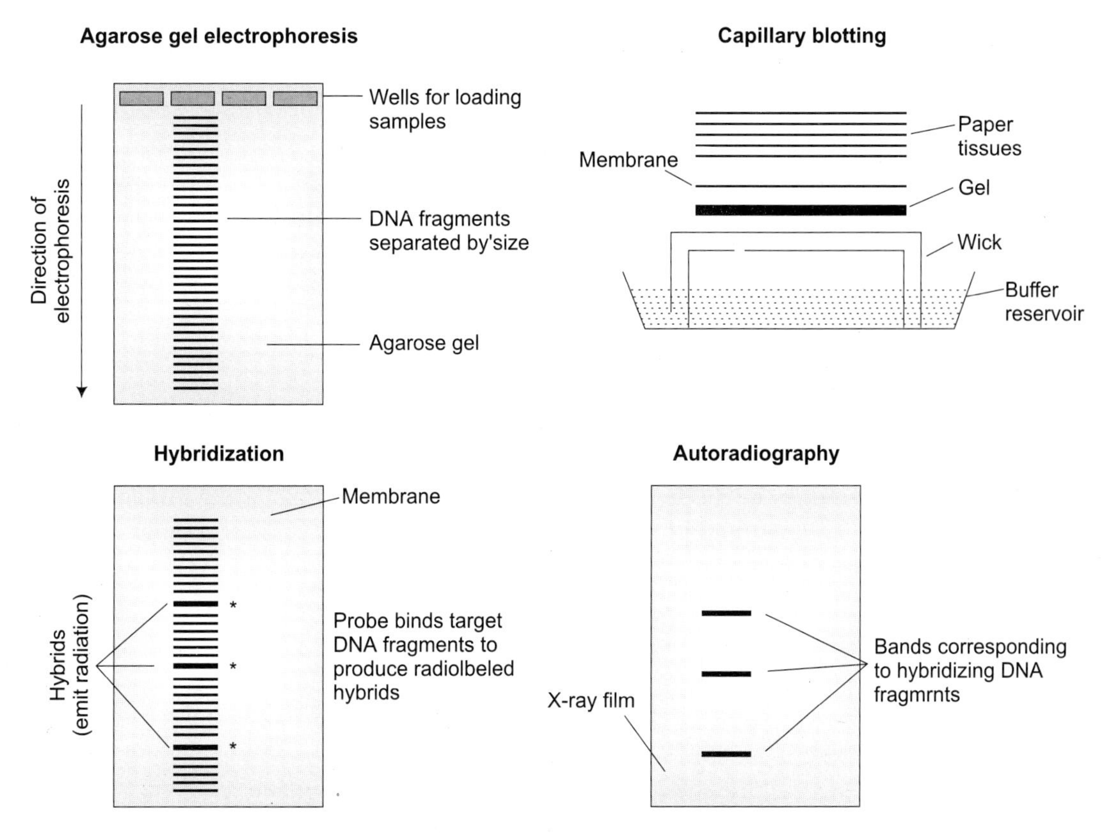

Fig. 41.1 Southern blotting.

The DNA fragments are then subjected to electrophoresis in an agarose gel. This separates them according to size the large ones near the top of the gel and the small fragments near the bottom. After electrophoresis, the gel is soaked in an alkaline solution that denture the fragments making them suitable for hybridization. The next stage involves transferring the dentured fragments out of the gel and onto a membrane made from nylon (or sometimes nitrocellulose) where they become accessible for analysis using a probe. Transfer is achieved by blotting. The gel is placed on a platform above a dish containing a buffer. A sheet of membranes is placed on top of the gel, covering it, and a stack of obsorbent paper tissues is placed on top of the membrane. The gel is blotted for several hours during which time the buffer in the reservoir is drawn, by capillary action, from the wick through the gel and the membrane into the stack of paper towels. The flow of buffer cause the DNA fragments to pass out of the gel and onto the membrane above, where they become attached. After several hours, a replica of the pattern of fragments in the gel forms on the membrane. At this point the blot is dismantled and the membrane containing the fragments is removed. The membrane is treated to firmly attach the DNA fragments either by backing it at 80° or by exposing it to ultraviolet radiation.

The next stage involves incubating the membrane with a labeled probe. This is known as hybridization and is carried out at a temperature and in a buffer that favor the formation of hybrids between the probe and fragments of DNA bound to the membrane whose sequence is complementary to the probe. The membrane is then washed with a buffer to remove probe is bound nonspecifically so that only labeled probe bound to target sequences remains. If the probe is labeled with radioactivity, the membrane is placed in the dark against a sheet of X-ray film. After several hours, development of the film reveals one or more dark bands which correspond t the position of the filaments bound by the probe. The length of the fragments can be calculated from their positive relative to marker DNA molecules of known length. In this way the structure of individual genes can be characterized in terms of the number of hybridizing fragments and their sizes. When a probe for a new gene is isolated one of the first experiments carried out is to use Southern blotting can also be used to reveal abnormalities in the structure of genes. For example, about 20% of individuals with the inherited bleeding disorder, hemophilia, show alterations in the structure of the gene for the blood clotting protein, Factor VIII. These alterations can be used to identify relatives who are carriers of hemophilia and may need genetic counseling.

NORTHERN BLOTTING

This technique is similar to Southern blotting but is used to analyze RNA rather than DNA. The DNA is every human cell contains copies of all of the genes present in the human genome. However, only about 15% are active in any particular cell. The remaining 85% are inactive. These are not transcribed and do not lead to the synthesis of protein. Furthermore, different genes are active in different cell types. For example, the genes that are active in a muscle cell are very different from those active red blood cells. The active genes reflect the very different protein compositions of these cells. Muscle cells contain proteins such as actin and myosin that are required for contraction and the genes for these proteins are very active. In contrast, in red blood cells the major protein present is hemoglobin. Consequently, the globin genes which encode the polypeptides that make up hemoglobin are very active. Northern blotting is used to identify which genes are active in different cell types. The method used for Northern blotting is similar to Southern blotting. The main difference is related to the starting material. For Northern blotting, RNA is isolated from cells. This contains messenger RNAs (mRNAs) derived from all the active genes in that cell type. The mRNAs are different sizes, depending on the size of the protein they encode. mRNAs that encode proteins present at high concentrations in a cell are also more abundant than those encoding proteins present at low concentrations. The RNA is separated by agarose gel electrophoresis. Each mRNA migrates to a position on the gel determined by its size, with the larger transcripts near the top and the smaller transcripts near the bottom. As was the case of Southern blotting, the gel is blotted and the membrane is hybridized with a probe specific for the mRNA being investigated. The membrane is then washed and exposed to X-ray film. Development of the film reveals usually a single band corresponding to the mRNA recognized by the probe. The position of the band relative to the top of the membrane can be used to estimated the length of the mRNA and the intensity or darkness of the band is a measure of how much of that mRNA was present in the original cells. Thus mRNAs of genes that are expressed at low levels appear as faint bands and very abundant mRNAs appear as dark bands. Northern blotting can, therefore, provide useful information about which genes are switched on in a cell, the level at which they are expressed and the sizes of the different mRNAs.

IN SITU HYBRIDIZATION

This technique is different from other hybridization methods in that it is used to detect nucleic acid sequences present in intact cells. The main use of in situ hybridization is to identify expression of specific mRNAs in individual cells present tissues containing a number of different cell types. For example, in situ hybridization can be used to show that insulin mRNA is produced only by the B cells of the pancreas. Thin slices of tissue, known as sections, are cut using an instrument called a microtome and are placed on a microscope slide. Probe is added to the cells on the microscopic slide and is taken up into the cytoplasm where it binds to target mRNA forming hybrids. These can be detected by the label on the probe and appear as areas of staining within cells which can be seen when the section is viewed under a microscope. Probes may be labeled in a number of ways. Radioactive isotopes, fluorescent molecules, enzymes that catalyze the formation of colored products and the steroid digoxigenin are all used.

A variation on the technique called **fluorescent in situ hybridization (FISH)** involves hybridizing probe to chromosomes in cells as a way to identifying the position of a gene on a chromosome. Cells undergoing metaphase contain chromosomes in a noncondensed form which have characteristic shapes. By hybridizing probe to these cells it is possible to determine the chromosomal locus of the hybridizing sequence.

DNA CLONING

The human genome is estimated to contain 50-100,000 genes. In DNA isolated from human cells, individual gene sequences are present in only very small amounts. DNA cloning is a powerful technique that allows specific DNA sequences to be separated from other sequences and copied so that they can be obtained in large amounts permitting detailed analysis or manipulation. An important use of DNA cloning is to isolate new genes allowing them to be investigated and characterized.

All DNA cloning experiments are based on the construction of recombinant DNA molecules. This involves joining different DNA molecules together. The DNA molecule to be cloned (often a fragment of human DNA containing a gene of interest) is inserted into another, usually circular, DNA molecule called a vector. The recombinant vector is introduced into a host cell, usually the bacterium *E. coli*, where it produces multiple copies of itself. When the host cell divided copies of the recombinant vector are passed on to daughter cells. Large amounts of the vector are produced which can be purified from cultures of the host cells and used for analysis of the foreign DNA insert.

Restriction Enzymes

The ability to join different DNA molecules together for cloning is dependent on the use of enzymes form bacteria called restriction endonucleases. These enzymes cut DNA molecules at specific sequences, usually of 4-8 bases. The sequences recognized are plindromes. This means that the sequence is the same reading 5'-3' on both strands. Each enzyme has a specific target sequence. For example the enzyme EcoRi, which is obtained from the bacterium *E. coli*, will cut any DNA molecule that contains the sequence GAATTC. Other restriction enzymes recognize different sequences. The cut made by restriction enzymes is usually staggered such that the two strands of the double helix are cut a few bases apart. This creates single-stranded overhangs called sticky ends at either end of the cut DNA molecules. Some restriction enzymes cut the DNA leaving the 5' end overhanging and other leave a 3' overhang. A few restriction enzymes cut both strands of the double helix at the same position creating what is known as a blunt end (Fig.41.2a). Two DNA molecules cut with the same restriction enzyme can be joined together by complementary base-pairing between the sticky ends (Fig. 41.2b). Although the molecules are joined by the sticky ends, they are not covalently linked. However, an enzyme called DNA ligase, which is found in all cells, can be used to catalyze the formation of a phosphosdiester bond between the two DNA molecules. Thus permanently recombining them. A number of systems exist for cloning DNA molecules based on the use of different types of vector. Each has its own individual characteristics and uses.

PLASMIDS

These are small, circular DNA molecules present in bacteria. They occur in addition to the main bacteria chromosome and can replicate autonomously. Plasmids are the commonest type of cloning vector used. The cloning procedure involves a number of steps (Fig. 41.3): first, the plasmid is digested with a restriction enzyme that cuts it is a single site converting it from a circular molecule into a linear molecule with sticky ends. The foreign DNA to be cloned is also digested with the restriction enzyme to produce the same sticky ends. When the plasmid and the foreign DNA are mixed, molecules of plasmid become joined top molecules for foreign DNA via their common sticky ends and circular recombinant plasmids are obtained. DNA ligase is then used to covalently join the two, the recombinant plasmid is introduced into host bacteria (usually *E. coli*) by a process called transformation. The transformed bacteria are spread on agar plates and bacterial colonies comprised of cell that have taken up the recombinant plasmid are grown. Individual colonies are isolated and culture in liquid medium. Large amount of plasmid can then be purified from the culture and the cloned DNA recovered for analysis.

Several features of plasmids make them especially suitable as cloning vectors Their small size (usually about 3 kbp) makes them easy to purify from bacterial cultures allowing the cloned DNA to be recovered easily. In addition, plasmids often contain genes encoding proteins that

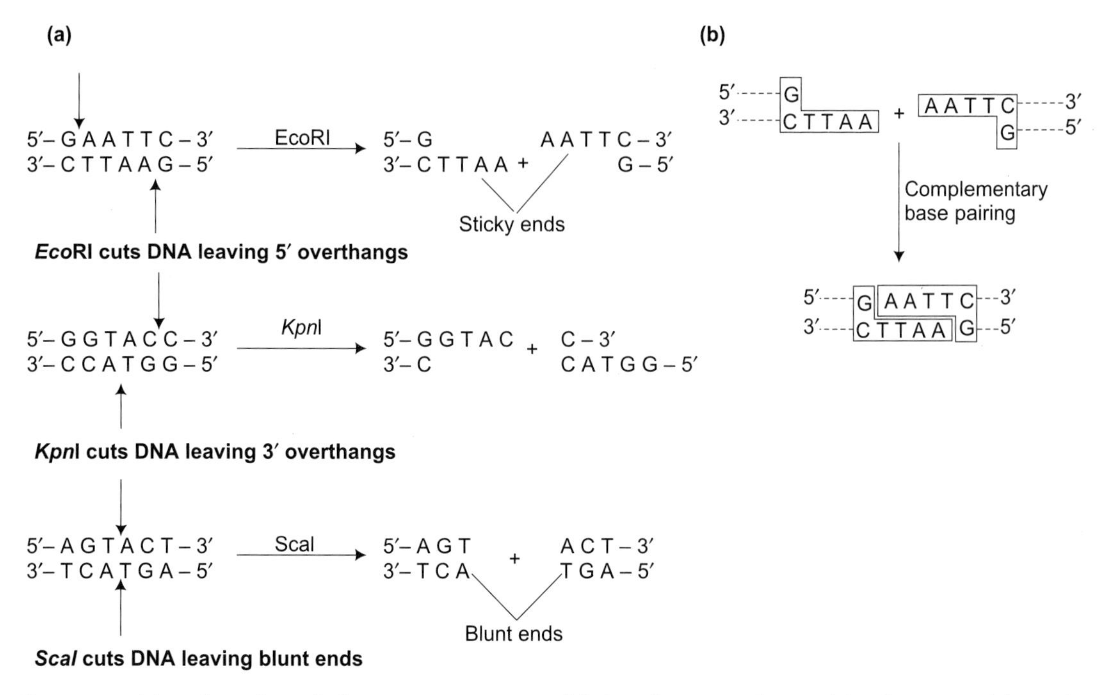

Fig. 41.2 (a) Action of restriction enzymes on DNA. (b) Complementary base-pairing between sticky ends.

make the bacteria resistant to antibiotics such as ampicillin and tetracycline. By growing colonies on agar plates containing antibiotic, it is possible to isolate bacteria that have taken up the plasmid during transformation because only these will be resistant to antibiotic and will be able to grow.

Plasmid cloning vectors were initially based on naturally occurring plasmids. These have gradually been replaced by improved vectors whose DNA sequences have been altered to include features useful for cloning.

One of the earliest plasmid vectors to be developed was pBR322. This plasmid contains two genes that confer resistance to the antibiotics ampicillin and tetracycline During cloning, foreign DNA is inserted into the tetracycline gene, thereby inactivating it. Transformed bacteria containing recombinant plasmid could therefore be identified by being resistant to ampicillin but not to tetracycline. Bacteria which had taken up plasmid that did not contain foreign DNA but had simply been relegated to itself could be identified be being resistant to both antibiotics.

The pBR322 plasmid was followed by the pUC series of vectors which allowed identification of colonies containing recombinant plasmid by a method called blue/white selection (Fig. 41.4). This method relies on the presence of a gene called lac Z which encodes the enzymes Β-galactosidase and is located on the plasmid at the point where the foreign DNA in inserted. Bacteria that contain the intact plasmid synthesize Β- galactosidase which acts on a synthetic substrate called X-gal (5-bromo 4-chloro-3-ndolyl-b-d-galactopyranoside) to produce a

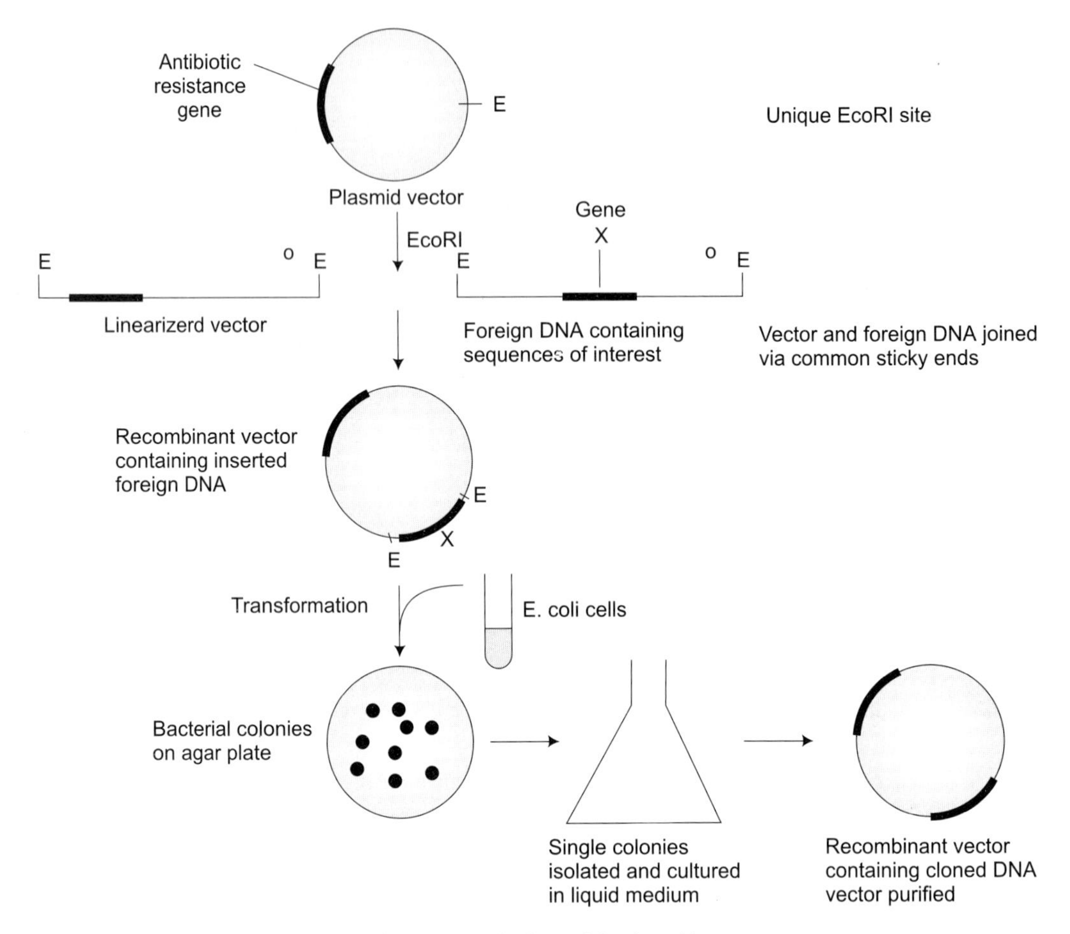

Fig. 41.3 Cloning with plasmids

colored product. When colonies are grown on agar plates containing X-gal they take on a blue color. However, when foreign DNA is inserted into the pUC plasmid, the lac Z gene is disrupted and B-galactosidase is no longer produced. As a result colonies containing recombinant plasmid remain white when grown on X-gal and are easily distinguished from colonies containing relegated vector which are blue.

Another useful feature of pUC vectors is that the sequences of part of the lac Z gene is modified to create a series of clustered restriction enzyme sites. This is called the multiple cloning site (MCS). Its purpose is to create extra flexibility during the cloning procedure by allowing the foreign DNA to be inserted at any one of several restriction sites (Fig. 41.5).

Other useful plasmid modifications include the presence of promoter sequences from bacteriophages inserted on either side of the MCS which allow in vitro transcription of the inserted foreign DNA by RNA polymerase. This feature is useful for producing RNA probes from cloned sequences. Some plasmids are modified to allow cloned sequence to be translated into protein. These are known as expression vectors.

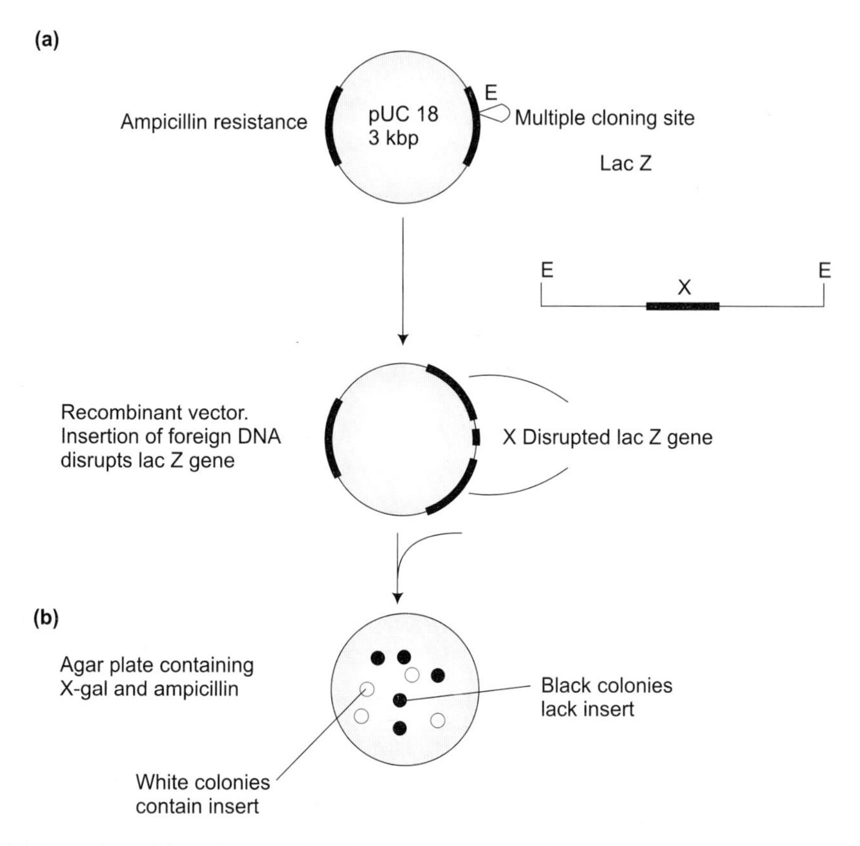

Fig. 41.4 (a) Insertion of forgeign DNA inactivates lac Z gene. (b) Recombinant colonies appear white on agar plates containing X-gal.

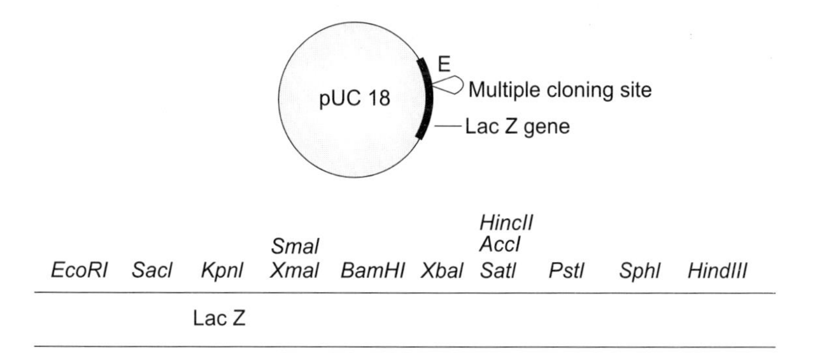

Fig. 41.5 Multiple cloning site of the pUC18 vector.

LAMBDA (b) PHAGE

Bacteriophages (phages) are viruses that infect bacteria. They consist of a nucleic acid geneome inside a protective protein coat called a capsid. The bacteriophage which infects *E. coli.* has a liner double-stranded DNA genome. The phage attaches itself to the surface of a bacterium and injected its DNA into the cell. In side the cell, the DNA is copied and capsid proteins are synthesized. The DNA is packaged into capsids and new phage particles are produced that are released by lysis of the infected cells.

Lambda phage has been adapted for use as a cloning vector. The central portion of the DNA, which is not essential for infection, is deleted leaving 5' and 3' fragments known as arms (Fig 41.6). The deleted region can be replaced be foreign DNA to produce recombinant phage DNA (Fig. 41.6). This is inserted in to phage capsids in vitro by a process called packaging which involves mixing the recombinant phage with a packaging extract containing phage capsid proteins and processing enzymes. Recombinant phage particles are produced which are highly efficient at infected *E. coli*. Infected cells are spread on an agar plate and produce a continuous sheet of bacteria called a lawn which contains small clear areas about the size of a pin head.

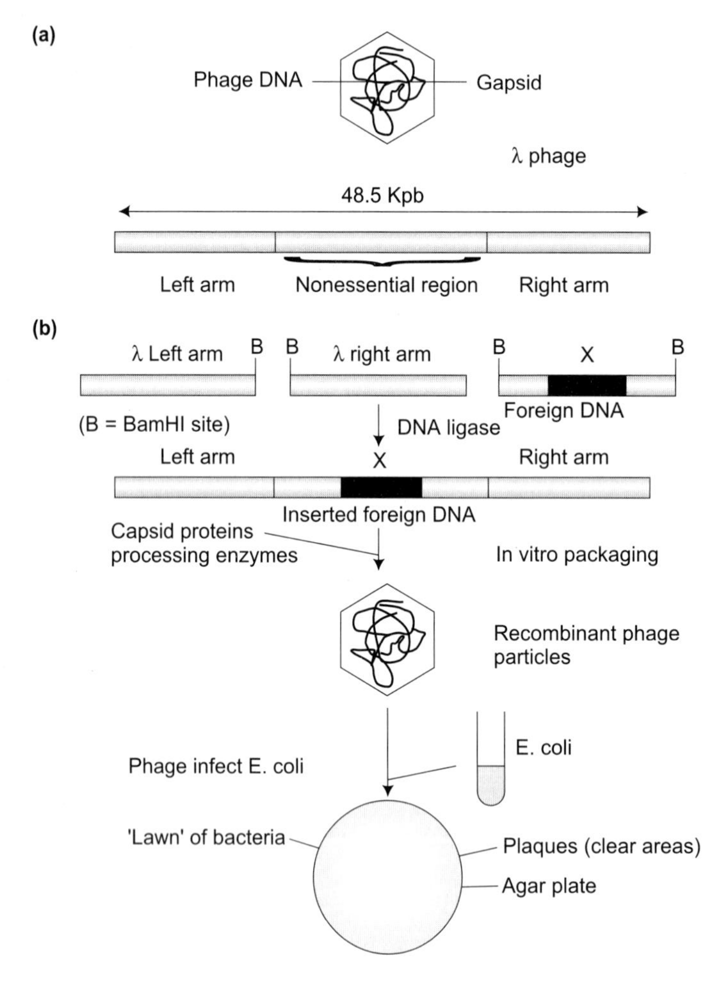

Fig. 41.6 (a) λ phage. (b) Use of λ phage as a cloning vector.

These correspond to areas of lysis produced by infection with phage and are known as plaques. Individual plaques can be isolated and used to generate large amount of cloned DNA by infection of fresh can be culture of *E. coli*.

The main advantage of as a cloning vector is that the size of the fragments that can be cloned is much larger than for plasmids. Lambda vectors accommodate fragments up to 25 kb as compared to less than 10 kb for plasmids. The ability to clone larger fragments has led to the development of the main use of vectors which is the construction of DNA libraries. A library is a collection of recombinant phage which together contain clones representative of all the DNA sequences present in the genome of an organism. Because they can accommodate relatively large fragments of DNA, many fewer clones are required to represent an entire genome than would be required if plasmids were used. Genomic libraries are constructed from DNA purified from cells which has been broken randomy into fragments of around 20 kbp either by digestion with a restriction enzyme or due to physical shearing by pipeting or sonication. The fragments are ligated to the arms and are cloned at random. Many thousands of plaques are produced, each of which contains of different cloned sequences. Plaques corresponding to a cloned sequence of interest can be identified by screening the library with a probe using a procedure called a plaque lift. This involves taking an agar plate containing plaques and laying a sheet of special nylon membrane on top of it. Some of each plaque adheres to the membrane and a replica of the pattern of plaques on the plate forms. The membrane is then treated with alkali to denature the DNA in the plaques and is hybridization to a DNA probe labeled with radioactivity. After washing away unbound probe, exposure of the membrane to X-ray film produces a series of block dots corresponding to the position of plaques containing the desired sequence. These can then be isolated from the agar plate and used to obtain the cloned DNA.

Libraries can also be produced using RNA. The enzyme reverse transcriptase is used to convert the DNA into complementary DNA (cDNA) which can then be cloned in the same way as for genomic libraries. Libraries made this way are called cDNA libraries and contain clones that are representative of the genes that are active in the cells used to isolate the RNA. Thus, a cDNA library from blood cells will have many different clones from one derived from lung cells or kidney cells because the active genes in each cell type will be different. cDNA libraries have the advantage that the cloned sequences do not contain introns. This greatly simplifies the characterization of cloned genes because the complete coding sequence of the gene may be present in a single clone. Expression libraries are a type of cDNA library in which the cloned sequences are translated into protein by the host bacteria. This allows the library to be screened using antibodies specific for the protein encoded by the cloned sequence.

COSMIDS

This type of vector combines features found in plasmids and phage. Cosmids contain all the normal features found in plasmids, including a MCS and genes conferring antibiotic resistance, but also include sequences found in called cos sequences. These occur at either end of the DNA

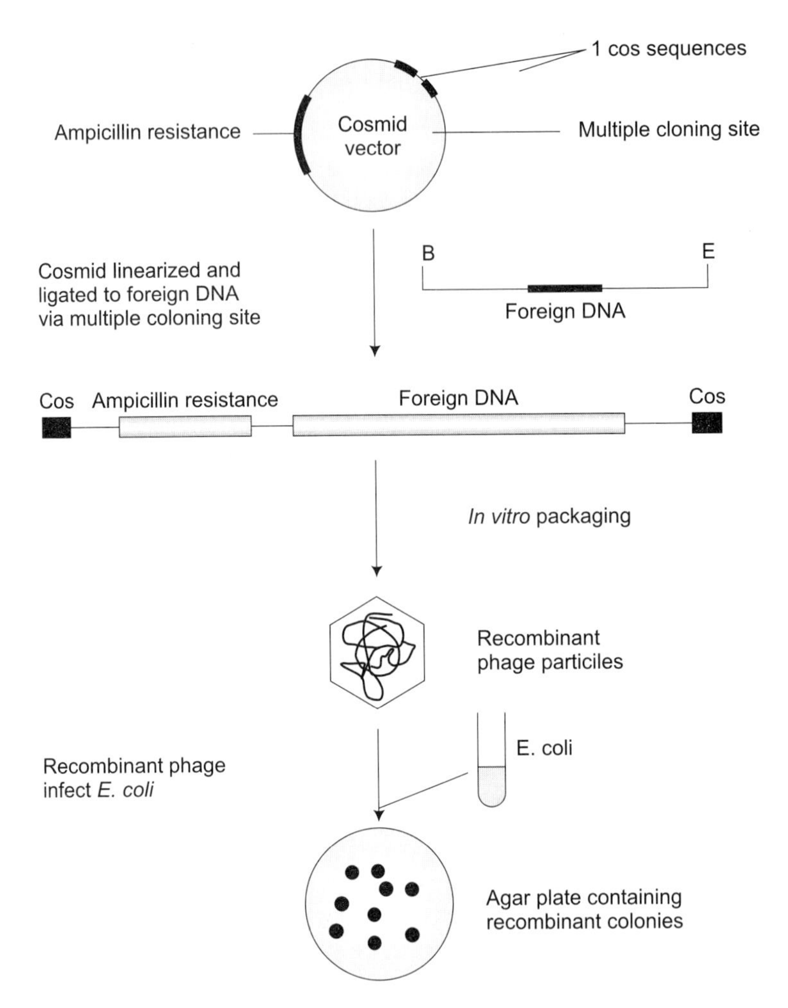

Fig. 41.7 Cloning with cosmid vectors.

molecules and are responsible for its insertion into the phage capsid. The presence of cos sites on cosmids allows them to be packaged into phage capsids. Cloning with cosmids combines features associated with the use of both and plasmids as cloning vectors (Fig. 41.7). Cosmid DNA is cleaved with a restriction enzymes and ligated to foreign DNA. The recombinant cosmid is then packaged into capsids and used to infect *E. coli*. Cosmids do not contain any genes and so do not form plaques after infection. Instead, infected cells grown on agar containing antibiotic and resistant colonies containing recombinant cosmids are obtained which can be propagated in the same way as plasmids. Cosmids have the advantage of being able to accommodate very large inserts. Because cosmids ae small, typically 8 kbp or less, and the capsid can accommodate up to 52 kbp, inserts of up to 44 kbp can be cloned.

YEAST ARTIFICIAL CHROMOSOMES (YAC)

These vectors were developed recently and represent a new approach to gene cloning which uses eukaryotic host cells with a vector that is replicated in the same way as a host cell chromosome. YACs contain all the essential features of a chromosome required for its propagation in a yeast cell including an origin of replication, a centromere to ensure segregation into daughter cells and telomeres to stabilize the ends of the chromosome (Fig. 41.8). Very large DNA molecules up to several hundred kbp can be cloned using YACs. This is significant because individual clones or large enough to encompass anentire mammalian gene. YACs are also used to construct maps of parts of the human genome by identifying clones containing adjacent regions of the genome. Two related types of vector are bacterial artificial chromosomes (BACs) and P1 artificial chromosomes (PACs) which have uses similar to those of YACs.

PLANT CLONING VECTORS

Most vectors for cloning with plants as the host organism are based on the Ti plasmid which occurs in a soil bacterium called *Agrobacterium tumefaciens*. The bacterim invades plant tissue and causes a cancerous growth called a crown gall. During infection part of the Ti plasmid called the T-DNA integrates into the plant chromosomal DNA. Cloning vectors based on the Ti plasmid use the ability of T-DNA to integrate to carry useful genes into the plant the plant genome; such genes may confer useful features to the plant such as resistance to disease.

APPLICATION OF DNA CLONING

Gene cloning is a powerful technique that has made important contributions to a variety of areas in biological research. These include:

Identification of genes with involvement in disease processes Cloning has led to the identification of defective genes that cause inherited disease such as hemophilia and muscular dystrophy and oniogenes and tumor suppressor genes that play am important role in tumor development. Characterization of these genes has greatly improved our understanding of disease processes.

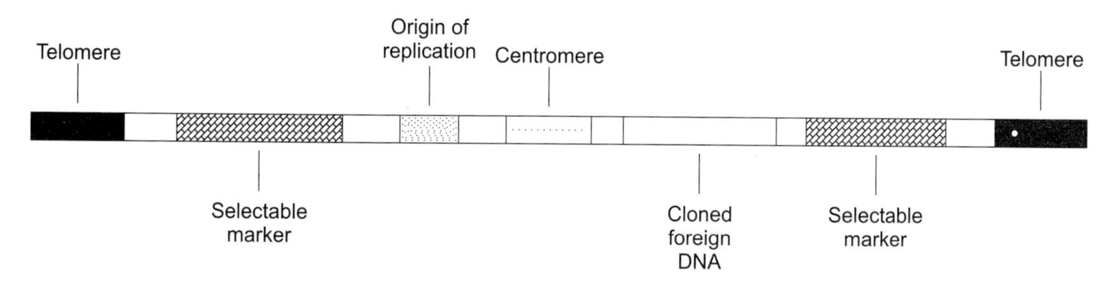

Fig. 41.8 Structure of a yeast artificial chromosome (YAC).

Genome mapping The identification and characterization of clones corresponding to adjacent regions on chromosomes is being used to construct maps which will allow the relative positions of genes to be determined.

Recombinant proteins Gene cloning is used to produce large amount of medically useful proteins such as insulin which is used in the treatment of diabetes and Factor VII which is used to treat hemophilia. Genes encoding useful proteins are cloned in expression vectors and introduced into a suitable host organism such as yeast which expresses the cloned sequence in large amount.

Genetically modified organisms Gene cloning can be used to transfer foreign genes into an organism creating transgenic plants and animals with modified characteristics.

POLYMERASE CHAIN REACTION

Principle

Polmerase chain reaction (PCR) allows specific DNA sequences to be copied or amplified over a million fold in a simple enzyme reaction. DNA, corresponding to genes or fragments or genes, can be amplified from samples or chromosomal DNA containing thousands of genes. Amplified DNA is used for the analysis or manipulation of genes.

The polymerase chain reaction (PCR) is a powerful and widely used technique that has greatly advanced our ability to analyze genes. Chromosomal DNA present in cells contains thousands of genes. This makes it difficult to isolate and analyze any individual genes PCR allows specific DNA sequences usually corresponding to genes or parts of genes, to be copied from chromosomal DNA in a simple enzyme reaction. The only requirement is that some of the DNA sequence at either end of the region to be copied is known. DNA corresponding to the sequence of interest is copied or amplified by PCR more than a million fold and becomes the predominant DNA molecule in the reaction. Sufficient DNA is obtained for detailed analysis or manipulation of the amplified gene.

Components

DNA is amplified by PCR in an enzyme reaction which undergoes multiple incubations at different temperatures. Each PCR has four key components:

TEMPLATE DNA. This contains the DNA sequence to be amplified. The template DNA is usually a complex mixture of many different sequences, as is found in chromosomal DNA, but any DNA molecule that contains the target sequence can be used. RNA can also be used or PCR by first making a DNA copy using the enzyme reverse transcriptase. Oligonucleotide primers. Each PCR requires a pair of Oligonucleotide primers. These are short single-stranded dna molecules (typically 20 bases) obtained by chemical synthesis. Primer sequences are chosen so that they bind by complementary base-pairing to opposite DNA strands on either side of the sequence to be amplified.

DNA POLYMERASE, A number of DNA polymerases are used for PCR. All are thermostable and can withstand the high temperatures (up to 100°C) required. The most commonly used enzyme is Taq DNA Oligonucleotide polumerase from Thermus aquaticus, a bacterium present in hot springs. The role of the DNA polymerase in PCR is to copy DNA molecules. The enzyme binds to single-stranded DNA and synthesizes a new strand complementary to the original strand. DNA polymerases require a short region of double-stranded DNA to get started. In PCR, this is provided by the oligonucleotide primers which create short double-stranded regions by binding on either side of the DNA sequence to be amplified. In this way the primers direct the DNA polymerase to copy only the target DNA sequence.

DEOXYNUCLEOTIDE TRIPHOSPHATES (dNTPs). These molecules correspond tot eh four bases present in DNA (adenine, guanine, thymine and cytosine) and are substrates for the DNA polymerase. Each PCR requires four dNTPs (dATP, dGTP, dTTP, dCTP) which are used by the DNA polymerase as building blocks to synthesize new DNA.

How the PCR works

PCR allows the amplification of target DNA sequences through repeated cycles of DNA synthesis (Fig. 41.9). Each molecule of target DNA synthesized acts as a template for the synthesis of new target molecules in the next cycle. As a result, the amount of target DNA increases with each cycle until it becomes the dominant DNA molecule in the reaction. During the early cycles, DNA synthesis increases exponentially but in later cycles, as the amount of target DNA is to copied increases and the reaction components are used up, the increase becomes linear and then reaches a plateau.

Each cycle of DNA synthesis involves three stages (denaturation, primer annealing, elongation) which take place at different temperatures and together result in the synthesis of target DNA.

Denaturation

The reaction is heated to greater than 90°C. At this temperature the double helix is destabilized and the DNA molecules separate into single strands capable of being copied by the DNA polymerase.

Primer annealing

The reaction is cooled to a temperature that allows binding of the primers to the single-stranded DNA without permitting the double helix to reform between the template strands. The process is called annealing. The temperature used varies (typically 40-60°C) and is determined by the sequence and the number of bases in the primers.

Extension

This stage is carried out at the temperature at which the DNA polymerase is most active. For Taq, this is 72°C. The DNA polymerase, directed by the position of the primers, copies the intervening target sequence using the single-stranded DNA as a template. A total of 20-40 PCR

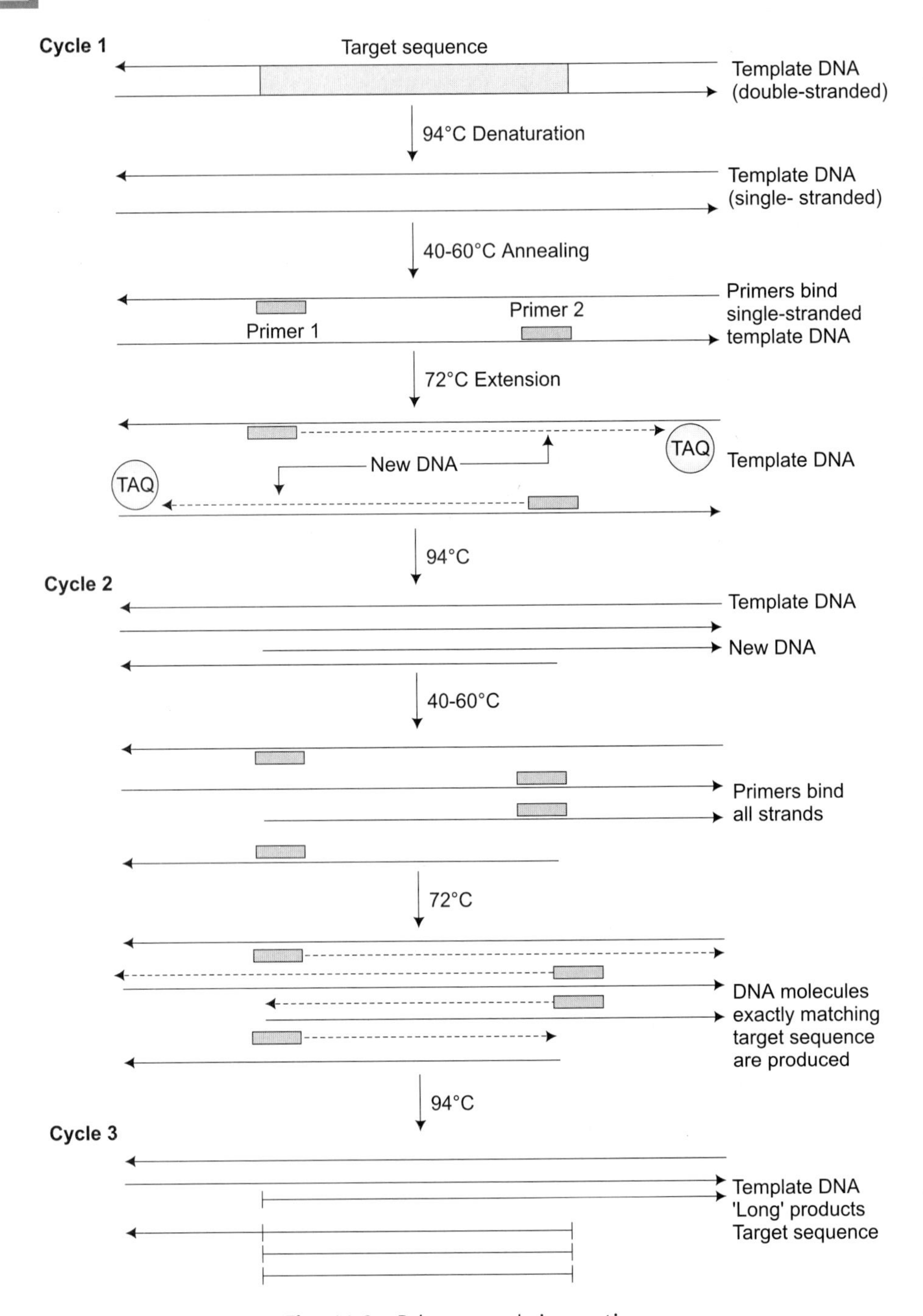

Fig. 41.9 Polymerase chain reaction.

cycles is carried out depending on the abundance of the target sequence in the template DNA. Sequences up to several thousand base pairs can be amplified. To deal with the large number of separate incubations needed, the PCR is carried out using a microprocessor controlled heating block known as thermal cycler. In the first cycle, DNA molecules are synthesized which extend beyond the target sequence. This is because there is nothing to prevent the DNA polymerase continuing to copy the template beyond the end of the target sequence. However in subsequent cycles, newly synthesized DNA molecules which end with the primer sequence act as templates and limit synthesis to the target sequence so that the amplified DNA contains only the target sequence.

Applications

PCR is used extensively as a research tool which has greatly improved our ability to study genes. Most studies in molecular genetics involve the use of PCR at some stage, normally as part of an overall strategy and in association with other techniques. For example, DNA amplified by PCR can be used for DNA sequencing, as a probe in Northern and Southern blotting, and to generate clones. PCR has applications in most areas of biology and medicine as well as in unexpected subjects such as anthropology and archaeology. It is also an important tool in the biotechnology industries.

PCR has made important contributions in many areas which include:

- Inherited diseases. These disorders are caused by gene mutations passed on from parents to their children. Examples include hemophilia and cystic fibrosis. PCR is used to amplify gene sequences which can then be screened for disease-causing mutations. The information obtained has dramatically improved our understanding of these disorders and has produced the important additional benefit of allowing carriers of the disorders to be identified.
- Cancer research. PCR has been widely used in studies of the role of genes in cancer. For example, mutations in oncogenes and tumor-supressor genes have been identified in DNA from tumors using PCR-based strategies. This has improved our understanding of how cancer develops.
- Forensic science. By amplifying repetitive sequences, PCR can be used to identify individuals from samples of their DNA. This is used to link individuals with forensic DNA samples from the scene of a crime. Analysis of variable sequences is also used in tissue typing to match organ donors with recipients and in anthropology to study the origins or races of people.
- Biotechnology. PCR has played an important role in the production of recombinant proteins such as insulin and growth hormone which are widely used as drugs and in the development of recombinant vaccines, such as that for hepatitis B virus.

DNA SEQUENCING

Basis of DNA Sequencing

Two methods have been developed to determine the nucleotide sequence of DNA molecules: the dideoxy chain termination method of Sanger and the chemical degradation method developed by Maxam and Gilbert. The chain termination method has now superseded the chemical method because it is more efficient and is simpler to perform.

The basis of the chain termination method is that the DNA molecule whose sequence is to be determined is copied in an enzyme reaction by a DNA polymerase. Modified nucleotide triphosphates are included in the reaction that cause termination of DNA synthesis randomly at each of the four bases where they occur in the template DNA. The overall effect is that a series of DNA molecules each one nucleotide longer than the next is synthesized which can be separated according to size by electrophoresis. The base sequence of the template DNA can then be determined by identifying the terminal base of each synthesized DNA molecule from the shortest to the longest.

Dideoxy Chain Termination Method

The DNA to be sequenced is called the **template.** It must be obtained in a purified form and must be homogenous; i.e, it must contain only DNA molecules with the same sequence. Purified plasmids containing cloned DNA and DNA produce by PCR are commonly used as templates. A requirement for sequencing by the dideoxy chain termination method is that the DNA must be present in single-stranded form capable of being copied by a DNA polymerase. Previously this was achieved by cloning the DNA into the phage vector, M13. When recombinant phage were used to infect *E. coli*, single-stranded copies of the cloned sequence were produced which could be used as templates for sequencing. More recently, single-stranded DNA for sequencing has been produced in a much simpler way without cloning by denaturation of the DNA template using heat or alkali. Several different DNA polymerases have been used for the sequencing reaction. These include part of the *E. coli* DNA polymerase I known as the **Klenow fragment,** a genetically modified DNA polymerase from the phage T7, called **Sequenase** and **Taq DNA polymerase** which is also used in the polymerase chain reaction.

All DNA polymerases require short regions of double-stranded DNA to initiate DNA synthesis on a single-stranded template. This is provided in sequencing reactions by the addition of short, single-stranded DNA molecules called **oligonucleotide primers** which are produced by chemical synthesis. The sequence of the primer chosen to be complementary to the template DNA such that it binds to it forming a short single-stranded region, thus determining the point at which the sequencing reaction is initiated. To sequence a DNA molecule four separate enzyme reactions are prepared. Each contains: template DNA in single-stranded form, DNA polymerase, primer, each of the four deoxynucleotide triphosphates (dNTPs) which are the building blocks for DNA synthesis and a modified nucleotide called a **dideoxynucleotide triphosphate (ddNTP).** There are four ddNTPs that correspond to the four bases in DNA.

They differ from dNTPs in that they lack hydroxyl group on the 3'carbon of the ribose sugar. They are incorporated into the growing DNA polynucleotide by the polymerase but their lack of 3'hydroxyl group prevents formation of a phosphodiester bond with the next nucleotide to be added. This prevents further elongation of the DNA polynucleotide being synthesized (Fig. 41.10). The ddNTPs thus act as specific inhibitors of DNA synthesis. Each of the four sequencing reactions contains a different ddNTP. Depending on which ddNTP is present, synthesis will terminate where that nucleotide is incorporated. At any point during DNA synthesis, the polymerase may incorporate a dNTP into the growing polynecleotide chain in which case synthesis continues or it may incorporate a ddNTP in which case chain elongation is blocked and synthesis ends at that position. The overall effect in that in each of the four reactions a series of DNA molecules of different lengths is produced each terminating at a position corresponding to the presence of that base in the template DNA (Fig. 41.11). For example, in the sequencing reaction containing ddATP a series of DNA molecules will be synthesized each of which ends in an A corresponding to the position of a T in the template.

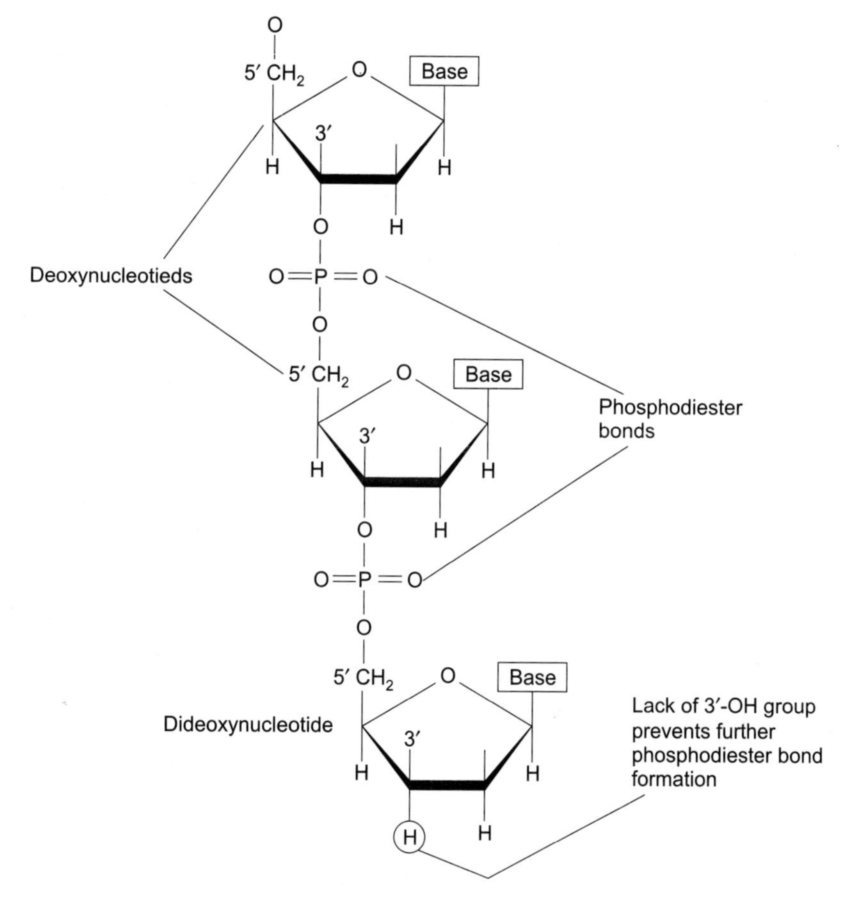

Fig. 41.10 Termination of DNA synthesis by dideoxynucleotides.

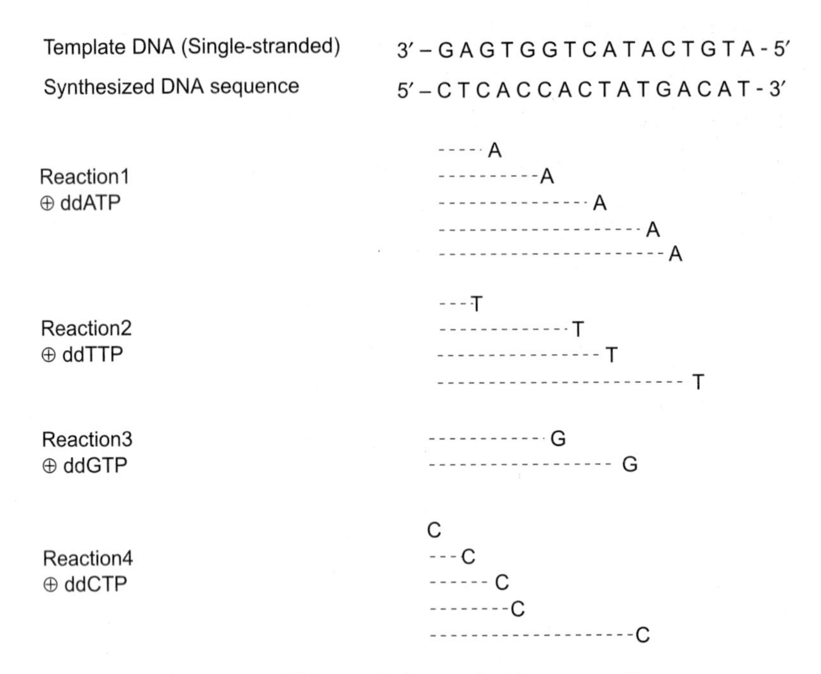

Fig. 41.11 Dideoxy chain termination sequencing.

When the sequencing reaction is complete the synthesized DNA molecules are separated according to size by electrophoresis on polyacrylamide gels with the four reactions run in adjacent lanes. By adding dNTP which contains radioactive phosphorus or sulfur to the sequencing reactions, the synthesized DNA becomes radioactive phosphorus allowing it to be detected by placing the polyacrylamide gel against a sheet of X-ray film. This is called autoradiography. When the X-ray film is developed a series of bands is visible in the four lanes which form a ladder. The sequence of the template DNA can be determined by identifying the smallest band followed by successively larger bands and assigning the terminal base in each from its lane on the gel (Fig. 41.12).

Automated DNA Sequencing

Although the basic principle of the sequencing reaction is unchanged, the technique has been developed to include semi-automated systems in which electrophoresis of DNA and detection and analysis of sequencing reactions is carried out by instruments controlled by computers. In this format, the four ddNTPs are labeled by the covalent attachment of a different colored dye. A single sequencing reaction is carried out and the products are a series of DNA molecules, each one nucleotide longer than the next and labeled with a different dye depending on the identity of the terminal base. The products are resolved by electrophoresis on polyacrylamide gels and are detected as they migrate past the end of the gel by a laser that excites the dye label to emit fluorescence which is then detected by the instrument. From the wavelength of the

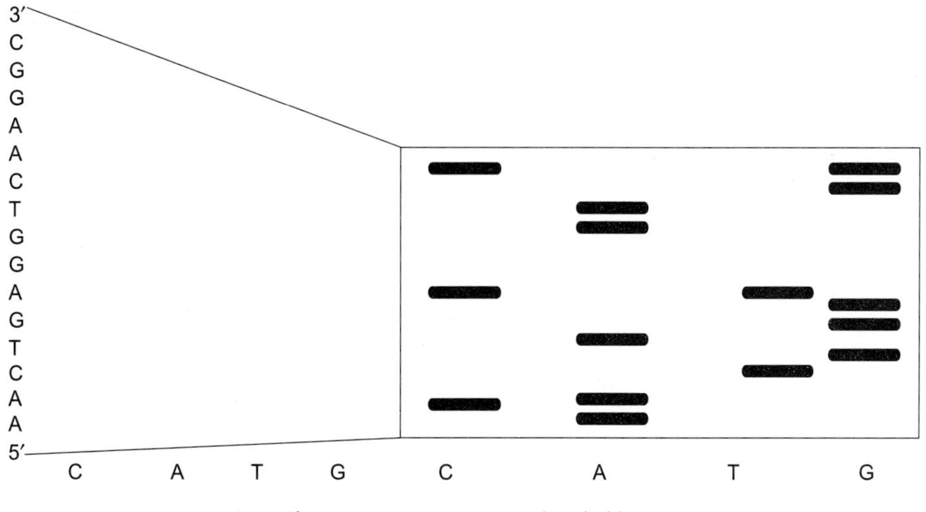

Fig. 41.12 DNA sequencing ladder.

emitted fluorescence the instrument identifies the dye and so the terminal base on each DNA species and build up a picture of the sequence of bases passing the detector. Automated sequencing systems represent a significant advance over conventional or manual sequencing approaches. Each sequencing reaction requires just one tube and one lane on the gel compared with four per sample required for manual sequencing. This feature couples with the automated detection of samples and analysis of data greatly increases the amount of sequence information that can be generated

Analysis of Sequence Data

DNA sequencing is one of the core techniques of molecular genetics and is widely used in research. The importance of the technique lies in its ability to determine the information content of DNA molecules. In recent years, the amount of DNA sequence information has expanded enormously to the point where it has become necessary to organize it in a series of databases. The main databases are called EMBL, which is based in Europ, and Genbank, based in the USA. These databases contain sequences of human genes and genes from other species generated by research projects worldwide. The information is freely available over the Internet and can be down loaded for analysis.

DNA sequencing generates a large amount of data. To aid the analysis of sequences, a series of software packages have been developed. These include the **University of Wisconsin GCG package** which allows important features in DNA sequences such as restriction enzyme sites, start codons, stop codons, open reading frames, intron-exon junctions, and promoter sequences to be identified. The programs also allow sequences to be compared with other sequences in the databases to look for similarities that suggest that genes are related.

The development of automated DNA sequencing systems has made it feasible to determine the entire DNA sequence of several important organisms. Already, the entire sequence of the genomes of *E. coli* (4.6×10 bp) and the yeast. S. cerevisiae (2.3×10 bp) has been determined. In addition, the human genome mapping project aims to identify and locate all human genes by determining the entire 3×10 bp sequence of the human genome.

Suggested Readings

Altenburg, E. 1973. *Genetics.* Oxford IBH. New Delhi. 489 pp.

Anonymous. 1970. *Manual on Mutation Breeding.* IAEA. Vienna. 237 pp.

Atlas, R.M. 1986. *Basic and Practical Microbiology.* Macmillan Pub1. Comp. N.Y. USA. 741 pp.

Avers C.J. *Genetics,* D.Van Nostrand Company. N.Y. USA. 659 pp.

Bailey, N.T.J. 1959. *Statistical Methods in Biology.* ELBS. Lond. 200 pp.

Bonner, D.M. 1963. *Heredity.* Prentice Hall of India. New Delhi. 112 pp.

Brock, T.K. and Madigan, M.T. *Biology of Microorganisms.* Prentice Hall. Englewood Cliffs. NJ. USA. 1994. 874 pp.

Brown, T.A. *Genetics. -A Molecular Approach.* Chapman And Hall. Lond. 1990. 387 pp.

Brown, T.A. 1990. *Gene Cloning.* Chapman and Hall. Lond. 286 pp.

Catcheside, D.G. 1977. *The Genetics of Recombination.* Edward Arnold. Lond. 172 pp.

Darlington, C.D. and Lacour, L.F. 1970. *The Handling of Chromosomes.* George Allen And Unwin LTD. Lond. 273 pp.

Drake, J.W. 1970. *The Molecular Basis of Mutation.* Holden Day. San Francisco. USA. 273 pp.

Freifelder, D. 1987. *Molecular Biology.* Jones And Bartlett Pub1. INC. Boston. USA. 883 pp.

Freifelder, D. 1987. *Microbial Genetics.* Narosa Publ. House. New Delhi. 601 pp.

Garber, E.D. 1974. *Cytogenetics.* Tata Mc Graw Hill Publ. Comp. Ltd. New Delhi. 258 pp.

Gardner, E.J., Simmons, M.J. and Snustad, D.P. 1996. *Principles of Genetics.* John Wiley and Sons. NY. USA. 649 pp.

Goodenough, U. 1990. *Genetics.* Holt, Rinehart, And Winston, NY, USA. 900 pp.

Hayes, W. 1965. *The Genetics of Bacteria and their Viruses.* ELBS. Lond. 925 pp.

Levine, R.P. 1962. *Genetics.* Holt, Rinehart, Winston INC. NY. USA.180 pp.

Levine, L. (ED). 1971. *Papers on Genetics. A book of Readings.* The C.V. Mosby Company. ST. Louis. USA. 497 pp.

Lewin. B. 1994. *Genes V.* Oxford UNIV. Press. Lond. 1292 pp.

Loewy, A.G. Siekevitz, P. 1974. *Cell Structure and Function.* Holt, Rinehart, and Winston Inc. NY. USA.506 pp.

Mandal, S. 1996. *Fundamentals of Human Genetics.* New Central Book Agency. Calcutta. 102 pp.

Poehlman, J.M. and Borthakur, D. 1972. *Breeding Asian Field Crops.* Oxford and IBH. Publ. CO. New Delhi. 375 pp.

Rothwell, V.M. 1976. *Understanding Genetics.* The Williams and Williams Company. Baltimore, USA. 486 pp.

Singh, B.D. 1983. *Plant Breeding.* Kalyani Publishers, Ludhiana. 506 pp.

Sinnott, E.W., Dunn, L.C. and Dobzhansky, T. 1962. *Principles of Genetics.* Mc Graw Hill Book Company. NY. USA. 460 pp.

SRB, A.M. and Owen R.D. 1967. *General Genetics.* W.H. Freeman and Comp. San Fransisco. USA. 561 pp.

SRB, A.M., Owen, R.D. and Edgar, R.S. (Eds.) 1970. *Facets of Genetics.* W.H. Freeman and Com. San Francisco. USA. 354 pp.

Stent, G.S. 1971. *Molecular Genetics. An Introductory Narrative.*W.H. Freeman And Company. Sanf Francisco. USA. 650 pp.

Strickberger, M.W. 1968. *Genetics* MacMillan Comp. NY. USA. 868 pp.

Swanson, C.P. 1965. *Cytology and Cytogenetics.* Macmillan and Co. Lond. 596 pp.

Swanson, C.P. Merz, T. and Young, W.J. 1967. *Cytogenetics.* Prentice Hall NJ. USA. 194 pp.

Tamarin, R.H. 1982. *Principles of Genetics.* Willard Grant Press. Boston. Mass. USA.732 pp.

Walker, J.M. and Gaastra, W. (EDS.). 1987. *Techniques in Molecular Biology.* Vol. 2. Croom. Helm. Lond. 332 pp.

Watson, J.D., Hopkins, N.H. Roberts, J.W., Steitz, J.A. and Weiner, A.M. 1988. *Molecular Biology of the Gene.* Benjamin/Cummings Publ. California USA. 1158 pp. 4th Ed.

Watson, J.D., Gilman, M. Witkowski, J. and Zoller, M. 1992. *Recombinant* DNA. W.H. Freeman and Company NY. USA. 626 pp.

Watson J.D., Baker T.A., Bell S.P., Gann. A., Levine M. and Losick, R. *Molecular Biology of the Gene.* 5th Edition. Pearson Education. Singapore. 2004. 750 pp.

Weaver, R.F. and Hedrick. P.W. 1989. *Genetics.*WMC Brown Publ. IOWA. USA. 575 pp.

White House, H.L.K. 1969. *Towards an Understanding of the Mechanism of Heredity.* ELBS. Lond. 447 pp.

References

Brink, R.A. 1958. Paramutation at the *R* locus in Maize. *Cold Spring Harbor Symp. Quant. Biol.* 23: 379–391.

Brink, R.A., Styles, E.D., Axtell, J.D. 1968. Paramutation: Directed genetic change. *Science,* 159: 161–170.

Catcheside, D.G. 1947. The P-locus Position Effect in Oenothera. *Jour. Genet.* 48: 31–42.

Demerec, M. 1941. Unstable genes in *Drosophila. Cold Spring Harbor Symp. Quant. Biol.* 9: 145–150.

Lewis, E.B. 1950. The phenomenon of Position Effect. *Adv. Genet.* 3: 73–116.

McClintock, B. 1951. Chromosome organization and genic expression. *Cold Spring Harbor Symp. Quant. Biol.* 16: 13–47.

Sambamurty, A.V.S.S. 1984. Instability at the CR-Locus in Rice (*Oryza sativa L.*). *Jour. of Cytol. and Genet.* 19: 67–79.

Sambamurty, A.V.S.S. 1994. Mutagenesis in rice. *Frontiers of Plant Science* (Ed). Dr. Irfan A. Khan. 471–477.

Suggested Readings

1) Loewy, A.G.: *Cell Structure and Function* Philadelphia: Somdey Cottage Publ. 1991. 980 pp.
2) Darnell, J., H. Lodish and D. Baltimore: *Molecular Cell Biology* Scientific American Books, NY. 1986. 1192 pp.
3) Alberts, B., B. Denis, J. Johnson, J. Lewis, M. Roff, A. Roberts and P. Walter: *Essential Cell Biology*. Garland Robertson Co., NY. 1997. 8000 pp.
4) Karp, G.: *Cell and Molecular Biology*. John Wiley & Sons, NY. 1996. 800 pp.
5) De Roberts: *Molecular Cell Biology* Saunders Co., NY. 1997. 1200 pp.
6) Watson, J.D. T.A. Baker, S.P. Bell, A. Gann, M. Levine & R. Losick: *Molecular Biology of the Gene* Pearson Education, Singapore. 2004. 732 pp.
7) Cooper, G.M.: *The Cell–A Molecular Approach*. ASM Press, Washington, USA. 1997. 650 pp.
8) Watson J.D., M. Gilman, I.J. Witkowsk, M. Zolle: *Recombinant DNA*. Second Edition. Freeman W.H. and Co., NY. 1992. 626 pp.

References

Abbas, A.K., A.H. Lichtman and J.S. Pober (1991). *Cellular and A Molecular Immunology*. Saunders. Philadelphia.

Abbas, A.K., K.M. Murphy and A. Sher (1996). Functional diversity of helper T lymphocytes. *Nature* **383:** 787-793.

Albers, K. and E. Fuchs (1992). The molecular biology of intermediate filament proteins. *Int. Rev. Cytol.* **134:** 243-279

Altman. S. (Ed.) (1978). *Transfer RNA*. MIT Press, Cambridge, USA.

Ameisen. J.C. (1996). The origin of programmed cell death. *Science* **272:** 1278-1280.

Amos, L.A. and W.B. Amos (1991). (1991). *Molecules of the Cytoskeleton*. Guilford Press. New York.

Anderson. L.K., H.H. Offenberg, W.M.H.C. Verkuijlen and C. Heyting (1997). RecA-like proteins are components of early meiotic nodules in lily. *Proc. Natl. Acad. Sci.* USA **94:** 6868-6873.

Anderson, S. *et al.* (1981). Sequence and organization of the BI human mitochondrial genome. *Nature* **290:** 457-465.

Anderson, W.F. (1984). Prospects of human gene therapy. *Science* **226:** 401-409.

Andre. I., A. Gonzalez. B. Wang, J. Katz, C. Benoist and D. Mathis (1996). Checkpoints in the progression of autoimmune disease: lessons from diabetes models. *Proc. Natl. Acad. Sci.* USA **93:** 2260-2263.

Andre, C., A. Levy and V. Walbot (1992). Small repeated sequences and the structure of plant mitochondrial genomes. *Trends in Genet.* **8:** 128-131.

Andrews, B. and V. Measday (1998) The cyclin family of budding yeast: abundant use of a good idea. *Trends in Genet.* **14:** 66-72.

Atkins, J.F and R.F. Gesteland (1996). A case for translation. *Nature* **379:** 769-771.

Attardi, G. and G. Schatz (1988). Biogenesis of mitochondria *Annu. Rev. Cell Biol.* **4:** 289-333.

Auerbach, C. (1962). *Mutation. Part 1. Methods*. Oliver and Boyd, Edinburgh.

Auerbach, C. (1976). *Mutation Research: Problems. Results and Perspectives*. Chapman and Hall, London.

Auerbach, C. and J.M. Robson (1946). Chemical production of mutations. *Nature* **157:** 302.

Auffray, C. and J.L. Srrominger (1987). Molecular genetics of the human histocompatibility complex. *Adv. Human Genet.* **15:** 197-247.

Avery, O. T., C.M. MacLeod and M. McCarthy (1944). Studies on the chemical nature of the substance inducing transformation by a deoxyribonucleic acid fraction isolated from *Pneumococcus* type III. *J. Exp. Med.* **79:** 137-158.

Baker, T.A. and S.H. Wickner (1992). Genetics and enzymology of DNA replication in *Escherichia coli. Annu. Rev. Genet.* **26:** 447-477.

Baltimore, D. (1970). Viral RNA-dependent DNA polymerase. *Nature* **226:** 1209-1211.

Barinaga, M. (1995). A new twist to the cell cycle. *Science* **269:** 631-632.

Barnes, D.E., T. Lindahl and B. Sedgwick (1993). DNA repair. *Curr. Opin. Cell Biol.* **5:** 424-433.

Barret, T., D. Maryanka, P.H. Hamlyn and H.J. Gould (1974). Non-histone proteins control gene expression in reconstituted chromatin. *Proc. Natl. Acad. Sci.* USA **71:** 5057-5061.

Bartel, D.P. and J.W. Szostak (1993). Isolation of new ribozymes from a large pool of random sequences. *Science* **261:** 1411-1418.

Barton, N.R. and L.B.S. Goldstein (1996). Going mobile: microtubule motors and chromosome segregation. *Proc. Natl. Acad. Sci. USA* **93:** 1735-1742.

Beadle, G.W. (1945). Genetics and metabolism in *Neurospora. Physiol. Rev.* **25:** 643-663.

Beadle, G.W. and B. Ephrussi (1937). Development of eye colors in *Dorsophila:* diffusible substances and their interrelations *Genetics* **22:** 76-86.

Beadle, G.W. and E.L. Tatum (1941). Genetic control of biochemical reactions in *Neurospora. Proc. Natl. Acad. Sci. USA* **27:** 499-506.

Beadle, G.W. and E.L. Tatum (1945). *Neurospora* II. Methods of producing and detecting mutations concerned with nutritional requirements. *Amer. J. Bot.* **32:** 724-740.

Beale, G.H. (1965). Genes and cytoplasmic particles in *Paramecium.* In P. Emmelot and O. Muhlbock (Eds.), "*Cellular Control Mechanisms and Cancer*". Elsevier, Amsterdam, pp. 8-18.

Beale, G. and J. Knowles (1978). *Extranuclear Genetics.* Edward Arnold, London.

Beardsley, T. (1991). Smart genes. *Sci. Amer.* Aug. 1991. pp. 86-95.

Becker, W.M. and D.W. Deamer (1991). *The World of the Cell* 2nd ed., Benjamin-Cummings, Redwood City, CA.

Beckwith, J.R. and D. Zipser, Eds. (1970). *The Lactose Operon.* Cold Spring Harbor Laboratory, New York.

Belling, J. (1933). Crossing over and gene rearrangement in flowering plants. *Genetics* **18:** 388-413.

Benda, C. (1902). *Die Mitochondria. Ergebn. Anat.* **12:** 743.

Benzer, S. (1962). The fine structure of the gene. *Sci. Amer. Jan.* 1962. pp. 70-84.

Benzer, S. (1973): Genetic dissection of behaviour. *Sci. Amer. Dec.* 1973. pp. 24-37.

Berg, P. (1991). Reverse genetics, its origin and prospects. *Biotechnology* 9: 342-344.

Bevilacqua, M.P. (1993). Endothelial-leukocyte adhesion molecules. *Annu. Rev. Immunol.* **11:** 767-804.

Bevilacqua, M.P. and R.M. Nelson (1993). Selectins. *J. Clin. Invest.* **91:** 379-387.

Beyer, E.C. (1993). Gap junctions. *Int. Rev. Cytol.* **137:** 1-38.

Bhattacharyya, M.K. *et al.* (1990). The wrinkled seed character of pea described by Mendel is caused by a transposon-like insertion in a gene encoding starch binding enzyme. *Cell* **60:** 115-122.

Birch, R.G. (1997). Plant transformation: problems and strategies for practical application. Annu. Rev. Plant physiol. *Plant Mol. Biol.* **48:** 297-326.

Birky, C.W. Jr. (1978). Transmission genetics of mitochondria and chloroplasts. *Annu. Rev. Genet.* **12:** 471-512.

Bittar, E.E. (Ed.) (1973). *Cell Biology in Medicine.* John Wiley, New York.

Bjorkman, O. and J. Berry (1973). High-efficiency photosynthesis. *Sci. Amer.* April 1973. pp. 80-93.

Bjorkman, P.J. and P. Parhem (1990). Structure. function and diversity of class I major histocompatibility complex molecules. *Annu. Rev. Biochem.* **59:** 253-288.

Blackbum, E.H. (1990). Telomeres: structure and synthesis. *J. Biol. Chem.* **265:** 5919-5921.

Blackbum, E.H. (1991). Telomeres. *Trends in Biochtm. Sci.* **16:** 378-381.

Blakeslee, A.F. (1934). New jimson weeds from old chromosomes. *J. Hered.* **25:** 80-108.

Boeke, J.D. and K.B. Chapman (1991). Retrotransposition mechanisms. *Curr. Opin. Cell Biol.* **3:** 502-507.

Bogorad, L. (1981). Chloroplasts. *J. Cell Biol.* **91:** 256s-270s.

Bolwell, G.P. (1993). Dynamic aspects of the plant extracellular matrix. *Int. Rev. Cytol.* **146:** 261-324.

Botchan, M. (1996). Coordinating DNA replication with cell division: current status of the licensing concept. *Proc. Natl. Acad. Sci. USA* **93:** 9997-10000.

Boulter, D., R.J. Ellis and A. Yarwood (1972). Biochemistry of protein synthesis in plants. *Biol. Rev.* **47:** 113-175.

Bourrett. R.B., K.A. Borkovich and M.I. Simon (1991). Signal transduction pathways involving protein phosphorylation in prokaryotes. *Annu. Rev. Biochem.* **60:** 401-441.

Branden. C. and J. Tooze (1991). Introduction to Protein Structure. Garland, New York.

Breathnach, R. and P. Chambon (1981). Organization and expression of eukaryotic split genes coding for proteins. *Annu. Rev. Biochem.* **50:** 349-383.

Bredt. D.S. and S.H. Snyder (1992). Nitric oxide, a novel neuronal messenger. *Neuron* **8:** 3-11.

Brenner, S. (1955). Tryptophan biosynthesis in *Salmonella typhimurium*. *Proc. Natl. A cad. Sci. USA* **41:** 862-863.

Brenner. S., F. Jacob and M. Meselson (1961). An unstable intermediate carrying information from genes to ribosomea for protein synthesis. *Nature* **190:** 576-581.

Bridges, C.B. (1916). Non-disjunction as proof of the chromosome theory of heredity. *Genetics* **1:** 1-52, 107-163.

Bridges, C.B. (1925). Sex in relation to chromosomes and genes. *Amer. Naturalist* **59:** 127-137.

Bridges, C.B. (1935). Salivary chromosome maps. *J. Hered.* **26:** 60-64.

Bridges. C.B. (1936). The bar 'gene', a duplication. *Science* **83:** 210-211.

Britten, R.J. and D.E. Kohne (1968). Repeated sequences in DNA. *Science* **161:** 529-540.

Broda, P.W. (1979). Plasmids. Freeman, Oxford.

Browder, L.W., C.A. Erickson and W.R. Jeffery (1991). *Developmental Biology*, 3rd ed. Saunders, Philadelphia.

Brown, D.D. (1973). The isolation of genes. *Sci. Amer. Aug.* 1973. pp. 20-29.

Brown. D.D. (1981). Gene expression in eukaryotes. *Science* **211:** 667-674.

Brown. G.W., P.V. Jallepalli, B.J. Huneycutt and T.J. Kelly (1997). Interaction of the S phase regulator Cdc18 with cyclin-dependent kinase in fission yeast. *Proc. Natl. Acad. Sci. USA* **94:** 6142-6147.

Brown, J.H. *et al.* (1993). Three-dimensional structure of the human class II histocompatibility antigen HLA-DRI. *Nature* **364:** 33-39.

Bryant. D.A. (1992). Puzzles of chloroplast ancestry. *Curr. Biol.* **2:** 240-242.

Bukhari, A.I., J.A. Shapiro and S.L. Adhya (Eds.) (1977). *DNA Inserting Elements. Plasmids and Episomes.* Cold Spring Harbor laboratory, New York.

Burke, D.T. (1991) The role of yeast artificial chromosomes in generating genome maps. *Curr. Opin. Genet. Dev.* **1:** 69-74.

Burke, T. *et al.* (1991). *DNA Fingerprinting: Approaches and Applications.* Birkhauser Verlag. Berlin.

Burley, S. K. (1996). Biochemistry and structural biology of transcription factor IID (TFIID). *Annu. Rev. Biochem.* **65:** 769-799.

Burnham, C.R. (1962). *Discussions in cytogenetics.* Burgesa, Minneapolis.

Buvat. R. (1969). *Plant Cells.* World University Library, London.

Cairns, J. (1963). The chromosome of Escherichia coli. *Cold Sp. Harb. Symp. Quant. Biol.* **28:** 43-45.

Cairns, J. (1966). The bacterial chromosome. *Sci. Amer. Jan.* 1966. pp 36-44.

Callan. H.G. (1982). Lampbrush chromosomes. *Proc. R. Soc. Lond. (Biol.)* **214:** 417-448.

Calvin. M. (1962). The path of carbon in photosynthesis. *Science* **135:** 879-889.

Campbel, A. (1962). Episomes. *Adv. Genet.* **11:** 101-145.

Capaldi, R.A. (1974). A dynamic model of cell membranes. *Sci. Amer. March* 1974. pp. 26-33.

Capaldi, R.A. (1990). Structure and function of cytochrome c oxidase. *Annu. Rev. Biochem.* **59:** 569-596.

Carpenter. A.T.C. (1979). Synaptonemal complex and recombination nodules in wild-type. *Drosophila melanogaster females genetics* **92:** 511-541.

Carvarelli. J. and D. Moras (1993). Recognition of tRNAa by aminoacyl-RNA synthetases. *FASEB J.* **7:** 79-86.

Catterall, W.A. (1995). Structure and function of voltage-gated ion channels. *Annu. Rev. Biochem.* **64:** 493-531.

Caveney, S. (1985). The role of gap junction in development. *Annu. Rev. Physiol.* **47:** 319-335.

Cech, T.R. (1986). RNA as an enzyme. *Sci. Amer. May* 1986. pp. 64-75.

Cech. T.R. (1990). Self splicing of group 1 introns. *Annu. Rev. Biochem.* **59:** 543-568.

Chambon, P. (1981). Split genes. *Sci. Amer. May* 1981. pp. 60-71.

Charlesworth, B. (1991). The evolution of sex chromosomes. *Science* **251:** 1030-1033.

Charbonneau, H. and N.K. Tonks (1992). 1001 protein phosphatases? *Annu. Rev. Cell Biol.* **8:** 463-493.

Chasan, R. and V. Walbot, (1993). Mechanisms of plant reproduction: questions and approaches. *Plant Cell* **5:** 1139-1146.

Chess, A. (1998). Expansion of the allelic exclusion principle? *Science* **279:** 2067-2069.

Clackson. T., H.R. Hoogenboom, A.D. Griffiths and G. Winter (1991). Making antibody fragments using phage display libraries. *Nature* **352:** 624-628.

Clayton D.A. (1991). Replication and transcription of vertebrate mitochondrial DNA. *Annu. Rev. Cell Biol.* **7:** 453-478.

Clowes. R.C. (1973). The molecule of infectious drug resistance. *Sci. Amer. April* 1973. pp. 18-27.

Coen, E.S. (1991). The role of homeotic genes in flower development and evolution. *Annu. Rev. Pl. Physiol. Pl. Mol. Biol.* **42:** 241-279.

Coen, E.S. and R. Carpenter (1993). The metamorphosis of flowers. *Plant Cell* **5:** 1175-1181.

Cohen, P. (1989). The sturcture and regulation of protein phosphatases. *Annu. Rev. Biochem.* **58:** 453-508.

Cohen, S.N. and J.A. Shapiro (1980). Transposable genetic elements. *Sci. Amer. Feb.* 1980. pp. 36-45.

Collins. K., T. Jacks. and N.P. Pavletich (1997). The cell cycle and cancer. *Proc. Natl. Acad. Sci. USA* **94:** 2776-2778.

Conawy, R.C. and J.W. Conaway (1993). General initiation factors for RNA polymerase II. *Annu. Rev. Biochem* **62:** 161-190.

Cotter. T. G., and M. Al-Rubeai (1995). Cell death (apoptosis) in cell culture systems. *Trends in Biotech.* **13:** 150-155.

Coux. O., K. Tanaka, and A. L. Goldberg (1996). Structure and functions of the 20S and 26S proteasomes. *Annu. Rev. Biochem.* **65:** 801-847.

Creighton. H.S. and B. McClintock (1931). A correlation of cytological and genetical crossing over in Zea mays. *Proc. Natl. Acad. Sci. USA* **17:** 492-497.

Creutz., C.E. (1992). The annexins and exocytosis. *Science* **258:** 924-930.

Crick., F.H.C. (1966). The genetic code III. *Sci. Amer.* Oct. 1966. pp. 55-62.

Crick, F.H.C. (1966). Codon-anticodon pairing: The wobble hypothesis. *J. Mol. Biol.* **19:** 548-555.

Crick. F.H.C. (1979). Split genes and RNA splicing. *Science* 204 264-271.

Darnell. J.E.. H.F. Lodish and D. Baltimore (1995). *Molecular Cell Biology*, 3rd ed., Scientific American Books, New York.

Das, A. (1993). Control of transcription termination by RNA binding proteins. *Annu. Rev. Biochem.* **62:** 893-930.

Davenport, C.B. (1913). Heredity of skin color in negro-white crosses. *Carneg. Inst. Wash. Publ., No* 554, Washington, D.C.

Davidson, J.N. (1965). *Biochemistry of Nucleic Acids.* Cox and Wyman, London.

Davis, L. I. (1995). The nuclear pore complex. *Annu. Rev. Biochem.* **64:** 865-896.

Davis, M.M.K. Calame, P.W. Early, D.L. Viant, R. Joho, I.L Weissman and L. Hood (1980). An immunoglobulin heavy chain gene is formed by atleast two recombinational events. *Nature* **283:** 733-739.

de Duve, C. and H. Beaufay (1981). A short history of tissue fractionation. *J. Cell. Biol.* **91:** 293s-299s.

DeFranco, A.L. (1993). Structure and function of the B cell antigen receptor. *Annu. Rev. Cell Biol.* **9:** 377-410.

Deisenhofer, J., O. Epp, K. Miki, R. Huber and H. Michel (1985). Structure of the protein subunits in the photosynthetic reaction centre of Rhodopseudomonas virdis at 3A resolution. *Nature* **318:** 618-624.

Deisenhofer, J. and H. Michel (1991). High-resolution structures of photosynthetic reaction centres. *Annu. Rev. Biophys. Chem.* **20:** 247-266.

Deisenhofer, J. and H. Michel (1991). Structures of bacterial phostosynthetic reaction centres. *Annu. Rev. Cell Biol.* **7:** 1-23.

Delbruck, M. and W.T. Bailey, Jr. (1946). Induced mutations in bacterial viruses. *Cold Sp. Harb. Symp. Quant. Biol.* **11:** 33-37.

Dennis, D. and W. Ray (1985). The genetic linkage map of the human X-chromosome. *Science* **230:** 753-759.

DePamphilis, M.L. (1993). Eukaryotic DNA replication: anatomy of an origin. *Annu. Rev. Biochem.* **62:** 29-63.

DePamphilis, M.L. (1993). Origins of DNA replication that function in eukaryotic cells. *Curr. Opin. Cell Biol.* **5:** 434-441.

Dessev, G.N. (1992). Nuclear envelope structure. *Curr. Opin. Cell Biol.* **4:** 430-435.

DeRobertis, E.D.P. and E.M.F. DeRobertis, Jr. (1987). *Cell Biology* 8th Edn. B. I. Waverly Pvt. Ltd., New Delhi.

DeVries, H. (1901 and 1903). *Die Mutationtheorie.* Veit, Leipzig (translated into English and reprinted 1909. Open Court, Chicago).

Dickerson, R.E. (1980). Cytochrome c and the evolution of energy metabolism. *Sci. Amer. March* 1980. pp. 136-153

Dickson, R.C., J. Abelson, W.M. Barnes and W.S. Reznikoff (1975). Genetic regulation: the lac control region. *Science* **187:** 27-35.

Diffley, J.E.X. and B. Stillmon (1990). The initiation of chromosomal DNA replication in eukaryotes. *Trends in Genet* **6:** 427-432.

Dillin, A. and J. Rine (1998). Roles for ORC in M phase and S phase. *Science* **279:** 1733-1737.

Dingwall, C., and R. Laskey (1992). The nuclear membrane *Science* **258:** 942-947.

Donovan, S., J. Harwood, L.S Drury and J.F.X. Diffley (1997). Cdc6p-dependent loading of Mcm proteins onto pre-replicative chromatin in budding yeast. *Proc. Natl. Acad. Sci. USA* **94:** 5611-5616.

Doolittle, W.F. and C. Sapienza (1980). Selfish genes, the phenotype paradigm and genome evolution. *Nature* **284:** 601-603.

Dover, G.S. and R.B. Flavell (Eds.) (1982). *Genome Evolution.* Academic Press, London.

Drake, J.W. (1970). *The Molecular Basis of Mutation.* Holden-Day Inc., San Francisco.

Drew, H., T. Takano, S. Tanaka, K. Itakura and R.E. Dickerson (1980). High-salt d(CpGpCpGp), a left handed Z DNA double helix. *Nature* **286:** 567-573.

Dulbecco, R. (1967). The induction of cancer by viruses. *Sci. Amer.* April 1967. pp. 28-37.

DuPraw, E.J. (1969). *Cell and Molecular Biology.* Academic Press, New York.

Duvick, D.N. (1965). Cytoplasmic pollen sterility in com. *Adv. Genet.* **13:** 1-56.

Dynlacht, B. D. (1997). Regulation of transcription by proteins that control the cell cycle. *Nature* **389:** 149-152.

Earnshaw, W.C. and J.E. Tomkiel (1992). Centromere and kinetochore structure. *Curr. Opin Cell Biol.* **4:** 86-93.

East, E.M. (1916). A Mendelian interpretation of variation that is apparently continuous. *Amer. Naturalist* **44:** 65.

East, E.N. (1916). Studies on size inheritance in *Nicotiana. Genetics* **1:** 164-176.

Echols, H. and M.F. Goodman (1991). Fidelity mechanisms in DNA replication. *Annu. Rev. Biochem.* **60:** 477-511.

Edelman, G.M. and K.L. Crossin (1991). Cell adhesion molecules: implications for a molecular histology. *Annu. Rev. Biochem.* **60:** 155-190.

Edge, M.D. *et al.* (1981). Total synthesis of a human leukocyte interferon gene. *Nature* **292:** 756-761.

Edger, B.A. and C.F. Lehner (1996). Developmental control of cell cycle regulators: a fly's perspective. *Science* **274:** 1646-1651.

Edidin, M. (1987). Rotational and lateral diffusion of membrane proteins and lipids: phenomena and function. *Curr. Top. Membr. Transp.* **29:** 91-127.

Edidin, M. (1990). Molecular associations and membrane domains. *Curr. Top. Membr. Transp.* **36:** 81-96.

Edlin, G. (1990). *Human Genetics: A Modern Synthesis.* Jones & Banlett Publ., Boston.

Elledge, S.J. (1996). Cell cycle checkpoints: preventing an identity crisis. *Science* **274:** 1664-1671.

Elledge. S.J. (1998). Mitotic arrest: Mad2 prevents sleepy from waking up the APC. *Science* **279:** 999-1002.

Ellis, R.J. (1974). The biogenesis of chloroplasts: protein synthesis by isolated chloroplasts. *Proc. 54th Meeting of Biochemical Society Transactions.* Vol 2, pp. 179-182.

Ellis, R.J. (1975). Chloroplast protein synthesis. *Nature* **254:** 13.

Ellis, R.J. (1991). Molecular chaperones. *Annu. Rev. Biochem.* **60:** 321-347.

Ellis, R.E., J.V. Yuan and H.R. Horvitz (1991). Mechanisms and functions of cell death. *Annu. Rev. Cell Biol.* **7:** 663-698.

Fantl, WJ., D.E. Johnson, and L.T. Williams (1993). Signalling by receptor tryosine kinases. *Annu. Rev. Biochem.* **62:** 453-481.

Fawcett. D.W. (1966). *The Cell: Its Organelles and Inclusions.* Saunders, Philadelphia.

Fincham, J.R.S. (1966). *Genetic Complementation,* Benjamin, New York.

Fincham, J.R.S. and G.R.K. Sastry (1974). Controlling elements in maize. *Annu. Rev. Genet.* **8:** 15-50.

Fishcer, L.M. (1982). DNA unwinding in transcription and recombination. *Nature* **299:** 105-106.

Fischer, A. and B. Malissen (1998). Natural and engineered disorders of lymphocyte development. *Science* **280:** 237-243.

Fisher, E.H. and E.G. Krebs (1955). Conversion of phosphorylase b to phosphorylase a in muscle extracts. *J. Biochem.* **216:** 121-132.

Flavell, R.A. (1982). The mystery of the mouse a globin pseudogene. *Nature* **295:** 370.

Forbes, D.J. (1992). Structure and function of the nuclear pore complex. *Annu. Rev. Cell Biol.* **8:** 495-527.

Forsburg, S.L. and P. Nurse (1991). Cell cycle regulation in the yeasts. *Saccharomyces cerevisiae and Schizosacharomyces pombe. Ann. Rev. Cell Biol.* **7:** 227-256.

Fraenkel-Conrat. H. and B. Singer (1975). Virus reconstitution: combination of protein and nucleic acid from different strains. *Biochem. et Biophys. Acta.* **24:** 540-548.

Fraenkel-Conrat, H. and R.C. Williams (1955). Reconstitution of active tobacco mosaic virus from its inactive protein and nucleic acid components. *Proc. Natl. Acad. Sci. USA* **41:** 690-698.

Frey-Wyssling, A. and K. Muhlethaler (1965). *Ultrastructural plant Cytology.* Elsevier, Amsterdam.

Friedberg, E. C. (1996). Relationships between DNA repair and transcription. *Annu. Rev. Biochem:* **65:** 15-42.

Frye, L.D. and M. Edidin (1970). The rapid intermixing of cell surface antigens after formation of mouse-human heterokaryons. *J. Cell Sci.* **7:** 319-335.

Gall, J.G. (1981). Chromosome structure and the C-value paradox. *J. Cell Bioi.* **91:** 3s-14s.

Garbers, D.L. (1989). Molecular basis of fertilization. *Annu. Rev. Biochem.* **58:** 729-742.

Garcia-Bellido, A., P.A. Lawrence and G. Morata (1979). Compartments in animal development. *Sci. Amer. Jan.* 1979. pp. 102-111.

Garrod, A.B. (1909). *Inborn Errors of Metabolism.* Oxford Univ. Press, Oxford.

Garrod, D.R. (1993). Desmosomes and hemidesmosomes. *Curr. Opin. Cell Biol.* **5:** 30-40.

Gehring, W.J. (1986). On the homeobox and its significance. *Bioessays* **5:** 3-4.

Geiger, B. and O. Ayalon (1992). Cadherins. *Annu. Rev. Cell Biol.* **8:** 307-332.

Gelfand, V.I. and A.D. Bershadsky (1991). Microtubule dynamics: mechanism, regulation, and function. *Annu. Rev. Cell Biol.* **7:** 93-116.

Germain, R.N. and D.H. Margulies (1993). The biochemistry and cell biology of antigen processing and presentation. *Annu. Rev. Immunol.* **11:** 403-450.

Gesteland, R.F. and J.F. Atkins, Eds. (1993). *The RNA World.* Cold Spring Harbor Laboratory Press. Cold Spring Harbor, New York.

Gesteland, R.F. and J.F. Atkins (1996). Recoding: dynamic reprogramming of translation. *Annu. Rev. Biochem.* **65:** 741-768.

Gianani, R. and N. Sarvetnick (1996). Viruses. cytokines, antigens, and autoimmunity. *Proc. Natl. Acad. Sci. USA* **93:** 2257-2259.

Gilbert, S.B. (1991). *Developmental Biology*, 3rd ed. Sinauer, Sunderland MA.

Gilbert, W. (1985). Genes-in-pieces revisited. *Science* **228:** 823-824.

Gilbert, W. and B. Muller-Hill (1966). Isolation of the lac repressor. *Proc. Natl. Acad. Sci. USA* **56:** 1891-1898.

Gilham. N.W. (1974). Genetic analysis of the chloroplast and mitochondrial genomes. *Annu. Rev. Genet.* **8:** 347-392.

Gilham, N.W. (1978). *Organellar Heredity.* Revan Press. New York.

Gilmour, R.S. and J. Paul (1973). Tissue specific transcription of the globin gene in isolated chromatin. *Proc. Natl. Acad. Sci. USA* **70:** 3440-3442.

Glick. B. and G. Schatz (1991). Import of proteins into mitochondria. *Annu. Rev. Genet.* **25:** 21-44.

Glover. D.M.. C. Gonzalez and J.W. Raff (1993). The centrosome. *Sci. Amer.* June 1993. pp. 62-68.

Golub. B.S. and D.R. Green (1991). *Immunology: A Synthesis.* 2nd ed. Sinauer, Sunderland, MA.

Goodman. M.F., S. Creighton, L.B. Bloom and J. Petruska (1993). Biochemical basis of DNA replication fidelity. *Crit. Rev. Biochem. Mol. Biol.* **28:** 83-126.

Goodnow, C. C. (1996). Balancing immunity and tolerance: deleting and tuning lymphocyte repenoires. *Proc. Natl. Acad. Sci. USA* **93:** 2264-2271.

Govindjee and Govindjee. R. (1974). The absorption of light in photosynthesis. *Sci. Amer.* **231:** 68-82.

Gray, D. (1993). Immunological memory. *Annu. Rev. Immunol.* **11:** 49-77.

Gray, M.W. (1989). Origin and evolution of mitochondrial DNA. *Annu. Rev. Cell Biol.* **5:** 25-50.

Gray. M.W. (1989). The evolutionary origin of organelles. *Trends in Genet.* **5:** 294-299.

Gray, M.W., P.J. Hanic-Joyce and P.S Covello (1992). Transcription. processing and editing in plant mitochondria. *Annu. Rev. Pl. Physiol. Pl. Mol. Biol.* **43:** 145-175.

Green, D.E. (1964). The mitochondrion. *Sci. Amer.* Jan. 1964, pp. 63-74.

Green, J. T. (1997). Programmed cell death in plant-pathogen interactions. *Annu. Rev. Pl. Physiol. Pl. Mol Biol.* **48:** 525-545.

Green, M.M. and K.C. Green (1949). Crossing over between alleles at the *lozenge* locus in *Drosophila melanogaster. Proc. Natl. Acad. Sci. USA* **35:** 586-591.

Greider, C. W. (1996). Telomere length regulation. *Annu. Rev. Biochem.* **65:** 337-365.

Grierson, D. (1991). *Plant Genetic Engineering.* (*Plant Biotechnoloy.* Vol. 1). Blackie. London.

Griffin, H.G. and AM. Griffin (1993). DNA sequencing: recent innovations and future trends. *Appl. Biochem. Biotech.* **38:** 147-159.

Griffith. F. (1928). The significance of pneumococcal types. *J. Hygiene* **27:** 113-159.

Grivell. L.A. (1983). Mitochondrial DNA. *Sci. Amer.*, March 1983. pp. 60-73.

Groner, Y.G., Monroy. M. Jacqet, and J. Hurwitz (1975), Chromatin as a template for RNA synthesis *in vitro. Proc Natl. Acad. Sci. USA* **71:** 194-199.

Gruenberg, J. and M.J. Clague (1992). Regulation of intracellular membrane transport. *Curr. Opin. Cell Biol.* **4:** 593-599.

Gruissem, W., A. Barken, S. Deng and D. Stem (1988). Transcriptional and post-transcriptional control of plastid mRNA in higher plants. *Trends in Genet.* **4:** 258-262.

Grunstein. M. (1992). Histones as regulators of genes. *Sci. Amer.* April 1992, pp. 68-74B.

Gualerzi, C.O. and C.L. Pon (1990). Initiation of rnRNA translation in prokaryotes. *Biochemistry* **29:** 5881-5889.

Guarente, L. and O. Bermingham-McDonogh (1992). Conservation and evolution of transcriptional mechanisms in eukaryotes. *Trends in Genet.* **8:** 27-31.

Guha. S. and S.C. Maheshwari (1967). Development of embryoids from pollen grains of *Datura in vitro. Phytomorphology* **17:** 454-461.

Gupta, P.K. (1989). *Cytology. Genetics and Evolution.* 5th ed., Rastogi Publications, Meerut.

Gupta, P.K. (1994). *Genetics.* 3rd ed., Rastogi Publications, Meerut.

Gupta, P.K. and P.M. Priyadarshan (1982). Triticale: present status and future prospects. *Adv. Genet.* **21:** 255-345.

Gupta, P.K. and A.K. Srivastava (1970). Natural triploidy in *Cynodon dactylon. Caryologia* **23:** 29-35.

Guthrie, C. and B. Patterson (1988). Spliceosomal snRNAs. *Annu. Rev. Genet.* **22:** 387-419.

Gutman, A. and B. Wasylyk (1991). Nuclear tergets for transcription regulation by oncogenes. *Trends in Genet* **7:** 49-54.

Hadjiolov, A.A. (1985). *The Nucleolus and Ribosome Biognesis.* Springer-Verlag, New York.

Hahn, S. (1993). Transcription: efficiency in activation. *Nature* **363:** 672-673.

Hall. A. (1998). Rho GTPases and the actin cytoskelcton. *Science* **279:** 509-524.

Hall. B.D. (1979). Mitochondria spring surprises. *Nature* **282:** 129-130.

Hanna-Rose. W. and U. Hansen (1996). Active repression mechanisms of eukaryotic transcription represson. *Trends in Genet.* **12:** 229-234.

Hanson, P.I. and H. Schulman (1992). Neuronal Ca^{2+}/calmodulin-dependent protein kinases. *Annu. Rev. Biochem.* **61:** 559-601.

Hardie, D.G. (1990). *Biochemical Messengers: Hormones, Neurotransmitters and Growth Factors.* Chapman and Hall, London.

Harper, K., and A. Ziegler (1997). Applied antibody technology. *Trends in Biotech.* **15:** 41-42.

Harriman, W.H., H. Volk, N. Defranoux and M. Wabl (1993) Immunoglobulin class switch recombinations. *Annu. Rev Immunol.* **11:** 361-384.

Harrison, S.C. and Aggarwal A.K (1990). DNA recognition by proteins with the helix-turn-helix motif. *Annu. Rev Biochem.* **59:** 933-969.

Hartwell, L.H. (1991). Twenty-five years of cell cycle genetics. *Genetics* **129:** 975-980.

Hartwell, L.H. and T.A. Weinert (1989). Checkpoints: controls that ensure the order of cell cycle events. *Science* **246:** 629-634.

Haselkom, R. (1992). Developmentally regulated gene rearrangements in prokaryotes. *Annu. Rev. Genet.* **26:** 113-130

Hayes, W. (1964). *The Genetics of Bacteria and their viruses.* John Wiley, New York.

Hearne, C.M., S. Ghosh and J.A. Todd (1992). Microsatellites for linkage analysis of genetic traits. *Trends in Genet* **8:** 289-293.

Heintz. N.H., L. Dailey, P. Held and N. Heintz (1992). Eukaryotic replication origins as promoters of bidirectional DNA synthesis. *Trends in Genet.* **8:** 376-380.

Helenius. A., I. Mellman, D. Wall and A. Hubbard (1983) Endosomes. *Trends in Biochem. Sci.* **8:** 245-250.

Henderson, P.J.F. (1993). The 12-transmembrane α-helix transporters. *Curr. Opin. Cell Biol.* **5:** 708-721.

Henderson, R., J.M. Baldwin, T.A. Ceska, F. Zemlin, E. Beckmann and K.H. Downing (1990). Model for the structure of bacteriorhodopsin based on high-resolution electron cryo-microscopy. *J. Mol. Biol.* **213:** 899-929.

Hendrick, J.P. and F.U. Hartl (1993). Molecular chaperone functions of heat-shock proteins. *Annu. Rev. Biochem.* **62:** 349-384.

Henikoff, S. (1990). Position-effect variegation after 60 years. *Trends in Genet.* **6:** 422-426.

Herschbach, B.M. and A.D. Johnson (1993). Transcriptional repression in eukaryotes. *Annu. Rev. Cell Biol.* **9:** 479-511.

Hershey, A.D. and M. Chase (1952). Independent functions of viral protein and nucleic acid in growth of bacteriophage. *J. Gen. Physiol.* **36:** 39-56.

Hershey, A.D. and R. Rotman. (1949). Genetic recombination between host range and plaque type mutants of bacteriophage in single bacterial cells. *Genetics* **34:** 44-71.

Hershko. A. and A. Ciechnover (1992). The ubiquitin system for protein degradation. *Annu. Rev. Biochem.* **61:** 761- 807.

Higgins, C.F. (1992). ABC transporters: from microorganisms to man. *Annu. Rev. Cell Biol.* **8:** 67-113.

Hill, T.M. (1992). Arrest of bacterial DNA replicatiol1. *Annu. Rev. Microbiol.* **46:** 603-633.

Hisatake, K., R.G. Roeder and M. Horikoshi (1993). Functional dissection of TFIIB domains required for TFIIB- TFIID-promoter complex formation and basal transcription activity. *Nature* **363:** 744-747.

Hoagland, M.B., M.L. Stephenson, J.F. Scott, L.I. Hecht and P.C. Zamecnik (1958). A soluble ribonucleic acid intermediate in protein synthesis. *J. Biol. Chem.* **231:** 241.257.

Hoagland, M.G. (1959). The present status of the adaptor hypothesis. *Brookhaven Symp. Biol.* **12:** 40.

Hoeijmakers, J.H. (1993). Nucleotide excision repair I: from *E. coli* to yeast. *Trends in Genet.* **9:** 173-177.

Hoeijmakers, J .H. (1993). Nucleotide excision repair II: from yeast to mammals. *Trends in Genet.* **9:** 211-217.

Hoey, T. (1997). A new player in cell death. *Science* **278:** 1578-1579.

Holland, P. (1992). Homeobox genes in vertebrate evolution. *Bioessays* **14:** 267-273.

Holley, R.W. *et al.* (1965). Structure of a ribonucleic acid. *Science* **147:** 1462-1465.

Holliday, R. (1964). A mechanism of gene conversion in fungi. *Genet. Res.* **5:** 282-304.

Holliday, R. (1989). A different kind of inheritance. *Sci. Amer.*, June 1989. pp. 60-73.

Holzman, E. (1976). *Lysosomes: A Survey.* Springer-Verlag. New York.

Horikoshi, M. *et al.*, (1992) Transcription factor TFIID induces DNA bending upon binding to the TATA element. *Proc. Natl. Acad. Sci. USA* **89:** 1060-1064.

Horisberger, J.D., V. Lemas, J.-P. Kraehenbuhl and B.C. Rossier (1991). Structure-function relationship of Na^+ K^+-ATPase. *Annu. Rev. Physiol.* **53:** 565-584.

Hortsch, M. and C.S. Goodman (1991). Cell and substrate adhesion molecules in *Drosophila. Immu. Rev. Cell. Biol.* **7:** 505-557.

Hotta. Y., M. Ito and H. Stem (1966). Synthesis of DNA during meiosis. *Proc. Natl. Acad. sci. USA* **56:** 1184-1191

Hatta, Y. *et al.* (1985). Meiosis specific transcripts or a DNA component replicated during chromosome pairing: homology across the phylogenetic spectrum. *Cell* **40:** 785-793.

Hoyt, M.A., A.A. Hyman and M. Bahler (1997). Motor proteins of the eukaryotic cytoskeleton. *Proc. Natl. Acad. Sci. USA* **94:** 12747-12748.

Hull, R. and H. Will (1989). Molecular biology of viral and non-viral retroelements. *Trends in Genet.* **5:** 357-359.

Hwang, L.H., L.F. Lau, D.L. Smith, CA. Mistrot, K.G. Hardwick, E.S. Hwang, A. Amon and A.W. Murray (1998). Budding yeast Cdc29: a target of the spindle checkpoint. *Science* **279:** 1041-1044.

Hyams, J.F. and C.W. Lloyd (1993). *Microtubules.* Wiley-Liss, New York.

Hyman, A.A and T.J. Mitchison (1991). Two different microtubule-based motor activities with opposite polarities in kinetochores. *Nature* **351:** 206-211.

Hynes, R.O. (1992). Integrins: versatility, modulation, an signaling in cell adhesion. *Cell* 69: 11-25.

Ingram, V.M. (1957). Gene mutations in human haemoglobi. *Nature* **180:** 326-328.

Ingram, V. (1965). *The Biosynthesis of Macromolecule.* Benjamin, New York.

Izquierdo, M. and D.A. Cantrell (1992). T-cell activation. *Trends in Cell Biol.* **2:** 268-271.

Jacks, T. and R. A. Weinberg (1996). Cell-cycle control and its watchman. *Nature* **381:** 643-644.

Jacks, T. and R.A. Weinberg (1998). The expanding role of cell cycle regulators. *Science* **280:** 1035-1036.

Jacob, F. and J. Monod (1961). Genetic regulatory mechanisms in the synthesis of proteins. *J. Mol. Biol.* **3:** 318-356.

Jacob, F. and E.L. Wollmann (1961). *Sexuality and Genetics of Bacteria.* Academic Press, New York.

Jacobson, M. (1991). *Developmental Neurobiology,* 3rd ed. Plenum Press, New York.

Janeway, C. and P. Travers (1994). *Immunology.* Current Science and Garland, London & New York.

Janknecht, R. and T. Hunter (1996). A growing coactivator network. *Nature* **383:** 22-23.

Janssens, FA. (1909). Spermatogenese dans les Batraciens V. La. theorie de la Chiasmatypie, Nouvelles interpretation des cineses de maturation. *Cellule* **25:** 387-411.

Jinks, J.L. (1964). *Extrachromusomal Inheritance.* Prentice-Hall. Englewood Cliffs, New Jersey.

Johannsen, W. (1903). Uber Erblickeit in Populationen und in reinen Linien. G-Fisher, Jena. (Excerpts have been reprinted in Peters, 1959).

John, B. and G.L.G. Milkos (1980). Functional aspects of satellite DNA and heterochromatin. *Intern. Rev. Cytol.* **58:** 1-114.

Johns, E.W. (1971). The preparation and characterization of histones In: Philips, D.M. (Ed.) *Histones and Nucleohistones.* Plenum Press, New York. pp. 246.

Jukes, T.H. (1983). Mitochondrial codes and evolution. *Nature* **301:** 19-20.

Kafer, E. (1958). An 8-chromosome map of *Aspergillus nidulans. Adv. Genet.* **9:** 105-146.

Kaguni, L.S, I.R. Lehman (1988). Eukaryotic DNA polymerase-primase: structure, mechanism and function. *Biochim. et Biophys. Acta* **950:** 87-101.

Kasha K.J. and K.N. Kao (1970). High frequency haploid production in barley (*Hordeum vulgare* L.). *Nature* **225:** 874-876.

Kasha, K.J. and G. Seguin-Swartz (1983). Haploidy and crop improvement. In : Swaminathan, M.S., P.K. Gupta and U. Sinha (Eds). *Cytogenetics of Crop Plants.* MacMillan India Ltd., Delhi. pp. 19-68.

Katagiri, F. and H-Hai Chua (1992). Plant transcription factors : present knowledge and future challenges. *Trends in Genet.* **8:** 22-27.

Kawashima, T., C. Berthet-Colominas, M. Wulff, S. Cusack and R. Leberman (1996). The structure of the *Escherichia coli* EF-Tu. EF-Ts complex at 2.5 A° resolution. *Nature* **379:** 511-518.

Kaziro, Y. *et al.* (1991). Structure and function of signal transducing GTP binding proteins. *Annu. Rev. Biochem.* **60:** 349-400.

Kendrew, J.C. (1961). The three-dimensional structure of protein molecule. *Sci. Amer.* June 1961. pp. 96-111.

Kenneth, W.J. and R.J. Ellis (1975). Protein synthesis in chloroplasts IV. Polypeptides of the chloroplast envelope. *Biochem. et Biophys. Acta.* **378:** 143-151.

Kessler, C. and V. Manta (1990). Specificity of restriction endonucleases and DNA modification methyltransferases-a review (edn. 3). *Gene* 92: 1-248.

Khorana, H.G. (1979). Total synthesis of a gene. *Science* **203:** 614-625.

Khush, G.S. (1973). *Cytogenetics of Aneuploids.* Academic press, New York.

Kim. S.H. *et al.* (1973). Three dimensional structure of yeast phenylalanine RNA: folding of the polynucleotide chain. *Science* **179:** 285-288.

Kim, SH., DP. Lin, S. Matsumoto. A. Kitazono and T. Matsumoto (1998). Fission yeast Sip 1: an effector of the. *Mad2-dependent* spindle checkpoint. *Science* **279:** 1045-1047.

King, R.C. (Ed.) (1975). *A Handbook of Genetics.* Vols 2-4. Plenum Press, New York.

King, RW.. R.J. Deshaies, J-M. Peters and M.W. Kirschner (1996). How proteolysis drives the cell cycle. *Science* **274:** 1652-1659.

Kirschner. M. (1992). The cell cycle: then and now. *Trends in Biochem. Sci.* **17:** 281-285.

Kjellen, L. and U. Lindahl (1991). Proteoglycans: structures and interactions. *Annu. Rev. Biochem.* **60:** 443-475.

Kleckner, N. (1981). Transposable elements in prokaryotes. *Annu. Rev. Genet.* **15:** 341-404.

Kim. Y., J. H. Geiger, S. Hahn, and P.B. Sigler (1993). crystal structure of a yeast TBP/TATA-box complex. *Nature* **365:** 512-520.

Klug. A. (1983). From macromolecules to biological assemblies. *Biosci. Rep.* **3:** 395-430.

Klukas, C.K. (1976). Non-histone proteins and transcription. *Nature* **263:** 545-546.

Knight, S.C. and A.J. Stagg (1993). Antigen-presenting cell types. Curr. Opin. *Immunol.* **5:** 374-382.

Kornberg. A. (1962). *Exzymatic Synthesis of DNA.* John Wiley, New York.

Kornberg. A. (1974). *DNA Synthesis.* Freeman, San Francisco.

Kornberg, A. (1980). *DNA Replication.* Freeman, San Francisco.

Kornberg, A. and T.A. Baker (1992). *DNA Replication*, Freeman, New York.

Kornberg, R.D. and A. Klug (1981). The nucleosome. *Sci. Amer.* Feb. 1981. pp. 52-64.

Kornberg, R.D and Y. Lorch (1992). Chromatin structure and transcription. *Annu. Rev. Cell Biol.* **8:** 563-587.

Kornberg, R.D. and J.O. Thomas (1974). Chromatin structure: oligomers of the histones. *Science* **184:** 865-868.

Kornfeld. S. and I. Mellman (1989). The biogenesis of lysosomes. *Annu. Rev. Cell Biol.* **5:** 483-525.

Korsmeyer, S.J (1995). Regulators of cell death. *Trends in Genet.* **11:** 101-105.

Kozak, M. (1992). Regulation of translation in eukaryotic systems. *Annu. Rev Cell Biol.* **8:** 197-225.

Kramer, A. (1.996). The. structure and function of proteins involved in mammalian pre-mRNA splicing. *Annu Rev. Biochem.* **65:** 367-409.

Krauss, N., W. Hinrichs, I. Witt, *et. al.*, (1993). Three-dimensional structure of system I of photosynthesis at 6A° resolution. *Nature* **361:** 326-331.

Krebs, H.A. (1970). The history of the tricarboxylic acid cycle. *Perspect. Biol. Med.* **14:** 154-170.

Kreis, T.E. and R.D. Vale, Eds. (1993). *Guidebook to Cytoskeletal and Motor Proteins.* Oxford University Press Oxford, UK.

Kuby, J. (1992). *Immunology.* W.H. Freeman, New York.

Kuhlbrandt, W. (1995). Many wheels make light work. *Nature.* **374:** 497-498.

Kuhlbrandt, W., Da N. Wang and Y. Fujiyoshi (1994). Atomic model of plant light-harvesting complex by electron crystallography. *Nature* **367:** 614-621.

Kuhlemeier, C., P. J. Green and N. Hai Chua (1987). Regulation of gene expression in higher plants. *Annu. Rev. Pl. Physiol* **38:** 221-257.

Kurland, C.G. (1972). Structure and function of the bacterial ribosomes. *Annu. Rev. Biochem.* **41:** 377.

Lake, J.A. (1981). The ribosome. *Sci. Amer.* Aug. 1981. pp 56-69.

Landsteiner, K. and P. Levine (1927). Further observations on individual differences of human blood. *Proc. Soc. Exp. Bio. Med.* **24:** 941-942.

Lasky, L.A. (1992). Selectins: interpreters of cell-specific carbohydrate information during inflammation. *Science* **258:** 964-969.

Lazarow, P.B. and Y. Fujiki (1985). Biogenesis of peroxisomes. *Annu. Rev. Cell Biol.* **1:** 489-530.

Leaver. C.J. *et al.*, (1988). Mitochondrial genome diversity and cytoplasmic male sterility in higher plants. *Phil. Trans. R Soc. Lond.* **B319:** 165-176.

Lederberg, J. (1947). Gene recombination and linked segregation, in *Escherichia coli. Genetics* **32:** 505-525.

Lederberg, J. and E.L. Tatum (1946). Novel genotypes in mixed cultures of biochemical mutants of bacteria. *Cold Sp. Harb symp. Quant. Biol.* **11:** 113-114.

Lehninger, A.L. (1965). *The Mitochondrion.* Academic Press New York.

Lehninger, A.L., D.L. Nelson and M.M. Cox (1993). *Principle of Biochemistry.* 2nd ed. Worth, New York.

Lerner, M.R., J.A. Boyle, S.M. Mount, S.L. Wolin and J.A Steitz (1980). Are SnRNPs involved in splicing. *Nature* **283:** 220-224.

Lerncr, R.A., S.J. Benkovic and P.G. Schultz (1991). At the crossroads of chemisry and immunology: catalytic antibodies. *Science* **252:** 659-667.

Lerner, R.A. and A. Tramontano (1988). Catalytic antibodies. *Sci. Amer.* March 1988. pp. 58-70.

Levings, C.S. III (1990). The texas cytoplasm of maize cytoplasmic male sterility and disease susceptibility. *Science* **250:** 942-947.

Levitzky, G.A. (1931). The karyotype in systematics. *Bull. Appl. Bot. Genet. Pl. Breed.* **27:** 220-240.

Lewin, Benjamin (1975). The nucleosome: subunit of mammalian chromatin. *Nature* **254:** 651-653.

Lewin, Benjamin (1997). *Genes VI.* Oxford University Press, to New York.

Lewin, R. (1983). How mammalian RNA returns to its genome *Science* **219:** 1052-1054.

Lewis. E.B. (1950). The phenomenon of position effect. *Adv. Genet.* **3:** 73-116.

Lewis, E.B. (1951). Pseudoallelism and gene evolution. *Cold Sp. Harb. Symp. Quanti Biol.* **16:** 159-172.

Lewis, E.B. (1993). Clusters of master control genes regulate the development of higher organisms. *Curr. Sci.* **64:** 640-649.

Lewis, KR. and B. John. (1963). *Chromosome Marker.* Chur-chill, London.

L Heritier Ph. (1958). The hereditary virus of *Drosophila. Adv. Virus Res.* **5:** 195-245.

Lilienfeld, F.A. (1951). H. Kiharn : Genonle analysis in *Triticum and Aegilops.* X. Concluding review. *Cytologia* **16:** 101-123.

Linder, M.D. and A.G. Gilman (1992). G proteins. *Sci. Amer.* Jan. 1992. pp. 36-43.

Lippincott-Schwartz, J. (1993). Bidirectional membrane traffic between the endoplasmic reticulum and Golgi apparatus. *Trends in Cell Bioi.* **3:** 81-88.

Long, E.O. and I.B. David (1980). Repeated genes in eukaryotes. *Annu. Rev. Biochem.* **49:** 727-764.

Lorimer, G.H. (1981). The carboxylation and oxygenation of ribulose-1, 5-bisphosphate: the primary events in photosynthesis and photorespiration. *Annu. Rev. Plant Physiol.* **32:** 349-383.

Lowy, D.R. and B.M. Willumsen (1993). Function and regulation of Ras. *Annu. Rev. Biochem.* **62:** 851-891.

Lucas, W.J. and S. Wolf (1993). Plasmodesmata: the intercellular organelles of green plants. *Trends in Cell Biol.* **3:** 308-315.

Luger, K., A.W. Mader, R.K. Richmond, D.F. Sargent and T.J. Richmond (1997). Crystal structure of the nucleosome core particle at 2.8 A° resolution. *Nature* **389:** 251-260.

Luna, E.J. and A.L. Hitt (1992). Cytoskeleton-plasma membrane interactions. *Science* **258:** 955-963.

Lundqvist, T. and G. Schneider (1991). Crystal structure of activated ribulose-1,5-bisphosphate carboxylase complexed with its substrate. ribulose-1,5-bisphophate. *J. Biol. Chem.* **266:** 12604-12611.

Luria, S.E. (1951). The frequency distribution of spontaneous bacteriophage mutants as evidence for the exponential rate of phage reproduction. *Cold Sp. Harb. Symp. Quont. Biol* **16:** 463-470.

Marians, K.J. (1992). Prokaryotic DNA replication. *Annu. Rev. Biochem.* **61:** 673-719.

Marx, J. (1987). Rice plants regenerated from protoplasts. *Science* **235:** 31-32.

Marx, J. (1992). Taking a direct path to the genes. *Science* **257:** 744-745.

Marx. J. (1993). Forging a path to the nucleus. *Science* **260:** 1588-1590.

Maxam. A.M. and W. Gilbert (1977). A new method for sequencing DNA. *Proc. Natl. Acad. Sci. USA* **74:** 560-564.

Maxwell. E.S. and M.J. Fournier (1995). The small nucleolar RNAs. *Annu. Rev. Biochem.* **35:** 897-934.

McCarthy. J.E.G. and C. Gualerzi (1990). Translational control of prokaryotic gene expresion. *Trends in Genet.* **6:** 78-85.

McClain, W.H. (1993). Transfer RNA identity. *FASEB J.* **7:** 72-78.

McClintock, B. (1956). Controlling elements and the gene. *Cold Sp. Harbor Symp. Quant. Biol.* **21:** 197-216.

McClintock. B. (1961). Some parallels between gene control systems in maize and in bacteria. *Amer. Naturallist* **95:** 265-277.

McClung. C.E. (1902). The accessory chromosomes-sex determinant. *Biol. Bull.* **3:** 43-84.

McGhee. J.D. and G. Felsenfeld (1980). Nucleosome structure. *Annu. Rev. Biochem.* **49:** 1115-1156.

McIntosh, J.R. and G.E. Hering (1991). Spindle fiber action and chromosome movement. *Annu. Rev. Cell. Biol.* **7:** 403-426.

McIntosh, J.R. and K.L. McDonald (1989). The mitotic spindle *Sci. Amer.* April 1989. pp. 48-56.

McKim. K.S., B.L. Green-Marroquin, J.J. Sekelsky, G. Chin. C. Steinbverg. R. Khodosh, and R.S. Hawley (1998). Meiotic synapsis in the absence of recombination. *Science* **279:** 876-878.

McKnight, S.L. (1991). Molecular zippers in gene regulation. *Sci. Amer.* April 1991. pp 54-64.

Melefors, O. and M.W. Hentze (1993). Translational regulation by mRNA/protein interactions in eukaryotic cells: ferritin and beyond. *Bioessays* **15:** 85-90.

Mendel, G. (1866). *Experiments in Plant Hybridization.* Originally published in the Brunn Natural History Society and the translation reprinted in 1965. Oliver and Boyd. Edinburgh. (Introduction and Commentary by R.A. Fisher).

Mercer. R.W. (1993). Structure of the Na^+, K^+-ATPase. *Int. Rev. Cytol.* **137C:** 139-168.

Mermall, V., P.L. Post, M.S. Mooseker (1998). Unconventional myosins in cell movement. membrane traffic, and signal transduction. *Science* **279:** 527-533.

Merrick. W.C. (1990). Overview: mechanism of translation initiation in eukaryotes. *Enzyme* **44:** 7-16.

Merrick, W.C. (1992). Mechanism and regulation of eukaryotic protein synthesis. *Microbiol. Rev.* **56:** 291-315.

Meselson M. and F.W. Stahl (1958). The replication of DNA in *Escherichia coli. Proc. Natl. Acad. Sci. USA* **44:** 671-682.

Meyer, R.R. and P. S. Laine (1990). The single-stranded DNA-binding protein of *Escherichia coli. Microbiol. Rev.* **54:** 342-380.

Miceli, M.C. and J.R. Parnes (1993). Role of CD4 and CD8 is T cell activation and differentiation. *Adv. Immunol.* **53:** 59-122.

Michel, F. and E. Westhof (1990). Modelling of the three-dimensional architecture of group I catalytic introes based on comparative sequence analysis. *J. Mol. Biol.* **216:** 585-610.

Miller, K.R. (1979). The photosynthetic membrane. *Sci. Amer.* April 1979. pp. 102-113.

Miller, L.J. and J. Marx, Eds. (1998). Apoptosis (special section). *Science* **281:** 1301-1326.

Miller, O.L. (1981). The nucleolus, chromosomes, and visualization of genetic activity. *J. Cell Biol.* **91:** 15s-27s.

Milstein, C. (1980). Monoclonal antibodies. *Sci. Amer.* April 1980 pp.66-74.

Mirkin, S.M. (1995). Triplex DNA structures. *Annu. Rev. Biochem.* **64:** 65-95.

Mitchell, P. (1961). Coupling of phosphorylation to electron and hydrogen transfer by a chemi-osmotic type of mechanism. *Nature* **191:** 144-148.

Mitchell, P.J. and R. Tjian (1989). Transcription regulation in mammalian cells by sequence-specific DNA binding proteins. *Science* **245:** 371-378.

Moncada, S., R.M. Palmer and E.A. Higgs (1991). Nitric oxide physiology, pathophysiology and pharmacology. *Pharmacel Rev.* **43:** 109-142.

Mondino, A., A. Khoruts and M.K. Jenkins (1996). The anatomy of T-cell activation and tolerance. Proc. *Natl. Acad. Sci. USA* **93:** 2245-2252.

Moore, P.B. (1995). Molecular mimicry in protein synthesized. *Science* **270:** 1453-1454.

Morgan; T.H. (1911). Random segregation versus coupling in Mendelian inheritance. *Science* **34:** 384.

Morris, S., S. Ahle and E. Ungewickell (1989). Clathrin-control vesicles. *Curr. Opin. Cell Biol.* **1:** 684-690.

Moss, S.E. (1995). Annexins taken to task. *Nature* **378:** 446-447.

Mount, S.M. (1996). AT-AC introns: An ATtACk on dogma. *Science* **271:** 1690-1692

Muller, H.J. (1916). The mechanism of crossing over II. Amer. *Naturalist* **50:** 284-305.

Muller, H.J. (1927). Artificial transmutation of the gene. *Sciences* **66:** 84-87.

Murray, A.W. (1989). The cell cycle as *cds2* cycle. *Nature* **342:** 14-15.

Murray, A.W. (1992). Creative blocks: cell cycle checkpoints and feedback controls. *Nature* **359:** 599-604.

Murray, A.W. and T. Hunt (1993). *The Cell Cycle.* Oxford University Press, Oxford, UK.

Murray, A.W. and M.W. Kirschner (1989). Dominoes and clocks: the union of two views of the cell cycle. *Science* **246:** 614-621.

Murray, A.W. and M.W. Kirschner (1991). What controls the cell cycle? *Sci. Amer.* March 1991. pp. 56-63.

Murray, R.K. *et al.* (1990). *Harper's Biochemistry.* 2nd ed. Appleton & Lange, Norwalk, CT (USA).

Nasmyth, K. (1996). At the heart of the budding yeast cell cycle. *Trends in Genet.* **12:** 405-412.

Nass, M.M.K. (1969). Mitochondrial DNA: advances, problems and goals. *Science* **165:** 15-35.

Nass, M.M.K.. S. Nass and B.A. Afzelius (1965). The general occurrence of mitochondrial DNA. *Exp. Cell Res.* **37:** 516-539.

Neale, D.B., K.A. Marshall and R.R. Sederoff (1989). Chloroplast and mitochondrial DNA are paternally inherited in *Sequoia semipervirens* D. Don Endl. *Proc. Natl. Acad. Sci. USA* **86:** 9347-9349.

Neher. E. (1992). Ion channels for communication between and within cells. *Science* **256:** 498-502.

Neher, E. and B. Sakmann (1992). The patch clamp technique. *Sci. Amer.* March 1992. pp. 28-35.

Newmeyer. D.D. (1993). The nuclear pore complex and nucleocytoplasmic transport. *Curr. Opin. Cell Biol.* **5:** 395-407.

Newport, J.W. and D.J. Forbes (1987). The nucleus: structure, function and dynamics. *Annu. Rev. Biochem.* **56:** 535-565.

Newton, K.J. (1988). Plant mitochondrial genomes : organization expression and variation. *Annu. Rev. Pl. Physiol. Pl. Mol Biol.* **39:** 503-532.

Nikaido, H. and M.H. Saier, Jr. (1992). Transport proteins if bacteria: common themes in their design. *Science* **258:** 936-941.

Nikolov, D.B. and S.K. Burley (1997). RNA polymerase II transcription initiation: A structural view. *Proc. Natl. Acad. Sci. USA* **94:** 15-22.

Nirenberg, M.W. (1963). The genetic code II. *Sci. Amer.* March 1963. pp. 80-94.

Nirenberg, M.W. and P. Leder (1964). RNA code words an, protein systhesis: the effect of trinucleotides upon the binding of sRNA to ribosiomes. *Science* **145:** 1399-1407.

Nirenberg, M.W. and J.H. Mathaei (1961). The dependence of cell-free protein syntheisis., in *E. coli* upon naturally occurring or synthetic polyribonucleotides. *Proc. Nat Acad. Sci. USA* **47:** 1588-1602.

Nishida, E. and Y. Gotoh (1993). The MAP kinase cascade is essential for diverse signal transduction pathways. *Trends in Biochem. Sci.* **18:** 128-130.

Nishimura, S., D.S. Jones and H.G. Khorana (1965). Studies on polynucleotides, XLVIII. The *in vitro* synthesis of a copolypeptide containing two amino acids in alternating sequence dependent upon DNA like polymer containing two nucleotides in alternating sequence. *J. Mol. Biol.* **13:** 302-324.

Noller, H.F. (1991). Ribosomol RNA and translation. *Annu. Rev. Biochem.* **60:** 191-227.

Nomura, M. (1969). Ribosomes, *Sci. Amer.* Oct. 1969. pp. 28.

Nomura, M. (1972). Assembly of bacterial ribosomes. *Fed. Proc.* **31:** 18.

Nomura, M., A.N. Tissieres and L.H. Lengyel (1974). *Ribosomes*. Cold Spring Harbor Laboratory, New York.

Nossal, G.J.V. (1992). Cellular and molecular mechanisms of B lymphocyte tolerance. *Adv. Immunol.* **52:** 283-331.

Novick, R.P. (1980). Plasmids. *Sci. Amer.* June 1980, pp. 102-127.

Novina, C.D. and A.L. Roy (1996). Core promoters and transcriptional control. *Trends in Genet.* **12:** 351-355.

Nudler, E., E. Avetissova, V. Markovtsov and A. Goldfarb (1996). Transcription processivity: protein-DNA interactions holding together the elongation complex. *Science* **273:** 211-217.

Olins, A.L. and D.E. Olins (1974). Spheroid chromatin units (v bodies). *Science* **183:** 330-331.

Olson, M. F., A. Ashworth and A. Hall (1995). An essential role for Rho, Rac, and Cdc42 GTPases in cell cycle progression through G_1. *Science* **269:** 1270-1272.

Olson. J.M and O.K. Pierson (1987). Evolution of reaction centres in photosynthetic prokaryotes. *Int. Rev. Cytol.* **108:** 209-248.

Olson, G.J. and C.R. Woese (1996). Lessons from an Archaeal genome: What are we learning from *Methanococcus jannaschii*? *Trends in Genet.* **12:** 377-379.

O'Malley' B.W., H.C. Towle and R.J. Schwartz (1977). Regulation of gene expression in eukaryotes. *Annu. Rev Genet.* **11:** 239-275.

Orgel, L. and F.H.C. Crick (1980). Selfish DNA: the ultimate parasite. *Nature* **284**: 604-607.

Orr-Weaver, T. L. (1995). Meiosis in *Drosophila*: seeing is believing. *Proc. Natl. Acad. Sci. USA* **92:** 10443-10449.

Orr-Weaver, T.L. and R.A. Weinberg (1998). A checkpoint on the road to cancer. *Nature* **392:** 223-224.

Ostro, M.J. (1987). Liposomes. *Sci. Amer. Jan.* 1987. pp. 102-111.

Oudet, P. M. Gross-Bel!i1rd and P. Chambon (1995). Electron microscopic and biochemical evidence that chromatin is a repeating unit. *Cell* **4:** 281-300.

Pabo, C.O. and R.T. Sauer (1992). Transcription factors: structural families and principles of DNA recognition. *Annu. Rev. Biochem.* **61:** 1053-1095.

Palade, G.E. (1953). The fine structure of mitochondria: an electorn microscope study. Histochem *Cytochem.* **1:** 118.

Palczewski. K. and J.L. Benovic (1991). G-protein-coupled receptor kinases. *Trends in Biochem. Sci.* **16:** 387-391.

Pante. N. and U. Aebi (1993). The nuclear pore complex. *J. Cell Biol.* **122:** 977-984.

Pardue. M.L. and J.G. Gall (1969). Molecular hybridization of radioactive DNA to the DNA of cytological preparations. *Proc. Natl. Acad. Sci. USA* **64:** 600-604.

Pardee. A.B. (1985). Molecular basis of biological regulatioa: origins of feedback inhibition and allostery. *Bioessays* **1:** 17-40.

Parijs. L. V. and A.K. Abbas (1998) Homeostasis and self-tolerance in the immune system: turning lymphocytes off. *Science* **180:** 243-248.

Parkinson. J.S. (1993). Signal transduction schemes of bacteria. *Cell* **73:** 857-871.

Parton. R.G.. P. Schrotz. C. Bucci and J. Gruenberg (1992). Plasticity of early endosomes. *J. Cell Sci.* **103:** 335-348.

Pastan. I. (1972). Cyclic AMP. *Sci. Amer.* Feb. 1972. pp. 97-105.

Paulingm. L., H.A. Itano. S.J. Singer and I.C. Wells (1949). Sickle cell anemia. a molecular disease. *Science* **110:** 543-548.

Pawson. T. and J. Schlessinger (1993). SH2 and SH3 domains. *Curr. Biol.* **3:** 434-442.

Pennisi. E: (1998). Cell division gatekeepers identified. *Science* **279:** 477-478.

Perutz. M.F. (1964). The hemoglobin molecule. *Sci Amer.* Nov. 1964. pp. 64-76.

Peters. J.A. (Ed.) (1959). *Classic Papers in Genetics.* Prentice-Hall, Englewood Cliffs. New Jersey.

Peterson. M.G. and Tjian (1992). Transcription: the tell-tail trigger. *Nature* **358:** 620-621.

Pollard. T.D., S.K. Doberstein and H.G. Zot (1991). Myosin-1. *Annu. Rev. Physiol.* **53:** 653-681.

Pontecorvo. G. (1958). *Trends in Genetic Analysis.* Columbia Univ. Press. New York.

Pontecorvo. G. and E. Kafer (1958). Genetic analysis based on mitotic recombination. *Adv. Genet.* **9:** 71-104.

Potrykus. I. (1991). Gene transfer to plants: assessment of published approaches and reuslts. *Annu. Rev. Pl. Physiol. Pl. Mol. Bio* **42:** 205-225.

Price, C.M. (1992). Centromeres and telomeres. *Curr. Opin. Cell Biol.* **4:** 379-384.

Ptashne. M. (1989). How gene activators work. *Sci. Amer.* Jan. 1989. pp. 40-47.

Ptashne, M. and A. Gann (1997). Transcriptional activation by recruitment. *Nature* **386:** 569-576.

Radman, M. and R. Wagner (1988). The high fidelity of DNA duplication. *Sci. Amer.* Aug. 1988. pp. 40-46.

Raff. M.C. (1994). Cell death genes: *Drosophila* enters the field. *Science* **264:** 668-669.

Rapoport, T.A. (1992). Transport of proteins across the endoplasmic reticulum membrane. *Science* **258:** 931-936.

Rapoport, T.A., B. Jungnickel and U. Kutay (1996). Protein transport across the eukaryotic endoplasmic reticulum and bacterial inner membranes. *Annu. Rev. Biochem.* **65:** 271-301.

Razin, A. and A.D Riggs (1980). DNA methylation and gene function. *Science* **210:** 604-610.

Reed, S.I. (1992). The role of p34 kinases in the G1 to S-phase transition. *Annu. Rev. Cell Biol.* **8:** 529-561.

Reinert, J. and H. Uroprung (1971). *Origin and Continuly of Cell Organelles.* Springer Verlag, Berlin.

Renkawitz, R. (1990). Transcriptional repression in eukaryotics. *Trends in Genet.* **6:** 192-197.

Rennie, J. (1993). DNA's new twists. *Sci. Amer.* March 1993. pp.88-96.

Reth, M. (1992). Antigen receptors on B lymphocytes. *Annu. Rev. Immunol.* **10:** 97-122.

Rhoades, M.M. (1946). Plastid mutations. *Cold Sp. Harb. Symp. Quant. Biol.* **11:** 202-207.

Rhoads, R.E. (1993). Regulation of eukaryotic protein synthesis by initiation factors. *J. Biol. Chem.* **268:** 3017-3020.

Rhodes, D. and A. Klug (1993). Zinc fingers. *Sci. Amer.* Feb. 1993. pp .56-59, 62-65.

Rich A and S. H. Kim (1978) The three dimensional structure of transfer RNA. *Sci. Amer.* Jan. 1978 pp. 52-62.

Richards. F.M. (1991). The protein folding problem. *Sci. Amer.* Jan. 1991. pp. 54-63.

Robertson, M (1988). RNA processing: the post-RNA world. *Nature* **335:** 16-18.

Robinson, C., R.B. Klosgen, R.G. Herrmann and J.B. Shackleton (1993). Protein translocation across the thylakoid membrane-a tale of two mechanisms. *FEBS Lett* **325:** 67-69.

Robinson, D.G. and H. Depta (1988). Coated vesicles. *Annu. Rev. PI. Physiol. Pl- Mol. Bioi.* **39:** 53-99.

Rochaix, J.D. (1992). Post-transcriptional steps in the expression of chloroplast genes. *Annu Rev. Cell Biol.* **8:** 1-28.

Rodnina, M.V., A. Savelsbergh, V.I. Katunin and W. Wintermeyer (1997). Hydrolysis of GTP by elongation factor G drives tRNA movement on the ribosome. *Nature* **385:** 37-41.

Roeder, G.S. (1990). Chromosome synapsis and genetic recombination: their roles in mitotic chromosomes segregation. *Trends in Genet.* **6:** 385-389.

Roeder, G.S. (1995). Sex and the single cell: meiosis in yeast. *Proc. Natl. Acad. Sci. USA* **92:** 10450-10456.

Rosenberg, M. and D. Court (1979). Regulatory sequences involved in the promotion and termination of RNA transcription. *Annu. Rev. Genet.* **13:** 319-354.

Rothman, J.E. (1994). Mechanisms of intracellular protein transport. *Nature* **372:** 55-63.

Rothman, J.E. and L. Orci (1990). Movement of proteins through the Golgi stack: a molecular dissection of vesicular transport. *FASEB J.* **4:** 1460-1468.

Rothman, J.E. and F.T. Wieland (1996). Protein sorting by transport vesicles. *Science* **272:** 227-234.

Rudnick, G. (1986). ATP-driven H^+-pumping into intracellular organelles. *Annu. Rev. Physiol.* **48:** 403-413.

Ruoslahti, G. (1988). Structure and biology of proteoglycans. *Annu. Rev. Cell Biol.* **4:** 229-255.

Sager, R. (1961). Genetic systems in *Chlamydomonas*. *Science* **132:** 1459.

Sager, R. (1972). *Cytoplasmic Genes and Organelles.* Academic Press, New York.

Sakes, M.E., J.R. Sampson and J.N. Abelson (1994). The transfer RNA identity problem: a search for rules. *Science* **263:** 191-197.

Sakmann, B. (1992). Elementary steps in synaptic transmission revealed by currents through single ion channels. *Science* **256:** 503-512.

Sambrook, J., E.F. Fritsch and T. Maniatis (1989). *Molecular Cloning: A Laboratory Manual,* 2nd ed. Cold Spring Harbor Laboratory Press, Cold Spring Harbor, New York.

Sancar, A. (1996). DNA excision repair. *Annu. Rev. Biochem* **65:** 43-81.

Sandler, L. and E. Novitski (1957). Meiotic drive as an evolutionary force. *Amer. Naturalist* **91:** 105-110.

Sanger, F. (1955). The chemistry of simple proteins. *Symp Soc. Exp. Biol.* **9:** 10-31.

Sanger, F. (1988). Sequences, sequences and sequences. *Annu. Rev. Biochem.* **57:** 1-28.

Sanger, F., S. Nicklen, and A.R. Coulson (1977). DNA sequencing with chain-terminating inhibitors. *Proc. Nail. Acad. Sci. USA* **74:** 5463-5467.

Sanger, F. *et al.* (1977). Nucleotide sequence of bacteriophage ϕ × 174. *Nature* 265: 687-698.

Sarabhai, A., A.D. W. Stretton, S. S\Brenner and A Bolle (1964). Colinearity of the gene with polypeptide chain. *Nature* **201:** 13-37,

Sarkar, N. (1997). Polyadenylation of mRNA in prokaryotes. *Annu. Rev. Biochem.* **66:** 173-197.

Sawin, K.E. and S.A. Endow (1993). Meiosis, mitosis and microtubule motors. *Bioessays* **15:** 399-407.

Schatz, D.G.. M.A. Oettinger and M.S. Schlissel (1992). V(D)J recombination: molecular biology and regulation. *Annu. Rev. Immunol.* **10:** 359-383.

Schatz, G. and B. Dobberstein (1996). Common principles of protein translocation across membranes. *Science* **271:** 1519-1524.

Schekman, R. and L. Orci (1996). Coat proteins and vesicle budding. *Science* **271:** 1526-1533.

Schimmel, P. (1993). GTP hydrolysis in proteins synthesis: two for Tu. *Science* **259:** 1264-1265.

Schindler, C. and J.E. Darnell, Jr. (1995). Transcriptional responses to polypeptide ligands: The JAK-Salt pathway. *Annu. Rev. Biochem.* **64:** 621-651.

Schmid, M. B. (1988). Structure and function of the bacterial chromosome. *Trends in Biochem. Sci.* **13:** 131-135.

Schmid, S.L. (1997). Clathrin-coated vesicle formation and protein sorting: An integraled process. *Annu. Rev. Biochem.* **66:** 511-548.

Schreurs, J., D.M. Gorman and A. Miyajima (1993). cytokine receptors: a new superfamily of receptors. *Int. Rev. Cytol.* **137B:** 121-155.

Schule, R. and RM. Evans (1991). Cross-coupling of signal transduction pathways: zinc finger meets leucine zipper. *Trends in Genet.* **7:** 311-381.

Schulz, G.E. (1993). Bacterial porins: structure and function. *Curr. Opin. Cell Biol.* **5:** 701-707.

Schutz, G. (1989). Control of gene expression by steroid hormones. *Interdisciplinary Sci. Rev.* **14:** 212-215.

Schwartz, R.H. (1993). T cell anergy. *Sci. Amer.* Feb. 1993. pp. 62-71.

Schwarz, M.A., K. Owaribe, J. Kartenbeck and W.W. Frank (1990). Desmosomes and hemidesmosomes: constitutive molecular components. *Annu. Rev. Cell Bioi.* **6:** 461-491.

Schwarz-Sommer, Z. *et al.* (1990). Genetic control of flower an development by homeotic genes in Antirrhinum *majus. Science* **250:** 931-936.

Shapeio, J.A. (Ed.) (1983). *Mobile Genetic Elements.* Academic Press, New York.

Shaplro, L., A.M. Fannon, P.D. Kwong, A. Thompson, M.S. Lehmann, G. Grubel, J.-F. Legrand, J. Als-Nielsen, D.R. Colman and W.A. Hendrickson (1995). Structural basis cell-cell adhesion by cadherins. *Nature* **374:** 327-337.

Sharp, L.W. (1934). *Introduction to Cytology.* McGraw Hill, New York.

Sherr, C. J. (1993). Mammalian Gl cyclin. *Cell* **73:** 1059-106.

Sherr, C. J. (1996). Cancer cell cycles. *Science* **274:** 1672-1677.

Shine, J. and L. Delgamo (1974). The 3′-terminal sequences of *Excherichia coli* 16S ribosome RNA: complementar nonsense triplets and ribosome binding sites. *Proc. Natl. Sci. USA* **71:** 1342-1346.

Shirai, T. and M. Go (1991). RNase-like domain DNA-directed RNA polymerase II. *Proc. Natl. Acad. Sci. USA.* **88:** 9056-9060.

Shortle, D., D. DiMaio and A. Nathans (1981). Direct mutagenesis. *Annu. Rev. Genet.* **15:** 265-294.

Sigurbjomsson B. and A. Micke (1974). Philosophy and accomplishment of mutation breeding. In "*Polyploidy and Induced Mutations in Plant Breeding*" IAEA, Vienna. pp. 303-343.

Simon, S. (1993). Translation of proteins across the endoplasmic reticulum. *Curr. Opin. Cell. Biol.* **5:** 581-588.

Simpson, G.G. (1966). *The Meaning of Evolution.* 2nd edn. Oxford & I.B.H., Calcutta.

Singer, S.J. (1990). The structure and insertion of integral proteins in membranes. *Annu. Rev. Cell Biol.* **6:** 247-296.

Singer, S.J. and G.L. Nicolson (1972).The fluid mosaic model of the structure of cell membranes. *Science* 175: 720-731.

Singer. B., R. Sager and Z. Remains (1976). Chloroplast genetics of chlamydomonas. *Genetics* **83:** 341-354.

Singh. L., L.F. Purdom and K.W. Jones (1980). Sex chromosome associated satellite DNA: evolution and conservation. *Chromosoma* **79:** 137-157.

Sinsheimer. R.L. (1959). A single-stranded deoxyribonucleic acid from bacteriophage $\phi \times 174$. *J. Mol. Biol.* **1:** 40-53.

Slayter, E. M. and H.S. Slayter (1992). *Light and Electron Microscropy.* Cambridge University Press, Cambridge, U.K.

So, A.G. and K.M. Downey (1992). Eukaryotic DNA replication. *Crit. Rev. Biochem. Mol. Biol.* **27:** 129-155.

Sollner-Webb. B. (1991). RNA editing. *Curr. Opin. Cell Biol.* **3:** 1056-1061.

Sollner-Webb. B. (1996). Trypanosome RNA editing: resolved. *Science* **273:** 1182-1183.

Solomon. M.J. (1993). Activation of the various cyclin/cdc2 protein kinases. *Curr. Opin. Cell Biol.* **5:** 180-186.

Sonneborn. T.M. (1959). Kappa and related particles in Paramecium. *Adv. Virus Res.* **6:** 229-356.

Southern, E.M. (1975). Detection of specific sequences among DNA fragments separated by gel electrophoresis. *J. Mol. Biol.* **98:** 503-517.

Stadler, L.J. (1928). Mutations in barley induced by X-rays and radium. *Science* **69:** 186-187.

Staiger, C. and J. Doonan (1993). Cell division in plants. *Curr. Opin. Cell Biol.* **5:** 226-231

Stebbins, G.L. (1950). *Variation and Evolution in Plants.* Columbia Univ. Press. New York.

Stebbins. G.L. (1970). *The Processes of Organic Evolution.* Prentice-Hall, New Delhi.

Stebbins, G.L.(1971). *Chromosomal Evolution in Higher Plants.* Edward Arnold. London.

Stein, G.S., F.S. Stein and L.J. Kleinsmith (1975). Chromosomal proteins and gene regulation. *Sci. Amer.* Feb. 1975. pp. 47-57.

Steitz, J.A. (1988). 'Snurps'. *Sci. Amer.* June, 1988. pp. 56-63.

Stem, B. and P. Nurse (1996). A quantitative model for the cdc2 control of S phase and mitosis in fission yeast. *Trend in Genet.* **12:** 345-350.

Stem, C. (1936). Somatic crossing over and segregation in *Drosophila melanogaster. Genetic* **21:** 625-730.

Stewart, A. (1990). The functional organization of chromosomes and the nucleus- a special issue. *Trends in Genet.* **6:** 377-379.

Stillman, B. (1996). Cell cycle control of DNA replication. *Science* **274:** 1659-1664.

Strycr, L. (1995). *Biochemistry,* 4th W.H. Freeman & Co. New York.

Sugiura, M. (1989) The chloroplast chromosomes in land plants. *Annu. Rev. Cell Biol.* **5:** 51-70.

Sulston, J. *et al.* (1992). The C. elegans genome sequencing project: a beginning. *Nature* **356:** 37-41.

Sutton, A. and R. Freiman (1997). The Caklp protein kinase is required at G_1/S and G_2/M in the budding yeast cell cycle. *Genetics* **147:** 57-71.

Suzuki, D.T. *et al.* (1989). *An Introduction to Genetic Analysis.* 4th edn. W.H. Freeman & Co., New York.

Swain, S.J. *et al.* (1991). Helper T-cell subsets: phenotype, function and the role of lymphokines in regulating their development. *Immunol. Rev.* **123:** 115-144.

Swaminathan, M.S. P.K. Gupta, U. Sinha (Eds.). (1983). *Cytogenetics of Crop Plants.* MacMillan India Pvt. Ltd., New Delhi.

Swanson, C.P. (1957). *Cytology and Cytogenetics.* Prentice Hall, Englewood Cliffs, New Jersey.

Swanson., C.P. (1971). *The Cell.* 3rd edit. Prentice Hall, New Delhi.

Swanson, C.P., T.F. Mertz and W.J. Young (1981). *Cytogenetics-The Chromosomes in Divsion, Inheritance and Evolution.* 2nd edn. Englewood Cliffs, Prentice-Hall, New Jersey.

Szathmary, E. (1992). What is the optimum size for the genetic alphabet. *Proc. Natl. Acad. Sci. USA* **89:** 26214-2618.

Szekely, M. (1978). Triple overlapping genes. *Nature* **272:** 492.

Tarn, W-Y. and J.A. Steitz (1996). Highly diverged U4 and U6 small nuclear RNAs required for splicing rare AT-AC introns. *Science* **273:** 1824-1832.

Tate, P.H., A.P. Bird (1993). Effects of DNA methylation on DNA-binding proteins and gene expression. *Curr. Opin. Genet. Dev.* **3:** 226-231.

Taylor, J.H. (1963). The replication and organization of DNA in chromosomes. In: Taylor, J.H. (Ed.). *Molecular Genetics.* Part I; Academic Press, New York. pp. 65-111.

Taylor, S.S., D.R. Knighton, J. Zheng, L.F. Ten Eyck and J.M. Sowadski (1992). Structural framework for the protein kinase family. *Annu. Rev. Cell Biol.* **8:** 429-462.

Temin, H.M. (1972). RNA directed DNA synthesis. *Sci. Amer.* Jan. 1972. pp. 24-33.

Temin, H.M. and S. Mizutani (1970). RNA dependent DNA polymerase in virions of Rous Sarcome Virus. *Nature* **226:** 1211-1213.

Tessman, I. (1965). Genetic ultrafine structure in the T4 rll region. *Genetics* **51:** 63-75.

Thommes, P. and U. Hubscher (1990). Eukaryotic DNA replication: enzymes and proteins acting at the fork. *Eur. J. Biochem.* **194:** 699-712.

Threadgold, L.T. (1976). *The Ultrastructure of Animal Cell.* Pergamon Press, London.

Tjio, J.H. and A. Levan (1965). The chromosome number of man. *Hereditas* **42:** 1-6.

Tryggvason, K. (1993). The laminin family. *Curr. Opin. Cell Biol.* **5:** 877-882.

Tsurimoto, T., T. Melendy and B. Stillman (1990). Sequential initiation of lagging and leading strand synthesis by two different polymerase complexes at the SV40 DNA replication origin. *Nature* **346:** 534-539.

Turner, B.M. (1991). Histone acetylation and control of gene expression. *J. Cell Sci.* **99:** 13-20.

Unwin, N. (1993). Nicotinic acetylcholine receptor at 9 A° resolution. *J. Mol. Biol.* **229:** 1101-1124.

Uptain, S.M., C.M. Kane, and M.J. Chamberlin (1997). Basic mechanisms of transcript elongation and its regulation. *Annu. Rev. Biochem.* **66:** 117-172.

Vallee, B.L., J.E. Coleman and D.S. Auld (1991). Zinc fingers, zinc clusters and zinc twists in DNA-binding protcin *domains. Proc Natl Acad Sci. USA* **88:** 999-1003.

Van der Rest, M. and R. Garrone (1991). Collagen family of Proteins. *FASEB J.* **5:** 2814-2823

Varki, A. (1994). Selectin ligands. *Proc. Natl. Acad. Sci. USA* **91:** 7390-7397.

Varmus, H. and R.A. Weinberg (1993). *Genes and the Biology of Cancer.* Scientific American Library, New York.

Vaux. D.L. and A. Strasser (1996). The molecular biology of apaptosis. *Proc. Natl. Acad. Sci. USA* **93:** 2329-2244.

Vogelstein, B. and K.W. Kinzler (1993). The multistep nature of cancer. *Trends in Genet.* **9:** 138-141.

Von Hippel, P.H. and T.D. Yager (1992). The elongation termination decision in transcription. *Science* **225:** 809.812.

Wadsworth P. (1993). Mitosis: spindle assembly and chromosome motion. *Curr. Opin. Cell Biol.* **5:** 123-128.

Wagner, R.P. and H.K. Mitchell (1964). *Genetics and Metobolism.* John Wiley, New York.

Wake, R.G. (1996). Tussle with a terminator, *Nature* **383:** 582-583.

Walker, 5.S., J.C. Reese, L.M. Apone and M.R. Green (1996). Transcription activation in cells lacking TAF_{IIS}. *Nature* **383:** 185-188.

Wallace, D.C. (1992). Diseases of the mitochondrial DNA. *Annu. Rev. Biochem.* **61:** 1175-1112.

Walton, K.M. and J.E. Dixon (1993). Protein tryosine phosphatases. *Annu. Rev. Biochem.* **62:** 101-120.

Walvot V. (1992). Strategies for mutagenesis and gene cloning using transposon tagging and T-DNA insertional mutagenesis. *Ann. Rev. Pl. Physiol. Pl. Mol. Biol.* **43:** 49-82.

Wang, A.H.J., G.J. Quigley, F.J. Kolpak, J.L. Crawford, J.H. van Boom, G. van der Merel and A. Cick (1979). Molecular structure of a left handed double helical DNA fragment at atomic resolution. *Nature* **282:** 680-686.

Wang, T.S.F. (1991). Eukaryotic DNA polymerases. *Annu. Rev. Biochem.* **60:** 513-552.

Warren, G. (1993). Membrane partitioning during cell division. *Annu. Rev. Biochem.* **62:** 323-348.

Watson, J.D. (1963). Involvement of RNA in the synthesis of proteins. *Science* **140:** 17-26.

Watson, J.D. and F.H.C. Crick (1953). Molecular structure of nucleic acids: a structure for deoxyribose nucleic acid. *Nature* **171:** 737-738.

Watson, J.D. *et al.* (1987). *Molecular Biology of the Gene.* 4th edn. Cummings, Menlopark, CA.

Watts, C. and M. Marsh (1992). Endocytosis: what goes in and how? *J. Cell Sci.* **103:** 1-8.

Weinberg, R.A. (1991). Tumor suppressor genes. *Science* **254:** 1138-1146.

Weinert, T. (1997). A DNA damage checkpoint meets the cell cycle engine. *Science* **277:** 1450-1451.

Weinhues, U. and W. Neupert (1992). Protein translocation across mitochondrial membranes. *Bioessays* **14:** 17-23.

Weintraub, H. (1990). Antisense RNA and DNA. *Sci. Amer.* Jan 1990. pp. 38-44.

Weintraub, H., J.G. Izant and R.M. Harland (1985). Anti-sense RNA as a molecular tool for genetic analysis. *Trends in Genet.* **1:** 22-25.

Weising, K., J. Shell and G. Kahl (1988). Foreign genes in plants: transfer, structure, expression and applications *Annu. Rev. Genet* **22:** 421-477.

West, S.C. (1992). Enzymes and molecular mechanisms of genetic recombination. *Annu. Rev. Biochem.* **61:** 603-640.

White, J.M. (1992). Membrane fusion. *Science* **258:** 917-924.

White. K., M. E. Grether, J M. Anrams, L. Young, K. Farrell and H. Steller (1994). Genetic control of programmed cell death in *Drosophila. Science* **264:** 677-678.

White, R.J. and S.P. Jackson (1992). TATA-binding protein: a central role in transcription by RNA polymerase I, II and III. *Trends in Genet.* **8:** 284-288.

White, T.J., N. Arnheim and HA. Erlich (1989). The polymerase chain reaction. *Trends in Genet.* **5:** 185-189.

Whitehouse, H.L.K. (1963). A theory of crossing over by means : of hybrid DNA. *Nature* **199:** 1034-189.

Whitehouse, H.L.K. (1965). *Towards an Understanding of the Mechanism of Heredity.* Edward Arnold, London.

Whitehouse, H.L.K. and P.J. Hastings (1965). The analysis of : genetic recombination on the polaron hybrid DNA model. *Genet. Rev.* **6:** 27-92.

Wilkins, M.H.F. (1965). Physical studies of the molecular structure of deoxyribose nucleic acid and nucleoprotein. *Cold Sp. Harh. Symp. Quant. Biol.* **21:** 75-88.

Willets, N and R. Skurray (1980). The conjugation system of F like plasmids. *Annu. Rev. Genet.* **14:** 41-76.

Williamson, B. (1977). DNA insertions and gene structure. *Nature* **270:** 295-297.

Williamson, B (1978). Split gene transcription. *Nature* **272:** 753.

Winter, G. and C. Milstein (1991). Man-made antibodies. *Nature* **349:** 293-299.

Woese. C.R (1967). *The Genetic Code: The Molecular Basis for Genetic Expression.* Harper & Row, New York.

Wolffe. A.P. (1994). Architectural transcription factors. *Science* **264:** 1100-1101.

Wolffe, A.P. (1997). Sinful repression. *Nature* **387:** 16-17.

Wood, R. D.. (1996). DNA repair in eukaryotes. *Annu. Rev. Biochem.* **65:** 135-167.

Wuthrich, K. (1989). Protein structure determination in solution by nuclear magnetic resonance spectroscopy. *Science* **243:** 45-50.

Yamamoto. T. *et al.* (1992). A bipartite DNA binding domain composed of direct repeats in the TAT-A box binding factor TFIID. *Proc. Natl. Acad. Sci. USA* **89:** 2844-2848.

Yamashita. S. *et al.* (1993). Transcription factor TFIIB sites important for interaction with promoter-bound TFIID. *Science* **261:** 463-465.

Yanofsky, C. (1960). The tryptophan synthetase system. *Bact. Rev.* **24:** 221-245.

Yanofsky, C., B.C. Carlton, J.R. Guest, D.R. Helinski and U. Henning (1964). On the colinearity of gene structure and protein structure. *Proc. Natl. Acad. Sci. USA* **51:** 266-272.

Yarus, M. (1982). Translational efficiency of transfer RNAs: uses of an extended anticodon. *Science* **218:** 646-652.

Yelton, D.E. and MD. Scharff (1981). Monoclonal antibodies: a powerful new tool in biology and medicine. *Annu. Res. Biochem.* **50:** 657-680.

Young. R.A. (1991). RNA polymerase II. *Annu. Rev. Biochem.* **60:** 689-715.

Youvan, D.C. and B.L. Marrs (1987). Molecular mechanisms of photosynthesis. *Sci. Amer.* June 1987. pp. 42-49.

Zawel L. and D. Reinberg (1995). Common themes in assembly and function of eukaryotic transcription complexes. *Annu. Rev. Biochem.* **64:** 533-561.

Zetka, M. and A. Rose (1995). The genetics of meiosis in Caenorhabditis elegans. Trends in Genet. **11:** 27-31.

Zillig. W. (1991). Comparative biochemistry of Archaea and Bacteria. *Curr. Opin. Genet. Devel.* **1:** 544-551.

Zimmerberg, J., S.S. Vogel and L.V. Chernomordik. (1993). Mechanisms of membrane fusion. *Annu. Rev. Biophys. Biomol. Struct.* **22:** 433-466.

Zinder. N.D. (1958). Transduction in bacteria. *Sci. Amer. Nov.* 1958. pp. 1-7.

Zinder. N.D. and J. Lederberg (1952). Genetic exchange in *Salmonella. J. Bacter.* **64:** 679-699.

Zouros, E. *et al.* (1992). Direct evidence for extensive paternal mitochondrial DNA inheritance in the marine mussel Mytilus. *Nature* **359:** 412-414.

Appendix 1: Experimental Techniques in DNA and RNA Analyses

WORKING WITH DNA AND RNA: THE TOOLS

Our knowledge of molecular biology and genetics has depended on the development of adequate research tools. Advances in knowledge of how nucleic acids work have generally been tied to the development of new methods. We discuss here some of these methods.

1. **Extraction and purification of DNA** The first requirement is a sample of DNA free of other cellular chemicals. The steps in the purification of DNA are shown here (Fig. A1.1). The aqueous solution in the final step is treated with RNAse to remove RNA. Proteins are then removed by use of denaturing solvents (usually phenol). By repeating the purification steps a number of times, a solution can be obtained that is virtually free of any components other than DNA.

Note that the solution of DNA obtained never consists of native DNA molecules of the length found in the cell. The purification process causes the DNA to break down into fragments of various (random) lengths. If the DNA has been handled gently during purification, the lengths of the fragments will be about one-hundredth of the length of the whole chromosome.

2. **Detecting the presence of DNA (Fig. A1.1 & A1.2)** There are several methods for detecting the presence of, DNA in a solution. One of the most widely used is by its absorption of ultraviolet radiation. DNA absorbs strongly ultraviolet radiation at a wavelength of 260nm. The absorption is due to the purino and pyrimidine bases. As seen in the Fig., double-stranded DNA absorbs less strongly than single stranded DNA. This is because the interaction between the bases on the opposite strands of the double-stranded DNA (hydrogen bonding) reduces the ultraviolet absorbance.
3. **Density-gradient centrifugation of DNA** DNA molecules vary in density, depending on their exact chemical composition. DNA molecules with higher content of guanine

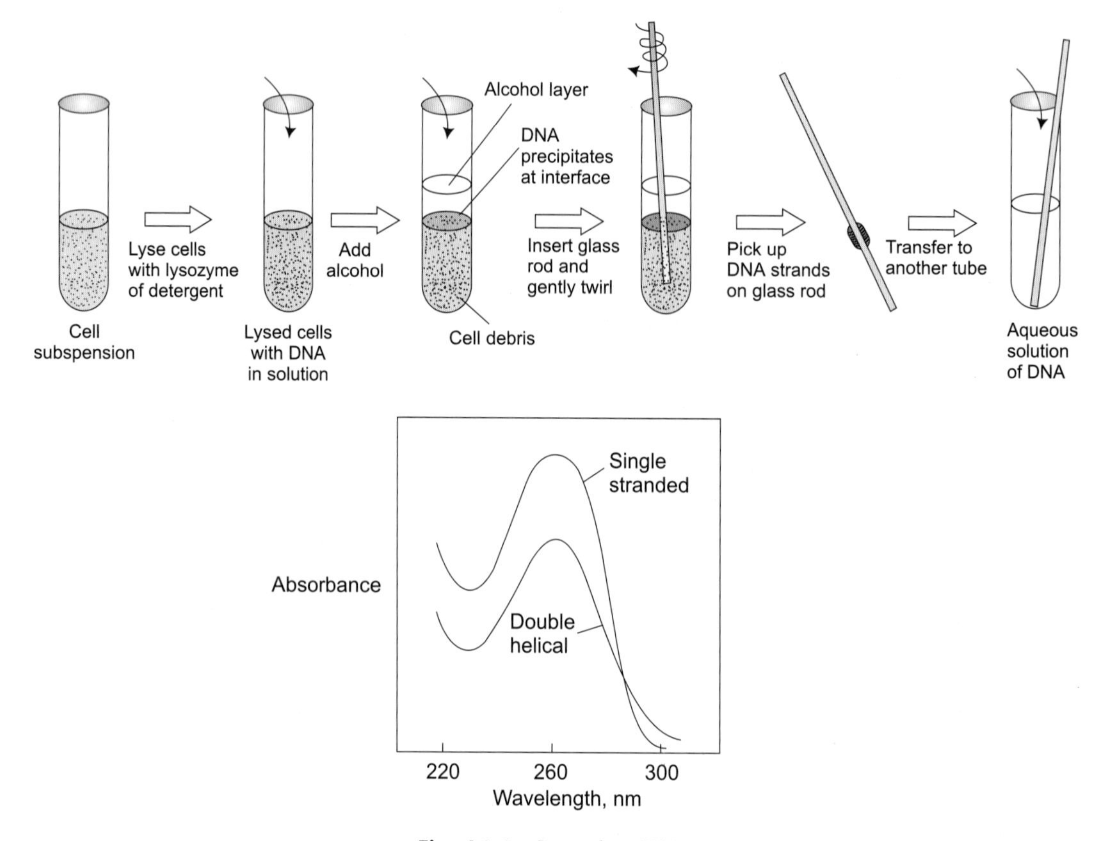

Fig. A1.1 Detecting DNA.

plus cytosine (GC) are denser than molecules with low GC. The density of DNA can be determined by centrifugation at very high speed in a gradient of cesium chloride (CsCL). The DNA solution is layered on top of a solution of CsCL forms and centrifuged at high speed for several days, until equilibrium is reached. The CsCL forms a density gradient from top to bottom of the tube and DNA molecules form band at appropriate densities. At equilibrium, the DNA molecules become positions in the gradient at position corresponding to their densities. Following the addition of ethidium bromide to make the DNA fluoresce (see below), observation of the centrifuge tube with ultraviolet radiation after the centrifugation reveals band of DNA. This method is called the buoyant density and separation of molecules of differing density.

4. **gel electrophoresis** One of the most wide-spread methods of studying nucleic acids is gel electrophoresis. Introduction of electrophoresis methods has revolutionized research on molecular genetics. Electrophoresis is the procedure by which charged molecules are allowed to migrate in an electrical field, the rate of migration being determined by the size of the molecules and their electrical charge. We discussed electrophoresis of proteins in Chapter 1. In gel electrophoresis, the nucleic acid is suspended in a gel, usually made

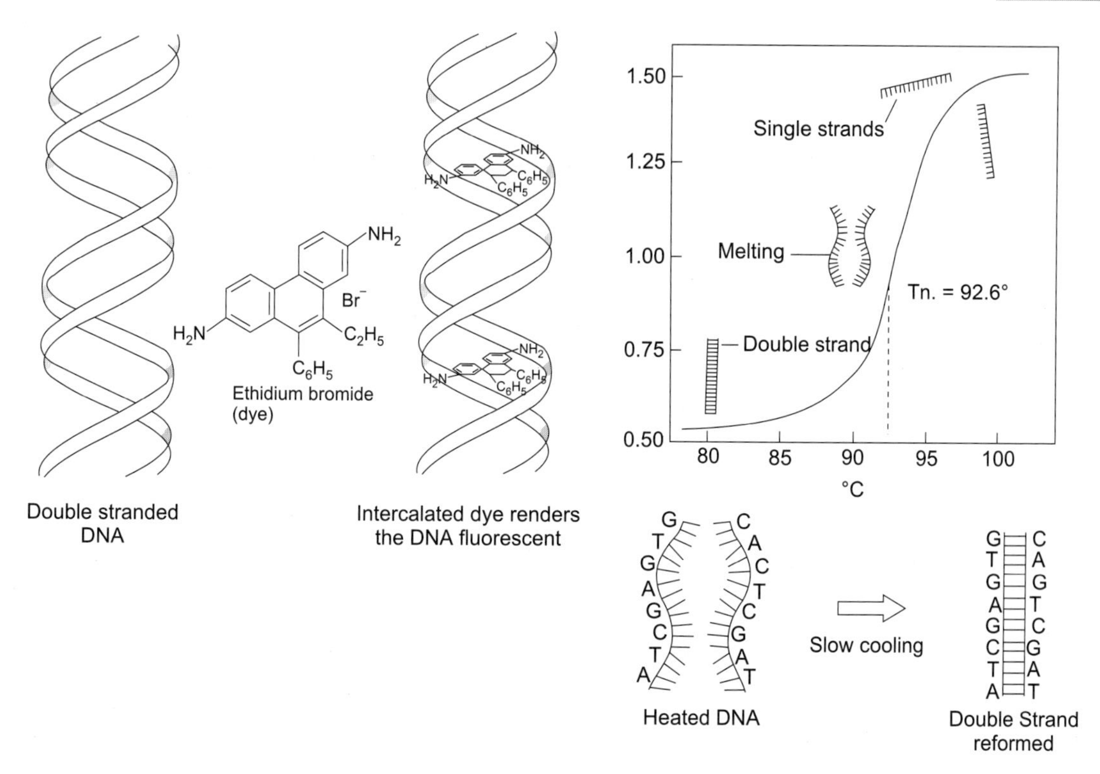

Fig. A1.2 Hydrogen bonding in DNA.

of polyacrylamide or agarose. The gel is a complex network of fibrils and the pore size of the gel can be controlled by the way in which the gel is prepared. The nucleic acid molecules migrate through the pores of the gel at rates depending upon their molecular weight and molecular shape. Small molecules, or compact, migrate more rapidly than large molecules. After a defined period of time migration (usually a few hours), the locations of the DNA molecules in the gel are assessed by making the DNA molecules fluorescent and observing the gel with ultraviolet radiation.

Shown here is a Fig. A1.3 an electrophoresis apparatus. The horizontal frame, made of Lucite plastic, holds the gel. The ends of the gel are immersed in buffer, which makes an electrical connection to the power supply (shown in fore-ground). The gel is observed after electrophoresis by use of ultraviolet radiation. In each lane, a mixture of DNA fragment had been applied. A computerized scanner can be used to locate the positions of the DNA bands.

5. **Detecting DNA by florescence** When nucleic acids are treated with dyes which are fluorescent and which are able to combine firmly with the nucleic acid chain, the nucleic acid is rendered fluorescent. The dye *ethidium bromide* is widely used to render DNA fluorescent because it combines tightly within the DNA molecules. Ethidium bromide interacts with double-stranded DNA. If the DNA is now observed with an ultraviolet source, it will be seen to fluoresce.

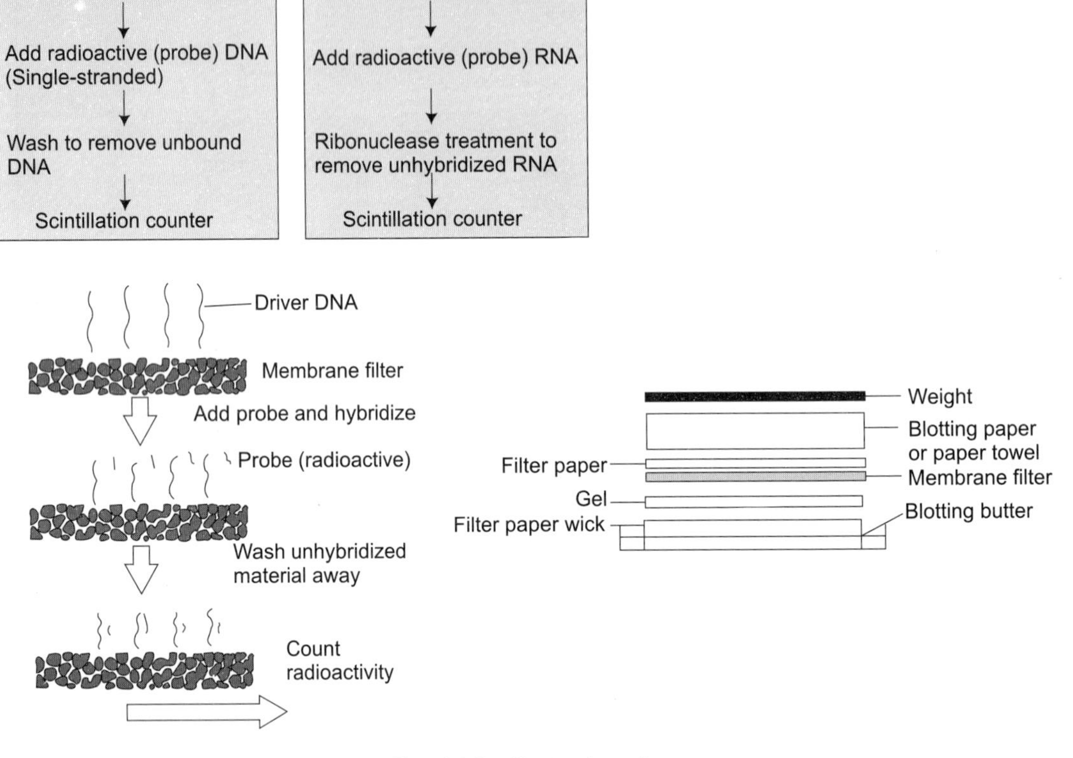

Fig. A1.3 Electrophoresis.

6. **Making nucleic acids radioactive** Radioactivity is widely used in nucleic acid research because radioactivity can be detected extremely tiny amounts. Radioactive nucleic acids can either be detected directly with Geiger or scintillation counters, or indirectly via their effects on photographic film (autoradiography). Autoradiography of radioactive nucleic acids is one of the most widely used techniques in molecular genetics because it can be applied to the detection of nucleic acid molecules during gel electrophoresis.
   (a) A nucleic acid can be made radioactive by incorporation of radioactive phosphate during nucleic acid synthesis. If radioactive phosphate is added to a culture while nucleic acid synthesis is taking place, the newly synthesized nucleic acid becomes radioactively labeled.

$P^{32}O$ - P^{32}-labeled nucleotides

P^{32} - labeled nucleic acid

(b) end-labeling of DNA that contains a free hydroxyl group at the 5′ position can be done, using radioactive ATP labeled in the third phosphate. The enzyme polynucleotide kinase specifically attaches the third phosphate of ATP to the free hydroxyl group at the 5′ end of the molecule. End–labeling is an extremely useful technique as it permits labeling of performed molecules. By tracing the radioactivity through subsequent chemical steps, the end of the molecules can be followed.

$$P^{32} - P \text{ adenosine} + HO - \text{deoxyribose} - DNA -$$
$$P^{32} - O - \text{deoxyribose -DNA} + ADP$$

7. **Effect of temperature on nucleic acids** As we have noted, double-stranded nucleic acid molecules are held together by large numbers of weak (hydrogen) bonds. These bonds break when the nucleic acid is heated, but he covalent bonds holding the polynucleotide chains together ate unaffected. As shown in part 2 above, double-stranded molecules how lower ultraviolet absorbance than single-stranded molecules. Therefore, if the ultraviolet absorbance of a nucleic acid solution is measured while it is being heated, the increase in absorbance when the double-stranded molecules are converted into single-stranded molecules will show the temperature at which strand separation occurs. Strand separation brought about by heat is generally called melting. The stronger the double-strands are held together, the higher will be the temperature of melting. Because guanine-cytosine base pairs are stronger than adenine-thymine base paris (three hydrogen bonds for GC pairs, only two for AT pairs), the higher the GC content the higher the melting temperature. The taxonomic significance of determining the G + C content of an organism's DNA is discussed in Chapter 18. The Fig. shows the change in absorbance at 260 nm when a solution containing double-stranded DNA is gradually heated. The mid-point of the transition, called Tm, is a function of the GC content of the DNA. If the heated DN is allowed to cool slowly, the double-stranded native DNA mat reform.
8. **Nucleic acid hybridization** By hybridization is meant the artificial construction of a double-stranded nucleic acid by complementary base pairing of two single – stranded nucleic acids. (These are side-to-side hybrids, not the end-to-end hybrids formed by genetic recombination or genetic engineering.). If a DNA solution that has been heated (see Above) is allowed to cool slowly, many of the complementary strands will reassociate and the original double-stranded complex reforms, a process called reannealing. The reannealing only occurs if the base sequences of the two strand are complementary. Thus, nucleic acid hybridization permits the formation of artificial double-stranded hybrids of DNA, RNA, or DNA-RNA. Nucleic acid hybridization provides a powerful tool for studying the genetic relatedness between nucleic acids. It also permits the detection of pieces of nucleic acid that are complementary to a single-stranded molecule of known sequence. Such a single-stranded molecule of known sequences is often called a probe. For instance, a radioactive nucleic acid probe can be used to locate in an unknown mixture a nucleic acid sequences complementary to the probe. Detection of

nucleic acid hybridization I usually done with membrane filters constructed of cellulose nitrate. Double-stranded DNA and RNA do not adhere to these filters, but single-stranded DNA and DNA-RNA hybrids do adhere.

Hybridization can also be done after gel electrophoresis. Blotting from the gel to a sheet of membrane filter material transfers the nucleic acid molecules, and the probe then added to the filter. The procedure when DNA is in the gel and RNA or DNA is the probe is often called a southern blot procedure, named for the scientist E.M. Southern, who first development it, When RNA is in the gel and DNA or RNA is the probe, the procedure is called a Northern blot. A Western blot (sometimes called an immunoblot involves protein- antibody binding rather than nucleic acids; the Fig. A1.3 also shows the use of a nucleic acid probe to search for complementary sequences in a mixture. The DNA fragments have been spread out by gel electrophoresis and then transferred to the membrane filter. The RNA probe, which is radioactively labeled, is allowed to reanneal to the DNA on the filter and its position is determined by autoradiography.

9. **Determining the sequence of DNA** Although the base sequences of both DNA and RNA can be determined, it turns out for chemical reasons that it is easier to sequences DNA. Even automatic machines are now available for determining the sequences of DNA molecules. Appropriate treatments are used to generate DNA fragments that end at the four bases, the ends of the fragments being radioactive. Then the fragments are subjected to electrophoresis so that molecules with one nucleotide difference in length are separated on the gel. His electrophoress procedure involves four separate lanes, one for fragments broken at each of the four bases of the DNA, adenine, guanine, cytosine, and thymine. The positions of these fragments are located by autoradiography and from a knowledge of which base is represented by each lane, the sequence of the DNA be read off.

Two different procedures have been developed to accomplish the above, called the Maxam-Gilbert and the Sanger dideoxy procedures. In the Maxam-Gilbert procedure for sequencing DNA the piece of DNA to be sequenced is made radioactive by labeling at the end with ^{32}P. The DNA is then treated with chemicals which break the DNA preferentially at each of the four nucleotide bases, under conditions in which only one break per chain is made. (thus, there are four separate test tubes prepared). After electrophoresis and autoradiography, the sequences can be read directly. The number of nucleotides in each fragment is shown on the Fig.

In the Sanger dideoxy procedure for sequencing DNA the sequence is actually determined by making a copy of the single-stranded DNA, using the enzyme DNA polymerase. This enzyme uses deoxyribonucleoside triphosphates as substrates and adds them to a primer. In the incubation mixtures (four separate test tubes) are small amounts of each of the dideoxy analogs of the deoxyribonucleoside triphosphates in radioactive form. Because the dideoxy sugar lacks the 3′ hydroxyl, continued lengthening of the chain cannot occur. The dideoxy analog thus acts as a specific chain termination reagent. Radioactive fragments of variable length are obtained, depending on the incubation conditions. Electrophoresis of these fragments is then carried out and the positions of the radioactive bands determined by autoradiography. By aligning the four dideoxynucleotide lanes and noting to its neighbor, the sequence of the DNA copy can be read directly from the gel.

Another approach based on the Sanger principle is to use fluorescent labels instead of radioactivity, one fluorescent color for each of the four bases. Then the electrophoresis can be done in one lane instead of four, with the fragments allowed to run off the bottom of the gel, where their fluorescence color is measured with a special laser fluorimeter. This procedure makes it possible to automate the sequencing process.

A major advantage of the Sanger method is that it can be used to sequence RNA as well as DNA. To sequence RNA, a single-stranded DNA copy is made (using the RNA as the template) by the enzyme reverse transcriptase. By making the single-stranded DNA in the presence of dideoxynucleotides, various sized DNA fragments are generated suitable for Sanger-type sequencing. From the sequence of the DNA, the RNA sequence is deduced by base-pairing rules. The Sanger method has been instrumental in rapidly sequencing ribosomal RNAs for use in studies on microbial evolution

For determining the DNA sequence of a long molecule, such as a whole gene, it is necessary to proceed in stages. First, the DNA is broken into small overlapping fragments and the sequence of each fragment determined. Using the overlaps as guide, the sequence of the whole molecule can be deduced.

Bonding, However, the base pairs must become unstacked vertically to allow for intercalation, so that the sugar-phosphate backbone is destroyed and the regular helical structure is destroyed. As we will see later, one consequence of intercalation is that the reading frame of the code can be changed ultimately resulting in formation of a faulty protein.

Examples of intercalating chemicals are the acridine dyes such as acriflavine, and acridine orange and ethidium bromide (see Nucleic Acids Box). Such compounds serve as useful tools in studying the structure and function of DNA and in visualizing DNA experimentally. Some of the planer molecules which intercalate into DNA are cancer-inducing agents, carcinogens, or mutation-inducing agents, mutagens. We will discuss carcingogens and mutagens later

Some antibiotics combine strongly and specifically with DNA. We will discuss later the details of how DNA-binding antibiotics act, but note here one impotent group of antibiotics, the actinomycins. These antibiotics not only intercalate into the double helix, but also have peptide side chains, which attach to the major groove, effectively inhibiting both DNA replication and transcription.

HYBRIDIZATION OF NUCLEIC ACIDS

Hybridization is the artificial construction of a double-stranded nucleic acid by complementary base pairing of two single-stranded nucleic acids. The procedure for constructing nucleic acid hybrids is shown in the Nucleic Acids Box. Both DNA: DNA and DNA: RNA hybrids can be made. There must be a high degree of complementarity between two single-stranded nucleic acid molecules if they are to fond a stable hybrid. In the most common use of hybridization in genetic engineering, one of the molecules is used as a radioactive probe to detect a specific nucleic acid sequence and formation of hybrids is detected by observing the formation of double-stranded molecules containing radioactivity.

One of the most common uses of hybridization is to detect DNA sequences that are complementary to mRNA molecules. Detection of DNA: RNA hybridiztion is usually done with membrane filters constructed of cellulose nitrate. RNA does not stick to these filters, but single-stranded DNA and DNA: RNA hybrids do. The single-stranded DNA is first immobilized on the filter and the radioactive RNA probe added. After appropriate incubation, the unhybridized RNA is washed out and the radioactivity still bound to the filter is measured.

INTERACTION OF NUCLEIC ACIDS WITH PROTEINS

Of great importance is the interaction of nucleic acids with proteins. Protein-nucleic acid interactions are central to replication transcription and translation and to the regulation of these processes. Two general kinds of protein-nucleic acid interactions are noted: nonspecific and specific depending upon whether the protein will attach anywhere along the nucleic acid or whether the interaction is sequence-specific. As an example of proteins that do not interact in a sequence-specific fashion, we mention the histones, proteins which are extremely important in the structure of the eukaryotic chromosome, although less significant in prokaryotes. Histones are relatively small proteins that have a high proportion of positively charged amino acids (arginine, lysine, histidine) (Fig. A1.4). DNA, as we have noted, is a polynucleotide, and has a high proportion of negatively charged phosphate groups. Since the phosphate groups are on the outside of the DNA double helix, DNA is a negatively charged molecule. Histones, because of their positive charge, combine strongly and relatively nonspecifically with the negatively charged DNA. In the eukaryotic cell there is generally enough histone so that all of the phosphate groups of the DNA are covered. Association of histones with DNA leads to the formation of nucleomes, the unit particles of the eucatyotic chromosome

There are also a number of proteins that interact with DNA in a sequence-specific manner. These interactions occur by association of the amino acid side chains of the proteins with the bases as well as with the phosphate and sugar molecules of the DNA. The major groove, because of its size, is an important site of protein binding (Fig. A1.5). In order to achieve specificity in such interactions, the protein must interact with more than one nucleic acid base, frequently several. We have already described a structure in DNA called an inverted repeat. Such inverted repeats are frequently the locations at which protein molecules combine specifically with DNA (Fig. A1.6). Proteins, which interact specifically with DNA, are frequently dimmers, composed of two identical polypeptide chains. On each polypeptide chain is a region, called a domain, which will interact specially with a region of DNA groove. A consideration of this type of interaction provides an explanation for the plaindromic nature of the base sequence at the point in the DNA where the protein interacts: in this way each of the

Ser - Gly - Arg - Gly - Lys - Gly - Gly - Lys - Gly - Leu -
Gly - LYs - Gly - Gly - Ala - Lys - Arg - His - Arg - Lys -
Val - Leu - Arg - Asp - Asn - lie - Gin - Gly - lie - Thr -
Lys - Pro - Ala - Ile - Arg - Arg - Leu - Ala - Arg - Arg -
Gly - Gly - Val - Lys - Arg - lie - Ser - Gly - Leu - lie -
Tyr - Glu - Glu - Thr - Arg - Gly - Val - Leu - Lys - Val -
Phe - Leu - Glu - Asn - Val - lie - Arg - Asp - Ala - Val -
Thr - Tyr - Thr - Glu - His - Ala - Lys - Arg - Lys - Thr -
Val - Thr - Ala - Met - Asp - Val - Val - Tyr - Ala - Leu -
Lys - Arg - Gin - Gly - Arg - Thr - Leu - Tyr - Gly - Phe -
Gly - Gly -

Fig. A1.4 Structure of histone protein. The positively charged amino acids are marked.

DNA strand (Fig. A1.7). Note, However, that the protein does not directly recognize the specific base sequence of the DNA. Rather it recognizes contact points such as electrostatic charges and hydrophobic interactions that are associated with specific base sequence.

Studies of several DNA binding proteins have revealed a common protein substructure named the helix-turn-helix, which is apparently critical for proper binding of many of these proteins to DNA (Fig. 5.7), the "helix-turn-helix" is so named because these DNA binding proteins are

C-C-A-G-G
G-G-T-C-C

and modification of this sequence results in methylation of two cytosines:

m

C-C-A-G-G
G-G-T-C-C

m

Note that a given nucleotide sequence can be a substrate for either a restriction enzyme or a modifications enzyme but not both. This

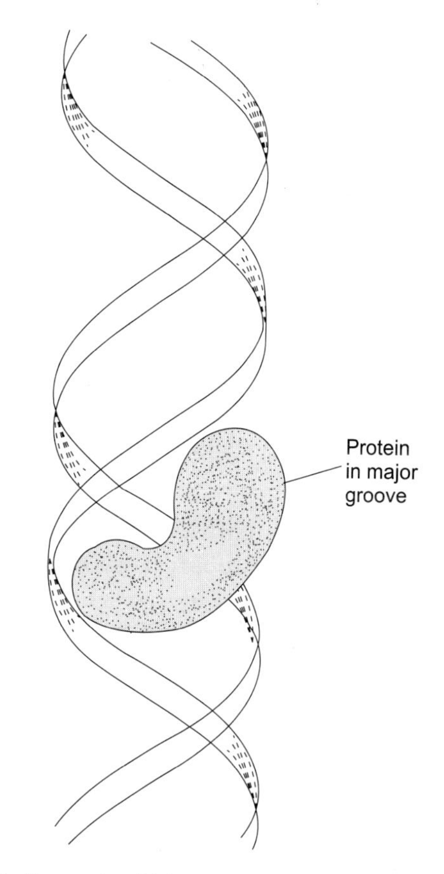

Fig. A1.5 Manner in which part of a protein molecule can fit into the major groove of the DNA double helix.

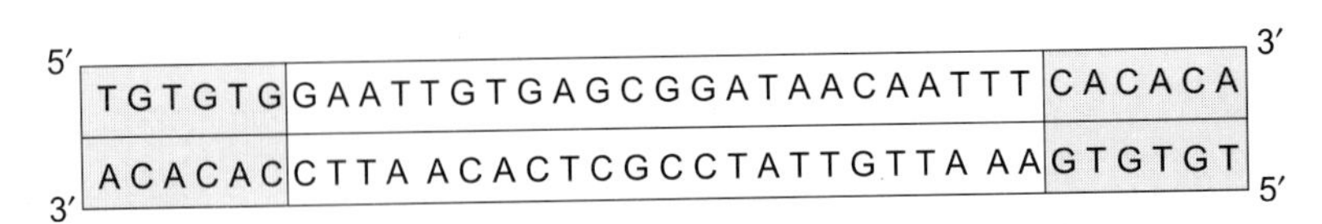

Fig. A1.6 Nucleotide sequence of the operator gene of the lactose operon. Nearby inverted repeats, which are sites in which the lac repressor makes contact with the DNA, are shown in shaded boxes.

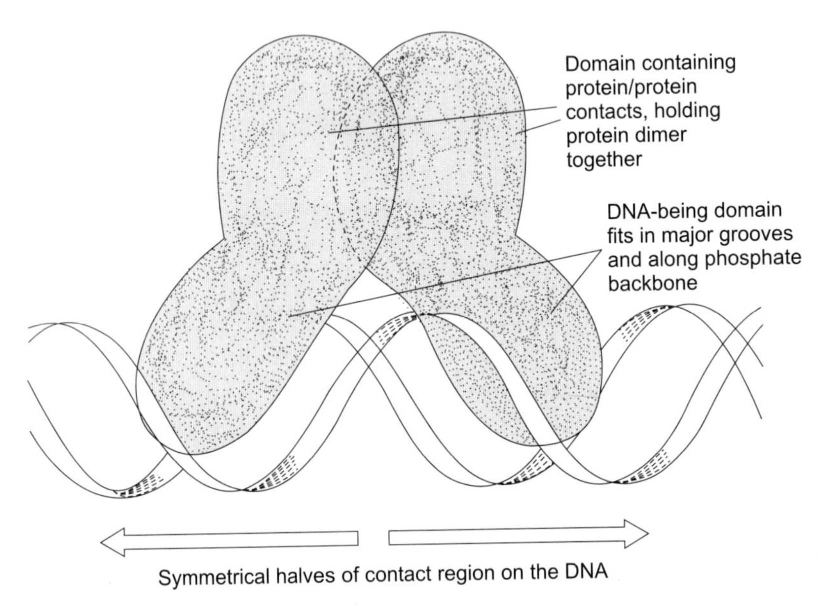

Fig. A1.7 A protein dimmer combines specifically with two sites on the DNA.

is because modification makes the sequence unreactive with restriction enzyme, and action of restriction enzyme destroys the recognition site of the modification enzyme.

Restriction enzymes are such important tools in modem molecular genetic research that they have become widely available commercially. A number of companies purify and market restriction enzymes of a variety of specificities. By referring to a catalog of base sequences, a research worker can generally obtain a restriction enzyme that will cut or near a particular site. On the other hand, if it is known that a particular piece of DNA has been cut by a particular restriction enzyme, they by reference to the base sequence which that restriction enzyme cuts, it is possible to deduce the base sequence around that site. This provides a powerful tool for studying DNA molecules.

RESTRICTION ENZYME ANALYSIS OF DNA

As noted, a DNA molecule can be cut at a specific location by a given restriction enzyme. Because the base sequences recognized by restriction enzyme are 4-6 nucleotides long, there will generally be only a limited number of such sequences in a piece of DNA. After cleaving the

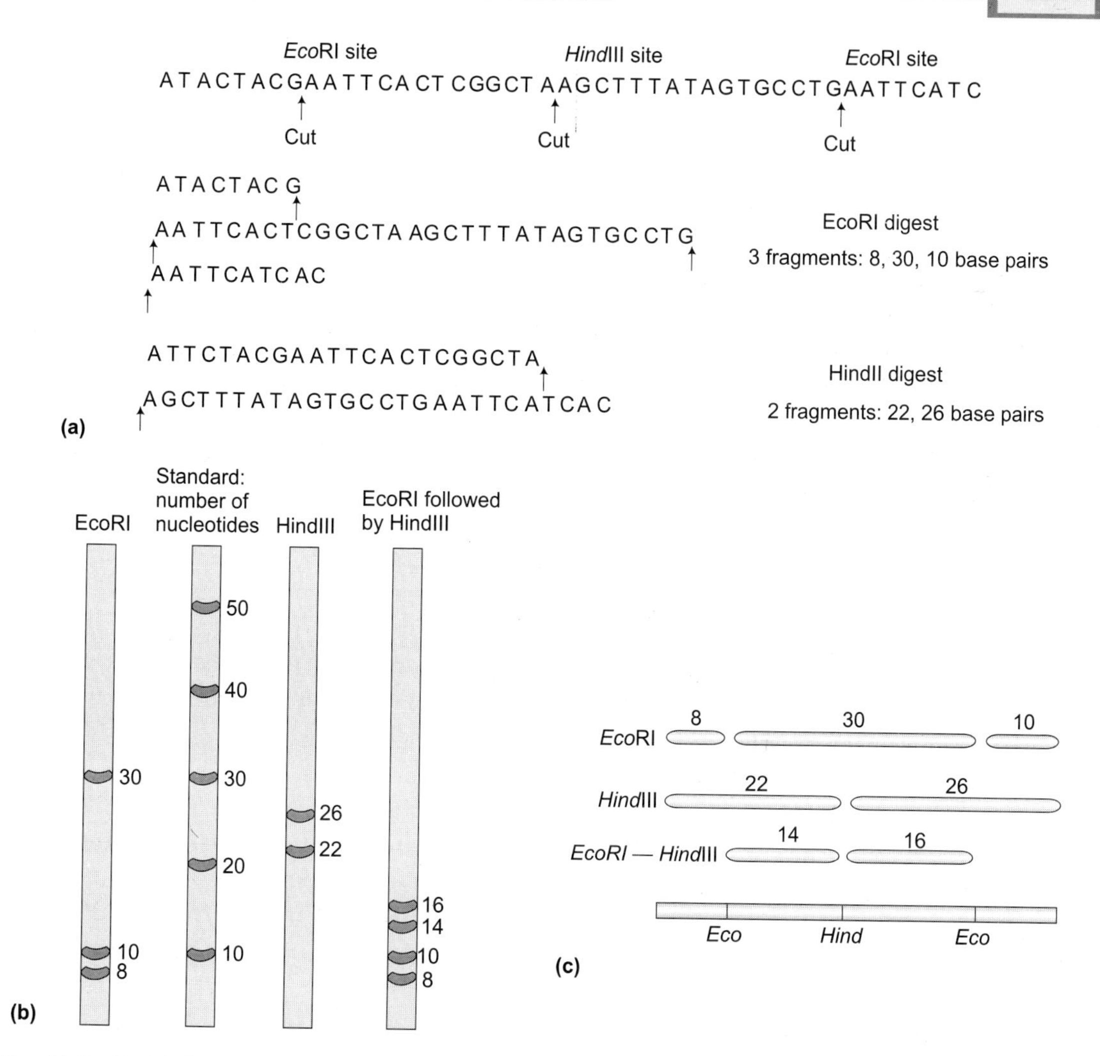

Fig. A1.8 Restriction enzyme analysis of DNA. (a) A 48 base sequence, with 2 EcoRI and 1 HindIII restriction sites. Note that the enzyme cuts double-stranded DNA. For simplicity, only a single strand is shown. (b) Results of electrophoresis of digest with each of the enzymes. Note that when the EcoRI digest is then digested with HindIII, the 30 base fragment is affected, showing it has a HindIII site. (c) By orientation of the overlapping framents, it is possible to deduce a restriction enzyme map of the DNA molecule.

DNA (Fig. A1.8a), the fragments can be separated by agarose gel electrophoresis, as shown in Fig. A1.8b, the distance migrated by any band of DNA in such a gel can be determined by calibrating the electrophoresis system with DNA molecules of known size. By judicious use of several restriction enzymes of different specificities, and by use of overlapping fragments, it is possible to construct a restriction enzyme map in which the positions cut by each of the several restriction enzymes can be designated (Fig. A1.8c).

Several procedures are now available for determining the base sequences of DNA molecules. In fact, automatic machines are available for sequencing DNA.

Details are presented in the Nucleic Acids Box. By successively determining the sequences of small overlapping fragments of DNA, it is possible to determine the sequence of very large pieces typically dimer and consist of a stretch of amino acids which form an A-helix secondary structure. This secondary structure is the so-called recognition helix, which is joined to a short stretch of three amino acids, the first of which is usually a glycine that functions to "turn" the protein, the other end of the "turn" is connected to a second helix which stabilizes the first by interacting hydrophobically with it. Recognition of specific DNA sequences occurs by a combination of noncovalent interactions including hydrogen bonds and van der Waals contacts between the protein and base pairs on the DNA. Interestingly, many different DNA binding proteins show the helix-turn-helix structure, including many repressor proteins such as the bacteriophage lambda repressor (Fig. A1.8b) and the lac and trp repressors of Escherichia coli. Although other structure types of DNA binding proteins exist, the helix-turn-helix is certainly a major class of bacterial DNA binding proteins.

Once a protein combines at a specific site on the DNA, number of outcomes can occur. In some cases, all the protein does is block some other process, such as transcription. In other cases, the protein is an enzyme which carries out some specific action on the DNA, such as RNA polymerase, which makes RNA using DNA as the template. However, one of the most interesting groups of such proteins is the restriction enzymes, enzymes which specifically cut DNA at site near where they combine. We discuss these interesting enzymes in the next section.

The complementary strands of a DNA molecule can be separated experimentally by chemicals or heat. Two complementary single strands can hybridize to form a stable double-stranded molecule. RNA can also hybridize with single-stranded DNA. Hybridization can be used to measure the degree of sequences homology of two single-stranded nucleic acid molecules and is an important tool in modern molecular genetics. Recognition also occurs between specific sequences in nucleic acids and certain proteins (DNA-binding proteins), and such protein-nucleic acid interactions affect the structure and function of nucleic acids.

RESTRICTION ENZYMES AND THEIR ACTION ON DNA

Organisms are occasionally faced with the problem of coping with foreign DNA, generally derived from viruses, that may derange cellular metabolism or initiate processes leading to cell death. Although a number of mechanisms for coping with foreign DNA exist, one of the most dramatic is that which results in its enzymatic destruction. The enzymes involved in the destruction of foreign DNA are remarkably specific in their action, an essential property if destruction of cellular DNA is to be avoided. One class of highly specific enzymes are called restriction endonuclease. Restriction enzymes combine with DNA only at sites with specific sequences of bases. Restriction enzymes have the unique property of making double-stranded breaks in DNA only at sequences, which exhibit two-fold symmetry around a given point (see Section 5.2), thus one restriction end nuclease of Escherichia coli, called EcoRI, has the following recognition sequence:

↓ ↓
5′ – G-A-A–T-T-C- – 3′
3′ – C-T-T–A-A-G- – 5′
↑

The cleavage sites are indicated by arrows, and the axis of symmetry by a dashed line. Nucleotide sequences with inverted repeats, such as are recognized by restriction enzymes, have been found to be widespread in DNA. The sequences recognized by restriction enzymes are relatively short, and frequently palinfromic. They are probably cleaved because the restriction enzymes are composed of identical subunits, each of which recognizes the sequence on a single chain. The significance of this specificity is that if the same sequence is found on each strand, such enzymes will always make double-stranded breaks, and such double-stranded breaks are not subject to correction by repair enzymes. This ensures that an invading nucleic acid will be destroyed.

Restriction enzymes are of great importance in DNA research, because they permit the formation of smaller fragments from large DNA molecules. Such fragments with defined termini, created as a result of the action of specific restriction enzymes, are amenanle to determination of nucleotide sequences, thus permitting the working out of the complete sequence of DNA molecules. A large collection of restriction enzymes has now been built up, which can be used in sequence determination. Recognition sequences for a few restriction enzymes are given in Table A1.1.

Another use of certain restriction enzymes is that they permit the conversion of DNA molecules into fragments which can be joined by DNA ligase. This enables laboratory researchers to create artificial genes.

An integral part of the cell's restriction mechanism is the modification of the specific sequences on its own DNA so that they are not attacked by its own restriction enzymes. Such modification generally involves methylation of specific bases within the recognition sequence so that the restriction nuclease can no longer act. Thus, for each restriction enzyme there must also

Table A1.1 Recognition sequences of a few restriction endonucloeases.

Organism	*Enzyme designation*	*Recognition sequence**
Escherichia coli	EcoRI	G↓AATTC
Escherichia coli	EcoRII	↓CCAGG (Not a palindromic sequence)
Haemophilus influenzae	HindII	GTPy↓PuAC
Haemophilus influenzae	HindIII	AAC↓CTT
Haemophilus hemolyticus	HhaI	GCG↓CC
Bacillu: subtilis	BsuRI	GG↓CC
Brevibacterium albidum	BalI	TGG↓CCA
Thermus aquaticus	TaqI	T↓CGA

*Arrows indicate the sites of enzymatic attack. Asterisks indicate the site of methylation (modification). G = guanine; C - cytosine; A – adenine, T - thymine, Pu - any purine; Py = any pyrimidine, Only the 5′ → 3′ sequence is shown.

be a modification enzyme, the two enzymes being closely associated. For Example, the sequences recognized by the EcoRII restriction enzyme (also see Table A1.1) is: of DNA. The sequences are now known for hundreds of genes, as well as for the complete genome of many viruses.

Restriction enzyme are cellular enzymes which recognize specific short base sequences in DNA and make single-stranded breaks at locations within the recognition sites. A restriction enzyme does not affect the cell that produces it because its own DNA is methylated at the recognition site by a modification enzyme specific for that site. Thus, restriction and modification enzymes constitute a specific system for protecting the cell's own DNA but destroying foreign DNA. Restriction enzymes are important research tools in modern molecular genetics.

Appendix 2: Membrane Function: Cell Adhesion and Cell Junctions

The cells in tissues are assembled together and remain in contact with the help of **extracellular matrix (ECM)**. These tissues in vertebrates include nerve, blood, lymphoid, epithelial and connective tissues. Among these tissues, at one extreme is the connective tissue in which cells are sparsely distributed and ECM is abundant bearing most of the mechanical stress. On the other extreme is the epithelial tissue, where cells are tightly bound together into sheets called epithelia and ECM is scanty consisting of a thin **basal lamina**, so that cells themselves bear most of the stress with the help of cytoskeleton. The cells in all these tissues are held together by **cell-cell adhesion**, which may or may not he associated with the formation of structures called '**cell junctions**. In early, stages of development, only cell-cell adhesions are displayed and no cell junctions are observed under the microscope, but in mature adult structures, cell junctions can be observed under the electron microscope. While the cell-cell adhesions have been studied through functional and biochemical texts, particularly, in embryonic tissues during development, the cell junctions have been studied by electron microscopy, particularly in the mature adult structures. In recent years, a unified picture of these cell adhesions and cell junctions has also emerged due to an understanding of their molecular basis. In this chapter, we first discuss the functional and biochemical aspects of cell-cell adhesion and then describe the structure and function of different types of cell junctions.

CELL SORTING AND DIFFERENTIAL CELL AFFINITY

Cell sorting is a phenomenon, which allows different types of cells to segregate and form specific regions of the developing embryo. Therefore, in recent years, a study of cell sorting helped in understanding the mechanism involved in the normal development of the embryo. Although the cell surface looks much the same in all cells, it actually differs in different cell types and facilitates cell sorting through cell-cell interactions. A definite evidence suggesting the presence of the phenomenon of "**cell sorting**" was actually provided by **Townes** and **Holtfreter** in the year 1955. They demonstrated that the dissociated cells of vertebrate embryonic organs can reaggregate and sort out to reconstruct semblances (structures looking alike) of the original structure. The amphibian tissues do become dissociated into single cells when placed in alkaline solution. Utilizing this feature, Townes and Holtfreter prepared single cell suspensions from each of the three germ layers (ectoderm, mesoderm, endoderm) soon after the neural tube had formed in amphibian embryos. It could be shown that when cells of different kinds were mixed, cells of same type aggregated together through the phenomenon of cell adhesion. The result of these experiments led to the following conclusions: (i) The reaggregated cells become spatially segregated in the reconstructed structures. (ii) The reaggregated cells reflected the embryonic positions; for instance, when ectodermal and mesodermal cells are mixed, they sort out and ectodermal cells are found at periphery, while mesodermal cells are found inside, so that epiderm envelopes the mesoderm (Fig. A2.1A); the mesoderm also migrates centrally with respect to endoderm or gut (Fig. A2.1B); similarly, when cells from three layers are mixed, the endoderm separates from ectoderm and mesoderm and is then enveloped by them; the neural plate cells also migrate to their corresponding position. These results were interpreted in terms of relative **positive affinity** of the inner surface of ectoderm for mesoderm, and its **negative affinity** for endoderm; mesoderm in its turn had positive affinity both for ectoderm and endoderm. (iii) The selective affinities of cells change during development; for instance, in blastula, all cells seem to have same affinity for each other but have high affinity for the extracellular matrix or ECM (hyaline layer) covering of the embryo and low affinity for the proteins inside the blastocoel. At the onset of gastrulation, a specific group of cells at the vegetal end of the blastula lose their affinity for neighbouring cells and for external extracellular matrix, but acquire affinity for protein fibrils inside the blastocoel. Consequently, the cells release their contacts with adjoining cells and migrate inside the blastocoel, where they form the skeleton (Fig. A2.2). These changes in affinities are accompanied with changes in adhesion properties and are extremely important during morphogenesis.

CELL ADHESION MOLECULES (CAMs)

Differentiation and morphogenesis during development of multicellular organisms involve a variety of interactions among different cells or between cells and its external environment. One of these interactions involves the phenomenon of selective **cell recognition** and **cell adhesion**, which may include adhesion among adjoining cells forming tissues, or between cells and the substratum leading to attachment of the cell. Cell adhesion is mediated by **cell adhesion molecules (CAMs)**, **substratum adhesion molecules** and **cell junction molecules**. These molecules excluding cell junctions) include the following four major families of receptors (i) cadherins, whose adhesion properties are dependent on Ca^{++} ions; (ii) members of Ig superfamily (they are Ca^{++} independent) including N-CAM (neural cell-cell adhesion molecules) and ICAM-1 (intercellular adhesion molecules, whose cell binding domains resemble those of

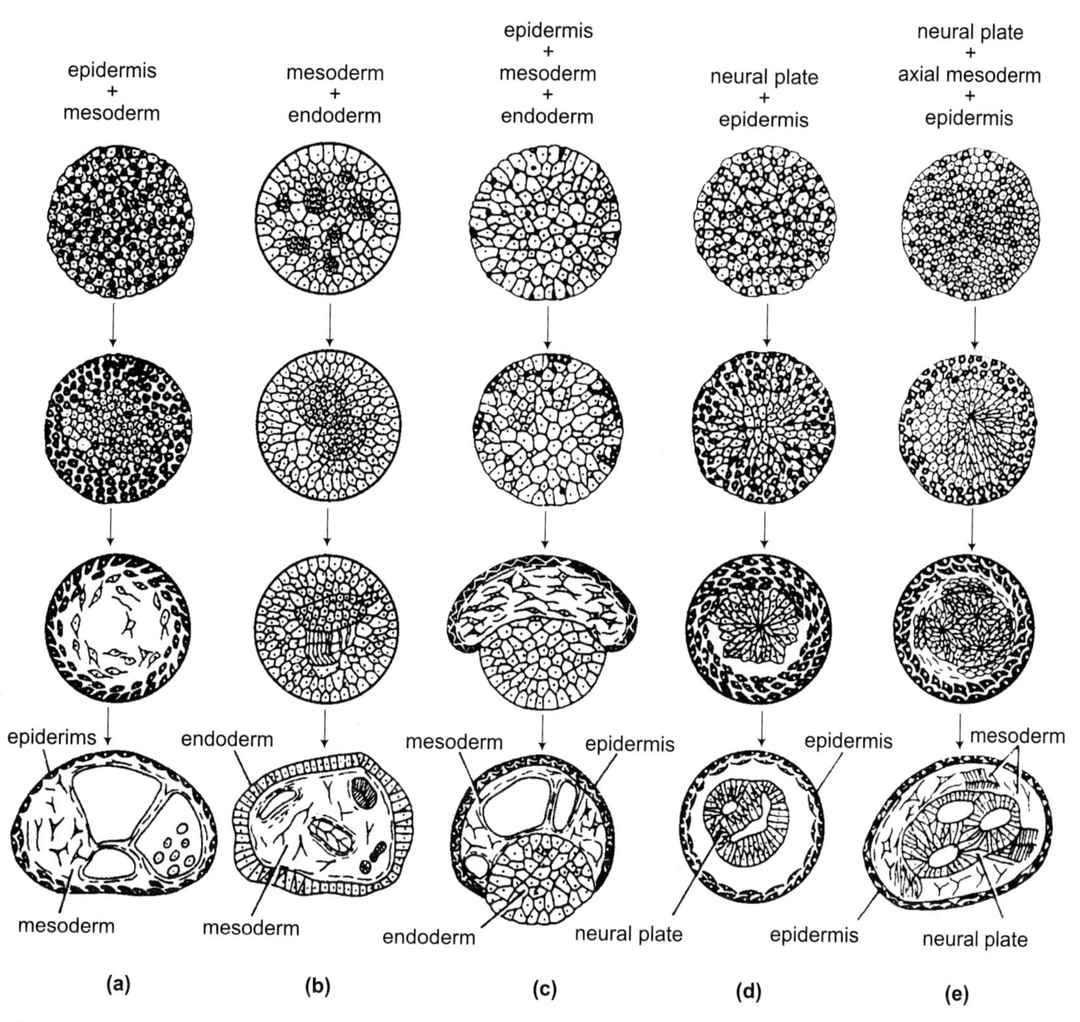

Fig. A2.1 Sorting out of the aggregates of embryonic cells and reoganization of their relationships (redrawn from Gilbert, 1994).

immunoglobulin molecules); (iii) selectins and (iv) integrins. The first three of these molecules mediate cell-cell adhesion, while integrins mediate cell-substratum adhesion.

Cadherins (Ca^{++} Dependent Cell-cell Adhesion)

Cadherins are calcium-dependent glycoproteins, which are crucial to the spatial segregation of cells leading to the organization of animal form. In mammalian embryos, following three classical cadherins are found (although 12 different types are known): (i) **E-cadherins** (epithelial cadherins, also called **uvomorulins** or L-CAM); (ii) **P-cadherins** (placental cadherins) and (iii) **N-cadherins** (neural cadherins, also called A-CAM). It has been shown that cells with a

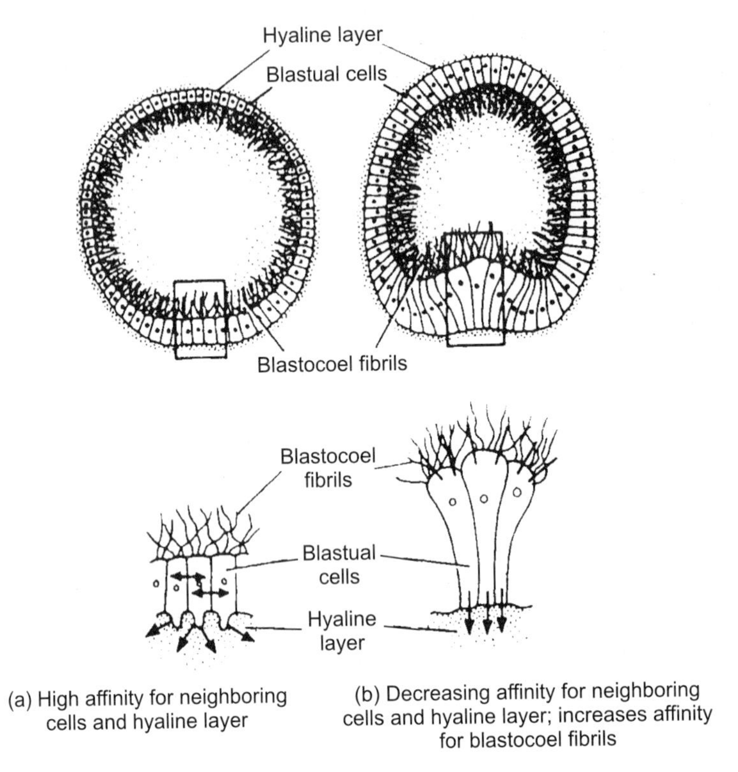

Fig. A2.2 Changes in cell adhesion properties of skeletal precursor cells with development (redrawn from Gilbert, 1994).

particular type of cadherins adhere preferentially to other cells expressing the same cadhcrins. This is called **homophilic binding**. Each cadherin has three major regions—an NH_2-terminal extracellular recognition domain of 113- amino acids (containing four major repeats) outside the cell, a transmembrane domain spanning the membrane and cytoplasmic domain within the cell The extracellular domain differs among cadherins and help in cell-sorting or cell recognition and is also responsible for Ca^{++} binding. The cytoplasmic domain mediates interaction with atleast three cytosolic proteins called α-, β- and **γ-catenins**. Although extracellular domain helps in **cell recognition**, but no cadherin mediated cell adhesion is observed in cells lacking either the cytoplasmic domain of cadherin or one of the three cytosolic proteins, the α-, β- and **γ-catenins** (Fig. A2.3). It is postulated that in the presence of Ca^{++}. the carboxy terminal cytoplasmic domain of a cadherin is connected to the **actin microfdaments** of the cytoskeleton through the catenins. The clustering of cadherins, catenins and actins in this manner leads to the formation of a belt-like region called **adherens-junction** or **zonulae adherens**, which is responsible for cell-cell adhesion. It has also been shown that the integrity of these junctions is regulated by protein-tyrosine phosphorylation.

Another example of cell-cell adhesion is represented by '**desmosomes**' (also called '**maculla adherens**'), which are spot-like sites of intercellular contact. In this case, the cell adhesion is

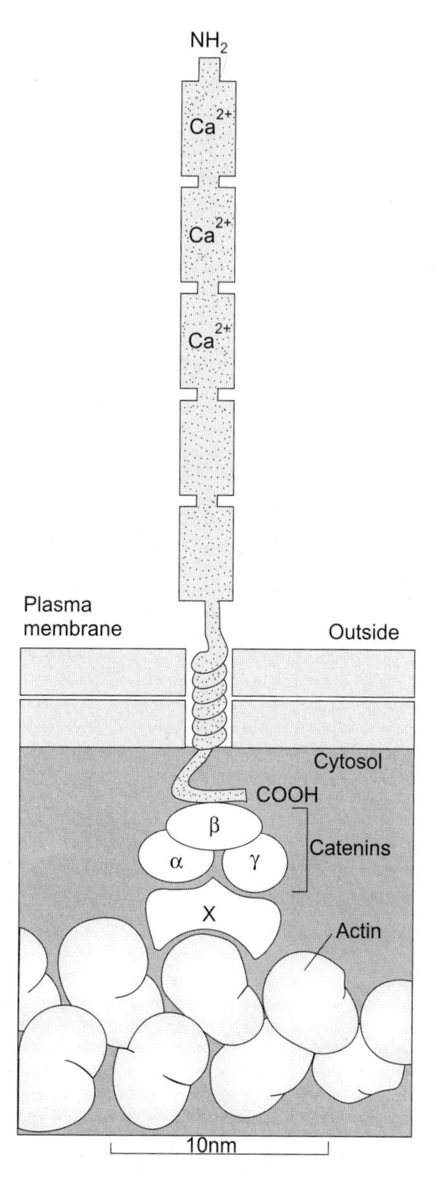

Fig. A2.3 Structure of a cadherin molecule with five extracellular domains, three of them having calcium binding sites, the farthest of which mediates cell-cell adhesion (redrawn from Alberts *et al., Molecular Biology of the cell,* 194).

mediated by three trans-membrane Ca^{++} dependent proteins (**desmoglein I, desmocollin I** and **desmocollin II**), each with four external extracellular domains identical to those found in classical cadherins. However, the cytoplasmic domains differ and are linked to intermediate filaments (not to actins as found in classical cadherins), through a cytoplasmic structure called the **desmosomal plaque**.

Cadherins are extremely important in establishment and maintenance of cell-cell interactions and their deficiency may lead to malignancy of cells or to other life threatening diseases like **pemphigus vulgaris**, in which epidermal cells lose adhesion, blister and fall off.

Immunoglobulin Superfamily CAMs (Ca^{++} independent Cell-cell Adhesion)

As the word 'superfamily' implies, these immunoglobulin related cell adhesion molecules (CAMs) are large and varied. They are the most, populous and represent a functionally diverse family of molecules present on the cell surface. (The designation 'superfamily' is used because immunoglobulins themselves constitute a family, and these immunoglobulin-related CAMs have structures resembling immunoglobulins, but do not represent a 'close' family). The structure of these CAMs resembles that of immunoglobulin molecule (Fig. A2.4) and it is considered likely that these CAMs later evolved into immunoglobulins during evolution. It is for this reason, that these glycoproteins are, called immunoglobulin superfamily CAMs'.

Neural cell adhesion molecules (N-CAM) are the most abundant and the best studied example of '**immunoglobulin superfamily CAMs**'. They are found in a variety of cell types including nerve cells and bind adjoining cells together by **homophilic binding** (like cadherins),

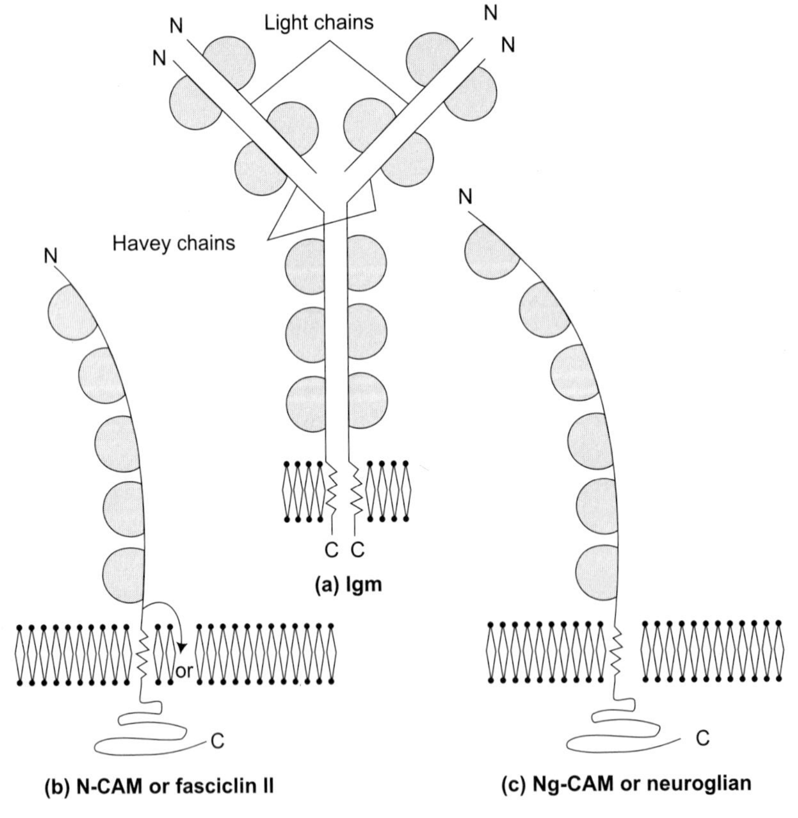

Fig. A2.4 Three members of the immunoglobulin superfamily (redrawn from Gilbert, 1994).

although there are other Ig-like CAMs, which use heterophilic binding. These other Ig like CAMs (involving heterophilic binding) include **intercellular adhesion molecules** (ICAMs), which are found on activated endothelial cells and bind to **integrins** (occurring on the surface of white blood cells) to help trapping of white blood cells sites of inflammation.

Selectins

'**Selectins**' (name used for the first time by **Bevilacqua** *et al*, 1991) represent a family of three cell surface glycoproteins, that contain lectin domains for recognizing cell-specific glycoproteins, (found on the surface of some cells) during regional inflammation. Therefore, the discovery of these selectins has given a meaning to the presence of cell-specific carbohydrates, which are utilized by leukocytes of cell-cell adhesion. The selectins actually provide for a mechanism of cell-cell (leukocycte-endothelial cell) adhesion, where **lectin** or **carbohydrate recognition domain (CRD)** in a selectin of one cell gecognizes a cell surface carbohydrate of another cell. (This is unlike CAMs, because although like CAMs. they facilitate) adhesion, but unlike CAMs, they encourage segregation of cell types). Following are the three known selectins (Fig. A2.5): (i) **L-selectin** or **leucocyte-selectin** (earlier also named as **homing receptor**, because it is involved in homing of lymphocytes back to the sites from which they were derived) is a 90 to 100kD antigen, confined to leukocyte cell surfaces, and synthesized constitutively, (ii) **E-selectin** or

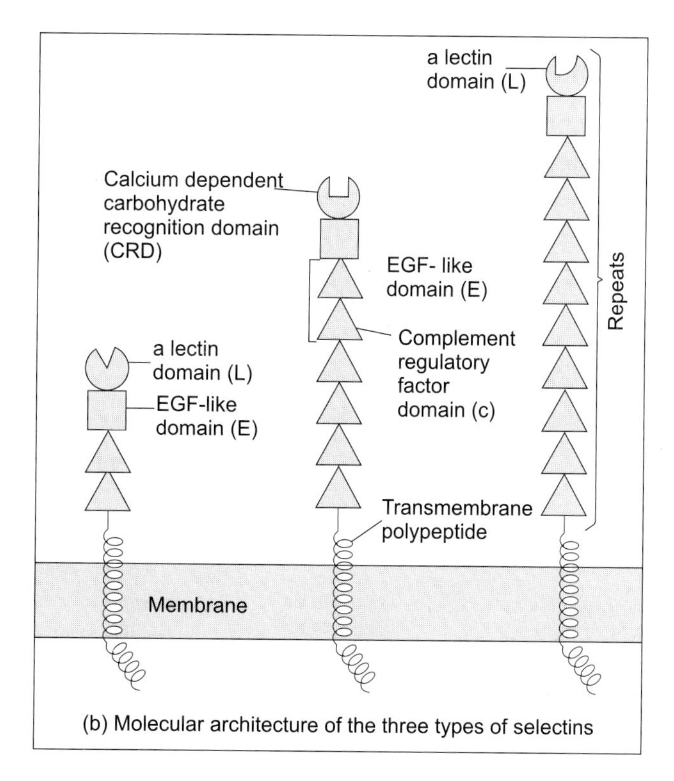

(b) Molecular architecture of the three types of selectins

Fig. A2.5 Three types of selectins.

endothelial-seletin (earlier also called **ELAM-1** or **endothelial leukocyte adhesion molecule**), located on endothelium, has a molecular mass of 95-155 kD and is synthesized within hours on induction of endothelial cells by inflammatory stimuli like interleukin-1 or tumor necrosis factor; this selectin functions as an adhesion molecule and binds neutrophils and monocytes to endothelial cell surface. (iii) **P-selectin** or **platelet-selectin** (earlier also called **GMP140** or **granule membrane protein** of molecular mass 140 kD) is also called **PADGEM** protein (platelet activation-dependent granule external membrane protein). It is also involved in binding of neutrophils of monocytes to endothelial cell surface, but is stored in platelet alpha granules or endothelial cell Weibel-Palade Bodies. It is induced within minutes (does not take hours as in case of E-selectin), by activation of cells with thrombotic stimuli.

Detailed studies of the above three selectins suggested excellent correspondence between the domain structure of these glycoproteins and the exon-intron structure of the genes encoding them. The domains in these selectins included the following: (i) a lectin domain (L) or carbohydrate recognition domain (CRD) for recognition of cell surface carbohydrate (CSC) , (ii) an epidermal growth factor (EGF) - like motif (E) juxtaoposd to lectin and (iii) a variable number of repetitive domains of glycoprotein C, specific to a selectin and similar to those in various complement binding proteins. It was shown that the overall similarity of L and E domains of the three selectins was high (65%). while the repetitive C domains were less conserved. The details of the three selectins are summarized in Table A2.1.

Integrins (Cell Receptors for ECM Molecules) Focal Adhesion and Hemidesmosomes

Integrins (so named by **Horwitz** *et al.*, 1986 and **Tamkun** *et al.*, 1986) are a class of transmembrane adhesion-receptor proteins, so called because they integrate the extracellular and intracellular scaffolds. They are the best characterized ECM receptors. The extracellular matrix (ECM) proteins on the substratum taking part in adhesion are substrate adhesion molecules and are represented by proteins like **fibronectin** and **vitronectin**, while the

Table A2.1 Some characteristic features of selectins.

Type	Location	Expression	Adherent cell types	Proposed function
1. L-selection (homing receptor)	Leucocytes (constitutive)	Decreases on cell activation	PLN endothelium; endothelium near inflammatory sites	Lymphocyte recirculation through PLN. neutrophil (+ other loucocytes) inflammation
2. E-selection (ELAM)	Endothlium (transcriptionally activated)	Increase on inflammation (within hours)	Monocytes, neutrophils, T-cell subsets (Th1, Th2)	Leukocyte inflammation
3. P-selection (GMP 140)	Platelets (α-granules); endothelium (Weibelpalade Bodies)	Increase upon thrombin activation histamine, etc. (within minutes)	Monocytes, neutrophiuls, T-cell subsets (Th1, Th2)	Leukocyte inflammation

intracellular scaffold is represented by **actin filament** bundles (stress bundles). These integrins are present in plasma membrane regions called **focal adhesions**, which are closely adherent to the substratum, (separated by only 10 nm-15 nm). The number of focal adhesions is inversely proportional to the rate of movement of the cell on the substratum (ECM).

CELL JUNCTIONS

Specialized junctions are found in all animal tissues among cells of a tissue and between cells and ECM to allow for cell-cell and cell-matrix contacts. They are particularly abundant and important in epithelia. Cell junctions are also found in plants and are described there as **plasmodesmata**. These cell-cell junctions and cell-matrix junctions are collectively called **cell junctions** and can be of the following three functional types: (i) **Occuluding junctions**, which seal cells together in an epithelial sheet preventing the passage of even small molecules from one side of the sheet to the other; (ii) **anchoring junctions**, which allow attachment of cells to their neighbouring cells or to ECM and (iii) **communicating junctions**, which allow the chemical or electrical signals to pass from one cell to another. These three types of junctions, and particularly the anchoring and communicating junctions are further classified into subclasses, depending upon the location and function of these cell Junctions (Table A2.2).

Cell Junctions in Animals

Occluding junctions (tight junctions). Occluding junctions or tight junctions between epithelial cells help formation of a sheet like structure, which functions as a selective barrier between fluids on either side of this sheet. For instance, the epithelial cells lining the mammalian small intestine or gut do not allow most of the contents in the cavity (lumen) to pass on to the other side of the epithelium, but do allow the transport of selected nutrients from the lumen across the epithelium sheet. After permeating through the connective tissue, these selected nutrients diffuse into blood vessels. This transcellular transport depends on two sets

Table A2.2 Different types of cell junctions.

(1) Occluding junctions (tight junctions)
(2) Anchoring junctions
  (a) Active filament attachment sites
    (i) cell-cell adherens junctions (e.g., adhesion belts)
    (ii) cell matrix adherens junctions (e.g., focal contacts)
    (iii) septate junctions (invenebrates only)
  (b) Intermediate filament attachment sites
    (i) cell-cell adheion (desmosomes)
    (ii) cell-matrix adhesion (hemi-desmosomes)
(3) Communicating junctions
  (a) gap junctions
  (b) chemical synapses
  (c) plasmodesmata (plants only)

of membrane bound **carrier proteins**: (i) carrier proteins confined to **apical surface** (surface facing the lumen) activity transport (Na^+—driven glucose symport) selected molecules from lumen into the cells and (ii) carrier proteins confined to **basolateral surface** allow passage of molecules from the cell to extracellular fluid by **facilitated diffusion** (by glucose carriers). In view of the different mechanisms involved in transport from apical surface and basolateral surface, the carrier proteins should not be allowed to migrate from one area to another, and the spaces between epithelial cells are sealed, so that the transported molecules do not diffuse back from intercellular spaces (Fig. A2.6). Permeability of different epithelia to small molecules differs, so that the epithelium lining the small intestine is 10,000 times more permeable to inorganic ions (e.g., Na^+) than the epithelium lining the urinary bladder.

The tight junctions have been shown to consist of an anastomosing network of strands, each composed of long rows of specific transmembrane proteins in each of the two interacting plasma membranes, which join to occlude the intercellular space (Fig. A2.7).

Cell Junctions in Plants (Fig. A2.8 & A2.9)

Plasmodesmata. The presence of cell wall in plants though eliminates the need for anchoring junctions to hold the cells in place, but the need for direct cell-cell communication remains.

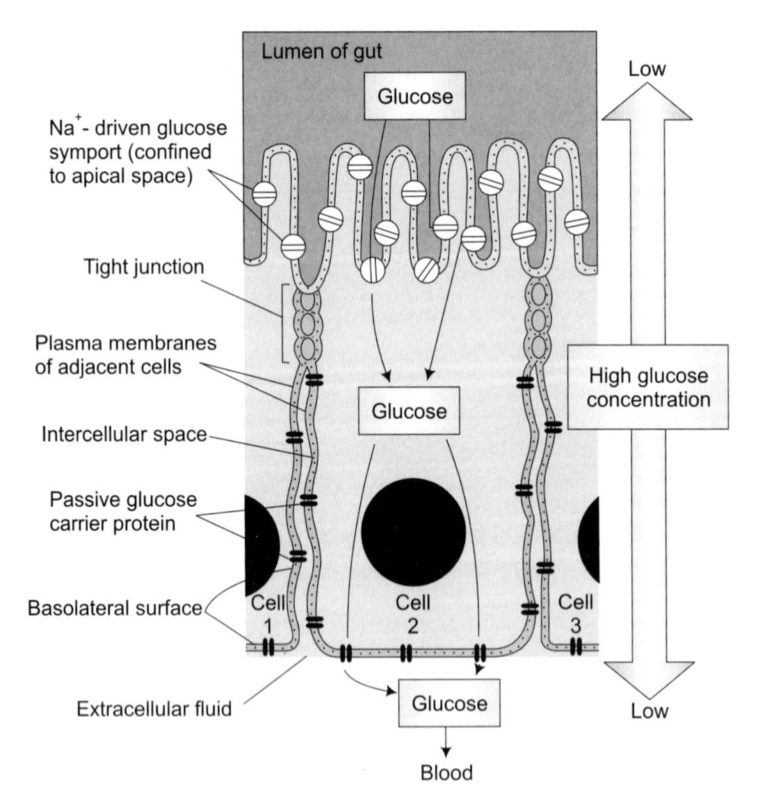

Fig. A2.6 The role of tight junctions in trnascellular transport (redrawn from Alberts *et al., Molecular Biology of the Cell,* 1994).

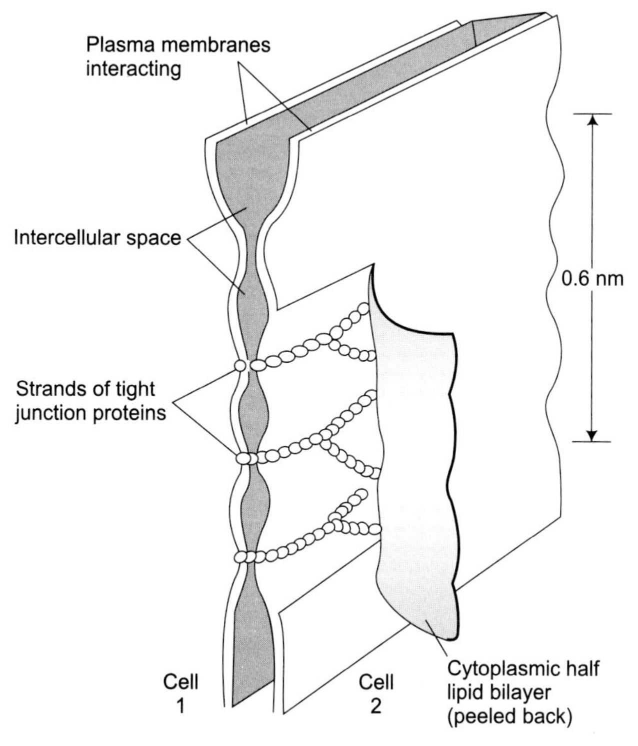

Fig. A2.7 A model of tight junction (redrawn from Alberts *et al., Molecular Biology of the Cell,* 1994).

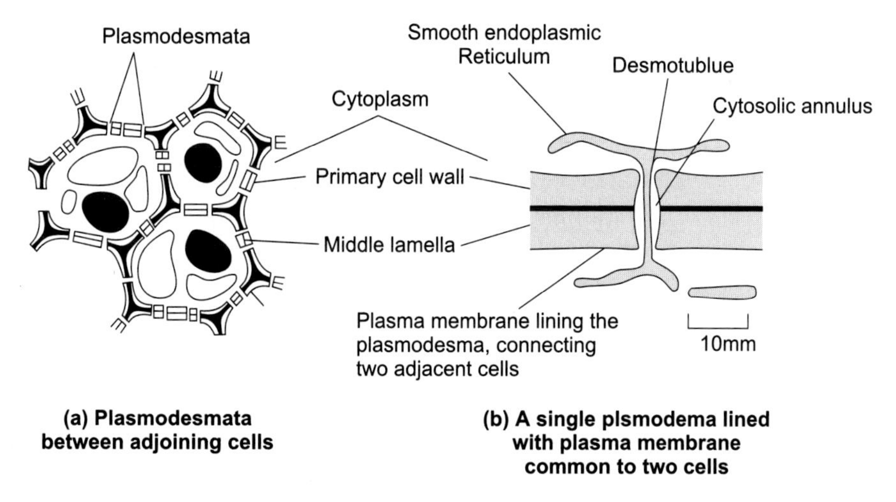

Fig. A2.8 Structure of plasmodesmata (redrawn from Alberts *et. al., Molecular Biology of the Cell,* 1994).

Only one type of cell junctions are known in plants and these are plasmodesmata (PD), which like gap junctions bring about coupling between cytoplasm of adjoining cells. The plasmodesmata form fine cytoplasmic channels (20-40 nm in diameter) between cells. Running

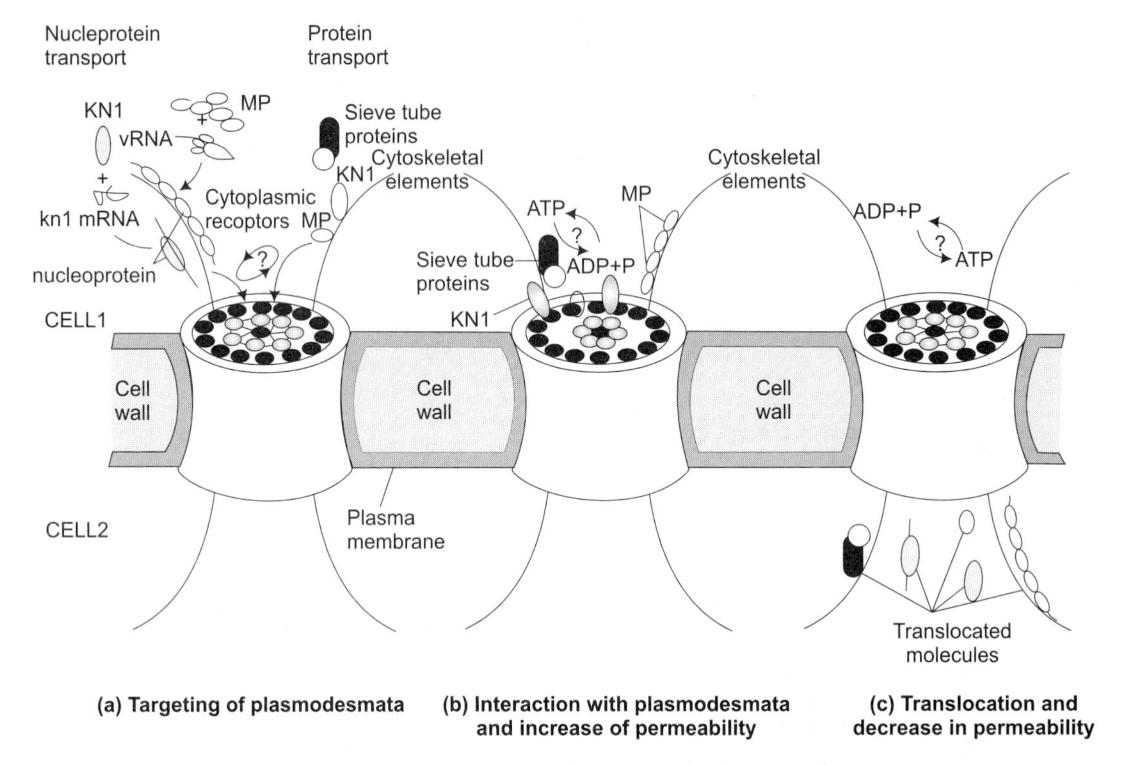

Fig. A2.9 Translocation of various nucleoproteins and proteins through plasmodsmata.

through the centre of this channel is a narrow cylindrical structure called **desmotubule**, which is continuous with the endoplasmic reticula of adjoining cells. Between the inner wall of plasmodesmata and the desmotubule, is a **cytosolic annulus** through which molecules can pass (Fig. A2.8). The plasmodesmata resemble gap junctions in many ways. However in contrast to these gap junctions in animals, plasmodesmata in plants arc clongated structures traversing the thick cell walls Each PD has (i) an outer sheath contiguous with the plasma membrane, (ii) a central core of endoplasmic reticulum (ER) and (iii) a collar or neck region.

Plasmodesmata (PD) are known for long and were initially assigned a passive role of permitting free movements of small metabolites and growth hormones (less than I kilodalton) between plant cells. Later, when it was found that large genomes of plant viruses could traverse from one cell to another, it was suggested that viruses alter the PD to allow transport of very large molecules like viral genomes. More recent work, however, suggests that plasmodesmata are not meant for a mere passive role as above, but are rather dynamic structures, rapidly altering their dimensions to allow transport of bigger molecules, like viral genomes and endogenous plant proteins. It has also been shown recently that the macromolecules track through the plant cytoplasm with the help of cytoskeleton, which functions as a major tracking system to the site of PD.

Movement proteins (MP) of plant viruses have been discovered, which operate an endogenous PD transport system. If MP is injected into single cells, the permeability of PD is increased within minutes of microinjection, confirming this role of MP has also been shown that PDs in different cell type in plants are diverse and therefore, respond differently to MP in different cell types. For instance, in tobacco leaves, in the absence of MP the plasmodesmata in trichome hair cell allow transport of molecules less than 7 kD and those in the mesophyll cells allow transport of molecules less than 1 kD. However, MP induced gating permits movement of dextrans (more than 20 kD) in mesophyll cells, but not in trichome cells, although MP itself (30 kD) moves in both cell types, It, has also been shown that a 30 kD MP mediates movement of a 68 kD reporter protein, when present in cis-orientation with MP, but not when present in trans-orientation (i. e., not transported unless attached with MP), This suggests that MP has a **transport signal** which dictates, transport irrespective of the size of the molecule to be transported. The four criteria for a molecule to be transported include size, shape, signal sequence an gating junction. The structure and behaviour of PD itself depends on cell/tissue type and the developmental stage.

Intercellular transport of RNA and, protein molecules through PD in maize has been inferred by locating the presence of RNA and, protein molecules encoded by specific genes. For Instance, during development, RNA transcribed *by kn 1* gene is not found in outermost L I layer of meristem, but the *knl* gene encoded **KNI (KNOTTED 1)** protein (a transcription factor) is found in this, layer suggesting that this must have been transported through PD from other cells, where. RNA is present and gets translated into protein. KN1 protein not only moves between mesophyll cells but also facilitates movement of dextrans and proteins larger than 20 kD. KN1 protein also selectively transports *knl* sense RNA but not antisense RNA between cells. Such regulation of the movement of molecules by regulation of the distribution and permeability of PD may actually regulate development and differentiation. Plasmodesmata are also implicated in the movement of numerous proteins (upto as large as 70 kD) from companion cells to the enucleate sieve elements of the phloem.

Another aspect relevant to the function of PD is the mechanism of the movement of molecules within the cytoplasm to reach PD. In this connection, co-localization of MP with microtubules and actin filaments and its involvement in transport through PD (as shown above) suggest that the cytoskeleton is used for tracking of macromolecules to PD. The cytoskeleton may also be involved in gating of PD. Actin filaments also traverse PD channels and actin may, act as sphincter at the neck region of PD. Additional cellular factors may interact with actin to generate an open/closed PD conformation thus regulating transport (Fig. A2.9).

MEMBRANE EXCITABILITY IN ANIMALS (NEUROTRANSMISSION AND ION CHANNELS (Fig. A2.10 & A2.11)

One of the important characteristics of all living systems is their ability to respond rapidly to stimuli such as sights, sounds, smells, etc. In animals, this will certainly involve movement of muscles and intercellular communications. These communications need to be very fast, certainly

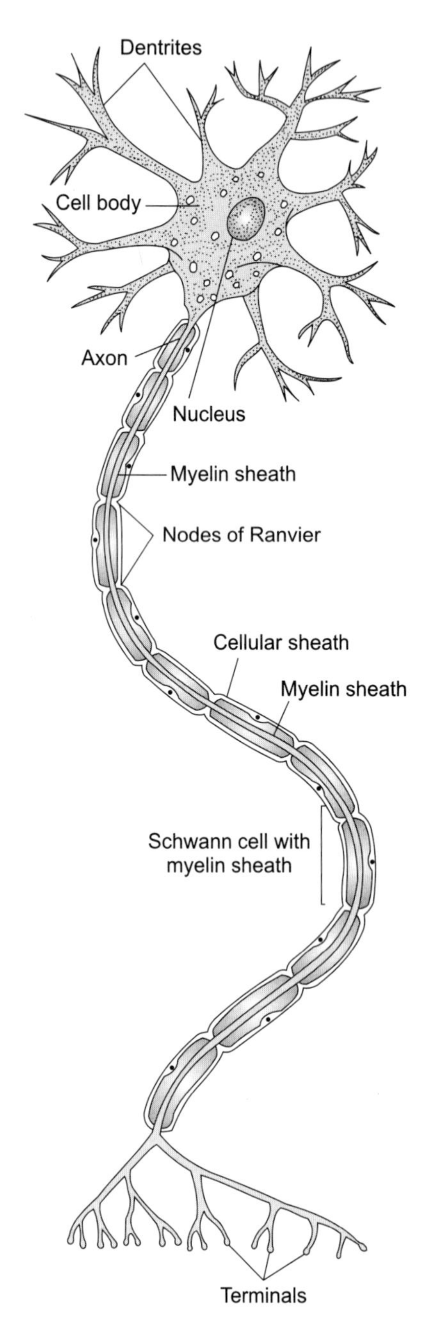

Fig. A2.10 Structure of a neuron showing three parts-cell body, axon and dendrites.

faster than the speed with which blood carrying hormones, or lymph flows in the circulatory system. Nerve impulses can move at a speed of 100 m/sec or more and thus fulfil the requirement of quick response to stimuli. These nerve Impulses are generated and transmitted through a complicated network connecting every part of the body to the brain, the latter itself having as many as 10^{12} cells.

The nervous system contains specialized **receptor proteins** in the membranes of excitable cells (cells which can be excited by stimuli). The stimulus causes a conformational change in these receptor proteins or causes a change in permeability of the excitable membranes. These changes in excitable membranes are propagated from one part of the cell to another and from one cell to another in a specific but reversible manner, carrying, information from one art of the organism to another. These changes involve regular transport of ions and substances called **neurotransmitters**. The progress in our knowledge in this area has been spectacular in recent years, and will be briefly discussed in this section. However, since transmission of signals in the nervous system involving transport of ions and neurotransmitters takes place through nerve cells or neurons, a brief account of their structure will be presented here, before discussing the mechanism involved in the transmission of nerve impulses (neurotransmission).

Neurons can be of three types–(i) **sensory neurons** which acquire sensory signals, (ii) **interneurons** connecting one neuron to another neuron and having the function of receiving signals from one neuron and pass it on to another neuron, and (iii) **motor neurons,** which pass signals from neurons to muscle cells, thus causing muscle movement. In some cases (lower organisms), interneurons are missing, and sensory neurons are connected directly to motor neurons. In still other cases (e.g., sea anemones), both interneurons and motor neurons are missing, so that sensory neurons are directly connected to muscle cells.

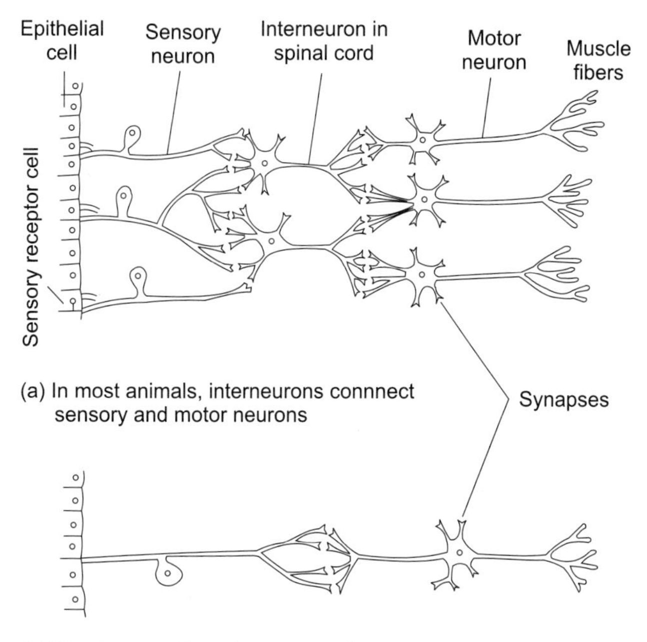

(a) In most animals, interneurons connnect sensory and motor neurons

(b) Direct connection of sensory and motor neurons

Fig. A2.11 Connections ('synapses') between sensory neurons, interneurons and motor neurons.

Neurons and Neuroglia (or Glial Cells)

Neurons and neuroglia are cell types unique to nervous system. while reception and transmission of nerve impulses are carried out by neurons, the protective and supportive functions are performed by glial cells (neuroglia = nerve glue). A neuron consists of three distinct regions–the **cell body** (soma), the **axon** and the **dendrites**. The cell body contains nucleus and the cell organelles (e.g. ER and mitochondria), but axon is simply an extension from the cell body, and so are the dendrites, except that the latter are much smaller in length and are branched, receiving nerve impulses and transmitting them to cell body. The axon itself is a long (sometimes upto one meter in length) thin structure, whose primary function is to carry nerve impulses from the cell body to cell extremities or termini. Axons of neurons outside the brain, are covered by two layers (outer **cellular sheath** and the inner **myelin sheath**) derived from **Schwann cells**. Gaps, between Schwann cells are called **nodes of Ranvier**, where no sheaths are found (Fig. A2.10). The axon terminates in structures called **synaptic terminals, synaptic knobs** and **synaptic bulbs**. The space between synaptic terminal of one neuron and a dendrite ending of an adjoining neuron is described as **a synapse** or **synaptic cleft**.

Ion Gradients for Transmission of Nerve Impulse or Action Potential

The nerve impulses, also called '**action potentials**', are transmitted from neuron to neuron in the form of transient changes in electrical potential differences (voltages) across the membranes of neurons (and other cells). These potential differences are generated by ion gradients. At the resting stage, a neuron is rich in K^+ (20 times nigher inside than outside) and poor in Na+ (9 times lower inside than outside) and Cl^- ions relative to extracellular fluid. This condition is maintained by Na^+, K^+-ATPase system, which sends out three Na^+ ions for every two K^+ ions imported. The potential difference at this resting stage is called '**resting potential**' and is approximately –60 mV (negative inside), at which no K^+ ions move across the membrane. It actually.

Cell-cell Communication at Synapse

A neuron transmits a signal to another neuron through a synapse and the same neuron may involve in several synapses (1 to 10,000). While in midbrain, there is only one synapse per post-synaptic cell, there may be 10,000 synaptic knobs impinging a single spinal motor neuron. In human forebrain also, the ratio of synapses to neurons is 40,000 : 1. The synapses may be **electrical synapses** or **chemical synapses**. In electrical synapses, the gap between presynaptic and postsynaptic cells is small, only 0.2 nm, but in chemical synapse this gap is 20 nm-50 nm wide. In electrical synapse, the arrival of action potential at presynaptic membrane leads directly to depolarization of the postsynaptic membrane, and a new action potential is initiated in the postsynaptic cell. On the other hand, in a chemical synapse, the action potential causes secretion of a chemical substance (called a **neurotransmitter**), by the presynaptic cell. This neurotransmitter has an affinity for the receptor located on postsynaptic membrane and therefore binds there, initiating a new action potential.

One of the most common examples of chemical synapse is **cholinergic synapse** involving acetylcholine as a neurotransmitter. There is a variety of other neurotransmitters which can be grouped as amino acids, catecholamines, peptides and gases (Table A2.3).

Acetylcholine as a neurotransmitter. In cholinergic synapses, at the synaptic knobs, there are synaptic vesicles containing acetylcholine molecules (approx. 1000 molecules/vesicle). When the action potential arrives at such a knob, special voltage-gated Ca^{++} channels open and Ca^{++} ions enter the synaptic knob, causing acetylcholine containing vesicles to attach and fuse with the knob membrane. The vesicles then open and release acetylcholine into the synaptic cleft (space

Table A2.3 A list or some neurotransmitters.

Sl no.	*Class of neurotransmitters*	*Neurotransmitters*
1.	Cholinergic agents	Acetylcholine
2.	Catechamines	Norepinephrine (noradrenaline), epinephrine (adrenaline), L-dopa dopamine, octopamine
3.	Amino acids and their derivatives	Gamma-aminobutyric acid (GABA) alanine, aspartate, cystathione glycine, glutamate, histamine, proline, serotonin, taurine, tyrosine
4.	Peptides	Cholecystokinin enkephalins and endorphins, gastrin, gonadotropin, neurotensin, oxytocin, secretin, somatostain substance P, thyrotropin releasing factor (TRF), Vasopressin, vasoactive intestinal peptide (VIP)
5.	Gases	Carbon monoxide (CO), nitric oxide (NO)

between presynaptic and postsynaptic membranes). Acetylcholine molecules then bind to acetylcholine receptor molecules located in the postsynaptic membrane and cause the opening of Na^+ and K^+ ion channels leading to the generation of new action potential in the postsynaptic neuron Acetylcholine receptor is itself (**nicotinic receptor** only; see later) an ion channel, which was the first ion channel to be purified and sequenced; it was also the first to be reconstituted in synthetic lipid bilayer and was the first for which electrical signal of a single open channel was recorded.

The fusion of acetylcholinc containing vesicles is facilitated by the presence of **synapsin-1**, a 75 kD protein located in the vesicle membrane. This protein binds calmodulin due to influx of Ca^{++} (calmodulin is regulated by Ca^{++}) and gets phosphorylated due to a kinase enzyme leading to the fusion of vesicles with the knob membrane.

Acetylcholine receptors found in postsynaptic membranes can be **muscarinic receptors** (stimulated by muscarine) or **nicotinic receptors** (stimulated by nicotine). The former (found in smooth muscles and glands) are transmembrane proteins, that interact with G proteins and the latter (found in sympathetic ganglia and motor end plates of skeletal muscle) arc ligand gated cation channels. **Nicotinic acetylcholine receptor** is a transmembrane glycoprotein of 270 kD molecular mass and consisting of five subunits α_2 $\beta\gamma\delta$. Each subunit includes five hydrophobic regions, which span the membrane in helical form. Each subunit has a binding site for acetylcholine. Nicotine locks the ion channels of nicotinic receptors in their open configuration. When acetylcholine (the ligand) binds nicotinic acetylcholine receptor, a conformational change opens the channels, which is equally permeable to Na^+ and K^+ the former rushing in and latter rushing out. Since the Nat gradient is steeper, Na^+ influx greatly exceeds K^+ efflux. The Na^+ influx depolarizes the membrane initiating an action potential. **Muscarinic receptors** are 70-kD glyeoproteins and are members of the **7-transmembrane segment** (**7-TMS**) family of receptors. Binding of acetylcholine to these receptors, through interaction with G proteins, leads to inhibition of **adenyl cyclase**, stimulation of **phospholipase** C and opening of **K^+ Channels.** Several toxins including insecticides/pesticides and nerve gases (used in chemical warfare) actually affect the, above acetylcholine receptors and lead to block in nerve impulses, stop breathing and cause death by suffocation.

After every cycle of synaptic, signal transmission, acetylcholine is degraded with, the help of **acetylcholinesterase** enzyme, to resensitize the acetylcholine receptor. The synaptic vesicles are formed again by endocytosis and restocked with freshly formed acetylcholine (acetylcholine enters through H+ pump with the help of acetylcholine transport protein; this H+ pump is a V-type-ATPase) to start another cycle of nerve impulse transmission.

The effect of Cl^- channels. The concentration of Cl^- ions is much higher outside the cells than in the cytosol. Therefore, when the channel opens, more Cl^- ions will move inside due concentration gradient and hyperpolarize the membrane (interior being already negatively polarized at the resting stage), unless the membrane potential stops such a flow. This leads to excitation of the cell.

G-protein linked receptors and enzyme-linked receptors. Many of the signaling molecules that are secreted by axon terminals (e. g neuropeptides) regulate ion channels only indirectly. These neurotransmitters or signaling molecules bind to receptors, which are linked to G-protein or an enzyme inside, initiating signal transduction pathway leading to the opening of certain ion channels. This action or signal transmission is slower and more complex and lasts longer.

Psychoactive drugs attack neurotransmitter-gated channels. Although the different neurotransmitter-gated ion channels have structural and functional similarity, there is sufficient diversity among these channels to allow designing of drugs having specificity for individual channels. This basic principle has allowed in practice to design drugs targeted against narrowly defined group of neurons and their synapses, thus influencing specific brain functions. '**Curare**' is a drug which blocks acetylcholine receptors on skeletal muscle, so that it can be used by a surgeon for keeping a muscle relaxed during operation. Most other drugs used for treatment of diseases like insomnia, anxiety depression and schizophrenia exert their influence at chemical synapses through binding transmitter gated ion channels. Two other drugs, '**valium**' and '**librium**' used as barbiturates and tranquilizers bind to GABA, receptors, potentiating the inhibitory effect of GABA, so that lower concentration of GABA is allowed to bind to receptors to open Cl^--channels. As more information about ion channels is becoming available, new generation of psycoactive drugs to alleviate the miseries of mental illness will become available.

Muscle contraction involves sequential activation of five different ion channels. When a nerve impulse 'stimulates a muscle cell to contract, the events take place in the following sequence involving five different ion channels: (i) When the nerve impulse as an action potential, reaches axon teminals and depolarizes the plasma membrane there, it transiently open voltage-gated Ca^{++} channels, so that Ca^{++} ions flow inside due to a concentration gradient. Increase in Ca^{++} ions in the cytosol leads to release of acetylcholine from synaptic vesicles: (ii) Acetylcholine binds to acetylcholine, receptors in the muscle, opening the Na^+ channels associated with receptors, allowing more Na^+ to enter the cytosol of muscle cell, causing localized depolarization. (iii) Localized depolarization opens voltage gated Na^+ channels permitting entry of more Na+ ions into the cytosol of muscle cell causing further depolarization; this, leads to opening of more voltage-gate Na^+ channels the action potential propagates involving the entire plasma membrane of muscle cell. (iv) The depolarization of entire plasma membrane of muscle cell activates voltage-gated Ca^{++} channels in specialized regions. This causes release of Ca^{++} within the cytosol; this then causes opening of gated Ca^{++} release channels of **sarcoplasmic reticulum (SR)** to release stored Ca^{++} ions, thus increasing Ca^{++} concentration in the cytosol. This causes myofibrils in the muscle to contract.

Appendix 3: Gene Therapy

During the 1970s, a young boy named David captured the attention of the American public as "the boy in the plastic bubble." The "bubble" was a sterile, enclosed environment in which the boy lived nearly his entire life. He required this extraordinary level of protection because he was born with a very rare inherited disease called *severe combined immunodeficiency disease* (SCID), which left him virtually lacking an immune system The bubble protected David from viruses or bacteria that might infect and kill him, but it also kept him from any direct physical contact with the outside world, including his parents. In approximately 25 percent of cases, SCID results from the hereditary absence of a single enzyme, adenosine deaminase (ADA), which catalyzes a reaction in the catabolic pathway that degrades purines. In the absence of the enzyme, the normal substrate for the enzyme (deoxyadenosine) accumulates to toxic levels, killing certain sensitive cells, including T lymphocytes, which provide cell mediated immunity.

Gene therapy is the process by which a patient is cured by altering his or her genotype. SCID is an excellent candidate for treatment by gene therapy for a number of reasons. First, there is no cure for the disease, which invariably proves fatal at an early age. Second, SCID results from an alteration in specific gene that has been isolated and cloned and thus is available for use in treatment. Finally, the cells that normally express the ADA gene are a type of white blood cell that can be easily removed from a patient, cultured in vitro, genetically modified, and then reintroduced into the patient by transfusion.

In 1990, a 4-year-olli girl suffering from SCID became the first person authorized by the National Institutes of Health and the Food and Drug Administration to receive gene therapy. In September 1990, the girl received a transfusion of her own white blood cells that had been genetically modified to carry normal copies of the ADA gene. It was hoped that the modified white blood cells would provide the girl with the necessary armaments to ward off future infections. Since white blood cells have a limited lifetime, the procedure must be repeated periodically to maintain the patient's immune capacity. At the time of this writing, the girl and another child who was later included in the program are both doing well. The results from this first venture into the realm of gene therapy indicate the treatment has been a success.

Intensive effort is currently focused on broadening the techniques to introduce foreign genes into cells. The white blood cells injected back into the children suffering from SCID were genetically engineered by infecting the cells with a retrovirus. carrying the normal human *ADA* gene. The virus had been genetically disabled so that it was incapable of replicating and producing viral progeny. It was still able, however to integrate its genome (and the extra *ADA* gene it carried) into the DNA of the host cell where the gene could serve as the base of operations for producing the needed enzyme. Their ability to integrate themselves into the genome makes retroviruses excellent vectors for gene therapy experiments. But, the use of retroviruses carries some risk. For example, retroviral integration occurs randomly within the genome, and 'thus, there is always the remote possibility that the virus will integrate at a site in the genome where it activates a cancer-producing oncogene, which could cause susceptible cells to become malignant. In addition, retroviruses are able to integrate their genes only into an actively dividing cell, which rules out their use on cells that no longer divide, such as neurons and muscle cells. A number of studies, such as the with cystic fibrosis, have been conducted using adenovirus as the gene vector. Since it doesn't integrate its genome into its host cell's DNA, adenovirus is not able to activate an oncogene and can also be used to infect a nondividing cell. But, adenovirus has disadvantages as well, including a tendency to be lost from the infected host cell.

Viruses are not the only way to get DNA into cells. Cells can be made to take up DNA from their environment (a procedure called *transfection*) by jolting them with an electric current, or fusing them with a liposome in which the DNA is encapsulated. Although these technique are not as efficient as the use of viral vectors, they are regarded as safer by some scientists, and their efficiency can be increased by combination with other approaches. For example, the gene to introduced into cells can be linked to a gene for drug resistance. Then, when the cells are grown in a medium containing the drug, only those cells that have taken up the DNA a able to grow and proliferate, thereby eliminating those cells that lack the foreign gene.

When gene therapy was first considered as a feasible possibility, inherited diseases, such as SCID cystic fibrosis, and familial hyper-cholesterolemia were the obvious candidates for treatment, since all such diseases could be corrected if a normal gene could be introduced into the cells of a affected tissues and organs. In facts, all three of these congenital diseases are the subjects of promising clinical trials. But the focus of gene therapy has shifted from these relatively rare diseases caused by a defect in a single gene to more common diseases that have a more complex basis, including cancer and cardiovascular disease. Let's take a brief look at the ways that gene therapy might be applied to both of these diseases.

CANCER

Clinical researchers have been attempting for several decades to apply the knowledge gained from research in molecular biology to the treatment of cancer. Unfortunately this effort has met with little success. With the advent of gene therapy a new list of protocols is currently being approved for clinical trials, and the research community is hoping for better results. Examination of one of these protocols will serve to illustrate the possibilities.

Herpes viruses contain a gene that encodes an enzyme that possesses a catalytic activity not found in mammalian cells. The enzyme is a type of thymidine kinase (called *HSV-TK*) that adds a phosphate group to the thymine base of a thymidine nucleotide. *HSV-TK* is, also able to use certain thymidine analogues (compounds of similar chemical structure) as substrates for phosphorylation, including a compound called ganciclovir (or acyclovir), which is commonly prescribed to fight herpes virus eruptions. Once the ganciclovir is phosphorylated, it enters the pathway for DNA synthesis, but cannot be incorporated into DNA: Instead, the compound blocks DNA synthesis leading to the death of the infected cell. Thus, in the presence of ganciclovir, the *HSV-TK* acts like as "suicide gene," leading to a cell's self-destruction.

The *HSV-TK* gene has been successfully used to destroy certain barin tumors in animals and currently being studied in limited clinical tests in humans. In these tests, disabled retroviruses that have been engineered to carry the *HSV-TK* gene are injected into the brain tumor mass of the patients, where they are taken up by a fraction of the cells. Since the neurons of the brain constitute a nondividing cell population, they are not susceptible to retrovirus infection and thus are not affected by the procedure (brain tumors are formed by malignant glial cells). Once the tumor cells have taken up the "suicide gene," ganciclovir can be injected into the body, leading to the destruction of the genetically modified cells. It has been found in tests with animals that, even if only a fraction of the tumor cells take up the *HSV-TK*–carrying retrovirus, the remainder to the tumor cells can be killed by toxic metabolites that leak from the infected cells (The so-called "bystander effect"). Whether or not this or other gene therapy protocols will prove useful against cancer remains to be seen.

CARDIOVASCULAR DISEASE

Congestive heart failure occurs as the heart gradually loses its ability to contract with the required force needed to circulate blood effectively. The force of contraction of the heart is normally increased by the hormone and neurotransmitter epinephrine, which binds to β-adrenergic receptors on the surface of the cardiac muscle cells. One of the apparent causes of congestive heart failure is the lack of the cells ability to respond to epinephrine. Experiments with animals indicate that the force of heart contraction is greatly increased when the animal's heart muscle cells contain extra copies of the gene that encodes the β-adrenergic receptor, which causes an increase in the number of epinephrine receptors on the surfaces of the cells. Although treatment of congestive heart failure in humans by transfection of heart cells with genes encoding epinephrine receptors remains a distant goal, the feasibility of such an approach has been demonstrated.

The primary cause of heart attacks is the occlusion of one or more coronary arteries by the build up of plaques. One of the common procedures employed to correct this condition is *balloon angioplasty*, in which a catheter is threaded into the blocked coronary artery. The tip of the catheter is then inflated like a balloon, Pushing the walls of the artery outward. Unfortunately, arteries opened by this procedure often become reblocked by the growth of smooth muscle cells in the wall of the vessel. Experiments with animals have indicated that proliferation of these cells can be blocked and the artery kept open using gene therapy. In one group of experiments the cells of the artery at the site of balloon angioplasty were infected by an adenovirus carrying the *HSV-TK* gene, followed by exposure to ganciclovir to kill the cells (as described above). In another group of experiments, proliferation of the smooth muscle cells was blocked by causing the cells to take up a gene that produced an antisense RNA that interfered with the translation of a cellular mRNA required for cell proliferation. Procedures of these types are likely to become routine in the practice of medicine in the following century.

ETHICAL CONSIDERATIONS

It is important to note that all of the procedures discussed above, as well as those considered for approval, involve the modification of *somatic cells*—those cells of the body that are not on the path to gamete formation. Modification of somatic cells only affects the person being treated, and the modified chromosomes cannot be passed on to future generations. This would not be the case if either the germ cells of the gonads or cells of the early embryo were modified, since these cells contribute to the formation of gametes. Thus far, the consensus remains that no studies involving the modification of human germline cells will be performed. Such studies would present risks for the genetic constitution of future generations and raise serious ethical questions about scientists tampering with human evolution.

PRIONS

Prions: Solving a Medical Mystery

In 1957, Carleton Gajdusek had been working as a visiting scientist in Australia, studying viral genetics and immunology. His interest in the medical problems of native cultures, had taken him to nearby New Guinea for what he expected would be a brief visit before returning home to the United States. Within a couple of days of his arrival in New Guinea, Gajdusek spoke with Vincent Zigas, a local physician, who told him of a mysterious disease that was responsible for up to half of the deaths among certain villagers living in the remote highlands of the island. The natives had named the disease kuru, meaning "shaking or shivering," because in the early stages, the victims exhibited involuntary tremors. Over the following months, the victims-who were primarily women and children-progressed through stages of increasing debilitation, dementia, and paralysis, which ultimately claimed their lives. Gajdusek decided to abandon his travel plans and remain in New Guinea to study the disease.

Upon hearing of the symptoms of the disease, Gajdusek had concluded that the people of the region were probably suffering from an epidemic of viral encephalitis. The disease was probably being spread through the population by the ritual practice of eating parts of the bodies of relatives who had died. Since it was the women in the villages who prepared the bodies and were must likely to engage in this form of cannibalism, they would be the ones who were at greatest risk of infection. Over the ensuing months, Gajdusek helped care for the sick villagers in a makeshift hospital, performed autopsies on those who died, and prepared tissue and fluid samples that were sent to laboratories in Australia. In one of his many letters to the outside world, Gajdusek wrote: "We had a kuru death and a complete autopsy. I did it at 2 a.m., during a howling storm in a native hut, by, lantern light; and I sectioned the brain without a brain knife. The sections revealed that victims of kuru were dying as the result of widespread degeneration of their brains.

The evidence began to mount that kuru is not a viral infection. Patients dying from kuru showed none of the symptoms that normally accompany central nervous system infections, such, as fevers, inflammation of the brain, and changes in the composition of cerebrospinal fluid. In addition, the best virology labs in Australia were unable to culture an infectious agent from the diseased tissue samples. Gajdusek began to consider alternate explanations for the cause of kuru. It was possible that the affected villagers were being exposed to some type of toxic substance in their diet. Blood analyses were performed in the hopes of finding elevated levels of trace metals or other common toxins, but no chemical abnormalities were found.

At one point, Gajdusek thought that kuru might be an inherited disease but he concluded from discussions with geneticists that this was very unlikely. For example, it would be practically impossible for an inherited disease (1) to be of such high lethality and apparently recent origin and attain such a high frequency in a population; (2) to manifest itself in individuals of such diverse age groups, from young children through older adults; (3) to strike an equal number of young boys and girls but strike 13 times as many adult women as men; (4) to appear in a person born in another area of the island who had moved into the affected population.

There seemed to be no reasonable explanation for the cause of kuru. Gajdusek even considered the possibility kuru that was a mental disease. "Since much suggests hysteria early in the disease... I cannot get the psychosis idea out of my mind. But the advanced classical parkinsonism and basal gangliar disorder that results in eventual death cannot easily by linked with psychosis-in spite of the role this illness plays in local sorcery, murder, warfare, etc."

William Hadlow was an American veterinary pathologist *who had* worked on a degenerative neurologic disease called *scrapie* that was common in sheep and goats. In 1959, Hadlow was visiting an exhibit in London sponsored by a British pharmaceutical company when he saw a display of neuropathologic specimens prepared by Carleton Gajdusek of a person who had died of kuru. Hadlow was struck by the remarkable resemblance between the abnormalities in the brains of kuru victims to those in the brains of sheep that had died from scrapie. Scrapie was known to be caused by an infectious agent; this been demonstrated by transmitting the disease to healthy sheep by injecting them with extracts prepared from sick animals. Because the agent responsible for scrapie was able to pass through filters that retarded the passage of bacteria, it was presumed to be a virus. Unlike other viral diseases, however, the symptoms of scrapie did not appear for months after an animal was infected with the pathogen, which, therefore, became known as a "slow virus." Hadlow concluded that kuru and scrapie were caused by the same type of infectious agent and published his speculation in a letter to the British medical journal *Lancet*. After reading the published letter and then speaking with Hadlow, Gajdusck was convinced that his first idea about kuru being an infectious disease was correct. After several years of work, Gajdusek was finally able to demonstrate that kuru could be transmitted from extracts of human tissue to laboratory primates. The incubation period between the inoculation of the animals is and the appearance of symptoms of the disease was nearly two years. Kuru became the first human disease shown to be caused by a *slow virus.*

Several years earlier, lgor Klatzo, an astute neuropathologist at the National Institutes of Health (NIH), had told Gajdusek that a rare *inherited* condition called Creutzfeldt–Jakob disease (CJD) produced brain abnormalities that resembled those of kuru. Three years after they showed that kuru could be transmitted from humans to animals, Gajdusek and his co-workers demonstrated, using extracts prepared from a biopsy of the brain of, a person dying from CJD, that CJD could also be transmitted to animals. There have also been numerous documented cases in which CJD was transmitted from one human to another during surgical procedures, such as corneal transplants, or in extracts of growth hormone prepared from the pituitary glands of cadavers.

How can an inherited disease, such as CJD be linked to the presence of an infectious agent? The answer to this question has been revealed over the past fifteen years, largely through the work of Stanley Prusiner and his colleagues at the University of California, San Francisco. prusiner began by studying the properties of the agent responsible for scrapie and soon arrived at two provocative conclusions. First, the agent was very small-much smaller than any known virus-having a total molecular weight of 27,000 to 30,00 daltons. Second the agent appeared to *lack a nucleic acid* component and to be composed *exclusively of protein.* This second conclusion was based on exhaustive treatments of infectious brain extracts with enzymes or other substances that would digest or destroy either proteins or nucleic acids. Protein-destroying treatments such as proteolytic enzymes or phenol, rendered the extracts harmless, whereas nucleic acid-destroying treatments including various types of nucleases and ultraviolet radiation had no effect on infectivity. The resistance of the scrapie agent to ultraviolet radiation compared to that of viruses is shown in Table A3.1. Prusiner named the agent responsible for scrapie, and presumably for kuru and CJD as well, *a prion, which stood for a proteinaceous infections particles.*

Table A3.1 Inactivation of small infectious agents by UV irradiation 254 nm.

Example	D_{37} *(J/m²)*
Bacteriophage T2	4
Bacteophage S13	20
Bacteriophage of $\Phi \times 174$	20
Rous sarcoma virus	150
Polyoma virus	240
Friend leukemia virus	500
Murine leukemia virus	1,400
Potato spindle tuber viroid	5,000
Scrapie agent	42,000

D_{37} is the dose of irradiation that permits 37 percent survival.

Reprinted with permission from S.B. Prusiner, Science 216: 140, 1982 Copyright 1982 American Association of the Advancement of Science.

The notion that an infectious pathogen consisted exclusively of protein was met with considerable skepticism, but subsequent studies by Prusiner and others have note produced any evidence to alter the original conclusion. It was shown in 1985 that the prion protein is encoded by a gene within the cells own chromosomes. The gene is expressed within *normal* brain tissue and encodes a protein of 254 amino acids designated PrP^{C} (standing for prion prtein celluler) whose function remains unknown. A modified form of the protein (designated PrP^{Sc}, standing for prion protein scrapie) is present in the brains of animals with scrapie. Unlike the normal PrP^{c}, the modified version of the protein accumulates within nerve cells forming aggregates that apparently kill the cells. Not only does PrP^{Sc} cause the degenerative changes in the brain that characterize scrapie it is also presumed to be the infectious agent capable of transmitting the disease from on animal to another.

Once its was discovered that scrapie could be traced to a modification of a normal gene product, it became possible to explain how a genetic disease, such as CJD, could be transmitted from one individual to another. Nearly all of the genes present in humans are shared with other mammals, and thus, there is a human version of PrP. Presumably, if this human gene were to become mutated in certain ways, it would lead to the production of a modified protein analogous in its activities to the modified sheep protein PrP^{Sc}. As expected analysis of DNA isolated from a number of human CJD patients has revealed the presence of specific mutations in the gene encoding PrP. In the past few years, the genetic analysis of susceptibility to prion diseases has depended upon mice that have been genetically engineered in a particular way. Two types of altered mice have been developed; one that lacks the PrP gene entirely (referred to as a PrP knockout mouse) and another that contains one or more copies of the mutated form of the human PrP gene (referred to as a PrP transgenic mouse.)

Given the fact that the PrP protein is produced normally in the brains (and other organs) of mice, the absence of the gene might be expected to cause dire consequences for the development or behavior of the PrP knockout mice. Despite this expectation, however, mice lacking the PrP gene show no ill effects. There are several reasonable explanations for this result including the possibility that the normal function of the PrP protein is replaced by another protein produced by a related gene; in other words, the mouse has a "backup" system that makes the PrP protein expendable. Regardless, mice that lack the PrP gene, and thus cannot make the PrPc protein, cannot be made to develop scrapie by injecting mouse scrapie prions into their brains. Thus, for a mouse to be susceptible to developing the disease, the animal must be able to produce the PrP protein from its own genes; it not enough to have the abnormal protein introduced into the body. This finding provides support for the hypothesis that the PrP protein is essential for propagation of the prion during an infection.

As noted above, studies have also been performed using transgenic mice; that is mice that have been genetically engineered so that they carry forgeign genes within their chromosomes. When a mutated human PrP gene is transferred to mice, the transgenic animals develop the same type of neuropathologic brain disease as that seen in humans. This experiment demonstrates that presence of a single mutated gene, which codes for a single abnormal protein, is sufficient to cause all of the symptoms associated with a devastating neurologic disease.

UNANSWERED QUESTIONS

The ides that an infectious disease can be caused by an agent consisting solely of a single protein remains controversial. Some biologists believe that the prion protein is associated with a small piece of nucleic acid that has yet to be discovered, others that the prion protein causes the individual to be susceptible to infection by a second agent; for example, a virus, that actually causes the disease. The development of the disease in mice transgenic for a mutant gene encoding the prion protein argues for the protein as the sole cause, but the finding would be greatly strengthened if it could be shown that brain extracts from the transgenic mice could, transmit the disease to normal, nontransgenic mice. To date, attempts to transmit the disease in this way have met with only limited Success, and the matter remains unsettled.

Another matter that remains unanswered is the mechanism by which the infectious agent is able to increase its numbers (replicate) within an infected individual, as clearly occurs, Replicil tion is generally attributed only to nucleic acids. How is it possible for a protein to produce more of itself? This unanswered question remains one of the major "sticking points' in the entire concept of the prion as an infectious agent. Prusiner and his colleagues have gathered evidence that suggests that the two versions of the PrP protein, PrP^{C} and PrP^{Sc}, differ in their three-dimensional structure (*conformation*), In other words, the same protein can exist in two different forms. According to their hypothesis, the protein would normally exist in the Pr^{PC} form. However, in humans or animals that develop prion diseases, formation of the PrP^{Sc} structure is favored and the abnormal protein accumulates. In the case of infectious prion diseases, such as kuru or scrapie, Prusiner suggests that "replication" begins when a scrapie version of the PrP protein binds to the normal PrP protein (or an unfolded version of the protein), causing a transformation of the normal protein into the modified form. Thus, if one PrP^{sc} molecule were to bind one PrP^{C}, the event would generate two PrP^{sc} molecules, which could then bind two more PrP^{C} molecules producing four PrP^{Sc} molecules, and so forth.

While prion diseases are very rare, other degenerative neural disorders, such as Alzheminer's and Parkinson's diseases, are very common. It is hoped that the study of prion diseases will prove useful in understanding the basis of these more familiar human conditions.

Cytological Techniques

There are two types of cytological preparations viz., *Squash* and *Smear* techniques. Squash preparations are meant for mitotic chromosomes or mitotic stages with root-tips. Smear preparations are meant for meiotic studies with the help of *flower buds.*

For karyotypic studies of mitotic chromosomes, pre-treatment of root tips with colchicine is used, an alkaloid obtained from the rhizomes of *Colchicum autumnale* (fam. Liliaceae). This alkaloid Colchicine, has the property of dissolving the mitotic *spindle* and hence the chromosomes which are divided, will be found free in the cell. Since there is no spindle, there will be no anaphase and telophase and hence the chromosomes lie spread out which are already divided ($2n \rightarrow 4n$). This cell will go into the next mitotic cycle with double the chromosome numbers. Thus a tetraploid cell is produced with double the chromosome complement, ($2n$ becomes $4n$). Colchicine is thus used in cytogenetic studies to induce *polyploidy.*

The chromosomes at metaphase, in a pretreated cell, is called *C-mitosis,* (C stands for *Colchicine*). A C-mitotic cell is used for karyotypic study for studying the morphology of the chromosomes at mitosis. The chromosomes can be studied end to end, their length, arm ratio (short arm/long arm), centromere position and morphology of satellite structures. Then an *idiogram* will be prepared for comparative study in different species or genera. The length measurements of chromosomes of a given species are constant as their number from generation to generation.

Karyotypic studies can be made with pachytene chromosomes also at meiosis, as in maize, tomato, etc. Recently karyotypic studies are made with different banding techniques like Giemsa banding, the Quinacrine banding, the R banding, etc. These banding techniques are quite refined and are used in animal and plant species especially in human karyotypes.

SQUASH PREPARATION FOR ROOT TIPS (MITOSIS)

Techniques

1. The fixed root tips are placed on a slide and carefully the tips of about 0.5 cm are cut with a blade and then they are transferred to a watch glass with 2% acetocarmine.
2. The root tips are boiled on the flame (spirit lamp) for 2 minutes.

3. Transfer the boiled root tips to 1N Hydrochloric acid (HCl), in a watch glass for 2-3 minutes.
4. Transfer the root tips to absolute alcohol or even 90 per cent alcohol, for 2-3 minutes.
5. Remove the root tips from the watch glass and place on a slide one root tip per each slide. Then put 2-3 drops of acetocarmine and place a cover slip neatly.
6. Warm only for 5 seconds on spirit lamp flame and press between folders of filter paper to drain off excess stain and also to spread the chromosomes. Pressing can be done with thumb or slight tapping with pencil or hard material.
7. If there are air bubbles, in the preparation put one drop of acetocarmine from the side of the coverslip, then warm and press between filter papers. The air bubbles disappear.
8. Now the slide is ready for examination under the compound microscope for detailed study.

Remember, the chromosomes should be stained deeply (orange red) and also well spread to identify the stages of mitosis. Observe the following stages of mitosis:

Prophase, Metaphase, Anaphase, Telophase and Cytokinesis.

SMEAR TECHNIQUE FOR MEIOSIS (FLOWER BUDS)

Normally for classwork always *Allium cepa* flower buds are given for the following reasons: (1) Flower buds are quite big and all the 6 stamens can be removed onto the slide very easily. (2) Chromosomes are large and are easily stainable with acetocarmine. (3) Well spread chromosomes are seen in all stages except pachytene. (4) Chromosome number is small (2n=16), i.e., 8 bivalents are seen clearly at diplotene, diakinesis and Metaphase I. Subsequent stages in Meiosis II are also seen quite clearly.

Technique

1. Place one flower bud (average size because if it is too small no divisional stages are seen except interphase and if too big only pollen grains are seen) on a slide and scoope out with a pair of fine needles all the six stamens and add 2 drops of acetocarmine.
2. With a smearing needle or with a needle squash the anther or crush the anther so that all the pollen mother cells will come out into the acectocarmine stain.
3. Carefully remove the *debris* with a fine needle by keeping the slide slanting (about 20 degree slope) so that all pollen mother cells go to a side of the stain and the debris which can be seen with a naked eye is removed by a needle.
4. Now put a clean coverslip from a side on the stain carefully by holding with a needle gently drop the coverslip.
5. Heat gently or warm for 2-3 seconds on a spirit lamp flame.
6. Squash or press the coverslip in between filter paper folds with a thumb. It can be also gently tapped with the back of needle or pencil.

7. If air bubbles are there, place a drop of acetocarmine from the side of the coverslip and gently warm and press gently again in between folds of filter paper.
8. Now the slide is ready for detailed observation of meiotic stages under the compound microscope.

Observe the following stages of meiosis:

Meiosis I	*Meiosis II*
Prophase-Leptotene, Zygotene, Pachytene, Diplotene, Diakinesis.	Anaphase II
Metaphase I	Telophase II
Anaphase I	Tetrads
Telophase I	

C-mitosis for karyotypic studies

Pre-treatment of Root tips with Colchicine.

Technique

1. Fresh root tips of *Allium* are cut and placed in a 0.5% colchicine for 2-3 hours in a petri dish.
2. Remove the colchicine treated root tips from the petri dish and wash thoroughly in running water for 3-5 minutes to remove the excess colchicine.
3. Transfer the washed root tips to 1 : 3 45% acetic acid:absolute alcohol fixative and keep it for 24 hours (From 10 AM to next day 10 AM).
4. Remove the fixed root tips from petri plate and wash in 90% alcohol for 2-3 minutes.
5. Transfer the washed root tips to 70% alcohol for storage. The root tips should be stored in refrigerator always. Whenever the root tips are needed for study, they can be used and put back in the refrigerator.
6. Now the root tips are ready for Squash technique for C-mitotic chromosomes. This technique is called C-*Mitosis* for Karyotypic studies of mitotic chromosomes in plants.

Note: Always fix either root tips or flower buds on a sunny day between 9 AM to 11 AM for maximum number of divisions.

EXPERIMENTS- IN CYTOGENETICS

1. Study of mitosis

Identification and drawings of different stages of mitosis.

2. Study of meiosis

Identify and draw different stages of meiosis.

3. Mitotic peak

Hourly fixation of root tips for 24 hours in a day and observe maximum no. of divisions. Draw a *graph, hours* on *X-axis* and per cent of divisions on *Y-axis.*

4. C-mitosis

Colchicine pretreated root tips. Study the morphology of chromosomes and prepare an Idiogram.

5. Chiasma frequency (meiosis)

Study of chiasma frequency per bivalent, and per cell at diakinesis and metaphase I. Obtain *chiasma terminalisation* per cell.

$$= \frac{\text{No. of chiasmata terminalised}}{\text{Total no. of chiasmata per cell}}$$

6. Meiosis in translocation heterozygotes

Observe No. of quadrivalents at diakinesis and mataphase.

Materials: *Rhoeo discolor,* Tetraploid maize (for *multivalents*), *Chlorolphytum, Peaonia, etc.*

SQUASH METHOD WITH FEULGEN STAINING

This is really a microchemical test for DNA. It depends on Schiff's reaction, which is a test for aldehydes, and the procedure is as follows:

1. The fixed material is washed in water for a few minutes.
2. It is then placed in normal hydrochloric acid at 60°C for a time depending on the type of fixative used. Following fixation in acetic-alcohol mixtures, 6–8 minutes is the optimum time. Since the time allows some latitude, it does not matter very much if the temperature fluctuates somewhat around 60°C. The procedure can be carried out, therefore, in an ordinary water-bath or even in a test tube in a beaker of water at 60°C. The acid should be at 60°C before the washed material is placed in it.
3. It is then convenient to pour the acid with its suspended material into a flat dish so that the material can be picked up with needles or a flat-ended instrument similar to that which can be used to raise free hand sections. The hydrolysis affects the cell walls and the middle lamellae so that the material is rather soft at this stage. This is one of the functions of the hydrolysis for it facilitates the separation and flattening of cells at a later stage. It means, however, that the material must be handled gently until it is on the slide and if anthers are being used they should be whole ones otherwise many cells are lost in suspension.
4. The material is transferred from the acid to water for a few minutes.

5. It is then placed in the Feulgen stain which is a decolourised acid solution of leuco-basic fuchsin (see fixatives).
6. The material is left in this solution in a corked tube in the dark for one to a few hours. Small 2 in. by 1 in. tubes are suitable for this purpose. At room temperature the material "goes off" after about 6 hours. in this solution. Consequently, the material should be treated as above on the day the preparations are to be made. In a deep-freeze unit, however, feulgen-stained material keeps indefinitely in 45% acetic.
7. After 1 hour or so in the Feulgen stain, the material is removed and it is prepared by the squash method described earlier. Since the material is already stained, however, it can be tapped out in 45% acetic acid but it can also be prepared in an acetic stain to intensify the colouration and to show up clearly the cytoplasm and the cell wall. This has the additional advantage of producing a preparation which will not fade whereas Feulgen alone tends to fade.

The basis of this technique is that mild hydrolysis liberates aldehydes from the DNA, which is a constant and confined constituent of the chromosomes. In effect, therefore, the hydrolysed material is tested for aldehydes which give a *pale magenta* colouration. RNA does not liberate stainable aldehydes on hydrolysis.

THE ACETIC STAINS

Three acetic stains are commonly used. These are *acetocarmine, acetic orcein* and *acetic lacmoid.* A solution of iron acetate in 45% acetic acid can be used in conjunction with acetocarmine. All these are used in the same way and in the manner described earlier. Acetic orcein gives the best results because it is more specific than the other stains in most cases. It does not give satisfactory results with mature pollen grains and similar material, however, because it tends to stain cuticularised walls heavily. Its other disadvantage is that it precipitates considerably on keeping and there seems to be considerable variation in the quality of the stain in different batches—even those from the same supply source. A good batch should show bluish when made up, a pure orcien is red in colour.

As a general rule both orcein and lacmoid should be used in solutions as concentrated as they can be. This is not the case with acetocarmine, the use of which involves more individual judgement, and the best concentration varies with the material. This stain keeps very well, however, and once made up it can be kept for years without taking any special measures. It does not colour mature pollen grain walls and it is worth learning how to use this stain. It is unquestionably the connoisseur's stain.

When using this stain, it is convenient to have three solutions at hand, a concentrated solution of acetocarmine, 45% acetic acid and a solution of iron acetate in 45% acetic acid, one the colour of a dry sherry is a convenient concentration.

THE PREPARATION OF SOLUTIONS

The following solutions can be used in the study of chromosomes at mitosis. They interfere with the organisation of the spindle and thus facilitate the study of chromosome morphology. Colchicine is the most effective spindle inhibitor. Following colchicine treatment, the chromosomes may become even more contracted than they are during normal metaphase and very few normal anaphases or later stages are seen. These can be observed following the other treatments which also have less effect on chromosome contraction.

	Inhibitor	*Concentration*	*Treatment time (hours)*
1.	Colchicine	0.01–0.05% aqueous	4–6
2.	para-dichlorobenzene	Saturated aqueous	4
3.	alpha-bromonaphthalene	Saturated aqueous	4

FIXATIVES

The following fixatives are suitable:

		Material and relative quantities by volume		
	Fixative	*Absolute alcohol*	*Glacial acetic acid (45%)*	*Chloroform*
1.	Acetic-alcohol (three-in-one)	Three parts (3)	One part (1)	—
2.	Carnoy	Six parts (6)	One part (1)	three parts

Since the acid and the alcohol react, these fixatives cannot be stored and they should be used the day they are made up. Chloroform is often included when fatty globules in the cytoplasm tend to be troublesome.

STORATIVES

Fixed material not required for staining immediately can be left in the fixative or else stored in 70–95% alcohol or in the following mixture (by volume):

Glacial acetic acid	*70% Alcohol*	*50% Glycerol*
3	6	3

ACETIC STAIN

(i) Weigh sufficient solid (carmine, orcein or lacmoid) for a 2% solution, (2gm in 100 c.c acetic acid (45%)).

(ii) Place the stain with the appropriate volume of 45% acetic acid in a flask with a vertical condenser.

(iii) Reflux all day, glass beads or pumice will help to prevent "bumping".

(iv) Allow to cool and filter.

Acetic-orcein may also be prepared as follows:

The standard solution is 1% in 45% acetic acid. American synthetic orceins need greater dye concentration. Because of deterioration in dilute acid the stain is best kept as a 2.2% stock solution in glacial acetic acid which can then be diluted to 45% as required.

FEULGEN STAIN

(i) Weigh 1 gm of basic fuchsin and dissolve by pouring 200 ml of boiling distilled water over it.

(ii) Shake well, allow the solution to cool to 50°C and filter.

(iii) Add 3 gm of potassium metabisulphite ($K_2S_2O_5$).

(iv) Add 30 ml normal HCl to the filtrate (normal HCl can be made by taking 87.3 ml of concentrated analar HCl and making this up to 1000 ml with distilled water).

(v) Add 3 gm of potassium metabisulphite ($K_2S_2O_5$)

(vi) Place the solution in a tight-stoppered bottle in the dark for 24 hours. This leads to the bleaching of the solution by the sulphur dioxide. (SO_2).

(vii) Add 0.5 gm of decolourising powdered carbon (animal charcoal).

(viii) Shake well and filter through a coarse paper.

(ix) Adjust the pH to 3.6 by adding normal NaOH.

(x) The stain is now ready for use. It should be *colourless* or a very pale yellow. It should be kept in a tight-stoppered bottle in the dark when not in use. A dark glass bottle is a suitable container.

Preparation of SO_2 Water

5 c.c. normal HCl

5 c.c. $K_2S_2O_5$ 10%

100 c.c. distilled water.

Preparation of an Albuminized Cover-slip

Place a small spot of glycering albumen on the cover-slip and smear it over the surface with a finger. A very thin film is sufficient. In fact, after smearing an attempt should be made to wipe off the albumen with a clean finger. The amount that remains following such an attempt is sufficient. The cover-slip is then passed through a flame until it begins to "smoke". This should not be overdone for the albumen then turns a brown colour. The cover-slip is then ready for use. They should be prepared immediately before use or else kept in a dust free chamber.

HYDROXYQUINOLINE PRE-TREATMENT FOR KARYOTYPE STUDY

8-hydroxyquinoline (HQ) is used as 0.004M-0.004M% aqueous solution.

Preparation

The solubility of HQ in water is poor. Take 500 c.c. or 1000 c.c. of distilled water in a glass-stoppered bottle and add powdered HQ of a quantity accurately weighed which when dissolved gives a solution of the required strength. Keeping the bottle at 40°C for over an hour aids in the dissolution of the chemical in water. In such a case the solution needs to be cooled to the room temperature before use.

PRETREATMENT IN HQ

1. Take fresh root tips or stem tips in about 10 c.c. of HQ in a small specimen tube.
2. Keep the specimen tube in a refrigerator at about 16-20°C for about 3 or 4 hours.

Lower temperatures induce abnormalities. Centromere constrictions and nucleolus organiser region become exaggerated and hence the centromeres and the nucleolus organisers become easily observable.

HQ induces chromosome contraction like colchicine but unlike it, it is not a spindle inhibitor. Metaphase blocking and wide spreading of chromosomes do not occur.

ACETO-ORCEIN SQUASHES

Aceto-orcein gives better results than aceto-carmine after pretreatment with hydroxyquinoline, paradichlorobenzene or alpha-bromo naphthalene.

1. After pretreatment in HQ wash material in distilled water.
2. Hydrolyse material by warming in a test tube for 4–5 seconds in a mixture of 2% aceto orcein (9 parts) and 1 N HCL (1 part). This can be repeated three to four times.

   For grass roots some more HCl (1N) can be added after the first treatment. Keeping the material in the warm mixture for about 30 minutes improves hydrolysis.
3. Squash as usual in 1 % aceto orcein.

CHROMATOGRAPHY

Chromatography techniques are used to separate compounds on the basis of their relative affinities for absorption onto a solid matrix. They are particularly useful for separating individual proteins, amino acids or nucleotides.

There are several different chromatography methods, distinguished by the nature of the solid matrix. In paper chromatography, this matrix is a strip of filter paper. A sample of the compounds to be analysed is placed at one end of the paper strip and eluted by soaking an aqueous or organic solution (the solvent) along the strip. As the solvent soaks along the paper it carries the compounds with it, but at different rates depending on the relative affinities of the compounds for absorption onto the matrix.

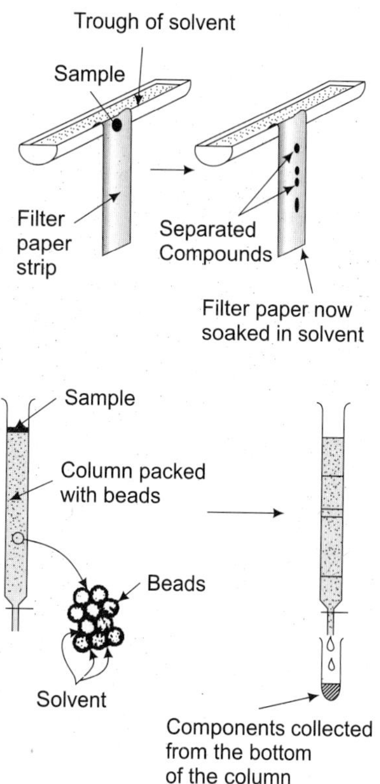

Chargaff's method for analysing the base composition of DNA involved, first breaking the molecule into its component nucleotides by treatment with acid or alkali. The sample was then eluted along a paper strip with any of several solvents (for example an n-butanol-water mixture) and the resulting spots cut out. The pure nucleotides were recovered from the paper by soaking in an aqueous solution, and their concentrations measured.

Chromatography is now more routinely carried out in a glass column packed with tiny beads, composed of substances such as cellulose or agarose, immersed in the solvent. The sample is layered on to the top of the column and eluted by passing through more solvent. The separated compounds are then collected as they drip out of the bottom of the column.

ELECTROPHORESIS, ULTRACENTRIFUGATION AND ULTRAVIOLET SPECTROSCOPY

These three techniques, which were in their infancy when used by Avery, Macleod and McCarty to analyse the transforming principle, are now important methods regularly employed in research projects concerning DNA.

Electrophoresis. It is defined as the movement of charged molecules in an electric field. DNA molecules, like proteins and many other biological compounds, carry an electric charge, negative in the case of DNA. Consequently, when DNA molecules are placed in an electrical field they will migrate towards the positive pole. In an aqueous solution the rate of migration of a molecule depends on two factors: its shape and its electrical charge, meaning that most DNA molecules will migrate at the same speed in an aqueous electrophoresis system. Avery's group subjected the purified transforming principle to electrophoresis and demonstrated that its migration rate is the same as that of a sample of pure DNA.

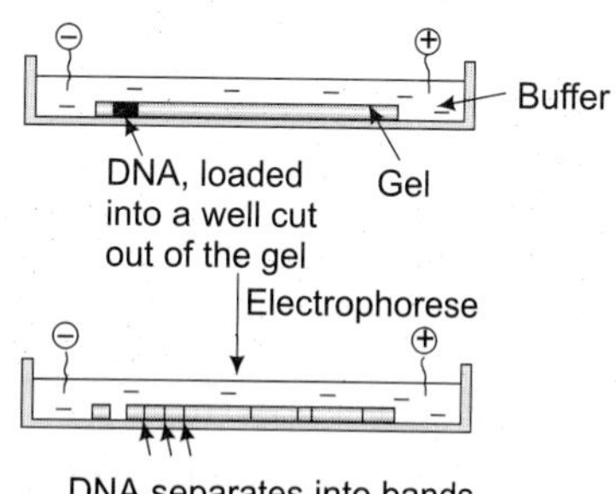

Nowadays, electrophoresis of DNA is usually carried out in a gel made of agarose, polyacrylamide or a mixture of the two. In a gel, the migration rate of a macromolecule is influenced by a third factor, its size. This is because the gel comprises a complex network of pores through which the molecules must travel to reach the electrode. The smaller the molecule, the faster it can migrate through the gel. Gel electrophoresis will, therefore, separate DNA molecules according to size.

Gel electrophoresis is also routinely used to separate proteins of different molecular masses.

Ultracentrifugation. The ultracentrifuge, invented by Svedberg in 1925, allows samples to be subjected to intense centrifugal forces, up to several hundred thousand xg. Cells, cell components and macromolecules sediment during ultracentrifugation at a rate dependent on their size, shape, density and molecular weight. Avery showed that the transforming principle sedimented in the ultracentrifuge at a rate similar to that of pure DNA.

Two versions of ultracentrifugation are now important in studying DNA. The first is called *velocity sedimentation analysis*, a procedure that involves measuring the rate at which a macromolecule or particle sediments through a dense solution, often of sucrose, whilst subjected to a high centrifugal force. The rate of sedimentation is a measure of the size of the molecular or particle (although shape and density also influence the rate) and is expressed as a sedimentation coefficient.

The second technique is called **density gradient centrifugation.** A density gradient is produced by centrifuging a dense solution (usually of caesium chloride), as a high centrifugal force will pull the caesium and chloride ions towards the bottom of the tube. Their downward migration will be counterbalanced by diffusion, so a concentration gradient will be set up, with the CsCl density greater towards the bottom of the tube. Macromolecules present in the CsCl solution when it is centrifuged will form bands at distinct points in the gradient, the exact position depending on the buoyant density of the macromolecule.

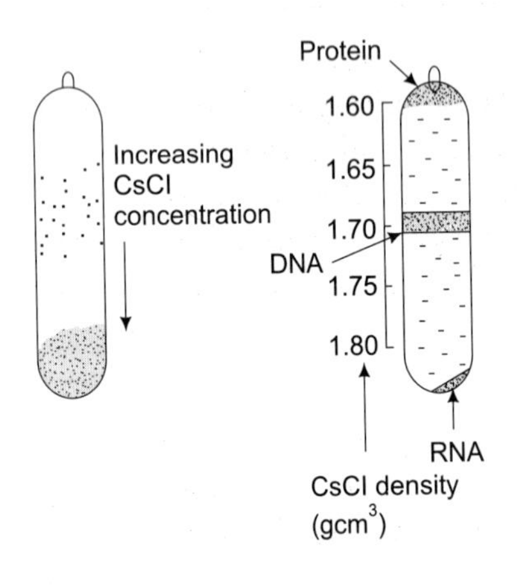

DNA has a buoyant density of about 1.7 g cm^3 and will, therefore, migrate of the point in the CsCl density is also 1. 7 gcm^3, Poteins have much lower and RNA somewhat higher buoyant densities.

Ultraviolet spectroscopy. Spectroscopy involves analysis of substances by the spectra they produce. The spectrum of light emitted or absorbed by a substance is characteristic of the substance. For example, DNA strongly absorbs ultraviolet radiation with a wavelength of 260 nm; proteins on the other hand have a strong absorbance at 280 nm. Avery showed that the ultraviolet absorbance spectrum of the transforming principle is the same as that for pure DNA.

Ultraviolet spectroscopy is often used to check that samples of DNA obtained from living cells are pure and do not contain protein or other contaminants. As the amount of ultraviolet radiation absorbed by a solution of DNA is directly proportional to the amount of DNA in the solution, the technique can also be used to determine the concentration of DNA in a sample.

Questions

Genetics 1

1. (a) Discuss the chromosome theory of inheritance and the parallelism between Mendel's postulates and the behaviour of chromosomes at meiosis. 4

   (b) What is a linkage group? How many linkage groups are there in:

   (i) Bacteria (ii) Human beings

   (iii) Onion? 2

   (c) In a trihybrid cross in *Drosophila* between a ♀ heterozygous for recessive (i) *b*-black body, (ii) *cn*-cinnabar eyes, (iii) *vg*-vestigial wings with a recessive homozygous male the following progeny was obtained:

   (I) + + + 39, (II) b + + 416, (III) + cn + 42, (IV) + cn vg 402, (V) b cn + 1, (VI) b cn vg 48, (VII) b + vg 50, (VIII) + + vg – 2

   1. hich are the non-crossover classes?
   2. Which are the double crossover classes?
   3. What kind of linkage occurs in the female?
   4. Which gene is in the middle?
   5. Calculate the map distance between the genes.
   6. Is there interference?
   7. What is the coefficient of coincidence? 7

   (d) Devise a method of non disjunction of chromosomes in human female gametes that would give rise to Klinefelter and Turner's syndrome following fertilisation by a normal gamete. 2

2. (a) Explain the terms. Attempt any *ten*.

   (i) Transgenic organism
   (ii) Nullisomic
   (iii) Reverse genetics
   (iv) Housekeeping genes
   (v) Polar mutation
   (ix) Hemizygous
   (x) Phenocopy
   (xi) Snurps
   (xii) Antisense RNA
   (xiii) Sense strand of DNA

(vi) Pseudodominance
(vii) Holandric genes
(viii) Test cross ratio
(xiv) Carcinogen
(xv) Penetrance. 10

(b) What do the following abbreviations stand for? (Attempt any *four*)
(i) YAC
(ii) RFLP
(iii) HUGO
(iv) IBPGR
(v) SBI
(vi) CAP
(vii) SCE. 5

Or

(c) Give the important contributions of any two scientists:
(i) Barbara McClintock
(ii) G.W. Beadle and E. Tatum
(iii) S. Benzer
(iv) A. Hershey and M. Chase. 5

3. (a) How do induced mutations differ from spontaneous mutations: Describe the mutagenic effects of:
(i) Base analogues
(ii) Tautomerism
(iii) Alkylating agents
(iv) UV radiations. 8

(b) How do somatic mutations differ from germinal mutations? Why are the products of somatic mutations of commercial importance in plants and not in animals? 2

(c) Describe Bruce Ames test. Of what value is this test? 3

(d) Describe the role of SOS genes and Rec A protein in DNA repair. 2

4. (a) Give an account of the achievements of Genetic Engineering Research. 8

(b) Describe any *two* of the following techniques and give their importance in genetic rescarch:
(i) DNA sequencing
(ii) Polymerase chain reaction
(iii) Southern blotting
(iv) Molecular hybridisation
(v) Gene cloning

(c) If you were interested in isolating a c DNA for human haemoglobin, why would you choose a c DNA library established from m RNA isolated from RBC? While if you wanted to isolate the gene for albumin, why could you choose a genomic library established from any human tissue?

5. (a) What genetic criteria distinguish a case of extra chromosomal inheritance from (i) a case of Mendelian autosomal inheritance (ii) a case of sex linked inheritance? 4

(b) Distinguish between any three of the following:
(i) Euchromatin and Heterochromatin
(ii) Exonuclease and Endonuclease
(iii) Nucleosome and Spliceosome
(iv) Template DNA and Primer DNA

(v) Pericentric inversion and paracentric inversion

(vi) Multiple alleles and polygenes. 6

(c) What are the genetic disorders? Describe the symptoms and the cause of *either* Phenylketonuria or cystic fibrosis. Have these genes been located and cloned? Can they be cured by gene therapy? 5

6. (a) Write notes on any *four*:

(i) Banding techniques

(ii) Turner syndrome

(iii) Heterosis

(iv) Hardy-Weinberg Principle

(v) $\phi \times 174$

(vi) Second messenger

(vii) Organelle genomes. 10

(b) Why are allopolyploids also called amphidiploids? With the help of a suitable example discuss the role of polyploidy in speciation *or* plant breeding. 5

7. (a)Discuss the salient features of the genetic code. How was it deciphered? To what extent is the code:

(i) degenerate, and

(ii) universal?

How can degenerate nature of the code and wobbling in the codon-anticodon pairing help to tolerate changes in the nucleotide sequence without bringing about a change in the amino acid sequence? 5

(b) Discuss why synthesis of proteins is of great importance to geneticists. 5

(c) What is transcription? Describe how it is (i) initiated, (ii) terminated in prokaryotes.

Where would chain elongation during protein synthesis be stopped if mutations were to occur in the genes of the following proteins: (i) EF–G; (ii) EF–Tu; and (iii) Peptidyl Transferase? 5

8. (a) Give an illustrated account of DNA replication in prokaryotes. 6

(b) What is gene regulation? Why is it assumed to be more complex in eukaryotes than in prokaryotes? At what different levels is it regulated? 6

(c) Thirty per cent of the nucleotides from locust DNA are As. What are the values for:

(i) T

(ii) G + C

(iii) G

(iv) C? 3

Genetics 2

1. Answer the following:

(a) Which of the following statements are *True* or *False*?

(i) RNA Polymerase activity of eukaryotic cells is found only in the nucleus.

(ii) During RNA formation, the RNA chain is extended in the $3' \rightarrow 5'$ direction.

(iii) The addition of an incoming ribonucleoside triphosphate to a growing RNA chain occurs by nucleophilic attack on its phosphate.

(iv) Heterogenous nuclear RNA is the precursor of *m* RNA and *r* RNA, but not of *t* RNA.

(v) The codon AVG specifies the termination of the coding region of *m* RNA molecule. $2^1/_2$

(b) Which of the following protein factor(s) is/are involved in binding aminoacyl-*t* RNA molecules to ribosomes?

(i) EF–G (ii) EF–Tu

(iii) EF–Ts (iv) elF–2

(v) EF–1. $2^1/_2$

(c) Explain the terms:

Phenocopy; Mulatto; Holandric gene; Cybrid; and RFLP. $2^1/_2$

(d) Distinguish between test cross and backcross. $2^1/_2$

(e) Comment upon the statement, "Mutations are fountain-heads" of evolution. 5

2. (a) Bridges and Olbrycht crossed fruit-flies with rough eyes with gene echinus, *ec*, to flies with scute, *sc*, and crossveinless, *cv* genes. These genes are recessive to their corresponding normal genes and are sex-linked. The hybrid females were backcrossed to males carrying three recessive genes and the following results were obtained:

echinus (+ *ec* +)	–	810
scute, crossveinless (*sc* + *cv*)	–	828
scute, echinus (*sc ec* +)	–	62
crossveinless (+ + *cv*)	–	88
scute (*sc* + +) –		89
Normal wild type (+ + +)	–	0
scute, echinus crossveinless (*sc ec cv*)	–	0
Total Flies	=	1980

From the above, diagram the cross., and identify the phenotypes and genotypes of F_1 and F_2. Calculate the recombination frequencies between the genes. Map the genes in their expected sequence. 10

3. (a) What is cell cycle? Discuss the mechanism of DNA replication. Bring out the differences between the process of DNA replication in eukaryotes and the prokaryotes. 10

(b) Write a critical account of the recombinant DNA technology. 5

4. (a) Give a brief account of the basic methods of crop improvements with special reference to maize. 5

(b) Discuss the role of haploids in genetics and crop improvement. 5

(c) Define any *five* of the following:

(i) Hybrid vigour (ii) Emasculation

(iii) Pure line (iv) Recurrent hybridisation

(v) Allopolyploidy (vi) Reading frame. 5

5. (a) Briefly mention the principles of human pedigree analysis. Give the example of a well known pedigree showing the inheriance of simple Mendelian traits in man. 5

   (b) Give a brief account of the symptoms and the causes of phenylketonuria, thalassemia, and sickle-cell anaemia in man. 5

   (c) Discuss the role of genetic counselling in human inheritance. 5

6. (a) What is somatic cell genetics? What possible roles does it play in the field of inheritance? 5

   (b) Write short notes on any *four* of the following:

      (i) Replica plating
      (ii) Southern blotting
      (iii) Transposons
      (iv) Maternal influence
      (v) Reverse transcription. 10

7. (a) Discuss briefly the Operon model of gene regulation. Compare and contrast the genetic regulation in prokaryotes and eukaryotes. 10

   (b) Discuss briefly the experimental evidences to prove that DNA is the genetic material. 3

   (c) Explain briefly the colinearity of genes and proteins. 2

8. (a) Compare and contrast any *five* of the following terms:

      (i) Translocation and Inversion
      (ii) B-DNA and Z-DNA
      (iii) Transition and Transversion
      (iv) Complementary and Epistatic genes
      (v) Pedigree method and Bulk method of selection
      (vi) Allelomorphs and Gynandromorphs. 10

   (b) Write a brief account of sex determination in plants and animals. 5

Genetics 3

1. Attempt any *ten* of the following:

   (i) Why can't auxotrophs grow in minimal media?
   (ii) Mendel failed to discover linkage. Why?
   (iii) What is the significance of chromosomal banding technique?
   (iv) Why are restriction endonucleases called 'biological scissors'?
   (v) Why is inversion repeat called 'crossing over suppressor'?
   (vi) Certain governments have put restrictions on genetic engineering research, why?
   (vii) In a case of disputed parentage, the blood group of mother is B, her father has group O and her husband has group AB. Is her child with blood group O legitimate? Why?
   (viii) In what circumstances should a genetic counsellor advise a pregnant woman to get abortion on seeing her aminocentesis report?

(ix) How many barr bodies are present in individuals with–

(a) Turner's syndrome,

(b) Kleinfelter's syndrome?

(x) What is the cause of the disease phenylketonuria in humans?

(xi) Why is *Neurospora* considered to be useful in the study of genetics?

(xii) How are Gynandromorphs formed in *Drosophila*? $10 \times 1^1/_2 = 15$

2. (i) In *Drosophila* three genes, forked (*f*), outstretched (*od*), and garnet (*g*) are present in one linkage group. Wild-type females heterozygous at all three loci were crossed to wild-type males. The following data were obtained for the males:

57	Garnet, outstretched
419	Garnet, forked
60	Forked
1	Outstretched, forked
2	Garnet
439	Outstretched
13	Wild-type
9	Outstretched, garnet, forked
1,000	

(a) Which gene is in the middle? $1^1/_2$

(b) What is the percentage of double cross overs? $1^1/_2$

(c) Calculate the map-distance. 3

(d) How much is the coefficient of coincidence and interference? 3

(ii) Discuss the experiment which gave cytological proof of crossing over. 5

(iii) Why doesn't crossing over between two genes loci exceed 50%? 1

3. Mention the scientist/s associated with the following concepts and briefly explain the concepts. Attempt any *four.*

(a) Teminism

(b) Chromosomal balance theory

(c) Wobble hypothesis

(d) Base equivalence rule

(e) Semi-conservative replication of DNA

(f) Heterosis 15

4. (i) How is inducible system different from repressible system? With the help of diagrams, explain positive control of *lac* operon. 8

(ii) Enumerate the characteristic features of the Genetic Code. Discuss Khorana's contribution in deciphering the genetic code. 7

5. Differentiate between any *five*:
   (a) Z DNA and B DNA
   (b) Euchromatin and Heterochromatin
   (c) Prokaryotic mRNA and Eukaryotic mRNA
   (d) Pure-line selection and Mass selection
   (e) Autopolyploidy and Allopolyploidy
   (f) Introns and Exons
   (g) Transition and Transversion. $3 \times 5 = 15$

6. (i) What are the objectives of a wheat breeding programme? Name *two* dwarf varieties of wheat which were introduced from Mexico. How did the hexaploid wheats originate? 7

   (ii) How are the different components of nucleosome arranged? 3

   (iii) Two non-allelic genes govern the comb shape in fowls. The genes for Rose comb (r) and Pea comb (p) together produce Walnut comb. Both genes in homozygous condition produce single comb. When a Rose was crossed with Walnut it produced offspring in the ratio of 15 walnut: 15 rose: 5 pea: 5 single. Determine the genotype of the parents and the progeny.

   (iv) How is *Triticale* artificially synthesised from its parents? Give the full form of the crop-improvement organisations given below:

   (a) SBI (b) CPRI
   (c) NBPGR (d) IRRI.

   Where are these located? 2

7. (i) Which chemical mutagens were discovered during World War II ? Discuss the molecular basis of action of alkylating agents. 6

   (ii) What are thymine dimers and how are they formed?

   Or

   Why is DNA polymerase I called a repair enzyme? 3

   (iii) *List* the proteins that unwind DNA during *in vivo* DNA replication. With the help of labelled diagrams, describe the synthesis of Okazaki Fragments. 6

8. (i) With the help of diagrams *only* show the process of translation in prokaryotes.

   Or

   Discuss the different steps involved in transfer of insulin gene from rat to *E. coli*. 7

   (ii) How is DNA-RNA hybridisation brought about *in vitro?* 4

   (iii) What could be the possible role of T_1-plasmid and SV 40 virus in genetic engineering research? 2

   (iv) The *Pseudomonas* strains created by Anand Chakrabarty and co-workers are often referred to as superbing. Why? 2

9. (i) What is Hardy-Weinberg equilibrium? Which genetic mechanisms disturb this equilibrium in a natural population? 5

(ii) Explain the mechanism of genetic drift. 4

(iii) Discuss the inheritance of plastid colour in *Mirabilis jalapa*. 4

(iv) What are oncogenic viruses? What is the genetic material of AIDS virus? 2

Genetics 4

1. (a) What is the difference between epistasis and dominance? How does epistasis modify the dihybrid ratios? 5

   (b) What is crossing over? At what stage of meiosis does it take place? How does it explain recombination between linked genes?

   What are the advantages and disadvantages of rocombination? 4

   Or

   Describe the Holliday model of recombination.

   (c) In corn colourless aleurone (c) is recessive to coloured (C), shrunken endosperm (sh) is recessive to full (Sh), waxy endosperm (wx) is recessive to starchy (Wx). When a plant heterozygous for 3 pairs of genes was crossed to a homozygous recessive for all the 3-pairs of genes the following progeny was obtained:

(a) Coloured shrunken starchy	2538
(b) Colourless full waxy	2708
(c) Coloured full waxy	116
(d) Colourless shrunken starchy	113
(e) Coloured shrunken waxy	601
(f) Colourless full starchy	626
(g) Coloured full starchy	4
(h) Colourless shrunken waxy	2

   (i) Map the genes wherein all the non-crossover and double crossover classes. 6

2. (a) Describe some of the techniques of plant cell and tissue culture and genetic engineering which are of use in plant breeding. How do they help in overcoming the difficulties encountered by traditional plant breeding methods? 5

   (b) Why is it necessary to collect wild relatives of cultivated plants for the improvement of the latter? 2

   (c) Describe Beadle and Tatum's experiment to prove one gene-one enzyme hypothesis? Why was it modified to one gene-one polypeptide concept. 5

   (d) You have isolated a *Neurospora* mutant that can't make amino acid Z, Normal *Neurospora* cells make Z from a precursor molecule *x* by a pathway involving three intermediates a, band c, x–a–b–c–z. How would you establish that mutant contains a defective gene for the enzyme that catalyse $b \rightarrow c$ reaction. 3

3. (a) Explain the terms (any *ten*):

   (i) Oncogene

   (ix) Test cross ratio

(ii) Nucleosome
(iii) Split genes
(iv) Shuttle vector
(v) Second messengers
(vi) Aneuploid
(vii) Transposon
(viii) Clone
(x) Multiple alleles
(xi) Central dogma
(xii) Auxotroph
(xiii) C-DNA
(xiv) DNA dependant DNA polymerase
(xv) Hot spots
(xvi) Multiple alleles 10

(b) What do the following abbreviations stand for (any *two*): 5

(i) T and F gene
(ii) PCR
(iii) ARC
(iv) SCE
(v) CSAR
(vi) IBPGR
(vii) RFLP

Or

Give the salient contributions of any *two* of the following scientists:

(i) H.J. Muller
(ii) J. Cairns
(iii) Erwin Chargaff
(iv) Watson and Crick
(v) H. Temin and D. Baltimore.

4. (a) Write notes on any *four*:

(i) Lyon's hypothesis
(ii) Replica plating
(iii) Criss-cross inheritance
(iv) Transcription factors
(v) Genetic counselling
(vi) Cloning vectors
(vii) Down's syndrome 10

(b) Describe any *two* of the following techniques and give their use in genetic research:

(i) RFLP
(ii) Southern blotting
(iii) PAGE
(iv) Amniocentesis

Or

What are banding techniques? Why are they important to geneticists? 5

5. (a) What is gene regulation? Describe the regulation of *Lac* operon in *E. coli* under positive and negative regulation. 5

(b) What are inborn errors of metabolism? Describe the symptoms and cause of either Sickle Cell Anaemia or Thalessemia. Have the genes been responsible for them being mapped (located and cloned)? Can they be cured by gene therapy? 5

(c) What do you understand by colinearity of genes and proteins? 5

Or

Describe Hardy-Weinberg principle.

6. (a) How does electromagnetic radiation differ from particulate radiation? Which of the two has greater genetic significance and why? Compare the type of genetic changes induced by them. 7

   (b) Distinguish between (i) Nonsense, (ii) Missense, (iii) Frameshift mutations and compare their effects on polypeptide synthesis. Which of them has the most drastic effect? 5

   (c)On the basis of ABO blood group *alone*, should a type AB woman never investigate a paternity suit? For the evidence to be most incriminating what should be the ABO type of her child? If the woman were type O would she be more or less likely to find ABO evidence in her favour than type AB? 3

7. (a) What are the different type of RNAs found in a Eukaryotic cell? Where are they located in the cell? Comment upon their structure and function. 6

   (b) Why do RNA viruses have small genomes? 1

   (c) Discuss the salient features of genetic code. How does one t-RNA recognise multiple codons? Is the universal nature of the genetic code of significance in gene cloning examples. 5

   (d) How will you distinguish between: 3

   (i) Nuclear genome;

   (ii) Mitochondrial genome?

8. (a) Distinguish between any *four*:

   (i) Pure line selection and mass selection;

   (ii) Transcription and translation;

   (iii) B DNA and Z DNA;

   (iv) Exons and introns;

   (v) Aneuploid and euploid;

   (vi) Pleiotropy and polygeny. 10

   (b) What are the different types of chromosomal aberrations? Which one has the most drastic effect in homozygous conditions? What different types of chromosomal abnormalities are found associated with cancer? 5

   Or

   What do you understand by cotransduction and cotransformation? In a series of transduction experiments involving genes A, B, C, D; the following combinations were found to cotransduce frequently A and B, C and D, A and D. Deduce the gene order.

Genetics 5

1. (a) Red flowers result from the genotype *WW*, white flowers from *ww*, and pink flowers from *Ww*. The flower shape will be straight if the genotype CC or Cc is present, whereas a genotype cc causes the flower to curl. The genes for flower colour and flower shape segregate independently of each other. What phenotypic ratio will result from a cross between two plants with pink flowers, both of which are heterozygously straight-shaped? 5

(b) In a series of crosses between plants of various heights to a 20 cm plant, the following results were obtained in the F_1 generations:

(i) 4 cm × 20 cm → All 12 cm

(ii) 8 cm × 20 cm → All 14 cm

(iii) 12 cm × 20 cm → All 16 cm

(iv) 16 cm × 20 cm → All 18 cm

Offer an explanation for the inheritance of height in the above crosses. Predict the genotypes of plants of each height. 5

(c) Prove that crossing over takes place at the 4-strand stage. 5

2. (a) Explain the following terms briefly (Attempt any *six*):

(i) Nucleosome
(ii) Replicon
(iii) Constitutive heterochromatin
(iv) Holandric gene
(v) Introgressive hybridisation
(vi) Cloning
(vii) Protooncogene
(viii) Z-DNA
(ix) Plasmid
(x) Introns
(xi) Epistasis
(xii) Map unit
(xiii) Barr body
(xiv) Marker gene
(xv) Base analogue
(xvi) Sister chromatid exchange
(xvii) Killer trait in *Paramecium*
(xvii) Reverse transcriptase 12

(b) How do acridines induce mutations? What does the use of acridines tell us about the genetic code? 3

3. (a) Write short notes (Attempt any *three*):

(i) Bacterial genome;
(ii) Giant chromosomes;
(iii) Repetitive DNA;
(iv) Cis-trans test;
(v) Hybrid vigour. 12

(b) What are the consequences of inversions?

Or

What are the traits which a plant breeder would wish to incorporate into a crop variety? 3

4. Distinguish between (Attempt any *five*):

(i) Maternal effect and maternal inheritance;
(ii) DNA polymerase and RNA polymerase;
(iii) Polygenes and pleiotropic genes;
(iv) Nonsense mutations and missense mutations;
(v) Splicesome and replisome;
(vi) Backcross and test cross;
(vii) Nucleus and nucleoid;
(viii) TATA box and Pribnow box. 15

5. (a) Attempt any *three*:
   (i) Why do isolated chloroplasts and mitochondria function only for brief periods?
   (ii) What is meant by criss-cross pattern of inheritance?
   (iii) What is the experimental evidence for the existence of *m*-RNA?
   (iv) Why did Gregor Mendel not detect linkage?
   (v) What are restriction endonucleases? What use can they be put to? 9

   (b) Discuss any *two* experiments which proved that DNA replication is by the semi-conservative method.

   Or

   Briefly describe the repair mechanisms which enable the DNA molecule to correct errors and damages that are likely to occur? 6

6. (a) What are inducible and repressible operons? 4
   (b) What are the *two* methods of selection practised in sexually reproducing crops? 4
   (c) Describe the Hardy-Weinberg principle. 4
   (d) An enzyme present in an organism has a polypeptide chain of 192 amino acid residues. It is coded for by a gene having, 1,440 base pairs. Explain the relationship between the number of amino acid residues in this enzyme and the number of nucleotide pairs in its gene.

   Or

   Explain the different meanings of the term transformation. 3

7. (a) What is the explanation offered for the origin of *Triticum aestivum*? 5
   (b) (i) What is chromosome banding technique? What are its applications? 3, 2
   (ii) Using suitable examples explain the significance of genetic counselling. 5

   Or

   What is genetic code? How was it deciphered? 10

8. (a) Describe the following techniques and their applications (Attempt any *three*):
   (i) RFLP;
   (ii) Double-cross method of breeding;
   (iii) PCR;
   (iv) Replica plating technique;
   (v) Somatic cell gene therapy. 9

   (b) What are the principal contributions of (Attempt any six):
   (i) Charles, Yonofsky;
   (ii) Bruce Ames;
   (iii) Erwin Chargaff;
   (iv) F. Sanger;
   (v) N. Borlaug;
   (vi) Barbara McClintock;
   (vii) Arthur Kornberg;
   (viii) Severo Ochoa;
   (ix) A.E. Garrod;
   (x) Lalji Singh;
   (xi) E.M. Meyerowitz;
   (xii) S. Benzer. 6

Life Cycles

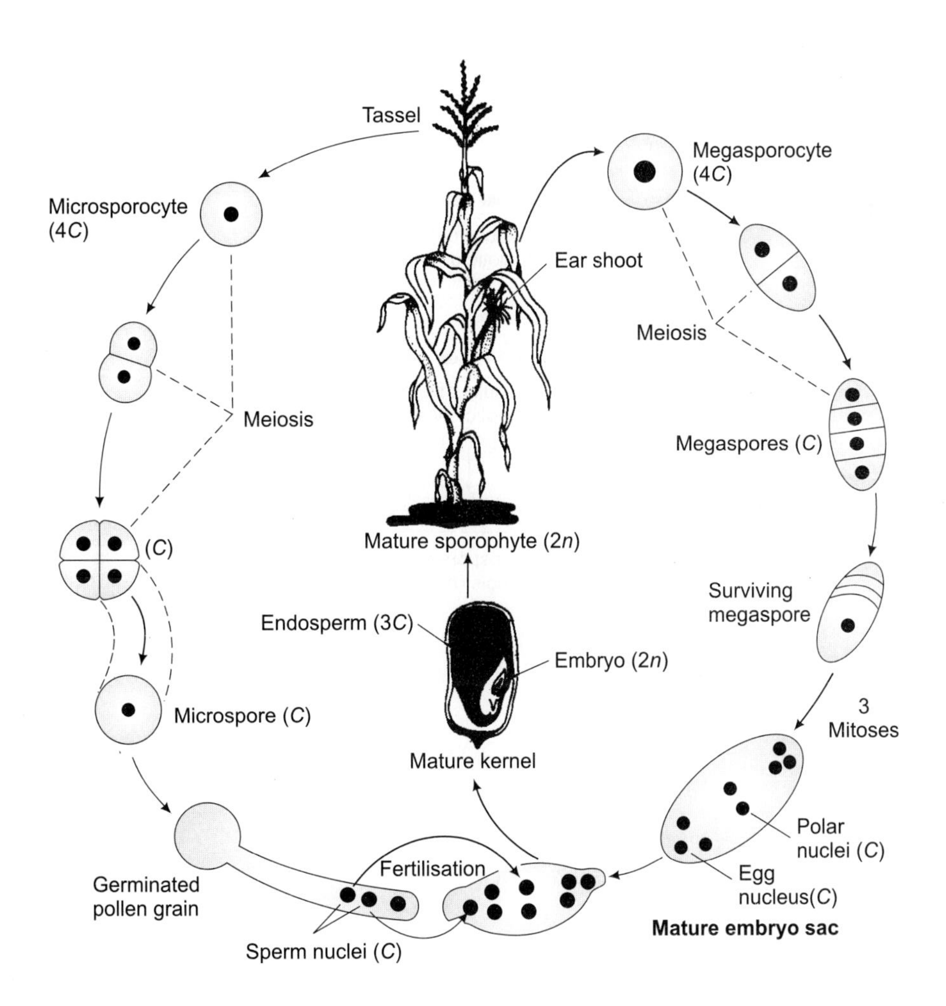

Corn Plant (Maize)

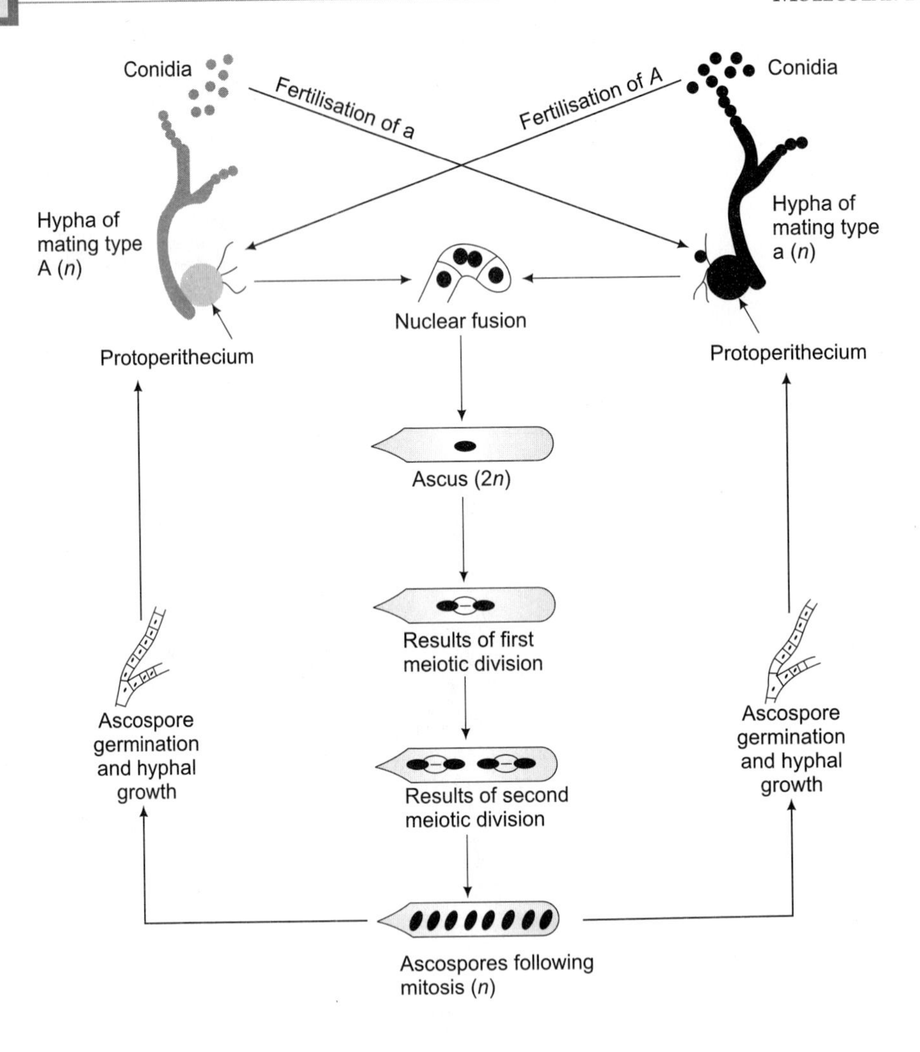

Neurospora erassa

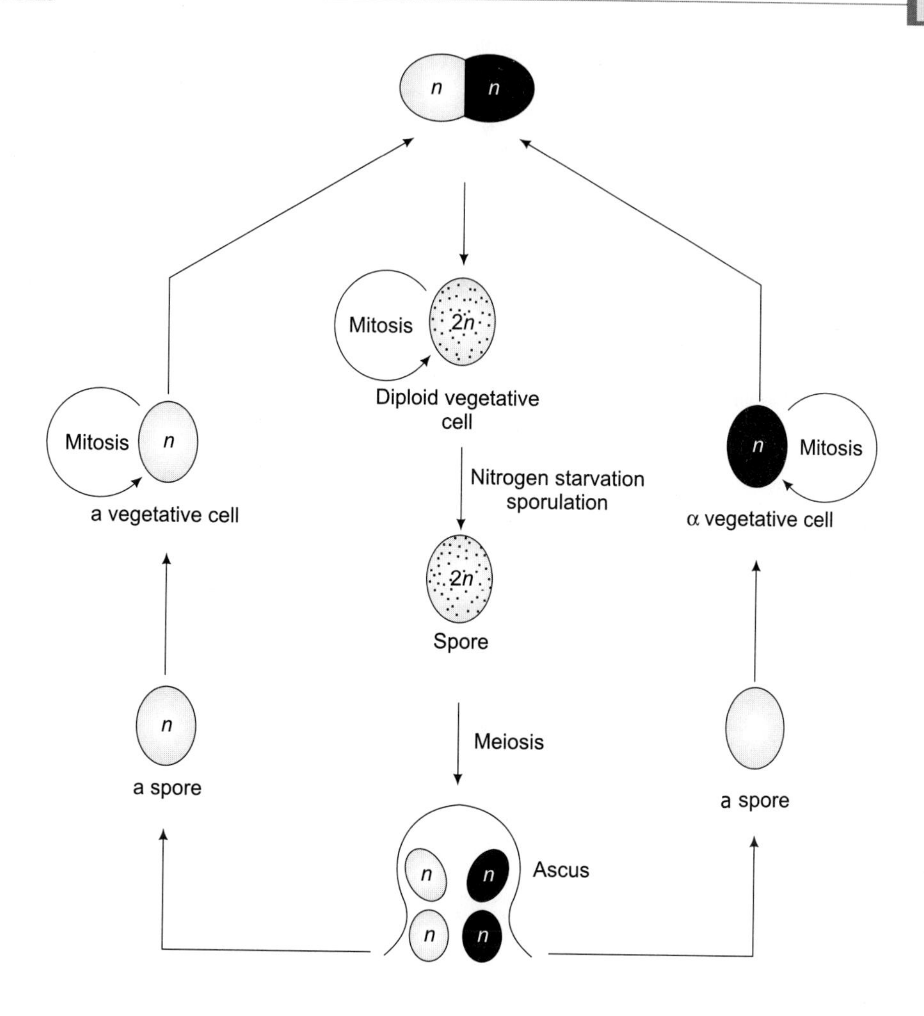

Yeast (*Saccharomyces cerevisiae*)

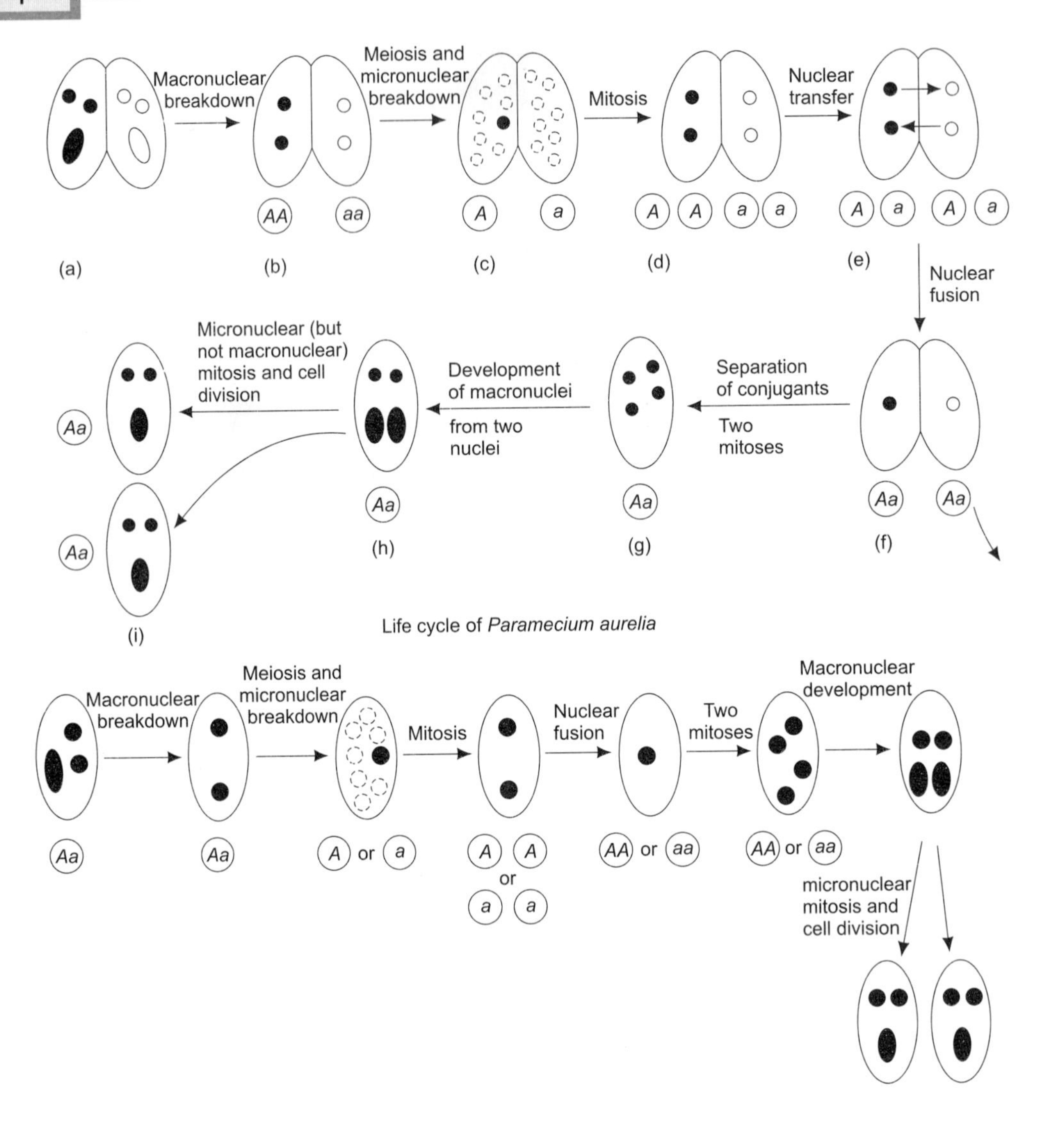

Autogamy in *Paramecium aurelia*

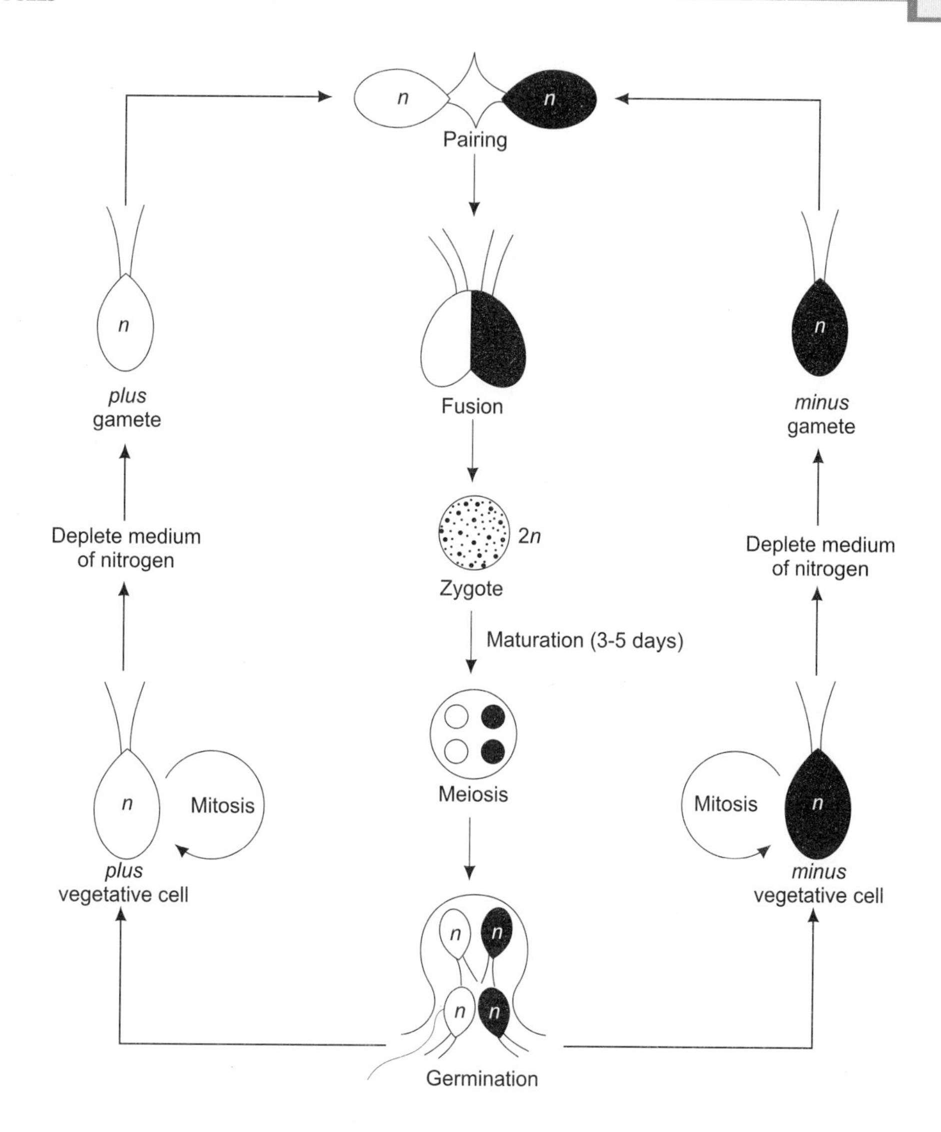

Chlamydomonas reinhardi as it occurs in the laboratory

Genetic Terms and Geneticists

Some Nobel Laureates in Genetics (Physiology/Medicine)

Name	*Year*	*Nationality*	*Cited for*
Thomas Hunt Morgan heredity	1933	USA	Discovery of the way that chromosomes govern
Hermann J. Muller	1946	USA	X-ray inducement of mutations
George W. Beadle	1958	USA	Genetic regulation of biosynthetic pathways
Edward L. Tatum	1958	USA	
Joshua Lederberg	1958	USA	Bacterial genetics
Severo Ochoa	1959	USA	Discovery of enzymes that synthesise nucleic acids
Arthur Kornberg	1959	USA	
Francis H. C. Crick	1962	UK	Discovery of the structure of DNA
James D. Watson	1962	USA	
Maurice Wilkins	1962	UK	
Francois Jacob	1965	French	Regulation of enzyme biosynthesis
Andre Lwoff	1965	French	
Jasques Monod	1965	French	
Robert W. Holley	1968	USA	Unravelling of the genetic code
H. Gobind Khorana	1968	USA	
Marshall W. Nirenberg	1968	USA	
Max Delbriick	1969	USA	Viral genetics
Alfred Hershey	1969	USA	
Salvador Luria	1969	USA	
Renato Dulbecco	1975	USA	Tumor viruses
Howard Temin	1975	USA	Discovery of reverse transcriptase
David Baltimore	1975	USA	
Werner Arber restriction	1978	Swiss	Discovery of sequencing of, and mapping with
Hamilton Smith	1978	USA	Endonucleases
Daniel Nathans	1978	USA	
Walter Gilbert	1980	USA	Techniques of sequencing DNA
Frederick Sanger	1980	UK	
Paul Berg	1980	USA	Pioneer work in recombinant DNA

LAND MARKING EVENTS IN THE HISTORY OF GENETICS

1856 Gregor Mendel, a monk at the Augustinian monastery of St. Thomas in Brunn, Austria (now Brno, Czechoslovakia), begins breeding experiments with the garden pea, *Pisum sativum.*

1859 Darwin publishes on the *Origin of Species.*

1860 T. A. E Klebs introduces paraffin embedding.

1861 L. Pasteur disproves the theory of spontaneous generation of microorganisms.

1865 Gregor Mendel presents the results and interpretations of his genetic studies on the garden pea at the Brunn Society for the study of Natural Science at their monthly meeting held on February 8 and March 8.

1866 G. Mendel's *Versuche uber Pflanzenhybriden* (*Experiments on Plant-Hybridisation*) is published and ignored.

F. Miescher isolates nucleoprotein.

1869 F. Galton publishes *Hereditary Genetics.* He emphasises the genetic basis of intelligence and is the first to compare mental traits in twins.

1870 W. His invents the microtome.

1873 A. Schneider gives the first account of mitosis.

1875 O. Hertwig concludes from a study of the reproduction of the sea urchin that fertilisation in both animals and plants consists of the physical union of the two nuclei contributed by the male and female parents.

E. Strasburger describes cell division in plants.

F. Galton draws attention to the use of twin studies for elucidating the relative importance of hereditary and environmental influence upon behavior.

1876 O. Bütschli describes nuclear dimorphism in ciliates.

1877 H. Fol reports watching the spermatozoan of a starfish penetrate the egg. He was to see the transfer of the intact nucleus of the sperm into the egg where it became the male pronucleus.

E. Abbe begins to publish important contributions to the theory of microscopic optics.

1878 W. Kuhne coins the word *enzyme.*

1879 W. Flemming studies mitosis in the epithelium of the tail fin of Salamanders. He shows that nuclear division involves a longitudinal splitting of the chromosome and a migration of the sister chromatids to the future daughter nuclei. He also coins the term *chromatin.*

1881 E.G. Balbiani discovers "cross-striped threads" within the salivary gland cells of *Chironomus* larvae. However, he does not realise that these are polytene chromosomes.

R. Koch devises the methods used to this day for isolating pure cultures of bacteria.

1882 W. Flemming discovers lampbrush chromosomes and coins the term *mitosis.*

1883 E. van Beneden studies meiosis in a species of the round worm, *Ascaris*, which (fortunately) has a diploid chromosome number of only four. He shows that the gametes contain half as many chromosomes as the somatic tissues and that the characteristic somatic number is re-established at fertilisation. He also describes fertilisation in mammals. W. Roux suggests that the filaments within the nucleus which stain with dyes are the bearers of the hereditary factors.

1887 A. Weismann postulates that a periodic reduction in chromosome number must occur in all sexual organisms.

1888 T. Boveri describes the centriole.

W. Waldeyer coins the word *chromosome* for the filaments referred to by Roux (1883).

1889 F. Galton introduces the quantitative measurement of populations into genetics.

1890 R. Altmann reports the presence of "bioblasts" within cells and concludes that they are "elementary organisms" that live as intracellular symbionts and carry out processes vital to their hosts. Later (1898) C. Benda names these organelles *mitochondria.* E. von Behring shows that blood serum from previously immunised animals contains factors that are specifically lethal to the organisms used for the immunisation. These factors are now called antibodies.

1896 E.B. Wilson publishes *The Cell in Development and Heredity.* This influential treatise distils the information gained concerning cytology in the half century since Schleiden and Schwann put forth the cell theory.

1898 T. Boveri describes chromatin diminution in *Ascaris megalocephala.*

1899 M.W. Beijerinck demonstrates that tobacco mosaic disease is due to a self-reproducing agent that will pass through bacterial filters and can neither be seen with light microscope nor grown upon bacteriological media. He proposes that this *virus* is a self-reproducing subcellular form of life, which represents a hitherto unknown class of organisms.

1900 H. de Vries, C. Correns, and E. Tschermak independently rediscover Mendel's paper. Using several plant species, de Vries and Correns had performed breeding experiments that parallelled Mendel's earlier studies and had independently arrived at similar interpretations of their results. Therefore, upon reading Mendel's publication they immediately recognised its significance. W. Bateson also stresses the impotance of Mendel's contribution in an address to the Royal Society of London.

K. Pearson develops the chi-square test.

K. Landsteiner discovers the blood agglutination phenomenon in man.

P. Ehrlich proposes that antigens and antibodies bind together because they have structural complementarity.

1901 H. de Vries adopts the term *mutation* to describe sudden, spontaneous, drastic alterations in the hereditary material of *Oenothera.*

T.H. Montgomery studies spermatogenesis in various species of Hemiptera. He concludes that maternal chromosomes only pair with paternal chromosomes during meiosis.

K. Landsteiner demonstrates that humans can be divided into three blood groups; A, B, and C. The designation of the third group was later changed to O.

E. von Behrin wins the Nobel Prize for his studies on antiserum therapy.

1902 C.E. McClung notes that in various insect species equal numbers of two types of spermatozoa are formed; one type contains an "accessory chromosome" and the other does not. He suggests that the extra chromosome is a sex determinant, and he next argues that sex must be determined at the time of fertilisation, not just in insects, but perhaps in other species (including man).

T. Boveri studies the development of haploid, diploid, and aneuploid sea urchin embryos. He finds that in order to develop normally, the organism must have a full set of chromosomes, and he concludes that the Individual chromosomes must carry *different* essential hereditary determinants. W.S. Sutton advances the chromosome theory of heredity, which proposes that independent assortment of gene pairs stems from the behaviour of the synapsed chromosomes during meiosis. Since the direction of segregation of the homologues in a given bivalent is independent of those belonging to any other bivalent, the genes they contain will also be distributed independently.

F. Hofmeister and E. Fischer propose that all proteins are formed by the condensation of amino acids bound through regularly recurrent peptide linkages.

1902-09 W. Bateson introduces the terms *genetics, allelomorph, homozygote, heterozygote*, F_1, F_2, and *epistatic* genes.

1903 W. Waldeyer defines *centromeres* as chromosome regions with which the spindle fibers become associated during mitosis.

1906 W. Bateson and R.C. Punnett report the first case of linkage (in the sweet pea).

1907 R.G. Harrison cultures fragments of the central nervous systems of frogs in haemolymph and observes the outgrowth of nerve fibers. In so doing, he invents tissue culture.

1908 G. H. Hardy and W. Weinberg, working independently, formulate the so-called Hardy-Weinberg law of population genetics.

1909 G. H. Shull advocates the use of self-fertilised lines in production of commercial seed corn. The hybrid corn programme which created an abundance of foodstuffs worth billions of dollars.

A.E Garrod publishes *Inborn Errors of Metabolism*, the earliest study of the biochemical genetics of man (or any other species).

F. A. Janssens suggests that exchanges between nonsister chromatids produce chiasmata. C.C. Little initiates a breeding programme that produces the first inbreds strain of mice (the strains now called DBA). W. Johannsen's studies of the inheritance of seed size in self-fertilised lines of beans lead him to realise the necessity of distinguishing between the appearance of an organism and its genetic constitution. He invents the terms *phenotype* and *genotype* to serve this purpose, and he also coins the word *gene*.

C. Correns and E. Bauer study the inheritance of chloroplast defects in variegated plants, such as *Mirabilis jalapa* and *Pelargonium zonate.* They find that the inability to form healthy chloroplasts is in some cases inherited in a non-Mendelian fashion.

H. Nilsson Ehle puts forward the multiple factor hypothesis to explain the quantitative inheritance of seed-coat colour in wheat.

1910 T.H. Morgan discovers white eye and consequently sex linkage in *Drosophila. Drosophila* genetics begins.

W. Weinberg develops the methods used for correcting expectations for Mendelian segregation from human pedigree data under different kinds of ascertainment applied to data from small families.

P. Rous shows that injection of a cell-free filtrate from chicken sarcomas induces new sarcomas in recipient chickens.

1911 T.H. Morgan proposes that the genes for white eyes, yellow body, and miniature wings in *Drosophila* are linked together on the X chromosome.

W.R.B. Robertson points out that a metacentric chromosome in one orthopteran species may correspond to two acrocentrics in another and concludes that during evolution metacentrics may arise by the fusion of acrocentrics. Whole arm fusions are called *Robertsonian translocations* in his honour.

1912 A. Wegener proposes the Continental Drift concept.

F. Rambousek suggests that the "cross-striped threads" within the salivary gland cells of fly maggots are chromosomes.

T.H. Morgan demonstrates that crossing over does not take place in the male of *Drosophila melanogaster.* He also discovers the first sex-linked lethal.

1913 Y. Tanaka reports that crossing over does not take place in the female of *Bombyx mori.* In this species the female is the heterogametic sex.

W.H. Bragg and W.L. Bragg demonstrate that the analysis of X-ray diffraction patterns can be used to determine the three-dimensional atomic structure of crystals.

A.H. Sturtevant provides the experimental basis for the linkage concept in *Drosophila* and produces the first genetic map.

1914 C.B. Bridges discovers meiotic nondisjunction in *Drosophila.*

C.C. Little postulates that the acceptance or rejection of transplanted tumors has a genetic basis.

1915 F. W. Twort isolates the first filterable bacterial virus.

R.B. Goldschmidt coins the term *intersex* to describe the aberrant sexual types arising from crosses between certain different races of the gypsy moth, *Lymantria dispar.*

J.B.S. Haldane, A.D. Sprunt. and N.M. Haldane describe the first example of linkage in vertebrates (mice).

1916 H.J. Muller discovers interference in *Drosophila.*

1917 F. d'Herelle coins the term *bacteriophage* and develops methods for assaying virus titre.

O. Winge calls attention to the important role of polyploidy in the evolution of angiosperms.

C.B. Bridges discovers the first chromosome deficiency in *Drosophila.*

1918 R. Spemann demonstrates that a living part of an embryo can exert a morphogenic stimulus upon another part, bringing about its morphological differentiation (embryonic induction). He thus discovers and names the *organiser.*

H. J. Muller discovers the balanced lethal phenomenon in *Drosophila.*

1919 T. H. Morgan calls attention to the equality in *Drosophila melanogaster* between the number of linkage groups and the haploid number of chromosomes.

C.B. Bridges discovers chromosomal duplications in *Drosophila.*

1920 A.F. Blakeslee. J. Belling, and M.E. Farnham describe trisomics in the Jimson weed, *Datura stramonium.* H.Kniep describes and correctly interprets the incompatibility system in *Schizophyllum commune* and related basidiomycetes.

1921 F.G Banting and C.H. Best isolate insulin and study its physiological properties.

R.B. Goldschmidt publishes the first genetic analysis and discussion of the evolutionary implications of industrial melanism.

C.B. Bridges reports the first monosomic (haplo-4) in *Drosophila.*

A.F. Blakeslee, J. Belling, M.E. Farnham, and A.D. Bergner discover a haploid *Datura*

1923 C.B. Bridges discovers chromosomal translocations in *Drosophila.*

R. Feulgen and H. Rossenbeck describe the cytochemical test which currently is most used for DNA localisation.

O.Winge discovers the first case of monosomy in a plant (wheat).

T. Svedberg builds the first ultracentrifuge.

A.E. Boycott and C. Diver describe "delayed" Mendelian inheritance controlling the direction of the coiling of the shell in the snail *Limnea peregra.*

A.H. Sturtevant suggests that the direction of coiling of the *Limnea* shell is determined by the character of the ooplasm, which is, in turn, controlled by the mother's genotype.

The XX-XY type of sex determination is demonstrated for certain dioecious plants: for *Elodea* by J.K. Santos, for *Rumex* by H. Kihara and T. Ono, and for *Humulus* by O. Winge.

1925 C.B. bridges completes his cytogenetic analysis of the aneuploid offspring of triploid *Drosophila,* and he defines the relations between the sex chromosomes and autosomes that control sexual phenotype.

E.M. East and A.J. Mangelsdorf propose the first satisfactory interpretation of the phenomenon of self-sterility in flowering plants.

A.H. Sturtevent analyses the *Bar* phenomena in *Drosophila* and discovers position effect.

F. Bernstein suggests that the ABO blood groups are determined by a series of allelic genes.

T.H. Goodspeed and R.E. Clausen produce an amphiploid in *Nicotiana.*

1926 E.G. Anderson establishes that the centromere of the X chromosome of *Drosophila* is at the end opposite the locus of *yellow.*

S.S. Chetverikov initiates the genetic analysis of wild population of *Drosophila.*

J.B. Sumner isolates the first enzyme in crystalline form and shows it (urease) to be a protein.

A.H. Sturtevant finds the first inversion in *Drosophila.*

R.E. Clausen and T.H. Goodspeed describe the first analysis of monosomics in a plant (*Nicotiana*).

1927 K.M. Bauer reports that the rejection of skin grafts does not occur when skin is transplanted from one monozygous twin to another.

J. Belling proposes that interchanges between nonhomologous chromosomes result in ring formation at meiosis.

J.B.S. Haldane suggests that the genes known to control certain coat colours in various rodents and carnivores may be evolutionarily homologous.

J. Belling introduces the aceto-carmine technique for staining chromosome squashes.

B.O. Dodge initiates genetic studies on *Neurospora.*

H.J. Muller reports the artificial induction of mutations in *Drosophila* by X-rays.

1928 L.J. Stadler reports the artificial induction of mutations in maize, and demonstrates that the dose-frequency curve is linear.

F. Griffith discovers type-transformation of pneumococci. This lays the foundation for the work of Avery, MacLeod, and McCarthy (1944).

L.F. Randolph distinguishes supernumerary chromosomes from the normal chromosomes of the plant cell. He calls the normal ones "A chromosomes" and the supernumerary ones "B chromosomes".

E. Heitz introduces the term *heterochromatin.*

1929 A. Fleming discovers penicillin.

C.D. Darlington is the first to suggest that chiasmats function to hold homologues together at meiotic metaphase I and so ensure that they pass to opposite poles at anaphase I.

1930 R.E. Cleveland and A.F. Blakeslee demonstrate that the peculiar patterns of the transmission of groups of genes in various *Oenothera races* result from a system of balanced lethal and reciprocal translocation complexes.

R.A. Fisher publishes *The Genetical Theory of Natural Selection.*

K. Landsteiner receives a Nobel Prize for his studies in immunology.

1931 C. Stern, and independently H.B. Creighton and B. McClintock, Provide the cytological proof of crossing over.

S. Wright publishes *Evolution in Mendelian Populations*. This work together with Fisher's book and a series of papers by J.B.S. Haldane published between 1930 and 1932 under the general title *A Mathematical Theory of Natural and Artificial Selection* constitute the mathematical foundation of population genetics.

C.D. Darlington suggests that chiasmata can move to the ends of bivalents without breakage of the chromosomes. This process of *terminalisation*, as he called it, is now known to occur in some species but not in others.

B. McClintock shows in maize that if a segment of a chromosome has become relatively inverted, individuals heterozygous for such an inversion often show reversed pairing at pachynema.

1932 M. Knoll and E Ruska describe the prototype of the modem electron microscope.

1933 T.S. Painter initiates cytogenetic studies on the salivary gland chromosomes of *Drosophila.* H. Hashimoto works out the chromosomal control of sex determination for *Bombyx mori.* A.W.K. Tiselius reports the invention of an apparatus which permits the separation of charged molecules by electrophoresis.

B. McClintock demonstrates in maize that a single exchange within the inversion loop of a paracentric inversion heterozygote generates an acentric and a dicentric chromatid.

T.H. Morgan receives a Nobel Prize for his development of the theory of the gene.

1934 M. Schleisinger reports that certain bacteriophages are composed of DNA and protein.

P.L' Heritier and G. Teissier experimentally demonstrate the disappearance of a deleterious gene from populations of *Drosophila melanogaster* maintained in population cages for many generations.

A. Folling discovers phenylketonuria, the first hereditary metabolic disorder shown to be responsible for mental retardation.

H. Bauer postulates that the giant chromosomes of the salivary gland cells of fly larvae are polytene B. McClintock shows that the nucleolus organiser of *Zea mays* can be split by a translocation and that each piece is capable of organising a separate nucleolus. She thus sets the stage for the later demonstration (1965) that the genes for rRNA are present in multiple copies.

1935 J.B.S. Haldane is the first to calculate the spontaneous mutation frequency of a human gene.

F. Zernicke describes the principle of the phase microscope.

G.W. Beadle and B. Ephrussi and A. Kuhn and A. Butenandt work out the biochemical genetics of eye pigment synthesis in *Drosophila* and *Ephestia*, respectively.

W.M. Stanley succeeds in isolating and crystallising the tobacco mosaic virus.

C.B. Bridges publishes the salivary gland chromosome maps for *Drosophila melanogaster.*

H. Spemann receives a Nobel Prize for his studies on embryonic induction.

1936 J. Schultz notes the relation of the mosaic expression of a gene in *Drosophila* to its position relative to heterochromatin.

T. Caspers son uses cytospectrophotometric methods to investigate the quantitative chemical composition of cells.

J.J. Bittner shows that mammary carcinomas in mice can be caused by a virus-like factor transmitted through the mother's milk.

A.H. Sturtevant and T. Dobzhansky publish the first account of the use of inversions in constructing a chromosomal phylogenetic tree.

C. Stern discovers somatic crossing over in *Drosophila.*

R. Scott-Moncrieff reviews the inheritance of plant pigments. The major part of this work was done by a group of English geneticists at the John Innes Horticultural Institute, and these early workers established that gene substitution resulted in chemical changes in certain flavanoid and carotenoid pigments.

1937 T. Dobzhansky publishes *Genetics and the Origin of Species.*

A.F. Blakeslee and A.G. Avery report that colchicine induces polyploidy.

T.M. Sonneborn discovers mating types in *Paramecium.*

F.C. Bawden and N.W. Pirie show that the tobacco mosaic virus, although being made mostly of protein, also contains a small amount (about 5%) of RNA.

P.A. Gorer discovers the first histocompatibility antigens in the laboratory mouse.

H. Karstrom points out that the synthesis of certain bacterial enzymes is stimulated when the substrates of these enzymes attack are added to the medium. He coins the term "adaptive enzymes" for these and differentiates them from the "constitutive enzymes" that are always formed irrespective of the composition of the medium.

1938 B. McClintock describes the bridge-breakage-fusion-bridge cycle.

T.M. Sonneborn discovers the killer factor of *Paramecium.*

M.M. Rhoades describes the mutator gene *Dt* in maize.

H. Slizynska makes a cytological analysis of several overlapping Notch deficiencies in the salivary gland X chromosomes of *Drosophila melanogaster* and determines the band locations of the *w* and *N* genes.

1939 E.L. Ellis and M. Delbruck perform studies on coliphage growth, which mark the beginning of modern phage work. They devise the "one-step growth" experiment, which demonstrated that after the phage adsorbs onto the bacterium, it replicates within the bacterium during the "latent period." and finally the progeny are released in a "burst".

P. Levine and R.E. Stetson discover maternal immunisation by a foetus carrying a new blood group antigen inherited from the father. This antigen is subsequently identified with the human Rh blood group system as the cause of erythroblastosis foetalis.

A. W. K. Tiselius and E.A. Kabat demonstrate that antibodies belong to the gamma class of serum globulins.

1940 W. Earle establishes the strain L permanent cell line from a C3H mouse.

1941 G.W. Beadlle and E.L. Tatum publish their classic study on the biochemical genetics of *Neurospora* and promulgate the one gene-one enzyme theory.

J. Brachet and T. Caspersson independently reach the conclusion that RNA is localised in the nucleoli and cytoplasm and that a cell's content of RNA is directly linked with the protein synthesising capacity. A.J.P. Martin and R.L.M. Synge develop the technique of partition chromatography and use it to determine the amino acids in protein hydrolyzates.

A. H. Coons, H. J. Creech and R.N. Jones develop immunofluorescence techniques to demonstrate the presence of antibody-reactive sites on specific cells.

1942 R. Schoenhejmer publishes *The Dynamic State of Body Constituents* and describes the use of isotopically tagged compounds in metabolic studies. He introduces the concepts of "metabolic pools" and the "turnover" of the organic compounds in cells.

L.J. Stadler and F.M. Uber demonstrate that the effectiveness of ultraviolet light in inducing mutations corresponds in corn pollen to the absorption spectrum of nucleic acid.

G. D. Snell sets out to develop highly inbred strains of mice to study the genes responsible for graft rejection.

1943 The production of radioisotopes is begun at the X-10 reactor at Oak Ridge, Tennessee.

A. Claude isolates and names the microsome fraction and shows that it contains the majority of the RNA of cells.

S.E. Luria and M. Delbruck initiate the field of bacterial genetics when they demonstrate unambiguously that bacteria undergo spontaneous mutation.

1944 O.T. Avery, C.M. MacLeod, and M. McCarty describe the pneumococcus transforming principle. The fact that it is rich in DNA suggests that DNA and not protein is the hereditary chemical.

T. Dobzhansky describes the phylogeny of the gene arrangements in the third chromosome of *Drosophila pseudoobscura* and D. *persimilis.*

E.L. Tatum, D. Bonner, and G.W. Beadle use mutant strains of *Neurospora crassa to work* out the intermediate steps in the synthesis of tryptophan.

1945 R.R. Humphrey demonstrates that the female is the heterogametic sex in urodeles.

E.B. Lewis discovers the cis-trans position effect in *Drosophila.*

R.D. Owen reports that in cattle dizygotic twins are born with, and often retain throughout life, a stable mixture of each other's red blood cells. This chimerism, which results from vascular anastomoses within the chorions of the foetuses, provides the first example of immune tolerance.

1946 A. Claude introduces tissue fractionation techniques based upon differential centrifugation and works out methods for characterising the fractions biochemically.

Genetic recombination in bacteriophage is demonstrated by M. Delbruck and W.T Bailey and by A.D. Hershey.

J. Lederberg and E.L. Tatum demonstrate genetic recombination in bacteria.

C. Auerbach and J.M. Robson report that mustard gas induces mutations in *Drosophila.*

J.A. Rapoport demonstrates the mutagenic effectiveness of formaldehyde in *Drosophila.*

J. Oudin develops the gel-diffusion, antigen-antibody precipitation test that bears his name. Nobel prizes are awarded to H.J. Muller, for his contributions to radiation genetics, to J. oudin.

1953 J.D. Watson and F.H.C. Crick unravel the double helical structure of DNA. Published in *Nature* June 1953.

Linus Pauling unravels *a*-helix structure of protein.

1955 R.H. Pritchard studies the linear arrangement of a series of allelic adenine-requiring mutants of *Aspergillus.* He confirms Pontecorvo's conclusion that crossing over can occur between different alleles of the same gene, provided they are characterised by mutations at different subsites.

C. de Duve and four colleagues describe intracellular vesicles that contain hydrolytic enzymes and name them *lysosomes.*

P. Grabar and C.A. Williams devise the technique of immunoelectrophoresis to analyse complex mixtures of antigenetic molecules.

1956 H.B. Kettlewell studies industrial melanism in the Peppered Moth. He demonstrates that moths that are conspicuous in their habitats are indeed eaten by birds more often than inconspicuous forms.

F. Jacob and E.L. Wollman are able experimentally to interrupt the mating process in *E. coli* and show that apiece of DNA is inserted from the donor bacterium into the recipient. Groups led by S. Ochoa and A. Kornberg succeed in the *in vitro* enzymatic synthesis of polymers of ribonucleotide and deoxyribonucleotides, respectively.

J.H. Tjio and A. Levan demonstrate that the diploid chromosome number for man is 46. C.O. Miller and his co-workers isolate and determine the chemical structure of kinetin, a substance promoting cell division in plants.

T.T. Puck, S.J. Cieciura, and P.I. Marcus succeed in growing clones of human cells *in vitro.*

G.E. Palade and P. Siekevitz isolate ribosomes.

M.J.Moses and D. Fawcett independently observe synaptonemal complexes in spermatocytes.

Gierer and G. Schramm and H. Fraenkel-Conrat demonstrate independently that a chemically pure nucleic acid, namely tobacco mosaic virus RNA, is infectious and genetically competent.

1957 S.A. Berson and R.S. Yalow report the first use of the radioimmunoassay procedure for the detection of insulin antibodies developed by patients in response to the administration of exogenous insulin.

J.H. Taylor, P.S. Woods and W.L. Hughes are the first to use tritiated thymidine in high resolution autoradiography in experiments that demonstrate the semiconservative distribution of label during chromosome replication in *Vicia faba.*

V.M. Ingram reports that normal and sickle-cell haemoglobin differ by a single amino acid substitution. A. Todd receives the Nobel Prize for his studies on the structure of nucleotides and nucleosides.

1958 F. Jacob and E.L. Wollman demonstrate that the single linkage of *E. coli* is circular and suggest that the different linkage groups found in different *Hfr* strain result from the insertion at different points of a factor in the circular linkage group which determines the rupture of the circle.

F.H.C. Crick suggests that during protein formation the amino acid is carried to the template by an adaptor molecule containing nucleotides and that the adaptor is the part which actually fits on the RNA template. Crick thus predicts the discovery of transfer RNA complexes.

H.G. Callan and H.G. MacGregor demonstrate that the linear integrity of chromatids of amphibian lampbrush chromosomes is maintained by DNA, not protein.

M. Okamoto and, independently, R. Riley and V. Chapman discover genes that control the pairing of homologous chromosomes in wheat.

J.R. Raper and co-workers work out the two-locus, multiple allelic structure of the A and *B* incompatibility factors of *Schizophyllum commune.*

F.C. Steward, M.O. Mapes, and K. Mears succeed in rearing sexually mature plants from single diploid cells derived from the secondary phloem of roots of the wild carrot, *Daucus carota.* They conclude that each cell of the multicellular organism has all the ingredients necessary for the formation of the complete organism.

M. Meselson and F.W. Stahl use the density gradient equilibrium centrifugation technique to demonstrate the semiconservative distribution of density label during DNA replication in *E. coli.*

Nobel prizes are rewarded to G.W. Beadle, E.L. Tatum, and J. Lederberg for their contributions to genetics and to F. Sanger for his contributions to protein chemistry.

1959 J. Lejeune, M. Garder and R. Turpin show that Down syndrome is a chromosomal aberration involving trisomy of a small telocentric chromosome.

C.E. Ford, K.W. Jones. P.E. Polani, J. C. de Almeida, and J. H. Briggs discover that females suffering from Turner syndrome are XO.

P. A. Jacobs and J.A. Strong demonstrate that males suffering from Klinefelter syndrome are XXY.

S.J. Singer conjuctes ferritin with immunoglobulin to produce a labelled antibody, which is readily recognised under the electron microscope.

R.L. Sinsheimer demonstrates that bacteriophage *f* ′ 174 of *E. coli* contains a single-stranded DNA molecule.

F.M. Burnet improves Jerne's selective theory of antibody formation by suggesting that the antigen stimulates the proliferation of only those cells that are genetically programmed to synthesise the complementary antibodies.

G.M. Edelman resolves immunoglobulin G into heavy and light chains.

A. Limia-de-Faria demonstrates by autoradiography that heterochromatin replicates later than euchromatin. M. Chevremont.S Chevremont-Comhaire, and E. Baeckeland demonstrate DNA in mitochondria using a combination of autoradiographic and Feulgen-staining techniques.

K. McQuillen. R.B. Roberts, and R.J. Britten demonstrate in *E. coli* that ribosomes are the sites where protein synthesis takes place.

E. Freese proposes that mutation can occur as the result of single, base-pair changes in DNA. He coins the terms transitions (where a purine is replaced by the other purine or a pyrimidine by the other pyrimidine) and *transversions* (where a purine is replaced by a pyrimidine and vice versa).

C. Pelling finds selective labelling of puffed regions of polytene chromosomes after they are incubated in a nutrient solution containing ^{3}H uridine.

R.H. Whittaker suggests the grouping of organisms into five kingdoms: the bacteria, the eukaryotic micro-organisms, animals, plants, and fungi.

S. Ochoa and A. Kornberg receive Nobel Prizes for their studies on the *in vitro* synthesis of nucleic acids.

1960 P. Nowell discovers phytohemagglutinin and demonstrates its use in stimulating mitoses in human leukocyte cultures.

H.E. Huxley and G.L. Zubay develop the negative staining procedure for electron microscopy of subcellular particles.

P. Siekevitz and G.E. Palade describe the synthesis of secretory proteins on a membrane bound ribosomes. P. Doty, J. Marmur, J. Eigner, and C. Schildkraut demonstrate that complementary strands of DNA molecules can be separated and recombined.

G. Barski, S. Sorieul, and F. Cornefert report the first successful *in vitro* hybridisation of mammalian cells.

U. Clever and P. Karlson experimentally induce specific puffing patterns in polytene chromosomes by injecting *Chironomus* larvae with ecdysome.

P.B. Medawar and F.M. Burnet receive a Nobel Prize for their studies on immunological tolerance.

1961 F. Jacob and J. Monod suggest that ribosomes do not contain the template responsible for the orderly assembly of amino acids. They propose that instead each DNA cistron causes synthesis of an RNA molecule of limited life-span which harbours the amino acid sequence information in its nucleotide sequence. This molecule subsequently enters into temporary association with a ribosome and so confers upon it the ability to synthesise a given protein. This messenger RNA is subsequently demonstrated by S. Brenner, F. Jacob, and M. Meselsones and by F. Gros, W. Gilbert, H. Hiatt, C.G. Kurland, and J.D. Watson.

M.F. Lyon and L.B. Russell independently provide evidence suggesting that in mammals one X chromosome is inactivated in some embryonic cells and their descendants, that the other is inactivated in the rest, and that mammalian females are consequently X chromosome mosaics.

V.M. Ingram presents a theory explaining the evolution of the four known kinds of haemoglobin, chains from a single primitive myoglobin-like haemeprotein by gene duplication and translocation.

B.D. Hall and S. Spiegelman demonstrate that hybrid molecules can be formed containing one single-stranded DNA and one RNA molecule, which are complementary in base sequence. Their technique opens the way to the isolation and characterisation of messenger RNAs.

S.B. Weiss and, T. Nakamoto isolate RNA polymerase.

G. von Ehrenstein and F. Lipmann combine messenger RNA and ribosomes from rabbit reticulocytes with amino acid-transfer RNA complexes derived from *E. coli.* Since this cell-free system synthesised a protein similar to rabbit haemoglobin, they concluded that the genetic code is universal.

F.H.C. Crick L. Barnett, S. Brenner, ana R.J. Watts-Tobin show that the genetic language is made up of three-letter words.

F. Jacob and J. Monod put forward the operon concept.

W. Beermann demonstrates that a puffing locus on a *Chironimus* Polytene chromosome is inherited in a Mendelian fashion.

M.W. Nirenberg and J.H. Matthaei develop a cell-free system from *E.coli* that incorporates amino acids into protein when supplied with template RNA preparations. They show that the synthetic polynucleotide, polyuridylic acid, directs the synthesis of a protein resembling polyphenylalanine.

M. Meselson and J.J. Weigle demonstrate in phage lambda that recombination involves breakage and reunion (but not replication) of the chromosome.

H. Dintzis shows that the direction of synthesis of the haemoglobin molecule is from amino to carboxyl terminus.

H. Moor, K. Muhlemhaler, H. Waldner, and A. Frey-Wyssling develop the first freeze-fracture procedure that permits ultrastructual observations impossible with conventional sectioning methods.

U.Z. Littauer shows that ribosomes contain only two high molecular weight species of RNA, with sedimentation values of l6S and 23S in bacteria and 18S and about 28S in animals.

C. Tokunaga demonstrates that the *engrailed* gene of *Drosophila melanogaster* causes a shift from one developmental prepattern to a different but related prepattern.

J.P. Waller and J.I. Harris find that bacterial ribosomes contain a large number of different proteins.

1962 H. Ris and W. plaut show by electron microscopy that chloroplasts contain DNA.

E. Zuckerkandl and L. Pauling calculate the approximate times of derivation of different haemoglobin chains from their common ancestors during eukaryotic evolution.

F.M. Ritossa discovers that the salivary gland chromosomes of *Drosophila buskii respond* to heat shocks by puffing.

The distinction between T and B lymphocytes is shown in publications by J.F.A.P. Miller, R.A. Good et al., and N.L. Warner et al.

R.R. Porter uses enzymes to cleave immunoglobulin molecules. He demonstrates that each molecule has two antigen binding portions (F_{ab}) and a crystallisable segment (F_c) that does not bind antigen. He shows that the heavy and light chains are present in 1:1 ratio and suggests the four chain model.

D.A. Rodgers and G.E. McClearn discover differences between mouse strains in alcohol preference.

U. Henning and C. Yanofsky show that amino acid replacements can arise from crossing over within triplets.

J.B. Gurdon reports that a normal fertile frog can develop from an enucleated egg injected with a nucleus from an intestinal cell. This experiment demonstrates that somatic and germinal nuclei are qualitatively equivalent.

Polyribosomes are discovered independently in three laboratories (by A. Gierer, by J.R. Warner, A Rich, and C.E. Hall, and by T, Staehelin and H. Noll).

A.M. Campbell proposes that episomes become integrated into host chromosomes by a crossover event resembling the exchanges that were previously reported between synapsed ring and rod-shaped chromosomes in eukaryotes.

Nobel prizes are awarded to J.D. Watson, F.H.C. Crick, and M.H.F. Wilkins for their studies on the structure of DNA and to M. F. Perutz and J.C. Kendrew for their studies on the structure of haemoglobin and myoglobin.

1963 B.B. Levine, A. Ojida, and B. Benacerraf publish the first paper on the immune response genes of guinea pigs.

R. Rosset and R. Monier discover 5S RNA.

T. Okamoto and M. Takanami show that mRNA binds to the small ribosomal subunit.

H. Noll, T. Staehelin, and F.O. Wettstein demonstrate the tape mechanism of protein synthesis.

J.G. Gall produces evidence that the lampbrush chromatid contains a single DNA double helix.

B.J. McCanhy and E.T. Bolton use their DNA-agar technique to measure genetic relatedness between diverse species of organisms.

E. Hadorn demonstrates allotypic differentiation in cultured imaginal discs of *Drosophila.*

J. Monod and S. Brenner publish the replicon model.

R. Sager and M.R. Ishida isolate chloroplast DNA from *Chlamydomonas.*

R.H. Epstein, R.S. Edgar, and their collaborators introduce the use of conditional lethal mutations in T_4 Phage for studying the action of indispensable genes.

J. Cairns demonstrates by autoradiography that the genophore of *Escherichia coli* is circular and that during its semiconservative replication a single Y-shaped, replicating fork proceeds from a starting point and generates two circular offspring genophores.

E. Margoliash determines the amino acid sequences for cytochrome c derived from a wide variety of species and generates the first phylogenetic tree for specific gene product.

L.B. Russel shows in the mouse that, when an X chromosome containing a translocated autosomal segment undergoes inactivation in somatic cells, the autosomal genes closest to the breakpoint are also inactivated. Thus the X inactivation spreads into the attached autosomal segment.

E. Mayr publishes *Animal Species and Evolution.* This volume provides a synthesis of modern ideas concerning the mechanism of speciation. and it has a profound influence on scientists working in this area.

1964 R.B. Setlow and W.L. Carrier and, independently R.P Boyce and P. Howard-Flanders describe the mechanism of excision repair in bacteria.

A.S. Sarabhai, A.O.W. Stretton, S. Brenner, and A. Bolle establish the colinearity of gene and protein product in the case of the protein coating the head of virus T_4 of *Escherichia coli.* C. Yanofsky, B.C. Carlton, J.R. Guest, D.R. Helinski, and U. Henning establish the colinearity of gene and protein product in the case of tryptophan synthetase for *Escherichia coli.*

M.S. Fox and M.K. Allen show that transformation in *Diplococcus pneumoniae* involves incorporation of segments of single-stranded donor DNA into the DNA of the recipient E.T. Menz, L.S. Bates, and O.E. Nelson show that the *opaque-2* mutation modifies the amino acid composition of the mature endosperm, resulting in a striking improvement in the nutritional quality of maize seed.

J.G. Gorman, V.J. Freda, and W. Pollack demonstrate that the sensitisation of Rh-negative mothers can be prevented by administration of Rh antibody immediately after delivery of their first Rh-positive baby.

D.J.L. Luck and E. Reich isolate mitochondrial DNA from *Neurospora.* They demonstrate subsequently (1966) that this DNA replicates by the classical semiconservative mechanism. G. Marbaix and A. Burny isolate a 9S RNA from mouse reticulocytes and suggest that it may be mRNA.

R. Holliday puts forth a model that defines a sequence of breakage and reunion events, which must occur during crossing over between the DNA molecules of homologous chromosomes.

J.W. Littlefield develops a method for selecting somatic cell hybrids utilising HGPRT and TK fibroblasts cultured on HAT medium.

D.D. Brown and J.B. Gurdon show that no synthesis of the l8S and 28S rRNAs occurs in *Xenopus* tadpoles homozygous for a deficiency covering the nucleolus organiser.

W.D. Hamilton puts forth the genetical theory of social behaviour.

W. Gilbert finds that nascent proteins bind to the large ribosomal subunit, as do the rRNAs.

1965 D.D. Sabatini. Y. Tashiro, and G.E. Palade show that the large subunit of ribosome attaches to the ER membrane.

L. Hayflick discovers that the *in vitro* life-span of human diploid cells in tissue culture is about 50 doubling cycles.

R. W. Holley and his colleagues determine the complete sequence of alanine transfer RNA isolated from yeast.

N. Hilschmann and L. Craig report that immunoglobulin molecules are made up of carboxyl-terminal segments, which are constant in their amino acid composition and amino terminal segments that are variable. This finding poses the problem of how a gene can code for those portions of the protein that vary in their amino acid compositions.

P. Karlson, H. Hoffmeister, H.Hummel, P. Hocks, and G. Spiteller determine the complete structural configuration of ecdysone.

S. Spiegelman, I. Haruna, I.B. Holland, G. Beaudreau. and D.R. Mills succeed in demonstrating the *in vitro* synthesis of a self-propagating infectious RNA (bacteriophage Q*b* of *E. coli*) using a purified enzyme (Q*b* replicase).

S. Brenner, A.O.W. Stretton, and S. Kaplan deduce that UAG and UAA are the codons that signal the termination of a growing polypeptide.

F.M. Ritossa and, S. Spiegelman demonstrate that multiple cistrons producing the ribosomal RNAs of *Drosophila* reside in the nucleolus organiser regions of each X and Y chromosome.

H. Harris and J.F. Watkins use the Sendai virus to fuse somatic cells derived form man and mouse and produce artificial interspecific heterokaryons.

A.J. Clark and A.D. Margulies report for *E. coli* that many mutants selected as deficient in recombination are also abnormally sensitive to ultraviolet radiation. This finding suggests that similar enzyme systems function both in repairing damaged DNA and in recombination.

F. Sanger, G.G. Brownlee, and B.G. Barrell describe a method for finger printing oligonucleotides from partially hydrolysed RNA preparations.

R. Rothman demonstrates that lambda phage has a specific attachment site on the *E. coli* chromosome. W.J. Dreyer and J.C. Bennett propose that antibody light chains are encoded by two distinct DNA sequences, one for the variable region and the other for the constant region. They suggest that there is only one constant region, but that the variable region contains hundreds of different minigenes.

F.Jacob, J. Monod, and A. Lwoff receive a Nobel Prize for their contributions to microbial genetics.

1966 B. Weiss and C.C. Richardson isolate DNA ligase.

M.M.K. Nass reports that mitochondrial DNA is a circular double-stranded molecule.

F.H.C. Crick puts forward the wobble hypothesis to explain the general pattern of degeneracy found in the genetic code.

J. Adams and M. Cappecchi show that N-formylmethionyl-t RNA functions as the initiator of the polypeptide chain forming on a ribosome.

W. Gilbert and B. Muller-Hill demonstrate that the lactose repressor of *E. coli* is a protein.

M. Ptashne shows that the phage lambda repressor is a protein and that it binds directly to the lambda DNA molecule.

H. Roller, K.H. Dahm, C.C. Sweely, and B.M. Trost determine the structural formula for the juvenile hormone of *Hyalophora cecropia.*

M. Waring and R.J. Brinen demonstrate that vertebrate DNAs contain repetitious nucleotide sequences. E.F.J. Van Bruggen and colleagues and J.H. Sinclair and B.J. Stevens independently demonstrate that the DNA of animal mitochondria is in the form of circles, each about 5 mm in circumference.

R.S. Edgar and W.B. Wood analyse the genetically controlled steps in the assembly of the T_4 bacteriophage.

V.A. McKusick publishes *Mendelian Inheritance in Man,* a catalogue listing some 1500 genetic disorders of *Homo sapiens.*

E. Terzaghi, Y. Okada, G. Streisinger, J. Emrich, M. Inouye, and A. Tsugita confirm that the genetic code is translated by the sequential reading of triplets of bases starting at a defined point in phage T_4 lysozyme.

H. Wallace and M.L. Birnstiel demonstrate that a nucleoate deletion in *Xenopus laevis* removes more than 99% of the rDNA.

R.C. Lewontin and J.L Hubby use electrophoretic methods to survey gene-controlled protein variants in natural populations of *Drosophila pseudoobscura.* They demonstrate that between 8 and 15% of all loci in the average individual genome are in the heterozygous condition. Using similar techniques H. Harris demonstrates the existence of extensive enzyme polymorphisms in human populations.

P. Rous receives the Nobel prize for his studies on oncogenic viruses.

1967 S. Spiegelman, D.R. Mills, and R.L. Peterson report the results of serial transfer experiments in which they select those *Qb* bacteriophage molecules that replicate most rapidly *in vitro.* As experiments in extracellular evolution progressed, the molecule became smaller as its replication rate increased. By the 74th transfer, the replicating RNA molecule was only 20% of its original length and was the smallest known self-duplicatng molecule.

H.G. Khorana and his coworkers use polynucleotides with known repeating di-and trinucleotide sequence to solve the genetic code.

K. Taylor, Z. Hradecna, and W. Szybalsi show that mRNA can originate from *both* DNA strands.

B. Mintz uses allophenic mice to demonstrate that melanocytes which provide colour to the fur of the mouse are derived from 34 cells which have been determined at an early stage in embryogenesis.

J.B. Gurdon transplants somatic nuclei into frog eggs at different developmental states. The synthesis of RNA and DNA of transplanted nuclei changes to the kind of synthesis characteristic of the host cell nucleus.

L. Goldstein and D.M. Prescott perform nuclear transplantations in a*moeba.* These show there are specific proteins that move from the cytoplasm to the nucleus, and these presumably control the nucleic acid metabolism of the nuclei they enter.

C.B. Jacobson and R.H. Barter report the use of amniocentesis for intrauterine diagnosis and management of genetic defects.

M. Goulian, A. Kornberg, and R.L. Sinsheimer report the successful *in vitro* synthesis of biologically active DNA. The template they presented to the purified *Escherichia coli* DNA polymerase was single-stranded DNA from *f* ´ 174.

M.L. Birnstiel reports the isolation of pure rDNA from *Xenopus laevis.*

T.O. Diener and W.B. Raymer show that the potato spindle tuber disease is caused by a viroid.

M.C. Weiss and H. Green develop the HAT selection procedure and localise the gene for thymidine kinase. This was the first use of somatic cell genetics to localise human genes. G. Wald receives the Nobel prize in physiology for his studies on the biochemistry of vision. This work include contributions to the biochemical understanding of inherited colour blindness.

1968 R.T. Okazaki and four colleagues report that newly synthesised DNA contains many fragments. These represent short lengths of DNA that are replicated in a discontinuous manner and then spliced together. J. Morgan, D.P. McKenzie, and X. Le Pinchon develop the concept of plate tectonics to explain continental drift.

Differential synthesis of genes for ribosomal DNA during amphibian oogenesis is reported by J.G. Gall and by D.D. Brown and I.B. Dawid.

M. Kimura proposes the neutral gene theory of molecular evolution.

H.O. Smith, K.W. Wilcox, and T.J. Kelley isolate and characterise the specific restriction endonuclease (Hind II).

D. Y. Thomas and D. Wilkie demonstrate recombination of yeast mitochondrial genes.

R.P. Donahue. W.B. Bias, Renwick, and V.A, McKusick assign the Duffy blood group locus to chromosome 1 in man. This is the first gene localised in a specific autosome.

J.A. Huberman and A.D. Riggs demonstrate that mammalian chromosomes contain serially arranged, independently replicating units each about 30 m long.

E.H. Davidson, M. Cripps, and A.E. Mirsky show that more than 60% of the RNA labelled during oogenesis in *Xenopus laevis* is synthesised during the lampbrush stage and stored during the remaining months of oocyte maturation. This RNA is presumably a long-lived mRNA stored for use in early embryogenesis.

O. Hess and G. Meyer report extensive studies on structural modification of the Y chromosome in various *Drosophila* species, which demonstrate that the Y chromosome contains genes that control stage-specific steps in the development of sperm.

S.A. Henderson and R.G. Edwards demonstrate that the number of chiasmata peroocyte declines with increasing maternal age in the mouse and that the number of univalents increases with age. If the same occurs in human female one would expect (as has been demonstrated) an increase in aneuploid offspring with advancing maternal age.

J.E. Cleaver shows that the repair replication of DNA is defective in patients with *Xeroderma pigmentosum.*

R.W. Holley, H.G. Khorana, and M.W. Nirenberg receive Nobel Prizes for discoveries concerning the interpretation of the genetic code and its function in protein synthesis.

1969 J. Abelson, L. Barnett, S. Brenner, M. Gefter, A. Landy, R. Russell, and J.D. Smith provide proof of the proposed mechanism of nonsense suppression by determining the actual nucleotide sequence of mutant tyrosine transfer ribonucleic acids.

The *in situ* hybridisation techniques for the cytological localisation of specific nucleotide sequences are developed by J.G. Gall and M.L. Pardue and by H. Johan, M.L. Brmoch and K.W. Jones.

B.C. Westmoreland, W. Szybalski, and H. Ris develop an electron microscopic technique for physically mapping genes in lambda phage. They photograph heteroduplex DNA molecules obtained by annealing of the *l* strand of one parent and the *r* strand of a second parent that has deletions, insertions, substitutions, or inversions.

R. Burgess, A.A. Travers, J.J. Dunn, and E.K. Bautz isolate and identify the sigma factor from RNA polymerase.

O.L. Miller and B.R. Beatty publish electron micrographs showing amphibian genes is the process of transcribing RNA molecules.

J.R. Beckwith (together with five associates) reports the isolation of pure *lac* operon DNA from *Escherichia coli.*

G.M. Edelman (together with five associates) publishes the first complete amino and sequence for human gamma G_1 immunoglobulin.

Y. Hotta and S. Benzer and W.L. Pak and J. Grossfield independently induces and physiologically characterise neurological mutants in *Drosophila.*

C. Boon find F. Ruddle correlate the loss of particular chromosomes from a somatic hybrid cell line containing both human and mouse chromosomes with the loss of specific phenotypic characters. This approach permits assignment of specific loci to certain human chromosomes.

R.E. Lockard and J.B. Lingrel purify the 9S RNA fraction obtained from polysomes of mouse reticulocytes and show that it directs the synthesis of mouse haemoglobin *b* chain. They thus confirm the suggestion of Marbaix and Burny (1964).

A. Ammermann reports that in the hypotrichous ciliate *Stylonychia mytilus* macronuclear anlage under endomitotic DNA replication to form polytene chromosomes. Subsequently the major portions of these are destroyed, and over 90% of the macronuclear DNA is degraded and excreted into the medium.

R.I. Huebner and G.I. Todaro put forward the oncogene theory.

H.A. Lubs describes a fragile site on the human X chromosome and shows that it is parent in mentally retarded males. Subsequent studies show that this locus (Xq27) is assorted with a common form of X-linked mental retardation.

M. Delbruck, S.E. Luria, and A.D. Hershey receive a Nobel prize for their contribution to viral genetics.

1970 B.M. Alberts and L. Frey isolate the protein product of gene 32 of phage T_4 and demonstrate that this protein binds cooperatively to single-stranded DNA. They suggest that proteins function to initiate unwinding of the DNA molecule so that replication can occur.

H.G. Khorana (together with twelve associates) reports the total synthesis of the an alanine tRNA from yeast.

J.Yourno, T, Kohno, and J.R. Roth succeed in fusing two bacterial enzymes into one large protein molecule that combines the functions of both. The enzyme fusion was accomplished by fusing the *his* D and *his* C genes in the histidine operon of *Salmonella,* using a pair of frameshift mutations.

D. Baltimore and H.M. Temin report the existence of an RNA-dependent DNA polymerase in two oncogenic RNA Viruses (Rauscher mouse leukemia and Rous fowl sarcoma).

M.L. Pardue and J.G. Gall demonstrated that pericentric heterochromatin is rich in repetitious DNA.

D.E. Wimber and D.M. Steffensen localise the 5S RNA cistrons on the right arm of chromosome 2 of *Drosophila melanogaster.*

T. Caspersson, L. Zech, and C. Johansson use Quinacrine dyes in chromosomal cytology and demonstrate specific fluorescent banding patterns in human choromosomes.

R.Sager and Z. Ramanis publish the first genetic map of non-Mendelian genes. This group of eight genes reside on a chloroplast chromosomes of *Chlamydomonas.*

R.T. Johanson and P.N. Rao induce premature chromosome condensation by fusing mitotically active cells with interphase cells *in vitro.*

1971 M.L. O'Riordan, J.A. Robinson, K.E. Buckton, and H.J. Evans report that all 22 pairs of human autosomes can be identified visually after staining with quinacrine hydrochloride. They demonstrate that the Philadelphia chromosome is a deleted chromosome 22. Y. Hotta and H. Stern characterise the DNA that is synthesised during meiotic prophase in the lily. Synthesis during zygonema represents the delayed replication of a small fraction of the DNA that failed to replicate during the previous S phase. The DNA synthesised during pachynema has the characteristics of repair replication.

S.H. Howell and Stern demonstrate that an endonuclease present in microspores reaches its highest concentration early in pachynema, the stage when crossing over is thought to occur.

B. Dudock, C. Di Peri, K. Scileppi, and R. Reszelbach present evidence for the phenylalanine tRNA synthetase recognition site being adjacent to the dihydrouridine loop.

C.R. Merril, M.R. Geier, and J.C. Petricciani infect fibroblasts cultured from a patient suffering from galactosemia with transducing lambda phage carrying the galactose operon. The cells then make the missing transferase and survive longer in culture than uninfected galactosemic cells.

RJ. Konopka and S. Benzer report recovery of induced clock mutants of *Drosophila.*

D.T. Suzuki, T. Grigliatti, and R. Williamson isolate a temperature-sensitive paralytic mutant of *Drosophila.*

J.E. Manning and O.C. Richards detect circular chloroplast DNA molecules in lysates of *Euglena* chloroplasts.

C. Kung induces and isolates behavioural mutants of *Paramecium aurelia* and shows that many mutants have electrophysiological defects in their plasma membranes.

K. Dana and D. Nathans use restriction endonucleases to cleave the circular DNA of simian virus 40 into a series of fragments and then deduce their physical order.

1972 P. Lobban and D. Kaiser develop a general method for joining any two DNA molecules, employing terminal transferase to add complementary homopolymer tails to passenger and vehicular DNA molecules. A.F. Zakharov and N.A. Egolma develop the BUDR labelling technique to produce harlequin chromosomes.

G.H. Pigott and N.G. Carr show that ribosomal RNAs from cyanobacteria hybridise with DNA from the chloroplasts of *Euglena gracilis.* This genetic homology provides strong to the theory that chloroplasts are the descendants of endosymbiotic cyanobacteria.

S.J. Singer and G.L. Nicholson put forth the fluid mosaic model of the structure of cell membranes.

B. Benacerraf and H.O. McDevitt show for the mouse that LIr genes are linked with the H_2 complex.

Y. Suzuki and D.D. Brown isolate and identify the mRNA for silk fibroin from *Bombyx mori*, and Suzuki, L.P. Gage. and Brown characterise the fibroin gene.

N. Eldredge nad S.J. Gould propose the punctuated equilibrium model to explain the tempo of major evolutionary changes.

R. Silber, V.G. Malathi, and J. Hurwitz discover RNA ligase.

D.A. Jackson, R.H. Symons, and P. Berg report splicing the DNA of SV40 virus into the DNA of the lambda virus of *E.coli.* They are thus the first to join the DNAs of two different organisms *in vitro.*

J. Mendlewicz, J.L. Fleiss, and R.R. Five demonstrate that manic-depressive psychosis is transmitted by a dominant gene located on the short arm of the X chromosome.

M.L. Pardue, E. Weinberg L.H. Kedes, and M.L. Birnstiellocate the histone genes on chromosome 2 or *Drosophila melanogaster.*

P.S. Carlson, H.H. Smith, and R.D. Dearing succeed in producing interspecific plant hybrids by parasexual means.

J. Hedgpeth, H.M. Goodman, and H.W. Boyer identify the nucleotide sequence in the DNA of coliphage lambda that is recognised by a specific endonuclease.

G.M. Edelman and R.R. Porter receive a Nobel Prize for their studies on the chemical structure of antibodies.

1973 D.R. Mills, F.R. Kramer, and S. Spiegelman publish the sequence for the 218 nucleotides in a replicating RNA molecule. The molecule is MDV-1, a variant derived from the RNA of Q*b* phage exposed to experimental selection and thus forced to undergo "extracellular evolution" to a shorter length.

S.H. Kim, G.J. Quigley, F.L. Suddath, A. McPherson, D. Sneden, J.J. Kim, J. Weinzierl, and A. Rich propose a three-dimensional structure for yeast phenylalanine transfer RNA.

R. Kavenoff and B.H. Zimm use a newly developed viscoelastic method for measuring the molecular weights of DNA molecules isolated from cells from different *Drosophila* species. They conclude that a chromosome contains one long molecule of DNA and that it is not interrupted in the centromere region.

P. Deheigh and C. Nitsch succeed in culturing haploid tomato plants directly from microspores.

W.G.Hunt and R.K. Selander analyse a zone of hybridisation between two subspecies of the house mouse using gel electrophoresis to trace the boundary.

P.J. Ford and E.M. Southern show for *Xenopus laevis* that different 5S RNA genes are transcribed in the oocyte than in somatic cells.

W. Fiers, W. M. Jou, G. Haegerman, and M. Ysebaert are the first to sequence gene coding for a protein (the coat protein of the male-specific phage MS2).

B.R. Roberts and B.M. Patterson report the preparation of a wheat germ cell-free system for the *in vitro* translation of experimentally supplied mRNAs.

A. Garcia-Bellido, P. Ripoll, and G. Morata report the developmental compartmentalisation of the wing disc of *Drosophila.*

S.N. Cohen, A.C.Y. Chang H.W. Boyer, and R.B. Helling construct the first biologically functional hybrid bacterial plasmids by *in vitro* joining of restriction fragments from different plasmids.

1974 J. Shine and L. Dalgamo show that the 3 terminus of *E.coli* 16S rRNA contains a stretch of nucleotides that is complementary to ribosome binding sites of various coliphage mRNAs. They suggest that this region of the 16S rRNA may play a base-pairing role in the termination and initiation of protein synthesis on mRNA.

1. Zaenen, N. van Larebeke, H. Teuchy. M. van Montagu, and J. Schell discover the tumor inducing plasmid of the crown gall bacterium.

K.M. Murray and N.E. Murray manipulate the recognition sites for restriction endonucleases in lambda phage so that its chromosome can be used as a receptor site for restriction fragments from foreign DNAs. Lambda thus becomes a cloning vehicle.

A. Tissieres, H.K. Mitchell, and U.M. Tracy find that heat shocks result in the synthesis of six new proteins in *Drosophila.* These are also synthesised by tissues that do not have polytene chromosomes. B. Dujon, P.P. Slonimski, and L. Weill propose a model for recombination and segregation of mitochondrial genomes in *Saccharomyces cerevisae.* According to it, mt DNA molecules are present in the zygote cell in multiple copies. These pair at random, and during any mating cycle a segment from one parent can exchange with that from a second parent mt DNA, yielding recombinant units.

R.D. Kornberg proposes that chromatin is built up of repeated structural units of 200 base pairs of DNA and two each of the histones H2A, H2B, H3 and H4. These structures, which are later called nucleosomes, are isolated by M. Noll, A. L. Olins and D.E. Olins publish the first electron micrographs of chromatin spreads from erythrocyte nuclei which show nucleosomes.

B. Ames develops a rapid screening test for detecting mutagenic and possibly carcinogenic compounds.

S. Brenner describes methods for inducing, isolating, and mapping mutations in the nematode *Caenorhabditis elegans.*

A. Claude. C. de Duve and G. Palade receive Nobel Prizes for their contributions to cell biology.

1975 G. Kohler and C. Milstein perform experiments with mouse cells, which show that somatic cell hybridisation can be used to generate a continuous "hybridoma" cell line producing a monoclonal antibody.

Molecular biologists from around the world meet at Asilomar, California, to write an historic set of rules to guide research in recombinant DNA experiments.

The NIH Recombinant DNA Committee issues guidelines aimed at eliminating or minimising the potential risks of recombinant DNA research.

M. Grunstein and D.S. Hogness develop the colony hybridisation method for the isolation of cloned DNAs containing specific DNA segments or genes.

D. Pribnow determines the nucleotide sequences of two independent bacteriophage T7 promoters, and compares these and other known promoter sequences to form a model for promoter structure and function.

E.M. Southern describes a method for transferring DNA fragments from agarose gels to nitrocellulose filters. The filters are subsequently hybridised to radioactive RNA and the hybrids detected by autoradiography.

W.D. Benton and R.W. Davis describe a rapid and direct method for screening plaques of recombinant *lgt* phages involving transfer of phage DNA to a nitrocellulose filter and detection of specific DNA sequences by hybridisation to complementary labelled nucleic acids.

F. Sanger and A.R. Coulson develop the first rapid method for determining the nucleotide sequences in DNA by primed synthesis with DNA polymerase. A second widely used method is described in 1977 by A.M. Maxam and W. Gilbert.

G. Morata and P.A. Lawrence show in *Drosophila* that the *engrailed* mutation allows cells of the posterior wing compartment to mix with those of the anterior compartment. Therefore, the normal allele of this gene functions to define the boundary conditions between the sister compartments of the developing wing.

B. Mintz and K. Illmensee inject XY diploid cells from a malignant mouse teratocarcinoma into mouse blastocytes, which then are transferred to foster mothers. Cells derived from the carcinoma appear in both somatic and germ cells of some F_1 males. When these are mated, some F_2 mice contain marker genes from the carcinoma. The experiments demonstrate that the nuclei of teratocarcinoma cells remain developmentally totipotent, even after hundreds of transplant generations during which they functioned in malignant cancers.

S.L. McKenzie, S. Henikoff, and M. Meselson isolate mRNAs for heat shock proteins and show that they hybridise to specific puff sites on the *Drosophila* polytene chromosomes.

L.H. Wang. P.H. Duesberg, K. Beemon, and P.K. Vogt locate within the RNA genome of the Roux sarcoma virus the segment responsible for its oncogenic activity.

R. Dulbecco. H. Temin, and D. Baltimore receive Nobel Prizes for their studies on oncogenic viruses.

1976 H.R.B. Pelham and R.J. Jackson describe a simple and efficient mRNA-dependent *in vitro* translation system using rabbit reticulocyte tysates.

R. V. Dippell shows in *Paramecium* that kinetosomes contain RNA (not DNA) and that RNA (not DNA) synthesis accompanies kinetosome reproduction.

N. Hozumi and S. Tonegawa demonstrate that the DNA segments coding for the variable and constant regions of an immunoglobulin chain are distant from one another in the chromosomes isolated from mouse embryos, but the segments are adjacent in chromosomes isolated from mouse plasmacytomas. They conclude that somatic recombination during the differentiation of B lymphocytes moves the constant and variable gene segments closer together.

W.Y. Chooi shows that ferritin-labelled antibodies raised against proteins (isolated from rat ribosomes) bind to the terminal knobs of fibers extending from Miller trees (isolated from the ovarian nurse cells of *Drosophila*). This observation proves that Miller trees are rRNA transcription units and shows that at least some ribosomal proteins attach to a precursor rRNA molecule before its transcription is completed. B.G. Burrell, G.M. Air, and C.A. Hutchinson report that phage *f* ´ 174 contains overlapping genes. Formal guidelines regulating research involving recombinant DNA are issued by the National Institutes of Health in the United States.

The first genetic engineering company is formed and named Genetech.

A. Efstratiadis, F.C. Kafatos, A. Maxam, and T. Maniatis are the first to enzymatically generate eukaryotic gene segments *in vitro*. They synthesise double-stranded DNA molecules that contain the sequences transcribed into the mRNAs for the alpha and beta chains of rabbit haemoglobin.

J.T. Finch and A. Klug propose that the 300 Å threads seen in electron micrographs of fragmented chromatin are formed by the folding of DNA-nucleosome filaments into solenoids.

1977 A. Knoll and E.S. Barghoorn find microfossils undergoing cell division in rocks 3400 million years old. This discovery pushes back the age of life on earth to the lower Archeaneon. C. Jacq, J.R. Miller, and G.G. Brownlee describe the presence of "pseudogenes" within the 5S DNA cluster of *Xenopus laevis* oocytes.

J.C. Alwine, D.J. Kemp, and G.R. Stark prepare diazobenzyloxymethyl (DBM) paper and describe methods for transferring electrophoretically separated bands of RNA from an agarose gel to the DBM paper. Specific RNA bands are then detected by hybridisation with radioactive DNA probes, followed by autoradiography. Since this method is the reverse of that described by Southern (1975), in that RNA rather than DNA is transferred to a solid support, it has come to be known as "northern blotting."

F. Sanger and eight colleagues report the complete nucleotide sequence for the DNA genome of bacteriophage *f* ´ 174.

M. Leffak, R. Grainger, and H. Weintraub show-that "old" histone octamers remain intact during DNA replication and that "new" octamer consist entirely of proteins synthesised during the time of replication. W. Gilbert induces bacteria to synthesise useful nonbacterial proteins (insulin and interferon).

R. W. Old, H.G. Callan, and K. W. Gross use labelled cloned histone genes from sea urchins to localise by *in situ* hybridisation, histone mRNAs being transcribed on the lampbrush chromosomes of salamander oocytes.

D.S. Hogness, D.M. Glover, and R.L. White report the occurrence of intervening segments in some of the 28S rRNA genes of *Drosophila melanogaster.* Introns are then described for genes that encode proteins, namely, the rabbit beta-globin gene (A. Jeffreys and R.A. Flavell) and the chicken ovalbumin gene (R. Breathnack, J.L. Mandel, and P. Chambon).

J.F. Pardon and five colleagues use neutron contrast matching techniques to demonstrate that in nucleosomes the DNA segment that attaches to the histone octamer is on the outside of the particle.

J. Sulston and H.R. Horvitz work out the postembryonic cell lineages for *Caenorhabditis elegans.* R.S. Yalow receives a Nobel Prize for developing the radioimmunoassay procedure.

1978 R.M. Schwartz and M.O. Dayhoff compare sequence data for a variety of proteins and nucleic acids from an evolutionary diverse assemblage of prokaryotes. eukaryotes, mitochondria, chloroplasts and their computer-generated evolutionary trees identify the times during evolution when protoeukaryotic organisms entered into symbiosis with mitochondria and chloroplasts (about 2 and 1 billion years ago, respectively).

W. Gilbert coins the terms *introns and exons.*

The world's first "test-tube baby" is born in Britain as a result of *in vitro* fertilisation and subsequent implantation of the embryo to the mother's uterus.

T. Maniatis, R.C. Hardison, E. Lacy, J. Lauer, C. O' Connell, D. Quon, G.K. Sim, and A. Efstratiadis develop a procedure for gene isolation, which involves construction of cloned libraries of eukaryotic DNA and screening of these libraries for individual sequences by hybridisation to specific nucleic acid probes.

D.J. Finnegan, G.M. Rubin, M.W. Young, and D.S. Hogness make detailed analysis of dispersed repetitive DNAs in *Drosophila.* Their discoveries represent a new beginning in the understanding of mutability, transposition, transformation, hybrid dysgenesis, and retroviruses in eukaryotes.

V.B. Reddy and eight colleagues publish the complete nucleotide sequence for simian virus 40 and correlate the sequence with the known genes and in RNAs of the virus.

Y.W. Kan and A.M. Dozy demonstrate the value of using restriction fragment length polymorphisms as linked markers for the prenatal diagnosis of sickle cell anaemia.

W. Arber, H.O. Smith, and D. Nathans receive Nobel Prizes for the development of techniques utilising restriction nucleases to study the organisation of genetic systems.

1979 J.G. Sutcliffe determines the complete 4,362 nucleotide pair sequence of the plasmid cloning vector pBR322.

J.C. Avise, R.A. Lansman, and R.O. Shade successfully use restriction endonucleases to measure mitochondrial DNA sequence relatedness on natural populations.

The National Institutes of Health relax guidelines on recombinant DNA to allow viral DNA to be studied.

B.G. Barrell, A.T. Bankier, and J. Drouin report that the genetic code of human mitochondria has some unique, non-universal features.

E.F. Fritsch, R.M. Lawn, and T. Maniatis determine the chromosomal arrangement and structure of human globin genes utilising recombinant DNA technology.

A. Wang, A. Rich, and their colleagues discover the Z form of DNA.

1980 L. Olsson and H.S. Kaplan produce the first human hybridomas that manufacture a pure antibody in laboratory culture.

The United States Supreme Court rules that genetically modified microorganisms can be patented. General Electric company, on behalf of A. Chakrabarty, obtains a patent for a genetically engineered microorganism capable of consuming oil slicks.

J.W. Gordon, G.A. Scangos, D.J. Plotkin, J. A. Barbose, and F.H. Rddle produce transgenic mice, A.R. Templeton provides a new theoretical framework for speciation by the founder principle.

M.R. Capecchi describes a technique for efficient transformation of cultured mammalian cells by direct microinjection of DNA into cells with glass micropipettes.

Nobel Prizes in Physiology and Medicine go to G.D. Snell, J. Dausset, and B. Benacerraf for their contributions to immunogenetics.

P. Berg, W. Gilbert, and F. Sanger receive Nobel Prizes in Chemistry for their contributions to the experimental manipulation of DNA.

1981 R.C. Parker, H.E. Varmus and J.M. Bishop demonstrate that the tumorigenic properties of the Rous sarcoma virus are due to a protein encoded by the *v-src* gene. Cells from various vertebrates contain a homologous gene, *c-src* genes differ in that *v-src* has an uninterrupted coding sequence, whereas *c- src* contains seven exons separated by six introns.

L. Margulis publishes *Symbiosis in Cell Evolution.* Here she summarises the evidence for the theory that organelles such as mitochondria, chloroplasts, and kinetosomes evolved from prokaryotes that lived as endosymbionts in the ancestors of modern day eukaryotes.

R. Lande proposes a new model of speciation based on sexual selection on polygenic traits. This model results in a revival of interest in sexual selection.

T.E. Wagner first demonstrates the successful transfer of functional genetic material between mammalian species; a rabbit blood protein is formed in mice.

J.D. Kemp and T.H. Hall transfer the gene of a major seed storage protein (phaseolin) from beans to the sunflower via a plasmid of the crown gall bacterium *Agrobacterium tumefaciens*, creating a "sunbean."

M.D. Edge assembles 514 paired nucleotides to make a synthetic gene for interferon, the longest gene to be pieced together in the laboratory so far.

The complete nucleotide sequence and genetic organisation of the human mitochondrial genome are worked out by a research group composed of S. Anderson, B.G. Barrell, F. Sanger, and eleven other scientists.

M.E. Harper and G.F. Saunders demonstrate that single copy genes can be mapped on human mitotic chromosomes utilising an improved *in situ* hybridisation technique.

J. Banerji, S. Rusconi, and S. Schaffner show that the transcription of the beta-globin gene is enhanced hundreds of times when this gene is linked with certain SV40 nucleotide sequences that they name "enhancer sequences".

M. Chalfie and J. Sulston identify among the touch-insensitive mutants of *Caenorhabditis elegans* five genes that affect a specific set of six sensory neurons.

T.R. Cech, A.J. Zaug, and P.J. Grabowski report that certain introns can function as ribozymes.

K.E. Steinbech, L. McIntosh, L. Bogorad, and C.J. Arntzen demonstrate that the resistance of a weed.

Amaranthus hybridus, to triazine herbicides is controlled by a chloroplast gene that encodes a polypeptide to which the herbicide binds. Resistant strains of the weed produce a modified gene product that fails to bind. Resistant strains of the weed produce a modified gene product that fails to bind triazine.

1982 Eli Lilly Internations Corporation is the first to market a drug made by recombinant DNA techniques.

The drug is human insulin, sold under the trade name "Humulin"

G. Scott and colleagues demonstrate that interferon produced by recombinant DNA techniques blocks virus action associated with the common cold.

E.P. Reddy, R.K. Reynolds, E. Santos, and M. Barbacid report that the genetic change which leads to the activation of an oncogene carried by a line of human bladder carcinoma cells is due to a single base substitution in this gene. The result is the incorporation of valine instead of lysine in the 12th amino acid of the protein encoded by the oncogene.

P.M. Bingharm, M.G. Kidwell, and G.M. Rubin show that P strains of *Drosophila* contain 30–50 copies per genome of a transposable P element. This is the cause of hybrid dysgenesis. Then A.C. Spradling and Rubin demonstrate that cloned P elements, when microinjected into *Drosophila*, embryos become integrated into germ line chromosomes and that p elements can be used as vectors to carry DNA fragments of interest into the *Drosophila* germ line.

J.G. Sutcliffe, R.J. Milner, F.E. Bloom, and R.A. Lerner report the isolation of identifier sequences unique to genes expressed in neural tissues.

A. Klug receives a Nobel Prize for the development of crystallographic electron microscopy and for structural elucidation of biologically important nucleic-acid complexes.

1983 E.A. Miele, D.R. Mills, and F.R. Kramer succeed in constructing the first recombinant RNA molecule, involving the insertion of a synthetic foreign decaadenylic acid into a variant of the RNA genome of a small bacterial virus called Q*b* via the action of the Q*b* replicase.

A bacterial enzyme called ribonuclease P, whose enzymatic activity depends on its RNA complement alone, is discovered by S. Altman, N. Pace and their colleagues. This is the first case in which an RNA molecule is shown to have such properties.

H.J. Jacobs (and six colleagues) report the presence of promiscuous DNA in the sea urchin. R. Doolittle and six colleagues show that the gene responsible for the oncogenic activity of the simian sarcoma virus is derived from a gene that encodes a growth factor normally synthesised by blood platelets.

I.S. Greenwald, P.W. Sternberg, and H.R. Horvitz demonstrate that the *lin*-12 mutant of *Caenorhabditis* functions as a developmental control gene.

B. McClintock receives the Nobel Prize for her discovery of transposable genetic elements.

1984 N.K. Jerne, G. Kohler, and C. Milstein receive the Nobel Prize in medicine for their contributions to immunology.

DEFINE THE FOLLOWING GENETIC TERMS

1. DNA
protein
nucleic acid
DNA
nucleoprotein
macromolecule
transforming principle
serotype
bacterial capsule
bacteriophage
radiolabelling
polymer
monomer
nucleotide
purine
pyrimidine
polynucleotide
phosphodiester bond
5′-terminus
3′-terminus
X-ray diffraction pattern
double helix
hydrogen bond
base pair
major and minor grooves
complementary

2. Gene Expression
gene expression
coding strand of DNA
operon
multigene family
homologous genes
pseudogene
discontinuous gene
exon
intron
central dogma
transcript
template
translation
messenger RNA
genetic code

3. Transcription
eukaryote
prokaryote

RNA polymerase
holoenzyme
core enzyme
upstream
downstream
transcription factor
promoter
consensus sequence
open promoter complex
leader segment
trailer segment
cruciform structure
stem-loop structure

4. RNA

ribosomal RNA
transfer RNA
stable RNA
turnover rate
ribosome
sedimentation coefficient
primary transcript
pre-rRNA
gene amplification
tRNA clover leaf
ribozyme
half-life
cap structure
polyadenylation
splicing
heterogeneous nuclear RNA
small
nuclear RNA
small nuclear RNA
small nuclear ribonucleoproteins
spliceosome

5. Genetic Code

R group
peptide bond
disulphydryl bridge
denaturation
renaturation
reading frame
codon
homopolymer
random heteropolymer
ordered heteropolymer
termination codon
initiation codon

6. Protein Synthesis

isoaccepting tRNAs
aminoacylation
anticodon
ribosome binding site
initiation complex
initiation factor
peptidyl site
aminoacyl site
elongation factor
translocation
polysome
release factor

7. Gene Expression

housekeeping gene
structural gene
regulatory gene
operator
catabolite repression
attenuation
upstream activating sequence
upstream repressing sequence
enhancer
homeobox

8. Replication or DNA

semiconservative replication
replication origin
replication fork
leading strand
lagging strand
Okazaki fragment
primer
primosome
concatamer

Role of following enzymes

DNA Polymerase I
DNA polymerase II
helicase
single-strand binding protein
primase
DNA ligase
DNA topoisomerase
DNA gyrase

9. Mutations

mutation
recombination
transposition
phenotype
genotype
wild-type
mutant
mutagen
mutagenesis
transition
transversion
silent mutation
lethal mutation
auxotroph
prototroph
minimal medium
temperature sensitive mutant
leaky mutant
back mutation
second site reversion
suppression
spontaneous mutation rate
base analogue
intercalating agent
tautomeric shift

heteroduplex
Holliday structure
chi form
transposable element
heterologous recombination

Eukaryotic Nucleus

diploid
haploid
repetitive DNA
Alu family
chromatin
histone
nucleosome
core octamer
linker DNA
30 nm chromatic fiber
metaphase chromosome
centromere
karyotype
chromatid
chromonemata
heterochromatin
euchromatin
telomere
mitosis
meiosis
homologous chromosomes
interphase
cell cycle
gap period
S phase

Organelle DNA

petite
granule
extrachromosomal gene
immunoscreening
protein forgeting
signal peptide
maturase
promiscuous DNA

Mendelism

monohybrid cross
self-fertilisation
cross-fertilisation
allele
dominant
recessive
segregation
gamete
homozygous
heterozygous
Punnett square
dihybrid cross
trihybrid cross
vegetative cell
linkage
crossing over

Linkage

partial linkage
autosome
sex chromosome
linkage group
map unit
centi Morgan
genetic marker
zygote
ascospore
replica plating
ascus
tetrad

Gene Mapping

donor bacterium
recipient bacterium
gene transfer
double recombination
selective medium
interrupted mating
abortive infection
contransduction
competence
contransformation

Gene Cloning

recombinant DNA technology
genetic engineering
gene cloning
cloning vector
recombinant DNA molecule
clone
restriction endonuclease
DNA ligase
blunt end
sticky end
microinjection
gene library
DNA sequencing
transcript mapping
HART and HRT
in vitro mutagenesis

GIVE THE CONTRIBUTIONS OF THE FOLLOWING GENETICISTS

T.H. Morgan
H.J. Muller
A.H. Sturtevant
H.K.K Whitehouse
P.J. Hastings
C. Stern
F. Sanger
H. Taylor
M. Delbruck

C.B. Bridges
B. McClintock
J.B.S. Haldane
A.E. Garrod
L.J. Stadler
M.M. Rhoades
T.M. Sonneborn
T. Dobzhansky
J.D. Watson
F.H.C. Crick
C.D. Darlington
A.F. Blakeslee
R.A. Fischer
E.L. Tatum
J. Lederberg
S. Ochoa
G.L. Stebbins
E. Wollman
N. Zinder
S. Luria
V. Mckusick
K. Mather
G. Mendel
R. Okazaki
H. Temin
F. Stahl
R.C. Punnett
W. Bateson
W. Johannsen
F. Griffith
F. Jacob
G.H. Jones
G. Karpechenko
H.G. Khorana
A. Kornberg
E. Lewis
S. Luria
M. Lyon
C.M. Macleod
M. Mc Carty
E. Chargaff
M. Nirenberg
F. Robertson
M. Strickberger
C. Swanson
B. Ames
C.R. Burnham
M.S. Swaminathan
E. Balbiani
G. Beadle
S. Benzer
S. Brenner
R.A. Brink
J. Cairns
H. Creighton
C. Darwin
H. De Vries
E. East
R. Emerson
D.F. Jones
W. Flemming
E. Freese
R. Gosling
G. Hardy
A. Hershey
P. Leder
K.R. Lewis
B.P. Pal
H.K. Jain
C.D. Darlington
R. Riley
R.A. Fischer

AGRICULTURAL RESEARCH INSTITUTES

IARI	Indian Agricultural Research Institute, New Delhi, India.
CTRI	Central Tobacco Research Institute, Rajahmundry, Andhra Pradesh.
CPRI	Central Potato Research Institute, Shimla.
SBI	Sugarcane Breeding Institute, Coimbatore, Tamil Nadu.
CTCRI	Central Fiber Crops Research Institute, Trivandrum, Kerala.
IIHR	Indian Institute of Horticultural Research, Hessarghtta, Bangalore, Karnataka.
CAZRI	Central Arid Zone Research Institute, Jodhpur, Rajasthan.
IGFRI	Indian Grassland and Fodder Research Institute, Jhansi, U.P.
NBPGR	National Bureau of Plant Genetic Resonance, IARI, New Delhi
CPCRI	Central Plantation Crops Research Institute, Kasargod, Kerala
CSSRI	Central Soil Salinity Research Institute, Karnal, Haryana.
CRRI	Central Rice Research Institute, Cuttack (Orissa).

CICR	Central Institute for Cotton Research, Nagpur, Maharashtra.
CIAE	Central Institute of Agricultural Engineering, Bhopal (M.P.).
VPAS	Vivekananda Parvatiya Anusandhan Shala, Almora, H.P.
IASRI	Indian Agricultural Statistics Research Institute, New Delhi.
FAO	Food and Agriculture Organisation, Rome.
IRRI	International Rice Research Institute, Los Banos, Phillipines.
CIMMYT	Centro International de Mejoramiento de Maizy Trigo. (Centre for Maize and Wheat Improvement, el Baton, Mexico).
CGIAR	Consultative Group for International Agricultural Research (Combined organisation of FAO, World Bank and United Nations Development Programme (UNDP)).
CIAT	International Centre for of Tropical Agriculture, Plamira, Colombia.
IITA	International Institute of Tropical Agriculture, Ibadan, Nigeria.
WARDA	West African Rice Development Association, Klonrovia, Liberia.
CIP	International Centre for Potato, Lima, Peru.
ICRISAT	International Crops Research Institute for Semi-Arid Tropics, Hyderabad, India.
ILRAD	International Laboratory for Research on Animal Diseases, Nairobi, Kenya.
IBPGR	International Board for Plant Genetic Resources, Rome, Italy.
ILCA	International Livestock Centre for Africa, Addis Ababa, Ethiopia.
ICARDA	International Centre for Agricultural Research in Dry Areas, Alippo, Syria.
JARI	Jute Agricultural Research Institute, Barackpore, West Bengal.
ILRI	Indian Lac Research Institute, Namkum, Ranchi, Bihar.

Glossary of Molecular Genetics

Abiogenesis. Origin of life on earth from non-living matter under reducing atmosphere when only methane, water vapours, ammonia and hydrogen were present in the environment. This view was given by Oparin who gave the theory of "*primary abiogenesis*" which can be summed up as "*abiogenesis at first and biogenesis ever after*". This is also referred to as "*chemical evolution of life on earth*".

Abiological (Abiotic) Evolution. Evolution of a mechanism which could synthesise macromolecules without involvement of living systems, during the origin of life.

A-chromosomes. A standard diploid or polyploid chromosome complement in a species.

Activation Reaction. A reaction using ATP as energy and conducted by an amino acyl synthetase to activate a specific amino acid. This reaction forms amino acid ~ AMP complex.

Activator-Dissociator (Ac-Ds) System. Originally reported by Mc Clinctock in 1956. Responsible for turning the expression of genes on or off. *Ds* gene located on chromosome 9 causes breakage at the site of its location. It functions in the presence of another gene *Ac* located on any other chromosome. Both Ac and Ds elements are transposable and regulate the activity of other genes. Ac-Ds system has been used for gene transfer in eukaryotes.

Activator RNA. A component of eukaryotic gene regulation which is the product of an integrator gene. It or its gene product complexes with a specific receptor gene.

Adaptor Hypothesis. The idea advanced by F.H.C. Crick in 1958) that there is a molecule that serves as an adaptor to match up amino acids on an mRNA template. This adaptor has one end that matches the mRNA and another end that is attached to the amino acid. This adaptor molecule is tRNA or sRNA.

Advanced Backcross Quantitative Trait Loci (QTL) Method. The method used in QTL mapping which integrates QTL analysis and variety improvement efforts.

Allelic-specific Probes. Short, labelled oligonucleotide probes used to distinguish single nucleotide differences in PCR products through hybridisation such that only a completely complementary probe will bind stably to the target sequence.

Allelic Exclusion. In immunoglobulin-producing cells only one member of each pair of alleles is involved in synthesis of immunoglobulin molecules. Thus, correct rearrangement of only one chromosome is required for generation of each expressed light or heavy chain gene.

Since DNA rearrangement of only one chromosome takes place, expression of the alternate allele is excluded.

Alloparapatric Speciation. New species arise through populations that are at first allopatric but later become parapatric before a completely effective reproductive isolating mechanism has evolved.

Allopatric (Geographical) Speciation. Speciation in which the evolution of reproductive isolating mechanisms occurs during physical separation of the populations.

Allopatric Species. Species that occupy regions that are geographically disjunctive (geographically separate).

Allotetraploid. A polyploid having whole chromosome sets from different species.

Allopolyploid. A polyploid formed by doubling of chromosome number in a dihybrid between two organisms with different genomes or by fusion of diploid gametes of such organisms.

Allozymes. Forms of an enzyme, controlled by alleles of the same locus, that differ in electrophoretic mobility. See *Isozymes*.

Alternative RNA Splicing. Different functional mRNA molecules are produced from the same primary transcript by differential removal of introns.

***Alu* Family.** A dispersed, intermediately repetitive DNA sequence found in the human genome about 300 nucleotides in size and repeated 10×10^5 times per genome. The name *Alu* comes from the restriction endonuclease that cleaves it. The repeats arise due to reverse transcription of RNAs into DNAs.

Amber Codon. The nonsense codon UAG.

Amber Suppressors. The mutant genes that code for tRNAs whose anticodons have been altered so that they can respond to UAG codons as well as or instead of to their previous codons.

Ames Test. A method which relies on reverse mutations of histidine-requiring (his^-) auxotrophs in *Salmonella typhimurium* to wild-type (his^+) auxotrophs. This method is widely used to test mutagenicity of environmental chemicals.

Amino Acid Activating Enzyme Alteration Suppression. Suppression of a mutation in a particular triplet, AUU (isoleucine) $\rightarrow$ UUU (phenylalanine) may take place by occasional attachment of isoleucine to $tRNA^{phe}$.

Aminoacyl-or A-site. The site on the ribosome to which the aminoacyl-tRNA attaches during translation.

Aminoacyl-tRNA Synthase. One of the family of enzymes responsible for aminoacylation.

Amniocentesis. A prenatal diagnostic technique for study of metabolic disorders by biochemically testing the cultured cells from the amniotic fluid. A foetus may be checked for any chromosomal abnormality also in early stages of development by karyotyping the cultured cells.

Amorphic (Antimorphic) Alleles. The recessive alleles that are not expressed in either homozygotes or heterozygotes compared to recessive alleles that are not expressed in heterozygotes. It is indeed equivalent to absence of it. Also known as *silent alleles.*

Amplification at Gene Level. A mutation at some site produces a new allele. Another mutation at a different site produces another new allele. Intragenic recombination produces another new allele which carries both the mutations. Thus, several *unisite mutant alleles* as a result of intragenic recombination can produce *two-site*, and *multisite mutant alleles.* Multisite mutant alleles have been reported to be actually present in natural populations. This is thus a mechanism for amplifying genetic variation at gene level.

Amplification at Genomic Level. Variation at gene level gets further amplified at genomic level. Independent assortment, crossing-over and random union of gametes are the mechanisms for amplification of genetic variation at genomic level.

Anagenesis. The evolutionary process whereby one species evolves into another without any splitting of the phylogenetic tree. See *cladogenesis.*

Ancient DNA. DNA of organisms which lived long ago and some of which have become extinct. Such a DNA is a treasure for understanding molecular evolution.

Ancillary Protein Factors. Proteins that are needed for initiation of transcription. These proteins recognise the sequence on DNA close to, or overlap with, the sequence bound by RNA polymerase itself.

Ancillary Sites. The sequence present between –30 and –40 position in prokaryotic or eukaryotic gene. In bacterial genes, it is known as cAMP-activated protein (CAP) site.

Annealing. Pairing of complementary single strands of DNA to form a double helix.

Antibody. A protein produced by B-lymphocytes (plasma cells) in response to a foreign antigen which is capable of combining with the antigen.

Anticoding Strand. The DNA strand that is complementary to the one that forms the template for transcribed mRNA. Also known as antisense strand.

Anticodon. The triplet of nucleotides in a tRNA molecule that is complimentary to and base pairs with a codon in an mRNA molecule.

Antileader. A nucleotide sequence in DNA upstream of the initiation codon of a gene which is transcribed but not translated. Antileader generally contains Shine-Dalgarno sequence which helps on binding of mRNA to the ribosome.

Antimutagen. An agent that has the ability to suppress mutagenicity of a mutagen.

Antimutator Gene. A gene that reduces the normal spontaneous mutation rate.

Antioncogene. Those genes whose protein products can prevent tumorigenic transformation of cells.

Antiparallel Strands. A term used to refer to the opposite but parallel arrangement of the two sugar-phosphate strands in double-stranded DNA; the $5' \rightarrow 3'$ orientation of one such strand is aligned along the $3' \leftarrow 5'$ orientation of the other strand.

Antisense DNA. A DNA strand that is complementary to the one used as a template during transcription.

Antisense Inhibition. A process where expression of a gene is inhibited with the help of an antisense RNA.

Antisense RNA. According to recent terminology, RNA transcribed on antisense DNA strand as a template. Also known as asRNA.

Antitermination. A process which causes RNA polymerase enzyme to continue transcription passed the terminator sequence, an event called *readthrough.*

Antitrialer. A nucleotide sequence in DNA downstream of the termination codon of a gene which is transcribed but not translated Antileader generally contains signal sequence for attachment of poly (A) tail to RNA polymerase II transcript.

Antizyme. A protein which renders some other enzyme unstable, e.g., in higher animals an antizyme renders enzyme ornithine decarboxylase unstable.

AP Endonucleases. Endonucleases that initiate excision repair at apurinic and apyrimidinic sites on DNA.

AP Sites. A purinic or apyrimidinic sites resulting from the loss of a purine or pyrimidine residue from the DNA.

Apoptosis. A physiological mechanism responsible for process of self-destruction of a cell. There is a controlled autodigestion of the cell through activation of endogenous enzymes. Apoptotic genes have been reported in mammals. Apoptosis plays a role in development. homeostasis and defence.

Artificial Gene Repressors. These are synthetic repressors of gene activity. These repressors are used in rational drug design.

Associative Intron. An intron which comes into being when two unrelated exons join by an unequal recombination between two genes. The introns would render this gene fusion event insensitive to a shift in the reading frame and allow a novel unequal recombination to succeed.

***att* B Site.** A 15-nucleotide long segment in bacterial genome having sequence 5′-GCTTTTTTATACTAA-3′ at which phage chromosome pairs for undergoing site-specific recombination.

***att* P Site.** A site on phage chromosome which is represented as P′OP site. P′ has 3 binding sites for integration protein whereas O and P have 2 sites each. Binding of integration protein molecules at these sites forms a structure, *intasome*, required for site-specific recombination between *E. coli* and phage l chromosomes.

***att* Sites.** These are the loci on a phage and the bacterial chromosome at which recombination integrates the phage into, or excises it from, the bacterial chromosome.

Attenuated Virus. Live viruses are allowed to live on a different host for a large number of generations. Through mutations, these viruses become attenuated which means the viruses are not able to cause disease in the original host. These attenuated viruses are used as vaccines. Through back mutations, the viruses may cause disease in the original host.

Attenuation. A mechanism by which expression of an amino acid biosynthetic operon (e.g., the *trp* operon) is regulated by the levels of the amino acid in the cell.

Attenuator. A stop site present within an operon which causes premature termination of transcription.

Attenuator Region. A sequence present downstream a gene which provides signal for termination of transcription and attachment of poly (A) tail, and thereby regulates gene activity.

Autonomous Controlling Element. In maize, there is an active transoposon with the ability to transpose (cf *non-autonomous controlling element*).

Autonomous Development. A type of development where a *Drosophila* imaginal disc from a mutant for eye colour develops into a mutant eye colour in wild-type host.

Autonomous Replicating Sequences (ARS). The sequences which serve as origin of replication and confers the ability to replicate the linked sequences.

Autotrophic. A living system which obtains complex organic materials from its surroundings and uses it as nutrition.

Auxotroph. An individual that, unable to carry on some particular synthesis, requires supplementing of minimal medium by some growth factor.

Avirulence Gene. Gene found in pathogens that elicits or fails to elicit the hypersensitive response in a plant carrying a defined resistance gene.

Back Mutation. A mutation that reverses the effect of a previous mutation by restoring the original nucleotide sequence.

Balanced Lethal System. An arrangement of dominant lethal alleles that maintains a heterozygous chromosome combination. Homozygotes for any lethal-bearing chromosome perish.

Balances Polymorphism. A type of genetic polymorphism in which a heterozygote is more adaptive than either homozygote. Natural selection maintains a gene even when it is lethal in homozygous state. In this case more than one forms of a gene are maintained by selection.

Balance in *Drosophila* Sex Determination. Sex in *Drosophila* is determined by the ratio of number of X chromosomes to number of sets of autosomes. X/A ratio of 1.0 is female, 0.5 is male, while 0.67 and 0.75 are intersexes, > 1.0 is superfemale and < 0.5 is supermale. Zygotes having X/A ratio < 0.33 and > 1.5 do not develop beyond embrynoic stage.

Balancing Selection. Heterozygotes when superior to the corresponding homozygotes in gene vigour or some specific component of fitness have a selective advantage. Both alleles will be maintained and a state of balanced polymorphism will be maintained.

Banded krait minor (Bkm) satellite DNA sequences. Repetitious DNA originally isolated from snake, *Bungarus fasciatus* which is localised exclusively in W chromosome of this snake. Bkm DNA-related sequences occur in the Y chromosome of the mouse. Bkm DNA-related sequences are in some way male determining.

Bank, Gene. A collection of recombinant DNA molecules containing inserts which together comprise the entire genome of an organism.

Basc. A stock of *Drosophila melanogaster* which carries five mutations—a semidominant marker Bar, a recessive marker *white-apricot*, and three overlapping inversions covering entire X chromosome. Also known as *Muller-5* stock, it is used to detect sex-linked recessive lethal mutations.

Base Analogue. A chemical compound that is structurally similar to one of the bases in DNA and which may act as a mutagen.

Base Pair. The hydrogen-bonded structure formed between two complementary nucleotides.

B cells. Plasma cells become B cells to form clones of lymphocytes. They are capable of secreting antibodies.

B-chromosomes. The chromosomes which are found in addition to the normal chromosomes. Also known as *supernumerary* or *accessory* or *extra chromosomes.*

b-galactosidase. The enzyme that splits lactose into glucose and galactose and coded by a gene, Z, in the *lac* operon.

b-galactoside acetyltransferase. An enzyme that is involved in lactose metabolism and encoded by a gene, *y*, in the *lac* operon.

b-galactoside permease. An enzyme that is involved in concentrating lactose in the cell and is coded by a gene, *a*, in the *lac* operon.

Bifunctional Vector (Plasmid). A DNA molecule able to replicate in two different organisms, e.g., in E. coli and yeast or *E. coli* and Streptomyces. These molecules are thus able to shuttle between the two alternative hosts and are, therefore, also known as shuttle vectors.

Bio-antimutagens. These are the suppressors of mutagenicity of mutagens by interfering with the process of mutagenesis.

Biochemical Basis of Heterosis. A hybrid between two types, each deficient in the ability to synthesise some chemical, grows normally since each type has a normal allele for the deficiency of the other. Consequently, the F_1 hybrid shows hybrid vigour or positive heterosis in comparison with the parents.

Biogenesis. The axiom that life originates only from pre-existing life.

Bioinformatic Tools. It includes computer software, etc. These help to collect, compile, analyse, process and represent the information in order to understand and navigate the genome, to look for gene, compare genomes and final information relevant to the stretch of sequence they want to study.

Biological Containment. One of the precautionary measures taken to prevent replication of recombinant DNA molecule in microorganisms in the natural environment. Biological containment involves the use of vectors and host organisms that have been modified so that they will not survive outside of the laboratory.

Biological Species Concept. In this species concept, emphasis is placed on "intrinsic" impediments to free exchange of genetic information between populations, i.e., emphasis is on reproductive isolation barriers.

Biotechnology. The use of living organisms, often but not always microbes, in industrial processes.

Biparental Zygote. The zygote which contains mitochondrial or chloroplast genes from both the parents. Two alleles for some cytoplasmic genes may thus occur in the zygote. For example, a *Chlamydomonas* zygote that contains cpDNA from both parents; such cells generally are rare.

Blood Group Antigens. Those antigens that are principally, but not exclusively, found on erythrocytes.

Blot. (i) As a verb, this means to transfer DNA, RNA or protein to an immobilising matrix such as DBM-paper, nitrocellulose or biodyne membranes. (ii) As a noun, it usually refers to the autoradiograph produced during the Southern or Northern blotting procedure.

Blunt End. An end or a DNA molecule, at which both strands terminate at the same nucleotide position with no single-stranded extension.

Bombay Phenotype. A blood group in which individuals lack H, A and B antigens.

Bottleneck Effect. A brief reduction in size of a population, which usually leads to random genetic drift.

Boundary between Abiological and Biological Evolution. Origin of genetic code seems to be a good criterion for distinguishing between abiological and biological evolution.

Boundary between Living and Nonliving. Scientists now believe in gradual origin of life. Origin of genetic code seems to be a good criterion for distinguishing between living and nonliving.

Branch Migration. The process in which a crossover point between two duplexes slides along the duplexes.

Britten-Davidson Model. A model that explains gene regulation in eukaryotes. See *gene regulatory mechanisms in eukaryotes* for details of the model.

C Region (Constant Region). A portion in light as well as heavy chains of an immunoglobulin molecule where the amino acid sequence is not variable.

CAP Site. A nucleotide sequence upstream of some bacterial genes and operons; the attachment point for the *catabolite activator protein.*

Cassette Mechanism of Mating Switch. The mechanism by which homothallic yeast cells alternate mating types. The mechanism involves two silent transposons (cassettes) and a region where these cassettes can be expressed (cassette player).

CAT Box. A conserved sequence found within the promoter region of the protein- encoding genes of many eukaryotic organisms. It has the canonical sequence GGPyCAATCT and is believed to determine the efficiency of transcription from the promoter.

Catenane. Interlocked circular DNA molecules that are the exclusive products of bacteriophage l integrative reaction *in vitro* when substrate is supercoiled molecule containing both attachment sites, attP and attB.

cDNA Cloning. A method of cloning the coding sequence of a gene starting with its mRNA transcript. It is normally used to clone a DNA copy of a eukaryotic mRNA. The cDNA copy, being a copy of a mature messenger molecule, will not contain any intron sequences and may be readily expressed in any host organism if attached to a suitable promoter sequence within the cloning vector.

cDNA Library. A library of clones that have been prepared from mRNA after conversion into double-stranded DNA.

Cell-mediated Immune Response. A type of immune response in which antigen interacts with certain cells.

Cell Transformation. The alteration in morphological and biochemical properties that occurs when an animal cell is infected by an oncogenic virus.

Cellular Oncogenes (C-*onc*). The genes found in the cells of a host organism that are derived from proto-oncogenes.

CEN (Centromere). A cloned eukaryotic centromere. Each CEN is given a number corresponding to the chromosome from which it was derived. This notation is confined to yeast molecular boilogy at the moment but, in principle, could be widely applied.

Central Dogma. The key hypothesis in molecular genetics, proposed by Crick in 1958, which states that DNA makes RNA and RNA makes protein.

Centromeric DNA Elements (CDEs). Yeast centromere sequence has three elements known as CDEs. CDEI is conserved 8 bp palindrome, CDEII is A/T rich region of variable length (78-86 bp), and CDEIII is conserved 25 bp sequence with palindromic characteristics. A 125 pDNA fragment is necessary and sufficient for complete centromeric function.

C Genes. The genes which code for the constant regions of immunoglobulin protein in chains.

Charged tRNA. A tRNA carrying an amino acid.

Charon Phages. A phage which contains two nonsense mutations that prevent phage from growing in E. coli strains that do not carry an appropriate nonsense suppressor mutation. This limits phage growth only to special laboratory bacterial strains and thus acting as a safety feature against spreading of undesirable strains.

Chaperone. Molecular chaperone is a protein that is needed for the assembly or proper folding of some other protein, but which is not itself a component of the target complex.

Chemical Cleavage DNA Sequencing. In this method of DNA sequencing, given by Maxam and Gilbert (1980), many uniform repeated lengths of DNA are isolated through restriction endonuclease digests which are then broken at various points by four chemical treatments. The fragments differing from one another by one nucleotide are then separated through gel-electrophoresis which are detected by autoradiography.

Chemoautotrophic. Systems which utilise simple organic molecules as nutrition. The system is thus environment-dependent.

Chimeric Gene. A recombinant gene having regulatory sequence from one gene and coding sequence from another gene.

Chimeric Plasmid. Hybrid, or genetically mixed, plasmid used in DNA cloning.

Chi (c) Site. Sequence of DNA at which the RecBCD protein cleaves one of the strands during recombination.

Chloroplast (cp) DNA. DNA present in chloroplasts. It is circular double stranded DNA (as in prokaryotes) and it replicates independently of the nuclear DNA by rolling circle method. Molecular weight = 55×10^6 to 97×10^6; Length = 419 mm; Size = 1.3×10^5 to 1.5×10^5,

Chromocenter. Pericentromeric heterochromatic regions of all the polytene chromosomes that are tightly paired such that all chromosomes appear to be fused in this region.

Chromogene. A heredity determiner in the chromosome, in contrast to determiner in the cytoplasm *plasmagene.*

Chromosome Imprinting. A phenomenon where, in some organisms, paternal X chromosome is inactivated in some or all the cells. Chromosome imprinting is obviously heritable but is germ line. DNA methylation is involved in this process.

Chromosome Jumping. A technique of isolated clones from a genomic library that are not contiguous but skip a region between known points on the chromosome. This is done usually to bypass regions that are difficult or impossible to "walk" through or regions known not to be of interest. Also known as *chromosome hopping.*

Chromosome Landing. In plants. Abundance or repetitive DNA makes chromosome walking impossible. Availability of high density physical maps based on markers such as RAPDs are found close enough to genes to be included on the same genomic clone. This method is analogous to *transposon tagging.*

Chromosome-mediated Gene Transfer. A method of gene transfer where a chromosome from a mammalian cell can be isolated by differential centrifugation of carefully broken cells, whose mitotic cycles has been arrested at metaphase by the drug colchicine, since metaphase chromosomes are relatively robust and are easy to be isolated. These isolated chromosomes are allowed to be taken up by the cells in culture.

Chromosome Puffs. During certain stages of development, diameter of specific sites of chromosome is greatly increased. These swellings indicate sites of gene activity. Also known as *Balbiani rings* or simply *puffs.*

Chromosome Walking. A technique for studying segments of DNA, larger than can be individually cloned, by using overlapping probes. Also known as *jumping libraries.*

Circularisation. A DNA fragment generated by digestion with a single restriction endonuclease will have complementary 5′ and 3′ extensions (sticky ends). If these ends are annealed and ligated the DNA fragment will have been converted to a covalently-closed circle or circularised.

Cis-position. When two mutations are located in the same chromosome.

Cis-*trans* Complementation Test. A genetic analysis that tests whether two mutations lie in the same or different genes, and which can also provide information on dominance and recessiveness. The test involves introducing the two mutated genes into a single cell, for example, by introducing an F' plasmid carrying one mutated gene into a recipient bacterium with a chromosomal copy of the second gene.

Cisternae. Apparent spaces within the endoplasmic reticulum whose profiles in electron micrographs suggest that cisternae are flattened vesicles.

Cistron. A segment of DNA specifying one polypeptide chain in protein synthesis. Under the concept of a triplet code, one cistron must contain three times as many nucleotide pairs as amino acids in the chain it specifies.

cl Repressor. One of the regulatory proteins responsible for controlling expression of l genes.

Cladogenesis. The evolutionary process whereby one species splits into two or more species. This is also known as *true speciation*. Also see *Anagenesis*.

Classical Phase of Gene Concept. A phase when gene was regarded as a hypothetical particle performing a specific function and recombining with other genes.

Class Switching. A change in the expression of the C region of an immunoglobulin heavy chain during lymphocyte differentiation.

ClB Method. A technique devised by Muller to rapidly screen in fruit-files recessive X chromosome lethal mutations. The ClB chromosome carries a recessive lethal (*l*), a dominant marker (B), and an inversion (crossover suppressor, C).

Clone. A population of identical cells, generally those containing identical DNA molecules.

Clone Library. A collection of clones that contain a number of different genes. See genomic library

Cloning Vector. A DNA molecule, capable of replication in a host organism, into which a gene is inserted to construct a recombinant DNA base paired.

Closed Promoter Complex. The initial complex formed between RNA polymerase and a promoter, in which the double helix is still completely base paired.

Closed Reading Frame. The reading frame contains termination codons that prevent its translation into protein.

Cloverleaf. A convenient two-dimensional representation of the structure of a tRNA molecule.

Cluster Clone Theory. Given by Bogorad in 1975. A prokaryote having a single duplicated strand of DNA has several genes. DNA strand breaks up into two parts containing different genes. Some of these genes are retained in nucleus and other DNA segments are converted into mitochondria through the acquisition of separate membranes. Neither the method nor the source of these membranes is explained.

Coacervate. Term given by Oparin. An aggregate of varying degrees of complexity, formed by interaction of two or more colloidal suspensions. Also called *microsphere* by Fox.

Code Dictionary. A listing of the 64 possible codons and their translational meanings (the corresponding amino acids).

Coding Strand. The DNA strand with the sequence complementary to transcribed mRNA (given U in RNA and T in DNA). Compare with *anticoding strand*. Coding strand is also called *sense strand*.

Codominance. The situation whereby both members of a pair of alleles contribute to the phenotype.

Codon. A triplet of nucleotides that code for a single amino acid.

Codon-Anticodon Misreading Suppression. A change in ribosomal configuration that enables codon-anticodon misreading to occur and thereby suppress some mutations through misreading.

codon-Anticodon Relationship. This is alignment on the ribosome of an anticodon of amino acid charged-tRNA molecule against its complementary codon on mRNA.

Codon Bias (Preference). The idea that for amino acids with several codons, one or a few are preferred and are used disproportionately. They would correspond with tRNAs that are abundant.

Cognate tRNAs. tRNAs that are recognised by a particular aminoacyl-tRNA synthetase.

Cohesive End. An end of a double-stranded DNA molecule where there is a single-stranded extension.

Colinearity. Refers to the fact that a gene and the polypeptide for which it codes are related in a direct linear fashion, with the 3′ end of the template strand of the gene corresponding to the amino-terminus of the polypeptide.

Col Plasmids. Plasmids that produce antibiotics (colicinogens) used by the host to kill other strains of bacteria.

Commaless Code. A genetic code in which successive codons are contiguous and not separated by noncoding bases or groups of bases.

Commensalism. An association in which one symbiont benefits and the other is neither harmed nor benefited.

Competitive Exclusion. The principle that no two species can coexist in the same place if there ecological requirements are identical. This is also known as *Gause principle.*

Complementary DNA (cDNA). DNA produced on a RNA template.

Complementary Genes. Genes which interact to produce an effect distinct from the effects of an individual gene. Complementary genes yield same mutant phenotype when present separately but when present together they interact to produce a wild-type phenotype. Two heterogygotes in F_2 progeny produce ratio 9 wild type: 7 mutant.

Complementary RNA. Synthetic RNA produced by transcription from a specific DNA single-stranded template.

Complementary Sequence. A nucleotide sequence that base pairs with another nucleotide sequence.

Complementation Map. A map developed from the complementation relationships between alleles, normally in a small segment of the chromosome.

Complementation Matrix. A tabular representation of complementation tests involving a number of phenotypically similar mutants.

Complete Dominance. Resemblance of F_1 with one of its parents.

Complex Gene. The gene or its protein product undergoes various rearrangements cleavages or modifications before a functional product is obtained.

Complexes in Eukaryotic Transcription Initiation. Comprises of five intermediate complexes containing distinct components involved in initiation of transcription of eukaryotic genes.

Complex Transcription Unit. When a primary transcript yields more than one mRNAs encoding different proteins as it has more than one poly (A) sites or splice sites.

Compound Gene. The genes in which coding sequences are separated by non-coding sequences. Also known as *split gene.*

Composite Transposon. A transposon constructed of two IS elements flanking a control region that frequently contains host genes.

Compound Antigen. Formation of unique serological determinant by combination of two common antigens.

C-onc. The cellular version of a gene carried by a transformation retrovirus.

Concatamer. A DNA molecule that comprises a series of smaller DNA molecules linked head-to-tail. Concatamers are formed during the replication of some viral and phage genomes.

Concept of Dominance. The concept that dominance is not a universal property of a gene. It is relative property of a gene to produce a particular phenotype in a particular genetic and environmental background.

Concerted Evolution. The ability of two related genes to evolve together as though constituting a single locus.

Condensation Reaction. A reaction in which a covalent bond is formed with loss of a water molecule, as in the addition of an amino acid to a polypeptide chain.

Conditional-lethal Mutant. A mutant that is lethal under one condition but not lethal under another condition.

Conditioned Dominance. Dominance affected by the presence of other genes or dominance affected by environmental variation.

Conjugative Plasmids. Bacterial plasmids which can be used to transfer genes to bacteria outside their own species.

Conjugative Transposons. Those transposons which are transferred between different bacterial cells through conjugation.

Consanguineous Mating. Mating between blood relatives. See *inbreeding.*

Consensus Sequence. A sequence of nucleotides most often present in a DNA segment of interest.

Constitutive Gene. A gene whose expression is not regulated. Its product is continuously synthesised by the cell whether or not substrate is present in the cell.

Constitutive Heterochromatin. Heterochromatin that surrounds the centromere.

Constitutive Secretion. The proteins which are continuously secreted from the bacterial cells.

Contig. Libraries of overlapping, contiguous clones.

Continuity of Life. Germ plasm is continuous between all descendent generations. It occurs through reproduction.

Continuity Replication. In DNA uninterrupted replication in the 5′ to 3′ direction using a 3′ to 5′ template.

Controlling Elements. These elements of maize are transposable units originally identified solely by their genetic properties. They may by autonomous (able to transpose independently) or nonautonomous (able to transpose only in the presence of an autonomous element).

Copy Number. Usually, the number of molecules of a plasmid contained in a single cell. Also, the number of copies of a gene, transposon or repetitive element in a genome.

Core DNA. The 146 bp of DNA contained on a core particle.

Core Octamer. The structure comprising two subunits each of histones H2A, H2B, H3, and H4, which forms the central component around which DNA is wound to form a nucleosome.

Corepressor. A small molecule that must bind to a repressor protein before the latter is able to attach to its operator site.

Cosmid. A hybrid plasmid that contains *cos* sites at each end. *Cos* sites are recognised during head filling of lambda phages. Cosmids are useful for cloning large segments (up to 45 kb) of foreign DNA.

***cos* Site.** The sequence with the l DNA molecule that forms single-stranded overhangs at the ends of the linear version of the genome.

C_0t. The product of DNA concentration and time of incubation in a reassociation reaction.

$C_0t_{1/2}$. Midpoint values on C_0t curves (C_0t values plotted against concentration of remaining single-stranded DNA) which estimate the length of unique DNA in a sample.

Co-transduction. The transfer of two or more genes on a single DNA molecule during transduction of a bacterium.

Co-transformation. The uptake of two or more genes on a single DNA molecules during transforming of a bacterium.

Co-translational Transfer. The process through membrane-bound polysomes synthesised proteins enter the endoplasmic reticulum when they are being synthesised. For entry of these proteins into the endoplasmic reticulum, leader sequence is required at the N-terminal end of the protein.

Cro Repressor. One of the regulatory proteins responsible for controlling expression of l genes.

Crown Gall. A tumor formed, usually, on the stems of broad-leaved plants when infected with *Agrobacterium tumefaciens* containing a Ti-plasmid. The bacterium is only necessary for the initiation of the tumor. The genome of the affected plant cells contains several copies of a segment of the Ti-plasmid (the T-DNA). Crown galls can be of two types, *octopine or nopaline,* depending on the type of Ti-plasmid which initiated the tumor. Whole plants can be regenerated from crown-gall tissue and some of these will contain the T-DNA.

Cruciform Structure. The cross-shaped structure that can arise by intramolecule pairing within a double-stranded DNA molecule that contains an inverted repeat.

Cryptic Plasmids. Plasmids to which phenotypic traits have yet not been ascribed.

Cryptic Satellite. A satellite DNA sequence not identified as such by a separate peak on a density gradient; that is, it remains present in main-band DNA.

Cryptic Variation. Differences among individuals in terms of some biochemical properties (e.g., denaturation) that are not revealed at morphological level or even electrophoretically.

Cryptomorphic Gene. The gene has a cryptic structure in that the ultimate active product or products are carried within the precursorial molecule. The active protein is released after enzymatic breakdown of the precursor followed by processing.

Curly-Lobe-Plum Drosophila. A balanced lethal system in chromosome 2 in *Drosophila.* Used to detect autosomal recessive lethals of chromosome 2.

C-value Paradox. The estimated number of genes in eukaryotes is much less than the amount of DNA present, e.g. haploid human genome has 2.8×10^7 bp and should contain 3×10^6 genes but the number estimated is 50,000. This anomalous situation is referred to as C-value paradox.

Cynobacteria. A simple prokaryote which in terms of its biochemistry and metabolism seems to occupy a middle position between anaerobes and eukaryotes because they can tolerate oxygen, make use of oxygen both metabolically in aerobic respiration and in synthetic pathways and can make unsaturated fatty acids by oxidative disaturation. Some cynobacteria under conditions can use H_2S instead of water releasing S instead of oxygen.

Cytohet. A genetic condition where a zygote contains in its cytoplasm genetically different mitochondria contributed by two parents. Thus the individual is cytoplasmically heterozygous.

Cytoplasmic Inheritance. Inheritance of characters whose governing genes are located in the cytoplasmic organelles lilke mitochondria, chloroplasts than in nucleus.

Cytoplasmic Male Recombination. When recombination is induced in males of *Drosophila melanogaster* under the influence of cytoplasmic factors.

Cytoplasmic Segregation and Recombination (CSAR). A process suggested to explain assortment and recombination of organelle genes.

Dark Repair. The repair of DNA damage by DNA-repairing enzymes that do not require light for their activation.

Darwinian Evolution. A theory of evolution proposed by Darwin which evolution in terms of natural selection.

Degeneracy. Refers to the genetic code and the fact that most amino acids are coded for by more than one triplet codon.

Delayed Segregation. Segregation occurs in F_3, rather than in F_2 since the phenotype of the individual is determined by the genotype of the mother. Example of delayed segregation is direction of coiling in snails.

Deletion Mapping. Mapping mutation by use of overlapping deletion mutants to determine whether or not a mutation includes the site of a mutant gene.

Demethylation Model. A model that explains role of undermethylation of cytosine in gene regulation and differentiation in eukaryotes. In undifferentiated cells DNA is fully or uniformly methylated at all sites that ever will be methylated. During development, sequence-specific proteins inhibit methylation during DNA replication, leading to methylation patterns specific for each tissue. Once demethylation events occur, the differentiated methylation patterns are clonally inherited as a result of maintenance methylation system.

Denominator Elements. Genes on the autosomes of *Drosophila* that regulate the sex switch (*sxl*) to the off condition (maleness). Refers to the denominator of the X/A genic balance equation. The elements of the sex determination located in the autosomes.

Deoxyribonucleic Acid (DNA). A usually double stranded, helically coiled, nucleic acid molecule composed of deoxyribose-phosphate "backbones" connected by paired bases attached to the deoxyribose sugar; the genetic material of all living organisms and many viruses.

Desmutagen. An agent which suppresses mutagenic activity of a mutagen by directly inactivating the mutagen by destroying oxygen radicals produced.

Developmental Homeostasis. The capacity of the developmental pathway to produce a normal phenotype in spite of developmental or environmental disturbances.

Developmentally-regulated Gene Rearrangements. A type of gene rearrangements that take place during differentiation of heterocysts in cyanobacteria or in mother cell of *Bacillus subtilis* during sporulation.

Dideoxynucleotide. A modified nucleotide that lacks the 3′-hydroxyl group and so prevents further chain elongation when incorporated into a growing polynucleotide.

Differential Processing. It refers to production of more than one functional mRNAs from one type of complex transcription unit through differential processing.

Differential Segments. Portions of chromosomes that do not pair in meiosis.

Dimorphic Gene. A gene whose protein product exists in two forms, inactive and active; active form is produced by cleaving a part of polypeptide chain of the inactive protein.

Direct DNA Repair. DNA repair that involves neither removal nor replacement of bases or nucleotides; rather the covalent modification in DNA is simply reversed.

Directed Gene Amplification. This is a type of gene amplification which is genetically programmed. This type of amplification occurs at some specific stage during development, e.g., programmed amplification occurs in rRNA genes of many amphibian oocytes.

Directional Selection. A type of selection that removes the individuals from one end of a phenotypic distribution and thus causes a shift in the distribution.

Direct Repeats. Two or more identical nucleotide sequences present in a single polynucleotide.

Discontinuous Replication. In DNA the replication in short 5′ to 3′ segments using the 5′ to 3′ strand as a template while going backwards, away from the replication fork.

Disruptive Selection. A type of selection that removes individuals from the centre of a phenotypic distribution and thus causes the distribution to become bimodal.

Divergent Transcription. The initiation of transcription at two promoters facing in the opposite direction, so that transcription proceeds away in both directions from a central region.

Diversifying Selection. Selection in which two or more genotypes have optimal adaptiveness in different subenvironments.

Divisive Introns. An intron which arises as a divisive element separating a previously contiguous exon into two halves.

D-loop. Configuration found during DNA replication of chloroplast and mitochondrial chromosomes wherein the origin of replication is different on the two strands. The first structure formed is a displacement loop, or D-loop.

DNA A Helicase. An *E. coli* protein which binds to replication origin site, Ori C. Twenty to thirty molecules are required for the formation of the initiation complex.

DNA B Helicase. The enzyme which causes unwinding of DNA double helix prior to the DNA replication in l phage and *E. coli*; this enzyme is localised in these systems by proteins lP and DNA C, respectively.

DNA-binding Protein. Any protein that attaches to DNA as a part of its normal function, e.g., histone, RNA polymerase, *lac* repressor.

DNA C Protein. An *E. coli* protein required to localise DNA replication proteins.

DNA Clone. A section DNA that has been inserted into a vector molecule, such as a plasmid or a phage chromosome, and then replicated to form many copies.

DNA Crosslinking. Interstrand thymines of DNA form dimers thus blocking replication.

DNA Fingerprinting. A technique developed by Jeffreys *et al.* in 1985 useful in establishing near perfect identity of an unidentified body. The technique is based on the fact that every human being is conferred with DNA variations in form of variable number of tandem repeats (VNTRs). DNA profile of an individual is also known as genetic signatures. Accuracy of this method is 1-in-75 billion error probability ratio.

DNA G Primase. An enzyme which synthesises primer required for l phage and *E. coli* DNA replication; this enzyme is localised in these systems by proteins lP and DNA C, respectively.

DNA Helix A. Right-handed helix; rotation/base pair 33.6° mean base pairs/turn 10.7; inclination of base to helix axis +19°; rise/base pair along helix axis 2.3 Å; pitch/turn of helix 24.6 Å, mean propeller twist +18°; glycosyl angle confirmation anti; sugar pucker conformation c3′-endo. This form of DNA occurs at high humidity.

DNA Helix B. Right-handed helix; rotation/base pair 38.0°; mean base pairs/turn 10.0; inclination of base to helix axis –1.2°; rise/base pair along helix axis 0.32 Å; pitch/turn of helix 33.2 Å; mean propeller twist +16°; glycosyl angle confirmation anti; sugar pucker conformation c1′-endo to C2′-endo. The most commonly found right-handed form of DNA.

DNA Helix Z. Left-handed helix; rotation/base pair -60°/2; mean base pairs/turn 12.0; inclination of base to helix axis -9°; rise/base pair along helix axis 3.8 Å; pitch/turn of helix 45.6 Å; mean propeller twist 0° glcosyl angle confirmation anti at (C), syn at G′, sugar pucker conformation G2′-endo at C, G1′ exo to C1′-exo at G.

DNA J. One of the two l phage proteins (the other being DNA K) required to free DNA B helicase (unwinding enzyme) from the specialised nucleoprotein (snup) complex.

DNA K. One of the two l phage proteins (the other being DNA J) required to free DNA B helicase (unwinding enzyme) from the specialised nucleoprotein (snup) complex.

DNA Glycosylases. Endonucleases that initiate excision repair at the sites of various damaged or improper bases in DNA.

DNA Gyrase. A type II topoisomerase found in *E. coli.*

DNA Ligase. An enzyme that repairs single-stranded discontinuities in double-stranded DNA molecules. In the cell, ligases are involved in DNA replication. Purified ligases are used in construction of recombinant DNA molecules.

DNA Looping. A mechanism of gene regulation which imagines that proteins bound with DNA at widely separated sites interact with each other with the intervening DNA looping or bending. It is the interaction between DNA-bound proteins, not the looping *per se* that regulates the gene expression.

DNA-mediated Gene Transfer. The method of gene transfer in which a purified DNA fragment carrying gene of interest is mixed with a carrier DNA and is precipitated out of solution with calcium phosphate. When target cells are incubated with the precipitate, there is increased uptake of DNA and some of the cells are transformed by the desired gene.

DNA Methylation Theory of Development. A theory put forward to explain development reprogramming by reprogramming of gametes prior to fertilisation which depends upon specific *de novo* methylation of chromosomal DNA.

DNA Modification. Methylation of adenine of prokaryotic DNA playing role in host-restriction and modification and cytosine in eukaryotic DNA playing role in maternal inheritance of chloroplast DNA, DNA replication, recombination, mutation, chromosome folding and packing, mammalian chromosome inactivation, conformational changes in DNA, etc. DNA methylation protects the host DNA from its restriction enzymes.

DNA Oozing. A gene regulatory mechanism according to which binding of a regulatory protein to its operator helps binding of another protein to adjacent sequences, which in turn helps another proteins bind next to it, and so on, until a procession of proteins has oozed out from the control sequence to the point where transcription is initiated.

DNA Polymerase I. The *E. coli* enzyme that completes synthesis of individual Okazaki fragments during DNA replication by degrading the primer and filling in the gaps. May also be helpful in the termination of DNA replication.

DNA Polymerase II. The main DNA replication enzyme of *E. coli.*

DNA Polymerase III. The enzyme required for *in vivo* DNA replication.

DNA-protein Interactions. Interaction between DNA and protein molecules for precise DNA transactions, viz, initiation of DNA replication, initiation of transcription, site-specific recombination, gene regulation.

DNA Puff. Cross band in a polytene chromosome swollen due to the relaxation of chromatin of banded region at the time of high activity. Also known as *chromosome puff.*

DNA Repair. The correction of nucleotide errors introduced during DNA replication or resulting from the action of mutagenic agents.

DNA Replication. The process by which a DNA molecule makes its identical copies.

DNA-RNA Hybridisation. When a mixture of DNA and RNA is heated and then cooled, RNA can hybridise (form a double helix) with DNA that has a complementary nucleotide sequence.

DNA sequencing. Determination of the order of nucleotides in a DNA molecule. Methods: (1) Chemical cleavage method of Maxam and Gilbert (1977), (2) Plus and minus method of Sanger *et.al.* (1978), and (3) Wandering Spot Analysis.

DNA Sliding. This mechanism of gene regulation imagines that a protein recognises a specific site on DNA and then moves (slides, tracks) along the DNA to another specific sequence where by interacting with another protein, it initiates transcription. This idea also imagines that the regulatory protein remains bound at the first site and the DNA is thread past or through the bound protein until the second critical site is encountered.

DNA Strand. One of the two DNA strands in a double-helical DNA molecule.

DNA Synthesis in vitro. Synthesis of DNA outside a cell (in a test tube) involving four types of triphosphate nucleosides. A template is used in artificial DNA synthesis.

DNA Topoisomerase. An enzyme that introduces or removes turns from the double helix by transient breakage of one or both polynucleotides.

DNA Topoisomerase Type I. Type I DNA topoisomerase rectifies super coiling of DNA by making a transient break in one strand of DNA.

DNA Topoisomerase Type II. Type II DNA topoisomerase relaxes negative supercoiling during transcription by introducing a transient double-stranded break in DNA.

DNA Tumor Virus. A virus containing a DNA genome and able to cause cancer after infection an animal cell, e.g., SV40, adenovirus.

DNA Twisting. A mechanism which explains gene regulation through binding of regulatory proteins to some altered form of DNA (left-handed or single stranded). Alternate belief is that regulatory protein has an enzymatic activity that alters DNA conformation and regulates gene expression.

Docking Protein. Responsible for attaching (docking) a ribosome to a membrane by interacting with a signal particle attached to a ribosome destined to be membrane bound.

Domain. Domain of a chromosome may refer either to a discrete structural entity defined as a region within which super-coiling is independent of other domains; or to an extensive region including an expressed gene that has heightened sensitivity to degradation by the enzyme DNase1.

Dominance Hypothesis. The theory the heterosis is caused by the masking of harmful recessive alleles by dominant alleles. See also *overdominance hypothesis.*

Dominance Modifiers. Those genes which through relative strength of action of the two alleles at a locus or action of other genes (epistasis) affect the dominance-recessiveness relationship.

Dosage Compensation. Any mechanism by which the effective dosages of sex-linked genes in organisms with an XX-XY or XX-XO mechanism of sex determination are made equal.

Dosage Effect. A type of inheritance in which two like dominant alleles have a stronger effect than a single one of the same type.

Dot Blotting. A blotting technique of DNA already cloned that eliminates the electrophoretic separation step. Autoradiographs reveal dots rather than bands on a gel, indicative of a cloned gene.

Double Digest. The product formed when two different restriction endonucleases act on the same segment of DNA.

Doubling Dose. The dose of a mutagenic substance to induce as many mutations as occur spontaneously in one generation.

Downgrading Regulation. A mechanism in dosage compensation where two X chromosomes in one sex decrease expression of its genes to equalise the expression of a single X chromosomes in the other sex. For example, in nematode *hypotranscription* exists.

Down Promoter Mutations. Mutations that decrease the frequency of initiation of transcription.

Downstream. Towards the 3′ end of a polynucleotide.

Drastic Amino Acid Substitutions. When amino acids having dissimilar charges, chemically reactive and non-reactive, or bulky and light side chain amino acids replace each other. These substitutions may have drastic effect on functioning of the protein.

Drosophila. A dipteran fruit-fly which was a central experimental organism in the development of classical genetics. Experiments on *Drosophila* led to the chromosome theory of inheritance by Morgan, Bridges, Sturtevant and Muller. *Drosophila* continues to be an important experimental organism in genetics.

Drug Resistance Factor. Duplex ring DNA plasmid or episome which confers upon its bacterial host resistance to one or more antibiotics.

Duplicate Loci. A pair of nonallelic loci having the same genetic meaning; two isoloci.

Duplicate Recessive Epistasis. Recessive alleles at either of the two loci can mask the expression of dominant alleles at the two loci resulting 9:7 ratio. Also called *complementary epistasis.*

Duplicate Genes. Either dominant or both dominant genes together produce the same phenotype to give 15:1 ratio in F_2 progeny.

Early Genes. Bacteriophage genes that are expressed during the early stages of infection of a bacterial cell; usually the products of early genes are involved in replication of the phage genome.

Ecdysone. A steroid hormone which triggers moulting in insect larvae. Ecdysone is considered to be an inducer of gene activity.

Effector Gene. The gene which drives transcription of another gene with the help of a promoter from another gene.

Effector Molecule. A molecule (a sugar, an amino acid or a nucleotide) that can bind to a regulator protein and thereby change the ability of the regulator molecule to interact with the operator.

Electroporation. A technique for transfecting cells by the application of a high-voltage electric pulse.

Elicitor. This is a specific signal molecule produced by the pathogen. Elicitors are identified by the host through receptors encoded by genes conferring resistance in plants.

Elicitor-receptor Model. A model proposed to explain gene-for-gene hypothesis. Accordingly, either primary virulence gene products or metabolites resulting from their metabolic activity are recognised by specific plant receptors encoded by disease genes.

Elicitor-suppressor Model. A model proposed to explain gene-for-gene hypothesis. Accordingly, pathogens are thought to produce general elecitors that initiate host defense unless a specific suppressor substance is also liberated by a particular pathogen. This scheme is understood to work where virulence rather than avirulence is dominant.

ELISA (Enzyme-linked Immunosorbent Assay). A sensitive immunological assay system which avoids both the hazards and expense of radioactive or fluorescence detection system. Two antibody preparations are used in ELISA. The primary antibody binds the antigen and is itself bound by the second, antiglobin, antibody. The antiglobin is linked to an enzyme, e.g., horseradish peroxidase, whose activity is easily monitored, for instance by colour change. The extent of the enzymatic reaction is then a quantitative indication of the amount of primary antibody or, indirectly, of antigen present.

Elongation Factors. The proteins that play an ancillary role in the elongation step during translation process molecule (EF-Ts, EF-Tu, EF-G).

Embiont. A heterotrophic protocell which through evolution of metabolic pathways became autotrophic.

End Labelling. The term describes the addition of a radioactively labelled group to one end (5′ to 3′) of a DNA strand.

Endocytosis. A process responsible for import of proteins into the cell.

Endomitosis. An increase in somatic DNA content which takes place within an intact nuclear envelope and gives rise to *endopolyploidy*.

Endonuclease. An enzyme that breaks phosphodiester bonds within a nucleic acid molecule.

Endoplasmic Reticulum. A double membrane system in the cytoplasm, continuous with the nuclear membrane and bearing numerous ribosomes.

Endopolyploidy. An increase in the number of chromosome sets caused by replication without cell division.

Endoreplication. Replication without separation of chromatids.

Endosymbiont. The organism that resides inside a host organism. The term also applies to those organisms that reside within the cells of an other organism.

End-product Inhibition. Describes the ability of a product of a metabolic pathway to inhibit the activity of an enzyme that catalyzes an early step in the pathway. Also known as *feedback inhibition.*

Engineered Vaccines. The vaccines prepared by recombinant DNA technology. These vaccines avoid dangers of incomplete inactivation. There are four steps involved in the preparation of vaccines by recombinant DNA technology. The first step is to identify the surface antigens of the parasite. Second step is to clone the gene for the antigen. Thirdly, gene is got expressed in *E. coli* or some other system. Finally, purified antigen is used as a vaccine.

Enhancer. A special type of eukaryotic regulatory sequence that can increase the rate of transcription of a gene located some distance away in either direction. They are mostly *cis-*acting but *transacting* ones are also known. They are effective whether lying upstream or downstream. They are active whether they lie in same or opposite polarity as the gene. They are effective regardless of the organism or the gene from which they are derived when attached to foreign DNA.

Enrichment. The process of increasing the proportion of mutant cells in a culture. Enrichment is used when the desired mutants cannot be selected directly. The technique is essentially the converse of selection. Conditions are established such that the desired mutants will not grow. Wild type (growing) cells are then killed by some physical or chemical treatment which does not affect non-growing (mutant) cells. Such methods include treatment with drugs such as penicillin (for bacteria) or nystatin (for fungi), inositol starvation, and heat shock.

Environmental Mutagenesis. Branch of genetics that deals with studies on the mutagenic effects of environmental agents using battery of test systems.

Epigenesis. A concept according to which development of an individual is epigenetic and genes act in epigenetic systems and adult structure develop from uniform embryonic tissues.

Epigenetic Landscape. An idea given by Waddington to explain the origin of epigenetic pattern where development is likened to balls rolling down a succession of valleys, called the epigenetic landscape. Depths and directions of this landscape are controlled by genes.

Epigenetic Modifications. Changes that influence the phenotype without altering the genotype. Also see *post-translational modifications*.

Epigenetic Pattern. The patterns involving spatial arrangement of tissues and organs are formed through a complex network of reactions and interactions influenced by genes.

Epimutations. Heritable changes in gene activity due to DNA modification.

Episome. A plasmid capable of integration into the host cell's chromosome. These are those particles that are added to a genome through an external source–not by mutation or rearrangement of the existing genome.

Epistasis. The ability of one gene of mask the phenotype derived from a second gene.

Epistatic Genes. The genes that suppress expression of other genes.

Error-catastrophy Model of Ageing. Mutations in the protein-synthesising apparatus have a cascading effect on for the synthesis of other proteins, a wrong enzyme will synthesize many more wrong proteins before it has degraded itself. Consequently, the activity of the cell gets so slow that it dies due to "error catastrophe".

Escherichia Coli. A species of human intestinal bacteria genetically well understood and often used in biological research.

Ethionine. A methionine analogue that lacks its methyl group and thereby inhibits most methyl transferases thus inducing gene expression in eukaryotes by undermethylation of cytosine.

Eviction, Gene. A method, originally developed in yeast molecular biology, which permits the retrieval of a chromosomal copy of any gene which has previously been cloned. Thus if the wild type copy of a gene has been cloned gene eviction may be used to obtain a mutant copy and vice versa. A feature of the gene eviction procedure is that DNA sequence adjacent to the gene of interest are retrieved at the same time and so the method may be used for *chromosome walking*.

Evolution. The transformation of an organism in a way that descendants differ from their predecessors.

Evolutionary Clock. This is defined by rate at which mutations accumulate in a given gene. By comparing amino acid sequence of the resultant proteins it is possible to discover how much the proteins and the DNA of various species differ from each other. This information is used to estimate the rate of evolutionary change.

Evolutionary (Phylogenetic) Species Concept. In this species concept, emphasis is laid on different evolutionary lineages.

Exchanges Reaction. The reaction involved in polypeptide elongation where the amino acyl located at A site is translocated to P site on 50S subunit of the ribosome. This step involves two protein factors Tu and Ts in prokaryotes.

Excision. Excision of phage or episome or other sequence describes its release from the host chromosomes as an autonomous DNA molecule.

Excision (Xi) Protein. A protein required for excision of phage chromosome from bacterial chromosome through site-specific recombination.

Excision Repair. Repair of DNA lesions by removal of the damaged segment and replacement with a newly synthesised corrected segment.

Exclusion Principle. The principle stating that two species cannot coexist at the same locality if they have identical ecological requirements.

Exocytosis. A process responsible for export of proteins from the cell to the outside.

Exon. One of the coding regions of a discontinuous gene. The sequence present in a gene which is complementary to that present in its mRNA.

Exon Length. A computer analysis of large sized random DNA sequences revealed that an exon has an upper limit of 600 nucleotides in most of the eukaryotic genes.

Exon Shuffling. A theory that supposes econs to contain discrete subunits of genetic information that may be "shuffled" to produce new functional genes.

Exon Theory of Genes. Split genes arise not by insertion of introns into unsplit genes but from combinations of primordial "minigenes" separated by spacers.

Exonuclease. An enzyme that sequentially removes nucleotides from the ends of a nucleic acid molecule.

Expression-Linked Copy (ELC). A copy of the gene which is active.

Expression Vector. A hybrid vector (plasmid) that expresses its cloned genes.

Expressivity. The degree of phenotypic expression within one phenotype under a variety of environmental conditions.

Expressivity Modifiers. Those genes which are responsible for variable expressivity of a phenotype.

Extended Anticodon Hypothesis. The structure of anticodon loop and the proximal anticodon stem are related to the sequence of anticodon. Thus anticodon is extended to (i) two nucleoside at the 5' end of the anticodon, (ii) three nucleosides of the anticodon, and (iii) five pairs of nucleosides in the anticodon stem. Extended anticodons are involved in translation efficiency.

Extension. The single-stranded DNA tail found at the end of a restriction fragment. Different restriction endonucleases generate extensions at either the 5′ or 3′ end of the DNA strand.

Extrachromosomal Gene. Any gene that is not carried by the cell's chromosome(s). For example, genes present on mitochondrial or chloroplast genomes, genes carried by plasmids.

Extranuclear Genes. The genes which reside in organelles such as mitochondria and chloroplasts outside the nucleus.

Extranuclear Inheritance. It is characterised by differences in reciprocal crosses, progeny shows characters of the female parent, no linkage with the nuclear genes. See *cytoplasmic inheritance.*

Favoured Mutations. They are missense mutations which affect less critical sites of polypeptide chain such that it does not change basic functional property of the polypeptide. Such mutations may, however, affect some kinetic properties of the protein/enzyme and may be more efficient under one environment or the other. Such mutations are favoured by the process of evolution.

Feedback Inhibition. A post-translational control mechanism in which the end product of a biochemical pathway inhibits the activity of the first enzyme of its pathway. Also known as *end-product inhibition.*

Fertility (F) Factor. The fertility factor in the bacterium *Escherichia coli*; it is composed of DNA and must be present for a cell to function as a donor in conjugation. Its presence confers donor ability (maleness). Also see *sex factor.*

Fidelity of DNA Replication. Deals with various process responsible for faithful replication of DNA through generations of dividing cells. The processes recognised in this process are base selection, proofreading and mispair correction.

First Eukaryotic Cell. Time of appearance of first eukaryotic cell is somewhat 1.5×10^6 million years ago. In terms of their biochemistry and metabolism, cynobacteria seems to occupy a middle group between anaerobes and the eukaryotes, because cynobacteria can tolerate oxygen, make use of oxygen both metabolically in aerobic respiration and in synthetic pathways, and can make unsaturated fatty acids by oxidative disaturation.

First Living Cell. The first living cell was probably a heterotroph which lacked chlorophyll. The atmosphere was then strongly reducing. The first living cell was also anaerobic and slow growing. It was enucleate, has genetic code, RNA replicase which functioned both as genetic material and enzyme. The first cell was delimited by a permeable membrane which was capable of withstanding abuse.

Fixed Allele. An allele for which all members of the population under study are homozygous, so that no other allele for this locus exists in the population.

5-azacytidine. A cytosine analogue with a nitrogen atom replacing the carbon atom at position 5 of pyrimidine ring thus unable to accept a methyl group. A short treatment with it produces foci of differentiated cells in mammalian cell line by undermethylation of cytosine.

5′ P Terminus. The end of a polynucleotide that terminates with a mono-, di- or tri-phosphate attached to the 5′ carbon of the sugar.

Flip-flop Model of Mating Switch. Assumes that two mating type alleles are present, one on each side of a common regulatory site. Mating switch is brought about by recombination within the common regulatory site of the mating type locus.

Fixity-of Species Doctrine. An old doctrine that argues that species were specially created and do not undergo any changes over time.

Flanking Regions. Non-coding DNA sequences on both ends of a mature gene.

Floating Rearrangements. When a chromosomal rearrangement is present within a species, it is called floating or *polymorphic.*

f^{Met} (N-formylmethionine). The modified amino acid carried by the $tRNA_f^{met}$ that initiates translation in bacteria.

Fokker-Planck Equation. An equation that describes diffusion processes. It is used by population geneticists to describe random genetic drift.

Foldback DNA. It consists of inverted repeats that have renatured by intrastrand reassociation of denatured DNA.

Foldback Elements. When DNA from any eukaryote is denatured and allowed to reanneal at low concentration, the first sequences to become double stranded are those that occurs as inverted repeats. The repeats may be immediately adjacent or separated by up to several thousand base pairs, and as the reaction is intramolecular, the structures formed are also called "*snap-back*" *elements.*

Footprinting. A technique used to determine the length of nucleic acid in contact with a protein. While in contact, the free DNA is digested. The remaining DNA is then isolated and characterised.

Forbidden Mutations. The mutations which affect the function of a gene drastically such that the individual cannot tolerate them, e.g., *frameshift mutations.* Such mutations do not accompany the process of evolution.

Forces of Evolution. Mutation, recombination, migration, random genetic drift, founder effect, bottlenecks, assortative mating, selection, etc., are regarded as forces of evolution.

Forward Mutation. Any change away from the standard (wild-type).

Frameshift Intergenic Suppressors. A mutation in a RNA gene suppresses mutagenic effect of a frameshift mutation.

Frameshift (Phaseshift) Mutation. A mutation that results from insertion or deletion of a group of nucleotides that is not a multiple of three, and which therefore changes the frame in which the altered gene is translated.

Free Radical Theory of Ageing. Free radicals which have unpaired electrons and which increase in concentration with age, cause ageing.

Frequency-dependent Selection. A selection whereby a genotype is at an advantage when rare and at a disadvantage when common.

Frozen-accident Theory. A theory given by Crick to explain the relationship between an amino acid and its codon. There was a chance selection of code words and then freezing of the selection of code words. Once this association took place, it perpetuated as such. This theory can explain particulate nature of the codon but not pattern of degeneracy. This theory was modified. Accordingly, choice of first codon was by accident but then choice of synonymous codons is by selection. This explains degeneracy of code.

Fully-methylated DNA Site. When both complementary strands of DNA carry a methyl group at a site. Also known as *homoduplex modified* DNA.

Functional Alleles. Alleles determined on the basis of complementation test.

Functional Genomics. It refers to the development and application of global experimental approaches to assess the gene function by making use of the information and reagents provided by structural genomics.

Functional Genomics Tools. These tools include: expressed sequence TAG's, serial analysis of gene expression, microarrays or DNA/RNA chip technology, proteomics and bioinformatics.

Functional RNAs. Different type of RNA species, viz., tRNA, rRNA and mRNA, which perform specific functions during protein synthesis. Also known as *nongenetic* RNAs.

Fusion of Species. Complete breakdown of reproductive isolating barriers between species. Consequently, some well-defined species merge into one species.

Gap Genes. Mutations in these genes lead to deletion of a group of several adjacent segments from final pattern.

GAL4. A sequence specific DNA-binding protein that activates transcription in yeast.

Galactose Upstream Activating Sequence (UAS_G). A piece of DNA recognised by amino domain of yeast GAL4 regulator.

G Proteins. These are guanine nucleotide-binding trimeric proteins that reside in the plasma membrane. When bound by GDP the trimer remains intact and is inert. When the GDP bound to the a subunit is replaced by GTP, the a subunit is released from the bg dimer. One of the separated units (either the a monomer or the bg dimer) then activates or represses a target protein.

Gene. A segment of DNA that contains biological ingromation and codes for an RNA and/ or polypeptide molecule. Recognised through its variant forms which transmit specification(s) from one generation to the next.

Gene Amplification. A process or processes by which the cell increases the number of a particular gene within the genome.

Gene Augmentation. A number of copies of a normal gene for the defective gene are introduced into the cell and are made to express at high level.

Gene Battery Model. Britten-Davidson model of gene regulation in eukaryotes is also known as gene battery model. A set of structural genes controlled by one sensor is termed as a battery.

Gene Chip Technology. The technology which permits analysis of thousands of genes at once. For this cheap and simple customised chips capable of deconstructing long segments of DNA have been manufactured. This technique enable biologists to screen huge chunks of animal and plant genomes in search of the genes that promote disease, the genetic switches that govern such biological phenomena as ageing, and the DNA codes that permit microorganisms to make antibiotics.

Gene Cloning. Insertion of a fragment of DNA, containing a gene, into a cloning vector, and subsequent propagation of the recombinant DNA molecule in a host organism.

Gene Concept. Understanding gene in terms of its structure and function.

Gene Conversion. In Ascomycete fungi a 2:2 ratio of alleles is expected after meiosis, yet a 3:1 ratio is sometimes observed. The mechanism of gene conversion is explained by repair of heteroduplex DNA produced by the Holiday model of recombination.

Gene Delivery. In gene therapy, this term refers to putting the new gene into correct target cells.

Gene Dosage. The number of times an allele is present in a particular genotype.

Gene-environment Interaction. A gene produces a particular effect only in the presence of a particular environment.

Gene Evolution. New genes in the genome evolve through gene duplication and diversification. At first the function of the second copy of the gene is knocked out through a *forbidden mutation* and becomes a *pseudogene.* Then with the help of new mutations the duplicate gene assumes a new function.

Gene Expression. The process by which the biological information carried by a gene is released and made available to the cell, through transcription possibly followed by translation.

Gene Family. A group of genes that has arisen by duplication of an ancestral gene. The genes in the family may or may not have diverged from each other.

Gene Flow. The movement of genetic factors (genes) within a population or from one population to another, resulting from the dispersal of gametes (such as pollen by wind) or zygotes (as by migration of fertile females, for example).

Gene-for-gene Hypothesis. This hypothesis states that for every gene for resistance in the host there is a corresponding and specific gene for virulence in the pathogen. A disease fully develops only if the resistance gene in the host is overcome by a corresponding and specific virulence gene in the parasite.

Gene-gene Interaction. Interaction between products of different genes leads to modification in phenotype or phenotypic ratios.

Gene Inhibition Therapy. For treatment of dominant negative loss-of-function mutations, and gain-of-function mutations, *gene augmentation* therapy is less powerful. Additional functional copies of a dominantly malfunctioning genes are unlikely to affect the phenotype. In principle, a valid approach to the treatment of such disease would be targeted correction (i.e., allele replacement) or gene knockout to remove the mutant allele. Targeted inhibition of gene expression can also take effect at the protein level, by the expression within a cell of genetically engineered antibodies (intrabodies) which bind to and inactive mutant proteins.

Gene Library. A large collection of cloning vectors containing a complete set of fragments of the genome of an organism.

Gene Manipulation. The formation of new combination of heritable material by the insertion of nucleic acid molecules, produced by whatever means outside the cell, into any virus, bacterial plasmid, or any other vector system so as to allow their incorporation into a host organism in which they do not naturally occur, but in which they are capable of continued propagation.

Gene Patenting. It is a sort of protection provided by a government of a country to discoverer of a new gene, genotype, a genetic strain, a gene test or a genetic procedure so that the detailed information can be declared publically.

Gene Regulation Theory of Ageing. Sequential changes in the expression of genes occur from the beginning of the life span of an organism. Until the attainment of reproductive maturity and then ageing results.

Gene Regulatory Circuitries. Study of evolution of genetic regulatory systems that are responsible for differences in gene expression at different developmental stages, in different tissues of the organism and in different species.

Gene Regulatory Mechanism in Eukaryotes. Transcription of a *producer gene* occurs only if at least one of its *receptors* was activated by forming a sequence-specific complex with *activator* RNA. The activator RNA is synthesised by *integrator genes* in response to signals by sensor genes that are sensitive to external/development signals.

Gene Regulatory Mechanism in Prokaryotes. They are based on presence or absence of substrate (inducer), end product (co-pressor), or product of a regulatory gene (repressor). Based on whether control is inducible or repressible, or whether control is positive or negative, several models of gene regulation in prokaryotes are conceived. Basically, gene regulation involves interaction between RNA polymerase and promoter to initiate transcription, repressor and operator thus blocking passage of RNA polymerase, and repressor and inducer to inactivate the repressor.

Gene Sequencing. The technique of determining the order of bases of DNA molecule which constitute a gene.

Gene Silencing. A process in which an endogenous gene is expressed by introduction of a related transgene. This approach has been used in crop improvement.

Gene Safety. In gene therapy, this term is used to emphasise that new gene introduced into target cells of the patient should not harm the cells, or by extension the organism.

Gene Shears. A genetically engineered molecule which can seek and destroy unwanted genetic messages within a cell. It may play role in prevention of plant and animal diseases, and destruction of viruses which invade human, plant and animal cells.

Gene Splicing. The process of removal of introns and joining of exons from heterogeneous nuclear RNA (hnRNA) to produce mature functional mRNA.

Gene Squelching. A gene is squelched when a transcription factor is over expressed and not only activate all its target genes, but sequences also, by protein-protein interactions, components of the basic apparatus of genes, it does not normally regulate, causing global down regulation.

Gene Synthesis. Chemical synthesis of a gene involving blocking of certain reactive groups of nucleotides and joining the same one by one. No template is used. For using this method, nucleotide sequence of the gene to be synthesised needs to be known.

Gene Synthesis Machines. These are the automated machines which can rapidly synthesise oligonucleotides chemically. This method is based on development of silica-based supports and stable deoxyribonucleoside phosphoramidites called **synthons.** There machines work under the control of microprocessor.

Gene Targeting. The process by which specified changes are introduced into the nucleotide sequence of a chosen gene.

Gene Therapy. It includes development of methods for curing genetic disorders by replacing a defective gene with a normal gene.

Gene Transfer. The passage of a gene or group of genes from a donor to a recipient organism.

Genetic Analysis. Deals with analysis of a gene in term of its (a) nature, i.e., nuclear or cytoplasmic, dominant or recessive, autosomal or sex linked (if applicable), (b) location on a particular chromosome, (c) position on the chromosome, and (d) nucleotide sequence.

Genetic Assimilation. Certain environmentally acquired characters could, through selection, become assimilated in the genotype. Genetic assimilation has experimental support.

Genetic Block. In biochemical genetics, a block in a step in the synthesis or segregation of a biochemical molecule brought about by a mutant gene.

Genetic Code. The rules that determine which triplet of nucleotides codes for which amino acid during translation.

Genetic Counselling. This is telling the parents of a child who suffers from a genetic defect the probability that another child born to them will also have that defect.

Genetic Death. Complete elimination of an allele from a population. It is difficult to eliminate an allele. According to an estimate, it needs 30,000 generations for genetic death of an allele.

Genetic Dissection. The use of recombination and mutation to piece together the various components of a given biological function.

Genetic Distance. The average number of electrophoretically detectable codon substitutions per gene that have accumulated in the population studied since they diverged from a common ancestor. There exists a clear-cut relationship between reproductive isolation and genetic distance.

Genetic Drift. Random change (in any direction) in gene frequencies due to sampling error.

Genetic Engineering. The use of experimental techniques to produce DNA molecules containing new genes or new combination of genes. It deals with isolation, synthesis, adding, removing or replacing genes in order to achieve permanent and heritable changes in diverse forms of life bypassing all reproductive barriers.

Genetic Fine Structure. The structure of the gene analysed at the level of the smallest units of recombination and mutation (nucleotides). Benzer used *cis-trans* complementation test to study fine structure of the gene.

Genetic Homeostasis. The tendency of populations under selection to regress toward the original mean.

Genetic Identity. The proportion of genes the products of which are not distinguishable by their electrophoretic behaviour.

Genetic Instability. In some instances genes mutate and back mutate with a very high frequency. Certain controlling elements are considered to be responsible for it.

Genetic Isolation. Isolation of a variety or species by genetic means, e.g., cross sterility in corn and cross incompatibility of amphidiploid species with parents or other species.

Genetic Marker. An allele whose phenotype is easily recognisable and which can, therefore, be used to follow the inheritance of its gene during a genetic cross.

Genetic Material. The chemical material of which genes are made, now known to be DNA in most organisms, RNA in a few.

Genetic Medicines. It is use of snippets of genetic material which can block the expression of defective genes. These blocks may be at transcription level (triplex approach) or at translation level (antisense approach). These techniques are yet to be perfected.

Genetic Mutagen. A gene or a DNA sequence that causes other genes to mutate, e.g., the *AC-Ds system* or the *Dt gene* in corn.

Genetic Polymorphism. The continued occurrence in a population of two or more discontinuous genetic variants in frequencies that cannot be accounted for by recurrent mutation.

Genetic Privacy. It deals with idea of regulating the collection, maintenance, use and dissemination of genetic information collected from individuals.

Genetic Prognosis. It illustrates some basic principles underlying genetic predictions in terms of probability of a particular genetic disorder to appear in an individual.

Genetic Risk Assessment. Quantitative estimate of probable impact on the gene pool of subsequent generations from a specific mutagenic exposure.

Genetic RNA. The RNA which acts as DNA, or the genetic material.

Genetic Screening. The testing of individuals for a particular genetic trait.

Genetic Switches. Different control mechanisms acts as switches which regulate different developmental phases of an organism.

Genetic Toxicology. A subdiscipline of toxicology which identifies and analyses the action of agents with toxicity directed towards the genetic material of the living systems.

Genetic Variance. The phenotypic variance due to the presence of different genotypes in the population.

Genetically-modified Food. By using genetic engineering techniques, it has been possible to have foods whose nutritional values are enhanced. Consumers demand safety of the genetically modified food. Also known as *genetic food.*

Genetics. A branch of life sciences that precisely understands the nature, structure, organisation, function, regulation, manipulation of the hereditary units called genes that are responsible for carrying out different life process and for transmission of biological properties from parents to offspring. Genetics also deals with finding out ways and means to use knowledge of genetic for welfare of mankind.

Genic Balance Theory. The theory of Bridges that the sex of a fruit-fly is determined by the relative number of X chromosomes and autosomal sets (X/A).

Genic Disharmony. When genes contributed by two parents fail to work together in hybrid or its progeny. It can be expressed at various stages of development and is associated with the process of nuclear metabolism, including DNA replication and transcription. Such individuals are at a selective disadvantage. It serves as a post-zygotic reproductive isolating mechanism.

Genic Explanation of Hybrid Vigour. Hybrid vigour is widely explained as the result of a dominance of linked genes. Many dominant genes are involved. Different inbreeds contribute different dominant genes, so that the F_1 has more than either parents.

Genic and Plasmagenic Interaction. Interaction between a plasma gene and a gene in the chromosome to produce an effect. For example, in *Paramecium*, plasmagene *kappa* and nuclear dominant gene *K* interact to produce a killer phenotype.

Genome. The entire genetic complement of a cell.

Genome Cutter. A peptide nucleic acid (PNA) "clamp", consisting of two predesigned 8-bp sequences of PNA-linked together, can bind strongly and sequence specifically to DNA, hence known as bis-PNA, protecting the binding region from methylation. After removal of this bis-PNA, methylated DNA can be cut quantitatively at restriction sites selected by the PNA clamp.

Genome Imprinting. A process that temporarily and erasably marks the genes passed on by females and males in different ways. Offspring that receives marked genes from their mothers are consequently different from those that receive the genes from their fathers.

Genome Organisation. The arrangement of the genetic material in the cells of an organism.

Genome Resetting Model. This model postulates that at least some of the DNA involved in species evolution should be functional. In particular, these sequences must be regulatory elements whose positions and sequences influence the development pathways in a coordinated manner.

Genomic Disease Model. This model postulates that at times an isolated population loses or fails to acquire a transposable element that otherwise has spread throughout the population, and as a consequence these isolated individuals on mating with the population lead to abnormalities that end up with sterility in F_1 and F_2 progenies, thereby establishing postzygotic reproductive isolation. And to avoid wastage of gametes, natural selection might then act to establish some prezygotic barriers. There are enough evidences in support of this model, especially in *Drosophila*.

Genomic Library. A collection of clones sufficing in number to include all the genes of a particular organism.

Genomics. It is a discipline of mapping sequencing and analysing genomes.

Genomic Tools. Genomic tools include genomic database, microorganisms, proteomics, etc.

Genophore. The chromosome (genetic material) of prokaryotes and virus.

Genotype-environmental Interaction. The interplay of a specific genotype and a specific environment, affecting the phenotype. The extent and nature of this interaction varies with each genotype and environment. The variance due to this interaction (V_1) is part of the total phenotypic variance of a trait.

Germ Line Sexual Development. A process that determines germ line sex of an individual.

Germ Line Theory of Antibody Diversity. According to this theory, the genome contains thousands of related but unique DNA sequences, each of which specifies one of the *V'* regions synthesised by an organism.

Gonadal Dysgenesis. A condition where the gonads in an individual are not clearly differentiated into a male or female sex. This condition is due to an X-linked recessive allele which does not allow differentiation of gonads to tests in 46 XY embryos.

Group I Introns. Self-splicing introns that do not require an external nucleotide for splicing. The intron is released in a lariat form.

Group II Introns. Introns that are not self-splicing. They need small guide RNAs or U RNAs for their splicing.

Group III Introns. Introns that have conserved sequences at their borders (GTGCGNY at 5′ end, and ATCHRYY(N)YYAY at 3′ end), similar to those in the eukaryotic nuclear genes. For example, *trnG, trnK, trnV, rpl2, rps12, rps16* have group III introns.

GT-AG Rule. Refers to the fact that with introns present in nuclear protein-coding genes, the first two nucleotides of the intron are GT and the last two AG.

Guide RNA (gRNA). An RNA molecule (or a part of one) which hybridises to eukaryotic mRNA and aids in the splicing of intron sequences. Guide sequences may be either external (EGS) or internal (IGS) to the mRNA being processed and may hybridise to either intron or exon sequences close to the splice junction.

Gyrase. An enzyme that relaxes supercoiling caused by unwinding of double helix of DNA.

Hairpin Loop. A double-helical region formed by base pairing between adjacent (inverted) complementary sequences in a single strand of RNA or DNA.

Half-chromatid Conversion. A type of gene conversion that is inferred from the existence of nonidentical sister spores in a fungal octad showing a non-Mendelian allele ratio.

Half-methylated Site. When only one strand of DNA carries a methyl group whereas the other complementary strand does not do so at a site. Also known as **heteroduplex modified DNA**.

Haplotype. It is the particular combination of alleles in a defined region of some chromosome, in effect the genotype in miniature. Originally used to described combinations of MHC alleles, it now may be used to describe particular combinations of RFLPs.

Hapten. A small molecule that acts as an antigen when conjugated to a protein.

HAT Medium. A medium containing hypoxanthine, aminopterin and thymidine which is used to select somatic hybrid cells.

Hayflick Limit. The observation that cells have a specific number of cell divisions that they can go through.

Heavy (H) Chain. One of the polypeptide chains present in every immunoglobulin molecule.

Heat-killed Viruses. These are used as vaccines because heat destroys pathogenic of virus but heat may even destroy antigenic property of the virus.

Heat Mutagenesis. Induction of mutations at above melting temperature by strand separation, depurination and chain breakage.

Heat-shock Elements. Two copies of a consensus sequence CNNGAANNTTCNNG present in *Drosophila* heat-shock protein genes, *hsp70*, required for its maximal transcription. Heat-shock transcription factors bind to these elements.

Heat-shock Genes. The genes which synthesise heat-shock proteins in response to heat shocks.

Heavy Chain Immunoglubulin Genes. Genes that determine heavy chain of immunoglubulin molecules.

Helicase. The enzyme responsible, during DNA replication, for breaking the hydrogen bonds that hold the double helix together.

Helix-turn-helix Motif. Configuration found in DNA-binding proteins consisting of a recognition helix and a stabilising helix separated by short turn.

Hermaphroditism. A condition in which an individual has both male and female sexes.

Heteroallele. The allele which arise due to base pair replacement in different mutons; those alleles between which recombination is theoretically possible.

Heterochromatic. Entire chromosomes or portions of the chromosomes that do not manifest the usual prophase-telephase transformations and appear to lack genes with major phenotypic effects.

Heterochromatin. The regions of a chromosome that appear relatively condensed and stain deeply with DNA-specific stains.

Heterocysts. In cynobacteria, nitrogenase enzymes are protected in specialised cells called heterocysts.

Heterodimer. A dimeric protein having two different polypeptides.

Heteroduplex. A base-paired structure formed between two polynucleotides that are not entirely complementary.

Heteroduplex DNA Model. A model that explains both crossing-over and gene conversion by assuming the production of a short stretch of heteroduplex DNA (formed from both parental DNAs) in the vicinity of a chiasma.

Heterogametic Sex. Sex producing gametes of two kinds with regard to sex determination.

Heterogeneous Nuclear RNA (hnRNA). The nuclear RNA fraction that comprises the unprocessed transcripts synthesised by RNA polymerase II.

Heterosis. (1) The increased vigour, growth, size, yield or function of a hybrid progeny over the parents that results from crossing genetically unlike organisms; (2) The increase in vigour of growth of a hybrid progeny in relation to the average of the parents.

Heterozygote Advantage. A selection model in which heterozygotes have the highest fitness.

High Density Molecular Maps. Morphological markers useful for constructing linkage maps are only in limited numbers. The molecular markers on the other hand are numerous and they provide capacity for complete coverage of nuclear, mitochondrial and chloroplast genomes. The saturated maps prepared with the help of molecular markers are known as high density molecular maps.

High frequency recombination (Hfr) Cell. A bacterium whose DNA molecule contains an intergrated copy of the F plasmid.

High Precision DNA Transactions. These are those genetic processors which involve interaction between DNA and proteins and between proteins and proteins. Such interactions are involved in site-specific recombination, initiation of DNA replication, initiation of transcription and regulation of gene action.

Highly Repetitive Sequence. When number of copies of a sequence in eukaryotic DNA is 10^6 to 10^7. These repeats are usually found in centromeric regions.

Histocompatibility. This refers to tissue compatibility. Tissues can be transplanted between genetically identical individuals without concern for immunological rejection whereas transplants between genetically non-identical individuals are usually rejected with time.

Histocompatibility Antigens. Antigens that determine the acceptance or rejection of a tissue graft.

Histocompatibility Genes. The genes that code for the histocompatibility antigens.

Histone. One of the basic proteins that make up nuclesomes and have a fundamental role in chromosome structure.

Histones as Genes Regulators. Histones not only are important basic proteins which impart structure to eukaryotic chromosome but also play important role in cell cycle of undergoing reversible modifications, such as acetylation of lysine, phosphorylation of serine and threonine and ubiquination of lysine.

Histone-dependent Activation. This is the first stage in activation of genes in which activator proteins at the upstream activator sequence (UAS) directly or indirectly cause the histone core particles to dissociate from TATA box and form a pre-initiation complex and thereby generate a base level of transcription.

Histone-independent Activation. This is the second stage in activation of genes in which activators stimulate the pre-initiation complex to maximise production of mRNA.

HU (Histone-like) Proteins. The proteins required to stimulate reaction leading to DNA replication in *E. coli*.

HLA Complex. Human leukocyte antigens complex comprises genes that control the synthesis of HLAs that are located on the surfaces of the leukocytes and which affect tissue compatibility in organ transplants and skin grafting. These antigens occur in quite a large number of types and are produced by at least four closely linked loci (in the sequence DBCA) an autosome 6.

Hogness Box. A nucleotide sequence that makes up part up of the eukaryotic promoter. Same as **TATA box.**

Holoenzyme. The version of the *E. coli* RNA polymerase that has the subunit composition $a_2bb's$ and is involved in efficient recognition of promoter sequences.

Homeo Box. A highly conserved segment of DNA found in all homeotic genes. This sequence may code for the part of a regulatory protein that binds to DNA.

Homeo Domain. The nucleotides of homeo box are translated into sixty amino acid peptide region called the homeo domain.

Homeotic Genes. The genes defined by mutations that convert one body part into another; for example, an insect leg may replace an antenna.

Homeotic Mutations. Mutations in which one developmental pattern is replaced by another. The genes in which the mutations occur appear to be involved in regulation of developmental patterns possibly by coding for of one flower whorl into a whorl of a different type.

Homoallele. The allele which arises due to base pair replacement in the same muton; recombination is theoretically not possible between homoalleles.

Homochromatography. A process in which solvent moves up the chromatagraphy medium separating the labelled oligonucleotide fragments according to the length. Shorter the fragment, faster the migration.

Homodimer. A dimeric protein having two identical polypeptides.

Homoduplex Modified DNA. DNA having methyl group on a particular adenine or cytosine of both strands included in a host specificity site. Also known as *fully methylated* DNA.

Homoduplex Unmodified DNA. DNA having no methyl group on a particular adenine or cytosine on both strands included in a host specificity site. Also known as *unmethylated* DNA.

Homosequential Species. The group of related species having identical gene arrangements in all their chromosomes as indicated by their banding patterns.

Homozygous Typing Cells (HTC). A technique used to type cells of an individual on the basis of class II HLA molecules. Lymphocytes of the individual to be typed are stimulated with a panel of cells each of which is homozygous for one of the known alleles. The individual carrying one or both alleles identical with a particular allele will not react with an individual. If an individual's cells respond to all known cells, it must carry a new allele.

Horizontal Resistance. The type of resistance which is governed by "minor" genes or polygenes which act additively together to control resistance which is non-specific in nature and does not reveal gene-for-gene hypothesis.

Hormone-receptor Protein Complexes. Specific receptor proteins which activate the transcription of specific genes.

Host-cell Reactivation. The repair of damaged DNA of viruses by host-cell repair mechanisms.

Host-controlled Modification. After DNA replication, the host methylates its own DNA with the help of its methylase enzyme. This enzyme is composed of polypeptide products of *hsdM* and *hsdS* genes in ratio 2:1.

Host-controlled Restriction. DNA synthesised in one cytoplasm that undergoes cleavage when it enters the host cytoplasm. The restriction endonuclease is made up of 3 different polypeptides, in ratio 2:2:1, synthesised by three genes, *hsdR*, *hsdM* and *hsdS*, respectively.

Host Genes. These are the genes that are responsible for resistance to diseases in the host. They are usually dominant. Resistance could be monogenic, oligogenic or polygenic depending upon whether it is controlled by one, a few or many genes. On the basis of effectiveness, resistance may be classified into *vertical* and *horizontal*.

Host-mediated Assay. A method in which microorganisms are introduced into a host organism (such as rat). Then the host is treated with the chemical whose mutagenicity is to be tested. After a period of time, the microorganisms are recovered and checked for mutations.

Host-parasite Specificity. Refers to ability of a pathogen to pathogenise distinct group of plants. It is known to operate at genus, species and genotypic level.

Host Specificity (*hsd*) Locus. It synthesises enzyme having methylase and/or restriction activity. It is composed of three genes in the order *hsdR* (responsible for restriction), *hsdM* (responsible for modification) and *hsdS* (required for recognition of host specificity site).

Host specificity (*hsd*) Promoters. The *hsd* locus has two promoters, P_1 and P_2. P_1 is located upstream of the first gene *hsdR* and is responsible for transcription of *hsdM*, hsd*R* and *hsdS*. P_2 is located upstream of the second gene *hsdM* and controls transcription of *hsdM* and *hsdS* genes.

Host Specificity Sites. The sites in host DNA where recognition and methylation takes place. The cleavage may take place elsewhere. Thus restriction enzymes recognises two sites.

Hot Dilute Soup. Belief that early earth was hot and first life originated in a hot solution containing simple organic compounds. Now this concept is changed to *cold dilute soup*.

Hotspot. It is a site at which the frequency of mutation (or recombination) is very much increased.

Housekeeping Genes. The genes whose products are required by the cell at all time, their activity is controlled by constitutive factors. Also see *constitutive genes*.

***hsdM* Gene.** The *E. coli* gene which encodes the DNA methylase involved in the modification system which prevents its DNA being attacked by its own restriction endonucleases. Host strains in gene cloning experiments usually have a functional hsdM gene facilitate gene transfers.

***hsdR* Gene.** The *E. coli* gene which encodes restriction endonuclease activity. The initials stand for host-specified defence restriction. Host strains for gene cloning experiments are often hsdR mutants so that the foreign DNA carried by the vector is not attacked by the host's restriction system.

***hsdS* Gene.** The *E. coli* gene which encodes host-specificity site recognition activity. The initials stand for host specified defence restriction. Host strains mutant for this gene produce enzyme that does not recognise host specificity site on DNA.

Humoral Immune Response. A type of immune response in which antigen stimulates the production of antibodies which react with specific antigens so as to destroy them.

H-Y Antigen. A histocompatibility factor determined by the Y chromosome. It is thought to be the major male-determining factor in mammals. In humans, the gene for the H-Y antigen is on the short arm of the Y chromosome.

Hybrid-arrested Translation (HART). A method used to identify the polypeptide coded by a cloned gene.

Hybrid Breakdown. Production of weak or sterile F_2 progeny by vigorous fertile F_1 hybrids.

Hybrid DNA. DNA whose two strands have different origins.

Hybrid Dysgenesis. The inability of certain strains of *D. melanogaster* to interbreed, because the hybrids are sterile (although otherwise they may be phenotypically normal).

Hybrid Inferiority. Hybrid is inferior to either of the parents.

Hybridisation *in situ*. Finding the location of a gene by adding specific radioactive probes for the gene and detecting the location of the radioactivity on the chromosome after hybridisation.

Hybridisation Kinetics. It is the speed with which a probe binds with complementary strands. It gives an estimate of number of copies of a sequence present.

Hybridisation Probing. A method that uses a labelled nucleic acid molecule to identify complementary or homologous molecules through the formation of stable base-paired hybrids.

Hybridoma. A cell hybrid between an antibody-producing B cell and a tumor cell, which divides indefinitely and produces a single antibody (monoclonal antibody) in culture.

Hybrid Plasmid. A plasmid that contains an inserted piece of foreign DNA.

Hybrid-released Translation (HRT). A method used to identify the polypeptide coded by a cloned gene.

Hybrid Resistance. The rejection of parental tissue grafts by the F_1 hybrids of two inbred strains.

Hybrid Screening. Radioisotope technique used to determine whether a hybrid plasmid contains a particular gene or DNA region.

Hybrid Vehicle (Vector). An episome or plasmid containing an inserted piece of foreign DNA.

Hybrid Vigour. Superiority of hybrid over the better parent in one or more traits. This is also known as *positive heterosis.*

Hydroxyapatite. A form of calcium phosphate that binds double-stranded DNA.

Hyperchromacity. The increase in optical density that occurs when DNA is denatured.

Hypermorph. An allele having an effect similar to but greater than that of the wild form.

Hypersensitive Response (HR). Also known as *systematic acquired resistance* (SAR). This is a non-host resistance which plant possesses to defend itself.

Hypersensitive Site for DNase I. A short region of chromatin detected by its extreme sensitivity to cleavage by DNAase I and other nucleases; probably comprises an area from which nucleosomes are excluded.

Hypervariable Loci. Loci with many alleles; especially those whose variation is due to variable members of tandem repeats.

Hypostatic Genes. The genes whose expression is suppressed by other genes.

Idiogram. A photograph or diagram of the chromosomes of a cell arranged in an orderly fashion. See *karyotype.*

Idiotypic Variation. Variation in the variable parts of immunoglobulin genes.

Idiotypic Reaction. The production of guanosine tetraphosphate (3′-ppGpp-5′) by the stringent factor when a ribosome encounters an uncharged tRNA in the A-site.

Illegitimate Recombination. Recombination between two double-stranded DNA molecules which have little orders of structure of many biomolecules.

Immunogenetics. Study of inheritance of those molecules which can trigger immune response, i.e., study of inheritance of antigens and antibodies.

Immunoglobulin (Ig). One of several types of antibodies secreted by derivatives of B lymphocytes (plasma cells) that protect an organism from an antigen.

Immunoglobulin Genes. The complex of genelets responsible for synthesis of immunoglubulins. These genes undergo DNA rearrangement at different levels before DNA acts as a template for transcription.

Incompatibility of Plasmids. The inability of certain types of plasmids to co-exist in the same cell.

Indirect and Labelling. It is a technique for examining the organisation of DNA by making a cut at specific site and isolating all fragments containing the sequence adjacent to one side of the cut; it reveals the distance from the cut to the next break(s) in DNA.

Induced Chromosome Break. Chromosome break caused by some agent usually external to the chromosome, such as radiations or chemicals. It can also be produced genetically, e.g., by the *Ac-Ds system* in maize.

Induced Mutations. Genetic changes produced by some physical or chemical agent or under changed growth conditions.

Inducer. A molecule that induces expression of a gene or operon by binding to a *repressor* protein and thereby preventing the repressor from attaching to its *operator* site. *Effectors* in inducible operons are known as *inducers.*

Inducible Control. In this case substrate acts to induce the production of the enzymes. Repression occurs in the absence of the substrate.

Induction. A chemical or physical treatment which results in excision from the host genome of the integrated form of a lysogenic phage, followed by the switch to the lytic mode of infection.

Industrial Melanism. Industrial melanism is the darkening of moths during the industrial revolution and is an interesting case study of natural selection. It illustrates the type of selection that can be caused by human intervention in natural systems. It is a spectacular example of very rapid evolution in nature.

Inert Gene. A gene which through apparent lack of mutation or effect in disturbing balance has been assumed to be inactive. Often associated with heterochromatin.

Informational Family. It has individual members that can differ markedly in sequence from one another although all are homologous and obviously show an ancient ancestry.

Informosome. A complex of mRNA and protein that protects mRNA from degradation.

Inhibitory Epistasis. One gene when dominant is epistatic to other but other when recessive is epistatic to the first. These two genes give 13:3 ratio is F_2 progeny.

Initiation Codon. The codon, usually but not exclusively 5′-AUG-3′, which indicates the point at which translation of an mRNA should begin.

Initiation Complex. The complex which comprises mRNA, a small ribosomal subunit, aminoacylated initiator-tRNA and initiation factors and which forms during the initiation stage of translation.

Initiation Factors. Protein molecules that play an ancillary role in the initiation stage of translation (IF1, IF2, IF3).

Initiation Site. The site on the DNA molecule where the synthesis of RNA begins.

Inosine. A newly discovered nucleotide which is found in third position in an anticodon (on tRNA) and can pair with A, U and C resulting in wobble base pairing. It is a deamination product of adenosine.

Insertional Inactivation. The technique in which foreign DNA is cloned into a restriction site which lies within the coding sequence of a gene in the vector. The insertion of foreign DNA at such a site interrupts the gene's sequence such that its original function is no longer expressed. This permits the detection of recombinant molecules following transformation.

Insertional Translocation. The insertion of a segment from one chromosome into another nonhomologous one.

Insertion Elements. Sequence of DNA that specify the location of insertion of different episomes into the genome.

Insertion Mutagenesis. Change in gene action due to an insertion event that either changes a gene directly or disrupts control mechanisms.

Insertion Sequences (IS elements). Small, simple *transposons*. See *transposable genetic element.*

***In situ* Hybridisation.** Hybridisation performed by denaturing the DNA of cells squashed on a microscope slide so that reaction is possible with an added single-stranded RNA or DNA; the added preparation is radioactively leballed and its hybridisation is followed by autoradiography.

Intasome. Nucleoprotein structure formed by l phage encoded integration (Int) protein and attP site required for pairing with *E. coli* attB site and site-specific recombination.

Integration Host Factor (IHF). A protein required for precise formation of an intasome. It is required both for integration and excision through site-specific recombination.

Integration (Int) Protein. Phage protein which binds with attP site on DNA to form a nucleoprotein structure called *intasome* necessary for integration and excision through site-specific recombination with bacterial chromosome.

Integrator Gene. A component of eukaryotic gene regulation which responds to the signals provided by *sensor gene.* These genes are transcribed to produce *activator* RNAs.

Interallelic Complementation. The change in the properties of a heteromultimeric protein brought about by the interaction of subunits coded by two different mutant alleles; the mixed protein may be more or less active that the protein consisting of subunits only of one or the other type.

Interbands. The region of chromosomes which are lightly stained. These regions mostly have unique DNA sequences.

Intercalation. Certain drugs or dyes, such as ethidium bromide, are able to insert into DNA or double-stranded RNA between adjacent base pairs. Molecules with this property are called *intercalating agents.* The binding of such dyes reduces the buoyant density of the DNA. The DNA duplex increases in length and, if the DNA is supercoiled, increasing concentrations of the dye first unwind the supercoils and then wind the molecule up again in the opposite sense.

Intercistronic Region. The distance between the termination codon of one gene and the initiation codon of the next gene. Also known as *intergenic spacers.*

Interferons. A heterogenous family of multifunctional cytokines that were originally identified as proteins responsible for the induction of cellular resistance to viral infection through cellular metabolic processes.

Interferon Regulatory Factors. These are the molecules that regulate the expression of the interferon genes by activation of kinases which phosphorylate substrate proteins called *signal transducers* and *activators of transcription* (STATs).

Intergenic (Extragenic) Suppression. A mutation at a second locus that apparently restores the wild-type phenotype to a mutant at a first locus. Intergenic suppressors are of 3 types, viz., *nonsense, missense and frameshift.*

Internal Control Signals. In this case, promoter sequences are located within a mature gene itself, e.g., genes coding for small pieces of RNA - tRNAs, 5S rRNA, etc., have internal control signals.

Internal Guide Sequence (IGS). In group I introns, a sequence pairing with exon sequences adjacent to both the 5' and 3' splice sites. The IGS usually begins with twenty nucleotide of the 5' splice site. IGS 5' exon pairing is important in both steps of RNA splicing.

Intragenic Suppression. When effect of a mutation in one gene is completely or partially suppressed by another mutation in the same gene.

Intron. A segment of DNA that is transcribed, but removed from within the transcript by splicing together the sequences (exons) on either side of it.

Invagination Theory. Given by Kaff and Mahler (1972). A prokaryotic cell having a flexible membrane evolves extensive invaginations that are derived from internal membranes of cell. Enzymes responsible for glycolysis and respiration became associated with these invaginations. The invaginated membranes lose contact with cell membranes and become converted into mitochondria.

Inversion. Alternation of the sequence of a DNA molecule by removal of a segment followed by its reinsertion in reverse orientation.

Inverted Repeats. The repeats that comprise two copies of the same sequence of DNA repeated in opposite orientation on the same molecule. Adjacent inverted repeats constitute a *palindrome.*

***In vitro* Mutagenesis.** Any one of several techniques used to produce a specified mutation at a predetermined position in a DNA molecule.

***In vitro* Transcription.** The specific, an accurate synthesis of RNA in the test tube on purified DNA preparations as template, so-called "*coupled system*" may be obtained from *E. coli* which carry out both mRNA synthesis and its translation into protein. For eukaryotes, separate cell-free systems have to be set up to demonstrate the activity of the three functionally distinct RNA polymerase complexes.

***In vitro* Translation.** The synthesis of proteins in the test tube from purified mRNA molecules using cell extracts containing ribosomal subunits, the necessary protein factors., tRNA molecules and aminoacyl tRNA synthetases. ATP, GTP, amino acids and an enzyme system for regenerating the nucleoside triphosphates are added to the mix. Prokaryotic translation systems are usually prepared from *E. coli* or the thermophilic bacterium *Bacillus stearothermophilus*. Eukaryotic systems usually employ rabbit reticulocyte lysates or wheat germ.

I-R System. This is a system of hybrid by dysgenesis in which when a female from a *reactive (R) strain mates* with a male from an *inducer (I) strain*, specific female (SF) sterility is induced. The reactive and inducer strains can be arranged from "strong" to "weak". Stronger the paternal strain, higher is the reduction of female sterility. Reactivity corresponds to a cytoplasmic state and inducer is a factor chromosomal in nature.

Isoacceptor tRNAs. Two or more tRNA molecules that are specific for the same amino acid.

Isoalleles. Alleles with similar phenotypic effect.

Isochromosome. A chromosome with two genetically and morphologically identical arms.

Isolating Mechanisms. Any structural, physiological, behavioural, or other features of an individual, or any geographical or geological barrier, that prevents individuals of one population from successfully interbreeding with those of other populations.

Isozymes. Different forms of the same enzyme which catalyse particular biochemical reactions during metabolism.

Jacob and Monod Model. The operon model of gene regulation in *lac* region of *E. coli.* This model explains mechanism of gene expression in prokaryotes.

Junctional Diversity. Variability in immunoglobulins caused by variation in the exact crossover point during V-J, V-D, and D-J joining.

Junk DNA. Same as *Selfish* DNA.

Jumping Gene. The gene which keeps on changing its position in a chromosome and also between the chromosomes in a genome.

Kappa Particle. A particle present in the cytoplasm of some Paramecia (killers) that secretes a substance that kills Paramecia not possessing the kappa particle. They contain their own DNA of length 0.4 microns.

Kinase. An enzyme which will remove or add a phosphate group to a protein or nucleic acid.

Kinetoplast. These are modified mitochondria located near to the base of flagella. A kinetoplast may be larger than a nucleus.

Kinetoplast DNA. DNA present in kinetoplasts. This DNA exists in form of a large number of interlocked "minicircles".

Kin Selection. The mode of natural selection that acts on an individual's inclusive fitness.

Klinefelter's Syndrome. A genetic disease in man due to the XXY karyotype. It produces sterile males with some mental retardation.

Konzak's Scanning Hypothesis. It pertains to translation initiation in eukaryotes. Accordingly, 40S ribosomal subunit associated with met-tRNA moves down the mRNA until it encounters the first AUG which in 90% of the cases occurs in form of consensus sequence PuNNAUGG, known as *Konzak's consensus sequence.*

***Lac* Operon.** The cluster of three structural gene that code for enzymes involved in utilisation of lactose by *E. coli.*

Lactose Repressor. The regulatory protein that controls transcription of the *lac* operon in response to the levels of lactose in the environment.

Lagging Strand. The strand of the double hilix which, during DNA replication, is synthesised in a discontinuous fashion in form of *Okazaki fragments.*

l protein. A phage protein required to localise replication proteins.

Lampbrush chromosome. A chromosome that has paired loops extending laterally, and occurs in primary oocyte nuclei; they represent sites of active RNA synthesis.

Late Genes. Bacteriophage genes that are expressed during the later stage of the infection cycle. Late genes usually code for proteins needed for synthesis of new phage particles.

Leader Peptidase. An enzyme used to cleave leader peptide from the preprotein to be secreted in bacterial cells.

Leader Peptide. In bacteria proteins destined to be secreted are synthesised as preproteins with N-terminal signal sequences sometimes termed leader peptides.

Leader Peptide Gene. A small gene within the attenuator control region of repressible amino acid operons. Translation of the gene tests the concentration of amino acids in the cell.

Leader Transcript. The untranslated segment of mRNA that lies upstream of the initiation codon on an mRNA molecule.

Leader Segment. The untranslated segment that lies upstream of the initiation codon on an mRNA molecule.

Leading Strand. The strand of the double helix which, during DNA replication, is copied in a continuous fashion.

Left Splicing Junction. The boundary between the right end of an exon and the left end of an intron.

Leucine Zipper. A structural motif found in several DNA-binding proteins.

Life. Any entity that is capable of making a reasonable accurate reproduction of itself and duplicate being able to produce the same task and subjected to low rate of alteration and these changes being heritable.

Light (L) Chain. One of the polypeptide chains present in every immunoglobulin molecule. There are two types of light chains: kappa and lambda.

Light (L) Chain Immunoglobulin Genes. Genes that determine light chain of immunoglobulin molecules. Two types of light chain genes known are *l* and *k*.

Linker DNA. The DNA that links nucleosomes together and which makes up the 'string' in the *beads-on-a-string model* for chromatin structure.

Lipsome-mediated Gene Transfer. A method of gene transfer which involves encasement of DNA in lipid bags, which can be made to fuse with protoplasts with the help of a peg. This technique has been used successfully for gene transfer in a number of plant species.

Live Vaccines. These are the vaccines which use large DNA viruses, such as *Vaccinia* virus as a biological delivery causing agent.

LOD Score Method. A technique (*logarithmic odds*) for determining the most likely recombination frequency between two loci from pedigree data.

Long Interspersed Nucleotide Element (LINE). Any of the long, (6 to 7 kb) repetitious (about 10^4 times), interspersed DNA sequence elements in nuclear genomes, the majority of which are retroposons. LINEs contain one or more open reading frames.

Long-term Gene Regulation. Gene regulation recognised in eukaryotes which operated during determination, differentiation, or more generally development.

Long Terminal Repeat (LTR). A sequence directly repeated at both ends of retroviral DNA.

Luxury Genes. The genes coding for specialised functions; their products are synthesised (usually) in large amounts in particular cell types.

Lyon Hypothesis. The hypothesis which states that in any given somatic cell of a female one X chromosome is active and the other is inactive.

Lysis. The bursting of a cell by the destruction of the cell membrane following infection by a virus.

Lysogeny. A state in which the genetic material of a virus and its bacterial host are integrated.

Maintenance Methylase. An enzyme which methylates a half-methylated site on DNA obtained after replication thus converting it into a fully-methylated site. This enzyme does not methylate unmethylated site.

Major Gene. A gene that may cause sufficiently large variation in the trait studied to be easily detected.

Major Groove. The larger of the two grooves that spiral around the surface of the double helix model for DNA.

Major Histocompatibility Antigens. These antigens act with high intensity and reject a graft fast.

Major Histocompatibility Complex. A group of highly polymorphic genes whose products appear on the surface of cells imparting to them the property of "self" (belonging to that organism). Some other functions are also involved.

Male Recombination Factors (MRFs). Genetic factors responsible for spontaneous recombination in males of *Drosophila melanogaster* where this phenomenon is normally known to be absent.

MAPMAKER. A computer package developed for detection and estimation of linkages. This package is based on the maximum likelihood method and is mostly used for construction of RFLP maps. MAPMAKER/EXP performs full multipoint linkage analysis for dominant, recessive and co-dominant (e.g., RFLP-like) markers in BC_1 backcrosses, F_2 and F_3 (self) intercrosses and recombination inbred lines. MAPMAKER/QTL is a companion programme to MAPMAKER/EXP which allows one to map genes controlling polygenic quantitative traits in F_2 intercrosses and BC_1 backcrosses relative to a genetic linkage map.

Marker. A locus or allele whose phenotype provides information about a chromosome or chromosomal segment during genetic analysis.

Masked mRNA. mRNA complexed with protein so that it is not translated or enzymatically degraded.

Master and Slave Hypothesis. This hypothesis explains C-value paradox. A cell has several copies of a gene, one of the copy being the master and the other copies being slave genes.

Master Regulators. The genes that respond most directly to X/A ratio. These regulators are located on the X chromosome in both fruitfly and nematode.

Maternal Effect. A nuclear gene product in the cytoplasm of egg determines the phenotype of the progeny.

Maternal Genes. The genes that are expressed during oogenesis by the mother. They function in nurse cells that surround the oocyte. Some of the maternal gene products behave as morphogens.

Maternal Inheritance. Phenotypic differences due to factors such as those in chloroplasts and mitochondria transmitted by the female gamete.

Mechanical Genome Incompatibility Model. This model postulates that species formation is an outcome of processes like unequal crossing over and gene conversion that alter the positions, numbers or sequences of clustered and dispersed repetitive DNAs.

Medical Genetics. A branch of genetics that deals with study of causes, identification, cure/ treatment of human disorders.

Meiotic Male Recombination. Recombination may rarely occur under natural conditions or may be induced with the help of a physical or a chemical agent in the meiotic germ line cells in males of *Drosophila melanogaster.*

Meristic Mutations. The mutations that alter the number of a structure within a floral whorl.

Merodiploidy. Temporary partial diploidy in bacteria due to the presence of transferred genes.

Messenger RNA (mRNA). A transcript of a protein-encoding gene which acts as a template for translation.

Messenger RNA (mRNA) Caps. Addition of m7G group(s) at 5′ end of the most of the eukaryotic primary transcript. These caps are of three types: (i) *Cap O.* Cap with a single methyl group, found in 100% cases, (ii) *Cap 1.* Methyl group may be present on penultimate base at 2′ O position of the sugar moiety, present in most cases, and (iii) *Cap 2.* methyl group may be present in the third base also at the 2′ O position of sugar, present in 10–15% cases.

Messenger RNA (mRNA) Decay. A process responsible for differential survival of mRNAs in the cytoplasm. Site specific endonucleases seem to control this process.

Metagon. An RNA necessary for the maintenance of mu particles in *Paramecium.*

Microinjection. Purified gene is directly injected into the nucleus of an animal cell through a micropipette.

Microprojectile Method. A method of gene transfer where gold or tungsten particles coated with DNA are carried by a macroprojectile and are accelerated into target cells where they penetrate the cell wall leaving DNA to be incorporated.

Microsatellites. A class of hypervariable loci consisting of tandemly repeated short (<10 bp) simple sequence of 100-200 bp length distributed in eukaryotic genome. These sequences serve as molecular markers for selection of agronomically useful traits and for preparing advanced high-density genetic linkage maps. Microsatellites, also called "*simple sequence repeats* (SSRs)", "*simple tandem repeats* (STRs)" or simply "*simple sequences* (SSs)" consist of head-to-tail tandem arrays of short DNA motifs (usually 1–5 bases). They are common component of eukaryotic genome but are almost absent from prokaryotes. Microsatellites are (1) highly variable due to a variable number of tandem repeat (VNTR)-type of polymorphism, (2) more or less evenly distributed throughout genome, and (3) probably nonfunctional and, therefore, selectively neutral.

Minichromosome. Chromosome of SV40 or polyoma is the nucleosoamal form of the viral circular DNA.

Minimal Life. Sequencing of the *Mycoplasma genitalium* genome represents the first complete molecule definition of minimal life with 482 genes. The genome is only 580 kb long.

Minisatellites. These are families of about 15 bp repeats forming 0.5 to 30 kb sequences distributed throughout the eukaryotic genome. These show allelic differences in the number of repeats called variable number of tandem repeats (VNTRs). The occurrence of many highly polymorphic DNA loci simultaneously is responsible for the specific DNA fingerprints of different individuals.

Minor Gene. A gene whose effect on a given trait is so small that it is not easily detectd.

Minor Groove. The smaller of the two grooves that spiral around the surface of the double helix model for DNA.

Minor Histocompatibility Antigens. These antigens act with low intensity and reject the graft slowly.

–25 Box. A component of the nucleotide sequence that makes up the prokaryotic promoter.

Misalignment Mutagenesis. Spontaneous mutation induction process in which the bases are correctly paired but the pairing occurs out of register.

Mismatch DNA Repair. A form of excision repair initiated at the sites of mismatched bases in DNA.

Missense Mutation. At alteration in a nucleotide sequence that converts a codon specifying one amino acid into a codon for a second amino acid.

Missense Intergenic Suppressor. It is a gene coding for a mutant tRNA able to respond to one or more of the missense mutations.

Mitochondrial (mt) DNA. In most eukaryotic cells mt DNA is double stranded and circular (as in bacteria) in structure. It replicates independently of nuclear DNA in semiconservative manner. Molecular weight = 9×10^6 to 10×10^6; length = 5 mm; size = 14,000 bp.

Mitochondrial DNA Diseases. Refers to human diseases associated with mutations in mitochondrial DNA. Diseases due to single deletions or duplications, multiple deletions or point mutations in mitochondrial DNA are known.

Moderately Repetitive Sequences. When number of copies of a sequence in eukaryotes is 10^3 to 10^5. These sequences are interspersed with unique sequences, and are also called dispersed repeats.

Modern Phase of Gene Concept. A phase where a gene is defined on the basis of *cis-trans* complementation test and is termed as a cistron which is made up of a large number of mutable sites (*muton*) which undergo recombination (*recon*).

Modes of Selection. Various modes of selection in natural populations are directional, disruptive, diversifying, balance, frequency-dependent (density-dependent), normalising, stabilising (centripetal).

Modification of DNA/RNA. Includes all changes made to the nucleotides after their initial incorporation into the polynucleotide chain.

Modified Bases. All those bases except the usual four from which DNA (T, C, A, G) and RNA (U, C, A, G) are synthesised; they result from post-synthetic changes in the nucleic acid.

Modifier (Modifying) Gene. A gene that affects or modifies the expression of another gene.

Molecular Drive. A molecular mechanism which explains evolution in multiple gene family. It includes unequal recombination, transposition and gene conversion.

Molecular Evolutionary Clock. A measurement of evolutionary time in nucleotide substitutions per year.

Molecular Farming. It is the use of transgenic plants as factories for producing specialty chemicals and pharmaceuticals.

Molecular Genetics. A branch of genetics that involves the study of the molecular nature of genes and gene expression.

Molecular Pharming. By transferring genes into animals it has been made possible to have pharmaceuticals secreted into milk, blood and urine which can be used for manufacture of drugs.

Molecular Tinkering. An evolutionary process responsible for production of a protein with new properties by joining together pieces of several different genes.

Molecular Zippers. These are the regulatory proteins which reveal the motifs which are composed primarily of amino acids that may not make direct contact with DNA. These motifs form three-dimensional scaffolds which steer the particular amino acid side chain of a regulatory protein into the grooves of double helical DNA where they can interact directly with DNA bases. These zippers may be of several types–steroid *receptors*, helix-turn-helix motifs, acid blobs, amphipathetic, helix-loop-helix motifs, zinc finger motifs, leucine zippers.

Monintron Gene. The gene which contains only one intron. Transcripts of such genes are not used for translation.

Monocistronic mRNA. An RNA molecule, mostly in eukaryotes, that contain information from only one cistron.

Monoclonal Antibody. A specific antibody produced by a hybridoma cell line against one specific antigen determinant.

Morphogen. A factor that induces development of particular cell types in manner that depends on its concentration.

Morphogenesis. Morphogenesis is the process of origin and development of a part, and organ, or organism. When this process is explained in molecular terms, it is known as *molecular morphogenesis.*

Morphological Species Concept. Organisms are classified in the same species if they appear similar.

Motifs. These are specific sequences found in some regulatory genes. These motifs code for conserved amino acid regions in proteins. The proteins containing these conserved amino acids bind to DNA to regulate structure proteins involved in development.

Multigene Family. A group of genes, possibly although not always clustered, that are related either in nucleotide sequence or in terms of function.

Multimarker Linkage Disequilibrium Mapping Method. A method developed for fine mapping QTL using dense marker map. The method compares the expected covariances between haplotype effects given a postulated QTL position to the covariances that are found in the data.

Multintron Gene. The gene that contains more than one intron. Transcripts of such genes are translated into proteins.

Multiple Sigma Factors. Various sigma factors are known to exist in *Escherichia coli* which have been discovered to operate under different conditions of heat shock, nitrogen starvation or chemotaxis and flagellar structure. In *Bacillus subtilis*, multiple sigma factors are involved in sporulation and in gene expression in phage SPO1 infecting *B. subtilis*.

Multiplicational Family. It consists of 10^1 to 10^4 copies of a gene, 80–100 nucleotides long. Repeat units are essentially identical.

Multisite Mutant Allele. A mutant differing from its wild-type form at two or more sites.

Multivalent. An association of more than two chromosomes whose homologous regions are synapsed by pairs.

mu Particles. Bacteria-like particles found in the cytoplasm of *Paramecium* that have the male-killer phenotype.

Mutagen. A chemical or physical agent able to cause a mutation in a DNA molecule.

Mutagenesis. Experimental treatment of a group of cells or organisms with a mutagen in order to induce mutations.

Mutant. A cell or organism with an abnormal genetic constitution.

Mutation. An alteration in the nucleotide sequence of a DNA molecule.

Mutation Theory of Evolution. de Vries advocated that a single mutation can lead to radical changes in an organism which can transform one species to another. Also known as *deVriesism*.

Muton. The smallest segment of DNA or subunit of cistron that can be changed and thereby bring about a mutation; can be as small as one nucleotide pair.

Natural Selection. The process in nature whereby one genotype leaves more offspring than another because of superior life history attributes such as survival of fecundity. Darwin gave natural selection as mechanism of evolution.

Nearest-neighbour Analysis. A technique of transferring radioactive atoms between adjacent nucleotides in DNA used to demonstrate that the two strands of DNA run in opposite directions.

Negative Control. Repression of an operator site by a regulatory protein that is produced by a regulator site.

N-end Rule. The life-span of a protein is determined by its amino-terminal (N-terminal) amino acid.

Neo-Darwinism. Term that refers to merger of classical Darwinian evolution with population genetics and thus leading to the synthesis of modern theory of evolution. Also see *synthetic theory of evolution.*

Neurogenetics. A branch of genetics that deals with studies on the manner in which genes specify the neuronal organisation of behaviour.

Neurospora. A pink mold, commonly found growing on old food. Being haploid it is very useful in genetic studies. Working on biochemical mutants of *Neurospora*, Beadle and Tatum put forward one gene-one enzyme hypothesis.

Neutral Alleles. The alleles whose differential contribution to fitness is so small that their frequencies change more due to generations than to natural selection.

Neutral Gene Hypothesis. The hypothesis that most genetic variation in natural populations is not maintained by selection.

Neutrality Theory of Protein Evolution. Rates of amino acid replacement in proteins and nucleotide substitutions in DNA during evolution may be approximately constant because the vast majority of such changes are selectively neutral.

New Codon Generation in Immunoglobulin Genes. Recombination at certain points between *V* and *J* sequences can generate new codons in the hypervariable regions. Recombination between different triplets can generate yet other new triplets.

Nick Translation. The ability of *E. coli* DNA polymerase I to use a nick as a starting point from which one strand of a duplex DNA can be degraded and replaced by resynthesis of new material. It is used to introduce radioactivity labelled nucleotides into DNA *in vitro.*

nif. The genetic notation for the genes involved in nitrogen fixation. The *nif* operon is a complex array of seventeen genes. The proteins encoded by the *nif* genes will fix atmospheric nitrogen (N_2) into ammonia (NH_4^+) and nitrate (NO_3^-). Many soil bacteria will fix nitrogen and there is much interest in manipulation of *nif* genes from bacteria to allow plants to fix nitrogen.

nod. The genetic notation for the genes involved in nodule formation. Also known as *nodule genes.*

Nonambiguous Code. Nature of genetic code where one codon codes for only one amino acid.

Nonautonomous Controlling Elements. Defective transposons that can transpose only when assisted by an autonomous controlling element of the same type.

Nonautonomous Development. When a *Drosophila* imaginal disc from a mutant for eye colour develops into normal eyes in wild-type host.

Non-Darwinian Evolution. A theory of evolution which considers natural selection as incompetent to account for arrival of the fittest. Saltation hypothesis, punctuation equilibrium model and evolution by random walk are in contrast to the concept of *Darwinian evolution*.

Non-disjunction. The failure of homologous chromosomes to separate at anaphase-I of meiosis.

Non-DNA Repair. Intrinsic features of the organism, like structure of the code, intergenic and intragenic suppression, diploidy, in addition to normal repair mechanisms.

Non-genetic RNA. The RNA which is found in association with DNA and does not act as genetic material. Also known as *functional* RNA, e.g., tRNA, rRNA and mRNA.

Non-histone Proteins. The proteins remaining in chromatin after the histones are removed. The scaffold structure is made of non-histone protiens.

Non-overlapping Codon. Nature of genetic code where one nucleotide is part of only one codon.

Non-parental Ditype (NPD). A spore arrangement in Ascomycetes that contains only the two recombinant-type ascospores (assuming two segregating loci).

Non-reciprocal Recombination. Complementary recombination products resulting from a crossing over event do not appear in 1 : 1 ratio (contrary to *reciprocal recombination*).

Non-repetitive DNA. The DNA which shows reassociation kinetics expected of unique sequences.

Nonsense Codon. One of the mRNA sequence (UAA, UAG, UGA) that signals the termination of translation.

Nonsense Intergenic Suppressor. It is a gene coding for a mutant tRNA able to respond to one or more of the nonsense codons.

Nonsense Mutation. An alteration in a nucleotide sequence that converts triplet coding for an amino acid into a termination codon.

Nontranscribed spacer. The region between transcription units in a tandem gene cluster.

Nopaline. A rare amino acid derivative which is produced by a certain type of crown-gall tissue. The genes responsible for the synthesis of nopaline are part of the T-DNA from a Ti-plasmid.

Normalising Selection. The removal by selection of all genes that produce deviations from the normal (= average) phenotype of a population.

Northern Blotting. A gel transfer technique used for RNA.

Novel Joint. A new DNA sequence and thus a new restriction fragment created during intrachromosomal amplification event.

Nu Body. Nucleosome; the subunits of chromatin produced during chromosome coiling in eukaryotes with roughly spherical shape, which is composed of a core octamer of histones and approximately 140 nucleotide pairs of DNA.

Nuclear Envelope. A layer of two membranes surrounding the nucleus. It is penetrated by nuclear pores and bound on the interior by the nuclear lamina.

Nuclease. An enzyme that degrades a nucleic acid molecule.

Nuclease Hypersensitive Site. Region of eukaryotic chromosome that is specifically vulnerable to nuclease attack because it is not wrapped as nucleosomes.

Nucleic Acid. Originally, the acidic chemical compound isolated from the nuclei of eukaryotic cells. Now, the polymeric molecules comprising nucleotide monomers: DNA and RNA.

Nucleic Acid Vaccines. Plasmid DNA encoding for an antigen of interest is injected into muscles. This results in sustained expression of the antigen and generation of an immune response. Several routes of nucleic acid vaccine administration are possible.

Nucleo-cytoplasmic Interaction. Activity of nuclear genes during development is limited by properties of the cytoplasm. Phenotype is regulated but not determined by cytoplasm. Genes determine a cell's potential; the cytoplasm determines whether or not that potential will be reached.

Nucleoid. The DNA-containing region within a prokaryotic cell.

Nucleolar Organiser. The chromosomal region around which the nucleolus forms; site of tandem repeats of the major rRNA gene.

Nucleolus. The region of the nucleus in which rRNA transcription occurs.

Nucleosome. The structure comprising histone proteins and DNA that is the basic organisational unit in chromatin structure.

Nucleotide Excision Repair (NER). DNA repair mechanism in eukaryotes which involves a large number of genes. A group of genes called *RAD* genes or *ERCC* (excision repair complementation competent) genes are involved in this process.

Null Allele. An allele whose protein product shows no histochemically detectable activity.

Null Mutation. A mutation which completely eliminates the function of a gene.

Numerator Elements. Genes on the X chromosome in *Drosophila* that regulate the sex switch (*sxl*) to the on condition (femaleness). Refers to the numerator of the X/A genic balance equation.

Ochre Codon. The nonsense codon UAA. One of the three nonsense codons that cause termination of protein synthesis.

Octad. An ascus containing eight ascospores, produced in species which the tetrad normally undergoes a post-meiotic mitotic division.

Octopine. A rare amino acid derivative which is produced by a certain type of crown-gall tissue. The genes responsible for the synthesis of octopine are part of the T-DNA from a Ti-plasmid.

Okazaki Fragment. One of the short segments of RNA-primed DNA that are synthesised during replication of the lagging strand of the double helix.

Oligogenes. One or a few genes governing the same qualitative character.

Oligonucleotide-directed Mutagenesis. An *in vitro* mutagenesis technique that involves the use of a synthetic oligonucleotide to introduce a predetermined nucleotide alteration into the gene to be mutated.

Oligonucleotide Ligation Assay (OLA). This assay is based on the ability of two dinucleotides to anneal immediately adjacent to each other on a complementary target DNA molecule.

Oligonucleotide Therapy. Use of oligoribonucleotides and oligodeoxynucleotides as the therapeutic agents. Three basic strategies are: antisense inhibition of expression, triple helix formation, and protein epitope targeting.

Oncogene. A gene which when carried by a retrovirus is able to cause cell transformation.

One Gene-One Enzyme Hypothesis. One gene controls the synthesis of one enzyme. This hypothesis is valid for only those enzymes or proteins that are made up of only one type of polypeptide.

One Gene-One Primary Function Hypothesis. One gene performs one specific primary cellular function.

Opal Codon. The nonsense codon UGA. One of the three nonsense codons that cause termination of protein synthesis.

Open Promoter Complex. The complex formed between *E. coli* RNA polymerase and a promoter, in which the double helix is partially unwound in readiness for the start of RNA synthesis.

Open Reading Frame (ORF). A series of codons with an initiation codon at the 5′ end but no termination codon. Often considered synonymous with 'gene' but more properly used to describe a DNA sequence which looks like a gene but to which no function has been assigned.

Operator. A nucleotide sequence element to which a repressor protein attaches in order to prevent transcription of a gene or operon.

Operator-constitutive Mutation (O^c). Mutation in the operator which leads to constitutive synthesis of all the gene products of the operon regardless of presence or absence of the substrate. The mutation affects activity of all the genes of the operon in *cis*-arrangement not in *trans*-arrangement.

Operon. A system of cistrons, operator and promoter sites, by which a given genetically controlled metabolic activity is regulated.

Opine. The general name given to rare amino acid and sugar derivatives found in crown-gall tumors.

O Protein. l phage protein which recognises replication origin site Ori l in phage l. Four or more molecules are used for the formation of initial complex.

Optimal Promoter. An optimal promoter is defined as a sequence consisting of the –35 hexamer, separated by 17 base pair from the –10 hexamer, lying 7 bp upstream of the start point.

Ori C. Replication origin of *E. coli.* It contains four 9 bp noncontiguous repeats in 240 bp stretch of DNA.

Ori l. Replication origin of phage l. It contains four 19 bp repeats each of which is an inverted repeat.

Origin of Species. The production of new types of organisms in descent from extant types. Natural selection is the process by which changes are favoured leading to the production of new types. The term was introduced by Charles Darwin.

Origin of Transfer (Ori-T). When gene transfer occurs through plasmids, one of the two strands of plasmid DNA is nicked at a site called Ori-T and this linear strand moves into the recipient bacteria.

Overdominance. Superiority of F_1 for one or more characters over both the parents.

Overdominance Hypothesis. The theory that heterosis is caused by the *Aa* genotype being superior to either the AA or the *aa* parent.

Overlapping Genes. Genes whose coding regions are overlapping such that a single nucleotide sequence produces more than one polypeptide.

Pair-rule Gene. A gene that influences the formation of body segments in *Drosophila.* The mutants have corresponding parts of pattern deleted in every other segment. The afflicted segments may be even-numbered or odd-numbered.

Palindrome. A sequence of DNA base pairs that reads the same on the completely other strands. For example, 5′-GAATTC-3′ on one strand, and 5′-CTTAAG-3′ on the other strand. Palindrome sequence are recognised by restriction endonucleases.

Panneutralist Theories. This theory believes in neutrality of most of the genetic variation and considers evolution as a random process. This is also referred to as *non-Darwinian evolution.* Also see *neutrality theory of protein evolution.*

Panselectionist Theories. Theories which recognise natural selection as the only guiding force of evolution.

Papovaviruses. A class of animal viruses with small genomes, including SV40 and polyoma.

Parallel Evolution. Evolution of similar characters separately in two or more lineages of common ancestry.

Paramutation. A term given by Brink to explain a mutation in which one allele in a heterozygous condition changes its partner allele permanently. In this allelic interaction, *paramutagenic allele* suppresses *paramutable allele* in the heterozygote.

Parapatric Speciation. Evolution of a taxon while it remains in contact with its ancestral population. Local selection pressures are strong enough to prevent homogenisation by interbreeding.

Paternal Disomy. A condition where both copies of a gene of a locus in a diploid are inherited from the father, e.g., in human patients with a disease known as Beckwith Weidermann Syndrome (BWS), both copies of a region on the short arm of chromosome 11 (11 p 15.5) are inherited from father.

Pathogen Genes. These are genes which are responsible for virulence/avirulence in pathogen. *Virulence* genes are usually inherited as recessive whereas *avirulence genes* are inherited as dominant. Virulence may be determined by one, a few or many genes like resistance in the host.

Patient Gene Therapy. A type of gene therapy in which cells with healthy genes may be introduced in the affected tissue, so that the healthy gene overcomes the defect without affecting the inheritance of the patient.

pBR 322. A particular type of artificial plasmid, frequently used as a vector for cloning genes in *E. coli*.

PCR (Polymerase Chain Reaction). A technique in which cycles of denaturation, annealing with primer, and extension with DNA polymerase, are used to amplify the number of copies of a target DNA sequences by more than 10^6 times.

PCR Mutagenesis. A method for mutation induction where primers are designed to mismatch with their target and conditions are chosen so that annealing is still permitted. Amplification introduces the mutation into the amplified product.

Pedigree Analysis. A method based on information from various generations of individuals to know the inheritance of a given trait. This method is largely used in those systems where crossing procedures are not applicable.

Penetrance. The extent to which a phenotype is expressed, measured as the proportion of individual with a particular genotype that actually display the associated phenotype.

Peptide Nucleic Acid (PNA). It has the ability to replicate itself and catalyse chemical reactions as RNA does. It is very simple molecule consisting of a peptide (polyamide) backbone composed of H-(2-amino ethyl) glycine units to which nucleobases are attached by carbonyl methylene linkers. It can serve as a template both for its own replication and for the formation of RNA from its subcomponents. A polycyticidine decamer of PNA can act as a template for oligomerisation of activated guanosine mononucleotides.

Peptidyl- or P-site. The site on the ribosome to which the tRNA attached to the growing polypeptide is bound during translation.

Peptidyl Transferase. The enzyme activity responsible for peptide bond synthesis during translation.

Permanent Heterosis. A hybrid condition maintained by a balanced lethal system so that *Oenothera lamarckiana* is a permanent hybrid of gaudensvelans complex.

Permissive Conditions. The conditions that allow conditional lethal mutants to survive.

Phage "Ghost". The phage protein shell left behind after the phage has injected its DNA.

Pharmacogenetics. The study of the relation between an individual's genotype and his reaction to various pharmaceutical agents.

Phasmid. A hybrid molecule formed *in vivo* between a plasmid containing multiple att sites of l and a bacteriophage l derivative. The formation of a plasmid or its breakdown to release the plasmid and phage is controlled by the site-specific recombination system of l. Phasmids can replicate as a plasmid (non-lytically) or as a phage (lytically). A number of sophisticated *in vivo* genetic manipulations can be accomplished using phasmids.

Phecomelia. A condition when pregnant women who are kept on drug thalidomide generally produced children with short limbs.

Phenocopy. An environmentally induced phenotype that resembles the phenotype produced by a mutation.

Phenocopying Abnormal. When a phenocopy resembles a known mutant.

Phenocopying Normal. When a phenocopy resembles a normal individual whereas non-phenocopy individual is mutant.

Philadelphia Chromosome. A translocation between the long arms of chromosomes 9 and 22, often found in the white blood cells of patients with chronic myeloid leukemia.

fX174. A bacteriophage which attacks *E. coli*. The nucleic acid within the viral particle is a single-stranded DNA circle of 5375 bases. The replicative form is a double-stranded circle and was the first complete DNA molecule to be sequenced.

Photoautotrophic. Those systems which can derive energy from sun and synthesise themselves most of their nutrition from simple inorganic molecules. Photoautotrophs evolved from chemoautotrophs. This great advance gave living systems independence from environment.

Photoreactivation. A light-induced reversal of ultraviolet light causing injury to cells. A large proportion of damaged cells are removed through light activated enzyme photolyase which cleaves the thymidine dimers.

Phyletic Speciation. The process of speciation caused by the gradual change in the genetic constitution of a population without the population splitting into demes and without any increase in the number of species produced by that population at any one time.

Phylogenetic Tree. A diagram indicating the supposed sequence of historical relationship linking together species or high taxa.

Physical Containment. The use of physical barriers to prevent the escape of genetically engineered organisms from the laboratory into the environment. Precautions employed include the maintenance of negative air pressure within the laboratory and the use of fume hoods with HEPA filters on their exhausts.

Physiological Suppressors. When a defect in one biochemical pathway is circumvented by another mutation.

Picornavirus. A small RNA virus.

Plasmagene. A self-replicating, cytoplasmically located gene.

Plasmid. A usually circular piece of DNA, primarily independent of the host chromosome, capable of replicating, often found in bacterial and some other types of cell.

Plastid Inheritance. Inheritance which is governed by chloroplast genes.

Pleiotropy. The ability of a single gene to produce a complex phenotype that consists of two or more distinct characteristics.

Plus-and-minus DNA Sequencing. In this method of DNA sequencing, given by Sanger *et al.* (1977), DNA polymerase-generated single stranded nucleotide are electrophoresed on polyacrylamide gels. Primer is elongated at the 3′ end. Method comprises plus series of four experiments where synthesis of new strand is halted at various points by insertion of dideoxy analogues. In the minus series of experiments, the oligonucleotide primer is elongated using only three of the deoxynucleoside triphosphates (e.g., dGTP, dCTP, dTTP are added, but dATP is missing).

P-M System. A system of hybrid dysgenesis where a female from maternal (M) strain when mated with a male from a paternal (P) strain, various germ line abnormalities characteristic of this syndrome are produced. In reciprocal crosses, these abnormalities are not observed.

Point Mutation. A mutation that results from a single nucleotide alteration in a DNA molecule.

Polyadenylation. The post-transcriptional addition of a series of adenin residues to 3′-end of a eukaryotic mRNA molecule.

Poly (A) Polymerase. The enzyme responsible for polyadenylation of a eukaryotic mRNA molecule.

Poly-A Tail. The initially long sequence of adenine nucleotides at the 3′ end of mRNA; added after transcription.

Poly-dA/poly-dT technique. A method of inserting foreign DNA into a vehicle by making 5′ poly-dA and 5′ poly-dT tails on the vehicle and foreign DNAs. Also feasible with poly-dG and poly-dC.

Polyacrylamide Gel Electrophoresis (PAGE). A method for separating nucleic acid or protein molecules according to their molecule size. The molecules migrate through the inert gel matrix under the influence of an electric field. In the case of protein PAGE, detergents such as sodium dodecylsulphate are often added to ensure that all molecules have a uniform charge.

Polyadenylation Signal. A sequence AAUAAA in pre-mRNA which defines the site of addition of poly(A) tail.

Polycistronic mRNA. An RNA molecule, mostly in prokaryotes, that contains information from more than one cistrons.

Polyclonal Antibodies. The antibodies synthesised against more than one antigen determinant.

Polydactyly. The occurrence of more than the usual number of fingers or toes.

Polygenes. Two to more different pairs of alleles, with a presumed cumulative effect that governs such quantitative traits as size, pigmentation, intelligence, among others. Those contributing to the traits are termed as *contributing* or *effective* alleles; those appearing not to do so are referred to as *noncontributing* or *noneffective alleles*.

Polyethylene Glycol (PEG) Stimulated DNA Uptake. A method of gene transfer which involves use of 15.25% PEG, which stimulates uptake of DNA by endocytosis without any gross damage to protoplasts.

Polymeric Genes. Two dominant alleles having similar effect when they are separate but produce enhanced effect when they come together resulting in 9:6:1 ratio.

Polypeptide Elongation. This process during protein biosynthesis starts when A site is occupied by an incoming amino acid tRNA and peptide bond is formed between $_f$met and the incoming amino acid. The amino acid-tRNA that binds to A site is specifically determined by the triplet codon of the mRNA that occupies A site.

Polypeptide Termination. This is a process in protein biosynthesis where, when complete, a polypeptide is released from the ribosome. Termination codons (UAA, UGA or UAG) serve as termination signals. In prokaryotes, three **released factors** (RF_1, RF_2 and RF_3), GTP and termination codons are required at this step.

Polyribosome (Polysome). An mRNA molecule in the process of being translated by several ribosomes at once.

Polytene Chromosome. Many-stranded giant chromosomes produced by repeated replication during synapsis in certain dipteran larval tissues. Synonym, *giant chromosome*.

Position Effect. A phenotypic effect dependent on a change in position on the chromosome of gene or group of genes. It may produce variegation, chimera, or a mosaic phenotype.

Positive Control. Activation of an operator site by a regulatory protein that is produced by a regulator site. Compare with *negative control.*

Postnatal Diagnosis. Diagnosis of a genetic disorder in an individual after birth.

Post-replicative Repair. A DNA repair process initiated when DNA polymerase bypasses a damaged area. Enzymes in the *rec* system are used.

Post-transcriptional Modifications. Changes in eukaryotic mRNA made after transcription has been completed. These changes include addition of cap and tail and removal of introns.

Post-translational Modifications. Changes in polypeptide after translation has been completed, such modifications may lead to change in phenotype. For example, post-translational modification lead to isozymes. Also see'*epigenetic modifications*.

Post-translational Steps. Steps in protein biosynthesis which occur after polypeptide chain has been released from the ribosome. In prokaryotes, these steps include removal of formyl group, the first methionine incorporated during initiation of translation and terminal tRNA.

Post-zygotic Isolating Mechanism. Any one of several mechanisms that keep populations reproductively isolated from each other even through fertilisation and hybrid zygotes may form. These hybrids are either sterile, nonviable, or so weak that they do not survive.

Preempter Stem. A configuration of leader transcript mRNA that does not terminate transcription in the attenuator-controlled amino acid operons.

Pre-meiotic Male Recombination. Recombination may rarely occur under natural conditions or may be induced with the help of a physical or a chemical agent in the premeiotic germ line cells in males of *Drosophila melanogaster*.

Pre-mRNA. The primary, unprocessed transcript of a protein-encoding gene.

Prenatal Diagnosis. Diagnosis of a genetic disorder in an individual before birth.

Pre-priming Complex. The structure, consisting of a series of proteins attached to the origin of replication, which initiates DNA replication in *E. coli*.

Pre-rRNA. The primary, unprocessed transcript of a gene or group of genes specifying rRNA molecules.

Pre-tRNA. The primary, unprocessed transcript of a gene or group of genes specifying tRNA molecules.

Pre-zygotic Isolating Mechanism. Any one of several mechanisms that keep populations reproductively isolated from each other by preventing fertilisation and zygote formation.

Pribnow Box. Relatively invariable six nucleotide DNA sequence TATAAT, located in eukaryotic promoters upstream (at position – 10) from the start codon.

Primary Cellular Function. Synthesis of a transcript - mRNA, tRNA or rRNA.

Primary Constriction. A construction which is determined by and associated with the centromere region.

Primary Non-disjunction. A type of non-disjunction which occurs anaphase I where homologous chromosomes fail to separate.

Primary Transcript. The immediate product of transcription of a gene or group of genes, which will subsequently be processed to give the mature transcript(s).

Primase. The RNA polymerase enzyme that synthesises the primer needed to initiate replication of DNA polynucleotide.

Primed Lymphocyte Typing (PLT). This is a technique to type cells of an individual for class II molecules of HLA complex. For this, a panel of cells is produced in which each cell of the panel is primed by one of the known class II molecules by being previously cultured with appropriate homozygous typing cells. These cells are then restimulated with cells of the individual to be typed. Then magnitude of the response is assessed.

Primer. A short oligoribonucleotide that is attached to a single-stranded DNA molecule in order to provide a site at which DNA replication or synthesis of Okazaki fragments takes place.

Primosome. A complex to two proteins, a primase and helicase, that initiates RNA primers on the lagging DNA strand during DNA replication.

Prion. An unusual infectious agent that appears to consist purely of protein with no nucleic acid.

Private Antigenic Determinants. The determinants that are unique to an antigen.

Probe. In recombinant DNA work, a radioactive nucleic acid complementary to a region being searched for in a restriction digestor genome library.

Processed Pseudogene. A pseudogene whose sequence resembles the mRNA copy of a parent gene, and which probably arose by integration into the genome of a reverse transcribed version of the mRNA.

Process Versus Discard Decision. This refers to the decision taken by pre-mRNA processing machinery whether to process a particular molecule in one cell type or to discard it in another cell type. There is no evidence for existence of such a mechanism for nuclear RNA.

Producer Gene. A eukaryotic structural gene that produces a pre-mRNA molecule which after going through some processing steps becomes mRNA. A producer gene may be under the control of several *receptor genes.*

Promiscuous DNA. The occurrence of the some DNA sequences in more than one cellular compartment. It suggests that DNA may have been exchanged between organelles, or between organelles and the nucleus.

Promoter. The nucleotide sequence, upstream of a gene, to which RNA polymerase binds in order to initiate transcription.

Proofreading. The ability of a DNA polymerase to correct misincorporated nucleotides as a result of its 3′ to 5′ exonuclease activity.

Propositus. The member of a family who first comes to the attention of geneticist. Usually phenotype of the propositus is unusual or exceptional in some way. Also known as *proband.*

Protamine. A low molecular weight protein, lacking prosthetic group but rich in arginine content, found in association with nuclear DNA in certain cells.

Proteinoids. Various amino acids in certain conditions of temperature and pH are referred to as "*proteinoids*" or "*thermal proteinoids*" which are thought to be involved in origin of life.

Protein Secretion. In eukaryotes, the proteins that are destined for secretion are initially targeted to the endoplasmic reticulum. This requires an N-terminal hydrophobic signal peptide which is recognised by a *signal recognition particle* (SRP). SRP consists of six proteins and a small cytoplasmic RNA (7S RNA) which is homologous to *Alu* element.

Protein Sorting. In eukaryotes, proteins can be targeted to any one of several able organelles, as well as to the nucleus. Targeted proteins must carry recognisable intracellular sequences or structures which allow them to be transported to the appropriate cellular compartment. Also known as *protein trafficking.*

Protein Targeting. It is a step in importing a protein from their site of synthesis to their site of function in which binding of proteins occurs with specific organelle membrane.

Proteome. Proteomes deal with field of research where the cell's proteins are looked at in order to define various processes in molecular terms compared to genome where study of DNA sequences is undertaken.

Protocell. Believed to have given rise to first living cell on the earth. It was thought to be "heterotrophic", i.e., it obtained complex organic molecules from surroundings. When complex organic molecules began to disappear, only those systems survived that evolved metabolic pathway to change simpler organic molecules to complex molecules. Thus auxotrophic mode of nutrition evolved from heterotrophic mode and origin of life can be traced back to auxotrophic protocells.

Proto-oncogene. A normal cellular gene that can be changed to an oncogene by mutation.

Protoplast Fusion. A technique for producing hybrids between two cells which would not normally mate. The two cells may belong to the same or different species. The cell walls are removed from the two parent cells to create protoplasts and then fusion of the two cell membranes is promoted, usually by the addition of polyethylene glycol and Ca^{2+} ions. These fusogenic agents cause proteins to migrate from certain regions of the two cell membranes and allow areas of naked phospholipid to fuse. Subsequent regeneration of the cell walls allows the propagation of the hybrid organism. If nuclear fusion does not follow cell fusion then *heterokaryons*, rather than diploid organisms, are produced. In some crosses, the genetic contribution of the two parents to the stable hybrid can be markedly unequal.

Prototroph. An organism that has no nutritional requirements beyond those of the wild-type and is, therefore, able to grow on minimal medium.

Provirus. The DNA copy of retroviral genome that is integrated into host chromosomal DNA.

pSC101. A small, non-conjugative plasmid which encodes tetracycline resistance. Its origin is not clear as it is now thought to have come from a contaminant in a transformation experiment. It was one of the first vectors used in genetic engineering. The pSC101 tetracycline-resistance gene was used in the construction of pBR322.

Pseudoalleles. Nonalleles so closely linked that they are often inherited as one gene, but shown to be separable by crossover studies. Peudoalleles are found to be allelic in complementation test but nonallelic in recombination test. Now this term is discarded.

Pseudoautosomal Inheritance. Pattern of inheritance shown by gene that are located on the pairing region of the X and Y chromosomes in individuals having XY mechanism of sex-determination.

Pseudodominance. The expression (apparent dominance) of a recessive gene at a locus opposite a deletion.

Pseudogene. A nucleotide sequence that has similarity to a functional gene but within which the biological information has become scrambled so that the pseudogene is not itself functional.

Public Antigenic Determinants. The antigens that carry determinants common in all the antigens.

Puff. A localised synthesis of RNA occurring at specific sites on giant chromosomes of Diptera. Also see *Balbiani ring, chromosome puff* and *DNA puff.*

Pulse-chase Experiment. These experiments are performed by incubating cells very briefly with a radioactively labelled precursor (of some pathway or macromolecule); then the fate of the label is followed during a subsequent incubating with a nonlabelled precursor.

Punctuation Codon. A codon that designates either the start or the end of a gene.

Quantitative Trait Loci (QTL). The loci which govern traits, such as height, that have a continuous pattern within a population that is typically determined by combined effect of a number of genes. There is a range of phenotypes differing in degree.

Quantitative Trait Loci (QTL) Meta-analysis. A meta-analysis, with respect to QTL analysis, is performed when we have data concerning different populations. It determines whether QTL identified for a given trait in one population correspond to those detected in other populations, whether QTL locations identified in one species correspond to QTL or other types of loci detected in corresponding regions of other species.

Quantum Sepeciation. The budding off of a new and very different daughter species from a semi-isolated peripheral population of the ancestral species in a cross-fertilising organism.

R_1, R_2, R_3, etc. The first, second, third, etc., generations following any type of irradiation to induce mutations.

R Loop. The structure formed when an RNA strand hybridises with its complementary strand in a DNA duplex, thereby displacing the original strand of DNA in the form of a loop extending over the region of hybridising.

R Plasmid. A plasmid containing one or several transposons that bear resistance genes.

Ramachandran Plots. Proteins have definite conformation which is determined by amino acid sequence.

Random Gene Amplification. A type of amplification that can occur randomly in the cell, e.g., in some type of cancer cells amplification is random.

Random Genetic Drift. Changes in allelic frequency due to sampling error.

Randomised Block Design. An experimental design which controls fertility variation in one direction only.

Rapid Lysis (*r*) Mutants. These mutants display a change in the pattern of lysis of *E. coli* at the end of an infection by a T-even phage.

Rare Base Pairing. Pairing between two purines or pyrimidines in DNA.

Readingframe. Sequence of triplet codons that are contained in any DNA/RNA sequence begnning at translation initiation codon.

Readingframe Lengths (RFLs). This is the piece of DNA in between two successive stop codons. Computer analysis concluded that an RFL had an upper limit of 600 nucleotides. This helped in defining *exon length.*

Readthrough. In the presence of a suppressor-tRNA, an amino acid can be inserted into a growing polypeptide chain in response to a stop codon. Termination is thus prevented and a longer than usual polypeptide synthesised. This process is known as readthrough and the protein product is a readthrough protein.

Reassociation of DNA. Pairing of complementary single strands to form a double helix.

recA. The product of the *rec*A locus of *E. coli* dual activities, activating proteases and also able to exchange single strands of DNA molecules. The protease-activating activity controls the SOS response; the nucleic acid handling facility is involved in recombination repair pathways.

***recBC*.** The genetic notation of an *E. coli* gene whose product mediates recombination. The enzyme encoded by the *recBC* gene is an ATP-dependent nuclease specific for double-stranded DNA. This enzyme will degrade linear DNA that has been taken up by transformation.

Receptor Element. A controlling element that can insert into a gene (making it a mutant) and can also exit (thus making the mutation unstable); both of these functions are nonautonomous, being under the influence of the regulator element.

Receptor Gene (site). A component of eukaryotic gene regulation with which a specific *activator* RNA molecule complexes. This interaction activates the receptor gene.

Recessive Epistasis. A condition where a homozygous recessive gene pair masks the effect of another gene.

Reciprocal Recombination. Complementary recombination products resulting from a crossing over event appear in 1:1 ratio (cf *non-reciprocal recombination*).

Reciprocal Translocation. The exchange of segments between two nonhomologous chromosomes.

Recoding. A programmed process where a minority of mRNAs in bacteria carry instruction that specify an alteration in how genetic code is to be read out, so that the meaning of the code words is altered.

Recoding Signals. Instruction present in mRNA responsible for programmed recoding.

Recognition Site. A sequence of bases within the promoter region that serves to recognise RNA polymerase molecule. See also *promoter* and *initiation site.*

Recognition System for Recombination in Immunoglobulin Genes. Analysis of the DNA sequence in the vicinity of several V_L and V_H, and J_L and J_H sequences shows a remarkable conservation of hepta- and decanucleotides. These conserved sequences are separated by a 10–11 or 21–23 bp non-homologous stretch of nucleotides. Presumably similar sequences are found in the vicinity of D regions of the heavy chain genes. An immunoglobulin heavy chain variable region gene is generated from three sequences of DNA, V_H, D and J_H. These conserved hepta- and decanucleotides in vicinity of V, D, and J genes from recognition system required for recombination.

Recombinant DNA. A DNA molecule created in the test tube by ligating together pieces of DNA that are not normally contiguous.

Recombinant DNA Technology. All the techniques involved in the construction, study and use of recombinant DNA molecules.

Recombinant Joint. The point at which two recombining molecules of duplex DNA are connected (the edge of the heteroduplex region).

Recombinant RNA. A term used to describe RNA molecules joined *in vitro* by T_4 RNA ligase. This technique may be used to join an RNA sequence to a phage Qb replicase template. The recombinant RNA molecule can be autocatalytically replicated by Qb replicase to produce large amounts of an RNA sequence of interest.

Recombinational Repair. Filling a gap in one strand of duplex DNA by retrieving a homologous single strand from another duplex.

Recombination Theory of Antibody Diversity. According to this theory, there are as many different light and heavy chain genes as many light and heavy chains a cell can produce. The diverse types of antibodies are produced as a result of recombination between light and heavy chains.

Recombinogen. An agent capable of inducing or enhancing recombination between linked genes.

Recon. The smallest segment of DNA or subunit of a cistron that is capable of recombination; may be as small as one deoxyribonucleotide pair.

***rec* System.** Several loci controlling genes (*recA*, *recB*, *rec*C, and others) involved in post-replicative DNA repair.

Redox Gene Regulation. A gene regulatory mechanism involving reductional-oxidation of a single conserved cysteine residue in the DNA binding domain of the regulatory proteins.

Redundant Gene. A group of genes which exhibit four properties namely, multiplicity, close linkage, sequence homology and related or overlapping phenotypic functions. Also known as *repetitive genes*. These genes form a *multigene family*.

Regulated Gene. A gene whose expression regulated. Its product is synthesised in the cell type, at a time and in the amount at which it is required.

Regulator(y) Gene. A gene that codes for a protein, such as a *repressor*, involved in regulation of expression of other genes.

Relaxed Plasmid. A plasmid whose replication is not linked to replication of the host genome, and which therefore can exist a multiple copies within the cell.

Release Factors. Proteins (RF1, RF2, RF3) in prokaryotes responsible for termination of translation and release of the newly synthesised polypeptide when a nonsense codon appears in the A site of the ribosome. Replaced by eRF in eukaryotes. Also known as *termination factors*.

Renaturation. The return of a denatured molecule back to its natural state.

Renner Complexes. Specific gametic chromosome combinations in *Oenothera*.

Repair. The correction of alterations in DNA structure before they are inherited as mutations.

Repeated Genes. Some genes may be present in a haploid genome in many copies. These copies may be identical or only similar.

Repetitive and Unique Sequence Hypothesis. Only small proportion of DNA is unique and is meant for genes carrying the genetic information and rest of DNA is repetitive and has some other functions like control of gene activity and serving as raw material for evolution.

Repetitive DNA. DNA made up of copies of the same nucleotide sequence.

Replica-plating. A technique used to make replicas of bacterial colonies from a master plate by placing a velvet covering wooden block over the master plate and making an imprint of the same on a fresh plate. This method is used to select particular mutants from a plate.

Replicase. An enzyme that unwinds double helical DNA.

Replication. (1) The process of copying; replication of DNA. (2) The equal incorporation of all combinations two or more times in an experimental design, which is then said to be replicated.

Replicating Form (RF). The double-stranded form of single-stranded DNA viruses (for example, fX174).

Replication Fork. The region of a double-stranded DNA molecule that is being unwound to enable DNA replication to occur.

Replication Origin. A site on a DNA molecule where unwinding begins in order for replication to occur.

Replication-defective Virus. The virus which has lost one or more genes essential for completing the infective cycle.

Replication Eye. A region in which DNA has been replicated within a longer, unreplicated region.

Replicator. A DNA segment which contains an origin of replication and is able to promote the replication of a plasmid DNA molecule in a host cell.

Replicon. A sequentially replicating segment of a nucleic acid, controlled by a subsegment known as a *replicator.* A single replicator is present in the bacterial "chromosome", whereas the chromosomes of eukaryotes bear large numbers of replicons in series.

Replisome. The DNA-replicating structure at the Y-junction consisting of DNA polymerase III enzymes and a primosome (primase and DNA helicase).

Reporter Gene. A gene whose phenotype can be assayed in a transformed organism, and which can be used in, for example, deletion analysis of regulatory regions.

Repressible Control. End product (co-repressor) acts to repress reproduction of the enzymes. Transcription is initiated in the absence of the end product.

Repressible Enzyme System. A coordinated group of enzymes, involved in a synthetic pathway (anabolic), is repressible if excess quantities of the end product of the pathway lead to the termination of transcription of the genes for the enzymes. These systems are primarily prokaryotic operons.

Repression. The ability of a bacteria to prevent synthesis of certain enzymes when their products are present; more generally, refers to inhibition of transcription (or translation) by binding of repressor protein to a specific site on DNA (or mRNA).

Repressor. A regulator molecule (protein) produced by a regulator gene that can combine with and repress action of an associated operator.

Resistance Transfer Factor (RTF). Infectious transfer part of R plasmids.

Resistant. Characteristic of a host plant such that it is capable of suppressing or retarding the development of a pathogen or other injurious factor.

Resolvase. Enzyme actively involved in site-specific recombination between transposons present as direct repeats in a cointegrate structure.

Restriction Digest. The results of the action of restriction endonuclease on a DNA sample.

Restriction Endonuclease. Any of a group of enzymes that break internal bonds of DNA at highly specific points.

Restriction Enzyme. One of a number of enzymes that break DNA molecules down by causing cleavage at specific points in the molecule; the points are determined by the base sequence.

Restriction Fragment Length Polymorphism (RFLP). Variations in banding patterns of electrophoresed restriction digests. RFLPs have certain traits—lack of dominance, multiple allelic forms, lack of pleiotropic effects on agronomic traits, co-dominance, no measurable effect on phenotype, no effect of environment.

Restriction Map. A physical map of a piece of DNA showing recognition sites of specific restriction endonuclease separated by lengths marked in number of bases. Also known as *cleavage map.*

Restriction Site. The base sequence at which a restriction endonuclease cuts the DNA molecule, usually a point of symmetry within a palindrome sequence.

Reteroregulation. The ability of a sequence downstream to regulate translation of an mRNA.

Retroposon. A transposon that mobilise via an RNA form; the DNA element is transcribed into RNA, and then reverse-transcribed into DNA, which is inserted at a new site in the genome. Such as *retrotransposon.*

Retrovirus. A viral retroelement whose encapsidated genome is made of RNA.

Reversal of Central Dogma. A phenomenon mostly observed in RNA viruses where reverse transcriptase uses RNA as a template to produce complementary DNA.

Reverse Genetics. The use of recombinant DNA techniques to investigate gene function, usually by introducing specific mutations *in vitro*, e.g., by site-directed mutagenesis, and often to its original site in that host, by homogenotisation or transplacement. It is the opposite of conventional genetics where mutations are introduced at random into the whole genome followed by a lengthy procedure to select for the desired mutant.

Reverse Mutation. Any change towards the standard (wild-type) by way of a second mutational event. Also known as *reversion* or *back mutation.*

Reverse Transcriptase. An enzyme that synthesises a DNA copy on an RNA template.

Reverse Transcription. Synthesis of a DNA copy on an RNA template.

Reverse Translation. A technique for isolating genes (or mRNAs) by their ability to hybridise with a short oligonucleotide sequence prepared by predicting the nucleic acid sequence from the known protein sequence.

Reversion. Reversion of mutation is a change in DNA that either reverse the original alteration (*true reversion*) or compensates for it (*second site reversion* in the same gene).

Revertants. Revertants are derived by reversion of a mutant cell or organism.

R Factors. Plasmids that carry genes that control resistance to various drugs. Also known as *R plasmids.*

Rho Factor. A protein that is required for termination of transcription.

Rho-dependent Terminator. A DNA sequence signalling the termination of transcription, termination required the presence of rho protein.

Rho-independent Terminator. A DNA sequence signalling the termination of transcription, termination does not require the presence of rho protein.

Ribonuclease. An enzyme that degrades RNA.

Ribonucleic Acid (RNA). A single-stranded nucleic acid molecule, synthesised principally in the nucleus from deoxyribonucleic acid, composed of a ribose-phosphate backbone with purines (adenine and guanine) and pyrimidines (uracil and cytosine) attached to the sugar ribose. RNA functions to carry the "genetic message" from nuclear DNA to the ribosomes.

Ribosome Binding Site. The nucleotide sequence that acts as the site for attachment of a ribosome to an mRNA molecule.

Ribozyme. An RNA molecule that possesses catalytic activity.

Right Splicing Junction. The boundary between the right end of an intron and the left end of the adjacent exon.

Ring Chromosomes. A physically circular chromosome. Usually found in bacteria.

Ri Plasmid. An *Agrobacterium* plasmid effectively used as a vector for gene transfer in plant cells.

R Looping. The technique in which an RNA molecule is annealed to the complementary strand to a partially denatured DNA molecule. The formation of the RNA. DNA hybrid displaces the opposite DNA strands as a single-stranded bubble. These R-loops can be visualised under the electron microscope using the Kleinschmidt technique. It was the R-looping technique which first revealed the presence of introns in eukaryotic genes. See *D-looping.*

r^-m^-. Shorthand genetic nomenclature for a bacterial strain deficient in both the restriction and modification of DNA.

r^-m^+. Shorthand genetic nomenclature for a bacterial strain deficient in the restriction but not the modification of DNA. Such strains are useful hosts for the cloning of heterologous genes (r^+m^- strains would commit suicide by restricting their own, unmodified, DNA).

RNA-dependent DNA Polymerase. A group of enzymes that catalyse formation of DNA molecule from RNA template. These occur in some viruses (e.g., those that produce tumors). See also *reverse transcriptase.*

RNA Editing. A process by which nucleotides not coded by the gene are introduced at specific positions in an mRNA molecule after transcription.

RNA Phages. Phages whose genetic material is RNA.

RNA Polymerase. An enzyme capable of synthesising an RNA copy of a DNA template. In prokaryotes, there is only one type of RNA polymerase.

RNA Polymerase I. A type of eukaryotic RNA polymerase which transcribes rRNA genes.

RNA Polymerase II. A type of eukaryotic RNA polymerase which transcribes protein-encoding genes.

RNA Polymerase III. A type of eukaryotic RNA polymerase which transcribes small nuclear RNAs and tRNA genes.

RNA Polymerase Core Enzyme. In bacteria it consists of four subunits (a_2bb') and is capable of elongating already initiated transcription by holoenzyme.

RNA Polymerase Holoenzyme. The version of the *E. coli* RNA polymerase that has five subunits ($a_2bb's$) and is involved in efficient recognition of promoter sequences and is capable of initiation of transcription.

RNA Self-replication. The mechanism where RNA acts as a template for its own replication. This is considered to be second modification to the original central dogma. The first modification being ability of mRNA to act as a template for synthesis of DNA. RNA self-replication is known to exist in RNA phages such as R17, f2, MS2 and QB.

RNase. An enzyme that hydrolyses RNA.

RNase P. Enzyme that removes precursor sequences from the 5′ end of the pre-tRNA in eubacteria.

RNA-driven Hybridisation Reactions. The reactions which use an excess of RNA to react with all complementary sequences in a single-stranded preparation of DNA.

RNase D. The enyzyme responsible for processing of pre-tRNA by cleaving at the 3′ termini of the mature tRNA sequences.

RNase P. The enyzyme responsible for processing of pre-tRNA by cleaving at the 5′ termini of the mature tRNA sequences.

RNA Processing. Involves the steps which heterogenous nuclear RNA (pre-mRNA) transcribed by RNA polymerase II undergoes to produce a finished mRNA molecule. These steps involve (a) addition of a "cap" (5-m^7Gppp) at the 5′ end, (b) addition of a poly(A) tail at the 3′ end, (c) splicing of noncoding sequences (introns), and (d) methylation of one out of every 400 adenines present.

RNA Replicase. A polymerase enzyme that catalyses the self-replication of single-stranded RNA.

RNA Splicing. The removal of large noncoding sequences (introns) from the primary RNA transcript followed by rejoining of the nonadjacent coding sequences (exons) to produce the functional mRNA.

RNA Transcript. An RNA copy of a gene.

Rolling Circle Replication. A model of DNA replication that accounts for a circular DNA molecule producing linear daughter double helices which later circularise.

R_0t. The product of RNA concentration and time of incubation in an RNA-driven hybridisation reaction.

Runaway Plasmid Vector. A plasmid vector which at low temperature (30°C) is present in a moderate number of copies per cell because of some control on plasmid replication but at a higher temperature (35°C) all control on plasmid replication is lost and number of copies per cell increases continuously.

Runaway Replication. Uncontrolled replication of plasmid molecules which multiply within the host building up to several thousand copies. This usually halts cell division but is a very powerful way of amplifying a gene product. Plasmid vectors which are temperature sensitive for a copy number control element have been constructed as runaway replication vectors.

S-adenosylmethionine (SAM or AdoMet). A molecule which donates its methyl group to cytosine or adenine during DNA methylation.

Salivary Chromosomes. The giant chromosomes in the cells of the salivary glands of larval flies and other dipterans such as mosquitoes and *Drosophila* characteristically each shows a highly specific pattern of contrasting bands. Also known as *giant chromosomes* or *polytene chromosomes*.

Satellite. A portion of the chromosome separated from the main body of the chromosome by secondary constrictions.

Satellite Association. All the chromosomes that have satellites tend to arrange themselves in a group during metaphase. This poses difficulty in assigning a particular number to a particular chromosome.

Satellite Chromosome. Chromosomes that seem to be additions to the normal genome.

Satellite DNA. DNA comprising of clustered repetitive sequences, so-called because it forms a satellite band in a density gradient.

sB Site. Host specificity site of *E. coli* B strain.

Scaffold. The eukaryotic chromosome structure remaining when DNA and histones have been removed; made from nonhistone proteins.

Scanning Hypothesis. Proposed mechanism by which the eukaryotic ribosome recognises the initiation region of an mRNA after binding the 5′ capped end of it. The ribosome scans the mRNA for the initiation codon.

scRNA. Any one of several small cytoplasmic RNA, molecules present in the cytoplasm and (sometimes) nucleus.

scRNPs. Small cytoplasmic ribonucleoproteins. scRNAs are associated with proteins.

Secondary Non-disjunction. A type of secondary non-disjunction which occurs during anaphase II when centromere fails to divide.

Secondary Response. The immunological response to a second challenge with a particular antigen. Secondary responses give rise to much higher levels of antibody in a much shorter time than do the primary responses.

Second Messenger. A molecule (cAMP) that mediates response of eukaryotic cells to altered extracellular environment.

Second Site Reversion. A second mutation that reverses the effect of a previous mutation in the same gene although without restoring the original nucleotide sequence.

Sedimentation Coefficient (S). A value used to express the velocity with which a molecule or structure sediments when centrifuged in a dense solution. Expressed as *Svedberg unit.*

Segmental Allopolyploid. An allopolyploid in which the combined genomes are homologous in many small segments throughout complement; crossing over may recombine material from different genomes.

Segment Polarity Genes. The genes concerned with controlling the number or polarity of body segments in insects. The *Drosophila* mutants have lost part of the 'P' compartment of each segment, and it has been replaced by a mirror image duplication of the 'A' compartment.

Segregation Distorter (SD). A factor that alters the segregation ratio in heterozygous *Drosophila*; SD^+ cells are eliminated during spermiogenesis in SD^+/SD males.

Selective Amplification. A type of amplification where a part of the genome has increased copy number. Selective amplification is a regulated developmental process occasionally used to increase the output of high demand gene.

Selector Genes. The genes which cause cells to commit themselves to a particular developmental direction. For example, a gene that influences the development of specific body segments in *Drosophila*; a homeotic gene.

Selfish DNA. DNA that appears to have no function and apparently contributes nothing to the cell in which it is found. Also known as *ignorant* DNA or *junk* DNA.

Self-splicing of Intron. In some lower eukaryotes, like *Tetrahymena*, an intron in pre-rRNA is self-spliced involving intramolecular recombination. Self-splicing occurs through transesterification mechanism.

Semi-conservative Replication. The mode of DNA replication in which each daughter double helix comprises one polynucleotide from the parent and one newly synthesised polynucleotide.

Semispecies. The populations that have acquired some but not yet all attributes of a species rank. These are border cases between species and subspecies.

Sense Codon (Sense Word). A codon specifying a particular amino acid in protein synthesis.

Sense DNA Strand. The strand of DNA which is used as a template during transcription. Also known as *coding strand.*

Sensor Gene. A eukaryotic regulatory gene that is sensitive to external/developmental signals.

Sequon. During protein secretion proteins with asparagine residues in the tripeptide motif An-Xaa-Ser/Thr are N-glycosylated with performed oligosaccharide units. These tripeptide motifs are known as sequon.

Serial Analysis of Gene Expression (SAGE). A technique that allows quantitative and simultaneous analysis of a large number of transcripts. It is based on the principle that a short nucleotide tag (9–10 bp) contains sufficient information to uniquely identify a transcript, and that concentration of short sequence tags allows efficient analysis of transcripts in a serial manner.

70S Initiation Complex. The complex formed at the second stage of initiation of polypeptide synthesis in prokaryotes. Two sites exist on 50S subunit–one for attachment of aminoacyl and the other for growing polypeptide chain attached to tRNA.

Sex Correction. Normal individuals with abnormal chromosome constitution. For example, normal individuals with chromosome constitution 46 + XX were produced from a loss of Y-chromosome from a 46 + XXY complement.

Sex Differentiation. Mechanisms that determine whether a developing embryo would be a male or a female.

Sexduction. Incorporation of bacterial chromosomal genes in the fertility plasmid, with subsequent transfer to a recipient cell in conjugation.

Sex Factor (F). An episome of bacteria which, when integrated into the chromosome, causes chromosome breakage at conjugation and facilitates genetic exchange; also exists autonomously in F^+ cell. Also known as *fertility factor* or *sex-particle*.

Sex Factor Affinity (SFA). A region on the bacterial chromosome that has a specific affinity for the sex factor. It is a region consisting of sex factor DNA left behind when the sex factor recombined out of the bacterial chromosome.

Sex-lethal. A gene in *Drosophila*, located on the X chromosome, that is a sex switch, directing development regulated by numerator and the denominator elements that act to influence the genic balance equation (X/A).

Sex Plasmid. Actually an episome, it is able to initiate the process of conjugation, by which chromosomal material is transferred from one bacterium to another.

Sex-promoting Alleles. Certain alleles *xa, xb, xc,* etc., concerned with sex determination. Females are always heterozygous, males homozygous or hemizygous.

Sex Reversal. Transformation of one sex into another.

Sex Switch. A gene in mammals, normally found on the Y chromosome, that directs the indeterminate gonads toward development as tests.

Sharon Phages. See *charon phages*.

Shine-Dalgarno Sequence. The prokaryotic ribosome-binding site AGGAGG located on mRNA just prior to AUG initiation codon which has complementarity with the 3′ end of 16S rRNA.

Short Interspersed Nuclear Element (SINE). A type of dispersed repetitive DNA, typified by the *Alu* sequences found in the human genome.

Short-period Interspersion. A pattern in a genome in which moderately repetitive DNA sequences of ~300 bp alternate with nonrepetitive sequences of ~1000 bp.

Short-term Gene Regulation. Gene regulation recognised in enkaryotes which operates in response to fluctuations in environment, changes in activities or concentrations of substrates, end-product or hormone levels. It is a feature of both developing and fully differentiated cells, or operates even when the cell is undergoing differentiation.

Shotgun Experiment. Cloning of an entire genome in the form of randomly generated fragments.

Shuttle Mutagenesis. In this process of mutagenesis, first a gene is cloned and then mutated through the use of transposons and the mutated gene is then transferred back in the original organism.

Shuttle Vector. A plasmid constructed to have origins for replication for two hosts (for example, *E. coli* and *S. cerevisiae*) so that it can be used to carry a foreign sequence in either prokaryotes or eukaryotes.

Sibling Species. Morphologically similar or identical populations that are reproductively isolated. These species may be sympatric or allopatric. These species may exit at the same time or at different times.

Sigma Factor. A proteinous component of RNA polymerase which is must for initiation of transcription. It is not required for elongation of the transcript.

Signal Codons. The codons which code for either start or stop signal in protein synthesis.

Signal Hypothesis. The major mechanism whereby proteins that must insert into or across a membrane are synthesised by a membrane-bound ribosome. The first thirteen to thirty-six amino acids synthesised, termed a *signal peptide*, are recognised by a *signal recognition particle* that draws the ribosome to the membrane surface. The signal peptide may be removed later from the protein.

Signal Molecule. Any molecule capable of giving indication to living cells to perform a specific biological function. Such a signalling phenomenon has been observed in *Agrobacterium*.

Signal Sequence. Protein sorting requires a specific amino acid sequence of the newly synthesised protein which interacts with membranes. These amino acid sequences are recognised by specific receptors located within the membrane. Also known as *transit peptide* or *leader sequence*.

Signal-sequence Receptors. The receptors which seem to determine the fate of the entry of different proteins into different membrane bound compartment.

Signal Transducers and Activators of Transcription (STATs). These are the proteinous factors which after phosphorylation by kinases move to the nucleus, bind to specific DNA elements and direct transcription of interferon genes whose protein products interface with viral replication.

Signal Transduction. The process by which a receptor interacts with a ligand at the surface of the cell and then transmits a signal to trigger a pathway within the cell.

Silencer. DNA sequence that helps to inhibit transcription, located at a long distance from the core promoter.

Silent Mutation. An alteration in a DNA sequence which dose not affect the expression or functioning of any gene or gene product.

Silent Sites. The sites in a gene describe those positions at which mutations do not alter the product.

Simple Gene. A continuous sequence in a nucleic acid that specifies a particular polypeptide or functional RNA.

Simple-sequence Family. The simple sequence family encompasses segments of DNA derived from 10^2–10^7 repetitions of a short fundamental sequence, generally 6–15 nucleotides in length. Degree of homology amongst the repeat units is often 80–100%.

Simple Transcription Unit. When a primary transcript produces only one type of mRNA.

Single-copy Plasmid. The plasmids that are maintained in bacteria at a ratio of one plasmid for every host chromosome.

Single-strand Assimilation. The ability of RecA protein to cause a single strand of DNA to displace its homologous strand in a duplex, that is, the single strand is assimilated into the duplex.

Single-strand Binding (ssb) Protein. One of the proteins that attaches to single-stranded DNA in the region of the replication fork, preventing reannealing of unreplicated DNA.

Site-directed Mutagenesis. It is construction of mutations at the predetermined site of a cloned DNA, precisely defining the nature of mutational change and then test functional effect of that mutation *in vivo* or *in vitro*.

Site-specific Recombination. A specialised recombination in which a plasmid is inserted into or is spliced out of a host chromosome. e.g., F-factor and *E. coli* chromosome.

sK Site. Host specificity site of *E. coli* K strain.

Smallest Eukaryote. *Ostreococcus tauri* is a photosynthetic picoeukaryote which is classified as green algae. Electron microscopy reveals their extreme simple ultrastructure. Mean length and width are 0.97 and 0.70 mm, respectively, and the DNA content per cell is 33.31 fg.

Smallest Know Gene. Microcin C7 (Mcc C7), a modified linear heptapeptide that inhibits protein synthesis in Enterobacteriaceae, consists of Acetyl Met-Arg-Thr-Gly-Asn-Ala-Asp. It is synthesised from a 21 bp open-reading-frame gene called *mccA*. This is one of the smallest genes so far identified.

Small Nuclear Ribonucleoproteins (snRNPs). Small nuclear ribonucleoproteins; components of the spliceosome, the intron-removing apparatus in eukaryotic nuclei.

Small Nuclear RNA (snRNA). The RNA component of the nucleus that comprises relatively small molecules thought to be involved in splicing and other transcript processing events.

Smart Transcription Factors. These are the proteinous transcription factors that are smart enough to make their contact with proteins though their phosphorylation to initiate transcription.

Solenoid. A higher order structure in a eukaryotic chromosome in which 100 Å fibers of nucleosomes are stacked with 5 or 6 nucleosomes/helix. H1 histone helps in solenoid formation which gives a package ratio of 1 : 50.

Somaclonal Variation. Variation produced in plants regenerated from tissue culture involving callus formation.

Somatic Cell Hybridizatyon. A technique that permits hybridisation between somatic cells of same or different species in tissue culture.

Somatic Cell Mutation and Recombination (SMART). A test which detects induction of mutations and recombi-nation in somatic cells (like eyes and wings of *Drosophila*). it is used to test mutagenicity of individual chemical compounds and mixtures.

Somatic Doubling. A disruption of the mitotic process that produces a cell with twice the normal chromosome number.

Somatic Hypermutation. The occurrence of a high level of mutation in the variable regions of immunoglobulin genes.

Somatic Diversity Theory. A theory to account for the high degree of antibody variability. It suggests that mutation of a basic immunoglobulin gene accounts for all the different types of immunoglobulins produced by B lymphocytes. See *germ-line theory*.

Somatic Mutation Theory of Ageing. Mutations-occur randomly and spontaneously destroying genes and chromosomes in post-mitotic cells during the life-span of an organism and this gradually increases the mutation load, decreases the production of functional proteins.

Somatic Pairing. The phenomenon of close association of homologues in somatic cells, as seen in dipteran *giant chromosomes*.

Somatic Sexual Development. A process that determines phenotypic sex of an individual.

S1 Nuclease. An enzyme that degrades specifically single-stranded molecules or single-stranded regions in predominantly double-stranded nucleic acid molecules.

SOS Box. The region of the promoters of various genes that is recognised by the LexA repressor. Release of repression results in the induction of the SOS response. SOS stands for "*save our souls*".

SOS Repair System. Repair systems (recA. uvr) induced by the presence of single-stranded DNA that usually occurs from post-replicative gaps caused by various types of DNA damage. The RecA protein, stimulated by single-stranded DNA, is involved in the inactivation of the LexA repressor, thereby inducing the response. This is also known as SOS *response*.

Southern Blotting. A method, first devised by E.M. Southern, used to transfer DNA fragments from an agarose gel to a nitrocellulose gel for the purpose of DNA-DNA or DNA-RNA hybridisation during recombinant DNA work.

Spacer DNA. Regions of nontranscribed DNA between transcribed segments. Term generally applies to immunoglobulin genes.

Specialised Gene. A gene whose activity is under the control of constitutive as well as regulatory factors.

Specialised Nucleoprotein Structures (snups). These are small deoxyribonucleoprotein particles that are involved in *high precision* DNA *transactions*.

Specialised Nucleoriboprotein Structures (snurps). These are small ribonucleoprotein particles that help to remove meaningless introns from the message issued by a cell's genes.

Specialised Transduction. A type of transduction where only a few bacterial genes are transferred because the phage has only specific sites of integration on the host chromosome.

Speciation. A process whereby, over time, one species evolves into a different species (*anagenesis* or *phyletic speciation*) or whereby one species diverges to become two or more species (*cladogenesis* or *true speciation*).

Species. (1) A group of actually or potentially interbreeding natural populations which are reproductively isolated from other such groups (Mayrs, 1910). (2) A largest and most conclusive reproductive community of sexual and cross-fertilising individuals which share a common gene pool (Dobzhansky, 1937). (3) A lineage evolving separately from others and with its own evolutionary role and tendencies (Simpson, 1949).

Specific-locus Test. A system for detecting recessive mutations in diploids. Normal individuals treated with mutagen are mated to tasters that are homozygous for the recessive alleles at a number of specific loci; the progeny are then screened for recessive phenotypes.

Specific Modifier. A gene that has the specific and perhaps exclusive function of modifying the expression of a gene at another locus.

Spheroplast. It is a bacterial or yeast cell whose wall has been largely or entirely removed.

Spherosomes. Vescicles in plant cells which appear to perform functions similar to those of lysosomes in animal cells.

Spiral Cleavage. The cleavage process in molluscs and some invertebrates whereby the spindle at mitosis is tipped in relation to the original egg axis.

Spliceosome. The protein-RNA structure believed to be responsible for splicing.

Splicing. The removal of introns from the primary transcript of a discontinuous gene.

Split Gene. A discontinuous gene. A gene in which coding sequences are separated by noncoding sequences.

Spontaneous Generation. The sudden, spontaneous origin of organisms from inert matter. The theory is now discredited.

Spontaneous Mutations. Genetic changes produced under normal growth conditions.

Spore-specific Repair. A type of DNA repair which in *Bacillus subtilis* restores the spore-specific UV photoproduct 5-thyminyl-5, 6-dihydrothymine into two residues in site leaving the DNA backbone intact.

Stable RNA. RNA molecules, such as rRNA and tRNA, but not mRNA, which are not subject to rapid turnover in the cell.

Stabilising Selection. A type selection that removes individuals from both ends of a phenotypic distribution divided by deviation of a sample of means.

Stable mRNA. An mRNA species that persists in time instead of being rapidly degraded. The stabilisation is probably the result of complexing with a protein.

Stacking. The packing of the flatish nitrogen bases at the center of the DNA double helix.

Stacking Forces. Weak hydrophobic forces between two adjacent base pairs of double helical DNA.

Stages in Evolutionary Divergence. According to Nie, there are five stage of increasing evolutionary divergence during speciation: (a) geographic species, (b) subspecies, (c)semi- or incipient species, (d) sibling species, and (e) full species.

Staggered Cuts. Cuts in duplex DNA are made when two strands are cleaved at different points near each other.

Start Codon. A codon which codes for initiation of protein synthesis, i.e., AUG.

Start Point. The position on DNA corresponding to the first base incorporated into RNA. Also known as *start site* or *startise*.

Stasipatric Speciation. Instantaneous speciation caused by polyploidy.

Statistical Proteins. In absence of a genetic code, proteins synthesised will have a random sequence of amino acids.

Stem-loop Structure. A lollipop-shaped structure formed when a single-stranded nucleic acid molecule loops back on itself to form a complementary double helix (stem), topped by a loop.

Step Allelomorphism. The occurrence of a series of multiple allelomorphs with overlapping effects which can supposedly be related to a linear order in the distribution of units of change within the gene.

Steps in Evolution of Organisms. Heterotrophic → Chemoautotrophic → Photoautotrophic → Photosynthesis → Respiration.

Sticky End. An end of a double-stranded DNA molecule where there is a single-stranded extension. Also known as *adhesive end*. Compare with *blunt end*.

Stochastic DNA Rearrangement. A type of DNA rearrangements involved in expression of certain genes in prokaryotes. These rearrangements take place without any need so that a small fraction of cell population carrying rearranged genes are always ready.

Stop Codons. The codons which provide signal for termination of polypeptide chain, viz., UAA, UAG and UGA.

Strand Displacement. It is a mode of replication of some viruses in which a new DNA strand grows by displacing the previous (homologous) strand of the duplex.

Strand Transfer Repair. A DNA repair mechanism in which a recombination-like process forms gaps opposite the induced lesion in a replicating template strand. These gaps are then filled in via strand transfer from the previously replicated normal strand.

Stringent Factor. A protein that catalyses the formation of an unusual nucleotide (guanosine tetraphosphate) during the stringent response under amino acid starvation.

Strong Promoter. A promoter whose copy number is under strict control and, therefore, has only one or two copies per chromosome, i.e., a low copy number plasmid, e.g., the F plasmid.

Structural Alleles. Alleles determined on the basis of recombination test.

Structural Gene. A gene that codes for an RNA molecule or protein other than a regulatory protein.

Structural Genomics. It represents an initial phase of genome analysis and involves the construction of high resolution genetic, physical and transcript maps of an organism.

Structural Hybridity. Heterozygosity for a chromosomal rearrangement.

Structural Proteins. The proteins that determine structure, form and properties of various parts of living cells.

Sub-clone. A method in which smaller DNA fragments are cloned from a large insert which has already been cloned in a vector.

Submetacentric Chromosome. A chromosome where the centromere is nearer one end than the other, resulting in the arms not being of equal length.

Subspecies. One of the several geographic or similar subdivisions between which interbreeding takes place at a reduced level.

Subvital Gene. A gene that causes the death of some proportion (but not all) of the individuals that express it.

Succession. The sequence of transient communities occurring in an area before the climax.

Sudden Speciation. New species evolve suddenly through individuals. It occurs through individuals. It is also known as speciation by saltation. Processes may be genetical (through macromutations) or chromosomal (through hybridisation and/or polyploidy). Plausibility of macromutations as a mechanism is less.

Supercoiling. Negative or positive coiling of double-stranded DNA that differs from the relaxed state.

Superfemale. An abnormal type, almost completely sterile, in *Drosophila* with an X chromosome/autosome set (X/A) ratio greater than 1, e.g., 3X/2A = 1.5.

Supermale. An abnormal type, almost completely sterile, in *Drosophila* with the X chromosome/autosome set (X/A) ratio less than 0.5, e.g., 1X/3A = 0.33.

Supergene. Advantageous grouping of genes within an inversion; the entire gene complex is inherited as a whole.

Supernumerary Chromosome. Chromosome present, often in varying numbers, in addition to the characteristic relatively invariable complement.

Super Suppressor. A mutation that can suppress a variety of other mutations; typically a nonsense suppressor.

Supplementary Genes. Two dominant genes when present together produce a novel phenotype. These genes yield 9:3:4 ratio in F_2 progeny.

Suppression. A mutation in a gene that reverses the effect of a previous mutation in a different gene. This term also refers to *rescue* of an allele by means of another gene.

Suppressor, Extragenic. Usually a gene coding a mutant tRNA that reads the mutated codon either in the sense of the original codon or to give an acceptable substitute for the original meaning.

Suppressor Gene. A gene that, when mutated, apparently restores the wild-type phenotype to a mutant of another locus.

Suppressor, Intragenic. A compensating mutation that restores the original reading frame after a frameshift.

Suppressor Rescue. A transferred gene is tagged with an *amber* suppressor gene to isolate a human oncogene.

Suppressor T (Ts) Cells. These T cells which suppress the immune response by interacting with the B cells.

Surrogate Genetics. A branch of genetics which deals with introduction of a manipulated DNA into a living nucleus where its expression can be monitored. Effects of directed DNA changes, changes in RNA transcript and changes in protein encoded by it can all be studied.

Survival of the Fittest. In evolutionary theory, survival of only those organisms best able to obtain and utilise resources (fittest). This phenomenon is the cornerstone of *Darwin's theory.*

Susceptible. Characteristic of a host plant such that it is incapable of suppressing or retarding an injurious pathogen or other factor.

S Value. The unit of measurement of a sedimentation coefficient. Also known as *Svedberg unit.*

Swivel. The place in DNA at which replication starts.

Symbiosis. Permanent or long-lasting association between two or more different species of organisms.

Symbiotic Theory. This theory of origin of eukaryotic cell was given by Marguilis in 1970. A prokaryotic cell having a flexible cell membrane ingested bacterial cell which evolved into mitochondria. Spirochetes became ingested and converted into nuclear fibers, centrioles and flagella of a primitive unicellular eukaryote. Blue green algae cells became incorporated into colourless unicellular eukaryotes to form a simple green flagellate. This is the most favoured theory.

Sympatric Speciation. Speciation in which the evolution of reproduction isolating mechanisms occurs within the range and habitat of the present species. This speciation may be common in parasites.

Synergistic Gene Expression. When two transcription activation factors together lead to higher rate of transcription than sum of the activation of either of the factors individually does.

Synonymous Codons. Different codons that specify the same amino acid.

Synteny. The occurrence of two or more genetic loci on the same chromosome. Depending on intergene distance (s) involved, they may or may not exhibit nonrandom assortment at meiosis.

Synthetic Theory of Evolution. The current extension of Darwin's theory of evolution, in which mutation, natural selection, and reproductive isolation play major roles.

Synthetic Vaccines. The vaccines which contain not the intact viruses but merely peptides that are constructed in the laboratory to mimic a very small region of virus's outer coat that can induce formation of antibodies capable of neutralising the virus.

TACTAAC box. A consensus sequence surrounding the lariat branch point of eukaryotic mRNA.

Tandem Repeats. Direct repeats that are adjacent to each other with no intervening DNA.

Targeted Gene Modification. In this method of mutation induction, genes are introduced by any one of the traditional gene transfer methods. This is followed by site-specific mutations.

Targeted Gene Replacement. In this method of gene replacement, homologous recombination replaces a resident gene with a foreign gene.

Targeted Gene Transfer. The gene transfer at homologous sites in the host genome.

Target Site Duplication. A sequence of DNA that is duplicated when a transposable element inserts, usually found at each end of the insertion.

Target Theory. A theory that predicts response curves based on the number of events required to cause the mutational phenomenon in a gene. This theory was used to estimate size of the gene.

TATA Box. A component of the nucleotide sequence located at position –25 that makes the prokaryotic promoter.

Tautomeric Shift. Reversible shift of proton position in a molecule. Based on nucleic acids shift between keto and enol forms or between amino and imino forms.

Tay-Sachs Disease. A degenerative brain disorder of infancy due to an autosomal recessive allele in the gene controlling the enzyme hexosaminidase A. The age of on-set of the disease is 4 to 6 months. Affected children show progressive mental deterioration, paralysis, deafness, blindness, and convulsions, leading to death usually between the ages of 3 and 5 years.

T-cells. Lymphocytes of the T (thymic) lineage; may be subdivided into several functional types. Then carry TcR (T cell receptor) and are involved in the cell-mediated immune response.

T-cell Receptors. Surface cells of T-cell that allow the T cells to recognise host cells that have been infected.

T-DNA. Complete Ti plasmid is not found in plant tumor cells, only 20kb DNA of Ti plasmid called T-DNA is found integrated in plant nuclear DNA. Genes of T-DNA are eukaryotic in origin and have been captured by Ti plasmid during evolution.

Telomerase. The enzyme that maintains the ends of chromosomes by synthesising telomeric repeat sequences.

Temperate Phage. A bacteriophage that is able to follow a lysogenic mode of infection.

Temperature-sensitive Mutation. An organism with an allele that is normal at a permissive temperature but mutant at a restrictive temperature.

Template. A pattern serving as a mechanical guide. In DNA replication, each strand of the duplex acts as a template for the synthesis of new double helix.

Terminal Redundancy. The repetition of the same sequence at both ends of, for example, a phage genome.

Terminalisation The movement of the chiasma toward the ends of the bivalent during meiosis.

Termination Codon. One of the three codons (5′-UAA-3′, 5′-UAG-3′ and 5′-UGA-3′ in the standard genetic code) that mark the position where translation of an mRNA should stop.

Termination Factors (TF). Proteins required to obtain release of newly synthesised polypeptide chain from tRNA. Also known as *release factors.*

Terminator Sequence. A sequence in DNA that signals the termination of transcription to RNA polymerase.

Terminator Stem. A configuration of the leader transcript that signals transcription termination in attenuator-controlled amino acid operons.

Testicular Feminisation. The creation of an apparent female phenotype in an XY individual as a result of an X-linked mutation.

Testis-determining Factor (TDF). General term for the gene determining maleness in human beings.

Thermoregulation of a Gene. Regulation of expression is controlled by temperature. Histone-like bacterial proteins seem to play role in this process.

30-nm Chromatin Fiber. The substructure of chromatin that consists of a possibly helix array of nucleosomes in a fiber approximately 30 nm in diameter.

30S Initiation Complex. The complex first formed during initiation of polypeptide synthesis in prokaryotes. It requires mRNA, 30S subunit of ribosome, a special initiating species of amino-acyl-tRNA and three initiation protein factors (IF_1, IF_2 and IF_3).

Thymine Dimer. Hydrogen bonding of two molecules of thymine by action of ultraviolet light.

Ti-plasmid. The large plasmid found in *Agrobacterium tumefaciens* cells and used as the basis for a series of cloning vectors for higher plants.

3′-OH Terminus. The end of a polynucleotide which terminates with a hydroxyl group attached to the 3′ carbon of the sugar.

Tissue Culture. A technique by which a full fledged individual can be developed, on a well defined medium, from a somatic cell or an individual sex cell.

Tolerable Mutations. The mutations which affect function of the gene to such a degree that they do not have a drastic effect on the individual and are tolerable. Such mutations accompany the process of evolution and play important role in speciation. Tolerable mutations are of two types—neutral and favourable.

Topoisomerase. An enzyme that can relieve (or create) supercoiling in DNA by creating breaks in one (type I) or both (type II) strands of the helical backbone.

Totipotency. The property of a cell (or cells) whereby it develops into a complete and differentiated organism to any and all adult cell types, as compared with a differentiated cell whose fate is determined.

Tracer. A radioactivity labelled nucleic acid component included in a reassociation reaction in amounts too small to influence the progress of reaction.

***Tra* Gene.** One of a group of genes carried by an F plasmid and coding for proteins involved in DNA transfer during conjugation.

Trailer Segment. The untranslated segment that lies downstream of the nonsense codon of an mRNA molecule. Also known as *train*.

Trans-acting. Referring to mutations of, for example, a repressor gene, that act through a diffusible protein product; the normal mode of action of most recessive mutations.

Transcribed Spacer. Part of an rRNA transcription unit that is transcribed but discarded during maturation; that is, it does not give rise to part of rRNA.

Transcriptional Element. DNA sequences in eukaryotic genes which control transcription of its gene.

Transcriptional Factors. Eukaryotic proteins that aid RNA polymerase to recognise promoters. Analogous to prokaryotic sigma factors.

Tyrosinosis. A human disease caused by metabolic defect and characterised by excretion of p-hydroxyphenyl pyruvic acid and tyrosine into urine.

U1-U7 RNA. A family of stable, small nuclear RNA molecules orginally found in rat Novikoff hepatoma cells but subsequently discovered in a range of mammals and other higher eukaryotes. The U RNAs are used in pre-mRNA processing.

mm (micrometer). 1×10^{-6} meter.

Ubiquitin. A heat-shock protein present in yeast and chicken cells, although apparently not present in *Drosophila*.

Underdominance. A phenotypic relation in which the phenotypic expression of the heterozygote is less than that of either homozygote.

Underwinding. Underwinding of DNA is produced by negative supercoiling (because the double helix is itself coiled in the opposite sense from the intertwining of the strands.)

Unicistronic mRNA. An mRNA molecule which contains information for synthesis of only one polypeptide chain.

Unidentified Reading Frame (URF). An open reading frame recognised from a DNA sequence for which no genetic function is known.

Unidirectional Replication. Movement of a single replication fork from a given origin.

Uninemic Chromosome. A chromosome consisting of one double helix of DNA.

Uniparental Inheritance. Inheritance pattern where the offspring have received certain phenotypes from only one parent. This inheritance pattern is due to transmission of DNA containing cytoplasmic particles.

Unique DNA Sequences. A length of DNA with no repetitive nucleotide sequences. Majority of the structural genes are unique DNA sequences. Also known as *single copy* DNA.

Unisite Mutant Allele. A mutant allele differing from its wild-type form at only one site.

Universal Codon. A triplet codon codes for same amino acid in diverse forms of life, viz., bacteria, wheat, *Drosophila*, man.

Universal Nucleotide. 1-(2′ deoxy-b-D-ribofuranosyl)-3-nitropyrrole, designated "M" is a universal nucleotide that maximises stacking while minimising hydrogen-bonding interactions without stearically disrupting a DNA helix. The universal nucleotide is used at ambiguous sites in DNA primers.

Unmethylated Site. When neither of the complementary strands of DNA carries methyl group at a site. Also known as *homoduplex unmodified* DNA.

Unstable Gene. A gene which mutates frequently.

Unstable Mutation. A mutation that has a slight frequency of reversion; a mutation caused by the insertion of a controlling element, whose subsequent exit produces a reversion.

Unusual Bases. Other bases, in addition to adenine, cytosine, guanine, and uracil, found primarily in tRNAs.

Unusual Chromosomes. Chromosomes which show adaptational forms of the normal chromosomes (e.g., lampbrush chromosomes, polytene chromosomes) or these may be permanently specialised structures (e.g., B chromosomes, sex chromosomes).

Unwinding Proteins. Proteins that bind to , and unwind, the DNA helix at the replicating fork.

Upgrading Regulation. A mechanism in dosage compensation where a single X chromosome in one sex elevates expression of its genes to equalise the expression of two X chromosomes in the other sex. For example, in *Drosophila* hypertranscription exists.

Up Promoter Mutations. Mutations in promoter that increase the frequency of initiation of transcription.

Upstream. Towards the 5′ end of polynucleotide represents sense strand of DNA.

Upstream Activating Sequence (UAS). This is a DNA segment required for activating transcription of eukaryotic genes.

U-RNP. A nuclear particle, consisting of one or two U-RNAs and several proteins, involved in splicing and other transcript processing events.

Use and Disuse Doctrine. The discredited idea that the more a structure is used, the more prominent it becomes in future generations, and that the less it is used, the less prominence it assumes in later generations. Also known as the doctrine of acquired characteristics.

U snRNPs. U-class small ribonucleoproteins that regulate sex-specific alternative splicing of transcripts required for sexual differentiation.

Variable-number-of-tandem Repeat (VNTR) Loci. Loci that are hypervariable because of tandem repeats. Presumably, variability is generated by unequal crossing over.

Variable Region. An immunoglobulin chain is coded by the V gene and varies extensively when different chains are compared, as the result of multiple (different) genomic copies and changes introduced during construction of an active immunogobulin.

Vector. A DNA molecule capable of replication into which a gene is inserted by recombinant DNA techniques.

Vectorial Discharge. A step involved in protein secretion in eukaryotes where a polypeptide is fed into endoplasmic reticulum lumen through Sec61 complex. Also known as *co-translational import.*

Vehicle. The host organism used for the replication or expression of a cloned gene or other sequence. (Compare with *vector*, the DNA molecule which contains the cloned gene.) The term is little used and is often confused with *vector.*

Vehicle Plasmid. A plasmid containing a piece of passenger DNA; used in recombinant DNA work.

Vertical Resistance. The type of resistance which is governed by "major" genes which have large effects and reveal gene-for-gene relationship.

Viral Retroelement. A virus whose genome replication process involves reverse transcription.

Virion. The genome of virus.

Viroids. Bare RNA particles that are plant pathogens.

Virulence. The extent to produce disease in any particular instance.

Virulent Bacteria. A disease causing bacteria leading to death of the host.

Virulent Phage. Refers to a bacteriophage that follows the lytic mode of infection and destroys its host bacterial cell.

V-J Joining. The joining of a variable gene and a joining gene in the first step of the formation of a functioning immunoglobulin gene.

V-onc. The version of an oncogene carried by a transforming retrovirus.

Wandering Spot Analysis. In this method of DNA sequencing, given by Sanger et al., DNA is labelled with ^{32}P. performed. Two-dimensionally separated nucleotides are autoradiographed. Partial digests are applied to cellulose acetate strip and on the basis of shift in mobility the results are inferred.

Watson-Crick Base Pairing. The normal base pairing, viz., A with T and G with C in DNA and A with U and G with C in RNA.

W Chromosome. Used for X chromosome where female is the heterogametic sex.

Western Blotting. A technique for probing for a particular protein using antibodies. See *Southern blotting.*

Whole Genome Amplification. A type of genome amplification which increases copy number of the complete genome in the cell.

Wobble Base Pairing. The pairing of mRNA codon with tRNA anticodon in which first two bases of a codon have normal pairing and the third base has abnormal base pairing.

Wobble Hypothesis. The partial or total lack of specificity in the third base of some triplet codons whereby two, three, or four codons differing only in the third base may code for the same amino acid.

X/A Ratio. Ratio between number of X chromosomes to number of sets of autosomes that determines sex in *Drosophila.*

Xanthophyll. A yellow-coloured compound $C_{40}H_{56}O_2$ found in plants.

X chromosome. The sex chromosome that is represented twice in the homogametic sex but only once in the heterogametic sex.

Xenia Effect. Direct effect of male gamete on tissues other than embryonic ones.

Xeroderma Pigmentosum. A genetic disorder in which the skin is extremely sensitive to sunlight and death usually occurs from skin cancer. It is inherited as an autosomal recessive condition.

X Inactivation. The genetic inactivation of all X chromosomes in excess of one, taking place on a random basis in each cell in an early stage in embryogenesis.

Xis Protein. A protein required in excisive site-specific recombination between l phage and *Escherichia coli* between prophage *attL* and *attR* sites. This protein is also required for formation of a localised nucleoprotein structure on *attR* DNA. Also known as *excisive protein*.

XO Condition. Having an X chromosome but no Y chromosome, as the female in poultry and some other birds.

Yeast Artificial Chromosome (YAC). A cloning vector comprising the structural components of a yeast chromosome and able to clone very large piece of DNA.

Yeast Centromere Plasmid (YCp). Plasmid which carries functional yeast centromere.

Yeast-episomal Plasmid (YEp). Many strains of yeast contain a 2 mm (6 kb) long plasmid that has no known function.

Yeast Integrative Plasmid (YIp). A yeast vector that relies on integration into a host chromosome for replication.

Yeast-replicating Plasmids (YRp). These plasmids contain autonomous replicating sequences (ARS) derived from yeast chromosome.

Y-junction. The point of active DNA replication where the double helix opens up so that each strand can serve as a template.

Z Chromosome. Used for the Y chromosome where female is the heterogametic sex.

Zero Time-binding DNA. DNA enters the duplex from at the start of a reassociation reaction; results from intramolecular reassociation of inverted repeats.

ZFY Gene. Originally believed to be the human male sex-switch gene, located on the short arm of the Y chromosome. ZFY stands for zinc finger on the Y chromosome.

Zinc Finger. A structural motif found in several DNA-binding proteins.

Z-DNA. The DNA in which sugar and phosphate linkage follow a zig-zag pattern. Such DNA has left handed double helical model.

Glossary of Genetic Terms

A Priori Probability. Probability determined by the nature or geometry of an event or a situation.

A (Aminoacyl) Site. The site on the ribosome occupied by an aminoacyl-tRNA just prior to peptide bond formation.

Acentric Fragment. A chromosome piece without a centromere.

Acrocentric. A chromosome whose centromere lies very near one end.

Active Site. The part of an enzyme where the actual enzymatic function is performed.

Adaptive Value. See fitness.

Additive Model. A mechanism of quantitative inheritance in which alleles at different loci add a fixed amount to the phenotype or add nothing.

Adenine. See purines.

Adjacent-1 Segregation. A separation of centromeres during meiosis in a translocation heterozygote such that homologous centromeres are pulled to the same pole.

Affected. Individuals in a pedigree that exhibit the specific phenotype under study.

Allele. Alternative form of a gene.

Allopatric Speciation. Speciation in which the evolution of reproductive isolating mechanisms occurs during physical separation of the populations.

Allopolyploidy. Polyploidy produced by the hybridisation of two species.

Allosteric Protein. A protein whose shape is changed when it binds a particular molecule. In the new shape protein's ability to react to a second molecule is altered.

Allotype. Mutant of the nonvariant parts or immunoglobulin genes that follows the rules of simple Mendelian inheritance.

Allozygosity. Homozygosity where the two alleles are alike but unrelated.

Alternate Segregation. A separation of centromeres during meiosis in a reciprocal translocation heterozygote such that balanced gametes are produced.

Aminoacyl-tRNA Synthetases. Enzymes that attach amino acids to their proper tRNAs.

Amphidiploid. An organism produced by hybridisation of two species followed by somatic doubling. It is an allotetraploid that appears as a normal diploid.

Anaphase. The stage of mitosis and meiosis where sister chromatids or homologous centromeres are separated by spindle fibers.

Aneuploids. Individuals or cells exhibiting an aneuploidy.

Aneuploidy. The condition of a cell or of an organism that has additions or deletions of whole chromosomes from the expected balanced number of sets.

Anticodon. The complementary sequence to a codon.

Antigen. A foreign substance capable of triggering an immune response in an organism.

Antimutator Mutations. Mutations of DNA polymerase that decrease the overall mutation rate of a cell or of an organism.

Antiparallel Strands. Strands as in DNA, that run in opposite directions.

Assortative Mating. The mating of individuals with similar phenotypes.

Autogamy. Nuclear reorganisation in a single *Paramecium* cell similar to the changes that occur during conjugation.

Autopolyploidy. Polyploidy in which all the chromosomes come from the same species.

Autotrophs. Organisms that can utilise carbon dioxide as a carbon source.

Autozygosity. Homozygosity in which the two alleles are identical by descent.

Auxotrophs. Strains of organisms that have specific nutritional requirements.

Bacillus. A rod-shaped bacterium.

Backcross. The cross of an F_1 hybrid with one of its parents.

Bacterial Lawn. A continuous cover of bacteria on the surface of the growth medium.

Balanced Lethal System. An arrangement of recessive lethals that maintains a heterozygous chromosome arrangement. Homozygotes for any lethal bearing chromosome, perish.

Balbiani Rings. The larger polytene chromosomal puffs. Generally synonymous with puffs. See chromosome puffs.

Barr Body. Heterochromatic body found in the nuclei of normal females but absent in the nuclei of normal males.

Binary Fission. Simple cell division in single-celled organisms.

Binomial Expansion. The terms generated when a binomial raised to a particular power is multiplied out.

Binomial Theorem. The theorem that gives the terms of the expansion of a binomial raised to a particular power.

Bivalents. Structures formed during prophase of meiosis I, consisting of the synapsed homologous chromosomes.

Bottleneck. A marked reduction in size of a population that potentially leads to genetic drift.

Breakage and Reunion Hypothesis. A model that suggests that breakage of homologous chromatids and their rejoining account for recombination.

Bubbles. Nucleic acid configuration relating to replication in eukaryotic chromosomes or the shape of heteroduplex DNA at the site or a deletion or insertion.

Buoyant Density of DNA. A measure of the density or lightness of DNA determined by the equilibrium point reached by the DNA in a density gradient.

Cancer. An informal term for a diverse class of diseases marked by abnormal cell proliferation.

Cancer Family Syndromes. Pedigree patterns in which unusually large numbers of blood relatives develop certain kinds of cancers.

Catalyst. A substance that increases the rate of a chemical reaction without itself being permanently changed.

Centric Fragment. A chromosome piece with a centromere.

Centrioles. Cylindrical organelles, found in eukaryotes (except in higher plants), that organise the formation of the spindle.

Centromere Markers. Loci located near their centromeres.

Centromeres. Constrictions in eukaryotic chromosomes in which the kinetochores lie.

Centromeric Fission. Creation of two chromosomes from one by splitting the centromere.

Chargaff's Rule. Chargaff's discovery that in the base composition of DNA, the quantity of adenine equalled the quantity of thymine and the quantity of guanine equalled the quantity of cytosine (equal purine to pyrimidine content).

Charon Phages. Phage lambda derivatives used as vehicles in recombinant DNA work.

Chiasmata. X-shaped configurations seen in tetrads during the later stages of prophase I of meiosis. They represent physical crossovers. (Singular: chiasma.)

Chimaeric Plasmid. Hybrid, or genetically mixed, plasmids used in recombinant DNA work.

Chimeras. See mosaics.

Chi-Square Distribution. The sampling distribution of the chi-square statistic. A family of curves depending on degrees of freedom.

Chloroplast. The organelle that carries out photosynthesis and starch grain formation in plants.

Chromatids. Two identical units (sister chromatids) held together at the centromere that, at prophase of nuclear divisions make up each chromosome. When the centromeres divide and the chromatids separate, each chromatid is then a chromosome.

Chromatin. The nucleoprotein material of the eukaryotic chromosome.

Chromomeres. Dark regions in eukaryotic chromosomes at meiosis or mitosis.

Chromosome Puffs. Diffuse, uncoiled regions in polytene chromosomes where transcription is actively taking place.

Cis. Meaning "near side of" and referring to geometric configurations of atoms, or mutants usually on the same chromosome.

Cis-Trans. Complementation Test. A mating test to determine whether two mutants on opposite chromosomes will complement each other. A test for allelism.

Cistron. Term coined by Benzer for the smallest unit that exhibits the cis-trans position effect. Synonymous with gene or locus.

ClB Method. A technique devised by Muller to rapidly screen fruit-flies for recessive X chromosome lethals. The ClB chromosome carries a recessive lethal (1), a dominant marker (b) and an inversion (crossover suppessor, C).

Clone. A group of cells arising from a single ancestor.

Coccus. A spherical bacterium

Coefficient of Coincidence. The percentage of observed double crossovers divided by the percentage expected.

Colicinogenic Factors. See col plasmids.

Col Plasmids. Plasmids that produce antibiotics (colciginogene) used by the host to kill other strains of bacteria.

Common Ancestry. The state of two individuals when they are blood relatives. When two parents have common ancestry, their offspring will be inbred.

Competence Factor. A surface protein that binds extracellular DNA and enables the cell to be transformed.

Complementarity. The correspondence of DNA bases in the double helix such that adenine in one strand is opposite thymine in the other strand and cytosine in one strand is opposite guanine in the other. This explains Chargaff's rule. (A–T; G–C pairs).

Complementation. The production of the wild-type phenotype by a cell or an organism that contains two mutant genes. If complementation occurs, the mutants are nonallelic.

Complete Linkage. The state in which two loci are so close together that alleles of these loci are virtually never separated by crossing over.

Complete Medium. A medium that is enriched to contain all of the growth requirements of a strain of organisms.

Component of Fitness. A particular variable in the lifecycle of an organism upon which selection acts.

Concordance. The amount of similarity in phenotype among individuals.

Condition lethal Mutant. A mutant that is lethal under one condition but not lethal under another condition.

Confidence Limits. A statistical term for a pair or numbers that predict, with a particular probability level, the region in which a particular parameter lies.

Conjugation. A process whereby two cells come in contact and exchange genetic material. In prokaryotes the transfer is a one way process.

Consanguineous. Mating between blood relatives.

Conservative Replication. A postulated mode of DNA replication where an intact double helix would act as a template for a new double helix.

Constitutive Heterochromatin. Heterochromatin that surrounds the centromere. See satellite DNA.

Constitutive Mutant. A mutant that is no longer under regulatory control but instead produces a fixed quantity of gene product.

Continuous Replication. In DNA, uninterrupted replication allowed in the 5' to 3' direction by a 3' to 5' template.

Continuous Variation. Variation measured on a continuum or distribution rather than in discrete categories (e.g. height in humans).

Copy Choice Hypothesis. An incorrect hypothesis that stated that recombination resulted from the switching of the DNA replicating enzyme from one homologue to the other.

Corepressor. The metabolite that when bound to the repressor (of a repressible operon) forms a functional unit that can bind to its operator and block transcription.

Correlation Coefficient. A statistic that gives a measure of how closely two variables are related.

Cot Values (Cot 1/2). The product of C_0, the original concentration of denatured, single stranded DNA and t, time in seconds, giving a useful index of renaturation. Cot 1/2 values are the midpoint values in cot curves—cot a values a plotted against concentration of remaining single-stranded DNA—and estimate the length of unique DNA in the sample.

Coupling. Allele arrangement in which mutants are on the same chromosome and wild-type alleles on the homologue.

Covariance. A statistical value measuring the simultaneous deviations of x and y variables from their means.

Criss cross Pattern of Inheritance. The phenotypic pattern of inheritance shown by traits controlled by X-linked recessive alleles in a diploid XY species.

Critical Chi-Square. A Chi-square for a given degree of freedom and probability level to which an experimental Chi-square is to be compared.

Crossbreed. Fertilisation between separate individuals.

Crossing Over. A process in which homologous chromosomes exchange parts by a breakage and reunion process. Chromatids take part in crossingover.

Cryptic Coloration. Coloration that allows an organism to match its background and hence become less vulnerable to predation.

Cytokinesis. The division of a cell into two daughter cells.

Cytoplasmic Inheritance. Extra-chromosomal inheritance controlled by no nonnuclear genomes.

Dauermodification. The persistance for several generations of an environmentally induced trait.

Degrees of Freedom. An estimate of the number of independent categories in a particular statistical test or experiment.

Deletion Chromosome. A chromosome with part deleted.

Denature. Loss of natural configuration (of a molecule) through heat or other treatment. Denatured DNA is single stranded.

Depauperate Fauna. A fauna, especially common on islands, lacking many species found in similar habitats.

Derepressed. The condition of an operon that is transcribing because repressor control has been lifted.

Development. The process of orderly change that an individual goes through in the formation of a structure.

Diakinesis. The final stage of prophase I of meiosis when chiasmata terminalise.

Dicentric Chromosome. A chromosome with two centromeres.

Dimerisation. The chemical union of two identical molecules.

Diploid. The state of having each chromosome in two copies per nucleus or cell ($2n$).

Diplotene. The stage of prophase I of meiosis in which chromatids appear to repel each other.

Directional Selection. A type of selection that removes individuals from one end of a phenotypic distribution and thus causes a shift in the distribution.

Disassortative Mating. The mating of individuals with dissimilar phenotypes.

Discontinuous Replication. In DNA, only interrupted replication allowed backward in 5' to 3' segments by a 5' to 3' template strand.

Discontinuous Variation. Variation that falls into discrete categories.

Discrete Generations. Generations that have no overlapping reproduction. All reproduction takes place between individuals in the same generation.

Dispersive Replication. A postulated mode of DNA replication.

Disruptive Selection. A type of selection that removes individuals from the centre of a phenotypic distribution and thus causes the distribution to become bimodal.

DNA-DNA Hybridisation. A technique in which, when DNA from the same or different sources is heated and then cooled, double helix configurations will reform at homologous regions. This technique is useful for determining sequence similarities and degrees of repetitiveness among DNAs.

DNA Ligase. An enzyme that closes nicks or discontinuities in one strand of double-stranded DNA by creating an ester bond between adjacent 3'OH and a $5'PO_4$ ends on the same strand.

DNA Polymerase. One of several classes of enzymes that polymerise DNA nucleotides by using single-stranded DNA as a template and that require a double helical primer.

DNA-RNA Hybridisation. A technique in which, when a mixture of DNA and RNA is heated and then cooled, RNA can hybridise (form a double helix) with DNA that has a complementary nucleotide sequence.

Dosage Compensation. The mechanism used in species with sex chromosomes to ensure that one sex does not suffer due to the different number of sex-linked alleles in the two sexes.

Double Helix. The structure of DNA that is made of two helices rotating about the same axis.

Double Reduction. The condition in polyploids in which a heterozygous individual produces homozygous gametes.

Doublesex. An allele that converts fruit-fly males and females into developmental intersexes.

Dyad. A centromere with two chromatids attached.

Elongation Factors (EF-T_s, EF-T_u, EF-G). Proteins necessary for the proper elongation and translocation processes during translation at the ribosome in prokaryotes.

Empirical Probability. Probability determined by observing a large number of relevant cases.

Endogenote. Bacterial host chromosome.

Endomitosis. Chromosomal replication without nuclear or cellular division that results in cells with many copies of each chromosome.

Enzyme. Protein catalyst.

Epistasis. The masking of the action of alleles of one gene by allele combinations of another gene.

Equational Division. The second meiotic division that is equational bacause it does not reduce chromosome numbers.

Euploidy. The condition of a cell or organism that has one or more complete sets of chromosomes.

Evolution. In Darwinian terms, a gradual change in phenotypic frequencies that results in a population of individuals better adapted to survive.

Evolutionary Rates. The rate of divergence between taxa, measurable as amino acid substitutions per million years.

Excision Repair. A process whereby cells repair certain kinds of mutations by the removal of the mutated DNA strand and replacement using the good strand as a template.

Exconjugant. Each of the two cells that separates after conjugation has taken place.

Exogenote. DNA that a bacterial cell has incorporated through one of its sexual processes.

Exon. A region of a gene that has intervening sequences (introns) and that is actually translated.

Exonucleases. Enzymes that digest nucleotides from the ends of polynucleotide molecules. They hydrolyse terminal phosphodiester bonds.

Expressivity. The degree of expression of a genetically controlled trait.

Eyes. Referring to the configuration of replicating DNA in eukaryotic chromosomes.

F-Pile. Sex pile. Hair-like projections of an F^+ or H^+r bacterium involved in anchorage during conjugation and presumably through which DNA passes.

Feedback Inhibition. A post translational control mechanism in which the end product of an enzymatic pathway inhibits the activity of the first enzyme of this pathway.

Fertility Factor. The plasmid that allows a prokaryote to engage in conjugation with and pass DNA into an F cell.

Fimbriae. Fringed. Referring to the surface of bacteria with pili (hair-like projections).

First-Division Segregation (FDS). The allele arrangement in spores of Ascomycetes with ordered spores that indicates no recombination between a locus and its centromere.

Fluctuation Test. At experiment by Luria and Delbruck that compared the variance in number of mutation between small cultures and subsamples of a large culture to determine the mechanism of inherited change in bacteria.

Founder Effect. Genetic drift observed in a population founded by small, nonrepresentative sample of a larger population.

Frameshift Mutation. An addition or deletion of nucleotides that causes the codon reading frame to shift.

Frequency Dependent Selection. Selection whereby a genotype is at an advantage when rare, and at a disadvantage when common.

G-Bands. Eukaryotic chromosomal bands produced by treatment with Giemsa stain.

Galactosidase. The enzyme that splits lactose into glucose and galactose (coded by a gene in the *lac* operon).

Galactoside Permease. An enzyme involved in concentrating lactose in the cell (coded by a gene in the *lac* operon).

Galetic Selection. The forces acting to cause differential reproductive success of one allele over another in a heterozygote.

Gametophyte. The stage of a plant life cycle that produces gametes (by mitosis). Alternates with a diploid, sporophyte generation.

Gene Conversion. In Ascomysete fungi, where a 2:2 ratio of alleles is expected after meiosis and a 3:1 ratio is sometimes observed. The mechanism of this gene conversion is explained by the Holliday model of recombination.

Gene Pool. All the alleles available among the reproductive members of a population from which gametes can be drawn.

Genetic Code. The linear sequences of nucleotides that specify the amino acids during the process of translation at the ribosome.

Genetic Fine Structure. The structure of the gene in relation to the number and size of the smallest units of recombination and mutation.

Genetic Load. The relative decrease in the mean fitness of a population due to the presence of genotypes that have less than the highest fitness.

Genetic Polymorphism. The occurrence together in the same population of more than one allele at the same locus, with the least frequent allele occurring more frequently than can be accounted for by mutation.

Genic Balance Theory. The theory of Bridges that stated that the sex of a fruit-fly is determined by the relative number of X chromosomes and autosomes.

Genome. The genetic complement of a prokaryote or virus. A haploid cell or gamete of a eukaryotic species.

Genotype. The genes that an organism possesses.

Germ Line Theory. A theory to account for the high degree of antibody variability. The germ line theory suggests that every B lymphoyte has all the genes for every type of immunoglobulin but only transcribes one. See somatic mutation theory.

Giemsa Stain. A complex of stains specific for the phosphate groups of DNA.

Group Selection. Selection for traits that would be beneficial to a population at the expense of the individual possessing the trait.

Guanine. See purines.

Gyandromorphs. Mosaic individuals having simultaneous aspects of both the male and female phenotype.

H-Y Antigen. The Histocompatibility Y-antigen, a protein found on the cell surfaces of male mammals.

Haploid. The state of having one copy of each chromosome per nucleus or cell (*n*).

Hemizygous. The condition of loci on the X chromosome of the heterogametic sex of a diploid species.

Heterochromatin. Chromatin that remains tightly coiled (and darkly staining) throughout the cell cycle.

Heritability. A measure of the degree to which the variance in the distribution of a phenotype is due to genetic causes.

Heteroduplex Analysis. Analysis in which, if double helix DNA is formed by strands from different sources, loops and bubbles identify regions where the two DNAs differ. This heterogeneous DNA is referred to as a heteroduplex. Electron microscopic observation of this DNA is a useful tool in recombinant DNA work.

Heterokaryon. A cell that contains two or more nuclei from different origins.

Heteromorphic Chromosomes. Chromosomes of which the members of a homologous pair are not morphologically identical (e.g., sex chromosomes).

Heterotrophs. Organism that requires an organic form of carbon as a carbon source.

Heterozygote. A diploid or polyploid with different alleles at a particular locus.

Heterozygote Advantage. A selection model in which heterozygotes have the highest fitness.

Hfr. High frequency of recombination. A strain of bacteria that has incorporated an F factor into its chromosome and can then transfer the chromosome during conjugation.

Histones. Arginine and lysine rich basic proteins making up a substantial portion of eukaryotic nucleoprotein.

Holoenzyme. The complete enzyme. Usually refers to RNA polymerase when indicating the core enzyme plus the sigma factor.

Homogametic. The sex with homomorphic sex chromosomes and which therefore only produces one kind of gamete in regard to the sex chromosomes.

Homologous Chromosomes. Members of a pair of essentially identical chromosomes that synapse during meiosis.

Homomorphic Chromosomes. Morphologically identical members of a homologous pair of chromosomes.

Hybrid DNA. DNA whose two strands have different origins.

Hybrid. Offspring of unlike parents.

Hypostatic Gene. A gene whose expression is masked by an epistatic gene.

Hypotheses Testing of. Statistical methods for determining the probability that a data set fits a particular hypothesis about it.

Idiogram. A photograph or diagram of the chromosomes of a cell arranged in an orderly fashion.

Idiotypic Variation. Variation in the variable parts of immunoglobulin genes.

Immunity. The ability of an organism to resist infection.

Immunoglobulins. Specific proteins produced by derivatives of B lymphocytes that protect an organism from antigens.

Inbreeding. The mating of genetically related individuals.

Inbreeding Coefficient, F. The probability of autozygosity.

Inbreeding Depression. A depression of vigour or yield due to inbreeding.

Incestuous. A mating between blood relatives who are more closely related than the law of the land allows. Incomplete Dominance. The situation in which both alleles of the heterozygote influence the phenotype.

Independent Assortment. Mendel's second rule describing the independent segregation of alleles of different loci.

Inducible System. A system, in which a coordinated group of enzymes is involved in a catabolic pathway, is inducible if the metabolite upon which it works causes transcription of the genes controlling these enzymes. These systems are primarily prokaryotic operons.

Induction. Regarding temperate phage, the process of causing a prophage to become virulent.

Industrial Melanism. The darkening of moths during the recent period of industrialisation in many countries.

Initiation Codon. The mRNA sequence AUG, which specifies methionine, the first amino acid used in the translation process.

Initiation Complex. The initiation complex of translation consisting of the 30S ribosome subunit, mRNA, N-formyl methionine tRNA, and three initiation factors.

Initiation Factors (IF1, IF2, IF3). Proteins required for the proper initiation of translation.

Insertion Sequences. Regions of homology between host chromosomes and plasmids that allow the latter to synapse with the former and become inserted into the host chromosome by a crossover.

Inside Marker. The middle locus of three linked loci.

Intercalary Heterochromatin. Heterochromatin, other than centromeric heterochromatin, dispersed through eukaryotic chromosomes.

Intergenic Suppression. A mutation at a second locus that apparently restores the wild-type phenotype to a mutant at a first locus.

Interphase. The metabolically active, nondividing stage of the cell cycle.

Interrupted Mating. A mapping technique that disrupts a bacterial conjugation after specified time intervals.

Intersex. An organism with external sexual characteristics that have attributes of both sexes.

Intervening Sequences. Sequences of DNA within a gene that are transcribed but later removed prior to translation. See intron.

Intragenic Suppression. A second change within a mutant gene that results in an apparent restoration of the original phenotype.

Intron. A length of DNA that makes up an intervening sequence.

Inversion. The replacement of an internal section of a chromosome in the reverse orientation.

In vitro. Biological or chemical work done in the test tube (literally, "in glass") rather than in living systems.

Iojap. A locus in corn that produces variegation.

Ionising Radiation. Radiation, such as X-rays, that causes atoms to release electrons and become ions.

Isochromosome. A chromosome with two genetically and morphologically identical arms.

Kappa Particles. The bacteria-like particles that give a *Paramecium* the killer phenotype.

Karyotype. The chromosome complement of a cell.

Kinetochores. The chromosomal attachment points for the spindle fibers, located within the centromeres.

lac Operon. The inducible operon including three loci involved in the uptake and breakdown of lactose.

Lampbrush Chromosomes. Chromosomes of amphibian oocytes having loops suggestive of a lampbrush.

Leader. The length of mRNA from the 5' end to the initiation codon.

Leader Peptide Gene. A small gene within the attenuator control region of repressible amino acid operon; Translation of the gene tests the content of the amino acid whose operon is being regulated.

Leader Transcript. The mRNA transcribed by the attenuator region of repressive amino acid operons. The transcript is capable of several alternate stem-loop structures dependent on the translation of a short leader peptide gene.

Leptotene. The first stage of prophase I of meiosis where chromosomes become distinct.

Linkage Groups. Associations of loci on the same chromosome. In a species. there are as many linkage groups as there are homologous pairs of chromosomes.

Locus. The position of a gene on a chromosome. Used synonymously with gene (Plural: loci).

Lyon Hypothesis. The hypothesis that suggests that the Barr body is an activated X chromosome.

Lystate. The contents released from lysed cell.

Lysis. The breaking open of a cell by the destruction of its wall or membrane.

Lysogenic. The state of a bacterial cell that has an integrated phage in its chromosome.

Mapping. The study of the position of genes of chromosomes.

Mapping Function. The mathematical relationship between measured map distance and actual recombination frequency.

Marker. A locus whose phenotype provides information about a chromosome or chromosome site segment during genetic analysis.

Maternal Inheritance. Extrachromosomal inheritance controlled by non-DNA cytoplasmic substances.

Mean. The arithmetic mean, or the sum of the data values divided by the sample size.

Mean Fitness of the Population. The sum of the fitnesses of the genotypes of a population weighted by their proportions; hence, a weighted mean fitness.

Meiosis. The nuclear process that results, in diploid eukaryotes, in gametes or spores with only one member of each original homologous pair of chromosomes per nucleus.

Merozygote. A partially diploid bacterial cell arising from one of the sexual processes.

Metacentric Chromosome. A chromosome with a centrally located centromere.

Metafemale. A fruit-fly with an X/A ratio greater than 1.0.

Metagon. An RNA necessary for the maintenance of mu particles in *Paramecium.*

Metamale. A fruit-fly with an X/A ratio below 0.5.

Metaphase Plate. The plane of the equator of the spindle into which chromosomes are manipulated at metaphase.

Metaphase. The stage of mitosis or meiosis in which spindle fibers are attached to kinetochores and the chromosomes are positioned in the centre of the cell.

Microtubules. Hollow cylinders made of the protein tubulin and making up, among other things, the spindle.

Mimicry. A phenomenon in which an individual of one species gains an advantage by looking like individuals of a different species.

Minimal Medium. A culture medium for microorganisms that contains the minimal necessities for growth of the wild type.

Missense Mutations. Mutations that change a codon for an amino acid to a codon for a different amino acid.

Mitochondrion. The eukaryotic cellular organelle in which the krebs cycle and electron transport reaction take place.

Mitosis. The nuclear division producing two daughter nuclei identical to the original nucleus.

Monohybrids. Offspring of parents that differ in only one characteristic. Usually implies heterozygosity at a single locus under study.

Monosomic. A diploid cell missing a single chromosome.

Mosaics. Individuals made up of two or more different cell lines.

mRNA. Messenger RNA. The basic function of the nucleotide sequence of mRNA is to determine the amino acid sequence in proteins.

Mu Particles. Bacteria-like particles found in the cytoplasm of *Paramecium* that cause the mate-killer phenotype.

Multihybrid. An organism heterozygous at numerous loci.

Multinomial Expansion. The terms generated when a multinomial raised to a particular power is multiplied out.

Mutants. Alternate alleles to the wild type. The phenotypes produced by alternate alleles.

Mutation. The process by which a gene or chromosome changes structurally and the end result of this process.

Mutational Load. Genetic load, caused by mutation, that brings deleterious alleles into a population.

Mutation Rate. The proportion of mutants per cell division in bacteria or single-celled organisms or the proportion of mutants per gamete in higher organisms.

Mutator Mutations. Mutations of DNA polymerase that increase the overall mutation rate of a cell or of an organism.

Muton. A term coined by Benzer for the smallest mutable site within a cistron.

Natural Selection. A process whereby one genotype leaves more offspring than another genotype.

Nearest-neighbour Analysis. A technique of transferring radioactive atoms between adjacent nucleotides in DNA that demonstrated that the two strands of DNA run in opposite directions.

Negative Interference. The phenomenon whereby a crossover in a particular region enhances the occurrence of other apparent crossovers in the same region of the chromosome.

Neo-Darwinism. The merger of classical Darwinian evolution with population genetics.

Neutral Gene Hypothesis. The hypothesis that suggests that most genetic variation in natural populations is not maintained by selection.

Nickase. An enzyme that nicks one strand of double-stranded DNA during DNA replication presumably to allow torsion to be released.

Nondisjunction. The failure of a pair of homologous chromosomes to separate properly during meiosis.

Nonhistone Proteins. The proteins remaining in chromatin after the histones are removed. The scaffold structure is made of nonhistone proteins.

Nonparental Ditype (NPD). A spore arrangement in Ascomycetes that indicates a four-strand, double crossover between two linked loci.

Nonparentals. See recombinants.

Nonrecombinants. In mapping studies, offspring that have alleles arranged as in the original parents.

Nonsense Codon. One of the mRNA sequences (UAA, UAG, UGA) that signals the termination of translation.

Nonsense Mutations. Mutations that change a codon for an amino acid to a nonsense codon.

Normal Distributions. Any of a family of bell-shaped curves defined on the basis of the mean and standard deviation.

Nuclear Transplantation. The technique of placing a nucleus from one source into an enucleated cell.

Nuclease. One of several classes of enzymes that degrade nucleic acid. *See* endonucleases and exonucleases.

Nucleolus. The globular, nuclear organelle formed at the nucleolus organiser.

Nucleolus Organiser. The chromosomal location of the ribosomal RNA genes around which the nucleolus forms.

Nucleoprotein. The substance of eukaryotic chromosomes consisting of proteins and nucleic acids.

Nucleosomes. Arrangements of DNA and histones forming regular spherical structures in eukaryotic chromatin.

Nucleotide. Subunits that polymerise into nucleic acids (DNA or RNA). Each nucleotide consists of a nitrogenous base, a sugar, and one or more phosphates.

Null Hypothesis. The statistical hypothesis that states that there are no differences between observed and expected data.

Nullisomic. A diploid cell missing both copies of the same chromosome.

Nutritional-requirement Mutants. *See* auxotrophs.

Okazaki Fragments. Segments of newly replicated DNA produced during discontinuous DNA replication.

Oogenesis. The process of ovum formation in female animals.

Operator. A DNA sequence that is recognised by a repressor protein or repressor-corepressor complex. When the operator is complexed with the repressor, transcription is prevented.

Peptidyl Site. The site on the ribosome occupied by the peptidyl tRNA just prior to peptide bond formation.

P_1. Parental generation.

Pachytene. The stage of prophase I of meiosis where chromatids are first distinctly visible.

Paracentric Inversion. An inversion that does not include the centromere.

Paramecin. A toxin liberated by "killer" *Paramecium.*

Parental Ditype. A spore arrangement in Ascomycetes that indicates no recombination between two linked loci.

Pascal's Trianlge. A triangular array made up of the coefficients of the binomial expansion.

Passenger DNA. DNA incorporated into a plasmid to form a hybrid plasmid.

Path Diagram. A modified pedigree showing only the direct line of descent from common ancestors.

Pedigree. A representation of the ancestry of an individual or family. A family tree.

Penetrance. The normal appearance in the phenotype of genetically controlled traits.

Peptidyl Transferase. The enzyme responsible for peptide bond formation during translation at the ribosome.

Pericentric Inversion. An inversion that includes the centromere.

Permissive Temperature. A temperature at which temperature sensitive mutants are normal.

Petite Mutations. Mutations of yeast that produce small, anaerobic like colonies.

Phenocopy. A phenotype that is not genetically controlled but that looks like a genetically controlled phenotype.

Phenotype. The observable attributes of an organism.

Phosphodiester Bond. Diester bond linking nucleotides together (between phosphoric acid and sugars) to form the nucleotide polymers DNA and RNA.

Photoreactivation. The process whereby dimerised pyrimidines (usually thymine dimers) are restored by an enzyme requiring light energy (deoxyribodipyrimidine photolyase).

Pile (Fimbriae). Hair-like projections on the surface of bacteria.

Plaques. Clear area on a bacterial lawn caused by cell lysis due to viral attack.

Plasmid. A genetic particle that can exist independently in a cell's cytoplasm without the ability to integrate into the host chromosome.

Plastid. A chloroplast prior to the development of chlorophyll.

Point Mutations. Mutations that are single changes in the nucleotide sequence and that consist of a replacement, addition, or deletion of a base pair.

Poky Mutations. Mutations in *Neurospora* that produce a petite phenotype.

Polarity. Referring either to an effect seen in only one direction from a point of origin or to the fact that linear meiotic (such as a single strand of DNA) have ends that differ from each other. Polarity means directionality.

Polarity Gene. A gene in mitochondrial DNA with alleles that are preferentially found in daughter mitochondria after recombination between mitochondria.

Poly-A-Tail. A sequence of adenosine nucleotides added to the 3' end of eukaryotic mRNAs.

Polygenic Inheritance. Quantitative Inheritance. (Inheritance of a trait by more than one gene).

Polynucleotide Phosphorylase. An enzyme that can polymerise diphosphate nucleotides without the need for a primer. The *in vivo* function is probably in its reverse role, as an RNA exonuclease.

Polyploids. Organisms with whole chromosome sets greater than two.

Polysome. The configuration of several ribosomes simultaneously translating the same mRNA. Polyribosome.

Polytene Chromosome. Large chromosome consisting of many chromatids formed by rounds of endomitosis followed by synapsis.

Position Effect. An alteration of phenotype caused by the shifting of genes.

Positive Interference. When the occurrence of one crossover reduces the probability that a second crossover will occur in the same region.

Postreplicative Repair. A DNA repair system initiated when DNA polymerase bypasses a damaged area. Uses enzymes in the *rec* system.

Posttranscriptional Modifications. The changes in eukaryotic mRNA made after transcription has been completed. These changes include additions of caps and tails and removal of introns.

Preemptor Stem. A configuration of leader transcript that does not terminate transcription in attenuator-controlled amino acid operons.

Pribnow Box. Relatively invariant sequence of 7 nucleotides in DNA that signal the start of transcription.

Primer. In DNA replication, a length of double-stranded DNA that continues as a single-stranded template leaving a 3′-OH end.

Probability. The expectation of the occurrence of a particular event.

Probability Theory. The conceptual framework concerned with quantification of probabilities.

Product Rule. The rule that states that the probability of the occurrence of independent events is the product of their separate probabilities.

Progeny Testing. Breeding of offspring to determine their and their parents' genotypes.

Prokaryotes. Organisms that lack true nuclei.

Promoter. A region of DNA that signals the initiation of transcription to RNA polymerase.

Proofread. Technically, to read for the purpose of detecting errors for later correction. DNA polymerase has 3′ to 5′ exonuclease activity, which it uses during polymerisation to remove nucleotides it has recently added. This is a correcting ability to remove errors in replication and is referred to as proofreading.

Prophage. A temperate phage integrated into a host chromosome.

Prophase. The initial stage of mitosis or meiosis in which chromosomes become visible and the spindle apparatus forms.

Proplastid. Mutant plastids that do not grow and develop into chloroplasts.

Propositus (Proposita). The person through whom a particular pedigree was discovered.

Prototrophs. Strains of organisms that can survive on the minimal medium.

Pseudoalleles. Alleles that are functionally but not structurally allelic.

Pseudodominance. The phenomenon in which a recessive allele shows itself in the phenotype when only one copy of the allele is present as in hemizygous alleles or alleles opposite deletions.

Punnett Square. A diagrammatic representation of a particular cross used to determine the progeny of this Cross.

Purines. Nitrogenous bases of which guanine and adenine are found in DNA and RNA.

Pyrimidines. Nitrogenous bases of which thymine is found in DNA, uracil in rRNA, and cytosine in both.

Quantitative Inheritance. The mechanism of genetic control of continuous variation.

Quaternary Structure. Of a protein, the association of polypeptide subunits to form the final protein.

R Plasmids. Plasmids that carry genes that control resistance to various drugs.

Ram Mutants. Referring to ribosomal ambiguity (RAM). Ribosomal mutants that allow incorrect tRNAs to be incorporated into the translation process.

Random Genetic Drift. Changes in allelic frequency due to sampling error.

Random Mating. The mating of individuals in a population such that the union of individuals with the trait under study occurs according to the product rule of probability.

Random Strand Analysis. Mapping studies in organisms that do not retain all the products of meiosis in a recoverable form.

Realised Heritability. Heritability determined by response to selection.

Recessive. A trait that does not express itself in the heterozygous condition.

Reciprocal Cross. Testing of the role of parental sex on a phenotype by repeating a particular cross with the phenotype of each sex reversed as compared to the original cross.

Reciprocal Translocation. A chromosomal configuration in which the ends of two homologous chromosomes are broken off and become attached to the nonhomologues.

Reciprocity. In relation to recombination, the conservation of the total amount of genetic material while allowing changes in the arrangement of alleles.

Recombinants. In mapping studies, offspring with allelic arrangements made up of combinations of the original parental arrangements.

Recombination. The non-parental arrangement of crossing over.

Recon. A term coined by Benzer for the smallest recombinable unit within a cistron.

rec System. Several loci controlling genes (recA, recB, recC, and others) involved in postreplicative DNA repair.

Regression to the Mean. A phenomenon of polygenic traits in which the offspring of extremes tend toward the population mean.

Regulator Gene. A gene primarily involved in control of the production of another gene's product.

Release Factors (RF-1, RF-2, RF-3). Proteins responsible for proper termination of translation and release of till newly synthesised polypeptide when a nonsense codon appears in the A site of the ribosome.

Renner Complexes. Specific gametic chromosome combinations in *Oenothera.*

Repetitive DNA. DNA X containing copies of the same nucleotide sequence.

Replica Plating. A technique to rapidly transfer microorganism colonies to numerous petri plates with different media.

Replication. The process of copying.

Replicons. A replicating genetic unit including the site for the initiation of replication.

Repressible System. A system in which a coordinated group of enzymes is involved in a synthetic pathway if excess quantities of the end product of the pathway lead to the termination of transcription of the genes for the enzymes. These systems are primarily prokaryotic operons.

Repressor. The protein product of a regulator gene that acts to control transcription of inducible and repressible operons.

Reproductive Isolating Mechanisms. Environmental, behavioural, mechanical and physiological barriers that prevent two individuals of different populations from producing viable progeny.

Reproductive Success. The unit of natural selection that is measured as the relative production of offspring by a particular genotype.

Repulsion. Allele arrangement in which each homologue has mutant and wild-type alleles.

Resistance Transfer Factor. A plasmid that confers on its host the simultaneous resistance to several antibiotics.

Restriction Endonucleases. Endonucleases that recognise certain DNA sequences and cleave that DNA. Thought to protect cells from viral infection; useful in recombinant DNA work.

Restrictive Temperature. A temperature at which temperature-sensitive mutants display the mutant phenotype.

Reverse Transcriptase. An enzyme that can synthesise single-stranded DNA by using RNA as a template.

Reversion. The return of a mutant to the wild-type through the process of a second mutational event.

RHO. A protein that is involved in the termination of transcription and release of the transcript at the terminator sequence.

Ribosomes. Organelles at which translation takes place. Made up of two subunits consisting of RNA and proteins.

RNA Phages. Phages whose genetic material is RNA. They are simple phages known.

RNA Polymerase. The enzyme that polymerises RNA by using DNA as a template. (Also known as transcriptase or RNA transcriptase).

Robertsonian Fusion. Fusion of two acrocentric chromosomes at the centromere.

RNA Replicase. A polymerase enzyme that catalyses the self-replication of single-stranded RNA.

Rolling Circle Replication. A model of DNA replication that accounts for a circular DNA molecule producing linear daughter double helixes.

rRNA. Ribosomal RNA. RNA components of the subunits of the ribosomes.

Satellite DNA. Highly repetitive eukaryotic DNA primarily located around the centromeres. Satellite DNA usually has a different buoyant density than the rest of the cell's DNA.

Scaffold. The eukaryotic chromosome structure remaining when DNA and histones have been removed; made from non histone proteins.

Screening Technique. A technique to determine the genotype or phenotype of an organism.

Secondary Oocytes. The cells formed by meiosis I in female animals.

Secondary Spermatocytes. The products of the first meiotic division in male animals and which undergo the second meiotic division.

Secondary Structure. Of a protein, the flat or helical configuration of the polypeptide backbone.

Second-Division Segregation (SDS). The allele arrangement in spores of Ascomycetes with ordered spores that indicates a crossover between a locus and its centromere.

Segregation, Rule of. Mendel's first principle describing how genes are passed from one generation to the next.

Segregational Load. Genetic load caused when a population is segregating less fit homozygotes under heterozygote advantage.

Selection Coefficients. The sum of forces acting to prevent reproductive success of a genotype.

Selective Medium. A medium that is enriched with a particular substance to allow the growth of particular strains of organisms.

Self-Fertilisation. Fertilisation in which the two gametes are from the same individual.

Semiconservative Replication. The mode by which DNA replicates.

Sex Controlled Traits. Traits that appear more often in one sex than to another but are neither sex linked, sex limited, or sex influenced.

Sexduction. A process whereby a bacterium gains access to and incorporates foreign DNA brought in by a modified F factor during conjugation.

Sex Influenced Traits. Traits controlled by alleles that show a different dominance-recessiveness relationship depending on the sex of the heterozygote.

Sex Limited Genes. Autosomal genes whose phenotypes are expressed in only one sex.

Sex Linked. The inheritance pattern of loci located on the sex chromosomes (usually X chromosome in XY species). Also refers to the loci themselves.

Sex Ratio Phenotype. A trait in *Drosophila* where females produce mostly, if not only, daughters.

Sexual Selection. The forces acting to cause one genotype to mate more frequently than another genotype.

Siblings (SIBS). Brothers and sisters.

Sigma Factor. The protein that gives promoter-recognition specificity to the RNA polymerase core enzyme.

Skew. A distortion of the shape of the normal distribution toward one side or the other.

Snurp. Small nuclear RNP (Sn RNP) present in nucleus, which are associated with specific protein molecules.

Scyrps. Small cytoplasmic RNP (Sc RNP) present in cytoplasm.

Somatic Doubling. A disruption of the mitotic process that produces a cell with twice the normal chromosome number.

Somatic Mutation Theory. A theory to account for the high degree of antibody variability. The somatic mutation theory suggests that mutation of a basic immunoglobulin gene accounts for all of the different types of immunoglobulins produced by B lymphocytes.

Spacer DNA. Regions of nontranscribed DNA between transcribed segments, as in the numerous spacer regions in the nucleolus organiser.

Speciation. A process whereby, over time, one species evolves into a different species or where one species diverges to become two or more species.

Species. A group of organisms belonging to the same species because they are capable of interbreeding to produce fertile offspring.

Spindle. The microtubule apparatus that controls chromosome movement during mitosis and meiosis.

Spirillum. A spiral bacterium.

Sporophyte. The stage of a plant life cycle that produces spores by meiosis and alternates with the gametophyte stage.

Stabilising Selection. A type of slection that removes individuals from both ends of a phenotype distribution and thus maintains the same mean of the distribution.

Standard Deviation. The square root of the variance.

Standard Error of the Mean. The standard deviation divided by the square root of the sample size. It is the standard deviation of a sample of means.

Statistics. Measurements of attributes of a sample from a population; denoted by Roman letters.

Stem Loop Structures. Structures formed nucleic acid a loops back on itself to form complementary double helixes (stems) topped by the loops. Lollipop-shaped structures.

Stochastic. A process with an indeterminate or random element as compared to a deterministic process that has no random element.

Stringent Factor. A protein that catalyses the formation of two unusual nucleotides during the stringent response under amino acid starvation.

Stringent Response. A translational control mechanism of prokaryotes that represses tRNA and rRNA synthesis during amino acid starvation.

Structural Allele. Mutant alleles that have changes at identical base pairs.

Submetacentric Chromosome. A chromosome whose centromere lies between the middle and the end, but closer to the middle.

Subtelocentric Chromosome. A chromosome whose centromere lies between the middle and the end, but closer to the end.

Sum Rule. The rule that states the probability of the occurrence of one of several of a group of mutually exclusive events is the sum of the probabilities of the individual events.

Supergenes. Close physical association of several loci that usually control related aspects of the phenotype.

Suppessor Gene. A gene that, when mutated, apparently restores the wild-type phenotype to a mutant of another locus.

Survival of the Fittest. In evolutionary theory, survival of only those organisms best able to obtain and utilise resources (fittest). This phenomenon is the cornerstone to Darwin's theory.

Svedberg Unit. A unit of sedimentation during centrifugation. Abbreviated as in 50S.

Sympatric Speciation. Speciation in which the evolution of reproductive isolating mechanisms occurs within the range and habitat of the parent species. This speciation is common in parasites.

Synapsis. The point by point pairing of homologous chromosomes during zygotene or in certain dipteran tissues prior to endomitosis.

Synaptinemal Complex. A proteinaceous complex that mediates synapsis during zygotene and breaks down shortly thereafter.

Synteny Test. A test that determines whether two loci belong to the same linkage group by observing concordance in hybrid cell lines.

Synthetic Medium. A chemically defined substrate upon which microorganisms are grown.

Target Theory. A theory that predicts response curves based on the number of events required to cause the phenomenon. Used to determine that point mutations are single events.

Tautomeric Shift. Reversible shifts in proton position in a molecule. Bases in nucleic acids shift between keto and enol forms or between amino and imino forms.

Telocentric Chromosome. A chromosome whose centromere lies at one end.

Telophase. The terminal stage of mitosis or meiosis in which chromosomes uncoil the spindle breaks down and cytokinesis usually occurs.

Temperate Phage. A phage that can enter into lysogeny with its host.

Temperature Sensitive Mutant. Mutants that are normal at a permissive temperature, but mutant at a restrictive temperature.

Template. A pattern serving as a mechanical guide. In DNA replication, each strand acts as a template for the synthesis of a new double helix.

Terminator Sequence. A sequence in DNA that signals to RNA polymerase the termination of transcription.

Terminator Stem. A configuration of leader transcript that signals transcription termination in attenuator controlled amino acid operons.

Tertiary Structure. Of a protein, the further folding beyond the secondary structure as well as the formation of disulfide bridges between cysteines.

Test cross. The cross of an F_1 hybrid female with a male homozygous recessive organism.

Testing of Hypotheses. The determination of whether to accept or reject a proposed hypothesis based on the likelihood that the hypothesis is correct.

Tetrads. The configuration made of four chromatids first seen in pachytene. There is one tetrad-bivalent per homologous pair of chromosomes

Tetranucleotide Hypothesis. Hypothesis, based on incorrect information, that DNA could not be the genetic material because its structure was too simple-repeating subunits containing one copy of each of the four DNA nucleotides.

Tetraploids. Organisms with four whole sets of chromosomes.

Tetratype (TT). A spore arrangement in Ascomycetes that indicates a single crossover between two linked loci.

Theta Structure. An intermediate structure formed during the replication of a circular DNA molecule.

–Thiogalactoside Acetyltransferase. An enzyme that is involved in lactose metabolism and encoded by a gene in the *lac* operon.

Three Point Cross. A cross involving three loci.

Thymine. See pyrimidines.

Trailer. The length of mRNA from the nonsense codon to the 3' end (or, in prokaryotes, from a nonsense codon to the next initiation codon).

Trans. Meaning "across" and referring to geometric configurations of atoms or mutants usually on different homologous chromosomes.

Transcription. The process whereby RNA is synthesised from a DNA template.

Transduction. A process whereby a cell can gain access to and incorporate foreign DNA. The new DNA is brought in by a viral particle.

Transfer Operon (TRA). Sequence of loci that impart the male (F-pili producing) phenotype on a bacterium. The cell can then transfer its genes to another bacterium.

Transformation. A process whereby prokaryotes take up DNA from the environment and incorporate it into their genomes.

Transformer. An allele in fruit-flies that converts chromosomal females into sterile males.

Transition Mutation. A mutation in which a purine/pyrimidine base pair is replaced by a base pair in the same purine/pyrimidine relationship.

Translation. The process of protein synthesis wherein the primary structure of proteins is determined by the nucleotide sequence in RNA.

Translocase (EF-G). Elongation factor necessary for proper translocation at the ribosome during the translation process.

Translocation. A chromosomal configuration in which part of a chromosome become attached to a different chromosome.

Transversion. A mutation in which a purine replaces a pyrimidine or vice versa.

Trihybrid. An organism heterozygous at three loci.

Triploids. Organisms with three whole sets of chromosomes.

tRNA. Transfer RNA. Small RNA molecules that transfer amino acids to the ribosome for polymerisation.

Two Point Cross. A cross involving two loci.

Two Strand Double Crossovers. Double crossovers that occur in only two of the four chromatids of a tetrad.

Type I Error. In statistics, the rejecting of a true hypothesis.

Type II Error. In statistics, the accepting of a false hypothesis.

Type Species Concept. The concept that organisms that are morphologically similar belong to the same species.

Uninemic Chromosome. A chromosome consisting of one double helix of DNA.

Unique DNA. A length of DNA with no repetitive nucleotide sequences.

Unusual Bases. Other bases, in addition to adenine, cytosine, guanine and uracil found primarily in tRNAs.

Uracil. See pyrimidines.

Variance. The average squared deviation about the mean of a set of data.

Variegation. Patchiness. A position effect caused when particular loci are contiguous with heterochromatin.

Vehicle Plasmid. A plasmid containing a piece of passenger DNA forming a hybrid plasmid, used in recombinant DNA work.

Virion. A virus particle.

Viroids. Bare RNA particles that are plant pathogens.

Wild-type. The phenotype of a particular organism as first seen in nature.

Wobble. When the third position of an anticodon is not as closely constrained as the other positions (wobbles) and thus allows additional complementary base pairing.

X Linked. See sex linked.

X-ray Crystallography. A photographic technique, using X-rays, to determine the atomic structure of molecules that have been crystallised.

Y Linked. Inheritance pattern of loci located on the Y chromosome. Also refers to the loci themselves.

Zygotene. The stage of prophase I of meiosis in which synapsis occurs.

Zygotic Induction. When a prophage that is passed into an F cell during conjugation becomes virulent.

Zygotic Selection. The forces acting to cause differential mortality of an organisms at any stage in its life cycle (other than gametes).

Index

About the Author

Dr. A.V.S.S. Sambamurty was a Senior Reader (Retired) in Sri Venkateswara College, Delhi University, New Delhi for three decades.

He obtained his B.Sc., M.Sc. and Ph.D. degrees from Andhra University, Waltair and has vast teaching and research experience. He has authored 20 textbooks in Botany and published 25 research papers in Rice cytogenetics.

Other books published by him with Narosa Publishing House are:

- ***Ecology***
  (First Edition 1999, Second Edition 2006)
- ***Genetics***
  (First Edition 2000, Second Edition 2005)
- ***Molecular Genetics***
  (2007)